Communications in Computer and Information Science 569

Commenced Publication in 2007
Founding and Former Series Editors:
Alfredo Cuzzocrea, Dominik Ślęzak, and Xiaokang Yang

More information about this series at http://www.springer.com/series/7899

Fuling Bian · Yichun Xie (Eds.)

Geo-Informatics in Resource Management and Sustainable Ecosystem

Third International Conference, GRMSE 2015
Wuhan, China, October 16–18, 2015
Revised Selected Papers

Editors
Fuling Bian
Wuhan University
Wuhan
China

Yichun Xie
Eastern Michigan University
Ypsilanti
China

ISSN 1865-0929 ISSN 1865-0937 (electronic)
Communications in Computer and Information Science
ISBN 978-3-662-49154-6 ISBN 978-3-662-49155-3 (eBook)
DOI 10.1007/978-3-662-49155-3

Library of Congress Control Number: 2015958555

Printed on acid-free paper

This Springer imprint is published by SpringerNature
The registered company is Springer-Verlag GmbH Berlin Heidelberg

Preface

The third Annual International Conference on Geo-Informatics in Resource Management and Sustainable Ecosystem (GRMSE 2015) was held in Wuhan, Hubei, China, during October 16–18, 2015. GRMSE aims to bring together researchers, engineers, and students working in the areas of geo-informatics in resource management and sustainable ecosystems. GRMSE 2015 featured a unique mix of topics on smart cities, spatial data acquisition, processing and management, modeling and analysis, and recent applications in the context of building healthier ecology and resource management.

We received a total of 321 submissions from various parts of the world. The Technical Program Committee worked very hard to have all papers reviewed before the review deadline. The final technical program consisted of 101 papers. There were four keynote speeches and five invited sessions. All the keynote speakers are internationally recognized leading experts in their research fields, who have demonstrated outstanding proficiency and have achieved distinction in their profession. The proceedings are published as a volume in Springer's *Communications in Computer and Information Science* (CCIS) series. Some excellent papers were selected and recommended for a special issue of the journal *Sustainability*. We would like to mention that, owing to the limitation of the conference venue capacity, we were not able to include many fine papers in the technical program. Our apologies to these authors.

We would like to express our sincere gratitude to all the members of the Technical Program Committee and the organizers for their enthusiasm, time, and expertise. Our thanks also go to the many volunteers and staff members for the long hours and hard work they generously gave to GRMSE 2015. We are very grateful to Wuhan University and Eastern Michigan University for their support in making GRMSE 2015 possible. The generous support from the International School of Software and School of Remote Sensing and Information Engineering, Wuhan University, and Eastern Michigan University is greatly appreciated. Finally, we would like to thank all the authors, speakers, and participants of this conference for their contributions to GRMSE 2015.

November 2015

Fuling Bian
Yichun Xie

Organization

Third Annual International Conference on Geo-Informatics in Resource Management and Sustainable Ecosystem GRMSE 2015

http://www.grmse2015.org/
October 16–18, 2015
Wuhan, Hubei, China

Organizer

International School of Software, Wuhan University, China

Co-organizer

School of Remote Sensing and Information Engineering, Wuhan University, China
Eastern Michigan University, USA

Publisher

Advisory Committee

Chenghu Zhou	Institute of Geographic Sciences and Natural Resources Research, Chinese Academy of Sciences, China
George Christakos	San Diego State University, USA, Zhejiang University, China, Shanghai Center at SIVA, China
Hui Lin	Institute of Space and Earth Information Science (ISEIS), The Chinese University of Hong Kong, SAR China
Xingfa Gu	Institute of Remote Sensing Applications China Academy of Sciences, China Academy of Sciences, China

Organizing Committee

Honorary Chair, Tom Venner	College of Arts and Sciences, Eastern Michigan University, USA

Honorary Chair, Xiaohui Cui	International School of Software, Wuhan University, China
Honorary Chair, Fuling Bian	Wuhan University, China
Honorary Chair, Zhiguo Huang	Wuhan University, China
Honorary Chair, Yichun Xie	Eastern Michigan University, China
Honorary Chair, Jianya Gong	Wuhan University, China
Honorary Chair, Weimin Bi	Wuhan University, China
Co-chair, Elisabeth Morgan	Eastern Michigan University, USA
Co-chair, Xinyue Ye	Kent State University, USA
Co-chair, Xiaoliang Meng	Wuhan University, China
Richard Sambrook	Eastern Michigan University, USA
Hugh Semple	Eastern Michigan University, USA
Wei Wei	Xi'an University of Technology, China
Secretary, Siyu Fan	Eastern Michigan University, USA
Secretary, Lin Sun	Wuhan University, China
Secretary, Quan Li	Wuhan University, China

Scientific Committee

Co-chair, Yichun Xie	Eastern Michigan University, China
Co-chair, Fuling Bian	Wuhan University, China
Co-chair, Jianya Gong	Wuhan University, China
Co-chair, William Welsh	Eastern Michigan University, USA
Wenzhong Shi	The Hong Kong Polytechnic University, SAR China
Guobin Zhu	Wuhan University, China
Zongyao Sha	Wuhan University, China

Editorial Committee

Co-chair, Xiaohui Cui	International School of Software, Wuhan University, China
Co-chair, Yichun Xie	Eastern Michigan University, China
Co-chair, Fuling Bian	Wuhan University, China
Co-chair, Jianya Gong	Wuhan University, China
Co-chair, Xinyue Ye	Kent State University, USA

Technical Program Committee

YiChun Xie	Eastern Michigan University, China
Xinyue Ye	Kent State University, USA
Ruihong Huang	Northern Arizona University, USA
Qunying Huang	University of Wisconsin-Madison, USA
Wenwen Li	Arizona State University, USA
Yingru Li	Aubern University, USA
Haifeng Liao	University of Idaho, USA

Cheng Hu	Wuhan University, China
Pariwate Varnakovida	Prince of Songkla University, Thailand
Manfred F. Buchroithner	Technische Universität Dresden, Germany
Anthony Stefanidis	George Mason University, USA
Chaowei Yang	George Mason University, USA
Xiaoxiang Zhu	Technische Universität München, Germany
Matt Rice	George Mason University, USA
Jianjun Bai	Shaanxi Normal University, China
Yongmei Lu	Texas State University, USA
Alberta Albertella	Technische Universität München, Germany
F. Benjamin Zhan	Texas State University, USA
Huamin Wang	Wuhan University, China
Edwin Chow	Texas State University, USA
Lin Liu	University of Cincinnati, USA
Shuqiang Huang	JiNan University, China
Weihua Dong	Beijing Normal University, China
Mengxue Li	University of Maryland, USA
Wenwen Li	Arizona State University, USA
André Skupin	San Diego State University, USA
Alan Murray	Arizona State University, USA
Mike Worboys	The University of Maine, USA
Shuang Li	Wuhan University, China

Contents

Smart City in Resource Management and Sustainable Ecosystem

Smart City in Resource Management and Sustainable Ecosystem

Assessing Governance Issues of Underground Utility As-Built Records

Di Wu(✉) and Xueqing Zhang

Department of Civil and Environmental Engineering,
The Hong Kong University of Science and Technology, Clearwater Bay, Kowloon,
Hong Kong SAR, China
{dwuaf,zhangxq}@ust.hk

Abstract. Insufficient management to underground utilities often appears as frequent and disorganized street works, unduly prolonged excavation period and underground properties damage. A centralized utility information system can facilitate utility information to share, update and exchange and has irreplaceable advantages in managing underground space. In order to fully utilize a utility information system, the fundamental component –utility as-built records must be complete, accurate and reliable. However, currently, the data quality of as-built records in many cities remain unsatisfactory, which also obstructs the process of building “Smart City”. The purpose of this paper is to investigate governance issues with as-built records in worldwide cities, in order to provide insight to current manage mechanism. The findings indicate that proper governance and management should be carried out to improve the data quality of as-built records in the long run.

Keywords: Underground utilities · As-built records · Data quality · Legal framework

1 Introduction

Underground utilities play an important part to the residents and serve as veins and arteries of the city. As the size of the city expands, the number of utilities under public roads began to explode. For example, in Hong Kong, there are around 47 km of diverse underground utility services laid under each kilometer of public road, and the density can be much higher in some downtown areas.

A centralized utility information system is often considered as an efficient instrument towards these problems. Generally, three data sources constitute the necessary database for the information system: data from utility permitting process, data from utility surveying, and as-built records from utility owners - the last one is often chosen as the main data source due to its larger coverage and rich information compared with the former two [1].

In order to make these systems fully utilized, the key components are the data stored in the system; otherwise, erroneous information would only bring about a “garbage in,

F. Bian and Y. Xie (Eds.): GRMSE 2015, CCIS 569, pp. 3–8, 2016.
DOI: 10.1007/978-3-662-49155-3_1

garbage out" situation. Nevertheless, the skyrocketing amount of underground utilities engenders the proliferation of heterogeneous utility information. Moreover, the contents, formats, attributes, accuracy, and data quality of these information varies. Marvin and Slater [2] estimated that only half of the utility as-built records in the US are accurate; Anspach [3] maintained the word "as-built" had already lost its meaning, since it seldom reflected the actual condition of buried facilities.

As-built records have irreplaceable roles in areas and phases such as urban planning and project designing. In the long run, the trend is to gradually improve the data quality of as-built records to meet satisfactory accuracy level, yet currently it seems there stands a considerable gap. In regarding of this situation, this paper takes a preliminary step by investigating the governance issues about as-built records.

2 Research Approach

First, we conducted a comprehensive literature review process covering ordinances and practices in several worldwide metropolises or countries, i.e. Hong Kong, Shanghai, Sydney, Melbourne, London, New York, United Kingdom, Vancouver and Singapore. More than 300 publications - government report, academic papers, legislative ordinances, codes of practices, technical manuals, etc. - related to management of urban utilities were acquired via local authorities' official publication, webpages, and online databases, and were scrutinized with care.

Second, in-depth interviews with relevant government officers, engineers and experts were conducted. Additionally, we carried out a citywide investigation on exiting as-built records in Hong Kong. The findings in this two-step process are discussed in following sections.

3 Management Methodologies and Functions of As-Built Records

A common procedure for new utility installation under public road contains three stages. We can divide these cities into two groups based on the procedures and control methods. One group includes cities where authorities have direct influences in the use of underground space; for the other group, local authorities' jurisdiction usually do not cover reviewing plans about detailed alignments of utilities, hence they cannot intervene in detailing requirement of as-built records. Table 1 selected several noteworthy management methodologies and functions of as-built records in various cities.

During occupation stage, project owners and designers gather preliminary information, conduct route selection, adjust and confirm utility alignment plans, and obtain approval from related authorities. In this period, cities where authorities have direct influences in the use of underground space (e.g. Vancouver) would use as-built information to refer to the status quo of underground space, and control the utility planning and positioning via planning permits issuance.

Table 1. The management methodologies and functions of as-built records

Management methodologies for utilities	Cities	Functions
Occupation stage: Approval for occupation of underground space		
Utility plan, design, existing as-built records and excavation are incorporated in one application process under control of the centralized authority–Utilities Management Branch. Conformity with this authorities' requirements and standards on planned utility alignments and depth are necessary to obtain approval from city council.	Vancouver	Utility plans and existing as-built records are used to establish requirements of utility design, alignment and excavation issues; assess utility application plan; grant permit to undertake street-works.
Excavation stage: Use damage-prevention methods before excavation		
Excavators are required to use one-call services before commencement of excavation or directly contact nearby property owner to avoid possible excavation damage	Sydney, Melbourne, New York	Use one-call services or other exchange platforms, color-coding, web-based information systems to prevent possible excavation damage by providing notification service to registered members
Completion stage: Management of as-built records after work completion		
Inspections, which are not only limit on issues relate to construction activities and traffic management, but also on compliance with planned utility depth and alignment, space allocation, are carried out by one or several authorities in charge of planning, transportation and construction. Penalties or compensations are imposed once offences of inappropriate space occupation, misleading as-built information is observed.	Vancouver	Final record drawings are stored and managed by Utilities Management Branch after submission by utility companies
Besides stored in utility undertakings' individual systems, as-records are required to be submitted to the city after completion.	Shanghai	Establish citywide GIS-based utility information database and system consolidating as-built records to provide GIS support for projects planning and route selection
Records are only stored in individual utility undertakers' mapping systems and can be circulated via "One-Call" center, Electronic Mark Plant Circulation system (EMPC), trade associations or organizations with similar functions.	Singapore, Sydney, New York, London, Hong Kong	Circulate as-built records among registered members in one-call centers, EMPC, etc.

During excavation stage, one-call center or Electronic Mark Plant Circulation system (EMPC) combines functions of damage prevention and existing records' inquiry to prevent damage during excavation however, EMPC itself will not store any data of other companies [4].

Besides as-built records, the other common approach for project owners and contractors to obtain utility information is Subsurface Utility Engineering (SUE) [5]. Countries and regions such as Australia, Malaysia, Canada, and UK kept pace with ASCE standard CI/ASCE 38-02 and published standards and guidelines on the accuracy of utility information. The major contribution from these practices is the stipulation of ranked quality levels for utility information based on the accuracy.

From the literature review on legislative frameworks, we observed a dilemma in established city utility information systems that no qualified bodies would be held responsible or accountable for the data quality of input as-built records. Thus, the specification and accuracy varies between agencies; and some agency provides plan data rather than as-built records. Logical administrative procedures and practices to enforce order and update of utility information is missing.

4 Current Data Quality of As-Built Records in Hong Kong

In addition to the literature review on worldwide practices, we also carried an investigation on existing data quality of as-built records in Hong Kong and on utility undertakings' expectation and definition of reliable as-built records. Prior to any possible improvement, if their definition to reliable as-built records remains the same level to current as-built records, it is highly possible that they favor maintaining status quo.

The investigation was conducted by Hong Kong University of Science and Technology and Highways Department of Hong Kong in 2014 sampling as-built records from 15 major utility undertakings (UU) randomly. As-built records from all the 15 major UU in Hong Kong have attributes of length and number. However, only 5 UU have dimension attribute; 4 UU recorded vertical position; and only 3 UU recorded service status of buried utilities. All the utility undertakings have established GIS systems to manage their as-built records. At present, there are six software vendors in the market, providing GIS services to these 15 major UU.

Although, the existing overall data quality of as-built records remains satisfactory and heterogeneous, from the interview with several practitioners in utility industries, many respondents have been aware of the consequences of uncharted or substantially deviated utilities and potential benefits of reliable as-built information. An inarguable fact is that for utility owners, high data quality usually implies additional expenses on the surveying, ipso facto, increasing data quality inevitably suggests purchasing additional apparatuses to locate and record the actual position.

5 Conclusions

In order to establish a three dimensional, comprehensive and accurate database and platform that enables all the relevant communities to assess detailed alignment and disposition, the as-built records should be reliable and up-to-date, which also meets the requirement for an integrated process of planning and management of citywide underground utilities.

If the government wants to establish a centralized utility information system to inventory installations of underground utilities insofar as efficient managing underground occupancies could be achieved, effective measures should be adopted to increase the quality of as-built records, which often come as the prime data source for the system. Based on the findings in literature review and in-depth interviews, there are three obstacles that must be surmounted with special attention:

(1) Information systems must be used in accompanied with corresponding procedures and legislations. Government should consider to consider an independent and centralized authority that dedicates to managing underground utilities and supervising quality of as-built records.
(2) Financial and human resources consumption is the prime obstacle that hinders the implementation of high quality as-built data. The expense for increasing as-built accuracy could be too substantial to hinder the improvement of as-built records. So the government might consider publishing incentive and compensatory policies to stimulate utility undertakings to adopt advanced technologies.
(3) Major concerns from utility owners about information safety must be allayed to encourage them to share necessary information fully and deeply. Government should draft legislation to safeguard utility information so as to encourage utility owners to consider more reliable as-built records.

To summarize, although mapping and surveying practices like SUE have been proved effective in generating accurate utility information, these results are still project-based with a relatively low spatial coverage; we should be aware that there is no single silver bullet or panacea that will solve the problems of managing underground utilities once for all.

Acknowledgements. This study is financially supported by the General Research Fund (project number: 623113) of the Research Grants Council, the Government of the Hong Kong Special Administrative Region.

References

1. Quiroga, C., Pina, R.: Utilities in highway right-of-way: data needs and modeling. In: Transportation Research Board 82nd Annual Meeting, p. 21. Transportation Research Board of National Academies, Washington (2003)
2. Marvin, S., Slater, S.: Urban infrastructure: The Contemporary conflict between roads and utilities. Prog. Plann. **48**, 247–318 (1997)
3. Anspach, J.H.: The Case for a National Utility "As -Builting " Standard. http://content.asce.org/files/pdf/10Asbuiltstandard.pdf.

4. Li, S., Cheng, H.: Develop a GIS-based Electronic Mark Plant Circulation System in a collaborative and end-user computing approach. www.esrichina-hk.com/events/presentation/MTRC_UtilityInfo.pdf
5. American Society of Civil Engineers: Standard Guideline for the Collection and Depiction of Existing Subsurface Utility Data (CI/ASCE 38-02). American Society of Civil Engineers, Reston (2002)

A Simple Approach for Guiding Classification of Forest and Crop from Remote Sensing Imagery: A Case Study of Suqian, China

Ni Wang[1,2(✉)], Taisheng Chen[1], and Shikui Peng[2]

[1] Chuzhou University, Fengle Road 1528, Chuzhou 239000, People's Republic of China
wangni2009@163.com

[2] College of Forest Resources and Environment, Nanjing Forestry University, Longpan Road 159, Nanjing 210037, People's Republic of China

Abstract. Basic scientific research of land cover in Suqian is fundamental to ensure the sustainability of land resource management. The major study is to monitor the confusable land cover types according to the remote sensing, and the integrated information. It supplies a new direction for integrating date sets and improves the monitor on land cover. It presents a simple fusion approach for integrating time series of the MODIS Vegetation Index products and Landsat TM data. The fusion supplies the prior probability to distinguish forest with crop for guiding supervised classification which served in monitoring forest quantities-increasing in future. The entire operation just uses primarily the fusion method from the fuzzy mathematics to achieve various kinds of information with some simple parameters. However, the fusion is a spatial feature classification conduced remote sensing training mask data blending the advantages of the phonological information, the feature characteristics and the spatial-temporal data.

Keywords: Forest and crop cover masks · Feature extraction · Fusion operator · Classification · Sustainability

1 Introduction

Knowledge of the changes in land uses and present land cover is crucial to be able to determine which areas require more attention from conservation and restoration programs. The construction of informational society and ecological city is the object of urban development and harmonious society. It established completing the classification and evaluation of land using status by taking advantage of all kinds of geographic information resources, which has a very important significance to establish a type of ecological and livable city, and to realize the sustainable development of land using. In view of China's state policy, collecting and classifying the geographic space information of land using, analyzing and evaluating the regional ecological environment, providing the suggestion and guidance for land using and regulating by 3S technology, and then achieving the development of land using is starting point of this paper [1–4].

F. Bian and Y. Xie (Eds.): GRMSE 2015, CCIS 569, pp. 9–21, 2016.
DOI: 10.1007/978-3-662-49155-3_2

In recent years, the forestry development in Jiangsu Province of China has made considerable progress. In order to improve efficiency and expand the extent of these investigations, remote sensing technology (RS) has begun with utilizing the characteristics by monitoring the covering forest resources, which is used to the real outside survey data with interpretation of supplementary Thematic Mapper (TM) imagery. Because remote sensing technology can decrease a lot of manpower, material and costs for the economic and social development on monitoring forest dynamically.

Firstly, extracting the boundary of forest is the principal task which should be operationalized by withdrawing the real location of forest from the existing data fully. The time-series high spatial resolution satellite data sets have been studied to quantify forest in an early form [5]. The Landsat products for Earth observation have provided invaluable information on the Earth's surface characteristics over the past four decades [6]. The analysis time has been chose by the phenological key period of every vegetation types [7, 8].

Secondly, high resolution satellite data has not only coarser temporal resolution which can easily be confounded efforts of rainy climate and cloud cover contaminated, but also it has high costs that not any researcher could afford [9–12]. The homogeneity of time series of satellite images is crucial when they has been studied abrupt or gradual changes from vegetation types via remote sensing data.

The Moderate Resolution Imaging Spectroradiometer (MODIS) Vegetation Index products can solve the above limitations of Landsat data sets. Hansen et al. [10] and Potapov et al. [13] uses the regression tree to integrate Landsat and MODIS imagery, which can monitor deforestation in Africa and North America except the problem of mixed sub-pixels. Gao et al. [14] introduces the Spatial and Temporal Adaptive Reflectance Fusion Model (STARFM) to blend Landsat and MODIS data. It can generate synthetic Landsat-like imagery with a spatial resolution of 30 m on a daily basis. Hansen et al. [10] demonstrates that can be used regional/continental MODIS to derive forest cover products from calibrating Landsat data for exhaustive high spatial resolution mapping of forest cover and clearing in the Congo River Basin. However, such data-fusion approaches are often based on spatially integrating reflectance observations and specifically being designed by the sub-pixel ranges of the coarse-spatial-resolution [15, 16]. But it is arguable due to high data costs and the difficulties of combining disparate resolution data.

This paper presents a fusion approach based on the fuzzy mathematics. That explains that it derive the masks of forest and crop locations from time series of the MODIS Vegetation Index products and Landsat TM data on the same which is addressed an antecedent feature extraction classification method. We are aware of only few studies about the problem of land-cover class spatial variability [17]. Feature extraction is implemented as a fast algorithm for segmenting TM imageries utilized homogeneous regions according to elements of the neighboring pixel like brightness, the texture, the color and the saturation [18]. However, the fusion approach focuses on the intersection operation of forest and crop masks and TM spatial feature classification which equivalently corrects the weight or likelihoods of the forest or crop cover types. The method provides a way of thinking by remote sensing, which serves in evaluating forest quantities-increasing in future.

2 Algorithm

In Fig. 1 we present an overview of the processing steps implemented in this methodology, which are described in detail below. The aim of the proposed procedure is to provide the guidance of classification and users' interpretation. To the Landsat data, the plants have the difficulty to be discerned due to Landsat spatial and temporal resolution limitations. First of all, it should prepare the dataset of MODIS products in 2006, which covers the same lands with Landsat images and the phenology data. At the same time, these images need the pre-processing procedure and atmospheric correction operation to decrease the cloud and shadow contamination. Secondly, it reconstructs the high-quality NDVI time series derived from the MODIS Vegetation Index Products by the filter method. Thirdly, the two sets of forest and crop cover masks must extract respectively with one of which derived from analyzing the time series of the phenology and other segments. Finally, the two sets of data are operated by implementation of fusion operator to extract samples included information as much as possible. It can serve for the further classification.

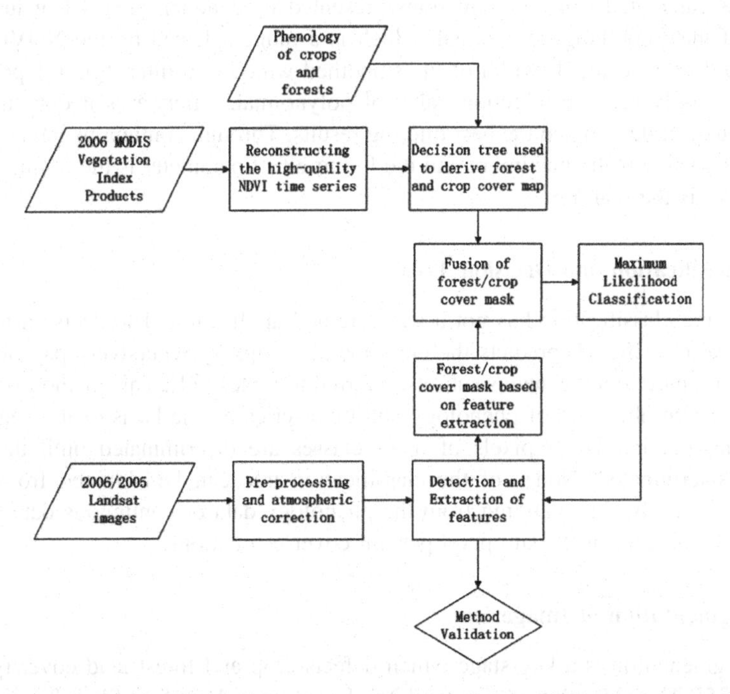

Fig. 1. It presents an overview of the processing steps. First of all, the dataset of MODIS products and Landsat images in 2006 have done the pre-processing procedure and atmospheric correction operation. Secondly, it reconstructs the high-quality MODIS Vegetation Index Products. Thirdly, it extracts the two sets of forest and crop cover masks which derived from analyzing the time series of the phenology and other segments. Finally, the two sets of data are operated by implementation of fusion operator to extract samples included information as much as possible.

2.1 Reconstructing NDVI Time-Series Data

The methodology smoothes noises out in NDVI time-series. Specifically the noises can be caused primarily by cloud contamination and atmospheric influence. Although the most often-used NDVI dataset are 16-day Maximum Value Composite (MVC) products [19], these still include a lot of noises caused by cloud contamination, atmospheric influence, and bi-directional effects. The prototype of the filter method is proposed by Savitzky and Golay [20] and improved by Chen et al. [21] as a simplified least squares-fit convolution for smoothing and computing derivatives of consecutive values. The general equation can be given as follows:

$$Y_j^* = \frac{\sum_{i=-m}^{i=m} C_i Y_j}{2m+1} \tag{1}$$

Yj* and Yj are represented the resultant and the original NDVI value. Ci is the coefficient for the its NDVI values of the smoothing window which can be obtained directly from Steinier et al. [22] as a corrected version of Savitzky and Golay's work [20]. It is calculated from the equations presented by Madden [23]. Ci includes the degree of the smoothing polynomial, which is a range selected by the NDVI observation, and m. m is a half-width of the smoothed window to filter 2 m + 1 points.

Now it only sets the selection value of polynomial. Afterwards it does not stop iteration until it determined the best filtering results. Through continuous iteration until new NDVI value is the maximum and the fitting-effect parameter is the minimum, the new NDVI is the real Y_j*.

2.2 Classification and Decision Tree

Decision tree classification has got a wide studied application. The decision tree is a hierarchical classifier. It predicts the class memberships by recursively partitioning a data set into more homogeneous subsets, referred to nodes [24]. This method is carried out with the combination of phenology and time-series on the basis of the vegetation time series curves. These pixels of other classes are discriminated until the tree's growth is terminated. Nodes of the classification tree can be obtained from either categorical data by summarizing from the phenology data or continuous data by performing the percent of the sub-pixel percent cover estimations.

2.3 Segmentation of Imageries

Image segmentation is a key stage which defects crop and forest land cover types in images [25]. The algorithm is a general one based upon Mumford-Shah function [26, 27]. The function is summarized by the following equation.

$$E(\mu, K) = \sum_{i=1}^{n} \|\mu(i) - g(i)\|^2 + \lambda l(K) \tag{2}$$

K is a discerned factor of the dataset. The boundary of K is a set of pixel edges which separate the regions and its length (K) is a number of edges of the Landsat data. u(i) is an expectation of g(i) from every segmented region which includes the information of region A, values of pixels DN (such as brightness, DN et al.) and texture T. Continued iterations cannot stop dividing and merging regions of homogeneous objects until the value of the function t(A, DN, T) equals to a threshold λ. The smaller the λ is, the more fragmented images we will get.

As λ is man-made, the result of segmentation is based on the nature of the expert experiences. Due to 30 m resolution of TM, an appropriate λ can be selected to achieve the fine segmentation without any redundant information.

2.4 The Fusion of the Cover Mask

In order to obtain the training data for the further classification, it should adequately use the guidance of the forest and crop cover masks. If a region of interests of forest covers is not a real forest mask, it will greatly affect the accuracy because of the false ROIs. To decrease the probability of false circumstances, this method is a traditional intersection of the two dataset which is defined based on the fuzzy mathematics.

$$St = \left\{ t \middle| \Phi S_{MODIS}(P, NDVI_i) \bigcap S_{TM}\left(B_m, DN_m, T_m, NDVI_j\right), i \in \Phi S_{MODIS}, j \in \Phi S_{TM}, t > 0 \right\} \quad (3)$$

t is the dataset of results St. St includes crossing values of S_{TM} and S_{MODIS}. S_{TM} is the TM vegetation cover masks derived from phenology of vegetations P and time-series NDVI values $NDVI_i$. S_{MODIS} is derived from brightness B_m (m is an original pixel of TM), the values of DN DN_m, the texture data Tm and the NDVI values $NDVI_j$. St is blending information to fill with some missing information and reduce a data volume of higher spatial resolution and costly time series data such as TM. It generates the class labels independently for each Landsat training data as a dependant variable and the Landsat cover masks as an independent variable. If the area of one sample from MODIS fully covers that of Landsat, the data of Landsat is selected in St. Contrary to the rule, the data of MODIS is selected so that TM data has different objects in the same spectrum. The overlap parts can be selected to reduce the length of the sample and make the forest or crop cover mask precise likelihoods if the extent of two fusion samples is these ones.

3 Materials and Methods

3.1 Study Area

The test area covers a surface of about 8555 km^2 located at northwest of Jiangsu Province, which is known as a 'pure land' and 'greatest oxygen bar' in the east of China (Fig. 2). It is a representative forestry and agricultural ecological city in Jiangsu. The main crops in Suqian are winter wheat, spring corn and summer rice, etc. in rotation.

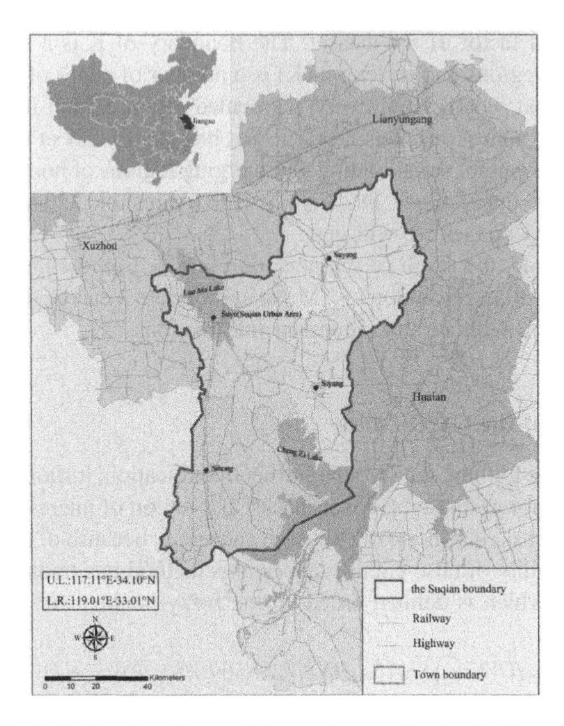

Fig. 2. This is the location figure of the study area, Suqian in Jiangsu province.

3.2 Satellite Data and Pre-Processing

There are four scenes of Landsat satellite data to cover all areas of Suqian. TM data are available in 185 km × 185 km per scene defined in a Universal Transverse Mercator projection, 50 N, and the datum is WGS-84. The dates of TM data sets show in Table 1. The images were atmospherically corrected using the FLAASH atmospheric correction algorithm [28].

Table 1. Landsat data used in this study: Due to the humid climate during the flourish stage of vegetation in rich soil in Jiangsu, Landsat data are selected by being visually identified small cloudy and little shadowy areas. Although the TM data come different time, they have little influence to the further processing because they are some difficult changes of places where trees are planted during one year.

Acquisition	Path/row
09/09/2006	120/37
09/09/2006	120/36
12/08/2005	121/36
12/08/2005	121/37

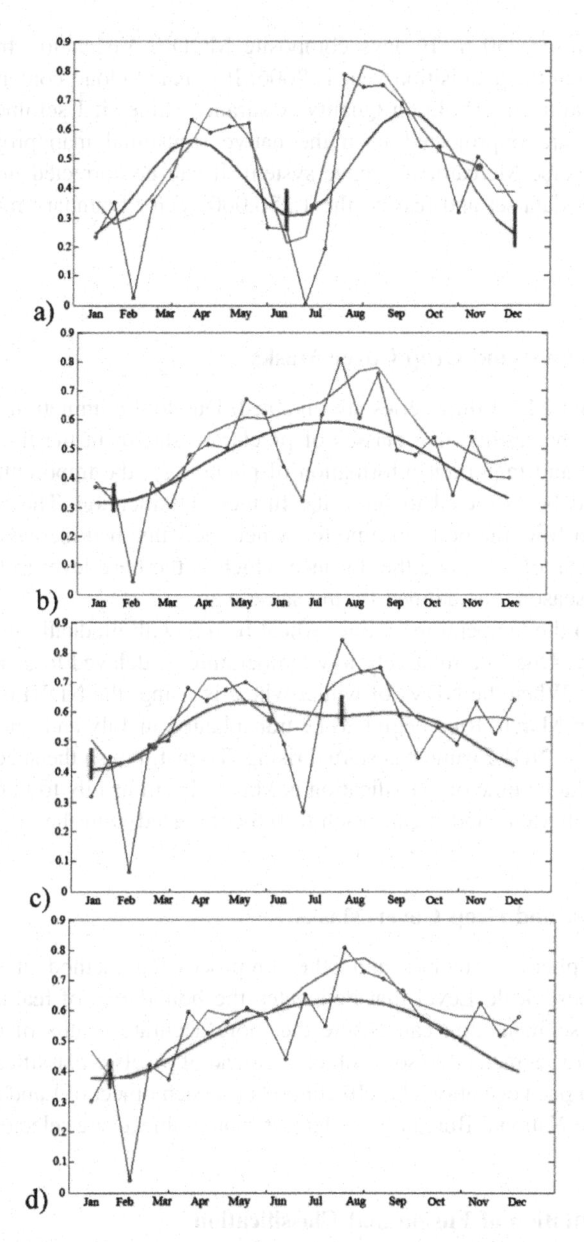

Fig. 3. The NDVI time-series curves of 4 vegetation types of land cover are generated using the SG method. The rule curve is depicted by hard lines with big dots, and the hard line with small dots delegates the original NDVI time-series curve. Seasonal fit is smoothy lines. (a) two crop annually (mainly are winter wheat and summer rice in rotation in Suqian); (b) broadleaf forest; (c) two crop mixed with forest; (d) evergreen forest.

The twenty-three 250 m 16-days composite MODIS Vegetation Index Products (MOD13Q1) when the acquisition year is 2006. It selected cloud contaminated pixels by fixing an arbitrary threshold on Quality Assurance values if Usefulness Index > 3. MODIS images are re-projected from the native sinusoidal map projection to the Universal Transverse Mercator reference system. It can be corrected into registration with the Landsat data set and resized the 1:250,000 vector boundary map.

4 Results

4.1 MODIS Forest and Crop Cover Masks

In Fig. 3, the final NDVI time-series are obtained. Due to the limitation of the spectral resolution, only the results of 3 classes of pixels are shown in the figure. Using SG analysis method and important information of phenology, the important phenological parameters could be extracted to serve the further classification. The beginning of a season is defined from the point in time for which the value had increased by a certain value. It sets 20 % of the value the distance which is the base level to the maximum. The end of the season was defined by the same way.

According to the temperature, winter wheat begins with gradually turning green to joint in February. Due to a relatively low temperature, it delayed to grow than that in the past slightly. When the NDVI of winter wheat is rising, the NDVI of forest begins to growing up in March. Rice begins to be transplanted in July and gradually to tiller and joint when the NDVI value has started rising. To distinguish the spectrum of forest and crop, the suitable time of classification is March. From in July to in October, it has maintained a high value and it can reach to 0.6 contrasted with the crops.

4.2 TM Forest and Crop Cover Masks

After the atmospheric correction and other preprocessing method, it should ideally choose the highest Scale Level that delineates the boundaries of features as well as possible. Good segmentation can ensure the more accurate results of classifications. Since TM data are segmented to some objects instead of pixels, a classification approach has been used to per-vegetation class likelihoods for each object of Landsat acquisition. According to the National Bureau Investigation, some objects are selected as the ROIs.

4.3 Implementation of Fusion and Classification

Figure 4 shows the result of the fusion operator, and it is the forest and crop cover samples respectively. The procedure generates training data independently for each Landsat vegetation types. In Table 2, it is the comparison of the Landsat cover masks operated with the MODIS cover masks. The results of the masks what we wanted are composited above the methodology described. This figure reveals that most of these samples fit closely with the boundary of some land cover classes. The patterns of human settlement and road infrastructure through the masks are excluded.

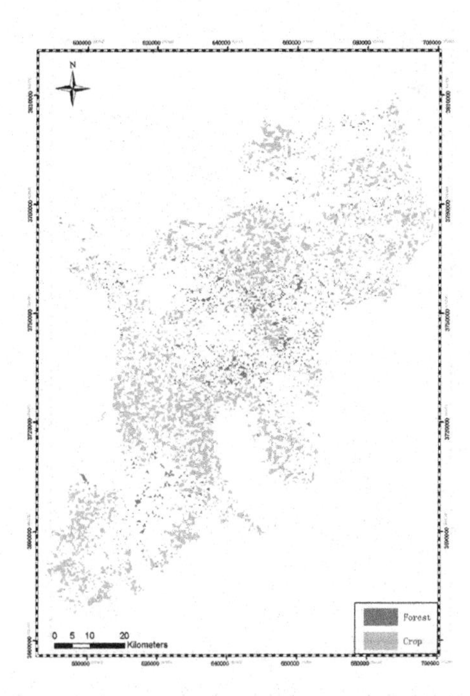

Fig. 4. The result of the fusion operator is derived from MODIS and Landsat forest or crop cover masks. The procedure is based on the raster mask data which is represented the precise likelihoods of land cover types. Then the result has been transferred to the vector mask data showed above.

Table 2. Confusion matrix indicated the samples extraction

	Original cover	Landsat forest cover	Landsat crop cover	Percent agreement
Original cover	/	18120	18726	/
MODIS forest cover	3110	838	/	26.95 %
MODIS forest cover	3437	/	2450	71.28 %
Percent agreement	/	4.62 %	13.08 %	/

In circumstance of the software ENVI, every part of the training samples is classified by the MLC classifier. Mentioned in Sect. 4.2, the pre-classification based on the feature extraction is used to compare with the method. The results are shown in Fig. 5 and the comparison of these contributions is shown in Table 3. We use training samples to add with the experimental area. Those are chosen as ROIs according to the visual inspection.

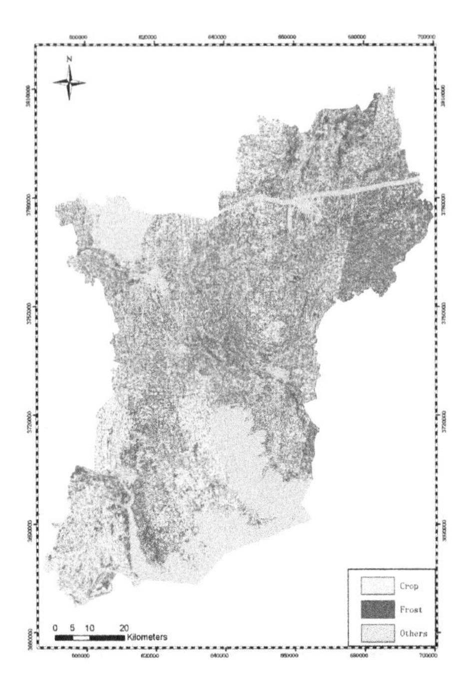

Fig. 5. Landsat derived classification result by the fusion cover mask of the Landat scene covering overall Suqian. The vegetation classification image was based on the MLC approach.

Table 3. The comparison of the accuracy assessment result from MLC and pre-classification

Class	Forest	Crop	/
Forest	73.97 %/66.72 %	24.84 %/33.08 %	/
Crop	24.89 %/30.04 %	74.41 %/69.95 %	/
Overall Accuracy	/	/	97.17 %/93.35 %
Kappa	/	/	0.8912/0.6915

Remark 1: The samples data derive the classification confusion matrix, and the overall accuracy and the Kappa coefficient are assessed by 3 classes (Crop, Forest, Others). The front number respects the result of MLC and the later one is pre-classification, which the percentage of each class is labeled as the percentage of the class.

Viewing the precision, the accuracy of the forest and crop classification achieves more than 70 % which shows a fine result. Comparing with the pre-classification, it plays a guiding role of the method.

5 Discussion

By relying on MODIS NDVI time series as the reference information, several limitations of the distinguishing procedures has been addressed, as the Landsat data allowed for vegetation masks with the fusion. The Landsat dataset derive vegetation cover masks from successfully the segmentation. Image segmentation is based on a preview of the results what you wanted. Certain texture could bring the bias, so it is the key stage to select an appropriate λ for segmentation of TM images. The MODIS products derive the vegetation masks with the phenology information. The filter is a simple and robust method comparing with the BISE algorithm, Gauss and Flourier methodology, which only sets a window size to derive the seasonal information. The masks served the further classification of Landsat data.

Finally, the maximum likelihood classification is based on the vegetation mask derived from the fusion operator. We select the validation from segmentation compared with the pre-classification. When the training data is labeled from the cover masks data, we have considered the center pixels on the samples. These pixels can bring some influence to the results of classification. The purpose of this paper found a simple, robust and economically affordable method to serve the earlier stage of the higher classification, which improves the phenomenon having the same spectrum from different field categories.

6 Conclusion

Basic scientific research of land cover in Suqian is fundamental to ensure the sustainability of land resource management. This paper presents a multi-resolution methodology to extract forest and crop masks from the scale of Landsat data, covering 8552 km2 of Suqian. Typical Landsat scale studies use a time series of good quality image to class many kinds of vegetations for one given year or decade. This can decrease the high resolution data costs and the difficulty of combining multi-temporal Landsat acquisitions. The paper describes a method to supply the basis for the supervised classification by processing multi-resolution and time-series imageries per year. Initial results indicate that Suqian has the tremendous changes in rural and urban greening. Also the area of forest has increased. However, the fusion of vegetation cover masks, integrating phenology, texture, brightness, NDVI etc., provides the higher accuracy greatly for the effective guidance of the classification and saves the manpower and costs. The training data from the new method provides the sample likelihoods at a certain extent which can be used to class.

The method is a simple and robust one that it is an operational alternative for large area vegetation cover to monitor at data-lacking satellite imageries. The full operation is just primarily an operation of the input data contained various kinds of information with simple parameters. When various types of vegetation had no start time, this method might represent a feasible approach. Thanks to minimize data costs, it is the better way to ensure accuracy monitoring for mining the information from imageries and improving the operating method.

Acknowledgements. This study is supported partly by National Natural Science Foundation of China (41201485), Scientific Research Startup Foundation of Chuzhou University (2012qd17) and the Doctorate Fellowship Foundation of Nanjing Forestry University and the Graduate Education Innovation Project of Jiangsu Province.

References

1. Chen, B., Zhang, F.: Trend and priority areas in land use research of China. Geogr. Res. **30** (1), 1–9 (2011)
2. Liu, F., Zhang, H.: Sustainability assessment of land use in main agricultural production regions in China. J. Nat. Resour. **27**(7), 1138–1153 (2012)
3. Liu, Q., Chen, L.: Comprehensive evaluation and spatial partition of sustainable utilization of land in Chang-Zhu-Tan region. Trans. Chin. Soc. Agric. Eng. (Transactions of the CSAE) **29**(6), 245–253 (2012)
4. Zeng, W.: Discussion on application of 3S in process of land use sustainable development. Sci. Surveying Mapp. **37**(2), 191–193 (2012)
5. Skole, D., Tucker, C.: Evidence for tropical deforestation, fragmented habitat, and adversely affected habitat in the Brazilian Amazon:1978–1988. Science **260**, 1905–1910 (1993)
6. Cohen, W., Goward, S.: Landsat's role in ecological applications of remote sensing. Bioscience **54**, 535–545 (2004)
7. Kang, S., Running, S.W., Lim, J., Zhao, M., Park, C., Loehman, R.: A regional phenology model for detecting onset of greenness in temperate mixed forests, Korea: an application of MODIS leaf area index. Remote Sens. Environ. **86**, 232–242 (2003)
8. Fensholt, R.: Earth observation of vegetation status in the Sahelian and Sudanian West Africa: comparison of terra MODIS and NOAA AVHRR satellite data. Int. J. Remote Sens. **25**, 1641–1659 (2004)
9. Asner, G.P.: Cloud cover in Landsat observations of the Brazilian Amazon. Int. J. Remote Sens. **22**, 3855–3862 (2001)
10. Hansen, M.C., Roy, D.P., Lindquist, E., Adusei, B., Justice, C.O., Altstatt, A.: A method for integrating MODIS and Landsat data for systematic monitoring of forest cover and change in the congo basin. Remote Sens. Environ. **112**, 2495–2513 (2008)
11. Helmer, E.H., Ruefenacht, B.: Cloud-free satellite image mosaics with regression trees and histogram matching. Photogram. Eng. Remote Sens. **71**, 1079–1089 (2005)
12. Yang, H., Tong, X.: Distribution information extraction of rubber woods using remote sensing images with high resolution. Geomatics Inf. Sci. Wuhan Univ. **39**(4), 411–416 (2014)
13. Potapov, P., Hansen, M.C., Stehman, S.V., Loveland, T.R., Pittman, K.: Combining MODIS and Landsat imagery to estimate and map boreal forest cover loss. Remote Sens. Environ. **112**, 3708–3719 (2008)
14. Gao, F., Masek, J., Schwaller, M., Hall, F.: On the blending of the Landsat and MODIS surface reflectance: predicting daily landsat surface reflectance. IEEE Trans. Geosci. Remote Sens. **44**, 2207–2218 (2006)
15. Gu, X., Han, L., Wang, J., Huang, W., He, X.: Estimation of maize planting area based on wavelet fusion of multi-resolution images. Trans. CSAE **28**(3), 203–209 (2012)
16. Guo, W., Ni, X., Jing, D., Li, S.: Spatial-temporal patterns of vegetation dynamics and their relationships to climate variations in qinghai lake basin using MODIS time-series data. J. Geogr. Sci. **24**(6), 1009–1021 (2014)
17. Zhukov, B., Oertel, D., Lanzl, F., Reinhackel, G.: Unmixing-based multisensor mul tiresolution image fusion. IEEE Trans. Geosci. Remote Sens. **37**, 1212–1226 (1999)
18. Chen, Q., Zhou, C., Luo, J., Ming, D.: Fast segmentation of high-resolution satellite images using watershed transform combined with an efficient region merging approach. In: Klette, R., Žunić, J. (eds.) IWCIA 2004. LNCS, vol. 3322, pp. 621–630. Springer, Heidelberg (2004)

19. Holben, B.: Characteristics of maximum value composite images from temporal AVHRR data. Int. J. Remote Sens. **6**, 1271–1328 (1986)
20. Savitzky, A., Golay, M.J.E.: Smoothing and differentiation of data by simplified least squares procedures. Anal. Chem. **36**, 1627–1639 (1964)
21. Chen, J., Jönsson, P., Tamura, M., Gua, Z., Matsushita, B., Eklundh, L.: A simple method for reconstructing a high-quality NDVI time-series dataset based on the Savitzky-Golay filter. Remote Sens. Environ. **91**, 332–344 (2004)
22. Steinier, J., Termonia, Y., Deltour, J.: Comments on smoothing and differentiation of data by simplified least squares procedure. Anal. Chem. **44**(11), 1906–1909 (1972)
23. Madden, H.: Comments on the Savitzky –Golay convolution method for least-squares fit smoothing and differentiation of digital data. Anal. Chem. **50**(9), 1383–1386 (1978)
24. Breiman, L., Friedman, J.H., Olshen, R.A., Stone, C.J.: Classification and regression trees. Chapman and Hall/CRC, Boca Raton (1984)
25. Funck, J.W., Zhong, Y., Butler, D.A., Brunner, C.C., Forrer, J.B.: Image segmentation algorithms applied to wood defect detection. Comput. Electron. Agric. **41**, 157–179 (2003)
26. Koepfler, G., Lopez, C., Morel, J.M.: A multiscale algorithm for image segmentation by variational methods. SIAM J. Numer. Anal. **31**(1), 282–299 (1994)
27. Morel, J.M., Solimini, S.: Variational Methods in Image Segmentation. Birkhauser, Boston (1995)
28. Adler-Golden, S.M., Matthew, M.W., Bernstein, L.S.: Atmospheric correction for short-wave spectral imagery based on MODTRAN4. Proc. SPIE **3753**, 61–69 (1999)

Landscape Changes and Ecological Effects on Dali City, Yunnan Province

Huan Yu[1(✉)], Tiancai Zhou[1], Ainong Li[2], Guangbin Lei[2], and Rongxiang Du[1]

[1] College of Earth Sciences, Chengdu University of Technology, Chengdu 610059, Sichuan, China
yuhuan0622@126.com

[2] Institute of Mountain Hazards and Environment, Chinese Academy of Sciences, Chengdu 610041, Sichuan, China

Abstract. Landscape change will cause flow and change of energy, material and nutrient between different units of land, and can cause regional ecological processes and systems functional changes. Based on the past two decades land cover changes of Dali city, the study applies ten types of landscape indices on class level and landscape level, analyzes the landscape changes of bare soil, grass land, water area, plough land, forest land, industrial land, and human habitation from year 1990 to 2010, then explores the various types of spatial and temporal evolution of the landscape, the regional landscape effects on the ecosystem, and provides a scientific reference for understanding the trend of ecological environment and promoting the coordinated development of regional economy and environment.

Keywords: Dali City · Spatial-Temporal Evolution · Landscape Index · Ecological Effects

1 Introduction

The change of land use / land cover is an important part of global environmental changes and is one of the main driving forces, which not only brings great changes in landscape structure, but also influences the cycle of material and the flow of energy in landscape, and has profound effects on regional biodiversity and important ecological processes [1]. Analyzing the change processes, laws and driving factors of regional land use, exploring the causes and mechanism of landscape pattern, knowing the influence of landscape pattern evolution on ecological system can provide the scientific basis for understanding the change trend of regional ecological environment, exploiting the natural resources reasonably, protecting the ecological environment and promoting the coordinated development of regional economy and the environment [2–4].

Dali is an important central city in western area of Yunnan Province, which is the station of ancient Silk Road and the Old Tea-trading Route, its location advantage is obvious [5]. In the process of reform and opening up and the western development, Dali develops industry and tourism, which brings the comprehensive development of the second industry and the third industry. At the same time, Dali also faces the high intensity

F. Bian and Y. Xie (Eds.): GRMSE 2015, CCIS 569, pp. 22–31, 2016.
DOI: 10.1007/978-3-662-49155-3_3

of human disturbance, which makes the landscape and ecological environment take place change greatly and present more stringent requirements to the protection of ecological environment. In addition, thick fog and haze suddenly appears a spread in the metropolis of China in recent years, the 'Escape from the city' and 'with quality living' has once again become the theme of the times. With this background, Dali will become the refuge for "climate refugees" in China [6]. Therefore, Dali must faces the inevitable problems of high intensity development in the future, and the contradiction between human development and environmental protection will be particularly evident. In order to coordinate the ecological environment protection and the economy development, advanced and scientific technologies are in urgent need. Based on the principle of ecology and environment science, this study adopt remote sensing and geographic information technology to analyses the response mechanism of ecological environment and landscape structure changes, and to provide decision support for regional sustainable development.

2 Study area

Dali is located in the central west of Yunnan Province, covering an area of 29459 km^2, within the longitude 98°52' ~ 101°03', latitude 24° 40' ~ 26°42' (Fig. 1). It belongs to the low latitude plateau monsoon climate, and riches in natural resources. Dali is located in the YunGui Plateau, the terrain become lower from northwest to southeast. Purple soil accounts for 31.75 % of the total area and red soil accounts for 27.7 %. There are

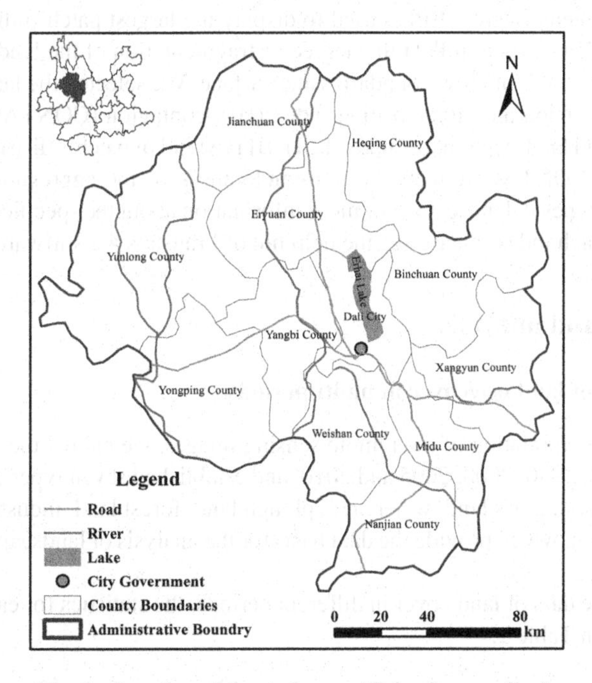

Fig. 1. Location of the study area (Color figure online).

eight lakes distribute in the region: Erhai, Tianchi, Cibi, West Lake, East Lake, Sword Lake, West Sea and Qinghai Lake. The rivers are belonging to the four major river systems of Jinsha River, Nu River, Lancang River and Red River [7].

3 Methods

The study used Landsat5 TM remote sensing images in different phases as data source (1990, 2000, 2005, and 2010), and adopted visual interpretation method to obtain the land cover of the study area, which provided the data basis for analysis of landscape pattern evolution. According to the regional characteristics, the landscape types were classified into seven categories: bare soil, grass land, water area, plough land, forest land, industrial land, and human habitation.

Moreover, the landscape pattern index is easy to understand and its ecological significance is clear, which can highly summarizes the information of landscape pattern [8], which can be classified as patch level index, class level index and landscape level index.

Considering the research objectives, class level index and landscape level index were chosen as the main indices to reflect the ecological effects. We selected the class type level index of class area (CA), percentage of landscape (PLAND), patch number (NP), patch density(PD), largest patch index(LPI), landscape shape index (LSI). The CA is used to define the area of each landscape type; PLAND shows the proportion of various landscape composition; NP is used to reflect the overall patch number of each landscape type; PD reflects the fragmentation degree of the landscape, also reflects the landscape spatial heterogeneity degree; LPI is used to display the largest patch' influence on the whole class; LSI is used to reflect the degree of fragmentation of the landscape, and is calculated on the basis of class boundaries and surface. We selected the landscape level index of interspersion and juxtaposition index (IJI), contagion (CONTAG), the patch number (NP) and landscape shape index (LSI). IJI is a kind of patch's distribution across the landscape; CONTAG reflects the non-randomness or the aggregation degree of different patch types in landscape. For more information about the specific meaning and algorithms of each index, please see the help file of Fragstats 4.2 software.

4 Results and analysis

4.1 Analysis of land use/cover in multi-periods

Based on the interpretation of the remote sensing images, we gained the land cover of the study area in 1990, 2000, 2005 and 2010, and established seven types of landscapes including bare soil, grass land, water area, plough land, forest land, industrial land, and human habitation, which provide the data basis for the analysis of landscape change and ecological effect.

Based on the data of land cover in different periods, the statistics for each landscape type is shown in Table 1.

Table 1. Statistics of dali landscape area in each period (km^2)

Class	1990	2000	2005	2010
Bare Soil	26.30	26.31	25.94	26.13
Grass Land	4792.63	4769.65	4767.79	4771.49
Water Area	396.26	403.19	403.03	403.41
Plough Land	5110.83	5071.07	5036.01	5024.30
Forest Land	17831.27	17826.41	17823.80	17805.07
Industrial Land	0.54	4.18	4.23	4.23
Human Habitation	356.41	413.44	453.45	479.79

It can be drawn from the statistical result that forest land, plough land and grass land are the main landscape types in Dali. Three landscape types account for more than 97 % of the total area, forest accounts for about 62 %, plough land accounts for about 18 %, and grass land accounts for about 17 %.

From the dynamic statistic data, we can see the forest land and plough land continues to decline over the past 20 years, and human habitation increases constantly. Industrial land has a sharply rising from 1990 to 2000, and keeps stable after that. The grass land is decreases in general, but has an increase from 2005 to 2010. The change of landscape mainly takes place among plough land, human habitation, grass land and forest land.

4.2 Class level index analysis

Based on remote sensing image classification results, class level landscape pattern index were calculated under Fragstats 4.2 software, the statistical result is shown in Table 2.

Table 2 shows that the CA index of water continues rising, that is to say, the area of water is steadily increased in the period from 1990 to 2010. This is consistent with the background of Nanjian County, Yangbi County, Yongping County and Eryuan county respectively building large water conservancy and hydropower project in recent years; the rising of PLAND index shows that the influence of water landscape on the entire regional landscape is more and more important; NP and PD index decrease, the disappearance of small patches and construction of reservoir may be the main reasons; LPI index changes a little from 1990 to 2010, which means the large patch water keeps stable; the rising of LSI index suggests that the water landscape patch shape is more complex. Because the value of LSI index is not affected by patch area, the increase of patch length is the main reason.

Table 2. Statistics of class level landscape pattern index in different periods

Index	Year	Forest Land	Grass Land	Plough Land	Human Habitation	Water Area	Bare Soil	Industrial Land
CA	1990	1783127	479263	511083	35641	39626	2630	54
	2000	1782641	476965	507107	41344	40319	2631	418
	2005	1782380	476779	503601	45345	40303	2594	423
	2010	1780507	477149	502430	47979	40325	2613	423
LPI	1990	53.127	3.494	0.561	0.038	0.868	0.003	0.000
	2000	53.022	3.491	0.545	0.066	0.868	0.003	0.002
	2005	53.021	3.511	0.487	0.122	0.868	0.003	0.002
	2010	52.952	3.534	0.484	0.127	0.868	0.003	0.002
LSI	1990	215.09	360.67	314.85	127.99	47.37	45.92	5.71
	2000	213.19	359.04	316.33	126.28	47.62	45.94	11.20
	2005	212.83	358.90	318.32	127.88	47.74	45.83	11.25
	2010	212.87	359.04	319.59	129.32	47.73	45.67	11.25
NP	1990	17799	59548	39988	6830	1762	713	12
	2000	17856	59075	40096	6482	1732	712	33
	2005	17850	59114	40169	6617	1725	711	34
	2010	17888	59128	40224	6764	1714	697	34
PD	1990	0.6242	2.0884	1.4024	0.2395	0.0618	0.0250	0.0004
	2000	0.6262	2.0718	1.4062	0.2273	0.0607	0.0250	0.0012
	2005	0.6260	2.0731	1.4087	0.2321	0.0605	0.0249	0.0012
	2010	0.6273	2.0736	1.4107	0.2372	0.0601	0.0244	0.0012
PLAND	1990	62.5346	16.8078	17.9238	1.2499	1.3897	0.0922	0.0019
	2000	62.5175	16.7273	17.7843	1.4499	1.4140	0.0923	0.0147
	2005	62.5084	16.7207	17.6614	1.5903	1.4134	0.0910	0.0148
	2010	62.4427	16.7337	17.6203	1.6826	1.4142	0.0916	0.0148

The CA index of human habitation continues to rise, which suggests that the area of human habitation is increased in the period from 1990 to 2010; furthermore, the increase of PLAND index indicates that the impact of human habitation on the regional landscape is more and more obvious; NP index decreases from 1990 to 2000, with the area increased background, which illustrates that the development of human habitation is more and more concentrated, and the city process is significantly. While NP index increases from 2000 to 2010, which suggests that the urbanization process is very obvious during this period, both the human habitation area and the number of patches are increased; PD index shows the same rule as NP index, which verifies the above analysis; LPI index shows a rapid increasing trend, which indicates that the impact of human habitation landscape increasing and has more and more important role in the ecological environment; LSI index shows the same rule as NP index and PD index, which further validates the rule and trend of regional city construction.

The CA index of industrial land is rising during 1990 to 2010, especially there is a sharp increase from 1990 to 2000, which indicates regional industry is fast developed in earlier ten years, and keeps stable in recent ten years; In addition, the rising of PLAND index illustrates that the influence of industrial land on the whole regional landscape has grew; NP and PD index show the same rule as PLAND index, which indicates that the industrial land both in area and number of patches are increased in the period of 1990 to 2010. The change amount before and after 2000 is obviously different, which is because the development mode of economy in Dali is changed. The industrial production is the main income for people before 2000, but green tourism has become the main method of regional economy development after that. The LPI index of industrial land keeps stable, which indicates that the large size of industrial land doesn't change; LSI index shows the same rule as NP index and PD index, which further validates the rule and trend of the industrial land changes.

The CA index of forest land is continuously lowing in the period of 1990 to 2010, which indicates that the forest resources has been destroyed during this period; PLAND index shows the same rule as CA index, which indicates that the influence of forest land on the whole regional landscape has decreased, and the ecological environment is worsened. The phenomenon is consistent with the fact that human habitation and industrial land are built in large numbers; the NP index decreases from 1990 to 2005, mainly because the destroy of forest resources, thus forming the trend that both the area and the number of patches is decreased; while NP index is rising from 2005 to 2010, under the background that the area is decreased, the fragmentation is aggravated, which can be concluded by human disturbance; PD index shows the same rule as NP index, which further validates the above analysis; the law of LPI index also validates the forest resources has been threatened from 1990 to 2010, also indicates that the forest ecological environment is under challenge; LSI index shows the same rule as PD and NP index, which further validates the specific changes of forest landscape.

The CA index of the grass land is continuously decreasing during the period of 1990 to 2000, and is slightly reduced from 2000 to 2005, which suggests that the area of grassland reduces a little during this period, and also indicates that the ecological environment protection policy begins to take effect. On the contrary, the CA index increases slightly after 2005, this result reflects that the area of grassland has increased, indicating that the grass resources are effectively protected and the grass land area is increasing at this time, which is consistent with the background that the government vigorously carry out afforestation, and returning farm land to grass land in recent years; PLAND index and CA index show the same rule, indicating that the influence of grass land continues to decrease before 2005 and becomes increasing after 2005; NP index shows a decreasing trend from 1990 and 2000, with the area and the number of patches decreased background, we can conclude that the grassland ecological environment is under threat and challenge. On the contrary, the NP index is rising from 2000 to 2010, under the background that the area increased, we can conclude that it is mainly because the activities of returning farm land to grass land; PD index shows the same law as NP index, which further validates the above analysis; LSI index shows the same rule as PD and NP index, which further validates the complex degree of grass land patch shape is decreased first and then increased.

The CA and PLAND index of bare soil show the same rule, the change trend is not obvious and the amount is not big, in other words, the change of bare soil landscape is not great in decades, which is related to the distribution of bare soil at high altitude, where environment is relatively stable; NP and PD index decline just a bit, which indicates that the fragmentation of bare soil landscape has been slowed, the phenomenon suggests a good response to the ecological environment; LPI index keeps stable, which suggests that the effects of bare soil on the whole regional landscape still can't be ignored, also validates the bare soil is stable; LSI index shows the same law as NP and PD index, which further validates the specific changes of bare soil landscape.

The CA index of plough land is decreased sharply from 1990 to 2010, indicating the area of plough land decreases during the period, which is caused by the construction of human habitation and industrial land, and the application of green policy such as returning farm land to grass land; PLAND decreases during the period of 1990 to 2010, which indicates that the influence of plough land on the whole regional landscape is more and more smaller, the change of plough land is obvious, which put forwards a challenge to the food production; NP and PD index are rising, which indicates the fragmentation degree of plough land becomes more and more serious; LPI index is decreased, indicates that the influence of plough land on the whole regional landscape is reduced, and the plough land resources are under threat; LSI index shows the same law as PD and NP index, which further proves the specific changes of plough land landscape and the complexity of patch shape of plough land is increasing.

For the CA index, human habitation, industrial land and plough land show a significant negative correlation rule from 1990 to 2010, which indicates there must be a certain mutual conversion between plough land and the other two kinds of landscapes. As the main landscape type, forest land decreasing means the protection of the ecological environment still should be concerned. For the PLAND index, forest land, grass land, plough land are the main landscape types, at the same time, the importance of human habitation and industrial land are increasing. For the NP and PD index, although the grass land area is less than the forest land, but the NP index and PD index of grassland is much higher than that of forest land, which means the spatial distribution of forest land is more concentrated and the spatial distribution of grass land is more cracked. For the LPI index, the forest land landscape is different with other types, which further verifies the main status of forest landscape. As for LSI index, the change of each type landscape is different in different periods, and the complex of each landscape type shows specific rules: the grass land is the highest, followed by plough land, forest land, human habitation, water area, bare soil, while the industrial land is the lowest. This reflects the spatial and temporal distribution characteristics of the landscape types in Dali during the past 20 years.

4.3 Landscape level index analysis

Based on remote sensing image classification results, landscape level landscape pattern index were calculated under Fragstats 4.2 software, the statistical result is shown in Table 3.

Table 3. Statistics of landscape level landscape pattern index in different periods

Year	NP	LSI	IJI	CONTAG
1990	126652	237.45	46.44	64.89
2000	125986	236.77	47.02	64.71
2005	126220	237.22	47.31	64.6
2010	126449	237.76	47.51	64.5

As shown in Table 3, The rising of IJI index illustrates that the landscape types is becoming complex in this area, the mixed distribution of different landscape types has intensified. In addition, the index CONTAG also reflects the same ecological significance. The NP and LSI index shows a decreasing trend during 1990 to 2000, which indicates the number and shape complex of regional landscape patch is decreased. However, in the periods of 2000 to 2010, they show the contrary rules. This is mainly due to the regional development background. 1980s and 1990s is a period of rapid development in Chinese economy, the extensive mode of economic development causes great destruction and waste of resources, lack the awareness of environmental protection is one of the main reasons for ecological environment quality serious declining. In the later time of 1990s, a series of ecological restoration work and sustainable development mode has been applied to Dali city construction. The state government try best to create a national famous tourist resort with Dali unique characteristics, and a series of policies such as 'Dali city overall urban planning', 'Dali scenery scenic area overall planning' have been introduced in recent years. The new development mode and intensive ecological restoration work has reduced the regional ecological environment deterioration to some extent. Considering the ecological environment vulnerability of Dali city, there is still a large space for economic restructuring and changing the economic growth pattern, in order to achieve better and faster development of this region.

5 Conclusion

Study applies the theory and method of landscape ecology to analyze the spatial temporal variations of landscapes (bare soil, grass land, water area, plough land, forest land, industrial land, and human habitation) in Dali from 1990 to 2010, and explores the dynamic characteristics of all kinds of landscapes. The results show that the area of water landscape increasing from 1990 to 2010, which indicates that the impact of water on the regional landscape is more and more important; the area of human habitation has been increasing, and shows a trend of urbanization, which has shown more and more important role in the ecological environment. Forest is the main landscape types in the study area, and the area of forest is decreased, which indicates that the influence of forest land on the whole regional landscape has decreased, also indicates that the forest ecological environment is under threat and challenge. The area of grass land decreases significantly before 2000, indicates that the importance of grass land landscape on the whole regional landscape has reduced, fortunately, grass land decreasing situation has been eased after 2000, which is mainly benefited from a series of activities such as returning farm land to grass. Industrial land shows the same law as human habitation, which indicates the industrial land plays a more and more important role in the regional environment; The area of plough land is decreased from 1990 to 2010, and the fragmentation degree is increased, which has a certain relationship with the process of urbanization and green policy such as returning farmland to forest or grass.

The analysis results show that 2000 is a turning point year for the ecological environment change. This shows that the national implementation of the western development strategy plays an important role to improve the regional economic development and ecological environment restoration. This also reflects the concept change of the Chinese economic development, changing emphasis on the speed of development in the past, to pay more attention to the quality and benefit of development now. With the development of economy, saving the resources, protecting environment, conserving land resources are put on a more prominent strategic position to achieve sustainable development of population, resources and environment. In order to achieve better and faster development of the regional economy, Dali should continue to build a mode of circular economy with itself characteristics, changing from the economy of high pollution and high consumption to the green tourism economy, and realizes the coordinated development of regional economy and environment at last.

Acknowledgment. This study is supported by the National Natural Science Foundation of China (grant no. 41101174), China Postdoctoral Science Foundation (grant no. 2013M540700), Special Grants for Postdoctoral Research Projects in Sichuan Province (grant no. SZD013), Lead Strategic Project of the Chinese Academy of Sciences (grant no. XDB03030507), Young and Middle-aged Excellent Teacher Training Program of Chengdu University of Technology.

References

1. Zou, X.P., Qi, Q.W., Xu, Z.R., et al.: Analysis of land use/ cover changes and its landscape ecological effects in nujiang watershed. J. Soil Water Conserv. **19**(5), 147–151 (2005)
2. Ge, F., Li, W.: Review of landscape change and its ecological impacts. J. Ecol. Environ. **17**(6), 2511–2519 (2008)
3. Fan Y J.: Study on dynamic of land use in three gorges reservoir area based on RS and GIS D. Beijing: China. M.S. Thesis, Chinese Academy of Sciences (2000)

4. He Y S.: Three gorges project and the sustainable development. Beijing: China. Ph.D. Dissertation, Chinese Academy of Social Sciences (2002)
5. Wang, K., Qin, C.X.: Study on the Development Mode of Modern Service Industry Based on Ecological Civilization. J. Kunming Univ. Sci. Technol. **11**(3), 75–79 (2011)
6. Gao, X.T.: Dali: Search for the ideal country. J. Chin. Country Geogr. **5**, 45–71 (2014)
7. http://baike.baidu.com/view/161491.htm?from_id=11161519&type=syn&fromtitle=Dalistate&fr=Aladdin
8. Turner, M.G., Gardner, R.H., O'Neill, R.V.: Landscape ecology in theory and practice: pattern and process. Springer, New York (2001)

Urban Traffic Operation Pattern and Spatiotemporal Mode Based on Big Data (Taking Beijing Urban Area as an Example)

Chao Sun[1(✉)], Yu Deng[2,3], Botao Tang[1], and Shaobo Zhong[1]

[1] Department of Engineering Physics/Institute of Public Safety Research, Tsinghua University, Beijing 100084, China
slayergod@163.com, zhongshaobo@gsafety.com

[2] College of Architecture and Urban Planning, Tongji University, Shanghai 200092, China

[3] Key Laboratory of Regional Sustainable Development Modeling, CAS, Beijing 100101, China

Abstract. An analysis of urban traffic operation pattern and spatiotemporal mode is an important basis to solve the problems of traffic congestion, emergency and extreme weather. Traditional studies on the urban traffic operation pattern and spatiotemporal mode usually are restricted by issues as poor time effectiveness, large space scale and coarse time granularity of traffic flow data, thus this essay choose to use the urban traffic speed data based on floating vehicle trajectory to dissect the urban traffic operation pattern and spatiotemporal mode in Beijing in a multi-dimensional and fine granularity. Differences of features in weekdays and weekends are also compared. This paper reports that "two-peak" mode is obvious in the urban traffic condition. Besides, the morning peak of weekends is postponed to 11-12 am, and the night peak appears shorter in 5 pm compared to weekdays. Finally, four modes of traffic and its driving mechanism are concluded.

Keywords: Traffic flow · Impact factors · Traffic speed data · "Two-peak" mode · Tide mode

1 Introduction

With urbanization and economic development fastened in recent years, traffic congestion has become an urgent issue to solve for urban economy and social development in China. Beijing, as the capital of the country with the largest population and fastest economy development, is now suffering from traffic congestion a lot, and traffic speed in Beijing now is also one of the slowest in the world. In 2008, the amount of social time delay and extra fuel consumption due to traffic congestion occupies between 0.5 % and 2.5 % of GDP of Beijing [1]. Main reasons of traffic congestion in Beijing include discordance of road network and urban planning layout, unreasonable road design, unreasonable arrangement of bus stops as well as entrances and exits of ring roads, and large amount of vehicles. The amount of vehicles has doubled from 2000 to 2007.

F. Bian and Y. Xie (Eds.): GRMSE 2015, CCIS 569, pp. 32–47, 2016.
DOI: 10.1007/978-3-662-49155-3_4

In order to deal with the stressful condition of traffic congestion in cosmopolitans, both domestic and foreign scholars have studied the issues from various perspectives.

Firstly, studies have been made from the perspective of urban traffic operation: Chen Xin has studied the method of evaluation and optimization of urban traffic network, and conducted field study in Gongyi, Henan [2]. The optimized solution was later applied in the city. Mao Tao used the main street road network in Xi'an as an example to analyze the static and dynamic efficiency of urban traffic network [3]. Ibrahim and Hall provided regression analysis on how the highways in Mississauga, Canada are affected by different weather elements, and they discovered that the average speed of rainy days may have a 5–10 km/h reduction, and the effect of snowy days can be larger, with a 38–50 km/h reduction [4]. Knapp and Smithson analyzed that average reductions range from 16 % to 47 % for different storm events in Iowa State during winter storms [5]. Keay and Simmonds reported that the traffic reductions resulted from rain-fall and other weather variables during wet days were 1.35 % in winter and 2.11 % in spring [6]. Smith investigated the impact of rainfall at varying levels of intensity on freeway capacity and operating speeds [7].

Secondly, researches have been made from the internet perspective of traffic flow structured by vehicle positioning based on urban GPS vehicle data. Bin Jiang analyzed the road network structure of Gavle, an eastern city in Sweden, based on taxi driving data from GPS, and received traffic flow allocation in different roads and hotspot information of urban traffic [8]. Guo Jing used the driving data collected from the 4th Ring Road expressway and main street floating vehicles in five weekdays in Beijing to accomplish the assessment of urban traffic condition, and he also provided related suggestions to deal with traffic congestions [9]. Gou Xirong used the data of taxi GPS and road network in Kunming to finish the work of data matching and dynamic visualization of data traffic, and he also analyzed the spatiotemporal allocation pattern of traffic condition in Kunming. In the essay, the mostly used map-matching algorithm and GPS data processing method are introduced in detail [10]. Jayakrishnan and Mahmassani used the method of on-line generation and real-time evaluation to convince that real-time simulation of a traffic network can predict future conditions and thus help design and implement more effective traffic operations [11, 12].

Existing studies are usually restricted by issues as poor time effectiveness, large space scale and coarse time granularity of traffic flow data, thus in this essay we choose to use the urban traffic speed data based on floating vehicle trajectory (three weekends and three weekdays) to dissect the urban traffic operation pattern and spatiotemporal mode in Beijing in a multi-dimensional and fine granularity using analysis of time series and spatiotemporal statistics. The feature differences between weekdays and holidays are also compared to provide strong theory support for solving congestion issues, adjusting urban road traffic security contingency plan and setting comprehensive traffic manage and control measures.

2 Areas Studied and Data Source

2.1 Brief Introduction of the Areas Studied

Currently, the areas inside the 3rd Ring Road of Beijing are capable of political, economic, scientific, educational, medical, financial functions. The overlap of urban functions on the one hand creates a dual economic structure between the urban area and the areas around; on the other hand, the condition of agglomeration economy has reached its extreme in the areas. Institutions, national sectors and functional organizations are intensely located in the urban areas, while residential areas are outside the 4th Ring Road, which causes a large cost of commute and directly influence the urban traffic condition. As supervised, the morning peak of weekdays witness 1.6 times as many cars driving into the 2nd Ring Road area as that driving outside, whereas twice as many driving outside during the evening peak. Every day, 5.5 million vehicles are on the road of Beijing [13], and floating population inside the 6th Ring Road is as high as 30.33 million (pedestrians are not included). According to statistics, an average commute time for people in Beijing is 97 min in 2014 [14], which lead to heavy burden on urban load and road traffic. Based on this, the government has conducted a series of measures including massive underground construction, real-time management and control, vehicle restrictions based on license plate number, congestion prompt and limited driving time for special vehicles.

2.2 Data Source and Management

Based on the trajectory data of floating vehicles, the driving speed of all floating vehicles can be received and an average speed is used in this essay. The data cover more than 470 main roads in Beijing and are collected every 5 min, so that real-time road network condition can be reflected. The advantage of data is the high spatiotemporal resolution ratio and accuracy, and the data are collected from 1 Jan, 2015 to 6, Jan 2015, which consist of three vacation days and three workdays. In order to guarantee the credibility and accuracy, all data used in analysis of traffic operation mode, spatiotemporal pattern and comparison between days are mean of corresponding periods of time.

2.3 Research Method

Auto-correlative indexes of space measurement include Moran Index, Geary Coefficient and G-statistics, which can all be used to study data of the whole areas and one specific unit of area. The standardized statistics of the general index can be used to judge if the aggregation of attribute value of the areas studied is notable from a mathematical perspective.

The formula of Moran I is shown as follows:

$$I = \frac{n \sum_{i=1}^{n} \sum_{j=1}^{n} W_{ij}(x_i - \bar{x})(x_j - \bar{x})}{\sum_{i=1}^{n} \sum_{j=1}^{n} W_{ij} \sum_{i=1}^{n} (x_i - \bar{x})^2} = \frac{\sum_{i=1}^{n} \sum_{j=1}^{n} W_{ij}(x_i - \bar{x})(x_j - \bar{x})}{S^2 \sum_{i=1}^{n} \sum_{j=1}^{n} W_{ij}} \tag{1}$$

In which, x_i represents the attribute value of unit i, I refers to the calculated Moran Index, i and j represent two different space units, and n is the amount of units in the whole area studied.

$$S^2 = \frac{1}{n}\sum_{i=1}^{n} (x_i - \overline{x})^2 \tag{2}$$

$$\overline{x} = \frac{1}{n}\sum_{i=1}^{n} x_i \tag{3}$$

The mathematical expectation of I:

$$E(I) = -\frac{1}{n-1} \tag{4}$$

With an increasing sample size n, E(I) will approach 0.

The result of calculation is usually between -1 and 1. When the result is positive, there is positive aggregation between the attribute values of the objective areas studied, which means the distance of space location is short and probable interactions may exist between the areas with similar attribute values. When the result is negative, it indicates that there is negative aggregation between the attribute values of the objective areas studied. When the result is zero, there is no aggregation between the attribute values. When the result is close to the mathematical expectation of I, then there is no interaction between the units of areas, and the distribution is random.

For the result of calculation I, the standardized statistic Z can be used to identify the significant level of the aggregations between the spatial area units. The formula of calculation is shown below:

$$Z = \frac{I - E(I)}{\sqrt{Var(I)}} \tag{5}$$

The two general statistics can reflect the overall situation of the attribute values of spatial area units in the areas studied, however, the level and the position of the attribute values under the abnormal aggregation cannot be gained; therefore, the local statistics are applied to analyze the data. The research shown below emphasize on applying local Moran index.

The local Moran index is defined as:

$$I_i = \frac{n(x_i - \overline{x})\sum_{j=1}^{n} W_{ij}(x_j - \overline{x})}{\sum_{i=1}^{n} (x_i - \overline{x})^2} \tag{6}$$

The standardized statistics are:

$$Z(I_i) = \frac{I_i - E(I_i)}{\sqrt{Var(I_i)}} \tag{7}$$

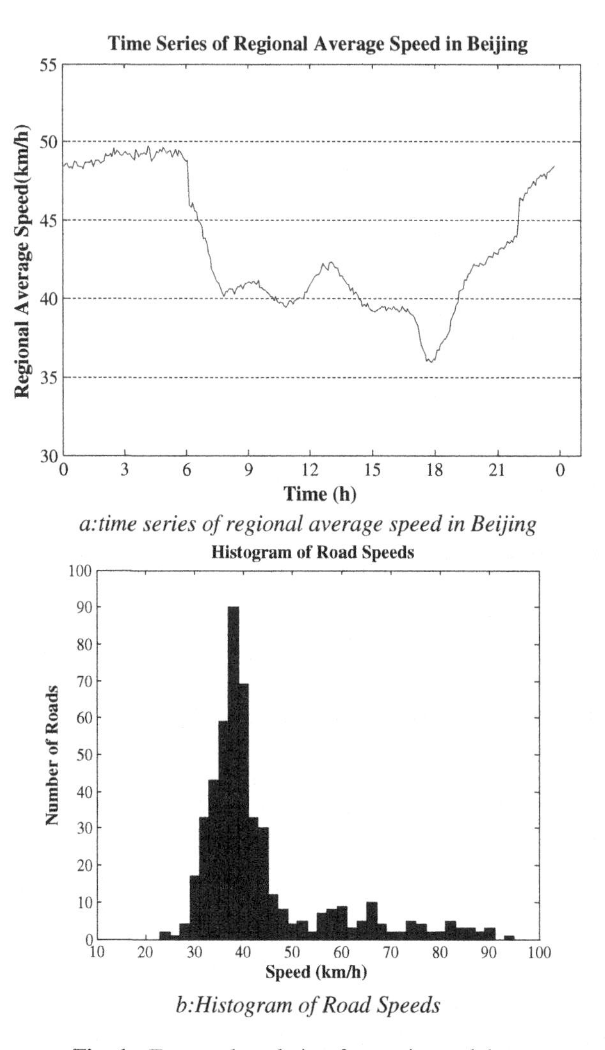

a:time series of regional average speed in Beijing

b:Histogram of Road Speeds

Fig. 1. Temporal evolution feature in weekdays

3 Analysis of Spatiotemporal Mode in Weekdays

3.1 The Spatiotemporal Evolution Pattern of Traffic Capacity

According to Fig. 1, the temporal evolution feature throughout a day shows an apparent "two-peak" mode in Beijing urban areas, in other words, there is a marked feature of morning peak and evening peak. From 12 am to 6 am, the average speed in the road network is nearly 50 km/h; after 6 am, the average speed drops drastically until the morning peak is formed at 7:30, when the average speed maintains at 40 km/h. It is until 11 am that the average speed begins to rise slowly. After 1 pm, the average speed declines again and lasts to the evening peak after 5 pm. It reaches the lowest—35 km/h

—at 6. After that, it rises and exceeds 40 km/h at 8 pm. The normal distribution curve for average speed in the road network is shown in the following figure, and the axis is at the point less that 40 km/h, while few roads have an average speed over 60 km/h.

The spatial distribution features of driving speed in the study area, shown in Fig. 2, reflects that ring roads and mainly roads into the city have a rather slow driving speed, for instance, the 2nd ring road, the northern and eastern part of the 3rd ring road, the Beijing-Tibet Highway, Xueyuan Road and Chang'an Avenue. We then analyze the spatial distribution features of morning and evening peaks, which is shown in the figure as well. During morning peak, areas with slow speed appear around Xuanwumen, Xinjiekou, Xizhimen, Jianguomen, Guangqumen, Guang'anmen, Fuxingmen and Deshengmen. During the evening peak, such areas still exist and expand to roads of more directions as well as side roads, which to some extent proves the basic analysis mentioned above that the average speed of evening peak is obviously slower than that of morning peak.

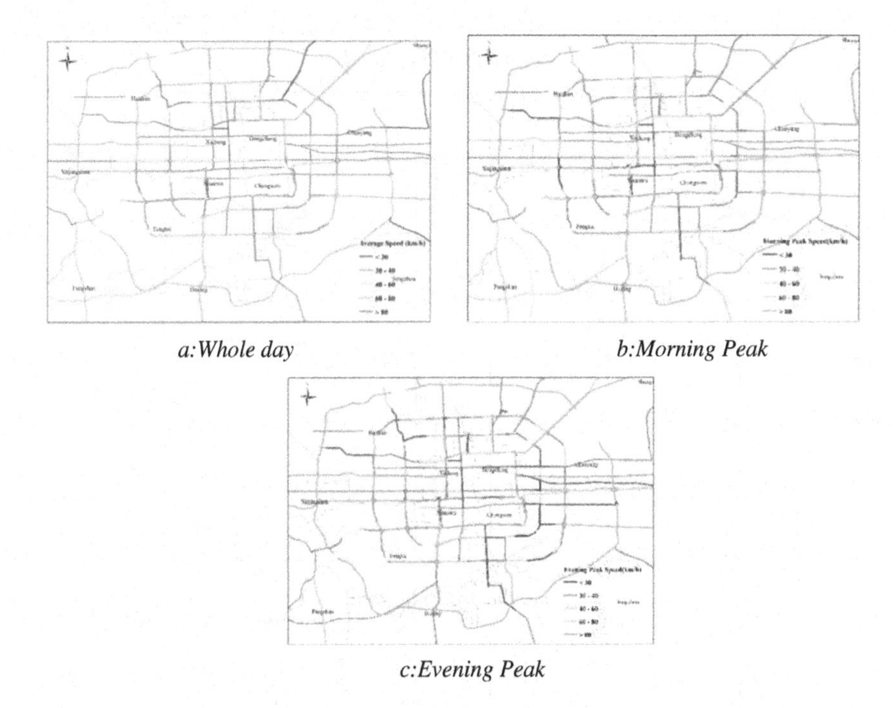

a:Whole day *b:Morning Peak*

c:Evening Peak

Fig. 2. Spatial distribution features of driving speed

In Fig. 3, a further analysis of distinctions of driving speed on difference directions of road is made, using the ring roads as an example. Both directions on different parts of the 2nd ring road have significant "two-peak" patterns, yet the change rules of different parts are various. Take the northern part as an example, the speed on the side to the west is much slower than that to the east during the morning peak, while the situation is the opposite during the evening peak, thus a "tide" phenomenon can seen

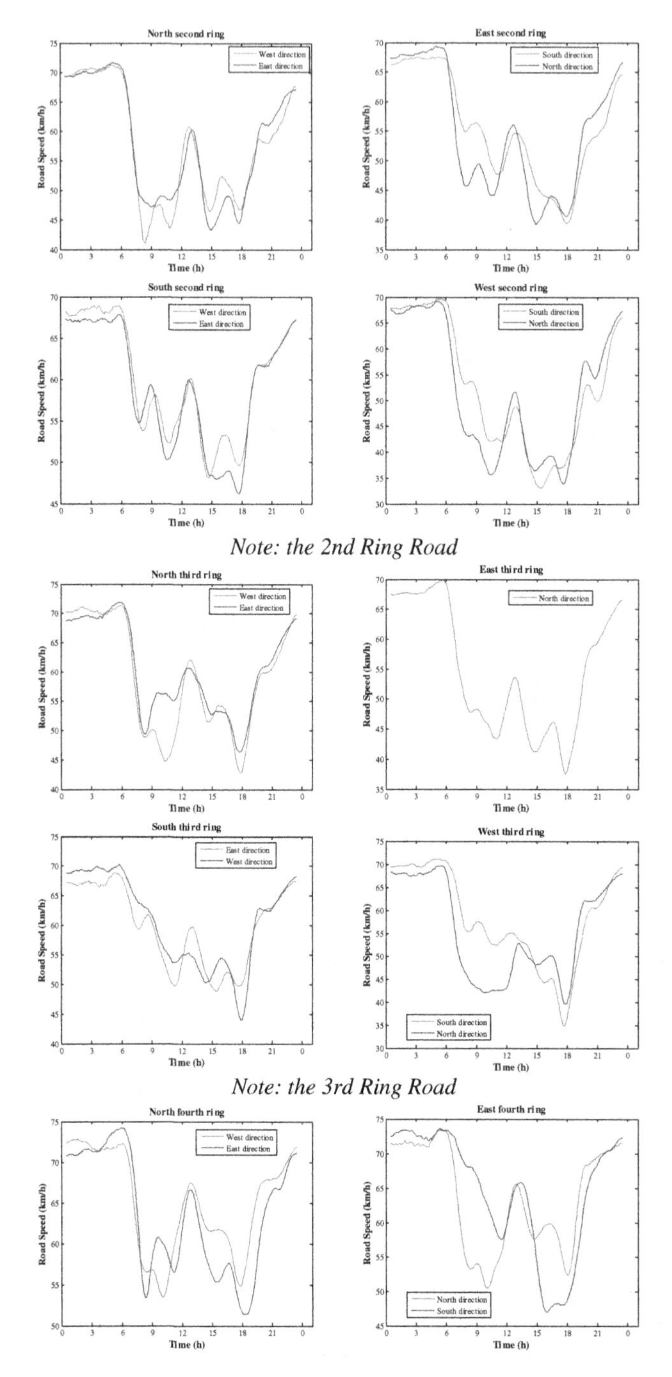

Fig. 3. Average speed of 24 h on a two-way ring road

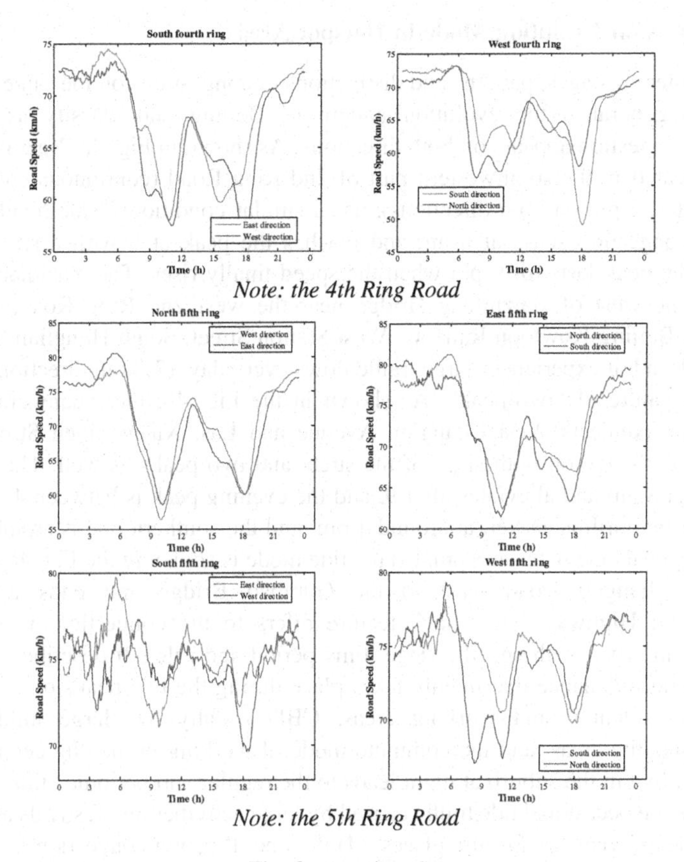

Fig. 3. (continued)

here, as well as on the western part of the 2nd ring road. Differently, the eastern part always sees a slower speed on the side to north than that to south, while the two directions on the southern part are largely similar. All parts of the 3rd ring road has the "two-peak" features, and the western part has the tide pattern while the other parts have similar variation trend on both directions. All parts of the 4th ring road has the "two-peak" features, and the eastern and western parts have the tide pattern while the other parts have similar variation trend on both directions. All parts of the 5th ring road has the "two-peak" features, but the average speed is much faster than the other ring roads mentioned. In all, all ring roads have the "two-peak" mode while it is easy for the tide mode to appear on the eastern and western part due to fact that most labor markets are located in the northern part of the city.

3.2 Analysis on Evolution Mode in Hotspot Areas

Based on different stages, patterns and distinctions of congestions on the same road, we can further generalize the evolution pattern of Beijing and classify as follows: (1) all-day congestion appears on both directions; As shown in Fig. 4a, Yaziqiao Road, which is located in the southwestern part of 2nd Ring Road (continuous congestions from 8 am to 22 pm on both directions), has a similar condition for both sides of the road—the congestion starts at 6 am and reaches the peak at 7 with a speed below 20 km/h. The peak lasts till 9 pm when the speed finally rises. The Yaziqiao Road is located to the west of Baizhifang Bridge near the west 2nd Ring Road, and as a junction for Beijing-Kowloon Railway, West Station Street, South Honglian Strret and Lianhuahe Road, it experiences large traffic flows every day. (2) both directions of road have same feature of "two-peak". As shown in the Fig. 4b, the Xuanwumen Inner Street, which connects West Chang'an Avenue and East Xuanwumen Street, has a similar speed change on both sides of the street and two peaks as well. The morning peak starts at 8 am and alleviates after 9, and the evening peak is between 4 and 6 pm (the side to the north is earlier, at around 4 pm, and the southern one is around 6 pm). The evening peak alleviates at 8 pm. (3) the tide mode is shown in the Fig. 4c using the example of Jianguo Road—east to the Guomao Bridge and ends up at old Beijing-Yushu Highway. The "tide" feature refers to the congestion on only one direction of the road with specific stage, time period and rules. It describes a kind of spatiotemporal imbalance that mainly takes place during the two peaks on main roads connecting residential and working areas. CBD usually has large influence on peripheral satellite areas, and the commute mode of working in the city center during daytime and live in the suburb at night leads to the regular and periodic "tide" mode in traffic. Also, the occasional tide traffic caused by large activities and festivals also exists due to residents' visits to specific places. (4) the one-direction congestion is shown in the Fig. 4d with the example of Shuanglong Road—which lies between east 3rd Ring Road and 4th Ring Road. The congestion on the side to the west starts from 8 am and ends at 8 pm, while the east side barely suffers from congestion. The reason is that Shuanglong Road is a vital part for Beijing-Harbin Highway, which brings much pressure on the specific side of the road.

3.3 Regional Differentiation Characteristics and Effect Mechanism

Through the recognition of local Moran I index, similar characteristics can be seen shared by the areas with morning and evening congestions in Beijing, which mainly locate inside the 3rd Ring Road, especially the north-south streets are in continuous congestion in the morning peak. Moreover, the congested streets located outside the 4th Ring Road mainly are roads into the city as Daxing City Road, Tongzhou City Road, as Fig. 5 shows below. The traffic operation pattern and congestion mode in Beijing are the results of several prospects: firstly, early as 2005, Beijing's overall urban planning determined the strategic goal of restricting the increase of downtown population and enhance regulation of land used as well as the core economic function including finance, commerce and trade. Such plan results in the enhancement of economic

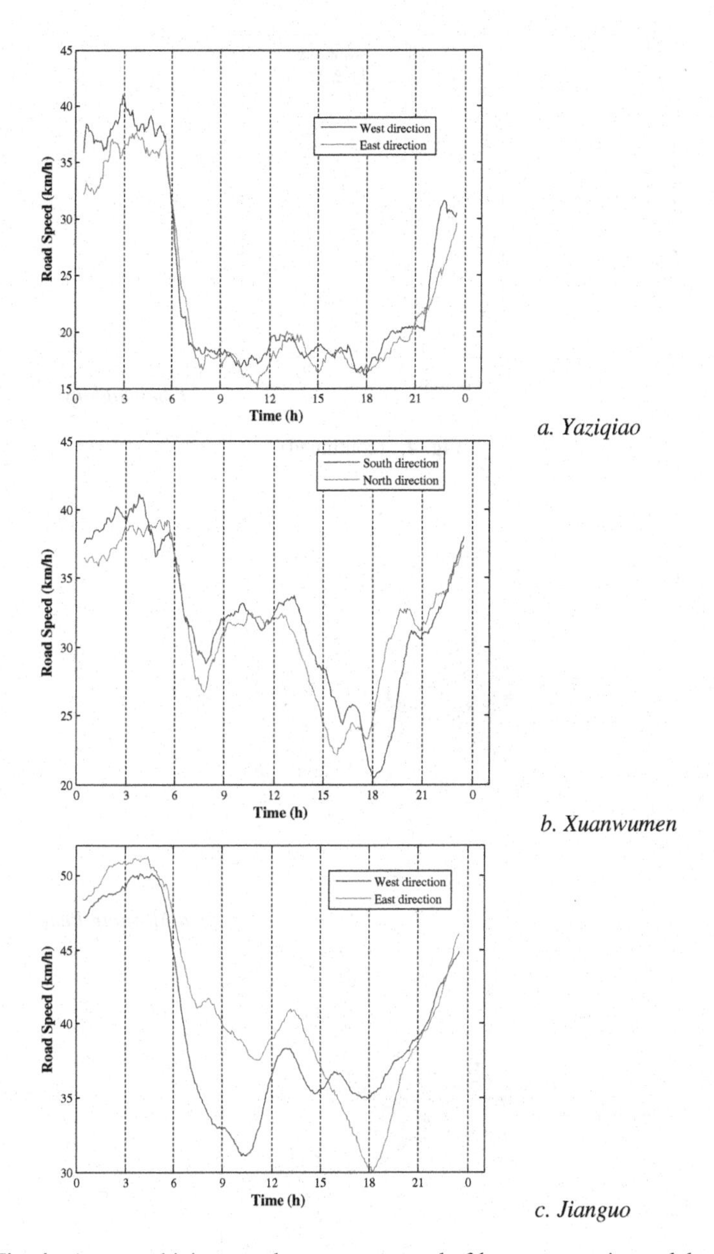

Fig. 4. Average driving speed on two-way road of hotspot areas in weekdays

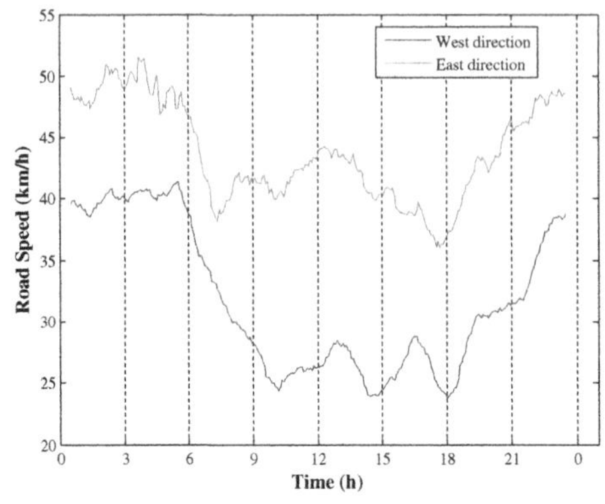

d. Shuanglong

Fig. 4. (continued)

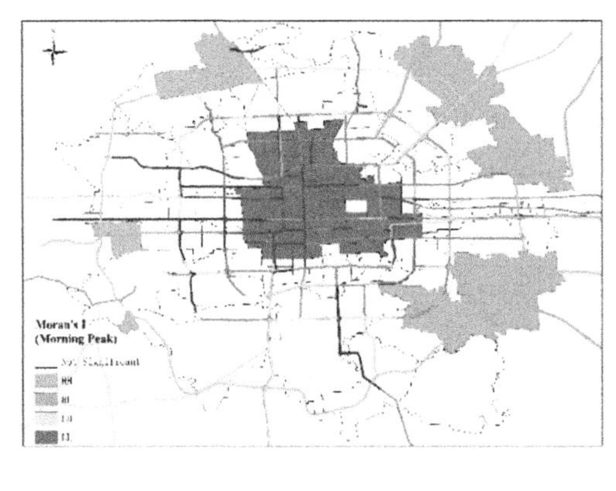

a:Morning Peak

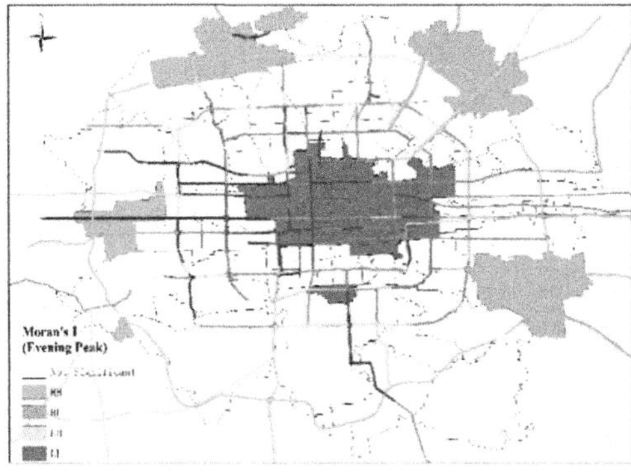

b:Evening Peak

Fig. 5. Congestion pattern from local Moran I

function, the growth in industry and the increase in the density of building land. In addition, as to the plentiful residential areas built outside the city, the total population

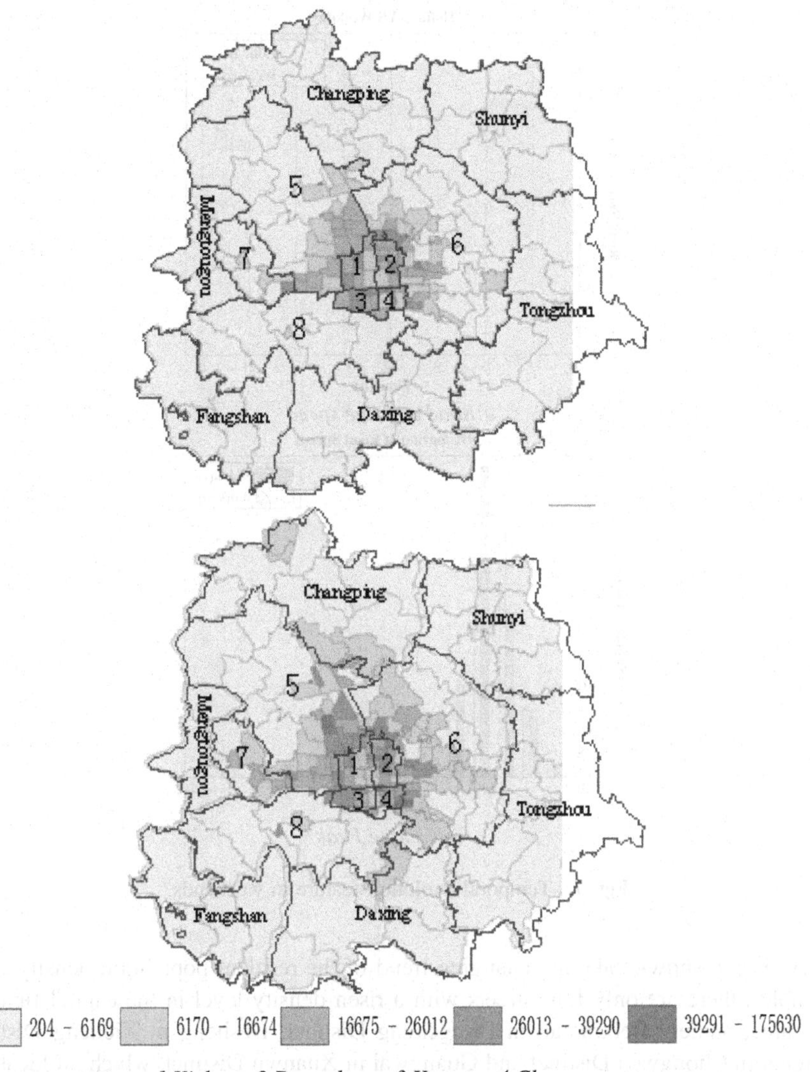

1.Xicheng;2.Dongcheng; 3.Xuanwu;4.Chongwen;
5.Haidian; 6.Chaoyang; 7.Shijingshan; 8.Fengtai;

Fig. 6. Total population distribution (Left 2000, Right 2010)

of Wangjing, Tiantongyuan and Huilongguan is equivalent to a medium-sized city. However, due to the lack of forecast and evaluation to the traffic effect, corresponding industrial planning and commercial activities, the cities turn out to become 'sleeping cities'. Therefore, an outward movement of living functions facilitates the separation of workplace and residence and tidal traffic phenomena, which causes traffic problem to deteriorate.

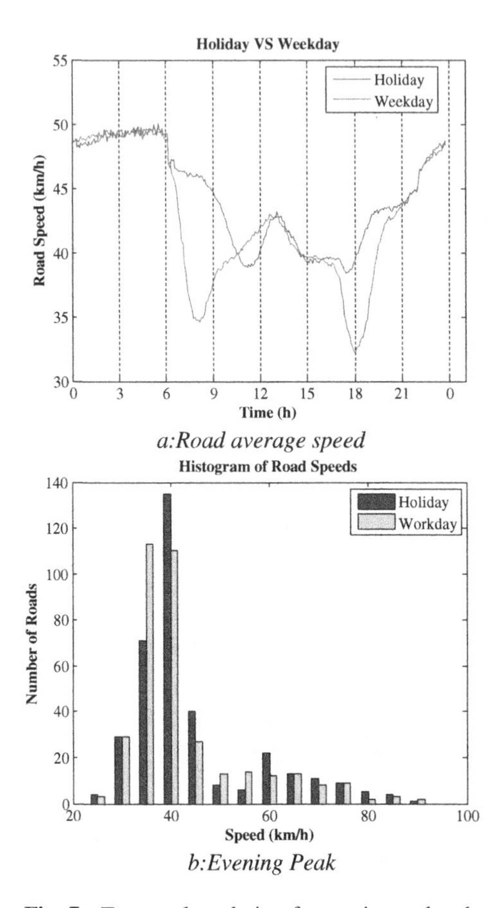

Fig. 7. Temporal evolution feature in weekends

As Fig. 6 shows (take the changing trend of the resident population density as an example), there are only four streets with a risen density level in the capital function core area, namely Beixinqiao in Dongcheng Distinct, Desheng in Xicheng Distinct, Longtan in Chongwen Distinct and Guangwai in Xuanwu Distinct, which all locate on the edge of the second ring road. In addition, the level of population density in the Stadium Road and Qianmen Streets decreased, in which the Qianmen Streets leveled down for two degrees. Since the government introduced the general plan for restoration improvement in the Qianmen District, the work has been accomplished gradually by having large buildings built and large number of residents directed to move out of the district. The city function expansion area, which includes 9 streets of Haidian District, 12 streets of Chaoyang District and several streets in Shijingshan and Fengtai District, now becomes the main area with a rising population density level. In other areas, districts as Changping and Huilongguan has their population density leveled up, and the rest remain stable. Thus it can be seen that most residents are moving outward and have formed new aggregation areas. Plus, there is no necessary traffic corridor established between the city center and outward residential areas, and only the east 3rd Ring

Road, east 4th Ring Road, west 3rd Ring Road and west 4th Ring Road are working as arterial traffic, which worsens the traffic congestion issues.

In addition, Beijing is now in the period of fast development of private transportation, with a ratio of private transportation usage higher than some of the cosmopolitan cities in developed countries, thus the conflict between the demand of private transportation and the supply of road traffic has become more serious. Till 2010, the vehicle inventory in Beijing has reached 4.809 million, which occupies 5.2 % of domestic inventory. Related policies have been introduced by the government of Beijing, from the vehicle restrictions based on license plate number to the car registration lottery, mainly to limit the inventory of private transportation in Beijing as well as the ratio of trip by private transportation. However, the policies forced citizens to buy more than one car to break the limitation, thus actually add the inventory. Moreover, there is a lack of foresight during the formulation of policies. For instance, the government considers the automobile industry as the breakthrough of modern manufacturing business, while gaining large amount of economic development, the easing policy of vehicle usage also has caused the sharp increase of number of cars.

4 Discussions

The Fig. 7 illustrate the contrast features of time-order evolution between driving speed during vacations and that of weekdays, and the "two-peak" mode can apparently be seen from the traffic operation during vacations in Beijing. Different from the weekday morning peak, which starts at 7:30 and ends at 11, morning peak during vacation is postponed to 11-12 am, and the night peak appears shorter in 5 pm. In Fig. 7, the contrast features of the normal distribution curves for average speed in both vacations and weekdays are shown. The axis of the data for vacation days locates at the point of 40 km/h, which is much higher than that of the workdays, and the frequency to the left of axis declines while that to the right rises.

Affected by family visits, shopping, self-driving tour and the cancellation of traffic controls during vacations, main transportation junctions, tourist attractions, business districts, main national roads, travel routes and four highways connected Beijing and Tibet, Chengde, Hong Kong, Macau and Kaifeng all suffer from large traffic pressure, while roads in nearby places may also witness traffic congestions. Also, some temporal traffic control measures may also influence social traffic during vacations. As shown in the figures, comparing the spatial distributions between the traffic speed of vacation days and that of weekdays, tourist attractions as Xiangshan Mountain, the Summer Palace, the Palace Museum and the Beihai Park, other areas for large activities, and business districts will become hotspot areas with large traffic.

5 Conclusions

(1) The temporal evolution feature of the areas studied throughout a day shows an apparent "two-peak" mode in Beijing urban areas, in other words, there is a marked feature of morning peak and evening peak. The spatial distribution feature of the

driving speed in the areas studied shows a slow speed on the ring roads and entrance route into Beijing. In general, morning peaks and evening peaks can be seen on every sector of the ring roads, while a "tide" phenomenon appears in the vehicle flows on the eastern part and the western part of the ring roads.

(2) According to different traffic speed in various time periods and directions, four modes of traffic are concluded: with same modes on both directions, congestion is an all-day feature; with similar modes on both directions, peaks occur in mornings and nights; tide mode; and one-direction congestion. The four modes have provided reliable evidence for setting comprehensive traffic control and management measures to solve various types of congestion issues based on specific regional conditions.

(3) Emphasize the importance of real-time report and alert of traffic condition, especially value the utility of apps on mobile phones. When serious congestion and extreme weather take place, relevant government departments must decisively take actions as temporary traffic control and restriction on visitor number in tourist attractions. Meanwhile, the congestion during vacations is different from the usual congestions which happen because demands cannot be met efficiently, and the vacation congestions usually appear as instant peaks. Such situation requires a larger effort of traffic dispersion by traffic management departments, which include technical supervision, on-site dispersion and alert beforehand.

Acknowledgment. The authors would like to thank the support of the National Natural Science Foundation of China (Study on Pre-qualification Theory and Method for Influences of Disastrous Meteorological Events, Grant No. 91224004), the youth talent plan program of Beijing City College (Study on Semantic Information Retrieval of Decision Analysis of Emergency Management for Typical Disastrous Meteorological Events, Grant No. YETP0117) and the National Natural Science Foundation of China (Key Scientific Problems and Integrated Research Platform for Scenario Response Based National Emergency Platform System, Grant No. 91024032) and Key Program of the National Natural Science Foundation of China, No. 71433008 and Key Research Program of the Chinese Academy of Sciences, No. KZZD-EW-06-04.

References

1. Xu-xuan, X., Shi-qiu, Z., Ru, Y., et al.: The social cost of transportation congestion in Beijing. China Popul. Resour. Environ. **21**(1), 28–32 (2011)
2. Chen, X.: A Dissertation Submitted in Partial Fulfillment of the Requirements for the Degree of Doctor of Philosophy in Management. Huazhong University of Science and Techology (2005)
3. Mao, T.: Research on Efficiency Evaluation of Urban Traffic Network Based on Analytical Perspective of Space—Taking Xi'an as an Example. Chang'an University (2013)
4. Ibrahim, A.T., Hall, F.L.: Effect of adverse weather conditions on speed-flow-occupancy relationships (1994)
5. Knapp, K.K., Smithson, L.D.: Winter storm event volume impact analysis using mul tiple-source archived monitoring data. Transp. Res. Rec. J. Transp. Res. Board **1700**, 10–16 (2000). Transportation Research Board of the National Acad-emies, Washington DC
6. Keay, K., Simmonds, I.: The Association of rainfall and other weather variables with road traffic volume in Melbourne, Australia. Accid. Anal. Prev. **37**(1), 109–124 (2005)

7. Smith, B.L., et al.: An Investigation into the Impact of Rainfall on Freeway Traffic Flow. Transportation Research Board, Washington, D.C (2004)
8. Jiang, B.: Street hierarchies: a minority of streets account for a majority of traffic flow. Int. J. Geogr. Inf. Sci. **23**, 1033–1048 (2009)
9. Jing, G.: Floating Car Data-based Measures and Approach to the Assessment of Road Traffic Operations for Beijing. Beijing Jiaotong University (2006)
10. Xi-rong, G.: Study of Spatial and Temporal distribution of the Urban traffic Based on GPS Floating car data. Kunming University of Science and Technology (2013)
11. Jayakrishnan, R., Mahmassani, H.S., Hu, T.-Y.: An evaluation tool for advanced traffic information and management systems in urban networks. Transp. Res. **2**(3), 129–147 (1994)
12. Mahmassani, H.S.: Dynamic network traffic assignment and simulation methodology for advanced system management applications. Netw. Spat. Econ. **1**:**2**(3), 267–292 (2001)
13. Beijing Municipal Bureau of Statistics. Beijing Statistical Yearbook (2000–2013). China statistics Press, Beijing (2013)
14. De-sheng, L.: 2014 China Labor Market Development Report: Working Time during the process of becoming a high-income country. Beijing Normal University Press (2014)

A Study of the Methods for the Division of the Boundary Between Urban and Rural Areas Based on Multiple Conditions: A Case Study of Xi'an

Lei Fang[1,2](✉) and Yingjie Wang[1]

[1] State Key Lab of Resources and Environmental Information System, IGSNRR, CAS, Beijing 100101, China
{fangl.13b,wangyj}@igsnrr.ac.cn
[2] University of Chinese Academy of Sciences, Beijing 100049, China

Abstract. Based on summarizing the research on the division of urban and rural areas, this paper defines the attributes of "cities" and "villages". By using the method based on multi-layer condition, in combination with cumulative percentage and other mathematical methods, adopting the spatial comparison method by using Google Earth satellite images, the optimal threshold is decided through multi-condition screening and then the definitional standards are confirmed. This paper divides the urban and rural areas by selecting Xi'an city as the research area. The divided results compare to the results concluded by the National Bureau of Statistics of the People's Republic of China. It is found that the urbanization rate is reduced by 3.24 %. Through overlapping comparison of satellite remote sensing images, it is proven that the feasibility and sophistication of the methods for the division of urban and rural areas through screening multiple conditions.

Keywords: Boundary division between Urban-Rural area · Connotation · Multi-conditional judgment · Cumulative percentage · Satellite remote sensing image

1 Introduction

Since the 1990s, China's urbanization progress has entered a fast growth period. Among various problems caused by "the fast urbanization", the blurred boundary between urban and rural areas is a basic problem.

Since the founding of the People's Republic of China, China has continuously changed the standards of dividing urban and rural areas and statistical caliber calculating the urban population, lacking a consistent definition and standards to calculate urban region and urban population. Due to the inconsistent standards on defining urban setting, urban region and population in China, the city concept used in China is different from international urban definitions [1, 2]. In particular, after the 1980s, many Chinese cities and regions have changed administrative regions. Due to unreasonable and inconsistent

F. Bian and Y. Xie (Eds.): GRMSE 2015, CCIS 569, pp. 48–59, 2016.
DOI: 10.1007/978-3-662-49155-3_5

urban and rural division, the urbanization level calculated based on urban population statistics could not truly reflect the actual development and its features. It is theoretically and realistically significant to explore a set of fast, effective and accurate indicator system aligning with the regional development and technical methods for urban-rural division.

Targeted at the above problems, this paper defines the "cities" and "villages" based on the connotation of urban and rural areas. Starting from the different spatial features and attributes between cities and villages, this paper selects indicators from the geospatial, economic, demographic and social fields based on multiple urban and rural attributes. The threshold is decided through such methods as linguistic screening, cumulative percentage, and satellite sensing image comparison method, and then urban-rural boundary division standard is decided. Also, the feasibility of method and indicator system is tested by selecting Xi'an as a research area.

2 The Concept Definition and Attributes of Cities and Villages

2.1 The Differences and Definition of Cities and Villages

The gist of urban-rural boundary division is why and how the boundary between cities and villages divides. The first task of solving this problem is to understand the connotation and attributable characteristics of cities and villages, as well as their differences, and then to find the ways to divide urban and rural areas.

Different organizations and research areas define the connotation of "cities" and "villages" differently. Scholars in the world give different definitions on cities and villages.

Urban geographical scientists believe that cities are residential areas based on non-agricultural population and reaching a certain population scale. Cities are a comparably permanent large settlement, emphasizing urban spatial features. Geologists believe that cities are a macro-phenomenon happened on earth, with certain spatial, regional and comprehensive characteristics. Cities are a centralized region gathering population engaging in the secondary and tertiary industry. Cities are investment points and clusters of national economic space and labor population [4]. Since the 1930s, the research on rural and urban areas in the academic community has generally focused on the respective economic functions of urban and rural areas. It's generally believed that rural areas are less-populous places engaging in agricultural production. By contrast, cities are populous places engaging in non-agricultural production. Cities are an economic, political, transportation, cultural and information center in a certain region. According to the Third Article of the Regulations on Statistical Classification of Urban and Rural Areas approved by the National Bureau of Statistics in 2008 (the Regulations), these Regulations divide Chinese regions into urban and rural areas based on Chinese administrative division, targeted at the communities under the jurisdiction of Committee of Residents and Committee of Villages, and also based on the actual construction. Actual construction refers to public facilities, residential buildings and other facilities that have been built or plan to build.

Through summarizing the literature, it could be concluded that the essential differences between urban and rural areas lie in that: cities and towns are large and centralized residential settlements based on industrial, commercial and other non-agricultural economic activities in city and township forms. Cities are normally a political, economic and cultural center in a certain region. By contrast, rural areas are based on agricultural economic activities featured as decentralization. The differences between urban and rural areas reflect in population distribution, cluster scale, transport, infrastructures and building aspects. These different attributes provide a theoretical basis for selecting indicators in this study.

2.2 Perception of Urban and Rural Attributes and Principles for Selecting Indicators

Urban and rural attributes are comprehensive indicators reflecting urban and rural differences in an area or region, quantifying the urban and rural division in this area or region. Previously, scholars classified urban and rural areas by urban functions. Among such classification, the division by urban economic and landscape functions is a valuable angle. In most of countries, the division of urban and rural space is based on economic and landscape functions. However, such division is subjective of nature and depended on some methods and indicators. Chinese scholars investigate and explore functional space division from different research purposes, scales, regions and angles. They propose different concepts, such as spatial function [6, 7]; regional function [8]; dysfunction [9] and specific function [10].

This study finds out the differences among different urban and rural definitions by analyzing different definitions. The indicators for assessing urban and rural attributes are selected based on the following principles: (1) Comprehensive principle; (2) Special principle; (3) Available principle; (4) Operable principle.

2.3 Establishing Indicator System

Following the above principles, this paper selects five indicators (i.e. Committee of Residents census block, population density in census block, census community density in census block, POI density, and road network density) from population, economic and social factors. Firstly, through linguistic screening, the census units containing the name "Committee of Residents" are selected as absolute urban areas. Among five indicators (i.e. Committee of Residents census block, population density in census block, census community density in census block, POI density, and road network density), if all these conditions are satisfied, such areas are 100 % urban areas. If four indicators are satisfied, such areas are 80 % of urban areas. If three indicators are satisfied, such areas are considered as 60 % of urban areas. According to this standard, the census unit is less satisfied the conditions, its urbanization level is lower (Table 1).

Table 1. Classification of urbanization attribute indicators.

Principles	Notes on urban and rural attributes
Linguistic screening	The census units containing the name "Committee of Residents" or "Committee of Villages" are selected as absolute urban areas.
Conditions to judge census blocks	The density of permanent resident population based on census block
	The density of census community is calculated based on census blocks
	POI density is calculated based on census blocks.
	Road network density is calculated based on census blocks.

3 Methods for the Division of the Boundary Between Cities and Villages

3.1 Technical Route

The technical route is illustrated in Fig. 1 for dividing urban and rural areas based on multiple conditions. Through literature review and the field visit of research areas, as

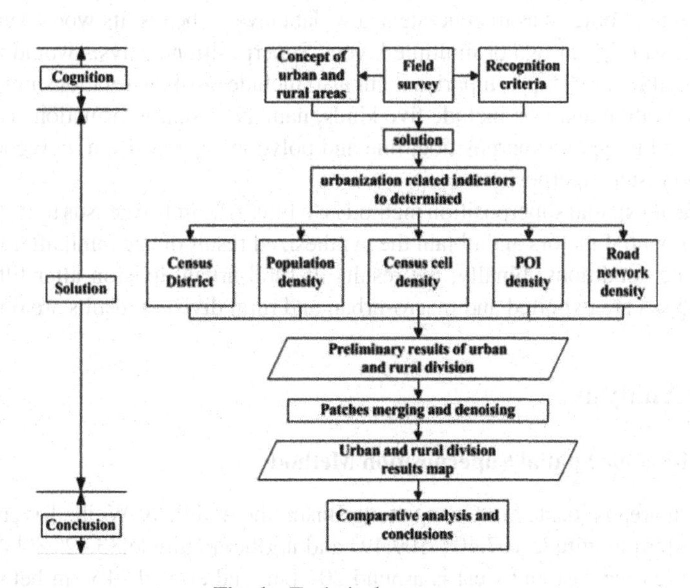

Fig. 1. Technical route map

well as the availability of data on five indicators (i.e. Committee of Residents census block, population density in census block, census community density in census block, POI density, and road network density), after the threshold decided, the result map of preliminary division of urban and rural areas is generated. Using Arcgis software, polygons are combined and de-noising is processed. The result compares with the result made by the National Bureau of Statistics (Fig. 1).

3.2 Deciding the Threshold to Judge the Boundary Between Cities and Villages

Combining quantitative and qualitative methods, this paper confirms city and town division standards. Through random sampling and removing data abnormalities, the histogram of urban attribute frequency is generated. According to the statistical results of sample data, the indicator thresholds are found and the range interval of urban and rural boundary based on multiple conditions are obtained preliminarily.

Regarding cumulative percentage, after grouping by an indicator, number of data distributed in each group is called frequency or number of times), while the percentage of each group in the sum of all percentages is called percentage or proportion. For the purpose of statistical analysis, we need to observe the sum of frequency of data above or under a certain value, called cumulative percentage or the total of a frequency. Cumulative upward refers to cumulative from smaller value to larger value, otherwise, called cumulative downward. The total cumulative value is 100 %.

3.3 Multi-Factor Spatial Superposition Method

Superposition analysis is one of the most common methods to extract hidden spatial information. Superposition analysis in geographic information system is to overlap data layer related to a topic so as to generate a new data layer. The results would synthesize all attributes of original two or multiple layers. Superposition analysis would not only include spatial relationship comparison, but also include attribute relationship comparison. Superposition analysis include five kinds, namely, visual information superposition, point and polygon superposition, line and polygon superposition, polygon superposition and raster superposition.

Multi-factor spatial superposition method is to filter researched census units by using multiple influential factors and obtain the synthesized result of each unit after overlapping multiple conditions. Finally, the results of rural-urban division after filtered by multiple factors are exported and macro-urban and rural division results are obtained.

4 Case Analysis

4.1 Multi-Factor Spatial Superposition Method

The research area is located in Guangzhong Basin, the middle of Weihe River Valley, between eastern longitude 107.40°~109.49° and northern latitude 33.42°~34.45°. The length between the east and west is around 204 km, and around 116 km between the south and north. The size is 9983 m^2, including 1066 m^2 of city area. By the end of 2012, Xi'an had 9 districts (i.e. Xincheng, Beilin, Lianhu, Yata, Weiyang, Baqiao, Yanliang, Lintong and Zhangan districts) and 4 counties (i.e. Zhouzhi, Lantian, Huxian and Galing counties). It had 176 streets, villages, and towns, including 89 street offices, 40 towns and 47 villages (Fig. 2).

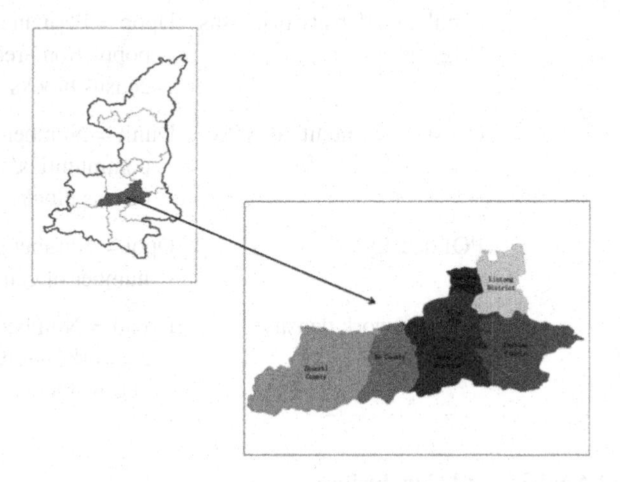

Fig. 2. The study area map

4.2 Data Preparation and Processing

Data used in this research include census block and community units under Xi'an jurisdiction (including 3,713 census block unit and 35,368 census community units). Data also include satellite remote sensing images, navigation road network data and POI data covered under Xi'an jurisdiction. Data processing includes attribute data pre-processing and spatial data pre-processing. Attribute data processing includes statistical analysis by using EXCEL and SPSS tools. Spatial data is processed by using AGCGIS and ENVI to normalize remote sensing images and vector data. Finally, image data and vector data are counted to match up with the results.

In this study, the urban-rural division method is based on the perception of urban-rural meanings, along with urban-rural attribute indicators. Main research unit is the smallest unit according to administrative division, namely urban-rural classification standards under Committee of Residents and Committee of Villages. Each selected indicator data is normalized to eliminate dimension influence. Selected data are summarized in Table 2:

Table 2. Urban and rural division indicators

Principles	Indicators	Data source and note
Linguistic Screening	Committee of Residents census block	The 6th Census Data in 2010
Conditions	population density in census block	Dpop = Permanent resident population/areas of census blocks
	Census community density in census block	Dunit = Number of census communities/number of census blocks
	POI density	Dpoi = Number of POI/ number of census blocks
	Road network density	Droad = Number of road network/number of census blocks

4.3 Empirical Analysis and Conclusions

Calculating City-Village Attribute and Deciding Indicator Standards.
(1) Extraction based on linguistic meaning "Committee of Residents" census blocks

According to the file name in vector data on Xi'an census blocks, "Committee of Residents" census block units are selected firstly. These census blocks are 100 % of urban areas because their name has urban attribute.

Overlapping remoting sensing images expands the scope of main city zones. It is found that "Committee of Residents" census blocks cover urban centers, aligning with the characteristics of urban areas.

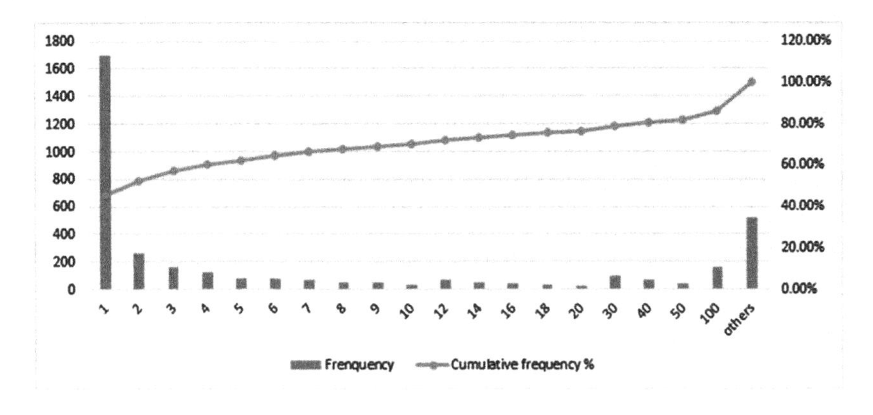

Fig. 3. The histogram of population density in census areas

(2) Decided by population density threshold in census blocks

Using cumulative percentage map, the population data in Xi'an census block are analyzed. According to 19 grades of cumulative percentage (50, 150, 350, ……, 5000, 10000, 3000), the histogram is generated (Fig. 3).

The frequency of population density and cumulative value in census blocks are calculated based on Xi'an census blocks. The population density = 1100, 1200, 1300 are selected at the turning points in the Figure and population density map is generated. The different population density overlapping results are overlapped with Google Earth images. It is found that the building coverage and population density = 1200 in Xi'an main city zones is close to the urbanization development scope. Thus, population density = 1200 is selected as indicators to judge urban census blocks.

(3) Deciding the threshold of other indicators

Following the above methods, the thresholds of census community density, POI density and road network density indicators are obtained. Given the above analysis, the urban-rural division standards are listed in Table 3 based on the turning points on the cumulative percentage chart and multi-factor spatial superposition method.

Table 3. Multi-conditional standard for screening the boundary between urban and rural areas

Indicator	Meaning	Features	Threshold
Committee of Residents census block	Meaning (Committees of Residents, Communities etc.)	Directly consider as cities and towns	Considered as the scope of cities and towns
Population density in census block	Population	Maxvalue: 616305 Min value: 0 Meanvalue: 6478.19	1200
Census community density in census block	Number of census communities in census blocks	Max value: 2133.96 Min value: 0 Mean value: 26	5
POI density	Service infrastructure	Max value: 1787.61 Min value: 0 Mean value: 58	10
Road network density	Transportation infrastructure	Max value: 60480.1 Min value: 0 Meanvalue: 3706.17	3500

Empirical Results. According to the division standard in Table 3, the research area is divided into urban and rural areas by using GIS software. The results of urban and rural division and grading are shown in the following Figure.

As shown in Fig. 4, it could be found that the following areas meeting five conditions, which are main city zones in Xian, and main satellite towns in Zhouzhi county, Hu County, Lantian County, Lintong County, Yanliang County and Gaoling County. With diffusing city/township centers, less and less areas meet the conditions, the level of urbanization becomes lower and lower.

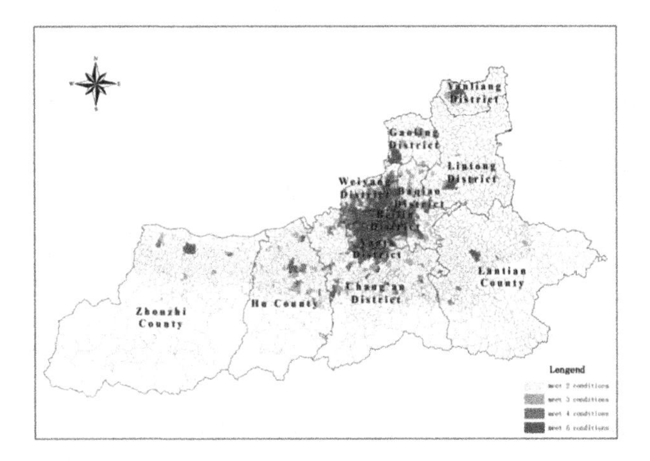

Fig. 4. Superposition results.

The data unit (i.e. surveyed data on census blocks) would report to higher administrative units by following a bottom-to-up approach. Census community shall be a closed and complete region without duplication and omission. They are spatial polygon data with a full coverage. Thus, it is important to keep the integrity of census community

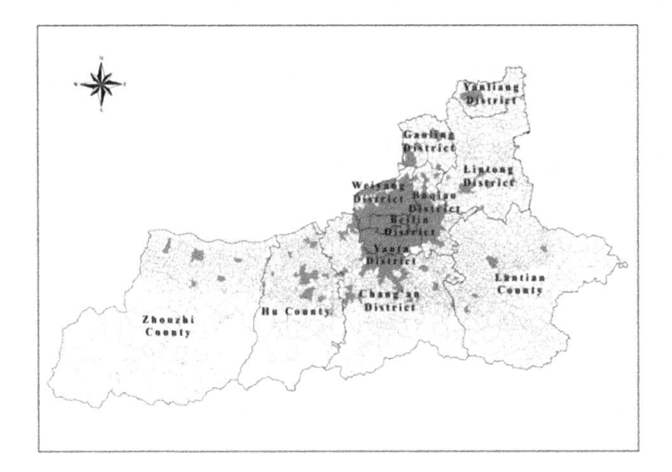

Fig. 5. The division of urban and rural areas based on multiple conditions.

data. The final result of urban and rural division is presented in Fig. 5 through making up missed parts in main city zones, linking the periphery of main city zones, processing discrete polygon image objects of the border of cities and villages, de-noising process of discrete polygon objects of rural are a sand patch processing of surrounding areas of satellite towns.

Compare the Division Results by National Bureau of Statistics. The above result of urban-rural division based on multiple conditions compares with the result made by the National Bureau of Statistics in 2012. The urbanization level is presented in Table 4.

Table 4. Comparing the statistical methods and the research results.

Discrepancy indicators	Division results (cities and towns) by the Plan	Division results by the National Bureau of Statistics (cities and towns)
Number of census blocks	1257	1370
Total area of cities and towns	957.34 km^2	1579.21 km^2
Population	5429354	5698970
Urbanization Rate	65.36 %	68.6 %

From Fig. 6, it could be found that the urban-rural division based on multiple conditions is smaller than the division by the National Bureau of Statistics in 2012. In the main city zones in Xi'an (i.e. Weiyang District, Lianhu District, Xincheng District, Yanta District and Baqiao District), the urban attributes are evident with many overlapped parts. So, the screening result is good. The city scope in other

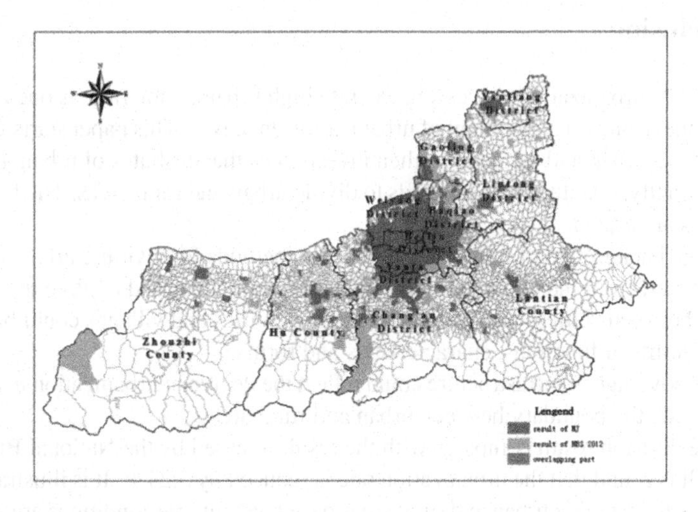

Fig. 6. Overlapping the research results with the division results by the National Bureau of Statistics

suburbs and counties are overlapped, and so these areas are considered as urban areas as well.

The division results based on multiple conditions are overlapped with the division by the National Bureau of Statistics in 2012. As illustrated in Fig. 7, Houzhenzi Village Committee in the south of Zhouzhi County and the south of Chang'an district have census blocks different from the division by the National Bureau of Statistics. Adopting the satellite image superposition method, it could be found that Houzhenzi Village Committee and Dasixin Village are mountainous areas, with limited urban areas. Their infrastructure could not reach urbanization level. Thus, they are not urban areas according to the conditions proposed in this study.

Fig. 7. Satellite image of a typical sample area

5 Conclusion

In China's fast urbanization process, inaccurate high urbanization ratio is one of issues to hamper the healthy development of urbanization endeavor. This paper starts from the connotation of urban and rural areas, then investigates the attributes of urban and rural areas, and finally, concludes the methods to divide urban and rural areas. The following conclusions are drawn:

(1) In the background of fast urbanization, the methods for dividing urban and rural areas must start from the connotation of urban and rural areas and explore the essence difference between "cities" and "villages". As such, cities and villages could be distinguished in terms of functions, characteristics and forms.

(2) The coverage of infrastructure could reflect the degree of urbanization in a region and fast divide the boundary between urban and rural areas.

(3) The division result compares with the result released by the National Bureau of Statistics. It is found that the urbanization rate is reduced by 3.24 %. It is illustrated that the methods for dividing urban and rural areas based on multiple conditions are accurate than current prevailing methods.

This study is significant because it firstly explores the connotation of "cities" and "villages" from linguistic perspective, and then establishes the indicators and methods to fast divide urban and rural areas screened by multiple conditions and based on urban construction attributes. This method could provide a more scientific technological support for government authorities to calculate urbanization ratio and formulate urbanization strategies and policies in the process of integrating urban and rural areas.

Acknowledgments. Foundation: Under the auspices of National Key Technology R&D Program of China (No. 2012BAI32B07)

References

1. Wu, F., Webster, C.J.: Simulation of land development through the integration of cellular automata and multicriteria evaluation. Environ. Plan. B **5**(1), 103–126 (1998)
2. Zhou, Y., Si, Y.: Establishing the concepts of physical urban area in China. J. Geog. Sci. **04**, 289–301 (1995)
3. Zhou, Y.: Urban Geographic Exploration. Zhou Yixing Collection. Commercial Press, Beijing (2001)
4. Wang, Y.A.: Study on China's Criteria for Urban-Rural Division. Dissertation from Southwest Jiaotong University (2012)
5. Jin, X., Wang, H., Tu, J., Zhang, Z.: To assess comprehensive urban environmental quality with multi-factor space superposition assessment method. Admin. Technol. Environ. Monit. **03**, 561–570 (2009)
6. Chen, W., Duan, X., Chen, J.: Methods for dividing spatial development functional zones. J. Geog. Sci. **S1**, 53–58 (2004)
7. Xie, G., Lu, C., Zhen, L.: The purpose, progress and method for regional spatial functional zone division. Geog. Sci. Res. **03**, 561–570 (2009)
8. Jiang, A.: Main functional zone: new frontier and new practice of regional development. Theory Dev. Res. **02**, 14–17 (2007)
9. Wang, C., Zhao, H., Sun, G.: Optimizing main functions to develop functional region division in counties: a case study of Shangyu. Geog. Sci. Res. **03**, 481–490 (2010)
10. Liu, C., Zeng, J.: Regional spatial supply-demand model and spatial structural optimization route selection – scientific basis for functional zone construction. Econ. Geol. **01**, 26–30 (2009)
11. Ni, X., Yang, Q., Liu, X.: Geographical simulation system: cellular automation and multi-agent. Scientific Press, Beijing (2007)
12. Ni, X., Yang, Q., Liu, X.: Knowledge mining and plan simulation based on CA urban evolution. China Sci. D Volume Earth Sci. **37**(9), 1241–1251 (2007)
13. Zhu, C., Shi, W.: Spatial Analysis Modeling and Theory. Scientific Press, Beijing (2006)

Empirical Analysis of the Impact Which the Shared Resources Have on the Competitive Advantages of the Logistics Enterprise in Logistics Industry Clusters

Rong Zhou(✉)

Wuhan Polytechnic, Wuhan, China
93h999@163.com

Abstract. The paper proposes seven theoretical hypotheses after firstly putting forward the composition of the shared resources of the logistics industry cluster and then analyzing theoretically the impact that the shared resources have on the competitive advantages of the logistics enterprise in logistics industry clusters. To prove these hypotheses, the researchers prepare questionnaires and collect the questionnaire data from 201 enterprises of the logistics industry clusters. In this paper, the researchers analyze these data by using such soft wares as SPSS for windows (version 17.0) and AMOS7.0, etc. and make the conclusion that location resources, brand resources, specialized division of labor and cooperation resources, supporting institutions resources and order resource have positive impacts on the competitive advantages of the logistics enterprises, while material resources are only partially supported as to their impacts on the competitive advantages for logistics companies, and collective learning and knowledge sharing resources do not have obvious impacts on the competitive advantages of logistics businesses.

Keywords: Logistics industry cluster · Shared resource · Competitive advantage · SEM (structure equation model)

1 Introduction

In recent years, due to the growing study of the industrial clusters and the emergence of the research into the companies' competitive advantages within the industrial clusters, the perspective of researchers have extended from the enterprises' internal resources to external resources which are still within the boundaries of the clusters. One of the earliest researchers of the cluster's shared resources is Molina-Morales [1], who named them shared resources and believed that these resources include assets, capabilities and their related expertise. Comparing the performance of the Spanish tile industry clusters' internal enterprises and external ones, he found out the correlation between the clusters' shared resources and the competitive advantages of the clusters' internal companies. Geng [2] discovered that there exist, outside the cluster enterprises and within the boundaries of clusters, six shared resources, namely the collective reputation, the smooth flow of inter-enterprise resource exchange and combination of channels, the high degree

F. Bian and Y. Xie (Eds.): GRMSE 2015, CCIS 569, pp. 60–69, 2016.
DOI: 10.1007/978-3-662-49155-3_6

of mutual trust between enterprises, the collective learning and knowledge sharing networks, the close interactive atmosphere of competition and cooperation between enterprises, and the active participation and support of the local organizations. In addition, he deemed that the clusters shared resources have a close relationship with the competitive advantages of firms in clusters. However, there are significant differences among scholars in the research of the elements of the clusters shared resources. What's more, the existing studies are all based on the research into the manufacturing clusters and the high-tech industrial clusters. So the research on the elements of logistics clusters' shared resources based on the industries remains to be deepened. The object of this paper is the empirical research on the composition of the shared resources within the logistics industry clusters and its correlation with the competitive advantages of the clusters' logistics enterprises.

2 The Composition of the Shared Resources in the Logistics Industry Clusters

On the basis of previous studies and the analysis of the organizational structure of the logistics industry clusters, this paper holds the opinion that the shared resources of the logistics industry clusters contain seven kinds of resources, location resources, material resources, brand resources, collective learning and knowledge-sharing channel resources, specialized division of labor and cooperation resources, supporting organization resources, and order resources.

Location resources are the combination of the supplying of the logistics demands and logistics facilities within the region of logistics industry clusters. The supplying of the logistics demands refers to the needs of economic activities in the region to logistics activities. And the supplying of the logistics facilities refers to the transportation network and the availability of cheap lands. Material resources mean the building of a public information platform and public logistics infrastructure within the logistics industry clusters. Brand resources are not only the signs that distinguish the logistics industry clusters with other clusters, and the representatives of the potential competitiveness and profitability of enterprises in the clusters, but also the good reputation accumulated through the long-term standardized operations, the good quality products and the comprehensive services of the enterprises in the clusters. Collective learning and knowledge sharing channel resources refer to the knowledge diffusion and overflow channels resulting from geographical proximity and concentration. Specialized division of labor and cooperation resources is due when logistics companies, under the leadership of the integrated logistics service providers, jointly provide their customers with integrated logistics services and enjoy the increased labor productivity and flexible specialization capacity brought about by the division of labor and cooperation. Supporting organization resources are made up of the related organizations providing value-added logistics services and additional services, including financial institutions, customs, industry and commerce, taxation and other government agencies, research enterprises, high-tech logistics enterprises, logistics consulting firms, and even catering, accommodation and enterprises providing auto parts and auto repair services. Order resources mean through following the trading rules and management methods of the unified management departments inside the clusters, all economic entities interface and coordinate with each other, compete equally, and allocate efficiently resources so as to become a harmonious environment.

3 Theory Analysis and Hypothesie Proposal

3.1 Impact of Location Resources on the Enterprises Competitive Advantage

Firstly, since the costs would be too high for the logistics enterprises to search for the favorable location conditions, if the land resources can be shared within the clusters, the costs to search land would be greatly reduced. What's more, because the joining of more enterprises can help sharing the costs of land use, the logistics enterprises' costs of land use would be considerably cut inside the clusters. Secondly, the more advantageous the locations of clusters are, the more the logistics enterprises can ensure the smoothness of conducting their logistic businesses, and the more they can improve the timeliness and availability of logistics services and also improve the customer service capabilities of logistics. Last but not the least, favorable location resources mean a large number of big industrial enterprises and retail enterprises are close to the clusters, which can bring plenty of client resources to clusters' logistics enterprises and thus improve their marketing capabilities. In summary, this paper proposes the following hypothesis.

H1: Location resources have a marked positive impact on the competitive advantages of the logistics enterprise in industry clusters.

3.2 Impacts of Material Resources on the Enterprises Competitive Advantages

First, a public logistics information platform for public clusters can reduce the costs of the enterprises in this regard and enhance their abilities to control costs. Second, the use of information systems to integrate customers, logistics companies and suppliers can allow the flow of logistics to achieve the best purpose and economy. Furthermore, public information platform can bring together logistics need information of various enterprises both inside and outside this region and make progress the marketing capabilities of the logistics enterprises. Last, due to the sharing of the logistics infrastructure in the clusters, logistics enterprises can reduce investment in logistics infrastructure. In summary, this paper proposes the following hypothesis.

H2: Information platform resources have a marked positive impact on the competitive advantages of the logistics enterprise in industry clusters.

3.3 Impacts of Brand Resources on the Enterprises Competitive Advantages

For the cluster logistics enterprises, the role of the logistics clusters' brand resources is manifested specifically in following aspects: First of all, they can enhance the overall image of the logistics enterprises settled in the logistics park. Most logistics enterprises have limited comprehensive strengths, so it is difficult to establish their own brands. However, through sharing the clusters' brand advantages, these enterprises have reduced their advertising costs to a large extent and accordingly the operating costs, thus improving their abilities to control costs. Next, compared with the brands of individual logistics enterprises, cluster brands have smaller risk coefficient, higher values and more sustainable brand effect, which would help enhancing customers' desire and loyalty to buy the logistics products and services in the park, and enable the individual logistics enterprises inside the clusters to attain sustainable brand advantages. Finally, good cluster brands mean good atmosphere of cooperation and institutional environment. They can also encourage enterprises to innovate and enhance their innovation capability. In summary, this paper proposes the following hypothesis.

H3: Brand resources have a marked positive impact on the competitive advantages of the logistics enterprise in industry clusters.

3.4 Impacts of Collective Learning and Knowledge Sharing Channel Resources on the Enterprises Competitive Advantages

Cluster enterprises' collective learning and knowledge sharing networks, resulting from cooperation and geographical closeness, enable logistics enterprises to observe and imitate their practice to learn the relevant knowledge in cooperation with other enterprises, and to stride cross organizational boundaries to obtain the knowledge and skills of other organizations and supplement the inadequate internal knowledge and skills. Cluster enterprises can also have access to information on market and management, etc. through networks, which greatly reduces the costs of obtaining knowledge and information and expands the ranges of getting these things. Thus they can enjoy the benefits brought by the spillover and spread of the knowledge, technical know-how, information and other elements in the cluster. These benefits can make the logistics enterprises to have more market intelligence and to learn more knowledge and skills. They can also help logistics enterprises improving business methods and business processes, imitate and learn more advance and appropriate management ideas and management models, so as to improve management efficiency. In summary, this paper proposes the following hypothesis.

H4: Collective learning and knowledge sharing channel resources have a marked positive impact on the competitive advantages of the logistics enterprise in industry clusters.

3.5 Impacts of Specialized Division of Labor and Cooperation Resources on the Enterprises Competitive Advantages

Specialized division of labor and cooperation resources in the logistics industry clusters influence the competitive advantages of logistics enterprises from two aspects:

On the one hand, the division of labor and collaboration between the logistics enterprises make enterprises to concentrate on their core businesses and do good to the accumulation of business experience and the improvement of the service efficiency. The effective collaboration of each unit enhances the abilities of companies to quickly adapt to environmental changes and boost their profitability. In addition, knowledge, information and skills circulate and deliver through the relationship networks established by way of cooperation and division of labor, which can improve the innovation capacity of enterprises. Their good relationships, built up during their cooperation, provide a good foundation for the integration of the enterprises' external resources.

On the other hand, in the process of cooperation and division of labor, competitions will take place inevitably owing to the similarity of businesses. Under the pressure of competitions, logistics enterprises will try their best to surpass others and obtain the differential advantages which can distinguish them with others. So they will improve continuously their logistics services and introduce innovative measures. Companies' capabilities of customer services and innovative abilities can develop subsequently. In summary, this paper proposes the following hypothesis.

H5: Specialized division of labor and cooperation resources has a marked positive impact on the competitive advantages of the logistics enterprise in industry clusters.

3.6 Impacts of Supporting Institution Resources on the Enterprises Competitive Advantages

The supporting institutions can provide logistics network design, needs analysis, personnel training and other services for logistics enterprises to help them increase their technical and service capabilities, and cultivate their creative spirits and innovative ideas. These institutions also offer facilities for logistics companies to achieve a one-stop logistics management. What's more, the settlement of banks and other financial institutions provides convenient conditions for the financial settlement of logistics companies, and the development of logistics finance. Because the proximity of distance between each other increase exchanges of them, which help financial institutions have a better understanding of the operational status and asset sizes of logistics enterprises, and build gradually trust mechanism to reduce the corporate finance difficulties of the logistics enterprises. In summary, this paper proposes the following hypothesis.

H6: Supporting institution resources has a marked positive impact on the competitive advantages of the logistics enterprise in industry clusters.

3.7 Impacts of Order Resources on the Enterprises Competitive Advantage

The logistics industry clusters usually have a central management agency whose responsibilities are to carry out day-to-day management of the entire logistics industry clusters through the development of a unified business management system and trading rules. In this well-ordered environment, the disorderly and vicious competition between logistics enterprises can be minimized to ensure the equal status of the main bodies of the transaction, to maintain the unity of the market and to reduce the operating costs. Besides this can improve the efficiency of services and resource integration, modify the operating environment of the entire cluster, and thus realize the co-amplification effects of industry gathering. In summary, this paper proposes the following hypothesis.

H7: Order resources have a marked positive impact on the competitive advantages of the logistics enterprise in industry clusters.

4 Empirical Research

Based on the above hypothesis, this paper carries out data analysis of the relevant hypothesis by using the software AMOS7.0. Firstly we created a measurement scale for logistics industrial clusters' shared resources. Secondly according to the convenience of the questionnaires distribution and the development state of logistics industry which were decided by using subjective perception and Linkert7 level scale scoring method, we selected Wuhan, Shenzhen, Zhejiang and Hangzhou to distribute our questionnaires, with the senior management staff of the enterprises as our main subjects. This research distributed totally 709 questionnaires and 238 of them were returned (returning rate is 33.6 %). Then we screened the questionnaires that were returned and obtained 201 valid questionnaires (valid returning rate is 28.3 %).

4.1 Scale Reliability and Validity Analysis[1]

The assessment of this research's reliability is mainly undertaken from two aspects, reliability of individual indicators and reliability of overall factors (also called C.R.). Reliability is assessed by using Cronbach's α coefficient. α coefficient over 0.7 indicates that the measure has acceptable reliability. In the confirmatory factor analysis, we will adopt R2, which means the square of the load coefficient of the standardized factors, as the reliability index of the individual variables. This paper will use the standard proposed by Tabachnica and Fidell [3], that is, R2 values can be declared well if they are more than 0.3. The overall reliability of the various factors can be measured by the reliability indexes of certain individual questions. We also call it CR (composition reliability).

To analyze validity, we adopt here the combination of two methods: EFA (exploratory factor analysis) and CFA (confirmatory factor analysis)[1]. We make confirmatory factor analysis of them, choosing x2/df value, RMSEA, GFI, AGFI, CFI, NFI, IFI and NFI value as the fit index. x2/df value should be less than 5, RMSEA value should be less than 0.1 (RMSEA < 0.05, perfect fit; $0.05 <$ RMSEA < 0.08, just fit and within the acceptable range; $0.08 <$ RMSEA < 0.10, moderate fit; RMSEA > 0.10, indicating that the model fits very bad, bad fit), and the values of the other fit indexes such as GFI, AGFI, NFI, IFI, CFI should be over 0.90, the model can be regarded as good fit [4], the closer the value is to 1, the fitter the model is.

In general, the exploratory factor analysis should use different samples with that of confirmatory factor analysis. In this study, the total number of samples is randomly split into two groups, of which 130 sample data are given to the first group, and the second group gets 71 sample data. The first set of sample data is used in the exploratory factor analysis and the confirmatory factor analysis use both the two sets of sample data.

After the above test of reliability and validity, we build a structure equation model.

[1] Because of limited space, this paper omits the EFA and CFA process.

4.2 The Initial Structure Equation Model

We set seven exogenous latent variables as the explanatory variables in the initial structure model. They are location resources, material resources, brand resources, collective learning and knowledge sharing resources, specialized division of labor and cooperation resources, supporting institutions resources and order resources. We set an endogenous latent variable as the explained variables. It is the competitive advantages of logistics enterprises. We then set 26 exogenous observable variables to measure the seven exogenous latent variables, and also set five endogenous observable variables to measure the endogenous latent variable. The initial model is as shown in Fig. 1.

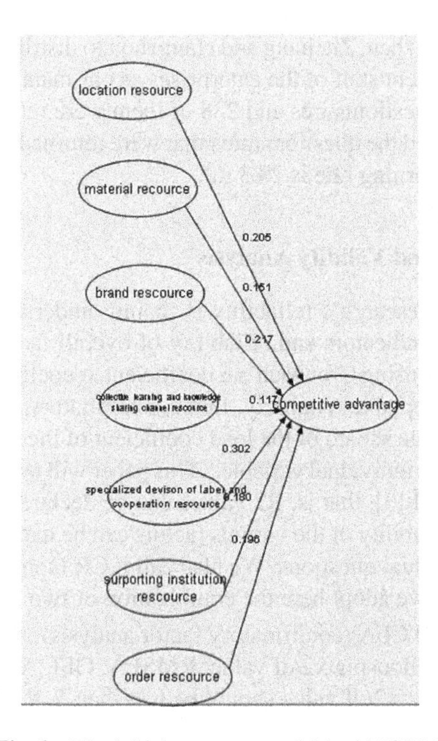

Fig. 1. The initial structure model in AMOS7.0

(1) Model identification. This model has a total of 31 measuring indexes and so 496 data points according to the t-rule. It will also estimate 31factor loadings, 31 error variances of the measure indexes, 21correlation coefficients among the factors, 7 path coefficients and 1 residual of endogenous latent variable. The model has totally to estimate 91 parameters to satisfy the necessary condition for model identification.

(2) The initial model fitting. After importing the sample data and computing these data with AMOS7.0, we gain the computing data of initial structural equation models. The computation results show that x2/df value is 1.601; RMSEA value is 0.055, less than 0.1; GFI, CFI values are separately bigger than 0.9; AGFI, NFI values are close to 0.9. In addition, in the 34 path coefficients, except a small number of path coefficients do not

reach the standards[2], the CR values corresponding to the rest of the path coefficients are greater than the reference value of 1.96, with statistical significance level of $p \leq 0.05$. The degree of model fitting is acceptable. However, two path coefficients, material resources → competitive advantages and collective learning and knowledge sharing channel resources → competitive advantages, do not meet the standards. Therefore, the model needs modifying.

Table 1. Model calculation result

The hypothesized path	Standardized path coefficient	Whether to support the hypothesis
competitive advantages (η) ← location resources (ξ1)	0.204 (2.857)***	support
competitive advantages (η) ← material resources (ξ2)	0.149(1.796)*	partial support
competitive advantages (η) ← brand resources (ξ3)	0.215 (3.072)***	support
competitive advantages (η) ← specialized division of labor and cooperation resources (ξ5)	0.314 (4.973)***	support
competitive advantages (η) ← supporting institution resources (ξ6)	0.183 (2.160)**	support
competitive advantages (η) ← order resources (ξ7)	0.194 (2.305)**	support
The fitting index	x2/df = 1.599 GFI = 0.912 AGFI = 0.892 CFI = 0.926 NFI = 0.872 RMSEA = 0.055	

Remarks: *p < 0.1, **p < 0.05, ***p < 0.01 Brackets for the C.R. value

[2] The standards for this paper which judge whether the path parameters support the corresponding hypothesis is: the P value of path coefficient less than 0.05 for support the hypothesis; the P value of path coefficient higher than 0.05 and less than 0.1 for partial support the hypothesis; the P value of path coefficient above 0.1 for deny the hypothesis.

(3) Model modification. After the comparison of the C.R. values of two paths which are not up to the standards, we select the path with smaller value (the path of collective learning and knowledge sharing channel resources → competitive advantages) and cut it out. Then we re-import the data and compute once again. The computing results show that x2/df is 1.599, RMSEA is 0.055, and the values of AGFI, NFI increase a little. After cutting out the path of collective learning and knowledge sharing channel resources → competitive advantages, the parameter of the path of material resources → competitive advantages improves and the model's fitting degree also improve a little, compared with the first time. We then conclude that the modified model have passed the test. The path coefficients and the fitting parameters of the model's final modification can be found in Table 1.

5 Conclusion

It can be seen from the results that H1, H3, H5, H6 and H7 are all verified by those data, and H2 partially verified, while H4 is denied by the model. Therefore, the vast majority of the shared resources of the logistics industry clusters have positive impacts on the logistics industries' competitive advantages. Location resources, brand resources, specialized division of labor and cooperation resources, supporting institutions resources and order resources have marked positive impacts on the competitive advantages of the logistics enterprises within the logistics industry clusters, among which the specialized division of labor and cooperation resources have the greatest impact on the competitive advantages.

Material resources are partly supported by the positive effects of the competitive advantage. The reasons may be the following two: first, many logistics industry clusters just gather those logistics companies together and don't committed to construct the information platform, or do not play its role of information publishing and distribution. Second, a public information platform can only achieve its maximum advantages when combined with the information system of the enterprises themselves. But many small-scale logistics enterprises have not built their own information system, thus the advantages of public information platform can not be maximized.

The impacts of collective learning and knowledge sharing channel resources on logistics enterprises' competitive advantages has not been proved by the models and so are excluded from the research. The reason for this may be that logistics industry cluster formation mechanism is different from spontaneous generation mechanism of other manufacturing industry cluster. Most of the logistics industrial clusters are led by the government formation. Furthermore China's logistics industry cluster formation and development time is not long, so mutual trust mechanism of the group interior enterprise is weaker than the spontaneous formation of a lot of manufacturing industry clusters. In such circumstances, compared to other industrial cluster, an important way of the collective learning and knowledge sharing which is the informal exchanges is greatly reduced. The exchange of information between enterprises mainly rely on the formal exchanges between the cooperation enterprises. So the impact of exchange of information and knowledge on the enterprise competitive advantages is enhanced by the specialized division of labor and cooperation. Possible the impact which the collective learning and knowledge sharing resources have on the enterprise competitive advantages is covered by the impact which specialization division of labor and cooperation have.

References

1. Molina-Morales, F.X.: European industrial districts: influence of geographic concentration on performance of the firm. J. Int. Manag. **7**, 277 (2007)
2. Geng, S.: Study on competitive advantage of firms in industrial cluster based on the shared rescource view. ZheJiang University (2005)
3. Tabachnick, B.G., Fidell, L.S.: Using Multivatiate Statistics. Pearson/Allyn and Bacon, Boston (2007)
4. Hou, J., Wen, Z.L., Cheng, Z.J., Zhang, L.: Structure Equation Model and Its Application, p. 7. Education Science Press, Beijing (2004)

The Research on the Dual-Channel Supply Chain Coordination with Compensation Strategy When the Costs of Production Are Changing

Boya Weng(✉) and Cancan Zhao

School of Safety and Environmental Engineering, Capital University of Economics and Business, Beijing, China
bettyweng@126.com, zhaocann@cueb.edu.cn

Abstract. This paper studies a dual-channel supply chain model which contains a manufacturer and a retailer. By using the game theory, we analysis and comparison the dual-channel supply chain when the costs of production is changing. We investigate the optimal price of the supply chain under centralized decision and decentralized decision making. We also design a compensation strategy which considering changes in the costs of production in order to both manufacturer and retailer are in win-win situation. Finally, this paper use the example demonstrated the effectiveness of the compensation strategy for the dual-channel supply chain coordination.

Keywords: Dual channel · Supply chain · Changes in production costs · Compensation strategy

1 Introduction

The developments of electronic commerce bring the market a lot of opportunities. The ministry of commerce had predicted that the e-commerce transactions will be more than 18 trillion Yuan in China in 2015 [12]. As a result, the rapid development of dual channel supply chain has made it attractive for large number of brick and mortar firms, and there have been a lot of largest examples which run the dual-channel supply chain and succeeded [11], such as Nike Company in the United States. However, it's difficult for some manufacturers to coordinate the relationship between the traditional channels and electronic channels. The development of electronic channels makes the retailer's profits are exploited which lead to traditional retailers believe online direct marketing channels are becoming a competitor [10]. Such approach also creates the potential for channel conflicts. In the issue, coordinate the relationship between the electronic channels and traditional channels plays a significant role in the dual-channel supply chain coordination.

Dual-channel supply chain coordination has been broadly studied in the literature [1–6]. It's not hard to find existing research mainly focused on decision making about the price and quantity of the dual channel supply chain under the stable condition. Generally the literature does not consider the costs of production [1, 2], or suppose

F. Bian and Y. Xie (Eds.): GRMSE 2015, CCIS 569, pp. 70–82, 2016.
DOI: 10.1007/978-3-662-49155-3_7

production costs does not change [3–6], such as Zhao and Xu [1] introduced a model when a manufacturer use electronic channel and traditional retail channel to sell products at same time, established a demand function from the perspective of consumer utility which demand change depend on price change. They also designed a contract that the wholesale price plus electronic channel price, in order to coordinate the channel conflict, and improve the revenue sharing contract. Ai et al. [2] introduce a two layers of the supply chain competition model in the case of considering environmental information factors, so as to provides the basic theory for further build the double channel cooperation and incentive mechanism. Yan et al. [3] established a dual-channel supply chain coordination stackelberg model under service-price sensitive demand conditions, study the two coordinate ways: the coordination between the upstream and downstream nodes and the coordination between the traditional channel and electronic channel. Yan et al. [4] compared the non integrated channels and the integration channels, points out that the channel integration strategy which share profits can make both online sale channels and traditional channels profitable.

Besides, there are a few literature takes the changes in production costs into consideration [7, 8]. But these studies did not analyzed and discussed the coordinate way to solve the channel conflicts problem. Huang et al. [7] analyzed when the demand and production costs are disrupted at same time, the influence on centralized and decentralized dual-channel supply chain pricing and production decisions. They found optimal pricing and optimal production decisions in different range of distribution, but article did not elaborate how to coordinate the dual-channel supply chain after distribution occurred. Wang [8] introduced pricing strategy of dual-channel supply chain when considering the manufacture costs increase and decrease, he analyzed the range of manufacturing costs which affect the production plan. But the paper didn't mention how to coordinate the influence.

Our paper addresses these limitations by studying the compensation strategy when the costs of production are changing in the supply chain. Our objective in this paper is to design a compensation mechanism to make decentralized decision profits and centralized decision profits equal. And ensure that both manufacturer and retailer can obtain a benefit, realize the pareto optimality.

2 Problem Statement

Our basic model includes one manufacturer and one retailer, the hypotheses are as follow:

(1) Both sides of the decisions are rational;
(2) The two sides share same information;
(3) The two sides follow the master-slave game. The manufacturer is leader, the retailer is follower. First of all, the manufacturer decide the wholesale prices and the sales price of electronic channel, then according to the price declares by the manufacturer, the retailer set the sales price of traditional channel.

Preliminaries:

w: the wholesale price;
c: the unit production cost of the manufacturer;
P_e: The sales price of electronic channel;
P_r: The sales price of traditional channel;
d_r: Traditional channel demand;
d_e: Electronic channel demand;
a: total market demand, $a > 0$;
b: the substitution effect coefficient between electronic channel sales product and traditional channel sales product, $0 < b < 1$;
f: the market demand at traditional channel, $0 < f < 1$;
1-f: the market demand at electronic channel.

According to the literature of Ai et al. [2] and Chen [9], the demand functions for the traditional and electronic channel, respectively, can be expressed as:
Traditional channel demand:

$$d_r = fa - P_r + b(P_e - P_r). \tag{1}$$

Electronic channel demand:

$$d_e = (1 - f)a - P_e + b(P_r - P_e). \tag{2}$$

3 Model Structure

3.1 The Centralized Decision-Making

When the production costs reduced, the manufacturer will not produce fewer products, and this phenomenon may touch off oversupply, resulting in a decline in sales price. The manufacturer cannot sell all the products, and incur the processing costs. Similarly, when the production costs increases, the manufacturer will not produce more products, and this phenomenon may touch off demand exceeds supply, resulting in a sale prices rise. The manufacturer cannot meet all the requirements, and incur the shortage costs [13]. As a result, changes in the costs of production will affect the sale price, and then change the best decision in the whole supply chain.

When the costs of production are changing, the total profit is:

$$\begin{aligned}\pi_1^c = (P_r - c - \Delta c)[fa - P_r + b(P_e - P_r)] + (P_e - c - \Delta c)[(1 - f)a - P_e \\ + b(P_r - P_e)] - \mu_1(d_e + d_r - d_e^* - d_r^*)^+ - \mu_2(d_e^* + d_r^* - d_e - d_r)^+\end{aligned}.$$

Among them, the third is when manufacturer produce more products may incur the processing cost. The fourth is when manufacturer produce fewer products may incur the shortage costs. And both costs can't exist at the same time.

μ_1: Unit producing costs caused by the manufacturer producing one more products.

μ_2: Unit shortage costs caused by the manufacturer producing one less products, max $\{\mu_1, \mu_2\} < c$.

The Production Costs Reduce Under the Centralized Decision-Making. The dual-channel supply chain's optimal profit is:

$$\pi_1^c = (P_r - c - \Delta c)[fa - P_r + b(P_e - P_r)] + (P_e - c - \Delta c)[(1 - f)a - P_e + b(P_r - P_e)] - \mu_1(d_e + d_r - d_e^* - d_r^*)^+. \tag{3}$$

We obtain the optimal solution by solving the first-order condition, $\frac{\partial \pi_1^c}{\partial p} = 0$. Thus, the optimal sales price of electronic channel is: $P_e = \frac{2bc + 2b\Delta c + 2b\mu_1 + a - fa + c + \Delta c + \mu_1 + ab}{4b + 2}$, the optimal sales price of traditional channel is:

$$P_r = \frac{ab + c + 2bc + fa + \Delta c + 2b\Delta c + \mu_1 + 2b\mu_1}{4b + 2}. \tag{4}$$

Considering when the costs of production reduce, the electronic channel and traditional channel total demand cannot less than the total demand when the costs of production constant, we can get: $\Delta c \leq -\mu_1$.

The Production Costs Increase Under the Centralized Decision-Making. The dual-channel supply chain's optimal profit is: $\pi_1^c = (P_r - c - \Delta c)[fa - P_r + b(P_e - P_r)] + (P_e - c - \Delta c)[(1 - f)a - P_e + b(P_r - P_e)] - \mu_2(d_e^* + d_r^* - d_e - d_r)^+$, similar to the reduction of cost, we calculate and obtain the optimal sales price of electronic channel is: $P_e = \frac{2bc + 2b\Delta c - 2b\mu_2 + a - fa + c + \Delta c - \mu_2 + ab}{4b + 2}$, the optimal sales price of traditional channel is:

$$P_r = \frac{ab + c + 2bc + fa + \Delta c + 2b\Delta c - \mu_2 - 2b\mu_2}{4b + 2}. \tag{5}$$

Considering when the costs of production increase, the electronic channel and traditional channel total demand cannot exceeds the total demand when the costs of production constant, we can get: $\Delta c \geq \mu_2$.

Thus, the total profit of the dual-channel supply chain under the centralized decision-making is:

$$\pi_1^c = \begin{cases} \dfrac{\begin{array}{c}-4ab\Delta c + 8\Delta cb\mu_1 + 4b\Delta c^2 + 8cb\Delta c + 4b\mu_1{}^2 - 4abc + a^2b + 4bc^2 + 4c\Delta c \\ 2\mu_1{}^2 + 2c^2 + a^2 - 2ac - 2fa^2 + 2\Delta c^2 + 2f^2a^2 - 2a\Delta c + 4\Delta c\mu_1\end{array}}{4+8b}, & \Delta c \le -\mu_1 \\ \dfrac{4bc^2 - 4abc + a^2b - 2a^2f + 2a^2f^2 + 2c^2 + a^2 - 2ac}{8b+4}, & -\mu_1 < \Delta c < \mu_2 \\ \dfrac{\begin{array}{c}-4ab\Delta c - 8\Delta cb\mu_2 + 4b\Delta c^2 + 8cb\Delta c + 4b\mu_2{}^2 - 4abc + a^2b + 4bc^2 \\ + 4c\Delta c + 2\mu_2{}^2 + 2c^2 + a^2 - 2ac - 2fa^2 + 2\Delta c^2 + 2f^2a^2 - 2a\Delta c - 4\Delta c\mu_2\end{array}}{4+8b}, & \Delta c \ge \mu_2 \end{cases}. \tag{6}$$

3.2 Decentralized Decision-Making

We assume that all the loss caused by the changes in the cost of production undertakes by the manufacturer alone. The manufacturer's profit is:

$$\pi_2^c = (w - c - \Delta c)[fa - P_r + b(P_e - P_r)] + (P_e - c - \Delta c)[(1 - f)a - P_e + b(P_r - P_e)] - \mu_1(d_e + d_r - d_e^* - d_r^*)^+ - \mu_2(d_e^* + d_r^* - d_e - d_r)^+. \tag{7}$$

The retailer's profit is:

$$\pi_3^c = [fa - P_r + b(P_e - P_r)](P_r - w). \tag{8}$$

The Production Costs Reduce Under the Decentralized Decision-Making. The manufacturer's profit is:

$$\pi_2^c = (w - c - \Delta c)[fa - P_r + b(P_e - P_r)] + (P_e - c - \Delta c)[(1 - f)a - P_e + b(P_r - P_e)] - \mu_1(d_e + d_r - d_e^* - d_r^*)^+ \tag{9}$$

We obtain the optimal solution by solving the second-order dynamic game on Eqs. (8) and (9), the optimal sales price of electronic channel is:

$$P_e^* = \frac{\begin{array}{c}2\mu_1 + 2a - 4fab - 2fab^2 + 8\Delta cb + 10\Delta cb^2 + 2\Delta c - 2fa + 8bc + 6ab + 10cb^2 + 6ab^2 + 2c \\ + 8\mu_1 b + 4cb^3 + 2ab^3 + 10\mu_1 b^2 + 4\mu_1 b^3 + 4\Delta cb^3\end{array}}{20b^2 + 8b^3 + 16b + 4}. \tag{10}$$

The optimal wholesale price is:

$$w^* = \frac{\mu_1 + fab + 3\Delta cb + 2\Delta cb^2 + \Delta c + fa + 3bc + ab + 2cb^2 + ab^2 + c + 3\mu_1 b + 2\mu_1 b^2}{4b^2 + 6b + 2}. \tag{11}$$

The optimal sales price of traditional channel is:

$$P_r^* = \frac{4\Delta cb^2 + 4cb^2 + 2ab^2 + 4\mu_1 b^2 + 2ab + 4fab + 4bc + 4\mu_1 b + 4\Delta cb + \mu_1 + \Delta c + 3fa + c}{4 + 12b + 8b^2}. \tag{12}$$

Considering when the costs of production reduce, the electronic channel and traditional channel total demand cannot less than the total demand when the costs of production constant, we can get: $\Delta c \leq -\mu_1$.

The Production Costs Increase Under the Decentralized Decision-Making. The manufacturer's profit is:

$$\pi_2^c = (w - c - \Delta c)[fa - P_r + b(P_e - P_r)] + (P_e - c - \Delta c)[(1 - f)a - P_e + b(P_r - P_e)] - \mu_2(d_e^* + d_r^* - d_e - d_r)^+. \tag{13}$$

Similar to the reduction of cost above, we calculate and obtain the optimal wholesale price is:

$$w^* = \frac{-\mu_2 + fab + 3\Delta cb + 2\Delta cb^2 + \Delta c + fa + 3bc + ab + 2cb^2 + ab^2 + c - 3\mu_2 b - 2\mu_2 b^2}{4b^2 + 6b + 2}. \tag{14}$$

The optimal sales price of electronic channel is:

$$P_e^* = \frac{\begin{array}{c}-2\mu_2 + 2a - 4fab - 2fab^2 + 8\Delta cb + 10\Delta cb^2 + 2\Delta c - 2fa + 8bc + 6ab + 10cb^2 \\ 6ab^2 + 2c - 8\mu_2 b + 4cb^3 + 2ab^3 - 10\mu_2 b^2 - 4\mu_2 b^3 + 4\Delta cb^3\end{array}}{20b^2 + 8b^3 + 16b + 4}. \tag{15}$$

The optimal sales price of traditional channel is:

$$P_r^* = \frac{4b^2c + 2ab^2 - 4\mu_2 b^2 + 4\Delta cb^2 + 4bc + 2ab - 4\mu_2 b + 4fab + 4\Delta cb + c + \Delta c - \mu_2 + 3fa}{8b^2 + 12b + 4}. \quad (16)$$

Considering when the costs of production increase, the electronic channel and traditional channel total demand cannot more than the total demand when the costs of production constant, we can get: $\Delta c \geq \mu_2$.

Thus, the total profit of the dual-channel supply chain under the decentralized decision-making is:

$$\pi_4^c = \begin{cases} \dfrac{\begin{array}{c} 4fabc + 4fab\Delta c - 4fab\mu_1 + 22bc^2 + 16c^2b^2 + 7c^2 + 22\mu_1{}^2 b + 16\mu_1{}^2 b^2 + 8a^2b + 4a^2 - 8a\Delta c - 8fa^2 - 8ac \\ + 4a^2b^2 - 16ab^2\Delta c + 4bc\mu_1 + 2fac + 2\Delta cfa + 32\Delta cb^2\mu_1 + 32\Delta cb^2c + 44\Delta cb\mu_1 + 44\Delta cbc - 2\mu_1 fa - 8a^2bf \\ + 6f^2a^2b + 2\mu_1 c + 14\mu_1\Delta c + 22\Delta c^2 b - 16ab^2c - 24ab\Delta c - 24abc + 16\Delta c^2b^2 + 14\Delta cc + 7\Delta c^2 + 7f^2a^2 + 7\mu_1{}^2 \end{array}}{16 + 48b + 32b^2}, & \Delta c \leq -\mu_1 \\ \dfrac{\begin{array}{c} -16ab^2c + 16c^2b^2 + 4a^2b^2 + 8a^2b + 22bc^2 - 24abc + 4fabc + 6f^2a^2b - 8a^2bf \\ 2fac - 8fa^2 - 8ca + 7c^2 + 4a^2 + 7f^2a^2 \end{array}}{48b + 32b^2 + 16}, & -\mu_1 < \Delta c < \mu_2 \\ \dfrac{\begin{array}{c} 2fac - 4bc\mu_2 + 2\mu_2 fa + 6f^2a^2b - 2\mu_2 c - 8a^2bf + 4fabc + 4fab\Delta c - 14\mu_2\Delta c + 7\mu_2{}^2 + 22\Delta c^2 b + 16\Delta c^2b^2 \\ + 14\Delta cc + 7\Delta c^2 + 7f^2a^2 + 16\mu_2{}^2 b^2 + 44\Delta cbc - 44\Delta cb\mu_2 + 32\Delta cb^2c - 32\Delta cb^2\mu_2 + 2\Delta cfa - 24abc - 24ab\Delta c \\ - 16ab^2c - 16ab^2\Delta c - 8ca + 8a^2b + 22bc^2 + 16c^2b^2 + 7c^2 + 22\mu_2{}^2 b + 4a^2 - 8a\Delta c - 8fa^2 + 4a^2b^2 + 4fab\mu_2 \end{array}}{32b^2 + 48b + 16}, & \Delta c \geq \mu_2 \end{cases} \quad (17)$$

From the Eqs. (6) and (17), we can find that if fa > c, the centralized decision-making total profits is always greater than the decentralized decision-making total profits. According to the Eqs. (12), (4) and (16), (5), we can find that if fa > c, the optimal sales price of traditional channel under decentralized decision-making is always greater than the optimal sales price of traditional channel under centralized decision-making. Thus, the retailer is not willing to accept the profit allocation scheme under centralized decision-making, because retailer can gain more under decentralized decision-making.

4 The Compensation Strategy

In order to solve the conflict put forward in Chap. 3, we set a compensation strategy: the manufacturer supply certain proportion electronic channel order to the retailer as compensation (increased demand for traditional channel). At the same time, the manufacturer requires a certain order transfer fee to ensure its profit. And we also realized that the order transfer fee can not hinder the retailer's optimal pricing.

The proportion of the order that the manufacturer supply the retailer set as $\theta (0 < \theta < 1)$. The order transfer fee is: $e = (P_r - w)d_r + (P_e - w)\theta d_e - E(E \geq 0)$, E is bigger than zero which ensures that e cannot exceed the retailer's optimal pricing.

The retailer's profit is:

$$\pi_5 = (P_r - w)d_r + (P_e - w)\theta d_e - e. \tag{18}$$

Substituting (1) and (2) into (18), we obtain the reaction function by solving the first-order condition, $\frac{\partial \pi_5}{\partial P_r} = 0$. The reaction function is: $P_r = \frac{fa + bP_e + w(1+b) + \theta b(P_e - w)}{2+2b}$, and we can get demand function:

$$d_r = \frac{fa + bfa + bP_e - w - 2wb - \theta bP_e + \theta bw + b^2P_e - wb^2 - \theta b^2P_e + \theta b^2w}{2+2b},$$

$$d_e = \frac{2a + 2ba - 2fa - bfa - 2P_e - 4bP_e - b^2P_e + wb + wb^2 + \theta b^2P_e - \theta b^2w}{2+2b}.$$

4.1 The Optimal Pricing of Electronic and Traditional Channel Under the Costs of Production Changes

According to the Δc to choose the manufacturer's profit function, then substituting the manufacturer's profit function into the demand function and solve two equations simultaneous. We can get the optimal sales price of traditional and electronic channel, the optimal wholesale price and the optimal demand of traditional and electronic channel. We can find that the retailer's optimal sale price under compensation strategy is the same as the retailer's optimal sale price under centralized decision-making. In light of the actual situation, the wholesale price which retailer purchase products must less than the electronic channel sales price. Thus $P_e > w$, the constraint is: $2bc - ab + c + fa - a < -(2b+1)(\Delta c - \mu_2)$.

4.2 The Total Profits of Dual-Channel Supply Chain When Meet the Constraint

Base on the result in Sect. 4.1, we can get the total profits of dual-channel supply chain:

$$\pi_1^c = \begin{cases} \dfrac{-4ab\Delta c + 8\Delta cb\mu_1 + 4b\Delta c^2 + 8cb\Delta c + 4b{\mu_1}^2 - 4abc + a^2b + 4bc^2 \\ 4c\Delta c + 2{\mu_1}^2 + 2c^2 + a^2 - 2ac - 2fa^2 + 2\Delta c^2 + 2f^2a^2 - 2a\Delta c + 4\Delta c\mu_1}{4+8b}, & \Delta c \le -\mu_1 \\ \dfrac{4bc^2 - 4abc + a^2b - 2a^2f + 2a^2f^2 + 2c^2 + a^2 - 2ac}{8b+4}, & -\mu_1 < \Delta c < \mu_2 \\ \dfrac{-4ab\Delta c - 8\Delta cb\mu_2 + 4b\Delta c^2 + 8cb\Delta c + 4b{\mu_2}^2 - 4abc + a^2b + 4bc^2 \\ + 4c\Delta c + 2{\mu_2}^2 + 2c^2 + a^2 - 2ac - 2fa^2 + 2\Delta c^2 + 2f^2a^2 - 2a\Delta c - 4\Delta c\mu_2}{4+8b}, & \Delta c \ge \mu_2 \end{cases} \tag{19}$$

According to the Eqs. (6) and (19), both total profits under the compensation strategy and under the centralized decision-making are equal. The result shows that the compensation strategy realizes the pareto optimality.

4.3 The Order Transfer Fee When Meet the Constraint

In order to satisfy the actual conditions, the manufacturer and retailer have to ensure its own profits under the compensation strategy more than the profits under the decentralized decision-making. We can get the order transfer fee condition: (the specific expressions of A,B,C,D,E,F,G,H,I,J,K…etc., see the appendix.)

$$\begin{cases} \dfrac{(Aa+Ba^2+C)\theta+D}{8(2b+1)(-1-b+\theta b)(2b^2+3b+1)} \le e \le \dfrac{(Ga^2+Ha+F)\theta^2+(Ja^2+Ka+I)\theta+E}{16(2b+1)^2(2\theta b^2+\theta b-1-2b^2-3b)^2(b+1)}, \Delta c \le -\mu_1 \\ \dfrac{(A_2a+B_2a^2+C_2)\theta+D_2}{8(2b+1)(-1-b+\theta b)(2b^2+3b+1)} \le e \le \dfrac{(G_2a^2+H_2a+F_2)\theta^2+(J_2a^2+K_2a+I_2)\theta+E_2}{16(2b+1)^2(2\theta b^2+\theta b-1-2b^2-3b)^2(b+1)(1+3b+2b^2)}, \Delta c \ge \mu_2 \\ \dfrac{(A_3a^2+B_3a+C_3)\theta^2+(D_3a^2+E_3a+F_3)\theta+G_3}{8(2b+1)^2(-1-b+\theta b)^2(1+b)} \le e \le \dfrac{(H_3a^2+I_3a+J_3)\theta^2+(K_3a^2+L_3a+M_3)\theta+N_3}{16(2b+1)^2(-1-b+\theta b)^2(b+1)}, -\mu_1 < \Delta c < \mu_2 \end{cases} \quad (20)$$

5 Numerical Example

In this section our objective is to illustrate our results with the help of a selected numerical example. The parameters are a = 1000, f = 0.3, b = 0.5, c = 20 Yuan, $\mu_1 = 1$ Yuan, $\mu_2 = 2$ Yuan. When the proportion of the order that the manufacturer supply the retailer set as: $\theta = 0.4$, we can obtain (Table 1):

Table 1. Order transfer fee under compensation strategy

Changes in the costs of production	Order transfer fee is more than	Order transfer fee is less than
−5	41654.76923	45062.10256
−4	41487.82692	44871.53526
−3	41321.25641	44681.42308
−2	41155.05769	44491.76603
−1	40989.23077	44302.5641
0	41036	44303
1	41036	44303
2	41129.23077	44302.5641
3	40963.77564	44113.81731
4	40798.69231	43925.52564
5	40633.98077	43737.6891

When we choose the order transfer fee as 42000 Yuan, we can get:

Table 2. The profits of manufacturer and retailer under compensation strategy

Changes in the costs of production	The manufacturer profits under decentralized decision-making	The retailer profits under decentralized decision-making	Order transfer fee	The manufacturer profits under compensation strategy	The retailer profits under compensation strategy
−5	120840	3360.67	42000	121185.23	6422.77
−4	120403.75	3337.04	42000	120915.92	6208.58
−3	119968.33	3313.5	42000	120647.08	5994.92
−2	119533.75	3290.04	42000	120378.69	5781.81
−1	119100	3266.67	42000	120110.77	5569.23
0	118670	3266.7	42000	119630.77	5569.23
1	118670	3266.7	42000	119630.77	5569.23
2	117800	3266.67	42000	118670.77	5569.23
3	117367.08	3243.38	42000	118403.31	5357.19
4	116935	3220.17	42000	118136.31	5145.69
5	116503.75	3197.04	42000	117869.77	4934.73

The Table 2 shows that the manufacturer and retailer profits under the compensation strategy were higher than the profits under decentralized decision-making. And we realized that the manufacturer and retailer profits are changed with the order transfer fee. The more order transfer fee has to pay, the fewer profits retailer can get. So it is important for the manufacturer have a discussion with the retailer, to draw up a satisfied order transfer fee, so as to arrange the distribution of profits reasonable.

6 Conclusions

Base on the dual-channel supply chain model which manufacturer is leader, by comparing when the costs of production is changing the dual-channel supply chain profits under centralized decision-making and the dual-channel supply chain profits under decentralized decision-making, we figure out when $fa > c$, the traditional channel profits under decentralized decision-making is higher than the profit under centralized decision-making. In order to coordinate the conflicts and find the optimal total profit, we introduced a compensation strategy: the manufacturer supply certain proportion electronic channel order to the retailer as compensation. At the same time, the manufacturer requires a certain order transfer fee to ensure its profit. Through this compensation strategy, both sides profit are not damaged, and can achieve the total profits optimal.

In this paper, we design conditions are completely rational, so in practical application, it's inevitably encounter loopholes. There are still a lot of problems that need to be studied. For example how to coordinate the dual-channel supply chain when the customers demand present exponential function distribution.

Acknowledgement. The corresponding author of this paper is Boya Weng. This research is supported by the Collaborative Innovation Center of Academy of Metropolis Economic and Social Development of Capital University of Economics and Business, by the 2014 Youth Foundation of Capital University of Economics and Business, and by the Research Project of Capital University of Economics and Business under grant No. 2014XJQ001. These supports are greatly acknowledged.

Appendix

1. When $\Delta c \leq -\mu_1$: $A = -2(2b+1)^2[(fb-2b-2+2f)(c+\Delta c)+2\mu_1(b-1)(f-1)]$
$B = (2b+1)(-2b^2+2f^2b^2+f^2b-4b+4bf-2-2f^2+4f)$
$C = -(2b+1)^2[6\Delta cb(c+\mu_1)+4(bc\mu_1+c\Delta c+\mu_1\Delta c)+ (3b+2)({\mu_1}^2+c^2+\Delta c^2)]$
$D = -(2b+1)^2(b+1)(2c\Delta c+2\mu_1\Delta c+c^2+4\mu_1 c-$
$4\mu_1 fa-2fac+f^2a^2-2\Delta cfa+\Delta c^2+{\mu_1}^2)$
$E = 3(b+1)^2(2b+1)^4(-fa+c+\Delta c+\mu_1)^2$
$F = -b(5b+4)(2b+1)^4(\mu_1+\Delta c+c)^2$
$G = 49152b^{11}c+16(3f^2-2)b^6+16(4f+6f^2-7)b^5+8(20f+5f^2-19)b^4$
$+4(-6f^2+36f-25)b^3+(56f-21f^2-32)b^2+4(2f-f^2-1)b$
$H = [32(4-3f)(\Delta c+c)-96\mu_1 f]b^6+[384(\Delta c+\mu_1)-320f(c+\Delta c+\mu_1)]b^5+$
$16(28-25f)(c+\Delta c+\mu_1)b^4+16(16-15f)(\mu_1+\Delta c+c)b^3$
$+2(36-35f)(\mu_1+\Delta c+c)b^2+8(1-f)(\Delta c+c+\mu_1)b$
$I = (146b^2+160b^5)(c^2+\Delta c^2)+38\Delta c^2 b+288{\mu_1}^2 b^3+8c\Delta c$
$J = 32(-3f^2+1)b^6+16(9-18f^2-4f)b^5+$
$8(-38f^2+33-28f)b^4+4(-28f^2+63-76f)b^3+2(66+9f^2-100f)b^2+$
$2(16-32f+11f^2)b+2(f-1)^2$
$K = (\Delta c+c+\mu_1)[64(3f-2)b^6+64(11f-8)b^5+32(33f-26)b^4$
$+64(13f-11)b^3+4(91f-82)b^2+ 4(21f-20)b+8(f-1)]$
2. When $\Delta c \geq \mu_2$: $A_2 = 2(2b+1)^2[(2b-fb-2f+2)(\Delta c+c)+2\mu_2(b+1)(f-1)]$
$B_2 = (2b+1)(-2b^2+2f^2b^2+f^2b-4b+4bf-2-2f^2+4f)$
$C_2 = (2b+1)^2[6b\Delta c(c-\mu_2)-4(bc\mu_2+\mu_2\Delta c+\mu_2 c+$
$c\Delta c)+(2+3b)(\Delta c^2+{\mu_2}^2+c^2)$
$D_2 = -(b+1)(2b+1)^2[4\mu_2(fa-c)-2(fa\Delta c-c\Delta c+\mu_2\Delta c+cfa)$
$+(f^2a^2+{\mu_2}^2+\Delta c^2+c^2)]$
$E_2 = 3(b+1)^3(2b+1)^5(fa-\Delta c+\mu_2-c)^2$
$F_2 = -b(\Delta c-\mu_2+c)^2(b+1)(5b+4)(2b+1)^5$
$G_2 = 32(3f^2-2)b^8+16(-20+8f+21f^2)b^7+32(16f+13f^2-31)b^6$
$+8(-96+104f+21f^2)b^5+2(-258+352f-37f^2)b^4+(-95f^2+328f-204)b^3$
$+(-33f^2-44+80f)b^2-2(f-1)^2b$

$H_2 = -2b(b+1)(2b+1)^5(\Delta c - \mu_2 - c)(3bf - 4b - 4 + 4f)$
$I_2 = (\Delta C^2 + {\mu_2}^2 + C^2)(64b^8 + 416b^7 + 1120b^6 + 1648b^5 + 1460b^4 + 802b^3 + 268b^2$
$+ 50b + 4) + (\Delta cc - \mu_2\Delta c - \mu_2 c)(128b^8 + 832b^7 + 2240b^6 + 3296b^5 + 2920b^4$
$+ 1604b^3 + 536b^2 + 100b + 8)$
$J_2 = 64(1 - 3f^2)b^8 + 16(24 - 8f - 54f^2)b^7 + 32(31 - 20f - 49f^2)b^6$
$+ 16(90 - 89f^2 - 84f)b^5 + 4(-384f - 151f^2 + 321)b^4 + 2(-7f^2 + 360 - 516f)b^3$
$+ 4(62 + 23f^2 - 102f)b^2 + 2(24 + 17f^2 - 44f)b + 2(f - 1)^2$
$K_2 = 4(b+1)^2(2b+1)^5(\Delta c - \mu_2 + c)(-2b + 3bf - 2 + 2f)$
3. When $-\mu_1 < \Delta c < \mu_2$: $A_3 = b(2b+1)(2f^2b^2 - 2b^2 + 4fb + f^2b - 4b - 2 - 2f^2$
$+ 4f)$
$B_3 = -2bc(2b+1)^2(-2b + fb - 2 + 2f);$
$C_3 = -bc^2(3b+2)(2b+1)^2$
$D_3 = -2(2b+1)(b+1)^2(2f^2b - b + 2f - 1 - f^2)$
$E_3 = 4c(b+1)^2(2b+1)^2(f-1)$
$F_3 = 2c^2(b+1)^2(2b+1)^2$
$G_3 = (4a^2b^4 + 12a^2b^3 + 13a^2b^2 + 6a^2b + a^2)f^2$
$(-24ab^3c - 12abc + 26ab^2c - 8ab^4c - 2ca)f + 4b^4c^2 + c^2 + 6bc^2 + 12b^3c^2 + 13b^2c^2$
$H_3 = b(2b+1)(6f^2b^2 - 4b^2 + 3f^2b - 8b + 8bf - 4f^2 - 4 + 8f)$
$I_3 = -2bc(2b+1)^2(3bf - 4b - 4 + 4f)$
$J_3 = -bc^2(5b+4)(2b+1)^2$
$K_3 = -2(-2b^2 + 6f^2b^2 + 3f^2b - 4b + 4bf - 2 - 2f^2 + 4f)(b+1)(2b+1)$
$L_3 = 4c(b+1)(2b+1)^2(3bf - 2b + 2f + 2)$
$M_3 = 2c^2(b+2)(b+1)(2b+1)^2$
$N_3 = 3(4fab^2 + 3fab + fa - c - 3bc - 2b^2c)(-c - 3bc - 2b^2c + fa + 3fab)$

References

1. Zhao, L., Xu, J.: Contract design for coordination conflict of dual channels supply chain. J. Manage. Sci. **22**, 61–68 (2014)
2. Ai, X., Tang, X., Ma, Y.: Performance of forecasting information sharing between traditional channel and e-channel. Chin. J. Manage. Sci. **11**, 12–21 (2008)
3. Yan, N., Huang, X., Liu, B.: Stackelberg game models of supply chain dual-channel coordination in e-markets. Chin. J. Manage. Sci. **15**, 98–102 (2007)
4. Yan, R., Wang, J., Zhou, B.: Channel integration and profit sharing in the dynamics of multi-channel firms. J. Retail. Consum. Serv. **17**, 430–440 (2010)
5. Tsay, A.A., Agrawal, N.: Channel conflict and coordination in the e-commerce age. J. Prod. Oper. Manage. **13**, 93–110 (2004)
6. Xu, G., Dan, B., Xiao, J.: Price discount model for coordination of dual-channel supply chain under e-commerce. J. Syst. Eng. **27**, 344–350 (2012)

7. Huang, S., Yang, C., Yang, J.: Pricing and production decisions in dual-channel supply chains with demand and production cost disruptions. J. Syst. Eng. Theor. Pract. **34**, 1219–1229 (2014)
8. Wang, J.: A study on the price decisions of the dual-channel supply chain. Southwest Jiaotong University, pp. 11–31 (2008)
9. Chen, Z.: The study of operation cooperation mechanism on the traditional channel combined with e-channel. University of Electronic Science and Technology of China, pp. 30–43 (2006)
10. Dan, B., Xu, G., Zhang, X.: A compensation strategy for coordinating dual-channel supply chains in e-commerce. J. Ind. Eng. Eng. Manage. **26**, 124–130 (2012)
11. Huang, F.: Research on Price Competition and Coordination in Dual-channel Supply chain in asymmertric Demand Information. South China University of Technology, pp. 1–16 (2012)
12. Modern Business: China's e-commerce market have huge development space. J. Mod. Bus. **1**, 60–61 (2011)
13. Xu, M., Qi, X., Yu, G., Zhang, H.: Coordinating dyadic supply chains when production costs are disrupted. J. IIE Trans. **38**, 765–775 (2006)

GIS-Based Commute Analysis Using Smart Card Data: A Case Study of Multi-Mode Public Transport for Smart City

Yuyang Zhou[1(✉)], Lin Yao[1], Yu Jiang[1,2], Yanyan Chen[1], and Yi Gong[1]

[1] Beijing Key Laboratory of Traffic Engineering, Beijing University of Technology, Beijing 100124, China
zyy@bjut.edu.cn

[2] School of Transportation Engineering, Tongji University, Shanghai 200092, China

Abstract. This paper utilized the one-week smart card data (SCD) and the control passenger flow survey to analyze the commute travel time and the passenger flow distribution in the multi-mode public transport. To research the commute pattern of the central business district (CBD), there were three large-scale residence communities selected for the survey areas. Based on SCD in the double ticket system, the average travel time and the passenger volume were estimated through clustering "the alighting time" and filling none value. As a result, the visualization of the station attraction and the travel time under multi-mode public transport was presented through the application of GIS. The analysis of the commute pattern is aimed to make a scientific guidance for commuters on the traffic model choice as well as provide a quantitative basis for the development of the smart city.

Keywords: Smart card data · Multi-mode public transport · Travel time · Station attraction · Smart city

1 Introduction

With the accelerating process of urbanization, the excessive passenger flow has put enormous pressure on the transit network as well as brought much inconvenience to commute at the peak time. In order to improve the heavy traffic situation, the metro, the ordinary bus, the Bus Rapid Transit (BRT) and the public bicycle are merged into the integrative traffic network constantly. As each mode is independent and interactional with others, the analysis of smart card data (SCD) under multi-mode public transport is quite essential to offer effective information service for travelers.

The analysis of the route choice under multi-mode public transport depends, to a large degree, on the data of questionnaire surveys. Zhang [1] studied the interactions among the three modes of morning commuting: transit, driving alone and carpool. The difference of the factors affect the network was achieved by the nested logit mode. Mishra et al. [2] suggested measures to decide connectivity from the graph theory which could be an indicator to evaluate and quantify transit service in the multi-mode

F. Bian and Y. Xie (Eds.): GRMSE 2015, CCIS 569, pp. 83–94, 2016.
DOI: 10.1007/978-3-662-49155-3_8

public transport networks. Chen and Xu [3] described the multi-mode network by considering the typical multi-mode transportation, the topology theory and the state-augmentation technique. Owen and Levinson [4] proposed an accessibility-based model to aggregate commute mode share. Hu et al. [5] developed a nested logit choice model and a travel time reliability factor to evaluate the influences of integrated multi-mode transit information service.

The research about SCD mainly focused on the data of one ticket transit system which obtain the boarding station and time by model estimation. Munizaga and Palma [6] presented a mathematical method for estimating an OD matrix from the bus smart card with an application of two one-week datasets. Kusakabe and Asakura [7] applied the bus smart card to derive the relationships among traveler behavior which was difficult to achieve a person trip survey. Chen et al. [8] proposed a method to identify the boarding stops of passengers by matching the data of smart card, intelligent dispatching system and automatic fare collection system. Wu [9] compared five methods of public transport passenger acquisition to check where passengers get on and off bus and the way of transferring by SCD in one ticket system.

The geographic information system (GIS) has been applied to public transit information system with great functions of spatial analysis and decision-making. But the related researches focus relatively lessin the multi-mode transport system. Long [10] combined the bus smart card data, household travel survey and the parcel- level land use map to identify commuting trips and job-housing places. Yan and Bian [11] employed the GIS-SDA to analyze the spatial distribution characteristics of settlement agglomeration. The nearest neighbor analysis was used to study the clumped pattern of spatial distribution of settlement. Zhao et al. [12] identified three principal dimensions: spatial, temporal and sequential patterns based on an analysis of daily travel-activity scenario which was implemented in a commercial GIS platform.

Combined with the big data and the intelligent service, the speedy development of the smart city has brought great convenience to people. Branchi et al. [13] analyzed matrix to find the different elements affecting the smart city environment. Li et al. [14] proposed a strategy based on the data mining and cloud computing to support technologies for the smart city. Yuan [15] presented scheduling mode of cloud computing virtual resource and the method of urban sensor network. The application of government linked data is helpful for the improvement of the smart city information system.

However, the above methods are obtained from the theoretical models which combined with the data of the questionnaire survey and one ticket system. On one hand, there are differences not to be neglected between the actual travel behaviors and the theoretical models. On the other hand, SCD from the double tickets system which is carried out in Beijing adds the record of boarding stops and mileage. This paper analyses the travel characteristics basing on the trip chain under the multi-mode public transport. The applying rules of SCD are established by the pretreatment of the data. Then the station attraction and the travel time are calculated through clustering "the alighting time" and filling none value. The processing on SCD contributes to increasing the accuracy in the public traffic information loaded by GIS as well as provide commutes with reference travel. Besides, taking full advantage of SCD plays a positive role in promoting the development of the smart city.

2 SCD Characteristics

2.1 The Data Category and Structure

The research is based on the one-week data of SCD and the actual measurement of the control passenger flow time. The research time of SCD is between 6:00 a.m. and 12:00 a.m. from January 7th, 2015 to January 12th, 2015. The survey of the controlpassenger flow time are April 27th, 2015 at Tongzhou Beiyuan Station (Subway Line Batong) and May 4th, 2015 at Huilongguan Station (Subway Line 13). The research of interest (ROI) data contains 7 bus lines and 5 metro lines with 150216 dates processed. Among them, 89960 are from the ordinary bus and BRT, 60256 from the metro. What is more, there are 24 net-sides of public bikes calculated for travel time in this research. The original data from buses contains 6 fields: "smart card code", "vehicle code", "running code", "boarding station code", "alightingstation code", "alighting time", and that from the metro contains 7 fields: "smart card code", "line code", "boarding station code", "boarding time", "alighting station code" and "alighting time".

2.2 The Data Characteristics

The Data fromthe Transit IC Card System. Since December 28th, 2014, Beijing public transportation has carried out the double tickets system. In this system, the bus stop code is not arranged in order. It is determined by the mileage from one stop to the starting stop. Therefore, the same bus stop code may correspond to several bus stops. The starting stop code from the same line is fixed in SCD.

The alighting time is equal to the boarding time at the same bus stop. Hence it is practicable to fill the data of boarding time by calculating the average value of the alighting time at the same bus stop. After the determination of the vehicle code, the travel time between the stops and the number of people at the station can be further calculated through the running code.

In the multi-mode public transport network, SCD recorded from the transit IC card system is the same to the one from the metro automatic fare collection (AFC) system. Hence tracing SCD number can acquire the owner's traffic information including the trip mode, route choice, origin destination and the transfer pattern.

The Data from the Metro AFC System. The travel time between the metro stations is basic fixed due to the less affection from the road condition. Therefore, it is regarded that the time interval calculated through SCD is equal to the running time between the metro stations.

In order to avoid the severe pressure from large passenger flow volume, some metro stations in Beijing with enormous passenger flow volume take measures to control the passenger crowding at the on-peaktime. On account of the short running time by the metro, it is essential to survey the travel delay time affected by the control of passenger flow volume.

The Data from the Public Bicycle System. The travel distance by bicycle means the distance from the rent net-sides to the return net-sides. On the basis of the fifth urban comprehensive transport survey, 57 % of the travel distance from cyclists concentrate on 2 km. And the average cycling time is 20 min. The paper calculates the cycling travel distance at a speed of 15 km/h.

3 Data Preprocessing and Model Building

3.1 The Data Pretreatment

It is necessary to preprocess the bus smart card data for the analysis of the travel time and the station attraction. There are several principles to be followed in data pretreatment:

1. Screen one code from "the vehicle code" in order to guarantee the data information come from the same vehicle.
2. Screen "the running code" successively in order to guarantee the preprocess data come from the same run.
3. Cluster "the alighting time" from the same run at the same stations. Eliminate outliers so as to calculate the average alighting time for the prediction of the bus arrival time.
4. Identify the travel directionby judging the sign of the travel time in order to guarantee the processed data from the same direction.

3.2 The Model Building

It is essential to calculate the average travel time and the number of people boarding and off the bus for analyzing the travel time between stations and the station attraction. 15 min is regarded as the unit for interval (Fig. 1).

The Data Calculation from the Transit IC Card System. The calculation procedure of the average travel time between the bus stations is presented as follow:

1. Select "the vehicle code" and "the alighting station code" through screening the data. Then sort the "alighting time" in ascending sort order. Suppose "the alighting station code" is n, the number of stations is N, the code of data is i,"the alighting time" is D_b^{ni} and the time interval is $\left\{d_b^{ni}\right\}_1^{N-1} = \left\{D_b^{n2} - D_b^{n1}, D_b^{n3} - D_b^{n2}, \ldots D_b^{nN} - D_b^{n(N-1)}\right\}$.
2. Cluster "the alighting time". If $d \leq 72s$ [16], keep D_b^{ni}, else, eliminate it. Suppose M is the number of kept D_b^{ni}.
3. Calculate the bus arrival time $t_b^n = \frac{1}{M}\sum_{i=1}^{M} D_b^{ni}$, the travel time between stops $t_b = \frac{1}{N}\sum_{n=2}^{N}\left|t_b^n - t_b^{n-1}\right|$, M is the number of metro stations and the average travel time $\overline{t_b}$ which is the mean value of t_b in 15 min.
4. If $M \leq 3$, estimate $\overline{t_b}$ by calculating the speed of bus from the last stop.
5. The calculation procedure flow diagram is shown below:

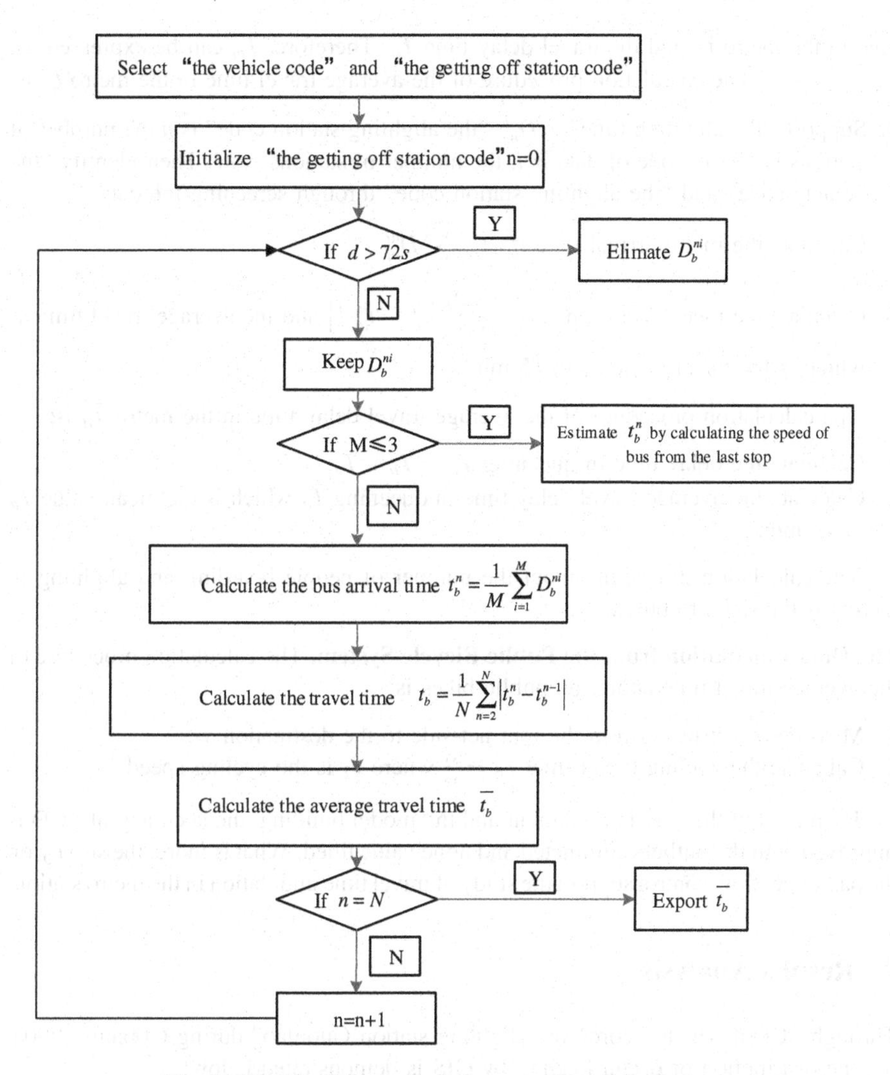

Fig. 1. The travel time calculation procedure flow diagram

The calculation procedure from the number of people alighting is shown below:

1. Screen "the alighting time" and " the alighting stop code". Then select the time and stops for calculation.
2. Count the number of data i.
3. The computing method about the number of people boardingis the same to alighting

The Data Calculation from the Metro AFC System. Suppose the travel time in the metro is $\overline{t_m}$. The time when passenger controlled in the metro pass the entrance of continuous barrier is T_b, and that of the ticket gate is T_g. The travel delay time $\bar{T}_d$. The average travel time between the metro stations $\overline{T_m}$ is divided into two parts: the travel time in the metro $\overline{t_m}$ and the travel delay time $\overline{T_d}$. Therefore, $\overline{T_m}$ can be expressed as: $\bar{T}_m = \overline{t_m} + \bar{T}_d$. The calculation procedure of the average travel time in the metro $\overline{t_m}$ is:

1. Suppose "the alighting time" is D_m^{ni}, "the alighting station code" is n,the number of stations is N, the code of data is i, the number of stations is M. Then identify "the vehicle code" and "the alighting station code" through screening the data.
2. Calculate the metro arrival time $t_m^n = \frac{1}{M}\sum_{i=1}^{M} D_m^{ni}$.
3. Calculate the metrotravel time $t_m = \frac{1}{N}\sum_{n=2}^{N} \left|t_m^n - t_m^{n-1}\right|$ and the average travel time $\overline{t_m}$ which is the mean value t_m in 15 min.

The calculation procedure of the average travel delay time in the metro $\bar{T}_d$ is:

1. Calculate the delay time in queening $T_d = T_b - T_g$.
2. Calculate the average travel delay time in queening $\bar{T}_d$ which is the mean value $\bar{T}_d$ in 15 min.

The calculation procedure about the number of people boarding and alighting in metros is the same to buses.

The Data Calculation from the Public Bicycle System. The calculation procedure of the average travel time about the public bikes is:

1. Measure the distance from the rent net-side to the destination s_2.
2. Calculate the cycling travel time $T_6 = \frac{s_2}{v_2}$ where v_2 is the cycling speed.

By means of the data pretreatment and the model building, the accuracy of SCD is improved with the outliers eliminated and none value filled. What is more, the survey on the passenger flow control support the study of travel time undulation in the metro station.

4 Results Analysis

Through SCDof which record "the alighting station Guomao" during 6:00a.m.-9:00a.m., the distribution of origin loading by GIS is demonstrated below:

The "Guomao" metro station is a transfer station (between Line1 and Line10) near the CBD in Beijing with 30000 people per hour at the peak time. It is full of economic vitality, modernization and internationalization. The metro station has three exits in all and the north-west exit, the north-east exit, the south-west exit are used for analysis. The Figs. 2 and 3 illustrate that in a travel for more than 10 km, Tongzhou, Tiantongyuan, Huilongguan are the concentration origins. The great transportation demand tends to cause traffic jams, increased travel time and some other traffic problems. Therefore, the selected lines crossing the dense area is shown in Table 1, and the structure of SCD is presented in the Table 2.

As shown in Fig. 4, most of the travel time concentrated on 30.01-60.00 min. Thetravel time by bus has a lager variation due to different periods than by the metro. Based on the 150216 travelers who commute to CBD during the week, the average

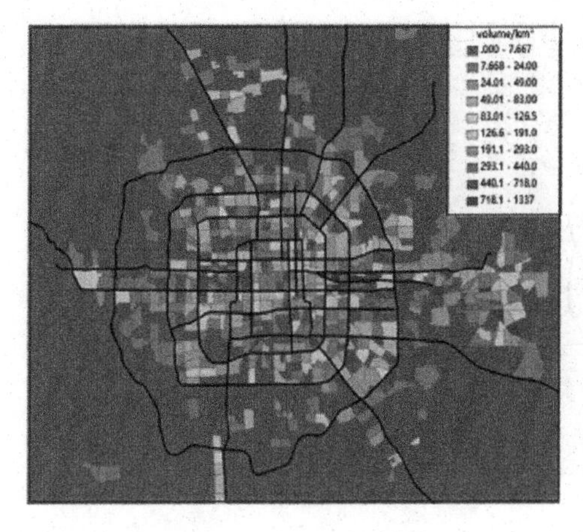

Fig. 2. The distribution of origin

Fig. 3. Three residential areas commute pattern [10].

Table 1. Line code for analysis

Public transit mode	Line code for analysis
metro	Metro Line Batong- Metro Line 1(Tuqiao-Guomao) Metro Line 13-Subway Line 10(Huilongguan-Guomao) Metro Line 1(Pingguoyuan-Guomao)
Ordinary bus	Bus NO.91(Fangzhuangbeikou-Dabeiyaonan) Bus NO.974(Xiaowujiqiaobei-Dabeiiyaonan)
BRT	Bus NO.647(Damazhuang-Dabeiyaodong)
Public bike	The rental net-side at a range of 3.5 km to CBD
Bus-metro	Bus NO.557-Metro Line 13-Metro Line10(XiaoshahecunGuomao) Bus NO.560-Metro Line 13- Metro Line10(Shigezhuangcun-Guomao) Bus NO.583- Metro Line 6- Metro Line 10(Changying-Guomao) Bus NO.363- Metro Line 1(Qingqingjiayuan-Guomao)

Table 2. The structure of SCD

NO.	**Survey time**	**Line code**	**Station code**	$\bar{T}$**(min)**	*i*
…	…	…	…	…	…
5503	7:00–9:00	363	1	40.33	325
5504	7:00–9:00	363	64	33.04	300
5505	7:00–9:00	363	75	32.31	286
5506	7:00–9:00	363	93	31.15	212
5507	7:00–9:00	363	166	28.26	260
5508	7:00–9:00	363	196	26.41	243
…	…	…	…	…	…

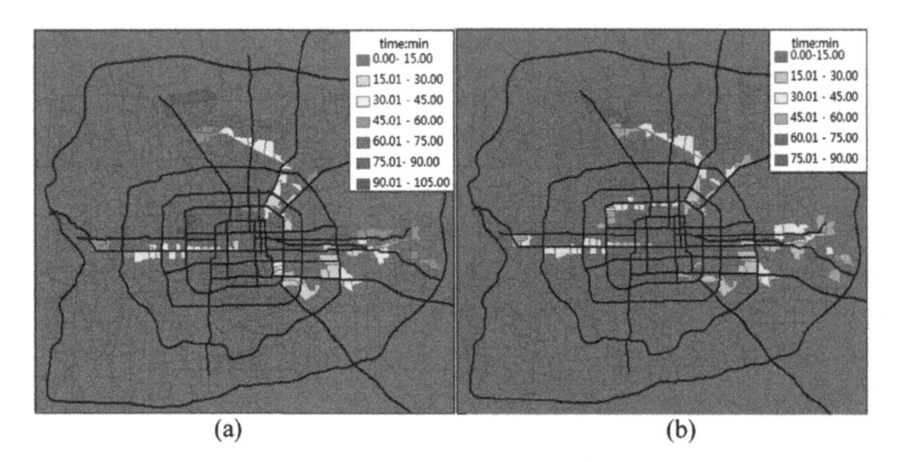

Fig. 4. The average travel time (a) peak time (6:00-9:00); (b) off-peak time (9:01-12:00)

travel time is 41 min at on-peaktime, and 24 min at the off-peaktime. What is more, the standard variance of the travel time is 4.56 min at peak time, and 4.73 min at the off-peaktime.

The Fig. 5 illustrates that a majority of travelers to CBD are commuters and most of them come from Tongzhou. At peak time, the average number of travelers at the metro station is 360, and 144 from the BRT station, 91 from the ordinary bus station.

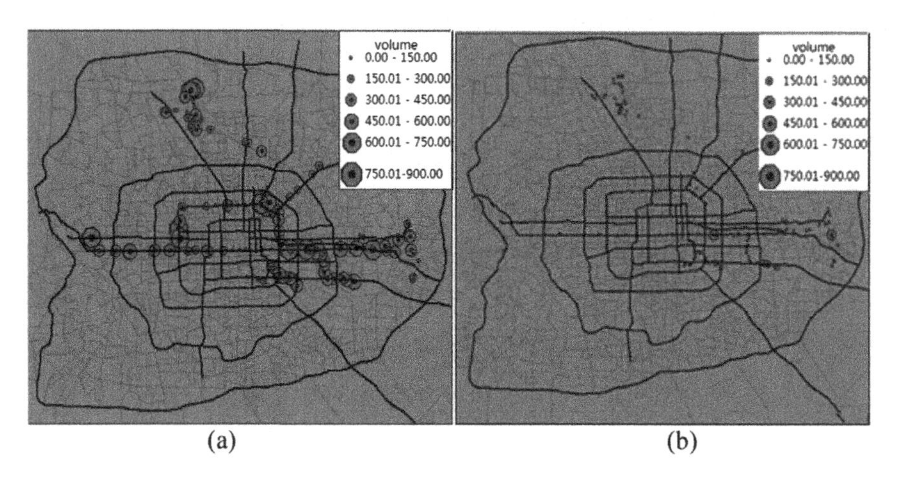

Fig. 5. The average number of travelers (a) peak time (6:00-9:00); (b) off-peak time (9:01-12:00)

Based on the Figs. 4 and 5, the land use has a great impact on the station attraction. As the traffic demand increases, optimizing the traffic resource in CBD and the residential communities play an important role in avoiding the traffic congestion.

The Fig. 6 indicates the relationship between the travel time and the distance to CBD by BRT. The increase in the travel time of the BRT from Damazhuang mainly causes after it drives down the Jingtong Expressway which is one of the urban express way use for the BRT. Though there is the bus lane on the express way, the traffic congestion near the CBD still causes great impacts on the travel time. Therefore, only improving the traffic condition in the vicinity of the CBD can manage the express way in an effective way.

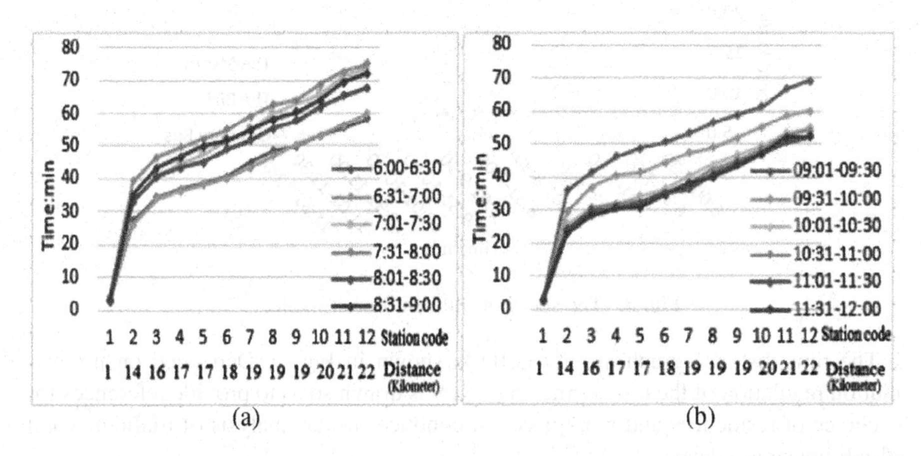

Fig. 6. The travel time by BRT (a) at peak time (6:00-9:00); (b) at off-peaktime (9:01-12:00)

As shown in the Fig. 7, the travel time by the metro is steadier than by the BRT. The significant growth of the travel time is between 8:01-9:00 when the commutes increase sharply in the metro. On account of the short running time by metro, the control of the passenger flow volume and the variation in the pedestrians' density may have an obvious influence on the travel time. Therefore, the station attraction have a significant impact on the variation of the travel time.

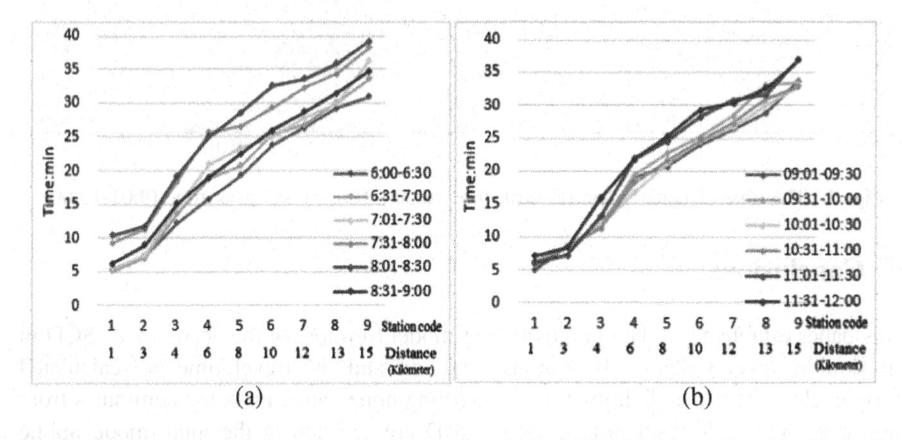

Fig. 7. The travel time by the metro (a) peak time (6:00-9:00); (b) off-peaktime (9:01-12:00)

Figure 8 manifests that the average travel speed at the peak time is 26.2 km/h by the metro, 18.8 km/h by the BRT, 11.1 km/h by the ordinary bus and at the off-peaktime, 28.5 km/h by the metro, 22.6 km/h by the BRT, 15.6 km/h by the ordinary bus.The ordinary bus is the most obvious transit mode affected by the road condition.

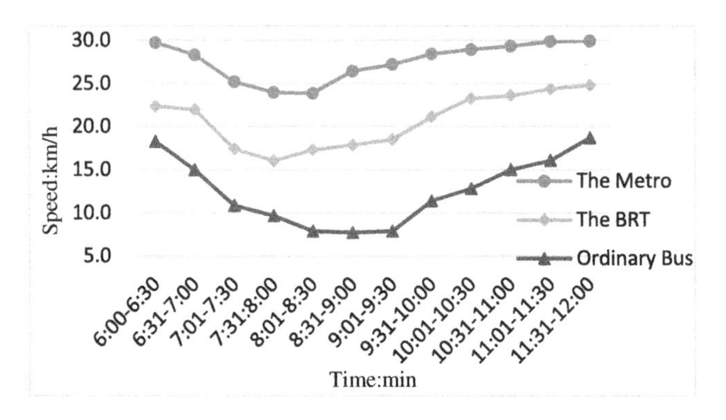

Fig. 8. The speed of different transit modes

The time-distance graph around CBD is shown in Fig. 9. Depended on the distribution regulation of the travel time, the Fig. 9 is drawn so as to provide references for the choice of residences and workplaces. It conduces to the analysis of traffic demand and job-housing balance.

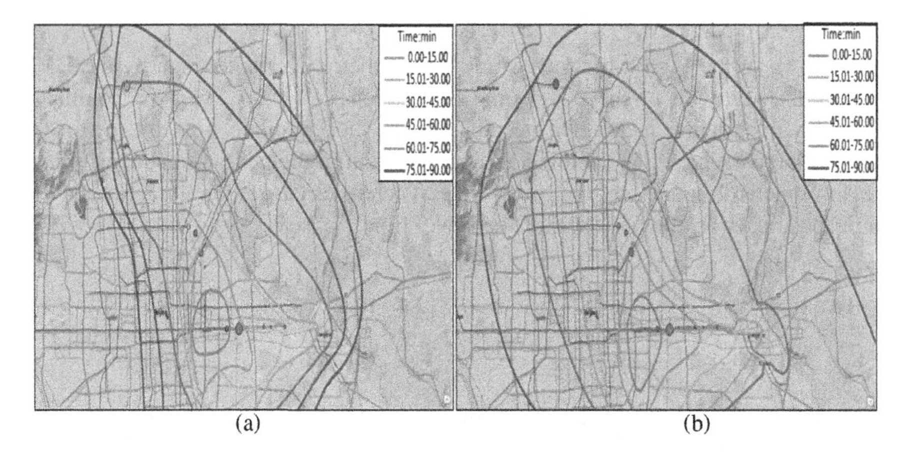

Fig. 9. The time-distance graph (a) peak time (6:00-9:00); (b) off-peak time (9:01-12:00)

5 Conclusions

This paper establishes a data preprocessing model to improve the accuracy of SCD in the double ticket system. The station attraction and the travel time are calculated through clustering "the alighting time" and filling none value. First, the commutes from the three large-scale residential areas to CBD are selected in the multi-mode public transport including the metro, the ordinary bus, the BRT and the public bicycle. Then the one-week SCD including 150216 dates is used to calculate the station attraction and the travel time. Through the data preprocessing model,the average travel time is 41 min at the peak time, and 24 min at the off-peak time. At the peak time, the average number of travelers is 360 at the metro station, 144 at the BRT station, and 91 at the ordinary bus station. The average travel speed at the peak hour is 26.2 km/h by the metro, 18.8 km/h by the BRT, 11.1 km/h by the ordinary bus while at the off-peak time, 28.5 km/h by the metro, 22.6 km/h by the BRT, and 15.6 km/h by the ordinary bus. Next the visualization of the commute patterns is presented through the application of GIS. Finally, the time-distance graph around CBD is made through calculating the speed of different traffic modes.

In an increasingly developed multi-mode public transport today, much more work need to be done to fully utilize SCD combining with GIS. The study of the commute pattern provides travel reference for commutes to arrange time more reasonably as well as assists the selection of the residents and workplaces. Through analyzing and applying SCD, the smarter data services can not only contribute to the analysis of job-housing balance, but also accelerate the progress of the smart cities and transportation.

Acknowledgment. Financial support for this research was obtained through the National Natural Science Foundation of China (NNSFC) project: The research of connectivity and accessibility matrix optimization model for the urban transport network (No.51208014); and Beijing Municipal Education Commission, the General Program of Science and Technology: Journey time estimator under traffic uncertainties.

References

1. Zhang, H.M.: Modeling multi-modal morning commute in a one-to-one corridor network. Transp. Res. Part C Emerg. Technol. **19**(2), 254–269 (2011)
2. Mishra, S., Welch, T.F., Jha, M.K.: Performance indicators for public transit connectivity in multi-modal transportation networks. General Inf. **46**(7), 1066–1085 (2012)
3. Chen, C., Xu, Y., Fu, X.: Travel Time reliability appraisal for urban multi-model public transport network. Comput. Commun. **28**(4), 27–30 (2010)
4. Owen, A., Levinson, D.M.: Modeling the commute mode share of transit using continuous accessibility to jobs. Transp. Res. Part A: Policy Pract. **74**(4), 110–122 (2015)
5. Hu, H., Teng, J., Gao, Y., Zhou, X.: Research on travel mode choice behavior under integrated multi-mode transit information service. China J. Highway Transp. **22**(9), 87–92 (2009)
6. Munizaga, M., Palma, C.: Estimation of a disaggregate multimodal public transport origin-destination matrix from passive smartcard data from Santiago. Transp. Res. Part C Emerging Technol. **24**(9), 9–18 (2012)
7. Kusakabe, T., Asakura, Y.: Behavioral data mining of transit smart card data: a data fusion approach. Transp. Res. Part C Emerging Technol. **46**(9), 179–191 (2014)
8. Chen, S., Chen, Y., Lai, J.: An approach on station ID and trade record match based on GPS and IC card data. J. Highway Transp. Res. Develop. **29**(5), 102–108 (2012)

9. Wu, M.: The Exploration of Data Processing and Analysis Application on the Bus Intelligent Card. Beijing Jiaotong University, Beijing (2010)
10. Long, Y., Yu, Z., Cui, C.: Identifying commuting pattern of Beijing using bus smart card data. Acta Geographica Sinica **67**(10), 1339–1352 (2012)
11. Yan, Q., Bian, Z.: Study on distribution patterns of settlements based on GIS-SDA. geography and geo-information. Science **24**(3), 57–61 (2008)
12. Zhao, Y., Shi, X., Guan, Z.: An analytical framework for travel-activity pattern analysis-a GIS approach. Acta Scientiarum Naturalium Universitatis Sunyatseni **49**(z1), 43–47 (2010)
13. Branchi, P., Fernandez-Valdivielso, C., Matias, I.: Analysis matrix for smart cities. Sustainability **6**(1), 61–75 (2014)
14. Li, D., Yao, Y., Shao, Z.: Big data in smart city. Geomatics Inf. Sci. Wuhan Univ. **39**(6), 631–640 (2014)
15. Yuan, Y.: Key Technologies for Smart City Information System. Wuhan University, WuHan (2012)
16. Guo, J.: The Method Confirming the Station of bus IC Card Passengers and its Application. Southeast University, Nan Jing (2006)

Urban Disaster Comprehensive Risk Assessment Research Based on GIS: A Case Study of Changsha City, Hunan Province, China

Chaolin Wang, Shaobo Zhong(✉), Qianying Zhang, and Quanyi Huang

Department of Engineering Physics/Institute of Public Safety Research, Tsinghua University, Beijing 100084, China
zhongshaobo@tsinghua.edu.cn

Abstract. Existing urban disaster risk assessment methods didn't consider the hazard, vulnerability, capacity synthetically, as well as the typical characteristics of city facing disasters. So they were difficult to meet the needs of emergency preparedness, mitigation and response of the city specific disasters. This paper, taking Changsha City, Hunan Province, China as an example, analyzes disaster risk features of the city taking history disasters, population, economy and social development into consideration from Hazard, Vulnerability and Capacity. And then a model, geospatial visual urban disaster comprehensive risk assessment based on GIS, is established, in which the indicator weights are calculated using an improved gray correlation method. A regionalized disaster risk map of Changsha City is acquired, which can be used for comprehensive emergency management of the city.

Keywords: Urban disaster · Comprehensive risk assessment · GIS · Gray correlation

1 Introduction

Nowadays, with the process of urbanization accelerating, coupled with the rapid economic development, the type and frequency of urban disasters are increasing. On May 14, 2013, a heavy rainfall occurred in Hunan Province, which resulted in a total of 61 million people and 35.76 million hectares crops affected in Changsha, Zhuzhou and other 16 counties. And it resulted in 2688 houses collapsed, and the direct economic loss of 263 million RMB in the whole province. On January 13, 2011, the Xinawan Hotel, in Yuelu District, Changsha suffered from severe fire. 10 people were killed and 4 people were wounded. This fire burned more than 150 square meters and caused economic loss of 602 thousand RMB. The security situation is much more serious. Not only there are more type and higher frequency of disasters, and the degree of hazard is also growing, which is a great threat to human lives and property. Therefore, urban disaster comprehensive risk assessment, the ability of emergency command and disaster prevention and mitigation need be paid more attentions.

F. Bian and Y. Xie (Eds.): GRMSE 2015, CCIS 569, pp. 95–106, 2016.
DOI: 10.1007/978-3-662-49155-3_9

Many researchers have been focused on disaster risk assessment. Zou et al. [1] established the flood risk model based on variable fuzzy theory according to flood, disaster-inducing environment and disaster bearers, which took Jingjiang diversion area as a study case, and calculated the risk degree and risk grade of every evaluation unit. Combing the semi-quantitative model with fuzzy hierarchy analysis method, Wang et al. [2] put forward a GIS-based spatial multi-criteria method for flood risk assessment of the Dongting Lake, and generated the index and comprehensive risk regionalized map. Both methods above have certain practical value, but they were just for a single disaster risk assessment, which could not well provide an appropriate foundation for the urban comprehensive emergency command and regional disaster prevention and mitigation. Yin et al. [3] studied the evaluation index system and comprehensive urban disaster risk assessment model of Shanghai City. But the evaluation index weights for some indicators were calculated in terms of experts scoring, which was much subjective and had a certain impact on the results of the assessment.

This paper analyzes the urban disaster comprehensive risk from three aspects of, hazard, vulnerability and capacity, and establishes the comprehensive assessment system. Then the assessment model based on historical statistical data is established with using an improved gray correlation method which overcomes the disadvantage that the difference between indicator weights is obvious [4]. Finally, a case study of urban comprehensive risk assessment for Changsha City is proposed through this model and the spatial analysis function of ARCGIS 10.2.2 software.

2 Materials and Methods

2.1 Overview of the Study Area

Changsha is located in south-central China, east central of Hunan Province, the lower reaches of the Xiang River, and the western edge of Changliu Basin. It ranges about 230 km from east to west, and about 88 km from south to north. In 2013, it covers an area of 11816 km^2, the urban area is 1909.9 km^2, and the built area is 325.51 km^2. Changsha belongs to Xiang River watershed, a subtropical monsoon climate. Furong District, Tianxin District, Yuelu District, Kaifu District, Yuhua District, Wangcheng District, Changsha County, Ningxiang County and Liuyang County are under the jurisdiction of Changsha. There are 7.04 million resident people, and the population density is 596 people per km^2 according to sixth census. GDP reaches 715.3 billion RMB.

2.2 Data Source

Data used in this paper includes evaluation indicators data and geographic information data.

(1) Evaluation indicators data: collected from 2014 Hunan Statistical Yearbook, 2014 Changsha Statistical Yearbook including GDP, numbers of beds in hospital, construction area and total area of each district/county; collected from sixth census in 2010 including numbers of people in education situation (primary school or less,

junior high school, senior high school, college/junior college or more) and population in different ages; collected from Changsha Department of Civil Affairs and Changsha Administration of Work Safety including the disaster statistical data containing the attributes of disaster type, location, losses of life and economy.

(2) Geographical information data: acquired from National Geomatics Center of China including Changsha administration map extracted from the county-level administrative map, which is vectorized by ARCGIS 10.2.2 software according to the API of Baidu Map for district boundary data of main city. The regional overview of Changsha is shown in Fig. 1.

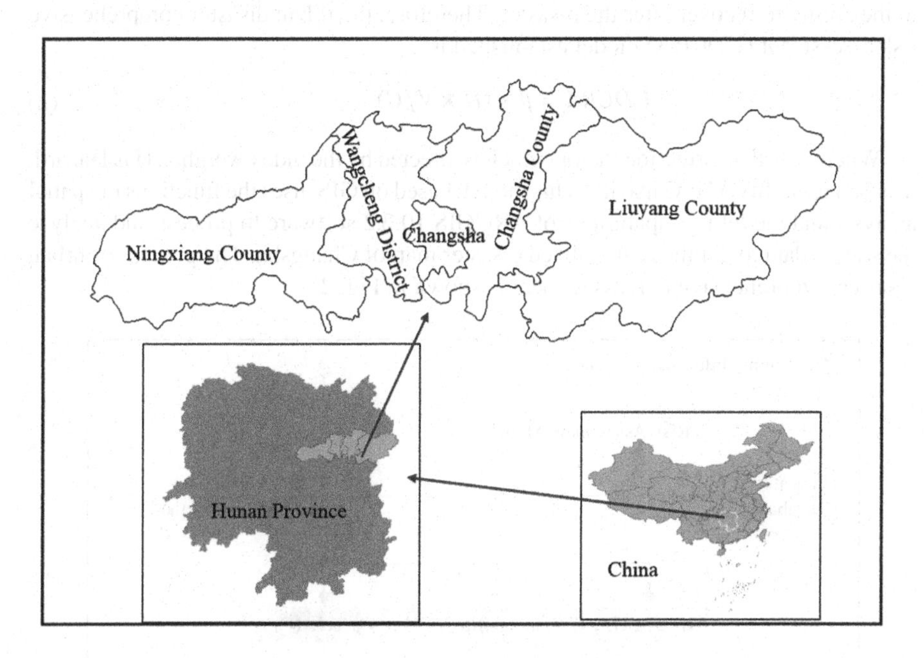

Fig. 1. Regional overview of Changsha

2.3 Urban Disaster Comprehensive Risk Assessment Model

Currently, some methods such as fuzzy comprehensive evaluation, AHP, gray prediction method, comprehensive risk index method are commonly used in urban disaster comprehensive risk assessment [5]. The risk assessment is aimed at calculating the value of risk by combing some indicators and some relevant indexes. The conceptual model of disaster risk is different according to different understanding and definition, which can be divided to three aspects [6]:

(1) The risk is defined as the probability of certain conditions loss from a point of risk themselves;

(2) The risk is defined as the probability of hazard factors from a view of hazard;
(3) The risk is defined as the results of hazard, exposure and vulnerability from a view of the definition of disaster risk system theory.

This paper establishes the risk assessment model based on the risk definition of United Nations International Strategy for Disaster Reduction, as shown:

$$Risk = (Hazard(H) \times Vu\ln erability(C))/Capacity(C) \quad (1)$$

Here: Hazard refers to the possibility of the occurrence of a potential threat; Vulnerability refers to the extent of exposure to the hazard and affordability; Capacity refers to the ability to recover after the disaster. Therefore, the urban disaster comprehensive risk assessment (UDCRA) model established is:

$$UDCRA = \beta \times (H \times V/C) \quad (2)$$

Where, β is the correction factor, which is affected by the index weights; H is Hazard; V is Vulnerability; C is Capacity. This model, based on GIS, uses the functions of spatial analysis and modeling capabilities of ARCGIS 10.2.2 software to process and analyze the source data to get the regionalized disaster map of Changsha. The process of urban disaster comprehensive risk assessment is shown in Fig. 2.

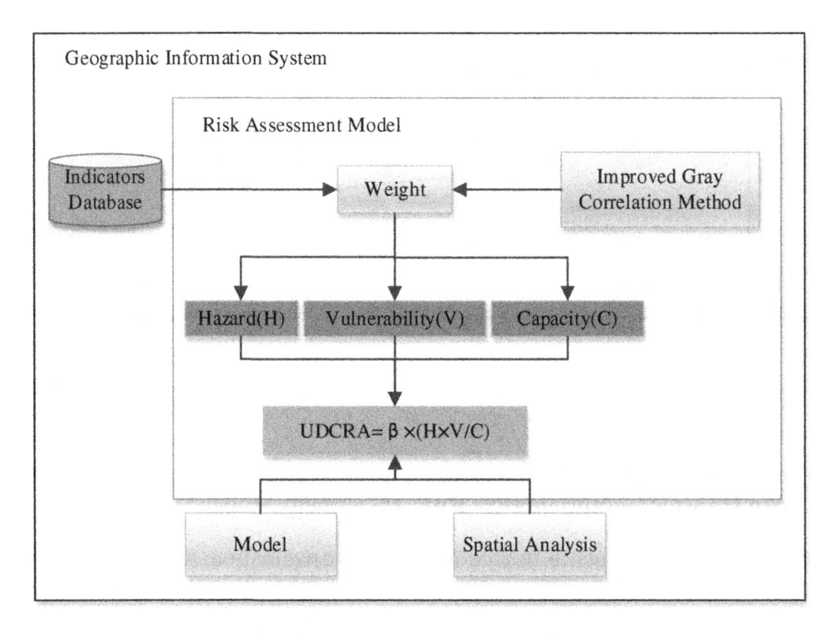

Fig. 2. Process of urban disaster comprehensive risk assessment

The indicators are normalized, which is to convert the data into dimensionless data. The formula is:

$$X'_{ij} = \frac{X_{ij}}{\sum_{j=1}^{m} X_{ij}} \tag{3}$$

Where X_{ij} is the original data, X'_{ij} is the normalized data. This method can effectively solve the problem of the multiple variation of the same index, and make the evaluation result more accurate.

Based on urban administration map, the normalized indicators data is added into personal geodatabase in ARCGIS 10.2.2 software. According to formula (2), the model (Fig. 3) for urban disaster risk assessment is established by using Conversion Tools (Feature to Raster), Spatial Analysis Tools (Weighted Overlay and Raster Calculate) and Toolbox Model. Take $\beta = 9.149$, and the formulas for H, V, C are as follows:

$$H = \sum h_i \times w_i \tag{4}$$

$$V = \sum v_i \times w_i \tag{5}$$

$$C = \sum c_i \times w_i \tag{6}$$

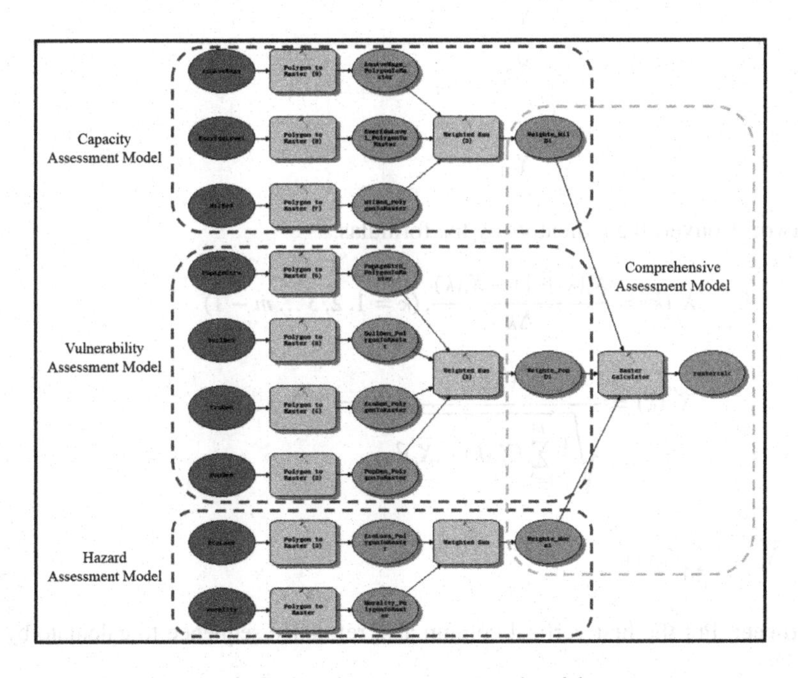

Fig. 3. Model of urban disaster comprehensive risk assessment

Where, h_i, v_i, c_i refers to the normalized value of H, V, C corresponding to secondary indicator, w_i refers to the weight of each indicator.

2.4 Calculation of Indicator Weight

It's difficult to avoid the influence of subjective factors when traditional methods are employed on calculating indicator weights. However, gray correlation method converts the data which is chaos and disorder into obvious regularity by some methods (such as weighted cumulative method, genetic factor accumulation), and solves uncertainties. The method is based on the historical statistics, and it can avoid the influence of subjective factors that are scored by experts, improving the accuracy of the evaluation results. But the difference of weight between indicators is small with this method. Therefore, the method is improved in this paper, and the improved calculation process is as follows [4, 7]:

Step one: Establish the original data table (Table 1), Y_i reflects the system behavior characteristic, called reference sequence, which is typical place for the study of urban disasters. X_i influence the system behavior, called comparison sequence, which is indicators for evaluating.

Table 1. Original data.

	X_1	X_2	...	X_n
Y_1				
Y_2				
...				
Y_m				

Step two: Convert the indicators by the formulas:

$$X_i'(k) = \frac{X_i(k+1) - X_i(k)}{\Delta k}, (k = 1, 2, 3 \ldots m - 1) \quad (7)$$

$$X_i''(k) = \frac{X_i'(k)}{\sqrt{\frac{1}{n}\sum_{i=1}^{n}(X_i(k) - \overline{X_i})^2}}, (k = 1, 2, 3 \ldots m) \quad (8)$$

Where $\overline{X_i} = \frac{1}{n}\sum_{i=1}^{n} X_i(k)$.

Step three: Put the first vertical column as a reference sequence to calculate by the formula:

$$\Delta_{ij}(k) = \left|X_i''(k) - X_j''(k)\right|, (k = 1, 2, 3 \ldots m) \quad (9)$$

Get the absolute difference sequence, then find the $\Delta_{\min}$ and $\Delta_{\max}$. Calculate the correlation coefficient by the formula:

$$\zeta_{ij}(k) = \frac{\Delta_{\min} + \rho\Delta_{\max}}{\left|X_i''(k) - X_j''(k)\right| + \rho\Delta_{\max}}, (k = 1, 2, 3 \ldots m) \tag{10}$$

Where ρ is discrimination coefficient, and $\rho \in (0, \infty)$. The larger the value, the smaller the resolution. Generally, $\rho \in (0, 1)$, which only affects the size of the correlation coefficient, does not affect the incidence order, so, usually, take $\rho = 0.5$.

Gray correlation, r_{ij}, is the number reflecting the degree of correlation between reference sequence and comparison sequence, which is calculated by the formula:

$$r_{ij} = \frac{1}{m}\sum_{k=1}^{m}\zeta_{ij}(k) \tag{11}$$

Step four: Following the above steps, change the sequence of letters, calculate all pairwise correlation to obtain the correlation matrix, $R_{n \times n}$. The diagonal elements of $R_{n \times n}$ are 1 s.

The indicators weight is calculated by the formula:

$$w_i = \frac{\frac{1}{n}\left(\sum_{j=1, j \neq i}^{n} r_{ij}\right)}{\sum_{i=1}^{n}\frac{1}{n}\left(\sum_{j=1, j \neq i}^{n} r_{ij}\right)} \tag{12}$$

3 Case Study

3.1 The Establishment of Evaluation Indicator System

Disaster risk assessment involves many indicators [8], so it's particularly important to select appropriate indicators that can well reflect the disaster risk level. According to the formation mechanism of urban disaster risk and referring extensively to domestic and international evaluation indicator selection principles, the evaluation indicator system of the UDCRA model includes three types of primary indicators: Hazard, Vulnerability and Capacity [9].

Hazard. Hazard, which can be divided into natural hazards and artificial hazards, is the root of urban disasters. In some cities, artificial hazards generally can result in more losses and threats. City waterlogging, mine accidents, traffic accidents and fire are most common in Changsha City. Disaster-related Mortality Rate (Mortality) and Disaster-related Economic Loss (EcoLoss) [10], are taken into consideration to fully represent the amount, frequency and severity of hazards in Changsha City.

Vulnerability. Vulnerability refers to the extent a city suffer from injury and damage, which is the bearers' exposure, sensitivity and tolerance to disaster, or the character of a particular bearer for a certain hazard to suffer from injury and loss in some disaster-inducing environment, at a certain socio-political, economic, cultural background [11]. Population Density Index (PopDen) [12], Economic Density Index (EcoDen), Building Density Index (BuilDen) [13] and Population Age Structure Index (PopAgeStru), are taken into consideration.

Capacity. Capacity refers to the ability of society and government to prevent and response to disasters, which is a composite of predictions, defense, rescue and recovery to disasters of city, reflecting the overall capacity level. Million Beds (MilBed), Emergency Education Level (EmerEduLevel) [14] and Annual Average Wage (AnnAveWage) are considered.

Detailed explanations and calculation of the weights of each indicator are shown in Table 2.

Table 2. Detailed explanation and weight of each indicator.

Indicator I	Indicator II	Explanation	Weight
Hazard(H)	Mortality	The ratio of each district annual death tolls and the total number of deaths tolls	0.097922
	EcoLoss	The ratio of each district economy losses due to disaster and its total GDP	0.106736
Vulnerability(V)	PopDen	The ratio of each district total population and its area	0.121372
	EcoDen	The ratio of each district total GDP and its area	0.116119
	BuilDen	The ratio of each district building area and its total area	0.120809
	PopAgeStru	The ratio of each district people who older than 65 or less than 14 and its total population	0.117297
Capacity(C)	MilBed	Number of beds per ten thousand people	0.111696
	EmerEduLevel	The ratio of each district above 6 year old reached junior high school and its total population	0.102968
	AnnAveWage	The annual average wage level of each district	0.105082

3.2 Urban Disaster Comprehensive Risk Assessment

A personal geodatabase is created in ARCGIS 10.2.2 software, which the normalized indicators data is added in. The projected coordinate system is Beijing_1954_3_Degree_GK_CM_114E in this paper. And the process is as follows.

Hazard. Display the layers of Mortality and EcoLoss which reflect the relative quantity and degree of hazard in a region together. The larger the indicator value, the more the relative amount of hazards, the greater the risk. According to formula (4) and Hazard Assessment Model in Fig. 2, the distribution map (Fig. 4) of hazard in each district of Changsha City is obtained in ARCGIS 10.2.2 software.

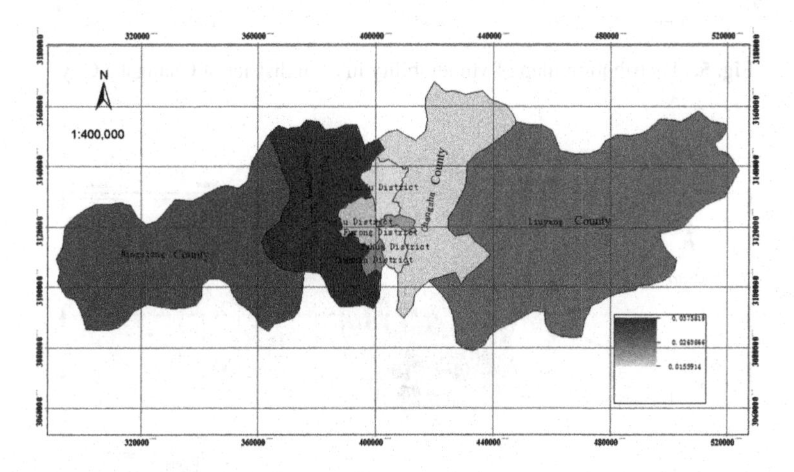

Fig. 4. Distribution map of hazard in each district of Changsha City

Vulnerability. PopDen and PopAgeStru reflect the distribution of people in cities, the greater the PopDen and PopAgeStru, the loss of life caused by more serious when disaster occurs; Similarly, EcoDen and BuilDen reflect social development, the indicator is greater, the more advanced a society, property damage caused by the disaster is bigger. According to formula (5) and Vulnerability Assessment Model in Fig. 2, the distribution map (Fig. 5) of vulnerability in each district of Changsha City is obtained in ARCGIS 10.2.2 software.

Capacity. MilBed and EmerEduLevel represent the level of society response to rescue at the time of disaster, the higher, the stronger the rescue capacity, the smaller the loss. AnnAveWage reflects the resilience after disaster, the higher, the stronger the resilience. According to formula (6) and Capacity Assessment Model in Fig. 2, the distribution map (Fig. 6) of capacity in each district of Changsha City is obtained in ARCGIS 10.2.2 software. In this figure, the greater the indicator, the stronger the disaster prevention and mitigation, the smaller the risk.

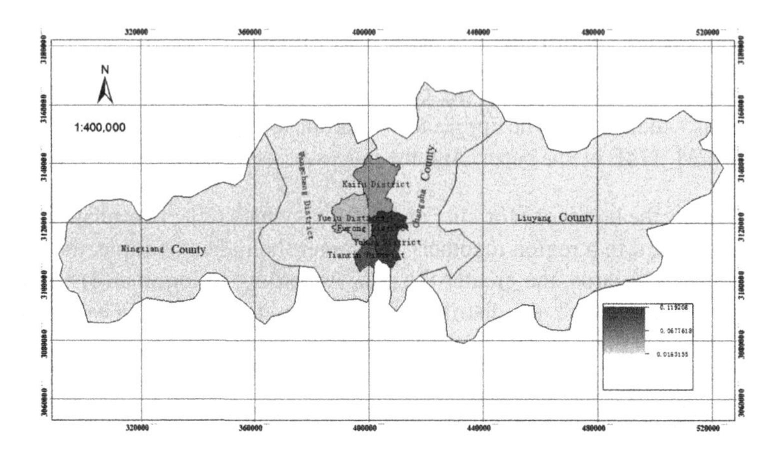

Fig. 5. Distribution map of vulnerability in each district of Changsha City

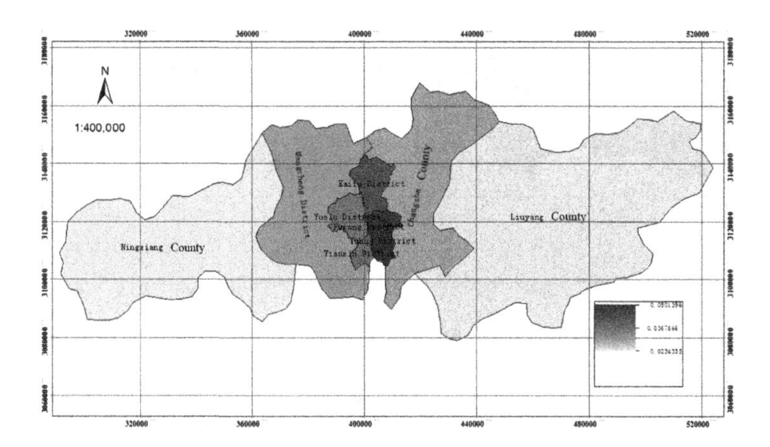

Fig. 6. Distribution map of capacity in each district of Changsha City

UDCRA. Based on the distribution map of hazard, vulnerability and capacity in each district of Changsha City, the risk is assessed by raster spatial computing according to formula (2) and Comprehensive Assessment Model in ARCGIS 10.2.2 software, and the disaster comprehensive risk map (Fig. 7) in each district of Changsha City is obtained. The risk increased gradually with the color from light to dark.

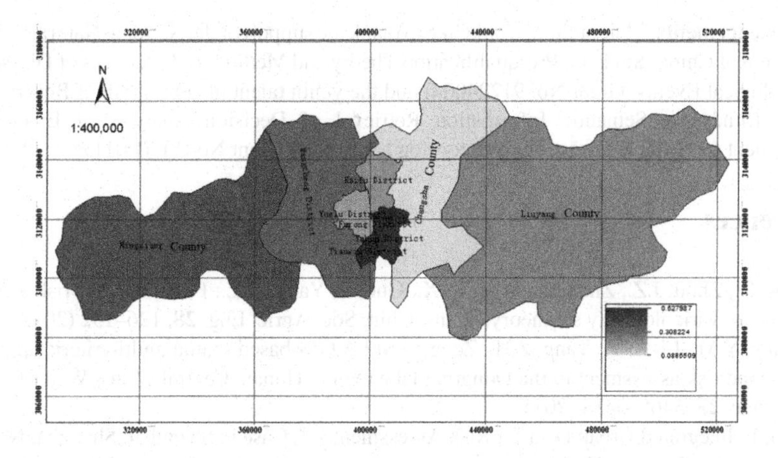

Fig. 7. Disaster comprehensive risk in each district of Changsha City

3.3 Results and Discussions

According to the disaster comprehensive risk map in each district of Changsha City, the highest risk areas are Furong District, Tianxin District and Yuhua District, Ningxiang County and Wangcheng District followed. Higher risk areas which have high population density and prosperous economic zone, account for 2.15 % of the total area of Changsha City. The damage is severe once disaster occurs. The Capacity of Furong District is strong, but due to its poor Hazard and Vulnerability, the comprehensive risk is very high compared to others. As for Ningxiang County, the Hazard level is high, and the Capacity is worse, resulting in its relatively high comprehensive risk.

4 Conclusions

This paper applied the disaster risk system theory and the spatial analysis functions of GIS to improve the existing comprehensive risk assessment model. An evaluation indicator system was established of Changsha City, Hunan Province, China, in which the indicator weights were calculated using an improved gray correlation method. This method overcame the disadvantage that the difference between indicator weights was not obvious. The regionalized disaster risk map, which could be used for comprehensive emergency management of the city, was acquired through ARCGIS 10.2.2 software, finding that: The areas that disaster risk was relatively high accounted for 27.05 % of total area of Changsha City, and the population in this area accounted for 31.98 % of total population. The Capacity of Tianxin District, Ningxiang County and Liuyang County should be stronger. The Hazard level of Wangcheng District was highest. So, its ability of disaster forecasting and warning, safety monitoring and hidden investigation should be paid more attention to lower it. The places where the Vulnerability was higher should be set some shelters and kept emergency supplies.

Acknowledgements. The authors would like to thank the support of the National Natural Science Foundation of China (Study on Pre-qualification Theory and Method for Influences of Disastrous Meteorological Events, Grant No. 91224004) and the youth talent plan program of Beijing City College (Study on Semantic Information Retrieval of Decision Analysis of Emergency Management for Typical Disastrous Meteorological Events, Grant No. YETP0117).

References

1. Zou, Q., Zhou, J.Z., Zhou, C., Song, L.X., Guo, J., Yang, X.L.: Flood disaster risk analysis based on variable fuzzy set theory. Trans. Chin. Soc. Agric. Eng. **28**, 126–132 (2012)
2. Wang, Y.M., Li, Z.W., Tang, Z.H., Zeng, G.M.: A GIS-based spatial multi-criteria approach for flood risk assessment in the Dongting lake region, Hunan, Central China. Water Resour. Manage. **25**, 3465–3484 (2011)
3. Yin, J.: Integrated Urban Disaster Risk Assessment—A Case in Shanghai. Shanghai Normal University, Shanghai (2008)
4. Cui, J., Dang, Y.G., Liu, S.F.: An improved approach for determining weights of attributes in decision making based on grey incidence. Chin. J. Manage. Sci. **16**, 141–145 (2008)
5. Liu, Y.Q., Shen, Y.P., Chen, Y.W.: Reviewing and commenting on the methods of risk assessment. In: 13th Annual Conference of System Engineering Society of China, Hong Kong, pp. 223–234 (2004)
6. United Nations: Risk awareness and assessment, in Living with Risk. ISDR, UN, WMO, Asian Disaster Reduction Centre. Geneva (2002)
7. Chen, X.M., Luo, G.Y.: Grey system analysis and evaluation of slope stability based on experience. Chin. J. Geotech. Eng. **21**, 638–641 (1999)
8. Tie, Y.B., Tang, C.: Establish the evaluation system of urban disaster emergency response capability. Urban Prob. **6**, 76–79 (2005)
9. Yuan, H.Y., Huang, Q.Y., Su, G.F., Fan, W.C.: Theory and Practice of Key Technologies of Emergency Platform System, pp. 101–111. Tsinghua University Press, Beijing (2012)
10. Cardona, O.D.: Indicators of Disaster Risk and Risk Management: Program for Latin America and the Caribbean: Summary Report. Inter-American Development Bank, Washington, DC (2005)
11. Shang, Y.R.: Vulnerability study — the new development of synthetized study on natural disasters. Areal Res. Dev. **19**, 73–77 (2000)
12. Peduzzi, P., Dao, H., Herold, C., Mouton, F.: Assessing global exposure and vulnerability towards natural hazards: the Disaster Risk Index. Nat. Hazards Earth Syst. Sci. **9**, 1149–1159 (2009)
13. Birkmann, J.: Risk and vulnerability indicators at different scales: applicability, usefulness and policy implications. Environ. Hazards **7**, 20–31 (2007)
14. Yin, J., Yin, Z.E., Xu, S.Y., Chen, Z.L., Wang, J.: Disaster risk theory and risk management method. J. Catastrophology **24**, 7–11, 15 (2009)

Synergetic Analysis and Assessment Study on Vital Area of Chengdu Plain Farmland Resource System

Chengyi Huang[1], Langji Deng[2(✉)], Conggang Fang[3], Weizhong Zeng[1], Meixiu Zhou[1], Ruoheng Tian[1], and Fashuai Qin[4]

[1] College of Management, Sichuan Agricultural University, Chengdu 611130, China

[2] College of Resources, Sichuan Agricultural University, Chengdu 611130, China
auh6@sicau.edu.cn

[3] Chengdu Land and Resources Information Center, Chengdu 610072, China

[4] College of Water Conservancy and Hydropower Engineering, Sichuan Agricultural University, Yaan 625014, China

Abstract. This thesis chooses the farmland resource system as the research object. Based on its attributive characters and multi-disciplinary theories such as systematology and synergetics, as well as the characteristics of Chengdu Plain which is one of the Coordinating Urban-Rural development experimental areas in China, the analysis of order parameter and the computation of synergy model give a response to the major problems: whether the farmland resource system of Chengdu Plain and relevant subsystems of the three major systems are synergic; what is the state and degree of synergy of the system. The results indicated that the three major attributes and the entire degree of order of the system developed in a trend of asynergy in the period from 2002 to 2008, and there were obvious area differences. Meanwhile, the degree of synergy between the farmland resource system of Chengdu Plain and the relevant subsystems of land resource system, socioeconomic system and eco-environment system was a negative value ranging between −0.1 and −0.01. Therefore the farmland resource system of Chengdu Plain and the relevant subsystems of land resource system, socioeconomic system and eco-environment system in regional area are in an asynergic and inharmonious state to a certain extent.

Keywords: Farmland resource system · Synergy · Degree of order · Chengdu plain

1 Introduction

With the rapid development of economic society, population growth and accelerated urbanization process, the cultivated land resource, especially those around the cities and towns, are occupied substantially, which severely aggravate the tense human-land relationship and the original orderly and harmonious spatial pattern. Meanwhile, due to the rigidity of food and agricultural product demand, the agricultural product has to rely

F. Bian and Y. Xie (Eds.): GRMSE 2015, CCIS 569, pp. 107–115, 2016.
DOI: 10.1007/978-3-662-49155-3_10

on the intensive cultivation, as well as substantial chemical fertilizer and pesticide for guaranteeing the yield, but it may result in a series of resource and environmental problems, such as soil degradation, agricultural land environmental pollution. Seen from the natural constitution, the cultivated land resources consist of paddy field, dry land, vegetable plot, etc. And the cultivated land resource system is formed by paddy field, dry land, vegetable plot, as a natural resource system. However, owing to the existence of absolute subject of cultivated land resource development, utilization and protection, namely human activities, it endows the economic and society attribute in complete sense, including all the development, protection and utilization activities, as well as the influences, and as a result, it turns to be the complicated and dynamic system closely related to the human activities formed by the mutual interlacing, mutual effect, interaction and interdependency of numerous complicated factors, such as nature, society, economy, ecological environment, etc. which is co-formed by the biotic factors, abiotic factors, and their interaction [1–3]. When the cultivated land resource system and land resource system, social and economic system, and the ecological environment system within certain range cooperate and integrate with each other, and develop towards benign, healthy and orderly direction, the basic production and life demands will be guaranteed, and it will promote the comprehensive coordination and sustainable development of economic, society, resource and eco-environment. Or human society may fall into the resource depletion, environmental degradation, and vicious circle of man-earth relationship [4–7]. Therefore, in the great context of urban and rural overall development, how to protect and develop cultivated land resources scientifically and reasonably and to coordinate the relationship between resource utilization, social economic development and ecological environmental protection turns to be the primary task of current urban and rural overall development. On that basis, with the Chengdu Plain core of the nationwide urban and rural overall development demonstration plot as an example, the deep analysis of the synergic relationship, synergic state and synergic degree among the cultivated land resource system, social and economic system and ecological environment system will be of great realistic significance.

2 Study Area and Data Source

2.1 Overview of Study Area

Chengdu Plain core area locates in the west of Sichuan Basin, about 103°21′08″–104°25′14″ in long east longitude and 30°14′55″–31° 03′44″ in northern latitude, and it mainly refers to the plain area in the south of Guanghan and Shifang County. As for the administrative region, it covers 14 counties and cities, including Chengdu, Jingyang District of Deyang, Guanghan, etc. The area is about 7162.70 km^2, covering Chengdu, and part of Deyang [8]. The land leans towards the southeast from the northwest, about 500 to 750 m above the sea, with an average slope about 3 % to 4 % [9]. The land is flat, the soil is fertile parent material, and the soil types mainly include paddy soil and purple soil. In the region, the cultivated land resource is 491016.68 hm^2, accounting for 36.75 % of the total area. The cultivated soil layer is deep, with moderate character and excellent natural conditions, and it has always been the main manufacturing base of agricultural products, such as food, oil plants, vegetables, etc. In 2011, the GDP of the research area was about 736.83 billion Yuan, the permanent resident population was 14.47 million, and the per capita gross domestic product was 51000 Yuan. With a high urbanization rate, it is the pilot region for comprehensive reforms of overall urban and rural development in China.

2.2 Data

In this research, the related data of land resource is from the survey data of land utilization and alteration in Sichuan; the related data of social economy is from Statistical Yearbook of Sichuan and Statistical Yearbook of Chengdu; in 2002 and 2008, related data of cultivated land fertility index and soil pollution in the research area was from the research results of Liu Yinghua, etc. [10, 11]. SHDI and SHEI are obtained through calculating the change data of land utilization in Sichuan Province.

3 Computing Method and Model

The objective of the construction of collaborative analysis model for cultivated land resource system mainly lies in that: the collaboration between related subsystems of the three systems, including cultivated land resource system and land resource, social economy and ecological environment, shall be analyzed and fully grasped according to the ‘contribution’ to the overall order degree made by the order degree between the key order parameters determining the three parameters of cultivated land resource, quality and spatial arrangement [11]. According to the related evaluation model of collaborative degree [12–15], model for evaluating the collaboration of cultivated land resource system in the research region is constructed by combining the practical and various characteristics of order parameters of cultivated land resource system.

3.1 Order Degree Evaluation Model of Cultivated Land Resource

The calculation model of the quantitative attribute, quality attribute, and order degree $\eta_1(\lambda_1)$ of the spatial arrangement attribute of cultivated land resource system is [11]:

$$\eta_1(\lambda_1) = \sqrt[m]{\prod_{i=1}^{m} \eta_{1m}(\lambda_{1m})} \tag{1}$$

or

$$\text{or } \eta_1(\lambda_1) = \sum_{i=1}^{m} \theta_1 \eta_{1m}(\lambda_{1m}),\text{式中，} \theta_1 \geq 0, \sum_{i}^{m} \theta_1 = 1 \tag{2}$$

The calculation model of the order degree $\eta(\lambda)$ of the key order parameter of cultivated land resource is [11]:

$$\eta(\lambda) = \sqrt[3]{\prod_{j=1}^{3} \eta_j(\lambda_j)} \tag{3}$$

or

$$\eta(\lambda) = \sum_{j=1}^{3} \theta_j \eta_j(\lambda_j), (\theta_j \geq 0, \sum_{i}^{3} \theta_j = 1) \tag{4}$$

In which: j = 1, 2, 3

3.2 Collaborative Evaluation Model of Cultivated Land Resource System

The evaluation model of collaborative degree SC_λ of the cultivated land resource system is [11]:

$$SC_\lambda = \pm \sqrt[4]{\prod_{j=1}^{3} \left|\eta_j^1(\lambda_j) - \eta_j^0(\lambda_j)\right| \left|\eta^1(\lambda) - \eta^0(\lambda)\right|} \; (j = 1, 2, 3) \tag{5}$$

3.3 Determination of Order Degree and Collaborative Evaluation Standard for the Cultivated Land Resource in Chengdu Plain

The constituents of the land resource system are quite complicated, with numerous influence factors. Therefore, when determining the order degree and collaborative evaluation standards, related research achievements [11, 16–18] are referred to.

4 Result and Analysis

4.1 Order Degree and Analysis of Cultivated Land Resource System in Chengdu Plain

In order to reflect the collaborative relationship between related sub-systems in the regional land resource system and land resource system, socioeconomic system and ecological environment system, comprehensive analysis of the order degree and collaborative characteristics of the regional cultivated land resource system is carried out from the perspective of time and space by combining the related calculation result.

(1) Characteristic analysis of the order of cultivated land resource in Chengdu Plain based on the time dimension.

Figure 1 reflects the coordination and cooperation expressed by the order degree characteristic of the overall system between the internal related sub-systems of cultivated land resource system of Chengdu Plain and land resource system, socioeconomic system and ecological system from 2002 to 2008. Over the past six years, the order

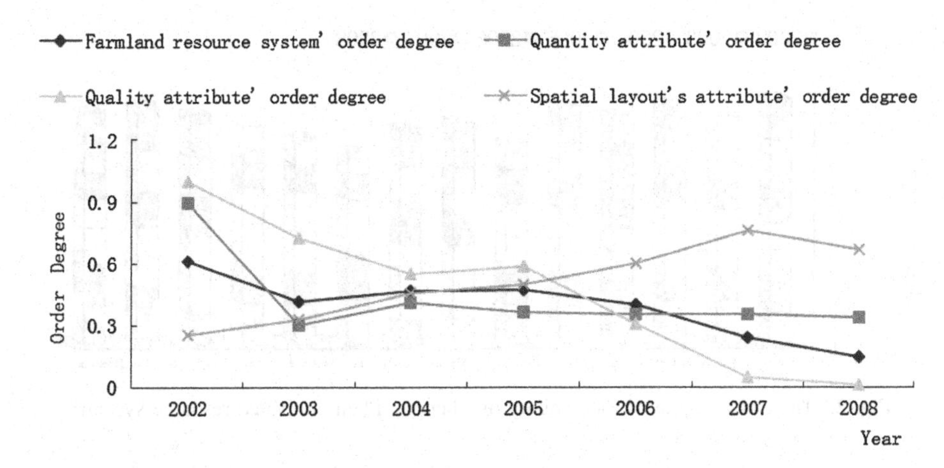

Fig. 1. The order degree of Chengdu Plain cultivated resource system

degree of the cultivated land resource system fluctuated between 0.6 and 0.1, displaying a constant and rapid declining trend, and the entire system gradually develops from the 'unordered' trend to the 'orderly' trend. From 2002 to 2003, since the order degree of the quantity attribute and quality attribute of the system declines at the same time, the order degree of the entire cultivated land resource system occurred in a declining trend, and the system also transformed from the 'orderly' state to 'basically orderly' state; from 2003 to 2005, the order degree of the system quantity and spatial distribution were improved differently. Meanwhile, the order degree of the quality attribute declined slowly, and owing to its impact, the overall order degree of the cultivated land resource system increased constantly; from 2005 to 2008, due to the dual influence of declining quality attribute and fluctuating spatial arrangement attribute of the cultivated land resource system, the overall order degree of the cultivated land resource system declined gradually, and the system developed towards the 'unordered' state.

(2) Characteristic analysis of the order of cultivated land resource in Chengdu Plain based on the space dimension.

It can be reflected by Fig. 2 that from 2002 to 2008 that the orderly varying spatial difference of the cultivated land resource system was relatively evident. In which, the order degree of cultivated land resource system in the center of the plain and next to Wenjiang and Shuangliu County surrounding Chengdu, Guanghan City in the northeast of the Plain varies greatly, displaying a declining trend. Moreover, besides the obvious decline in 2007 and 2008, the order degree of cultivated land resource system in Pixian County within the region in the rest years varied gently. In Pengzhou, north of the Plain, as well as Pujiang and Xinjin, etc. in the southwest of the Plain, the order degree of the cultivated land resource increased gradually year by year from 2002 to 2008. In addition, other regions of the Plain, the fluctuation of the order degree of cultivated land resource system was obvious, occurring in an increasing or declining trend in different years.

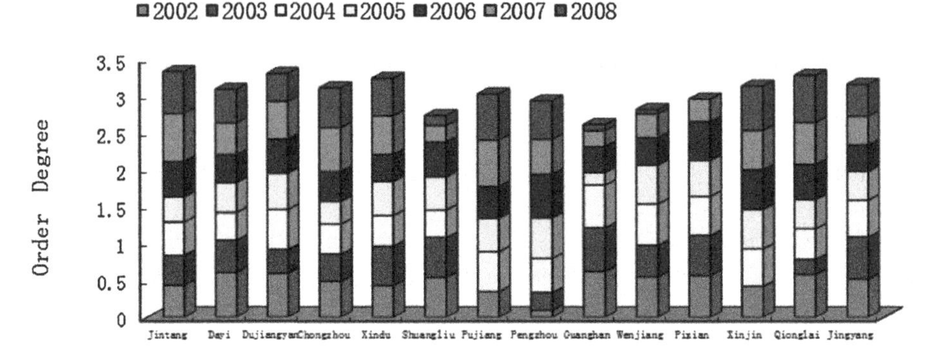

Fig. 2. The order degree of each county of Chengdu Plain cultivated resource system

4.2 Characteristic Analysis of the Collaboration of Cultivated Land Resource System in Chengdu Plain

(1) Characteristic analysis of the collaboration of cultivated land resource system in Chengdu Plain based on the time dimension.

It was displayed by Fig. 3 that from 2002 to 2008, the collaboration between sub-systems of the cultivated land resource system and land resource, socio-economic system and ecological environment system varied between −0.1 and −0.01, suggesting that the system was not in collaborative state. From 2003 to 2005, since the order degree of the system quantity, quality and spatial distribution of the land resource system varied relatively gentle, and meanwhile, the order degree of the quantity attribute, spatial distribution attribute and overall system was improved in the fluctuation. Due to the impact, the overall collaboration of the cultivated land resource system increased constantly, and the overall system gradually developed from un-collaborative state to collaborative state;

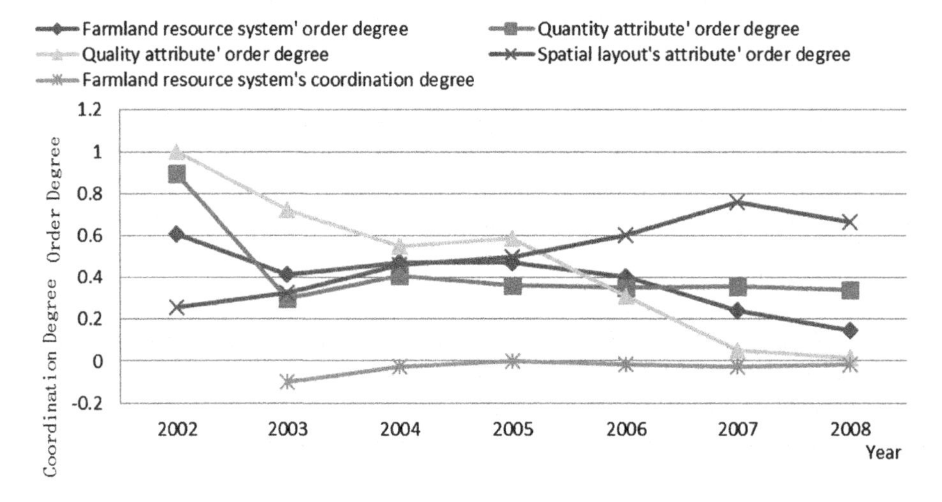

Fig. 3. Chengdu Plain farmland resource system's coordination degree

from 2005 to 2007, due to the gradually decreasing order degree of quality attribute and overall order degree, together with the stable variation of order degree of the quantity attribute, the overall collaboration of cultivated land resource system declined gradually; from 2007 to 2008, the order degree variation of the quantity attribute, quality attribute, spatial arrangement and overall system was steady, and as a result, the collaboration of cultivated land resource system increased a little.

(2) Characteristic analysis of the collaboration of cultivated land resource system in Chengdu Plain based on the space dimension.

In Fig. 4, it is reflected that from 2002 to 2008, the spatial difference of the collaboration of the cultivated land resource was relatively evident. In Dujiangyan, Pengzhou and Chongzhou, north of the plain, the collaboration of the cultivated land resource system displayed evident fluctuation characteristics, occurring in an increasing trend. In which, the cultivated land resource system of Dujiangyan in 2004, Pengzhou in 2004 and Chongzhou in 2007 displayed evident collaborative state. In Qionglai, Dayi and Pujiang in the southwest of the Plain, as well as Xinjin in the couth of the Plain, the collaboration of the cultivated land resource system was improved. In which, the cultivated land resource system of Pujiang in 2007, Qionglai in 2007 and Xinjin in 2008 displayed evident collaborative state; in Jintang, Guanghan, Jingyang, etc. in the northeast of the Plain the collaborative of the system declined gradually. However, the cultivated land resource system of Jintang in 2006 and 2007, as well as Xindu in 2007 displayed certain collaborative state. The four countries in the center of the plain and next to the surrounding of Chengdu, including Wenjiang, Shuangliu, Bixian and Xindu, the collaboration of the cultivated land resource system was not evident. Besides the collaborative state of Xindu in 2007, the collaboration of cultivated land resource system in the rest countries displayed certain declining trend.

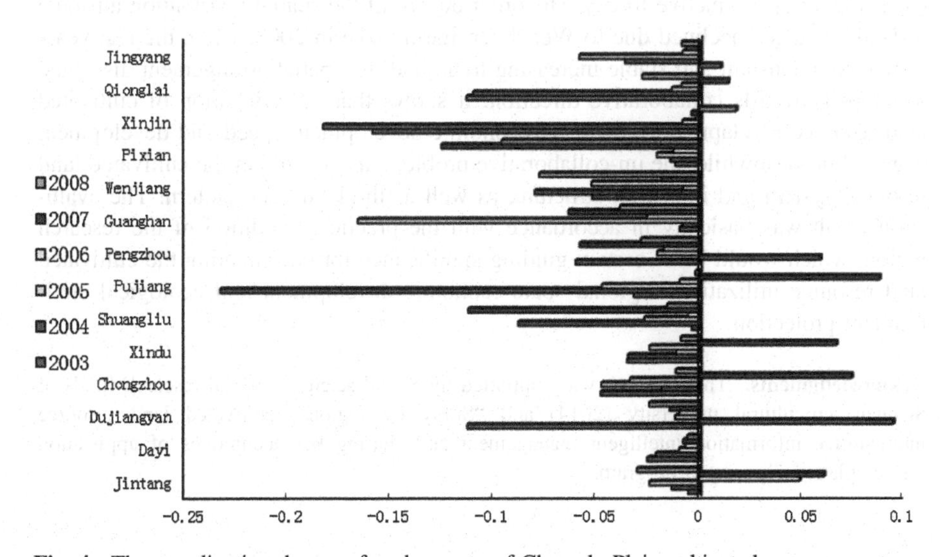

Fig. 4. The coordination degree of each county of Chengdu Plain cultivated resource system

5 Conclusion and Discussion

In this paper, the simulation of coupling relationship between numerous complicated factors impacting the cultivated land resource system, such as nature, society, economy and eco-environment, etc. is analyzed with the order parameters from the characteristics of the three attributes, including quantity quality and spatial distribution of cultivated land resource system, and the collaborative evaluation model of regional cultivated land resource system shall be established on the basis of the systematics and collaboration theory, so as to realize the quantitative exploration of collaborative relation between the sub-systems of the cultivated land resource system and land resource system, socio-economic system and ecological environment system.

In conclusion, the sub-systems of the cultivated land resource system and land resource system, socio-economic system and ecological environment system in Chengdu Plain displayed certain degrees of different collaborative state, with evident time and spatial difference characteristics. From 2002 to 2008, the order degree of the quantity attribute of regional cultivated land resource system experienced the variation process from 'relatively orderly to basically orderly to unordered' state, while the quantity attribute developed towards the un-collaborative direction, which actually reflected that the cultivated resource system, population sub-system of the socio-economic system and other sub-systems of the land resource system were connected and coupled, presenting increasingly tense human-land relationship and structural proportion. In the corresponding period, the order degree of the quality attribute of cultivated land resource system declined constantly, and the annual average reduction rate reached 16.42 %, while the quality attribute developed towards un-collaborative trend drastically, reflecting that influence of socio-economic development on the utilization of cultivated land resource, yield capacity and environmental quality in the current level of productive forces. The order degree of the spatial distribution attribute of land resources declined due to Wenchuan Earthquake in 2008, while the rest years displayed a constant and stable increasing trend and the spatial arrangement distribute developed towards collaborative direction. It shows that the utilization of cultivated land resource is adaptive to the socio-economic development speed and development degree, but meanwhile, the un-collaborative problem also occurs in the cultivated land resource system and industrial structure, as well as the landscape pattern. The evaluation result was basically in accordance with the practical condition of the research region, which would be of certain guiding significance for coordinating the cultivated land resource utilization, regional socio-economic development and ecological environment protection.

Acknowledgments. This research was supported by Social sciences special research funds of Sichuan agricultural university (2014) and "Study on regional cultivated land resource multi-source information intelligent management and sharing key techniques of application service platform" research program.

References

1. Chen, H.: Introduction to Systems Engineering. Higher Education Press, Beijing (2006)
2. Yu, Z., Qiu, J., Wang, J.: Methods Analysis and Realization of Sustainable Land Utilization System. Chinese Agricultural Science and Technology Press, Beijing (1998)
3. Zhu, X., Cai, Y., Jiang, W.: Self-organized character of land use system. Resour. Sci. **19**(2), 62–67 (2005)
4. Hanna: Science, technology and future sustainability. Future **1,** 27 (1997)
5. Hofmann, N., Filose, G., Schofield, M.: The loss of dependable agricultural land in Canada. Rural Small Town Can. Anal. Bull. **21**(6), 3–16 (2005)
6. Tisdell, C.: Economic indicators to assess the sustainability of conservation farming projects: an evaluation. Agric. Ecosyst. Environ. **57**, 2–3 (1996)
7. Hubbard, K.G., Flores-Endoza, E.J.: Relating United States crop land use to natural resources and climate change. J. Clim. **8**, 329–335 (1995)
8. Tan, T., Wang, C., Li, B.: Pollution and evaluation of Pb in soil in Chengdu Plain. Resour. Environ. Yangtze Basin **14**(1), 71–75 (2005)
9. Li, T., Zhang, S., Gan, W.: Temporal-spatial distribution characteristics and influence factors of soil pH in the Chengdu Plain. J. Sichuan Agric. Univ. **24**(3), 313–318 (2006)
10. Li, Y.: Study on Spatio-temporal Variation of Soil Fertility of Chengdu Plain. Master's Degree Thesis of Sichuan Agricultural University, 5 (2004)
11. Huang, C.: Sysergetic Analysis and Adjustment Mechanisms Study on Chengdu Plain Cultivated Land Resource System. Doctor's Degree Thesis of Sichuan Agricultural University, 6 (2011)
12. Yan, L., Suyi, H., Ke, L.: Basic characteristics of material flow, energy flow and information flow in large-seale systems. Adv. Syst. Sci. Appl. **8**(2), 335–341 (2008)
13. Long, Y., Huang, S., Zhang, H.: Approach to synergy among material flow, energy flow and information flow. J. Chem. Ind. Eng. **57**(9), 2135–2139 (2006). (China)
14. Huang, C., Deng, L., Fang, C.: Study on cooperativity evaluation model of farmland resources system. Adv. Inf. Sci. Serv. Sci. **12**(4), 434–443 (2012)
15. Meng, Q., Han, W.: Study on overall coordination degree model research of the composite system. J. Hebei Normal Univ. (Nat. Sci. Edit.) **23**(2), 177–179 (1999)
16. United Nation: Indicators of Sustainable Development Framework and Methodology, New York (1996)
17. Zang, L.: Research on the Optimal Allocation of Land Use Structure in Western Jilin. Doctor's Degree Thesis of Jilin University, 6 (2010)
18. He, B.: Research on Regional Cultivated Land Resource Security—Take Jiangsu Provence as an Example. Doctor's Degree Thesis of Nanjing Agricultural University, 6 (2009)

Spatial-Temporal Monitoring of Urban Growth: A Case in Kunming, Southwest China

Min Liu[1,2], Zhiming Zhang[1], Hongli Zhang[1], Mingyu Yang[1], Ding Song[3], and Xiaokun Ou[1(✉)]

[1] Institute of Ecology and Geobotany, Yunnan University, 2 Northern Green-Lake Road, Kunming 650091, China
{minliu14,xkou}@ynu.edu.cn

[2] Yunnan Key Laboratory of International Rivers and Transboundary Eco-Security, Yunnan University, 2 Northern Green-Lake Road, Kunming 650091, China

[3] Kunming University of Science and Technology, No. 727 South Jingming Road, Chenggong 650500, China

Abstract. With the rapid growth of urban population and economic development, the urban growth has also accelerated dramatically. The paper monitored the urban growth of Kunming by detecting the land use change after supervised classification and analyzing urban expansion rate and intensity index in 1974–2013. The result shows the urban has experienced rapid expansion. Moreover, since 1992 the spatial extension has speed up. The main source of land expansion were farmland, woodland and grassland. And the urban expansion is expanding rapidly to the southeast, northwest and northeast with the old city as the core in Kunming. The urban growth is mainly affected by the natural terrain, economy, population and administrative factors. The study summarized the regularity of expansion and the driving force factors of the city growth, and provide a basis theory for future urban healthy development and provide experience for relevant government and scholars.

Keywords: Urban expansion · Land use change · Remote sensing · Dynamic monitoring · Kunming city

1 Introduction

LUCC (Land Use/Cover Change) research has become hot topics of global change, in where the urban growth (sprawl) is the main content [1–6]. The contemporary urbanization differs markedly from historical patterns of urban growth in terms of scale (large-scale), rate (high speed), location (Europe and America transferred to Asia and Africa) and form (expansion of concentric circles to complex forms), especially concentrated in developing countries [7, 8]. China as the world's fastest growing economy since reform and opening up in 1970s, experienced a rapid urbanization process with a urbanization rate of 17.16 % in 1974 to 53.7 % in 2013 [9]. Urbanization process not only is the concentration of population and economy, especially a geographical space change process, and the urban expansion is a significant characteristics [10, 11].

F. Bian and Y. Xie (Eds.): GRMSE 2015, CCIS 569, pp. 116–127, 2016.
DOI: 10.1007/978-3-662-49155-3_11

With the urban growth, there are great changes in urban land use structure with sharp increase of urban land and a large reduction of non urban land resources such as farmland, which cause more serious problems of the conflict between people and the environment. In recent years, the research on urban land use is getting more and more attention, mainly concentrated in the urban growth pattern change and driving force analysis [12–17]; urban expansion morphology and growth pattern [20–22]; the environment and global change effects of urban growth [23–25]; and urban development simulation research [26–31]. The combination of RS and GIS is a effective way to dynamic monitoring and simulating urban growth, because it can grasp the dynamic land space with specific, rapid and quantitative [14, 18, 19, 32]. However, the current domestic urban expansion research case is mainly concentrated in the eastern developed cities, few in middle-west underdeveloped cities [10].

Kunming as the economic, political and cultural center of Yunnan province in southwest China. Its urbanization process is very significant [33]. In recent years, with the upsurge of LUCC research, some researchers have carried out a series of related research work on urban growth feature and driving force of Kunming [35]. The landuse change in the process of urbanization have rarely been studied, which are either lack of large spatial-temporal scale or direct analysis of land use change in urban area [35–37]. In this context, in order to understand the urbanization characteristics and development mechanism in underdeveloped cities in Western China, Kunming city was selected as a typical sample. The paper used supervised classification to extract the land use of the city based on images of 1974, 1992, 2002, 2013 year, to explore the urban growth of Kunming during 40 years, and also attempt to analyze the specific features and driving factors of urban expansion for the plateau city. The study can effectively guide and control the urban growth of Kunming city and will have positive practical significance in reasonably guiding the regional planning and controlling the scale of urban land. It will provide scientific support for scientific and reasonable land use and other related urban development.

2 Study Area

Kunming city (102°10′–103°40′E, 24°23′–26°33′N; 1890 a.s.l.) is the capital and the only one mega-city of Yunnan province in southwestern China, located in the central of Yunnan province. Situated in a lake basin, surrounded by mountains on three sides, the city enjoys a mild, temperate climate and is known as the "City of Eternal Spring", which has experienced rapid urban construction and expansion since 1992. The city began to receive more attention after 1999 when it held the International Horticulture Exposition and was recognized as an international radiation center for South and Southeast Asia. The master plan 2008–2020 of Kunming city identified the main direction of the city development, it is more obvious that the trend of the city development toward the south of Kunming. In this paper, the study area focused on the main city of Kunming (Wuhua district, Panlong district, Guandu district, Xishan district), Chenggong district and Jinning county (Fig. 1).

Fig. 1. The location of study area.

3 Materials and Methods

3.1 Data Source

The remote sensing image data of this study were derived from The Geospatial Data Cloud platform (http://www.gscloud.cn/). Four Landsat imageries of MSS, TM, OLI-TIRS, ETM+ in 1974, 1992, 2002 and 2013 were adopted. Table 1 shows the specific information of data source.

Table 1. The information of remote sensing image data.

Satellite type	Imaging date	Data type	Bands	Spatial resolution	Cloud amount
Landsat1-3	1974.01.20	MSS	4	79 m	0
Landsat4-5	1992.08.16	TM	7	30 m	0
Landsat7	2002.10.07	ETM+	7	30 m	0
Landsat8	2013.04.20	OLI-TIRS	11	30 m	0.01

3.2 Image Data Processing

This research adopted the Landsat images from many periods and multiple sensors, firstly, the four remote sensing images were geo-referenced to the Universal Transverse Mercator (UTM) coordinate system using ERDAS IMAGINE software package, secondly, geometric correction and image fusion and other preprocessing were made. Finally, in order to match the spatial resolution, the images were resampled to 79 m as the Landsat 1–3 images.

3.3 Land Use Classification

Considering the spatial resolution of the acquired satellite images and the focus of the study on land use change and urban expansion, According to current land use classification (GB/21010-2007) and prior researcher's land classification standard, and combining remote sensing image features and land use status of the study area, land use in this study was subdivided into urban, forest land, agriculture land, grass land, water and other land [33]. The current Google Earth image resolution of the study area is up to 0.27 m enough to replace the field sampling to examine the accuracy, so we used Google Earth image data as a reference for obtaining the ground data of training samples and testing samples. The maximum likelihood classifier of supervised classification in IDRISI was used for land use classification, fifty sampling plots for each land use class were randomly selected to divide into training data and testing data for classification and assessing the classification accuracy. After classification, the corresponding merger and other post processing were carried out, and the precision of Kappa coefficient method was used.

The standard false color composite images are used in the process of classification. The phenomenon of different object with the same spectra characteristics and same object with different spectrum often appears in the process of image recognition, which will affect the classification. So we make the original 6 kinds of landuse for detailed classification to set up the corresponding interpretation signs, according to the texture, color and other information of images. For example, urban was divided into white and blue built up area, water can be divided into blue water, black water, etc. Table 2 shows the main features of the interpretation in the study area and 1992 TM fusion image as an example.

3.4 Urban Growth Monitoring

Urban growth mainly contains area and space change of built-up land. Analysis of land use change is a good way to understand the overall trend of urban sprawl [35, 36]. In this study, the urban expansion rate and the urban expansion intensity are used to reflect the dynamic of urban growth, and the urban spatial expansion rate reflects the average expansion rate, which was calculated by comparing the urban area change of four period in the study area. Its calculation formula is:

$$V = (B - A)/T \tag{1}$$

V is the urban expansion rate; A is the urban area at the beginning of the study; B is the urban area in the end of the study; T is the time interval.

Table 2. Various types of land use interpretation symbol of TM remote sensing image in study area.

Land use	Description	Interpretation instructions(RGB-432)	Image feature
Urban	Including urban and rural residents, transportation facilities and other construction sites (such as industrial area, airport, etc.).	Mainly located in the inner cities, residential areas are light blue, hard court is bright white.	
Agric ulture	Including cultivated land, paddy field, vegetable field, etc.	Mainly has two kinds: slope cropland and paddy field. The former general distributed in the slope on the mountain; the latter distributed in the flat area of the water margin or the edge of the city.	
Forest	Namely, woodland, parks and protective green space, etc.	Mainly distributed in a certain elevation of the mountain, and present a dark red or red.	
Grass	Shrub forest, grass and sparse trees	Generally mixed in forest land, a small amount distributed in the river and the plain area.	
Water	Rivers, lakes, reservoirs, pits and tidal flats.	Mainly has main rivers, lakes and reservoirs and shows blue or bluish black	
Other land	Bare soil, bare rock, gravel and other construction built land	Including the construction land, bare land, gravel, etc. Mainly distributed in the mountains and nearby rivers , few distributed in nearby the built up area.	

Urban expansion intensity index is an important indicator to reflect the change of urban space, which can be quantitatively shows the urban expansion rate and degree [34]. Its calculation formula is:

$$R = (B - A)/A \times (1/T) \times 100\%. \quad (2)$$

R is the urban expansion intensity index; A is the urban area at the beginning of the study; B is the urban area in the end of the study; T is the time interval.

4 Results

4.1 The Land Use Change in the Process of Urbanization

The land use classification of the four periods is shown in Fig. 2. The overall accuracy is respectively: 80.58 % (1974), 84.30 % (1992), 86.48 % (2002), 89.82 % (2013), which all meet classification interpretation standards.

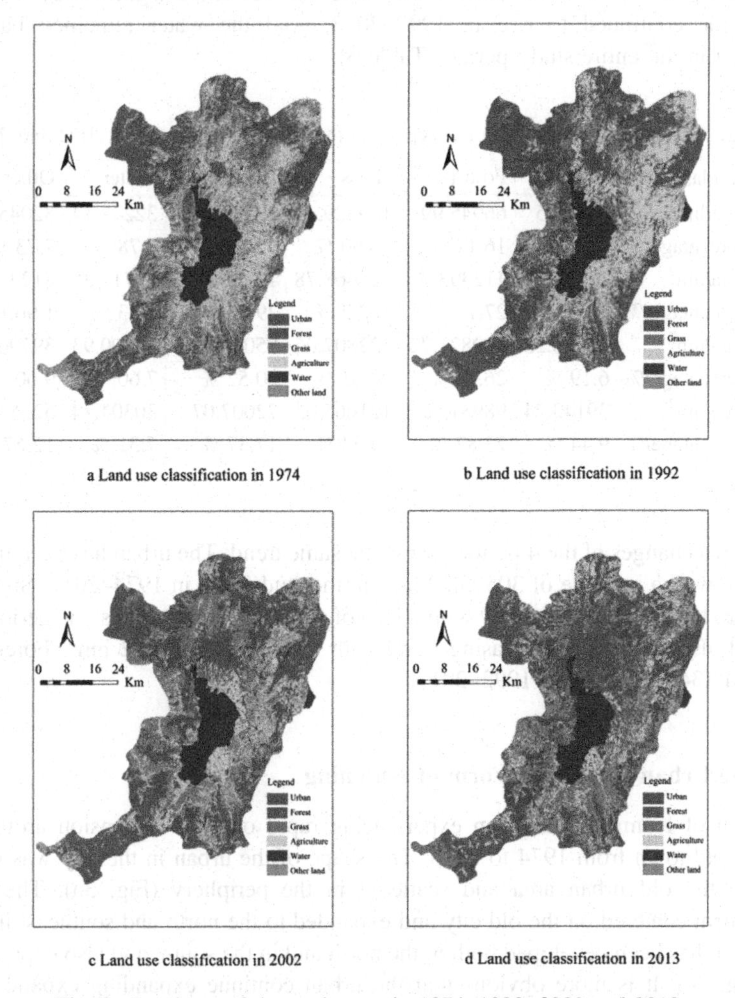

a Land use classification in 1974

b Land use classification in 1992

c Land use classification in 2002

d Land use classification in 2013

Fig. 2. Land use of the study area in 1974, 1992, 2002 and 2013.

Four image interpretation results shows there are significant urban expansion and intense land use/cover change in the study area, which reflects in four landuse of urban, agriculture, forest and grass land. During 1974–2013, the urban had substantial increase and the proportion of increased by 7.38 %. In particular, the growth of the last two periods was outstanding with an increase of 3.23 % in 1992–2002 and an increase of 3.25 % in 2002–2013. However, farmland is experiencing sustained and rapid reduction. It was reduced by 15.31 % in recent 40 years from 1974 to 2013. The proportion reduced largest with a decrease of 9.31 % in 1992-2002, followed by 2002–2013 years with a decrease of 3.15 %. The decrease rate of grassland was relatively slow with a decrease of 4.26 % during the last 40 years, and the largest decline in 1974–1992, followed by 1992–2002. Forest had increased during 1974–1992, and decreased by 3.3 % in 1992–2013. Other land use decreased significantly in 1974–1992, while continued to rise in 1992–2013. And the water remained basically unchanged in the entire study period (Table 3).

Table 3. The comparison of land use types of study area in 1974, 1992, 2002 and 2013.

Land use class		Urban	Forest	Grass	Agriculture	Water	Other land
1974	area/hm^2	8546.95	66945.99	139329.45	135510.45	32277.14	32045.7
	percentage/%	2.06 %	16.14 %	33.60 %	32.68 %	7.78 %	7.73 %
1992	area/hm^2	12272.19	112393.7	124468.78	123684.64	33713.35	8123.02
	percentage/%	2.96 %	27.11 %	30.02 %	29.83 %	8.13 %	1.96 %
2002	area/hm^2	25668.25	109821.72	122807.06	85067.36	31500.93	39790.35
	percentage/%	6.19 %	26.48 %	29.62 %	20.52 %	7.60 %	9.60 %
2013	area/hm^2	39142.24	98981.78	121665.03	72007.07	30304.74	52554.81
	percentage/%	9.44 %	23.87 %	29.34 %	17.37 %	7.31 %	12.67 %

The area changes of the 4 periods show the same trend. The urban has been in rapid expansion with a increase of 30595.29 hm^2 in the study area in 1974–2013. Similarly, agriculture land decreased rapidly with a lose of 63503.37 hm^2 in the same period. The grassland also showed a decreasing trend with a lose of 17664.42 hm^2. Forest land decreased 13411.91 hm^2 in 1992–2013.

4.2 The Urban Expansion Form of Kunming

The urban of Kunming has been experiencing rapid outward expansion around the center of old town from 1974 to 2013. The shape of the urban in the city was mainly based on the old urban area and scattered in the periphery (Fig. 3a). The urban development centered on the old city and expanded to the north and southeast in 1992 (Fig. 3b). It has been developed both in the north and in the south, and also expanded to west (Fig. 3c). It is more obvious that the urban continue expanding expand to the southeast in 2013, the northeast expansion also was led by the new airport construction (Fig. 3d). To the area of urban expansion, the city area of Kunming was 8546.95 hm^2 in 1974, and 39142.24 hm^2 in 2013, which was 4.5 times of 1974.

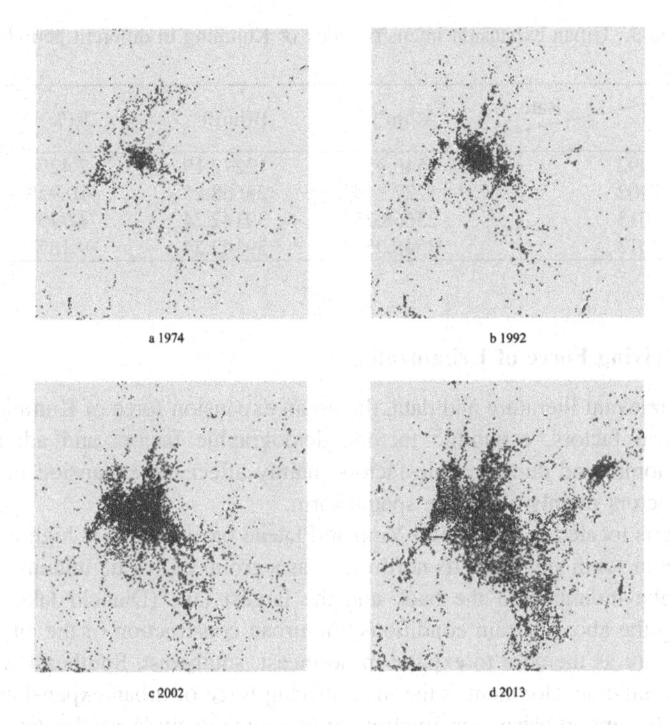

Fig. 3. Spatial distribution of urban (built up area) in 1974, 1992, 2002 and 2013.

4.3 Urban Expansion Rate and Intensity

Table 4 shows the urban spatial expansion rate from 1974–2013. Overall, the average expansion rate was 784.49 hm^2/ year during 1974–2013. And fastest extension at 1339.61 hm^2/year was happened in the period of 1992–2002, followed by the period of 2002–2013 at 1224.91 hm^2/year.

Table 4. The rate of urban expansion.

Time \ Rate	A/hm²	B/hm²	v/hm²
1974-1992	8546.95	12272.19	206.96
1992-2002	12272.19	25668.25	1339.61
2002-2013	25668.25	39142.24	1224.91
1974-2013	8546.95	39142.24	784.49

At the same time, expansion intensity of 4 periods showed a consistent trend with the extension rate. The urban expansion intensity index in study area was 9.18 % in the whole study period and maximum of 10.92 % during 1992–2002. Followed by 2002–2013 with a rate of 4.77 % (Table 5).

Table 5. Urban expansion intensity index of Kunming in different periods.

Time \ Rate	A(hm^2)	B(hm^2)	R(%)
1974-1992	8546.95	12272.19	2.42%
1992-2002	12272.19	25668.25	10.92%
2002-2013	25668.25	39142.24	4.77%
1974-2013	8546.95	39142.24	9.18%

4.4 The Driving Force of Urbanization

Refer to the relevant literature and data, the urban expansion force of Kunming mainly include natural factors, economic factors, demographic factors and administrative factors. Economic and demographic factors mainly affect the expansion rate, natural and policy factors mainly affect the spatial form.

Kunming is located in the central Yunnan Plateau basin, where is high in the north and east, low in south and west, Its northwest has a group of low mountains, hills, with a ribbon snake mountain in the back and the largest lake (Dianchi lake) in south. Restricted by the above terrain conditions, the urban construction of the city only can take the old city as the core to expand in northeast, southwest, Southeast with "Star" pattern. Economic development is the main driving force of urban expansion. There is ample funds to support urban construction, to promote the city's accelerated expansion with high-speed economic development. During the period of 1992–2002, the conduct of "99 World Expo of Kunming" led the economic development of the city and promoted the urban development by leaps and bounds. Population growth will increase the demand of urban land, which will directly result the increase of residential land with rising demand for the traffic and public services and have a impact on the size and shape of the city development. In 1974, the population of Kunming was 1.09 million [37], and the resident population was 6.43 million in 2010 (the sixth national census of Yunnan province). So that, the population expansion of 1974–2013 in Kunming directly lead to the expansion of urban area. On the other hand, the expansion of the city is largely dependent on the government policy, which not only guide the development of the city, but has a certain extent for the blind expansion of the city and make the urban development more scientific and reasonable. The master plan (1996–2010) was proposed to form the space structure with the second ring area as core to develop four vice center in each direction. Subsequently, the master plan (2008–2020) puts forward the idea of the construction of modern new Kunming, which will implement "4 zones with 1 lake and 4th ring road with 1 lake" around the Dianchi Lake to expand the scope of the original city (Wuhua, Panlong, Guandu, Xishan) to develop in the Chenggong, Jinning, Songming. These administrative policies are consistent with the direction of urban space expansion, and play a key role in the expansion form of the city in a large extent.

5 Conclusion

Supervised classification was used to extract and analyze the land use of Kunming in 1974–2013, and analyze the urban expansion rate and strength to monitoring dynamic the urban growth feature in relatively long time scale (nearly 40 years). The results shows that the city is experiencing rapid and dramatical urban space expansion. The expansion of the urban area in 2013 was 4.5 times in 1974, it was the most rapid during the period of 1992–2002 years, followed by the period 2002–2013. Land use change and urban expansion rate and intensity also reflect the above results. The urban expansion makes the corresponding landuse pattern changes. It is mainly reflected in other land use transform into a new urban land. Through the analysis of the land use classification map in 4 periods, the urban expansion is relatively slow and the land mainly from farmland and grassland in 1972–1992 due to the economic and demographic factors. In 1992–2002, urban sprawl was relatively quickly, mainly due to "99 World Expo of Kunming" driving the development of regional economy, population growth and urban planning policy factors, the land mainly from farmland which has disappeared largely, and some from the forest land and grassland. During 2002–2013, the urban sprawl was also relatively fast by the economic development and population growth. It develop a new district of Chenggong affected by the terrain and the government's macro planning policies. The main source of expansion is still farmland, forest and grassland. There has consistent results with the previous researches, the urban expansion led to a sharp loss of farmland and woodland, and ecological landscape fragmentation serious [33, 36]. And the urban spatial form development shows that it is mainly influenced by the terrain and government policy.

The paper has monitored the urban growth dynamic characteristics of Kunming in large spatial-temporal scale, to investigate the historical evolution of urban growth, and explore its mechanism. The study will provide the scientific theory and guidance for city planning, at the same time, it can be used for the construction of ecological and livable city construction for government, planners and ecological experts. However, China will still be in the stage of rapid urbanization, especially in the western cities, where locate the early stage of rapid urbanization.

Further study of urban growth drivers is urgent need, our next work will simulate the future development trend of the city and optimize the structure of urban space to provide important support for the urban development.

References

1. Nunes, C., Auge, J.I.: Land-use and land-cover change (LUCC): Implementation strategy (1999)
2. Turner, B.L., Skole, D.L., Sanderson, S., et al.: Land-use and land-cover change. Science/Research Plan. J. Global Change Report (Sweden) (1995)
3. Turner, B.L., Robbins, P.: Land-change science and political ecology: similarities, differences, and implications for sustainability science. Annu. Rev. Environ. Resour. **33**, 295–316 (2008)

4. Benjamin, S.P., Clarke, K., Findley, J.E., et al.: Geography for a changing world: a science strategy for the geographic research of the US Geological Survey, 2005–2015. US Department of the Interior, US Geological Survey (2005)
5. Rindfuss, R.R., Walsh, S.J., Turner, B.L., et al.: Developing a science of land change: challenges and methodological issues. Proc. Natl. Acad. Sci. U.S.A. **101**(39), 13976–13981 (2004)
6. Zong, W.: Research on Land Use/Land Cover Change and Driving Forces Mechanism in Coastal Zone of Shanghai. East China Normal University, Shanghai (2012). (in Chinese)
7. Shen, L., Cheng, S., Gunson, A.J., et al.: Urbanization, sustainability and the utilization of energy and mineral resources in China. Cities **22**(4), 287–302 (2005)
8. Ramalho, C.E., Hobbs, R.J.: Time for a change: dynamic urban ecology. Trends Ecol. Evol. **27**(3), 179–188 (2012)
9. State Statistics Bureau, China's urbanization rate statistics (1949–2013). (in Chinese). http://wenku.baidu.com/
10. Lin, M., Shi, Y., Chen, Y., et al.: A study on spatial-temporal features of construction land expansion in Changsha urban area. Geogr. Res. **226**(2), 265–274 (2007). (in Chinese)
11. Li, B.: Dynamic Monitoring and Simulation of Urban Spatial Expansion Using Multi-Sources Remote Sensing Data. Zheiiang University, Hangzhou (2012). (in Chinese)
12. Sudhira, H.S., Ramachandra, T.V., Jagadish, K.S.: Urban sprawl: metrics, dynamics and modelling using GIS. Int. J. Appl. Earth Obs. Geoinformation **5**(1), 29–39 (2004)
13. Dietzel, C., Herold, M., et al.: Spatio-temporal dynamics in California s Central Valley: Empirical links to urban theory. Int. J. Geog. Inf. Sci. **19**(2), 175–195 (2005)
14. Zhang, X., Liu, M., Meng, F.: Expansion of urban construction land in shanghai city based on RS and GIS. Resour. Environ. Yangtze Basin **15**(1), 29–33 (2006). (in Chinese)
15. Kuang, W., Liu, J., Shao, Q., et al.: Spatio-temporal patterns and driving forces of urban expansion in Beijing central city since 1932. J. Geo-information Sci. **11**(4), 428–435 (2009). (in Chinese)
16. Ren, W., Jiang, D., Dong, D., et al.: A study on remote sensing monitoring and evolution characteristics of urban expansion of Chengdu city. J. Gansu Sci. **26**(2), 15–21 (2014). (in Chinese)
17. Cui, X., Wang, Y.: Social and human driven mechanism of urban land expansion-case study of city of Shanghai. China Civ. Eng. J. **S2**, 306–310 (2012). (in Chinese)
18. Liu, L., Wang, Y., Chen, G., et al.: Research of urban change monitoring based on multi-source remote sensing data: a case of Zhengzhou city. Areal Res. Dev. **29**(1), 136–140 (2010). (in Chinese)
19. Chen, S., Pu, X.: Monitoring and analysis of urban expansion of Chengdu city based on RS. Resour. Dev. Market **9**, 779–781 (2012). (in Chinese)
20. Jiao, L., Wu, S.: Analyzing the characteristics of the expansion of the metropolises in China from 1990 to 2010 using self-organizing neural network. Geomatics Inf. Sci. Wuhan Univ. **12**, 009 (2014). (in Chinese)
21. Zhou, X.: Research on Urban Space Shape Evolution of Kunming. Chongqing Uniwersity, Chongqing (2008). (in Chinese)
22. Zhu, Y., Yao, S., Liyujian: On the Urban Spatial Evolution in the Process of Urbanization in China. Geogr. Territorial Res. **16**(2), 12–16 (2000). (in Chinese)
23. Herold, M., Goldstein, N.C., Clarke, K.C.: The spatiotemporal form of urban growth: measurement, analysis and modeling. Remote Sens. Environ. **86**(3), 286–302 (2003)
24. Grimm, N.B., Faeth, S.H., Golubiewski, N.E., et al.: Global change and the ecology of cities. Science **319**(5864), 756–760 (2008)
25. Bart, I.L.: Urban sprawl and climate change: a statistical exploration of cause and effect, with policy options for the EU. Land Use Policy **27**(2), 283–292 (2010)

26. Batty, M., Xie, Y., Sun, Z.: Modeling urban dynamics through GIS-based cellular automata. Comput. Environ. Urban Syst. **23**(3), 205–233 (1999)
27. Zeng, H., Yu, H., Guo, Q.: Dynamic model construction and simulation study of town landuse for Longhua area, Shenzhen city. Acta Ecologica Sin. **20**(4), 545–551 (2000). (in Chinese)
28. Barredo, J.I., Kasanko, M., McCormick, N., et al.: Modelling dynamic spatial processes: simulation of urban future scenarios through cellular automata. Landscape Urban Plann. **64** (3), 145–160 (2003)
29. Oluseyi, O.F.: Urban land use change analysis of a traditional city from remote sensing data: The case of Ibadan metropolitan area, Nigeria. Humanity Soc. Sci. J. **1**(1), 42–64 (2006)
30. Shen, X.: Urban Sprawl and Dynamie Simulation of Shanghai. East China Normal University, Shanghai (2010). (in Chinese)
31. Yang, B., Fang, Y., Feng, H., et al.: Construction land expansion in urbanization process of Dongguan city. J. Geo-information Sci. **11**(5), 684–690 (2009). (in Chinese)
32. Li, X., Yang, G., Zhang, X., et al.: Application of multiscal sementation to the detection of urban expansion. J. Mt. Res. **3**, 005 (2014). (in Chinese)
33. Zhou, X., Wang, Y.: Spatial–temporal dynamics of urban green space in response to rapid urbanization and greening policies. Landscape Urban Plann. **100**, 268–277 (2011)
34. Xu, H.: Remote sensing information extraction of urban built-up land based on a data-dimension compression technique. J. Image Graph. **10**(2), 223–229 (2005). (in Chinese)
35. Cai, H., Li, J.: Comparison of the urban area expanding of Kunming supported by ERDAS software. Geomatics Spat. Inf. Technol. **29**(4), 72–75 (2006). (in Chinese)
36. Feng, S.: Studies on 32 years Land use cover change and its drivering forces in Kunming suburbs. TroPical Botanical Garden Chinese Aeademy of Science, Xishuangbanna (2009). (in Chinese)
37. Li, H.: Research on temporal and spatial evolution and driving forces of urban construction land expansion based on GIS in Kunming. Yunnan University of Finance and Economics, Kunming (2012). (in Chinese)

A Kind of Vibratory Isolation Algorithms Based on Neural Network

Shuqing Li[1], Na Zhang[1(✉)], ZhiFei Tao[2], Jianliang Li[1], LiangLiang Wang[1], and Lei Ma[2]

[1] Tianjin University of Science and Technology, Tianjin 300222, China
Zhangna428@yeah.net

[2] China Petroleum Group Oriental Geophysical Exploration Co., Ltd., Hebei 072750, China

Abstract. Vibration isolation technology makes a significant effect in the high-precision instruments field, however, the anti-interference technology at low-frequency and ultra-low frequency becomes the bottleneck of high-precision instrument development obstructively. The regular vibratory used in oil and gas exploration has a good effect on controlling the interfering signals above 6 Hz, but it doesn't work well under 6 Hz. However, the Low-frequency excitation for hydrocarbon detection become a hotspot. In this paper, a hybrid vibration isolation method is proposed to suppress the interfering signal 6 Hz below and to improve the accuracy of the controllable vibratory excitation signal. A neural network (NN) with unique non-linear approximation capability is adopted to identify the vibration system and a NN predictive controller takes active control for the vibration systems. A simulation model is established using MATLAB/SIMULUNK. The simulation results showed that the proposed NN-based hybird isolation method can suppress the interference signals magnitude down by more than 92 % for 3–6 Hz interference signals, which put forward a novel effective anti-interference method for low-frequency vibration applications.

Keywords: Active vibration isolation · Controllable vibratory · Neural network predictive control algorithm · MATLAB/SIMULUNK simulation

1 Introduction

From the beginning of the 1990s, the researchers came from Sweden and the former Soviet Union (Russia) proposed that the low-frequency signals(1–3 Hz) from natural seismic activity may be one reflection of reservoir, and can be used for directing hydrocarbon indicator (DHI), especially in the reservoir gas detection.

In recent years, the high-yield oil and gas flow were drilled from the volcanic rocks at Daqing Oilfield, Shengli Oilfield, Liaohe Oilfield and Karamay Oilfield in China. Most exploration experts would abandon when they encounter the volcanic rocks due to the difficulty of exploration at past. With the development of volcanic exploration, it becomes an important exploration targets instead of traditional sandstone exploration. As the transferring speed and the impedance changes fast, seismic energy transfer to the

F. Bian and Y. Xie (Eds.): GRMSE 2015, CCIS 569, pp. 128–136, 2016.
DOI: 10.1007/978-3-662-49155-3_12

underground deeply becomes difficult with low noise ratio. To compare the transferring effect of high frequency and low frequency, a vibratory is used to generate low frequency signal and field validation test was carried out. The imaging results shown that the objective layer is more clear when low frequency signal is applied contributed by the low-frequency characteristics of wavelengths [1–4]. The recorded field images were shown in Fig. 1.

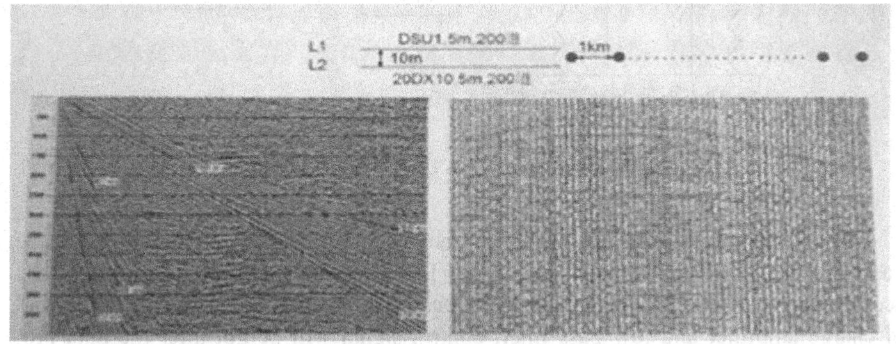

a,The recorded near-surface seismic waves at low frequency excitation signal
b,The recorded near-surface seismic waves at high frequency excitation signal

Fig. 1. The images of field verification test

But the biggest challenge using low-frequency signal excitation is the anti-interference technology. Compared with passive vibration isolation, active vibration isolation has better adaptive performance, in particular, the advantages of ultra-low frequency vibration isolation is more obvious [5]. Adaptive feed forward control based on the filtering LMS algorithm is the main method of controlling active vibration isolation system [6], however, this method is based on linear control theory, the components (such as the electromagnetic actuator, the magnetostrictive actuator) in actual system introduce a significant non-linear, so the method is difficult to achieve a good effect. Therefore, the nonlinear control method based neural networks have received increasing attention [7–9].

The nonlinear neural network model applied to the neural network model predictive control can predict the future performance of the system; the controller calculates the control input and control input signal within the specified time to achieve the optimal performance of the system [10]. In this study, using the neural network model predictive control method, taking electromagnetic exciter as a initiative executing agency. A model of the active vibration control system and conducted a simulation study is designed.

2 Controlling System Schematic

Active vibration control system structure is shown in Fig. 2. The main components of the control system with Cantilever Beam, Vibration Platform, Sensor, Charge Amplifier, Control Systems, Actuator and Power Amplifier, etc.

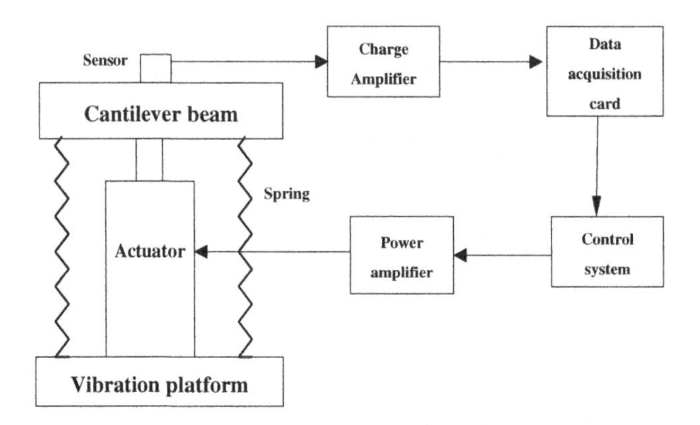

Fig. 2. The structure of active vibration control system

The interference signals caused by the environment and people activities will be passed to the cantilever through the ground [11]. The vibration signal acquired by the acceleration sensor, then, the signal will be passed to the control system after the work of charge amplifier and low pass filter, control signal will be got after the data processing of the control system, the actuator obtain the amplified signal by the power amplifier and drive the actuator to generate a force for suppressing vibration disturbance, so as to achieve the purpose of the vibration attenuation.

3 Mathematical Model of Vibration System

Restricted the Vibration interference signal mainly come from the vertical vibration, in this paper, we build the model based on single degree of freedom [12, 13]. Kinetic models established in the vertical direction is shown in Fig. 3.

In Fig. 3, m represents the mass of cantilever, x represents the vibration displacement of the cantilever, y represents the displacement of the foundation vibration, f represents the force generated by the actuator, c and k represent the spring stiffness and damping respectively.

Equations of motion can be listed according to Fig. 3.

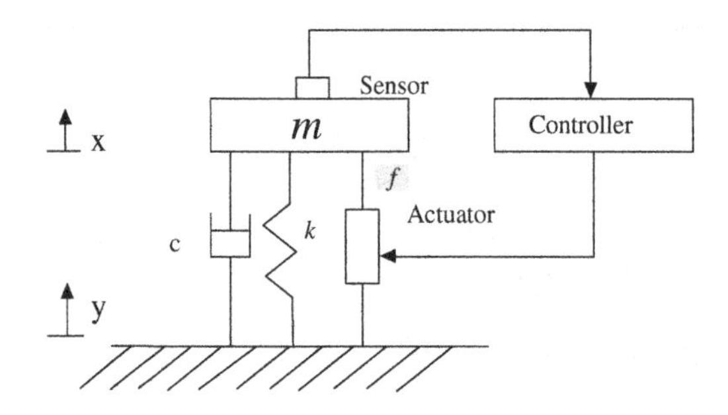

Fig. 3. The dynamics model of active vibration control system

$$m\ddot{x}(t) + c\dot{x}(t) + kx(t) = c\dot{y}(t) + ky(t) + f(t) \tag{1}$$

Combining the Eq. (1) as obtained to take the Laplace transform can express as:

$$(ms^2 + cs + k)X(s) = (cs + k)Y(s) + F(s) \tag{2}$$

Without applying control, the corresponding transmissibility equation can be expressed as:

$$G(S) = \frac{X(s)}{Y(s)} = \frac{cs + k}{ms^2 + cs + k} \tag{3}$$

4 Neural Network Predictive Control Algorithm

For the active vibration control, many control methods are widely used in the field of industrial control, the main control method comprising: PID control, adaptive control, fuzzy control, neural network control, etc. The paper uses the neural network predictive control algorithm, the characteristic of this method lies in the following three aspects [10]:

(a) The controller uses neural network model can predict the system response for all possible control signals.
(b) To achieve the optimal performance of the system, the optimization algorithm is choose to calculate the control signal.
(c) Training the model of neural network system is offline, we can choose any kind of algorithm with batch processing approach as training methods.

Model predicts mainly composed of two parts: the first part is to build a neural network model (System Identification); The second part is to predict the future performance of the system with this model.

4.1 System Identification

To simulate the dynamics of the system, neural network training is necessary. Using the prediction error between system outputs and neural network outputs as neural network training signal, the training process is shown in Fig. 4.

4.2 Predictive Control

Model predictive control algorithm based on Receding Horizon Technique (RHT). Neural network model prediction responds within the specified time, using numerical

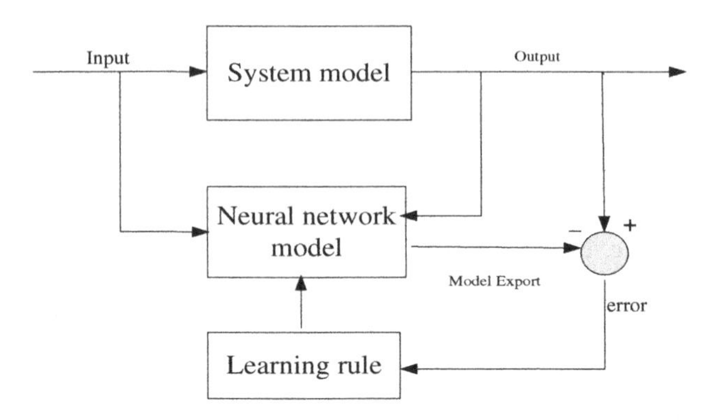

Fig. 4. The process of neural network training

optimization algorithms to predict for determining the control signal, optimal performance of the function can be expressed as [10]:

$$J = \sum_{j=N_1}^{N_2} (y_r(t+j) - y_m(t+j))^2 + \rho \sum_{j=1}^{N_u} (u'(t+j-1) - u'(t+j-2))^2 \quad (4)$$

Where, N_1, N_2,..., N_u, expressed as a range of variables, u' expressed as a test control signal, y_r expressed as a expected response, y_m expressed as a neural network model response, size ρ value reflects the distribution of the sum of control gain's squares.

The process of model predictive control is shown in Fig. 5.

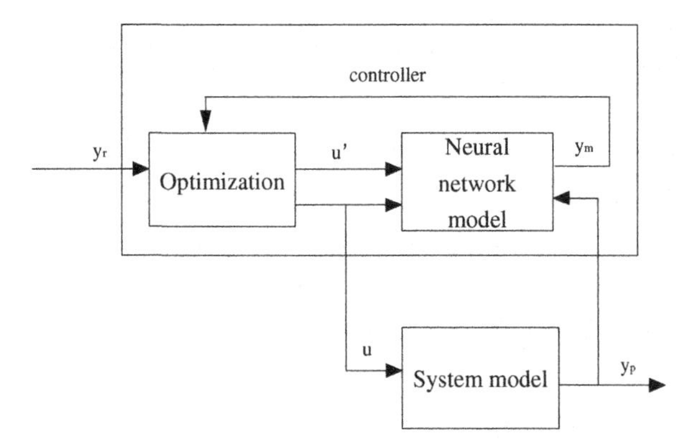

Fig. 5. Process of model predictive control

5 The Simulation of Active Vibration Isolation System

According to the mathematical model of vibration system and transfer function, the simulink simulation model of active vibration isolation system based on neural network predictive control can be made in the MATLAB environment [14]. As shown in Fig. 6.

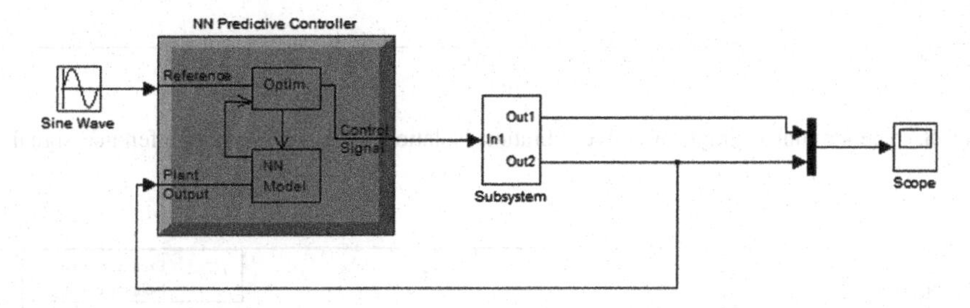

Fig. 6. The simulation model of active vibration isolation system

Wherein the model subsystem shown in Fig. 7:

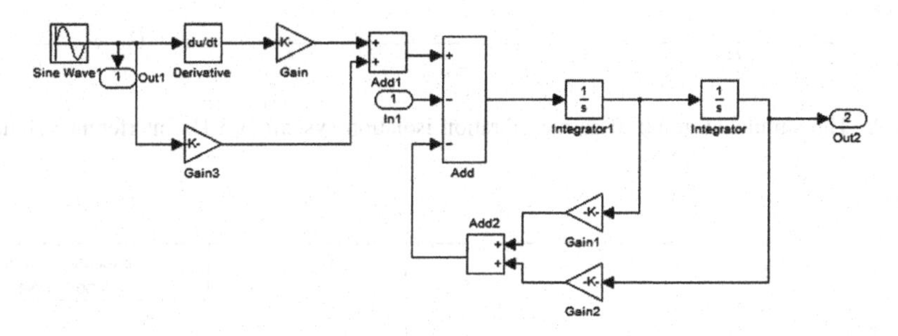

Fig. 7. Subsystem model

In this paper, vibration system parameters are: $m = 2.8 \times 10^4$ kg, the stiffness of spring $k = 2.1144 \times 10^6$ N/m, damping is $c = 13760$ N.S/m.

Set the frequency of 3–6 Hz, the vibration amplitude of the interference of 2 mm, the simulated graph of active vibration isolation system is shown in Figs. 8, 9, 10 and 11.

Equations of the isolation efficiency is $\eta = \frac{x - x'}{x}$, wherein, η is the isolation efficiency, x is the amplitude of interference signal, x' is the amplitude after isolation, we can calculate the isolation rate according to the value from the simulated graph. The results are shown in Table 1.

Simulation curve from Figs. 7, 8, 9 to 10 show that the vibration isolation efficiency decreases with the reduced frequency. The hybrid vibration isolation efficiency can reach 96 % at 6 Hz, the hybrid vibration isolation efficiency is still able to reach 92 % at

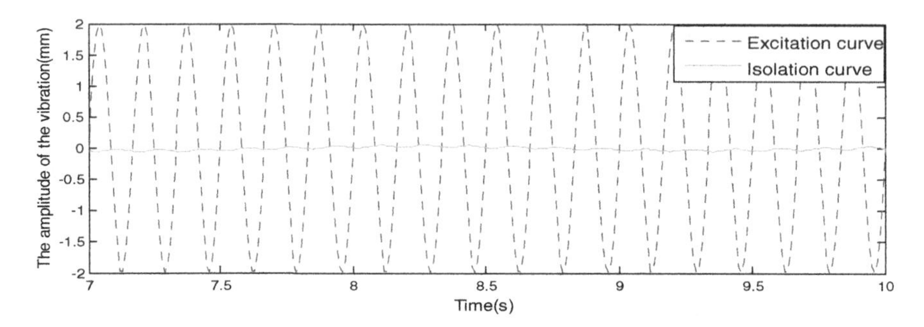

Fig. 8. The simulated graph of active vibration isolation system at 6 Hz interference signal

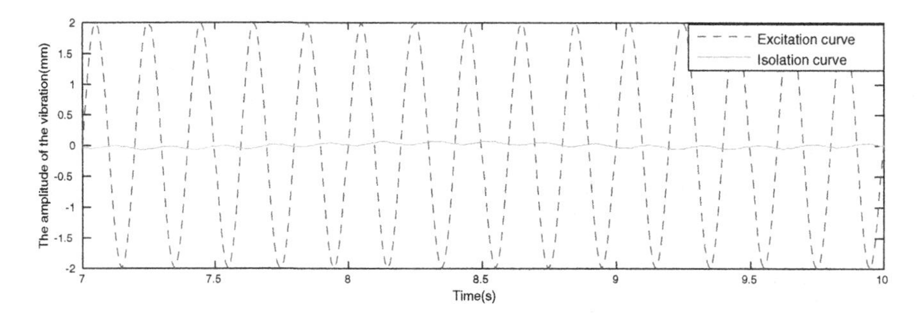

Fig. 9. The simulated graph of active vibration isolation system at 5 Hz interference signal

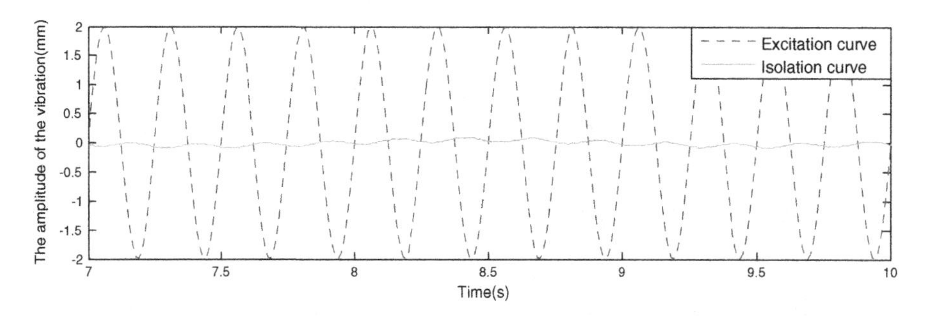

Fig. 10. The simulated graph of active vibration isolation system at 4 Hz interference signal

3 Hz. According to the Table 1, the NN-based hybird isolation method can suppress the interference signals magnitude down by more than 92 % for 3–6 Hz interference signals, which has a good isolation effect.

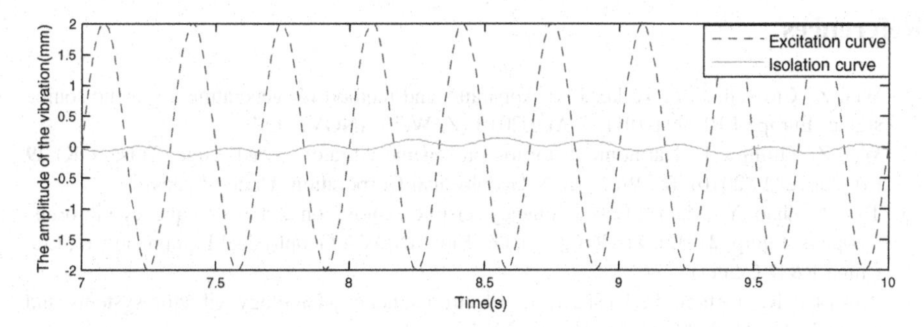

Fig. 11. The simulated graph of active vibration isolation system at 3 Hz interference signal

Table 1. The isolation efficiency at 3–6 Hz

f/(Hz)	x/(mm)	x'/(mm)	η
6	2	0.08	96 %
5	2	0.10	95 %
4	2	0.12	94 %
3	2	0.16	92 %

6 Conclusion

In seismic exploration, the excitation signal generated by vibratory with low-frequency passes downward deeply. We can get more abundant geological messages and can determine the status of oil and gas storage effectively. The vibratory use hybrid vibration isolation technology. A low-frequency micro-vibration control system with a electromagnetic exciter as initiative executing agency is established based on the interaction mechanism between vibratory and the earth. The predictive control algorithm based on neural network improve the control accuracy and suppress the low-frequency interference signals evidenced by the simulation results. The system can decrease the amplitude of interference signals by 92 % at 3 Hz, and the ability of restraining interference vibrations is satisfied. It has high reference value to improve the accuracy of the vibratory excitation signal. The proposed method for vibratory provide a theoretical basis for vibration isolation system installment.

Acknowledgement. The research work is supported financially by National 863 Project (Grand Number: 2012AA061201).

I would like to express my gratitude to Dr. Xiaoyan Chen and Dr. Zhigang Qu who helped me during the writing of this thesis.

References

1. Wei, Z., Crowell, J.M., Teske, J.E.: Apparatus and method for generating a seismic source signal. Europe EP1999490B1, 7 Aug 2013 (Z. Wei – INOVA Ltd.)
2. Wei, Z., Phillips, T.: Harmonic distortion on seismic vibrators. Lead. Edge (Tulsa, OK) **29** (3), 256–261 (2010). (Z. Wei – ION Geophysical Corporation, United States)
3. Tao, Z., Zhao, Y., Ma, L.: Low-frequency seismic exploration and low-frequency vibroseis. Geophys. Equip. **21**(02), 71–76 (2011). (Z. Tao – INOVA Geophysical Equipment Co., Ltd. China R&D Center)
4. Zowarka, R., Uglum, J., Upshaw, J.: Electromagnetic seismology vibrator systems and methods. US US8699302 B2, 15 Apr 2014
5. Zhou, J., Jiang, S.: Sensing technology and application. Central South University Press, pp. 60–65 (2003)
6. Fuller, C.R., Elliott, S.J., Shaffer, S., Nelson, P.A.: Active Control of Vibration, pp. 185–217. Academic Press, Second printing. London (1997)
7. Riyanto, T.B., Lazuardi, A., Kenko U.: DSP based RBF neural modeling and control for active noise cancellation. In: Proceedings of the 2002 IEEE International Symposium on Intelligent Control, Canada, pp. 460–466 (2002)
8. Snyder, S.D., Tanaka, N.: Active control of vibration using a neural network. IEEE Trans. Neural Netw. **6**(4), 819–828 (1995)
9. Kraft, L.G., Pallotta, J.: Real-time vibration control using CMAC neural networks with weigh smoothing. In: Proceedings of the American Control Conference, Chicago, pp. 3939–3943 (2000)
10. Zhang, D.: MATLAB neural network application design, pp. 246–247. Machinery Industry Press, Beijing (2009)
11. Heertjes, M., deGraaff, K., vanderToorn, J.-G.: Active vibration isolation of metrology frames: a modal decoupled control design. J. Vib. Acoust. **127**, 223–233 (2005)
12. Chen, K.T., Chou, C.H., Chang, S.H., Liu, Y.H.: Intelligent active vibration control in an isolation platform. Sci. Direct **69**(12), 1063–1084 (2008)
13. Tjepkema, D., vanDijk, J., Soemers, H.M.J.R.: Sensor fusion for active vibration isolation in precision equipment. J. Sound Vib. **331**, 735–749 (2012)
14. Wang, W.: Study mechanical vibration analysis of MATLAB/SIMULINK simulation. Mod. Electron. Technol. **24**, 46–48 (2006)

Research on Classification and Coding of Government Information Resources for Urban Area Development

Wensheng Zhou[1(✉)], Shenglei Zhang[2], and Haichuan Shi[3]

[1] School of Architecture, Tsinghua University, Beijing, China
zwsbj@163.com
[2] Chinese Society for Urban Studies, Beijing, China
[3] School of Surveying and Mapping, Wuhan University, Wuhan, China

Abstract. Among studies on urban area development, information classification is one of the important aspects. This paper analyzes the current situation and problems on the simulation of urban area development and puts forwards the concept of government information resources for urban area development under the era of big data. After analyzing related research at home and abroad and the specific needs for urban area development simulation, principles and methods suitable for classification of government information resources for urban area development is suggested. Based on the above studies, the classification system of government information resources for urban area development is constructed, which includes urban fundamental geographic information urban thematic information and special information about urban simulation system. Furthermore, the paper discusses its data coding scheme. Finally, an application system based on the above research is developed for simulating distribution of urban land and traffic loads of Wuhan.

Keywords: Urban area development · Information resource · Classification · Coding

1 Introduction

As a process necessary for development of human beings, urbanization has been sped up since 1996 in China, where urban population increases by approximately 20 million on an annual basis and annual urbanization rate have grown by nearly 1.5 % on average. By late 2014, there had been 0.74 billion urban population in China in which urbanization rate was up to 54.77 %. However, while rapid urbanization has been promoting fast economic growth and industrialization, it has brought about some problems, such as over centralized population in large and first-tier cities, especially excessive congestion, traffic jam, serious pollution, too high housing price, heavy burden over residents' life and poor living standards as well as increasingly serious "urban diseases". In some small or medium-sized cities and small towns, there are neither impetus for economic development nor sufficient competitive industries, so they aren't attractive for transferring population. Although it has been densely populated in many urbanized rural areas, urban functions are rather incomplete. Thus, how to

F. Bian and Y. Xie (Eds.): GRMSE 2015, CCIS 569, pp. 137–146, 2016.
DOI: 10.1007/978-3-662-49155-3_13

coordinate development of urban areas so that sustainable development may be achieved in such areas hasn't only become an import part of current research concerning unprecedented strategies, but also one of the most active research fields in terms of geography [1].

As an essential part of research on urban sustainable development, urban simulation orienting towards urban development may abstractly reflects spatial phenomena and processes of urban areas from the perspective of mathematics. Covering land conservation, resource conservation, environment-friendliness, transportation demand management, water saving and low-carbon construction, it is not only a powerful means for understanding, simulating and predicting changes to urban spatial phenomena, but also highly helpful for scientifically managing and planning urban systems. Furthermore, it may provide feasible technical support for implementing urban policies, developing and evaluating schemes for urban planning.

Support of databases is indispensable for large-scale systematic research for simulating and predicting urban development, while databases shall be constructed based on data standards. According to "GB/T21063.4—2007 Government Information Resources Catalog System Part 4 - Government Information Resources Classification" released in September 2007 [2], this set of classification system belongs to a government information resources classification system orienting towards public services divided into 21 and 132 primary and secondary categories, including comprehensive government affairs, economic management, national land resources, energies, industries, transportation, postal services, information industry, urban construction and environmental protection. In a National Key Technology Research and Development Program of China during the "10th Five-Year Plan", information about sustainable development is divided into fundamental geography, resources, environment, population and social economy, disaster, another six categories and 29 classes [3]. After summing up existing research work, Jianbang et al. proposed a program for categorizing and coding information about resources and environment. Concerning this program, information was divided into categories, classes, sub-classes, class I, class II and class III, among which there were 5 categories, 36 major classes and 209 minor classes [4].

Above research findings may be used as references by urban areas for constructing and modeling databases, but can't fully accommodate their needs in this respect. Thus, it is also necessary to classify and build the models according to demands for urban simulative and predictive research.

2 Government Information Resources for Urban Area Development

Government information resources refer to information generated or used by government in performing functions. In a narrow sense, they mean the information resources government agencies generate, acquire, utilize, transfer, store and dispose of to perform functions. In a broad sense, they are collectively known as all information resources produced inside government or impacting government activities in spite of their generation outside government [5].

Government information resources for urban area development are constructed to provide sufficient information resources necessary for sustainable development of such areas, including urban planning, urban management, construction, landscaping, transportation and environment, which may be helpful for related departments to investigate simulation and prediction of sustainable development in urban areas.

Government information resources for urban area development are mainly utilized for examining simulation and prediction of their development. To simulate and predict development of urban areas, a wide range of contents are covered, ranging from macroscopic simulation of land-use, population changes and migration, traffic, atmospheric environment and water environment to microscopic simulation of construction environment. To carry out these simulations, support from much spatial and non-spatial data with different sources, formats and semantics is necessary. These data are multi-source heterogeneity, distributed, poly-semantics and great amount.

Multi-source Heterogeneity: Government data about urban development include information about various spatial, natural, functional and social attributes as well as their connections. Collected from different different professional fields and functional departments with distinct techniques, they are presented and stored in multiple forms and formats, so they are highly multi-source heterogeneous like typical spatial data.

Distributed Management and Maintenance: Generally, government data for urban development are physically stored, used and maintained in different geographic locations. For instance, fundamental geographic data are maintained by the Surveying and Mapping Department, while cadastral data were kept by the Bureau of Land and Resources or the Real Estate Bureau. These data may be distributed through the internet.

Multi-scale Characteristics: Owing to different research objects and objectives, geographic phenomena shall be usually observed and examined at different scales. Geospatial data are spatially unique that they have characteristics of spatial scales. Meanwhile, geographic phenomena change with times, apparently characterized by time scale. In general, topography is presented by fundamental urban data in the form of scale, so a common object may present characteristics of multiple scales in these different-scale topographic maps. Various remote sensing images with different resolution are presented at multiple scales as well.

Spatial Correlation: To develop government data in urban areas, efforts shall not be only made to convey information about geo-spatial locations, but also relationships among spatial elements, mainly including proximity and connectivity. Based on such topological relationships, various operations may be realized for spatial data. For example, traffic simulation may be analyzed based on the shortest path.

Massive Data: Including multi-source and multi-scale spatial data of a whole urban area, government data for urban development may be collected from different industries with various techniques and presented in rich forms, including vector and raster data about maps, remote sensing data, DEM data, data about 3D models, multimedia data and those about society, economy and environment. In addition, there is a growing amount of data reflecting dynamic changes to urban areas as an era of big data is approaching. Therefore, there is a huge amount of government data for urban development.

3 Classification System and Coding

3.1 Classification Principles

Apart from basic standard and regular principles of information research field, following criteria shall be also met in classifying government information resources for urban area development [6]:

(1) Scientific
Classification shall be based on specific scientific and reasonable evidences, and its results may scientifically describe or reflect architecture of data resources supporting research on urban development.
(2) Systematic and Complete
On the whole, data classification system shall be generic and inclusive that may incorporate all existing government data for urban development and potential data that may be generated in the future. Data attributes and connections may be reflected relatively completely by classification. Thus, a rational scientific classification system may come into being that a specific position may be made available to each dataset during classification.
(3) Expandable
Data classification system needs to be established to meet growing demands of research on urban development for category and quantity of information. For this purpose, adequate categories shall be set during classification to preliminarily retain sufficient space, in order that classification system may be constantly expanded with the increase of information, so as not to disturb the classification system that has been established.
(4) Practical
In addition for being helpful for data organization and management, data classification shall also pay attention to and reflect users' general habits of searching and retrieving data. During classification, names shall be used according to practices of different specialties if practical.
(5) Compatible
Government data for urban development shall be classified by referring to pertinent international, national or industrial criteria or consistent with them whenever possible. To be exact, fundamental geographic data shall be classified based on existing national standards as far as possible.

3.2 Classification System

Architecture of government information resources classification system for urban described according to the formula as follows [7]:

$$\sum \min\left(\mathrm{TL}, \sum\left|\mathrm{NN}_{\mathrm{li}} - \frac{\sum_{i=1}^{\mathrm{nL}} \mathrm{NN}_{\mathrm{li}}}{\mathrm{nL}}\right|, \frac{1}{\mathrm{R}}\right) \tag{1}$$

Where
TL = number of layers of classification system
NNl i = number of categories of the *i*th node of Layer i
nL = number of nodes on Layer L
R = magnitude of nodes

From description of Formula 1, it may be known that government information resources for urban area development shall be classified into suitable layers, which shall be neither too rough nor detailed. Granularity shall guarantee overall coordination and equilibrium, while nodes shall be arranged according to magnitude of each category, in order to highlight the information users are concerned about. Therefore, architecture of categories needs to be properly arranged during practical classification, and layers shall be divided pursuant to actual needs, in order that there could be an equivalent number of categories on each layer.

A government data classification system may be constructed according to above classification principles and requirements by referring to government information resources classification criteria and other related standards [8–10]. It is divided into categories, major classes, divisions and minor classes, among which categories include urban fundamental geographic information, urban thematic information and special information about urban simulation system, as shown in Table 1.

(1) Urban Fundamental Geographic Information
As basis of government information resources, urban fundamental geographic information is about geographic locations, relationships and attributes of spatial elements. It is mainly classified according to GB/T 13923-2006 national standard for classification [11]. In this classification system, it is divided into 8 classes, including fundamentals, waters, residential areas and facilities, transportation, pipelines, boundaries and government areas, landscape, soil and vegetation. Geography of foreign areas and national administrative areas isn't involved in studying simulation of urban development. In consideration of needs for statistical units on different-scale maps during simulation analysis, boundaries and government areas are replaced by government areas and statistical units, where foreign areas and national administrative areas are deleted, whereas spatial statistical units are added. Besides, seven minor classes are arranged in this category, including cities, districts, streets (offices), residents' committee, land mass, building blocks and parcels. For convenient unified management of DEM and DOM data, raster data is added as a category, where DEM and DOM data belong to divisions. Therefore, urban fundamental geographic information is classified into 9 major classes, 49 divisions and 303 minor classes.

(2) Urban Thematic Information
Urban thematic information refers to that conveying content of all urban fields based on urban fundamental geographic information. Based on corresponding national and industrial classification standards for sectors of national economy, it

Table 1. Government information resources classification system for urban development

Categories	Classes	Divisions
Urban fundamental geographic information	Location basis	Surveying control points, foundations of mathematics
	Waters	Rivers, ditches, lakes, reservoirs, marine elements, elements of other waters, water conservancy and ancillary facilities
	Residential areas and facilities	Residential areas, industrial mining and the facilities, agriculture and the facilities, public services and the facilities, places of interests, religious facilities, scientific observation stations, other buildings and the facilities
	Transportation	Railways, inter-city roads, urban roads, village roads, road structures and ancillary facilities, water transport facilities, courses, air transportation facilities and other transportation facilities
	Pipelines	Power transmission lines, communication lines, gas, main water delivery pipelines, urban pipelines
	Government areas and statistical units	Provincial administrative areas, regional administrative areas, prefectural administrative areas, township-level administrative areas, other areas, statistical units of urban space
	Landscape	Contour lines, altitude lettering points, contour lines of waters, underwater lettering points, natural landscape, artificial landscape
	Vegetation and soil	Farmland and forestry land, urban green land, soil
	Raster data	DEM, DOM
Urban thematic information	Population	General situation of population, components of labor forces, cultural composition, living conditions
	Resources	Climatic resources, land resources, water resources, mineral resources and tourist resources
	Economic growth	Integrated economy, rural economy, industrial economy and other economy
	Environment	Environmental pollution, environmental deterioration, natural disasters

(*Continued*)

Table 1. (*Continued*)

Categories	Classes	Divisions
	Planning and construction	Urban planning, urban construction
	Dynamic big data	Monitoring, social network, active sensing, industrial operation
Special information about urban simulation system	Simulation of land for construction	Land use, planning and control, control factors
	Simulation of traffic load	Analytic elements
	Simulation of architectural energy consumption	Agricultural information, agricultural energy consumption
	Simulation of aquatic environment	Lots, aquatic environment
	Simulation of vegetation and carbon sequestration	Areas, carbon emissions

may be categorized into 6 major classes (including population, resources, economy, environment, planning and dynamic big data), 22 divisions and 90 minor classes. Dynamic big data are made up of monitored data (concerning environment monitoring, traffic monitoring and video monitoring), data about social network (e.g. data about micro blogs and forums), data about active sensing (temperature, humidity, PM2.5 data and cellphone location), data about industrial operation (such as IC cards, utilities, business approval, taxi trajectories from GPS, mobile communications, finance, logistics, supermarket shopping and medical data). These data are rarely covered in traditional data management and analysis. With the approaching of big data era, people may obtain increasingly rich, detailed, real-time and cost-effective data that are associated with each other, in examining simulation of urban development, in order to conduct more complicated and meticulous urban research on a larger scale [12].

(3) Special Information about Urban Simulation System
Special information about urban simulation system means that used for research for simulating and predicting urban development other than urban fundamental geographic information and urban thematic information. In this paper, simulation system involved mainly includes simulation of land for construction, traffic load, agricultural energy consumption, water environment, vegetation and carbon sequestration. Surely, adjustments will be made to such information accordingly with constant expansion of urban simulation.

3.3 Data Coding

Government information resources for urban area development may be coded according to GB/T 13923-2006 “Basic Geographic Information Classification and Codes”, namely a national standard of the People’s Republic of China. The code is made up of six bits, so another bit is added before the original code as a specialty code in consideration of unifying coding of the entire information resource system and represents three categories of information in the classification system, where 1, 2 and 3 are used as specialty classification codes for urban fundamental geographic data, urban thematic data and special data about urban simulation system. The classification coding structure of government information resources for urban area development are shown in Fig. 1.

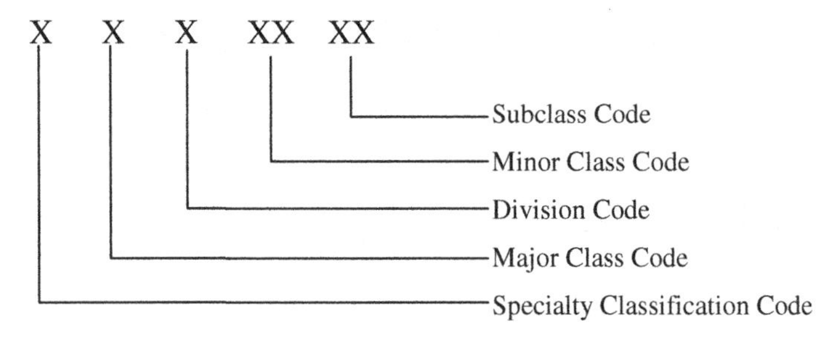

Fig. 1. Classification coding structure diagram

Fig. 2. A predicting and simulating system for traffic load

4 Application of Classification System

Based on above research, corresponding standard was established for database construction. According to this standard, a database orienting towards simulation and prediction of urban development was initially constructed. Then, a system was developed to analyze the distribution of urban land and traffic loads of Wuhan (Fig. 2). Practices have proven that government information resources classification system for urban areas is of great referential value for studying government information resources classification system for urban areas.

5 Conclusion

With constant deterioration of human residential environment, regional sustainable development has become an important issue for the whole world. The approaching of a big data era has made good opportunities available for quantitatively examining regional sustainable development. Under this background, it is undoubtedly that references may be provided for research on sustainable development of Chinese urban areas by investigating government information resources classification system for urban development. After related research work is analyzed, government information resources for urban area development are defined. Meanwhile, government information resources classification system for urban area development and coding schema are proposed, which lay a foundation for China to systematically and constantly perform quantitative studies to simulate urban development. Surely, this classification system will be further improved and optimized on a non-stop basis as research is constantly deepened.

Acknowledgments. This research has been funded by National Key Technology R&D Program (NO. 2012BAH24B00) and National Natural Science Foundation of China (NO. 51178236).

References

1. Lu, D., Fan, J.: The rise and effects of regional sustainable development studies in China. Bull. Chin. Acad. Sci. **27**(3), 290–300 (2012)
2. National Standard of the People's Republic of China. Government Information Resources Catalog System Part 4: Government Information Resources Classification (GB/T 21063.4-2007) (2007)
3. He, J., Li, X.: Research on the ontology of geographic information classification and coding. Geogr. Geo-Inf. Sci. **18**(3), 1–7 (2002)
4. He, J., Li, X., Bi, J., et al.: Research on classification and coding of resources and environment information and its association with ontology. Geomatics World **1**(5), 6–11 (2003)
5. Zhang, W.: Information System Integration Technology. Publishing House of Electronics Industry, Beijing (2002). (in Chinese)

6. Liu, R., Jiang, J.: Research on classification principles and methods for geographic information - a case of classification of fundamental geographic data. Sci. Surv. Mapp. **28** (S1), 84–87 (2004)
7. Wu, S.: Ways for optimizing website information classification and the application. Nanjing Forestry University, Nanjing (2009)
8. Shunbao, L., Lin, J.: Study on classification system of data for earth system science. Progress Geogr. **24**(6), 93–98 (2005)
9. Chen, Y.: Study of coding system of the urban fundamental geographic information. Urban Geotech. Invest. Surv. **12**(1), 15–17 (1997)
10. Yan, Z.: Standard Guideline of Urban Geographic Information System. Science Press, Beijing (1998). (in Chinese)
11. National Standard of the People's Republic of China. Classification and Code of Fundamental Geographic Information Elements (GB/T 13923-2006) (2006)
12. Zhang, X.: Urban planning opportunity, challenge, and thinking in big data era. Planners **20** (8), 38–42 (2014)

Spatial Data Acquisition through RS and GIS in Resource Management and Sustainable Ecosystem

Consistency Verification for GML Data Based on DOM

Xiaoli Gao[1(✉)], Haixia Li[1], Tingguang Yan[1], Zhencai Cui[1], Jiyu Yu[1], and Yehua Sheng[2]

[1] Shandong Water Polytechnic, Rizhao, Shandong, China
wishuluck@126.com

[2] Key Laboratory of Virtual Geographic Environment, Nanjing Normal University, Nanjing, China

Abstract. GML schemas are metadata files, which define the structure, content and restriction of GML instances. As a kernel of the GML parser, consistency verification decides whether GML documents are consistent with the relevant application schemas. In order to parse GML data more effectively and accurately, an algorithm based on DOM was developed as to how GML consistency can be verified. Furthermore, some primary user-defined methods and the homologous regular expression technology, involving in this algorithm, were discussed in detail. Experimental results show that the consistency verification algorithm is efficient.

Keywords: GML · Schema-based parser · DOM · Consistency verification · RegExp · HRegExp

1 Introduction

As the encoding standard for geographic information, GML documents have to meet certain grammar specification to guarantee the correctness of physical structure and logical structure, thus, which can be interpreted, process and sharing [4]. GML parsing is the basis of geographic information storage, compress, transformation, conversion, index, query and share, etc., which is related to studying and application of GML tightly [5]. Almost all the studies related to GML, cannot be separated from GML parsing. GML schema-based parser should be versatile. It is able to preserve the properties of adaptive and self-expansion [6].

The integrative grammatical and semantic database, addresses the issue of how to describe simple data type and complex data type in schemas, and lays the foundation for GML grammar validation and GML grammatical and semantic parsing [1].

The GML grammar validation is the most basic and important foundation of GML schema-based parser, and its purpose are to verify the correctness and the validity of

This work is sponsored by the Special Water Resources Projects Aided by Special Fun of Water Resources Department of Shandong Province (No. sdw200709027) and the Provincial Water Conservancy Science Research and Technology Promotion Project of Shandong Province (No. SDSLKY201304).

F. Bian and Y. Xie (Eds.): GRMSE 2015, CCIS 569, pp. 149–158, 2016.
DOI: 10.1007/978-3-662-49155-3_14

the GML data that based on the two aspects: validity and consistency, to ensure the GML data is available.

Consistency verification is also known as the validity verification. It is an inference that whether the GML instances and the GML patterns can match in the data structure, which mainly for the effective judgment that whether elements, attributes, and data structure of the GML instances conform with the corresponding defined requirements of GML application schemas. When an inconsistent error occurs, it can prompt the wrong position and the wrong type information, etc.

Normative GML documents should comply with the requirements of two aspects: To be well-formed and to be valid.

The well-formed GML document is known as the legal GML document, whose content and structure must comply fully with the GML basic grammar rules. Such as the start tag and the end tag must appear in pairs, and the structure of the document must be a hierarchical tree structure, etc.

The GML instance document that has completely passed the consistency test is referred to as consistent (or effective) GML document. Such kind of document must be well-formed, firstly, which also calls for its content and structure shall abide by the provisions of the relevant application schemas or DTDs [7].

The consistent GML document has more restrictions than a well-formed GML document, and has to follow a stricter specification. A well-formed GML document cannot achieve the standard of consistency. However, consistent GML document will meet the requirements of well-formed. From this point, consistency is the proper subset of the legality [3]. In functional implementation of GML consistency verification, the DOM technology mainly used as the GML parsing scheme.

2 The Specification of DOM API

DOM is the abbreviation of Document Object Model, proposed by the World Wide Web Consortium (W3C), was an application programming interface (API) for XML and HTML documents [8].

When parsing GML data using DOM, GML document is organized into a hierarchy tree structure (called a DOM tree), and the DOM tree is loaded into the memory. In DOM tree, all elements are organized as the tree nodes. DOM parser supplies a series of API methods, provides abundant supports for manipulating tree nodes, such as dynamic traversal, retrieving, adding nodes, removing nodes, modifying nodes, etc. Moreover, DOM parser can establish, update, or access the GML document structure and GML data style [7].

DOM offers the following four basic API classes.

- Document class represents the entire GML document. The document object is actually the root of the DOM tree, which is the entrance leads to access data, or other elements of the DOM tree.
- Node class, on behalf of an abstract node of the DOM tree, is the parent class of many other classes.

- NodeList class, providing the abstract definition of an ordered list of nodes, each node appears as an item with its index value in the list.
- NamedNodeMap class, defines the operational methods set. The methods in the set are used to process elements in an unordered collection of nodes. The instance object is primarily used to access an attribute node, or to describe the corresponding relations between a set of nodes and their names.

The working mechanism of DOM tree is shown in Fig. 1.

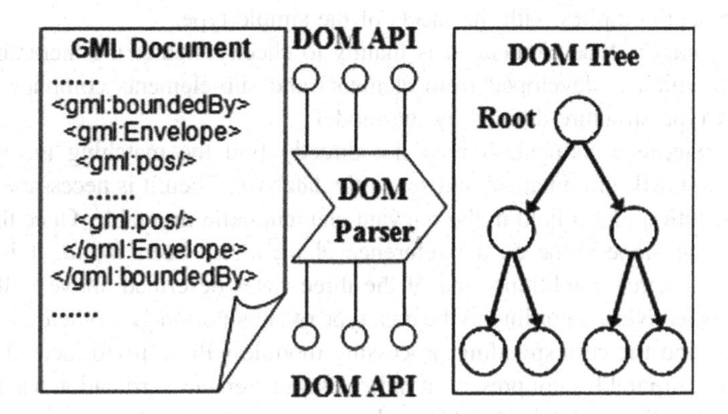

Fig. 1. The working mechanism of GML parsing using DOM.

3 The Realization Principle of Consistency Verification

When the data entries in a GML document are manipulated using DOM, first organize the collections of independent elements, attributes, text, data entity, and so on, into a DOM tree structure in the memory, and make each node represent an embedded object in the GML document. Secondly, a series of API functions are provided; they support the application to access the contents of the DOM tree through the interface functions. So many kinds of operations are allowed to apply to the nodes on a DOM tree, and then process the results of operations, maps the results indirectly to the corresponding documentation [2]. Finally, by complying with the requirements of the application, API functions allow programmers to modify the DOM, and decide whether to save changes back to the GML document.

Consistency validation process depends largely on the basis of the DOM tree and the integrative GML grammatical and semantic database, and the recursive method is utilized in the implementation. It is important in order to inspect whether the contents of a data instance match the corresponding records in the integrative GML grammatical and semantic database. The contents of a data instance are made up of many aspects in each data entity, including the organizational structure, data type, and range of domain [7].

Once found that elements do not agree with the schema definition, the wrong type and the error position will be recorded, so that the verified conclusions will be reported to users at the end of the consistency validation process.

In view of the different element types, the consistency validation behavior will change.

- For the built-in standard simple type element. It is mainly to check whether the value scope of each element complies with the provision of the standard type.
- For the restricted simple type element. It is mainly to check whether the value of each element complies with the facets of the simple type.
- For the complex type element. It is mainly to check whether the hierarchical tree structure which is developed from elements and sub-elements complies with the complex type structure defined by the model.
- For the reference element. It may not directly find the matching record in the integrative GML grammatical and semantic database. Then it is necessary to check the substitution group field in the relevant element definition table. Once the record that has the same name as the reference element has been found, it is time to perform the three conditions, one of the three cases described above will be processed respectively according to the data type of substitution group field is simple or complex, and the corresponding processing module will be performed. If the substitution group field is not present, it indicates that element verification has failed, so the error handling module should invoke.

4 Implementation Algorithm of Consistency Verification

GML parsing relies on DOM trees mapping in memory. The tree data structure belongs to the nonlinear hierarchical structure, and its definition is recursive. Because the recursion has characteristics of concise, delicate, function perfectly, short code, efficient performance, so recursive algorithm was designed in the implementation for consistency verification.

The function of the method named consistency verification is to implement dynamic analysis and check on the specified GML elements. A detailed algorithm description is depicted below.

Input Description: Input objects consist of the specified GML instance document to be verified, the integrative grammatical and semantic database of GML core schemas, and the integrative grammatical and semantic database of GML application schemas.

Function Description: The core function of the algorithm is to check consistency of the specified element. This function returns TRUE on success in consistency verification, or FALSE on failure in consistency verification. When the specified element takes its failure in consistency verification, the method named registerError will be invoked immediately. The function of this method is to record the error location, the error type, the wrong code, etc.

Output Description: Output result is the conclusion through verification test, or error messages caused by verification failures.

4.1 Description of Consistency Verification Method

The pseudo code of the key process in the method called consistency verification is described as follows [1].

```
boolean consistencyVerification(Element checkedElement)
{ //Define a variable to store value returned by method
  boolean verifiedResult=true;
  if ( dataTypeOf(checkedElement)∈standard simple type)
    verifiedResult=
            VerifyStandardSimpleType(checkedElement);
  //Error occurs when verifies standard simple type
  if (verifiedResult==false)
    registerError(); //Invokes the registerError method
  if (dataTypeOf(checkedElement)∈restricted simple type)
    verifiedResult=
        VerifyRestrictedSimpleType(checkedElement);
  //Error occurs when verifies restricted simple type
  if (verifiedResult==false)
  registerError(); //Invokes the registerError method
if (dataTypeOf(checkedElement)∈complex type)
{//To concatenate checkedElement and its child elements
destinationConcatenate=
    generateConcatenateStr(checkedElement);
//To gain the value of  rule field from database
sourceRule=getRuleFromDB(DBFile,checkedElementType);
//To convert the parameter into a regular expression
sourceRegExp=regularization(sourceRule);
//Using regular expression as a template
boolean isMatch= Pattern.matches(
    sourcePattern,destinationConcatenate);
if (isMatch==false) {//Pattern match suffers a failure
  registerError(); //Invokes the registerError method
  verifiedResult=false;  //Assign false to variable
} else
  verifiedResult=true;  //Assign true to variable
//Realize the verification function using recursive
for (eachSubElement∈checkedElement.getChildNodes())
if ((consistencyVerification(eachSubElement))==false) {
  registerError(); //Invokes the registerError method
  verifiedResult=false; //Assign false to variable
} }//To complete the verification of complex type
return verifiedResult; //Returns value of the variable
}  //To stop consistencyVerification method process
```

4.2 Illustrations of Custom Methods

Some primary user-defined methods involving in the method of consistency verification are given in Table 1.

Table 1. Important functions referenced in consistency verification module.

NO.	Method Prototype	Function description
1	String dataTypeOf(Element checkedElement)	To determine the data type of parameter checkedElement
2	void registerError( )	To record and store the error location, the error type, the wrong code, etc.
3	boolean verifyStandardSimpleType(Element checkedElement)	To check whether the parameter of checkedElement complies with the definition of standard simple type according to the records in the integrative grammatical and semantic database
4	boolean verifyRestrictedSimpleType(Element checkedElement)	To check whether the parameter of checkedElement complies with the definition of restricted simple type according to the records in the integrative grammatical and semantic database
5	String generateConcatenateStr(Element checkedElement)	To construct a string expression that is comprised of the element specified by the parameter of checkedElement and its child elements
6	String getRuleFromDB(String DBFileName, Element checkedElement)	To gain the homologous regular expression of a complex type element specified by the parameter of checkedElement from the database called DBFileName
7	String regularization(String sourceRule)	To convert the homologous regular expression into the corresponding regular expression, To prepare for the operation of pattern matching

Specially as is pointed, in generate ConcatenateStr() method, after producing a string expression, pattern matching will be carried out on the expression and the corresponding regular expression of a complex type. The next step is to decide whether the structure of the element specified by the parameter of checkedElement consistents with the structure of data type defined in the pattern files.

4.3 Regular Expression and Homologous Regular Expression

Regular expressions (abbreviated as RegExp) provide a concise and flexible means to identify strings of text, such as particular characters, words, or patterns of characters. RegExp can express the syntactic structure and nesting relationship perfectly, so we select it to describe GML model information defining in complex types. There are numerous advantages in describing and modeling regular language with RegExp. For example, it is easy for people to understand and to use RegExp; it is easy for computers to parse and to process RegExp, and RegExp can describe complicated things in a simple form [5], etc.

A regular expression is composed of some branches; and a branch is composed of some pieces. The Piece can be seen as the basic independent element of RegExp, which consists of atoms and quantifier.

Primary quantifiers and operators used in RegExp is given in Tables 2 and 3 respectively.

Table 2. List of the quantifiers in RegExp.

No.	Quantifier	Description of usage
1	?	Matches the preceding element zero or one time
2	*	Matches the preceding element zero or more times
3	+	Matches the preceding element one or more times
4	{n}	Preceding element can only occur n times
5	{n,}	Preceding element occurs at least n times
6	{n, m}	Preceding element occurs at least n times and not more than m times
7	No quantifier	Element can only occur one time

Table 3. List of the operators in RegExp.

No.	Operator	Description of usage
1	$\alpha\beta$	Sequence operator, matches token α firstly, then token β
2	$\alpha \mid \beta$	Choice operator, matches either token α or token β
3	(α)	Token α is treated as grouping

The above quantifiers and operators can be brought together to form arbitrarily complex expressions. In order to enhance describing capacity of RegExp for parsing the syntactic and semantic information, some grammar rules are extended. The improved regular expression is called homologous regular expression(abbreviated as HRegExp) in this paper [1].

The following syntax rules were added in HRegExp:

1. For an element in the complex type model, HRegExp decorates the element name with a pair of characters (“<”, “>”) as the delimiters.
2. For an element reference to the complex type model, HRegExp decorates the referenced name with a pair of square brackets (“[”, “]”) as the delimiters.
3. For a named model group in the complex type model, HRegExp decorates the referenced name of the group with a pair of tokens (“[#”, “#]”) as the delimiters.
4. All sub elements in a “sequence group” must be divided by the blank character.
5. All sub elements in a “choice group” must be separated by vertical bar(“|”).
6. All sub elements in a “all group” must be separated by slash(“/”).

7. Uses multi-pairs of nested parentheses("(", ")") to reflect the syntax structure of the nested complex type model.
8. Named model group will conform to the above definition.

Therefore, it can be seen that HRegExp adds some special delimiters for four different types of complex type models, to depict additional semantic information in GML models.

5 Example Test for Consistency Verification Algorithm

The consistency validation program for checking GML data was developed with C#. It can help users to judge whether a GML document is coherent or not [7].

A concrete example test has been laid on. The result of verification as showed in Fig. 2, it indicates that the checked data segment does not meet the requirement of consistency.

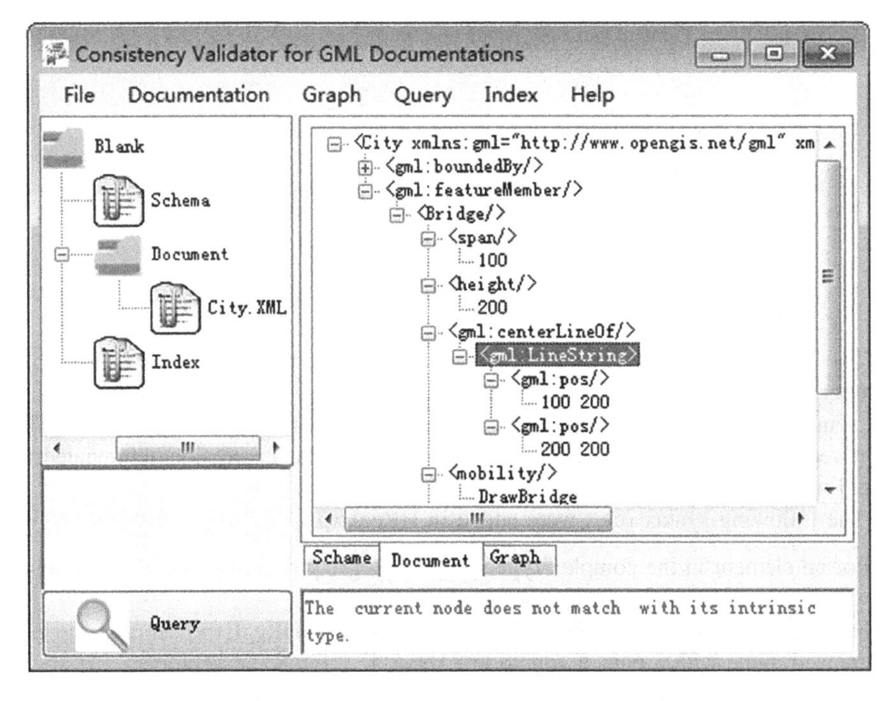

Fig. 2. Example test for a GML data segment file.

The test data comes from a document titled "City.XML", which is a GML data segment file describing the characteristics of the bridge element. Its content is shown as follows.

```
<gml:featureMember>
  <Bridge>
    <span>100</span>
    <height>200</height>
    <gml:centerLineOf>
      <gml:LineString>
        <gml:pos>100 200</gml:pos>
        <gml:pos>200 200</gml:pos>
      </gml:LineString>
    </gml:centerLineOf>
    <mobility>DrawBridge</mobility>
    <spans> <Gorge/> </spans>
  </Bridge>
</gml:featureMember>
```

A considerable amount of GML data instances were selected to verify the correctness and reliability of the program. The test results show that: The consistency verification algorithm can verify the consistency of GML data effectively, and the algorithm performance is higher. The detected errors are accurate and reliable [1].

6 Conclusion

DOM parsing technologies based on objects are adapted in the implementation of GML grammar validation, which is contained within GML schema parsing. An algorithm for GML data's consistency verification using recursion technique is built and implemented.

Experiments show that the algorithm relies on the integrative GML grammatical and semantic database, which can make a correct judgment about the GML data consistency, to resolve errors and eliminate hidden dangers effectively for GML application data. The algorithm has the relatively high accuracy and fast execution efficiency.

References

1. Gao, X.: Research on universal GML schema parsing based on grammatical and semantic database. Dissertation for The Degree in M.A.Sc (in Chinese), pp. 11–29 (2006)
2. Lake, R., Burggraf, D.S., Trninic, M., Rae, L.: Geography Mark-UP Language (GML). John Wiley & Sons Ltd, USA (2004)
3. Gao, X., Cui, Z., Jia, N., Xiao, H., Zhang, S.: Design and implementation of essential algorithms for parsing GML schemas. In: 6th International Conference on Intelligent Human-Machine Systems and Cybernetics, vol. 2, pp. 284–287. IEEE Computer Society CPS Press, USA (2014)
4. Lake, R.: The application of geography markup language (GML) to the geological sciences. J. Comput. Geosci. **31**, 1081–1094 (2005)

5. Gao, X., Li, H., Zhang, S., Sheng, Y.: Research on HRegExp applied to GML parsing. In: 5th International Symposium on Computational Intelligence and Design, vol.2, pp. 214–217. IEEE Computer Society CPS Press, USA (2012)
6. Walmsley, P.: Definitive XML Schema. Prentice Hall PTR, Upper Saddle River (2002)
7. Open Geospatial Consortium Inc.: Geographic information-Geography Markup Language Version3.3". http://www.opengeospatial.org/standards/gml
8. The World Wide Web Consortium: XML Technology. http://www.w3.org/standards/xml

Algorithm of Trawler Fishing Effort Extraction Based on BeiDou Vessel Monitoring System Data

Shengmao Zhang[1,2], Bailang Yu[1], Qiaoling Zheng[2], and Weifeng Zhou[2(✉)]

[1] Key Laboratory of Geographic Information Science, Ministry of Education, East China Normal University, Shanghai 200241, China
[2] East China Sea Fisheries Research Institute, Chinese Academy of Fishery Sciences, Ministry of Agriculture of the People's Republic of China, Shanghai 200090, China
zhwfzhwf@163.com

Abstract. Performing statistical computations for traditional fishing effort takes much time and effort, and the macro fishing effort cannot be accessed immediately. Through the Beidou satellite vessel position monitoring system, the position, time, speed and other information of vessels can be got and used to data mining. In this paper, the speed threshold of each vessel's fishing state is obtained by the statistics of navigational speed. And fishing state points can be judged by the speed threshold and heading deviation. Via the correction of filtering window, the fishing area grid is calculated by the cumulative fishing time. The cumulative fishing is the product of the cumulative fishing time and the vessel power, such as kW•h. This method has the characteristics of real-time, large-scale, fast and high resolution, which can provide good service in fishery resources protection.

Keywords: Beidou satellite · Vessel monitoring system · Fishing effort · Heading · Speed

1 Introductions

The European Union uses Vessel Monitoring System (VMS) to monitor fishing activities. It has been applied to fishing vessels exceeding 18 m in length [1, 2] as from 1 January 2004 and fishing vessels exceeding 15 m in length [3] as from 1 January 2005. It is mainly used to fisheries management [4, 5]. VMS data have time, speed and heading information. In recent years, there are research on Fishing state judgment [6], fishing effort estimate [7, 8], resource distribution analysis [9, 10], and retrospective fishing boat [11, 12] etc. Fishing effect, the work done in fishing, is an important parameter for researching the variations of fishery resources [13, 14]. The conventional calculation method considers several elements, such as the number of vessels, power, and number of days.

More than fifty thousand fishing vessels had installed the terminal unit of Beidou satellite positing system. After these fishing services carrying out for nine years, the system have recorded billions historical cruising data for each vessel, including the

F. Bian and Y. Xie (Eds.): GRMSE 2015, CCIS 569, pp. 159–168, 2016.
DOI: 10.1007/978-3-662-49155-3_15

time, position, speed, direction, rate of turn, etc. These data can be analyzed deeply by big data mining technology.

2 Materials and Methods

2.1 Data Sources

BeiDou Vessel Monitoring System Data comes from BeiDou data service center which mainly includes ship-card number, latitude, longitude and time. The vessel position information of latitude and longitude has a temporal resolution of 3 min and a spatial resolution of 10 m. Data are managed by SQL Server, because of its superior performance in spatial data's management and analysis. The spatial relationships between geometry instances can be determined and the vessel position can be saved with geometry data type.

Fishing types, vessel powers, vessel names and other information are mainly from official materials which are published by Marine Fisheries Service of every province. By matching official data with BeiDou Vessel Monitoring System Data, the types of 3333 vessels are determined. Most of them are trawling and gill-netting. There are 2212 trawlers, accounting for 66 % of the total. In this paper, it is focus on trawler.

2.2 Methods

Every point position of trawlers includes speed, course, time and other information. With that information, the state of trawlers can be determined. Whether the state is fishing or not can be judged by Formula (1). When the speed and course are in the range of threshold value, the trawler is in the state of fishing.

$$P = F(v, d)\ V_{\min} \leq v \leq V_{\max}\ D_{\min} \leq d \leq M_{\max} \tag{1}$$

In Formula (1), V_{min} and V_{max} are the threshold range of speed. D_{min} and D_{max} are the threshold range of course deviation for fishing states.

There may be many trawlers in one fishing grid. One trawler in fishing is divided into many nets. In general, a net lasts a few hours. When a net is over, the next net will get under way after some interval of time. Every net is composed of several discrete point of vessel position. Hence, Formula (2) can be used to calculate the cumulative fishing as kW•h in one grid.

$$Z_i = \sum_{i=0}^{p} \sum_{j=0}^{m} \sum_{k=1}^{n} (P_{i,j,k} - P_{i,j,k-1}) * W \tag{2}$$

In Formula (2), Z_i is the cumulative fishing effort of one fishing grid, in unit of kW•h. $P_{i,j,k}$ and $P_{i,j,k-1}$ are the time of adjacent two position points of one trawler. The difference between $P_{i,j,k}$ and $P_{i,j,k-1}$ is the length of time, and W is the power of trawler. The first summation is the cumulative fishing for one net. The second summation is the cumulative fishing of one trawler in a period of time for several nets. The third summation is the cumulative fishing of all trawlers in one fishing grid.

Distance inverse weight method is used to generate the thematic map of variation tendency. As long as the positions of trawlers are intensive enough, the thematic map of trend surface can be generated by interpolation, which can assist the trend analyses of global change in resource.

$$Z_0 = \frac{\sum_{i=1}^{n} \frac{Z_i}{d_i^r}}{\sum_{i=1}^{n} \frac{1}{d_i^r}} \tag{3}$$

In Formula (3), r is specified exponent. d_i is the distance between control point i and point o. With the increase of distance between forecasted points, the weight of impacts of control points on forecasted points decrease exponentially. Z_o is the estimated value of Point o. n is the number of control points used in the estimation. Z_i is the Z value of Control Point i.

3 State Division of Trawlers

In order to analyze a trawler of Zhejiang Province whose Beidou ship-card number is 300585 (hereafter, it is named Trawler 300585) and report time is Oct 1st, 2013, the time period from 0 a.m. to 24 p.m. is divided into 9 sections. There exist 3 different states. Stage A and stage C is dropping anchor. Stage B and stage D to stage H is fishing. Stage I is sailing. The general process of fishing is as follows. The trawler sails to one fishing ground. Then it casts net in a short-time sailing and begins to tow net. When the quantity of fish in the nets reaches a certain amount, Trawler is slow down and net are pulled.

The course of trawler (azimuthal angle) is the horizontal contained angle between the target direction line and the line drew by centering on the vessel position and lining up from the north direction of it. The course value ranges from 0 to 360°. When varying near 0° or 360°, the variation of course changes largely. To further analyze the actual variation of course, a deviation calculation is performed. The course deviation is the difference between two neighboring report times, that is, the later report time and the former report time. The positive value reacts turning clockwise and the negative value reacts turning anticlockwise. The deviations mainly change near 0°, because the interval of report time in Beidou data is 3 min and the course fluctuation cannot change largely in that short time. Referring to Fig. 1 and according to the statistics of several trawlers, the course deviations of operating states are set as D_{min} and D_{max}. They are −50° and 50°.

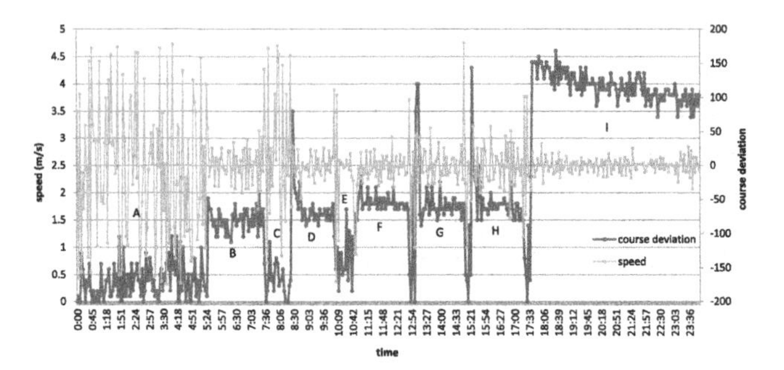

Fig. 1. Distribution map of trawler speed

A navigation path of Trawler 300585 was plotted on Oct 1st, 2013, as illustrates in Fig. 2. Comparing Figs. 1 and 2 an intuitive judgment can be made. In Fig. 2, the points of fishing and sailing are obvious, while the points of anchoring and low speed are covered by the points of fishing and sailing.

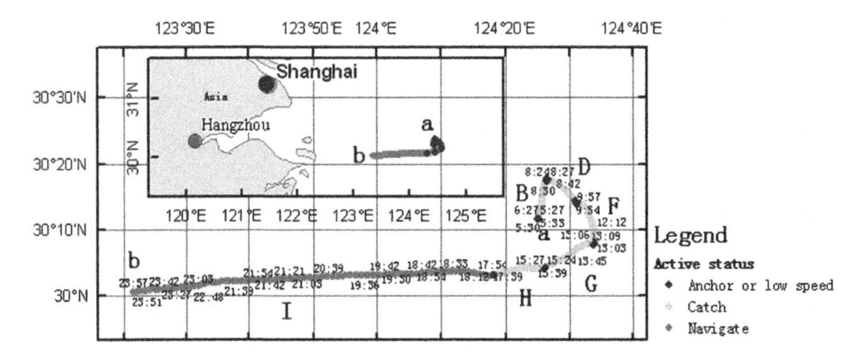

Fig. 2. Tracks of trawler 300585

Through analysis, the path of the trawler from 0:00 to 23:57 is from Point a (30°12′ N, 124°25′E) to Point b (30°6′N, 123°22′E). Point a is in the state of anchoring from 0:00 to 5:27. Segment B, D, F, G and H are in the state of fishing from 5:27 to 17:33. Segment I is in the state of sailing from 17:33 to 23:57.

4 State Judgments of Trawlers

4.1 Determining Threshold by the Statistics of Trawler Speed

Under the circumstances of operating with different power, placing trawl in different water depth and catching different kinds of fish, the speed of trawlers are different. When judging the states of trawlers, Speed threshold should be set for trawlers in fishing. Figure 3 illustrates the quantity of position points changing over the speed with

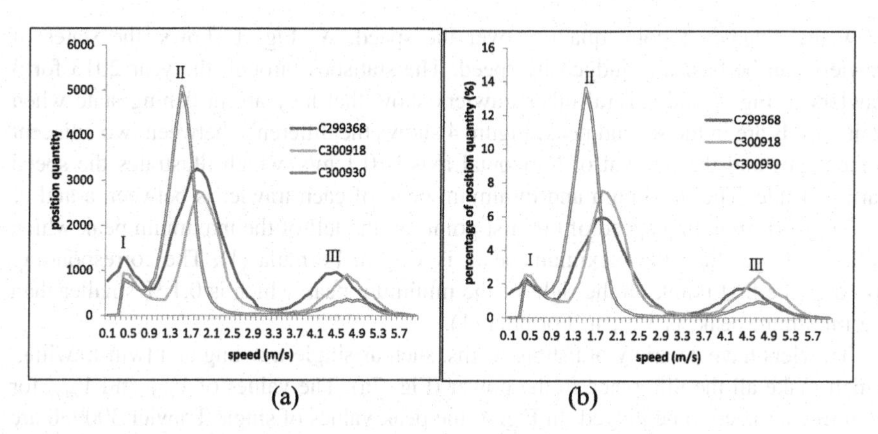

Fig. 3. Position record changing with speed in 2013. (a) Position quantity changing with speed (b) Percent of position quantity changing with speed.

performing statistical analysis on position points of twin Trawler 299368, twin Trawler 300918 and single Trawler 300930 in 2013. Since the trawlers anchoring in port also send the position data, large amounts of data with value of 0 m/s are recorded. These data make no influence on judging the states of trawlers, so they are ignored from the statistics.

Through the statistics of vessel position data for a long time, characteristics can be revealed and fishing states can be judged. The speeds of trawlers in Fig. 3 mainly show 3 peaks. The first peak, I, shows that the trawler is in a low speed (e.g. arriving at port, dropping anchor, drifting, proceeding at a slow speed). The second peak, II, shows that the trawler is in fishing state. The third peak, III, shows that the trawler is in sailing. There are striking differences among the quantities of vessel positions. As a positon point obtained by every 3 min, the quantity of points in 10 days to 1 year is 4800 to 170880. In Fig. 3, the position number is 54408 of Trawler 299368, 36716 of Trawler 300918 and 37341 of Trawler 300930.

In order to unify the data of trawlers into the same order of magnitude, marking the total of points with the speed in a range from 0.1 to 6 m/s as 100 %, The percentage of total are used to substitute for the quantity. Figure 3(b) is expressed as a percentage to

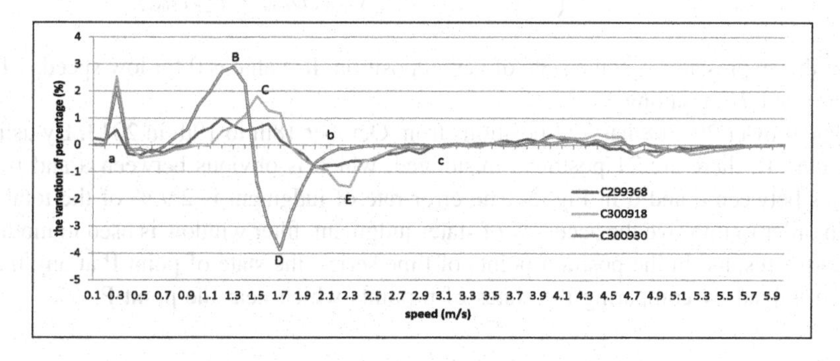

Fig. 4. The variation of percentage in 2013

show the change of point quantity over the speed. As Fig. 1 shows, the states of trawlers can be basically judged by speed. The statistics through the year 2013 for 3 trawlers in Fig. 3 and several other trawlers show that they are in fishing state when their speeds are in the second peak. Figure 4 shows the difference between two adjacent percentages and the interval of horizontal axis is 0.1 m/s, which illustrates the speed variation rate. The maximum and minimum peak of each trawler is between a and c.

The corresponding speed of the first point, on the left of the maximum peak which is 0.1 % larger than the maximum peak, is V_{min} in Formula (1). The corresponding speed of the first point, on the right of the minimum peak which is 0.1 % smaller than the minimum peak, is V_{max} in Formula (1).

Trawlers have a variety of fishing forms, such as single-trawling and twin-trawling, which make all the difference to the power (Fig. 7a). The values of V_{min} and V_{max} for each trawler need to be judged. In Fig. 4, the peak values of single Trawler 300930 are at point B and point D, and the velocity threshold are near point a and point b. The peak values of twin Trawler 299368 and 300918 are at point C and point E, and the velocity threshold are near point a and point c.

In the annual statistics of many trawlers whose speeds are between 0.1 m/s and 0.3 m/s, their position numbers are 0. Hence, there exist exceptions in the statistics. For example, the peak value at point A of Trawler 300918 (Fig. 4) is higher than its peak value at point C. Because the percentage of speed between 0.1 m/s and 0.3 m/s is 0, the change rate will become much larger when the percentage of speed in 0.4 m/s is more than 0. Therefore when the threshold is calculated, the peak value must be captured as when it is larger than 0.4 m/s.

4.2 Modifying the State of Trawlers

In Formula (4), the state of each vessel position can be judged by the combination of speed and course deviation. The threshold of velocity and course deviation can be obtained by the statistical analysis in the preceding paragraphs.

$$S = F(v,d) = \begin{cases} 0 & v \leq V_{\min}, D_{\min} \leq d \leq D_{\max} \\ 1 & V_{\min} < v \leq V_{\max}, D_{\min} \leq d \leq D_{\max} \\ 2 & v \leq V_{\max}, D_{\min} \leq d \leq D_{\max} \end{cases} \quad (4)$$

In this expression, S is the state of vessel position. Its value is 0 for low speed, 1 for fishing, and 2 for sailing.

The trawler 299368 has 724 positions from October 18th to 19th in 2013. By using Formula (4), there are 21 positions misjudged, which is obvious between a and b as well as between c and d in Fig. 5. The error rate of judgment is 2.9 % of the total.

In order to improve the accuracy of states judgment, filter window is used to modify the above results. In the position points of time series, the state of point P at any time can be judged by comparing the states of n points which near the point P.

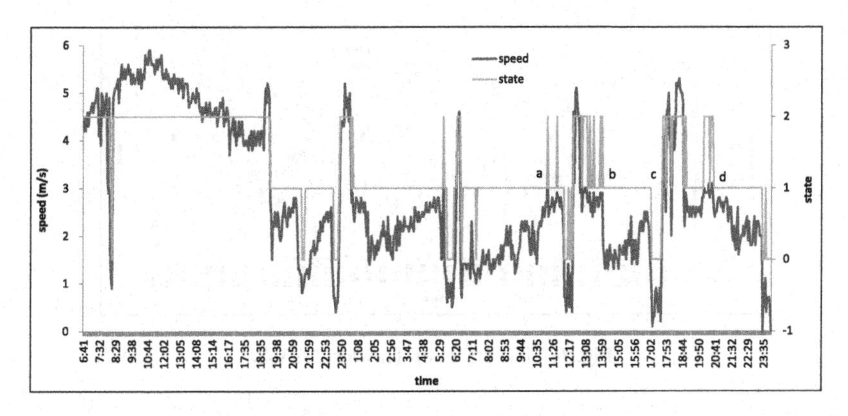

Fig. 5. The states of Trawler 299368

$$S_{front} = F(S_1, S_2, S_3 \ldots S_{m-1})\ S_{back} = F(S_{m-1}, S_{m-2}, S_{m-3} \ldots S_n)\ S_x \in S_1 \ldots S_{m-1}, S_m, S_{m+1} \ldots S_n \quad (5)$$

In Formula (5), S_x belongs to the collection ($S_1 \ldots S_m \ldots \ldots S_n$) which is the states of n points. The state of point P_m is S_m. The most frequency state in [S_1, S_{m-1}] which is earlier than the state of P_m can determine the state S_{front}. The most frequency state in [S_{m+1}, S_n] which is later than the state of P_m can determine the state S_{back}. If there are two states in the same frequency, S_{front} or S_{back} can be determined by the states of the points which is closer to P_m.

$$S_m = F(S_{front}, S_m, S_{back}) = \begin{cases} S_m & S_{front} \neq S_m, S_m \neq S_{back}, S_{front} \neq S_{back} \\ S_m & S_{front} = S_m \text{或} S_m = S_{back} \\ S_{front} \text{或} S_{back} & S_f = S_b \end{cases} \quad (6)$$

There are four situations in judging state S_m.

If S_{front}, S_m and S_{back} are different from each other, S_m is kept in the original state.
If S_{front} is the same as S_{back}, S_m is modified to state S_{front} and S_{back}.
If S_{front} is not the same as S_{back} but the same as S_m, S_m is kept in the original state.
If S_m is not the same as S_{front} but the same as S_{back}, S_m is kept in the original state.

After being modified by filter window, states still have 5 misjudgments, where n takes 3 and m takes 1. The error rate is 0.7 % of the total, 724 points, as Fig. 6 illustrates. It is lower than the error rate by only using threshold method in Sect. 4.1. There are 3 stages in trawl fishing. Stage A is casting; stage B is trawling and stage C is pulling. In this paper, we use the time of trawling (stage B) as the time of each net.

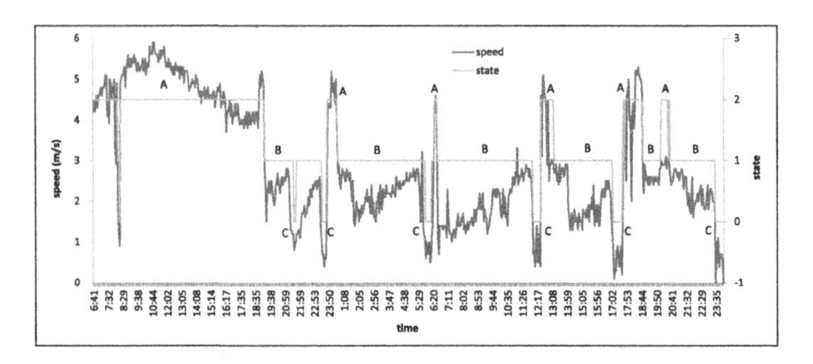

Fig. 6. The states of Trawler 299368

5 Calculation of the Fishing Effort

In 2013, there are 1443 trawlers whose quantity of vessel position is more than 4800 in Xiangshan Harbor. The differences of their power are relatively large (Fig. 7a). Therefore time multiples power as the fishing effort.

With the combination of speed and course deviation, the points in the state of fishing are extracted by Formulas (4) to (6). There are 410949 points in fishing extracted from 1443 trawlers on Oct 10th 2013. These points are putted into a $0.1° \times 0.1°$ grid according to the Formula (2) and calculate the cumulated fishing (kW•h) in each lattice.

The points in Fig. 7b are the center of lattices and the different colors denote different cumulated fishing (kW•h). As the fishing targets are moving fish, the neighboring fishery resources will be affected when we conduct fishing in one area. The stronger the fishing effort is in one area, the faster the neighboring fishery resources decrease. The fishing effort of several adjacent points will prone to a trend surface of resource impact in the surrounding area. The values of lattices are interpolated by Formula (4) and the interpolation figure of Oct 10th 2013 is generated. It reflects the distribution trend surface of trawlers fishing effort from Xiangshan Harbor. As the Fig. 7(b) shows, the strongest fishing effect these days is in Yushan Fishing Area, and then there are Zhoushan Fishing Area and Zhouwai Fishing Area.

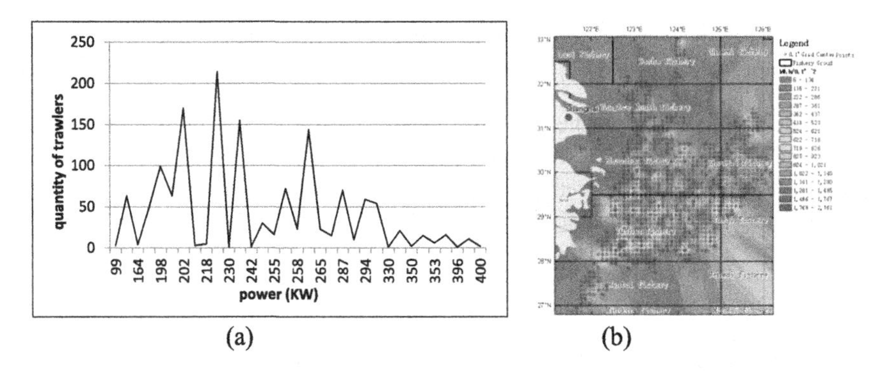

Fig. 7. Calculating the fishing effort (a) Fishing boat power (b) Fishing effort of Xiangshan

6 Conclusion

Via the analysis of speed and course characteristics, the fishing states can be judged and the operating rules can be obtained. According to the statistics of speed for a long period of time, the speed threshold of fishing states can be acquired for each trawler. With the combination of the above speed threshold and course deviation, the position in the state of fishing can be extracted. Then these position data are modified by filter window and calculate the cumulative fishing time according to the state of each trawler in one lattice.

The cumulative fishing is the product of the cumulative fishing time and the vessel power, such as kW•h. The lattice figure of the cumulative fishing is made, and its trend chart generated by interpolation.

In the next phase of work, we will analyze the other types of fishing, explore the methods of distinguishing different types of trawlers by speed and course and produce the complex chart of fishing effect for several fishing types. Analyzing the change rules of cumulative fishing in space and time and comparing advantages and disadvantages with the traditional method, we will improve our method further. Automatic mapping of cumulative fishing for fishing grid will be realized as a business service through programming. The automatic acquisition and business service of cumulative fishing effort will be regarded as a reference and guidance to fishery management and fishing condition analysis.

Acknowledgments. Thanks for the data provided by Shanghai Ubiquitous Navigation Technologies Ltd. The work is Funded by open research Funding program of KLGIS (KLGIS2015A06), Yangtze River Delta joint research project from Shanghai science and Technology Committee (15595811000) and the central level Public Welfare Scientific Research Institute of basic scientific research business fee special funds project (East China Sea Fisheries Research Institute 2014T13).

References

1. Walker, E., Bez, N.: A pioneer validation of a state-space model of vessel trajectories (VMS) with observers' data. Ecol. Model. **221**(17), 2008–2017 (2010)
2. Bensow, R.E., Larson, M.G.: Residual based VMS subgrid modeling for vortex flows. Comput. Methods Appl. Mech. Eng. **199**(13–16), 802–809 (2010)
3. Zhong, S., et al.: Guidance compliance behaviors of drivers under different information release modes on VMS. Inf. Sci. **289**, 117–132 (2014)
4. Detsis, E., et al.: Project catch: a space based solution to combat illegal, unreported and unregulated fishing: Part I: vessel monitoring system. Acta Astronaut. **80**, 114–123 (2012)
5. Ram-Bidesi, V., Tsamenyi, M.: Implications of the tuna management regime for domestic industry development in the Pacific Island States. Mar. Policy **28**(5), 383–392 (2004)
6. Joo, R., et al.: Optimization of an artificial neural network for identifying fishing set positions from VMS data: an example from the peruvian anchovy purse seine fishery. Ecol. Model. **222**(4), 1048–1059 (2011)

7. Fock, H.O.: Estimating historical trawling effort in the German Bight from 1924 to 1938. Fish. Res. **154**, 26–37 (2014)
8. Cicuendez Perez, J., et al.: The efficiency of using remote sensing for fisheries enforcement: application to the Mediterranean bluefin tuna fishery. Fish. Res. **147**, 24–31 (2013)
9. Papaioannou, E.A., et al.: Using indicators based on primary fisheries' data for assessing the development of the German Baltic small-scale fishery and reviewing its adaptation potential to changes in resource abundance and management during 2000–09. Ocean Coast. Manag. **98**, 38–50 (2014)
10. Fock, H.O.: Fisheries in the context of marine spatial planning: defining principal areas for fisheries in the German EEZ. Mar. Policy **32**(4), 728–739 (2008)
11. Zhang, S., et al., Method of trawling tracing based on beidou vessel monitoring system data, pp. 20–28 (2014)
12. Zhang, S., Wang, X., Zhou, W.: Offshore fishing aquatic products traceability based on vessel monitoring system. Comput. Dev. Appl. **4**, 16–19 (2014)
13. Bastardie, F., et al.: Effects of fishing effort allocation scenarios on energy efficiency and profitability: an individual-based model applied to Danish fisheries. Fish. Res. **106**(3), 501–516 (2010)
14. Guillemot, N., et al.: Effects of fishing on fish assemblages in a coral reef ecosystem: from functional response to potential indicators. Ecol. Ind. **43**, 227–235 (2014)

Spaceborne Multispectral Image Compression by Exploiting Temporal Correlation

Shigao Li(✉) and Liming Jia

School of Mathematic and Computer Science, Wuhan Polytechnic University, Wuhan 430023, Hubei, China
sg51@163.com

Abstract. Earth observation satellites usually scan the ground at a fixed period to capture remote sensing images. It's no doubt that there exists a strong correlation between images obtained at a small interval. This paper is devoted to the compression of spaceborne multispectral images and investigating the coding gain obtained by exploiting temporal correlation. To exploit temporal correlation, a temporal compensation (TC) scheme based on rate-distortion optimization (RDO) is proposed to remove redundancies between two adjacent-period multispectral images along temporal direction and a wavelet-based coding method is used to encode residue images. Experimental results indicate that the TC-based method produces significant improvement compared to those coding schemes of only exploiting spectral and spatial correlation.

Keywords: Motion compensation · Rate-distortion optimization · Image compression · Multispectral image compression · Wavelet-based image compression

1 Introduction

Earth observation satellites are extensively applied in administration of resources and environment. In the last decade, many countries have launched many observation satellites installed various sensors. A typical kind of sensors is multispectral CCD camera to produce multispectral images. At present, the resolution of spaceborne images is getting higher and higher. How to transmit vast image data by wireless channel becomes a challenge. Therefore, many related works are devoted to multispectral image compression [1–7].

Generally, there exists a considerable spectral correlation as well as spatial correlation for multispectral images. The spectral correlation can be removed by some spectral transforms. How to perform efficiently the transforms in the spectral domain gets the key of compression methods and many different methods are thus developed. A typical spectral transform for multispectral image compression is Karhunen-Loeve Transform (KLT) [8]. In general, KLT is first carried out in spectral domain followed by a 2D spatial

This work was supported by the Chinese Natural Science Foundation (61201452).

F. Bian and Y. Xie (Eds.): GRMSE 2015, CCIS 569, pp. 169–176, 2016.
DOI: 10.1007/978-3-662-49155-3_16

transform coding such as discrete wavelet transform (DWT). Due to the efficiency of KLT and DWT in removing the correlations, those compression methods based on KLT and DWT achieve good performance [1–3]. It is no exaggeration that if only spectral correlation and spatial correlation are taken into account, these methods have almost reached the performance limit.

In general, earth observation satellites scan the ground at a fixed period to capture remote sensing images. In a short time, ground landscape hardly changes. As a result, there exists a strong correlation between images obtained at a small interval. Being similar to video, the temporal correlation can thus be exploited to improve spaceborne image compression. So far, no researchers take the temporal correlation into account in spaceborne image compression. As far as I'm concerned, there is a key factor impeding temporal correlation utilization, i.e. complexities. It's of higher computational burden to exploit the temporal correlation. Moreover, we need more storage on satellites to save the images obtained in the last period. However, with the development of hardware technologies, it seems no longer to be a problem.

In according to features of multispectral spaceborne images, this paper investigates a compression method for multispectral spaceborne images. Due to the good performance of DWT for image compression, the wavelet-based coding method is still used for residue images. To reduce the time complexity, big images are split into tiles with fixed size and tiles are encoded one by one. A tile-based match module is carried out to find the best reference tile from the image obtained in the last period to predict the current tile.

In those vegetation regions, the correlation is very weak between images obtained in two adjacent periods. Compensation won't help. It's wise to encode those regions without temporal compensation. Therefore, we design an adaptive RDO-based temporal compensation in this paper.

2 Temporal Correlation Analysis

Theoretically, the temporal correlation is very high. However, many factors impair the correlation. In this section, we present a quantitative comparison between temporal correlations and spectral correlations. We adopt the dataset of IKONOS launched in 1999. The resolution of multispectral imagery of IKONOS is 4-m. We extracted four 512×512 tiles from IKONOS images.

The Table 1 gives the correlation coefficients. The first two columns present the temporal correlation coefficients between the two blue components obtained in the two periods. The first column is the result after local block-based registration while the second column indicates the result by applying tile-based registration. The last 3 columns are the correlation coefficients between the blue component and other 3 components respectively. As shown in Table 1, the correlation is very strong after the process of local registration. However, the correlation is weak if only tile-based registration is carried out.

Table 1. Comparision of correlation coefficients.

Image tiles	Temp. vorr. Coe.1	Temp. corr. Coe.2	Spectral corr. Coe.1	Spectral corr. Coe.2	Spectral corr. Coe.3
IKONOS-1	0.9496	0.6946	0.9791	0.9437	0.5697
IKONOS-2	0.9274	0.5664	0.9820	0.9542	0.8307
IKONOS-3	0.9721	0.7701	0.9870	0.9272	0.4654
IKONOS-4	0.9740	0.7244	0.9779	0.9336	0.3898

3 The Proposed Scheme Based on Adaptive TC

To reduce the complexity, we encode images based on tiles. To exploit temporal correlation, we store images obtained in the last period on-line as reference images. To compress efficiently the current images, we should find corresponding regions in the reference images. Instant positions of satellites can be exploited to give a searching region in reference images. The matching of the first tile may be time-consuming. However, if we obtain the match result of a tile, it'll be easy to carry out the match for its subsequent tiles. In addition, many methods can be used to finish the registration base on tiles. For instance, C.D. Kuglin presents a matching method based on the Fourier transform [8].

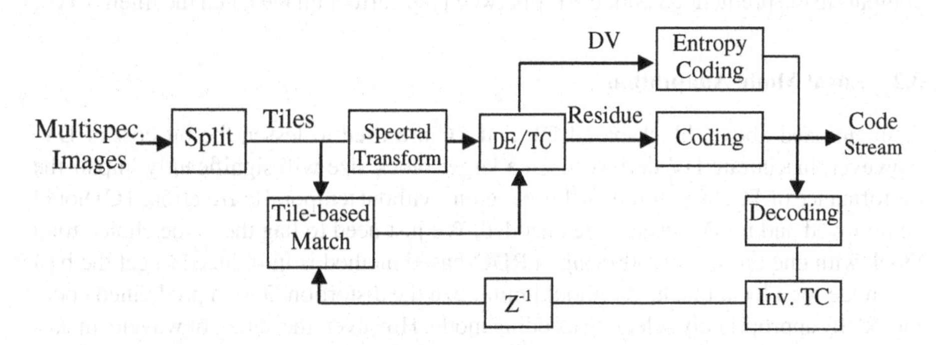

Fig. 1. Block Diagram of multispec. images based on TC.

As shown in Table 1, spectral correlation is stronger than temporal correlation. Therefore, we still introduce spectral transform prior to the temporal compensation. Due to the good performance of KLT, we adopt KLT to exploit spectral correlation in the following. And the temporal compensation will be applied to the images generated by KLT. The proposed scheme is shown as Fig. 1. Reference images (or tiles) are used to predict the current images (or tiles) and the generated prediction residues are then encoded. In the following, we'll give the design of these modules.

3.1 Adaptive TC

As analyzed in Sect. 2, to remove the temporal redundancies, local registration should be first carried out. This procedure is similar with motion estimation in video compression. In this paper, we call it displacement estimation (DE). And the local math result is displacement vector (DV). To reduce the bit-rate, a DE procedure with a fixed block size is carried out. Images (or tiles) produced by spectral transform are split into blocks with fixed size. For a block, the corresponding block can be found within a given range in reference images and the DV (u, v)can be generated by minimizes the Sum of Absolute Differences (SAD) *as*

$$(u, v)_{m,n} = \underset{u,v}{\operatorname{argmin}} \sum_{i=0}^{I-1} \sum_{j=0}^{J-1} \left| I_t(m \times I + i, n \times J + j) - \tilde{I}_{t-1}(m \times I + i + u, n \times J + j + v) \right| \tag{1}$$

where $\tilde{I}_{t-1}$ is the reconstruction version of the image obtained in the last period, the subscript t is the time, I and J denote the size of a block indexed by m and n. A temporal compensated image (TCI) is then produced as $\tilde{I}_{t-1}(x + u_{x,y}, y + v_{x,y})$.

Experiments show that there is little correlation between high-frequency components of the current image and the compensated image. An adaptive filter is thus applied to TCI to generate a filtered TCI in the DE/TC module in Fig. 1. The design of the adaptive filter is detailed in [9]. The temporal compensated difference (TCD) is obtained by computing the prediction residue error between the current image I_t and the filtered TCI.

3.2 Local Mode Adaptation

As mentioned above, block-based DE and TC are used to lessen the bit cost of DV. However, inaccurate DV derived from a larger block size will significantly impair the performance of TC. In addition, in those regions without temporal correlation, TC should be removed and no DV need to be encoded. We just need to flag the mode choice for a block with one bit. In the following, a RDO-based method is introduced to get the best TC mode for a local block. We should minimize the distortion D for a predefined code-rate R* by appropriately selecting a coding mode. However, the nature of wavelet makes it a very complicated dependent optimization problem to get the best modes for all blocks. In other word, the mode of a block is affected by its neighbor blocks. In the following, an iterative process with greedy strategy is carried out to get a suboptimal solution.

Rate-Distortion Model. However, to get the distortion D for a coding mode at a given bit-rate R* is a strenuous task. We can get the accurate D by performing a coding procedure. However, it's a very big computational burden. A practical method is to construct an analytical R-D model for wavelet-based coding methods.

In [10], a relationship between of peak signal to noise ratio (PSNR) of reconstruction images and the entropy of differential images was constructed as

$$PSNR = \frac{a}{CR} + b \times DEntropy + c \tag{2}$$

where *CR* is the compression ratio, a, b and c are const coefficients obtained by data fitting. If the intensity resolution of samples is NB bits, by using the define of PSNR, we can rewrite (2) as

$$D(R) = (2^{NB} - 1)^2 \times 10^{-\frac{1}{10} \times \left(\frac{a \times R}{NB} + b \times DEntropy + c\right)}. \tag{3}$$

Iterative Solution for Dependent Optimization. The mode selection described above can be seen as a labeling problem with regular sites and discrete labels [11]. In this labeling problem, we have $S = \{(i,j) | 1 \le i \le M, 1 \le i \le N\}$ corresponds to the set of M×N spatial blocks and the label set $L = \{1,2\ldots m\}$ corresponds to the set of prediction modes. We should assign a label from the label set L to each sites in S.

In order to simplify the problem, we transfer the mode selection problem into two binary decision problems. First, the images are segmented into the blocks with the size of 8x8. Each block site is assigned a label in {0, 1}. The label '0' denotes the mode without TC and the label '1' means exploiting the mode with TC in the blocks. After getting the labeling map with 8 × 8 blocks, a new binary decision problem whether to merge 4 blocks to a 16 × 16 block in terms of minimizing rate-distortion cost should be solved. Correspondingly, the label '0' denotes no merging and the label '1' means adopting TC mode with block-size 16 × 16.

We can model this label filed as a Markov random filed and solve this dependent optimization problem by minimizing the following potential function

$$U(f,R) = \sum_{i \in S} J(f_i, R) + \sum_{i' \in N_i} J(f_{i'}, R) \tag{4}$$

where J is the cost as (3) for the block at site i and f_i is the label, i.e. the prediction mode. N_i is the neighbor block sites of the i-th block site. For a 512 × 512 tile, there are 4096 blocks with the size of 8×8. Although an exhaustive searching method can leads to the optimal solution, its computational cost makes it impractical. So, we use an iterative method called iterated conditional modes (ICM) which uses the greedy strategy. An iterative solution that closely approximates the optimal strategy consists of starting from a possible solution. The algorithm sequentially updates the label for each site by minimizing (4). The process is repeated until no more improvements are possible or a maximum number c of iterations is reached. Experimentally it has been found that this method converges to a local minimum into 4–6 iterations. We can use the iterative method to solve the merging problem. However, experiments have proved that the improvement by using the method is negligible.

3.3 Implementation of TC

With the R-D model and the iterative method, we can give the overall algorithm of TC with the following steps. We assume the DVs with block-size 8×8 and 16 × 16 are obtained.

First, the DVs with block-size 8×8 are adopted to optimize the adaptive filter for a tile and the filtered TCI are generated.

Second, for each block, the TC-based coding mode and the coding mode without TC with the block-size of 8×8 are evaluated by using the model as (2) and the one with minimum D is selected.

Three, the iterative procedure as illustrated in the last subsection is performed based on the initial solution obtained in the last step.

And then, a merging procedure is applied for every four adjacent blocks with size 8×8 if the TC-based coding mode with size 16×16 leads to a decrease of distortion.

Finally, for those blocks with TC mode, a new filter is optimized and a new filtered TCI are generated.

4 Simulation and Analysis

This section designs several experiments to investigate the compression method based on TC. Since images applied in remote sensing are compressed with very high fidelity, we only investigate the compression with high fidelity. We obtained a group of multi-spectral images of two adjacent periods from the website 'http://www.isprs.org/data/ikonos_hobart/default.aspx'. The dataset originates from the satellite of IKONOS. We extract a part of images with the size of 1024×1024 which contains urban and hill areas as Fig. 2. The intensity resolution is 11 bit.

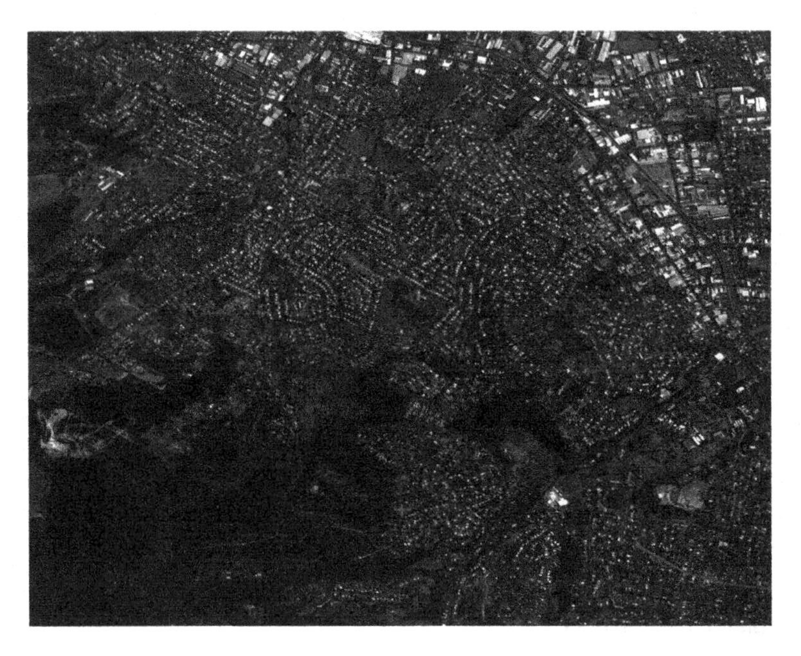

Fig. 2. The test image extracted from IKONOS.

We compare the proposed method to the traditional method based on KLT and JPEG2000. To simulate the TC procedure, we first compress and decompress the first time-phase images by using the traditional method. The reconstructed images are used to predict the second time-phase images as reference images. The proposed adaptive TC scheme is based on overlapped block TC with ALF and RDO-based local prediction. Residue images generated by TC are encoded by using the JPEG2000. The 9/7 wavelet coding mode is used in JPEG2000. A simple block matching method with integral pixel accuracy based on the dissimilarity criterion of SAD is adopted in DE procedure. To lower the code-rate of side information, the DPCM and arithmetic coding method are used to encode the DV. We split the extracted images into 512 × 512 tiles and encode them one by one.

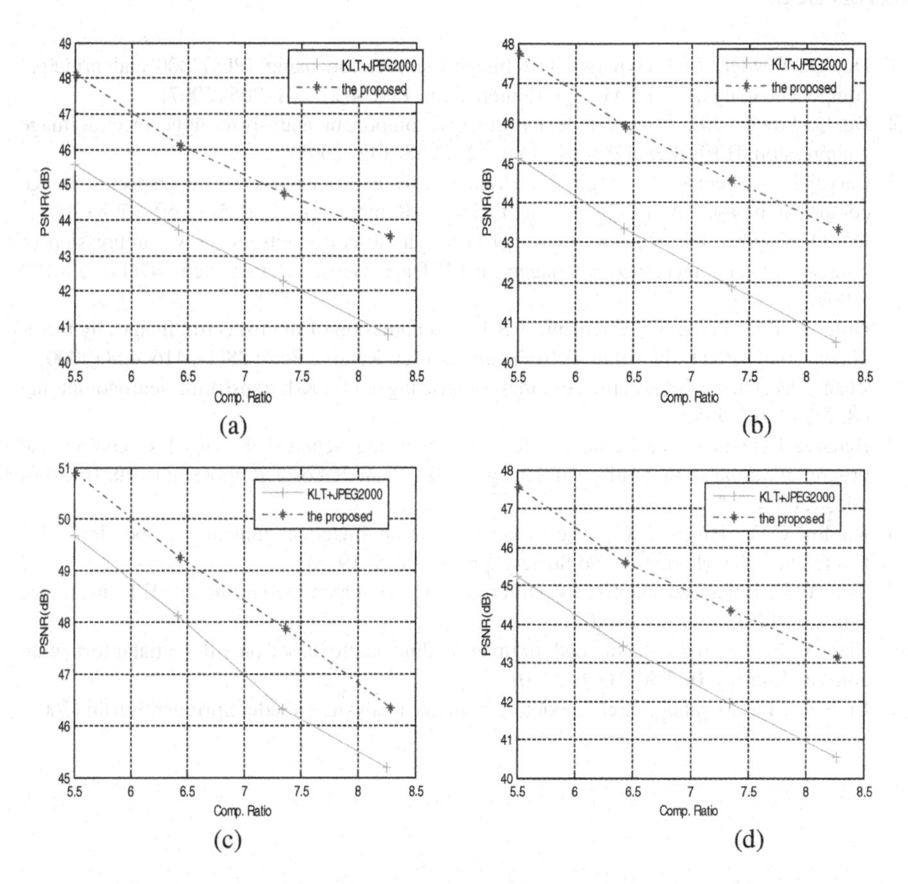

Fig. 3. Comparsion of compression results. **a–d** are the results of 4 tiles respectively.

We present the compression results of 4 tiles for IKONOS in Fig. 3. In Fig. 3, the PSNR denotes average PSNR of all components of multispectral images. The horizontal axis denotes the compression ratio. As shown in Fig. 3, for dataset IKONOS, the proposed adaptive TC scheme obtains the better PSNR. Its coding gains range from 1.5 to 3 dB compared to the method based on KLT and JPEG2000.

5 Conclusions

In this paper, we investigate the coding gain obtained by exploiting temporal correlation for the compression of spaceborne multispectral images. To exploit temporal correlation, an adaptive temporal compensation scheme is proposed. A block-based prediction mode selection by using rate-distortion optimization model is designed to cater to local features of images. Experimental results show that the proposed scheme can provide significant coding gain compared to the coding scheme of only exploiting the spatial and spectral correlations.

References

1. Du, Q., Fowler, J.E.: Hyperspectral Image compression using JPEG2000 and principal component analysis. IEEE Geosci. Remote Sens. Lett. **4**(2), 201–205 (2007)
2. Du, Q., Fowler, J.E.: Low-complexity principal component analysis for hyperspectral image compression. IEEE Signal Process. Lett. **12**(2), 38–142 (2005)
3. Carvajal, G., Penna, B., Magli, E.: Unified lossy and near-lossless hyperspectral image compression based on JPEG 2000. IEEE Geosci. Remote Sens. Lett. **5**(4), 593–598 (2008)
4. Du, Q., Fowler, J.E.: On the impact of atmospheric correction on lossy compression of multispectral and hyperspectral imagery. IEEE Trans. Geosci. Remote Sens. **47**(1), 130–132 (2009)
5. Dragotti, P.L., Poggi, G., Ragozini, A.R.P.: Compression of multispectral images by three-dimensional SPIHT algorithm. IEEE Trans. Geosci. Remote Sens. **38**(1), 416–428 (2000)
6. Zhanga, D., Chena, S.: Fast image compression using matrix K-L transform. Neurocomputing **68**, 258–266 (2005)
7. Benazza-Benyahia, A., Pesquet, J.-C.: Vector-lifting schemes for lossless coding and progressive archival of multispectral images. IEEE Trans. Geosci. Remote Sens. **40**(9), 2011–2025 (2002)
8. Kuglin, C.D., Hines, D.C.: The phase correlation image alignment method. In: IEEE Conference on Cybernetics and Society, pp. 163–165 (1975)
9. Yoo, Y.-L.: Enhanced adaptive loop filter for motion compensated frame. IEEE Trans. Image Process. **20**(8), 2177–2188 (2011)
10. Tian, X., Li, T.: Prediction method for image coding quality based on differential information entropy. Entropy **16**, 990–1001 (2000)
11. Li, S.Z.: Markov random field modeling in image analysis, 3rd edn. Springer, Berlin (2009)

A Decision Tree Classification Method Combining Intensity and RGB Value for LiDAR Data

Piyuan Yi(✉), Peng Tong, and Yingjun Zhao

National Key Laboratory of Remote Sensing Information and Image Analysis Technique, Beijing Research Institute of Uranium Geology, Beijing 100029, China
yipiyuan@163.com

Abstract. Airborne light detection and ranging (LiDAR) has played an important role in obtaining spatial information. But most existing LiDAR data classification algorithms mainly based on elevation and need more manual participation. Compared to these algorithms, we emphasize the use of intensity, RGB and echo number, and put forward a decision tree classification method. Before using this method, the intensity value must be calibrated first, and the RGB usually assigned from orthophoto. Then the experiment show that classification work can be completed with high accuracy while reducing manual workload. In addition, it was found intensity information is useful in target detection.

Keywords: Classification · Intensity · RGB · Echo number · Decision tree

1 Introduction

Airborne light detection and ranging (LiDAR) systems are mainly used to obtain the 3D coordinates of objects on the Earth surface for the generation of detailed elevation and topographic models. It has become an effective means of acquiring 3D information in China recent years [1, 2].

Most algorithms of feature extraction and object recognition using airborne LiDAR data mainly rely on the 3D geometric information of the laser points [3, 4]. In addition to containing 3D information, but also some other information recording, such as intensity, echo number, scanning angle.

The intensity values correspond to the backscattered radiometric information of objects. Most LiDAR systems can record 4 echoes of a laser pulse and give the corresponding intensity values. So a laser pulse produces one single echo or multi-echo pulses consist of first, last and intermediate echoes for the different backscattering characteristics and morphological structures of objects [5]. There are also full-waveform LIDAR systems that can digitize the shape of return signal [6, 7]. Many studies investigated the use of LiDAR intensity data for land cover classification and object recognition [8, 9]. In these studies, some algorithms were put forward to calibrate the intensity value, such as normalization, geometric calibration and radiometric correction [10, 11]. The main intensity processing methods are normalization and physical model-driven approach [12].

F. Bian and Y. Xie (Eds.): GRMSE 2015, CCIS 569, pp. 177–189, 2016.
DOI: 10.1007/978-3-662-49155-3_17

Application based on combining LiDAR data with optical images is also a research focus field. For example, extraction of multilayer vegetation coverage [13], urban vegetation mapping [14], building detection [15], land use classification [16], coastal and estuarine habitat mapping [17]. In these research, the data was processed separately and then analyzed the results comprehensively.

The main purpose of this study was to full use the intensity, echo number and RGB to develop a new approach for LiDAR data classification. The RGB value of laser points are assigned by the optical image and this is a main difference between with the research above. Based on analyzing the reflection characteristics, a decision tree method was put forward to recognize various ground objects and it obtained good results.

2 Data Pre-processing and Analysis

2.1 Data Collection

The study area is located in Nanning City, Guangxi Province, Southern China. The LiDAR data was acquired by using Optech ALTM Gemini system. The main parameters of flight plan are as shown in Table 1 and final point density within the study area is 2.6 points per square meter. The system also has a DIMAC RGB camera that can get digital photograph synchronously. For the camera focus length is 55 mm, spatial resolution of images is 10 cm, and then orthophoto can be produced based on the DEM (Digital Elevation model) generated by LiDAR points. The orthophoto and flightlines of the study area are as shown in Fig. 1. It can be seen the study area has a feature of high complexity for there are many kinds of ground objects, such as buildings, river, road, bridge, power lines, trees, sod and so on. So it's a good sample to evaluate the effect of classification method.

Table 1. Main parameters of flight plan

Parameter	Specification
Laser wavelength	1064 nm
Range capture	Up to 4 range measurements, including 1, 2, 3, and last returns
Flight Height (Altitude)	850 m
Pulse frequency	100 kHz
Scan frequency	35 Hz
Scan width (FOV)	$\pm 25^0$

Fig. 1. Orthophoto and flightlines of study area

2.2 Pre-processing LiDAR Point Cloud

Pre-processing of LiDAR points mainly contain intensity calibration and RGB value assignment.

According to the theoretical formulas of laser backscattering [18], assuming the target surface was a diffuse component, backscattering coefficients related to the reflectance of objects and incidence angles, atmospheric attenuation, and ranges between the laser sensor and objects [19]. So LiDAR intensity was normalized before using it [20]. The calculation formula is:

$$I_{(R)normalized} = I \cdot R^2 / \left(R_S{}^2 \cdot \cos\alpha\right) \quad (1)$$

where $I_{(R)normalized}$ is the value of the normalized intensity, I is the raw intensity, R is the distance between the point and the LiDAR sensor, R_s is the reference distance and a is the angle of incidence [21].

"Color laser point" is to combining it with the spectral information from the camera. The main principle is to assign the RGB value of the image pixel to the corresponding laser point according to the coordinate. For the average laser point distance is about 0.4 m while the orthophoto spatial resolution is 0.1 m, So each point can get a corresponding pixel value.

All above work were completed on the ENVI LiDAR + IDL platform. The effect of results are as shown in Figs. 2 and 3.

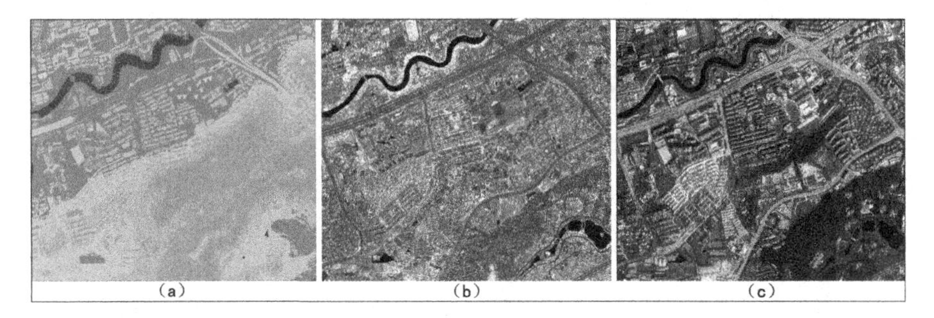

Fig. 2. Display of the study area laser points based on different attributes: (a) elevation, (b) intensity, (c) RGB value

Fig. 3. 3D view of the study area laser points based on RGB values

2.3 Data Analysis

The recording information of LiDAR data mainly depend on the laser wavelength and attributes of ground objects. So the function of different recording information in classification was analyzed first.

3D coordinates: Utilization of 3D information, especially the elevation, is the basis of mainly current classification algorithms. In this paper, we also use some existing methods to process the 3D information, such as extracting buildings based on height and area.

Intensity: Each material is inconsistent with the reflection of the laser pulse, so according to the characteristic of echo intensity data, objects of different materials can be distinguished in theory.

The echo intensity of various objects in study area is analyzed statistically and the results are as shown in Table 2. Water has a strong absorption function for 1064 nm laser, so the intensity value is almost 0. Trees have a big intensity value range from 54 to 72 for the influence of canopy shape, leaf density and some other factors. The echo intensity values of some different objects are the same, such as real turf and red roof, blue roof and plastic course, white roof and trees. This shows that the feasibility of the object classification only based on the intensity value is not good.

Table 2. Intensity statistics of some ground objects

Objects	Intensity	Main medium component
River	0	Water
Trees	54–72	Branch, leaves
Asphalt road	12–30	Asphalt, pebble
Red roof	70–79	Red paint, iron sheet
Grey roof	19–24	Concrete, asbestos tile
Blue roof	35–39	Iron, blue lacquer
White roof	43–58	Unknown
Roof solar panels	200–400	Crystalline silicon material
Real turf	75–80	Grass, sand grains
False turf	12–18	Polyethylene plastics
Plastic course	34–37	Rubber, sand grains

Echo number: ALTM Gemini LiDAR systems can record 4 echoes of a laser pulse and give the corresponding intensity values. So a laser pulse produces one single echo or multi-echo pulses consist of first three and last echoes for the different backscattering characteristics and morphological structures of objects. Statistical analysis show that some objects have single echo, such as roof, road, turf, bare land, square, some trees with thick foliage and so on, while most trees, shrub, power lines, street lamps, cranes and sides of buildings have multi-echo. It was think multi-echo of building sides mainly caused by the salient parts of building (such as balcony, window) when the incidence laser beam are approximate vertical to the ground. Statistics show that most of these laser points have incident angle values between 1^0 and 4^0, few are 10^0.

RGB value (Spectral information): The principle of RGB value assignment has been introduced. For influence of some buildings' geometric distortion and shadow, the RGB assignment of a few laser points is not correct and this will be avoided in the next step of data processing.

Based on the above analysis, it is feasible, combining the advantages of different laser point cloud attributes, to realize the high accuracy classification.

3 Classification Experiment

3.1 Method

The general thinking is to full use of some existing algorithms, and further combined echo number, intensity, RGB value by setting the threshold decision condition, progressively extract laser point subsets of different kinds objects. After many analysis and experiments, a decision tree classification method is formulated as shown in Fig. 4.

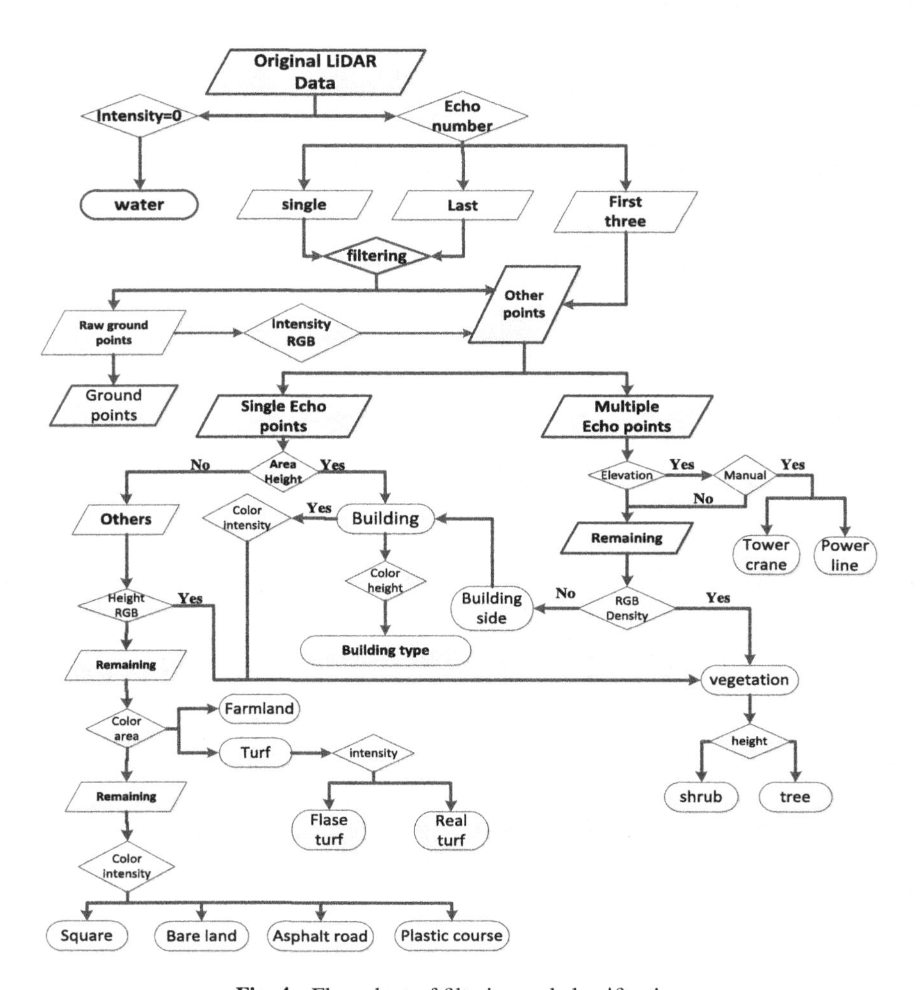

Fig. 4. Flow chart of filtering and classification

The process can be realized by using Terrasolid software. By setting the steps and relevant parameters in the macro operation, LiDAR data will be classified according to this process.

3.2 Classification Procedure

Using the decision tree method, the experimental data was classified with high accuracy.

Filtering—Ground Points Extraction. Although there are many classification methods, but it's primary to isolate ground points. Filtering of Lidar Point Clouds means to remove non-ground laser points and extract the digital terrain/elevation model (DTM/DEM) [22]. For ground points are from the lowest points, the data was firstly classified by echo number, then the single and last echo points were separated out to do filtering by using an iterative encryption triangulation algorithm [23]. Some error points can be further removed taking color and intensity values as reference.

Building Extraction. Laser points on building roofs almost have single echo, while few points of building sides have multi-echo. So the data was divided into two categories: single and multiple echo points. Using ground points as reference, by setting the parameters as height, area, maximum top angle (for inclined roof), maximum point interval (distance of nearest roof scan point), the building points were extracted from data set of single point [24].

As shown in Fig. 5, it was found that there are two types of errors by checking the extraction results. First is the garden vegetation, such as trimming holly, mainly for it's high leaf density, having a certain height and continuous distribution. These points can be removed by RGB value. Second is overpass, mainly due to bridge deck has a certain width, area and height. This part of the data is difficult to remove because it's similar with many other objects in RGB, intensity values. For the length of the overpass road more than 100 meters which is longer than building, so most overlap points can be

Fig. 5. Building extraction results and errors

extracted through to do neighborhood search by setting a certain radius range, step size, but there are still some residues can only be manually modified.

Finally, the total number of building points was 2563952, there are 201836 points of overpass and vegetation among them, so the correct rate was 92.12 %.

Other Single-echo Points Classification. After extraction of buildings, the points with a certain height in the remaining contain vegetation, a small number of building side points, while other low points mainly contain square, cement, playground, turf, bare land and few other objects.

By taking the ground points as reference and setting a certain height value, the points above ground 1 meter were extracted and then the building laser points can be filtered out by RGB value as shown in Fig. 6. Other vegetation points can be further classified to low vegetation (shrubs) and tall trees according to height. Through artificial interpretation, there are also few error points in vegetation category which corresponding some artificial facilities such as street lamp.

Fig. 6. Building extraction results and errors

As shown is Fig. 6, the remaining laser points within 1 meter distance to ground contain many objects, such as road, turf, square and so on. The height difference of these features is not so high, so it can not be classified only by the height informa-

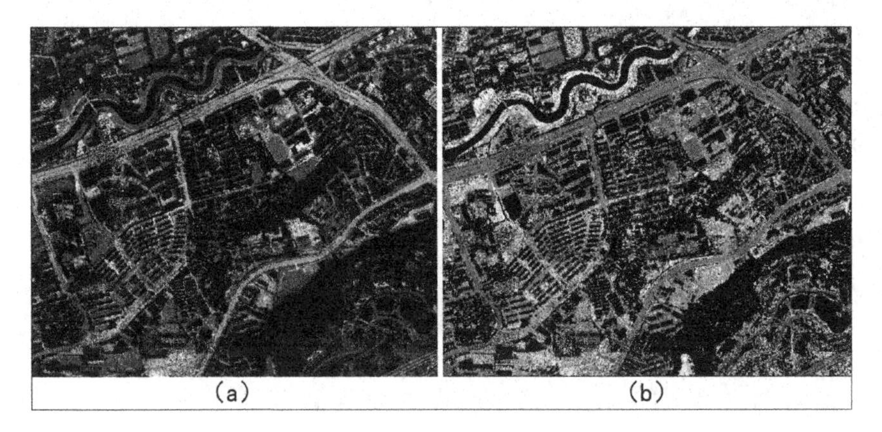

Fig. 7. Display of remaining low points based on different attributes: (a) RGB value, (b) intensity

tion. But as shown in Fig. 7, it was think the data has a good separability based on RGB and intensity.

For some ground objects' intensity value range has overlap, such as asphalt road and false turf, so the data was first classified according to RGB and green turf points were extracted. As shown in Fig. 8, these points can be further classified to real and false turf according to intensity for real one has higher value.

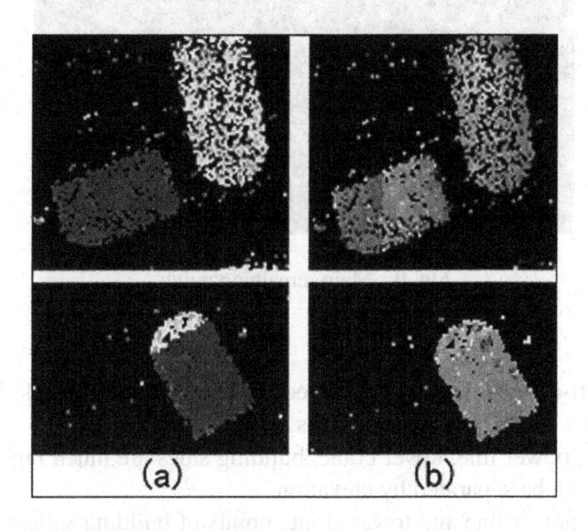

Fig. 8. Comparison of real and false turf (Color figure online)

On the other hand, because some real turf in residential areas is located in the shadow area, so the RGB assignment is wrong, and it can't be extracted only by RGB. But the intensity value of real turf and road has a big difference, so these points can be further extracted according to intensity. In this process, there are also few other objects' points which have the same intensity with real turf would be classified wrongly, such as bare land, zebras, these few points can be removed by RGB. The whole number of turf points is 578636 while the error is 3667, the correct rate is 99.36 %.

For the bare land, it has khaki color and high intensity value, so it was also mainly extracted by RGB and intensity. There are some error discrete points which are mainly distributed on road edge and footpath in residential area and this may be mainly due to the presence of a thicker dust cover. The number of these discrete points is 269568 while total number of bare land is 284530.

Now as shown in Fig. 9, the main remaining points mainly belong to roads after extracting of vegetation, turf and bare land. Some isolated points were removed firstly, and then different types of road were classified by using RGB and intensity. First, the plastic course is extracted according to color. Second, several square were extracted mainly by area. Finally, the rest is mainly asphalt road.

Fig. 9. Main remaining points

Process of Multi-echo Points. Multiple-echo points mainly include building sides, power lines, tower crane, street lamps, trees, shrubs.

Among them, power line, tower crane, building sides are much higher than others, so these objects can be separated by elevation.

Compared to power line and tower crane, points of building salient parts are more discrete, so it can be separated by density. But power line and tower crane has little difference in RGB and intensity, so it only can be further distinguished by manual operation. The results are as shown in Figs. 10 and 11.

Fig. 10. Points of power lines

Finally, most of the remaining multi-echo points are vegetation, and it can be almost extracted by RGB. There still few tree points which in the shadow area with wrong RGB value, for the other residual points are discretely distributed, so these can be extracted by density. Then the results can be merged with the previous single-echo points classification.

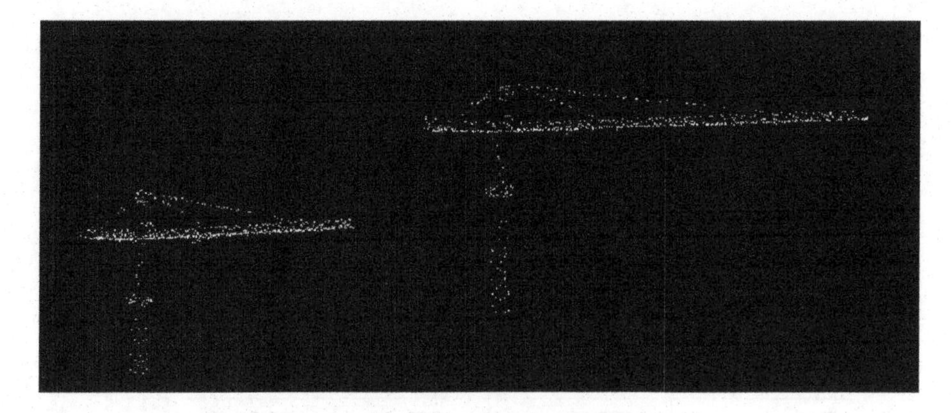

Fig. 11. Points of tower cranes

3.3 Results

Through the above work, laser point classification and extraction of seven typical features are accomplished and some results are as shown in Fig. 12.

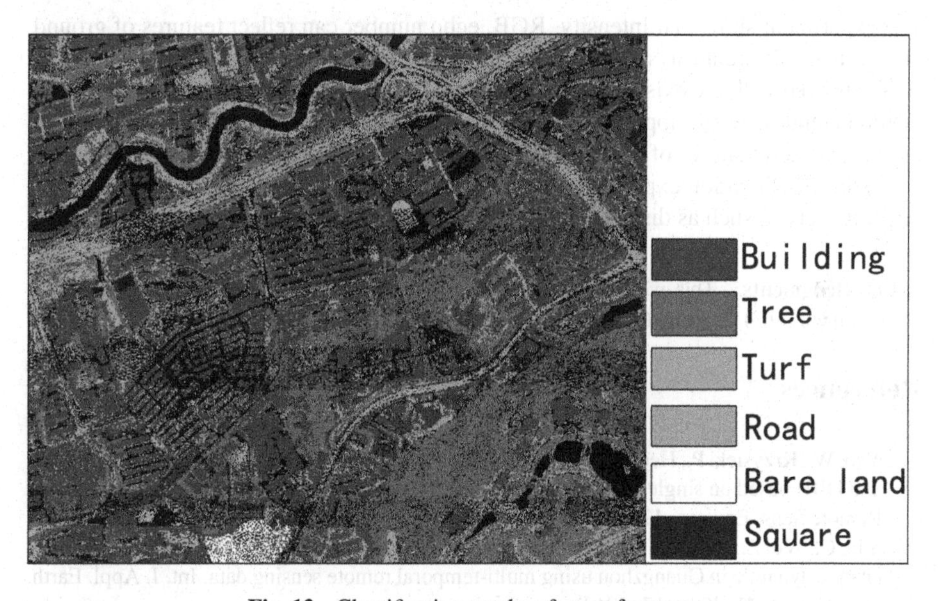

Fig. 12. Classification results of some features

The statistics of correct rates are as shown in Table 3. Total number of laser points of the area is 13891257, except 9092287 points in Table 3 and 3567381 ground points, there are still 1231589 points remaining. These points were also analyzed. Some points with intensity value greater than 200 on the building, it is inferred as roof solar panels. But more points are not regular, so it's hard to classify, or must be manual processing.

Table 3. Statistic of classification results

	Total number	Wrong number	Correct rate
Building	2563952	201836	92.12 %
Vegetation	4442687	341818	92.31 %
Bare land	284530	269568	94.74 %
Turf	578636	3667	99.36 %
Plastic course	14875	1685	88.67 %
Road	1189825	59372	95.01 %
Square	19046	903	95.26 %
	9092287	878928	90.34 %

4 Conclusion

The experiment show that intensity, RGB, echo number can reflect features of ground objects from different ways, and can play important roles in classification and detection.

Compared with the existing algorithms mainly based on elevation, the decision tree method emphasizes the application of intensity and RGB, combining other attributes of laser points and features of objects, classification can be completed with high accuracy.

In the classification experiment, it was found that intensity information is useful in target detection, such as distinction of true and false turf.

Acknowledgments. This work was supported by National High Technology Research and Development (863) Program (2012AA121304).

References

1. Yao, W., Krzystek, P., Heurich, M.: Tree species classification and estimation of stem volume and DBH based on single tree extraction by exploiting airborne full-waveform LiDAR data. Remote Sens. Environ. **123**, 368–380 (2012)
2. Sun, C., Wu, Z., Lv, Z., Yao, N., Wei, J.: Quantifying different types of urban growth and the change dynamic in Guangzhou using multi-temporal remote sensing data. Int. J. Appl. Earth Obs. Geoinf. **21**, 409–417 (2013)
3. Zhang, K., Yan, J., Chen, S.C.: Automatic construction of building footprints from airborne LiDAR data. IEEE Trans. Geosci. Remote Sens. **44**(9), 2523–2533 (2006)
4. Dorninger, P., Pfeifer, N.: A comprehensive automated 3D approach for building extraction, reconstruction, and regularization from airborne laser scanning point clouds. Sensors **8**(11), 7323–7343 (2008)
5. Korhonen, L., Korpela, I., Heiskanen, J., Maltamo, M.: Airborne discrete return LIDAR data in the estimation of vertical canopy cover, angular canopy closure and leaf area index. Remote Sens. Environ. **115**(4), 1065–1080 (2010)

6. Mallet, C., Bretar, F.: Full-waveform topographic lidar: state-of-the-art. ISPRS J. Photogrammetry Remote Sens. **64**(1), 1–16 (2009)
7. Qin, Y., Li, B., Niu, Z., et al.: Stepwise decomposition and relative radiometric normalization for small footprint LiDAR waveform. Sci. China Earth Sci. **41**(1), 103–109 (2011)
8. Bork, E.W., Su, J.G.: Integrating LIDAR data and multispectral imagery for enhanced classification of rangeland vegetation: a meta analysis. Remote Sens. Environ. **111**(1), 11–24 (2007)
9. Dalponte, M., Bruzzone, L., Gianelle, D.: Fusion of hyperspectral and LiDAR remote sensing data for classification of complex forest areas. IEEE Trans. Geosci. Remote Sens. **46**(5), 1416–1427 (2008)
10. Yan, W.Y., Shaker, A., Habib, A., Kersting, A.P.: Improving classification accuracy of airborne LiDAR intensity data by geometric calibration and radiometric correction. ISPRS J. Photogrammetry Remote Sens. **67**(2), 35–44 (2012)
11. Wagner, W.: Radiometric calibration of small-footprint full-waveform airborne laser scanner measurements: basic physical concepts. ISPRS J. Photogrammetry Remote Sens. **65**, 505–513 (2010)
12. Donoghue, D.M.M., Watt, P.J., Cox, N.J., Wilson, J.: Remote sensing of species mixtures in conifer plantations using Lidar height and intensity data. Remote Sens. Environ. **110**(4), 509–522 (2007)
13. Han, W., Zhao, S., Feng, X., Chen, L.: Extraction of multilayer vegetation coverage using airborne LiDAR discrete points with intensity information in urban areas: a casestudy in Nanjing City, China. Int. J. Appl. Earth Obs. Geoinf. **30**, 56–64 (2014)
14. Ramdani, F.: Urban vegetation mapping from fused hyperspectral image and LiDAR data with application to monitor urban tree heights. J. Geogr. Inf. Syst. **5**, 404–408 (2013)
15. Chen, L., Zhao, S., Han, W., Li, Y.: Building detection in an urban area using LiDAR data and Quickbird imagery. Int. J. Remote Sens. **16**, 5135–5148 (2012)
16. Huang, C., Peng, Y., Lang, M., Yeo, I.-Y., McCarty, G.: Wetland inundation mapping and change monitoring using Landsat and airborne LiDAR data. Remote Sens. Environ. **141**, 231–242 (2014)
17. Chust, G., Galparsoro, I., Borja, A., Franco, J., Uriarte, A.: Coastal and estuarine habitat mapping, using LIDAR height and intensity and multi-spectral imagery. Estuar. Coast. Shelf Sci. **78**, 633–643 (2008)
18. Baltsavias, E.P.: Airborne laser scanning: basic relations and formulas. ISPRS J. Photogrammetry Remote Sens. **54**(2/3), 199–214 (1999)
19. Yoon, J.-S., Shin, J.-I., Lee, K.-S.: Land cover characteristics of airborne LiDAR intensity data: a case study. Geosci. Remote Sens. Lett. **5**(4), 801–805 (2008)
20. Mesas-Carrascosa, F.J., Castillejo-González, I.L., de la Orden, M.S., Porras, A.G.-F.: Combining LiDAR intensity with aerial camera data to discriminate agricultural land uses. Comput. Electron. Agric. **84**, 36–46 (2012)
21. Höfle, B., Pfeifer, N.: Correction of laser scanning intensity data: data and model-driven approaches. ISPRS J. Photogrammetry Remote Sens. **62**(6), 415–433 (2007)
22. Zhang, X.: The Theory and Methods of Airborne Light Detection and Ranging Technology. Wuhan University Press, WuHan (2007)
23. Axelsson, P.: DEM generation from laser scanner data using adaptive TIN models. In: International Archives of the Photogrammetry, vol. XXXIII(1), pp. 10–117 (2000)
24. Soininen, A.: Terrasolid. TerraScan User's Guide, 3 October 2011

GIS-Based Distribution and Land Use Pattern of the Monasteries in Guoluo Tibetan Autonomous Prefecture in China

Yi Xiao[1], Luo Guo[1(✉)], Sulong Zhou[1], Fen Li[2], and Bingsheng Wu[3]

[1] College of Life and Environmental Sciences, Minzu University of China, Beijing 100081, China
guoluo@muc.edu.cn

[2] Chinese Research Academy of Environmental Sciences, Beijing 100012, China

[3] National Institute of Education, Nanyang Technological University, Singapore 637616, Singapore

Abstract. Based on the 2013 remote sensing data, DEM and field survey, we researched the characteristics of spatial distribution of 66 monasteries in Guoluo Tibetan Autonomous Prefecture which is located in three rivers headwater region of Qinghai province in China. By use of GIS we analyzed the differences of land use types and landscape patterns between the monasteries and the settlements with in 4 km of area. The result indicated that the monasteries in Guoluo are mainly distributed in the elevation of 4000–4500 m, and most of them stand on a slope or a gentleslope. Compared with the land use type of the settlements, the monasteries showed a more stable land use pattern.

Keywords: Land use pattern · Monastery distribution · Guoluo Tibetan Autonomous Prefecture

1 Introduction

As a unique civilization located in the Tibetan plateau, Tibetans know how to live harmoniously with a vulnerable ecological environment. For the Tibetan - an all-people religious nationality, the religious belief not only becomes an important part for their material and spiritual life but also plays a significant role in mediating the relationship between mankind and nature as well as interpersonal relationships. A typical example is Tibetans' reverence for the divine (e.g. monasteries, holy mountains or sacred lakes), which is a tradition of ecological culture showing awareness of natural environment, behaviors of respecting biological and cultural diversities (Liu et al. 2013). The unique culture implicitly exerts positive impacts on ecological protection.

F. Bian and Y. Xie (Eds.): GRMSE 2015, CCIS 569, pp. 190–196, 2016.
DOI: 10.1007/978-3-662-49155-3_18

Land use and land cover change (LUCC) has been broadly discussed in the research on environmental change and sustainable development (Liu et al. 2002; Turner et al. 2007). It is also an important factor in the change of ecological environment (Drummond et al. 2012; Parton et al. 2004; Fairman et al. 2011; Simbay-Kabba and Li 2011; Tian et al. 2013; Foley et al. 2005). Land use in China has experienced dramatic changes due to economic shift from agriculture to industries and been formed as various spatial patterns (Liu et al. 2009). This study explores spatial dynamics of monasteries in Guoluo Tibetan Autonomous Prefecture. Remote sensing and geographic information technologies are adopted to develop a geodatabase of spatial locations and land use types of 66 monasteries. Ecological and ethnological are used to analyze landscape patterns. Through spatial and ecological analysis, method, this study aims to reveal the relationship between monasteries and surrounding settlements from temporal and spatial perspectives of land use and land cover changes.

2 Study Area and Methods

2.1 Study Area

Guoluo Tibetan Autonomous Prefecture is located in the southeast of Qinghai Province with a unique geographical position. It is connected to Gannan Prefecture of Gansu Province on the east side. It is adjacent to Aba and Ganzi Tibetan and Qiang Autonomous Prefecture on the south side. It is next to the Yushu Tibetan Autonomous Prefecture and the Haixi Autonomous Prefecture on the west and the north side respectively. It is also an important region because it is the origins of two rivers, Yellow River and Yangtze River. In Guoluo Prefecture, administrative divisions include the Maduo County, Maqin County, Dari County, Gande County, Jiuzhi County, Banma County. The total population in Guoluo Prefecture is about 14.94 million, and the majority (91.24 % of population) is Tibetans. Among the 30 minority autonomous prefectures, the natural environment and economic development in Guoluo is rigid.

2.2 Methods

Data Source. Landsat images are selected to classify land use types in this study. Images are taken from Landsat 8 satellites in 2013. Object-oriented classification method is used to extract various land-use types including farmland, woodland, grassland, water area, urban and rural resident land, and unused land. After classification, field survey is conducted and 200 points are chosen to validate the classification results. The in situ fieldwork shows that the Kappa index of remote sensing image are all above 0.75. The distribution of the monasteries in the research area is obtained from the State Bureau of religious affairs and later validated by field survey (Fig. 1).

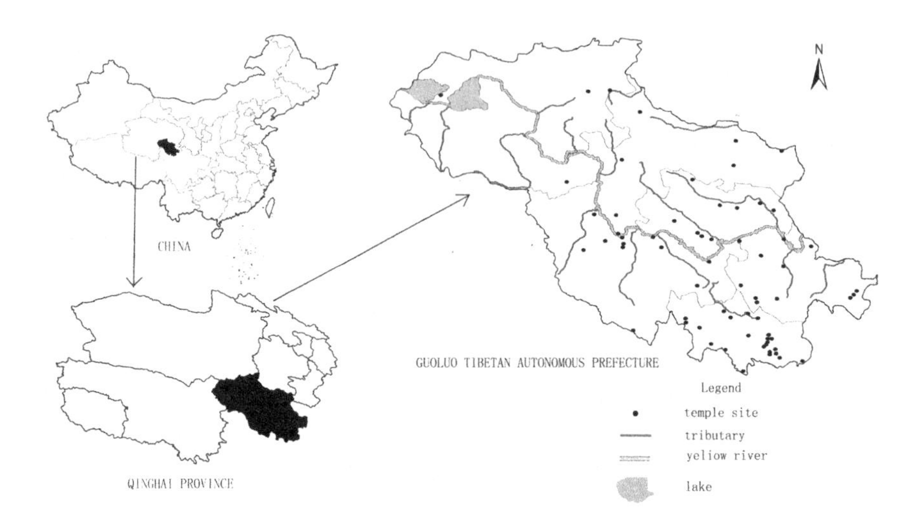

Fig. 1. The distribution map of the monasteries in Guoluo Prefecture

Analysis Method. Under the Spatial Analyst extension module of ArcGIS, the monastery points are used as the center and a 4-km buffer zone is created for each point. The land use data within the 4-km area is obtained from the classification results in 2013. Terrain factors such as slope, gradient, and aspect are created by 3D analyst module in ArcGIS.

3 Results and Analysis

3.1 The Characteristics of the Elevation Change in Spatial Distribution of Monasteries

Guoluo Prefecture is high in the northwest and low in the southeast. The areas between 4000–4500 m, 4500–5000 m, and 3500–4000 m are 65.56 %, 21.07 %, and 8.85 % of the total area respectively. The percentage of areas that are below 3500 m and above 5000 m is negligible. The average altitude is nearly 4500 m. The distribution of monasteries at different levels of altitude is categorized to 8 classes: 0–1000 m, 2500–3000 m, 3000–3500 m, 3500–4000 m, 4000–4500 m, 4500–5000 m, 5000–5500 m and >5500 m (Table 1). Table 1 reveals that the distribution density tends to increase first and then decrease. Between 3500 and 4000 m there are 22 monasteries, while 38 monasteries are located between 4000 and 4500 m, which the most monasteries are located. This region also occupies the largest percentage of the total area. The second largest region that are located between 4500 and 5000 m only has 5 monasteries. There is no monastery distributed above the altitude of 5000 m. It indicates a sharp decline in the number of monasteries with the raise of elevation. Therefore, the altitude of 5000 m can be regarded as the limit for distribution of monasteries in Guoluo Prefecture.

Table 1. The elevation class and monastery distribution in study area

Elevation (m)	Total area (km^2)	Area ratio (%)	Number of monasteries (n)	Proportion of monasteries (%)
0–1000	2812.1376	3.65	0	0
2500–3000	2.5515	0	0	0
3000–3500	344.7930	0.45	1	1.52
3500–4000	6809.6100	8.85	22	33.33
4000–4500	50454.7000	65.56	38	57.58
4500–5000	16212.8000	21.07	5	7.58
5000–5500	285.3140	0.37	0	0
>5500	33.3072	0.04	0	0
Total	76955.2133	100	66	100

3.2 The Characteristics of Slope Gradient and Slope Aspect of Monastery Distribution

According to the topographical characteristics of the prefecture, the slope is divided into four levels: 0–7° gently, 7–15° gentle slope, 15–25° slope, 25° above steep slope. As for slope aspect, there are nine categories including the horizon, the north, the northeast, the east, the southeast, the south, the southwest, the west and the northwest. Combine terrain data and monastery distribution map, and the monastery distribution on different slope gradient and aspect can be seen below (Tables 2 and 3).

As shown in Table 2, from the perspective of terrain gradient composition in Guoluo Prefecture, gently (0–7°, desirable or suitable to grazing, farming and human living) accounts for 40.04 %, gentle slope (7–15°) accounts for 26.39 %, slope (15–25°, applicable to agricultural production) accounts for 23 %, and steep slope which is not suitable for agricultural production and human living (equal to or more than 25°) accounts for 10.57 %. It can be seen that the gradient change in Guoluo Prefecture is small, and the land area suitable for grazing and farming accounts for a major proportion. In terms of the distribution density of monasteries, when the slope is between 0–25°, the distribution density of monasteries increases with the increase of the gradient. From low to high slope, the monastery distribution ratio is ordered as gently (7.58 %), gentle slope (27.3 %), and slope (51.5 %). When the slope is more than 25°, the distribution density of monasteries is obviously decreased. Therefore, in Guoluo Prefecture, most monasteries are mainly distributed in the slope, followed by the gentle slope, steep slope and gently.

According to the analysis of slope aspect, Guoluo Prefecture features the northwest slope, northern slope, west slope, southeast slope and east slope which accounts for 12.79 %, 12.25 %, 12.24 %, 12.23 %, and 12.05 % respectively. Given the monastery distribution ratio analysis, monasteries in Guoluo Prefecture are mostly distributed in the northeast slope, southeast slope, east slope, respectively, 12, 11, 10, followed by the northwest slope, north slope and west slope, respectively 9, 8 and 8 (see Table 3).

Table 2. The relationship between slope gradient and monastery distribution

Slopegradient (°)		Patch number (n)	Total area (km²)	Area ratio (%)	Number of monasteries	Proportion of monasteries
0–7	Gently	13076970	30811.5	40.04	5	7.58 %
7–15	Gentle slope	2506992	20306.6	26.39	18	27.27 %
15–25	Slope	2185542	17702.9	23.00	34	51.52 %
>25	Steep slope	1004228	8134.25	10.57	9	13.64 %

Table 3. The relationship between slope aspect and monastery distribution

Slope aspect (°)		Patch number (n)	Total area (km²)	Area ratio (%)	Number of monasteries	Proportion of monasteries
0–45	North slope	1163464	9424.06	12.25	8	12.12
45–90	Northwest slope	1214862	9840.38	12.79	9	13.64
90–135	West slope	1163118	9421.26	12.24	8	12.12
135–180	Southwest slope	1026031	8310.85	10.80	5	7.58
180–225	South slope	1125601	9117.37	11.85	3	4.55
225–270	Southeast slope	1161822	9410.76	12.23	11	16.67
270–315	East slope	1144999	9274.49	12.05	10	15.15
315–360	Northeast slope	992082	8035.86	10.44	12	18.18

3.3 The Characteristics of Land Use of Buffer Zones Around Monasteries

Buffer zones are built within a 4-km radius around 66 monasteries in order to analyze their land use pattern in 2013. The results show that most monasteries are mainly distributed in the grassland with high coverage and medium coverage, of which nearly 20 % monasteries are distributed in high coverage grassland types, nearly 64 % in medium coverage grassland types and nearly 16 % in low coverage grassland types. The difference of land use pattern around 66 monasteries is not significant.

4 Discussion and Conclusion

The study shows that the distribution limit of monasteries in study area is the altitude of 5000 m, and monasteries are mainly distributed in the elevation of 4000–4500 m. The slope is a key factor that affects the distribution of monasteries. The number of monasteries on the northeast (12), southeast (11), and east (10) side are the most. It is hardly to find any monastery at the high altitude and steep slope. Due to the lack of human activities, the higher latitude region is less disturbed so the ecological environment is relatively stable and healthy. Furthermore, the traditional ecological view in Tibet culture also deeply influences the Tibetan people. The core in Tibet culture is to develop consciousness of protecting the surrounding environment of the monastery and long-standing worship of nature toward holy mountains and sacred lakes. It then creates a stable ecological environment around the monasteries so the changes of land use types are slow at the origin of the Yellow River. The results from this study reflect these traditional Tibetan ecological views of monasteries and sacred sites have positive impacts on the ecological environment regarding regulation and protection.

Guoluo Prefecture belongs to the Yellow River source region and is a core area of three rivers headwater region characterized by its conservation status and unique ecological value. In this study, the traditional culture of ethnic minorities serves as a starting point and the traditional culture gathering point - monasteries as the research object. Based on remote sensing and geographic information technology, a contrast analysis is made regarding the superiority and stability of ecological and environmental conditions around the monastery distribution. This research result is not only quantitative verification of the positive influence of Tibetan traditional culture on ecological environment in the region, but also conducive to the government and the masses by providing a theoretical basis for resource management and sustainable development of ethnic minority areas of China so that they can learn valuable experience and knowledge from the traditional culture of ethnic minorities on the harmony between nature and human.

Acknowledgment. The work presented in this paper was supported by the National Natural Science Foundation of China (No. 31370480) and 111 Project (B08044). We also acknowledge all people who help us built the library.

References

Liu, L.Y., Liu, B., Li, J.Q., et al.: Ecological implication and significance of holy mountains in Tibetan area, northwest Yunnan, China. J. MUC (Nat. Sci. Ed.) **22**(2), 76–80 (2013). (in Chinese)

Liu, J.Y., Liu, M.L., Zhuang, D.F., et al.: Spatial pattern analysis on land use change of China in recent years. Sci. China (Series D) **32**(12), 1031–1040 (2002). (in Chinese)

IGBP Secretariat: GLP (2005) Science Plan and Implementation Strategy. IGBP Report No. 53/ IHDP Report No. 19, 2005, Stockholm, 64

Turner, B.L., Lambin, E.F., Reenberg, A.: The emergence of land change science for global environmental change and sustainability. PNAS **104**(52), 20666–20671 (2007)

Drummond, M.A., Auch, R.A., Arstensen, K.A., Ayler, K.L., et al.: Land change variability and human–environment dynamics in the United States great plains. Land Use Policy **29**, 710–723 (2012)

Parton, W.J., Tappan, G., Ojima, D., et al.: Ecological impact of historical and future land-use patterns in Senegal. J. Arid Environ. **59**(3), 605–623 (2004)

Fairman, J.G., Nair, U.S., Christoper, S.A., et al.: Land use change impacts on regional climate over Kilimanjaro. J. Geophys. Res. **116**(D3), 24–32 (2011)

Simbay-Kabba, V.T., Li, J.F.: Analysis of land use and land cover changes, and their ecological implications in Wuhan, China. J. Geogr. Geol. **3**, 104–112 (2011)

Tian, Y.C., Liang, M.Z., Ren, Z.Y.: Simulation of land use change and temporal-spatial heterogeneity of eco-risk in urban fringe. Res. Environ. Sci. **26**(5), 540–548 (2013). (in Chinese)

Foley, J.A., Defries, R., Asner, G.P., et al.: Global consequences of land use. Science **309**, 570–574 (2005)
Liu, J.Y., Zhang, Z.X., Xu, X.L., et al.: Spatial patterns and driving forces of land use change in China in the early 21st century. Acta Geogr. Sin. **64**(12), 1412 (2009). (in Chinese)

Parallelization of the Kriging Algorithm in Stochastic Simulation with GPU Accelerators

Lin Liu[1,2(✉)], Chonglong Wu[1], and Zhibo Wang[2,3]

[1] Information Technology Institude, China University of Geosciences, No. 388 Lumo Road, Wuhan, People's Republic of China
xlliu@ecit.cn
[2] Software Institute, East China Institute of Technology, Nanchang 430074, China
[3] International School of Software, Wuhan University, Wuhan 430079, China

Abstract. 3D realtime modeling places a heavy load on CPU. This paper presents a new method on 3D visualization in reservoir modeling system by using the computation power of modern programmable graphics hardware (GPU). The proposed scheme is devised to achieve parallel processing of massive reservoir logging data. By taking advantage of the GPU's parallel processing capability, moreover, the performance of our scheme is discussed in comparison with that of the implementation entirely running on CPU. Experimental results clearly show that the proposed parallel processing can remarkably accelerate the data clustering task. Especially, although data-transferring from GPU to CPU is generally costly, acceleration by GPU is significant to save the total execution time of data-clustering, and significantly alleviates the computing load on CPU.

Keywords: Kriging algorithm · Massive data · Stochastic simulation · Graphics processing unit (GPU)

1 Introduction

In the past few years, the 3D scientific visualization and acceleration has more application in numerous areas (the online game, the 3D scene wander, the flight simulation, virtual operation). In particular, 3D visualization of complex reservoir modeling is proved to be a highly competitive and important task.

In the reservoir modeling, the location and the shape of moving objects should be drew in real time. The refresh of the frames should not be noticed by users when the viewpoint changed [1]. The scene also should be redrew to adjust with the action of people. Especially, there are enormous of data should be processed to construct the complex reservoir modeling. Unfortunately, most of the modeling software can only sustain limited logging data [2]. The transformation of the model would be too slow to display. At the meantime, the information could lost if we reduce the logging data increasing with the size and dimension of data sets [3]. For the purpose of accelerating

F. Bian and Y. Xie (Eds.): GRMSE 2015, CCIS 569, pp. 197–205, 2016.
DOI: 10.1007/978-3-662-49155-3_19

the speed for massive data processing, many approaches for parallel data clustering have been proposed.

This paper presents an effective implementation scheme of 3D scientific visualization of the reservoir modeling system, in which each PC is equipped with a commodity programmable graphics processing unit (GPU). The proposed scheme is designed to achieve parallelly processing commoditization of modern GPUs, leading to a relatively low price per unit and rapid development of next generation processors.

2 Graphics Processing Unit and Cuda

GPU's amazing evolution on both computational capability and functionality extends application of GPU to the field of non-graphics computation, which is so-called general purpose computation on GPUs (GPGPU) [4, 5]. Design and development of GPGPU are becoming significant because of the following reasons:

(1) Cost-performance: Using only commodity hardware is important to achieve high computing performance at a low cost, and GPUs have become commonplace even in low-end PCs. Due to the hardware architecture designed for exploiting parallelism of graphics, even today's low-end GPU exhibits high-performance for data-parallel computing. In addition, GPU has much higher sequential memory access performance than CPU, because one of GPU's key tasks is filling regions of memory with contiguous texture data [6]. That is, GPU's dedicated memory can provide data to GPU's processing units at the high memory bandwidth.
(2) Evolution speed: GPU's performance such as the number of floating-point operations per second has been growing at a rapid pace [7].

Due to their highly parallel architecture, the programmable pixel pipeline of modern GPUs is capable of a theoretical peak performance that is an order of magnitude higher than CPU. An NVIDIA 7900 GTX 512 has a FLOPS rating of around 200 GigaFLOPS compared to a high-end PC, which is capable of around 10 GigaFLOPS. Furthermore, GPU performance has been increasing by a factor of 2 to 2.5 per year, which is faster than the increase in CPU performance as predicted by Moore's law. The high performance-to-cost ratio, rapid increase in performance and widespread availability of GPUs, which can deliver several times the performance of a single CPU, have propelled them to the forefront of high performance computing. The utility of GPUs has expanded beyond traditional graphics rendering.

nVIDIA's CUDA programming guide [8] estimates CUDA hardware to be approximately 1000 % faster than a Core2Duo. However SunlightLB's CUDA kernel achieved speed-ups of 149 % running on GeForce 8800GTS and 119 % on GeForce 8800GT in relation to Intel's Core2Duo. Hence, the port seems not to be programmed well enough to exhaust CUDA hardware's power. This is caused by non-optimized memory access patterns of SunlightLB's core simulation functions from CUDA's point of view. To achieve high performance on CUDA, it is very important for the GPU software to access data as big sequential blocks in GPU's D-RAM memory [9]. Traditional CPU software is not that dependent on block data access patterns, as the various caches of a modern CPU absorb this matter in a transparent way. If the desired algorithm is not adaptable to block access, the software must embed cache-like routines inside the GPU code for blockwise loading and unloading of input and output data between GPU's DRAM memory, and share memory located on the GPU itself. Then the algorithm can access randomly to data inside the shared memory space without major performance penalty. The CUDA program model is shown as Fig. 1.

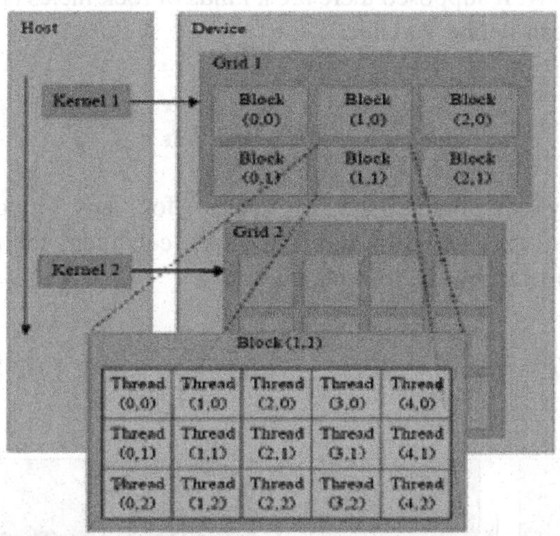

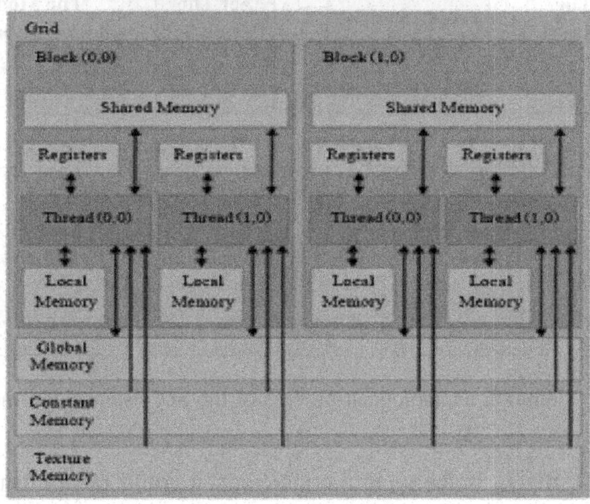

Fig. 1. The CUDA program model

3 The Methods of Geology Statistics

In the present spatial data interpolation method, the Kriging method is an optimal interpolation method, with an unbiased interpolated value and minimal estimation variance, most of the 3D Visualization is using the Methods as spatial data interpolation method [10]. By using the method of Kriging interpolation in Limited area, continuous reservoir data body can be obtained. Several forms of the Kriging interpolation method exist, such as the simple Kriging method, the ordinary Kriging method, the co-Kriging method, the stratified Kriging method, and the non-linear Kriging method, but each form has particular characteristics and it is suitable for a specific task. In this study, an ordinary Kriging interpolating approach was used to construct the data body. It supposed there are k kinds of rock facies $(s_1, s_2, \ldots, s_k)$ in the modeling area, we can define variable:

$$I(u) = \begin{cases} 1 & Z(u) \in s_k \\ 0 & Z(u) \notin s_k \end{cases} \tag{1}$$

The probability of being the k facies for any modeling point is: $P(I_k) = 1|Z(u_\alpha) = s_\alpha.\forall\alpha)\alpha$ could be the area which concludes the points, the probability could be calculated by the formula below:

$$P(I_k = 1|z(u_\alpha = s_\alpha, \forall\alpha) = E(I_k) + \sum_{\alpha=1}^{n} \lambda_\alpha[1 - E(I_k)]) \tag{2}$$

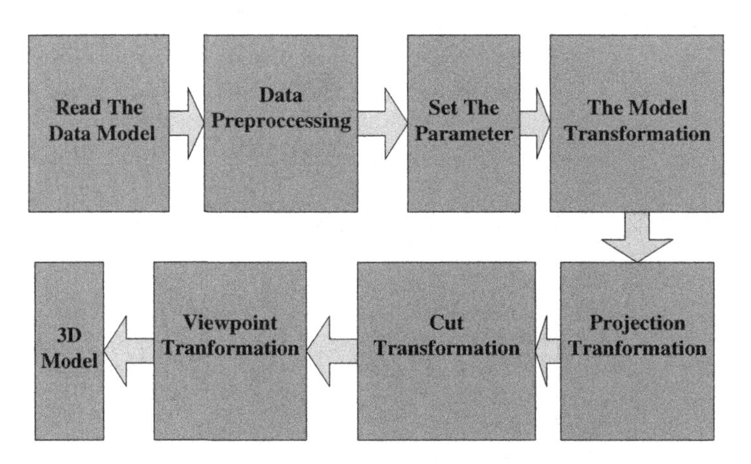

Fig. 2. Common technological process of the 3D visualization model

In this formula, the λ_α is the weight coefficient, it could be confirmed by Eq. (2).

Before the geology model can be drew on the computer, it should go through a series of coordinate transformation. The common technological process is shown in Fig. 2: Firstly, the mathematical description of the model can be preprocessed after reading it, set appropriate parameters such as length, width, etc., then set viewpoint to observe interested landscape. The description of how to observe the 3D model should be present after the construction. According to a series of coordinate transformation, the observation of 3D model can be observed in a appropriate position which is adapted to the viewpoint. In the observation process of 3D model, the observation way is up to the type of the projection transformation, different projection transformation get different

3D scene. The scene of transformation model is cut or zoom in the viewport transform which decides the whole 3 d model of the image on the screen.

3D visual modeling of Reservoir can be divided into three layers on macroscopic: the data interface layer, business logic disposal layer and human-computer interaction layer (as shown in Fig. 3 below). The format of the data set (such as file, database) can be translated in the interface layer, then it would be loaded in the business logic disposal layer for processing to establish reservoir data model. Finally, the model results can be shown in human-computer interaction layer. Among them, the business logic disposal layer and divided into three main steps to complete: the tectonic modeling, sedimentary facies modeling, property modeling (usually in a phased conditions).

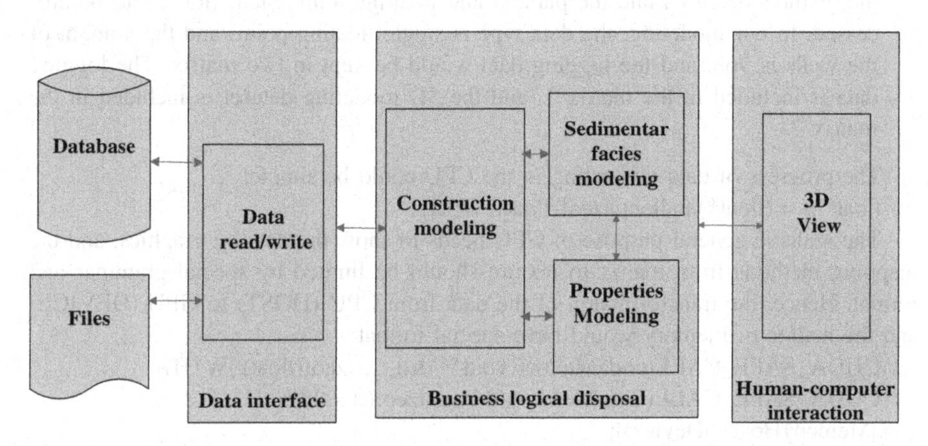

Fig. 3. The common technological process of the geology model

4 The Acceleration of 3D Modeiling by GPU

The scheme can be divided into two parts: the programme in the CPU or in the GPU.

A. *Element copy kernel function design*

The kernel function operation principle is that the CUDA program, which is designed to be a kernel, could be executed by sending to a grid. A number of blocks are contained in a grid, and several threads could be executed by every block.

The element copy kernel function declared below:

```
__global__ void kernel(float* d_1,float* d_2,float* d_3,float* d_4,float* d_5)
```

The function which is defined by global_would be a kernel function, and it would be invoked in the host computer. This part could be transformed from CPU to GPU for parallel processing. The logging parameters would be transformed to the GPU by kernel function. The kernel function could be invoked like this:

```
dim3 grid(5,5);
dim3 thread(BLOCKDIM_X,BLOCKDIM_Y);
kernel<<<grid,block>>>(d_1,d_2,d_3,d_4,d_5);
```

The grid.x*grid.y is the amount which would be sent, thread.x* thread.y is the amount of threads in every block.

B. *The code design of host in 3D visualization*

The processing results of logging data are not unique. It is not very complex to compare the results and discover the suited results, but the amount of the data involved in the modeling is enormous and a wide range of data could be processed. It would be unrealistic and low efficiency relying on the CPU only. Therefore, the logging data are considered to be arranged as matrix. The data should be mapped in the textures of GPU, and the parallel and floating-point calculation could be processed. In our modeling, the data type is single floating-point, and the amount of the wells is 968, and the logging data would be kept in two matrix. The logging data is included in the matrix 1, and the 3D modeling dataset is included in the matrix 2.

The progress of data generating in the CPU could be simple:

```
float* a = (float*)malloc(sizeof(float)*W*H)
```

The realistic general purpose of GPU needs to show the data by graphics, and the mapping methods from matrix to texture should be limited by special grammar and format. Hence, the transformation of the data from CPU (HOST) to GPU (DEVICE) and the malloc of memory would have special format.

```
CUDA_SAFE_CALL(cudaMalloc((void**)&d_1,sizeof(float)*W*H));
CUDA_SAFE_CALL(cudaMemcpy(d_1,a,sizeof(float)*W*H,
cudaMemcpyHostToDevice));
```

C. *The modeling speed test in CPU*

Due to the GPU data processing ability is far higher than the CPU, and the display work can be accomplished soon after the compute, we should assign calculation amount to the GPU as mush as possible, the CPU only need to translate the data to the GPU. The GPU would send the data to CPU after the parallel computing is finished and calculation results is accomplished. So we just need to record the speed of CPU modeling test.

5 The Results

The 3D visualization displays in the VS2005. The x- coordinate of reservoir modeling area is from 410 km to 470 km, the y- coordinate is from 450 km to 520 km. The CPU and GPU run in different equipments, which have the same price. The software and hardware environment is shown in Table 1: the GPU is NVIDIA GeForce 8400 M GS and the CPU is Intel Core2 Duo 1.5 GHz. The 3D modeling results display in the computer, and the running time is recorded in Table 2 and shown Fig. 4: the visualization was running for ten times in the CPU and GPU. The costing time in the CPU is from 44.786 ms to 72.556 ms, the costing time in the CPU is from 20.542 ms to

Table 1. The software and hardware environment

GPU	NVIDIA GeForce 8400 M GS
CPU	Intel Core2 Duo 1.5 GHz
RAM	2.0 GB
Operating system	Microsoft Windows XP Pro SP2
Programme environment	Microsoft Visual Studio 2005 Visual C ++
Software	NVIDIA CUDA Toolkit 2.0beta.NVIDIA CUDA SDK 2.0beta \NVIDIA CUDA Driver 2.0beta

Table 2. Comparison of 3D modeling runtime in cpu and gpu

Number	1	2	3	4	5
The time running on CPU/ms	56.245	67.458	44.786	60.395	72.556
The time running on GPU/ms	20.542	29.575	26.455	35.343	32.546
The ratio of CPU/GPU	2.738	2.28	1.692	1.708	2.229
Number	6	7	8	9	10
The time running on CPU/ms	70.37	65.735	58.368	58.465	57.35
The time running on GPU/ms	33.483	29.476	38.571	34.28	31.439
The ratio of CPU/GPU	2.23	1.951	2.23	1.513	1.824

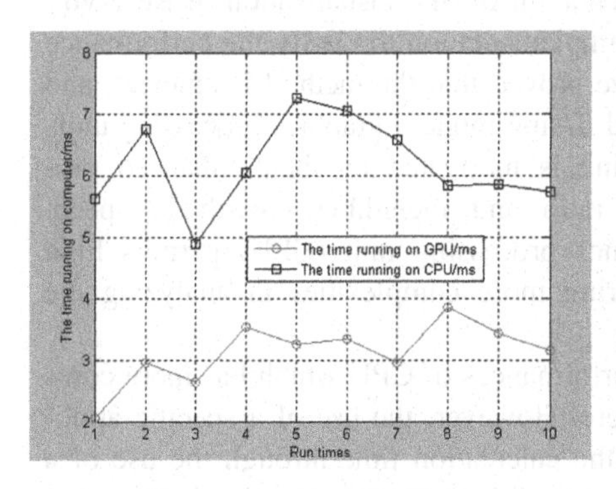

Fig. 4. Comparison of 3D modeling runtime in cpu and gpu

Fig. 5. The 3D modeling of restraint area

38.57 ms, the acceleration is obvious. The 3D modeling of restraint area is shown in Fig. 5. The 3D modeling of restraint area with boundary is shown in Fig. 6. The3D modeling of restraint area with 1st section is displayed in Fig. 7.

The results show that the general purpose is suited for the processing of mass data. The logging data has a huge amount in our model, and most of them are floating point data. The display speed would be slow if only relying on CPU. We presented the methods of parallel processing by GPU with the same price to share the calculated amount, and then the speed is accelerated. So, the general purpose of GPU could be an important and meaningful research.

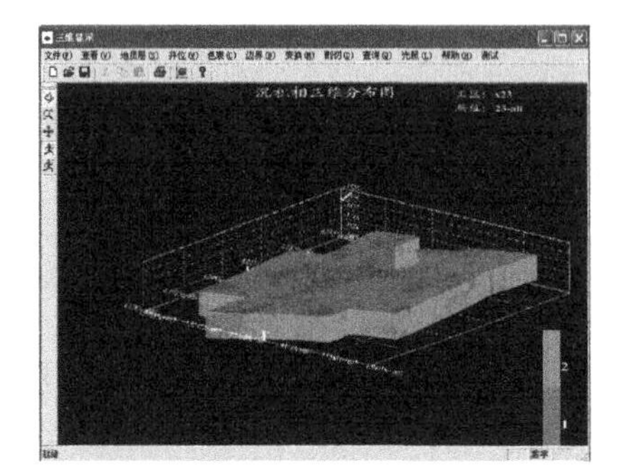

Fig. 6. The 3D modeling of restraint area with boundary

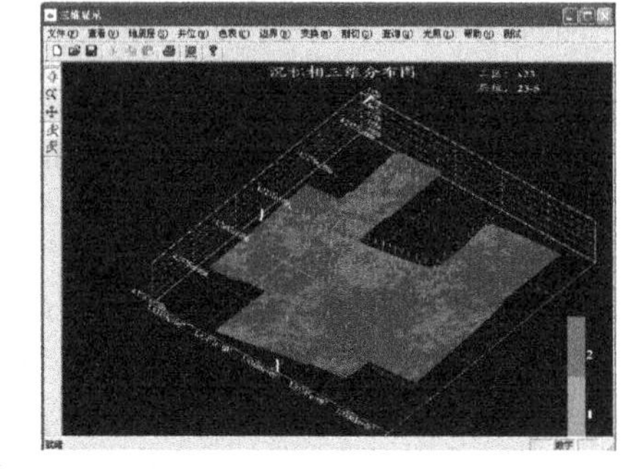

Fig. 7. The 3D modeling of restraint area with 1st section

6 The Conclusion and Future Work

This paper has discussed the GPU implementation of 3D Visualization in Reservoir Modeling System, compared with the CPU implementation to clarify the performance gain of GPU co-processing. The results have proved that the method is efficient, and the computing load of CPU is also alleviated. Using modern graphics processing units for no-graphics high performance computing is motivated by their enhanced programmability, attractive cost/performance ratio and incredible growth in speed. Although the pipeline of a modern graphics processing unit (GPU) permits high throughput and more concurrency, they bring more complexities in analyzing the performance of GPU-based applications.

In this study, we did not compare the performances of CPU which has more cores instead restricted our comparison to dual-core. However, the use of a specific application programming interface might prove the calculation time through the use of a multicore CPU rather than a dual-core CPU [11]. The multi-threading implementation with effective load balancing between CPU and GPU will be investigated in our future work.

Acknowledgment. Lin Liu thanks the support of Youth Fund of JiangXi Province (N0: GJJ14491).

References

1. Goovaerts, P.: Geostatistical modelling of uncertainty in soil science. Geoderma (2001)
2. Anctil, F., Mathieu, R., Parent, L.E., Viau, A.A., Sbih, M., Hessami, M.: Geostatistics of near-surface moisture in bare cultivated organic soils. J. Hydrol. **260**, 30–37 (2002)
3. Trendall, C., Steward, A.J.: General calculations using graphics hardware, with applications to interactive caustics. In: Proceedings of Eurogaphics Workshop on Rendering (2000)
4. Manocha, D.: General-purpose computations using graphics processors. Computer **38**(8), 85–88 (2005)
5. Manssen, M., Weigel, M., Hartmann, A.: Random number generators for massively parallel simulations on GPU. Eur. Phys. J. Spec. Top. (2012)
6. Moreland, K., Angel, E.: The FFT on a GPU. In: Proceedings of Graphics Hardware, San Diego (2003)
7. Zhang, E.Z., Jiang, Y., Guo, Z., Tian, K., Shen, X.: On-the-fly elimination of dynamic irregularities for GPU computing. In: ASPLOS 2011: Proceedings of the 16th International Conference on Architectural Support for Programming Languages and Operating Systems (2011)
8. Trendall, C., Steward, A.J.: General calculations using graphics hardware, with applications to interactive caustics. In: Proceedings of Eurogaphics Workshop on Rendering (2000)
9. Li, W., Wei, X., Kaufman, A.: Implementing lattice Boltzmann computation on graphics hardware. Vis. Comput. **19**(7–8), 444–456 (2003)
10. Houlding, S.W.: 3D Geoscience Modeling: Computer Techniques for Geological Characterization. Springer, Berlin (1994)
11. Spoerk, J., Bergmann, H., Wanschitz, F., et al.: Fast DRR splat rendering using common consumer graphics hardware. Med. Phys. **34**, 4302–4308 (2007)

Study on Spatial-Temporal Feature of Natural Disasters in Qinghai Province Based on GIS

Shurui Feng[1], Luo Guo[1(✉)], Jizhuoma Cai[1], and Yan Ai[2]

[1] College of Life and Environmental Sciences, Minzu University of China, Beijing 100081, China
guoluo@muc.edu.cn

[2] Chinese Research Academy of Environmental Sciences, Beijing 100012, China

Abstract. The main purpose of this research is to Qinghai Province as the research object. We use GIS to analyze distribution of natural disasters, such as hail, snow disaster, from 1950 to 2013. **Then, obtained the temporal and spatial distribution and law of natural disasters.** Results show that: (1) Qinghai occurred 15 times drought which reached 31.5 %, **between 1960 to 2010.** (2) The high incidence areas of drought are Xining City and Haidong area, and sub high incidence areas are southwest Yushu Tibetan Autonomous Prefectures and west Haixi Tibetan Autonomous Prefectures, and the relatively low frequency of disasters in other areas. (3) From the spatial distribution, the most areas of snow disaster concentrated in the six counties of Yushu Prefectures and Guoluo Prefectures. (4) From the time distribution, the most serious areas of snow disaster are Yushu and Guoluo Prefectures.

Keywords: Spatial-Temporal feature · Natural disasters · Qinghai province

1 Introduction

China is one of the most serious natural disasters **area** in the world. All kinds of serious natural disasters in the world which exist in our country, and these disasters have enormous damage and impact to our country [1, 2]. At present, researches on the professional assessment of disasters have two main ideas. One is establish the index system of comprehensive disaster situation to evaluate the intensity of regional disaster and classification. Another is using the temporal and spatial distribution of hazard factor to analyze spatial pattern of disasters [3]. Many scholars have published some articles to discuss normal disasters and sudden event aspect [4–7]. As part of the Tibetan Plateau, Qinghai Province is a sensitive area of climate change, and mainly in alpine and arid climate in China. It has special geographical position and severe natural environment conditions and makes it become a high and mixed incidence area of natural disasters in China. Qinghai has variety of natural disasters, and all kinds of meteorological disasters are higher frequency. According to the statistics data, since the 20th century, there are more than 160 times of a magnitude-5.0 earthquake over in **Qinghai**, it ranks first in the country's province [8]. Natural disasters seriously affect local economic development and herdsmen's living

F. Bian and Y. Xie (Eds.): GRMSE 2015, CCIS 569, pp. 206–212, 2016.
DOI: 10.1007/978-3-662-49155-3_20

standard. Qinghai Province is a typical area to research spatial- temporal feature and distribution of natural disasters in China. It has a great demonstration effect for our research and analysis in this aspect.

2 Study Area and Methods

2.1 Study Area

Qinghai Province is located in the northeastern of Tibetan Plateau. The study area lies between E 89°35′- 103°04′ N 31°9′-39°19′, and covering an area of approximately 69.67km2. The area is characterized by a cold and dry climate. And **has** numerous rivers, glaciers and lakes, and complex terrain and landform. The study area includes all or some of the counties from five Tibetan Autonomous Prefectures and a Mongolian Autonomous Prefectures, which are Yushu Prefecture, Guoluo Prefecture, Hainan Prefecture, and Huangnan Prefecture Haimen Prefecture, Haixi Prefecture and two cities, which is Geermu City and Xining city (Fig. 1). It has a total population of about 58.34 million.Qinghai is a typical area to research climate and ecological environment change in sensitive and frail area, and spatial-temporal feature and distribution of natural disasters [9].

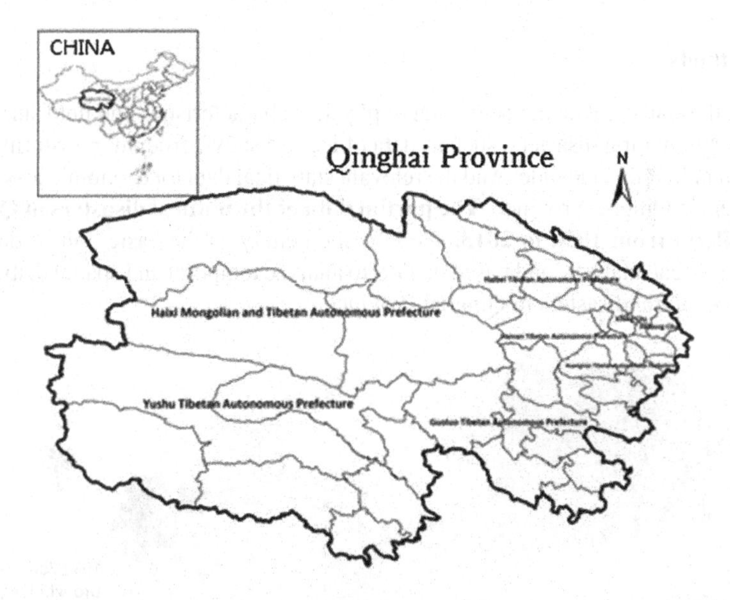

Fig. 1. Administrative map of Qinghai province

Drought and snow disaster are natural disasters which have the high frequency, large range and serious harm. Not only have a greater effect on agriculture, but also severe obstruct local economic development. At the same time, disasters restrict stable and sustainable development of animal husbandry in pastoral area. **The annual mean temperature in Qinghai, showing increasing trend from 1962 to 2013** (Fig. 2). Especially because of effects of climate change, since 1998, the annual mean temperature higher than reference value obviously, and the increase of temperature is faster from that point. It affects the natural disasters in Qinghai Province, particularly meteorological disasters [10, 11].

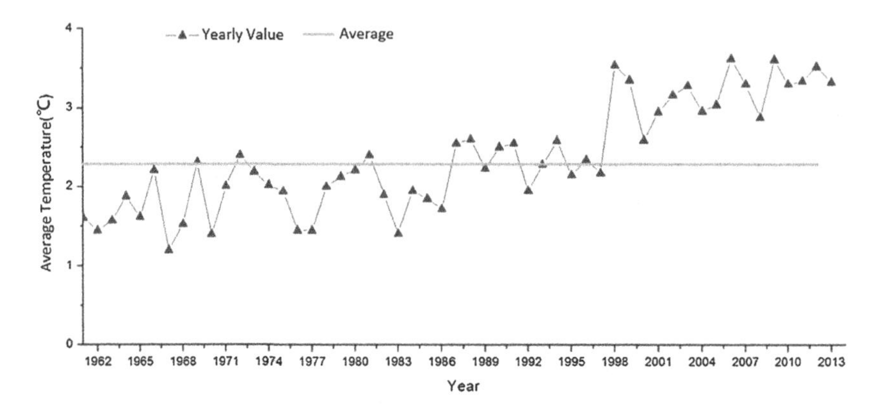

Fig. 2. Annual average temperature of Qinghai Province from 1962 to 2013

2.2 Methods

Based on the statistical data, vector data of physical characteristics and field survey, we obtain data of natural disasters, such as data of ice and snow, drought, rainstorm, earthquake and mountain landslide. And the relevant statistical data for economic loss caused by disasters in Qinghai province. **The partial data of the natural disasters of Qinghai from bulletin from 1950 to 2013.** We take the county as the basic unit of data, use spatial statistical analysis, and based on GIS to analyze temporal and spatial distribution and law of natural disasters in Qinghai Province.

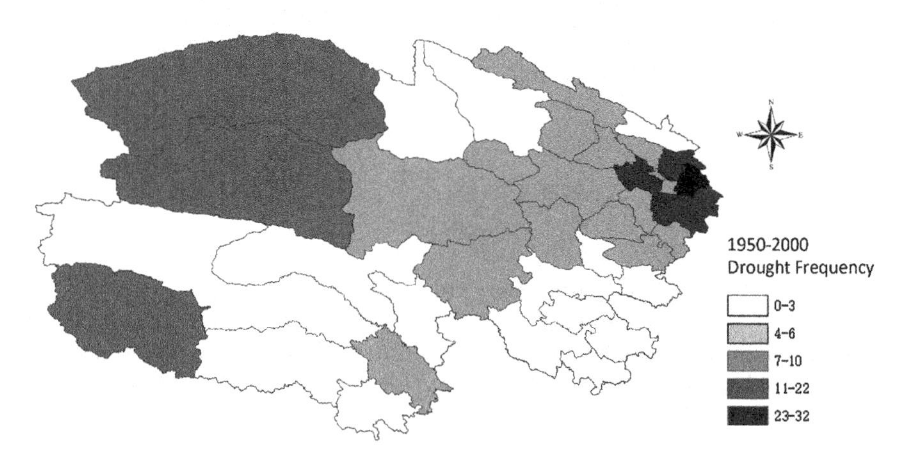

Fig. 3. Geographic distribution of drought disaster in Qinghai 1950–2000

3 Results and Analysis

3.1 The Temporal-Spatial Distribution of Drought Disaster

According to the statistical data of Qinghai Province in nearly 60 years, and combined with the local population distribution change, by using GIS, we get the spatial distribution of drought disaster in the study area (Fig. 3). The global warming leads to temperature gradually increased. Precipitation in Qinghai experienced a humid period in the 1980 s, to 1990 s it entered a drought period. This situation results cause severe drought year increased significantly, the disaster area expanded continuously, and the trend of drought still not been contain. From 1960 to 2010, there have been 15 years of drought disaster in Qinghai, the probability achieved 31.5 %. Among them, there were an extraordinary drought, severe drought twice and moderate or light drought six times. Light drought relatively more frequency of occurrences, and bring to serious economic losses in the local. The high incidence areas of drought are Xining City and Haidong Tibetan Autonomous Prefectures in the past 50 years, secondary high incidence areas of drought are southwest Yushu Tibetan Autonomous Prefectures and west Haixi Tibetan Autonomous Prefectures, the other place of the relatively low frequency of disasters. The population is mainly concentrated in the Xining, Haidong, Guoluo, Huangnan and southeastern Yushu in Qinghai. We combined with the affected population, area and probability to analyze, and the result is more serious drought disaster area is Xining and Haidong, the disaster mainly located in Ledu, Minhe, Xuhua, Huangyuan, Huangzhong, Huzhu.

3.2 The Temporal and Spatial Distribution of Snow Disaster

Snow disaster means snow for a long time and continue to melt in a certain region, resulting people had been frostbit, the electricity and transportation congestion, and livestock death. Qinghai Province is located in Tibetan Plateau. The annual average temperature is below 4 centigrade. The average altitude is above 2000 m. So, cold is the main climatic characteristic in Qinghai. The index of snow disaster in snow cover and duration as parameters to measured. Among them, the snow cover is an important parameter to analyze snow disaster distribution. Count distribution of snow disaster nearly 60 years we can be found: from the spatial distribution (Fig. 4), snowfall and snow cover are most; the most areas of snow disaster concentrated in the six counties of Yushu and Guoluo. The mountains of southern Qinghai about 3800 meters, so susceptible to cold air in the winter, plus a special environment of the high altitude and low pressure, caused more snowfall and snow cover in here. Especially in the Bayan Har Mountain and Tangula Mountain in southern Qinghai, where snow all the year, the average snow days can be reached 60-147 days, and the maximum snow cover can reached 90 cm. In the end, the least areas of snowfall and snow cover mainly concentrated in the warm zone of the western Qinghai where near the Yellow River, such as Qaidam Basin, Huangzhong Country and Huangyuan Country. In these areas the days of snow less than a month, and the maximum thickness of snow cover less 10 cm.

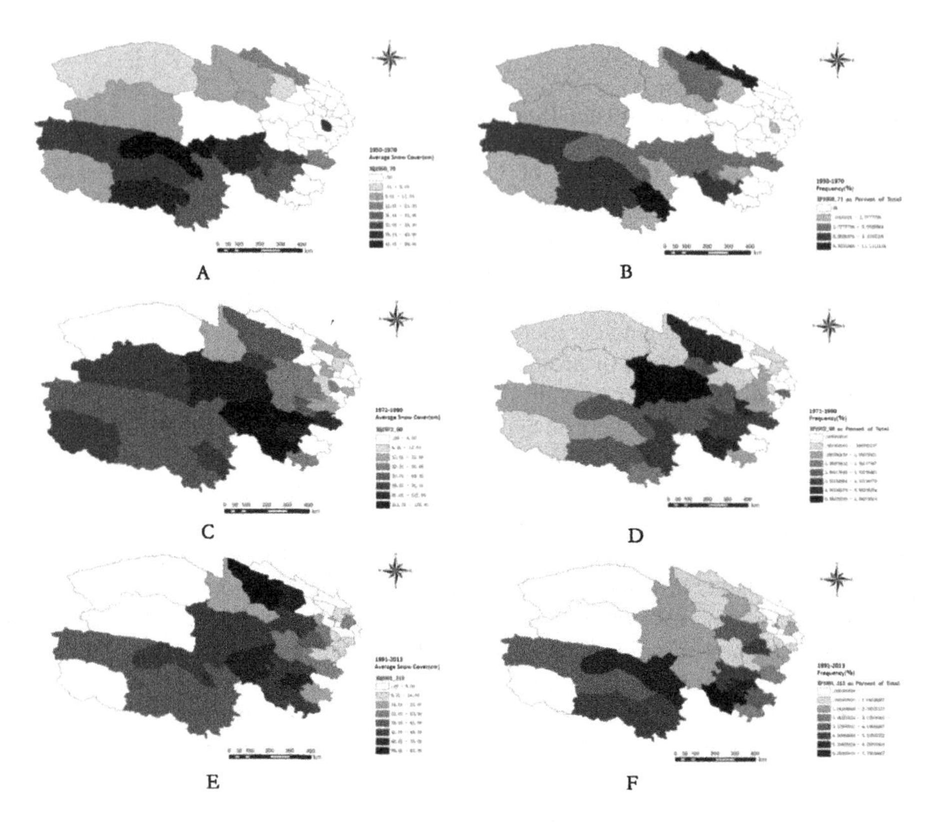

Fig. 4. Geographic distribution of snow disaster in Qinghai 1950–2013 (A, C, E represent depth for 1950–1970, 1971–1990, 1991–2013 and B, D, F represent frequency for 1950–1970, 1971–1990, 1991–2013)

From the time distribution, between 1950 and 1970, the biggest average thickness of snow cover area is Yushu, especially in Qumalai County. In the 20 years, the maximum thickness of snow cover can reach the 80 cm and extremely heavy disaster occurs frequently in this county. From the frequency of snow, Yushu County is the county where occurred the largest number of snow disaster in this period, and Zaduo County and Zhiduo County second only. Between 1971 and 1990, the severe area of snow disaster were transferred from the Yushu to Guoluo, the worst afflicted county are Maduo County, Machen County and Dari County. In these areas, the thickness of snow cover reached more than 100 cm and snow days overtake a month during the years. From the frequency of snow disaster, the highest number of snow disaster of county is Dulan County nearly the Guoluo and Tianjun County of northern **Qinghai**, the Dari County suffered snow disaster followed by number. Between 1990 and 2010, the extremely heavy snow disaster mainly focused on Guoluo and Qilian Mountain of northern **Qinghai**, and Yushu is still the most susceptible area of snow disaster, Dari County second only. By the end of 1995 to early 1996, the snowfall more than usual in southern Qinghai, it makes a heavy snow disaster in Yushu, and resulting in economic losses of more than 300 million. Because of snow disaster caused vast damage of environment in parts of Qinghai Province, the urgent problems are relying on science and technology to resist snow disaster and improve the ability of disaster prevention and mitigation.

4 Discussions and Conclusion

Qinghai Province is located in the northeastern of Tibetan Plateau, belongs to the plateau landform. Due to strong uplift and complex terrain of Tibetan Plateau, make the Qinghai has diverse climate types, also become a sensitive area of climate change. There **are a typical continental climate** in Qinghai province, the performance of the overall low average temperature, warm cool in summer and cold in winter. Annual mean temperature ranges between −5.76°Cand 9°C, and annual mean precipitation between 17 mm and 773 mm. there are cause the low temperature phenomenon, hail, snow disaster and frozen easy to appear. The 80–90 % annual precipitation concentrated in the May to September, and mainly focuses on eastern areas, resulting in uneven temporal and spatial distribution of rainfall. Therefore, lead to occurring continuous drought in spring and summer in eastern and western of Qinghai. At the same time, study area is easier to cause torrential rain and flood. Due to Qinghai is a bigger desertification area, and belongs to semi-arid area, determine the regional suffer invasion of sandstorm frequently. In addition, the terrain of study area is west high and east low, the average altitude more than 3000 m, the geological structure is complicated, the tectonic activity is extremely violent, the percentage of forest cover is lower, with its special geographical location and severe climate environment conditions, make it occurs in a variety of geomorphic disaster frequently, and become a high incidence area of earthquake. Natural disasters depend not only on natural conditions, but also have a close relation of some human factors. With the development of the economic and social, environment has been demolished, and climate changed greatly. Forest, land and mineral, etc., once been a broken widely, it is very difficult to recover in a short time, and soil and water loss severely, thus aggravate the variety of geological hazards. In the other hand, climate change and ecological environment have been destroyed when they reached some degree will promote ecological and biological hazards.

Based on geographic information technology We analyzed the temporal and spatial features of natural disasters in Qinghai Province that can take a theoretical basis to establish and improve monitor and manage system of disasters including climate, hydrology, earthquake monitoring, environmental protection and soil and water conservation in whole study area. It should be take measures to prevent and reduce artificial damages such as conversion of farmland to forest, forest conservation, windbreak, sand-fixation and biological control. Special attention will be given to strengthen the development, utilization and protection of water resources, These management measures can fundamentally reduce meteorological and geological disasters such as drought and desertification.

Acknowledgment. The work presented in this paper was supported by the National Natural Science Foundation of China (No. 31370480) and 111 Project (B08044). We also acknowledge all people who help us built the library.

References

1. Zhou, Y., Jin, X., Wang, Q., et al.: Comprehensive assessment of natural disaster risk for agricultural procuction in Guanzhong region based on GIS. Scientia Geographica Sinica **32**(12), 1464–1471 (2012)
2. La, B., Cheng, T.: Research on Chinese terrestrial Gale disaster based on GIS. Tibet's Sci. Technol. **8**, 59–62 (2008)
3. Liao, Y., Zhao, F., Wang, Z., et al.: Spatial pattern analysis of natural disasters in China from 2000 to 2011. J. Catastrophology **28**(4), 55–60 (2013)
4. Fu, Y., Li, F., Guo, G., et al.: Natural disasters in Qinghai province and characteristics analysis. Earthquake Res. Plateau **16**(4), 59–67 (2004)
5. Luo, P.: GIS-based risk evaluation model of meteorological disaster: a case study on Hail Disaster in Chongqing administrative region. J. Nat. Disasters **16**(1), 3–44 (2007)
6. Liu, Y., Li, L., Yan, L., et al.: Risk division of pasture drought in Qinghai province based on loss assessment. J. Glaciology Geocryology **35**(3), 681–686 (2013)
7. Sui, X., Yang, Z.: Characteristics of natural disasters in 12 counties along the yellow river in Qinghai province and vulnerability evaluation of hazard-bearing body, 19(3), 7–13 (2004)
8. Yang, L., Rong, Y.: The main types of naturel disasters in Qinghai province and strategy of disaster prevention and reduction. Catastrophology **16**(1), 78–83 (2001)
9. Qinghai Statistical Yearbook. National Bureau of Statistics of China (2013)
10. China Meteorological Disaster Yearbook. China Meteorological Press, vol. 11, pp. 161–166 (2010)
11. Wang, X., Wen, K.: China meteorological disaster grand ceremony (Qinghai volume), vol. 4 pp. 2–197 (2007)

Monitoring and Analyzing of Poyang Lake Wetland Land Use Change Based on RS and GIS

Yuanxuan Yang[1,2] and Zhigang Yan[1,2](✉)

[1] College of Environment and Spatial Informatics, China University of Mining and Technology, Xuzhou, Jiangsu, China
yyxcumt@163.com, zhg-yan@126.com
[2] Jiangsu Key Laboratory of Resources and Environmental Information Engineering, Xuzhou, Jiangsu, China

Abstract. Applying object-based image analysis and supervised classification to remote sensing images, this paper analyzes the land use and land cover change (LUCC) in low water period of Poyang Lake wetland from 2009 to 2014. Result shows that the wetland had restored in recent years. Area of lake and non-wetland increased remarkably due to natural factors and human activities; hydraulic projects are the main reason for the radical increase in non-wetland area from 2009 onwards; landscape indexes indicate the wetland landscape is in a lower level of fragmentation.

Keywords: Remote sensing · LUCC · Landscape pattern · Poyang lake wetland

1 Introduction

Land use/cover change is a key factor of regional ecological environment evolution. As a vital component of International Human Dimensions Programme on Global Environmental Change (IHDP) [1], land use change, which is affected by natural factors and socioeconomic factors, has become one of the driving forces of global change.

Wetland has significant influence on regulating climate, water conservation, soil degradation, concentration of pollutants and maintaining biodiversity. Study on wetland land use change can help to understand the impacts of human activity on ecological environment and is advantageous to the analysis of the wetland ecological environment evolutionary trend and direction. As the biggest freshwater lake in China, Poyang Lake has been a key and hot area of wetland science research. At present, most of land use change researches of Poyang Lake wetland dividing the wetland boundary in county-level or regard the wetland area as a wetland economic zone, but less paid attention to the natural wetland area itself [2]. In addition, the Poyang Lake key water-control project, which has started since December 2009, has a huge impact on Poyang Lake wetland, but fewer studies involved the recent period of time [3].

Supported by National Natural Science Foundation of China (Project No: 41271445).

F. Bian and Y. Xie (Eds.): GRMSE 2015, CCIS 569, pp. 213–221, 2016.
DOI: 10.1007/978-3-662-49155-3_21

This paper took multiphase satellite remote sensing images and hydrological data as basic data, combined object-oriented technique and ECOC SVMs for image classification, then output of land use changes, analyzed the changes of land use and landscape level to research the land use change of Poyang Lake wetland from 1996 to 2004. The research results can reveal ecological evolution regulation and serve for land use plan of Poyang Lake area.

2 Study Area

Poyang Lake is located in the north of Jiangxi province (28°24′–29°46′N, 115°49′–116°46′E). As one of the biggest flood water storage wetland areas, Poyang Lake sustained flood control safety and hydrologic cycle of the middle and lower reaches of the Yangtze River. The high water period of Poyang Lake is from April to September and the rest time is the low water period. Benefit from the periodic change of water level, Poyang Lake has broad bottomland wetland vegetation development and diverse wetland creatures, which attract a large number of winter migratory birds and make the Poyang Lake wetland become one of the world's most important wintering habitat for rare migratory birds [4, 5]. Biodiversity conservation and habitat services function are very significant in Poyang Lake wetland. The wetland also plays an important role in maintaining regional and national ecological security in this area.

In order to alleviate the contradiction between population expansion and shortage of farmland, there has been numerous reclamation activity in Poyang Lake wetland for a long time. Especially in the period between late 1950s to 70s, under the effect of "food comes first" policy, the reclaim land from lake activities rose toward a climax and posed a great threat to the wetland environment. After entering the 1980s, the damage of excessive reclamation in Poyang Lake came to be known, the reclaim land from lake activities was banned. The government and the society became aware of the necessity for the protection of wetland and began to implement the returning farmland to lake project after the catastrophic flood in 1998 [6, 7].

To prevent the Poyang Lake wetland degradation and make full use of ecological and economic benefits of Poyang Lake wetland, the state council approved the Poyang Lake ecological economic zone planning in 2009 and rose the construction of Poyang Lake ecological economic zone to the level of national strategy, putting forward the Poyang Lake water conservancy engineering construction program [8].

3 Materials and Methods

3.1 Research Data Sources

Remote sensing (RS) is increasingly valued as a useful tool for providing large-scale basic information. Remote sensing technology can partially replace the time-consuming and expensive ground surveys [9]. With relatively straightforward techniques, these data could provide important knowledge for wetland ecological studies, such as information on their surface area, number, and mutual distance. Besides detection of the land surface change can be investigated with the availability of long-term data. Remote sensing data used in the study were TM/ETM/OLI_TIRS images on 9 December 1996, 4 February 2003, 11 January 2009 and 14 March 2014, with low percent cloud cover in the study area. These images were collected on the low water period of Poyang Lake and had been corrected (including the geometric precision correction, image registration, image mosaic, image clipped and atmospheric scattering correction). This paper took research area scope based on the date of DEM data, diking data and hydrological data of Poyang Lake and the rule of the boundary inside the causeway and below the average highest water level [10].

3.2 Classifying and Information Extraction of Land Use Type

According to the "Ramsar Convention", "Standard of the China wetland resources survey and monitoring techniques" (China's State Forestry Administration, 2008), and the actuality of the Poyang Lake wetland, this paper classifies the study area as natural wetland (lakes, mud flat, meadows), constructed wetland (paddy fields, reservoirs and ponds, aquaculture region), and non-wetland.

(1) Remote sensing images feature extraction
Extracting reasonable remote sensing images feature can improve the accuracy and reliability of classification. NDVI (Normalized Difference Vegetation Index) is one of the most commonly used vegetation index, which can extract the vegetation information in remote sensing image effectively, and has been widely utilized in vegetation detection, crop yield estimation, and other fields. PCA (principal component analysis) is a kind of data dimension reduction, which can focus the useful information in original multiple wavelengths to the new image as little as possible, and compress image data effectively. The first four components of PCA can be used in SVM classify effectively [11]. This paper combined the data of PCA information, NDVI and original spectrum to classify.

(2) Image segmentation and merging
The phenomenon of different spectra characteristics with the same object and different object with the same spectra characteristics will affect the image classification, and object-oriented classification technology can reduce the influence in a certain degree [3]. ENVI FX software using multi-scale segmentation algorithm based on edge, combination of remote sensing image spectral feature and shape feature and texture information, and completing the object segmentation according to the specified thresholds. ENVI FX can use Full Lambda-Schedule algorithm for merging of plaques to solve the problem.

(3) Classification
Classify segmented image data using ECOC SVMs with Matlab 2014a. ECOC (Error correcting output code) is an efficient multi-class classification frame which needs less child separator but has fault tolerant capability than commonly used frame (1v1 SVM) [12]. Its performance is greatly influenced by the code matrix, so code matrix selecting is needed before classification [13]. ECOC SVMs can

effectively improve the classification precision of remote sensing image with higher efficiency.
Select more than 30 samples from each type of feature as the training samples and establish optimal code matrix; establish each SVM classifier and use cross validation method for parameter optimization. Then output the classification results of Poyang Lake remote sensing image.

(4) Mapping and analysis
Import the remote sensing image classification results to ArcGIS 10 software for mapping and statistical analysis of land use changes. The projection coordinate system of these data is WGS_84_UTM_zone_50 N. After all the images were turned into classified landscape patches, they were converted to the format of ArcGIS GRID, and input FRAGSTATS 4.2 to compute landscape metrics for the imageries captured in four dates mentioned above.

4 Results and Analysis

4.1 Land Use/Cover Change in Poyang Lake Wetland

According to the interpretation characteristics of remote sensing image, the study selected different types of training samples, calculated the spatial, spectral and texture information, the paper creates the landscape type maps of the Poyang Lake wetland with ArcGIS software (Fig. 1). The area and transition matrix of LUCC in the Poyang Lake wetland during 1996 to 2010 were calculated by Arcgis and shown in Tables 1 and 2.

Returning Farmland to Lake Project started after catastrophic flood happened in the Yangtze River in 1998. Small earth-fill dams were destroyed, farming were not allowed in diked marsh area and resident were required to relocate [6].

Paddy fields area decreased from 262 km^2 to 222 km^2 in the period from 1996 to 2009 because of the strict regulations. Most of these paddy fields were transformed into lakes and meadows according to Table 2. The speed of the project began to slow down in 2009 because sufficient area of paddy fields had been returned to lake and the figure was only 220 km^2 in 2014. Figure 1 demonstrates that the project achieved a remarkable result as the lake area in 2003, 2009, 2014 were much larger than that of 1996. New lakes areas were mostly transformed from meadows, mudflats and paddy area. In early March of 2014, large area precipitation occurred on the middle and lower reaches of the Yangtze River Valley, which made lake area slightly larger than former years.

The launch of Poyang Lake water conservancy engineering construction program directly generated large area of construction land, which is the main part of non-wetland. The figure for non-wetland area doubled between 2009 and 2014, reaching to 111 km^2.

The program and the returning farmland to lake project generated a large number of shallow water areas, providing suitable places for aquaculture. Because of regional economic development and industrial transformation, aquaculture water area had increased significantly since 2009. The figure reached at 83 km^2; besides library pond area also increased. Poyang Lake water conservancy engineering construction program had certain improvement effect on water quality (especially in the low water period) and wetland biodiversity. Besides, the program reduced the areas of meadows and mudflat which are the habitat of spiral shell, making great contribution to schistosomiasis prevention.

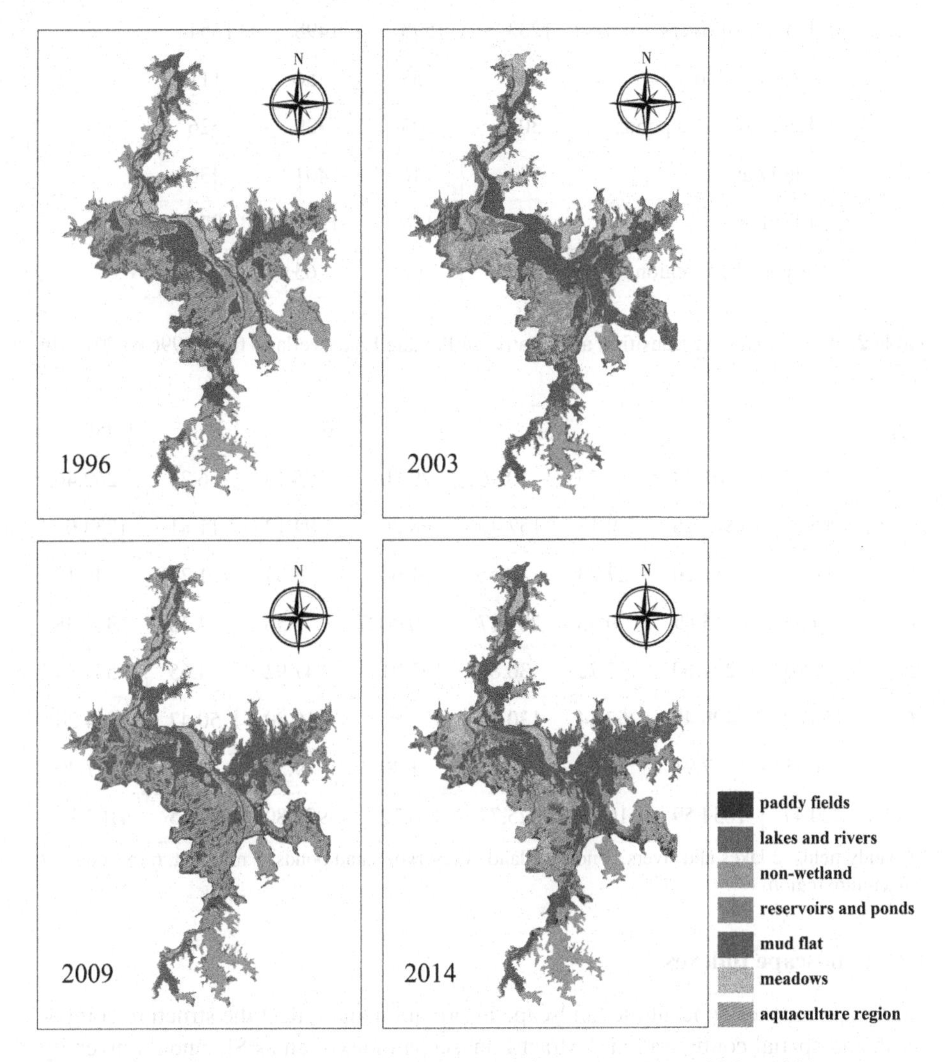

Fig. 1. Changes of land use of Poyang Lake wetland from 1996 to 2014

Table 1. Area change of Poyang Lake wetland land cover/use from 1996 to 2014 (in km^2)

Land cover/use type	1996	2003	2009	2014
Paddy fields	262	246	222	220
Lakes and rivers	1233	1598	1498	1654
Non-wetland	49	53	56	111
Reservoirs and ponds	303	283	329	626
Mud flat	534	746	401	436
Meadows	1684	1114	1533	972
Aquaculture region	37	62	63	83

Table 2. Land use/cover transition area matrix in Poyang Lake wetland from 1996 to 2014 (in km^2)[a]

Type	1	2	3	4	5	6	7	Total
1	42.02	61.49	7.35	10.96	29.20	105.19	6.24	262.46
2	15.03	857.26	9.13	157.40	88.26	94.12	11.81	1233.0
3	1.21	11.20	27.23	2.55	1.67	3.59	1.72	49.18
4	0.09	15.69	0.38	281.17	0.0027	1.56	4.48	303.38
5	9.90	294.60	7.72	30.89	137.91	47.92	4.65	533.59
6	150.88	406.45	57.14	130.56	177.87	710.90	50.47	1684.26
7	1.35	7.91	1.63	12.20	1.40	9.52	3.27	37.29
Total	220.47	1654.59	110.58	625.74	436.32	972.80	82.65	4103.15

[a] 1 paddy fields, 2 lakes and rivers, 3 non-wetland, 4 reservoirs and ponds, 5 mud flat, 6 meadows, 7 aquaculture region

4.2 Landscape Indexes

Landscape indexes concentrate landscape information and reflect the structure composition and spatial configuration. Extract Landscape index such as Shannon's diversity index (SHDI), landscape shape index (LSI), patch density (PD), Shannon's evenness index (SHEI) and Fractal dimension (FRAC) from multiphase data according to the characteristics of the wetland to reflect the dynamic changes of the Poyang Lake wetland landscape pattern (Tables 3, 4).

Table 3. Landscape indexes of Poyang Lake wetland

Time	PD	LSI	SHDI	SHEI
1996	0.7615	36.7033	1.4558	0.7481
2003	1.1632	40.0265	1.517	0.7796
2009	1.0538	39.6542	1.4551	0.7478
2014	0.7588	36.0452	1.572	0.8078

Table 4. Mean fractal dimension of each land cover/use type in Poyang Lake wetland

Land cover/use type	1996	2003	2009	2014
Paddy fields	1.0175	1.0334	1.0436	1.0235
Lakes and rivers	1.0317	1.0124	1.0142	1.0343
Non-wetland	1.0387	1.0436	1.0395	1.0532
Reservoirs and ponds	1.0358	1.0181	1.0325	1.051
Mud flat	1.0494	1.0257	1.0279	1.0272
Meadows	1.0426	1.036	1.0509	1.058
Aquaculture region	1.0164	1.017	1.0192	1.0381

PD and LSI of Poyang Lake wetland landscape increased at the initial phase and dropped afterwards. Landscape of paddy field, meadow, mud flat and lake changed dramatically. Landscape patches tend to be broken and the complexity increased at the first stage of the returning farmland to lake project. With the relative reduction of human activity on the paddy field and the ecosystem repair integration, the indexes had fallen and the figures of 2014 were slightly lower than those of 1996.

SHDI reflects the diversity of landscape types and distribution of complicated patch level, which is the embodiment of the plaques abundance. SHEI were used to determinate landscape structure on dominating landscape level of one or several kinds of landscape types. Landscape diversity in Poyang Lake wetland in 2014 was highest during the whole period, indicating the wetland ecosystem had been well protected. SHEI rose mainly because the lakes landscape area increased during low water period.

Fractal dimension is a valid indicator of patch edge complexity, which closely related to the interference of human activities. Table 4 shows that Fractal dimension value of natural wetland was significantly higher than that of artificial wetland. The reason is that the two were exposed to different level of human activity interference. Fractal dimension of the paddy field first increased then decreased and reached a peak in 2009. Mud flat were still notably affected by human activities in 1996, mainly because of reclamation activities had not fully been suspended, the mud flats were still being transformed into paddy fields in the early 90 s. In 1998, returning farmland to lake project launched, which made a great contribution to the decrease of patch fractal dimension of Mud flat.

The figure had fallen sharply and leveled off in recent years. The fractal dimension value of non-wetland remained steady from 1996 to 2009 but soared in 2014. The change showed the Poyang Lake water conservancy engineering construction program impacts regional landscape dramatically. The project generated a large number of shallow water areas, which transformed into aquaculture region as a result of regional economic development and industrial transformation. The figure for aquaculture region area increased rapidly after 2009.

5 Conclusion

The study analyzed the changes of land use and landscape patterns of Poyang Lake wetland in 1996, 2003, 2009 and 2014 using remote sensing techniques. Profound changes had taken place in land use and landscape patterns due to natural factors and human activities. Implementation of policy and projects played an important role in wetland restoration. There was a huge increase in lake area of 2003 comparing with the figure of 1996. Wetland had been well protected and restored with a lower degree of landscape fragmentation. From 2009 onwards, lake area in dry seasons tended to be more stable and more related to regional precipitation.

The interference of human activities is a major cause of ecosystem changes in Poyang Lake wetland after 2009. Due to the construction of water conservancy facilities and economic development, artificial wetland (reservoirs and ponds, aquaculture region) area increased and natural wetland (bottomland grassland) area decreased. Local government should improve water conservancy facilities construction, optimize the industrial structure and perfect the policy system to protect Poyang Lake wetland. There will be ongoing changes in the lake ecology. Further research is necessary for supporting wetland ecological restoration and exert the social and economic benefits in the future.

References

1. Turner II, B.L., Skole, D., Sanderson, S.: Land-use and land-cover change. Science/Research plan (1995)
2. Michishita, R., Jiang, Z., Xu, B.: Monitoring two decades of urbanization in the Poyang Lake area, China through spectral unmixing. J. Remote Sens. Environ. **117**, 3–18 (2012)
3. Dronova, I., Gong, P., Wang, L.: Object-based analysis and change detection of major wetland cover types and their classification uncertainty during the low water period at Poyang Lake, China. J. Remote Sens. Environ. **115**(12), 3220–3236 (2011)
4. Xia, S., Yu, X., Fan, N.: The wintering habitats of migrant birds and their relationship with water level in Poyang Lake, China. J. Resour. Sci. **32**(11), 2072–2078 (2010)
5. Liu, C., Tan, Y., Lin, L.: The wetland water level process and habitat of migratory birds in Lake Poyang. J. Lake Sci. **23**(1), 129–135 (2011)
6. Min, Q.: On the restoring lake by stopping cultivation to Poyang Lake and its impacts on flood mitigation. J. Lake Sci. **16**(3), 215–222 (2004)
7. Shankman, D., Liang, Q.L.: Landscape changes and increasing flood frequency in China's Lake Poyang region. J. Professional Geogr. **55**(4), 434–445 (2003)

8. Qi, S., Liao, F.: A study on the scheme of water level regulation of the Poyang Lake hydraulic project. J. Acta Geogr. Sin. **68**(1), 118–126 (2013)
9. Baker, C., Lawrence, R., Montague, C., Patten, D.: Mapping wetlands and riparian areas using Landsat ETM+ imagery and decision-tree-based models. J. Wetlands. **26**, 465–474 (2006)
10. Xie, D., Zheng, P., Deng, H.: Landscape responses to changes in water levels at Poyang Lake wetlands. J. Acta Ecologica Sinica. **31**(5), 1269–1276 (2011)
11. Tan, K., Du, P.: Hyperspectral remote sensing image classification based on support vector machine. J. Infrared Millim. Waves. **27**(2), 123–128 (2008)
12. Dietterich, T.G., Bakiri, G.: Solving Multiclass Learning Problems via Error-Correcting Output Codes. J. Artif. Intell. Res. **2**, 263–286 (1995)
13. Yan, Z., Du, P.: Generalization performance analysis of M-SVM s. J. Data Acquisition Process. **24**, 469–475 (2009)

The Security Management Information System of Subgrade and Pavement Based on Grid GIS

Ji Zhou[1,2](✉), Xiekui Zhang[1], Qiong Tian[2], Mingfang Chen[3], and Yongqin Rui[3]

[1] Guangxi University, Nanning 530004, Guangxi, China
hnkjxy_zhouji@163.com
[2] Hunan University of Science and Engineering, Yongzhou 425199, Hunan, China
[3] Changsha University of Science and Technology, Changsha 410001, China

Abstract. There are lots of factors produce an effect on subgrade and pavement, these influence may directly relate to the security of transportation. According to the influencing factors for the security of subgrade and pavement, the author suggests that we should build a security management information system based on Grid GIS. The paper introduces the characteristics and structure of the Grid GIS at first; Then mainly describes the security management framework of subgrade and pavement based on Grid GIS; Finally discusses the implementation and application of this security management information system. It has profound significance to build highway subgrade and pavement information system security management system based on Grid GIS.

Keywords: Grid GIS · Subgrade and pavement · Security · Management system

1 Introduction

China is one of the countries with a serious natural disasters in the world. The various natural disasters in china are widely distributed, taking place frequently, the damage and economic losses are also extremely serious. It has become an important factor influence us to build the harmonious society.

The highway construction usually includes several stages such as feasibility study, survey and design, construction and operation. There are different security factors in every stage, for example, we may encounter a variety of complex geological problems in construction, such as collapse, landslide, debris flow, karst and other special geological disasters, so we should try to avoid the adverse geological section and optimization design for the route at the stage of road route selection. In operation, we should set up hazard warning mechanism of the highway, apply the space monitoring technology to forecast a possible geological disasters and take effective preventive measures. We also

F. Bian and Y. Xie (Eds.): GRMSE 2015, CCIS 569, pp. 222–227, 2016.
DOI: 10.1007/978-3-662-49155-3_22

need to strengthen the Pre-maintenance technology of road, prevent premature pavement damage occur in the vehicle load or natural conditions, maintain the original design of the road, improve the working condition and prolong its service life.

China has paid great attention to disaster monitoring, forecast and control, the government invested a lot of money to solve these problems, it already acquired and accumulated a massive relevant data, such as satellite images and ground elevation, geological and seismological observations, hydrological data and so on. But the vast amounts of data dispersed form unique longitudinal information system resource islands, disaster data types monotonous, low availability, lack of data description, the information is incomplete, the existing disaster information system does not adapt to social development, and hinder the development of disaster science, its difficult to make decision due to lack of full and useful information in the disaster process.

The emergence of technology of Grid GIS provides the possibility to solve the issue mentioned above, by means of information technology to control a variety of resources of the road in "digital form", it can facilitate efficient storage of resources, query and use the resources, and we can full sharing of the computing resources, storage resources, data resources, information resources and expert resources, so we can greatly enhance the management of the road.

2 Grid Geographic Information System

Grid GIS is a kind of super processing environment which integrated with the distributed and heterogeneous computer, spatial data server, large storage system, geographic information system and virtual reality system by means of high-speed interconnection network, and become the formation of transparent virtual space information resources to customer. It is a seamless integration and collaborative processing system in the range of wide spatial information area.

In other words, Grid GIS is a basis platform and technical system combine with all space-related information resources, such as computing resources, storage resources, communication resources, GIS software resources, spatial information resources, spatial knowledge resources and get the access to achieve full connection to the Internet. is a huge super GIS server combine with the internet, which can realize the full sharing of various spatial resources, and eliminate the support technology of resource island. The core of Grid GIS is to realize sharing and collaborative work of various space information processing resources.

As the means and methods of spatial data organization, management, analysis and display, GIS is one of the essential technologies to solve information management of subgrade and pavement, involves collected, input, storage, retrieval, handling, analysis various spatio-temporal data and thematic data. which provide necessary information and knowledge support for highway management and decision making.

We should to address the following issues before building of highway subgrade and pavement information management. First of all, it needs to achieve effective storage and sharing of resources within the region and don't produce waste of storage resources. Secondly, based on its specific resource storage, it should to provides the effective resource query method, makes the information query, search and location of the resources more easy, for example, we can realize the visual information query of the highway.

3 Grid Geographic Information System Management System

Security management system of highway subgrade and pavement involves huge data, includes raster data, vector data and attribute data, etc. Grid GIS not only can be used to store and manage all types of subgrade and pavement deformation information, but also can be used to visualize query of various information of the highway and publish on web.

The application of Grid GIS in highway subgrade and pavement information management is to integrate the various information of the highway through various technical means, and realize the unified management of all kinds of information on the highway. By means of building of highway subgrade and pavement information management and using the query module, we can display all the static data such as the route curve data, pavement width, etc. and the dynamic data such as deflection, flatness, Settlement deformation curve, etc. to the costumer.

The user can full sharing of subgrade and pavement management information system when they are carrying out the work of subgrade and pavement disease about early warning, diagnosis and treatment. The resource in subgrade and pavement management information system contains the types of subgrade pavement survey method, disease detection technology, subgrade and pavement deformation prediction prediction technology, deformation monitoring, deformation analysis evaluation method and treatment measures and so on.

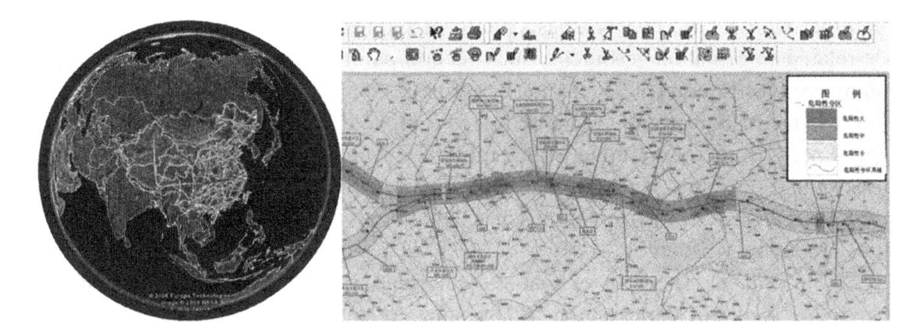

Fig. 1. The security management information system of subgrade and pavement based on grid GIS

As shown in Fig. 1, it is a map of China highway network on the left, and it is a specific highway in China on the right, we can use the technology of Grid GIS to build the security management information system of subgrade and pavement engineering.

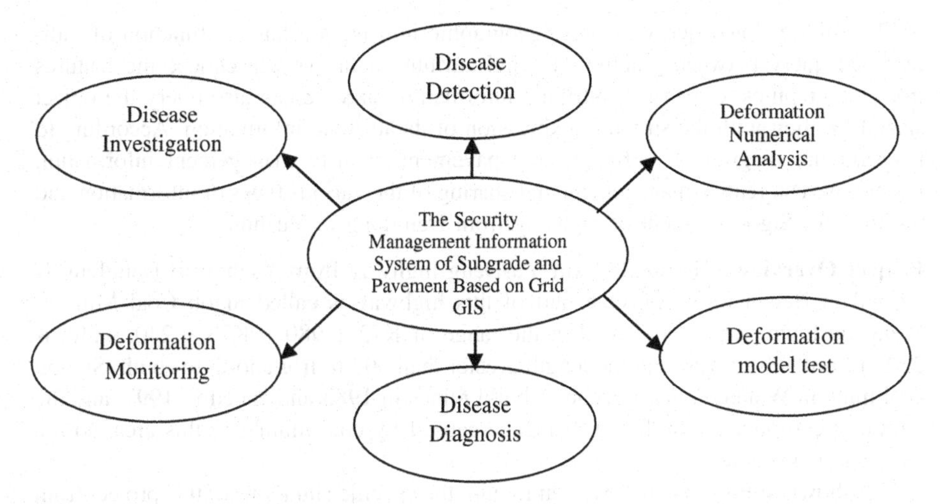

Fig. 2. The structure diagram of subgrade and pavement security management information system

As can be seen from Fig. 2, it is the structure diagram of subgrade and pavement security management information system, that is the Security management information system of subgrade and pavement based on Grid GIS is composed of disease investigation, disease detection, deformation numerical analysis, deformation monitoring, deformation model test, disease diagnosis and so on.

Grid GIS Integrated all the security information of highway subgrade and pavement in every stage into a system, including the feasibility study stage, survey and design phase, information construction stage, the operation stage and so on. It is beneficial to realizing full sharing the highway engineering security management of disaster forecasting method, monitoring technology, detection method, analysis and evaluation method, protection, management measures and other resources. Therefore, It is easy to become a strong function information management infrastructure, including the demand service capabilities, a variety of powerful data management and information integration processing capabilities.

4 The Example of Engineering Application

The security management system of subgrade and pavement is established by means of the Grid GIS, then we can build the deformation information management system of Highway subgrade and pavement based on the example of highway engineering from Yuzhou to Dengfeng. The main application of this system of its information storage and information to realize visual query function.

The bidirectional query of basic geographic information has the function of bidirectional query between graphic data and attribute data, we can choose the features from the graphics to query its attribute information, and we can also query the object spatial location with the structure expression of its attribute information. According to the structure diagram of subgrade and pavement security management information system, we can realize management and sharing of the various types of information and resource for Sigou segment of highway from Dengfeng to Yuzhou.

Project Overview: The 48.380 km Yu-Deng highway from Yuzhou to Dengfeng is located in Henan province. One part of this highway is called Sigou Coal Mine in Wangcun town, which is located in the range of K72 + 980 − K73 + 240, width is 230 m from east to west and the length greater than 300 m from north to south. Si Gou coal mine in Wangcun town began to build mines in 1983 and ended in 1997, and the terrain is complicated, surface subsidence caused by coal mining in this area, so the subgrade is instable.

As shown in Figs. 3 and 4, we can inquire the specific situations of this project from the deformation information management system of Yu-Deng Highway subgrade and pavement we builded before.

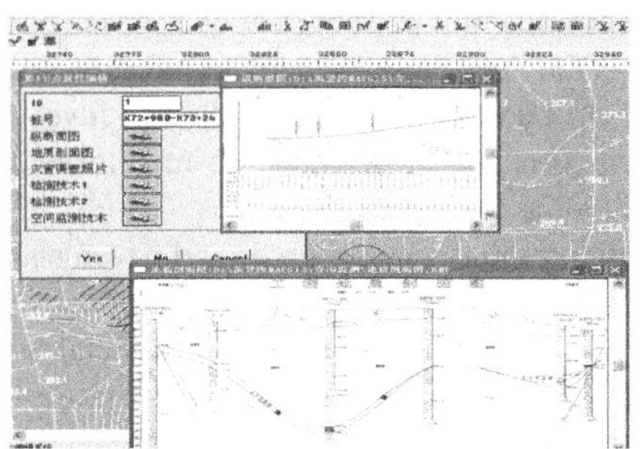

Fig. 3. Diagram of vertical section and geological section

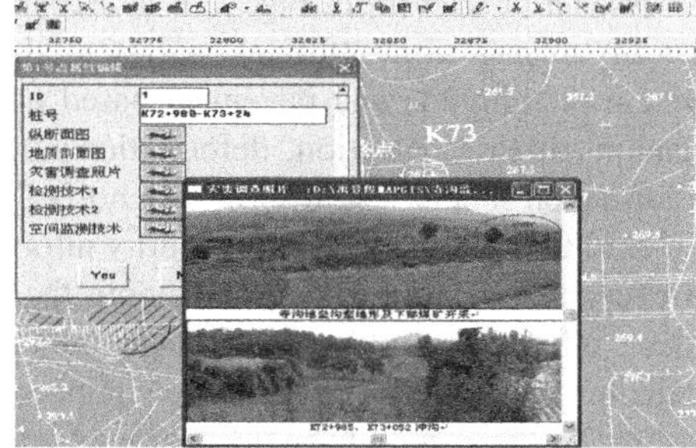

Fig. 4. Photos of disaster survey

From Figs. 5 and 6, we can also obtain large amounts of data of Field test and Settlement Monitoring in the deformation information management system of Yu-Deng Highway subgrade and pavement.

We can carry out Numerical simulation analysis of "activation" of mined area and Small scale model test, and Stored these information in the deformation information management system of Yu-Deng Highway subgrade and pavement the system, such as Figs. 7 and 8, we can inquire the data we need at any time.

Fig. 5. Field test

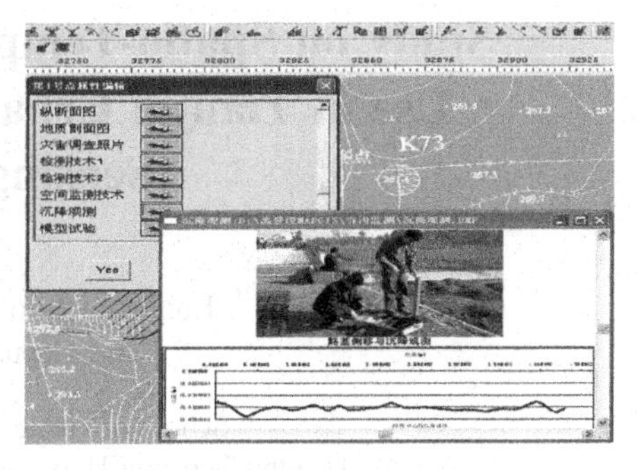

Fig. 6. Settlement Monitoring

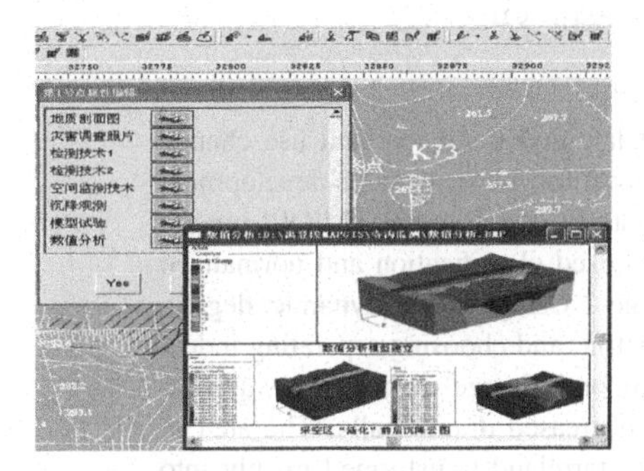

Fig. 7. Numerical simulation analysis of "activation" of mined area

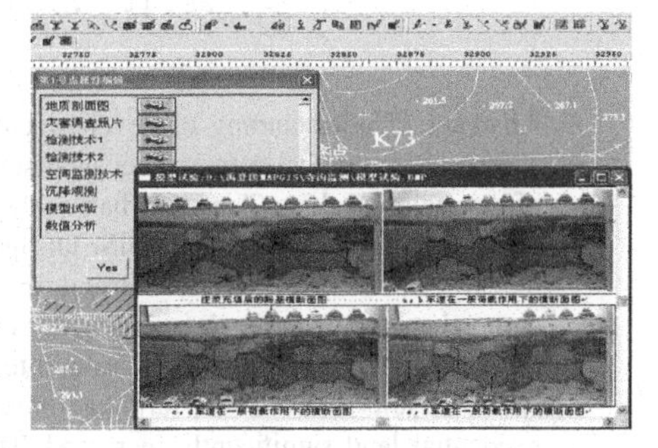

Fig. 8. Small scale model test

5 Conclusion

The Grid GIS has a powerful spatial analysis and statistical function that provides an effective tool for the standardization, networking, and spatial information management of subgrade and pavement deformation information management. It has realized the sharing of computing resources, storage resources, communication resources, GIS software resources, spatial information resources, spatial knowledge resources and so on.

It has profound significance to build highway subgrade and pavement information system security management system based on Grid GIS, and it will comprehensively improve the safety of Highway Engineering in our country and strengthen diagnosis and treatment of early warning ability and enhance information infrastructure service levels.

Acknowledgments. The research in this paper is funded by Guangxi University Postdoctoral Science Foundation and China Social Science Foundation (No. 14BJL093).

Study on Spatio-Temporal Change of Land Use in Tianjin Urban Based on Remote Sensing Data

Qiaozhen Guo[1(✉)], Lingchun Luo[1], Hongrui Zhao[2], Yingyang Pan[1], and Qixuan Bing[1]

[1] School of Geology and Geomatics, Tianjin Chengjian University, Tianjin 300384, China
gqiaozhen@tcu.edu.cn, llcchun@126.com

[2] Department of Civil Engineering, Institute of Geo-Spatial Information, Tsinghua University, Beijing 100084, China
zhr@tsinghua.edu.cn

Abstract. Understanding of development law and trend for land use change can provide effective data and decision support for the sustainable development of the region. Taking Tianjin Urban as the study area, Landsat TM/OLI images were used. Based on RS and GIS, unsupervised classification and normalized indexes were combined to interpret images. Using single dynamic degree, comprehensive dynamic degree, transfer matrix, and choosing separating index, diversity index, evenness index, spatio-temporal change of land use was analyzed. Results showed that farmland area decreased dramatically. The area of residential land significantly increased. The farmland transformed mainly into the residential land, which showed that rapid urbanization took up a large amount of farmland. The separation degree of residential land reduced. The growth of residential land was more concentrated, and expanded outward from the city center gradually. The comprehensive dynamic degree, diversity index and evenness index of land use decreased.

Keywords: Landsat · OLI · Normalized index · Landscape index

1 Introduction

Land use change has significant effects on the ecological environment [1]. Using GIS and RS to extract spatio-temporal information of land use change and analysis of land use dynamic change is the frontier problem and the important direction of the research [2]. With the rapid economic development of Tianjin, population growth and urbanization process accelerated the contradiction between land use and economic development. Therefore, research of spatio-temporal dynamic changes can provide basic data for the formulation of long-term land use planning scheme, and provide decision support for the sustainable development of the economy in Tianjin Urban.

Remote sensing technology provides data sources for land use, such as MSS [3], TM [3], ETM + [4], MODIS [5], SPOT [6], and so on. Remote sensing technology has a great potential to extract detailed land use information [7]. The methods of monitoring land use include image differencing, rationing, principle component analysis,

F. Bian and Y. Xie (Eds.): GRMSE 2015, CCIS 569, pp. 228–237, 2016.
DOI: 10.1007/978-3-662-49155-3_23

and post-classification comparison [8]. International researchers attach importance to land use, and international organizations and research institutions have formed a series of global land use database in the basic data construction [9]. It has been a long history for the study of land use change in China [10]. Many domestic scholars made outstanding contributions, such as Liu et al. who carried out "national resources remote sensing macro survey dynamic research" and established the Chinese Resources and Environment Database [11]. The spatio-temporal dynamic characteristics of land use in the last 5 years in China were analyzed quantitatively by Wang [12]. Taking Tianjin Urban as an example, using Landsat TM/OLI data, under the support of RS and GIS technology, combining unsupervised classification and normalized index, land use information was extracted in this paper. The spatio-temporal dynamic change of land use was analyzed.

2 Study Area and Data Source

Tianjin City is located in longitude of 116°43′E to 118°04′E and latitude of 38°34′N to 40°15′N, which is located in the northeast of the North China Plain. Tianjin City is located in the lower reaches of the Haihe River. Its wide is 117 km from east to west, and its length is 189 km from north to south. The length of geosphere is 1137 km long, and the length of coastline is 153 km. It is the largest coastal open city in Northern China. Tianjin Urban (Fig. 1) has a total of ten districts, Heping District, Hexi District, Hebei District, Hongqiao District, Nankai District, Dongli District, Xiqing District, Beichen District and Jinnan District.

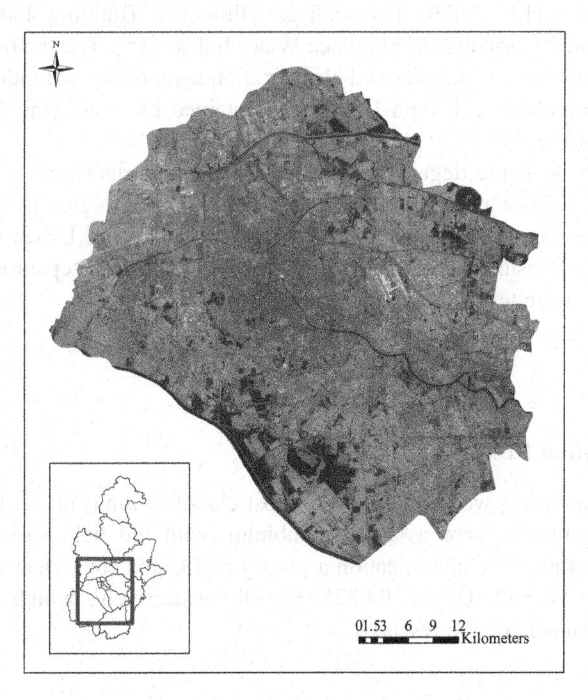

Fig. 1. Location of the study area

The data included remote sensing images, basic geographic data, statistical yearbook and the related data. Remote sensing images included two scenes of Landsat TM images with spatial resolution of 30 m in 2006 and 2010, one scene of Landsat OLI image with spatial resolution of 15 m in 2014.

3 Data Preprocessing and Research Method

3.1 Data Preprocessing

First of all, the bands were stacked. TM and OLI image have 7 and 11 bands respectively. Then the bands were combined. TM and OLI remote sensing image were selected 5, 4, 3 (red, green and blue) and 7, 5, 4 (red, green and blue) band combination respectively. The images were geometrically corrected and basic geographic data was converted projection. The vector boundary map of the study area was used to cut remote sensing images and the range of the study area was extracted. According to the classification principle of land classification system and the actual situation of Tianjin Urban, the land of the study area was divided into 5 categories, which were water, woodland, farmland, residential land and unused land.

3.2 Research Method

The normalized indexes used in this paper consisted of NDVI (Normalized Difference Vegetation Index) [13], NDBI (Normalized Difference Building Index) [14] and MNDWI (Modified Normalized Difference Water Index) [15]. The methods were used through a combination of unsupervised classification and normalized indexes. The land use change information in Tianjin Urban was obtained by overlaying NDVI, NDBI, MNDWI and unsupervised classification results.

The land use dynamic degree is one of the most important indicators to analyze land use change [16], which is divided into single dynamic degree [17] and comprehensive dynamic degree [18]. The landscape pattern of Tianjin Urban was added up and analyzed by choosing the average patch area [19], landscape separation index [20], landscape diversity index [21] and evenness index [22, 23].

4 Results

4.1 Classification Results

The classification results were obtained after post-classification (Fig. 2). The accuracies of classification results were assessed combining with the field survey and visual interpretation method. The classification accuracy and kappa coefficient were 96.00 %, 95.70 %, 96.09 % and 0.9436, 0.9375, 0.9398 respectively, which can meet the accuracy requirements of the research.

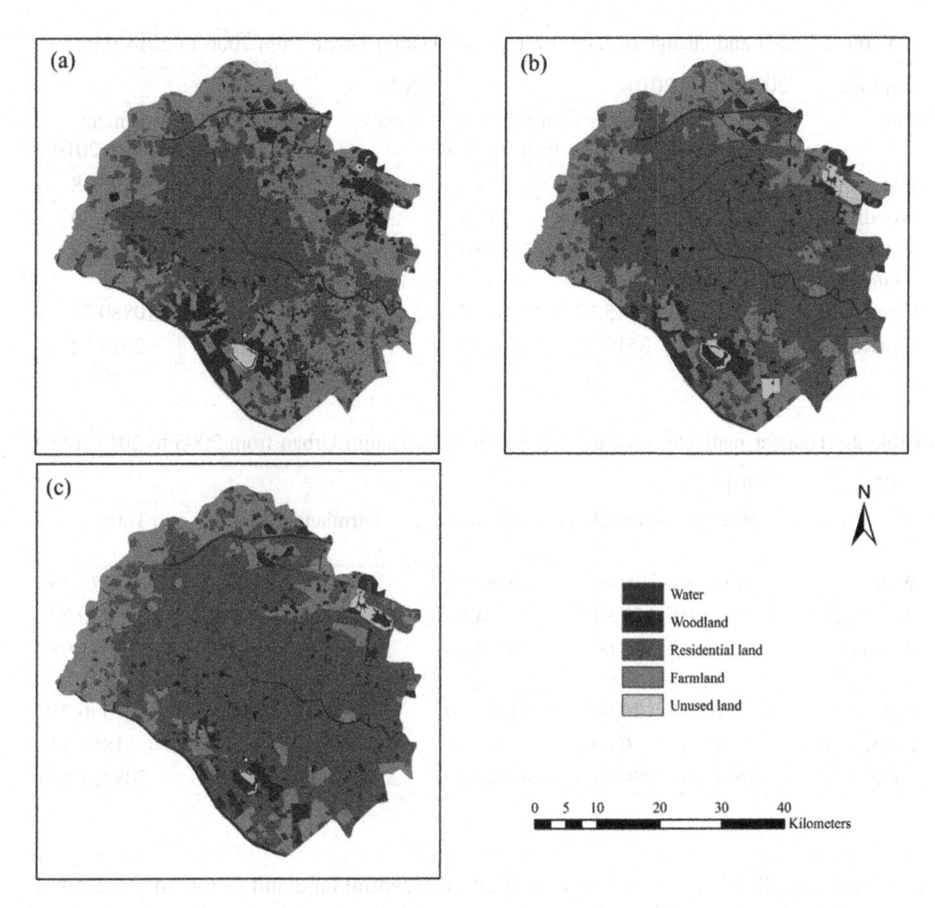

Fig. 2. Land use classification chart in Tianjin Urban: (a) 2006, (b) 2010, (c) 2014

4.2 Temporal Change

(1) The quantitative change of land use type

According to the interpretation results of remote sensing images, the changes of area and proportion of land types were obtained, which were as shown in Table 1.

As can be seen from Table 1, the area of residential land increased year by year, and it increased the most from 2006 to 2010, which increased 47576.04 hm^2. The area of water and farmland reduced year by year. From 2006 to 2014, water area reduced 9686.86 hm^2, while the decrement of farmland reached 52461.31 hm^2. The changes of woodland and unused land were gentle relatively.

(2) The transfer matrix of land use types

The transfer matrix of land use in Tianjin Urban from 2006 to 2014 was as shown in Table 2. As can be seen from Table 2, water mainly changed into residential land and

Table 1. Area and change of land use types in Tianjin Urban from 2006 to 2014 (hm^2)

Land use type	2006	2010		2014		
		Area	Increment than in 2006	Area	Increment than in 2006	Increment than in 2010
Water	25673.46	17938.98	−7734.48	15986.60	−9686.86	−1952.38
Woodland	440.22	512.16	71.94	296.37	−143.85	−215.79
Residential land	78861.98	126438.02	47576.04	141510.00	62648.02	15071.98
Farmland	101396.21	59815.19	−41581.02	48934.90	−52461.31	−10880.29
Unused land	1852.24	3519.75	1667.51	1496.23	−356.01	−2023.52

Table 2. Transfer matrix of land use types change in Tianjin Urban from 2006 to 2014 (hm^2)

2006	2014					
	Water	Woodland	Residential land	Farmland	Unused land	Total
Water	11593.94	10.90	10863.52	2469.23	735.86	25673.46
Woodland	11.01	137.57	104.55	187.09	0.00	440.22
Residential land	685.80	108.00	75584.16	2484.01	0.00	78861.98
Farmland	2767.34	39.90	54630.39	43668.64	289.94	101396.21
Unused land	928.51	0.00	327.37	125.93	470.43	1852.24
Total	15986.60	296.37	141510.00	48934.90	1496.23	208224.10

farmland; woodland also mainly changed into residential land and farmland; residential land mainly transformed into farmland and water; farmland mainly transformed into residential land; unused land mainly changed into water.

From 2006 to 2014, transform amount from water to woodland, residential land, farmland and unused land was 10.9 hm^2, 10863.52 hm^2, 2469.23 hm^2 and 735.86 hm^2 respectively. At the same time, transform amount from woodland, residential land, farmland and unused land to water was 11.01 hm^2, 685.8 hm^2, 2767.34 hm^2 and 928.51 hm^2 respectively. However, water area reduced 9686.86 hm^2 totally. The transform amount from woodland to residential land and farmland is 104.55 hm^2 and 187.09 hm^2 respectively. And woodland area was decreasing and decrement was 143.85 hm^2. The conversion area of residential land was the greatest. From 2006 to 2014, the total amount transferred to residential land was 62648.02 hm^2, most of which came from farmland, followed by water and unused land. During this period, residential land was converted into water, woodland and farmland, but transform amount was lower. The conversion quantity was 54630.39 hm^2 from farmland to residential land, while the amount of conversion from residential land to farmland was only 2484.01 hm^2. It can be seen that farmland mainly changed into residential land.

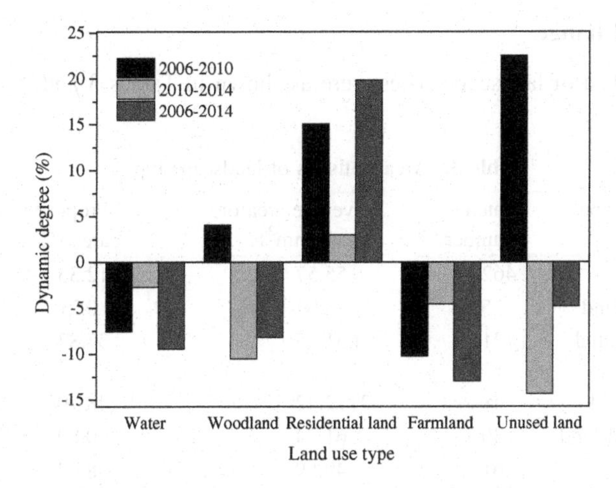

Fig. 3. Comparison of single dynamic degree in different period

The amount was 2767.34 hm^2 from farmland to water. Unused land mainly transformed into water, but the amount of unused land decreased and decrement was 356.01 hm^2.

(3) Dynamic degree of land use type

The single dynamic degree of land use type was calculated in different periods and the results were as shown in Fig. 3.

It can be seen residential land increased from 2006 to 2014 in Fig. 3, and other types kept descending trend. Single dynamic degree of water was negative value in each period, which indicated that it was in a decreasing trend. From 2006 to 2010, single dynamic degree of water was −7.53 %, while it was −2.72 % from 2010 to 2014, which showed that water has been declining. Single dynamic degree of woodland was −8.17 % from 2006 to 2014, 4.09 % from 2006 to 2010, −10.53 % from 2010 to 2014, which indicated it increased in the first stage and decreased progressively in the second stage. Single dynamic degree of residential land was 15.08 % in the first stage and 2.98 % in the second stage, which showed that the area of residential land increased in the whole and the rate of increase slowed down gradually. Single dynamic degree of farmland was −12.93 % in the whole research phase, −10.25 % in the first stage and −4.55 % in the second stage, which indicated that farmland decreased and the rate of decrease appeared to be accelerating. The changing trend of unused land was similar to woodland, and the area increased in the first stage and reduced slowly in the second stage.

Comprehensive dynamic degree of land use was 36.87 % from 2006 to 2014, 30.38 % from 2006 to 2010 and 14.82 % from 2010 to 2014. Comprehensive dynamic degree explained the annual change rate of land use, also explained the disturbance degree of human to the land use change. Among them, comprehensive dynamic degree in the first stage was two times more than the second stage, which indicated that land use change in the first stage was faster than in the second stage obviously.

4.3 Spatial Change

The statistics data of landscape types were as shown in Tables 3 and 4.

Table 3. Area statistics of landscape types

Year	Landscape type	Patch number	Average area of patch (hm^2)	Proportion of patch area (%)
2006	Water	462	55.57	12.33
	Woodland	5	88.04	0.21
	Residential land	121	651.75	37.87
	Farmland	48	2112.42	48.70
	Unused land	30	61.74	0.89
2010	Water	361	49.69	8.62
	Woodland	8	64.02	0.25
	Residential land	66	1915.73	60.72
	Farmland	50	1196.30	28.73
	Unused land	21	167.61	1.69
2014	Water	350	45.68	7.68
	Woodland	9	32.93	0.14
	Residential land	62	2282.42	67.96
	Farmland	53	923.30	23.50
	Unused land	5	299.25	0.72

Table 4. Landscape pattern indexes

Year	Index	Water	Woodland	Residential land	Farmland	Unused land	Total
2006	Separation degree	0.19	1.16	0.03	0.02	0.67	–
	Diversity	–	–	–	–	–	1.03
	Evenness	–	–	–	–	–	0.64
2010	Separation degree	0.24	1.26	0.01	0.03	0.30	–
	Diversity	–	–	–	–	–	0.2
	Evenness	–	–	–	–	–	0.13
2014	Separation degree	0.27	2.31	0.01	0.03	0.34	–
	Diversity	–	–	–	–	–	0.19
	Evenness	–	–	–	–	–	0.12

Through comparing the patch area proportions of residential land and farmland according to Table 4, the proportion of residential land increased year by year and farmland decreased year by year. It can be seen that diversity indexes were 1.03, 0.2 and 0.19 in 2006, 2010 and 2014 respectively. Among them, diversity index was the largest in 2006. The rapid urbanization led to diversity index of Tianjin Urban lower and lower. From Table 4, it can be seen that separation degree of woodland increased year by year and that of residential land and farmland were relatively stable. The reason was because the development of residential land focused on central city, whiles farmland always on the edge of central city and reduced year by year. Evenness decreased year by year which showed that distribution of land use types was more and more uneven.

5 Conclusions

In this paper, based on RS and GIS, the land use change information of Tianjin Urban was extracted using unsupervised classification and normalized index. The land use types of Tianjin Urban were mainly farmland and residential land. The total land area of Tianjin Urban was 208224.10 hm^2 in 2006, including water, woodland, residential land, farmland and unused land. Land use has been rapid changes from 2006 to 2014. The area of residential land was 141510 hm^2, increased 62648.02 hm^2; the area of farmland was 48934.90 hm^2, decreased 52461.31 hm^2 in 2014. The area of farmland and water gradually decreased year by year and the area of residential land increased rapidly during the period of research. The residential land was from water and farmland. The comprehensive dynamic degrees of two stages during the period of research were 30.38 % and 14.82 % respectively, and comprehensive dynamic degree of the whole research was 36.87 %, which indicated that land use change rate from 2006 to 2010 was larger than that from 2010 to 2014. Through analyzing separation degree, diversity, evenness, we can obtain the landscape diversity index of Tianjin Urban became lower and distribution of land use types was more and more uneven.

Acknowledgements. This research is financially supported by the Natural Science Foundation of Tianjin, China (Grants No. 13JCQNJC08600).

References

1. Badreldin, N., Goossens, R.: Monitoring land use/land cover change using multi-temporal Landsat satellite images in an arid environment: a case study of El-Arish. Egypt. Arab. J. Geosci. 7(5), 1671–1681 (2014)
2. Yang, Y.J.: Studies on land use/land cover of Wushan county based on RS and GIS. Southwest University (2009)

3. Vittek, M., Brink, A., Donnay, F., Simonetti, D., Desclée, B.: Land cover change monitoring using Landsat MSS/TM satellite image data over west Africa between 1975 and 1990. Remote Sens. **6**(1), 658–676 (2014)
4. Xian, G., Homer, C., Fry, J.: Updating the 2001 national land cover database land cover classification to 2006 by using Landsat imagery change detection methods. Remote Sens. Environ. **113**(6), 1133–1147 (2009)
5. Hulley, G., Veraverbeke, S., Hook, S.: Thermal-based techniques for land cover change detection using a new dynamic MODIS multispectral emissivity product (MOD21). Remote Sens. Environ. **140**(1), 755–765 (2014)
6. Cockx, K., Voorde, T.V.D., Canters, F.: Quantifying uncertainty in remote sensing-based urban land-use mapping. Int. J. Appl. Earth Obs. Geoinf. **31**(9), 154–166 (2014)
7. Beykaei, S.A., Zhong, M., Zhang, Y.: Development of a land use extraction expert system through morphological and spatial arrangement analysis. Eng. Appl. Artif. Intell. **37**, 221–235 (2015)
8. Khalifa, I.H., Arnous, M.O.: Assessment of hazardous mine waste transport in west central Sinai, using remote sensing and GIS approaches: a case study of Um Bogma area. Egypt. Arab. J. Geosci. **5**(3), 407–420 (2012)
9. Zhao, W.W.: International comparison of land use research. J. Earth Environ. **1**(3), 249–256 (2010)
10. Yu, X.X., Yang, G.S.: The advances and problems of land use and land cover change research in China. Progress Geogr. **21**(1), 51–57 (2002)
11. Liu, J.Y.: Study on national resources & environment survey and dynamic monitoring using remote sensing. J. Remote Sens. **1**(3), 225–230 (1997)
12. Wang, S.Y.: Study on land use/land cover change based on geo-spatiotemporal database in China. Institute of Remote Sensing Applications, Chinese Academy of Sciences (2002)
13. Rouse, J.W., Haas, R.H., Schell, J.A., Deering, D.W.: Monitoring Vegetation Systems in the Great Plains with ERTS, vol. 351, p. 309. Nasa Special Publication, Washington, D.C. (1974)
14. Zha, Y., Gao, J., Ni, S.: Use of normalized difference built-up index in automatically mapping urban areas from TM imagery. Int. J. Remote Sens. **24**(3), 583–594 (2003)
15. Xu, H.Q.: A study on information extraction of water body with the modified normalized difference water index (MNDWI). J. Remote Sens. **9**(5), 589–595 (2005)
16. Wang, A.Z., Zhang, G.B., Zheng, J., Zhao, J.J.: Analysis on land use change in Xinxiang city. Res. Soil Water Conserv. **15**(1), 163–165 (2008)
17. Wang, S.Y., Zhang, Z.X., Zhou, Q.B., Wang, C.Y.: Study on spatial-temporal features of land use/land cover change based on technologies of RS and GIS. J. Remote Sens. **6**(3), 223–228 (2002)
18. Zhao, D.B., Liang, W., Yang, Q.K., Liu, A.L.: Analysis of dynamic landuse changes of past 30 years in the hilly area of Loess plateau. Bull. Soil. Water Conserv. **28**(2), 22–26 (2008)
19. Li, Y.J.: Spatio-temporal changes analysis of land use in Pingdu county on RS and GIS. Shandong University (2008)
20. Chen, L.D., Fu, B.J.: The ecological significance and application of landscape connectivity. Chin. J. Ecol. **15**(4), 37–42 (1996)

21. Guo, Q.Z., Jiang, W.G., Li, J., Chen, Y.H., Yi, W.B.: Evolvement of urban landscape pattern and its driving factors in Haidian district, Beijing from 1985 to 2006. Urban Environ. Urban Ecol. **21**(1), 18–21 + 25 (2008)
22. Romme, W.H.: Fire and landscape diversity in subalpine forests of Yellowstoin national park. Ecol. Monogr. **52**(2), 199–211 (1982)
23. Wang, X.L., Xiao, D.N., Bu, R.C., Hu, Y.M.: Analysis on landscape patterns of Liaohe delta wetland. Acta Ecologica Sinica **17**(3), 317–323 (1997)

Research on the Capture Effect for RFID Tag Anti-Collision Algorithm

Chengshun Xu and Di Lu(✉)

Harbin University of Science and Technology, Harbin, China
xuchengshun2013@163.com, ludizeng@hrbust.edu.cn

Abstract. In the passive RFID system, the backscatter powers of tags are affected by the path loss and cause the capture effect. The capture effect makes some tags hidden and reduces the efficiency of identification. In this paper, an improved anti-collision algorithm capture effect tags optimization grouping (CEOG) is presented. The novel algorithm analysis the captures effects of RFID, then adopts Chebyshev estimation and group the capture effect optimization tags. By the theory and simulations, the CEOG make the system throughput exceed 60 %.

Keywords: RFID · Anti-collision algorithm · Capture effect · Tags grouping · Throughput

1 Introduction

Radio frequency identification (RFID) is a non-contact automatic identification technology with inductive coupling. Compared with the traditional identification technology, RFID technology has the advantages of fast transmission speed, high recognition accuracy and corrosion resistance [1].

In RFID system, the tags collision will happen when multiple tags send data at the same time, so the anti-collision algorithm is the key of solving this problem. At present, anti-collision algorithm is classified into aloha-based or tree –based. Aloha-based algorithm is a random algorithm while tags transmit their data to readable slots randomly. It is efficiently and widely used in passive tags identification system, including slotted aloha (SA), framed slotted aloha (SFA), and dynamic framed slotted aloha (DFSA). Tree-base is a deterministic algorithm which is based on the binary tree, it is accurately but long latency, including binary splitting tree (BST), query tree algorithm (QTA). In recent years, Lei Zhu et al. propose an optimal framed aloha (OFA) based anti-collision algorithm which adopts the adaptive markov decision process theory to solve the tags collision problem. Yuan-Cheng Lai et al. propose a dynamic blocking adaptive binary splitting (DBA) which can prevent the newly-arriving tags. Xin-Qing Yan et al. propose a successive scheme binary query tree (SS) which apply to new nodes added problem [2–4]. Although these algorithms can well improve the efficiency of tags identification, they do not consider the influence of the capture effect phenomenon to the system throughout in the RFID. Capture effect is a phenomenon that weak signal is hidden by

F. Bian and Y. Xie (Eds.): GRMSE 2015, CCIS 569, pp. 238–245, 2016.
DOI: 10.1007/978-3-662-49155-3_24

strong signal, and tags will be leakage when they answer reader request command, it will affects the system identification efficiency seriously [5].

The anti-collision algorithm under capture effect are few, Le-Xun Xu proposed a dynamic framed allotted aloha under capture effect (CE-DFSA) which establish an average acquisition probability mathematical model to achieve optimal capture ratio, and it improves the system throughput effectively. Qian Yang et al. propose a capture aware and tag population estimation protocol (CAPTE) which estimates the number of tags by idle slot number without affected, and it reduce the computational complexity [6]. But these algorithms are based on the average capture effect probability model and do not consider changing capture effect probability while tags are distributed in different frames. To solve this problem, an improved anti-collision algorithm capture effect tags optimization grouping (CEOG) dynamic framed slotted aloha is proposed. It can correct the number of collision slots and success slots by grouping the tags with capture effect optimization, and the system throughput is increased 60 %.

2 Tag Identification under Capture Environment

In DFSA, there are three types of tag C_s (the number of successful slot), C_e (the number of empty slot) and C_c (the number of collision slot), and the tag identification process is a random binomial distribution. We denote L is the length of previous frame, and n is the number of tags. Without capture effect, they are defined in (1), (2) and (3).

$$C_e(N,n) = L\left(1 - \frac{1}{L}\right)^n \tag{1}$$

$$C_s(L,n) = n\left(1 - \frac{1}{L}\right)^{n-1} \tag{2}$$

$$C_c(N,n) = L\left[1 - \left(1 - \frac{1}{L}\right)^n - \left(\frac{n}{L}\right)\left(1 - \frac{1}{L}\right)^{n-1}\right] \tag{3}$$

Where C_e, C_s, and C_c will be effect by capture effect, and change into C_e', C_s' and C_c' as follow:

$$C_e' = C_e \tag{4}$$

$$C_s' = C_s - p_c C_s \tag{5}$$

$$C_c' = C_c + p_c C_s \tag{6}$$

So, capture effect probability equation:

$$p_c = \sum_{i=1}^{k} p_{succ}(i) p_{cap}(i) \tag{7}$$

In Eq. (7), $p_{succ}(i)$ is the probability of successful slot, and $p_{cap}(i)$ is the capture effect probability that i tag success in the previous frame, and system throughput under capture effect is

$$T(N,n) = \frac{C'_s t_s}{C'_e t_e + C'_s t_s + C'_c t_c} \tag{8}$$

Where t_e is the empty slot time, t_s is the success slot time, and t_c is the collision slot time.

3 Capture Effect Tags Optimization Grouping Protocol

In DFSA, adjusting the length of the current frame is important to improve the system throughput by estimated tags number, so advanced tag estimation algorithms need to be proposed.

3.1 Tag Estimation under Capture Effect

It existed many tags number estimation algorithms which are based on LowerBound (LB), Schoute, Chebyshev and Bayes [7–10]. Among them, Chebyshev and Bayes are more accuracy, but Bayes need prior information which is too difficult to obtain, so we give a new tag estimation based on Chebyshev as

$$\varepsilon = \min_{(\hat{n},p_c)} \left\| \begin{pmatrix} C_e \\ C_s \\ C_c \end{pmatrix} - \begin{pmatrix} N_e \\ N_s \\ N_c \end{pmatrix} \right\| \tag{9}$$

Where N_e is the calculation of empty slot, N_s is the calculation of success slot, and N_c is the calculation of collision slot. We search a ε to makes (9) minimum, and then obtain the capture probability p_c and estimated tags number $\hat{n}$.

Figure 1 shows the estimation error $e = |(\hat{n} - n)/n|$ under capture effect of Lower-Bound (LB), Schoute, Chebyshev and Bayes, where we denote previous frame $L = 128$, capture probability $p_c = 0.5$. When tags range from 1 to 400, estimation error of LB and Schoute are worse than proposed method and Bayes. As tags number increasing, the estimation error increases simultaneously, and the result indicate that the proposed method is more accurate and below 0.05.

3.2 Tags Optimization Grouping under Capture Effect

In RFID, the backscatter powers of tags are affected by the distance to the reader with the different characterization of objects, so some tags will be hidden and lose to reader

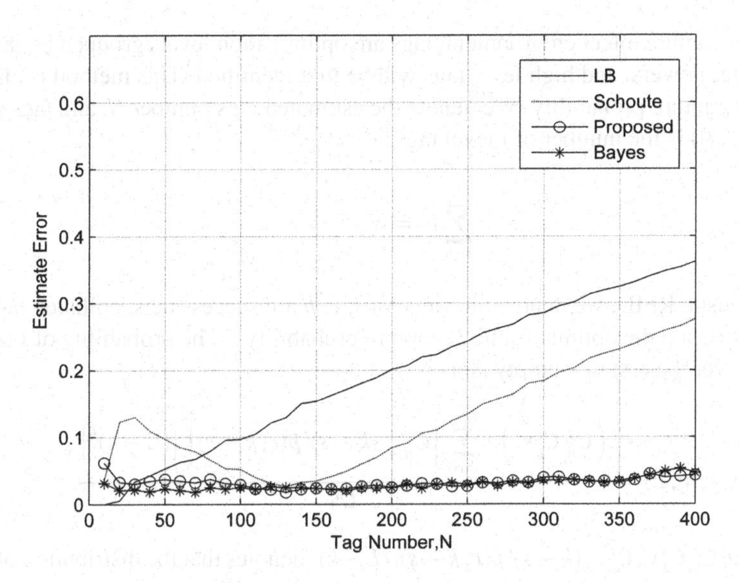

Fig. 1. Tags number estimation error under capture environment

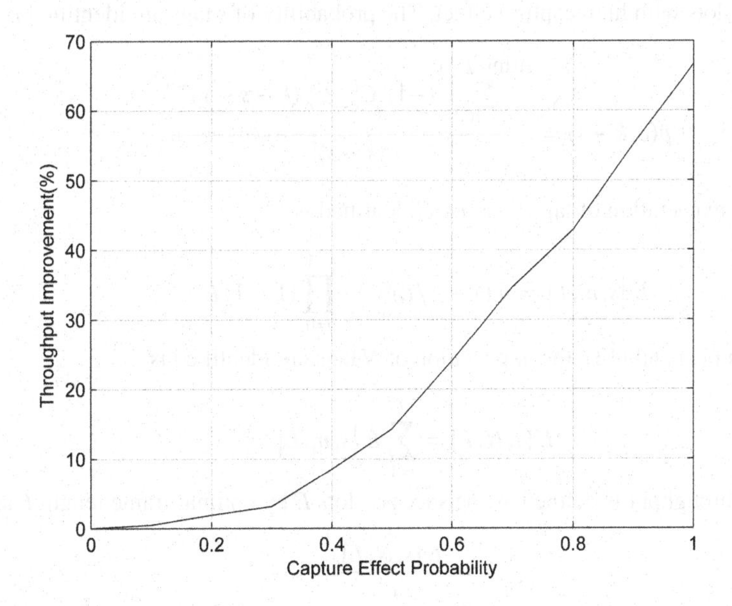

Fig. 2. Throughput improved of capture effect probability

by capture effect. The CE-DFSA and CAPTE can work well, but because of average capture probability model and do not consider changing probability in different frame as Fig. 2.

Figure 2 shows that the system throughput increases as the capture probability increases.

Under capture effect environment, tags are optimization levels grouped by different backscatter powers, and high level tags will be first identified. This method is effective to reduce capture probability. We denote the estimated tags number N, and tags groups l, $n_i(i \in [1, l])$ is the number of i level tags as

$$\sum_{i=1}^{l} n_i = N \tag{10}$$

In realistic RFID, we denote the value of s, c, h are success tags, collision tags, and hidden tags, and the optimal frame L, capture probability i. The probability of s tags are identified for the rest of k empty slots is

$$p(s, k, n_i, L) = \frac{\left(C^s_{n_i} C^s_k s!\right) \cdot \sum_{c=0}^{n_i - s} \left[C^c_{n_i - s}(k - s)^c \bar{p}(c, k - s)\right] \left[(L - k)^h\right]}{L^{n_i}} \tag{11}$$

Where $C^s_{n_i} C^s_k s!$, $C^c_{n_i - s}(k - s)^c \bar{p}(c, k - s)$, $(L - k)^h$ denotes that the distribution of s tags are identified, c tags are collision for the rest of $k - s$ empty slots, and h tags are hidden in $L - k$ slots with high capture effect. The probability of s tags are identified is

$$\bar{p}(c, k - s) = \frac{\sum_{v=0}^{\min(c, k-s)} (-1)^v C^v_{k-s} C^v_c (k - s - v)^{c - v}}{(k - s)^c} \tag{12}$$

Then the expectation of tags level i are identified is

$$E\left[s, n_i, L\right] = n_i(1 - 1/L)^{n_i - 1} \cdot \prod_{j=0}^{i-1} (1 - 1/L)^{n_j} \tag{13}$$

Based on (11) and (13), the expectation of N tags are identified is

$$E\left[s, N, L\right] = \sum_{i=1}^{l} E\left[s, n_i, L\right] \tag{14}$$

System throughput is defined as the success slots E by optimal frame length L as

$$T = \frac{E\left[s, N, L\right]}{L} \tag{15}$$

Where $\tilde{n}_i$ and $\tilde{l}$ can be obtained by searching the optimal solutions to system throughput T.

$$\left[\hat{n}_i, \hat{l}\right] = \arg\max_{(\tilde{n}_i, \tilde{l})} T \tag{16}$$

4 The Procedure of CEOG

The proposed algorithm apply to the passive tags identification system, and its procedure as follows.

Step 1. Initializing the register of reader, and let denote previous frame length $L = 128$, then prepare for identification.
Step 2. The reader send query order and wait for tags activation. After tags require, RN16 code are send by reader and check with them. If they are matching, tags will be successful identified, or wait for the next cycle.
Step 3. The reader record previous frame occupy status, and use our proposed method to estimated tags number N.
Step 4. If tags number is zero, finish the identification cycle, or tags are capture effect optimal grouped and go to step 2 to recycle.

5 Simulation and Performance Analysis

The performance of the proposed CEOG is simulated by Monte Carlo method. I consider the following metrics:

(1) Communication channel model is ideal capture effect model, capture effect will be happen while different capture levels tags select the some slot.
(2) Tags will send their ID if they are successful identified. It can well induces system overhead by empty slots and collision slots.

The system throughput equation is

$$T = \frac{n_s \cdot t_s}{n_e \cdot t_e + n_s \cdot t_s + n_c \cdot t_c} \tag{17}$$

Where t_e, t_s, t_c are the length of empty slots, success slots and collision slots.

Figure 3 shows that the system throughput increases by group number, and it searching times increases faster. Considering algorithmically complex and system throughput improvement, the value of group number often is set between 3 and 6, and in this paper, $l = 3$.

Figure 4 shows that CEOG makes system throughput better than CAPTE and CE-DFSA, and it is below 0.6 while tags number increase 0 to 200. Results indicate the CEOG can well deal with the capture effect probability changing, and the high level capture tags are identified before lower level capture tags, so it can effective reduce the number of lower power tags hidden and improve the system throughput.

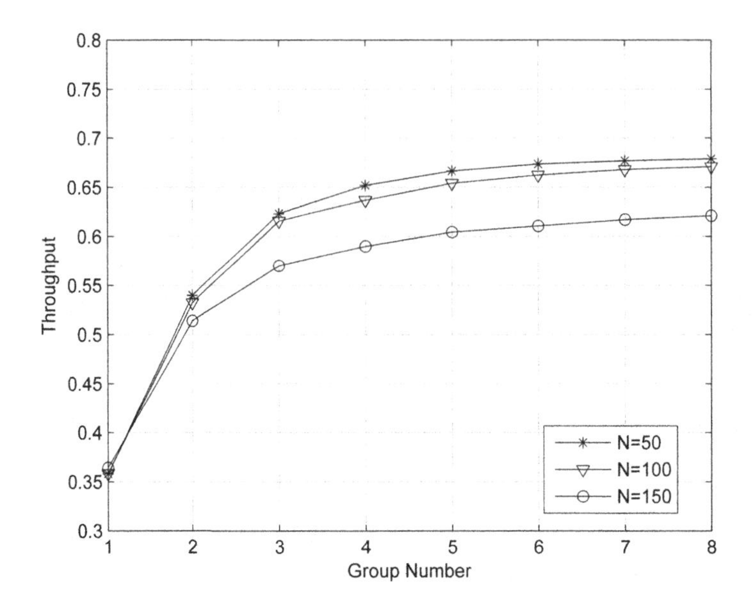

Fig. 3. Throughput of tags grouped level

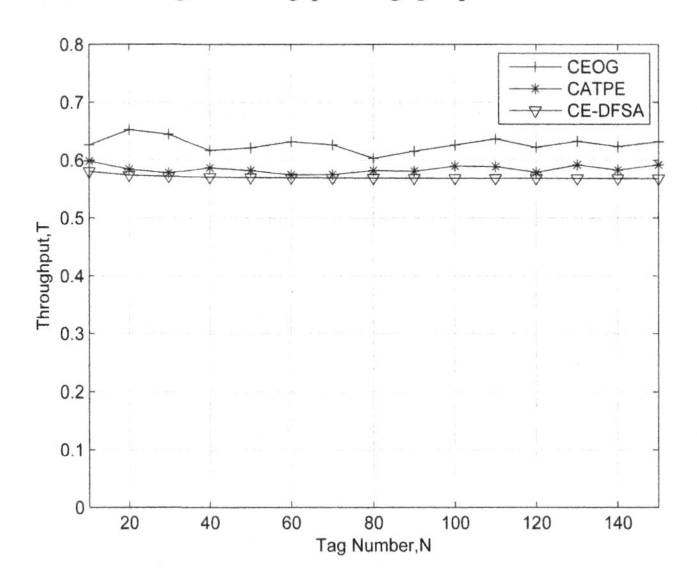

Fig. 4. Throughput improvement of the CEOG under capture effect

6 Conclusion

In this paper, we propose a CEOG algorithm for solving tags anti-collision under capture effect which is compatible with DFSA protocol. Analytical and simulation results show that our proposed method can effectively identify hidden tags and improve system throughput.

References

1. Finkenzeller, K.: RFID Handbook: Radio-Frequency Identification Fundamentals and Applications, 3rd edn, pp. 6–9. John Wiley and Sons, England (2010)
2. Zhu, L., Yum, T.S.P.: Optimal framed aloha based anti-collision algorithms for RFID systems. IEEE Trans. Commun. **58**, 3583–3592 (2010)
3. Lai, Y.C., Hsiao, L.Y., Lin, B.S.: An RFID anti-collision algorithm with dynamic condensation and ordering binary tree. Comput. Commun. **36**, 1754–1767 (2013)
4. Yan, X.Q., Liu, Y., Li, B., Liu, X.M.: A memoryless binary query tree based successive scheme for passive RFID tag collision resolution. Inf. Fusion **22**, 26–38 (2015)
5. Maguire, Y., Pappu, R.: An optimal Q-algorithm for the ISO 18000-C RFID protocol. IEEE Trans. Autom. Sci. Eng. **6**(1), 16–24 (2009)
6. Yang, X., Wu, H., Zeng, Y., et al.: Capture-aware estimation for the number of RFID anti-collision algorithm based on dynamic frame length ALOHA. IEEE Trans. Autom. Sci. Eng. **6**(1), 9–15 (2009)
7. Schoute, F.C.: Dynamic frame length ALOHA. IEEE Trans. Commun. **31**(4), 565–568 (1983)
8. Vogt, H.: Efficient Object Identification with Passive RFID Tags. In: Mattern, F., Naghshineh, M. (eds.) PERVASIVE 2002. LNCS, vol. 2414, pp. 98–113. Springer, Heidelberg (2002)
9. HaiFeng, W., Zeng, Y.: Bayesian tag estimate and optimal frame length for anti-collision ALOHA RFID system. IEEE Trans. Autom. Sci. Eng. **7**(4), 963–969 (2010)
10. Chen, W.T.: An accurate tag estimate method for improving performance of an RFID anti-collision algorithm based on dynamic frame length ALOHA. IEEE Trans. Autom. Sci. Eng. **6**(1), 9–15 (2009)

Design of Glass Defect Inspection System Based on Wavelet Transform

Qin Xu[1] and Haitao Zhang[2(✉)]

[1] College of Mathematical and Information Engineering, Anhui Science and Technology University, Fengyang 233100, China
xuqincher@163.com

[2] College of Mechanical Engineering, Anhui Science and Technology University, Fengyang 233100, China
toby_sh@163.com

Abstract. The study on glass defect inspection system is very important to glass industry. It can improve the accuracy of inspection and the quality of product as well as reduce the labor intensity of workers and the enterprise cost. Based on OpenCV Library, an inspection system of glass defects is designed. The Wavelet Transform is used to image preprocessing and the library function is applied to implement image gray, filtering, binarization and contour recognition. The results show that the system realizes high efficient inspection and offers data for the further study.

Keywords: OpenCV · Glass defect · Wavelet transform · Image recognition

1 Introduction

With the rapid development of automobile, architecture, new energy and other industries, the quantity and quality demand for glass keeps growing. However, the defects like bubbles and cracks seem unavoidable for the reason of the technology and the cost. These defects have serious impact on the transparency, heat resistance and mechanical property of glass. In most traditional glass factories, the glass is still inspected by artificial naked eyes. In this way, the labor intensity of workers is large but the accuracy is low. Some small defects might be ignored [1]. These would increase the burden of workers and risk of deep processing of glass. To avoid such situations, many applications on the inspection of glass defect based on the visual inspection technology are developed.

OpenCV is an open-source, cross platform visual library. It includes image transform, wave filtering, edge inspection, histogram, geometric transformation, mathematical morphology processing and other commonly used image processing methods. It has the characteristics of light weight, high efficiency, multi interface and complete algorithm and could run on multiple platforms [2, 3]. With the image processing technology of OpenCV, the system could realize image acquisition and processing and provides the valid image data for the subsequent processing. Such methods have great advantages, especially in the high speed processing system [4, 5].

F. Bian and Y. Xie (Eds.): GRMSE 2015, CCIS 569, pp. 246–253, 2016.
DOI: 10.1007/978-3-662-49155-3_25

In this paper, an open-source image processing algorithm library based on OpenCV is proposed and the glass defect inspection system is developed. The system is tested with three typical kinds of glass defects. The results show that the system has the characteristics of high efficiency and accuracy.

2 The Overall Design of the System

Glass defect is mainly divided into bubbles, cracks, inclusions, stones etc. [6]. However, the defects of glass do not have fixed shape and size as well as the obvious difference of gray scale and details. So the relatively stable statistical approach must be used to describe various morphological characteristics of the defects, for example, the inclination, centrifugal rate, length-wideth ratio, stretch, filling degree, edge straightness, equilibrium degree of horizontal extension, equilibrium degree of vertical extension, vergence and the connectivity leading rate of defects. In engineering application, the method of glass defect inspection is generally to recognize the defect contour, according to which the feature of defects would be analyzed and recognized.

The identification system of glass defect is mainly composed of image acquisition, image processing and defect recognition. The basis of image acquisition part is the transparency of glass. The part gets the raw image of glass defects mainly through the CCD camera and other image acquisition equipment at strong light source [7, 8]. The image processing part obtains the binary image through the graying, WT, threshold filtering and binarization to provide data for defect recognition. Defect recognition is mainly to find out the defect contour, according to binary image. Then the characteristic parameters of defects would be calculated. The statistical characteristic parameter would be used to make the identification and the recognition of defects. This paper uses the existed glass defect image, therefore, the image processing and the method for identification of defects would be emphatically researched and discussed. The basic software process of image processing and recognition is shown as Fig. 1.

3 Image Processing and Defect Recognition

3.1 Image Graying

The image pretreatment includes two parts, graying and filtering. Generally, the color format of image got by digital camera is Lab or RGB. However, in glass defect inspection system, the valid information is only the contour of defect. The existence of color would greatly increase image pretreatment time. Graying is the compressing process for the color images by gray scale mapping function. Graying could remove the useless reflected information of color. It could improve the processing speed of the histogram analysis and the binarization processing, which could realize the high efficiency and the accuracy of image recognition. OpenCV library has special function cvCvtColor which would converse the image to grayscale. The conversion process is shown as Fig. 2.

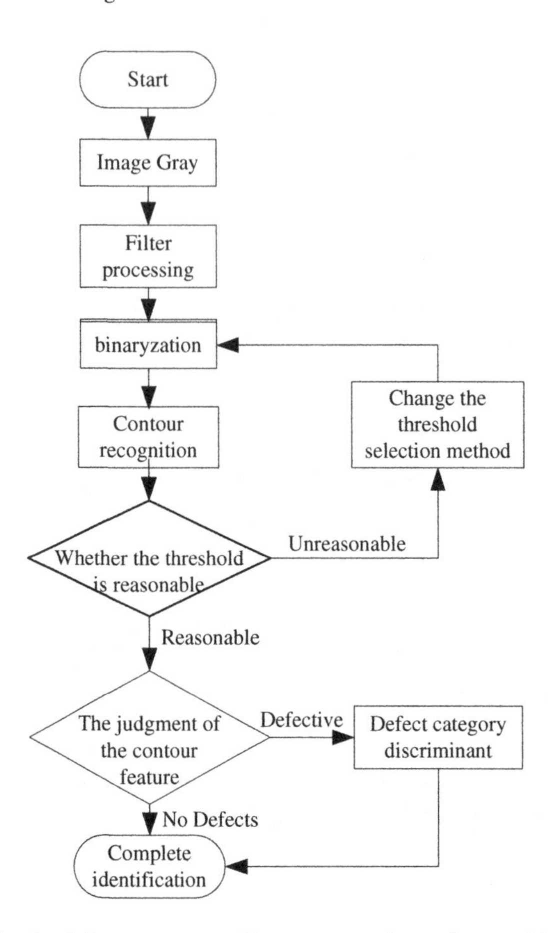

Fig. 1. Software process of image processing and recognition

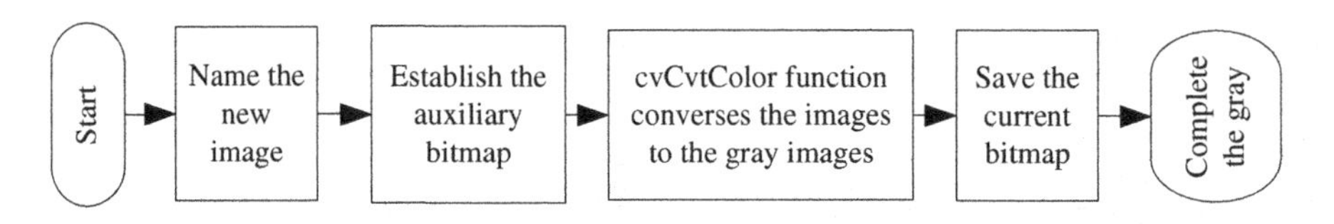

Fig. 2. Gray image processing

3.2 Image de-Noising Based on WT

Image de-noising mainly has three methods: (1). Gaussian noise generated by random thermal motion of electron photosensitive components and parts; (2). Poisson noise generated by photovoltaic conversion; (3). Grain noise generated by photographic process. On account of the error accumulation, image noise easily causes anamorphose in subsequent process. So image de-noising is the most important part in this paper.

OpenCV library offers many image de-noising functions. The most common ones are the CV_MEDIAN for median filtering and CV_GAUSSIAN for Gaussian filtering. The median filtering uses median of gray value of all the pixels in neighboring window as gray value of the selected pixel. This algorithm is simple and could effectively protect marginal information. Through Fourier transformation, the noise located on high-frequency zone, and the signal located on low-frequency zone. According to these characteristics, Gaussian filtering de-noise with low-pass filter (LPF). The algorithm could achieve good effect, when restrains the noise which obeys the normal distribution. However, these two algorithms are liable to destroy important signal singular point like cracks or corners.

Wavelet transform has good characteristic of time-frequency localization and multi-resolution. It can keep the whole contour, signal singular point and edge features. So the wavelet transform has been widely used in the glass defects image processing system. Continuous wavelet transform (CWT) formula is as follows

$$W_f(a,b) = \langle f, \Psi_{a,b}(x) \rangle = \int_{\infty}^{\infty} f(x)\Psi_{a,b}(x)dx = \frac{1}{\sqrt{a}}\int_{\infty}^{\infty} f(x)\Psi\left(\frac{x-b}{a}\right)dx \quad (1)$$

In which, $\Psi_{a,b}(x)=\frac{1}{\sqrt{a}}\Psi\left(\frac{x-b}{a}\right)$ is basic wavelet. With the scale and displacement of basic wavelet, wavelet transform window of different sizes and positions could be acquired. Basic wavelet has good attenuation, which satisfies

$$\int_{\infty}^{\infty} \Psi(t)dt = 0 \quad (2)$$

$$C_{\psi} = \int_{\infty}^{\infty} \frac{\Psi(s)}{s} ds < \infty \quad (3)$$

In the numerical calculation application of glass defect inspection, the scale factor and position factor are discrete. Thus, the discrete wavelet transform is widely used, the basis wavelet is

$$h_{m,n}(x) = \frac{1}{\sqrt{a_0^m}} h\left(\frac{1}{a_0^m} x - nb\right) \quad (4)$$

OpenCV library offers two-dimensional discrete wavelet transform function DWT and inverse transform function IDWT. The fundamental principle is dividing noise signal into odd sequence $s_o(k)$ and even sequence $s_e(k)$. According to correlations of the two sequences, it can be predicted even sequence $d(k)$ instead of odd sequence

$$d(k) = s_o(k) - P \bullet s_e(k) \quad (5)$$

Update even sequence with $c(k)$

$$c(k) = s_e(k) + U \bullet d(k) \tag{6}$$

Then acquire sequence $d(k)$ with glass defect information and sequence $c(k)$ with general information.

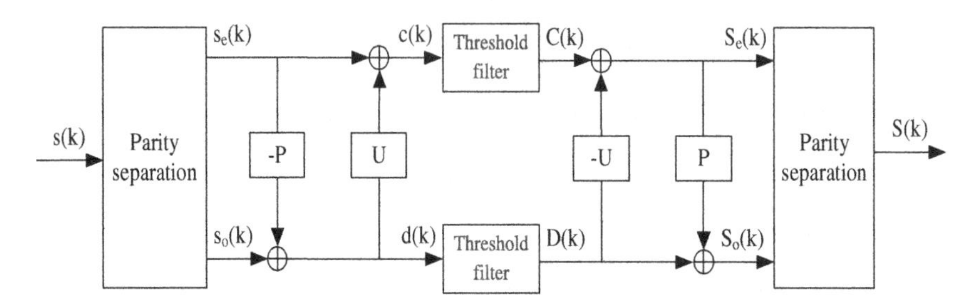

Fig. 3. Wavelet de-noising process

For the glass defect information, the wavelet coefficients of feature point and singular point like contour information have high amplitude and small quantity, located in high-frequency zone; and the wavelet coefficients of noise signal (white noise) have low amplitude and large quantity, located low-frequency zone. Through setting proper threshold value for different zones the system could keep glass defect information, while complete de-noising. After that, the new sequences $C(k)$ and $D(k)$ could be acquired. The inverse discrete wavelet transform can be represented as follows:

$$S_e(k) = C(k) - U \bullet D(k) \tag{7}$$

$$S_o(k) = P \bullet S_e(k) + D(k) \tag{8}$$

$$S(k) = S_e(k) + S_o(k) \tag{9}$$

where, the $S(k)$ represents the de-noised image, the whole process of Wavelet de-noising is illustrated in Fig. 3.

3.3 Binarization of Image and Image Contour Inspection

Binarization is the most critical step in the image recognition process and the basis of image contour inspection. It selects the gray image with 256 brightness levels through appropriate threshold. Then, the images could map into only two brightness levels, black and white. The image contour could be recognized with the boundary of black and white blocks in binary image. The threshold of the binarization will directly influence the precision and accuracy of the system, thus the effectiveness of the method should be verified repeatly. The OpenCV library contains the cvAdaptiveThreshold function which could make the self-adaptive binarization for the image. However, this function has poor effects in terms of the glass defect recognition. In this paper, the otsu method is adopted to recognize the defects, the realization of the method is as follows.

```
for (i=0;i<256;i++)
  ZFT[i]=0;
for (i=0;i<Img->imageSize;i++)
ZFT[(BYTE)Img->imageData[i]]++; %statistical image gray value%
Gs2=Mean_Gs( ); %calculate the average value of gray%
do
 { Gs1=Gs2;
   Cal_Gs_Low( ); %calculate the sum of the low gray level group%
   Mean_Gs_low( ); %calculate the average value of low gray level group%
   Mean_Gs_High( ); % calculate the average value of high gray level group%
   Gs2=New_Gs( ); % get the new estimated value of threshold%
 }while(Gs1!=Gs2);
```

Firstly, It can be found the contour by cvFindContours function. Then, the cvDrawContours function is used to draw the contour. The cvCvtSeqToArray function will then be used to extract each contour point set in order to make the analysis of statistical features and to determine the type of defects. It is important to note that the identified contour might have errors like unclose, isolated point, for improper binarization threshold selection. Then the threshold needs to be changed to seek a better result.

3.4 Image Recognition Results

In order to verify the effectiveness of glass defect inspection system which is developed based on OpenCV function library, glass photos with typical defects such as bubbles, stones and impurities are adopted as the test images. They were input into the developed software separately. Through gray, binarization and contour recognition, the binary images and the contour recognition graphs would be get.

Figure 4-1 is the glass image processing effect with bubble defects, Fig. 4-2 represents the glass image processing effect with impurities defects, and Fig. 4-3 is the glass image processing effect with Impurities. From the comparative analysis of resulting picture (b) of binarization and the original image (a), the binarization could process the originals very well and offer the effective data for contour recognition. From the comparative analysis of the contour recognition image (c) and the original image (a), the algorithm can identify the defect contour accurately and effectively. In the subsequent analysis, the type of defects could be identified by the analysis the morphological characteristics of defects according to the contour points.

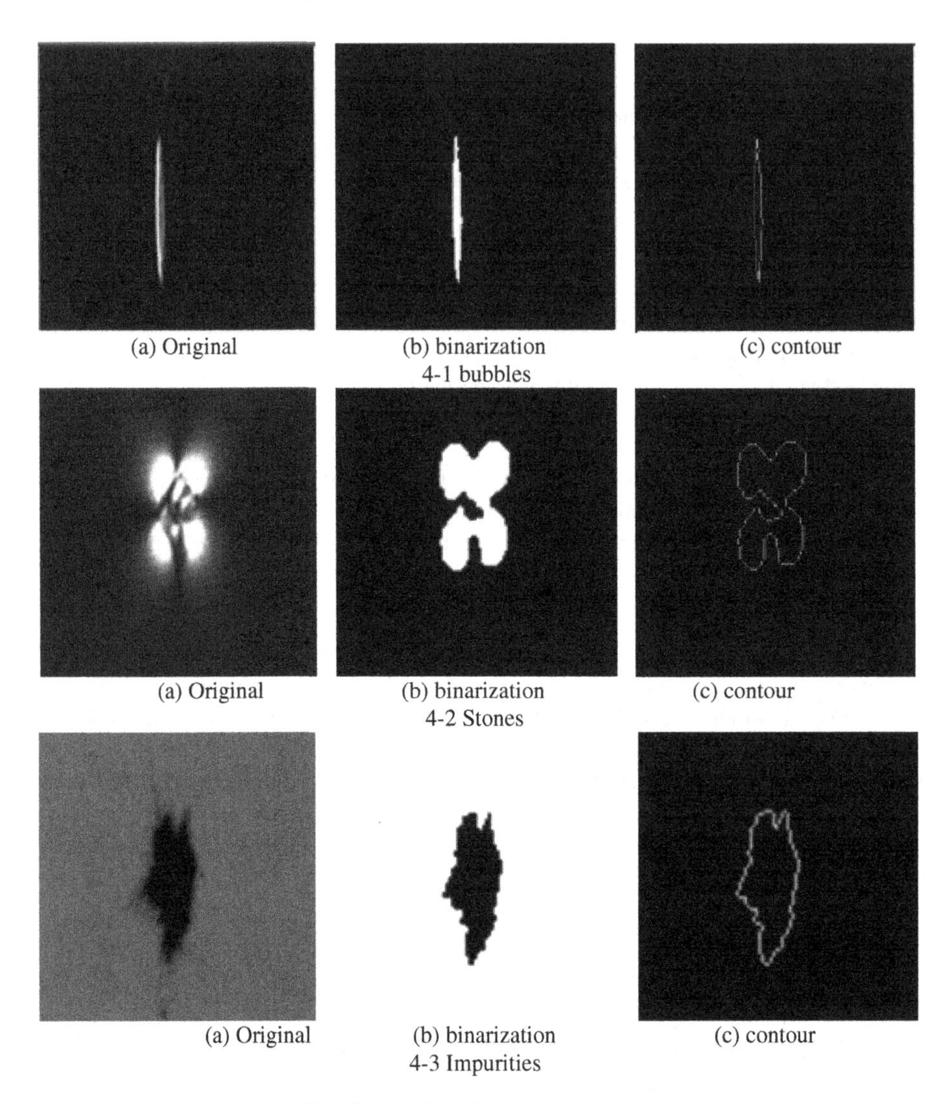

Fig. 4. Results of image recognition

4 Conclusions

In the glass defect recognition inspection, image processing is an effective method for image analysis and feature recognition. OpenCV image processing algorithm library has been used in this paper. The image processing based on wavelet transform and the defects recognition software of the glass defects inspection system have also been developed. The algorithm has been tested by the glass image with typical defects. The results showed that the defects could be identified by the glass defects inspection system effectively. The system has comprehensive functions, high efficiency and strong transportability. It also has good application value in the development of the similar systems of glass defect recognition.

Acknowledgement. This research was supported by University Natural Science Foundation of Anhui Province (No. KJ2013B075), Foundation of Anhui Science and Technology University (No. ZRC2013338).

References

1. Huang, Q.K., Zhou, Y.C.: A method based on watershed algorithm for core particles image segmentation. In: 3rd IEEE International Conference on Computer Science and Information Technology, vol. 3. IEEE (2010)
2. Zhang, Y., Duanquan, X.: Improvement on watershed algorithm of OpenCV and its application in cell image segmentation. J. Comput. Appl. **32**(1), 134–136 (2012)
3. Qin, X., Yu, H., Wen, Z., Qiao, W.: Image segmentation base on OpenCV. Sci. Technol. Inf. **7**, 39–41 (2011)
4. Blayvas, I., Bruckstein, A., Kimmel, R.: Efficient computation of adaptive threshold surfaces for imagine binarization. In: Proceeding of the 2001 IEEE Computer Society Conference Computer Vision and Pattern Recognition (2001)
5. Liu, Y.: Face inspection system design based on OpenCV. In: Proceedings of the 9th International Symposium on Linear Drives for Industry Applications. Lecture Notes in Electrical Engineering (2013)
6. Shopa, P., Sumitha, N., Patra, P.S.K.: Traffic sign inspection and recognition using OpenCV. In: 2014 International Conference on Information Communication and Embedded Systems, ICICES (2014)
7. Liu, Z., Guo, Z., Li, C.: The Study and Realization of glass defect image segmentation. J. Zhongyuan Univ. Technol. **18**(7), 26–28 (2007)
8. Yu, B.-Y., Wang, Z.-B., Jin, Y.: Research on new method of glass-defect inspection, Transducer and Microsystem Technologies **27**(1), 183–187 (2008)

Research on Relationship Between Vegetation Coverage and Height/Slope in Chongqing

Yanying Chen[1(✉)], Yangsheng You[2,3], Yunhui Tang[1], and Jianping Zhang[1]

[1] Chongqing Institute of Meteorological Sciences, Chongqing 401147, China
chenyanying1618@163.com
[2] College of Civil Engineering, Chongqing University, Chongqing 400045, China
youyangsheng@126.com
[3] Key Laboratory of New Technology for Construction of Cities in Mountain Area, Ministry of Education/Chongqing University, Chongqing 400030, China

Abstract. In order to clarify and quantify the relationship between vegetation coverage and topography in Chongqing, MODIS-250 m 16 days NDVI from 2010 s to 2013 s was used to synthesize monthly and seasonal NDVI enlarged 100 times. Then the NDVI was divided into 6 height districts and 5 slope districts. The relationship between NDVI and height/slope was analyzed and the equations of NDVI and height/slope were established. The results showed as following: (1) Different month and season have obvious effect on the relationship between NDVI and height/slope. The value of NDVI in months 6-9 and summer continued to increase with the increase of altitude. While NDVI increased firstly then decreased with the increase of height in months 10-12 and 1-5, seasons of spring, autumn and winter. The turning height of the NDVI value decreased in cold month/season and increased in warm month/season. The trend was relatively obvious and simple with increase of slope that was the NDVI monotonically increased with slope increasing. (2) The increasing trend of NDVI with height/slope increasing gradually decreases. (3) There was good binomial relationship between NDVI and height and most of these passed correlation coefficient test (P<0.001). There was binomial and exponential relationship between NDVI and slope and most of the equations passed correlation coefficient test (P<0.001).

Keywords: Chongqing · Normalized difference vegetation index (NDVI) · Topography (height/slope) · Relationship

1 Introduction

Mountains and hills are the main topographical in Chongqing, and mountainous areas mostly in south and north of Chongqing account 75.8%, hills mostly in western and central in Chongqing account 18.2%, ground and hyper located in the Yangtze and Jialing coast account 6%. Owing to complex terrain, climate types diverse in Chongqing. Affect by complex terrain and climate types, vegetation distribution in space is diversification.

F. Bian and Y. Xie (Eds.): GRMSE 2015, CCIS 569, pp. 254–266, 2016.
DOI: 10.1007/978-3-662-49155-3_26

Influence of topography on vegetation can be analyzed from different angles, such as the impact on landscape [1, 2], the impact on spatial and temporal distribution of vegetation [3, 4], the impact on vegetation coverage [5, 6], the impact on the community and biological reserves [7].

NDVI is an important indicator to describe vegetation coverage and vegetation growth [8, 9], usually used to monitor vegetation growth, estimate productivity and yield [10, 11], assess vegetation coverage [12], further more used to assess Ecological environmental. Most vegetation distribute in the area of complex terrain that not only directly affect the distribution of vegetation but also vegetation partially reflective bands thus affecting the results of remote sensing. Yao Chen etc assessed impact of terrain on different vegetation indices, the results is it can't be ignored that impact of terrain on vegetation index [13]. Jiang Hong etc build a topography-adjusted vegetation index based on vegetation Index algorithm, further more he monitor and assess vegetation status in complex terrain more accurately [14].

2 Research Methods and Main Content

2.1 Data Processing and Methods Introduction

In this paper, the official website of NASA MODIS data of MOD13Q1 that is synthetic 16d global vegetation index products with 250 m resolution. Data time is from January 1-2010 to December 31-2013, the study area spans h27v05, h27v06 two bands, a total of 184 view images.

Used MRTSwath, the MODIS image was splicing, reprojection and format conversion. Then image of Chongqing was cut out with GIS, and ultimately get MODIS/NDVI series including 92 scene data with projection system of Albers / WGS84. In order to avoid the influence of more clouds, in the GIS software monthly NDVI was the maximum data. Seeing Eq. (1):

$$NDVI_m = \max(NDVI_n, NDVI_{n+16}) \tag{1}$$

In formula (1), $NDVI_m$ is monthly NDVI, and m is the num of month, m=1, 2, ……, 12, $NDVI_n$ is NDVI data after projection and conversion that is the first phase NDVI of synthesis a month, and $NDVI_{n+16}$ is the second phase NDVI of synthesis a month.

$$NDVI_S = \frac{1}{3} \times \sum_{i=1}^{3} NDVI_m \tag{2}$$

Basic data of synthesis monthly NDVI is 16 days so that month NDVI can't contain exactly each day of a month. Because vegetation growth is a gradual process, and NDVI trend does not appear mutation under natural state, there is little effect on monthly NDVI missing a few days of NDVI. Synthetic monthly NDVI can reflect vegetation growth trends. Then season NDVI is obtained with monthly NDVI as basic data. Season NDVI is average of the used month because that represents the average state of vegetation, seeing Eq. (2).

In formula (2), $NDVI_S$ is season NDVI, s from 1-4 represent spring, summer, autumn and winter seasons, $NDVI_m$ is monthly NDVI, m is 3–5 in spring, 6–8 summer, 9–11 autumn, 12,1–2 winter.

The monthly and seasonal NDVI were divided into different areas according height/slope [15]. There are ≦ 400 m, 400–800 m, 800–1200 m, 1200–1500 m, 1500-2000 m and> 2000 m 6 districts by height, and ≦ 5°, 5-15°, 15-25°, 25-35° and> 35° 5 zones by slope.

2.2 Research Content

Chongqing is located in the eastern of Sichuan Basin, belonging to transition zone of the Tibetan Plateau and the Yangtze River Basin. Chongqing covers an area of 82403 km^2. Chongqing highest is elevation of 2796.8 m, the minimum altitude of 73.1 m [16]. The climate type is humid subtropical monsoon climate, average temperature of 17.8°C, annual rainfall of 1000 ~ 1450 mm. Vegetation zones are subtropical evergreen broadleaf forest, most of which located in area with higher elevation or slope greater than 15° or relatively high elevation areas. With altitude increasing, vegetation presents vertical zonal characteristics, from bottom to top as follows: evergreen broad-leaved forest (baseband) - evergreen broad-leaved and deciduous broad-leaved forest zone - subalpine coniferous forest (including deciduous broad-leaved and needle leaf forest zone).

Analysis found that vegetation distribution is affected by topography in Chongqing. Firstly, vegetation mainly distribute in height (slope) larger areas, secondly, with height (slope) increases, NDVI is bigger. This paper, based on MODIS-250 m NDVI series, relationship between NDVI and height/slope will be analyzed from a statistical point of view, and formula of them establish.

3 Analysis on NDVI Response to Terrain in Chongqing

3.1 Trends of Monthly Average NDVI in Height/Slope

Analysis found there are some differences in monthly NDVI between adjacent height/slope areas from January 2010 to December 2013. In Tables 1 and 2, monthly NDVI ranges of 2010-2013 in height / slope area were given.

Seeing from Table 1, there is difference of monthly NDVI trend with height. Max/minimum NDVI increased with height increase under 2000/1500 m area in January, while decreased with height increasing in areas above 2000/1500 m. Because NDVI affected by clouds circumstances vary in areas, so the maximum/minimum trends fluctuated with height increasing in month 2 and 3. Max/minimum NDVI increased with height increase under 1500/1200 m area in April, while decreased above 1500/1200 m. Max/minimum NDVI increased with height increase under 2000 m area, while NDVI presents essentially the same, slightly decrease or slightly increased three trends above 2000 m area in month 5–9. Max NDVI increased under 2000 m area in October, while decreased with height increasing above 2000 m. Minimum NDVI trends are basically the same as the maximum, but at 400 m and below 400–800 m, decreases 0.2 with increasing altitude. Maximum/minimum of NDVI increased under 1500 m/2000 m in month s 11 and 12, while decreased with height increasing in areas above 1500 m/2000 m. Table 2 shows range of monthly NDVI in height.

Table 1. Monthly NDVI average range on each height district in 2010 s-2013 s

Months	≦400 m	400-800 m	800-1200 m	1200-1500 m	1500-2000 m	>2000 m
Jan	19-43.3	36.1-54.1	47.3-58.1	42.3-56.6	42.2-60.2	31.6-56
Feb	40.4-53.8	51.2-58.4	42.7-59.3	35.8-56.8	31.6-61.3	32-63.4
Mar	28.2-54.7	31.1-59	32.4-59.4	33.2-49	35.9-59.8	34-54.9
Apr	39.4-60.1	44-66.8	47.3-71.2	46.9-72.9	46.7-71.9	44.4-62.1
May	55.8-69.5	56.7-73.8	63.4-81.5	68.6-81.8	71.3-84.5	69.3-84.5
Jun	56.5-70.5	58.1-75.8	60.9-79.1	64.6-83.7	68.1-87.2	68.6-87.9
July	65.3-77	64.8-79.9	68-85.6	69.6-85.6	70.4-86.9	68.4-86.5
Aug	64.8-70.2	64.8-79.9	68.9-84.3	72.7-87.4	75.5-89.2	76.5-88.2
Sep	47.7-69.7	55.5-78.9	64.5-82.3	68.8-84.8	71.7-86.7	71.9-87.5
Oct	54.5-64.9	54.3-73.6	63.3-78.1	64-80.7	65.7-80.9	64.2-75.6
Nov	41.3-58.8	44.4-65.3	52.2-65.6	56.4-72.3	59.4-70.7	55.7-63.6
Dec	41.3-49.5	44.4-60.5	52.2-64.8	57.7-64.9	59-64.1	54.2-57.7

Table 2 shows the monthly average NDVI trend was more single with slope increasing, maximum/minimum NDVI increased with increasing slope in 1–12 months.

Table 2. Monthly NDVI average range on each slope district in 2010 s-2013 s

Months	≦5°	5-15°	15-25°	25-35°	>35°
Jan	22.4-46	38.1-54.6	44.2-57	47.4-57.5	51.2-59.2
Feb	44.6-54.6	49.6-57.9	44.1-59.6	37-60.2	35.8-62.1
Mar	33.9-55.1	35.2-58.7	37.6-59.7	39.5-60	41.7-61.5
Apr	38.5-60.7	45-58.1	48.2-71	50.1-72.8	52.4-73.8
May	54.5-60.7	60.3-75.5	64.8-78.8	68.9-78.9	69.9-82.9
Jun	55.9-72.3	59.6-77.3	63-80.5	66.5-83.7	67.2-84.4
July	65.1-76.9	67.2-80.7	68.7-82.9	70.2-83.7	70.3-84.4
Aug	63.7-75.5	67.4-80.9	70.4-83.2	72.5-85.6	72.2-86.2
Sep	60.1-73.6	59.2-79.8	65.5-82.4	68.7-84.6	68.3-84.9
Oct	51.9-66	58.2-72.8	62-76	64.3-78.3	65-79.5
Nov	40.1-60.2	48.1-67.1	53.6-68.9	58.1-69.7	60.3-69.4
Dec	45.8-50.6	52.5-60.6	55.3-63	56.3-62.8	57.6-62.4

Seen from Fig. 1 left, in 6–9 months, NDVI present significantly different at different heights, and NDVI continued to increase with height increasing. In months 1–5 and 10–12, the trends of NDVI were complex. NDVI increased first and then decreased with

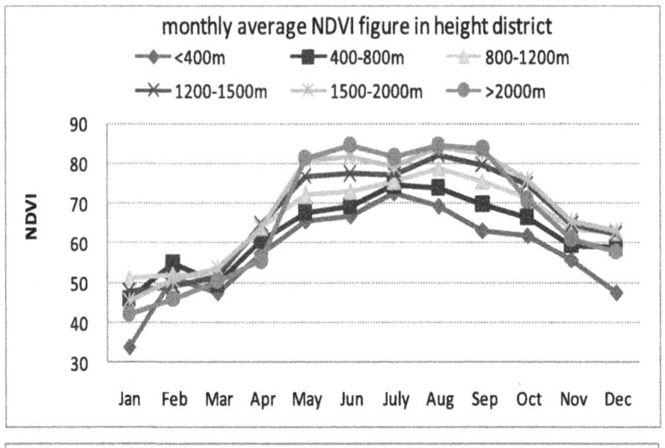

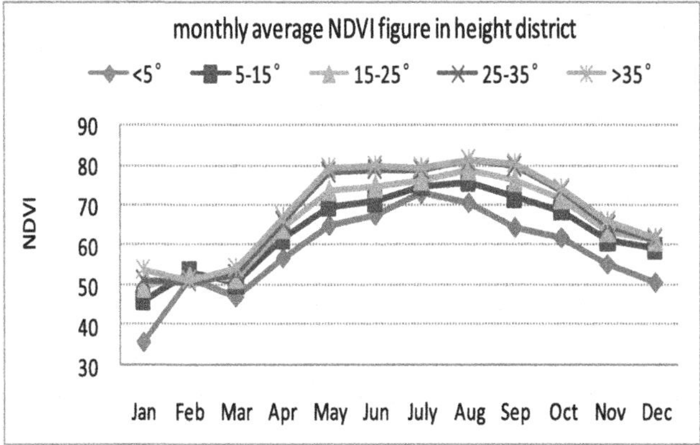

Fig. 1. Monthly average NDVI figure of height/slop district in 2010 s-2013 s

height increasing in months 1, 3–5 and 10–12 which turning height appeared in 1200 m in month 1 and in 2000 m in months 3–5 and 10–12, then decreased with height increasing above the turning height.

Seen from Fig. 1 right, except to February, NDVI trend clearly increased with slope increasing.

3.2 Analysis on Seasonal Average NDVI in Area Classed with Height and Slope

Figure 2 shows the seasonal average NDVI graph in height/slope area respectively.

Seen from Fig. 2 left, NDVI in spring and autumn increased firstly and then decreased with height increasing which turning point was 2000 m. NDVI increased continuously in summer, was complex in winter, with increasing trend below 1200 m and then showing fluctuations above 1200 m. Seen from Fig. 2 right, NDVI in all season increased monotonically with slope increasing.

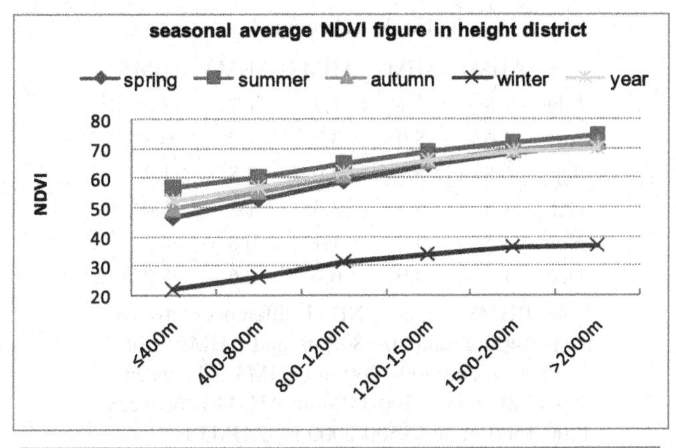

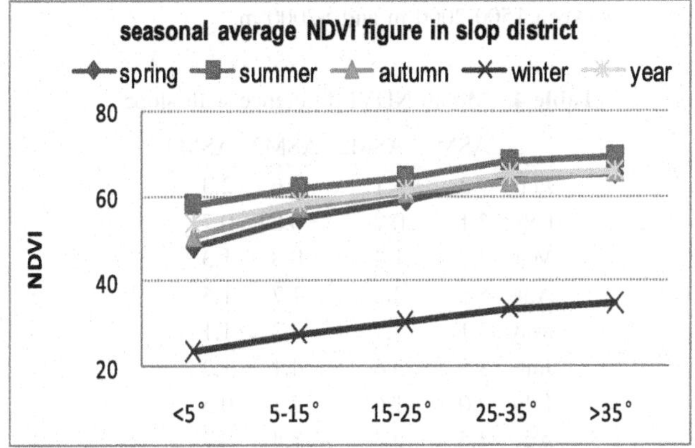

Fig. 2. Seasonal average NDVI figure of height/slop district in 2010 s-2013 s

3.3 The Relationship Analysis Between NDVI and Terrain

By the foregoing and articles [15] analysis, found NDVI and height/slope have goof correlation and regularity.

NDVI difference in adjacent area decreased with height/slope increasing. Tables 3 and 4 are month NDVI difference with height and slope, and Tables 5 and 6 are season NDVI difference with height and slope.

Table 3. Month NDVI difference with height

	AHM1	AHM2	AHM3	AHM4	AHM5
Jan	12.2	5.3	−3.6	−2.2	−3.4
Feb	3.4	−2.8	−2.7	1.4	−4.8
Mar	2.3	2.0	−0.3	2.2	−3.3
Apr	3.5	3.2	0.7	−1.0	−8.2
May	2.0	4.5	4.8	4.2	0.4
Jun	2.6	3.6	4.3	4.5	2.7

(Continued)

Table 3. (*Continued*)

	AHM1	AHM2	AHM3	AHM4	AHM5
July	1.8	1.1	1.7	1.7	2.5
Aug	4.6	5.0	3.3	1.9	0.3
Sep	6.8	5.7	4.0	2.8	1.2
Oct	4.5	5.1	3.3	1.1	−5.0
Nov	3.5	4.2	1.1	0.8	−4.3
Dec	11.3	4.0	−0.4	0.6	−5.2

Note: RHM1 is month NDVI difference between areas ≦400 m and 400-800 m, and AHM2 is of 400-800 m and 800-1200 m, AHM3 is between 800-1200 m and 1200-1500 m, AHM4 is between 1200-1500 m and 1500-2000 m, AHM5 is between 1500-2000 m and >2000 m.

Table 4. Month NDVI difference with slope

	ASM1	ASM2	ASM3	ASM4
Jan	9.3	3.1	1.6	2.3
Feb	3.1	0.2	−0.8	0.7
Mar	2.7	1.4	0.8	1.3
Apr	5.5	2.6	1.7	1.3
May	5.1	4.1	4.2	1.1
Jun	3.5	3.6	4.1	1.2
July	2.0	1.6	2.1	0.5
Aug	4.5	2.8	2.4	0.2
Sep	6.4	3.8	3.2	0.5
Oct	6.1	3.3	2.4	0.6
Nov	6.4	3.1	2.3	1.7
Dec	7.8	2.3	0.3	0.7

Note: RSM1 is month NDVI difference between areas ≦5° and 5-15°, and ASM2 is of 5-15° and 15-25°, ASM3 is between 15-25° and 25-35°, ASM4 is between 25-35 and >30°.

Seen form Tables 3, 4, 5 and 6, in height area, the difference becomes negative at turning height in months 1–5, 10–12, and all season, while in months 6–9, the difference remains positive with height increasing. NDVI difference remains decreasing in all height area most of which became negative at turning height 2000 m. In slope area, the NDVI difference decreased and remained positive in all area.

In order to choose the best relationship and follow the law of vegetation distribution, binomial relationship was selected to express NDVI and height, most formulas passed the test of 0.001. NDVI and slope follow power relationship. Samples of NDVI and height are 6, NDVI and slope is 5. The correlation coefficient of $P_{\alpha=0.001}$ was 0.925

Table 5. Month NDVI difference with height

	AHS1	AHS2	AHS3	AHS4	AHS5
Spring	5.3	4.0	2.7	2.3	−3.3
Summer	4.9	2.3	1.9	1.4	−0.5
Autumn	7.5	4.9	3.3	2.7	−0.7
Winter	6.7	0.6	−1.4	0.6	−3.5

Note: RHS1 is season NDVI difference between areas ≦400 m and 400-800 m, and AHS2 is of 400-800 m and 800-1200 m, AHS3 is between 800-1200 m and 1200-1500 m, AHS4 is between 1200-1500 m and 1500-2000 m, AHS5 is between 1500-2000 m and >2000 m.

Table 6. Month NDVI difference with height

	ASS1	ASS2	ASS3	ASS4
Spring	6.0	3.4	3.2	1.3
Summer	3.6	2.0	2.0	0.3
Autumn	7.0	3.5	3.6	1.1
Winter	5.0	1.9	0.3	1.8

Note: RSS1 is season NDVI difference between areas ≦5°and 5-15°, and ASS2 is of 5-15° and 15-25°, ASS3 is between 15-25° and 25-35°, ASS4 is between 25-35 and >30°

Table 7. The formulas of NDVI and height/slop

Month	R1	Formulas of NDVI and height	R2	Formulas of NDVI and slope
Jan	0.9141○	$y = -0.00001\ h^2 + 0.032\ h + 28.96$	0.9960※	$y = 35.91\ s^{0.114}$
Feb	0.4845	$y = -0.000002\ h^2 + 0.005\ h + 48.83$	0.8597○	$y = 49.74\ s^{0.018}11$
Mar	0.9413※	$y = -0.0000035\ h^2 + 0.011\ h + 39.52$	0.9915※	$y = 43.98\ s^{0.039}$
Apr	0.9960※	$y = -0.000009\ h^2 + 0.022\ h + 46.61$	0.9990※	$y = 50.17\ s^{0.062}$
May	0.9850※	$y = -0.000003\ h^2 + 0.017\ h + 56.76$	0.9783※	$y = 59.07x^{0.069}$
Jun	0.9944※	$y = -0.0000011\ h^2 + 0.012\ h + 60.18$	0.9592※	$y = 61.49x^{0.057}$
July	0.9959※	$y = -0.000001\ h^2 + 0.005\ h + 69.44$	0.9618※	$y = 69.39x^{0.027}$

(*Continued*)

Table 7. (*Continued*)

Month	R1	Formulas of NDVI and height	R2	Formulas of NDVI and slope
Aug	0.9999※	$y = -0.000004\ h^2 + 0.018\ h + 63.14$	0.9920※	$y = 67.78x^{0.044}$
Sep	0.9997※	$y = -0.000005\ h^2 + 0.022\ h + 55.99$	0.9945※	$y = 60.75x^{0.066}$
Oct	0.9820※	$y = -0.0000072\ h^2 + 0.023\ h + 53.9$	0.9980※	$y = 58.49\ s^{0.062}$
Nov	0.9948※	$y = -0.0000069\ h^2 + 0.022\ h + 45.50$	0.9985※	$y = 49.16\ s^{0.074}$
Dec	0.9839※	$y = -0.000011\ h^2 + 0.033\ h + 37.61$	0.9793※	$y = 48.27\ s^{0.067}$

Note: R1 is correlation coefficient of height and NDVI, R2 is of slope and NDVI. ※ notes passed the test of 0.001, O notes passed the test of 0.01, no dark notes not passed test. Y represents NDVI, h height and s slope.

0.951when number of samples is 6 and 5 respectively, 0.834 and 0.935 when α is 0.01. (seeing Table 7).

Formulas of NDVI and height in months and season were built. February formula do not passed inspection because of NDVI fluctuating with height, January passed test of 0.01, formulas of rest month and all seasons have passed the test of 0.001. The formulas of NDVI and slope passed the test of 0.01 in February, and passed the test of 0.001 in rest of months.

The above analysis shows there were good the presence relationship between NDVI and height/slope, Figs. 3 and 4 gives graph of NDVI and height/slope.

Seen from Fig. 3, seasonal formula of NDVI and height is binomial relationship and correlation coefficients above 0.925 that passed test of 0.001. Formula showed height under a certain number, NDVI increased with height increasing, but the increasing trend decreased, then NDVI decreased with height increasing reaching the certain height. The certain turning height is affected by temperature of season which is high temperature the turning height number is big and low temperature turning height decrease.

Following Fig. 4, formula of NDVI and slope present power relationship, and NDVI increased with slope increasing. Because samples selected to fit relationship between NDVI and slope only reflects the monotonically increasing, power function is good now to describe NDVI and slope.

The NDVI trends in height/slope reflect not only the vegetation vertical zone, also influence from meteorological conditions, topographical conditions and human activities. NDVI increases rapidly with height increasing below 200 m /1500 m or 2°area. This mainly reflects land use affected by human activities. Most low lying areas has complex land cover type and is the main place of human activity, so that most areas is urban land, residential villages, agricultural and industrial land, thus the NDVI is low in there. In these low lying areas, rapid reduction of human activities result in rapid

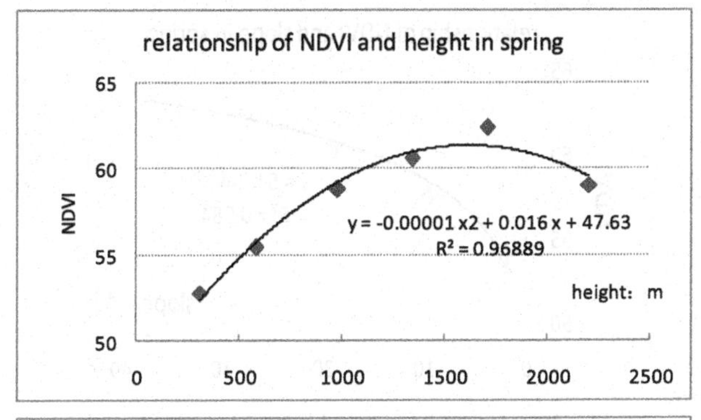

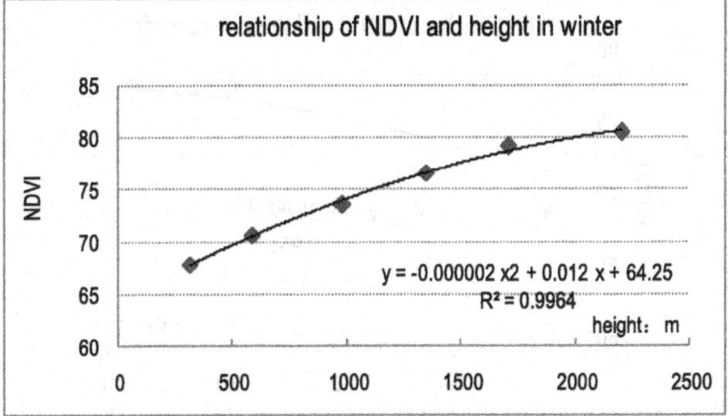

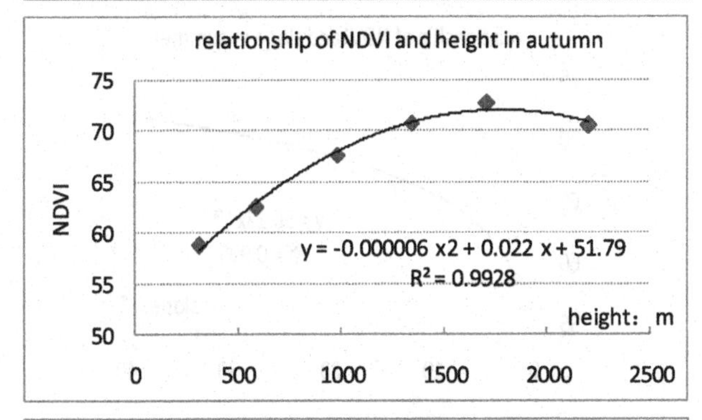

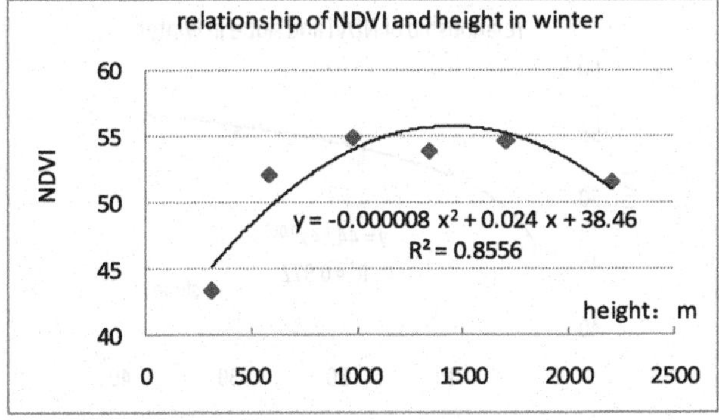

Fig. 3. The relationship figure between NDVI and height

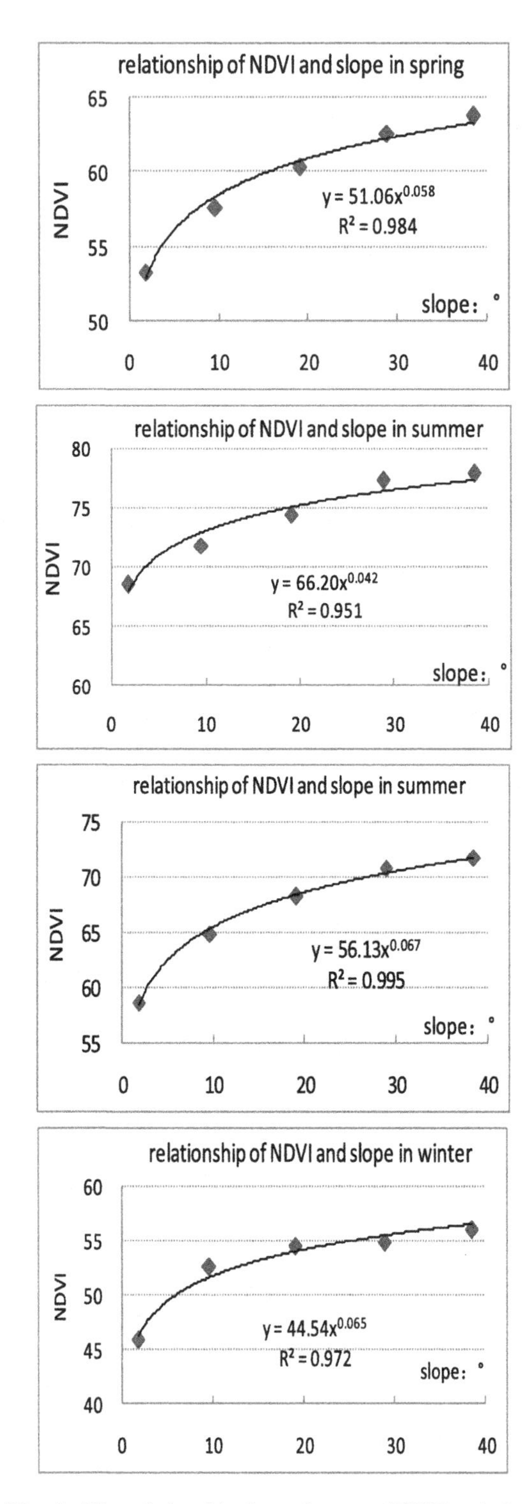

Fig. 4. The relationship figure between NDVI and slope

increasing of vegetation cover rate, which lead to significant increase of NDVI with height/slope increasing. In most areas between 1200 m/1500 m and 1500 m/2000 m or above 20°, land use proportion of human activities quickly reduce, that duce to vegetation cover affected more less by human factors. While meteorological and topographical conditions affection on vegetation cover is greater, NDVI increasing rate reduce with height/slope increasing. With height/slope increasing, natural conditions are more unfavorable for vegetation growing, when height above 1500 m/2000 m, vegetation types change with environment changes. Although more green area, NDVI is still smaller or even decline because vegetation leaf characters changed by monitoring results changed.

4 Summary

The relationship between NDVI and height/slope is significantly affected by month and season. NDVI in 6–9 months and summer continuously increase with height/slope increasing, in months 1–5, 10–12 and spring, autumn, winter, NDVI trends was complex which increase firstly then decrease. Turning height relates to temperature of month and season. Value of height is bigger when temperature high, is smaller when temperature low, and the biggest height is 2000 m around. Relationship between NDVI and slope is clear and single, NDVI increase with slope increasing.

Formulas between NDVI and height/slope show that NDVI and height follow good binomial relations, most of which passed 0.001 test only of Jan passed test of 0.01 and of Feb did not pass inspection. The relationship between NDVI and slope follow power function, in addition to the Feb passing by 0.01, the rest month and season passed the test of 0.001.

NDVI trend with height/slope and their relationship show NDVI is affected by multiple factors. In low lying or small slope areas, mainly influenced by human activity, in high or steep areas, mainly affected by natural factors that is affected by month and season.

The NDVI trends with height is related to season, which are supported in theory, but this results still need a detailed study that is supported by temperature and NDVI trend in height, so that further improve the support theory.

Acknowledgments. This paper was supported by famers fund projects from Ministry of Science and Technology [2013GB24160637] and Key found of Chong Meteorological Bureau [ywgg-201411].

References

1. Chen, J.J., Yi, S.H., Qin, Y., et al.: Responses of Alpine grassland landscape in the source region of Shule River Basin to topographical factors and frozen ground type. Chin. J. Appl. Ecol. **25**(6), 1599–1606 (2014)

2. Weng, N.Y., Liu, K., Wang, J.W.: A study of relationship between spatial vegetation pattern and terrain factors based on GIS techniques. Bull. Soil Water Conserv. **34**(1), 232–236 (2014)
3. Tong, X.W., Wang, K.L., Yue, Y.M., Liao, C.J., et al.: Trends in vegetation and their responses to climate and topography in northwest Guangxi. Acta Ecologica Sinica **34**(12), 3425–3434 (2014)
4. Xu, Q., Ren, Z.Y., Yang, R.: The spatial and temporal dynamics of NDVI and its relation with climatic factors in Loess Plateau. J. Shaanxi Normal University **40**(1), 82–87 (2012). Natural Science Edition
5. Cai, H., He, Z.W., An, Y.L., et al.: Correlation Intensity Of Vegetation Coverage And Topographic Factors in ChiShui Watershed Based on RS And GIS. Earth Environ. **42**(4), 518–524 (2014)
6. Liu, Y., Fu, B.J.: Topographical variation of vegetation cover evolution and the impact of land use/cover chang in the Loess Plateau. Arid Land Geography **36**(6), 1097–1102 (2013)
7. Fan, Y.Q., Zhou, G.M., Shi, Y.J., et al.: Effects of terrain on stand structure and vegetation carbon storage of Phyllostachys edulis Forest. Scientia Silvae Sinicae **49**(11), 177–183 (2013)
8. Zhang, M.W., Zhou, Q.B., Chen, Z.X., et al.: Supervision and testing on the growth of winter wheat grown based on MODIS EVI time sequence. Chin. J. Agricultural Resour. Regional Plann. **28**(2), 29–33 (2007)
9. Jiang, D., Wang, N.H., Yang, X.H., et al.: Principles of the interaction between NDVI profile and the growing situation of crops. ACTA Ecoloica Sinica **22**(2), 247–252 (2002)
10. Wang, R., Xia, W.T., Liang, T.G., et al.: Spatial and temporal dynamic changes of net primary product based on MODIS vegetation index in Gannan grassland. ACTA Pratacuiturae Sinica **19**(1), 201–210 (2010)
11. Huang, K., Liu, Z., Yang, L.F.: Evaluation of winter wheat productivity in Huang-Huai—Hai region by multi—year graded MODIS-NDVI. Trans. Chin. Soc. Agric. Eng. (Trans. CSAE) **30**(2), 153–161 (2014)
12. Gu.J, Li.X, Huang. C.C.L. Land Cover Classification Based on Time-series MODIS
13. NDVIData in Heihe River Basin. Advances in Earth Science, 25(3), 317–326 (2010)
14. Yao, C., Huang, W., Li, X.H.: Evaluation of topographical influence on vegetation indices of rugged terrain. Remote Sens. Technol. Appl. **24**(4), 496–502 (2009)
15. Jiang, H., Mao, Z.Y., Wang, X.Q.: A topography-adjusted vegetation index (TAVI) and its application in dynamic forest monitoring. J. Bering Forestry Univ. **33**(5), 8–12 (2011)
16. Chen, Y.Y., Tang, Y.H., Zhang, J.P., et al.: Response of vegetation index based on MODIS to topographic factors in Chongqing. Chin. J. Agrometeorology **33**(4), 587–594 (2012)
17. Sun, F., Xie, S.Y., Li, Z.Q.: Development of the characteristic agriculture in the context of topographic and geomorphologic diversity in Chongqing. J. Chongqing Three Gorges Univ. **27**(132), 120–122 (2011)

Real-Time Land Information Survey System Based on GPS/GIS/PDA

Cuiying Zhang(✉), Xuexiang Yu, and Xingwang Zhao

School of Geodesy and Geomatics, Anhui University of Science and Technology, Huainan 232001, China
vicky8453@126.com

Abstract. With the accelerating process of urbanization in China, the demand for urban land is increasing rapidly; in some city with a large population, shortage of land, intensive land use is the core problem of sustainable development in the future. According to the difficulties of land surveying, recording, a land survey terminal system is developed by integrating GPS, GIS, GPRS and PDA, providing effective data for land use evaluation and planning. The results of application experiment in a certain regional of Huainan city has shown that it can high-precision, real-time collect the land attribute and spatial data and share the data with land evaluation service. The land survey terminal system realizes the automation and networking and it effectively solves the existing problems of land survey in field.

Keywords: GPS/GIS/PDA integration · Land survey · Data communication · Working map

1 Introduction

In the process of urbanization in our country, the phenomenon of land extensive use is very serious, which results in imbalance between supply and demand of urban land. It is become a bottleneck problem of the sustainable development of city of our country. To intensive use of urban land resources, it is necessary to clearly understand the present situation of the current urban land use, analyzes the reasons of extensive use, and making land evaluation. There are two main methods to acquire land information, one is making accurate judgments for project related information such as area, and the other is land survey, to obtain the ownership of the land information, and so on other information. Therefore, experts and scholars in the field of urban land intensive use have carried out a series of relevant researches, it can be seen that survey methods such as the remote sensing (RS), Global Navigation Satellite System (GNSS), total station and drawing in the field [1–5]. And GNSS technology is used in the related research mostly. Using remote sensing to obtain and update land information has some limitations, such as the accuracy is low, versatility not strong, and so on [1].

While, the pattern of total station, drawing in the field, has some drawbacks, such as increasing the difficult because of lack control points, great work because of field survey content variety. And the most critical is the separation of spatial information and attribute information, which is not conducive to the indoor processing and data

F. Bian and Y. Xie (Eds.): GRMSE 2015, CCIS 569, pp. 267–276, 2016.
DOI: 10.1007/978-3-662-49155-3_27

management [6]. So, a kind of portable data acquisition terminal system is presented in this paper. The terminal system use GPS technology as a means of survey method, and integrating GPS, GIS and GPRS in PDA terminal. In field survey, we build the topological relationship between attribute information and geographic entity based on survey data in terminal system. Finally, the survey data information will be real-time sending to land evaluation service center by used of GPRS wireless transmission technology, providing useful data for land evaluation and planning.

2 Land Survey System Design

Land information acquisition terminal system for intensive use is an important component of land resources intensive use evaluation system. It can achieve fast acquisition, editing, mapping, data transmission of land use information, providing timely, accurate and detailed land information for urban land use management and evaluation. In order to meet much of land use information real-time collection and transmission, we developed an embedded land use survey system by integrating GPS/GIS/GPRS based on VC#.NET and windows CE operation system, that uses PDA as for mobile platforms to collection land information by GPS technology. The system includes four modules, namely data collection, data editing analysis, data transmission and measurement instruments integrated.

2.1 Data Collection and Storage Module Design

The system wills real-time collect the attribute data, spatial data and topological information. Among them, the attribute data including land classification, land owners, administrative, cadastral number and other information; the spatial data mainly including three-dimensional coordinates(X,Y,Z), surveying by GPS-RTK. Topological information is mainly the relationship between the points position, which used to represent geographic entity. During field data collection, we first open the working map of surveyed area, then connect the measuring equipment(GPS receiver), and send relevant instructions to the GPS receiver based on PDA terminal to control it to automatic collect data and transport to terminal system. The data is stored in the PDA which used to map editing, etc. Finally, we entry the point attribute information.

Land use information is stored in form of a SQL Server Express, including fundamental geographic elements, land use elements, land ownership elements, and grid elements, etc. In order to efficient, we need to classify the land. According to the classification and coding principles, the land use information is divided into specialty class,service class, first class, second class, third class and four class, the code of which is made up of ten digits. And its structure is as follows (See Table 1).

Where, XX represents the two digitals, X represents the one digital. Specialty class code is set to two digitals, such as the code of fundamental geographic elements is ‘10’, etc. service class is set to two digitals, space filled with ‘0’. For example, land use business code of ‘01’. The codes of first class and second class are both set to two digitals, the third class and four class are one digital. For the fundamental geographic

Table 1. Database structure of land use information.

Elements code	Elements category
XX	specialty class
XX	service class
XX	first class
XX	second class
X	third class
X	four class

elements, it is coded based on "Classification and Code of Fundamental Geographic Information (GB/T 13923-2006)", then, the land use information database is built up. Such as, the code of '310000'and '320000'are stand for inhabitant area, industrial and mining facilities respectively.

2.2 Data Editing and Analysis

Data management of different types land is on the PDA, aiming at achieving a series of operations including drawing, editing and deletion, which mainly reflects in two aspects: (1) the investigators can record land information using field editing module in the survey stage. The boundary point coordinate captured from GPS is used to plotting in the working map, by the editing module to represent the topological relationships between points; the land data is stratified on the working map with different colours to meet categorized management requirements. User can set layer attributes for each points and control each layer is or not displayed in the map. (2) Using the data management module can edit the attributes of boundary point in field survey, such as land owners, administrative, cadastral number and other information. This module also includes saving the point information to SQL database and dynamically query, update database, etc.

2.3 Data Communication Module

The data acquisition terminal system of land intensive use can high-speedily transmit land information to land evaluation service center. The realization of information transmission is through the internetwork. So we can use the GPRS function of SIM card provided by the PDA terminal to access the internet, then a connection to land evaluation service center is established.

Meanwhile, land evaluation service center need to connect the internet via network cable, sharing the land use information. The system need to share data with project information, boundary point attributes, boundary line attributes, registration information, etc. The detailed steps are as follows: (1) setting the rover's IP address and port for connecting to land evaluation service center; (2) land evaluation service center responds and ask for the user ID and password; (3) rover sends the user ID and password to log in the service; (4) if it is valid, the rover will real-time transmit the land information to the land evaluation service center; (5) if land evaluation service center has some requirements, it can also send the relevant data or instruction that control carry out.

In the network programming, the land evaluation service center is treated as the server, mobile terminal is a client. According to the requirements and characteristics of the project, the client needs to establish a connection using to send and receive messages form server. And the server can communicate simultaneously with multiple clients.

To achieve the above functions, the system must meet the following conditions:

(1) the server must be able to establish contact with multiple clients; (2) the server must be able to asynchronous read data from a client and can send the message to the client at any time; (3) the client must be able to asynchronous read message from the server and can send message to the server at any time.

The process of intensive use of land acquisition terminal system is shown in Fig. 1.

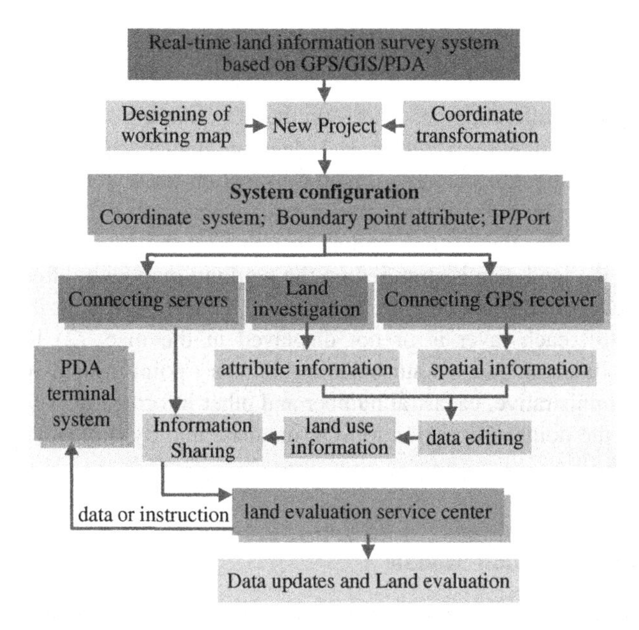

Fig. 1. The flow chart of intensive use of land acquisition terminal system

3 Key Technologies of Terminal System

3.1 GPS Positioning and Location

The coordinate of boundary point is obtained from GPS receiver via access to the Continuous Operational Reference System (CORS). In the system, PDA is used to control the operations of GPS receiver. For example, we can send an instruction to GPS receiver on the PDA, which obtains the survey results by GPS receiver.

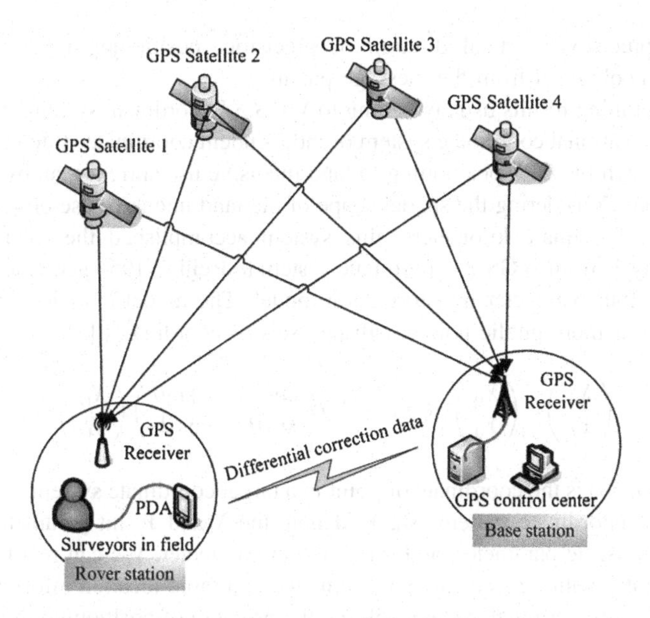

Fig. 2. GPS-RTK survey mode for land data survey

GPS-RTK survey mode is as show in Fig. 2. PDA with integrated GPS is through Bluetooth and serial cable; PDA operating system (such as windows mobile system) provides the corresponding I/O drivers and communication module. So the operation, such as serial port configuration, sending, receiving, and serial closure, can be implemented by calling the module on the PDA.

NMEA 0183 is a combined electrical and data specification for communication between marine electronics such as echo sounder, sonars, GPS receivers and many other types of instruments. It has been defined by, and is controlled by, the National Marine Electronics Association. Based on the NMEA protocol, this effectively transfers all GPS information from the receiver to the PC computer or PDA, and other equipment, including Global Positioning System Fix Data ($GPGGA), GPS Satellites in View ($GPGSV), GPS DOP and Active Satellites ($GPGSA), etc.

As an example, a Global Positioning System Fix Data has the form: $GPGGA, <1>, <2>, <3>, <4>, <5>, <6>, <7>, <8>, <9>,M, <11> ,<12> *hh.

Where: $GPGGA is the sentence Identifier, ',' is the field separator, <l> refers to the UTC time, <2> refers to the latitude, <3> refers to the either character N or character S, <4> refers to the longitude, <5> refers to the either character E or character W, <6> refers to the fix quality, <7> refers to number of Satellites, <8> refers to the horizontal dilution of precision, <9> refers to the altitude, <10> refers to the height of geoid above WGS84 ellipsoid, <11> refers to time since last DGPS update, <12> refers to the DGPS reference station id, *hh is the checksum.

A set of sentences transmitted by PDA in response to received messages, and storage in buffer memory. For ensuring the reliability of GPS data, we use a dynamic data chain table to maintain the received GPS data by identification and judgments the integrity of the received message. Then the UTC time, station position, horizontal dilution of precision, vertical dilution of precision, positioning mode, and other information, is obtained from the message parsing.

GPS positioning results usually belong to WGS-84 coordinate system, but land use planning is in National coordinate system or independent coordinate system. Therefore, we need to match the GPS positioning to land intensive use current map by coordinate transformation. Considering the spatial scope of the land intensive use project is small, commonly in 10 square kilometers, this system accomplished the calculation and accuracy analysis from WGS-84 coordinate system to Beijing 1954 geodetic coordinate system using four-parameter transformation model. The model is as follow, which is based on two or more public points with two sets of coordinates [8].

$$\begin{pmatrix} X_T \\ Y_T \end{pmatrix} = \begin{pmatrix} X_0 \\ Y_0 \end{pmatrix} + (1 + \mathrm{K}) \begin{pmatrix} \cos\theta & -\sin\theta \\ \sin\theta & \cos\theta \end{pmatrix} \begin{pmatrix} X_G \\ Y_G \end{pmatrix} \quad (1)$$

Where (X_T, Y_T) is the coordinate of point in a target coordinate system, and (X_G, Y_G) is in WGS-84 coordinate system; X_0, Y_0 denote the X and Y shit parameters, respectively; K is the scale parameter, and θ is rotation parameter. We can real-time get the point coordinates with a precision of 1–3 cm, and can rapid location information using the data acquisition terminal system, solving the problem of positioning difficult in the field.

After receiving the real-time location information, we will display it on the working map based on dynamic refresh technology to update the map information. At the same time, establishing the GIS topology relationship based on the recorded point classification, land owners, cadastral number and other information. Through the above operation, we can realize the seamless combination of GPS and GIS, which benefit of survey and management of spatial information and attribute information.

3.2 Designing of Working Map

In order to realize the seamless combination of spatial information and attribute information, we adopt the ESRI shape format map as an embedded working map. The land current map-spot is stored with ESRI shape format in data acquisition. But intensive use map of the land planning is usually edited and processed based on AutoCAD or CASS software, so it results that the shape format map cannot be directly imported to this kind of software. If you want to make full use of existing map data to record the attribute information and update the map, you must convert the AutoCAD map into GIS format in the terminal system. In the process of map transformation, it is necessary to follow these rules [6]: (1) Retaining the original layer manner. If a layer includes various types of geometry features, such as point, line, and polygons, we will further subdivide them according to geometry type. (2) Convert the ground objects into the layer of shape format map, and allow setting it display or not. (3) Importing the attribute data from other data sources into attribute table of shape format map.

The designing of working map is a precondition of surveying, the contents of which includes main ground object and landform in the land survey area. Then we can store and update the point information on the working map, which is beneficial to improve the working efficiency.

3.3 Expression of Land Data

In order to show the real ground object, we can build the map-cells based on the discrete spatial location information from the GPS, which is consists of geographic entities and layer object. The geographic entity object includes graphic data and characteristics data, the former includes points, line and area, the later includes coordinate, color, filling, etc. We can develop it according to the geometric meaning of dot, line, and area to show the real roads, houses, forest, and so on. And the layer object is a set of geographic entities with the same property fields. So we can save the common information of geographic entities to the same layer, such as layer color, land category, avoiding repeated store [7]. For example, we get the coordinate of feature point from GPS-RTK, and then activate the menu 'boundary point survey' to generate the map spot. Finally, input the boundary point serial, type and other information, saving the map spot to GIS map-cells. Then we can browse, attribute edit the graphic in GIS environment.

4 Application Experiment

To test the system performance, we take a certain regional in Huainan city as the investigating area; we survey the land use status which has been changed using the data acquisition terminal system. The detailed operating processes are as following: (1) convert the AutoCAD map into GIS shape format, and import it into your PDA as a working map; (2) setting the system parameters, such IP address and serial port to connect the land evaluation service center; (3) connecting the GPS receiver by Bluetooth, when the GPS solution with ambiguity fixing, we begin to survey the location information. The survey results of boundary point of No.023312301 are shown in Figs. 3 and 4. Like that, we get the remaining boundary points information in turn until it forms a closure of map spot.

(4) transmitting the land use information to the land evaluation service center used to land evaluation and planning. Figure 5(a) shows that how we select the data category from the drop-down list. Then click the button 'Loading information', you can to transmit a single or all boundary points to service center, as shown in Fig. 5(b).

Figure 6 shows the project information, boundary points coordinate and the topological relationship among them received from terminal system. On the one hand, the received data is stored to the SQL server; on the other hand, it uses to update the map spot of the changed area in real time. The updated map spot is marked with a blue box in the Fig. 7(b).

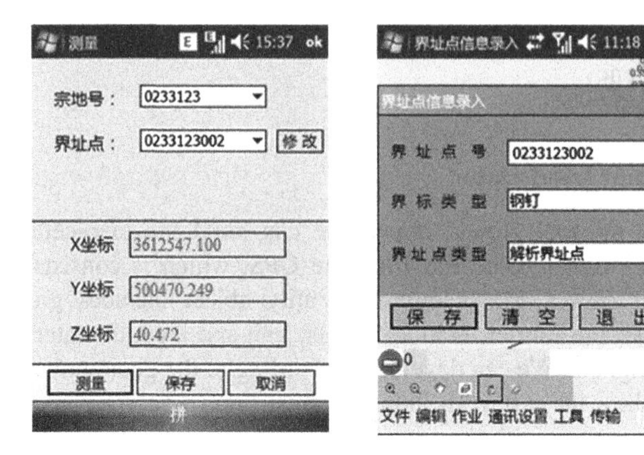

Fig. 3. Boundary point survey

Fig. 4. Information input for boundary point

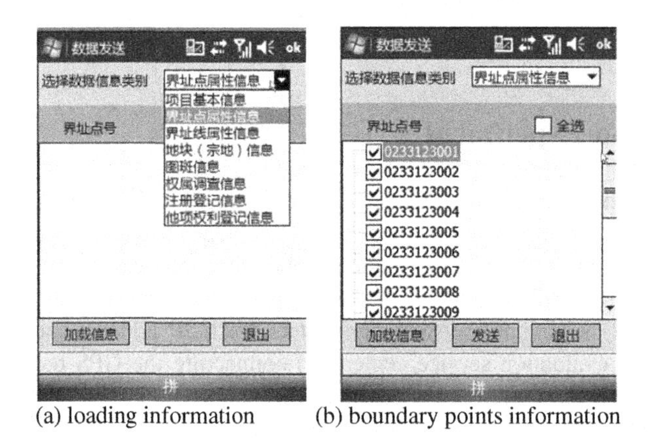

(a) loading information (b) boundary points information

Fig. 5. Data transmission

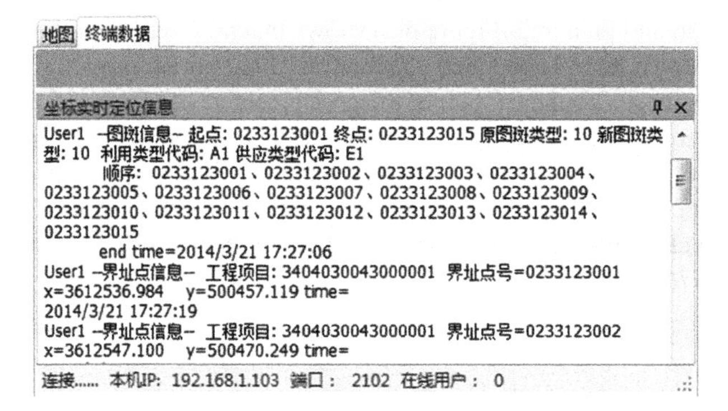

Fig. 6. Data receiving from the PDA terminal system

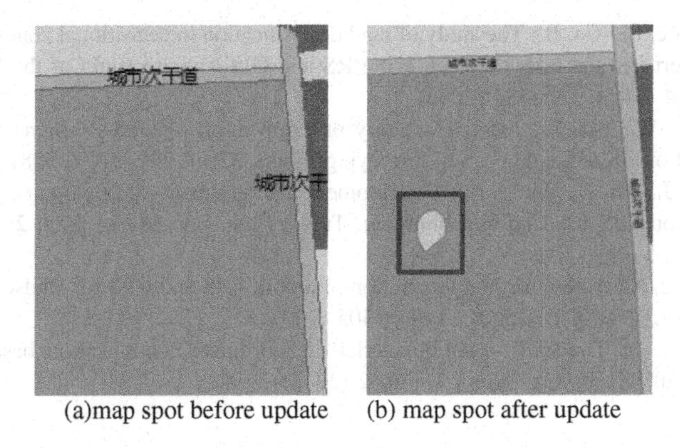

(a)map spot before update (b) map spot after update

Fig. 7. Updating the map information on the server

5 Conclusion

Land survey will provide valuable basis data for land evaluation and planning. As a technology of land survey, GPS-RTK is playing an important role in data acquisition. And with the further development of information technology, automation, information, real-time and networking of land information acquisition is imperative. This paper introduces the key problems of GPS/GIS/PDA integration for embedded data acquisition system, and then a mobile data acquisition mode was proposed in mobile environment, which has realized the rapid data acquisition and real-time data sharing of land use information. It greatly reduces the labor intensity and complexity of indoor processing, improve the operation efficiency. Of course, the integration of multiple techniques is not mature, there are many theoretical and technical problems need to further study, such as the compatibility and stability, which is work in the future.

Acknowledgments. This work is partially supported by National Natural Science Foundation of China (No. 41474026) and Anhui Provincial Natural Science Foundation (No. 1308085QD71 and 1408085QD72) and Anhui province land and resources science and technology project under Grant (No. 2011-K-22).

References

1. Liu, Y., Pei, Z., Wu, Q., Guo, L., Zhao, H., Chen, X.: Land use/land cover classification based on multi-resolution remote sensing data. In: Computer and Computing Technologies in Agriculture V, pp. 340–350. Springer, Heidelberg (2012)
2. Zhao, M., Li, Q., Feng, J., Chen, L., Li, H.: Evaluation of the urban land intensive use and it's regional differences in Shaanxi province based on GIS. In: Bian, F., Xie, Y., Cui, X., Zeng, Y. (eds.) GRMSE 2013 Part II. CCIS, vol. 399, pp. 435–444. Springer, Heidelberg (2013)
3. Kelarestaghi, A., Jeloudar, Z.J.: Land use/cover change and driving force analyses in parts of northern Iran using RS and GIS techniques. Arab. J. Geosciences **4**, 401–411 (2011)

4. Zhang, L., Yue, L., Xia, B.: The study of land desertification in transitional zones between the MU US Desert and the Loess Plateau using RS and GIS—a case study of the Yulin region. Environ. Geol. **44**(5), 530–534 (2003)
5. Li, X., Yang, W., Ma, L., Jiang, H.: Study on application of GPS-PDA in land changing survey based on 3S technology. Sci. Surveying Mapp. **33**(6), 209–210 (2008)
6. Jia, W., Liu, J., Yu, L., Wang, M.: Development and application of field survey technology based GPS and GIS for land consolidation. Trans. Chin. Soc. Agric. Eng. **25**(5), 197–201 (2009)
7. Tian, G., Tong, X.: A new method integration of mobile GIS and GPS for landscaping survey. J. Tongji Univ. (Nat. Sci.) **35**(10), 1400–1405 (2007)
8. Dong, J., Yue, J.: The transformation model of coordinate centralization based on robust estimation. Bull. Surveying Mapp. **7**, 39–42 (2012)

Drought Monitoring and Analyzing by Remote Sensing Based on MODIS Data in Heilongjiang Province of 2009

Dongping Wu[1(✉)], Lixia Jiang[2], Jianwei Ma[3], and Yayong Sun[3]

[1] Heilongjiang Province Flood Control and Drought Relief Office, Harbin 150001, China
fxbzgb@163.com

[2] Heilongjiang Province Institute of Meteorological Science, Harbin 150001, China

[3] China Institute of Water Resources and Hydropower Research, Beijing 100038, China
mjw147258369@126.com, yifei12645@163.com

Abstract. Heilongjiang is China's first major grain-producing province, which is typical in dry farming. Drought is one of the major natural disasters in the agricultural production area. Soil moisture is an important index of drought monitoring. Remote sensing has advantages such as a wide range of observation, high instantaneity and low cost, which can be widely applied to soil moisture monitoring. In this paper, we analyze the relationship among soil moisture and NDVI (Normalized Difference Vegetation Index) and LST (Land Surface Temperature), and built remote sensing estimating model of soil moisture based on NDVI and LST. According to this model, we dynamically monitored the drought in Heilongjiang province from April to June in 2009 based on NDVI and LST product of MODIS data and the ground measured soil moisture. The monitoring results show that the drought in Heilongjiang Province gradually relieved at first stage, then worsened significantly and eventually reached the mitigation, and the development of drought exhibits good agreement with the practical situation. The research indicates that the soil moisture monitoring method based on satellite remote sensing and the ground measured soil moisture can be applied to large-scale drought monitoring in Heilongjiang province.

Keywords: MODIS · Soil moisture · Heilongjiang province · Drought

1 Introduction

Drought is one of the major natural disasters with high frequency and wide range of effect which has caused large losses in China; due to the vast and complex terrain, local or regional drought appears almost every year. According to statistics, in recent 10 years, drought disasters occur frequently in China, expanding the scope of impact and increasing losses significantly. Continuous and extreme drought occurs occasionally which poses a serious threat to food security, water security and ecological safety in China. It has become one of the main factors restricting the economic development of China [1].

Soil moisture is an important index of drought monitoring. The traditional method of soil moisture monitoring is based on the point measurement of stations, which can

F. Bian and Y. Xie (Eds.): GRMSE 2015, CCIS 569, pp. 277–285, 2016.
DOI: 10.1007/978-3-662-49155-3_28

only receive a small amount of data. Coupled with the human, material, financial and other constraints, it is difficult to quickly and timely get access to the information of soil moisture, making a wide range of drought monitoring and assessment lack of timeliness and representability. Drought monitoring by remote sensing method has such advantages as wide range of monitoring, high spatial resolution, real-time information collection and good business application, can make up the disadvantages of ground observing system such as high cost, low space coverage and measurement delays. It can provide timely and efficient decision support services for disaster reduction departments at all levels. With the development of satellite remote sensing technology, the practicability of drought monitoring model by remote sensing of practical is higher and higher, and remote sensing has become an important method.

In the late 1980 s, research on soil moisture monitoring by remote sensing got quickly and full development, involving ground remote sensing, aviation remote sensing and satellite remote sensing; using band from visible, (near, middle and far) infrared, and thermal infrared to microwave remote sensing, increasingly diversification of monitoring [2]; Main methods include thermal inertia method, vegetation index method, temperature-vegetation index method, evapotranspiration method based on energy balance and microwave remote sensing method. Microwave remote sensing method is considered to be the best way to monitor soil moisture, but it is easily influenced by surface roughness and vegetation. So the wide application is subject to certain restrictions. In addition, the accuracy needs to be improved [3].

The feature space method of NDVI-LST combines vegetation index and surface temperature. It has more specific biological and physical significance, a mature retrieval method of soil moisture monitoring by remote sensing. Based on this, the paper selects NDVI (Normalized Difference Vegetation Index) and LST (Land Surface Temperature) products of MODIS data, and combined with ground measurement data, based on characteristics space of NDVI-LST, the paper constructs a soil moisture retrieval model, and applying the model for dynamically monitoring drought in Heilongjiang province from April to June in 2009.

2 Research Area and Data

2.1 Research Area

Heilongjiang is located in the North-East with an area of about 460,000 km^2. The geographic range is 43° 25' ~ 53° 33' N, 121° 11' ~ 135° 05' E. Terrain is mainly hills and Plains. It is an important grain production base of China with arable land of about 160,000 km^2. The average annual precipitation of Heilongjiang province is 400 ~ 650 mm, and precipitation space distribution has large differences. The average annual precipitation line is roughly parallel with warp. Differences between east and west are significant while North-South differences are not obvious. The time distribution of precipitation presents obvious monsoon features. Summer precipitation accounts for 65 % of annual precipitation while winter precipitation only accounts for 5 %, and spring and autumn distribution account for 13 % and 17 % respectively. Average annual temperature is −4.0 ~ 5.0°C, gradually declining from Southeast to Northwest, with certain zonal features. Annual variation in temperature is unimodal distributed with minimum temperature in January and maximum temperature in July. Due to the heterogeneous distribution of rainfall, widespread drought in Heilongjiang province often appears.

2.2 Remote Sensing Images

The data of this paper comes from geospatial data cloud of Computer Network Information Center of Chinese Academy of Sciences (http://www.gscloud.cn/), including 10-days synthesis NDVI products (MODND1T) and 10-days synthesis LST products (MODLT1T) of China in 1 km resolution based on Terra-MODIS data. Among them, the NDVI product is the calculation of bi-directional surface reflectance through atmospheric correction algorithm. The reflectivity data has undergone the severe treatment on water, cloud, aerosol and cloud shadow mask. LST products are made from the infrared emissivity calculated by the split window algorithm through MODIS 31, 32 band, and after geographic correction, radiometric calibration, cloud mask, atmospheric temperature and water vapor contour correction. NDVI product is the maximum value of 10 days and LST product is the average value of 10 days.

According to the drought development in Heilongjiang province in 2009, the paper has collected NDVI and LST product of 9 periods from April to June 2009 (where the third 10-day period of May, the range of time of NDVI and LST product is from May 21, 2009 to May 31; other products are in the range of 10 days).

2.3 The Measured Data of Ground Moisture

Ground measurement of soil moisture is an important basis for determining soil moisture retrieval model. The paper collects observation data about soil moisture of 32 stations from April to June 2009 of Heilongjiang province (8th, 18th, 28th/month). According to the amount of soil moisture in each station, convert soil absolute moisture content to relative soil water content of 0-20 cm and get soil moisture data of 9 periods. Soil moisture distribution is shown in Fig. 1.

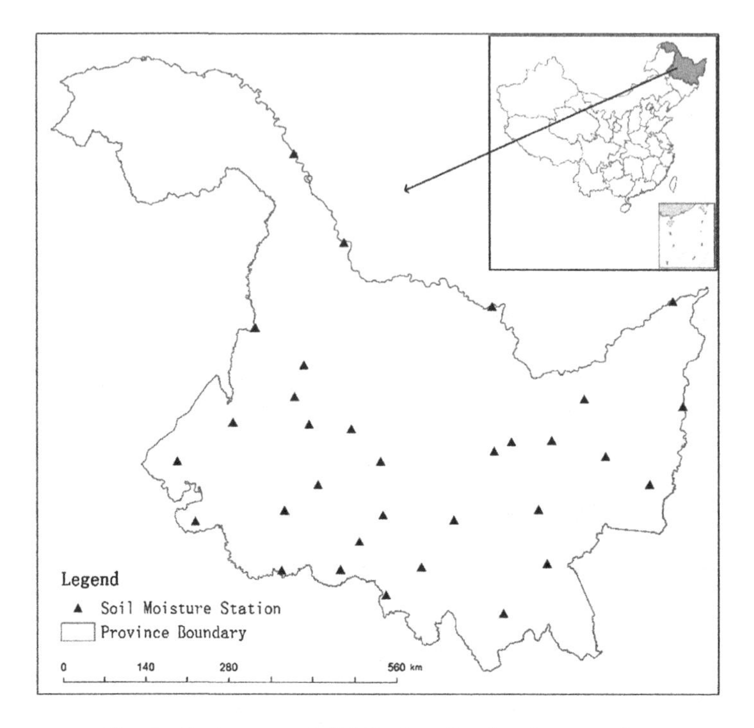

Fig. 1. Study area and distribution of soil moisture station

3 Soil Moisture Inversion and Drought Monitoring

Carlson et al. used a SVAT model to study the relationship between soil moisture, vegetation cover and surface temperature, and proved that soil moisture (SM) [4] can be estimated from LST and FVC):

$$SM = \sum_{i=0}^{i=2} \sum_{j=0}^{j=2} a_{ij} FVC^{(i)} LST^{*(j)} \tag{1}$$

In which a_{ij} is the polynomial coefficient, LST^{*} is the normalized surface temperatures,

$$LST^{*} = \frac{LST - LST_{o}}{LST_{s} - LST_{o}} \tag{2}$$

LST is surface temperature and LST_s the maximum surface temperature while $LST_o$is the minimum surface temperature. In the practice, vegetation cover is difficult to get, but has certain relations with the NDVI. Carlson et al. describes the relationship of "feature space" existing between soil moisture, NDVI and LST, as shown in Fig. 2 [4].

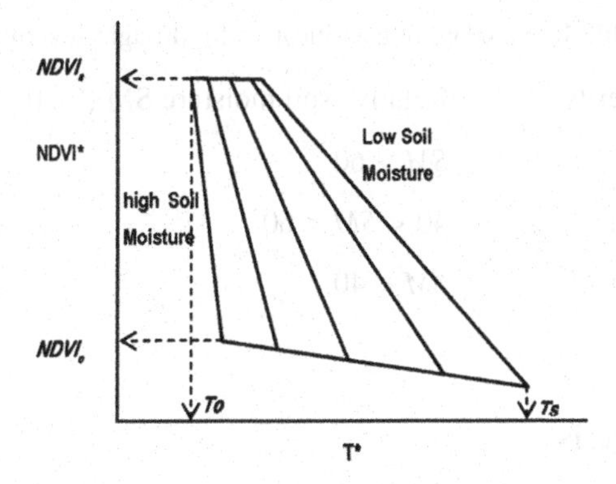

Fig. 2. NDVI-LST space

Lambin et al. from the perspective of evaporation, transpiration, vegetation coverage, more clearly illustrates and analyzes the feature space [5]. Based on feature space of NDVI-LST, during the research on soil moisture, Sandholt et al. discovered there are many contours in NDVI-LST characteristic spaces [6]. On this basis, using MODIS data and ground observation, the paper studies on inversion of soil moisture content in real time. Through a large number of experiments and analysis, based on normalized difference vegetation index and multiple regressions of surface temperature, we can better inverse soil moisture [7–10].

$$SM = \sum_{i=0}^{i=2} \sum_{j=0}^{j=2} a_{ij} NDVI^{*(i)} LST^{(j)} \tag{3}$$

In which, $NDVI^* = \frac{NDVI - NDVI_o}{NDVI_s - NDVI_o}$ normalized NDVI, $NDVI_s$ is the maximum NDVI, $NDVI_o$ is the minimum NDVI, *SM*is the average soil moisture of 0–20 cm.

Specific technical process includes the following steps:

(1) Model data collection: if corresponding time (day) has ground observation, calculate the distance between each ground site and each remote sensing pixel to look for the corresponding remote sensing pixel, and then integrate site code, time, corresponding vegetation index, surface temperature, and site soil moisture ion into an article record for documents.

(2) After the completion of model data collection, according to the formula (3), it sets up soil moisture retrieval model. Then according to the soil moisture retrieval model, the paper uses surface temperature and vegetation index products to inverse soil moisture distribution.

(3) On the basis, combined with land distribution and rating standard of drought severity in Heilongjiang province (as shown in Table 1), land drought classification and statistics is further conducted.

Table 1. Standard of classification for drought severity

Drought severity	Relative soil moisture SM (%) (0-20 cm)
Wet	$SM > 60$
Mild drought	$40 < SM < 60$
Severe drought	$SM < 40$

4 Results Analysis

Using the feature space method of NDVI-LST, we carried out inversion of a total of 9 periods of the soil moisture in Heilongjiang province from April 1 to June 30 in 2009, and classified drought severity according to land distribution and Table 1 (further divided into two types: wet and dry). Finally, we got drought map of arable land in Heilongjiang province from April to June of 2009, as shown in Fig. 3 (a)-(I). From drought monitoring inversion results of April 1 to April 10 in 2009 (Fig. 3 (a)), drought mainly occurred in the southwest; subsequently, drought area gradually reduced (Fig. 3 (b), (c)); into May 2009, drought in Heilongjiang province began to develop significantly, especially in late May 2009, most areas of Heilongjiang province appeared drought (Fig. 3 (f)), mainly in the southeast of Heilongjiang. After entering June, drought in Heilongjiang province began to ease. Influenced by rainfall, the drought-hit area had been significantly reduced by the end of June 10. Since then, drought-hit areas further reduced. From the monitoring results from June 20 to 30 of 2009, most of the arable land in the province was in the normal state.

In order to further analyze the development of drought in Heilongjiang province in 2009, the paper calculated the area of drought farmland of different periods (Fig. 4). From April 1, to April 10 of 2009, drought area of Heilongjiang province was 12592 km^2; from April 11 to April 30, drought-hit area decreased, respectively 2199 km^2 (April 11 to April 20) and 1411 km^2 (April 21 to April 30); While entering May 2009, drought suffered remarkable development, drought areas of the early and mid-May were 2037 km^2 and 9417 km^2 respectively; after the late May 2009, drought area was 46029 km^2, about 30 % of the crop land in the province; after entering June, drought area decreased; in late June 2009, the drought-hit area was 4464 km^2, 3 % of crop land in the province, at this time, the province's drought was basically eradicated. The whole drought development matches the actual development (the worst period of the drought in late May).

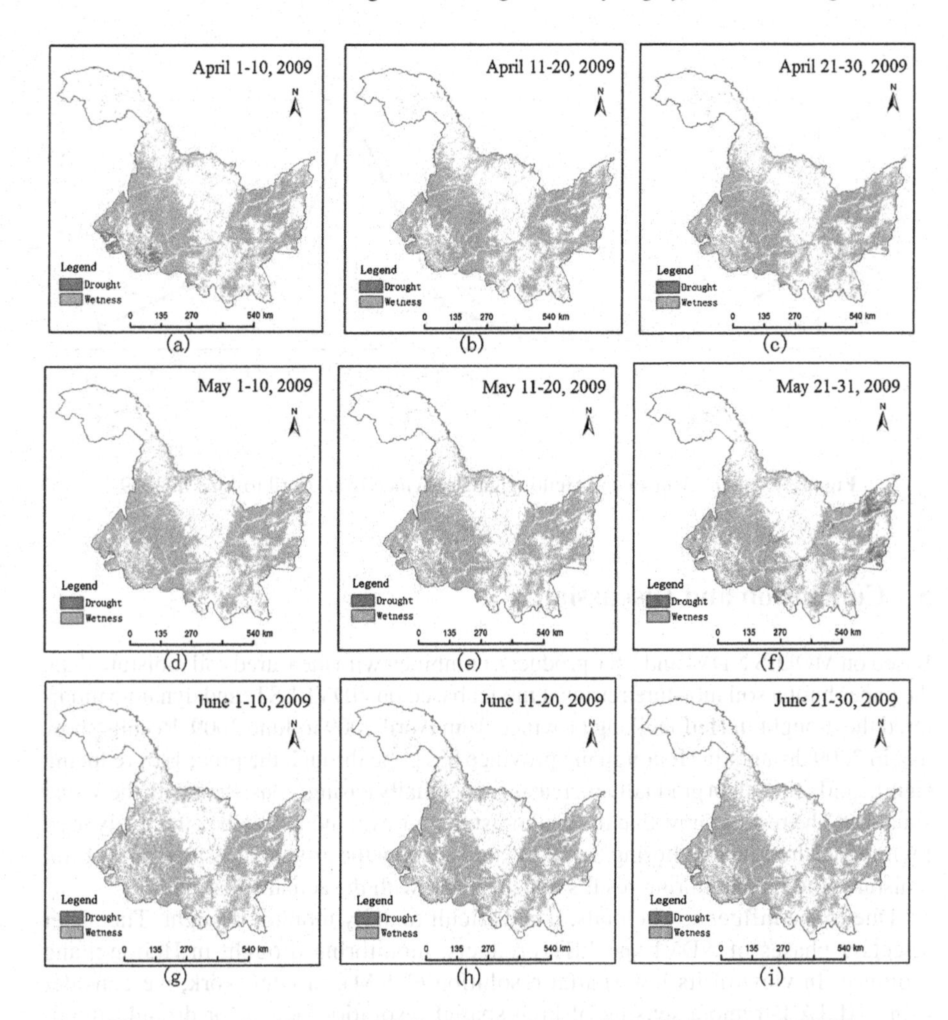

Fig. 3. Drought crop map in Heilongjiang province from April to June in 2009

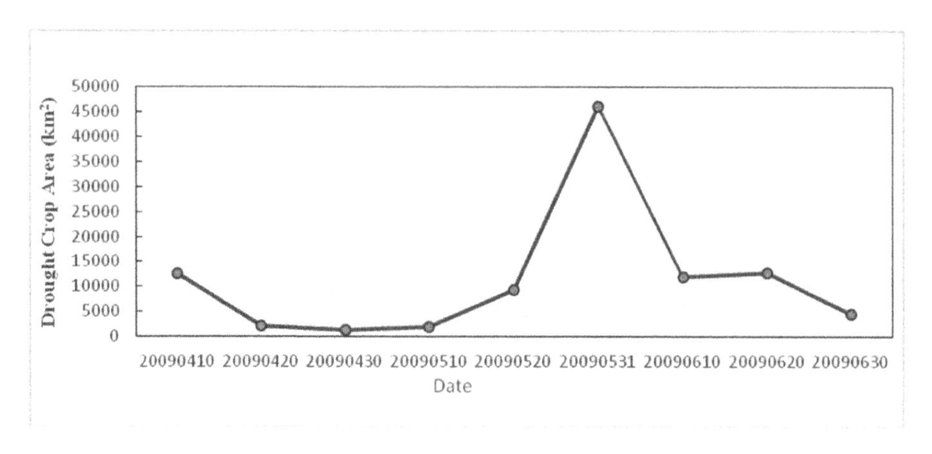

Fig. 4. Drought crop area in Heilongjiang province from April to June in 2009

5 Conclusion and Discussion

Based on MODIS NDVI and LST products, combined with measured soil moisture data, the paper built a soil moisture retrieval model based on NDVI-LST, and dynamic monitored the drought in Heilongjiang province from April 2009 to June 2009. Results show that in 2009 drought in Heilongjiang province has gone through the process of reducing significantly first, then gradually increasing, eventually easing, consistent with the actual situation. The results show that the soil moisture retrieval model can be effectively used for regional drought monitoring for taking into account the ground measured data of soil moisture, and its monitoring results matches well with the actual situation.

Due to the influence of clouds, it is difficult to daily monitor drought. The paper selects 9 phases of NDVI and LST products, monitoring drought in Heilongjiang province. In view of its low spatial resolution (1 KM), in later work, we consider using HJ-1A/B remote sensing of high spatial resolution image for drought monitoring by remote sensing. To further increase the frequency of monitoring, we should consider using radar data to compensate for the lack of optical remote sensing data. Radar has the all-day and all-weather imaging capabilities free from the effect of clouds and rains, ensuring access to remote sensing data. Based on optical and radar data fusion, the paper carries out the inversion of soil moisture, absorbs the advantages of data to establish a more effective drought monitoring model, thus providing more accurate and timely monitoring which is also another point for research.

Acknowledgements. The MODIS NDVI and LST data set is provided by International Scientific & Technical Data Mirror Site, Computer Network Information Center, Chinese Academy of Sciences. (http://www.gscloud.cn).

Fund Project: State 863 projects (2002AA2z4251)

References

1. Qi S.: Drought monitoring models with remote sensing and spatial-temporal characteristics of drought in China. In: Graduate University of Chinese Academy of Sciences(Institute of Remote Sensing Application Chinese Academy of Science) (2004) (In Chinese)
2. Zhiming, L., Bai, Z., Ming, Y., et al.: Some research advances and trends on soil moisture and drought monitoring buy remote sensing. Adv. Earth Sci. **4**, 576–583 (2003). (In Chinese)
3. Feng, G., Jiemin, W., Chengquan, S., et al.: Advances in study on microwave remote sensing of soil moisture. Remote Sens. Technol. Appl. **2**, 97–102 (2001). (In Chinese)
4. Carlson, T., Gillies, R., Perry, E.: A method to make use of thermal infrared temperature and NDVI measurements to infer surface soil water content and fractional vegetation cover. Remote Sens. Rev. **9**, 161–173 (1994)
5. Lambin, E.F., Ehrlich, D.: The surface temperature-vegetation index space for land cover and land-cover change analysis. Int. J. Remote Sens. **17**(3), 463–487 (1996)
6. Sandholt, I., Rasmussen, K., Andersen, J.: A simple interpretation of the surface temperature/vegetation index space for assessment of soil moisture status. Remote Sens. Environ. **79**, 213–224 (2002)
7. Wang, L., Qu, J.J., Zhang, S., et al.: Soil moisture retrieval using EOS MODIS and ground measurements in the eastern China. Int. J. Remote Sens. **28**(6), 1413–1418 (2007)
8. Hao, X.: Estimation of live fuel moisture and soil moisture using satellite remote sensing. George Mason University (2006)
9. Wang, L.: Remote sensing techniques for soil moisture and agricultural drought monitoring. George Mason University (2008)
10. Soriano, M.: Estimation of soil moisture in the southern United States in 2003 using multi-satellite remote sensing measurements. George Mason University (2008)

Effect of Landform in Loess Hilly-Gully Region on the Dynamic Change of Land Use in Qingyang City

Aihong Gai(✉), Qian Lu, Peijie Yan, Renzhi Zhang, and Huihui Kang

College of Resource and Environment Science, Gansu Agricultural University, Lanzhou 730070, China
gaiah@gsau.edu.cn

Abstract. By applying GIS and RS technologies and using Landsat TM/ETM data, the study, from the angle of geomorphology, analyzes the land use changes in 1990, 2000 and 2006 in Qingyang City. The study results are as follows: The proportion of the grassland, water body, cropland and unused land decreased while that of the forest land and construction land increased. The cropland area in divisions II and IV increased while in the other geomorphological units decreased. The water body area in division III decreased but increased at varying degrees in other divisions. The comprehensive dynamic degree of land use in the later period of the study was higher than that in the earlier period and the dynamic degree of single land type was characterized by the same trend. The land use change map indicated that in divisions I and V the conversion of cropland to forest land took up the highest proportion of all the land change types, that the land use change varied in different landforms and that the distribution and shift of the land use change hot areas were closely related to the distribution characteristics of human activity intensity and government policies.

Keywords: GIS · Landuse · Hilly-Gully landform · Dynamic change · Qingyang city

1 Introduction

Land is the nature-economy-society complex ecosystem which is composed of temperature, soil, hydrology, topography, geology, biology and the outcome of human activities. Human being's transforming and using land changes the environment, one of the results is that land use and land cover(LUCC) are changed [1, 2]. Land use and land cover are the result of the interaction between nature and human activities [3, 4]. Since the IHDP and IGBP jointly launched the LUCC research program in 1995 [5–7],many researches have been conducted by numerous scholars on land use classification system, land use security pattern, land use ecological zoning, optimal allocation of land use, analysis of current land use, LUCC and its driving force, sustainable land use and the ecological effect of LUCC [8–11]. As an important natural factor, geomorphology is not only a significant background for land but also plays a role in the distribution pattern of regional land use [12, 13]. The research on the effect of geomorphology on land use change has always been an important subject of land use and land cover

F. Bian and Y. Xie (Eds.): GRMSE 2015, CCIS 569, pp. 286–297, 2016.
DOI: 10.1007/978-3-662-49155-3_29

[14–16]. It directly affects substance flow and energy transformation on land, which in turn deeply affects the way land is used and to some extent decides the direction and speed of local land use change [17]. The analysis of regional geomorphological features for land use can enrich LUCC research content and is also of great significance to study the role of geomorphology in land use change.

At present, with the development of 3S technology, elevation, gradient, slope direction, degree of relief, terrain index, comprehensive topography index are used as the major indicators for topographic gradient analysis using spatial analysis function of GIS [18, 19]. Sun et al. [21] and Bu et al. [22] have applied GIS technology to study the effect of elevation, gradient, slope direction, relief amplitude and the change rate of slope on the poverty belt around Beijing-tianjin Cities and on the land use dynamic in Shenzhen City, and discussed the terrain gradient effect in land use change and its causes. However, most of the researches were conducted in western China with the focus on the analysis of single factor and few researches have been conducted to study the effect of special geomorphology type in western China on the spatial-temporal change of land use.

Qingyang City, located in the east end of Gansu Province neighboring Shananxi and Ningxia Provinces, belongs to the losses Gully region along the middle - lower reaches of the Yellow River. The action of natural forces has formed the current geomorphology with valley, plain, plateau, ravines, hill ridges, mountains and slope. In recent years, with economic development and rapid urbanization, huge change has taken place in land use in Qingyang City. This paper, from the angle of geomorphology, analyzes the spatial-temporal characteristics of land use change over the past 30 years from 1990–2006 in Qingyang City with the aim to provide decision making basis for land use control and management.

2 Materials and Methods

2.1 Overview of the Study Area

Qingyang city (106°20′E-108°45′E Long. and 35°15′N-37°10′N Lat.), located in the eastern Gansu Province and the west end of the Loess Plateau, belongs to the inland area along the middle-lower reach of the Yellow River. It's surrounded by Yangquan Mountain in the north, Liupan Mountain in the west and Ziwuling Mountain in the east with the middle and south areas lower than the other three sides, forming a dustpan in shape and hence called "the basin of eastern Gansu Province". Its relative elevation is 1204 m. The city covers an area of 27,119 square kilometers with a total population of 2.61 million including 2.2329 million of agricultural population. The City consists of 1 urban city, namely Xifeng City and 7 rural counties, namely Huanxian County, Huachi County, Qingcheng County, Zhenyuan County, Ningxian County, Zhengning County and Heshui County including 146 townships (Fig. 1). The City has a dry, mild continental climate with more rainfall and higher temperature in the south than in the north. The total static-storage of groundwater body resources is about 4.339 billion m3 and the dynamic-storage of groundwater body resources 37.14 million m3. The city is rich in petroleum and natural gas. Since 1978, the population in the City has kept growing

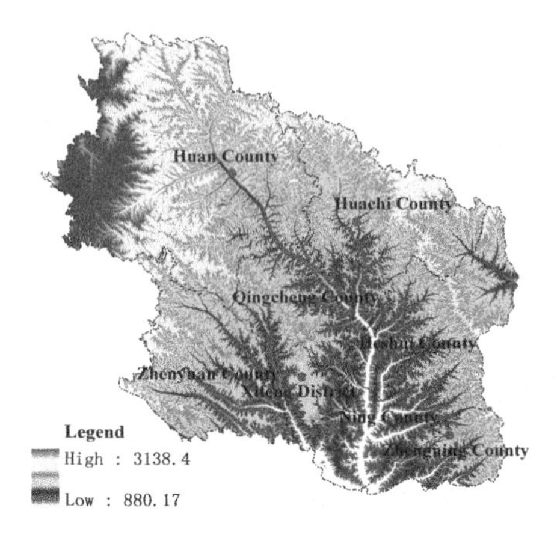

Fig. 1. Geographical distribution map of Qingyang city (Color figure online)

and accompanying the population growth is the growth of GDP which increased from 142 RMB Yuan per person on average to 6585 RMB Yuan from 1978 to 2007.

2.2 Data Sources and Processing

The TM images of bands 7, 6, 5, 4, 3, 2, 1 taken at three periods in 1984, 1996, 2001 were adopted. The images in each period contained 9 scenes and the spatial resolution of all the images was 30 m. Then the images were classified and interpreted using ERDAS IMAGE 9.2, MapGIS K9 and ArcGIS 9.3 and the land use type maps of target areas in 1990, 2000 and 2006 were obtained. To process the images, 1:50,000 topographic map was used for the geometric correction of the 27 scenes of images. Then the images were intensified and combined together. The mosaics at the three periods were cut and spliced based on the digitized contour of Qingyang City. The processed images presented in this paper were transverse Mercator projection. Based on the GPS fixed position data, pictures and the records about land use change obtained in field investigation, evaluating template was established. The images at the three periods were classified and then combined according to the land types. The small map spots were combined, modified and then their precision was verified using the precision verification point. The general interpretation precision reached 86 %. According to the land use classification system established by the Ministry of Land and Resources of People's Republic of China for the second nationwide land survey in 2007, the land use in the study area was classified into 6 types, namely, cropland, forestland, grassland, construction land, water body, unused land. The current land use situation maps in the study area at the three periods were obtained using human computer interactive interpretation technology. The image data were provided by the data sharing infrastructure of earth system science (www.geodata.cn), one of the programs built by national infrastructure of science and technology (Fig. 2).

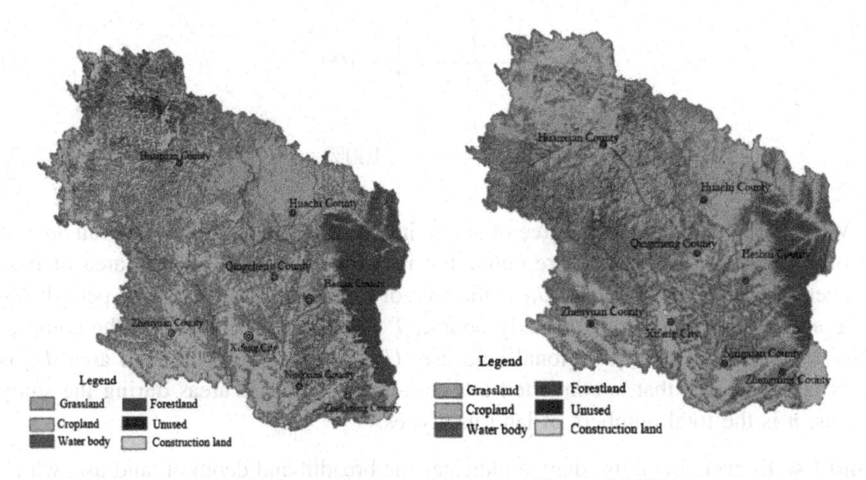

Fig. 2. Land use in study area in 1990 and 2006

2.3 Methods

Landform Classification. According to the geomorphologic features and their contributing factors in the study area, the whole study area was classified into five types of geomorphological divisions, namely, losses tableland division (I), broken loess tableland-mountain ridge division (II), loess Gully-mountain ridge division (III), loess hilly area (IV), bedrock hill area (V) (Fig. 3).

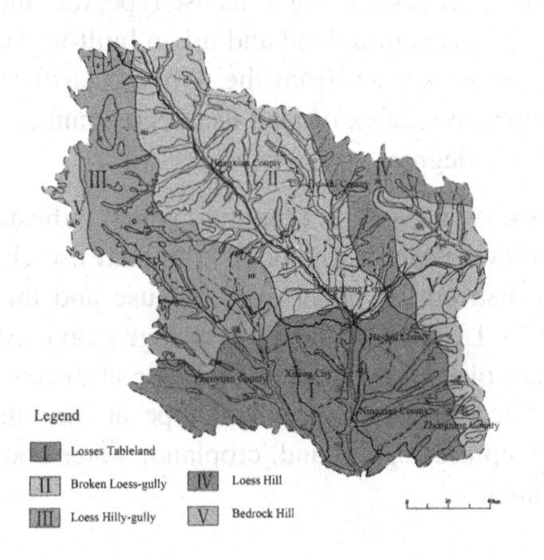

Fig. 3. Landform patterns of Qingyang City

The Dynamic Degree of Land Use. The dynamic degree of land use indicates the change intensity of regional land use (model (1), (2)), revealing the hot spot for land use change research [23, 24].

$$L_i = \frac{U_i + U_{i-}}{U_{i1} + U_{i2}} \times \frac{1}{T} \times 100\% \quad (1)$$

$$L = \frac{\sum_{i=1}^{n} \sum_{j=1}^{n} U_{ij}}{U} \times \frac{1}{T} \times 100\% \ldots (i \neq j) \quad (2)$$

Where, L_i is the dynamic degree of single land use of type i; U_{i+} is the total area of the other types of land which are converted into type i. U_{i-} is the total area of type i converted into the other types; U_{i1} is the area of type i at the earlier study period; U_{i2} is the area of type i at the later study period; T is the study period; L is the comprehensive dynamic degree of regional land use; U is the total regional land area; U_{ij} is type i land use area that is converted into non-type i land use areas during the study periods. n is the total numbers of land use types.

Land Use Degree. Land use degree indicates the breadth and depth of land use, which reflects not only the natural property of land use, but also the degree at which human activities affect land system. Wang et al. [25] put forward the classification principle of land use degree and land use quantitative expressions as follows:

$$P = 100 \times \sum_{i=1}^{n} A_i \times C_i \ldots\ldots P \in [100, 400] \quad (3)$$

Where, P is the comprehensive index of land use degree in the study area; A_i is the class i classification index of land use degree in the study area; C_i is the classified area percentage of class i land use degree; n is the classification number of land use degree. At present, the land is classified, according to its use type, into unused land, forest land, grassland and water body, agricultural land and urban built-up land. The grading index is set to 1, 2, 3, 4. It can be known from the above classification principle and the formula that the comprehensive index of land use degree ranges between 100 and 400, which reflects the land use degree.

Land Use Change Map. As it combines the map which indicates the spatial structure of land use with that showing land use process, the land use change map can present visual spatial-temporal distribution pattern of land use and thus has become a very important research tool for LUCC. The land use change map used in this paper consists of double digits. The tens place indicates the land type at the earlier period of land use change and the ones place represents the land type at the later period of land use change. 1, 2, 3, 4, 5, 6 represent grassland, cropland, water body, forest land, unused land and construction land.

3 Results and Analysis

3.1 The Analysis of Spatial-Temporal Characteristics of Land Use Change

During the study period, the areas of different types of land use showed different characteristics in terms of change number in the study area (Table 1). In division I, the

Table 1. Chang of land use quantity in Qingyang city during 1990–2006 (hm^2)

Division I	Division II	Division III	Division IV	Division V	The whole city	Division I
Grassland	−160204	−107216.88	−90000.96	−131031.43	−42010.88	−10151.1
Cropland	−112492.1	109311.12	−157973.92	28476.96	−84600.16	454511.26
Water body	250631.2	9791.76	−3655.92	17719.6	10441.28	−310.31
Forestland	122876.4	10734.16	2919.84	24248.16	6876.4	462713.3
Unused	−41948.7	−113558	−56807.6	−11316.56	−6183.84	−37698.57
Construction land	3462.4	−29650	−8295.68	3214.64	1894.4	50801.17

area of forest land, water body and grassland dropped while that of the cropland and construction land increased from 1990 to 2000; from 1990 to 2006, the area of grassland, cropland and unused land decreased while that of forest land and construction land increased; from 2000 to 2006, the area of cropland, water body and unused land decreased while that of grassland, forest land and construction land increased. In divisionII, the area of cropland, water body, unused land and construction land dropped while that of the grassland and forest land increased from 1990 to 2000; From 1990 to 2006, the area of grassland, unused land and construction land decreased while that of cropland, forest land and water body increased; from 2000 to 2006, the areas of grassland and unused land decreased while that of cropland, water body, forest land and construction land increased. Generally, in division III, the area of grassland, cropland, water body, construction land and unused land dropped while that of the forest land increased. From 1990 to 2000, the area of cropland, water body, construction land and unused land decreased; from 2000 to 2006, the area of grassland, cropland and unused land decreased while that of water body, forest land and construction land increased. Generally, in division IV, the area of grassland and unused land dropped while that of the cropland, water body, forest land and construction land increased; from 1990 to 2000, the area of cropland and water body decreased while that of grassland, forest land, unused land and construction land increased. In divisionV, the area of grassland, cropland and unused land dropped while that of water body, the forest land and construction land increased from 1990 to 2006, within which from 1990–2000 the area of cropland, water body and forest land decreased while that of grassland, unused land and construction land decreased; from 2000 to 2006, the area of grassland, cropland and unused land decreased while that of water body, forest land and construction land increased. In each geomorphological unit, the cropland had the largest reduced area at the three study periods, followed by grassland; in the earlier study period, cropland had the largest increased area while in the later study period the forest land had the largest increased area.

3.2 Spatial-Temporal Difference in Dynamic Degree Change of Land Use

The dynamic degree of land use in Qingyang City varied from geomorphologic unit to geomorphological units (Table 2). In division I, the comprehensive dynamic degree of

Table 2. Dynamic of Land Use in Qingyang City during 1990-2006 (%)

	Division I	Division II	Division III	Division IV	Division V	The whole city
Grassland	31.70	24.79	12.66	29.82	31.90	19.64
Cropland	18.14	14.60	9.10	28.64	30.45	56.56
Water body	18.71	25.12	1.32	5.25	2.75	35.50
Forestland	58.72	25.60	38.57	16.43	9.13	13.53
Unused	39.92	49.53	3.89	42.16	41.24	57.06
Construction land	63.39	18.23	1.43	19.04	39.25	27.63
Comprehensive dynamic degree of land use	13.27	14.70	8.58	3.91	3.70	1.81

land use in the later study period was 1.2 time more than that in the earlier study period; the dynamic degree of each single land type was ranked as: construction land > forest land > unused land > grassland > water body > cropland. In division II, the dynamic degree in the later study period remained the same as that in the earlier study period and the dynamic degree of each single land type was ranked as: unused land > forest land > water body > grassland > construction land > cropland. In division III, the dynamic degree of land use in the earlier study period was 1.5 time more than that in the later study period; the dynamic degree of each single land type was ranked as: forest land > grassland > cropland > unused land > construction land > water body. In division IV, the dynamic degree of land use in the later study period was 1.1 time more than that in the earlier study period. In divisionV, the dynamic degree of land use in the earlier study period remained the same as that in the later study period; the dynamic degree of each single land type was ranked as: unused land > construction land > grassland > cropland > forest land > water body. Generally, the dynamic degree of all geomorphological units in Qingyang City was higher in the later study period than in the earlier study period with only the exception of divisionV.

Generally, the comprehensive dynamic degree of land use in Qingyang City and all the geomorphological divisions in the later study period was higher than in the earlier study period and this was true of the dynamic degree of single land type. Divisions I and II in the earlier study period were the hot spots of land use change which shifted to division I in the later study period. The distribution and shift of hot spot of land use change were connected to the distribution of human activity intensity and to the government policy on land use. Division I was the most populous area, which demanded for more construction land and naturally the land use changed most intensively; the hot spot shifted to the other divisions around Qingyang municipality as a result of implementing the policy of "returning cropland to forest land".

3.3 The Spatial-Temporal Difference in Land Use Degree Change

From Table 3, it can be seen that the land use change in Qingyang City could roughly be divided into two periods. The earlier study period was the period for development while the later study period was the period for adjustment. The land use degree varied from geomorphological unit to geomorphological unit: Division I had the highest land use degree with the comprehensive index of land use degree higher than the average one in the city by 54.94, 56.76, 1.57 in 1990, 2000 and 2006 respectively. Following division I was division II whose comprehensive index of land use degree was higher than the average one by 38.09, 21.78, 28.59 in 1990, 2000 and 2006 respectively. DivisionV had the lowest land use degree whose comprehensive index of land use degree was higher than the average one by 32.97, 23.53, 49.54 in 1990, 2000 and 2006 respectively. The geomorphic difference in land use was closely related to land use structure. For division I, most of the land was for urban construction and agricultural use and therefore the forest land took up a high proportion. The comparison of the comprehensive index of land use degrees in the earlier and later study periods with the average one in Qingyang City indicated that the difference in land use degree was widening. The land use in divisions I and II developed most rapidly, and this was mainly due to the fact that all the types of land in the two divisions were converted into urban construction land on large scale. The other divisions were in adjustment period as large quantity of cropland was converted into forest land which led to low land use degree. The land use degree ranked in descending order in Qingyang City from 1990 to 2006 was as follows: division III > division V > division I > division IV > division II. From 1990 to 2000, the land use degree in all the divisions decreased in varying degree except division I. From 2000 to 2006, the land use degree in all the divisions decreased in varying degree except division II.

Table 3. Comprehensive Index of Land Use in Qingyang City during 1990-2006

Year	Division I	Division II	Division III	Division IV	Division V	The whole city
1990	250.22	233.37	242.01	225.18	228.25	123.81
2000	252.04	217.06	235.19	207.69	218.81	229.31
2006	196.85	223.87	197.38	188.64	145.74	232.73
1990–2000	1.82	−16.31	−6.82	−17.49	−9.44	105.5
2000–2006	−55.19	6.81	−137.81	−19.05	−73.07	3.42
1990–2006	−53.37	−9.5	−144.63	−36.54	−82.51	108.92

3.4 Analysis of Land Use Change Map

From 1990 to 2006, the changed land area in Qingyang City reached 422,696.3 hectares, accounting for 15.59 % of the total land area of Qingyang City. During the study period, the changed land in division I was 207,738 hectares, representing 7.66 % of the total land area in this division. The conversion of cropland into forest land took up the largest proportion of land conversion, followed by that of grassland into forest land (Table 4). The total area of changed land use in division V was 76,774.56 hectares,

Table 4. Map Types of Major Land Use Change in Qingyang City during 1990–2006(amount of change: hm^2; accumulative proportion: %)

Type	Map	14	24	21	52	
Division I	Amount of Change	55001.25	85367.3	9949.5	35570.25	
	Proportion	7.42	11.51	1.34	4.80	
	Accumulative proportion	7.42	18.93	20.28	25.07	
Division V	Map	14	24	54	34	
	Amount of change	14439.28	57310.88	1757.52	89.04	
	Proportion	3.89	15.45	0.47	0.02	
	Accumulative proportion	3.89	19.34	19.82	19.84	
Division III	Map	25	26	21	24	
	Amount of change	14836.4	1032.32	25917.7	271.44	
	Proportion	5.87	0.41	10.26	0.11	
	Accumulative proportion	5.78	6.19	16.44	16.55	
Division II	Map	25	26	21	24	
	Amount of change	19352.25	5224.5	74180.3	22.5	
	Proportion	2.33	0.63	8.95	0.00	
	Accumulative proportion	2.33	2.96	11.91	11.92	
Division IV	Map	24	25	26	21	
	Amount of change	6916.5	6563.25	101.25	27675	
	Proportion	1.32	1.26	0.02	5.30	
	Accumulative proportion	1.32	2.58	2.60	7.90	
The whole city	Map	14	24	54	34	
	Amount of change	97463.25	154422	11191.5	2279.25	
	Proportion	3.59	5.69	0.41	0.08	
	Accumulative proportion	3.59	9.29	9.70	9.78	
Type	Map	26	31	32	53	51
Division I	Amount of Change	5339.25	1788.75	5796	587.3	3424.5
	Proportion	0.72	0.24	0.78	0.08	0.46
	Accumulative proportion	25.79	26.03	26.82	26.90	27.36
Division V	Map	12	52	23	26	35
	Amount of change	20.72	1369.36	40.32	329.44	8.32
	Proportion	0.01	0.37	0.01	0.09	0.00
	Accumulative proportion	19.85	20.22	20.23	20.70	20.70
Division III	Map	35	36	34	51	31
	Amount of change	124.16	256.96	22.24	38709.4	5838.9
	Proportion	0.05	0.10	0.01	15.32	2.31
	Accumulative proportion	16.60	16.70	16.71	32.03	34.34
Division II	Map	35	36	31	51	54
	Amount of change	2079	132.75	5528.3	43713	13.5
	Proportion	0.25	0.02	0.67	5.27	0.56
	Accumulative proportion	12.17	12.36	13.03	18.30	18.30
Division IV	Map	35	36	31	32	15
	Amount of change	427.5	72	951.75	27	3390.75
	Proportion	0.08	0.01	0.18	0.01	0.65
	Accumulative proportion	9.42	9.43	9.61	9.62	10.27
The whole city	Map	46	16	31	32	16
	Amount of change	146.25	119277	1347.75	5388.8	13959
	Proportion	0.01	4.40	0.05	0.20	0.51
	Accumulative proportion	9.79	14.19	14.24	14.44	14.95

accounting for 2.83 % of the total changed land area in this division. The total changed land area in division III was 87,923.52 hectares, accounting for 4.24 % of the total changed land area in this division where the conversion of unused land into forest land was the major type of conversion followed by that of cropland into grassland. The total changed land area in division II was 22,6912.5 hectares, accounting for 8.37 % of the total changed land area in this division. The total changed land area in division IV was 53,633.25 hectares, accounting for 1.98 % of the total changed land area in this division where the conversion of cropland into grassland was the major type of conversion followed by that of cropland into forest land. From 1990 to 2000, the conversion of cropland into the construction land made up the largest percent land conversion in the whole Qingyang City. In division I, the conversion of cropland into the construction land made up the largest percent land conversion. In division V, the conversion of water body into the forest land made up the largest percent land conversion. In division III, the conversion of grassland into the forest land made up the largest percent land conversion. In division II, the conversion of unused land into the construction land took up the largest proportion of the land conversion. In losses division IV, the conversion of cropland into the unused land and that of water body into construction land made up the largest percent of the land conversion. From 2000 to 2006, the changed land in Qingyang City reached 176,847.75 hectares, accounting for 6.52 % of the total changed land area in Qingyang City. The conversion of cropland into the grassland and that of cropland into forest land were the major types of land conversion, which was mainly contributed to the policy of "returning cropland into forest land and grassland".

4 Conclusion and Discussion

Based on geomorphological type, quantitative analysis was made of land use dynamic degree, land use degree and land use change map from 1990-2006 using RS and GIS technologies. The study results were as follows:

(1) From 2000 to 2006, the land use change in the study area was characterized by the decrease in cropland and unused land areas and the increase in forest land and construction land areas with biggest change taking place in cropland and forest land.
(2) The grassland, cropland and forest land were the major land use types in the study area. During the study period, the land use change trend of each geomorphological unit remained the same as the general trend of the whole study area but there were still some differences in land use change between different geomorphological units. The area of forest land, grassland and water body in division I had the biggest change; the area of construction land and unused land in division II was larger than the other geomorphological units. Division III had the biggest change in cropland use.
(3) In division I and division V, the grassland and cropland had the biggest dynamic degree; the forest land, construction land and unused land had the biggest dynamic degree in 1 and division II. The comprehensive index of land use degree

in divisions I and II was on the rise during 1990–2000 and 2000–2006 while the comprehensive index of land use degree in other geomorphological units was declining during the study period.

(4) Most of the land use change took place in division II and division I with the conversion of cropland into grassland and cropland into forest land being the major types of land use change. The characteristics was also true of the other geomorphological units only with the conversion of unused land into forest land being the major type of land use in division III.

(5) The above characteristics of land use change could be attributed to the economic development and government policy. With rapid economic development, the population moved into the low and plain areas, resulting in the reduction of cropland area and increase in the construction land area, which was particularly true of the losses tableland division. In addition, affected by the government policy of "returning cropland into forest land and grassland", large quantity of steep sloping cropland was converted into forest land and grassland, reducing the cropland area and increasing the areas of grassland and forest land.

(6) The difference in the comprehensive land use degree and in the land use change map in different study periods in the study area was related not only to the implementation of environmental protection policy, but also to economic growth rate. However, the factors that affect land use change do not stop just at policy and economic development. Population and natural conditions such as temperature, precipitation and geomorphology also play a significant role. This paper just discussed the geomorphological causes of land use change in Qingyang City and the other natural and human causes remain to be discussed.

References

1. Liu, J., Gao, J., Geng, B., Wu, L.: Study on the dynamic change of land use and landscape pattern in the farming-pastoral region of Northern China. Res. Environ. Sci. **20**(5), 148–154 (2007)
2. Ademola, K.B.: Random and systematic land-cover transitions in Northern Ghana. Agric. Ecosystems Environ. **113**(1–4), 254–263 (2006)
3. Liu, Y.Y., Wen, Q., Cui, W.G., et al.: Analysis on landscape spatial patterns of different topography on county scale-a case study of Liling city, Hu'nan Province. Res. Soil Water Conserv. **16**(3), 89–94 (2009)
4. Min, J., Zhang, A.L., Gao, W., et al.: The comparison of driving mechanism of rural-urban land conversion in different physiognomies of Hubei Province. Res. Sci. **31**(7), 1125–1132 (2009)
5. Turner, B.L.II., Skole D., Sanderson S., et al.: Land Use and Land Cover ehang(LUCC) Implementation Strategy. IGBP Report No.48 and HDP Report No.10. Stockholm: IGBP (1999)
6. Lambin, E.F., Banlies, X., Boekstael. N., et al.: Land Use and Land Cover hange: Implemeniation Strategy. IGBP Report No.48 and IHDP Report No.10, Stockholm: IGBP (1999)

7. Chen, B.M., Liu, X.W., Yang, H.: Review of most recent progresses of study on land use and land cover change. Prog. Geogr. **22**(1), 22–29 (2003)
8. Zhu, H.-Y., Li, X.-B., He, S.-J., et al.: Spatio-temporal change of land use in Bohai Rim. Acta Geographica Sinica **56**(3), 253–260 (2001)
9. Chen, Q.-Y., Yin, C.-L., Chen, G.-H.: Spatial temporal evolution of urban morphology and land use sorts in Changsha. Scientia Geographica Sinica **27**(2), 273–280 (2007)
10. Tang, H.-J., Wu, W.-B., Yang, P., et al.: Recent progresses of Land Use and Land Cover Change (LUCC) models. Scientia Geographica Sinica **64**(4), 456–468 (2009)
11. Yue, S., Zhang, S., Yan, Y.: Impacts of land use change on ecosystem services value in the Northeast China Transect (NECT). Scientia Geographica Sinica **62**(8), 879–886 (2007)
12. Zhang, F.F., Qi, S.H., Shu, X.B., et al.: Study on the relationship between land use spatial patterns and topographical factors for mountainous region: taking Jiangxi Province as an example. Geo-inf. Sci. **12**(6), 784–790 (2010)
13. Liang, F.C., Liu, L.M.: Analysis on distribution characteristics of land use types based on terrain gradient: a case of Liuyang city in Hunan province. Res. Sci. **32**(11), 2138–2144 (2010)
14. Zhong, D.Y., Chang, Q.R.: Landscape patterns of land utilization under different geomorphologic types in loess Hilly-Gully region. Bull. Soil Water Conserv. **32**(3), 192–197 (2012)
15. Cui, B., Li, X., Jiang, G., et al.: Study on land use/cover in mountain area based on the DEM) taking the Qinghai Lake Basin as an example. J. Nat. Sources **16**, 871–880 (2011)
16. Wenfeng, Gong, Li, Yuan, Wenyi, Fan: Analysis on land use pattern changes in Harbin based on terrain gradient. Trans. Chin. Soc. Agric. Eng. **29**(2), 250–260 (2013)
17. Gao, H.-J., Zhang, C.-Q., Zhang, F.-T.: Spatial-temporal Patterms of Land Use Change in Guizhou Province of terrain Gradient. SiChuan Agric. Univ. **33**(1), 62–70 (2015)
18. Wang, X., Zheng, D., Yuancun, Y.: Land use change and its driving forces on the Tibet Plateau during 1990-2000. Catena **72**(1), 56–66 (2008)
19. Gong, W., Yuan, L., Fan, W.: Analysis on land use pattern changes in Harbin based on terrain gradient. Trans. Chin. Soc. Agric. Eng. **29**(2), 150–260 (2013)
20. Kong, F.H., Nakagoshi, N.: Spatial-temporal gradient analysis of urban green spaces in Jinan. China. J. Landscape Urban Plan. **78**(3), 147–164 (2006)
21. Sun, P.L., Xu, Y.Q., Wang, S.: Terrain gradient effect analysis of land use change in poverty area around Beijing and Tianjin. Trans. Chin. Soc. Agric. Eng. **30**(14), 277–288 (2014)
22. Bu, X.G., Wang, Y.L., Shen, C.Z., et al.: Influence of landforms of on the land use dynamics in Shenzhen City. Geogr. Res. **28**(4), 1011–1021 (2009)
23. Li, J., Ren, Z.Y.: GIS-based evaluation of ecological security of the loess plateau in Northern Shaanxi Province. Res. Sci. **30**(5), 732–736 (2008)
24. Zuo, T., Su, W., Ma, J., et al.: Ecological security of land evaluation in the three gorges reservoir area of Chongqing for water and soil erosion. J. Soil Water Conserv. **24**(2), 74–78 (2010)
25. Wang, S.-Y., Liu, J.-Y., Zhang, Z.-X., et al.: Analysis on spatial-temporal features of land use in China. Acta Geographica Sinica **56**(6), 631–639 (2001)
26. Xu, L., Zhao, Y.: Forecast of land use pattern change in Dongling district of Shenyang: an application of Markov Process. Chin. J. Appl. Ecol. **4**(3), 272–277 (1993)

Ecological and Environmental Data Processing and Management

A Comparison of Different Methods for Studying Vegetation Phenology in Central Asia

Yonggang Ma(✉), Xinmin Niu, and Jie Liu

Xinjiang Academy of Science and Technology Development Strategy, KeXue 1 Street, Urumqi 830011, China
thank5151@163.com

Abstract. The global inventory modeling and mapping studies (GIMMS) data has been extensively used to extract vegetation phenological data globally or regionally for phenological trend analysis. However, most of preview researches focus on the phenological change based on individual phenological metrics extracted method and the potential difference is less discussed. To compare the difference and identify the character vegetation phenology change, we use two phenological extract methods (threshold method and inflection point method) to calculate two series of phenological data for detecting the phenological change based on 25 years of satellite-derived Normalized Difference Vegetation Index (NDVI) in Central Asia. The Mann-Kendall trend analysis was used to examine the change trend of start of season (SOS), end of season (EOS) and length of season (LOS). Different phenological spatial pattern are conducted. There is an unexpected consistent distribution in detecting the significant change zone between methods. The most significant change was found in agriculture zones. The result also indicated that Vegetation phenology in central Asia does not change overall.

Keywords: Vegetation phenology · Remote sensing · Central Asia

1 Introduction

Remote sensing data can contribute to better understanding of vegetation phenology and its relations with the climate system. Many kings of Long time series remote sensing data, e.g., the GIMMS, Spot vegetation and Moderate Resolution Imaging Spectroradiometer (MODIS), are commonly used and proved successfully to derive phenological metrics. However, Among the different kinds of data, the GIMMS was still extensively used for monitoring vegetation phenology dynamic with its longest temporal range.

Lots of methods were introduced by preview researches and a series of friendly tools were developed for extraction of the phenological data. I.e.YU used the GIMMS data by Derivation and Threshold Approach to examine the seasonal vegetation response to the recent climatic variation on the Mongolian steppes [1, 2]. Kariyeva used the threshold method to analysis the relationship between phenological metrics and climate factor in Central Asia [3, 4]. However, different extract methods may product different phenological result [5]. The conclusion concerning Central Asia phenology derived from GIMMS

F. Bian and Y. Xie (Eds.): GRMSE 2015, CCIS 569, pp. 301–307, 2016.
DOI: 10.1007/978-3-662-49155-3_30

need to be treat carefully [6]. Consequently, We use two different physical-based methods (the threshold method and inflection point method) to extract phenological data in Central Asia. The Mann-Kendall trend analysis was used to detect the phenological change trend. The purpose was to examine the difference of methods in phenological pattern and monitor the regional vegetation phenology change trend.

2 Data and Methodology

2.1 Data

The GIMMS dataset compiles NDVI images acquired by the advanced very high resolution radiometer (AVHRR) sensor aboard National Oceanic and Atmospheric Administration (NOAA) satellites [7]. The database is composed of 15 day composites from July 1981 to December 2006. The composite images are obtained by the maximum value compositing (MVC) technique, which minimizes the influences of atmospheric aerosols and clouds. Total 25 years of data have been covered by five different satellites: NOAA-7, 9, 11, 14, 16 and 17. The NDVI images are obtained from AVHRR channel 1 and 2 images, which correspond respectively to red (0.58 to 0.68 mm) and infra-red wavelengths (0.73 to 1.1 mm). For this study, the dataset for special range (30–60 N, 45–98 E) has been used.

2.2 Phenological Metrics Extract

There are many methods to detect specific phenological events such as the start, end, and length of the growing season (SOS, EOS, and LOS). The most common methods include threshold method and inflection point method. The TIMESAT [8, 9] and PhenoSat [10–12] software based these two methods were used commonly to extract metrics of vegetation phenological metrics. An adaptive Savitsky-Golay smoothing filter was applied because it maintains distinctive vegetation time-series curves and minimizes various atmospheric effects. For the purpose of this study, SOS EOS and LOS were extracted from GIMMS by TIMESAT with a dynamic threshold of 20 %. Meanwhile, the PhenoSat was also implemented from GIMMS data to obtain the second phenological dataset, which aim at examine the dissimilarity from different phenological extracted methods.

2.3 Trend Change Detect

Once the time of SOS, EOS and LOS have been retrieved, a statistical analysis was conducted. A Mann–Kendall test is performed to determine if trends were present in the time series [13]. Then, confidence intervals at 95 % and 99 % were estimated to reject the null trend hypothesis and six classes of trend change are extracted for reflecting the overall changing trend.

3 Result

3.1 Comparison Between Methods

As expected, there is a clear north-south and low-high gradient of the SOS metric for both methods (Fig. 1), with a growing season starting earlier in the southern zone (blue areas) and progressively later in more northern zone (green, yellow and red areas).

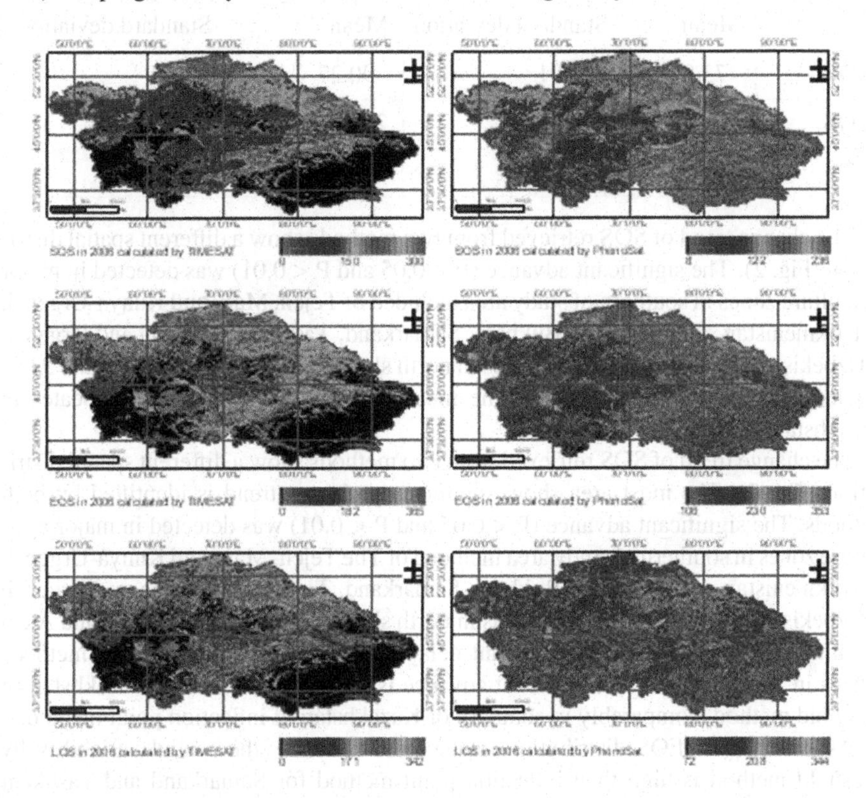

Fig. 1. The difference of phenological metrics between threshold method and inflexed point method (Color figure online)

The EOS shows an inconsistent distribution. the latest EOS of threshold method located in the south of Kazakhstan and earlier dormancy spread at the eastern and northern edges of study area, but the EOS of inflection point method keep the gradient similar as SOS. The LOS of threshold method shows a stronger spatial variation than the LOS of inflection point method. Moreover, there is no phenological metrics detected in the desert area (karakum desert, Kyzyl-Kum desert and Taklimakan desert) in the result of threshold method. Globally, the SOS, EOS and LOS of threshold method show a more clearly zonal transition, higher value and less spatial heterogeneity than those of inflection point method. Thus, threshold method conduct a lower standard deviation value of SOS, EOS and LOS than inflection point method. It is remarkable that the mean value of SOS and EOS of threshold method always smaller than those of inflection point method (Table 1). It mean no matter how to adjust the threshold value for threshold method, the value range of SOS, EOS and LOS for both methods would hardly keep consistent concurrently.

Table 1. Summary of phenological metrics retrieved from different method

	Threshold method		Inflection point method	
	Mean	Standard deviation	Mean	Standard deviation
SOS	73.86	54.91	90.27	39.87
EOS	209.50	140.52	280.25	50.43
LOS	135.64	96.07	189.98	59.42

The change trend of SOS retrieved from two methods show a different spatial distribution (Fig. 2). The significant advance ($P < 0.05$ and $P < 0.01$) was detected in major agriculture zones in southern of study area included of Tejen, Mary and Kunya-Urgench in Turkmenistan, as well as the Bukhara, Samarkand, Tashkent, Fergana and Urgench in Uzbekistan; on the contrary, the oasis in north slope of Tianshan Mountain and Aksu in China show a significant delay. The most difference of two methods locates in Kazakhstan.

The change trend of SOS retrieved from two methods show a different spatial distribution (Fig. 2). The most area show significantly change trend is identified by both methods. The significant advance ($P < 0.05$ and $P < 0.01$) was detected in major agriculture zones in southern of study area included of The Tejen, Mary and Kunya-Urgench in Turkmenistan, as well as the Bukhara, Samarkand, Tashkent, Fergana and Urgench in Uzbekistan; on the contrary, the oasis in north slope of Tianshan Mountain and Aksu in China show a unexpected significant delay. The most difference of two methods locates in Kazakhstan. The significant advance is found in northern of Kazakhstan of threshold method, comparably in southern of Kazakhstan of inflection point method.

As observed in EOS distribution image from methods, the spatial variability by threshold method is high than inflection point method for Samarkand and Tashkent where the significant advance trend change have been monitored by both methods. However, some region show significant delay in the result of threshold method, such as the north slope of Tianshan Mountain, but not perform the similar trend in the result of inflection point method.

LOS is effected by the distribution and intensity of change trend of SOS and EOS. The region of the Tejen, Mary and Kunya-Urgench, Bukhara, Samarkand, Tashkent, Fergana and Urgench with significant advanced SOS, generally show a significant increased LOS for both methods. The result may indicate the range and intensity of SOS is stronger than EOS.

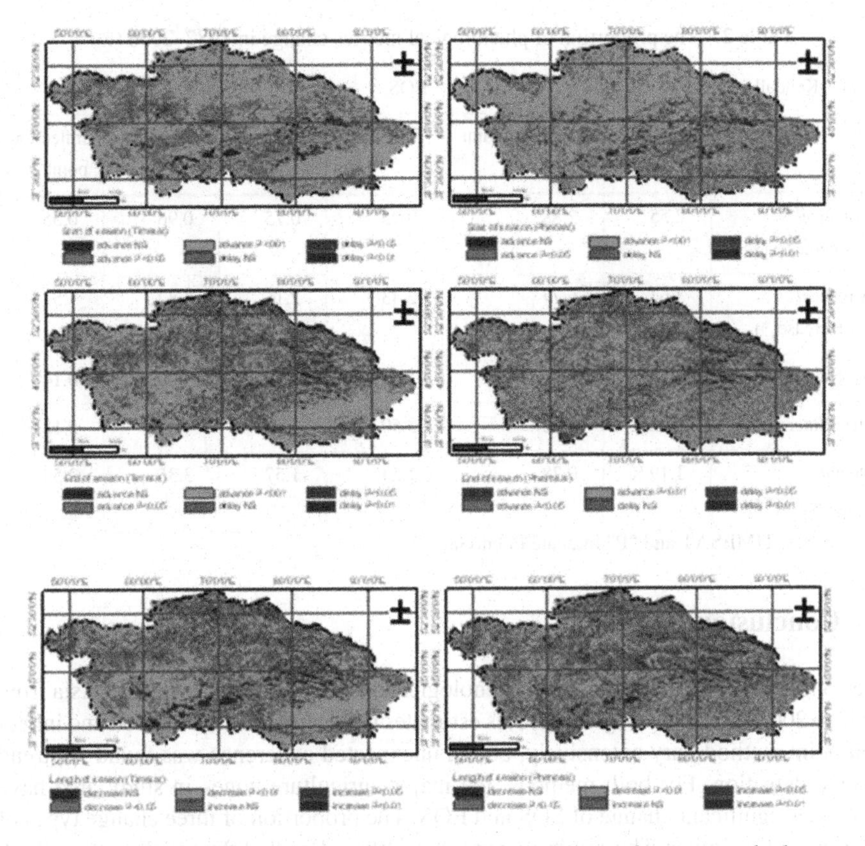

Fig. 2. The difference of phenological metrics trend change between methods

3.2 Phenological Metrics Trend Analysis

The proportion of change trends of SOS, EOS and LOS derived from GIMMS NDVI data is shown in Table 2. According to the level of significance and direction of change, five classes were summarized. Excluded of some slightly difference between two methods, strong coherence exists concerning basic pattern. About 90 % area does not change significantly for SOS, EOS and LOS. For SOS, the proportion of advance (sum of $P < 0.01$ and $P < 0.05$) significantly is larger than the area of delay significantly; as for EOS, the proportion of delay (sum of $P < 0.01$ and $P < 0.05$) significantly is small than the area of delay significantly; as for LOS, excluded of 90 % of no significance, 8 % increase and 2 % decrease.

Table 2. The proportion of phenological metrics change in 1982–2006 (%)

Mann-Kendall trend	SOS		EOS		LOS	
	Threshold	Inflection point	Threshold	Inflection point	Threshold	Inflection point
Advance/ decrease(**)	2.55	2.88	0.95	0.75	0.96	0.55
Advance/ decrease(*)	3.64	6.77	2.43	2.16	1.82	1.54
No change	90.59	88.11	91.58	91.59	89.00	90.14
Delay/increase(*)	2.02 %	1.46	3.50	3.93	4.87	5.18
Delay/ increase(**)	1.19 %	0.78	1.54	1.57	3.35	2.59

"T" indicate TIMESAT and "P" indicate PhenoSat

4 Conclusion

We conducted a comparison of two phenological extract methods for Central Asia from 1982 to 2006. The phenological metrics estimates between threshold method and inflection point method vary extensively, but an unexpected coherence was found for trend change detection. For both methods, the major agriculture zones in study area have showed a significant change of SOS and EOS. The proportion of three change types of metrics is also similar. The result suggests that, although with different algorithms, both methods are considerably effective for monitoring vegetation phenological change.

Acknowledgements. This work was financially supported by the Natural Science Fund of Xinjiang Uyghur Autonomous Region (NO. 2015211B028) and Xinjiang Youth science and technology innovation talents training project (NO. 2014721037).

References

1. Yu, F., Price, K.P., Ellis, J., Shi, P.: Response of seasonal vegetation development to climatic variations in eastern Central Asia. Remote Sens. Environ. **87**, 42–54 (2003)
2. Yu, F., Price, K.P., Ellis, J., Kastens, D.: Satellite observations of the seasonal vegetation growth in Central Asia: 1982–1990. Photogramm. Eng. Remote Sens. **70**, 461–469 (2004)
3. Kariyeva, J., van Leeuwen, W.J.D.: Environmental drivers of NDVI-based vegetation phenology in Central Asia. Remote Sens. **3**, 203–246 (2011)
4. Kariyeva, J., van Leeuwen, W.J.D., Woodhouse, C.A.: Impacts of climate gradients on the vegetation phenology of major land use types in Central Asia (1981–2008). Front. Earth Sci. **6**, 206–225 (2012)

5. White, M.A., de Beurs, K.M., Didan, K.: A continental phenology model for monitoring vegetation responses to interannual climatic variability. Glob. Change Biol. **15**, 2335–2359 (2009)
6. Schwartz, M.D., Hanes, J.M.: Continental-scale phenology: warming and chilling. Int. J. Climatol. **30**, 614–1626 (2010)
7. Tucker, C., Pinzon, J., Brown, M.: An extended AVHRR 8-km NDVI dataset compatible with MODIS and SPOT vegetation NDVI data. Int. J. Remote Sens. **26**, 4485–4498 (2002)
8. Jönsson, P., Eklundh, L.: Seasonality extraction by function fitting to time-series of satellite sensor data. IEEE Trans. Geosci. Remote Sens. **40**, 1824–1832 (2002)
9. Jönsson, P., Eklundh, L.: Timesat—a program for analyzing time-series of satellite sensor data. Comput. Geosci. **30**, 833–845 (2004)
10. Rodrigues, A., Marcal, A.R.S., Cunha, M.: PhenoSat– a tool for vegetation temporal analysis from satellite image data. In: Proceedings of the 2011 6th International Workshop on the Analysis of Multi-temporal Remote Sensing Images (Multi-temp), pp. 45–48 (2011)
11. Rodrigues, A., Marcal, A.R.S., Cunha, M.: Phenology parameter extraction from time-series of satellite vegetation index data using PhenoSat. In: IEEE International Geoscience and Remote Sensing Symposium, pp. 4926–4929 (2012)
12. Rodrigues, A., Marcal, A.R.S., Cunha, M.: Monitoring vegetation dynamics inferred by satellite data using the PhenoSat tool. IEEE Trans. Geosci. Remote Sens. **51**, 2096–2104 (2013)
13. Hirsch, R.M., Slack, J.R.: A nonparametric trend test for seasonal data with serial dependence. Water Resour. Res. **20**, 727–732 (1984)

Pollution Source Manage and Pollution Forecasting Platform Based on GIS

Gang Gou[1], Lianzi Feng[1], Yanqi Zhao[1(✉)], and Lingli Huang[2]

[1] College of Computer Science and Technology, Guizhou University, Guiyang 550025, Guizhou, China
1784788577@qq.com

[2] Huaxi People's Procuratorate, Guiyang 550025, Guizhou, China

Abstract. To strengthen the capabilities that environmental protection departments monitor and manage pollution sources, and improve the ability of predicting and analyzing water pollution, we use embedded component library ArcGIS Engine to develop a GIS-based sources of pollution monitoring and pollution forecasting platform which is based on geospatial databases. The following functions based on spatial properties, such as sewage disposal information inquiry, real-time monitoring of sewage, correlation analysis river pollution and dynamic simulation of water pollution were achieved. The results show that system is running the implementation and visualizations of above functions can well achieve the intended purpose.

Keywords: ArcGIS engine · Pollution forecasting · Dynamic simulation

1 Introduction

The eighteen report of the Communist Party of China points out that the ecological environment is one of the most important factors in the construction of a well-off society in China. However, the water pollution accident in the industry of our country has aroused public concern, and also become a major obstacle to the sustainable development of society and economy. According to the relevant laws and regulations of the state, the industrial enterprises have set up a sewage disposal management system, in order to monitor and manage enterprise sewage treatment and disposal. Currently most sewage monitoring is managed by the enterprises themselves. But due to the uneven of polluted water management ability of the enterprises, it can easily cause poor sewage treatment and harm to public health and social stability. Therefore, in order to reduce the various problems that enterprise sewage mismanagement may lead, environmental protection and other relevant departments hope to manage the monitoring of enterprises sewage concentrated, monitor the sewage discharge unified. For enterprises in the excessive discharge of pollutants, we must handle in time, thereby reducing the damage caused by pollution. In order to improve the efficiency of sewage management, the environmental protection department also hopes to improve the data visualization. Through the analysis and other means, we can get the key information. Through statistical analysis of a large number of water pollution accidents, it shows that the cause of water pollution is not only enterprises excessive discharge of

F. Bian and Y. Xie (Eds.): GRMSE 2015, CCIS 569, pp. 308–317, 2016.
DOI: 10.1007/978-3-662-49155-3_31

sewage, but also a large part of disclosure of toxic and harmful substances in the process of transportation. For all types of sudden water pollution accident, the environmental protection department hopes to quickly understand the situation of water pollution and accurately grasp the trend, so they can quickly take measures to maximize the protection of public safety. The most critical is how to improve the predictive analysis capabilities of water pollution accidents on the existing foundation. Therefore, the prediction of water pollution research has very important significance.

Geospatial data plays an important role in the realization of the sewage information query and water pollution prediction analysis. For weak ability in spatial data processing, traditional management information system can not very well provide solution for dealing with spatial data. With the rapid development of geographic information system (GIS), GIS is considered as the best technology to solve the problem related to spatial data. Using GIS technology, you can easily achieve the management of spatial data, to provide visual information display platform for the pollution enterprises. Its powerful spatial data processing capabilities provides important technical support for sewage information query and water pollution prediction analysis.

Based on in-depth study of geographic information system, water quality model, spatial database, and other related knowledge, this study uses ArcGIS Engine component library for secondary development to implement the pollution sources monitoring and pollution prediction platform which is combined with the actual needs of the related departments of environmental protection.

2 System Analysis and Design

2.1 System Analysis

The spatial data involved in the system mainly includes the point features of the pollution sources, the line features of the river, the surface features of river basin, elevation DEM (Digital Elevation Model) and satellite images. Among them, the sources of pollution, rivers and water are stored in SHP format, elevation DEM and satellite images stored in TIF and JPG format. The spatial data coordinate system is not unified. For example, water basin uses the plane coordinate system Beijing_1954, and the DEM uses the latitude and longitude coordinates GCS_WGS_1984.

GIS has excellent data storage capacity, which can be good for the spatial and attribute data centralized management, and transfer format of spatial data in different coordinate system and match coordinate system. Based on the spatial database, this research analysis all the functional requirements, which core is GIS technology. The research is focused on the management, query and analysis of spatial data.

2.2 System Logic Design

The system uses C/S architecture, the use of three-tier structure, namely the presentation layer, business logic layer and data access layer. The logical structure of the system is shown in Fig. 1. Presentation layer is the system client which is the interface interacted with the system. Business logic layer encapsulates the business logic to

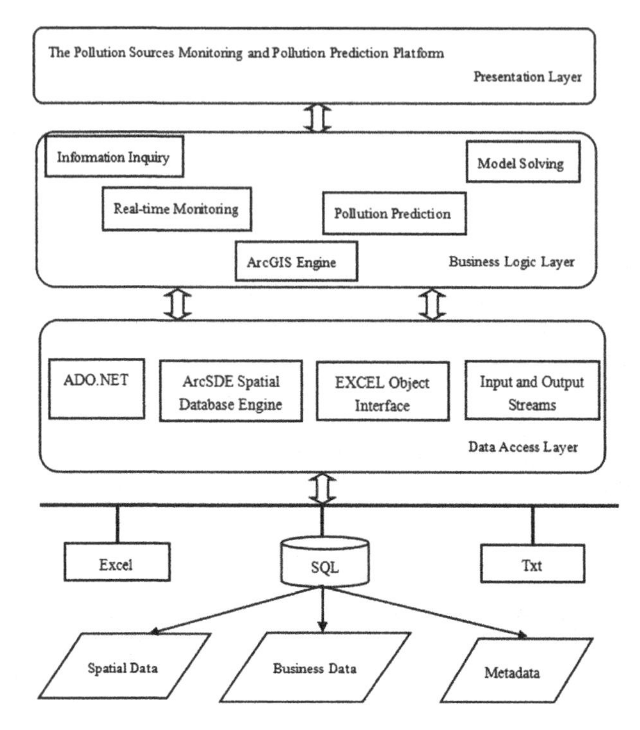

Fig. 1. System logic structure diagram

provide services to the presentation layer, such as information inquiry and pollution forecasting analysis. The data access layer encapsulates the attribute data access interface ADO.NET and the spatial data access interface ArcSDE, which provides data access services for the business logic layer.

2.3 System Function Design

The pollution sources monitoring and pollution prediction platform build GIS application system based on the geographical spatial information as the main research object, which provides a wealth of spatial query and analysis functions, and achieve visualization, and provides data information is true, reliable and intuitive for managers. System function structure is shown in Fig. 2.

Sewage Discharge Information Query. According to the enterprise sewage discharge monitoring data, we can analysis which enterprises sewage containing specific pollutants and whether the excessive emissions. Then we marked out the related enterprises on the map to provide intuitive data information for managers.

Real-Time Monitoring of Sewage. Real-time monitoring of the sewage is a unified monitoring management for enterprise sewage discharge monitoring data, which can be expressed in graphical form. And it can display real-time monitoring fluctuations in the data visually.

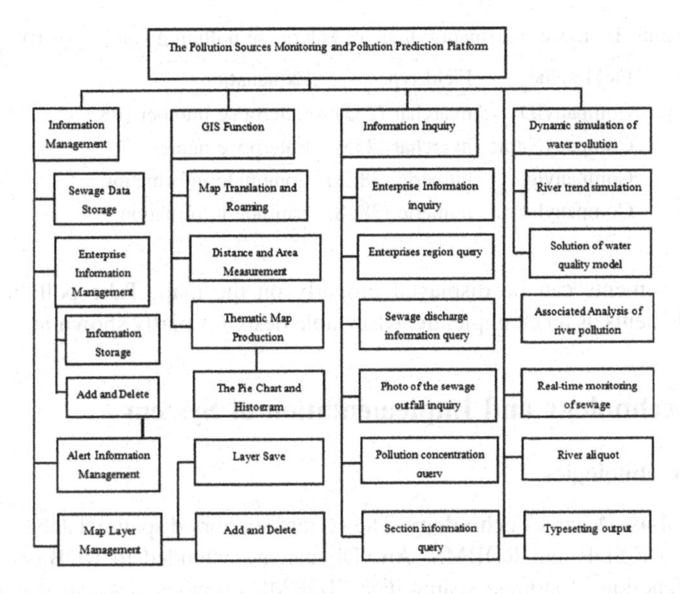

Fig. 2. System function structure diagram

Thematic Map. Thematic maps can express enterprise information in multiple fields as the form of graphics on the map. It can directly reflect a phenomenon in geographic distribution.

Correlation Analysis of River Pollution. First we select a section of the contaminated river, and set the search range. Then the associated pollution enterprises marked on the map. So managers can query and analyze the causes of the accident.

Dynamic Simulation of Water Pollution. For the river of water pollution, we make pollution diffusion simulation. Based on concentration of pollutants in the river dynamic displayed as the depth of the color, so that managers intuitively understand the pollutant concentrations and trends for some time in its river.

2.4 System Database Design

According to the functional design, the system must deal with a large amount of space and attribute information, so it is very important to build a suitable database. System database design mainly includes two aspects: one is business database design, the other is the spatial database design.

Business database design is to meet the needs of relevant data processing, design related forms, convenient data query.

Spatial database design is to achieve the storage and management of spatial data. Spatial data is a layer in the form of performance in the map, such as point, line, surface and other factors. Each element is corresponding with a basic table and a number of related system tables. The system table records the spatial interpretation of the geometric figures,

Table 1. Basic information table of enterprise pollution sources (part)

Field name	Field type	Explanation
CompanyID	nvarchar (255)	Enterprise number (PK)
CompanyName	nvarchar (255)	Enterprise name
Companylgt	numeric (38, 8)	Longitude information
Companylat	numeric (38, 8)	Latitude information

so that the elements can be displayed properly on the map. Take pollution source enterprise elements as an example, the basic table design is partly shown in Table 1.

3 Key Technology and Implementation of System

3.1 Key Technologies

Spatial Database Access Technology. The system supported spatial database is jointly built by the ArcSDE and RDBMS. ArcSDE can be extended to RDBMS, so as to realize the function of storage spatial data. RDBMS provides physical storage as the form of a table. ArcSDE enhances the ability to interpret spatial data in RDBMS, and stores data in a binary form in RDBMS. Thus the system database can store, query and manage the spatial and attribute data unified.

Pollution Simulation. The water pollution prediction problems mainly include the establishment, solution and visualization of water quality model. The key of forecast analysis is to solve the water quality model, so we can obtain the concentration of pollutants at different times in different sections. The other is to simulate the moving situation of pollutants in the river.

Water Quality Model and Its Solution. This paper selects the typical one-dimensional water quality model, and its differential equation is shown in the formula (1):

$$\frac{\partial C}{\partial t} + u\frac{\partial C}{\partial x} = E\frac{\partial^2 C}{\partial x^2} - KC \tag{1}$$

In the formula, C is the concentration of pollutants (mg/L), t is contamination occurred time (h), u is the river flow rate (km/h), x is the diffusion distance (km), E is pollutant diffusion coefficient (km^2/h), K is the pollutant attenuation constant (h^{-2}).

Formula (1) does implicit four point difference, specific alternatives as follows: Before the moment of j the different is the partial derivatives of t; after the moment of h the different is the partial derivatives of x; after the moment of j + 1 the different is the second-order partial derivatives of x. Thus, the original differential equation turn into a differential equation, such as formula (2).

$$\frac{C_i^{j+1} - C_i^j}{\Delta t} + u\frac{C_i^j - C_{i-1}^j}{\Delta x} = E\frac{C_{i+1}^{j+1} - 2C_i^{j+1} + C_{i+1}^{j+1}}{\Delta x^2} - \frac{1}{2}K\left(C_i^{j+1} + C_{i-1}^{j+1}\right) \tag{2}$$

To sort it, and get formula (3),

$$\alpha_i C_{i-1}^{j+1} + \beta_i C_i^{j+1} + \gamma_i C_{i+1}^{j+1} = \delta_i (i = 1, 2, 3, \ldots, n) \tag{3}$$

Among them,

$$\alpha_i = -\frac{E}{\Delta x^2}, \beta_i = \frac{1}{\Delta t} + \frac{2E}{\Delta x^2} + \frac{K}{2}, \gamma_i = -\frac{E}{\Delta x^2}, \delta_i = C_i^j \left(\frac{1}{\Delta t} - \frac{u}{\Delta x}\right) + C_{i-1}^j \left(\frac{u}{\Delta x} - \frac{K}{2}\right)$$

When i = 1, it is substituted into the formula (3), $\alpha_i C_{i-1}^{j+1}$ is a constant, convert and sort it, and get formula (4),

$$\beta_1 C_1^{j+1} + \gamma_1 C_2^{j+1} = \delta_1' = \delta_1 - \alpha_1 C_1^{j+1} \tag{4}$$

When i = n, it is substituted into the formula (3), $C_{n+1}^{j+1} = 2C_n^{j+1} - C_{n-1}^{j+1}$, convert and sort it, and get formula (5),

$$\alpha_n' C_{n-1}^{j+1} + \beta_n C_n^{j+1} = \delta_n \tag{5}$$

Among them, $\alpha_n' = \alpha_n - \gamma_n$, $\beta_n' = \beta_n + 2\gamma_n$

Thus the formulas (3), (4) and (5) form a one-dimensional linear equations of water quality model, which is written in a matrix form, as shown in formula (6). $C_{n+1}^{j+1} = 2C_n^{j+1} - C_{n-1}^{j+1}$.

$$\begin{bmatrix} \beta_1 & \gamma_1 & 0 & \cdots & \cdots & \cdots & 0 \\ \alpha_2 & \beta_2 & \gamma_2 & \ddots & \ddots & \ddots & \vdots \\ \vdots & \ddots & \ddots & \ddots & \ddots & \ddots & \vdots \\ \vdots & \ddots & \ddots & \ddots & \alpha_{n-1} & \beta_{n-1} & \gamma_{n-1} \\ 0 & \ddots & \ddots & \ddots & 0 & \alpha_n' & \beta_n' \end{bmatrix} \begin{bmatrix} C_1^{j+1} \\ C_2^{j+1} \\ \vdots \\ C_{n-1}^{j+1} \\ C_n^{j+1} \end{bmatrix} = \begin{bmatrix} \delta_1' \\ \delta_2 \\ \vdots \\ \delta_{n-1} \\ \delta_n \end{bmatrix} \tag{6}$$

Using Thomas algorithm to solve this three-diagonal matrix of formula (6), you can get the approximate value of the original problem at discrete points.

Movement of Pollutants in the River. Pollutants in the river are moving with the direction of the river, so the problem is to study the movement of water in the river. First, the pollutants move to the river course in the shortest path. When in the river course, it move as the average flow rate of rivers. When the river walked, according to the river network topology it will select a associated river to move until reaching the outlet. Contaminants move use a special logo to simulate its trajectory. When move to the next river section nodes, we will re-rendering the concentration changes of pollutants in the river, so as to achieve dynamic simulation results.

3.2 System Implementation

Sewage Discharge Information Query. Sewage discharge information query can improve the efficiency of the manager, and timely process the illegal enterprises. It plays an important role in reducing pollution. As shown in Fig. 3, the result of the query that select enterprise of the discharge of sewage containing heavy metal lead and excessive emissions display on the map.

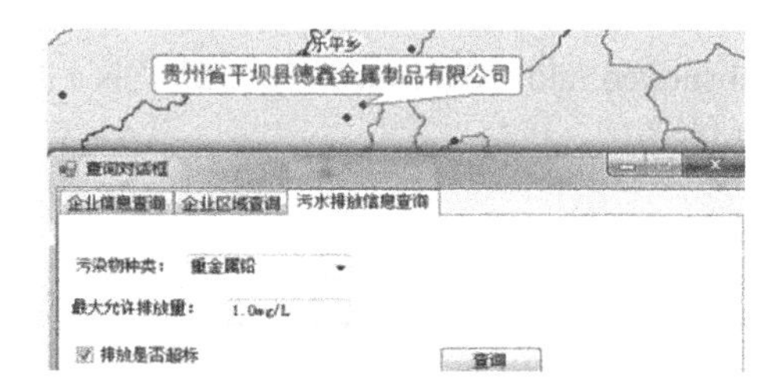

Fig. 3. Sewage discharge information query graph

Real-Time Monitoring of Sewage. You can choose any pollution source enterprise, and simulate the fluctuation of the sewage discharge monitoring data in the form of graphics. As shown in Fig. 4, it is shown that lead content of heavy metal pollutants in the sewage of the enterprises in the 60 h. In the process of monitoring, if the data exceeds the set threshold, the sound and light alarm will warn people.

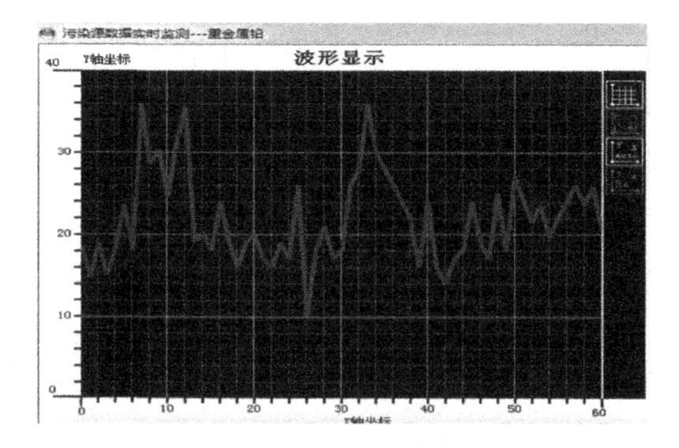

Fig. 4. Monitoring chart of heavy metal lead in sewage of a company (part)

Thematic Map. Thematic maps can visually express and present the data. There are bar charts, pie charts and other types. Take the pie chart as an example, as shown in Fig. 5, it shows the enterprise information of the percentage of employees, net profit and fixed assets in the total amount.

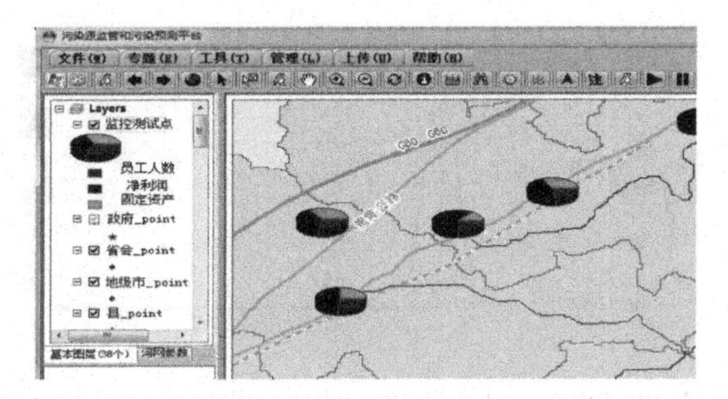

Fig. 5. The percentage of employees, profits and fixed assets of enterprises

Correlation Analysis of River Pollution. The correlation analysis of river pollution can help to investigate the causes of the pollution, and provide reliable basis for the sources of pollution. As shown in Fig. 6, we select a section of the polluted river, and set 800 range. Then carried out correlation analysis of pollution, we can find out the pollution enterprises surrounding river, and mark the associated information on the map.

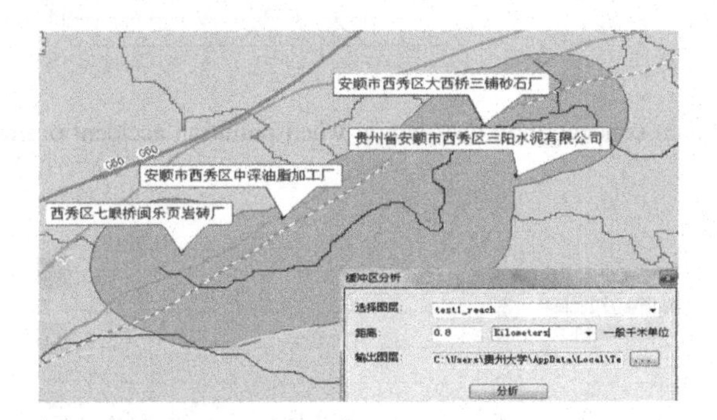

Fig. 6. Correlation analysis of river pollution

Water Pollution Simulation. The simulation object is Hongfeng Lake in Guizhou. We can arbitrarily select river in the map, input parameters of the simulation. We assume that the river flow velocity is 5, and diffusion coefficient of river section is 0.2, and time step is 0.8, and the space step is 1, and continued pollutants emission is 1, and the initial concentration of pollutants is 12, then modeling predict pollutant diffusion process within 2 h. According to a certain water quality evaluation standard, the results can be divided into five types of different colors to be displayed. The simulation results are shown in Figs. 7, 8, and 9. During the simulation, the contaminants to migrate with the river, the concentration of pollutants at different times in each section of river displayed graphically. We can clearly see the maximum peak of pollutant concentration in a period and the corresponding distance. In the biochemical effects of water, pollutants concentration gradually decreased. The diffusion of pollutants is in agreement with the actual conditions. The numerical solution obtained by simulation and the analytical solution obtained by Matlab programming are basically same.

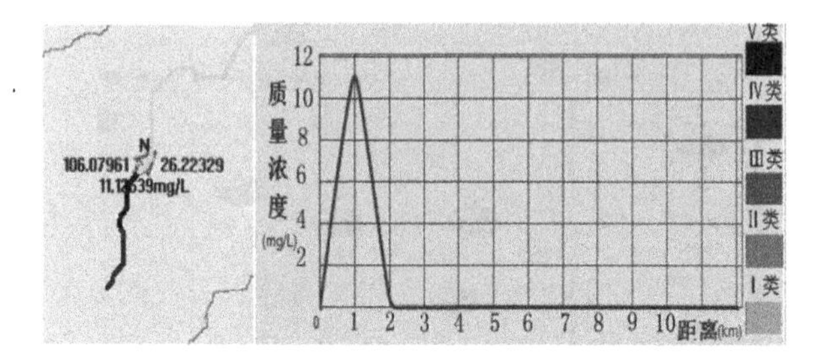

Fig. 7. Pollutant concentration distribution when pollution accident occurred 0.2 h

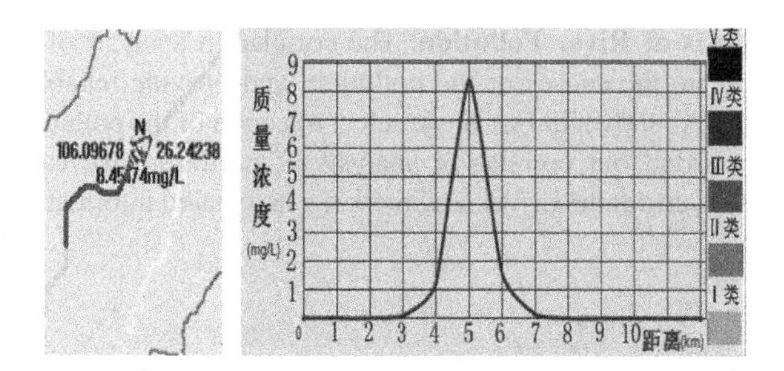

Fig. 8. Pollutant concentration distribution when pollution accident occurred 1.0 h

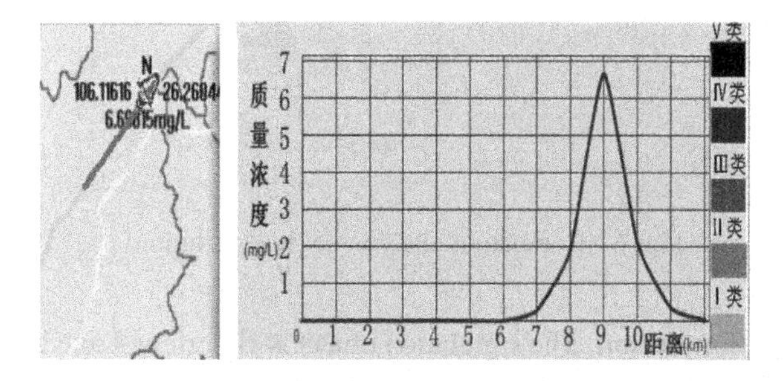

Fig. 9. Pollutant concentration distribution when pollution accident occurred 1.8 h

4 Conclusion

Combined the GIS technology with enterprise management of sewage and water pollution prediction analysis for the first time, this study establish the pollution sources monitoring and pollution prediction platform. It is the improvement and perfection of traditional enterprise sewage system function and innovation on the traditional management model. The system has achieved goals that are the environmental protection department to the enterprise sewage unified monitoring and the forecast analysis for water pollution. But the system still exist some shortcomings, such as lack of consideration for complex situations. For those rivers with large flow, relatively wide and deep, we need to use a high-dimensional water quality model for predicting water pollution. Therefore, the improvement of the function of water pollution prediction and analysis will be the direction of further research.

Ackownledgements. This work was supported in part by them:

Guizhou Province Department of Science Research Project, (Qian NY [2013] 2013)

Agricultural Resources and Environment Information Service Platform in Guizhou Key Technology Research and Development

Base Research Project in Guizhou Province Department of Education, (JD2014226)

Tunbao Culture Digitization project planning

Guizhou university teaching reform key projects, (JG2013021)

To Pressure and Successful Experience of the Computer Practice Teaching Mode to Stimulate Interest in Guizhou Province Department of Science Research Project, (Qian NY [2013] 3078);

Agricultural Resources and Environment Information Service Platform in Guizhou Key Technology Research and Development; Base Research Project in Guizhou Province Department of Education, (JD2014226).

We give our thanks to them!

References

1. Xu, J., Xie, X., Fu, Y., Li, Y.: Sudden water pollution accident simulation and visualization of time and space. China's Rural Water Conservancy and Hydropower (2012)
2. Zhao, K.: Maanshan Yangtze river section of sudden water pollution warning system research based on GIS. Anhui University of Technology, Anhui (2010)
3. Xu, S., Wang, S., Bi, Z.: Design architecture of logistics information system based on GIS technology. Comput. Eng. Des. **31**(6), 1259–1263 (2010)
4. Liang, L., Wang, D., Wang, F.: The SWAT model and its application study. J. China Inst. Water Resour. Hydropower Res. **5**(2), 125–130 (2007)
5. Xu, S., Yang, L.: One-dimensional river water quality diffusion of time and space dynamic simulation. J. Geogr. Geographical Inf. **26**(5), 41–43 (2010)
6. Liu, S.: A one-dimensional water quality model of simple simulation of contaminant distribution in the river. Water Transp. Manag. **27**(4), 33–34 (2005)
7. Page, B., Wohlgemuth, V.: Advances in environmental informatics: integration of discrete event simulation methodology with ecological material flow analysis for modelling eco-efficient systems. Procedia Environ. Sci. **2**, 696–705 (2010)
8. Liang, H.: Urban environmental information system development and application based on GIS. Graduate Sch. Chin. Acad. Sci. (Inst. Remote Sens. Appl.) (2005)
9. Ormsby, T., Napoleon, E., Burke, R., Groessl, C., Feaster, L.: Getting to know ArcGIS desktop: basics of ArcView, ArcEditor, and ArcInfo (2008)
10. Chai, J.: Sewage outfall into the river management research progress at home and abroad. Yangtze River Sci. Res. Inst. (2014)

Monitoring Health of Artificial Robinia Pseudoacacia Changes in the Yellow River Delta by the Analysis of Multiyear NDWI Data

Ling Yao[1,2(✉)], Qingsheng Liu[3], and Gaohuan Liu[3]

[1] Harbin Institute of Technology Shenzhen Graduate School, Shenzhen, China
yaol@lreis.ac.cn
[2] Center for Assessment and Development of Real Estate, Shenzhen, China
[3] Institute of Geographic Sciences and Natural Resources Research, Chinese Academy of Sciences, Beijing, China
{liuqs,liugh}@lreis.ac.cn

Abstract. While remote sensing and geographic information systems have been used successfully to classify forest health using recent image, applying this process to older images is problematic because contemporaneous field data are not available to measure the accuracy of the classification of historical images. Data ranges of normalized difference water index(NDWI) were established for each Robinia Pseudoacacia health class using a contemporary image and field data by sequential cluster analysis. These ranges were used to separate Landsat Thematic Mapper (TM) images acquired from 1999 to 2007 into a series of health-class maps, By applying cross-tabulation procedures to pairs of classified images, we can see how the Robinia Pseudoacacia health class of each pixel in the images of the study area had changed over time. The resulting maps provide a look back at forest conditions of the past and can be used to identify areas of special interest. Further analysis carried out between environmental factors(soil salt, soil texture, soil type, DEM, groundwater depth and groundwater salinity)and the Robinia Pseudoacacia health led to identify the likely causes of these negative trends.

Keywords: Robinia pseudoacacia health · Remote sensing · Spatial analysis · Change detection

1 Introduction

Artificial Robinia Pseudoacacia is an important part of ecosystem in the Natural Reserve of Yellow River Delta. Its area reached 12000 hm^2 in 1999, which is the biggest artificial Robinia Pseudoacacia forest in the northeastern plain of China [1]. These forests have great aesthetic value and have become the tourist attraction in the Yellow River Delta. But the dieback or dead of Robinia Pseudoacacia were noted at some areas of Yellow River Delta in the late 1990's, which may be caused by intrinsic vegetation processes (over mature), land-use and/or other human-induced changes (e.g., pollution stress, oil exploitation, forest fire, deforestation), and natural disasters (e.g., insect infestation, storm tide), till 2007 its area is only 3182 hm^2. Forest dieback and recovery are key

F. Bian and Y. Xie (Eds.): GRMSE 2015, CCIS 569, pp. 318–326, 2016.
DOI: 10.1007/978-3-662-49155-3_32

processes in forest ecosystem dynamics. Recovery is the reestablishment of a new forest stand or the regeneration of a partially disturbed stand following a previous disturbance. From a management perspective, natural forest diebacks are often considered to cause unforeseen loss of forest biomass or to decrease the actual or potential value of forest stands [2]. However, from an ecological perspective, a natural disturbance is merely a cyclical stage of forest destruction–creation dynamics [3]. The dominant natural disturbance agents in European forests are storm events (53 %), fires(16 %) and biotic factors (e.g., pest, 16 %) [2].

There is a critical need to measure, monitor, and predict the dynamic changes and health status of artificial Robinia Pseudoacacia forest. Over the last few decades of remote sensing history, numerous digital change detection techniques have been developed for use with Landsat and other satellite images [4]. Using time series of remote sensing data both continuous and subtle (associated with forest degradation or recovery) as well as discontinuous and sudden (e.g., clear-cuts,wind-throws) forest change phenomena can be assessed,quantified and monitored [5]. The Landsat satellites provide a unique, continuous record of earth observations at ample spatial and spectral resolution since 1972 and, hence, are well-suited for long-term forest change analyses [6, 7]. Using this archive, there has been success in relating the relative changes in Landsat spectral response to the severity of aspen defoliation [8].

In this study, we had two main goals. The first was to quantify and map the spread of the dieback over space and time. This was accomplished by interpreting remotely sensed data to locate Robinia Pseudoacacia stands on images of different years. A GIS was used to enhance the images, classify the Robinia Pseudoacacia by health, and quantify the areas. The second component of this study was to explore the data to identify forest health patterns and relationships among environmental site factors and Robinia Pseudoacacia decline. Hazard rating maps would assist in the placement of biologic controls, would guide the implementation of best management practices of Robinia Pseudoacacia forests, and could focus land preservation efforts throughout the region.

2 Materials and Methods

2.1 Study Area

The study site is located in the Modern Yellow River Delta of Dongying city, Shandong province, China. Soil texture is sandy loam, and the product of interaction of sea and land and Yellow River. The area has a monsoon climate of the warm-temperate zone. Four seasons can be distinguished: a dry cold winter, a dry warm spring, a humid hot summer and an autumn with temperate conditions. The average total sunshine is 2190 ~ 2380 h per year. The average annual temperature is 12.4 degree centigrade and the lowest temperature is minus 20.2 degree centigrade. The frost-free period lasts 211 days per year. The average annual precipitation is 530 ~ 630 mm, of which 70 % is rainfall during summer. Average annual evapo-transpiration is 1900 ~ 2400 mm, about 3.6 times of the precipitation. Because of the flat terrain and short history of the territory, this area is serious in soil salinization and alkalization except some farm land with good drainage conditions. Usually the ration of salt content is above 1 %. The rate of mineral content of groundwater (1 ~ 2 m under the surface) is above 10 g/l [9].

Nearly pure and relatively integrated artificial Robinia Pseudoacacia forest land in Gudao forestry, nearby Machang, nearby Xianhe, nearby the old course of Yellow River were selected as the experimental area(Fig. 1).

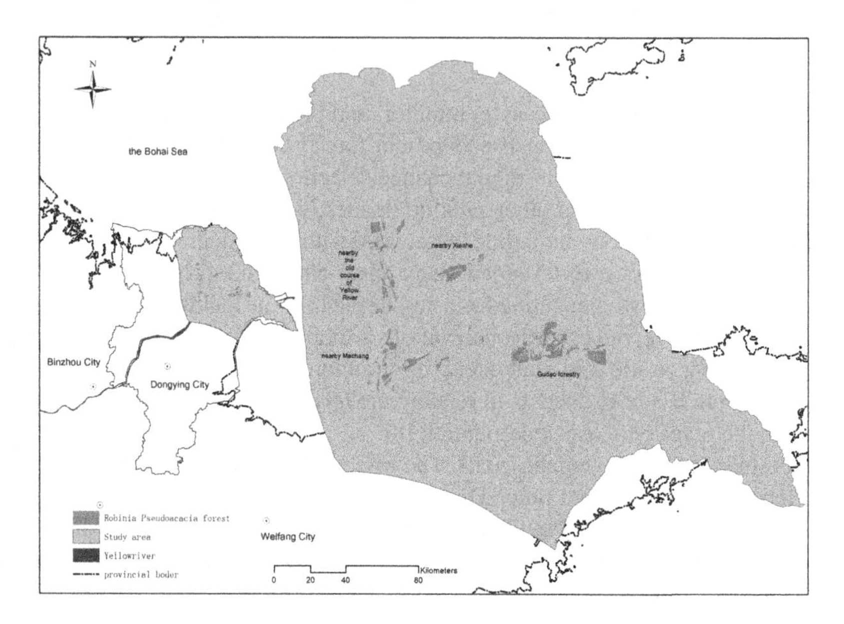

Fig. 1. Study area in the Yellow River Delta

2.2 Satellite Data Preparation

This study used five Landsat TM/ETM images spanning eight years. Each image was captured in the summer, prior to local leafout, to minimize the effect of foliar reflectance from deciduous trees interspersed among Robinia Pseudoacacia. Image dates were acquired on 25 June 1999, 6 June 2001, 11 May 2003,8 May 2005 and 14 May 2007.

The images had been corrected geometrically and radiometrically such that they have the highest level of achievable geolocation accuracy and radiometric consistency. A root mean square error is controlled within 0.5 pixels or less. Registration process is basically the same as geometric precision correction, RMSE (root mean square error) is generally controlled within 0.3 pixels, individual pixels of 0.5. Nearest neighbor re-sampling was used to create the new image.

Since there were many serious forest diebacks which were unable to show the reflection characteristics of Robinia Pseudoacacia, it is unlikely to extract the complete Robinia Pseudoacacia forest area and all health categories from the recent Landsat TM image. So we decided to use an earlier image of 1999 to get all Robinia Pseudoacacia forested pixels at the beginning of the study period. The supervised classification was applied to the images of 1999 to extract Robinia Pseudoacacia forest. A Robinia Pseudoacacia mask was created to remove all non-Robinia Pseudoacacia pixels from the study, and then the mask was applied to the image acquired on 14 May 2007. For comparing analysis, we need to get the same Robinia Pseudoacacia area between 1999 and 2007. The Yiqianer forest and the northeast of Gudao forest have been replaced by reed meadow and dry land crops in 2007, so these two forest areas will be removed.

2.3 Robinia Pseudoacacia Health Classification Using the 2007 TM Image

Several image enhancement transforms were applied to the Robinia Pseudoacacia forest portions of the 2007 image, including Normalized Difference Vegetation Index (NDVI), Soil Adjusted Vegetation Index (SAVI), Modified Soil Adjusted Vegetation Index (MSAVI), normalized difference water index (NDWI) and Tasseled Cap Transform (TCT). A cluster analysis was performed on each image to create five unique, non-overlapping classes.

The USFS Crown Condition Rating Guide (CCGR) was used by ground crews to measure the health of Robnia Pseudoacacia stands at 40 locations across the study area [10]. The NDWI transform produced the most accurate classification of Robnia Pseudoacacia health, with an overall accuracy of 82.5 % for the 30 m buffer site accuracy measurement, followed by the VIS for Modified Soil Adjusted Vegetation Index (72.5 %), Normalized Difference Vegetation Index (62.5 %), Soil Adjusted Vegetation Index (57.5 %). The Tasseled Cap transform could not separate the health and light dieback, which provided no advantage over classifying an image with no transform.

The single-band image is divided into five categories by segmenting threshold of NDWI using a decision tree classifier. Larger index values correspond to healthier vegetation, and the first four classes were labeled Healthy, Light dieback, Moderate dieback, Severe dieback. The fifth (lowest) class was labeled others(dry land corps). The sofeware ENVI 4.5 supported all the processing.

2.4 Classification of the Earlier TM Images

The Robinia Pseudoacacia mask was applied to the pre-processed images acquired from 1999, 2001, 2003 and 2005 to eliminate non-Robinia Pseudoacacia forest pixels from further study. Since the acquisition date for each scene images were not the same, weather conditions were different, biomass vegetation varied each year, which resulted in vegetation reflectance values were not the same. LeMarie proved that the same year of vegetation NDVI value decreased in his study [11]. So using the same threshold range to divide Robinia Pseudoacacia forest category is infeasible. We got new Robinia Pseudoacacia forest health classes maps for each year based on the health classification ranges using different level-slicing of NDWI (Table 1).

Larger index values correspond to healthier vegetation, and the first four classes were labeled Health, Light dieback, Moderate dieback, and Severe dieback. Statistics for the number of pixels used to calculate the percentage of each classification image.

Table 1. The health classification ranges of the clusters created after the image was transformed by the vegetation indices NDWI

	Health	Light dieback	Moderate dieback	Severe dieback
1999	>0.129	0.073 ~ 0.129	0.002 ~ 0.073	-0.972 ~ 0.002
2001	>0.136	0.067 ~ 0.136	-0.008 ~ 0.067	-0.167 ~ -0.008
2003	>0.071	0.010 ~ 0.071	-0.046 ~ 0.010	-0.169 ~ -0.046
2005	>-0.028	-0.099 ~ -0.028	-0.172 ~ -0.099	-0.359 ~ -0.172
2007	>0.0196	-0.0440 ~ 0.018	-0.110 ~ -0.046	-0.292 ~ -0.112

2.5 Temporal and Spatial Analysis

These ranges were used to separate Landsat Thematic Mapper (TM) images acquired from 1999 to 2007 into a series of health-class maps. An important aspect of the Robnia Pseudoacacia decline is the change in health, by location, over time. This can be examined by comparing the health of each Robinia Pseudoacacia pixel in two classified images captured at different times. Health classes were assigned numerical values from 1 to 4, with Health = 1; Light dieback = 2; Moderate dieback = 3; and Severe dieback = 4. Next, By applying cross-tabulation procedures to pairs of classified images, it is possible to construct a transition map that indicates how the Robinia Pseudoacacia health class of each pixel in the images of the study area has changed over time. The visual comparison of maps is useful for identifying the dominant trends and changes between images, but it is a subjective exercise that may miss subtle variations in the data.

In an effort to quantify map comparisons, Minnick proposed the use of a Coefficient of Areal Correspondence (Ca) which could be used to evaluate and compare the correspondence between natural patterns and surfaces (Unwin, 1981). The Ca has a possible range of 0, where the distribution of pixels are completely separate, to + 1, where pixels from both phenomena overlap each other completely. The Coefficient of Areal Correspondence formula is:

$$C_a \equiv \frac{x_{ii}}{x_{i+} + x_{i-} - x_{ii}} \tag{1}$$

where xii is the number of pixels that were in health class i in both years; xi + is the row total, or number of pixels that were in health class i in year one; and x + i is the column total, or number of pixels that were in health class i in year two.

The above comparison of Robinia Pseudoacacia health by pixel provides a progressive view of the data between any two times. While this is useful information, it may present a misleading picture of the changes in Robinia Pseudoacacia health among multiple time spans. Pixels that were Light dieback in both 1999 and 2001, and pixels that were Light dieback in both 2001 and 2003, can belong to completely different pixel sets. A more revealing method of analyzing the changes in health over time is to examine the health of the pixels in each of the four images, or times, involved in this study.

We use transition matrix of 1999 and 2001 with the transition matrix of 2003, 2005 and 2007 to get a transfer matrix. In the new class in a five-phase image pixel represents a single pixel of health categories, such as 11111 pixel represents that the five phases are healthy, 13244 pixels represents that 1999 was healthy, 2001 was moderate dieback, 2003 was light dieback, 2005 was severe dieback or death, and 2007 had the same health status with 2005.

We use six major trends to describe Robinia Pseudoacacia forests health change over time. These trends represent 68 % of the entire study area, the trend does not belong to the five trends as other class. Figure 2 depicts six trends of the study area. We calculated the statistical number of pixels trend throughout the study area and the four sub-regions (Table 2).

Table 2. The dominant trends and changes of Robinia Pseudoacacia health over time

Trends	Description	The study area		The old course of yellow river	
		Ha	%	Ha	%
1	The combination of healthy and light dieback in five images	545.2	15.5	64.3	8.9
2	The combination of healthy and light dieback in four images	534.7	15.2	66.4	9.2
3	The combination of healthy and light dieback in former two or three images, moderate and severe dieback in subsequent two or three images	432.7	12.3	59.0	8.2
4	The combination of moderate and severe dieback at least in four images	661.3	18.8	225.7	31.2
5	The combination of moderate and severe dieback in former two or three images and the combination of healthy and light dieback in subsequent two or three images	207.5	5.9	86.4	11.9
6	Other combinations	1136.2	32.3	221.2	30.6
Total		3517.7	100	689.04	100

Nearby Xianhe		Nearby Machang		Gudao Forest	
Ha	%	Ha	%	Ha	%
81.7	18.7	161.9	23.5	238.4	14.3
106.1	24.2	137.8	20	223.4	13.4
9.7	2.2	28.9	4.2	336.8	20.2
34.5	7.9	105.4	15.3	295.1	17.7
68.9	15.7	31.7	4.6	20	1.2
137.1	31.3	223.3	32.4	553.6	33.2
438.03	100	689.1	100	1667.4	100

The first trend represents the combination of healthy and light dieback in five images, these Robinia Pseudoacacia forest pixels represent a relatively good condition and which are stable. The second trend is the combination of healthy and light dieback in four images, which did not show any obvious decrease. Robinia Pseudoacacia in these areas did not exhibit any decline in health throughout the study. The third trend represents the forest was healthy and light dieback in former two or three images, moderate and severe dieback in subsequent two or three images. These areas' healthy condition is on recession, showing

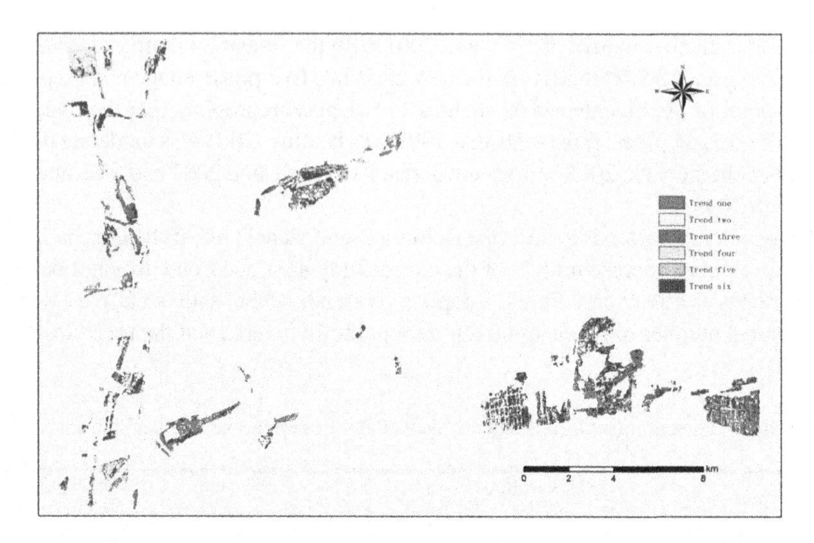

Fig. 2. Describing six major trends of Robinia Pseudoacacia health over time

Robinia Pseudoacacia hardly returned back to health. One dominant trend, or transition pattern, in the overall study area is the fourth trend, which represented the combination of moderate and severe dieback in at least four years, These pixels represented Robinia Pseudoacacia with consistently bad health throughout the study period. The fifth trend represents the forest was moderate and severe dieback in former two or three images, healthy and light dieback in subsequent two or three images, These areas represent the health of Robinia Pseudoacacia are from bad to good.

Overall, from 1999 to 2007, Robinia Pseudoacacia Forest Health declined. Health and light dieback categories basically showed a decreasing trend, moderate dieback rate remained unchanged, severe dieback category showed an increasing trend as a whole. Severe dieback ratio change is also the largest and the only one area of the increased percentage categories.

The healthy level variation span is quite large, in this study only about 45 % of Robinia Pseudoacacia changed between adjacent pixels, which those pixels belong to the three combinations: Health and light dieback, light dieback and moderate dieback, moderate dieback and severe dieback.

3 Result and Discussion

Many standard statistical procedures assume that populations are randomly distributed over space [12]. In the field of ecology the random distribution assumption frequently does not apply. Many events are related to adjacent phenomena such as, the heterogeneous landscape structure, surrounding land cover types, or occurrences of nearby pest infestations [13]. Spatial autocorrelation is the property of nearby points to take on values that are more similar or less similar to each other than would occur randomly [14]. Spatial autocorrelation is the property of nearby points to take on values that are more similar or less similar to each other than would occur randomly [14]. This phenomenon impacts the Robinia Pseudoacacia forests in this study. Robinia Pseudoacacia of different health classes are not randomly distributed over the area but are found in clustered patterns. So the chi-square test was used to explore relationships between Robinia Pseudoacacia health and soil salinity. Landscape features were derived from existing layers of digital data. Digital soil salinity data was generated by geostatistical analyst. GIS was used to calculate the frequency of occurrence for each combination of landscape features and health class.

According to the spatial and statistical analysis of site conditions and Robinia forest health category, we explored the factors of the Robinia forest dieback or death. The research showed that: Soil salinity, soil texture, elevation, water depth and salt content of underground water were closely related to the health condition of Robinia forest.

Through the above analysis the suitable soil conditions for growth of Robinia Pseudoacacia in the Yellow River Delta are: The Robinia Pseudoacacia grows well in medium soil; has resistance to a certain salinity, can grow healthily from 0.4 % to 0.6 % salt content of the soil; the water level is too high, causing the mud, affecting the health of Robinia Pseudoacacia forest; the lower groundwater salinity is, the better the Robinia Pseudoacacia forest's health situation, the more compatible groundwater salinity with the growth of Robinia Pseudoacacia is 0-6 g/l.

4 Conclusion

While there has been a general decline in Robinia Pseudoacacia health throughout much of the region between 1999 and 2007, there have been substantially different patterns of change in specific locations in study area. This suggests that several factors may be involved in the decline. The spatial aspects of this study are important because an examination of health changes over time and space may help to reveal trends and patterns in the data that can contribute to a greater understanding of the complex interactions between forest and insect communities. Transition maps can be used to consolidate information from a series of similar images. By applying a cross-tabulation procedure to sequential pairs of images, one combined image can be produced that captures changes in health over time. These transition maps, and underlying matrices, can be inspected to identify dominant patterns or trends in the data and locate where these patterns occur.

The examination of the forests at the landscape level has helped to identify several spatial attributes of the Robinia Pseudoacacia decline. It is hoped that this research will be useful to scientists studying the decline of Robinia Pseudoacacia. This information may also guide the resource management efforts of forest professionals and the land acquisition objectives of conservation organizations. Several areas of additional research have been identified that may lead to new insights into the interactions between the Robinia Pseudoacacia and the site conditions.

Acknowledgment. This research work was jointly supported by a grant from the National Natural Science Foundation of China (Project No. 40771172) and a grant from National Key Technology R&D Program of China (Project No. 2008 BAC34B06) and a grant from Innovative

Program of The Chinese Academy of Sciences (Project No. kzcx2-yw-308 and 066U03003SZ) and a grant from Special Project of Water Body Contamination Control and Curement of China (Project No. 2008ZX07526-007).

References

1. Wang, S.M.: Quicken steps in greening in the Yellow River Delta. Land Greening, Beijing **3**, 18 (2002)
2. Schelhaas, M.J., Nabuurs, G.J., Schuck, A.: Natural disturbances in the European forests in the 19th and 20th centuries. Global Change Biol. **9**, 1620–1633 (2003)
3. Rull, V.: Sustainability, capitalism and evolution-nature conservation is not a matter of maintaining human development and welfare in a healthy environment. EMBO Rep. **12**, 103–106 (2011)
4. Coppin, P., Lambin, E., Jonckheere, I., Nackaerts, K., Muys, B.: Digital change detection methods in ecosystem monitoring: a review. Int. J. Remote Sens. **25**, 1565–1596 (2004)
5. Kennedy, R.E., Cohen, W.B., et al.: Trajectory-based change detection for automated characterization of forest disturbance dynamics. Remote Sens. Environ. **110**(3), 370–386 (2007)
6. Cohen, W.B., Goward, S.N.: Landsat's role in ecological applications of remote sensing. Bioscience **54**, 535–545 (2004)
7. Kennedy, R.E., Yang, Z., et al.: Detecting trends in forest disturbance and recovery using yearly Landsat time series: 1. landtrendr — temporal segmentation algorithms. Remote Sens. Environ. **114**(12), 2897–2910 (2010)
8. Hall, R.J., Fernandes, R.A., Hogg, E.H., Brandt, J.P., Buston, C., Case, B.S., Leblanc, S.G.: Relating aspen defoliation to changes in leaf area derived from field and satellite remote sensing data. Can. J. Remote Sens. **29**, 299–313 (2003)
9. Liu, G.H., Drost, H.J.: Atlas of the Yellow River Delta. The publishing House of Surveying and Mapping, Beijing, pp.23 (1997)
10. Yao, L., Liu, G.H., et al.: Remote sensing monitoring the health of artificial Robinia Pseudoacacia Forest. Geomatics Inf. Sci. Wuhan Univ. **35**(7), 863–866 (2010)
11. LeMarie, M., van der Zaag, P., et al.: The use of remote sensing for monitoring environmental indicators: the case of the incomati estuary, Mozambique. Physics Chemistry Earth, Parts A/ B/C **31**(15–16), 857–863 (2006)
12. Reich, R.M.: Geils, B.W.: Review of spatial analysis techniques. In: Liebhold, A.M.: Barrett, H.R. (eds.) Proceedings: Spatial Analysis and Forest Pest Management, Mountain Lakes, Virginia, USA, 27–30 April 1992, pp. 142–149 USDA Forest Service General Technical Report NE-175 (1992)
13. Legendre, P.: Spatial autocorrelation: trouble or new paradigm? Ecology **74**(6), 1659–1673 (1993)
14. Cliff, A.D., Ord, J.K.: Spatial autocorrelation. Pion, London Congalton RG a review of assessing the accuracy of classifications of remotely sensed data. Remote Sens. Environ. **37**, 35–46 (1991)

Design of Power System Parameters Collection Device Based on Wireless Transform Technology

Liming Wang(✉), Xiaoling Yan, and Dingyuan Yang

Electric and Information College, Naval University of Engineering,
Wuhan 430033, Hubei, China
Yanxl0213@163.com

Abstract. With the rapid development of electronic technique in ships, electric propulsion warship had been paid high attention. Power parameters' real-time detection of warship exerts an active influence to power quality and warship power system's safe running. The article designs a power parameter collection device based on WIFI wireless transform technology. This device uses voltage clamp and current sensor to collect three-phase voltage and current data of ship power system. Through AD module, the data are turned into digit data and transferred to control terminal by WIFI module. By experiment, the device was confirmed that it can complete precise data acquisition at real-time.

Keywords: Electric propulsion · Frequency control · Current sensor

1 Introduction

In recent years, the development of warship gradually trends to large-scale and automatization. More and more high power alternating current machines are used in ships. With the widespread application of electrical energy, it supplies not only to usual lighting load, but also to every auxiliary engine, communication navigation equipment and all kinds of warning devices. The warship electric propulsion technology is paid high attention to at home and abroad and develops rapidly. Because of the increase of power consumption equipment device, the problem of power quality becomes more and more serious. Power parameters' detection is the basis of heightening power quality, so its position is also very important.

Aiming at power parameters' detection, the article designs a parameter collection device based on WIFI wireless transform technology. This device applies STM32F103 as processor and embedded system. It can monitor warship power parameters while it's running and transfer them to control center at real-time by WIFI. Wireless transform technology makes real-time supervisory control system installation and maintenance easier and the monitoring can be in progress at any time and place. The collection device not only enhances the observability of ship power system which is helpful to heighten the security, responsibility and stability of ship power system, but also makes power quality better, cuts down the costs of working and maintenance, and lengthens the electric equipment's service life.

F. Bian and Y. Xie (Eds.): GRMSE 2015, CCIS 569, pp. 327–335, 2016.
DOI: 10.1007/978-3-662-49155-3_33

2 The Entirety Design of Collection Device

The entirety design of the power parameter collection device is like Fig. 1. The original voltage and current parameters in ship power system are captured by voltage clamp and current sensor. They are then transferred to wireless data acquisition and transport module by AD unit. At last, the parameters will be sent to control center through WIFI module. The device applies STM32F103 as process to construct the system configuration. ARM processer corresponds with peripheral AD module through SPI interfaces, and through SDIO interfaces with WIFI module. The AD reset and sample are controlled by ARM through I/O interfaces and EXTI interrupt indicates the finish of sampling.

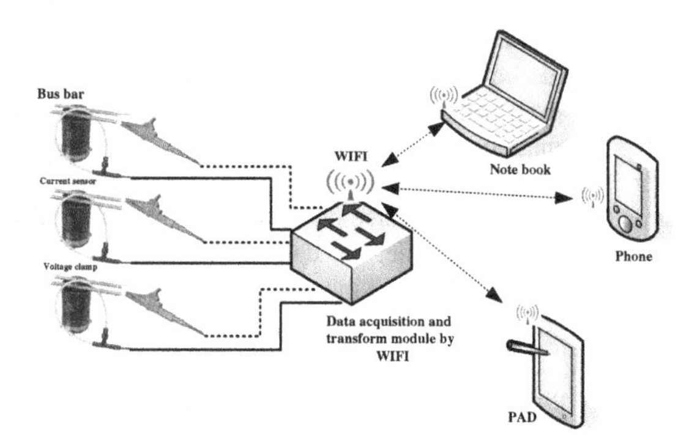

Fig. 1. Collection device diagrammatic sketch

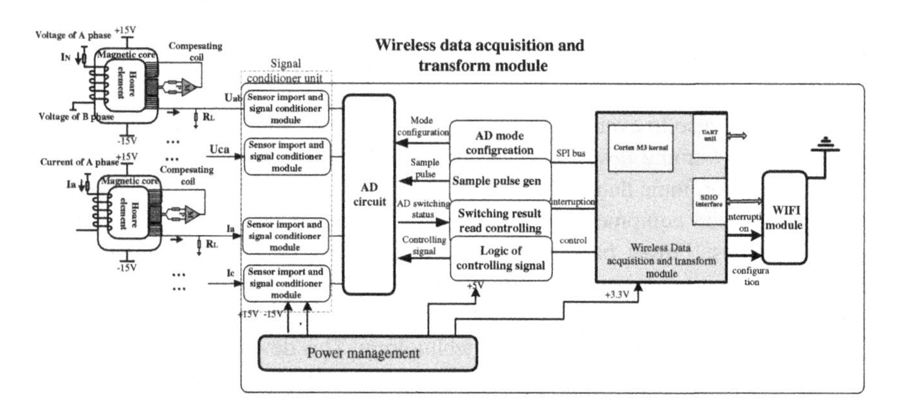

Fig. 2. The architecture of power parameters collection device

The function architecture of power parameter collection device is like Fig. 2. From Fig. 2, it can be known that the device mainly completes parameters' converter and transform by data acquisition and transport modules. The power module supplies three kinds of voltages:15 V, 5 V and 3.3 V. These voltages are separately provided to sensor importing and conditionermodule, controller and wireless module. The power parameters will go through AD conversion after passing signal conditionermodule. The micro-processor is a bridge between AD conversion and wireless transport. It is controlled by interrupt and GPIO, it gets AD conversion value by SPI bus, corresponds with WIFI module by SDIO interface, and its UART serial communication interfaces are used to debug the program.

2.1 Power Management Unit

From Fig. 2, it can be known that the device running needs three categories of voltage: 15 V, 5 V and 3.3 V. These three kind voltages are supplied by power management unit. 5 V is the device's source voltage. +15 V and -15 V are produced by E0515 and supplied electricity to operational amplifier LM258. 3.3 V is important for the device. In consideration of device's low power consumption, and in order to decrease voltage ripple, the device chooses low dropout linear regulator (LDO). By comparing different parameters of LDO, the device chooses LM117 to supply 3.3 V voltage. Its LDO power supply circuit is Fig. 3. In Fig. 3 C3, C44, C45, C46 are I/O smoothing capacitance.

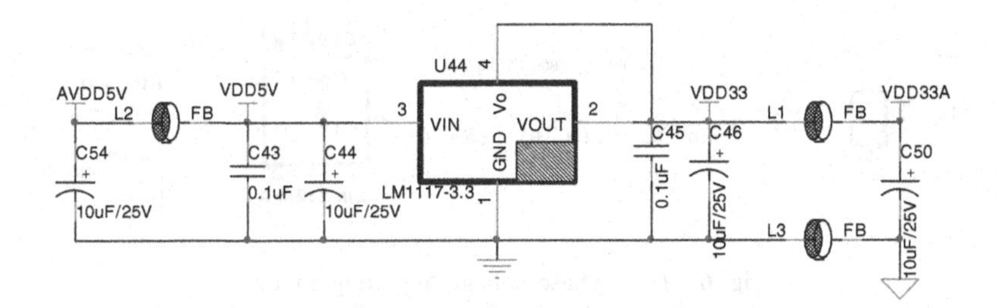

Fig. 3. LDO power supply circuit

The source power of WIFI is provided by VDD33I and its on-off is controlled by STMF103 PB9 which can control the start and finish of WIFI. The VDD33I circuit is like Fig. 4.

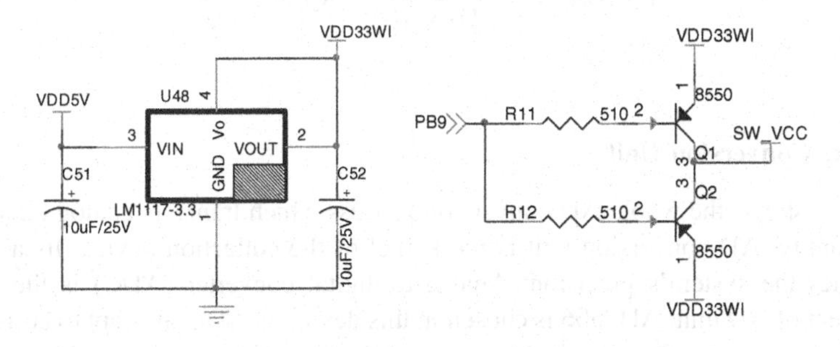

Fig. 4. VDD33WI circuit

2.2 Signal Conditioner Unit

After the amplitude of external signals of voltage and current falls to a reasonable value to detection device by Hoare sensor, if they need to be imported to AD7606, they have to go through conditioner circuit. The conditioner circuit is shown as Figs. 5, 6.

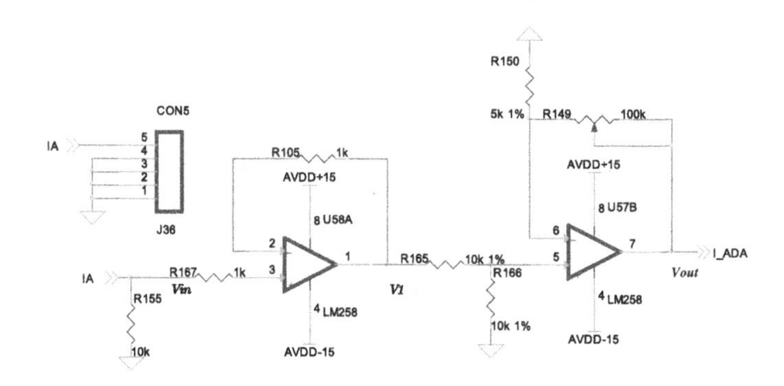

Fig. 5. A phase current importing circuit

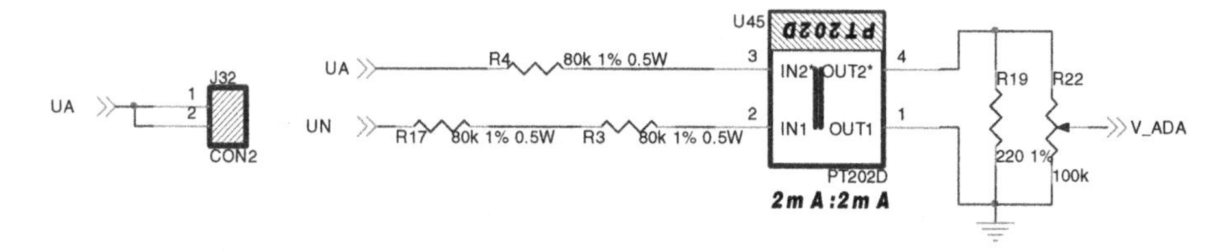

Fig. 6. Three phase voltage importing circuit

From Fig. 6, Voltage can be turned to current signal by passing three 240 K resistances, and then it will go through a 2mA : 2mA instrument transformer. The output value V_AD can be adjusted to fit to range ability of AD7606 by changing the resistance value of R22. The output value can be obtained by relationship as follow.

$$V\_ADA = \frac{U_{AN}}{3 \times 80 \times 10^3} \times R_{19} \times \frac{R_{22}}{100K}$$

2.3 AD Conversion Unit

The Fig. 2 shows the whole AD conversion process which transfers analog quantity to digit quantity. AD conversion unit is the kernel of the collection device. Its accuracy determines the system's precision. Analog-to-digital converter (ADC) is the central component of AD unit. AD7606 is chosen in this device as ADC. It is apt to correspond with microprocessor. The circuit diagram of AD conversion unit is like Fig. 7.

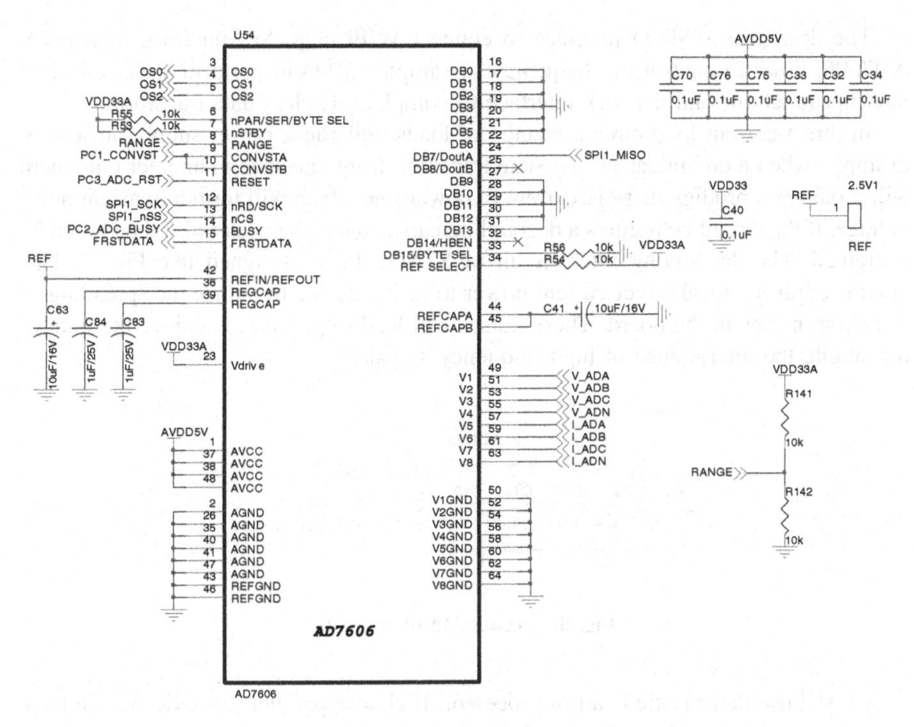

Fig. 7. Circuit diagram of AD7606

In AD circuit, the hardware circuit is designed to uses serial interface mode. It connects to DOUTA pin by MISO interface of SPI. This connection can complete data transmission. When CONVST pin is pulled to high level from low level by input signal, this signal starts to be sampled and converting starts too. AD7606 uses 5 V as its source supply. It can handle bipolar input signal of ± 10Vor ± 5 V and all channels can sample data at 200 kps speed. The clamping protective circuit can bear ± 16.5 V voltage. $\overline{PAR}$/SER/BYTE SEL connects high level, DB15 connects GND. RANGE pin decides the analog input range. When it connects high level, the input range is ± 10 V, otherwise, the range is ± 5 V. RESET is reset pin which is valid at low level. After the component powers on, it will receive a reset pulse. CS is chip selection pin which can start serial data transmission when it is at low level.

2.4 Micro Controller Unit

The power parameter collection device involves STM32F103 as its micro-controller produced by Mouser Electronics, Inc. the kernel of this microcontroller is ARM Cortex-M3 processor. Its basic frequency can reach 72 MHz. This chip integrates many peripheral interfaces such as timer, 7-channels DMA, CAN bus, ADC analog-to-digit converter, SPI interface, I^2C bus interface, SDIO, USART interface and GPIO etc.

The design uses SDIO interface to connect WIFI chip, SPI interface to connect AD7606, timer to control the frequency of sample, GPIO to perform start, reset and interrupt functions and network interface to complete wireless data transform.

In direct-current loop circuit, change of loads will cause power supply noise. For example, when a circuit carries out state transition from one state to another one, there will produces a prodigious peak current in power line which will form a transient noise voltage. If the circuit configures a decoupling capacitance, the peak noise then could be restrained. The decoupling filter circuit of STM32103 is designed like Fig. 8. This circuit supplies a local direct-current power to active device to reduce the spreading of switching noises in the board. These noises will lead to ground to achieve the goal of restraining the interference of high frequency signals.

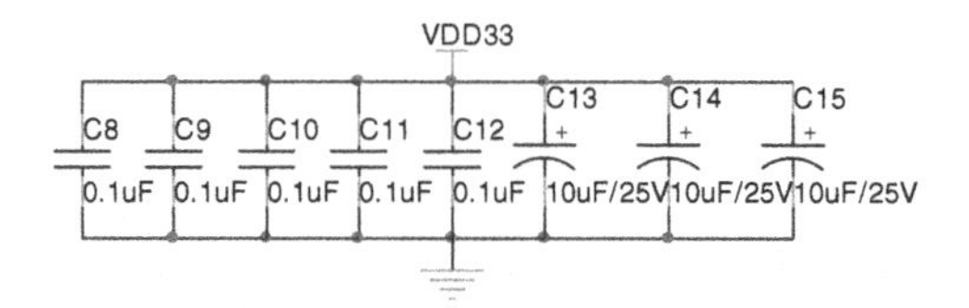

Fig. 8. Decoupling filter circuit

Crystal oscillator is the heart of processor. If choose parallel capacities which have the same resistance with load capacities, we can get the resonance frequency of crystal oscillator. The oscillating circuit is usually designed to link crystal oscillate to the both ends of inverting amplifier. In crystal oscillator circuit, there will exist two capacitances that one end connects the crystal oscillator, and the other end connects ground. The resistance of these two capacitances is equal to load capacitance's. The resonant circuit of STM32F103 is Fig. 9.

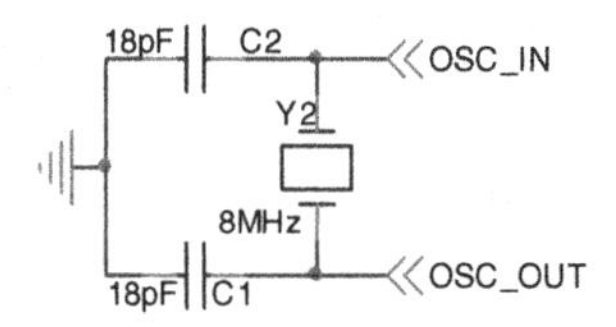

Fig. 9. Resonant circuit of STM32F103

2.5 WIFI Communication Unit

The WIFI wireless communication unit chooses WM-G-MR-09 module produced by USI Company. This module is packaged by 88W8686 chip of Marvell Company. This chip is compatible with protocol of IEEE802.11a/b/g and it also packages memory circuit, antennal loading coil circuit, and etc. If designing on WM-G-MR-09 module, according to requirement, there at least needs jobs as follow: configure its pins needs, design power supply circuit, and plan interface circuit.

Because the device system is real-time supervisory control system, the data transform needs to be real-time when working. So the wireless can't to be allowed in sleep state. As a result, the input pin of sleep pin 35(SLEEP_CLK) is at open circuit state. The WM-G-MR-09 module supports two interface modes——G-SPI and SDIO. Its interface mode can be decided by pin 23(IF_SEL_1) and pin 24(IF_SEL_2). In our system, we choose SDIO interface mode to correspond with STM32F103, so pin 23 (IF_SEL_1) and pin 24(IF_SEL_2) are not connect pull-down resistor. The pins relating to PDA and GPIO are in open circuit state because PDA and GPIO functions are not involved.

Because WM-G-MR-09 is a type of low voltage and high speed processing chip, the turbulence of power-supply will produce big disturb for chip's normal running. This disturb may lead to endless loop, system halt situations, and etc. therefore, the supply power voltage to chip needs to be filtered which adds 1uF and 0.1uF decoupling capacitor on every power supply positions. These positions are also added high frequency magnetic bead. These steps can not only filter harmonic component, but also avoid mutual interference between two power supply circuits. The design diagram of WM-G-MR-09 is Fig. 10.

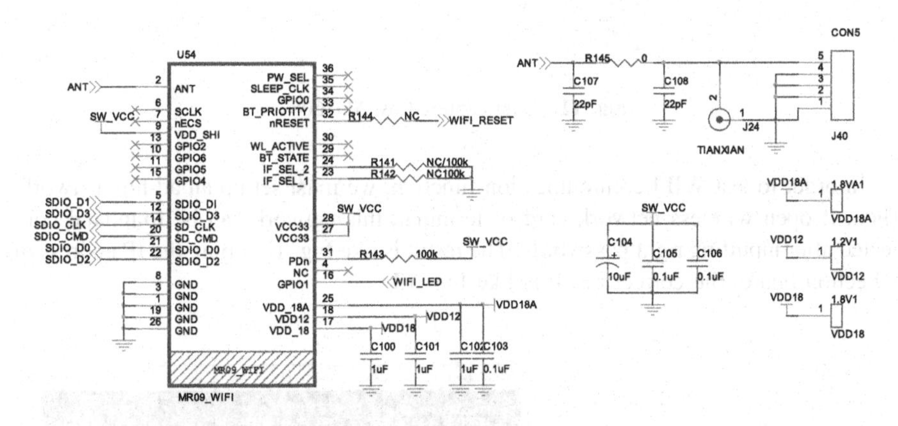

Fig. 10. Circuitous philosophy of WM-G-MR-09

The G-SPI and SDIO of WM-G-MR-09 share the common pin. Its mode can be determined by IFSEL1 and IFSEL2 pins. Because the system uses SDIO interface, the two pins' state is NC. Pin 6(SCLK) and pin 7(ECSn) are used to choose start mode. If SCLK is empty and ECSn links a 100 KΩ resistance, the system starts from SPI EEPROM interface; if SCLK and ECSn are both empty, it starts from host's interface bus. The start mode of this device chooses the second mode. Moreover, the 5, 12, 14, 20, 21, 22 of SDIO pins must be corresponding with STM32, and air wire interface needs 50Ω impedance.

3 System Test of Device

After the design is finished, the system needs test. Firstly, it needs to verify whether the device can truly collect voltage and current data. This test can be completed by Keil's on line debugging facility. The data captured by AD7606 are shown as Fig. 11.

Watch 1

Name	Value	Type
AD7606	0x20000082 &AD7606	struct <untagged>
CHn	0x20000082 &AD7606	short[8][32]
[0]	0x20000082 &AD7606	short[32]
[0]	0x3FFF	short
[1]	0x3FFF	short
[2]	0x3FFF	short
[3]	0x3FFF	short
[4]	0x3FFF	short
[5]	0x0000	short
[6]	0x0000	short
[7]	0x0000	short
[8]	0x0000	short
[9]	0x0000	short
[10]	0x0000	short
[11]	0x0000	short
[12]	0x0000	short
[13]	0x0000	short
[14]	0x0000	short
[15]	0x0000	short
[16]	0x0000	short
[17]	0x0000	short
[18]	0x0000	short
[19]	0x0000	short
[20]	0x0000	short
[21]	0x0000	short
[22]	0x0000	short
[23]	0x0000	short
[24]	0x0000	short

Fig. 11. Data captured by AD7606

In order to test WIFI communication function, we must set up an ad-hoc network. Then, if open wireless network card of terminal, this network can be foundand connected after inputting right passwords. The result is like Fig. 12 if ping the IP address of collection board, the correct result is like Fig. 13.

Fig. 12. Connect successfully diagram

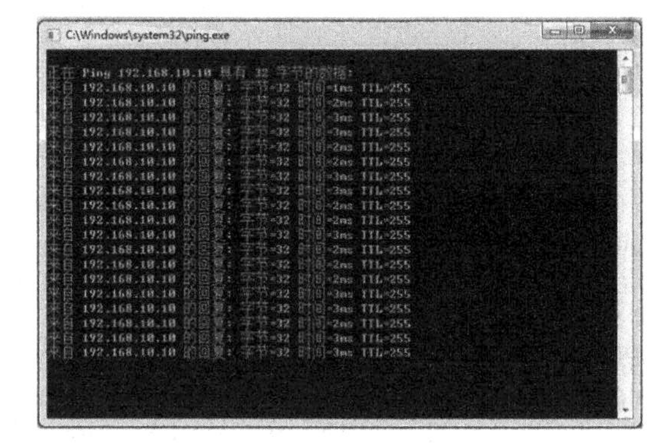

Fig. 13. Result of ping

If connect successfully, the terminal can receive the power parameters from collection device through WIFI by TCP/UDP debugging software. The data are shown as Fig. 14.

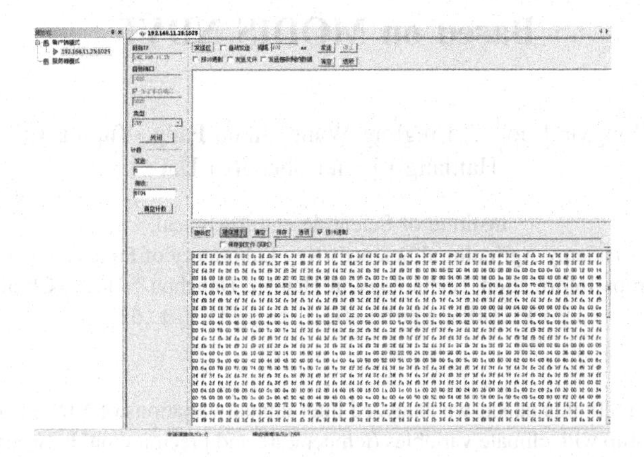

Fig. 14. Data received by TCP/UDP debugging software

4 Conclusion

In order to heighten the power quality of warship power system and gather its power parameters, the article designed a data collection device based on WIFI wireless transform technology. This device mainly completes the job of data collection and transform which includes data acquisition, AD converting and WIFI transform. By test, the device is confirmed that it can send warship power parameters to control terminal at real time and it satisfies design requirement.

References

1. Wu, F.: Study on the Method of Harmonic Control in Marine Electric Propulsion System. Wuhan university of technology (2009)
2. Xu, X.: Research on Marin Combned Electrical Power Plant and Electric Network Power Quality. Shanhai Maritime university (2007)
3. Mindykowski, J.: New challenges in power quality assessment on ships. In: Proceedings of 2006 International Marine Electrotechnology Conference and Exhibition, 6-24 Shanghai (2006)
4. Mindykowski, J.: Assessment of Electric Power Quality in Ship System Fitted with Converter Subsystem. Gdansk: Press of Shipbuilding and Shipping Ltd (2003)
5. IEEE802.11 WORKING GROUP. Draft Supplement to STANDARD FOR Telecommunications and Information Exchange between Systems-LAN/MAN Specific Requirements (2003)
6. Zhou, X.: The Design of Wireless Sensor Network Node. DongNan University (2010)

NDVI, Temperature and Precipitation Variables and Their Relationships in Hainan Island from 2001 to 2014 Based on MODIS NDVI

Hongxia Luo(✉), Lingling Wang, Jihua Fang, Yuping Li, Hailiang Li, and Shengpei Dai

Institute of Scientific and Technical, Chinese Academy Agricultural Sciences/Key Laboratory of Practical on Tropical Crops Information Technology in Hainan, Danzhou 571737, China
{120081008, zishi-010, fangdil}@163.com

Abstract. In this paper, we analyzed inter-annual variations of NDVI and their relationship with climate variables (temperature and precipitation) between 2001 and 2014 in Hainan island. Temporal response characteristics of NDVI-temperature and NDVI-precipitation was analyzed in spring, summer, autumn and winter based on the MODIS NDVI data and daily temperature and precipitation data from 2001 to 2014. The result indicated that the monthly and seasonal NDVI increased significantly over the study period. The mean monthly NDVI reached maximum value in August, and the mean monthly NDVI value was 0.785. The appearance of the largest mean monthly NDVI trend (February) lagged behind that of the largest temperature (May) trend by three months. The response of vegetation NDVI to temperature was more pronounced than to precipitation in the whole year. The maximum response of NDVI to the variation of temperature on the whole had not lag, while the maximum response of NDVI to the variation of precipitation had a lag of about 48 days.

Keywords: MODIS NDVI · Temperature · Precipitation · Lag time · Hainan island

1 Introduction

Vegetation is the natural link of soil, atmosphere and moisture on the earth, and it indicated environment and global changes [1–3]. Studying the inter annual and seasonal changes of NDVI and their relationship with climate is critical for understanding the mechanisms of climate-derived variations in vegetation activity, carbon and hydrological cycles [4]. Among them, temperature and precipitation are the main factors to describe climate conditions, and can affect vegetation growth in an obvious manner [5, 6]. Many scholars studied the correlation between NDVI and climate. For example, Guo et al. (2008) [7] indicated that NDVI changes were significantly correlated with both temperature and precipitation in north China. Luo et al. (2009) [8] reported that there were strong correlations among NDVI, precipitation and temperature. Chuai et al. (2012) [9]

F. Bian and Y. Xie (Eds.): GRMSE 2015, CCIS 569, pp. 336–344, 2016.
DOI: 10.1007/978-3-662-49155-3_34

analysed changes in NDVI, temperature and precipitation, and performed correlation analyses of NDVI, temperature and precipitation for eight different vegetation types in Inner Mongolia. The result indicated that for NDVI correlated quite differently with temperature and precipitation, with obvious seasonal differences.

In this paper, we used the MODIS NDVI datas from 2001 to 2014 to explore inter annual variations of monthly and seasonal NDVI and their relationships with precipitation and temperature in Hainan island.

2 Materials and Methods

2.1 Study Area

Hainan extends from19°20′N -20°10′N, and 108°21′E -111°03′E. Land area of Hainan is about 35000 Km2. The study area is surrounded by the sea (Fig. 1). The Wu Zhishan mountains is located in the central part. The northern part of the study area is close to the Guangdong province. The interior region is lined with some rivers, such as Nandu, Tongtian, and Wanquan rivers. The major types of vegetation included monsoon forest, rain forest evergreen broad leaved forest coniferous forest and mangrove. The study area is characterized by a temperate subtropical monsoon climate. Air temperature spatially increase from north to south with a mean annual value of 22 to 26°C. Precipitation varies greatly within and between years, with 70-90 % of total precipitation occurring between May and October. The mean annual precipitation decreases from 2200 mm in the east to 900 mm in the west.

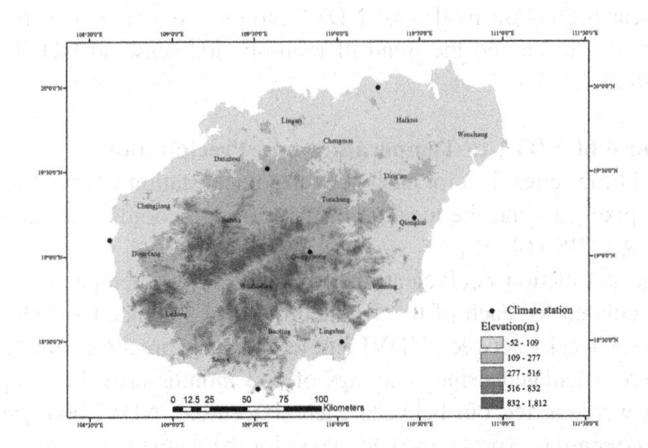

Fig. 1. Study area and location of 7 climate stations used

2.2 Data Source

The NDVI dataset used in this study is MODIS NDVI 16-day composite grid data (MOD13Q1) in HDF format. The spatial resolution of MODIS NDVI dataset obtained from NASA's Earth Observing System from 2001 to 2014 was 250 km × 250 km. Detail documenting the MODIS NDVI compositing process and Quality Assessment

Science Data Sets(QASDS) can be found at NASA's MODIS web site [10]. For this study area, two tiles of the MODIS NDVI product image(h28/v06, h28/v07) were required to cover the entire region.

NDVI data were re-projected from a sinusoidal to WGS84 projection and mosaicked, using a nearest neighbor resampling routine. Monthly NDVI data were obtained for each pixel form a predefined compositing period to represent the current period. This approach is based on the logic that low-value observation are either erroneous or have less vegetation vigor for the period under consideration [11, 12].

We extracted the regional data that cover Hainan using the ARCGIS-10.0 software. Precipitation is a direct factor influencing vegetation primary productivity, but other climatic indicators, such as solar radiation, temperature, and wind, also have impact on vegetation productivity [13]. In order to analyze the relationship between vegetation biomass and climatic features, the precipitation, temperature were selected to explore the correlation with NDVI.

The climatic datasets including daily mean temperature and daily precipitation data were obtained form the Notional Meteorological Center of China; the data include 7 meteorological stations during the period 2001–2014 in Hainan.

2.3 Method

2.3.1 Monthly and Seasonal NDVI, Temperature and Precipitation

Monthly and seasonal NDVI trends and their relationships with climatic changes variables were discussed by four seasons: spring(from March to May), summer(from June to August), autumn(from September to November) and winter(from December to February). Linear regression model for NDVI and climate changes were estimated in this paper. We also discussed the trend of monthly and seasonal NDVI, temperature and precipitation.

2.3.2 Response of NDVI to Temperature and Precipitation

First, the NDVI time series data for each meteorological station were extracted from the mean of 3 × 3 pixels around the location of the stations acoording to the geographical position using ArcGIS [14, 15].

Second, the correlation coefficients between NDVI and temperature and precipitation were calculated for each of the 7 meteorological stations. Considering the time lag of about 1–12 weeks between NDVI and temperature/precipitation, the correlation coefficients were calculated using time lags of 0–3 months [16]. In this paper, correlation analysis were carried out between each sixteen-day NDVI and precipitation of previous 0–5 sixteen-days (previous 0–80 days) for the whole year, spring(from March to May), summer(from June to August), autumn(from September to November) and winter(from December to February). The maximum values were chosen as the correlation coefficients for each station [15, 17].

3 Result

3.1 Trends of Monthly NDVI and Climatic Changes

The magnitude of monthly NDVI and its change over time are important indicators of the contribution of vegetation activity in different months to total annual plant growth [18]. In Hainan, the mean monthly NDVI reached maximum value in August, and the mean monthly NDVI value was 0.785. From February to September, the mean monthly NDVI continued to increase. And the mean monthly NDVI was rather small from December to March. In the 14 years, the trends of monthly NDVI showed positive values except in January, indicating that NDVI increased throughout almost the year over the study period. The maximum trend value was observed in May, three months ahead of the maximum mean monthly NDVI; whereas the minimum was observed in January (Fig. 2a). That means plant growth peaks in summer, while the maximum mean monthly NDVI appeared in spring.

The monthly temperature value increased at first and decreased latter with the increase of the months, while a higher value was observed in summer and a lower values in winter. The monthly temperature values were more than 18°C. That is, Hainan island is an area with clearest feature of tropical climate. The trends of monthly

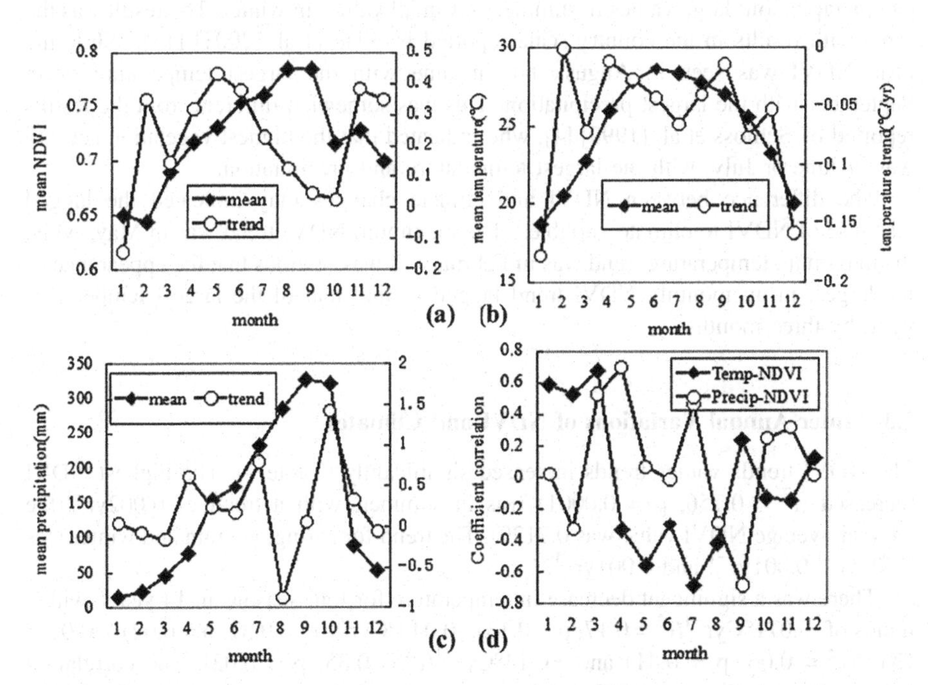

Fig. 2. Seasonal changes in monthly NDVI and climatic variables over the 14 years from 2001 to 2014 in Hainan. (a) 14-year averaged monthly NDVI and its trends. (b) 14-year averaged monthly mean temperature and its trends. (c) 14-year averaged monthly mean precipitation and its trends. (d) correlations between monthly NDVI and temperature and between monthly NDVI and precipitation

temperature showed negative trend values (Fig. 2b). The largest temperature decline occurred in January with an annual rate of 0.1774°C, indicating that temperature in this month decreased by 2.5°C from 2001 to 2014.

For monthly precipitation, a highest value was observed in September and a lowest value was observed in January. Monthly precipitation showed a pattern similar to that of NDVI: a higher value was observed in autumn and lower value winter (Fig. 2c). The trends of monthly precipitation showed positive values except in February, March, August and December. The minimum was observed in August, whereas the largest precipitation rise occurred in October with a rate of 14.3 mm, indicating that precipitation during this month increase by 200 mm in the past 14 years.

The correlation between monthly climatic and NDVI were shown in (Fig. 2d). However, the correlation indicated a large fluctuation: A positive correlation between NDVI and precipitation was observed for most of months; Nevertheless, the negative correlation between NDVI and temperature was observed for most of months. In March, both of NDVI-temperature and NDVI-precipitation correlations were positive.

3.2 Lagged Effect of Climate

In Hainan island, NDVI showed a consistent seasonal pattern with those of temperature and precipitation: large values in summer and small values in winter. The result was the same with results in the country scale reported by Piao et al. [2003] [18]. While the peak NDVI was seen in August, not in June with the largest temperature or in September with the largest precipitation. This was somewhat different from the results reported by Schloss et al. [1999] [4], who indicated that the highest vegetation activity was in June or July, with the largest temperature and precipitation.

The difference between NDVI and climate changes may indicated the lagged response of NDVI to climate variables. The maximum NDVI trend was in May, while the maximum temperature trend was in February. It was obvious that the appearance of the largest mean monthly NDVI trend lagged behind that of the largest temperature trend by three months.

3.3 Inter Annual Variations of NDVI and Climates

The NDVI trends values trends increased significantly (Table 1). The highest NDVI increased (r^2 = 0.756, p = 0.0001) was in summer, with a trend of $0.002yr^{-1}$(the 14-year average NDVI value was 0.7129). The trend for spring, autumn and winter was $0.002yr^{-1}$,$0.001yr^{-1}$ and $0.001yr^{-1}$.

There was a significant decrease in temperature for four seasons in 14 years, with a trends of -0.071°Cyr^{-1}(r^2 = 0.17, p = 0.13), -0.049°Cyr^{-1}(r^2 = 0.37, p = 0.02), −0.029° Cyr^{-1}(r^2 = 0.06, p = 0.41) and −0.14°Cyr^{-1}(r^2 = 0.35, p = 0.02). The correlation between seasonal mean NDVI and temperature for four seasons were -0.09, -0.66, -0.41 and 0.35. Precipitation had a increase trend in spring(r^2 = 0.07, p = 0.40), summer (r^2 = 0.01, p = 0.80), autumn(r^2 = 0.16, p = 0.20), but a slightly decreased in winter with a trend of −0.31mm yr-1(r^2 = 0.06, p = 0.41).

Table 1. The inter-annual changes of NDVI, temperature and precipitation in four seasons.

		Linear regression equations	r^2	p
NDVI	spring	y = 0.003x + 0.69	0.256	0.0648
	summer	y = 0.002x + 0.747	0.756	0.0001
	autumn	y = 0.001x + 0.731	0.129	0.2068
	winter	y = 0.001x + 0.651	0.069	0.361
Temperature	spring	y = −0.071x + 26.04	0.173	0.1381
	summer	y = −0.049x + 28.64	0.372	0.0205
	autumn	y = −0.029x + 25.46	0.057	0.0411
	winter	y = −0.145x + 20.84	0.352	0.0251
Precipitation	spring	y = 2.329x + 76.17	0.068	0.4
	summer	y = 0.808x + 224.9	0.007	0.8
	autumn	y = 9.446x + 175.5	0.162	0.2
	winter	y = −0.306x + 33.73	0.005	0.802

A relatively large NDVI trend was found in the middle of growing season(summer) with a rate of $0.002yr^{-1}$, following that in spring($0.003yr^{-1}$).

3.4 Response of NDVI to Temperature and Precipitation

In the whole year, the NDVI had higher correlation with temperature of previous 0–32 days, and had higher correlation with precipitation of previous 32–64 days (Fig. 3a). The NDVI had the highest correlation with temperature of previous 0 days, and had the highest correlation with precipitation of previous 48 days. The figure showed that the maximum response of NDVI to the variation of temperature on the whole had not lag, while the maximum response of NDVI to the variation of precipitation had a lag of about 48 days. With the increase of the previous days, the correlation between NDVI and temperature of previous 0–32 days first decreased gradually, and then the correlation between NDVI and temperature of previous 32–64 days increased. The correlation between NDVI and precipitation increased gradually with the increase of the previous days, but it was not obvious. The correlation between NDVI and temperature was higher than that between NDVI and precipitation of the corresponding periods, which showed that the effect of temperature seemed more significant than that of precipitation in Hainan island.

In the four seasons, the NDVI-temperature and NDVI-precipitation correlation were highest in autumn, and the response of NDVI to temperature was more significant than that to precipitation.

In spring, the NDVI-temperature and NDVI-precipitation correlation of previous 48 days were the highest correlation (Fig. 3b), and then decreased quickly with the increase of the previous days.

In summer, NDVI had the highest correlation with temperature of the same time and with precipitation of previous 80 days (Fig. 3c), the difference was that the correlation between NDVI and temperature continued to rise while the NDVI-precipitation

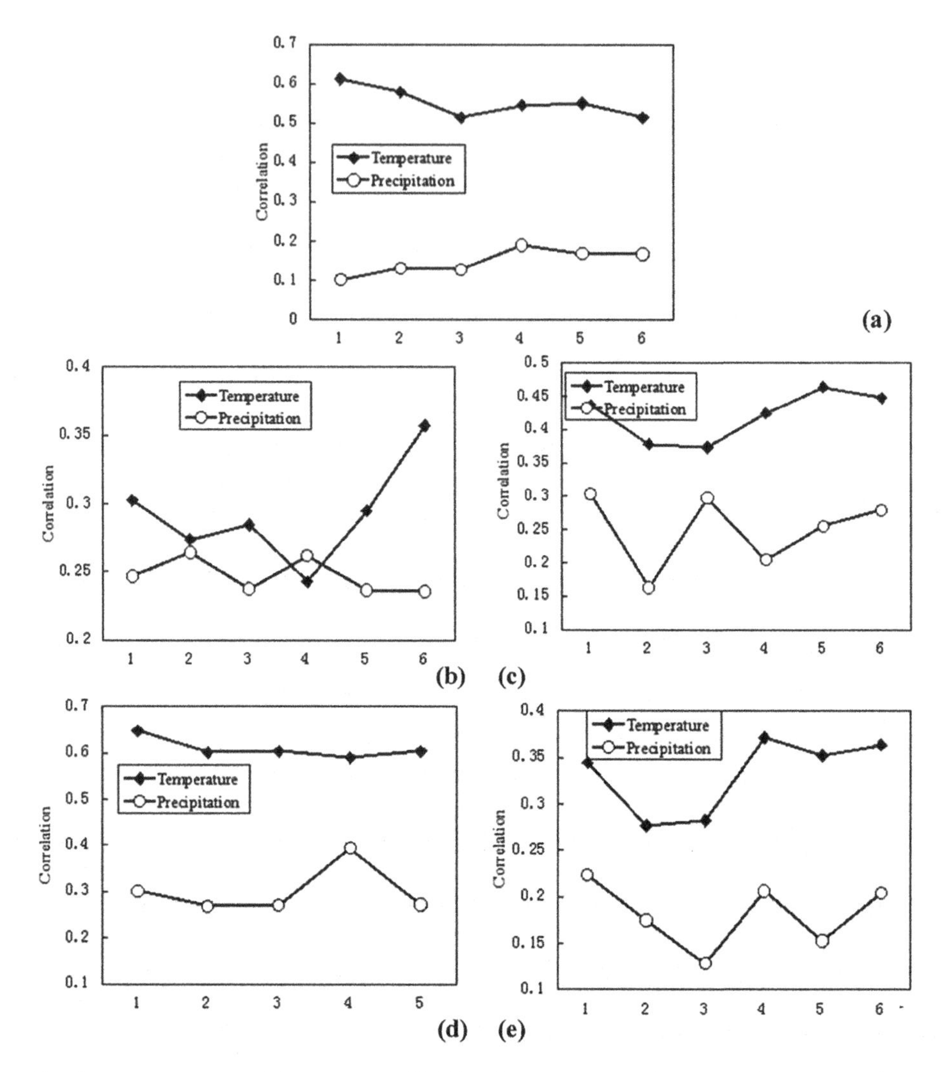

Fig. 3. Correlation coefficient between NDVI and temperature and precipitation. (a) the whole year. (b) spring. (c) summer. (d) autumn. (e) winter.

correlation of previous 0–32 days decreased gradually, then increased rapidly with the increase of the previous days.

In autumn, the NDVI-temperature and NDVI-precipitation correlation of previous 0 days were the highest correlation (Fig. 3d). And the NDVI- temperature correlation of previous 0–64 days, that showed the lag time of NDVI response to temperature was longer.

In Winter, the correlation between NDVI and temperature increased gradually, while that between NDVI and precipitation presents downtrend. The NDVI-temperature of previous 32 days was the highest correlation, while the NDVI-precipitation of previous 48 days was the highest (Fig. 3e).

4 Conclusion

(1) The monthly NDVI increased significantly from March to August. Monthly climatic changes and trends in the 14 years provided a clear illustration of NDVI trends. The maximum NDVI was observed in August, which was consistent with the findings of Piao et al. (2003) who conducted a study of monthly NDVI changes and trends from 1982 to 1999 in China. The value increase was attributed to the influence of adequate temperature and precipitation for vegetation growth in Hainan island. The unchanged increase in NDVI in October was attributed to the influence of Typhoon in this period with heavy rainfall and decrease temperature. Lower temperature and more precipitation inhibited vegetation growth.

(2) Negative correlation between NDVI and temperature in spring, summer and autumn was attributed to the special tropical monsoon climate in Hainan island with higher temperature. Because of clearest feature of tropical climate, the mean temperature in Hainan island was more than 18°C in the whole year. Monthly precipitation showed a pattern similar to that of NDVI: a higher value was observed in autumn and lower value winter. The largest precipitation rise occurred in October with a rate of 14.3 mm, indicating that precipitation during this month increase by 200 mm in the past 14 years.

(3) The difference between NDVI and climate changes may indicated the lagged response of NDVI to climate variables. The maximum NDVI trend was in May, while the maximum temperature trend was in February. It was obvious that the appearance of the largest mean monthly NDVI trend lagged behind that of the largest temperature trend by three months.

(4) In Hainan island, the response of vegetation NDVI to temperature was more pronounced than to precipitation, which is consistent with other results. The maximum response of NDVI to the variation of temperature on the whole had not lag, while the maximum response of NDVI to the variation of precipitation had a lag of about 48 days. In the four seasons, the NDVI-temperature and NDVI-precipitation correlation were highest in autumn, and the response of NDVI to temperature was more significant than that to precipitation.

Acknowledgments. The authors greatly thanks Li M.F for her helpful suggestions. The also thanks the NASA and the National Meteorological Center of China for satellite images and climatic data. This research was funded by the key laboratory of practical on tropical crops information technology fund project(rdzwkfjj014), Hainan application technology development and demonstration promotion(ZDXM2014082), and the natural science foundation of Hainan(20154184).

References

1. Cui, L.L., Shi, J.: Temporal and spatial response of vegetation NDVI to temperature and precipitation in eastern China. J. Geogr. Sci. **20**(2), 163–176 (2010)
2. Habib, A.S., Chen, X.L., Gong, J.Y.: Analysis of Sudan vegetation dynamics using NOAA-AVHRR NDVI data from 1982-1993. Asian J. Earth Sci. **1**(1), 1–15 (2008)

3. Zhang, G.L., Xu, X.L., Zhou, C.P., et al.: Responses of grassland vegetation to climatic variations on different temporal scales in Hulun Buir Grassland in the past 30 years. J. Geogr. Sci. **21**(4), 634–650 (2011)
4. Schloss, A.L.: Comparing global models of terrestrial net primary productivity(NPP): comparison of NPP to climate and the normalized difference vegetation index(NDVI). Global Change Biol. **5**, 25–34 (1999)
5. Fang, J.Y., Piao, S.L., He, J.S., et al.: Increasing terrestrial vegetation activity in China, 1982–1999. Sci. China **47**, 229–240 (2004)
6. Ji, L., Peters, A.J.: A spatial regression procedure for evaluating the relationship between AVHRR-NDVI and climate in the northern great plains. Int. J. Remote Sens. **25**, 297–311 (2004)
7. Guo, N., Zhu, Y.J., Wang, J.M., Deng, C.P.: The relationship between NDVI and climate elements for 22 years in different vegetation areas of Northeast China. Chin. J. Plant Ecol. **32** (2), 319–327 (2008)
8. Luo, L., Wang, Z.M., Song, K.S., et al.: Research on the correlation between NDVI and climatic factors of different vegetation in the northeast China. Xibei Zhiwu Xuebao **29**(4), 800–808 (2009)
9. Chuai, X.W., Huang, X.J., Wang, W.J., et al.: NDVI, temperature and precipitation changes and their relationship with different vegetation types during 1998–2007 in Inner Mongolia. China. Int. J. climatology. **33**, 1696–1706 (2012)
10. MODIS. MODIS Vegetation Index (MOD 13): Algorithm Theoretical Basis Document p. 26 of 29 (version 3) (1999). (http://modis.gsfc.nasa.gov/data/atbd/atbd_mod13.pdf)
11. Holben, B.N.: Characteristics of maximum-value composite images from temporal AVHRR data. Int. J. Remote Sens. **7**, 1417–1434 (1986)
12. Mao, D.H., Wang, Z.M., Luo, L., et al.: Intergrating AVHRR and MODIS data to monitor changes and their relationships with climatic parameters in northeast China. Int. J. Appl. Earth Obs. Geoinformation **18**, 528–536 (2012)
13. Okin, G.S., Murray, B., Schlesinger, W.H.: Degradation of sandy arid shrubland environ ments: observations, process modeling, and management implications. J. Arid Environ. **47** (2), 123–144 (2001)
14. Ding, M.J., Zhang, Y.L., Liu, L.S., et al.: The relationship between NDVI and precipitation on the Tibetan Plateau. J. Geophys. Sci. **17**(3), 259–268 (2007)
15. Nicholson, S.E., Davenport, M.L., Malo, A.R.: A comparison of the vegetation responses to rainfall in the Sahel and east Africa, using normalized difference vegetation index from NOAA AVHARR. Clim. Change **17**, 209–241 (1990)
16. Dehua, M., Zongming, W., Ling, L., Chunying, R.: Integrating AVHARR and MODIS data to monitor NDVI changes and their relationships with climatic paramenters in Northeast China. Int. J. Appl. Obs. Geoinformation **18**, 528–536 (2012)
17. Li, B., Tao, S., Dawson, R.W.: Relations between AVHRR NDVI and eco-climatic paramenters in China. Int. J. Remote Sens. **23**(5), 989–999 (2002)
18. Piao, S.L., Fang J.Y., Zhou, L.M., et al.: Interannual variations of monthly and seasonal Normalized Difference Vegetation Index (NDVI) in China form 1982 to 1999. J. Geophys. Res. **108**(D14), 4401–4413 (2003)

Dynamic Analysis of Oil Spill in Yangtze Estuary with HJ-1 Imagery

Yi Lin[1,2](✉), Jie Yu[1,2], Yuguan Zhang[1,2], Pengyu Wang[1,2], and Zhanglin Ye[1,2]

[1] College of Surveying and Geo-informatics, Tongji University, Shanghai 200092, China
linyi@tongji.edu.cn
[2] Research Center of Remote Sensing and Spatial Information Technology, Tongji University, Shanghai 200092, China

Abstract. This study was conducted to monitor oil spill changes in Yangtze estuary and to analyze their dynamic distribution by using HJ-1 charge-coupled device (CCD) imagery. First, the spectral response curve of the oil and other typical objects were analyzed to build the spectral feature space. Second, the classification algorithm of the polynomial kernel-based support vector machine (SVM) was studied to extract various levels of oil spill information including severe pollution, moderate pollution, and slight pollution. Third, the performance of the classification model was validated by comparison with other traditional approaches and the ground investigation data supported by the Shanghai Environmental Science Research Institute. Fourth, multi-temporal HJ-1 images were used to implement the classification with the polynomial kernel-based SVM algorithm. Finally, the oil-covered areas were calculated, the changes in spatial distribution were analyzed on the basis of the extracted results, and a statistical histogram was obtained. The results prove that the polynomial kernel-based SVM classification model has high accuracy with reliable performance for oil spill extraction. In addition, the dynamic analysis can be used to predict drifting trends and to provide important information for oil spill emergency response teams. Moreover, the HJ-1 satellite data can be applied to environmental monitoring.

Keywords: Oil spill · HJ-1 satellite images · Polynomial kernel · Support vector machine · Changing monitoring

1 Introduction

Developments in modern science and technology have enabled the oil industry to gradually become an important driving force for the global economy. However, seas, rivers, and other waterways for oil transportation are being polluted by thousands of tons of oil spills each year. The International Tanker Owners Pollution Federation (ITOPF) has reported that during 1970–2014, 5.74 million tons of oil were released into waterways due to oil tanker accidents worldwide [1]. In the coastal areas of China, about 2900 oil tanker spills occurred during the same period [2]. Such oil spills can

F. Bian and Y. Xie (Eds.): GRMSE 2015, CCIS 569, pp. 345–356, 2016.
DOI: 10.1007/978-3-662-49155-3_35

cause serious harm to the environment, economy, and society, particularly in coastline ports, reservoirs, and other environmentally sensitive areas. Oil spills in these areas pose direct hazards or potential threats to the security of human drinking water and food. In addition, oil spills can damage property, leading to economic losses particularly in the fishing and tourism industry [3–5].

The main methods of oil spill monitoring in current use include patrol ship/near surface drifters, closed circuit television (CCTV) systems, and airborne and satellite remote sensing. Because oil spills occur without warning and cause severe damage to the environment, such traditional methods cannot satisfy the requirement of oil spill monitoring in a large area for a short period. Specifically, patrol ship/near surface drifters lack mobility and are easily affected by the ocean environment, and the CCTV system is limited by the narrow range of the measuring instrument. Although airborne remote sensing method avoids such limitations, it is highly expensive, and its discontinuous scans introduce errors in spill monitoring of a wide area. Satellite remote sensing, however, is a comparatively effective method for monitoring oil spills on the global sea surface because it is dynamic and instantaneous and covers a wide region [6–9].

Among the applications of satellite remote sensing, synthetic aperture radar (SAR) is widely used because of its monitoring ability in all-weather conditions during the day and at night; however, this method is limited by its high cost of development [10–12]. Ultraviolet remote sensors are highly sensitive to oil spills; ultraviolet radiation is very high even in oil film thicknesses of less than 0.05 μm. However, false information is easily generated by the interference of external environmental factors such as solar flares, sea surface bright spots, and other biological phenomena [13, 14]. Visible and near-infrared (near-IR) satellite sensors are able to detect sea oil spills, estimate oil thickness, and identify oil species, particularly with the high-spectrum method; moreover, they offer large scale coverage, real-time processing, a short revisit period, and low cost [15–17].

Spectral analysis is an important procedure in the monitoring of oil spills by using satellite data. Oil reflectance decreases with an increase in oil film thickness at wavelengths of 0.700, 0.740, and 0.800 μm. Peaks appear at 0.580 and 0.700 μm, and reflectance occurs at the IR band. Spectral analysis can be used to extract oil information at 0.400–0.900 μm [18]. HJ-1 satellite data have four bands including blue (0.430–0.520 μm), green (0.520–0.600 μm), red (0.630–0.690 μm), and near-IR (0.760–0.900 μm). Such analysis detects important oil information and can be applied to oil spill monitoring. However, current classification algorithms used for oil spill detection are defective in classification accuracy and automaticity owing to limitations in spatial, spectral, and temporal characteristics in addition to environmental science expertise. To resolve such problems, this paper proposes an improved classification model and applies it to the HJ-1 satellite data. The model accuracy is evaluated by using ground survey data.

2 Study Area and Background

The study area is located in the Yangtze River estuary (Fig. 1). The main body of water, with an area of about 336 km^2, is situated on the south branch. The Yangtze estuary region belongs to a subtropical monsoon climate zone with prevailing northwest wind in winter and southeast wind in summer. The annual average temperature is 15.2–15.7 °C, with the lowest temperature occurred in January and the highest in July. This estuary is classified as a medium-strength tidal estuary. According to tide data observed by Xiuliujing tide level station, the water level difference is 4.62 m, the minimum water level difference is 0.17 m, and the average water level difference is 2.66 m. The coast of the study area includes many important economic and nature reserves; important spawning grounds and feeding grounds of major economic fish and shrimp species; Xisha wetlands, a multi-functional wetland ecology model district; and Dongfengxiasha reservoir, the source area of centralized drinking water.

On December 30, 2012, a ship carrying 400 tons of heavy fuel oil sank in the Yangtze River. Under the influence of tidal action, the oil began to affect Chongming Island on December 31 by causing severe pollution on the beaches and in the water. Statistics indicate that an 18-km-long stretch of tidal flat was polluted, with an area of 1.53 km^2.

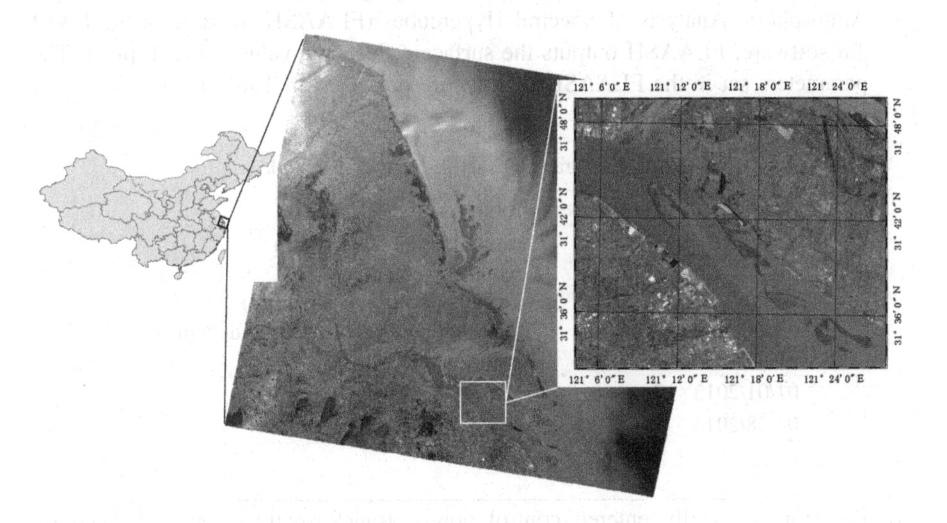

Fig. 1. The study area in the Yangtze River estuary

3 Data and Methods

3.1 Data Set

In this study, images acquired by the HJ-1 satellite on December 4, 2012 (before the oil spill accident), January 1, 2013 (one day after the accident), and January 28, 2013, were selected for oil detection in the Yangtze River estuary. The images are cloud-free in the research area. The HJ-1A/B CCD has visible bands at 430–520, 520–600, and 630–690 nm and one near-IR band at 760–900 nm with spatial resolution of 30 m. In addition, the environment monitoring data of water in the oil spill area employed Global Positioning System (GPS) location data and oil sample information.

3.2 Image Preprocessing

Conventional preprocessing procedures were performed for HJ-1 A/B CCD images. The procedure is shown in Fig. 2 and outlined below.

(1) First, radiometric calibration is conducted for each band. The digital number (DN) values are converted to sensor spectral radiance, L ($Wm^{-2}\ sr^{-1}\ \mu m^{-1}$). The radiometric calibration of HJ-1 data is performed by using Eq. (1):

$$L = \frac{DN}{gain} + bias. \tag{1}$$

where gain ($W^{-1}\ m^{2}\ sr\ \mu m$) is the calibration factor and bias ($Wm^{-2}\ sr^{-1}\ \mu m^{-1}$) is the calibration offset.

(2) Atmospheric correction is then performed by using the Fast Line-of-sight Atmospheric Analysis of Spectral Hypercubes (FLAASH) module in the ENVI 4.8 software. FLAASH outputs the surface reflectance value for each pixel. The parameters set in the FLAASH module are presented in Table 1.

Table 1. Parameters set in FLAASH module

Data	Visibility/km	Visibility/km
12/04/2012	9	Sensor altitude: 649.093 km Ground elevation: 3.4 km (average elevation of Shanghai) Atmospheric models: Mid-latitude winter Aerosol model: Urban
01/01/2013	10	
01/28/2013	8	

(3) By using manually entered control points, transformations are performed on separate images acquired on December 4, 2012, and January 28, 2013 to make the major features align with a second image captured on January 1, 2013. Accuracy better than one pixel is ensured; the Nearest Neighbor resampling method is used to ensure that the pixel difference percentage is low [19].

(4) Image subsets are created to acquire the image containing only the study area.

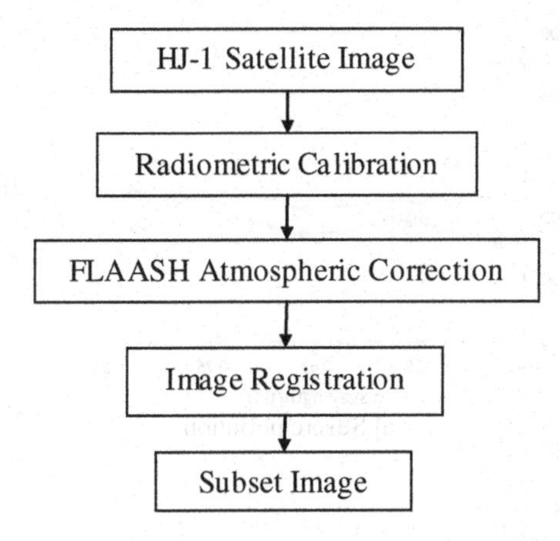

Fig. 2. Preprocessing procedures of HJ-1 satellite image

3.3 Spectral Response Curve of Water and Oil Spill Area

The spilled oil, a type of heavy non-persistent fuel oil, has higher levels of aromatic hydrocarbon compounds than those found in other oil types. Compared with other hydrocarbon components of oil, aromatic hydrocarbon compounds are soluble in water, easily spread, and difficult to degrade; moreover, their biological toxicity is high. It is difficult to distinguish the oil areas in RGB color images because after the oil dissolves in water, the inherent optical properties of the water are changed. However, oil can be distinguished in the spectral view [20, 21]. According to field investigation data, the oil spill was divided into four levels: severe pollution, moderate pollution, slight pollution, and water without pollution. The spectral curve of surface reflectance is presented in Fig. 3.

As shown in the figure, the surface reflectance of severe pollution, moderate pollution, slight pollution and water without pollution had the same tendency at three visible bands of B1 at 0.430–0.520 μm, B2 at 0.520–0.600 μm, and B3 at 0.630–0.690 μm. However, the effects of oil pollution caused a gradual increase in the reflectance value at different rates owing to the differences in pollution level; the largest difference appeared at the near-IR band B4 at 760–900 μm. On the contrary, non-polluted water showed the opposite tendency.

3.4 Support Vector Machine with Polynomial Kerne

Support Vector Machine (SVM) is a supervised learning model used for classification and regression analysis [22]. It has advantages of training small samples, nonlinear classification, and high dimensional pattern recognition. It effectively resolves the problems of dimensionality, over-learning, and other issues found in traditional learning methods.

The SVM algorithm adopts kernel function for nonlinear issues, which first maps the nonlinearity of input samples to high dimensional feature space. To a certain extent,

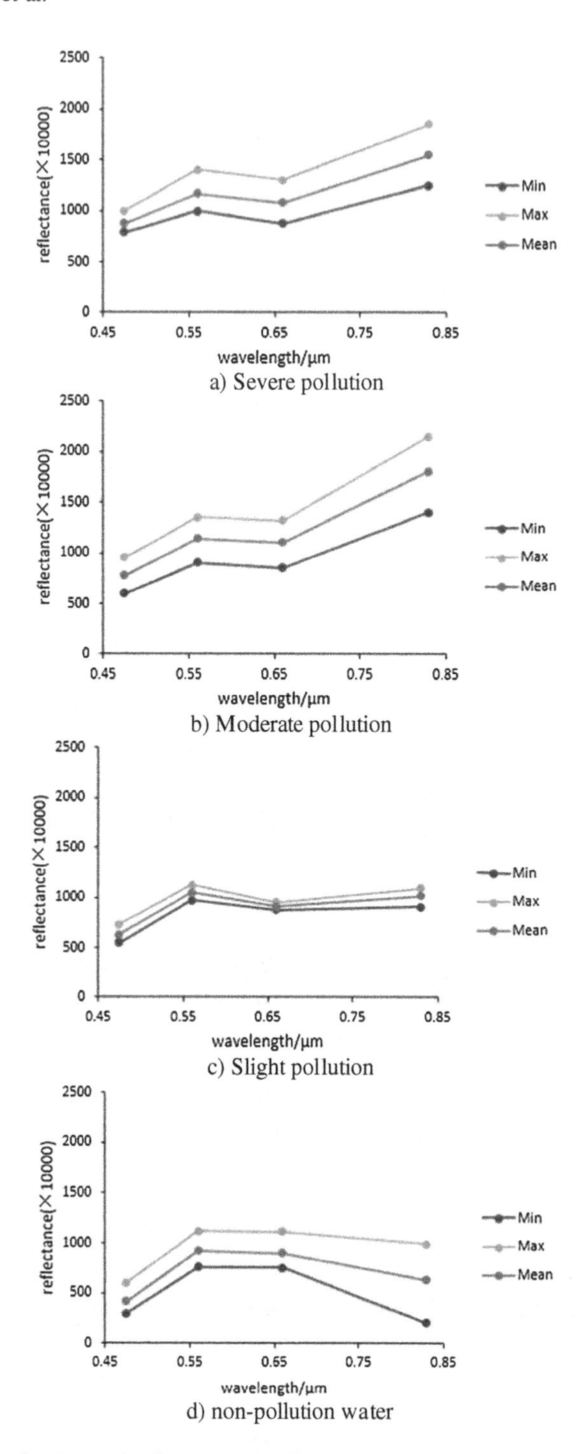

Fig. 3. Spectral reflectance of different oil spill pollution condition.

building a kernel function determines the complexity and accuracy of SVM decision. In SVM theory, selection of a model is equivalent to the selection of a kernel function. It is proposed that the kernel function allows the data that are not linearly separable in the original coordinate system to be mapped into a much higher dimensional space. The dimensionality can then be solved in the high-dimension feature space.

In this study, non-polluted water, severe pollution, moderate pollution, and slight pollution constitute a multiple feature space. The three-order polynomial kernel function Eq. (2) is applied as

$$K(x_i, x_j) = [(x_i, x_j) + 1]^3. \tag{2}$$

The classification procedure, shown in Fig. 4, is coded by MATLAB. First, the training set is selected according to the spectral features of the oil spill area, and the polynomial classification model is then constructed. Next, the optimal classification hyperplane is found, and the classifier is constructed. Finally, the pixel of the classified image is mapped into the feature space, and the ground object classification is completed.

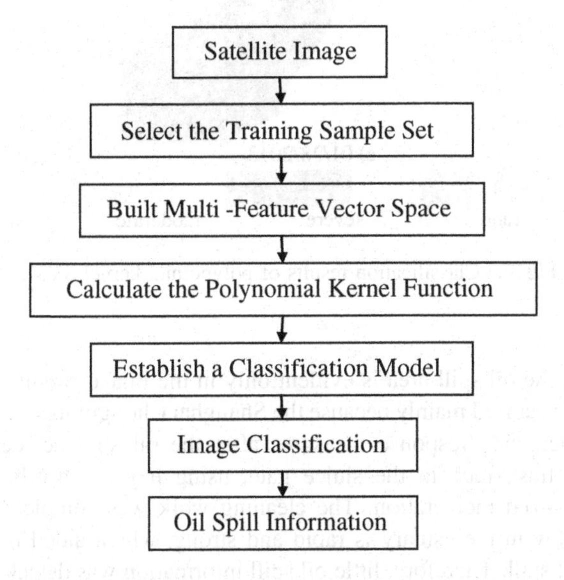

Fig. 4. Classification procedure of the Support Vector Machine (SVM).

4 Result and Accuracy Assessment

4.1 Classification Results

The classification results are given in Fig. 5.

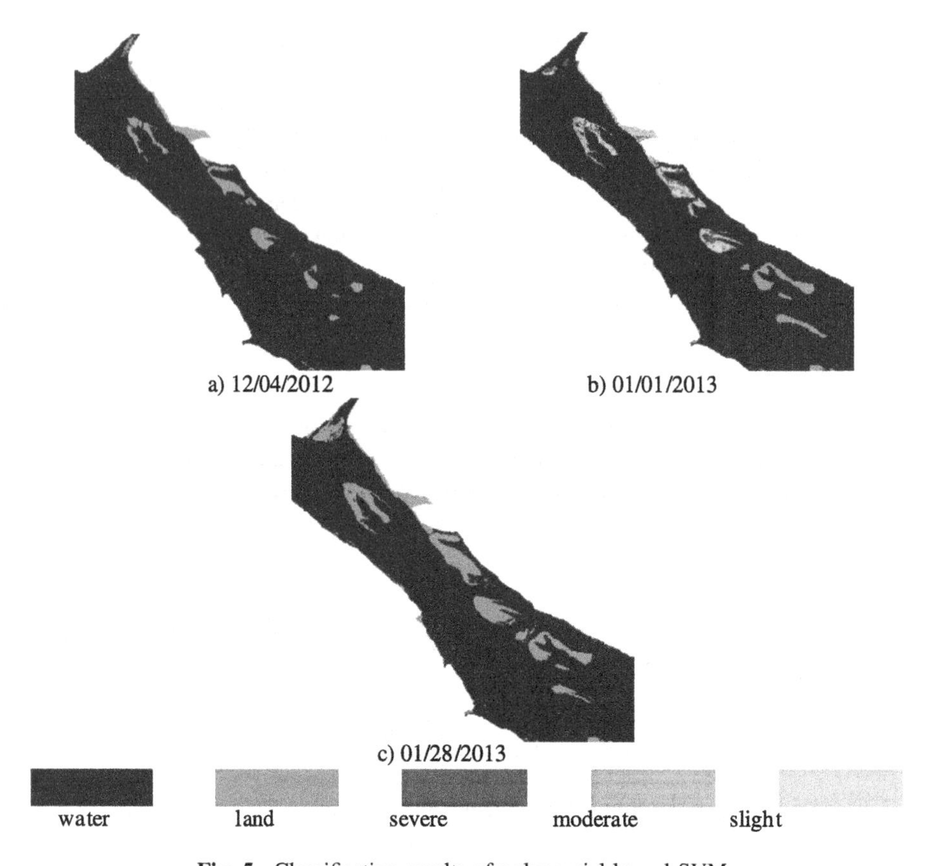

Fig. 5. Classification results of polynomial kernel SVM.

In the figure, the oil spill area is evident only in the image captured on January 1, 2013. This result occurred mainly because the Shanghai Chongming county government implemented emergency response measures after the oil spill accident that include installation of a trash rack in the sluice gate, using hay to absorbing the oil, and employing centralized incineration. The cleaning work was completed in one week. Moreover, the flow in the estuary is rapid and strong, which aided the diffusion and dilution of the oil spill. Therefore, little oil spill information was detected a month later on January 28, 2013.

We conducted vectorization for the classified image in ArcGIS and calculated the area of the various regions. The results are shown in Fig. 6.

The results of SVM classification showed that the total oil spill area was 3369.42 hm^2. The proportions of severe pollution, moderate pollution, and slight pollution areas were 22.074, 12.108, and 65.818 %, respectively.

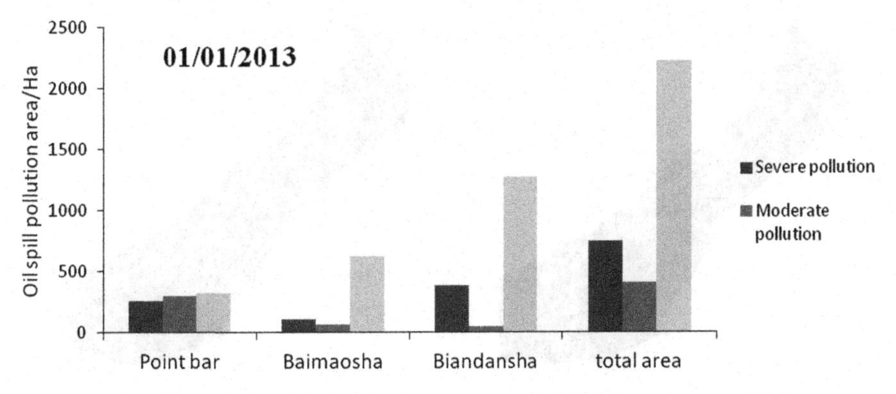

Fig. 6. Oil distribution, on January 1, 2013

4.2 Accuracy Assessment

Considering the real-time performance and the limitation of 30 m resolution, this paper adopted random sampling to verify the classification accuracy. By using ground survey, we evaluated the accuracy by calculating the confusion matrix and kappa coefficient. The confusion matrix of the classified image obtained on January 1, 2013, is shown in Table 2.

Table 2. Confusion matrix of polynomial kernel SVM classification results

Testing samples	Water	Severe pollution	Moderate pollution	Slight pollution	Land	Total
Water	1120	17	0	0	0	1137
Severe pollution	0	47	7	0	0	54
Moderate pollution	0	0	40	8	0	48
Slight pollution	0	17	13	126	1	157
Land	0	22	13	51	433	519
Total	1120	103	73	185	434	1915
Overall accuracy	99.22					
Kappa coefficient	0.8670					

In addition, to verify the effectiveness of the algorithm proposed in this study, we compared this algorithm with three other traditional algorithms: minimum distance method, maximum likelihood classification and neural network. The results are plotted in Fig. 7.

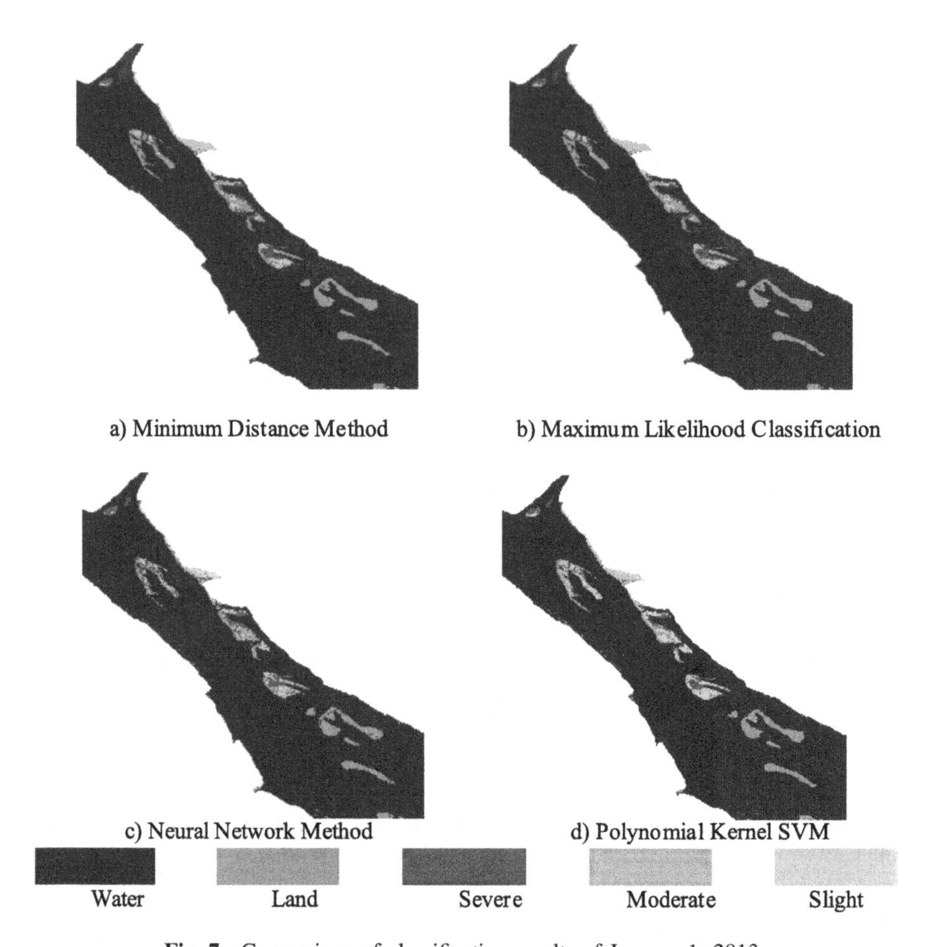

Fig. 7. Comparison of classification results of January 1, 2013.

The kappa coefficient and overall accuracy of different classification method are shown in Table 3.

Table 3. Comparison of accuracies of various classification methods

Classification method	Kappa coefficient	Accuracy
Minimum distance method	0.6586	77.28%
Maximum likelihood classification	0.7847	85.80%
Neural network method	0.6928	80.99%
Polynomial kernel SVM	0.8670	92.22%

The comparisons listed in Table 3 clearly show that the accuracy of the polynomial kernel SVM is the highest for effective extraction of oil spill information from HJ-1 satellite CCD images.

5 Conclusion

This study examines the monitoring of an oil spill by using HJ-1 satellite data, based on the algorithm of polynomial kernel SVM. We analyzed the availability of HJ-1A/B satellite CCD data for the oil spill, and a spectral feature space suitable for oil spill extraction was constructed on the basis of the spectral analysis of the oil spill at various pollution levels. The proposed polynomial kernel SVM classification model can generate results with a 92.22 % overall accuracy. Therefore, this method can be effectively used to obtaining oil spill information.

Acknowledgments. This study is a part of research project "Shanghai Innovation Action Plan" supported by Science and Technology Commission of Shanghai Municipality (STCSM) (Project ID: 13231203601). The environmental monitoring data was provided by the Environmental Science Research Institute of Shanghai and the HJ-1 satellite data was obtained from the China Centre for Resources Satellite Data and Application.

References

1. http://www.itopf.com/knowledge-resources/data-statistics/statistics/ (2015)
2. Lan, G.: Study on spectral information mining and application for oil spill remote sensing monitoring. Dalian Maritime University, Dalian (2012)
3. OECD: The environmental effects of freight. OECD work programme on trade and environment, Paris, France (1997)
4. Wang, Z., Liu, M.: Study on damage of marine ecosystem by oil spill. J. Waterway Harb. **29** (5), 367–371 (2008)
5. Li, Y., Zeng, W., Yu, Q., et al.: Adverse effect of marine oil spills on human health and ecosystem: a review. Asian J. Ecotoxicol. **6**(4), 345–351 (2011)
6. Yan, Y., Shang, H.: Tracking and monitoring oil spills by near-surface drifters. Ocean Technol. **26**(2), 17–20 (2007)
7. Liu, K., Yang, W.: The research progress of oil spill detecting technique. Saf. Health Environ. **7**, 25–29 (2012)
8. Yin, Q., Chu, X., Sun, X., et al.: Application and development trend of detecting techniques for oil slick from vessel. Ship Ocean Eng. **39**(5), 246–250 (2010)
9. Li, S.: Application of remote sensing for oil slicks detecting and its progress. Remote Sens. Inf. **2**, 53–56 (2004)
10. Tian, W., Bian, X., Shao, Y., et al.: On the detection of oil spill with China's HJ-1C SAR image. Aquat. Procedia **3**, 144–150 (2015)
11. Solberg, A.H.S., Brekke, C., Husoy, P.O.: Oil spill detection in radarsat and ENVISAT SAR images. IEEE Trans. Geosci. Remote Sens. **45**(3), 746–755 (2007)
12. Solberg, A.H.S., Storvik, G., Solberg, R., Volden, E.: Automatic detection of oil spills in ERS SAR images. IEEE Trans. Geosci. Remote Sens. **37**(4), 1916–1924 (1999)
13. Fingas, M.F., Brown, C.E., Mullin, J.V.: The visibility limits of oil on water and remote sensing thickness detection limits. In: Proceedings of the Fifth Thematic Conference on Remote Sensing for Marine and Coastal Environments, Environmental Research Institute of Michigan, Ann Arbor, Michigan, pp. 411–418 (1999)
14. Fang, S., Huang, X., Yin, D., et al.: Research on the ultraviolet reflectivity characteristic of simulative targets of oil spill on the ocean. Spectrosc. Spectr. Anal. **30**(3), 738–742 (2010)

15. Li, Y., Liu, B., Lan, G., et al.: Study on spectrum of oil film in ice-infested waters. Spectrosc. Spectr. Anal. **30**(4), 1018–1021 (2010)
16. Fingas, M., Brown, C.: Review of oil spill remote sensing. Mar. Pollut. Bull. **83**, 9–23 (2014)
17. Zhao, D., Gong, P.: The research of visual light wave-band feature spectrum of sea-surface oil spill. Remote Sens. Technol. Appl. **15**(3), 160–164 (2000)
18. Palmer, D., Boasted, G.A., Boxall, S.R.: Airborne multispectral remote sensing of the January 1993 Shetlands oil spill. In: Proceeding of the Second Thematic Conference on Remote Sensing for Marine and Coastal Environments: Needs, Solutions and Applications, ERIM Conference, Ann Arbor 2, pp. 546–558 (1994)
19. Zhang, Z., Han, J., Yu, T., et al.: Preprocess order of HJ Satellite Data. Geomat. Inf. Sci. Wuhan Univ. **38**(12), 1456–1459 (2013)
20. Lu, Y., Chen, J., Bao, Y., et al.: Using HJ-1 satellite CCD data for remote sensing analysis and information extraction in oil spill scenarios. Scientia Sinica Informationis **41**, 193–201 (2011)
21. Sun, P., Song, M., An, J.: Study of prediction models for oil thickness based on spectral curve. Spectrosc. Spectral Anal. **33**(7), 1881–1885 (2013)
22. Cortes, C., Vapnik, V.: Support-vector networks. Mach. Learn. **20**(3), 273–297 (1995)

A Time-Frequency Algorithm for Noisy ICA

Jing Guo[1(✉)] and Ying Deng[2]

[1] College of Electronics and Information Engineering, Southwest University, Chongqing 400715, China
poem24@163.com
[2] Chongqing College of Humanities, Science and Technology Chongqing, Chongqing, China

Abstract. The performance of standard algorithms for Independent Component Analysis (ICA) quickly deteriorates when the signals are contaminated by additive noise. In this paper, we propose an ICA approach exploiting the difference in the time-frequency (t-f) signatures of noisy signals to be separated. The approach uses a high-resolution t-f distribution to obtain the t-f matrices of mixed signals, then localizes the signal energy by Hough transform and obtains the estimated signals based on the diagonalization of a combined set of auto-term matrices. Furthermore, its performance is evaluated using the Signal-Noise-Ratio (SNR) as it is commonly employed to assess the ICA algorithms. Both the results of mathematical analysis and numerical simulations indicate that we could enhance the ICA performance by improving the input SNR or increasing the number of sampling points. The approach could increase the ICA robustness by spreading the noise power and localizing the source energy in the t-f domain.

Keywords: Independent component analysis · Noisy source · Time frequency distribution · Hough transform · SNR

1 Introduction

In many applications of ICA, a degree of background noise may be present. However, in practice, ICA seems to be quite difficult when noise is present [1, 2]. There have been a variety of attempts to derive algorithms that perform noisy ICA. Recently, solutions based on the bias removal techniques [3, 4], higher-order cumulant [5–7], maximum likelihood function [7–9], sparse code shrinkage [10, 11] and wavelet filtering [12, 13] have been proposed. Bias removal technique means that noise-free ICA methods are modified so that the bias due to noise is removed, or at least reduced.

This technique could be used for noisy ICA as well, if only we had measures of non-gaussianity which are immune to gaussian noise. A different approach to estimation of the mixing matrix is given by methods using higher-order cumulants only. Higher-order cumulants are unaffected by gaussian noise, and therefore any such estimation method would be immune to gaussian noise. However, their lack of robustness may be very problematic in a noisy environment. Another approach for estimation of noisy data is given by maximum likelihood estimation. Maximum likelihood methods could maximize the joint likelihood of the mixing matrix and the

F. Bian and Y. Xie (Eds.): GRMSE 2015, CCIS 569, pp. 357–365, 2016.
DOI: 10.1007/978-3-662-49155-3_36

realizations of the independent components. A problem with this algorithm is, however, that the computational complexity grows exponentially with the dimension of the data. Sparse code shrinkage means that we transform the data into a sparse, i.e., super-gaussian code, and then apply shrinkage on that code. But at present it still has some drawbacks such as considerable complexity of computation and inevitable loss of signal details. Wavelet filtering methods could localize information of the noisy signals in limited number of the wavelet coefficients according to the discrete wavelet transform. But, it has the shortcoming of complex computation and fixed number of decomposition layers in de-noising of inertial sensors.

The paper introduces a new noisy ICA approach exploiting the difference in the time-frequency (t-f) signatures of noisy sources to be separated. The approach firstly uses a high-resolution t-f distribution to obtain the t-f matrices of mixed signals, then localizes the source energy by Hough transform and finally obtains the estimated signals based on the diagonalization of a combined set of auto-term matrices. Furthermore, its separation performances are evaluated using the Signal-Noise-Ratio (SNR) as they are commonly employed to assess the ICA algorithms. Both the results of mathematical analysis and numerical simulations indicate that we could enhance the ICA performance by improving the input SNR or/and increasing the number of points sampled. In contrast to other noisy ICA approaches, the proposed approach allows the separation of contaminated signal with identical spectral shape but with different localization properties.

2 Data Model

We consider the following linear instantaneous signal model

$$x(t) = As(t) + n(t) \tag{1}$$

Where:

① $x(t) = [x_1(t), x_2(t), \cdots, x_m(t)]$ is a m-vector of the observations.

② $s(t) = [s_1(t), \cdots, s_i(t), n_{i+1}(t), \cdots, n_n(t)]$ has i-vector containing zero mean, non-stationary, mutually uncorrelated random sources and $n - i$-vector of the additive noise.

③ A is a $m \times n$ full rank mixing matrix with $m \geq n$.

The covariance matrices of vectors $\mathbf{s}(t)$ and $\mathbf{x}(t)$ are:

$$\mathbf{R_{ss}}(t, \tau) = \mathbf{E}\{\mathbf{s}(t+\tau)\mathbf{s}^*(t)\} = \mathrm{diag}[\rho_1(t, \tau), \cdots, \rho_n(t, \tau)] \tag{2}$$

$$\mathbf{R_{xx}}(t, \tau) = \mathbf{E}\{\mathbf{x}(t+\tau)\mathbf{x}^*(t)\} = \mathbf{A}\mathrm{diag}[\rho_1(t, \tau), \cdots, \rho_n(t, \tau)]\mathbf{A^H} + \delta(\tau)\sigma^2\mathbf{I}_m \tag{3}$$

Where $\rho_i(t, \tau) = \mathbf{E}\{\mathbf{s}_i(t+\tau)\mathbf{s}_i^*(t)\}, \delta(\tau)\sigma^2\mathbf{I}_m$ respectively denote the auto-covariance of $s_i(t), n(t)$, and the superscript $\mathbf{H}$ represents the conjugate transpose.

3 Separation Algorithm

Step I. Whitening the mixed signals: This aim of this step is to transform the mixing matrix **A** into a unitary matrix **U**. It is achieved by searching for a $n \times m$ whitening matrix **W** such that $\mathbf{U} = \mathbf{WA}$ and $\mathbf{WAA^H W^H} = \mathbf{I}_n$. The matrix **W** can be estimated by [14]

$$W = [(\lambda_1 - \hat{\sigma}^2)^{-1/2} h_1, \cdots, (\lambda_n - \hat{\sigma}^2)^{-1/2} h_n] \tag{4}$$

Where $[\lambda_1, \cdots, \lambda_m]$ is the eigenvalues of $R_{xx}[0]$ sorted in decreasing order, and $[h_1, \cdots, h_m]$ is the corresponding eigenvectors.

Step II. Constructing TFD of the whitened signals: The objective of this step is to obtain a set of different t-f localization characteristic of whitening signals. It is achieved by calculating the TFD whitening signals to localize the t-f properties. The literature [15] provides a high t-f resolution and the maximum cross-term reduction with the preferable diagonal or off-diagonal structure of TFD matrices in noisy ICA applications, and it is given by

$$GD_X(t,f) = \iint (e^{-|\tau|}\mathrm{sech}^2(v))^a \mathbf{x}(t+v+\frac{\tau}{2}) \times \mathbf{x}^*(t+v-\frac{\tau}{2}) e^{-j2\pi f\tau} dv d\tau \tag{5}$$

Where superscript $*$ denotes the complex conjugate. This TFD can be used to improve the noisy ICA performance by accurately estimating the instantaneous frequency of multi-component signals.

Step III. Hough Transform of $GD_{XX}(t,f)$: The objective of this step is to localize the source energy by Hough transform. The Hough transform is a technique which can be used to isolate features of a particular shape within an image by a voting procedure. The transformation can be expressed mathematically as [16]

$$\rho = x\cos(\theta) + y\sin(\theta) \tag{6}$$

Where ρ is the distance from some set origin of the image to the line, and θ is the angle corresponding to the angle at which that line is rotated from a set axis. Each line in the data space is mapped to a point (ρ, θ) in Hough space. For any given point (x, y), because there are an infinite number of lines that can pass through it, then the point (x, y) gets mapped to a sinusoidal curve in the Hough space. The GD-Hough (GDHT) of $X(t)$ is defined by

$$GDHT_X(f,g) = \int_{-\infty}^{\infty}\int_{-\infty}^{\infty} GD_X(t,v)\delta(v-f-gt)dtdv = \int_{-\infty}^{\infty} GD_X(t,f+gt)dt \tag{7}$$

The parameters corresponding to $\mathbf{x}(t)$ are $g = -\cot(\theta), f = \rho/\sin(\theta)$.

Step IV. Joint-diagonalization a set of GDHT matrices: The objective of this step is to jointly diagonalize a set of GDHT matrices and hence to retrieve the unitary matrix **U**. Given estimates W, we get

$$GDHT_{\mathbf{xx}}(f,g) = \mathbf{A} \times GDHT_{\mathbf{ss}}(f,g) \times \mathbf{A}^{\mathbf{H}} + \sigma^2 \mathbf{I}_m \tag{8}$$

Considering a low noise environment, that is,

$$GDHT_{\mathbf{xx}}(f,g) \approx \mathbf{A} \times GDHT_{\mathbf{ss}}(f,g) \times \mathbf{A}^{\mathbf{H}} \tag{9}$$

By pre and post multiplying $GHT_{\mathbf{xx}}(f,g)$ by $\mathbf{W}$, we have

$$\begin{aligned} GDHT_{\mathbf{zz}}(f,g) &\approx \mathbf{W} \times GDHT_{\mathbf{xx}}(f,g) \times \mathbf{W}^{\mathbf{H}} \\ &\approx \mathbf{WA} \times GDHT_{\mathbf{ss}}(f,g) \times \mathbf{A}^{\mathbf{H}}\mathbf{W}^{\mathbf{H}} \approx \mathbf{U} \times GDHT_{\mathbf{ss}}(f,g) \times \mathbf{U}^{\mathbf{H}} \end{aligned} \tag{10}$$

By calculating the eigenvalue of GDHT, Eq. (10) becomes

$$eig\{V_{zz}\} \approx eig(UV_{ss}U^H) \approx eig(V_{ss}) \tag{11}$$

According to the literature [17], if following inequality exists

$$if(\frac{\max(|eig(V_{zz}(f,g))|)}{\sum |eig(V_{zz}(f,g)|}) > \varepsilon \tag{12}$$

Then its corresponding (t,f) is a single auto-term position. The threshold value ε is bounded to the interval [0, 1]. Since $GDHT_{\mathbf{zz}}(f,g)$ is known, Eqs. (10), (12) shows that **U** may be obtained as a joint diagonalizing matrix of a set of whitened TFD matrices [18].

In summary, the separation algorithm is composed of the following steps.

① *Estimate the auto-correlation matrix R from data samples.*

② *Estimate the whitening matrix W form Eq.* (4).

③ *Construct K matrices by computing the GD-Hough distributions of $X(t)$ form Eqs.* (5) *and* (7).

④ *Choose auto-term t-f points based on expression* (12).

⑤ *Obtain the unitary matrix U as joint diagonalizer of the set* $\{GDHT_{ZZ}(t_i f_i).$ $|i = 1, \cdots, M\}$.

⑥ *Estimate the source signals as* $\hat{S} = U^H Wx(t)$.

4 SNR Analysis

The linear frequency modulation (LFM) signal is a classical non-stationary random signal. A LFM signal is shown in Eq. (13)

$$s(k) = c\ \exp(j2\pi(f_0 k + \frac{1}{2} g_n k^2) + j\phi_0) \tag{13}$$

$n(k)$ is additive gaussian noise with $R_n(k) = \sigma_n^2 \delta(k)$ and $A = 1$. The input SNR and the output SNR are respectively defined as then

$$SIR_{IN} = \frac{c^2}{\sigma_n^2} \tag{14}$$

$$SNR_{out} = \frac{\text{var}\{GDHT_S(f,g)\}}{\text{var}\{GDHT_{S+N}(f,g)\}} \tag{15}$$

If signal $x(k) = s(k) + n(k)$ is sampled at T points, we can obtain

$$\text{var}\{GDHT_S(f,g)\} = \frac{c^4 T^4}{4} \tag{16}$$

According to the literature [19],

$$\text{var}\{GDHT_{S+N}(f,g)\} \propto \frac{c^4 \sigma_n^2 T^3}{2} + \frac{\sigma_n^4 T^2}{2} \tag{17}$$

Hence, the output SNR is approximated by

$$SNR_{out} \propto \frac{c^4 T^4}{c^2 \sigma_n^2 T^3 + \sigma_n^4 T^2} = \frac{T^2 \times SNR_{in}^2}{T \times SNR_{in} + 1} \tag{18}$$

To estimate the separation performance of noisy ICA, an approximation to identify independent components one by one is designed as follows.

$$SNR_{out} \propto \left\{ \begin{array}{ll} T^2 \times \frac{1}{\frac{T}{SNR_{in}} + \frac{1}{SNR_{in}^2}} & if\ T\ is\ fixed\ \text{value} \\ SNR_{in}^2 \times \frac{1}{\frac{SNR_{in}}{T} + \frac{1}{T^2}} & if\ SNR_{in}\ is\ fixed\ \text{value} \end{array} \right\} \tag{19}$$

We can see from Eq. (19) that its output SNR is approximately directly proportional to the input SNR or the number of sampling points, respectively. Hence, we could enhance the separation performance of noisy ICA by improving the input SNR or/and increasing the number of sampling points.

5 Numerical Experiments

In this section, several experiments were conducted using the simulated and the real data to investigate the separation performance of the proposed algorithm. Three signals including a white gaussian noise, a speech signal and a color noise were taken from the web page of Prof. Andrzej Cichocki at the Laboratory for Advanced Brain

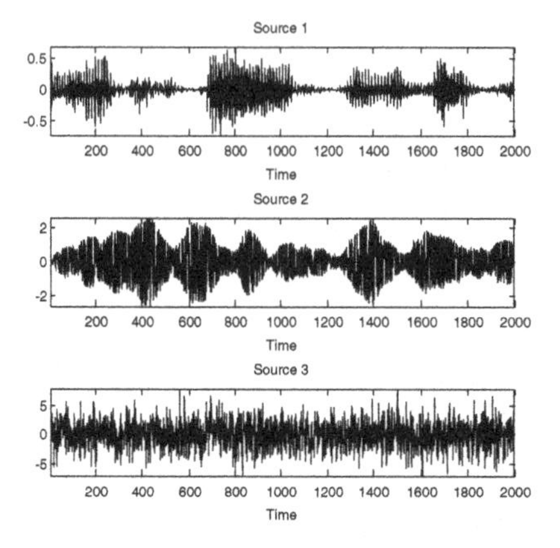

Fig. 1. The waveform of source signals

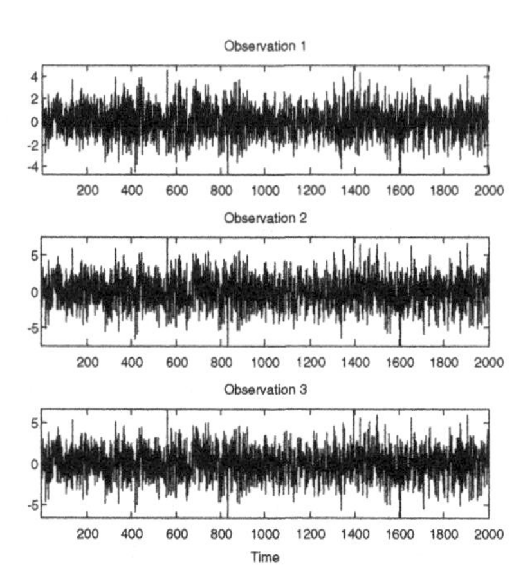

Fig. 2. The waveform of observed signals

Signal Processing, RIKEN Brain Science Institute, Tokyo, Japan. These signals were sampled at 2000 points and were mixed by the matrix A = [0.1509 0.8600 0.4966; 0.6979 0.8537 0.8998; 0.3784 0.5936 0.8216]. The plots of the source signals are shown in Fig. 1 and the observed signals are displayed in Fig. 2. Figure 3 shows the estimated signals. It is clear that the new algorithm works well in this case.

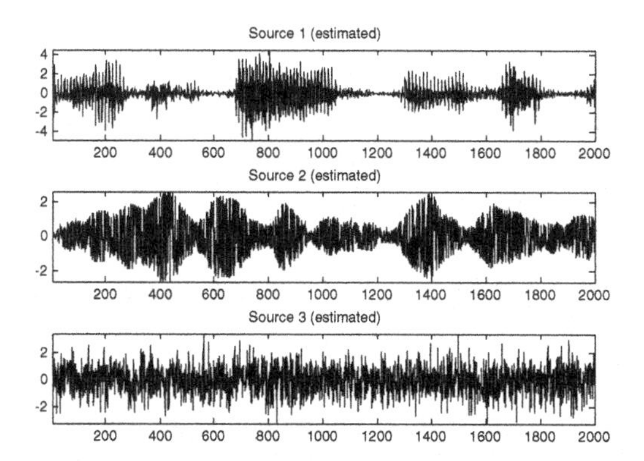

Fig. 3. The waveform of estimated signals

We used the SIR to compare the separation performance of several classical ICA algorithms and the proposed approach, and it is defined as

$$SIR_i = -10\log_{10}\{\frac{||\hat{s}_i(t) - s_i(t)||_2^2}{||s_i(t)||_2^2}\} \tag{20}$$

Where $\hat{s}_i(t)$ and $s_i(t)$ are the estimated signal and the source signal, respectively. *SIR* is a parameter measuring the similar level of the separated signal and the source signal. Table 1 shows the compared results evaluated by 50 Monte-Carlo simulations. Table 1 indicates that the proposed approach achieves the higher SIR (SIR = 30.8704) comparing with other methods.

Table 1. SIR of the multiple ICA algorithms

Algorithm	SIR	Algorithm	SIR
AMUSE	7.8052	FPICA	23.1289
EVD2	17.6694	FJADE	14.3204
SOBI	19.6233	NGFICA	19.8903
SONS	14.1091	Our algorithm	30.8704
FOBI	10.4241		

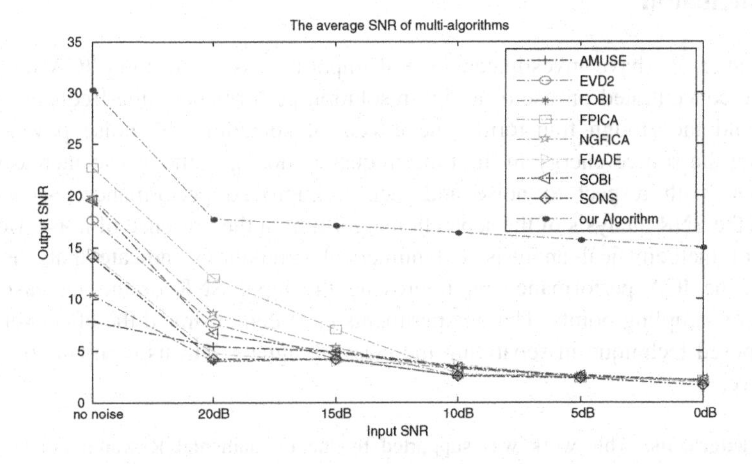

Fig. 4. The curves of the output SNR versus the input SNR

Figure 4 shows the mean output SNR of several ICA algorithms under the different input SNR in the presence of noise. It's obvious that the separation efficiencies degrade with the decrease of the input SNR. Figure 5 shows the influence of the sampling points on the output SNR in the presence of noise. It's obvious the higher the sampling points, the better the output SNR. Hence, we could enhance the separation performance of noisy ICA by improving the input SNR or/and increasing the number of sampling points.

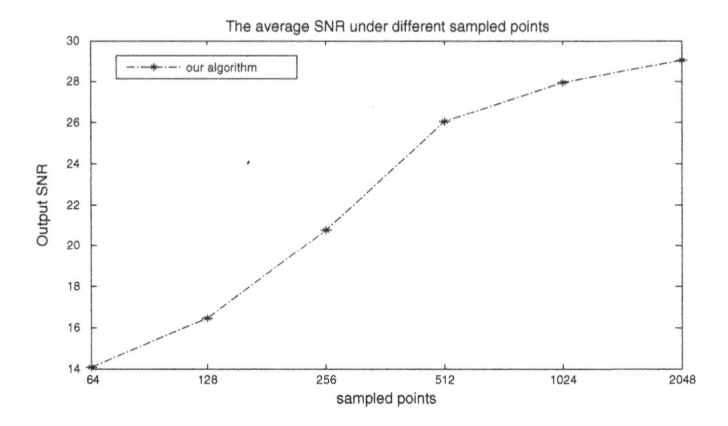

Fig. 5. The curves of the output SNR versus the number of sampling points

6 Conclusion

In the paper, we have investigated several important issues in noisy ICA problems. First, we concentrated on use of a high resolution performance time-frequency distribution and the Hough transform. The effects of spreading the noise power while localizing the source energy in the time-frequency domain amounts to increasing the robustness with respect to noise and, hence, improved performance. Second, we focused the SNR analysis of the separation algorithm in the presence of noise. Both the results of mathematical analysis and numerical simulations indicate that we could enhance the ICA performance by improving the input SNR or/and increasing the number of sampling points. These experiments have demonstrated the effectiveness of the proposed technique in separating the class of signals. But its computation is still expensive.

Acknowledgments. This work was supported by the Fundamental Research Funds for the Central Universities (No. XDJK2014C015), and the Doctoral Funds of Southwest University (No. SWU112056).

References

1. Robert, A., Herbert, B., Walter, K.: Speech and Audio Processing in Adverse Environments. Springer, Berlin (2008). GmbH & Co. K
2. Andrzej, C., Shunichi, A.: Adaptive Blind Signal and Image Processing: Learning Algorithms and Applications. Wiley Press, New York (2002)
3. Hui, T., Shu, W.: Noisy blind source separation based on adaptive noise removal. In: 10th World Congress on Intelligent Control and Automation, pp. 4255–4257. IEEE Press, China (2012)

4. Takahashi, Y., Takatani, T., Osako, K., et al.: Blind spatial subtraction array for speech enhancement in noisy environment. IEEE Trans. Audio Speech Lang. Process. **17**(4), 650–664 (2009)
5. Blanco, D., Mulgrew, B., Ruiz, D.P., et al.: Independent component analysis in signals with multiplicative noise using fourth-order statistics. Sig. Process. **87**(8), 1917–1932 (2007)
6. Karfoul, A., Albera, L., Birot, G.: Blind underdetermined mixture identification by joint canonical decomposition of HO cumulants. IEEE Trans. Sig. Process. **58**(2), 638–649 (2010)
7. Sepideh, H.S., Laurent, A., Mohammad, B.S., et al.: An efficient Jacobi-like deflationary ICA algorithm: application to EEG denoising. IEEE Sig. Process. Lett. **22**(8), 1198–1202 (2015)
8. Li, X., Zhong, W.D., Alphones, A., et al.: Channel equalization in optical OFDM systems using independent component analysis. J. Light Wave Technol. **32**(18), 3206–3214 (2014)
9. Adali, T., Anderson, M., Fu, G.S.: Diversity in independent component and vector analyses: identifiability, algorithms, and applications in medical imaging. IEEE Sig. Process. Mag. **31** (3), 18–33 (2014)
10. Jafari, M.G., Abdallah, S.A., Plumbley, M.D., Davies, M.E.: Sparse coding for convolutive blind audio source separation. In: Rosca, J.P., Erdogmus, D., Príncipe, J.C., Haykin, S. (eds.) ICA 2006. LNCS, vol. 3889, pp. 132–139. Springer, Heidelberg (2006)
11. Phatak, K., Jakhade, S., Nene, A., et al.: De-noising of magnetic resonance images using independent component analysis. In: 2011 IEEE Recent Advances in Intelligent Computational Systems, pp. 807–812. IEEE Press, India (2011)
12. He, Q., Su, S., Du, D.: Separating mixed multi-component signal with an application in mechanical watch movement. Digital Sig. Process. **18**(6), 1013–1028 (2008)
13. Muhammad, T.A., Wataru, M., Christopher, J.: Employing spatially constrained ICA and wavelet denoising for automatic removal of artifacts from multichannel EEG data. J. Sig. Process. **92**(2), 401–416 (2012)
14. Holobar, A., Févotte, C., Doncarli, C., Zazula, D.: Single autoterms selection for blind source separation in time–frequency plane. In: 11th European Signal Processing Conference, pp. 565–568. IEEE Press, France (2002)
15. Guo, J., Zeng, X., She, S.: Blind source separation based on high-resolution time-frequency distributions. Spec. Issue New Trends Sig. Process. Biomed. Eng. **38**(1), 175–184 (2012)
16. Chau, C.P., Siu, W.C.: Adaptive dual-point Hough transform for object recognition. Comput. Vis. Image Underst. **96**(1), 1–16 (2004)
17. Cédric, F., Christian, D.: Two contributions to blind source separation using time-frequency distributions. IEEE Sig. Process. Lett. **11**(3), 386–389 (2004)
18. Fadaili, E.M., Moreau, N.T., Moreau, E.: Nonorthogonal joint diagonalization/zero diagonalization source separation based on time-frequency distributions. IEEE Trans. Sig. Process. **55**(5), 1673–1687 (2007)
19. Bian, H.: The research of some problems in nonstationary signals time-frequency joint analysis methods and application. Ph.D. thesis, University of Electronic Science and Technology of China (2008)

Retrieval of Atmospheric Aerosol Optical Depth with an Improved Algorithm Over East China Sea

Yi Wang(✉), Jie Xiang, and Zeming Zhou

Institute of Meteorology and Oceanography, PLA University of Science and Technology, Nanjing 211101, China
wangyi_rsc@126.com

Abstract. Aerosol plays an important role in global climate change and environment depravation, which influences the earth radiation balance through direct forcing and indirect forcing. The uncertainty of aerosol in global climate prediction is one of the most important and difficult issue in the aerosol research field. With the development of earth observation satellites and their observated data, it's very important to improve the aerosol retrieval precision and the suitable algorithm. For the influence of the Asia continent, the remote sensing of aerosol distribution is complicated in this area. An iterative algorithm is presented in this study for simultaneous determination of the aerosol optical depth. Centered with the research of East China sea (especially over Bo and Yellow sea) aerosol optical depth distributing, our conclusion is listed as follows: With the MODIS data we implemented the aerosol retrieval with the improved algorithm and the AOD distribution over East China sea was analyzed. In total, there is the larger retrieval value over ocean due to the impact of the coastal turbid seawater and water leaving radiance, and as a result of lacking the imputing parameters in the area; there isn't the suitable arithmetic to improve the AOD retrieval quality yet.

Keywords: Aerosol Optical Depth (AOD) · Junge size distribution · MODIS · Retrieval

1 Introduction

Atmospheric aerosols are widely distributed in the atmosphere, which is very important trace elements in the atmosphere, and plays an important role inglobal and regional climate, air quality and environmental transmission of light in the atmosphere. With the increasingly frequent human activities, increasing industrial emissions, global atmospheric aerosol content increased significantly, and even cause serious air pollution, endangering human health and ecological balance. Climatic effects of atmospheric aerosols are considered a key factor in causing climate change uncertainty and more and more people's attention [1–5]. Currently, the use of remote sensing satellite data of

All authors have used the western naming convention, with given names preceding surnames.

F. Bian and Y. Xie (Eds.): GRMSE 2015, CCIS 569, pp. 366–373, 2016.
DOI: 10.1007/978-3-662-49155-3_37

aerosol optical parameters without geographical limits of environmental factors, can reflect a wide range of real-time dynamic aerosol information, with the incomparable superiority [6–12]. International research on the theory of satellite remote sensing of aerosols began in the mid-1970s, Chinese scientists study began in the mid-1980s. With the development of international new generation of Earth observation satellites, the current and future period of time, the main use of aerosol remote sensing study abroad MODIS multispectral information, the polarization information and POLDER multi-angle information to the inversion of MISR aerosol optical parameters.

Some studies show that in recent years, China climate change may be related to aerosol radiative effects, and therefore it is necessary to carefully study the radiative effect of aerosols in eastern China. The level of industrial development close to the area can be approximated that aerosol distribution and radiative properties of spatial inhomogeneity small relative to other regions. Therefore, this article tries to select the MODIS data of Terra and Aqua from 2000–2008 level and the corresponding auxiliary data, especially for the Chinese Yellow Sea and Bohai Sea coastal areas, the use of alternative Junge spectral distribution algorithm double NASA business normal spectral distribution, improving the efficiency of the iterative improvement over the ocean air AOD inversion algorithm, and on this basis to analyze research on AOD distribution of the yellow Sea and Bohai sea, some conclusions have practical value.

2 Theoretical Analysis

Main text paragraph (M_Text). In this paper, the study area selected in the Yellow Sea area 31.0°N~43.0°N and 117.0°E~129.0°E. Generating a query-based method for calculating and inversion tables are simplified method Wang and Gordon (1994) proposed that particle radiation weighted by double lognormal distribution patterns of multiple scattering by the same optical thickness of each mode Xia particle generation average approximate calculation. Troposphere aerosol model with a dual mode can be described in a lognormal distribution, i.e., the sum of the accumulation mode and the coarse mode, expressed as

$$\mathrm{n}(r) = \frac{dN(r)}{dr} = \sum_{j=1}^{2} \frac{dN_j(r)}{dr} \tag{1}$$

Each of these modes can be expressed as

$$\frac{dN(r)}{dr} = \frac{N}{(2\pi)^{1/2}\,\sigma \bullet 2.3r} \exp\left\{-\frac{1}{2\sigma^2}\left[\frac{\ln r - \ln r_m}{\ln(10)}\right]^2\right\} \tag{2}$$

Where N is the particle number density, r_m is the average radius, σ is the $\ln r$ standard deviation. Four kinds of modes and five kinds of small particles pattern composed of large particles, small particles of any kind of patterns and pattern combinations can represent a large particle aerosol models. Small models including the contribution to the process of gaseous nuclear accumulation mode and cloud-based condensation processes in large models include marine and dust particles.

The total radiation can be expressed as satellite observations

$$\mathrm{L}_{\lambda}^{\mathrm{c}}\left(\mu_s,\mu_v,\phi_v\right)=\eta\mathrm{L}_{\lambda}^{\mathrm{s}}\left(\mu_s,\mu_v,\phi_v\right)+(1-\eta)\mathrm{L}_{\lambda}^{\mathrm{l}}\left(\mu_s,\mu_v,\phi_v\right) \tag{3}$$

Which $\mathrm{L}_{\lambda}^{\mathrm{s}}\left(\mu_s,\mu_v,\phi_v\right)$ and $\mathrm{L}_{\lambda}^{\mathrm{l}}\left(\mu_s,\mu_v,\phi_v\right)$ are the small radiation pattern and the value of the large mode particles; η is the ratio between the pattern size, superscript c, s and l is calculated, small and large patterns, the goal is to obtain the inversion fit the observed data ratio and good examples of small-scale models and large patterns. Aerosol model is to select the following parameters to the minimum to achieve.

$$\varepsilon_{sl}=\sqrt{\frac{1}{n}\sum_{j=1}^{n}\left(\frac{\mathrm{L}^{\mathrm{mj}}\left(\mu_s,\mu_v,\phi_v\right)-\mathrm{L}^{\mathrm{cj}}\left(\mu_s,\mu_v,\phi_v\right)}{\mathrm{L}^{\mathrm{mj}}\left(\mu_s,\mu_v,\phi_v\right)+0.01}\right)^2} \tag{4}$$

Where $\mathrm{L}^{\mathrm{mj}}\left(\mu_s,\mu_v,\phi_v\right)$ and $\mathrm{L}^{\mathrm{cj}}\left(\mu_s,\mu_v,\phi_v\right)$ are in the radiation and the calculated value of the observation and the channel j, and the superscript m indicates the observed and the calculated value c. For small particles 4 and 5 large particle pattern mode, the field in the given set of relationships, the wavelength of 550 nm using a lookup table 5 kinds of different optical thickness receive $\mathrm{L}_{\lambda}^{\mathrm{s}}\left(\mu_s,\mu_v,\phi_v\right)$ and $\mathrm{L}_{\lambda}^{\mathrm{l}}\left(\mu_s,\mu_v,\phi_v\right)$, and even if the radiation value, any η value for the small and large particles synthetic particle model, we can use (3) the total optical thickness of 5 to calculate the radiation $\mathrm{L}_{550}^{\mathrm{c}}\left(\mu_s,\mu_v,\phi_v\right)$, the use of 550 channels by observing radiation values in these five small optical thickness for all modes and combinations of modes of large linear interpolation to obtain optical thickness. The best choice for η value, given a minimum ε_{sl} of two modes is selected residual aerosol mode.

7 spectral channels of MODIS for aerosol remote sensing center wavelength located 0.47, 0.55, 0.66, 0.86, 1.24, 1.64, 2.13 μm. Input to the algorithm is the input channel 7 MODIS observations of the solar spectrum in the cloudless instrument on 10 km × 10 km square area. Calculations show that, on the basis of a single pixel, because the value of the SNR is small, not the inversion of marine aerosol optical characteristics, so the inversion product is 10 × 10 km of pixel-based. Although to some extent at the expense of the resolution, but the SNR greatly increases and credibility of the product were given. Each channel satellite observations of radiation were calculated, namely the production of various types of patterns lookup table for retrieval of aerosol optical thickness at sea. The total content of aerosol consider several values (as 0.0, 0.2, 0.5, 1.0, 2.0, 3.0) for each mode, the optical thickness of 0.55 μm wavelength describe other values are linearly interpolated between. In 9 solar incidence angle is calculated, 16 and 15 satellite zenith azimuth radiation.

Ignoring the surface sun glitter and whitecaps in clear water, at the top of the atmosphere the total reflectance at a near-infrared wavelength λ in the single-scattering case can be written as

$$\rho_p\left(\theta_0,\phi_0,\theta,\phi\right)=\rho_m\left(\theta_0,\phi_0,\theta,\phi\right)+\frac{\omega_a\tau_a p_a\left(\theta_0,\phi_0,\theta,\phi\right)}{4\cos\theta_0\cos\theta} \tag{5}$$

where ρm (θ_0,φ_0,θ,φ) is the reflectance by atmospheric molecules (Rayleigh scattering), parameters ωa, τa, and are ρa (θ_0,φ_0,θ,φ) are the aerosol single-scattering albedo, aerosol optical thickness, and scattering phase function, respectively. Angles θ_0 and φ_0 are the solar zenith and azimuth angles, respectively. Likewise, θ and φ are the viewing zenith and azimuth angles.

For the Junge power-law size distribution, the wavelength dependence of an aerosol optical thickness between 0.4 and 1.1 μm is described in the Angstrom formula, which is written as

$$\tau(\lambda)=\beta\lambda^{-\alpha} \tag{6}$$

where α is the Angstrom coefficient and β is the turbidity factor. Combining Eqs. (5) and (6), the following expression can be obtained

$$\ln\frac{\rho_{p1}-\rho_{m1}}{\rho_{p2}-\rho_{m2}}=\ln\frac{\omega_{a1}\rho_{a1}}{\omega_{a2}\rho_{a2}}-\alpha\ln\frac{\lambda_1}{\lambda_2} \tag{7}$$

where the subscripts 1 and 2 denote two near-infrared wavelengths. The Angstrom coefficient α describes the relative spectral course of the extinction coefficient σ_e in

$$\sigma_e=\beta'\lambda^{-\alpha} \tag{8}$$

where β′ is the aerosol extinction coefficient at wavelength 1 μm. The formula is valid if the particle size distribution obeys the Junge power-law model. From Eq. (6), the ratio of the single scattering albedo in the two near-infrared bands is given by

$$\frac{\omega_{a1}}{\omega_{a2}}=\frac{\sigma_{s1}}{\sigma_{s2}}\left(\frac{\lambda_1}{\lambda_2}\right)^{\alpha} \tag{9}$$

where σ_s is the scattering coefficient. The phase function is expressed as

$$p_a(\varphi)=\frac{4\pi\beta_s(\varphi)}{\sigma_s} \tag{10}$$

where φ is the scattering angle and $\beta_s(\varphi)$ is the scattering function. The ratio of the phase function value in two near-infrared bands is given by

$$\frac{p_{a1}(\varphi)}{p_{a2}(\varphi)}=\frac{\sigma_{s2}\beta_{s1}(\varphi)}{\sigma_{s1}\beta_{s2}(\varphi)} \tag{11}$$

For the Junge power-law size distribution, the aerosol scattering function may be written as

$$\beta_s=0.4343C\left(\frac{\lambda}{2\pi}\right)^{-\alpha}\bullet\frac{1}{2}\eta(\varphi) \tag{12}$$

where C is a constant and $\eta(\varphi)$ is a dimensionless function that is used to describe the aerosol scattering direction. $\eta(\varphi)$ is related to the scattering angle, complex refractive index, and size parameter, which may be written as

$$\eta(\varphi) = \int_{x1}^{x2} (i_1 + i_2) x^{-(v+1)} dx \tag{13}$$

where x is the size parameter and i is the intensity distribution function. $\eta(\varphi)$ is insensitive to wavelength if the scattering angle is more than 4°. The scattering angle is always larger than 4° for the downward observations by satellite sensors. Thus, according to Eqs. (9), (11), and (12), we obtain

$$\frac{\omega_{a1} P_{a1}(\varphi)}{\omega_{a2} P_{a2}(\varphi)} = \frac{\eta_1(\varphi)}{\eta_2(\varphi)} \approx 1 \tag{14}$$

So, Eq. (7) can be modified by

$$\alpha = -\left(\ln \frac{\rho_{p1} - \rho_{m1}}{\rho_{p2} - \rho_{m2}}\right) \Big/ \ln \frac{\lambda_1}{\lambda_2} \tag{15}$$

According to the Mie scattering theory, the relationship between the exponent of the Junge power law ν and the Angstrom coefficient α is

$$\nu = \alpha + 2 \tag{16}$$

Deschamps et al. approximated the contribution of the aerosol on the right-hand side of Eq. (5) as the aerosol optical thickness, and further approximated it as $\lambda^{-\alpha}$, i.e.,

$$\begin{cases} \frac{\omega_a \tau_a P_a(\theta_0, \theta, \phi)}{4\mu\mu_0} \approx \tau_a \\ \tau_a \approx \lambda^{-\alpha} \end{cases} \tag{17}$$

Based on the simplification in Eqs. (17) and (15) could also be obtained. However, the aerosol optical thickness must be small enough to keep the approximation of Eq. (17) reasonable.

This article uses MODIS L1B data and the corresponding auxiliary data, the Yellow Sea area 10 km resolution aerosol optical thickness results. Inversion algorithm basic idea is to use MODIS L1B data through strict filtration cloud, land and water turbidity and flares like culling and other treatment, and to 10 km × 10 km for the parameter input window, based on various assumptions aerosol and surface radiative transfer calculation parameters established radiative transfer lookup table (LUT), and to find the table is based on the observation of the top of atmosphere radiation and radiation prior lookup table values are compared until the best fit so far [4, 5]. According to inversion algorithm, MODIS aerosol optical thickness has the average and optimal two solutions. The optimal solution is to take the minimum value as the inverse value of ε, the solution is the average value is taken as the average inversion of all ε values of less than 3 %, if ε is greater than 3 %, the averaged value ε minimum 3 value as the inversion of values. There are two kinds of

MODIS aerosol optical thickness suitable for the validation analysis to determine the best research area MODIS aerosol optical thickness data, the optimal solution and the average solution. In this article aerosol optical depth study of common 0.55, 0.66 μm visible channel for authentication was calculated. Input data include MODIS L1B calibration data: MOD021KM 1000 m resolution data, MOD02HKM 500 m resolution data, MOD02QKM 250 m resolution data and MOD03 positioning data sets. You also need some auxiliary data: (1) global daily snow-covered case, is used to determine cloud detection snow background. (2) Daily global sea ice conditions, to determine the background of the sea ice cloud detection. (3) NCEP global data every six hours reanalysis data, meteorological background field letter for inversion clouds and aerosols. (4) Weekly global sea surface temperature conditions, cloud cover and testing for Genting characteristics. (5) Daily global total ozone, aerosol effects for inversion.

3 Numerical Analyses

April 19, 2009, Shandong Marine Meteorological Observatory issued a gale warning, northeasterly winds are expected to gradually increase the Yellow Sea region, and the northern part of the Yellow Sea and Bohai Sea will storm the central region. Using the above method to the Yellow Sea area 31.0°N~43.0°N and 117.0°E–129.0°E between the conduct AOD inversion, can be atmospheric AOD instant distribution, due to the impact of the cloud area and invalid data exists inversion, following only select the four representative diagram representing the analysis. Among them, the figure goes blank area to cover and land areas or invalid data points. Dark colors tend to express more optical thickness is smaller, more tend to bright colors, said the larger optical thickness. Time 1 to 3 were 2008046, 2008048, 2008051, which is roughly 02 in adjacent days were 25, 0215, 0245 (UTC), the three almost the same time of day distribution of AOD Figure. The three figures together observations we find that the general distribution and trends of aerosols.

Figure 1, although most areas are blue, the entire Yellow Sea atmospheric visibility is better, there is only one AOD larger area in the southern Yellow Sea, but especially along the coast of Bohai Bay and land near where we can see it color lighter. Starting from Fig. 1, AOD large value area to move quickly to the northeast, indicating the presence of a fast-moving weather phenomena Recently this area; But in the depths of the Yellow Sea while moving large values of color throughout the region have gradually changed light, which may be caused by two reasons, one is due to the gradual demise of the weather phenomenon, in addition to the atmospheric mitigation and the role of deep-sea area. Figure 1 shows the distribution of aerosols 2009309 this day 0245. Seen from the figure, the type of the previous several figures, the AOD atmosphere landing area is smaller than the areas far from the sea and, in the northern region of the peninsula bias evident EASTERN an optical thickness of approximately elliptical extreme areas, indicating that the an area more obvious weather phenomenon makes visibility poor.

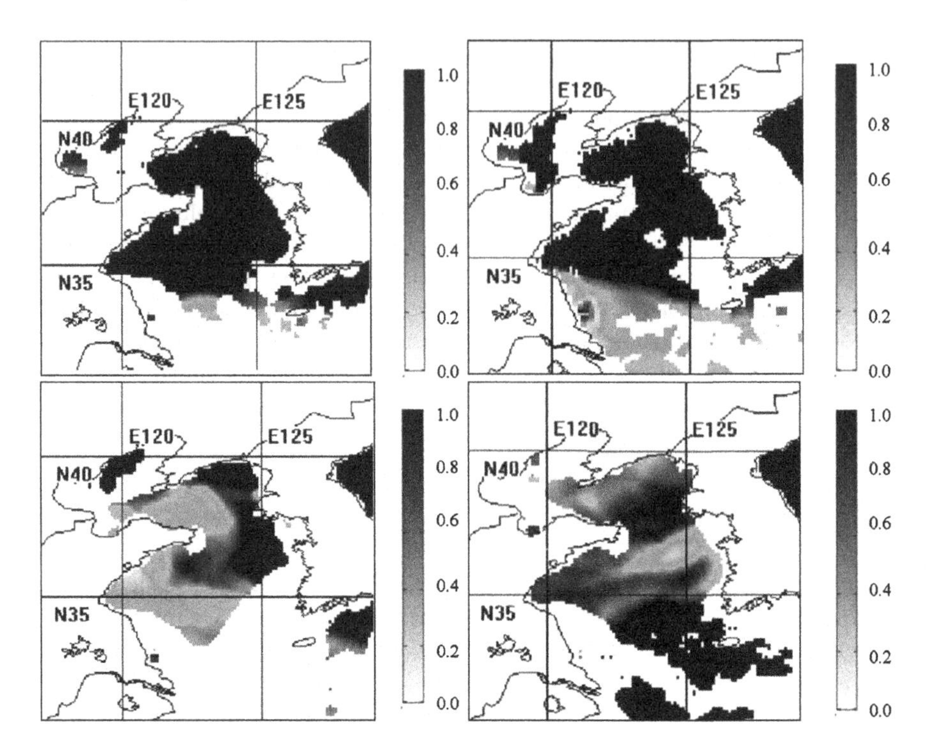

Fig. 1. (a) 2008046.0225 MODIS AOD; (b) 2008048.0215 MODIS AOD Add a descriptive label of the figure here; (c) 2008051.0245 MODIS AOD; (d) 2009309.0245 MODIS AOD

4 Conclusions

MODIS aerosol business in today's more advanced algorithm is an algorithm, which can be widely used in the global distribution of aerosols inversion, which makes it deviate local inversion accuracy, especially for rapidly changing and offshore China complex geography hydrological characteristics, its aerosol model and algorithm remains to be improved. For the establishment of new regional II water aerosols over global oceans based on the original MODIS aerosol retrieval algorithm on the inversion algorithm, extended MODIS aerosol model, will be the focus of future research, but also widely used MODIS aerosol products a necessary prerequisite. In this paper, the use of MODIS aerosol optical depth inversion Yellow Sea region, analyzed the aerosol distribution characteristics of the region and the weather combined with a case study of the origin and temporal distribution of marine aerosol for further research and monitoring of coastal marine gas sol provides a useful attempt methods.

References

1. WMO: Aerosol measurement procedures guidelines and reeommendations, No. 153. World Meteorological Organization, Geneva, pp. 24–37 (2003)
2. WMO: Strategy for the implementation of the global atmosphere water programme (2001~2007, No. 142. World Meteorological Organization, Geneva, pp. 43–45 (2001)
3. Holben, B.N., EeK, T.F., Slutsker, I., et al.: A federated instrument network and data archive for aerosol characterization. Remote Sens. Environ. **66**, 1–16 (2001)
4. Holben, B.N., Tanre, D., Smironv, A., et al.: An emerging fround-based aerosol climatology aerosol optical depth from AERONET. J. Geophys. Res. **106**, 12067–12098 (2001)
5. Bolin, Z., Wang, Q., Mao, J., et al.: Study on atmospheric aerosol and vapor with optical remote sensing. Chin. Sci. **10**, 951–962 (1983)
6. Luo, Y.F., Lv, D.R., Zhou, X.J., et al.: Analyses on the spatial distribution of aerosol optical depth over China in recent 30 years. Chin. J. Atmos. Sci. **26**(6), 722–730 (2002)
7. Dong, H., Liu, Y., Guan, Z., et al.: Validation of MODIS aerosol optical depth retrievals over East China Sea. J. Nanjing Inst. Meteorol. **29**(3), 335–342 (2007)
8. Xu, X.: Charaeteristies of aerosol optical properties in China. Nanjing University of Information Science & Technology, Nanjing **6** (2008)
9. Griggs, M.: Measurement of atmospheric aerosol optical thickness over water using ERTS-l data. Air Pollut. Contr. Assoc. **25**, 622–626 (1975)
10. Stowe, L.L., Carey, R.M., Pellegrino, P.P.: Monitoring the Mt. Pinatubo aerosol layer with NOAA/11 AVHRR data. Geophys. Res. Lett. **19**, 159–162 (1992)
11. Long, C.S., Stowe, L.L.: Using the NOAA/AVHRR to study stratospheric aerosol optical thickness following the Mt. Pinatubo eruption. Geophys. Res. Lett. **21**, 2215–2218 (1994)
12. Husar, R.B., Prospero, J.M., Stowe, L.L.: Characterization of tropospheric aerosols over the oceans with the NOAA advanced very high resolution radiometer optical thickness operational product. J. Geophys. Res. **102**, 16889–16909 (1997)

Study of Soil Water Retention in Horqin Sand Land

Shuxia Yao[1,2(✉)], Chuancheng Zhao[1,2], Suyun Wang[1], Yongli Liu[1], and Mansheng Li[1]

[1] Lanzhou City University, Lanzhou 730070, China
yaoshuxia@163.com
[2] Lanzhou Cold and Arid Regions of Environmental and Engineering Research Institute, Chinese Academy of Sciences, Lanzhou 730000, China

Abstract. Soil water retention curve (SWRC) of 0 ~ 100 cm at typical sandy lands of Horqin Sand Land, such as grassland, fixed dune and mobile dune, were measured by the laboratory methods, and the variation of this parameter along with sandy types and soil depths were analyzed. At the same time, combined with soil moisture diffusivity studied previously, the soil unsaturated hydraulic conductivity was calculated. The results showed that (1) The SWRC of 0 ~ 100 cm in each sandy lands could be described better by the Gardner empirical equation $\theta = Ah^{-B}$. Soil water holding capacity is decreased in the order: grassland > fixed dune > mobile dune; When the soil water potential changes, the soil water content of mobile dune changes fastest, followed by fixed dune, and grassland changes slowest. In addition, Soil water holding capacity and soil moisture change rate vary considerably along with the type of sandy land and soil depth. (2) Soil unsaturated hydraulic conductivity is smallest for mobile dune, followed by fixed dune, and for grassland is largest. And soil unsaturated hydraulic conductivity decreases generally with soil depth increased. (3) Soil organic matter content and the particle size of <0.05 mm are conducive to water holding capacity, but bulk density and the particle size of 2 ~ 0.1 mm are discouraged to water holding capacity.

Keywords: Soil water retention curve · Soil moisture diffusivity · Hydraulic conductivity · Water comparison content · Horqin Sand Land

1 Introduction

Unsaturated soil water movement parameters, including hydraulic conductivity, soil moisture diffusivity ($D(\theta)$) and soil water retention curve (SWRC), are the important parameters for quantifying soil water and solute movement and for modeling hydrogeologic processes (Do-Hun 2005). These parameters drive the flow of water in the soil-plant-atmosphere system, and hence control processes such as aquifer recharge or nutrient fluxes between soil and vegetation (Rubio 2008). The constitutive relationship between water content or degree of saturation and suction is called by many authors the soil water retention curve (SWRC) (Farimah et al. 2008). The SWRC contains vital information for deriving unsaturated soil property functions for the coefficient of

F. Bian and Y. Xie (Eds.): GRMSE 2015, CCIS 569, pp. 374–380, 2016.
DOI: 10.1007/978-3-662-49155-3_38

permeability, shear strength and volume change, and it can reveal the relationship between the amount of water in the soil and soil suction (Sillers et al. 2001). Data on SWRC are used in research and applications in hydrology, agronomy, meteorology, ecology, environmental protection, and many other soil-related fields. Reliable measurement of SWRC is essential for solving unsaturated flow problems (Farimah et al. 2008). Thus, knowledge of SWRC and its spatial distribution in various landscapes in Horqin Sand Land is of particular importance (Jia et al. 2006).

The Horqin Sand Land lies in a semi-arid area of eastern Inner Mongolia, in northern China. Due to the long-term influence of heavy grazing, land reclamation for agriculture, and extensive harvesting of fuelwood, this region has become one of the most severely decertified areas in China (Zuo et al. 2008). In recent years, desertification has produced distinctive mobile dune landscapes in this region. However, despite the importantly of soil water movement parameters, there was little studies have studied the characteristics of soil water retention curve and the unsaturated hydraulic conductivity in the Korqin sand land. In the present study, we used laboratory methods to measure the SWRC of 0 ~ 100 cm at three sandy land of Horqin Sand Land, such as grassland (GL), fixed dune (FD) and mobile dune (MD), and the variation of the parameter in different sandy type and different soil depth were analyzed. At the same time, combined with soil moisture diffusivity studied previously, the soil unsaturated hydraulic conductivity was calculated.

2 Materials and Methods

2.1 Study Site Description

This study was conducted in south-western Horqin Sandy Land, Inner Mongolia, China (42°55′ N, 120°42′ E; elevation approx. 360 m). The landscape in this area is characterized by sand dunes alternating with gently undulating lowland areas. The soils are sandy, light yellow and loose in structure. The climate is temperate, semi-arid and continental, receiving 360 mm annual mean precipitation. Scant rains and frequent winds between November to May makes this period both the major wind-erosion season (Li et al. 2005). The annual mean open-pan evaporation is about 1935 mm. The mean annual temperature is 5.8 ~ 6.4°°C, with a minimum mean monthly temperature of −12.6 ~ −16.8 °C in January and a maximum mean monthly temperature of 20.3 ~ 23.5 °C in July (Yao et al. 2013).

In this study, three typical sandy lands from Naiman County were selected for measurement SWRC, including: MD, FD and GL. For each sandy land, three sites similar to each other were selected to obtain a total of three replications. The basic and natural conditions for the nine sites listed in Table 1. In each site, three sampling points were set up to provide numerous replications.

2.2 Measurement of SWRC

SWRC was measured by pressure plate test (No. 1500 15 bar Pressure Extractor). For each sampling point, one soil profile with 1 m depth was excavated. For each soil

Table 1. Basic situations of each site

Site	Species	Coverage (%)	Type of disturbance (%)	Longitude	Latitude	Altitude (m)
GL 1	*Artemisia scoparia, Setarria viridis*	60	30 ~ 50	120.665	42.965	351
GL 2	*Artemisia scoparia, Setarria viridis*	80	/	120.713	42.938	357
GL 3	*Echinops gmelini Turcz, Artemisia scoparia*	75	30 ~ 50	120.661	42.956	353
FD 1	*Artemisia scoparia, Artemisia halodendron*	70	/	120.698	42.929	360
FD 2	*Artemisia scoparia, Setaria viridis*	80	/	120.692	42.929	368
FD 3	*Artemisia halodendron, Agriophyllum squarrosum, Echinops gmelini Turcz*	60	/	120.714	42.941	363
MD 1	*Agriophyllum squarrosum*	3	10 ~ 20	120.613	43.149	342
MD 2	*Agriophyllum squarrosum, Setaria viridis, Artemisia halodendron*	6	/	120.607	43.183	338
MD 3	*Agriophyllum squarrosum, Setaria viridis*	5	/	120.401	43.034	329

profile, five soil layers were designed: 0 ~ 20 cm, 20 ~ 40 cm, 40 ~ 60 cm, 60 ~ 80 cm and 80 ~ 100 cm. Soil sample was obtained from each soil layer by cutting ring (with high 1 cm, diameter 5 cm), which were brought back to the laboratory. The SWRC of each soil sample was measured by pressure plate test after saturated 24 h. The pressure was set up with 0.1, 0.3, 0.5, 1, 3, 5, 10 and 15 bar. The soil water content of each soil sample was obtained by formula (Wang et al. 2007): $V_i = [(W_i - W)/\rho]/V$. Where V_i ($cm^3 \cdot cm^{-3}$) is soil volumetric water content of the *i*-pressure; W_i (g) and W(g) is wet soil quality and dry soil quality for *i*- pressure, respectively; ρ ($g \cdot cm^3$) is density of water, which was taken with $\rho = 1$; V (cm^3) is the volume of cutting ring used in the test.

The $\theta \sim h$ curve was the curve of soil water content (θ) and soil water suction (h), namely SWRC. By SWRC or θ ~ h curve we can obtain soil water comparison content by formula $C(\theta) = -d\theta/dh$ (Shao and Horton 1996).

To reduce errors, SWRC of each soil sample was all the average of three replicates.

Combined with $D(\theta)$ (which was studied by Yao et al. (2014) and $C(\theta)$, soil unsaturated hydraulic conductivity ($K(\theta)$) was calculated by formula (1) (Chen and Shao 2002).

$$K(\theta) = D(\theta) \cdot C(\theta) \tag{1}$$

The primary statistical analysis was performed using version 11.5 of the SPSS software (SPSS Inc., Chicago, IL, USA).

3 Results

3.1 Soil Water Retention Curve

Studies have shown that, Gardner power function of the empirical formula $h{=}a\theta^{-b}$ is suitable for most of soil in China (Junhong et al. 2007). Where h (bar) is the soil water potential, θ $(\mathrm{g} \cdot \mathrm{g}^{-1})$ is the soil water content. From the formula of $h = a\theta^{-b}$ we obtained:

$$\theta = Ah^{-B} \tag{2}$$

Where A denote the level of water holding capacity, and the larger A, the stronger the water holding capacity. When A was constant, the larger B $(0 < B < 1)$, the curve closer to h axis, which indicated that B can reflect how fast the change of soil water when soil water potential change (Wang et al. 1999).

The SWRC of three typical sandy lands measured by the pressure plate test was suited formula (2) and with high correlation coefficients (Table 2). That was to say, the empirical formula of (2) can be better applied in Horqin Sand land.

Table 2. Estimate results of soil water retention curve and $C(\theta)$

Sandy land	Soil layer (cm)	A	B	Correlation coefficients	C(θ)
GL	0~20	0.104	0.050	$R^2 = 0.853$	$C(\theta) = 0.005h^{-1.050}$
	20~40	0.114	0.060	$R^2 = 0.954$	$C(\theta) = 0.007h^{-1.060}$
	40~60	0.108	0.065	$R^2 = 0.892$	$C(\theta) = 0.007h^{-1.065}$
	60~80	0.119	0.067	$R^2 = 0.894$	$C(\theta) = 0.008h^{-1.067}$
	80~100	0.117	0.070	$R^2 = 0.864$	$C(\theta) = 0.008h^{-1.070}$
	Mean	0.112	0.062	$R^2 = 0.891$	$C(\theta) = 0.007h^{-1.062}$
FD	0~20	0.036	0.105	$R^2 = 0.899$	$C(\theta) = 0.004h^{-1.105}$
	20~40	0.050	0.087	$R^2 = 0.952$	$C(\theta) = 0.004h^{-1.087}$
	40~60	0.036	0.105	$R^2 = 0.964$	$C(\theta) = 0.004h^{-1.105}$
	60~80	0.031	0.112	$R^2 = 0.995$	$C(\theta) = 0.004h^{-1.112}$
	80~100	0.036	0.107	$R^2 = 0.941$	$C(\theta) = 0.004h^{-1.107}$
	Mean	0.038	0.103	$R^2 = 0.950$	$C(\theta) = 0.004h^{-1.103}$
MD	0~20	0.027	0.322	$R^2 = 0.971$	$C(\theta) = 0.009h^{-1.322}$
	20~40	0.038	0.457	$R^2 = 0.976$	$C(\theta) = 0.017h^{-1.457}$
	40~60	0.031	0.352	$R^2 = 0.970$	$C(\theta) = 0.011h^{-1.352}$
	60~80	0.035	0.392	$R^2 = 0.983$	$C(\theta) = 0.014h^{-1.392}$
	80~100	0.036	0.402	$R^2 = 0.992$	$C(\theta) = 0.014h^{-1.402}$
	Mean	0.034	0.385	$R^2 = 0.978$	$C(\theta) = 0.013h^{-1.385}$

C(θ) was water comparison content, which referred the changes of soil water caused by changes of unit matrix potential, and hence, it equaled to the slope of SWRC (Wang et al. 1999). From formula (2), we obtained:

$$C(\theta) = -\mathrm{d}\theta/\mathrm{dh} = (\mathrm{A} \cdot \mathrm{B}) \cdot \mathrm{h}^{-(B+1)}. \tag{3}$$

Where $A{\cdot}B$ was the value of $C(\theta)$ when $h = -1 \times 10^5$ Pa.

It can be seen from Table 2 that SWRC of each soil depth at each sandy land can be suited better by Cardner model, with correlation coefficients at least 0.85. The value of *A* fitted for GL, FD and MD were 0.112, 0.038 and 0.034, respectively. This indicated that water holding capacity was highest in GL, followed by FD, and MD was lowest. For GL, FD and MD, The value of *B* were 0.062, 0.103 and 0.385, respectively. That was to say, when soil water potential changes, soil water of MD changed fastest, followed by FD, and it changed slowest in GL.

The change of SWRC with soil depth was also shown in Table 2. The trend of SWRC with soil depth was not consistent among the sandy lands. For example, *A* of GL was highest at 60 ~ 80 cm and 80 ~ 100 cm, but for FD and MD, it generally higher at 20 ~ 40 cm. For three sandy lands, *A* of 0 ~ 20 cm was lowest. This indicated that the water holding capacity of GL was higher at 60 ~ 100 cm, and it was higher at 20 ~ 40 cm for MD and FD; the water holding capacity of three sandy lands was lowest at 0 ~ 20 cm. The value of *B* at GL increased generally with soil depth increasing. But there was no rule for the change of *B* with soil depth at FD and MD, which mainly due to the frequently sandstorms activities of FD and MD, and have high degree of fragmentation of landscape, and hence led to the higher spatial heterogeneity at different soil layers of these two typical sandy lands (Yao et al. 2013).

Water comparison content ($C(\theta)$) of five soil depths at GL, FD and MD were calculated by formula (3), which were also shown in Table 2. It can be seen, the extent of soil water capacity released was lowest for GL, followed by FD, and it was highest for MD.

3.2 Unsaturated Hydraulic Conductivity

Form formulas (3) and (4), we obtained formula (4):

$$C(\theta) = -\mathrm{d}\theta/\mathrm{dh} = (\mathrm{A} \cdot \mathrm{B}) \cdot \mathrm{h}^{-(\mathrm{B}+1)} = \mathrm{B} \cdot \mathrm{A}^{-1/B} \cdot \theta^{(1+1/B)}, \tag{4}$$

$D(\theta)$ of these nine sites were studied previously by Yao et al. (2014). Combing $C(\theta)$ (Table 2) and $D(\theta)$, $K(\theta)$ was calculated by formula (1) and listed in Table 3.

Seen from Table 3, $K(\theta)$ of MD was lowest, followed by FD, and it was highest at GL. For each sandy land, $K(\theta)$ decreased generally with increasing soil layer.

Table 3. Estimating of unsaturated soil hydraulic conductivity $K(\theta)$

Soil layer (cm)	Sandy land		
	GL	FD	MD
0 ~ 20	$K(\theta) = 2.40 \times 10^{18} \times \theta^{21.04} \times e^{-6.75\theta}$	$K(\theta) = 5.60 \times 10^{12} \times \theta^{10.52} \times e^{-4.63\theta}$	$K(\theta) = 2.35 \times 10^{4} \times \theta^{4.10} \times e^{-3.56\theta}$
20 ~ 40	$K(\theta) = 2.96 \times 10^{14} \times \theta^{17.61} \times e^{-8.36\theta}$	$K(\theta) = 7.60 \times 10^{13} \times \theta^{12.47} \times e^{-5.17\theta}$	$K(\theta) = 5.80 \times 10^{2} \times \theta^{3.19} \times e^{-3.22\theta}$
40 ~ 60	$K(\theta) =4.05 \times 10^{13} \times \theta^{16.29} \times e^{-6.49\theta}$	$K(\theta) = 6.22 \times 10^{12} \times \theta^{10.52} \times e^{-4.01\theta}$	$K(\theta) = 6.59 \times 10^{3} \times \theta^{3.84} \times e^{3.71\theta}$
60 ~ 80	$K(\theta) = 5.65 \times 10^{12} \times \theta^{16.04} \times e^{-6.60\theta}$	$K(\theta) = 2.73 \times 10^{12} \times \theta^{9.90} \times e^{-3.97\theta}$	$K(\theta) = 2.01 \times 10^{3} \times \theta^{3.55} \times e^{-3.98\theta}$
80 ~ 100	$K(\theta) = 1.40 \times 10^{12} \times \theta^{15.25} \times e^{-6.62\theta}$	$K(\theta) = 3.58 \times 10^{12} \times \theta^{10.33} \times e^{-4.16\theta}$	$K(\theta) = 1.59 \times 10^{3} \times \theta^{3.49} \times e^{-3.60\theta}$

3.3 Influencing Factors of SWRC

Studies have showed that SWRC was affected mainly by soil texture, bulk density, organic matter content and temperature and other factors (Chen and Shao 2002; Shao and Horton 1996). From the soil physical-chemical properties (including bulk density, organic matter content, the particle size of 2 ~ 0.1 mm, 0.1 ~ 0.05 mm and <0.05 mm) studied by Yao et al. (2014), we found that the organic matter content and the particle size of <0.05 mm were conducive to water holding capacity, but bulk density and the particle size of 2 ~ 0.1 mm were discouraged to water holding capacity. Because the water holding capacity, organic matter content and the particle size of <0.05 mm were all decreased in the order: GL > FD > MD, but bulk density and the particle size of 2 ~ 0.1 mm were increased in the order: GL < FD < MD.

4 Conclusions

The SWRC of 0 ~ 100 cm in GL, FD and MD could be described better by the Gardner empirical equation $\theta = Ah^{-B}$. Soil water holding capacity is highest at GL, followed by FD, and it is lowest at MD. $K(\theta)$ of MD is lowest, followed by FD, and it is highest at GL. For each sandy land, $K(\theta)$ decreases generally with increasing soil layer. Soil organic matter content and the particle size of <0.05 mm are conducive to water holding capacity, but bulk density and the particle size of 2 ~ 0.1 mm are discouraged to water holding capacity.

Acknowledgments. This study was supported by the National Natural Science Foundation of China (No. 31300388; No. 41361013); Dr. Start-up Science Research Foundation of Lanzhou City University (No. LZCU-BS2013-09; No. LZCU-BS2013-12); Research Foundation of university in Gansu Province (No. 2013B-074) and Chancellor Research and Innovation Fund of Lanzhou City University (No. LZCU-XZ2014-12).

References

Chen, H.B., Shao, M.A.: A review on indirect methods for estimating unsaturated soil hydraulic properties. J. Basic Sci. Eng. **10**(2), 103–109 (2002). (in Chinese with English summary)

Do-Hun, L.: Comparing the inverse parameter estimation approach with pedo-transfer function method for estimating soil hydraulic conductivity. Geosci. J. **9**(3), 269–276 (2005)

Farimah, M., Katiá, V.B., Katsuyuki, K.: Laboratory hydraulic testing in unsaturated soils. Geotech. Geol. Eng. **26**, 691–704 (2008)

Jia, H.W., Kang, S.Z., Zhang, F.C.: One-parameter models of soil hydraulic parameters. Shui li Xue bao **37**(3), 272–277 (2006). (in Chinese with English summary)

Junhong, B., Wei, D., Baoshan, C., et al.: Water diffusion coefficients of horizontal soil columns from natural Saline-Alkaline Wetlands in a Semiarid area. Eurasian Soil Sci. **40**(6), 660–664 (2007)

Li, F.R., Kang, L.F., Zhang, H., et al.: Changes in intensity of wind erosion at different stages of degradation development in grasslands of Inner Mongolia. China J. Arid Environ. **62**, 567–585 (2005)

Rubio, C.M.: Applicability of site-specific pedotransfer functions and rosetta model for the estimation of dynamic soil hydraulic properties under different vegetation covers. J. Soils Sediments **8**(2), 137–145 (2008)

Shao, M.A., Horton, R.: Soil water diffusivity determination by general similarity theory. Soil Sci. **161**, 727–734 (1996)

Sillers, W.S., Delwyn, G.F., Noshin, Z.: Mathematical attributes of some soil water characteristic curve models. Geotech. Geol. Eng. **19**, 243–283 (2001)

Wang, M.B., Chai, B.F., Li, H.J., et al.: Soil water holding capacity and soil available water in plantations in the loess region. Scientia Silvae Sinicae **35**(2), 7–14 (1999). (in Chinese with English summary)

Wang, S.P., Zhang, Z.Q., Wu, J., et al.: Preliminary study on spatial variability of soil water retention function in a Chinese Pine (Pinus Tabliforms) Plantation. Res. Environ. Sci. **20**(2), 28–35 (2007)

Yao, S.X., Zhang, T.H., Zhao, C.C., et al.: Saturated hydraulic conductivity of soils in the Horqin Sand Land of Inner Mongolia, northern China. Environ. Monit. Assess. **185**(7), 6013–6021 (2013)

Yao, S.X., Zhao, C.C., Zhang, T.H., et al.: Soil moisture diffusivity in different habitats in Horqin Sand Land. Chin. J. Ecol. **33**(4), 867–873 (2014). (in Chinese with English summary)

Zuo, X.A., Zhao, H.L., Zhao, X.Y., et al.: Spatial pattern and heterogeneity of soil properties in sand dunes under grazing and restoration in Horqin Sand Land, northern China. Soil Tillage Res. **99**, 202–212 (2008)

Research on Meteorological Disasters of Guilin's Tourism

X.D. Bai(✉)

Guilin Meteorological Bureau, Guilin 541001, China
glbxd@126.com

Abstract. By research on meteorological disasters of tourism, it can provided meteorological services for the healthy development of Guilin tourism. Investigated of Guilin tourism projects and meteorological disaster, and analysis the meteorological disasters of tourism projects, tourism response to meteorological disasters and mitigation disaster prevention measures are put forward. Research suggests tourism of Guilin meteorological disasters mainly include: heavy rain, thunder and lightning, wind, drought, fog, haze, high temperature, low temperature, etc. Secondary disasters are: landslide, debris flow, flood, etc. Strengthen the research of tourism of meteorological disaster and understand its distribution characteristics and its influence. Strengthen the safety measures and management in the tourist attractions. Add some display meteorological information electronic display screen in the tourist attractions, by training to the tour guide of meteorological disasters and their defense knowledge, it can effectively avoid or alleviate the harm caused by meteorological disasters for tourism.

Keywords: Guilin tourism · Meteorological disasters · Defensive measures · Study

1 Introduction

Climate is a major conditions for the development of tourism in a region, it is must considered that weathered rain and snow or the temperature change are important issues to develop tourism activities. With the rapid development of economic society, tourism has become an important part of a modern economic life and cultural life. The influence of climate conditions of tourism is more and more attention, tourism climatology has become a new subject. Wu specifically written a textbook of <tourism climatology> [1]. Yang discusses the basic concept and research content of tourism climatology [2], Wei summarized the research of tourism climatology of our country [3], Many places have some introduction on study of tourist climate resources [4], Zhao has made a comprehensive assessment of tourism environment in oases of Xinjiang [5]. In those studies, scholars for the influence of climate conditions are given wide attention [6]. Climate impact ecosystem [7], affect human body comfort [8], affect human body health [9], affect the from all walks of life [10]. With the rapid development of social economy, meteorological disaster loss is more and more [11]. There are many kinds of meteorological disasters in our country. According to the Ge and Li [12] for the classification of meteorological disasters in our country, the various types

F. Bian and Y. Xie (Eds.): GRMSE 2015, CCIS 569, pp. 381–391, 2016.
DOI: 10.1007/978-3-662-49155-3_39

of meteorological disasters will cause adverse effect to Guilin tourism. Meteorological disasters around the tourism safety accidents often happen, the disaster prevention and mitigation of tourism is a long way; Meteorological disaster tourism research is attention by scholars widely, Lou and Wang [13] was studied the distribution features of the tourism meteorological disasters of Qinghai province. Liu and Yang research on the extreme climate events dangers of tourism development in our country [14, 15]. Cui et al. studied on non-engineering countermeasures of meteorological disaster prevention in large cities [16]. Yuan et al. research and application of rainstorm disaster risk assessment along the middle and lower reaches of Yangtze river [17]. Luo et al. discussion on common problems in risk assessment of lightning disaster [18]. Wu et al. [19] analysis on the relationship between Geo logical disasters and precipitation in Zhuanghe Region. Bai [20] analysis of characteristics of high temperature weather in Guilin area. Tourism activities is a new way of high consumption, need to provide a variety of services, meteorological service is also very important. Guilin is an international tourism resort, is also the area where the meteorological disasters often happened. Some of the geological disasters often occurred due to the influence of meteorological condition, derivative of meteorological disasters of Guilin tourism safety influence is very big also, it must be strengthen the research of meteorological disasters of Guilin's tourism, to provide meteorological service timely and efficiently for the healthy development of the tourism of Guilin. In this paper, based on field investigation and data analysis, author study of meteorological disasters of Guilin's tourism.

2 Guilin's Main Tourism Projects and Tourism Mode

Special geographical position, Guilin's mountains and rivers are beautiful, and KAST landscape development is complete, customs peaks, caves, rivers, vegetation, local customs, etc., the tourism resources is abundant. Guilin has developed into an international tourist destination. Existing tourism projects can be divided into: rivers and reservoirs, geological cave, plant flowers, ancient residence, national culture, rural holiday entertainment, landscape ecology, health, leisure, exercise and other types.

2.1 Rivers and Reservoirs

Led by the Li-river, Guilin landscape essence travel; Urban two lakes (the Li-river, Peach-river) and four lakes (Mulong lake, Gui lake, Rong lake, Shan lake) travel, drift on the Zi-river, Wupai-river, twelve beach, Yulong river, Longjing river, Jiuwu river, and Longji gorge. Travel reservoirs are Qingshi pool, Ban gorge, the heavenly lake, the Wuli gorge, etc., it has more than 200 reservoirs in the city, these area are both residents and visitors leisure and good place for fishing.

2.2 Geological Cave

Guilin KAST landscape development is complete, a lot of stone mountain have caves, exquisitely shaped cave, stalagmites, stone milk and stone curtain, rock flowers, stone forest, stone pillars, encrinite, etc. full of beautiful things in eyes. The most distinctive features of the tourist caves are: Seven star rock, reed rock, cap rock, silver rock, abundant fish rock, Jvlong rock, bodhisattva water rock, century glacier, lotus cave; The caves in Fubo hill, folded brocade hill, Nanxi hill, are also added the scenery for the tourist.

2.3 Historic Residence

Famous ancient villages and towns are: West street, Welfare, Xingping town, Xingping fish village, stone town, peace township, Daxv, Jiangtou village, Lipu green village, Wenshi village, Baishou village. The famous historic sites are: Jingjiang King Hours, Jingjiang King Mausoleum, The eighth route army offices, Wang Zhenggong statue, Lizhongren former residence, Baichonxi former residence, Chenhongmou former residence, Ling canal, Xiang river battle monument, Zhouwei temple, Hunan hall, Wen and Wu temple, Jinshan temple, Jianshan temple, Xvbeihong former home, Xiangshan temple.

2.4 National Culture

Guilin has Zhuang, Hui, Miao, Yao, Dong and other 36 ethnic minorities. The main ethnic cultural attractions are: Liusanjie view garden, Impression of Liusanjie, Totem ancient path, the peach orchard land out of the world, Butterfly spring, Yao village, Gold pit terrace, Dong village, etc.

2.5 Rural Entertainment

Guilin landscape gardens are: Yanshan botanical garden, Yu mount, West mount, Nanxi mount, fool oneself paradise, Rawmand paradise. Pastoral scenery mainly include: Nanshan Pasture, Longji terraced fields, Yulong Bridge, Gingkgo park, tallow tree sands, West hill peach blossom, etc.

2.6 Landscape Ecology

Guilin famous mountains are: solitary peak, The li-river ten famous mountains, Yu mount, West mount, Chuan mount, Yao mount, Elephant Trunk hill, Decai hill, Fubo hill, Nine horses mountain, moon hill, Mao'er Mountain, etc. Hot springs: Sishui, Xianjia, Jinzhong hill, Ziyuan, big west river. The waterfall landscape are: Ziyuan Baoding, Lingchuan Gudong, Longmen, etc.

2.7 Keeping in Good Health Holiday

In Yangshuo county as the center of leisure tourism has become the fashion, Foreign visitors often to Yangshuo for a period of time, keeping in good health holiday; Burning hot summer, people often choose the mountains where has a river for the weekend. For this, a lot of leisure holiday villa constructed on city surrounding, it provides conditions for holidays in Guilin. There are many foreign guests to Yongfu county where is a longevity township, to purchase house and free to live in a period of time.

2.8 Leisure Exercise

Mainly has the project such as mountain climbing, hiking, rafting. The mountains are green and water is elegant of Guilin and the surrounding, hundreds of classical hiking routes, attract public enthusiasm to participate in. Every weekend, a large number of travelers can go hiking, to enjoy the beautiful scenery of nature, both also exercise the body.

2.9 Tourism Modes

Guilin tourism pattern can be divided into: (1) Join the tour team, along with the travel service group, with a guide to lead; (2) scattered tour, not with group, oneself; (3) tour by drive oneself, enjoy the sight of nature along the drive. Guilin has become the back garden of Guangdong, Guangdong guest drive to Guilin at holiday has becoming formed; (4) Walking Travel. A group of travelers carry luggage and corresponding facilities, to the wild hiking. (5) Bike to tour, Guilin has opened some bicycle tour route, a few people rent the bicycle, the alternative route, is also very comfortable. (6) Leisure keeping in good health. In the better place to live down, relax oneself.

3 The Main Meteorological Disasters in Guilin and Its Impact on Tourism

Guilin is located in the northeast of Guangxi, a humid subtropical monsoon climate, Mild climate, abundant rainfall. But the drought and flood season is clear, it is more heavy rain at April to July, August to December, it is more drought; Spring South wind and wet weather, the thunderstorm winds, hail, rainstorm, flood, drought, high temperature, chilling injury, fog, haze, such as meteorological disaster will cause adverse effect to tourism. The classification of meteorological disasters and the impact can be seen in Table 1.

Table 1. Main kinds of meteorological disasters in Guilin

Kind	Weather phenomena	Direct and indirect damage	Tourism countermeasures
Flood	The intense rain, continuous heavy rain	Flash floods, river flood, water logging, city water, damaged roads, telecommunications, water supply, power supply, construction, crops, material; causing casualties. Secondary disasters are: geological disasters, disasters of agriculture and forestry, the spread of disease	Don't stay in rivers and low areas during the torrential rain, notice torrent and geological disasters, flood after pay attention to prevent the spread of disease
Drought	Little rain and long fine	Production and living difficulties, ecological environment destruction, agriculture, forestry and water disasters (plant diseases and insect pests, fire), geological disasters, land desertification	Pay attention to the field in the fire, drink plenty of water in the dry air weather
High temperature	Sunny hot temperature	People heatstroke, plant growth or wither, animal discomfort, ecological environment destruction, disease increased	Pay attention to avoid heat stroke, try to choose the shade for travel or walk
Cold damage	Cold, frost, snow and glaze	Cold, freezing in people and livestock, crops, economic trees, frozen can lead to power lines, communications lines, water pipes, gas pipes, icy roads lead to road traffic accidents increase	Attention to keep warm, wild field pay attention to prevent slippery, pay attention to the road traffic safety
Continuous rain	Damp, cold and rainy	Normal affect crop growth, the material easy to mildew, the body feel uncomfortable, cause disasters of agriculture and forestry, plant	Pay attention to prevent slippery, and the geological disasters, not to eat mould material

(*Continued*)

Table 1. *(Continued)*

Kind	Weather phenomena	Direct and indirect damage	Tourism countermeasures
		diseases and insect pests	
Severe convection	Thunderstorms, hail, wind, rain	Damage to buildings and crops, farmland, supplies, transportation, communications, electricity, water, gas and other facilities, human and animal casualties, traffic accidents. Disasters of agriculture and forestry, geological disasters	Note against hail, lightning, wind, rain, go out as far as possible to avoid strong convective weather
Else	The fog, haze	Affect traffic safety, body health, induced by a variety of diseases	Pay attention to traffic safety, pay attention to prevention

3.1 South Wind and Continue Rain Day

Water vapor content in the high incidence in the spring, the air is nearly saturated, wet everywhere, the room floor will be water, due to high humidity, slippery, easy to cause tourists wrestling. Continue rain day for a long time, caused the surface wetting, geological disasters easily happened. In April 2015, folded brocade hill rock collapse because of the early rain for a long time, major accidents caused casualties in Guilin.

3.2 Severe Convection Weather, Including Hail and Thunderstorm Winds, Thunderstorm

This kind of weather often accompanied, March to September are likely to happen. The field has great influence on the safety of the tourists, especially affected on the li-river cruising ship. The thunderstorm winds could cause major disasters. The li-river has occurred many times on the major water safety accident caused by strong wind.

3.3 Heavy Rain

April to July, Guilin is often heavy rain. Maximum rainfall reached 400 mm in one day. Especially Xiang-Gui along the railway is a heavy rain area. The boom in heavy rains often caused, li-river tour ships are suspend, low-lying scenic forced to close down by water; Heavy rains also often triggered floods, landslides and geological disasters. During the torrential rain of Guilin, the tourism project will be severely affected, is not only effect an outdoor tourism projects, indoor the tour will also be affected because of heavy rain is difficult to make tour occur. Storm floods also pose damage to water tourism facilities.

3.4 High Temperature

Mainly appear in the July to September, occasionally occur in May to June. Daily maximum temperature of 37 °C or more hot weather had a great influence on people outdoors, easily lead to heat stroke, especially wild people hiking, mountain climbing and sightseeing, people need more attention.

3.5 Dry

After the fall, less precipitation distinctness, it caused drought. Drought can affect the ecological environment around, causing the boat in the li-river had to reduce the distance because the water is little, also can cause a cruise ship ran aground, all water sports are affected because of water shortage; The ecological project also because drought is normal; Waterfall sightseeing, drift is affected the most serious project.

3.6 Chilling Injury

Including the icy roads, cold wave, frost and freezing. Because cold air activity, the main cause is adaptation for human; Icy roads are had a great influence on road traffic safety. Icy roads mainly appeared in the mountains and township of the north of Guilin city. Serious low temperature sleet freezing weather in January 2008, Guilin north and the mountains of all tourism projects was stopped, many roads can't normal traffic. Walking tour is not carried out in the wild.

3.7 The Fog

Fog can affects the visibility of the plane, and affect the road vehicle the driver's line of sight, are prone to accidents, affect the tourists in Guilin normal tourist activity.

3.8 Haze

Haze is caused due to large amounts of air pollutants concentration, of reduced visibility, it influences the atmospheric environment, affect human health, affected the choice of the foreign guests to Guilin tourism.

4 Tourism Meteorological Disasters Defense

Tourism meteorological disasters defense also should be done in case putting prevention first. Respectively according to different meteorological disasters, different tourism project, make different defense plans. All official foreign business tourism projects must formulate and improve all kinds of meteorological disaster emergency response measures. Tourism meteorological disasters defense is also a complicated project, in addition to make visitors pay attention to, park, tour guide and driver have very big responsibility, meteorological disaster early warning information timely and accurate transmission is also essential.

4.1 Tourist Attractions Need to Strengthen the Construction of Safety Facilities, and Check Regularly

Especially the buildings and facilities lightning protection, mountain road prevent slippery, must do to check every year, security fence to timely maintenance; Prone to collapse rock mountain, prone to landslides and debris flow area, to undergo screening on a regular basis, found that the problem should be handled in a timely manner. Install the sign on the easy place to a disaster facilities remind visitors at any time to pay attention to safety.

4.2 The Tourism Practitioners have to Strengthen Safety Knowledge and Skills Training

Tour guide must be familiar with his own route, prone to what kind of meteorological disasters, and know how to deal with. In the main tourist attractions to set up the weather information of electronic display, meteorological departments to strengthen disaster weather monitoring and forecast, meteorological information release at any time. Starting in 1990, Guilin tourism transport company every year please weather experts to work in the travel on the li-river course how to prevent against strong wind weather, the effect is very good, many effective to avoid the dangers of the strong wind.

4.3 Travel on Rivers and Reservoirs

Pay special attention to the monsoon rain floods, pay attention to prevent thunderstorm winds strong and convective weather. In the fog or haze weather, don't get lost. Drifting should pay attention to the depth of the water, avoid accident caused by water less.

4.4 Travel on Geological Cave

Be alert the heavy rain, rainy weather for a long time, when easy to cause the geological disasters such as landslides, rock falls. Rain also can cause mountain road slippery, easy to cause the guest wrestling. To observe whether may cause the cave pond because of storm water.

4.5 Travel in Former Residence, National Culture and Leisure Vacation Tour

Need to pay attention to the influence of high temperature. Pay attention to the lightning protection facilities construction sites, lightning disasters prevention, pay attention to the heavy rain and long time rain made the road very slippery.

4.6 Rural Entertainment and Landscape Ecological Tourism

Flood season to pay attention to the heavy rain, beware of flood disasters and geological disasters because of the rain. In the summer season to pay attention to prevent heat, in the winter should pay attention to chilling injury such as icy roads.

4.7 Walking Exercise to Tour

Tourists walk in the countryside, on the mountain path, rainfall, flash floods, high temperature, geological disasters, fog, haze, icy roads, thunderstorm, strong wind will affect tourists form, such as some can even be dangerous. Especially for tent accommodation in the field of tourists, more attention should be paid to choose the safety camp and safety zone. Attention to the wild in the fire in dry season.

4.8 Self-driving Trips are Often Driving Their Own Car, to Travel from Alone the Road

Because not local full-time driver, not understanding of the meteorological disaster in tourism destination, need careful driving more, had better want to strengthen consultation, pairing.

5 Brief Summary

(1) Guilin tourism projects can be divided into: rivers and reservoirs, geological rural residence of KAST, monuments, national culture, entertainment, landscape ecology, health, leisure and exercise such as 8 categories, many of all kinds of tourist attractions distribution in urban areas and various counties.

(2) Guilin tourism pattern can be divided into: join the travel team, travel by oneself at will, self-drive tour, hiking tourism, leisure, health, etc. Visitors can choose their favorite way for sightseeing, leisure, vacation, exercise, hiking, spa tourism, etc.

(3) Guilin tourism meteorological disasters are: storms, drought, high temperature and chilling injury, sounds, strong convection, fog, haze. By the meteorological causes of secondary disasters are: landslide, landslide, mud-rock flow, forest fires and the spread of disease.
(4) Tourism meteorological disasters defense to do first tourist attractions attaches great importance to the management department, improve the safety measures, and regular safety inspection of these safety measures; Second is to strengthen the safety knowledge training of tourism professionals, all tour guide must pass defense training, the basis of meteorological disasters in the disaster comes to take effective measures, minimize disaster losses.

References

1. Wu, W.Z.: Tourism Climatology. China Meteorological Press, Beijing (2001)
2. Yang, S.Y.: Several problems about tourism climatology. J. Guilin Tourism Coll. **16**(3), 24–27 (2005)
3. Wei, F.Y., Wang, J.C.: Tourism climate research in China in 2001–2007 were reviewed. J. Changchun Norm. Coll. **26**(6), 73–76 (2007). (Natural Science Edition)
4. Yang, B.H., Su, Z., Chen, G.L.: The influence of climatic conditions of Guangxi north gulf tourism evaluation. Tourism Forum **4**(4), 118–120 (2011)
5. Zhao, Y.F.: Comprehensive assessment of tourism environment in Oases of Xinjiang. Meteorol. Environ. Res. **4**(9), 1–5 (2013)
6. The national climate change assessment report writing committee: The national climate change assessment report. Science press, Beijing (2007)
7. Fu, G.B., Li, K.R.: Global warming and the research progress of wetland ecological system. Geogr. Res. **20**(1), 120–128 (2001)
8. Qin, W.J.: Guangxi tourism climate comfort analysis. Guangxi Meteorol. **24**(4), 50–51 (2003)
9. Zhang, Q.Y., jv, J.H., Wang, W.D., et al.: Climate warming on human health of shadow. Meteorol. Sci. Technol. **35**(2), 20–22 (2007)
10. Chen, J.G., Wang, X.F., Long, H., et al.: The influence of climate change on the main industry in Yunnan. J. Yunnan Norm. Univ. **42**(3), 1–20 (2010)
11. Liu, T., Yan, T.C.: Our country the main meteorological disasters and economic loss. J. Nat. Disasters **20**(2), 90–95 (2011)
12. Ge, J.X., Li, Z.C.: Our country the classification of meteorological disasters and disaster prevention and mitigation countermeasures. Disaster **20**(4), 106–110 (2005)
13. Lou, S.Z., Wang, Q.C.: Temporal variation characteristics of meteorological disasters analysis of tourism in Qinghai province. J. Qinghai Univ. **1**, 1–7 (2013). (Natural Science Edition)
14. Liu, Y.L.: Frequent extreme weather thing under the development of Chinese tourism industry thinking. Mod. Tourism **10**, 22–23 (2014)
15. Yang, S.Y., Hu, J.: The influence of meteorological disaster to our country tourism. Anhui Agric. Sci. **38**(13), 6977–6980 (2010)
16. Cui, X.Q., Jiang, H.R., Peng, Y.H.: Study on non-engineering countermeasures of meteorological disaster prevention in large cities. Meteorol. Environ. Res. **5**(9), 38–42, 46 (2014)

17. Yuan, H.M., Wang, X.R., Zhang, M., et al.: Research and application of rainstorm disaster risk assessment along the middle and lower reaches of Yangtze River. Meteorol. Environ. Res. **5**(10), 38–44 (2014)
18. Luo, M.Z., Jiang, X.H., Shi, L., et al.: Discussion on common problems in risk assessment of lightning disaster. Meteorol. Environ. Res. **5**(9), 28–32 (2014)
19. Wu, Q., Zou, J.H., Wu, W.J., et al.: Analysis on the relationship between geo logical disasters and precipitation in Zhuang he region. Meteorol. Environ. Res. **5**(8), 28–30 (2014)
20. Bai, X.D.: Analysis of characteristics of high temperature weather in Guilin area. In: Proceedings of the 2014 Technical Congress on Resources Environment and Engineering (CREE 2014), Hong Kong, 6–7 Sept 2014

Stand Height Estimation Based on Polarization Coherence Tomography

Xiange Cao, Jinling Yang(✉), Jianguo Hou, Jiwen Zhu, Weicheng Zhang, Jiang Liu, and Xianglai Meng

Shool of Surveying and Mapping Engineering, Heilongjiang Institute of Technology, Harbin 150050, China
{caoxiange,yangjinlingkm,houjg2006}@126.com
{1060968268,45379749,121281531,563827838}@qq.com

Abstract. With the information of interferometric coherence coefficient of different polarization states, Polarization Coherence Tomography (PCT) technology can reconstruct the vegetation vertical structure, it is an important development direction of SAR technology applications. This paper first introduced the principle of PCT, and then summarized vegetation vertical profile reconstruction process of PCT, and finally studied the applicable of PCT technology for stand height estimation with the simulation data and ALOS PALSAR data of Tahe region, the experiment results show that the estimation stand height of PCT is accurate, this technology has strong robustness, whenever choose which ALOS PALSAR data as the master image for registration, the effect on the estimation results are relatively small.

Keywords: Polarization coherence tomography · Stand height · ALOS PALSAR · Full polarization SAR

1 Introduction

Polarization interferometric synthetic aperture radar (POLInSAR) as a new kind of radar imaging technology, can be used to some important feature's remote sensing measurement, and also can be used in forest vegetation height and biomass detection, snow/ice thickness monitoring, urban land subsidence measurement, etc. [1]. According to the change of polarization interference phase we can extract important biological and geophysical parameters, this is extremely important to the remote sensing of vegetation coverage area; at the same time, the POLInSAR has an important features that it can based on the operation of single frequency's single baseline or multiple baseline sensor model conduct parameter estimation [2].

Cloude et al., first introduced the Polarization Coherence Tomography (PCT) method in 2006 [3], this method can reconstruct vegetation vertical distribution structure through the interferometric coherence coefficient of different polarization states, this method further expanded the theory and method of polarization interference SAR be used for vertical structure parameter inversion. Since then, Cloude et al., 2007, Hong Zhang et al., 2010, Haiqiang Fu et al., 2014, Wenmei Li et al., 2015, and many other

F. Bian and Y. Xie (Eds.): GRMSE 2015, CCIS 569, pp. 392–399, 2016.
DOI: 10.1007/978-3-662-49155-3_40

scholars have carried out a lot of research about PCT from algorithm, application and other aspects [4–7].

PCT technology has a wide application prospect, and so the research in this filed is vital significance. This paper based on the analysis of the principle of PCT and process steps, carried out the study through the simulation data and ALOS PALSAR data, and this study focus on the applicability of PCT technology on stand height estimation.

2 PCT Technology

PCT technology using scattering matrix of each pixel which observed from different location, through the effective combination of polarization and interference information to acquire object's space structure characteristic information [5]. The principle of PCT technology is shown in Fig. 1 [3].

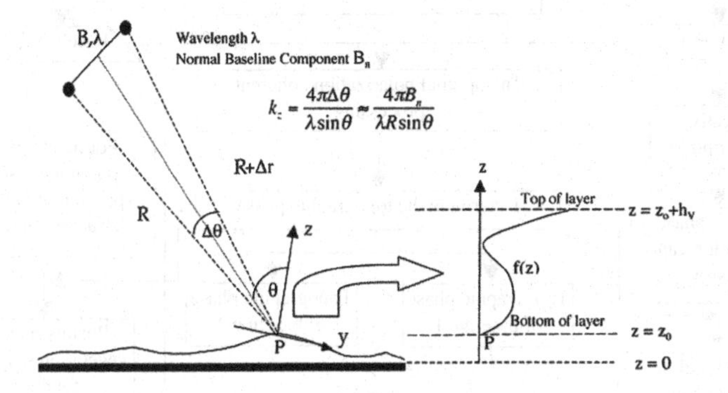

Fig. 1. Principle of PCT

The major measurement of PCT technology is the interference coherence $\tilde{\gamma}$ that related to volume scattering, the coherence can be reconstruct by a random distribution of volume scattering (e.g., scattering of vegetation coverage), as shown in formula 1 [8].

$$\tilde{\gamma}=e^{ik_z z_0}\frac{\int_0^{h_v} f(z)e^{ik_z z}dz}{\int_0^{h_v} f(z)dz}=e^{i\phi_0}\frac{\int_0^{h_v} f(z)e^{ik_z z}dz}{\int_0^{h_v} f(z)dz} \tag{1}$$

In formula, ϕ_0 is ground phase, z_0 is the bottom location of the scattering layer, $f(z)$ is vertical structure function, k_z is vertical wave number (related to interference baseline). The reconstruction of vertical structure function using every interference complex coherence of each pixel called PCT. $f(z)$ is a bounded function (from underlying surface to the top of vegetation canopy), so it can be expanded with Fourier-Legendre polynomial series. For n series vertical structure function $f(z)$, only need calculate n Legendre coefficients, the polarization coherence tomography was reduced to calculating these coefficients from interference data. The more expandable Legendre series, the more accurate of vertical structure function estimation.

3 PCT Process Steps

As a further extension of InSAR technology, in the process of data processing PCT technology also used two main algorithms of InSAR: high precision matching technology of SAR image and baseline estimates. The process of PCT technology reconstruct vertical structure is shown in Fig. 2 [3, 5].

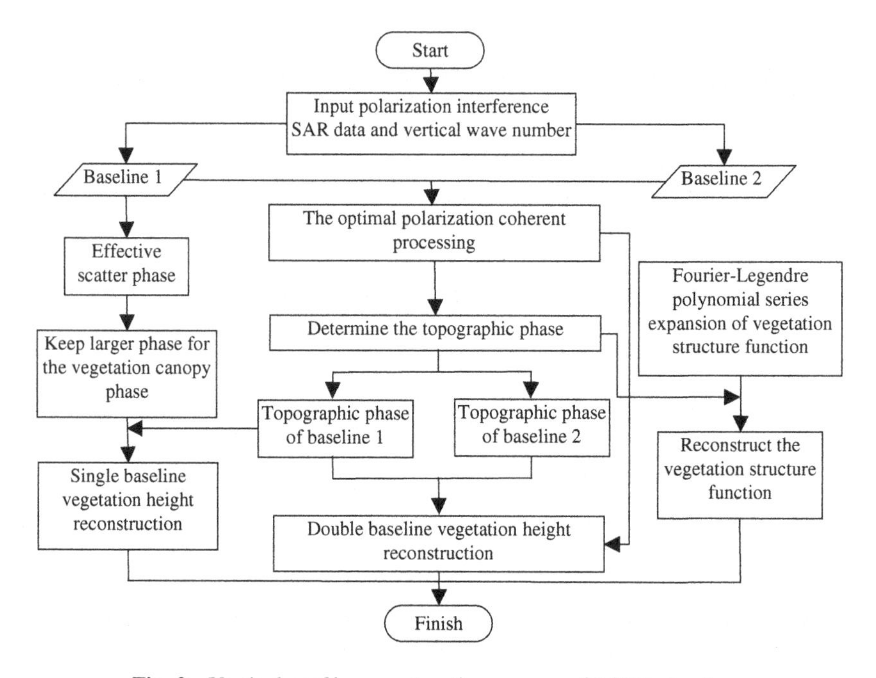

Fig. 2. Vertical profile reconstruction process of PCT technology

4 Stand Height Estimation Based on Simulation Data and ALOS PALSAR Full Polarization Data

This experiment by using the simulation data and ALOS PALSAR full polarization data of Tahe region to validate the accurate of stand height that estimated by PCT technology. The experimental data and the results are as follows:

4.1 Introduction of Simulation Data

The simulation data used in this experiment is generated by PolSARpro Simulator software provided by the European space agency (ESA), this software can generate simulation data by set up the parameters such as trajectory, incidence angle, center frequency, azimuth resolution, distance resolution, tree species, tree height, stand density. For simulation data, because the tree height can be manually set, so the tree height that input to model can be regarded as known values, and by comparing it to the estimation stand height of PCT technology then the accuracy can be analyzed.

4.2 Estimation Results of Simulation Data

This experiment generated the L-band SAR simulation data of 18 and 22 m respectively, the simulation data diagram as shown in Fig. 3, the estimation height map as shown in Fig. 4, height vertical profile as shown in Fig. 5, 3D view of height is shown in Fig. 6.

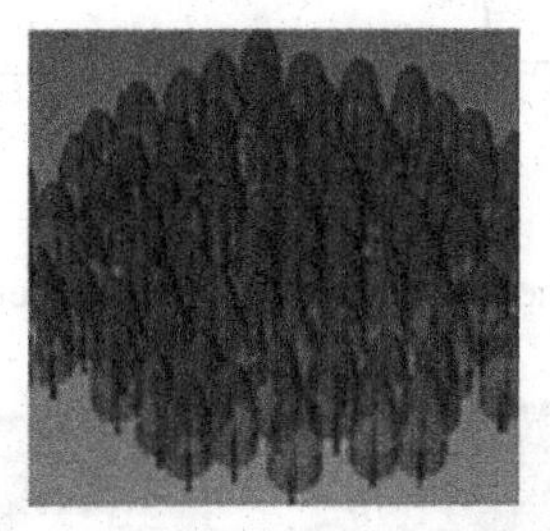

Fig. 3. L-band SAR simulation data diagram

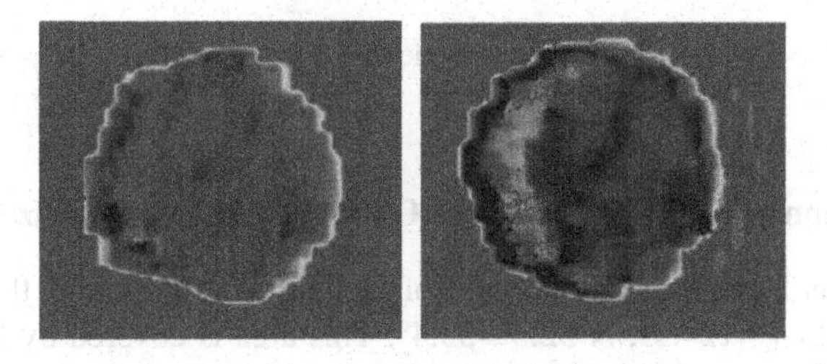

Fig. 4. Estimation height based on PCT

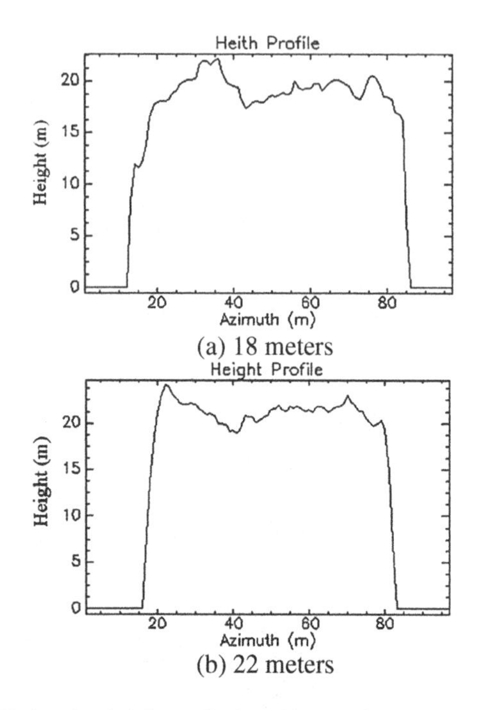

(a) 18 meters

(b) 22 meters

Fig. 5. Estimation height vertical profile, true height is 18 and 22 m

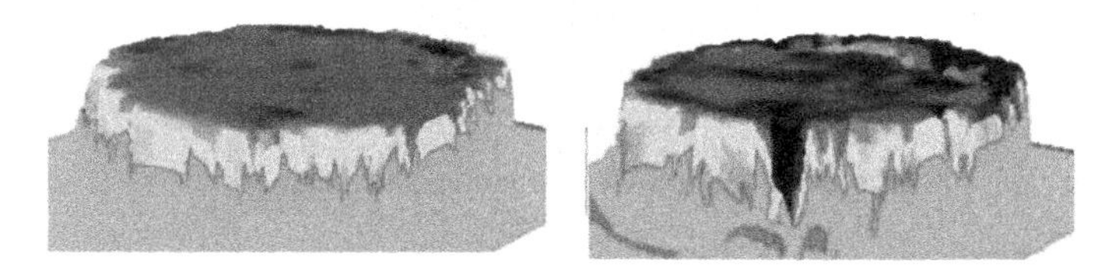

Fig. 6. 3D view of height

4.3 Introduction of Study Area and ALOS PALSAR Full Polarization Data

The study area is located in Tahe region of Heilongjiang province, the geographical position is: E 123.4°~124.9°, N 52.3°~53.4°. This area is covered by forest and with rich vegetation species.

ALOS satellite is the first satellite in the world with full polarization SAR and can obtain the L-band full polarization data, the data can be used for repeated trajectory interferometry [9]. The data used in this study was ALOS PALSAR full polarization data, and the data acquisition time is May 7, 2007, and November 7, 2007, details are shown in Table 1.

Table 1. Track and time of PALSAR data of Tahe region

Area	File name	Track	Time
China (Tahe region)	A0907325–033	421–1050	20071107
China (Tahe region)	A0907325–034	421–1050	20070507

At the same time, the subcompartment data of 2005 of the study area was also being collected, according to subcompartment data the average tree height of this region is about 20 m; Subcompartment data and ALOS PALSAR data are not the same period, because the stand height change of 2 years can be ignored, therefore, this experiment didn't consider the growth of tree height from 2005 to 2007 when analyzing of the estimation accuracy.

4.4 Estimation Results of ALOS PALSAR Data

By taking Tahe region's ALOS PALSAR data of May as master image and data of November as slave image for precise registration, after the process of interference, the estimation forest height based on PCT technology as shown in Fig. 7.

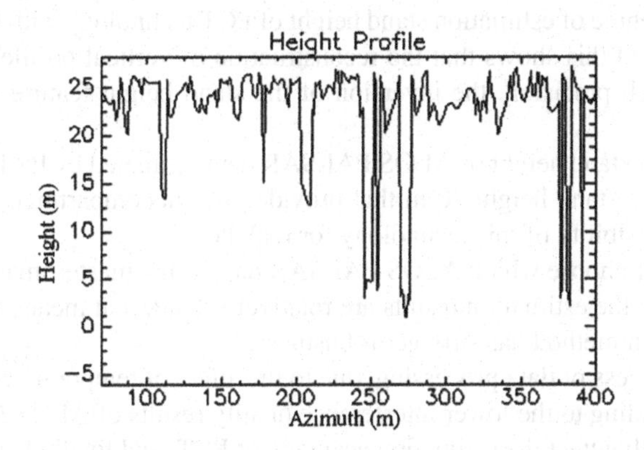

Fig. 7. Forest stand high profile estimated based on PCT technology

By taking Tahe region's ALOS PALSAR data of November as master image and data of May as slave image for precise registration, after the process of interference, the estimation forest height based on PCT technology as shown in Fig. 8.

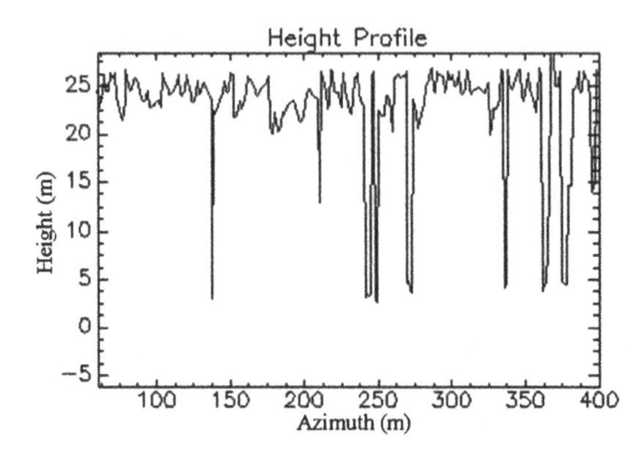

Fig. 8. Forest stand high profile estimated based on PCT technology

4.5 Analysis of the Stand Height Estimation Results

The stand height estimation results of L-band SAR simulation data and L-band ALOS PALSAR data show that:

(1) For simulation data, it can be seen from the estimation results of Figs. 5 and 6 that the match degree of estimation stand height of PCT technology with the input stand height is high, and this shows that the reconstruction of vertical profile by PCT technology has a high precision, the inversion of the stand height feature information is accurate;

(2) The forest stand height of ALOS PALSAR data estimated by PCT technology is close to reference stand height 20 m that provided by subcompartment data, and this illustrates the feasibility of this technology for real data;

(3) Whenever choose which ALOS PALSAR data as the master image for registration the effect on the estimation results are relatively small, that means that PCT technology estimation method has stronger robustness;

(4) In the process of data processing, due to the affect of temporal decoherence and other factors, leading to the lower interference quality results of ALOS PALSAR data, this will inevitably affect the estimation accuracy of PCT, and finally leads to the deviation between estimation results of height (showed in Figs. 7 and 8) and reference stand height provided by subcompartment data;

(5) Although there is a certain deviation between estimation stand height of ALOS PALSAR data and reference height, if eliminating the affect of temporal decoherence and other related factors, the estimate precision of PCT technology will be further improved.

The study results show that the PCT technology can effectively reconstruct the vertical structure of vegetation and also has a high estimation accuracy.

5 Conclusion

As a new technology of SAR, PCT technology can provide the spatial structure information of research object and has been developing rapidly in the last decade; PCT technology is an important developing direction of future SAR application technology. With the improvement of the SAR system, PCT technology will be further improved. Compared with abroad, in China, although in the aspect of SAR system and data processing we have made great progress, but research in PCT technology still needs to be further strengthened.

The forest stand height estimated by PCT technology has a great importance for the estimation of large areas forest biomass and volume and degree of density. The applicability of the PCT technology for stand height estimation still need further research, comparative analysis with other existing methods, the eliminating of temporal decoherence and related influence factors will be the next focus of further research.

Acknowledgments. This work was financially supported by Dr. Fund project of Heilongjiang Institute of Technology (2012BJ03).

References

1. Cloude, S.R., Papathanassiou, K.P.: Polarimetric SAR interferometry. IEEE Trans. Geosci. Remote Sens. **36**, 1551–1565 (1998)
2. Papathanassiou, K.P., Cloude, S.R.: Single baseline polarimetric SAR interferometry. IEEE Trans. Geosci. Remote Sens. **39**, 2352–2363 (2001)
3. Cloude, S.R.: Polarization coherence tomography. Radio Sci. **41**, 1–27 (2006)
4. Cloude S.R.: Multibaseline polarization coherence tomography. In: POLinSAR 2007 (2007)
5. Zhang, H., Jiang, K., Wang, C., Chen, X., Tang, Y.: The current status of SAR tomography. Remote Sens. Technol. Appl. **25**, 282–287 (2010)
6. Fu, H., Wang, C., Zhu, J., Xie, Q., Zhao, R.: A modified PolInSAR PCT method to invert vegetation vertical structure. Eng Surv. Mapp. **23**, 56–61.66 (2014)
7. Li, W., Chen, E., Li, Z., Ke, Y., Zhan, W.: Forest above ground biomass estimation using polarization coherence tomography and PolSAR segmentation. Int. J. Remote Sens. **36**, 530–550 (2015)
8. Fontana, A., Papathanassiou, K.P., Iodice, A., Lee, S.K.: On the performance of forest vertical structure estimation via polarization coherence tomography. http://ieee.uniparthenope.it/chapter/_private/proc10/22.pdf
9. Gan, T., Li, M.: The advanced land observing satellite-ALOS. Surv. Mapp. Jiangxi **67**, 11–15 (2007)

The Detection System Based on Machine Vision for the Process of Book Binding

Wenbin Bu[1], Fucheng You[1(✉)], Yue You[2], and Shangrong Rong[1]

[1] Beijing Institute of Graphic Communication, Beijing 102600, China
294140080@qq.com, bwb2015@163.com
[2] Huazhong University of Science and Technology, Wuhan 430074, China

Abstract. In the process of book binding, the detection for signatures is an essential step. In order to improve the speed of detection, and solve the problem of signatures position rotation when detecting. In this paper, to detecting signatures, the template matching method is used to detect signatures without page number, and for signatures with the page number, the method of OCR recognition is used to detect them. Found by experiment, the OCR identification method to detect the page number is a good one for the rotary signatures.

Keywords: Signatures · Detection · OCR recognition

1 Introduction

In the process of books and periodicals printing, book binding is essential. However, as mechanical failure and other various aspects error, unqualified signatures will appear for these reasons. Then lots of loss will be brought to the production. In order to improve the production efficiency, reduce production costs, it is very necessary to detect signatures in the process of book binding. The purpose of detection is finding out wrong signatures, blank signatures and reverse signatures. When the control system received the result of detection, it will take corresponding treatment measures to deal with it [1–3].

In order to solve the signatures chaos in the binding process, this paper proposes a solution. A system based on machine vision to prevent confusion will be designed with CCD camera, and using it to take real-time detection for the process of book binding through image processing related technologies and OCR identification technology. Once appears signatures chaos, this system will issued a corresponding alarm signal immediately [4, 5].

2 The System Design

The detection system block diagram is shown in Fig. 1.

The most striking feature of this system is that combining template matching with OCR recognition. Most of existing detection based on machine vision are based on

F. Bian and Y. Xie (Eds.): GRMSE 2015, CCIS 569, pp. 400–407, 2016.
DOI: 10.1007/978-3-662-49155-3_41

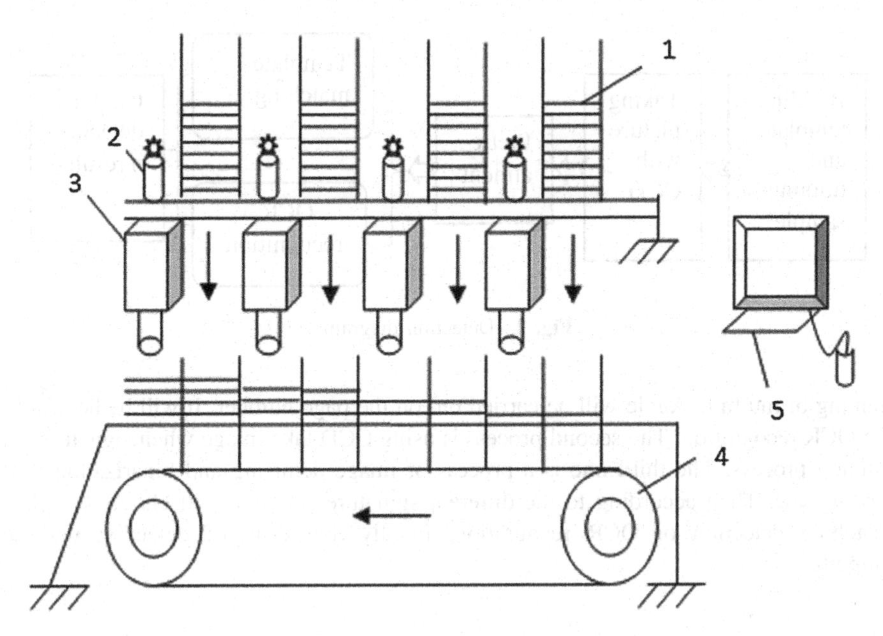

Fig. 1. Detection system block diagram. 1-Signatures; 2-Alarm; 3-CCD Camera; 4-Conveyor belt; 5-PC

graphic gray level or simply based on template matching. The faults of based on graphic gray level to detect, however, is that the detection of graphic gray level must be within the prescribed scope when detect the graphic grey value. Therefore, it must be neat when operators are stacking signatures in box, and folding also must be neat, or signatures may move on the relative position of the detection area, then the difference between the result of detection and the standard grey value will increase and bring fault detection; The inky depth of the same signatures is required to be consistent, otherwise the instrument will mistake the gray value of the signatures different which will cause fault detection, so as to increase the number of interrupt, and affecting work efficiency. If all printing graphics are small texts, and grey value of the signatures changed little in the area of the detection, the detection error rate will be extremely high. This may cause machine stop often, and it may be less suitable to use this detection method at this moment.

However, the detection rate of the method based on template matching is slow. The benefits of the combination of template matching with OCR are improving the detection rate in this system discussed in this paper. Because of the OCR has smaller amount of calculation than the template matching and has greater advantage for rotation. So for signatures with page number, this system will use the OCR for detection. And for covers which have no page numbers, system will use the template matching method to detect. The detection diagram of this system is shown in Fig. 2.

Creating templates and training sample are first process in this system. For signatures without page number, it is needed to modeling for the graphic. Otherwise, the

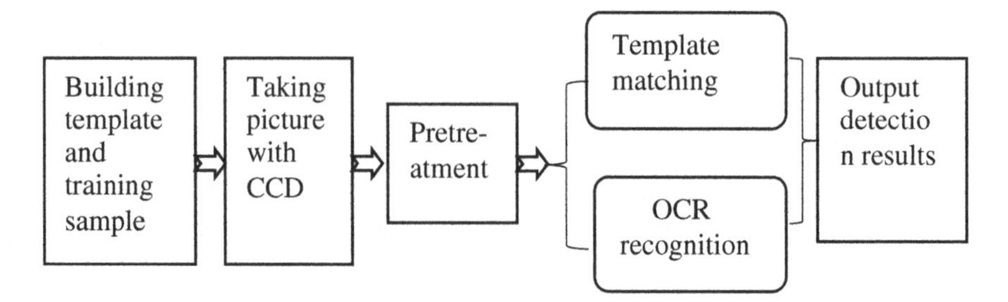

Fig. 2. Detection diagram

training of multiple Angle will be carried out on the page number. It will be helpful to the OCR recognition. The second process is using CCD take image when signatures in binding process. The third one is a process of image demising and binarization processing, etc. Then according to the different signatures, system will choose template matching detection or OCR recognition. Finally the result of detection will be outputted.

3 Algorithm

In this system, the template matching algorithm comes from the OpenCV library. The input image is represented as I, template is T and R is output image. The template T is smaller than input image I. If the size of I is W × H, and the size of T is w × h, than the size of R is (W – w + 1) × (H – h + 1). This method slides through image, compares the overlapped patches of size w × h against T using the correlation coefficient method and stores the comparison results in R. The summation is done over T and/or the I patch: $x' = 0 \ldots w-1$; $y' = 0 \ldots h-1$. Then we get R as follows.

$$R(x,y) = \sum_{x',y'} (T'(x',y') \bullet I'(x+x',y+y')) \quad (1)$$

where

$$T'(x',y') = T(x',y') - 1/(w \bullet h) \bullet \sum\nolimits_{x'',y''} T(x'',y'') \quad (2)$$

$$I'(x+x',y+y') = I(x+x',y+y') - 1/(w \bullet h) \bullet \sum\nolimits_{x'',y''} I(x+x',y+y') \quad (3)$$

The rest of this paper will mainly introduce the process of OCR recognition algorithm implementation. OCR recognition part is divided into two parts: Sample training and character recognition.

3.1 Sample Training

In order to be able to identify characters correctly, it is necessary to training the sample of the characters to be identified. Let the computer remember the inputting image and its corresponding characters. The more samples are trained, the higher identification accuracy will be achieved. The training of the sample can be divided into several steps: noise reduction, binarizing, determining the character boundary, making character image into small squares, character encoding, saving character and its coding.

At the beginning of training samples, to reduce the interference, the process of noise reduction is necessary to image. If the image is color images, we should make it to gray scale image firstly and do binary image processing. If the image is grayscale image, then binarization image can be obtained directly. Conventional algorithms will be used in noise reduction processing, graying and binarizing. Due to the size of the image is generally larger than that of character itself, removing some of the character image parts is necessary. Making the image just contain characters, namely, width and height of the image consistent with the width and height of the characters. In order to determine the boundary of the characters, we will do it in two steps. Firstly, making sure the upper and lower boundary of characters. Secondly, confirming the left and right boundary of the characters. Specific methods is shown in Fig. 3 below. At first we should get the Y direction histogram. To do this, we need gain the number of black pixels on the coordinates which are vertical with the Y axis.

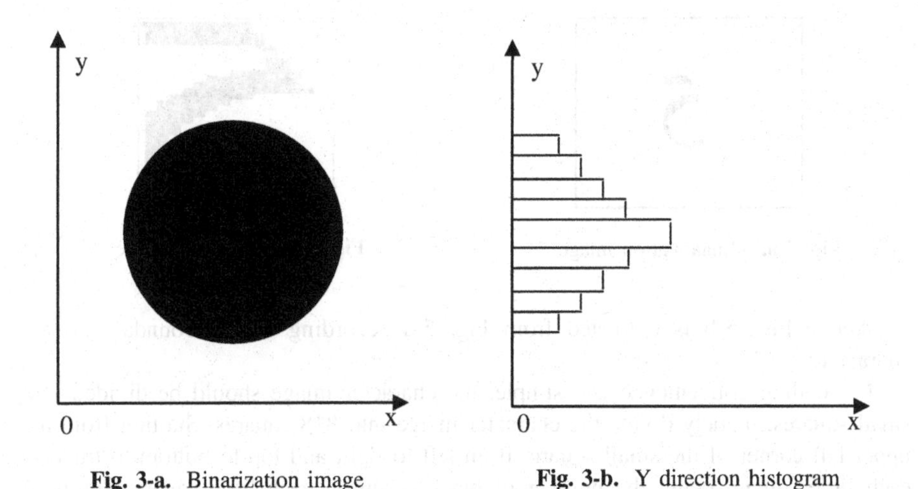

Fig. 3-a. Binarization image **Fig. 3-b.** Y direction histogram

After getting the Y direction histogram, according to the predetermined threshold in advance we can find the upper and lower boundaries of the histogram, namely upper and lower boundaries of the characters in the image.

The same method, as shown in the Fig. 4 below, after getting the X direction histogram, we can find the left and right boundaries of the histogram, namely left and right boundaries of the characters in the image.

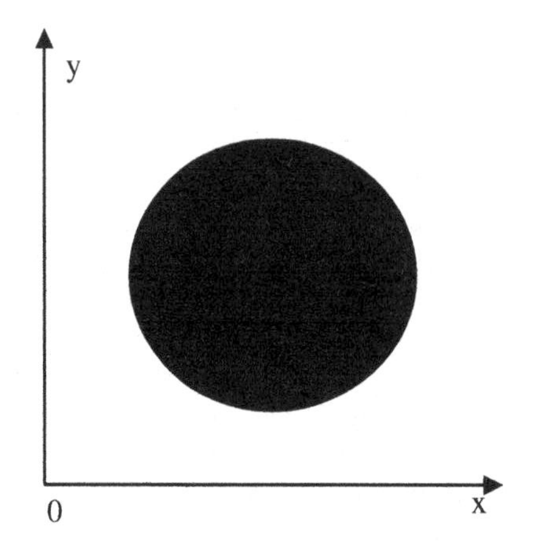

Fig. 4-a. Binarization image

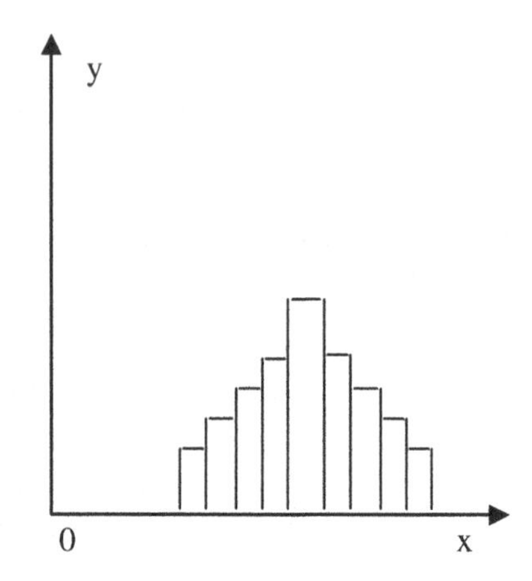

Fig. 4-b. X direction histogram

At this moment, we can get the whole boundary of the characters. According to the boundary of character, it is possible to determine the size and location of the ROI (region of interest). Then the character can be extracted from the image. As shown in the Fig. 5 below.

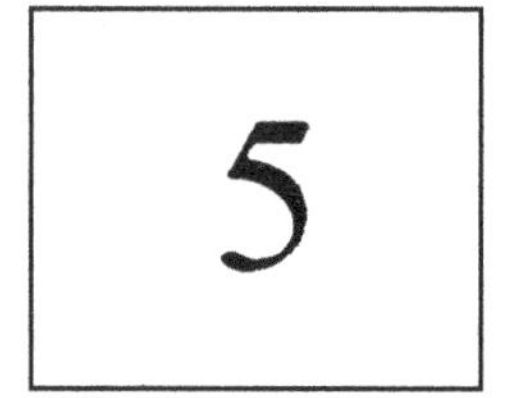

Fig. 5-a. Binaarization image

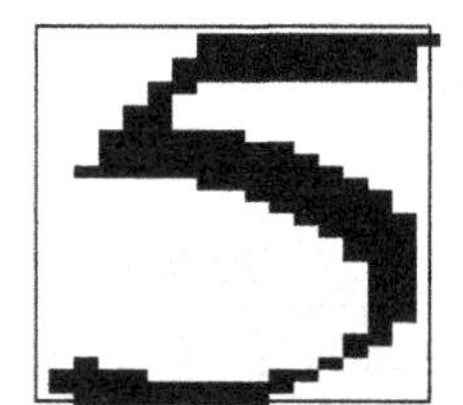

Fig. 5-b. Character image

Above Fig. 5-b is extracted from Fig. 5-a according to the boundary of the character.

For coding convenience and simple, the character image should be divided into small squares, namely divide the character image into 8*8 squares. Starting from the upper left corner of the small square, from left to right and top to bottom to traverse each little square, count the number of black pixels in each little square. Setting a threshold, when the number of black pixels in the square exceeds a certain value, encoded by this little square is 1, otherwise the code is 0. Finally we can get a 64-bit binary 01 series, a total of 8 bytes. By some of the above listed steps, so that a character image can be converted into a 8 bytes of digital information. Last stored characters and character encoding, so that we complete the encoding process.

3.2 Character Recognition

For the Character recognition, the previous process which is the same as the training before, it is need to encode the recognized character first, and then matching the encoded character with the stored encoded character. First we define the difference degree of the encoded character, and it will add one degree when there is a different binary which is a difference on squares. If the encoded character has the minimum different degree which gets from matching with the other character one by one, then the corresponding characters is the recognized character at last.

4 Experimental Analysis

As in the process of signatures detection, the OCR is used to detect the signatures with page number in this system. And for signatures without page number the system will use the template matching method. The sample training and recognition will be introduced in the following.

In order to improve recognition accuracy, we need to train sample as much as possible. Considering the signatures' position in the platform may change when it falling down. So it is necessary to train samples with multiple different angles. As shown in Fig. 6 below.

Fig. 6. All cases of rotations

The pictures above, we rotate the original image (a) both clockwise and counterclockwise at 3 (Fig. 6-b/6-h), 5 (Fig. 6-c/6-i), 7 (Fig. 6-d/6-j), 10 (Fig. 6-e/6-k), 12 (Fig. 6-f/6-m), 15 (Fig. 6-g/6-n) degrees, then training them respectively. So when signatures rotation occurs, the page number can be effectively recognized. Because of training all samples of different angle is unrealistic, and it will have a great deal of work. So in order to test whether the system can identify no training rotation angle, the following will take a test to images rotated both clockwise and counterclockwise at 2, 4, 6, 8 degrees.

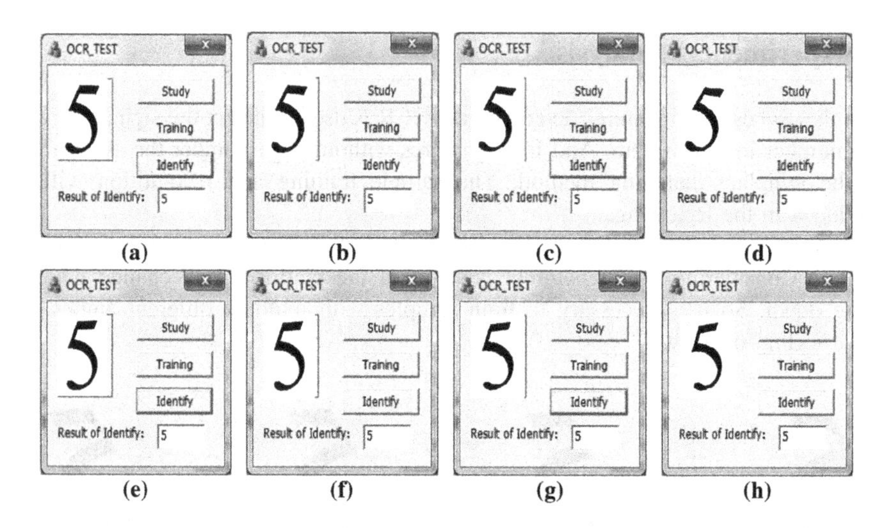

Fig. 7. Results of recognition

In Fig. 7, a, b, c, d are the results of recognition for clockwise 2, 4, 6, 8 degrees respectively. And e, f, g, h are the results of recognition for counterclockwise 2, 4, 6, 8 degrees respectively. The experimental results show that the character in the image untrained rotation angle can also be correctly recognized after a multi-angle training.

5 Conclusion

Experimental results show that using template matching and OCR recognition to detect signatures is feasible. The template matching method is used to detect signatures without page number, and adopting the method of OCR recognition to detect signatures with page number. This method compared with the method using template matching to detect has certain advantages. As the OCR recognition is easier than template matching method in dealing with the rotation image. Therefore, when detecting signatures, the system discussed in this paper could be helpful.

Acknowledgments. This paper is supported by Institute Level Key Project Funded by Beijing Institute of Graphic Communication (Ea201506) and Scientific Research Key Project of Beijing Municipal Commission of Education (KZ201210015015).

References

1. Wang Y., Li, D.: Design of a paper-checking instrument based on fuzzy pattern recognition. Sixth World Congress on Intelligent Control and Automation (2006). (in Chinese)
2. Huang, W., Wang, Y.-Z., Li D.-S.: College of Mechanical Engineering and Applied Electronic Technology. Beijing University of Technology, Beijing 100124, China. (in Chinese)
3. Yaofa, W.: Page checking technology in bookbinding mechanically, p. 19. Shanghai, Printing Technolygy (2008). (in Chinese)
4. Shengli, K.: Control points in book-binding quality inspection, p. 34. Shanghai, Printing Technolygy (2007). (in Chinese)
5. Huang, J., Li, L.: The solution of misplaced pages in book-binding, p. 05. Printing Technology, Shanghai (2000). (in Chinese)

Landscape Pattern Changes of Coastal Wetland in Nansha District of Guangzhou City in Recent 20 Years

Caige Sun[1,2,3], Kaiwen Zhong[2], Rubing Ge[1,3], Yu Zhao[1,3], Xulong Liu[2], and Tao Lin[1(✉)]

[1] Key Lab of Urban Environment and Health, Institute of Urban Environment, Chinese Academy of Sciences, Xiamen 361021, China
tlin@iue.ac.cn

[2] Key Lab of Guangdong for Utilization of Remote Sensing and Geographical Information System, Guangzhou Institute of Geography, Guangzhou 510070, China

[3] University of Chinese Academy of Sciences, Beijing 100049, China

Abstract. In this paper, 6 types of coastal wetland (River, Dike-pond, Lake, Tidal flat, Swamp, Shallow sea waters) was classified precisely through supervised method and decision trees based on DEM, depth contour, high tide level, low tide level, vegetation coverage and soil map. Landscape metrics including area metrics, shape metric, aggregation metric and diversity metric were used to analysis the coastal wetland changes on class-level and landscape-level in 1995, 2001, 2005, 2009 and 2015. The results showed that the landscape pattern changed greatly in recent 20 years, especially since 2009, the coastal wetland area decreased sharply, distribution of coastal wetlands is becoming more and more uneven and the fragmentation is becoming more and more serious. Our study will provide scientific supports for coastal wetland planning and development policy making.

Keywords: Coastal wetland · Landscape pattern · Decision trees · Nansha · Guangzhou

1 Introduction

Winding coastline gave birth to the rich coastal wetland resources. The coastal wetland refers to the place including intertidal beach, coastal lowland could be water inundated and shallow sea waters deep depth of less than 6 m and a depth of less than 6 m at low tide in shallow waters under the land-sea interaction. It is a complex natural ecosystem between land and sea. Studies showed that the coastal wetlands of China mainly includes 12 types [1–4]. They are shallow sea waters, subtidal aquatic layer, coral reefs, rock coast, intertidal silt sand beach, intertidal beach, intertidal salt marsh, mangrove swamps, coastal lagoon, coastal freshwater lakes, estuaries and delta wetland.

With the development of coastal economy, coastal wetland has a serious decline in scale and functions under the construction of land reclamation and port industry, especially in the last 20 years [5]. The destruction is difficult to recover in the short

F. Bian and Y. Xie (Eds.): GRMSE 2015, CCIS 569, pp. 408–416, 2016.
DOI: 10.1007/978-3-662-49155-3_42

term. The coastal wetland mainly has two distinct functions, one is to maintain biodiversity, conserve fishery resources, another is to provide aesthetic, recreational places. The former supports the living conditions of plant, animal and fishery resources on the continental shelf of the rare and endangered birds, the latter is the basis of coastal tourism. From this point of view, the ecological value of coastal wetland is far higher than the same area of the wetland ecosystem or inland marine ecosystem. So the study of coastal wetland has been the concern of many scholars [6–9].

In this study, 6 types of coastal wetland (River, Dike-pond, Lake, Tidal flat, Swamp, Shallow sea waters) would be classified precisely through supervised method and decision trees based on DEM, high tide level, low tide level, vegetation coverage and soil map. Landscape metrics including area metrics, shape metric, aggregation metric and diversity metric were used to analysis the coastal wetland changes on class-level and landscape-level in 1995, 2001, 2005, 2009 and 2015. The result has important scientific significance for protecting the coastal wetland resources and the reasonable development and utilization.

2 Study Area

The research area was Nansha District, at the center of Delta, which is located in the south Guangzhou City on the south coast of China. It comprises a larger region of pearl river alluvial plain, some hill plateau and islands. Since Nansha is the Pearl River Estuary, there are so many rivers, channel, sewage stream and dike-ponds in it. River flow is gently but the tide is obvious, with the average tidal range 2.4 m. Coastal wetland ecological resources in Nansha are rich. It is a large wetland ecosystem in typical subtropical tropical transition (Fig. 1).

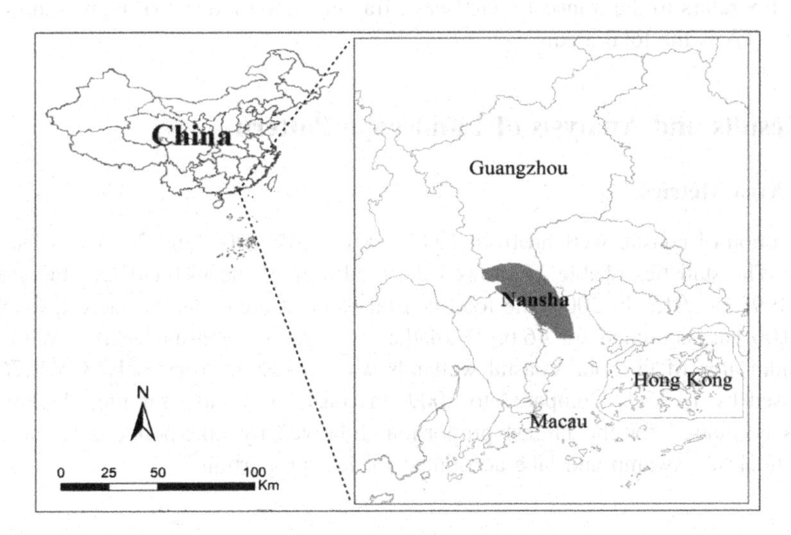

Fig. 1. Study area

3 Data and Methods

Five sets of Landsat TM images were used in this study. They were got on 30 December 1995, 1 March 2001, 23 November 2005, 2 November 2009 and 19 January 2015. The first four obtained by the landsat5 sensor and the last obtained by the landsat8 sensor. They were processed using ERDAS IMAGINE software, involved geometric correction, image clipping et al.

The preliminary classification of coastal wetlands was built by supervised classification method through the training area. Decision trees was also constructed based on DEM, depth contour, high tide level, low tide level, vegetation coverage and soil map to make the secondary classification [10, 11]. Compared with supervised classification, the combining methods will improve the accuracy of coastal wetland extraction and classification.

Landscape metrics of coastal wetland were also calculated. The vector data of transect 's coastal wetland were then converted to raster format at the pixel size of $1*1\ m^2$ using ArcGIS 10.1. To capture the spatial patterns, a suite of landscape metrics were quantified based on raster format by using FRAGSTATS 4.2 software [12]. The final class-level metrics included total class Area (CA), percentage of landscape (PLAND), fractal dimension (FRAC), contiguity index distribution (CONTIG), landscape shape index (LSI), patch density (PD) and aggregation index (AI). The final landscape-level metrics included total area (TA), patch density (PD), landscape shape index (LSI), aggregation index (AI), Shannon's diversity index (SHDI) and Shannon's evenness index (SHEI). At the same time, landscape fragmentation index (FN), belonging to landscape-level metrics, was calculated by TA and NP from formula (1) [13]:

$$FN = {}^{(NP\ -\ 1)}/_{TA} \tag{1}$$

where FN refers to the transect's landscape fragmentation index. NP is the number of patches. TA is the total area.

4 Results and Analysis of Landscape Pattern

4.1 Area Metrics

Distribution of coastal wetland from 1995, 2001, 2005, 2009, and 2015 was listed in Fig. 2. The statistics (Table 1) showed the total area of the wetland has fluctuations from 1995 to 2015. In 2009, the total coastal wetland area was the largest, reaching 36700.06 ha, accounted for 46.65 % of the total area of Nansha District. While the minimum area of the total coastal wetlands was appear in 2015, only 31471.76 ha, decreased by 14.25 % compared to 2009. In coastal wetland structure, shallow sea waters accounted for the largest proportion, followed by dike-pond, is again river, while tidal flat, swamp and lake accounted for less proportion.

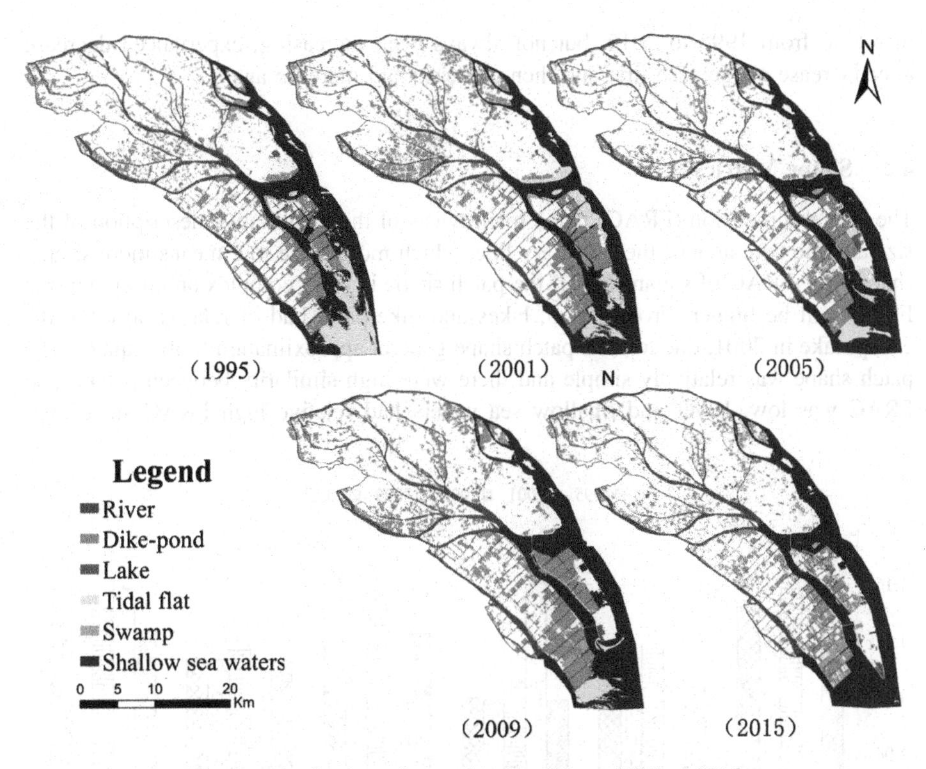

Fig. 2. Distribution of coastal wetland from 1995 to 2015

Table 1. Statistics of the coastal wetland area from 1995 to 2015 (Unit: ha)

	River	Dike-pond	Lake	Tidal flat	Swamp	Shallow sea waters	Total
1995	5767.3	10,747.08	147.08	4929.30	364.14	13249.22	35,204.12
2001	5577.58	12,252.18	101.44	3407.81	385.82	13625.85	35,350.68
2005	5632.92	9430.61	384.37	2202.31	592.51	14660.62	32,903.34
2009	6339.14	12,826.24	249.00	1499.33	560.19	15226.16	36,700.06
2015	6051.39	8824.29	392.93	94.04	719.73	15389.38	31,471.76

During 20 years of evolution, tidal flat area decreased furthest year by year. The areas were 4929.3, 3407.81, 2202.31, 1499.33 and 94.04 ha respectively in 1995, 2001, 2005, 2009 and 2015, accounting for 14.00, 9.64, 6.69, 4.09 and 0.30 % of the total coastal wetland respectively. The changing trend of shallow sea waters was on the contrary, which increased largest year by year. There are two reasons account for this phenomenon. One is the reclamation of construction land from tidal flat, especially on Longxue island, another is some tidal flats changes into shallow sea waters. The area of shallow sea waters was increased from 13,249 ha in 1995 to 15389.38 ha in 2015. Dike-pond area also reduced from 1995 to 2015. But it wasn't always in reducing, 2001 and 2009 have a substantial rebound. The area of river, lake and swamp has increased from 1995 to 2015, but not always been increasing, experienced decrease after increase or increase first and then decrease and increase again.

4.2 Shape Metrics

The fractal dimension (FRAC) is the main twists of the quantitative description of the size and the core area of the boundary line, which more close to 1 means more simple shape [14]. FRAC of square is 1. If the patch shape is more complex and irregular, the FRAC will be bigger. From Fig. 3, lakes and dike-pond had of relative low FRAC, except lake in 2001, due to their patch shape general approximation to the square. The patch shape was relatively simple and there were high similarity between patches, so FRAC was low. River and Shallow sea waters had relative high FRAC since their

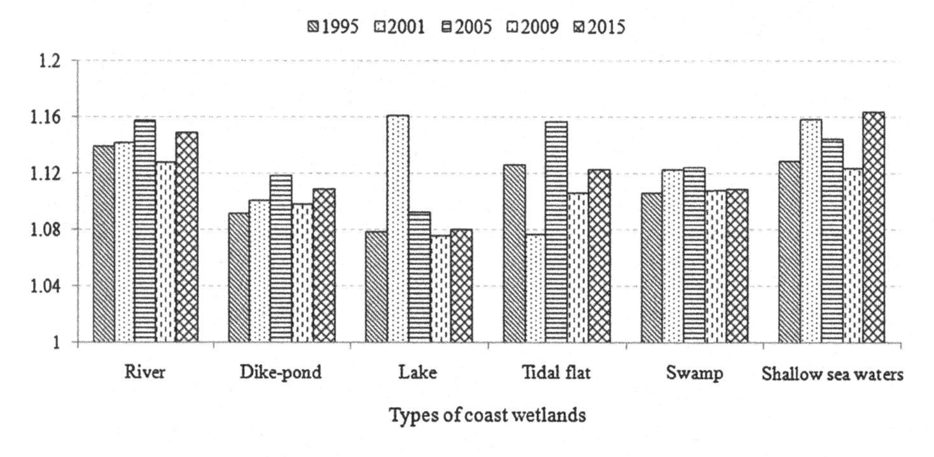

Fig. 3. Fractal dimension (FRAC) of different types of coastal wetlands

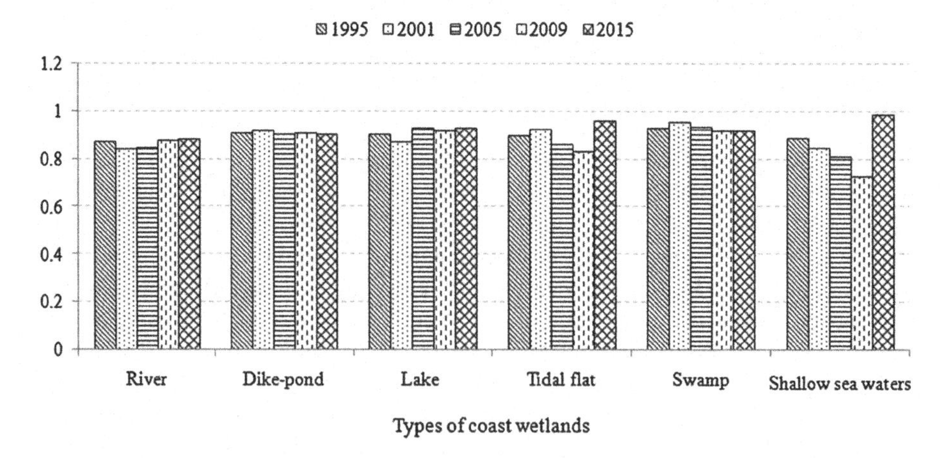

Fig. 4. Contiguity index distribution (CONTIG) of different types of coastal wetlands

complex landscape structure and irregular shape. The reason why lake in 2001 had high FRAC was because many artificial excavation ponds were made by the urban land development and construction.

Contiguity index distribution (CONTIG) was used to evaluation the spatial connectivity and proximity of patches, which also provides indicators for patch boundary line and the shape of the distribution. Its Range is 0 to 1. CONTIG is 0 for patch in only one pixel, while with the increasing connectivity and proximity of patches, CONTIG will close to 1. From Fig. 4, all the types of coastal wetland had relative high CONTIG, most of which above 0.8. But the CONTIG of shallow sea waters declined from 1995 to 2009 and increased a lot in 2015.

4.3 Aggregation Metrics

We select Patch density (PD), Landscape shape index (LSI) and Aggregation index (AI) to description aggregation. The values were listed in Table 2.

Table 2. Patch density (PD), Landscape shape index(LSI) and Aggregation index(AI) of coastal wetlands on class-leve land landscape-level from 1995 to 2015

Aggregation metrics	Year	Class-level						landscape-level
		River	Dike-pond	Lake	Tidal flat	Swamp	Shallow sea waters	The whole coastal wetlands
PD	1995	0.24	3.45	0.09	0.32	0.10	0.20	4.41
	2001	0.34	2.91	0.23	0.12	0.04	0.14	3.77
	2005	0.43	2.77	0.17	0.50	0.10	0.05	4.02
	2009	0.71	2.48	0.12	0.02	0.10	0.02	3.99
	2015	0.99	4.84	0.24	0.01	0.12	0.01	6.20
LSI	1995	32.86	49.33	6.80	14.16	9.65	15.19	46.04
	2001	32.12	44.17	16.66	8.63	6.92	13.97	46.28
	2005	33.73	48.08	9.43	16.44	9.79	8.60	45.15
	2009	32.16	41.69	7.53	4.44	8.71	7.71	41.51
	2015	35.77	63.31	10.83	3.85	7.88	7.04	52.92
AI	1995	97.90	97.67	97.60	99.06	97.73	99.38	98.55
	2001	97.91	98.05	98.66	96.19	98.49	99.44	98.62
	2005	97.82	97.57	97.84	98.35	98.19	99.69	98.62
	2009	98.04	98.20	97.92	98.47	98.37	99.61	98.77
	2015	97.76	96.68	97.51	98.52	98.71	99.76	98.46

Patch density (PD) is the number of patches in 100 ha, which can reflect the degree of landscape fragmentation by segmentation [15]. But it cannot reflect the size and spatial distribution information of the patches. From Table 2, Dike-pond had relative high PD on class-level, which indicated that it had high segmentation than other types. For the whole coastal wetlands, PD seemed more higher. The lowest was 3.77 in 2001, and the highest was 6.20 in 2015.

Landscape shape index (LSI) quantifies the degree of dispersion of patches. When there is only one patch in the landscape, and it is a square or nearly square, LSI is 1. With the discrete of patches, LSI gradually becomes large and there is no maximum limit. From Table 2, we can see dike-pond maintained the highest LSI, followed by river. For the whole coastal wetlands, LSI kept relative high level, but not higher than dike-pond. The lowest was 41.51 in 2009, and the highest was 52.92 in 2015.

Aggregation index (AI) quantifies the degree of aggregation of patches. When the fragmentation of a certain type of patch is maximized, AI is equal to 0, AI will increase with the increased degree of aggregation. When the patches aggregate into a coherent whole, AI is equal to 100. From Table 2, shallow sea waters had the highest AI, and the index showed an upward tendency from 1995 to 2015. This indicated that the patches of shallow sea waters were becoming more and more close. For the whole coastal wetlands, AI presented the trend that increased firstly but then decreased. The lowest was 98.46 in 2015, and the highest was 98.77 in 2015. From the overall perspective AI showed a negative correlation to LSI.

4.4 Diversity Metrics

We analysis diversity metrics just from landscape-level. Shannon's diversity index (SHDI) and Shannon's evenness index (SHEI) were chosen. Landscape fragmentation index (FN) was calculated by total area (TA) and number of patches(NP).

Shannon's diversity index (SHDI) quantifies landscape diversity based on the number and areal proportion of different patch types [16]. When the landscape consists of a single element, the landscape is homogeneous, SHDI is 0. With the increase of landscape types or with different types of landscape type distribution change to more balanced, SHDI will increase. If the landscape type is equal in proportion, SHDI will be the highest. From Fig. 5, we can see that SHDI of the whole landscape declined from

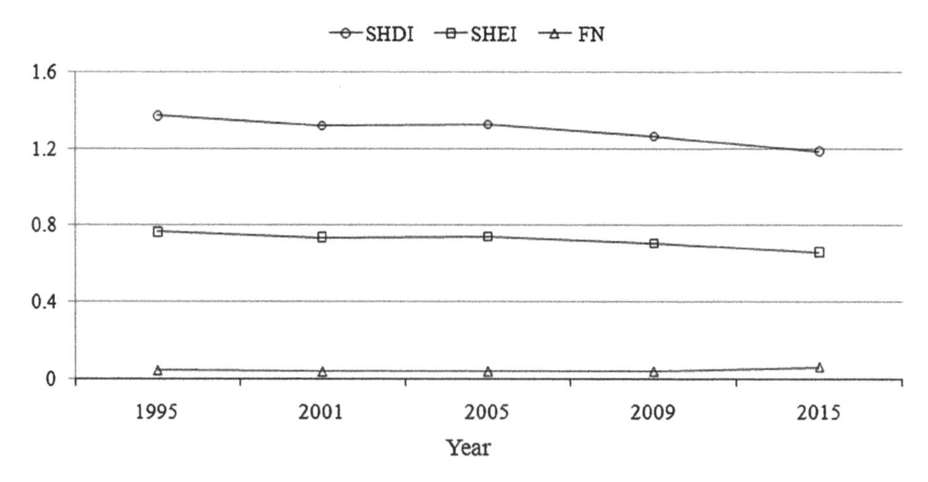

Fig. 5. Shannon's diversity index (SHDI), Shannon's evenness index (SHEI) and landscape fragmentation index (FN) of total coastal wetlands on landscape-level from 1995 to 2015

1.37 in 1995 to 1.18 in 2015, which means the difference in the proportion of coastal wetland types was increasing, especially during the period of 2009–2015.

Shannon's evenness index (SHEI) quantifies landscape control degree by a few main types of landscape, which similar to landscape dominance to some extent, with the range between 0 and 1. If the whole landscape is only one patch, SHEI is 0. When the landscape distribution is extremely uneven, SHEI will close to 0. When the landscape type distribution is very uniform, SHEI will be 1. The change trend of SHEI can be seen from Fig. 5. The whole landscape was composed of 6 elements from 1995 to 2015, but SHEI also showed the tendency of decline. The change range was between 0.66 in 2015 and 0.77 in 1995, which means the distribution of coastal wetland in Nansha district was always in a state of relatively uniform, while the relatively uniform state became more and more uneven over time.

Landscape fragmentation index (FN) quantifies landscape fragmentation based on the number of patches and the total area, which reflects the degree of human disturbance to the landscape. From Fig. 5, we can see that only a little changes in FN between 1995 and 2009, but there was a substantial increase from 2009 to 2015. They indicated the fragmentation of coastal wetland in Nansha district is becoming more and more serious.

5 Discussion and Conclusion

Through above analysis, area metrics showed that the total area of the wetland has fluctuations from 1995 to 2015, with the largest in 2009 and minimum in 2015. Shallow sea waters accounted for the largest proportion in coastal wetland structure, followed by dike-pond, is again river, while tidal flat, swamp and lake accounted for less proportion. Tidal flat area decreased furthest year by year while shallow sea waters was on the contrary. Shape metrics indicated river and shallow sea waters had complex landscape structure and the spatial connectivity of shallow sea waters increased a lot in 2015. Aggregation metrics showed dike-pond had the highest PD and LSI while shallow sea waters had the highest AI, which means dike-pond had high segmentation and dispersion while shallow sea waters had great aggregation. Diversity metrics indicated the difference in the proportion of coastal wetland types was increasing and the relatively uniform state became more and more uneven. The fragmentation of coastal wetland in Nansha district is becoming more and more serious.

Since 2009, the coastal wetland area decreased sharply, especially in the types of tidal flat and dike-pond. For the whole landscape-level, distribution of coastal wetlands is becoming more and more uneven and the fragmentation is becoming more and more serious. The structure is not balanced. It is high time to make the sustainable development route for coastal wetland resources. Whether the landscape pattern changes had caused some functional degradation of coastal wetland and what is the degradation mechanism are waiting to be further studied in the future.

Acknowledgements. This work was funded by Science Foundation for Youths of GDAS (qnjj201308), Major Special Project-the China High-Resolution Earth Observation System, Natural Science Foundation of Guangdong Province (2014A030313747 and S2013010013320) and Irrigation works Science and Technology Innovation Project of Guangdong Province (201514 and 2014-13).

References

1. Baidu. http://baike.baidu.com/view/4210524.htm
2. Liu, Y., Wu, D., Zeng, L., et al.: Evolvement of landscape pattern of the coastal Wetland in Zhuhai during 1988–2008. Trop. Geogr. **31**(2), 199–204 (2011)
3. Li, X., Wan, R., Lin, J.: Mapping of Coastal Wetland in Beibu bay by remote sensing. Res. Ecol. South China Sea. **8**, 28–36 (2010)
4. Wei, L., Wang, X.Q., Chen, Y.Z.: Change of Coastal Wetlands in Fuzhou City in recent 10 years. Wetland Sci. **9**(3), 251–256 (2011)
5. An, X.L.: Research on China Coastal zones III. J. Anhui Agric. Sci. **37**(4), 1712–1713 (2009)
6. Zhang, S.W., Yan, F.Q., Yu, L.X., et al.: Application of remote sensing technology to wetland research. Sci. Geogr. Sinca. **33**(11), 1406–1412 (2013)
7. Ci, H., Qin, Y., Yang, H., et al.: Study on extraction methods of coastal wetland information. Comput. Eng. Appl. **47**(33), 44–248 (2011)
8. He, D.J., Lin, L., You, W.B., et al.: Landscape pattern changes of coastal wetlands and its simulation in eastern Fujian Province. J. Fujian Coll. Forest. **33**(2), 97–105 (2013)
9. Gao, Y., Su, F.Z., Sun, X.Y., et al.: On changes in landscape pattern of coastal wetland around the pearl river estuary in past two decades. Trop. Geogr. **30**(3), 215–220 (2010)
10. Hou, M.H., Liu, H.Y., Zhang, H.B., et al.: Influences of topographic features on the distribution and evolution of landscape in the coastal wetland of Yancheng. Acta Ecol. Sinica. **33**(12), 3765–3773 (2013)
11. Zhang, L., Ma, H.C., Wu, J.W.: Utilization of LiDAR and tidal gauge data for automatic extracting high and low tide lines. J. Remote Sens. **2**, 405–416 (2012)
12. McGarigal, K., SA Cushman, E Ene.: FRAGSTATS v4: Spatial pattern analysis program for categorical and continuous maps. University of Massachusetts, Amherst, USA. Source http://www.umass.edu/landeco/research/fragstats/fragstats.html
13. Zheng, X., Fu, M.: Landscape spatial analysis technology and its application. Science Press, Beijing (2010)
14. Cao, Y.K., Wang, X.H.: The change of landscape pattern of land use based on fractal dimension and entropy-a case study of Xi'an city. Ground water. **34**(4), 152–154 (2012)
15. Du, P.J., Chen, Y., Tan, K.: Monitoring and analyzing wetland landscape pattern change and ecological security using remote sensing images: A case study of Jiangsu coastal wetland. Remote Sens. Land Resour. **26**(1), 158–166 (2014)
16. Su, S., Yang, C., Hu, Y., et al.: Progressive landscape fragmentation in relation to cash crop cultivation. Appl. Geogr. **53**, 20–31 (2014)

Analysis of MODIS Satellite Thermal Infrared Information Before and After M_S 6.5 Ludian Earthquake

Xiang Wen, Hua Zhang(✉), Bin Zhou, Huining Huang, and Yongdong Yuan

Earthquake Bureau of the Guangxi Zhuang Autonomous Region, Nanning 530022, China
{yaya997, huazhang1222, dztzb, bixirong1989}@163.com, 40024540@qq.com

Abstract. Continuous MODIS/Terra satellite remote sensing thermal infrared data of the 6.5-magnitude earthquake region in Ludian, Yunnan, from June to August 2014 were collected. Through cloud removing and other data treatments, the infrared data obtained during the optimal observation period from 5:00 a.m. to 7:00 a.m. of Beijing time were chosen for surface temperature retrieval. The relationship between the time-based evolution of anomalous surface temperature and anomalous space distribution, and the active fault before and after the earthquake was analyzed; the time and space relevance between the anomalous surface temperature and the latent heat flux changes was studied; and the influence of non-structural factors, such as landform and seasonal climate, on the anomalous surface is studied. Results suggested that: (1) one month before the occurrence of Ludian Earthquake, the epicenter showed anomalously increasing temperature of the thermal infrared, meaning that the anomalous temperature increase was related to the earthquake occurrence time; the significant temperature increase lasted for half a month before the earthquake, and the anomalous temperature increase reached the peak five to six days before the earthquake, but it decreased dramatically after the earthquake; (2) based on the hyperspectral data and the geological observation meteorological data, the influence of clouds and vegetation cover in the earthquake area was removed. The latent heat flux changes inverted showed a close time and space relevance with the anomalous surface temperature increase; (3) the analysis of the relevance with the landform and the seasonal climate factors suggested that there was structural "temperature increase" information of out-of-season changes before the earthquake; (4) the anomalous temperature increase developed in the shape of "X" from the epicenter to the conjugate fault. This coincided with the mechanical effect generated by the "L-shaped" asymmetric conjugated fault extended in the NW-SE direction of the advantageous distribution of the horizontal maximum principal stress and the NW ~ EW direction after the earthquake. Considering the influence of landform, seasonal climate and other non-structural factors on anomalous temperature increase, the author thought

Fund Project: This work was supported by the science-technology plan of Guangxi (Project number: 1377002, 1298005-2, 12426001)

F. Bian and Y. Xie (Eds.): GRMSE 2015, CCIS 569, pp. 417–433, 2016.
DOI: 10.1007/978-3-662-49155-3_43

that the temperature increase of the thermal infrared might be a short-term anomalous phenomenon before the earthquake.

Keywords: Surface temperature · Latent heat flux · Landform · Seasonal changes · Anomalous omens

1 Introduction

Anomaly of satellite remote sensing thermal infrared means that the geo-temperature obtained after the inversion of the energy information radiated by the earth atmosphere system contains the thermal information within the earth [1]. The time and space analysis of the satellite thermal infrared anomaly can not only help recognize the existing geological structures, but also analyze the active state of the geological structures according to the changing process of the infrared anomaly, and find the geological disaster omens. Since the 1990s, there have been a large number of earthquake scholars studying anomalousities of the geo-temperature, brightness temperature and long-wave radiation of the thermal infrared of the moderately strong structural earthquake before its occurrence, and attempting at revealing the corresponding relationship between these anomalousities and the time, space and intensity of the earthquake. Currently, remarkable achievements have been made in studying infrared anomaly and heat active state of the active fault before the earthquake, but scholars have also raised some difficult questions worth further discussion based on the current situations [2–7]. First is about the complex causes of the anomalous infrared radiation of surface features; second, natural factors, including landform, atmosphere situation, rock aquosity and electrical conductivity, vegetation growth, wind and rain, latitude and seasonal changes, can impede people's correct judgment of earthquake omens. Therefore, extraction of heat information genuinely related to the faulting activities based on an objective understanding of the disturbance of various non-earthquake factors is one of prerequisites to conduct earthquake prediction through the satellite infrared remote sensing technique. Based on the research findings of former scholars, this paper adopts M_S (Magnitude Scale) 6.5 Ludian Earthquake as an example. The temperature field changing images of the geological near the epicenter is obtained through inversion and based on the continuous MODIS satellite remote sensing infrared data. The relationship between the time evolution and the space distribution of geo-temperature anomaly before and after the earthquake, and the active fault is analyzed, and the time and space relevance between the geo-temperature and the latent heat flux anomaly is studied. Besides, the influence of the non-structural factors, including landform and climate, on anomalous geo-temperature is discussed.

2 Overview of the Research Area

An M_S 6.5 earthquake (27.08°N, 103.37°E) happened in Ludian near the Zhaotong Fault on August 3, 2014. The earthquake was quite active in the earthquake area. In history, there were 44 earthquakes higher than M_S 5 happening within 100 km, of which 37 ranged from $5M_S$ to $5.9M_S$; 5 ranged from M_S 6 to M_S 6.9; 2 ranged from M_S 7 to M_S 7.9. The maximum earthquake was the one happening on August 2, 1733, whose magnitude scale reached 7¾. This Ludian Earthquake has been the maximum one happening in Yunnan Province since 2000. In the disaster-stricken area, the focus is shallow, the population density is high, the buildings are vulnerable towards earthquake, and the landform is rugged, so rolling stones, landslides, debris flows and other secondary disasters happen frequently, causing serious casualties and property losses. The earthquake fault of Ludian Earthquake was Xiaohe Fault of the north-west direction in Baogunao Village and the north-west secondary strike-slip fault of the north-east Zhaotong-Ludian Fault. It belongs to the Xiaojiang Fault Zone. The fault rupture caused by the principal earthquake expanded on both sides, and might be impeded in the north-east direction, thus causing asymmetric conjugated fractures. The factures expanded from the north-west direction to the east-west direction. The after-shock was mainly concentrated in the conjugated intersection area. The earthquake structural chart is shown in Fig. 1.

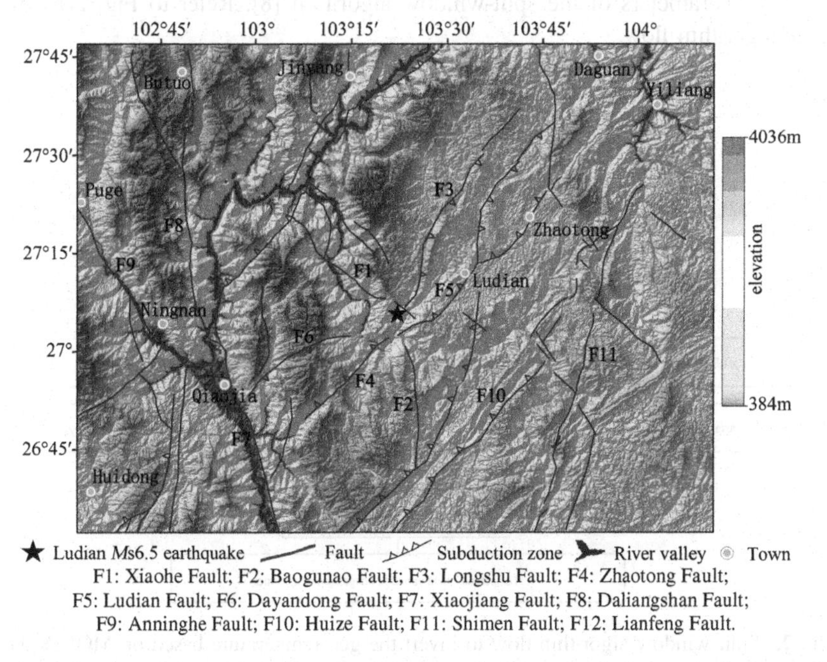

Fig. 1. Epicenter of M_S 6.5 Ludian Earthquake and the geological structural chart of the surrounding area

3 MODIS Data and Inversion Process

In recent years, the data for the short-term earthquake omens research based on the satellite remote thermal infrared have mainly come from GMS-5, AVHRR geostationary meteorological satellite and MODIS polar orbit. Comparing the three data sources, MODIS data are superior to GMS-5 and AVHRR. MODIS data of this paper mainly comes from the website of NASA, http://ladsweb.nascom.nasa.gov/data/search.html. MODIS image space resolution is 1km. According to the transmission characteristics of the thermal infrared radiation in the atmosphere, No. 31 and No. 32 wave section of the MODIS thermal infrared data are the most suitable for the inversion of the geo-temperature. Besides, after 5:00 a.m., the influence caused by the solar radiation is the smallest, which can best reflect the situation of the geo-temperature field. Therefore, this paper chooses data of the No. 31 and No. 32 thermal infrared wave section, and inverts the surrounding temperature field changing images according to the split-window algorithm put forward by MAO Kebiao [8]. The expression of the split-window algorithm is:

$$T_S = A_0 + A_1 T_{31} - A_2 T_{32} \tag{1}$$

where, Ts stands for the geo-temperature; T31 and T32 stands for the brightness temperature of the No. 31 and No. 32 wave section of MODIS, respectively; A0, A1 and A2 are parameters of the split-window algorithm [8]. Refer to Fig. 2 as to the specific algorithm flow.

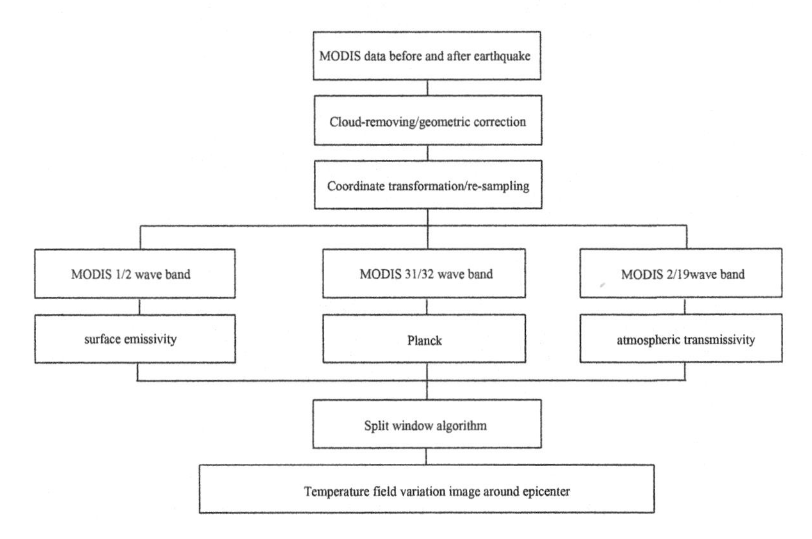

Fig. 2. Split-window algorithm flow to invert the geo-temperature based on MODIS data

During the remote-sensing image processing, the information obtained suffers the disturbance of the cloud noise due to climate reasons. This directly influences the inversion accuracy of the geo-temperature. Therefore, cloud removal is an important link of MODIS data pretreatment considering different characteristics of imaging and imaging principles of the thin and thick clouds, different cloud removing methods are adopted according to different cloud thickness. In terms of removing the thin cloud, due to the shrinking dynamic scope of images caused by the thin cloud, the detailed part of images is covered, thus reducing the resolution rate. At the same time, the noises of the thin cloud results in the strengthening of the image frequency spectrum in terms of low-frequency component. Therefore, the thin cloud is removed by enhancing the high-frequency component and inhibiting the low-frequency component of images. This paper adopts the "homomorphic filtering method" [9, 10] to remove the thin cloud. Based on the characteristic of the thin cloud generally occupying the low-frequency information in the frequency domain, this method transforms the remote-sensing images to the frequency domain to remove the low-frequency information, and enhance the results so as to enhance the surface features information under the cover of clouds. In terms of thick cloud removal, the image area (the target image area) influenced by the thick cloud cannot restore the surface features information of the covered area. Therefore, this paper adopts the "data substitution method" [10] to remove thick clouds. At the same time, if more than three days between 5:00 a.m. to 7:00 a.m. within 15 days are cloudless, this paper thinks that the data within the period of time have a small disturbance on the inversion accuracy of the geo-temperature, and can directly participate in the inversion calculation.

The thermal infrared geo-temperature observed by the satellite remote-sensing is very complex. Its spatial distribution is influenced by physical and chemical factors, including landform, territory and medium. Even under the normal meteorological conditions or the condition of no earthquakes, the infrared geo-temperature background is uneven. In terms of time, it is simultaneously influenced by meteorology and structural activities. The infrared anomaly before the earthquake often occurs along with the geo-temperature changes caused by meteorological factors. This increases the difficulty for the recognition of infrared anomaly before the earthquake. Currently, there has not yet been an agreement as to the infrared temperature increase mechanism. The extensive research area results in a greater difference of the underlying surface and more influencing factors. In terms of the current research level, the research cannot be exhaustive yet. However, if the research area can be limited to the surroundings of the active fault belt, the influencing factors can be simplified to some extent [11]. Therefore, this paper adopts the active fault belt (26°30′ ~ 28°N, 102°30′ ~ 104°30′E) near the epicenter as the research object, and MODIS data from June 3, 2014 to August 11, 2014 for geo-temperature inversion and analysis of the time evolution of thermal infrared anomaly before and after the earthquake. The anomalous images extracted are connected the active faults to explore the connection between the space distribution of thermal infrared anomaly and the active faults. The inversion results are shown in Fig. 3.

(1) The inversion images of the temperature field on June 3 show that there existed several points of temperature increase in the intersection area between Lianfeng Fault and Xiaohe fault, the south section of Longshu Fault, the central section of Dayandong Fault, and the central section of Anninghe Fault (Fig. 3a). On June 9, there were multiple heat source points near Lianfeng Fault, Xiaohe Fault, Dayandong Fault, Anninghe Fault and Ludian Fault (Fig. 3b). Later, the heat spreads to the surrounding faults. The interaction area between Lianfeng Fault and Xiaohe Fault, the north section of Longshu Fault, the south section of Zhaotong

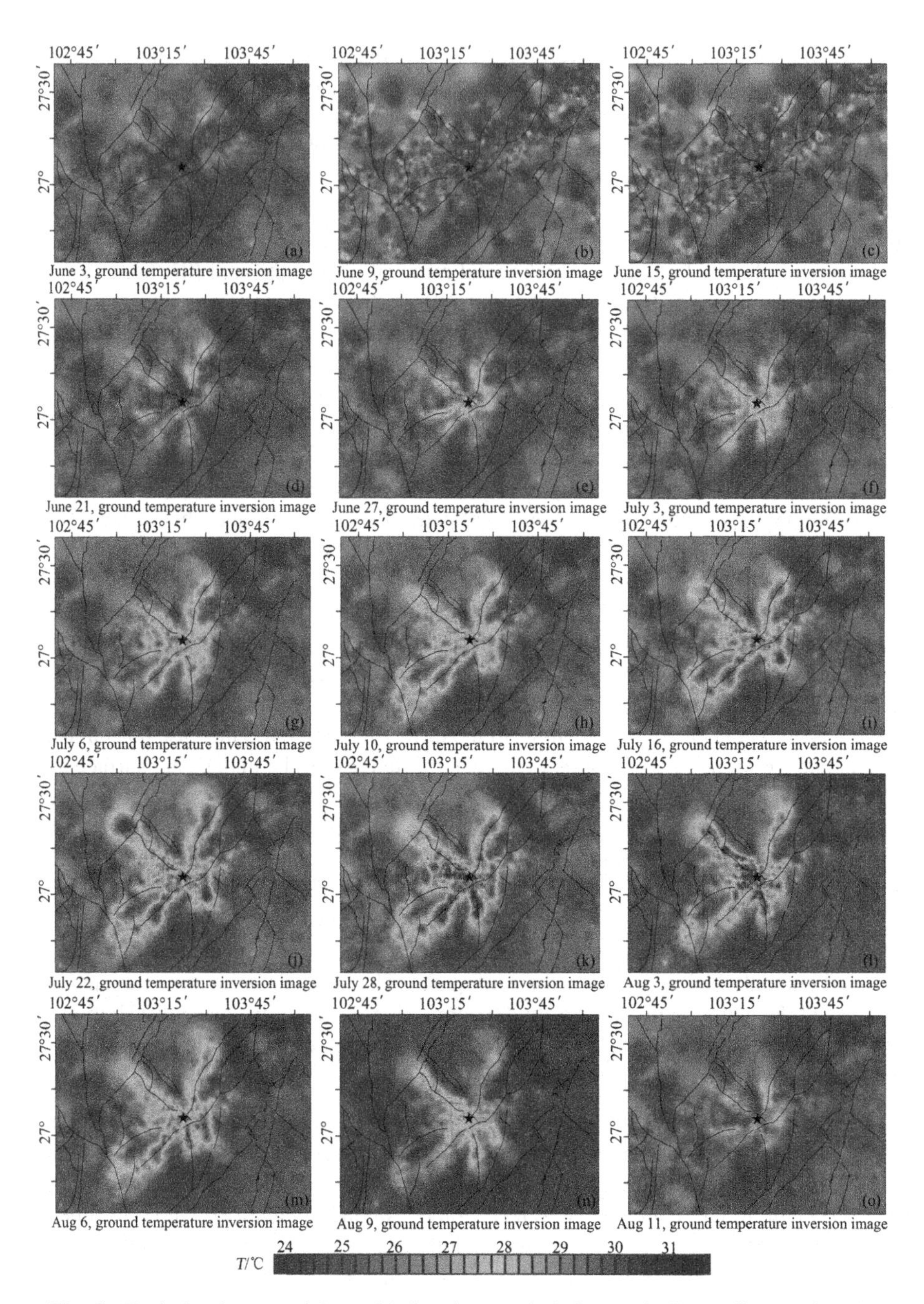

Fig. 3. Evolution images of thermal infrared anomaly before and after Ludian Earthquake

Fault and the north section of Ludian Fault shows the trend of temperature increase (Fig. 3c).

(2) Since June 16, the temperature had been increasing rapidly in Xiaohe Fault, the north section of Baogunao Fault, the south section of Longshu Fault, the north section of Dayandong Fault, the north section of Zhaotong Fault and the south section of Ludian Fault near the epicenter (Fig. 3d). The temperature field images on June 27 showed that the wide expansion of the heat source points. The anomalously temperature increasing area of the south section of Xiaohe Fault, the north section of Baogunao Fault, the south section of Longshu Fault, the north section of Dayandong Fault, the north section of Zhaotong Fault and the south section of Ludian Fault was yellowing gradually (Fig. 3e).

(3) Since July 6, the temperature anomalously increasing area of the epicenter and the surrounding areas had been expanding, yellowing and darkening. The anomalous temperature increase belts developed in the north-east direction, and were mainly distributed between 26°45′ ~ 27°30′N and 103° ~ 103°45′E (Fig. 3g). The temperature field images on July 10 showed that the anomalous temperature increasing region of Xiaohe Fault and Baogunao Fault near the epicenter developed in the north-west direction, intersected with the anomalous temperature increase belts of Zhaotong Fault, Ludian Fault and Longshu Fault, and formed a radial pattern (Fig. 3h). The anomalous temperature increase of the interacting belts of the north-east direction and the north-west direction on July 16 was more significant. A large area near the epicenter was red, and distributed in the shape of "X" and it still kept on developing (Fig. 3i). On July 28, the geo-temperature of the conjugated interaction area of the south section of Xiaohe Fault, the north section of Baogunao Fault, the south section of Longshu Fault, the north section of Dayandong Fault and the south section of Zhaotong Fault reached the peak value, and averaged at around 31 °C. However, the average temperature near Anninghe Fault, Huize Fault, Shimen Fault stayed around 26 °C(Fig. 3k).

(4) The temperature field images from the day the earthquake happened showed that the anomalous area was mainly concentrated near the south section of Xiaohe Fault, the north section of Baogunao Fault, the north section of Zhaotong Fault and the south section of Ludian Fault, and was distributed in the north-east direction; the geo-temperature on the south section of Zhaotong Fault and the north section of Ludian Fault showed a falling trend (Fig. 3l). After the breakout of the earthquake, the geo-temperature near the earthquake area decreased rapidly. The temperature field images on August 6 showed that the red temperature increasing area decreased, with the color fading. The heat source spread from the epicenter to the surroundings, and the anomalous distribution showed a "corolla pattern" (Fig. 3m). After August 11, the anomalous temperature increasing area near the epicenter basically disappeared (Fig. 3o). Refer to Table 1 for the specific description of the anomalous situations.

Table 1. Anomalous temperature increase statistics before Ludian Earthquake

Item	Anomalous date (MM/DD)	Temperature increase region	Temperature increase area /km^2	Temperature increase range/°C
1	06/15	Xiaohe Fault, the north section of Baogunao Fault, the south section of Longshu Fault, the north section of Dayandong Fault, the north section of Zhaotong Fault, the south section of Ludian Fault and point sources	About 0.3×10^5	1~2
2	06/21	The south section of Xiaohe Fault, the north section of Baogunao Fault, the south section of Longshu Fault, the north section of Dayandong Fault, the north section of Zhaotong Fault, the south section of Ludian Fault and point sources	About 0.3×10^5	1~3
3	06/27	The south section of Xiaohe Fault, the north section of Baogunao Fault, the south section of Longshu Fault, the north section of Dayandong Fault, the north section of Zhaotong Fault, the south section of Ludian Fault, and sheet distribution	About 0.3×10^5	2~3
4	07/06	The south section of Xiaohe Fault, the north section of Baogunao Fault, the south section of Longshu Fault, the north section of Dayandong Fault, the north section of Zhaotong Fault, the south section of Ludian Fault, and the belt distribution in the north-east direction	About 0.4×10^5	2~4
5	07/10	Xiaohe Fault, the north section of Baogunao Fault, Zhaotong Fault, the south direction of Ludian Fault, the south direction of Longshu Fault, the intersected development in the north-east and north-west direction, and the radial distribution pattern	About 0.5×10^5	2~4

(*Continued*)

Table 1. (*Continued*)

Item	Anomalous date (MM/DD)	Temperature increase region	Temperature increase area /km^2	Temperature increase range/°C
6	07/16	Xiaohe Fault, the north section of Baogunao Fault, Zhaotong Fault, the south section of Ludian Fault, Longshu Fault, the intersecting area between Zhaotong Fault and Xiaojiang Fault, the intersecting development in the north-east and north-west direction, and the X-shaped distribution pattern	About 0.7×10^5	3 ~ 5
7	07/22	Xiaohe Fault, the north section of Baogunao Fault, Zhaotong Fault, the south section of Ludian Fault, Longshu Fault, the north section of Dayandong Fault, the intersecting area between Lianfeng Fault and Xiaohe Fault, the intersecting area between Zhaotong Fault and Xiaojiang Fault, the intersected development in the north-east and north-west direction, and the X-shaped distribution pattern	About 0.7×10^5	3 ~ 5
8	07/28	The south section of Xiaohe Fault, the north section of Baogunao Fault, the south section of Longshu Fault, the north section of Dayandong Fault, the north section of Zhaotong Fault, the south section of Ludian Fault, the intersected development of the north-east and north-west direction, and the asymmetric conjugated distribution	About 0.6×10^5	4 ~ 6
9	08/03	The south section of Xiaohe Fault, the north section of Baogunao Fault, the north section of Zhaotong Fault, the south section of Ludian Fault and the north-west distribution pattern	About 0.5×10^5	4 ~ 5

4 Relationship Between the Latent Heat Flux and the Anomalous Geo-Temperature Before the Earthquake

During the earthquake development process, the focus keeps on conducting material and energy change with the external world. The heat energy transmission is the most important form of energy transmission. Most of the geological heat can be released through radiation, latent heat and sensible heat. Radiation is a relatively common energy transmission style. Any object in the nature can keep on sending heat in the form of magnetic wave as long as the temperature is above the absolute zero. The energy transmission style is called radiation. The original meaning of latent heat means the heat consumed when materials undergo phase change. In this paper, it refers to the heat taken away by the evaporating water in the form of evaporation heat. The sensible heat refers to the heat exchange of atmosphere in the form of turbulence, and is caused by the geo-temperature difference. Globally speaking, the energy radiated by the ground accounts for 41 %; the energy released by the latent heat accounts for 45 %; the energy released by the sensible heat accounts for 14 %. The geo-temperature rising phenomena before earthquakes has been observed by the meteorological stations and satellite remote sensing; while the geological latent flux heat before earthquakes has been just discovered recently. There have been research findings suggest that there will be different degrees of anomaly of latent heat flux and geo-temperature before inland earthquakes. To make it clear whether there were anomalous changes of latent heat flux before the earthquake, and whether anomalous changes of latent heat flux were well corresponded to anomalous geo-temperature, this paper adopts the remote-sensing thermal infrared data, hyper-spectrum and ground observed meteorological data of the No. 1, No. 2, No. 19, No. 31 and No. 32 wave section of MODIS from July 2014 to August 2014, and combines the characteristics of sparse geological vegetation before the earthquake. The dual-temperature difference dual-source model [12] was adopted to invert the sensible heat flux. Through the combination of the geological energy equilibrium equation, the latent heat flux near the epicenter can be obtained. The inversion results suggested that the latent heat flux value near the south section of Xiaohe Fault, the north section of Baogunao Fault, the south section of Longshu Fault, the north section of Dayandong Fault, the north section of Zhaotong Fault and the south section of Ludian Fault showed a significantly high value (Fig. 4d). Through the multi-wave-section cloud removing and the analysis of the vegetation coverage, it was found that the high-value anomalous area was slightly influenced by clouds, and was out of line with the vegetation coverage and growth situations of the area both in terms of time and space. This indicated that the anomaly was not caused by clouds or the vegetable of the underlying surface, but an anomalous reflection before the breakout of Ludian Earthquake. Further comparison between it and the geo-temperature anomaly sowed that the latent heat flux changes formed a mutually explanatory logic chain both in terms of time and space distribution (Figs. 3 and 4), and could more clearly reflect the anomalous heat phenomena before the outbreak of Ludian Earthquake.

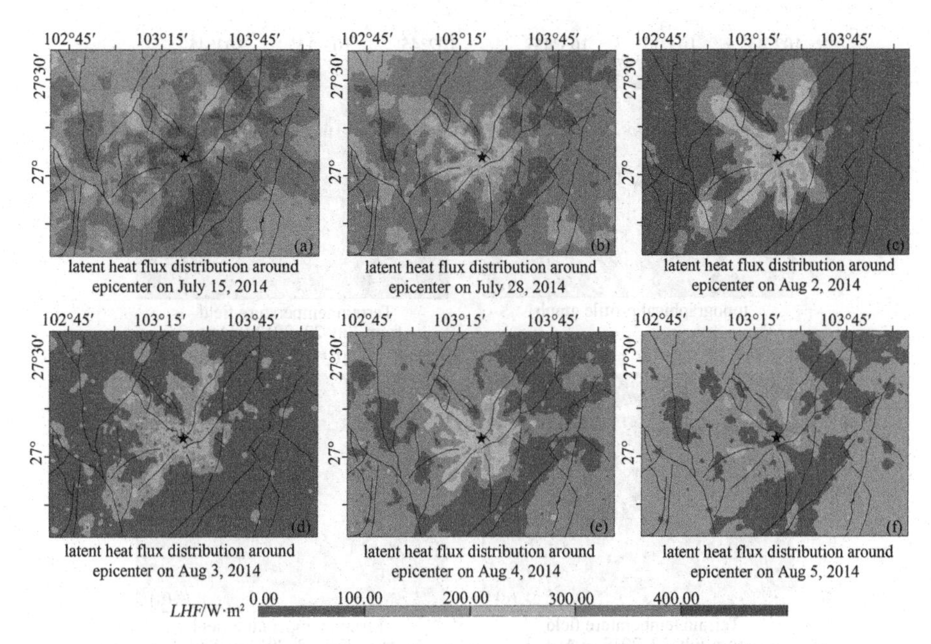

Fig. 4. Latent heat flux distribution before the breakout of Ludian Earthquake

5 Discussion

5.1 Relationship Between the Anomalous Geo-Temperature and the Landform in the Earthquake-Stricken Area

Landform is one of the disturbing factors for a correct judgment of anomalous earthquake omens. SHAN Xinjian et al. [13] thought that the overall background of the geo-temperature of the areas with a high elevation, such as highlands and mountains is relatively low; while the overall background of the geo-temperature of areas with a low elevation, such as basins, is relatively high, and that there is favorable corresponding relationship between the two. In particular, the since ravines feature a low terrain, an intersected drainage system and a high ground water content, the geo-temperature will be about 5 ~ 10 °C higher than that of the surrounding mountains. Within certain areas and periods, the geo-temperature background and the landform are stable. Through the tracking of the time evolution process and the comparison with the other years, the thermal infrared anomalous information can be effectively extracted. In this paper, the numerical elevation raster data near the earthquake-stricken area are overlapped with those of the thermal infrared geo-temperature field to study the dynamic varying relationship between the landform elevation difference and the heat anomaly near the active faults. The landform temperature field obviously showed that, compared with the non-earthquake periods in 2012 and 2013, the band-shaped river valleys and basins in the earthquake-stricken area before Ludian Earthquake showed obvious high temperature anomaly and featured a pink color; while the temperature increase in the area with a high elevation near the earthquake-stricken area was not significant and featured a

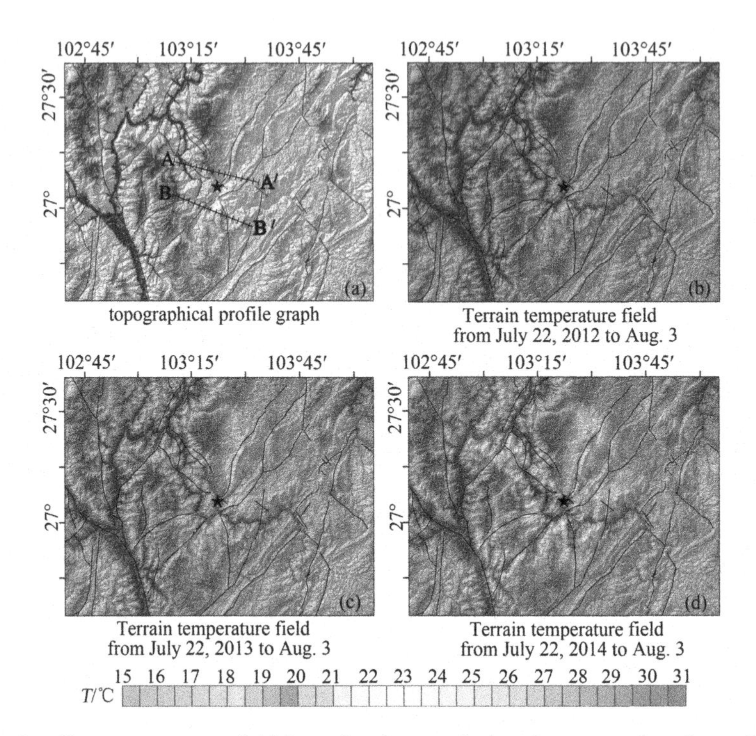

Fig. 5. Landform temperature field inversion images during the non-earthquake periods and before the earthquake near the earthquake-stricken area

yellow green (Fig. 5d). In order to obtain a more direct understanding of the phenomenon, the numerical elevation landform was used to draw two 40kmlong topographic profiles from the west to the east near the epicenter (Refer to A-A′ and B-B′ in Fig. 5a for the profile line position). The profile A-A′ goes across Xiaohe Fault, Longshu Fault and Ludian Fault, with an elevation difference of about 1,900 m. The profile B-B′ profile goes across Dayandong Fault, Zhaotong Fault and Baogunao Fault. The landform cutting was sharp on two sides, with an elevation different of about 2,300 m. Adopted the average value of the thermal infrared geo-temperature of the same-name pixel points on profile lines, namely A-A′ and B-B′, before the earthquake and during the non-earthquake period, to draw the geo-temperature profile (the red and blue curves in Fig. 6a and b). After comparing it with the landform profile (the black curves in Fig. 6a and b), it was found that, as the landform lifted, the infrared geo-temperature of the pixel points whose name was the same to the landform decreased. The two showed a favorable negative relevance. At the same time, the geo-temperature average curves of the same-name pixel points during the non-earthquake periods from July 22, 2013 to August 3, 2013 wed that the geo-temperature background was stable, and that the average foreshock geo-temperature was higher than that during the non-earthquake period by about 5∼6 °C. The landform factor relevance analysis suggested that the

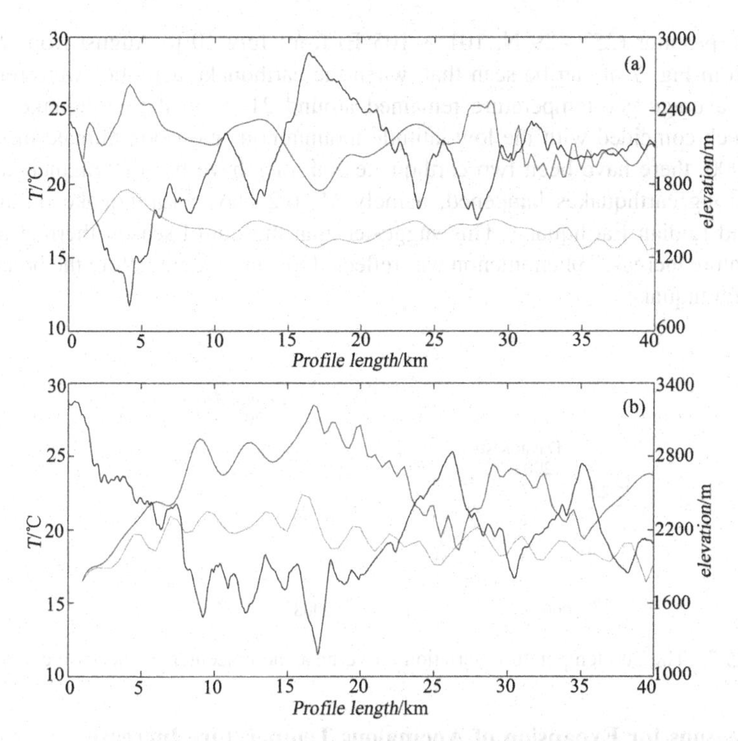

Fig. 6. Relationship between the profile landform elevation difference and the geo-temperature (a) Comparison of the landform elevation difference and the average geo-temperature of the A-A′ profile crossing Xiaohe Fault, Longshu Fault and Ludian Fault. The black curve stands for the elevation, the green curve stands for the average geo-temperature of the same-name pixel points during the non-earthquake periods from July 22, 2013 to August 3, 2013; the red curve stands for the average geo-temperature of the same-name pixel points before the earthquake from July 22, 2014 to August 3, 2014; (b) Comparison of the geological elevation difference and the average geo-temperature of B-B′profile crossing Dayandong Fault, Zhaotong Fault and Baogunao Fault. The black curve stands for the elevation, the green curve stands for the average geo-temperature of the same-name pixel points during the non-earthquake periods from July 22, 2013 to August 3, 2013; the red curve stands for the average geo-temperature of the same-name pixel points before the earthquake from July 22, 2014 to August 3, 2014.

geo-temperature near the earthquake-stricken area decreased along with the increase of elevation. There showed obvious foreshock structure "temperature increase" near the faults of the earthquake.

5.2 Relationship Between the Geothermal Anomaly and the Seasonal Climate Factor

Climate is another factor impeding the correct judgment of earthquake omens. Therefore, a correct understanding of the annual climate changes in the earthquake-stricken area is a necessary prerequisite to judge the infrared geothermal anomaly. This paper conducts a longitudinal comparison of the actually-measured average geo-temperature near the epicenter (25° ~ 29°N, 101° ~ 105°E) from July 10 to August from 2000 to 2014. From Fig. 7, it can be seen that, when the earthquake activities were relatively few, the average geo-temperature remained around 21 °C in the earthquake-stricken area, which coincided with the low-latitude mountainous monsoon climate in Ludian. Since 2000, there have been two earthquake activities have been strengthened twice. Later, strong earthquakes happened, namely M_S 6.2 Dayao Earthquake on July 21, 2003, and Ludian Earthquake. This suggested that the out-of-season thermal infrared "temperature increase" phenomenon was reflected to some extent before the breakout of Ludian Earthquake.

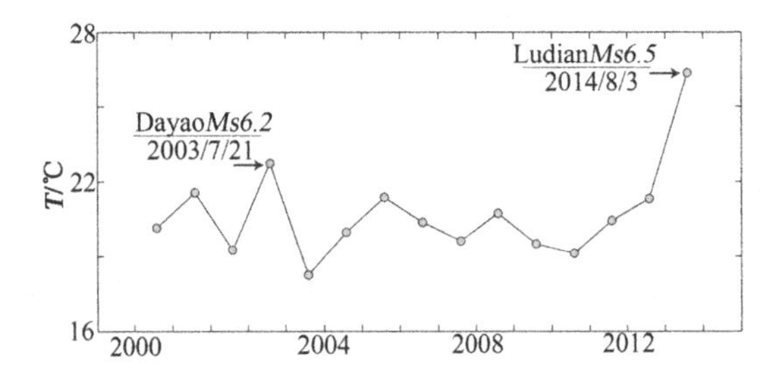

Fig. 7. The geo-temperature variation curve near the epicenter in successive years

5.3 Reasons for Expansion of Anomalous Temperature Increase from the Epicenter to the Conjugated Fault

Some scholars thought that, under the influence of stress field in certain area, the structural activity is a regional concept, instead of the independent activity of certain fault. Therefore, except that the fault in the principal earthquake controlled the foreshock, aftershock and most omens, the other faults in the area were also influenced by the stress field of the area. It could be the fault parallel to the principal earthquake fault or another group of fault conjugated and connected with it in terms of mechanical causes. Some fault zones in the mutually conjugating fault network become the anomalous distribution belts or concentration belts of earthquake omens. The phenomenon might exist in the long-and mid-term omens anomaly field. Its influence on the short-term omens anomaly field was clearer. The aftershock sequence of Ludian Earthquake showed asymmetric conjugated distribution in the east-west direction and north-west direction. Its length in the east-west direction and the north-west direction was about 17 and 22 km (Fig.8). According to the aftershock distribution and the principal earthquake positioning results, the earthquake fault was the conjugated intersection of Baogunao—Xiaohe Fault, Ludian-Zhaotong Fault and Longshu Fault. The fault in the conjugated intersection was located in the sensible position with complex stress state changes. The anomaly migrated from the intersection part to several relevant faults. This well explained the conjugated development phenomenon of the foreshock anomalous temperature increase from the surrounding of the epicenter to Baogunao—Xiaohe Fault and Ludian—Zhaotong Fault.

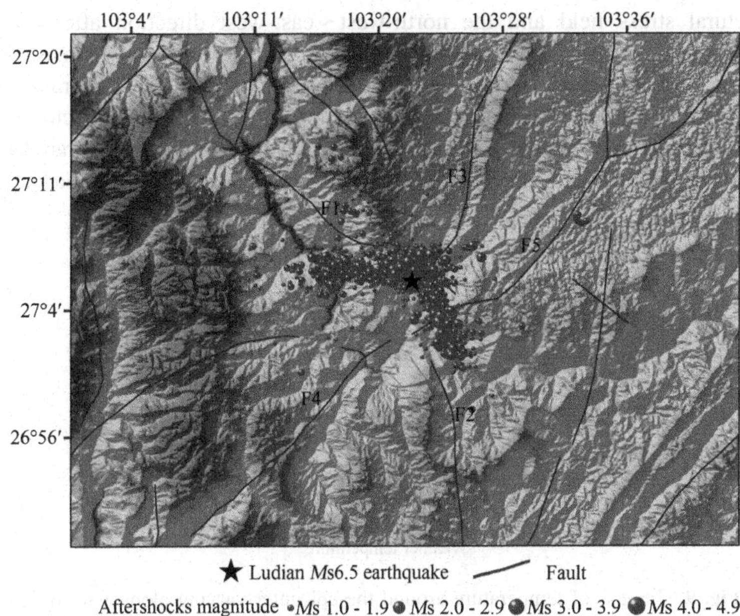

Fig. 8. M_S 6.5 Ludian Earthquake sequence epicenter distribution chart

5.4 Confirmation of Whether the X-shaped Anomalous Temperature Increase Belt is a Short-term Anomaly

Foreshock anomaly is an opposite idea to the normality. From the whole anomaly evolution process from June 3 to September 3, it can be seen that: First, in terms of temperature increase area, with the approaching of the earthquake, the temperature increase area significantly expanded. Second, in terms of temperature increase amplitude, in order to achieve a vivid analysis, the geothermal data were collected near the epicenter, and the comparison chart of the varying curve of temperature along with time was drawn based on the Ludian meteorological data (Fig. 9). It can be seen that the epicenter temperature inverted by the thermal infrared sensing was basically the same to the value actually measured by Ludian County Meteorological Bureau (http://lishi.tianqi.com/ludian/index.html). The temperature increase amplitude was not significant from June 3 to June 21, but gradually increased since June 27, and reached the peak five days before the earthquake. Third, in terms of the temperature increase region, the foreshock structural activities in the intersection area of Xiaohe Fault, the north section of Baogunao Fault, Zhaotong Fault, the south section of Ludian Fault, Longshu Fault and the north section of Dayandong Fault showed obvious thermal infrared anomalous temperature increase, and the anomaly developed in the north-east direction and the north-west direction, forming an "X" pattern (Fig. 3k). This was basically in line with the mechanical effect generated by the L-shaped asymmetric conjugated fault expanded by the north-west ~ south-east direction of the horizontal maximum principal stress of the structural stress field and the north-west ~ east-west direction aftershock. The foreshock latent heat flux changes and the anomalous geothermal increase showed consistence both in terms of time and space. Based on a full consideration of the influence of landform, seasonal climate and other non-structural factors on the anomalous temperature increase, the author thought that the thermal infrared temperature increase might be a foreshock short-term phenomenon of Ludian Earthquake.

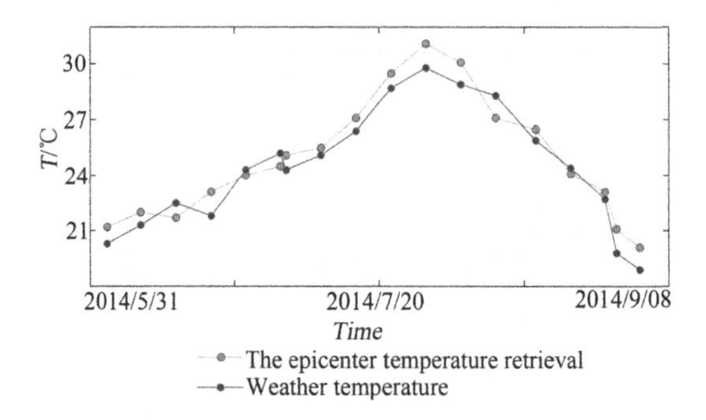

Fig. 9. Images of temperature around the epicenter varying along with time

6 Conclusions

The example of Ludian Earthquake suggested, though the geothermal anomaly generated by the crustal activities is hidden in the complex changes of the non-structural factors, including landform, seasonal climate, snows and rains. By analyzing their relationship with thermal infrared anomaly, the author removed the disturbance of non-structural factors at an attempt at obtaining information of crustal activities. The following points should be taken into consideration before using geo-temperature to obtain the thermal information of earthquake omens: First, the combination of the annual variation standard field and other research findings [14], and the comparison of the thermal field changes in recent years near the earthquake-stricken area could contribute to the removal of the influence of non-structural factors; second, the comprehensive treatment and analysis of fault data and data of the surrounding rocks could help study whether there are similar temperature increase phenomena in the same period and the same scope (such as the same structural partition or the same or similar structural belt). Third, the temperature increase mechanism of the thermal field should include strain temperature increase, friction-caused temperature increase and temperature increase caused by underground water changes. Based on the research findings of former scholars in terms of temperature increase mechanism [15], the temperature increase mechanism of the thermal field in the earthquake-stricken area can be further analyzed. At last, the thermal information related to the earthquake omens should be able to form a mutually explanatory logic chain with the findings (such as brightness temperature, long-wave radiation and meteorological data) obtained through the other methods. Most importantly, the space distribution of the thermal field should well coincide with the fault activity and its mechanical model. The latter acts as verification of research results. Only after verification, the research results could be more reliable.

Acknowledgement. I want to thank Association Professor CHEN Meihua for her patient guid ance and support of. My thanks also go to Ludian County Meteorological Bureau for providing the geothermal data products.

References

1. Qiang, Z.J., Kong, L.C., Wang, Y.P., et al.: Earth gas emission, infrared thermo-anomaly and seismicity. Chin. Sci. Bull. **37**(24), 2259–2262 (1992)
2. Ma, J., Wang, Y.P., Chen, S.Y., et al.: Discussion on the relationship between satellite thermal infrared information and fault activity. Prog. Nat. Sci. **15**(12), 1467–1475 (2005)
3. Guo, W.Y., Shan, X.J., Ma, J.: Discussion on the anomalous increase of ground temperature along the seismogenic fault before the Kunlunshan M_S 8.1 earthquake in 2001. Seismolog Geol. **26**(3), 548–556 (2004)
4. Guo, W.Y., Shan, X.J., Qu, C.Y.: Correlation between infrared anomalous and earthquakes in Tarim Basin. Arid Land Geogr. **29**(5), 736–741 (2006)
5. Guo, W.Y., Ma, J., Shan, X.J.: Researches on the yearly characteristic of IR brightness temperature and the different hypsography on the active faults. Prog. Geophys. **23**(5), 1437–1443 (2008)
6. Qu, C.Y., Shan, X.J., Ma, J.: Formation cause of thermal infrared high temperature belt along Honghe Fault and its relation to earthquake. Acta Seismol. Sin. **28**(1), 91–97 (2006)
7. Qu, C.Y., Shan, X.J., Ma, J.: Study on the methods for extracting earthquake thermal infrared anomaly. Adv. Earth Sci. **21**(7), 699–705 (2006)
8. Mao, K.B.: Study on land surface temperature retrieved based on MODIS data. Nanjing University (2004)
9. Liu, Y., Bai, J.W.: Research on the cloud removal method of remote sensing images. Geomatics Spat. Inf. Technol. **31**(3), 120–123 (2008)
10. Ren, H.: Study on the cloud processing with MODIS data and the retrieval of LST. Xidian University, China (2013)
11. Deng, Z.H., Wang, Y., Chen, M.H., et al.: Satellite infrared anomaly of several strong earthquakes in China mainland. Seismol. Geol. **25**(2), 327–337 (2003)
12. Wang, K.C., Zhou, X.J., Li, W.L., et al.: Using satellite remotely sensed data to retrieve sensible and latent heat fluxes. Adv. Earth Sci. **20**(1), 42–48 (2005)
13. Shan, X.J., Qu, C.Y.: Analysis on mechanism of thermal anomalies before the earthquake in infrared image. In: Proceedings of Remote Sensing Technology Forum and Annual Meeting of Association of Remote Sensing Application, pp. 279–285 (2004)
14. Chen, S.Y., Ma, J., Liu, P.X., et al.: A study on the normal annual variation field of land surface temperature in China. Chin. J. Geophys. **52**(9), 2273–2281 (2009)
15. Chen, S.Y., Liu, L.Q., Liu, P.X., et al.: Theoretical and experimental study on relationship between strees-strain and temperature variation. Sci. China Ser D-Earth Sci. **39**(10), 1446–1455 (2009)

Measurement Method Improvement of China's Potential Output: Based on the Synthetically Evaluation of Fuzzy Entropy

Rubin Chen(✉)

School of Economics and Commerce, South China University of Technology, Guangzhou Higher Education Mega Centre, Panyu District, Guangzhou 510006, China
Zsucrbnol@163.com

Abstract. Potential output and output gap are important reference variables in economy cycle fluctuation. According to the situation, that measurement methods of potential output are numerous and independent, this article establishes the improvement method based on synthetically evaluation of fuzzy entropy. This method could produce discourse domain for different measurement results, calculate membership function value with the method of local replacement iteration, and determine the best time series of output gap. Empirical results shows, that the output gap of 1992Q1-2013Q3, which is obtained by the improvement method, is better than the estimated values of HP filtering method, production function method, and Vector auto-regression in terms of predictive ability of inflation, the agreement degree of factual economic cycle, and stability.

Keywords: Potential output · Output gap · Fuzzy entropy · Measurement method

1 Introduction

Potential output and potential economic growth rate are important reference values (or baseline values) to analyze and judge macro-economy situation, and compare macro-economy variable [1]. Since the concept of potential output was put forward by Okun [2], it has caused interests of many scholars and institutions. Because it is impossible to observe the level of potential output directly, the study of measurement methods has become a focal point. Methods, which are based on different assumptions and economic theories, are springing up continually. It has greatly expanded the measurement scope [3]. Scholars in China have also estimated China's potential output, output gap, and economic growth rate with one or several measurement methods (Including [4–8]). Because of the limitation of measurement methods, and the difference of sample data, different scholars have various and even contradictory opinions on the measurement methods of China's potential output and output gap [9–15]. Then, whether or not we can simply tell that some certain method is inapplicable in out country? As we know, foreign mainstream methods are all based on the foundation of

F. Bian and Y. Xie (Eds.): GRMSE 2015, CCIS 569, pp. 434–446, 2016.
DOI: 10.1007/978-3-662-49155-3_44

certain reasonable hypothesis. Therefore, they have relatively strong logical reasoning process, and could explain some information of potential output. If simply to abandon a certain method, the only information contained by it could be ignored.

Based on such an assumption, this article uses research results of estimated value of China's potential output and output gap for reference, and tries to introduce synthetically evaluation of fuzzy entropy in order to make proper improvement. Different from the evaluation of certain measurement method of existed documents, this article is based on the evaluation of applicability of a certain measurement method, trying to produce new time series from estimated values measured by different measurement methods at the same time point. Eventually, it is to determine the best time series as China's final estimated value of potential output by synthetically evaluation of fuzzy entropy.

2 Rethinking on the Measurement of Potential Output

In order to get a reasonable improvement, it is necessary for us to rethink the measurement of potential output. Take the definition by neoclassical theory as an example. Potential output is the trend value of actual output. It is the output without any impact of unexpected financial and monetary policy. Actual output would fluctuate within a very small range of potential output, and output gap could be plus or minus [16]. The concept of actual output has clear denotation, and it is an important indicator of national economy accounting. It could be obtained by observation and statistics. As the trend value, potential output is estimated by the following measurement methods: statistical detrending techniques, estimation of structural relationships, semi-structural method, and etc. Because the quantity of measurement method is limited, and sample data adopted are discrete time series, therefore, this article only covers the discussion of discrete variables, not including continuous variable here. Roughly, the logic of measurement is as below:

Provide an actual output series.

$$Y = \{y_t | t \in [0, T]\} \tag{1}$$

Through N_1 methods

$$\Phi = \{\varphi_n | n \in N_1, 1 \leq n \leq N_1\} \tag{2}$$

Namely, $y_t \longrightarrow \varphi_n x_{tn}$ to get the measurement result of potential output.

$$X=\{x_{tn} | t \in [0,T], n \in N_1 \text{且} 1 \leq n \leq N_1\} \tag{3}$$

Then $\forall t \in [0, T]$, which could get a subset of X.

$$\tilde{X}_t=\{x_{tn} | x_{tn} \in X, n \in N_1 \text{且} 1 \leq n \leq N_1\} \subseteq X \tag{4}$$

It is a set calculated by N_1 different methods, and composed by t different estimated values at the same time point.

If the actual value $\bar{x}_t$ of potential output could be measured, then it is very simple for the assessment of the best estimated value in $\tilde{X}_t$. Namely, it's $\hat{x}_t = arg \min_{x_{tm}} \|x_{tm} - \bar{x}_t\|$, which represents the closest value to $\bar{x}_t$. However, according to the definition of potential output, the observable actual output y_t is potential output value $\bar{x}_t$ plus random fluctuation ε_t, $y_t = \bar{x}_t + \varepsilon_t$ (y_t is a known number). We know that an equation couldn't determine the only two unknown variables. Before additional information is provided, $\bar{x}_t$ is to be determined. Trend value couldn't be specifically observed and measured. It also couldn't be concluded by strict analysis method. Therefore, it has also been proved that an important aspect of potential output is the study of measurement method.

Under the condition of potential output actual value $\bar{x}_t$ to be determined, the above evaluation methods could not be used. With the consideration of potential output actual value $\bar{x}_t$ as an objective value of macro-economy, features of time series $\bar{X} = \{\bar{x}_t | t \in [0, T]\}$ could mostly explain all kinds of economic phenomena. In another word, if a time series of estimated value could describe features of economic operation better, it would have a better degree of fitting on the time series of actual value. Therefore, under the condition, that the actual value couldn't be known exactly, estimated values of different measurement methods could also be compared and assessed by providing indirect feature indicator system. Camba-Mendez and Rodriguez-Palenzuela [17] have raised three standards for the assessment of output gap measurement as below: the first is the forecast ability on inflation; the second is the consistency with historical economic cycle turning point; the third is the stability of estimation, namely the consistency between the estimated value of previous period and output gap re-estimated by new data. It is only for the assessment of one certain measurement method, however, this article is for the assessment of estimated values by different measurement methods based on it.

Therefore, take any element x_{ti} from $\tilde{X}_t$ to form a time series subset of X.

$$X_i = \{x_{ti} \in \tilde{X}_t | t \in [0, T]\} \tag{5}$$

Meanwhile, take the power set $\mathbb{T} = \{X_i, 1 \le i \le N_1\}$ of $X_i s$ as universe of discourse. Because potential output actual value $\bar{x}_t$ could not be precisely measured, its time series could be defined as fuzzy subset $\mathbb{A} = \{\bar{X}\}$, therefore, its mapping

$$\mu_A : \mathbb{T} \to [0, 1], X_i \to \mu_A(X_i) \in [0, 1] \tag{6}$$

It has determined the membership function from $\mathbb{T}$ to $\mathbb{A}$, and represent the degree of fitting of estimated value upon actual value. Obviously, $\mu_A(\bar{X}) = 1$.

$\forall X_{i_1}, X_{i_2} \in \mathbb{T}$, If $\mu_A(X_{i_1}) \le \mu_A(X_{i_2})$, then it means X_{i_2} is closer to actual value. Therefore, the best estimated value is

$$\hat{X} = arg \max_{X_i} \mu_A(X_i) \tag{7}$$

It is the final estimated value time series of potential output. By now, we have got the analysis frame of fuzzy synthetically evaluation of improved method.

3 Improved Measurement Method

3.1 Basic Assumption

This article is to improve the measurement method of potential output. The general idea is to take a time series from estimated value by different measurement method. Among all time series, the taken one could mostly represent economic features of actual value series of potential output. In this time series, the values of different time points could be the result of the same measurement method or different methods. From above mentioned analysis frame, we know that in order to get the best series, the only thing needs to be determined is membership function value $\mu_A(X_i)$. The determining methods of membership function of fuzzy set mainly include: fuzzy statistic method, assignment method, to take from existed "objects" and "dimensions", dualistic contrast compositor method, and etc. [18]. In this article, we take existed "objective" dimensions, realize fuzzy overall evaluation of estimated value by different methods, and eventually get improved measurement series of potential output.

A given assessment system is composed by assessed time series $\mathbb{T}$, assessment indicator system $\mathbb{C}$, assessment result set $\mathbb{V}$. Standardized matrices of eigenvalues of $\mathbb{T}$ to $\mathbb{C}$ and $\mathbb{C}$ to $\mathbb{V}$ are as below:

$$\mathbf{C} = (c_{ik})_{N_1 \times N_2}, i = 1, 2, \cdots, N_1; k = 1, 2, \cdots, N_2 \tag{8}$$

$$\mathbf{V} = (v_{kj})_{N_2 \times N_3}, k = 1, 2, \cdots, N_2; j = 1, 2, \cdots, N_3 \tag{9}$$

The membership degree matrix of $\mathbb{T}$ to $\mathbb{V}$ is

$$\mathbf{U} = (\mu_{ik})_{N_1 \times N_3}, i = 1, 2, \cdots, N_1; k = 1, 2, \cdots, N_3 \tag{10}$$

Among which, $\boldsymbol{U}_i = (\mu_{i1}, \mu_{i2}, \cdots, \mu_{iN_3})$ is the membership degree vector of the i time series on different assessment results, including constraint condition:

$$\sum_{j=1}^{N_3} \mu_{ij} = 1, u_{ij} \geq 0, \forall i = 1, 2, \cdots, N_1 \tag{11}$$

The constraint condition reflects the surjection from universe of discourse to membership degree. In another word, without exception, there is always corresponding assessment result for any element of universe of discourse.

Obviously, there are numerous vectors and membership degree matrices, which could meet the constraint condition. The assessment process for different time series is the process to determine the best membership matrix U.

3.2 Determination of the Best Membership Degree Matrix and the Best Time Series

In order to obtain the best membership degree matrix, fuzzy entropy is introduced and defined as below:

Definition 1: The uncertainty of any time series X_i belonging to assessment result set $\mathbb{V}$ is given by the following fuzzy entropy E_i, and the difference is given by weighted general Euclidean distance D_i.

$$E_i = -\sum_{j=1}^{N_3} \mu_{ij} \ln \mu_{ij} = -\boldsymbol{U}_i \boldsymbol{lnU}_i \tag{12}$$

When $u_{ij} = 0$, $u_{ij} \ln u_{ij} = 0$.

$$D_i = \sum_{j=1}^{N_2} \mu_{ij} d_{ij} = \sum_{j=1}^{N_3} \mu_{ij} \sum_{k=1}^{N_2} \left[\left(w_k \| c_{ik} - v_{kj} \| \right)^2 \right]^{1/2} = \boldsymbol{U}_i \boldsymbol{d}_i \tag{13}$$

In which, $\boldsymbol{lnU}_i = \left(\ln \mu_{i1}, \ln \mu_{i2}, \cdots, \ln \mu_{iN_3} \right)^T$, $\boldsymbol{d}_i = (d_{i1}, d_{i2}, \cdots, d_{iN_3})^T$, and w_k is the weight of No. k indicator.

According to the maximum entropy principle, the best membership degree matrix U makes information entropy maximized. In another word, the uncertainty is minimized, and the difference is also minimized. Because:

$$\max_U \left(-\sum_{i=1}^{N_1} E_i \right) \Leftrightarrow \min_U \left(\sum_{i=1}^{N_1} E_i \right) \tag{14}$$

With linear weighting, build double-objective programming question P.

$$\begin{cases} \min\limits_U \quad P = \sum\limits_{i=1}^{N_1} P_i = \sum\limits_{i=1}^{N_1} E_i + \sum\limits_{i=1}^{N_1} D_i = \sum\limits_{i=1}^{N_1} [\theta \boldsymbol{U}_i \boldsymbol{lnU}_i + (1-\theta) \boldsymbol{U}_i \boldsymbol{d}_i] \\ \qquad = \sum\limits_{i=1}^{N_1} \sum\limits_{j=1}^{N_3} \left[\theta \mu_{ij} \ln \mu_{ij} + (1-\theta) \mu_{ij} d_{ij} \right] \\ s.t. \quad \sum\limits_{j=1}^{N_3} \mu_{ij} - 1 = 0 \end{cases} \tag{15}$$

In which, $0 \leq \theta \leq 1$ is weighting factor.

According to Kuhn-Tucker conditional construct Lagrangian function,

$$L(\mu_{ij}, \lambda_i) = \sum_{i=1}^{N_1} \sum_{j=1}^{N_3} \left[\theta \mu_{ij} \ln \mu_{ij} + (1-\theta) \mu_{ij} d_{ij} \right] + \sum_{i=1}^{N_1} \lambda_i \left(\sum_{j=1}^{N_3} \mu_{ij} - 1 \right) \tag{16}$$

According to necessary condition of extreme existence, there are:

$$\frac{\partial L}{\partial \mu_{ij}} = \theta\left(\ln \mu_{ij} + 1\right) + (1-\theta)d_{ij} + \lambda_i = 0 \tag{17}$$

$$\frac{\partial L}{\partial \lambda_i} = \sum_{j=1}^{N_3} \mu_{ij} - 1 = 0 \tag{18}$$

Based on Eq. (17), then:

$$\mu_{ij} = \exp\left\{-\left[(1-\theta)d_{ij} + (\theta + \lambda_i)\right]/\theta\right\} = \exp\left[-\left(1+\frac{\lambda_i}{\theta}\right)\right]\exp\left[\left(1-\frac{1}{\theta}\right)d_{ij}\right] \tag{19}$$

Simultaneous Eq. (18), then

$$\exp\left[-\left(1+\frac{\lambda_i}{\theta}\right)\right] = 1\Big/\sum_{j=1}^{N_3}\exp\left[\left(1-\frac{1}{\theta}\right)d_{ij}\right] \tag{20}$$

Substitute back Eq. (17), then:

$$\mu_{ij}^{*} = \exp\left[\left(1-\frac{1}{\theta}\right)d_{ij}\right]\Big/\sum_{j=1}^{N_3}\exp\left[\left(1-\frac{1}{\theta}\right)d_{ij}\right] \tag{21}$$

It means the stable point is $\boldsymbol{U}_i^{*} = \left(\mu_{i1}^{*}, \mu_{i2}^{*}, \cdots, \mu_{iN_3}^{*}\right)$. It will be proved to be the minimum of all as below:

According to Definition 1, the uncertainty and difference of every taken time series is given respectively, which means they are independent from each other. Therefore, $P_1, P_2, \cdots, P_{N_1}$ are linear independent.

From $\forall \boldsymbol{U}_i^1, \boldsymbol{U}_i^2 \in \mathbb{U}_i$ and $0 \leq \alpha \leq 1$, it's obvious that

$$\sum_{j=1}^{N_3}\left[\alpha\mu_{ij}^1 + (1-\alpha)\mu_{ij}^2\right] = \alpha\sum_{k=1}^{N_3}\mu_{ij}^1 + (1-\alpha)\sum_{k=1}^{N_3}\mu_{ij}^2 = 1 \tag{22}$$

Namely, $\alpha\boldsymbol{U}_i^1 + (1-\alpha)\boldsymbol{U}_i^2 \in \mathbb{U}_i$, and $\mathbb{U}_i$ is a convex set. The following is to prove $\forall \boldsymbol{U}_i \in \mathbb{U}\ominus_i$, $P_i(\boldsymbol{U}_i)$ is a convex function [19].

From Eq. (15), we know $P_i(\boldsymbol{U}_i)$ gradient vector is

$$\nabla P_i(\boldsymbol{U}_i) = \left[\theta(\ln u_{i1} + 1) + (1-\theta)d_{i1}, \cdots, \theta(\ln u_{iN} + 1) + (1-\theta)d_{iN_3}\right] \tag{23}$$

Therefore, the Hesse matrix H_i of $P_i(\boldsymbol{U}_i)$ is

$$H_i = \begin{pmatrix} \theta/\mu_{i1} & 0 & \cdots & 0 \\ 0 & \theta/\mu_{i2} & \cdots & 0 \\ \vdots & \vdots & \ddots & \vdots \\ 0 & 0 & \cdots & \theta/\mu_{iN_3} \end{pmatrix} \tag{24}$$

From $0 \leq \theta \leq 1$ and $\mu_{ij} \geq 0$, we know

$$\det H_i = \frac{\theta^N}{\prod_{j=1}^{N_3} \mu_{ij}} \geq 0 \tag{25}$$

Namely, H_i is semi-positive definite. $P_i(\boldsymbol{U_i})$ is convex function. Since convex function meets the linearity, we know $\sum_{i=1}^{L} P_i(U_i)$ is still convex function. Goal programming P is convex programming, therefore $\boldsymbol{U_i^*}$ is the minimum point [20] among all.

Meet the requirement of $\hat{\mu}_{ij}^* = \max\left\{\mu_{ij}^* \middle| \mu_{ij}^* \in \boldsymbol{U_i^*}, j = 1, 2, \cdots, N_3\right\}$, which is corresponding to the assessment result element V_{j_i}, which is also the assessment result of the No. i time series. The assessment result set, which is produced by all assessed time series, is

$$V = \left\{V_{j_1}, V_{j_2}, \cdots, V_{j_{N_1}}\right\} \tag{26}$$

Because μ_{ij}^* is a normalized dimension, $\hat{\mu}_{ij}^*$ of different time series could be compared directly. Let $\hat{j} = \min\{j_1, j_2, \cdots, j_{N_1}\}$, which means the element of minimal subscript value in V is the highest assessment result. The corresponding no. i time series is also the highest one of assessment result. It is also the final time series of estimated value of potential output.

3.3 Choice of Calculation and Specific Calculation Steps

From the previous assumption, we know the assessed time series $\mathbb{T}$ is the assembly of different time points of different measurement methods. Obviously, measurement methods and time series are limited, so their combinations are also limited. $\mathbb{T}$ is universe of discourse of limited elements with supremum and infimum. Therefore, membership function μ_A has supremum and infimum, and the best time series combination $\hat{X} = arg \max_{X_i} \mu_A(X_i) = arg \sup_{X_i} \mu_A(X_i)$ is existed. For specific X, time series, which could be picked out, are as many as N^T. If all of time series of combination are directly calculated for membership function, the arithmetic amount will be bigger with longer sample period. Then its feasibility would be very low, and even impossible to get a result. In order to solve this problem, this article adopts sub-optimal solution approximation method to determine a solution as the measurement result value of potential output. Detailed steps are as below:

First step: Determine assessment indicator system $\mathbb{C}$, assessment result set $\mathbb{V}$, and eigenvalue normalized matrices **C** and **V** of $\mathbb{T}$ to $\mathbb{C}$, and $\mathbb{C}$, to $\mathbb{V}$, indicator weight w_k and weighting factor θ.

Assessment indicator system $\mathbb{C}$ could use the three standards of Camba-Mendez and Rodriguez-Palenzuela [17], namely indicator system composed by predictive ability on inflation, consistency judgment of economic cycle turning point, and stability, namely

$\mathbb{C} = \{C_1, C_2, C_3\}$ = {Predictive ability on inflation, consistency judgment of economic cycle turning point, and stability}

The eigenvalue determination of $\mathbb{T}$ to $\mathbb{C}$, and $\mathbb{T}$ to $\mathbb{V}$ could use "Thinking of Comparative Methods" by Ma and Wei [21] as reference. The eigenvalue of assessed time series to the first indicator is represented by Theil inequality coefficient (TIC, which is between 0 and 1, the smaller the better in predictive ability). The eigenvalue of second indicator is represented by the ratio of amount (which is forecasted by time series and in consistency with peak-number of basic cycle period) and peak-number of basic cycle (PR, which is between 0 and 1, the bigger the better). The eigenvalue of third indicator is represented by Pearson coefficient of variation (PCV, which is more stable if closer to 1). Namely

C = {TIC value, PR value, PCV value}

Assessment result set $\mathbb{V}$ have optimal, sub-optimal, sub-inferior, and inferior, namely

$\mathbb{V} = \{V_1, V_2, V_3, V_4\}$ = {optimal, sub-optimal, sub-inferior, and inferior}

According to eigenvalue features of indicator, the eigenvalue matrix of $\mathbb{C}$ to $\mathbb{V}$ is as below

$$\mathbf{V} = \begin{pmatrix} 0 & 1/3 & 2/3 & 1 \\ 1 & 2/3 & 1/3 & 0 \\ 1 & 2/3 & 1/3 & 0 \end{pmatrix} \qquad (27)$$

The first line and third line of matrix represent that eigenvalue is between 0 and 1. The closer it is to 0, the better, otherwise, the worse. The second line represents eigenvalue is between 0 and 1, the closer it is to 0, the worse, otherwise, the better. If normalize each line of the matrix, then we could get eigenvalue normalized matrix **V** of $\mathbb{C}$ to $\mathbb{V}$.

Because the importance of indicator is no difference, indicator weight w_k is the same, namely $1/N_1$. Here, it's 1/3.

The acceptation or rejection of uncertainty and difference are also important. Take $\theta = 0.5$ as weighting factor.

Second step: Choose combination value of initial calculation time series, calculate optimal membership degree matrix, and determine the combination of sub-optimal time series.

The time series combination of initial calculation includes:

Time series measured by different methods, totally N_1. Calculate TIC values, PR values, and PCV values for these N_1 time series. Then put them into Eq. (21), then we'll get the optimal membership function value μ_{ij}^*. Keep the time series with the biggest assessment optimal value, and abandon others.

Third step: Search substitute values to form new time series, and calculate optimal membership degree matrix, and abandon inferior values.

Based on the time series determined in the second step, replace local time series through graph observing, analyzing the influence of time point value on assessment eigenvalue, and etc. Because PR values are closely related with cycle division, try the best to replace local time series within a single cycle. Repeat the calculation process of second step, if it is worse than original time series, then continue to replace local values based on original time series, otherwise, abandon original time series and keep searching and calculating with a new time series.

Fourth step: Round robin of the third step. The quantity of round robin should be determined by actual needs. Eventually, take the sub-optimal solution of optimal time series as measurement value of potential output.

4 China's Potential Output Measurement Based on Improved Method

4.1 Different Measurement Methods and Results

Here, three commonly used methods by researchers in China are adopted. They are HP filtering method, production function method (PF), and Structural Vector auto-regression (SVAR).

GDP quarterly data (1992Q1-2013Q3)[1] issued by China's statistics department have been adopted. Quarterly GDP time series X_t and sample size $T = 87$ based on 1992 constant prices has been obtained. Adjust X_t and get logarithm by X-12 ARIMA season adjustment method. Obtain series Y_t for analysis. On capital stock estimation, obtain it by using base period capital stock estimation by Fan [22], and converting into the constant prices of 1992. Integrating the calculation result of perpetual inventory method by Shan [23] and Operation Department Research Team of People's Bank of China (2011) [24]. On employment estimation, because there is no public quarterly statistic data, we could obtain yearly employment data of 1992–2013, then calculate quarterly weight according to 2004Q1-2012Q4 quarterly new entrants to the labor force[2] issued by Ministry of Human Resources and Social Security, and eventually calculate 2004Q1-2012Q4 quarterly data. Meanwhile, take season adjustment with X-12ARIMA. On price level, obtain it by adopting quarterly GDP deflator. The price level of each quarter is quarterly GDP calculated by constant prices of 1992 to compare with last year's.

[1] National Statistic Bureau of China, http://data.stats.gov.cn/workspace/index?m=hgjd.

[2] Data are published by Ministry of Human Resources and Social Security of China on quarterly press conference..

Adopt EVIEWS7.1 HP as filtering tool, take $\lambda = 1600$ to calculate the measurement result time series of potential output based on HP filtering method. When adopting production function method, it's found that labor input is not obvious. Actually, An and Dong [25] have pointed out that China's economic growth is capital-driven type under the condition of labor surplus, therefore, this article is using production function of capital-driven to get total factor productivity (TFP) and potential value with HP filtering method, and eventually get measurement result time series of potential output based on production function method. Adopting similar method as Guo and Chen [26], build SVAR model with two factors of actual production and price level, and use long term constraint condition to calculate potential output measurement result time series based on SVAR method. Corresponding output gap (which is represented by the difference of logarithms of actual output and potential output) is shown as Fig. 1.

4.2 Calculate Value of Membership Function and Determine Sub-optimal Solution

Taking the method by Ma and Wei [21] as reference, calculate values of TIC and PCV (with 1994Q1-2013Q3 as comparison samples) for the above mentioned three time series. Taking BN decomposition method by Wang and Hu [27] as reference, get 12 base cycles (shown as Fig. 2) and calculate PR, therefore, eigenvalue matrix **C** is as below.

$$\mathbf{C} = \begin{pmatrix} 0.344424 & 0.500000 & 0.859624 \\ 0.340161 & 0.583333 & 0.891963 \\ 0.297116 & 0.083333 & 0.999798 \end{pmatrix} \tag{28}$$

According to definition 1, put into **C** and **V**, and get the matrix of d_{ij}, which is

$$\mathbf{d} = \begin{pmatrix} 0.359784 & 0.147346 & 0.369052 & 0.687685 \\ 0.317259 & 0.136374 & 0.398453 & 0.721910 \\ 0.559495 & 0.362757 & 0.415178 & 0.659830 \end{pmatrix} \tag{29}$$

Then, by Eq. (21), we get membership degree matrix as

$$\mathbf{u} = \begin{pmatrix} 0.253299 & 0.313253 & 0.250963 & 0.182486 \\ 0.264026 & \underline{0.316377} & 0.243436 & 0.176161 \\ 0.233797 & 0.284631 & 0.270094 & 0.211478 \end{pmatrix} \tag{30}$$

From above matrices, we know membership degrees of three assessment methods belong to sub-optimal level. However time series produced by production function method belongs to relatively higher sub-optimal level. Therefore, time series produced by production function method is chosen as sub-optimal series s_1, and it is also taken as the initial value of iteration calculation of searching method.

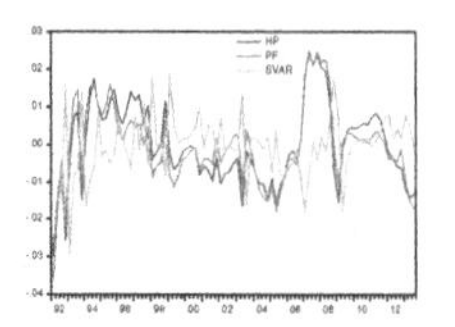

Fig. 1. Results

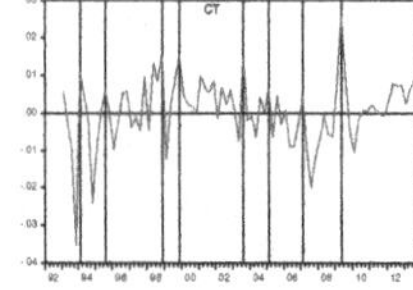

Fig. 2. Base cycles

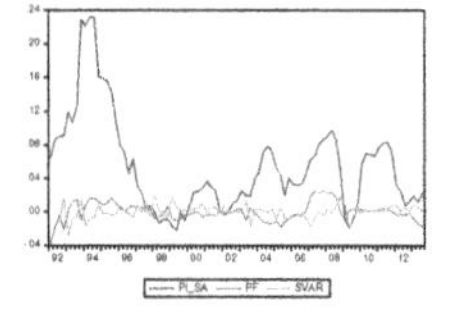

Fig. 3. GDP deflator

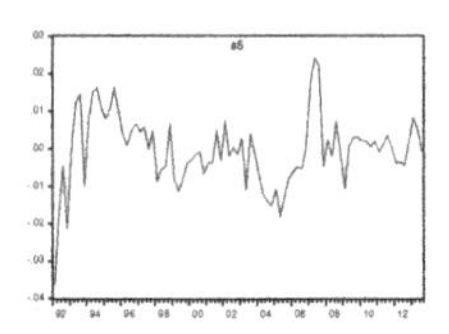

Fig. 4. Final results

4.3 Search Local Replacing Time Point Value, and Re-calculate Membership Degree Function to Compare with Sub-optimal Series

Firstly, search proper local substitute value. Comparing eigenvalues of three methods, we could find that SVAR has better ability and stability on inflation-predicting, therefore, we take local time-point value of SVAR estimated value as priority instead of HP estimated value. PF Estimated Value, SVAR Estimated Value, and GDP Deflator (With Season-Adjustment) Time Series are shown as Fig. 3.

Considering the estimated values of end-data by different methods have big differences, we use 2012Q4-2013Q3 time point data estimated by SVAR in the last base cycle to replace s_1, and get s_2. Correspondingly, we use SVAR estimated value calculated by 1994Q1-2013Q3 as comparison sample to replace corresponding for stable PCV value, and we get optimal membership degree value as below (Table 1):

From the result, we can see s_2 has a higher sub-optimal level, therefore, we abandon s_1 and use s_2 as the next iteration initial value.

Table 1. The first iteration membership degree function values

	Optimal	Sub-optimal	Sub-inferior	Inferior
s_1	0.264026	0.316377	0.243436	0.176161
s_2	0.249433	0.323920	0.248090	0.178556

4.4 Take Multiple-Iteration to Determine a Sub-optimal Time Series as Final Measurement Estimated Value

Observe Fig. 3 to get the next substitute value, and choose SVAR time-point, which is similar to the trend of GDP deflator, to substitute step by step and have iteration calculation (Fig. 1). Here, successively substitute 2010Q4-2011Q3, 2001Q3-2001Q4, 2002Q1-2002Q4, 2005Q4-2006Q2, 2007Q4-2008Q3, and successively produce three time series s_3, s_4 and s_5, and the optimal membership degree is as below (Table 2):

From above chart, we can see three iterations have all promoted sub-optimal membership value, and they are all better than the last time series. In this article, we choose s_5 as final output gap value calculated with the improved method.

Table 2. The multiple iteration membership degree function values

	Optimal	Sub-optimal	Sub-inferior	Inferior
s_3	0.249084	0.324009	0.248249	0.178658
s_4	0.253544	0.324568	0.245365	0.176523
s_5	0.241645	0.327881	0.250602	0.179872

5 Conclusion

During the process to measure China's potential output and output gap with improved methods, we found that the membership degree values are all sub-optimal for each measurement method, however, there are big differences for eigenvalues of three assessment indicators. From the membership result of three methods, we can see none of them could reach the optimal level, which also means each method could only partly reflect output gap information. PF method is better in describing economic cycle fluctuation, and the agreement degree with base cycle is more than half. This is because of its strong theoretical support. PF method takes capital and labor as endogenous variables, and is able to describe the impact caused by capital and labor input. SVAR method is better in the ability and stability of inflation forecast, therefore we take local time-point of SVAR to substitute as a priority. Although the introduced local value couldn't improve the level of membership degree to optimal level, the sub-optimal membership degree keeps getting stronger, which means through substitute and iteration, there are more information about output gap adding into new time series, which makes its uncertainty and difference smaller, therefore, it is better than original series.

References

1. Tian, Q.S.: Analyzing issues of macroeconomic situation. J. Soc. Sci. Guangdong **2**, 33–45 (2013)
2. Okun, A.M.: Potential GNP: its measurement and significance. In: Proceedings of the Business and Economics Statistics Section of the American Statistical Association (1962)
3. Cerra, V., Saxena, S.C.: Alternative Methods of Estimating Potential Output and the Output Gap-An Application to Sweden. International Monetary Fund, Washington (2000)
4. Liu, B., Zhang, H.Q.: The estimation of output gap in China. J. Fin. Res. **10**, 69–77 (2001)
5. Guo, Q.W., Jia, J.X.: China economic fluctuation: the impact of investment and total factor productivity shock. J. Manage. World. **7**, 22–28 (2004)
6. Zhang, H.W.: Output gap in China and to estimate the potential economic growth rate. J. Econ. Perspect. **8**, 44–49 (2005)
7. Shi, Z.X., Huang, H.M., Shi, Q.H.: Empirical studies of China potential GDP, fluctuation and inflation. J. World Econ. **8**, 34–41 (2004)
8. Zhao, Q.D.: Estimation of China output gap based on the Phillips curve. J. World Econ. **1**, 57–64 (2008)
9. Guo, Q.W., Jia, J.X.: Estimating potential output and the output gap in China. J. Econ. Res. J. **5**, 31–39 (2004)

10. Xu, Z.Y.: Estimating on China potential output and output gaps—by the method of Kalman filter. J. Quant. Tech. Econ. **12**, 3–15 (2005)
11. Zhang, C.S.: Estimating output gap: a multivariate dynamic model approach. J. Stat. Res. **7**, 27–33 (2009)
12. Zhao, Q.D., Gen, P.: Estimation method of state apace model using Bayesian Gibbs sampler and its application on estimating China's potential growth. J. Stat. Res. **9**, 55–63 (2009)
13. Yang, T.Y., Huang, S.F.: Estimating China's output gap based on wavelet denoising and quarterly data. J. Econ. Res. J. **01**, 115–126 (2010)
14. Zhen, T.G., Wang, X., Shu, N.: Real-time inflation forecasting and its applicability of Phillips curve to China. J. Econ. Res. J. **03**, 88–101 (2012)
15. Huang, T., Yang, J.L., Jiang, M.Q.: The effect of inflation, the output gap and Chinese inflation outlook in 12th Five-Year Plan period. J. Macroecon. 04, 3–9 + 38 (2013)
16. Yan, S.B., Zhang, L.C.: The definition and measurement method of potential output and output gap. J. Cap. Univ. Econ. Bus. **01**, 42–48 (2007)
17. Camba-Mendez, G., Rodriguez-Palenzuela, D.: Assessment criteria for output gap estimates. J. Econ. Model. **20**(3), 529–562 (2003)
18. Xie, J.J., Liu, C.P.: The method of fuzzy mathematics and its applications, 3rd edn, pp. 29–37. Huazhong University of Science and Technology Press, Wuhan (2006)
19. Gong, L.T.: Optimization in Economics, pp. 65–67. Peking University Press, Beijing (2000)
20. Qian, S.D., et al.: Operations Research, pp. 136–178. Tsinghua University Press, Beijing (2000)
21. Ma, W.T., Wei, F.C.: Quarterly output gap measure based on the new Keynes dynamic stochastic general equilibrium model. J. Manage. World. **8**, 39–65 (2011)
22. Fan, Q.: Details of perpetual inventory method and capital stock estimation of China from 1952 to 2009. J. Yunnan Univ. Fin. Econ. **3**, 42–50 (2012)
23. Shan, H.J.: Reestimating the capital stock of China: 1952–2006. J. Quant. Tech. Econ. **10**, 17–31 (2008)
24. Yang, G.Z., Li, H.J.: Potential output estimated with the production function method, the relationship between output gap and inflation: 1978–2009. J. Fin. Res. **3**, 42–50 (2011)
25. An, L.R., Dong, L.D.: Research on the potential economic growth and natural employment based on capital-driven theory. J. Quant. Tech. Econ. **02**, 99–112 (2011)
26. Guo, H.B., Chen, P.: The estimation and evaluation of China's output gap based on the SVAR mode. J. Quant. Tech. Econ. **05**, 116–128 (2010)
27. Wang, S.P., Hu, J.: Trend-cycle decomposition and stochastic impact effect of Chinese GDP. J. Econ. Res. J. **04**, 65–76 (2009)

Study on Spatial Data Acquisition and Processing Based on Vehicle-Mounted Mobile Measurement System

Yanmin Wang, Guannan Wei, Ming Guo(✉), Deng Pan, and Guoli Wang

Key Laboratory for Modern Urban Surveying and Mapping of National Administration of Surveying, Mapping and Geoinformation, Engineering Research Center of Representative Building and Architectural Heritage Database, Beijing University of Civil Engineering and Architecture, No.1 Zhanlanguan Road, Beijing, China
374453408@qq.com, guoming@bucea.edu.cn

Abstract. In order to achieve a collection of the three-dimensional scenes information of main streets automatically and quickly, vehicle-mounted mobile measurement system, which is consisted of sensors, vehicular laser scanner, global position system (GPS) receiver, inertial navigation system (INS) and high-precision computer, is a important means of record. Vehicular laser scanner rapidly scan street on both sides of the structure of the two-dimensional elevation information with the GPS and INS. In this way we can get high precision of urban architecture composed of spatial three-dimensional coordinates. Real time acquisition and processing of information can be realized.

Keywords: Vehicular laser scanner · Coordinate transformation · Space registration · Time registration · Image display

1 Introduction

As a high-tech field concept, the construction of "Digital City" has gradually become the focus of hot spots. "Digital City" mainly relies on 3S technology, positioning technology, remote sensing technology, virtual reality technology, integrated technology and geographic information technology [1]. It is widely used in urban planning and design, landscape architecture simulation, design of communication base station and net, and other fields. Acquisition of spatial data with high speed is a key technology of digital city engineering.

Nowadays the most common used methods for "Digital City" data collection are the ground-to-ground collection and air-to-ground collection [2]. The air-to-ground observation methods are mainly concentrated in high resolution, high spectrum of airborne and space borne and radar, it can collect the ground data in a large range and high efficient way, but data cannot be effectively collected for a region with trees, building shade and building facade, as shown in Fig. 1. In view of this situation, the ground observation can make up for the lack of the air-to-ground observation, collection the data of building facade and easy to be obstacles information timely, complement each other to form an effective air ground data collection, to build the three-dimensional space model of "digital city".

F. Bian and Y. Xie (Eds.): GRMSE 2015, CCIS 569, pp. 447–455, 2016.
DOI: 10.1007/978-3-662-49155-3_45

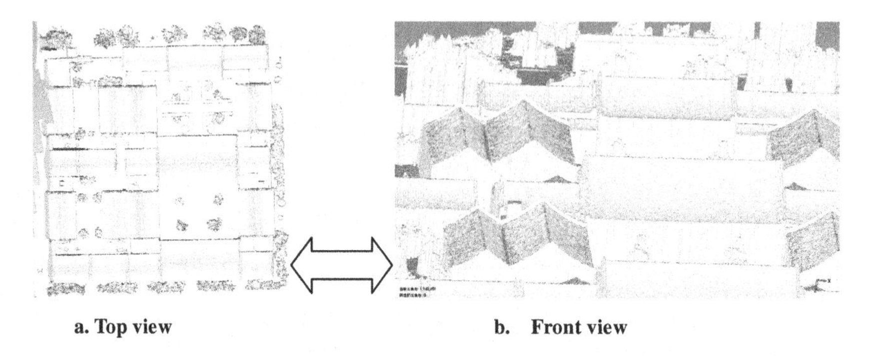

a. Top view **b. Front view**

Fig. 1. Air-to-ground surveying images

The ground data acquisition methods is the use of 3D laser scanner, collecting point cloud data of building models in a short period of time. In practical engineering, commonly used types are fixed 3D laser scanner and vehicular 3D laser scanner. Fixed 3D laser scanner is generally used in a small area of the data acquisition, high precision. But it is for the large-scale urban spatial data mining is low efficiency. With GPS technology, inertial navigation technology and CCD imaging technology, the features of vehicular 3D laser scanning system can play its portability, real-time positioning, convenient, and efficient urban space-data acquisition, rapid access to building space information and texture information.

In this paper, it depends on the working principle of 3D laser scanner and characteristics of the vehicular laser scanning system. This thesis researched the vehicular laser scanning system software and hardware structure, data collection and processing method. This article is based on the three-dimensional spatial data acquisition method. The principle is clear, and the vehicle-mounted mobile measurement system application example is given. It is believed that these results have important significations and practical values for this kind of engineering designs and applications.

2 The Composition of Vehicle-Borne Laser Scanning

In this research, as the carrier of building three-dimensional spatial information, the on-board motion platform is used to collect the urban information. The vehicle-borne laser scanning system consists of three main parts: The vehicular laser scanner is one of the most important parts in the digital image acquisition system [3]. It can provide position, distance, direction and polar coordinates information of the target points; GPS provide real-time location information for laser scanner; Inertial navigation system (INS) provide position, posture and other navigation information. They are important equipments in improving the accuracy of the vehicle-borne navigation. The latter two also often called position and orientation system.

2.1 Basic Principle of the System

The main tasks of this system are data collection and data processing part. During the vehicle running, the system runs stabilization and acquires data fully and accurately.

Under the control of each sensor in the vehicle-borne computer system features, such as synchronous collection construction in the vehicle running on one side of the 3D spatial data and attribute information, and stored in the onboard computer. It is convenient to the follow-up data processing on the integration process.

The entire vehicle scanning system is controlled by the vehicle-borne computer, through the GPS second pulse signal to synchronize. First of all, we need to install the scanner, GPS and inertial navigation on the vehicle. Through the strict calibration, it can get the vehicle relatively fixed geometry relationship between each sensor. After the installation is complete, we use precision grade sub-millimeter joint arm scanner to measure the relative position between each sensors, calculate its relative position parameters for the subsequent data processing to provide relatively accurate calibration of the original information. For the quality of the data in testing, the vehicle should move in a constant low speed. We regard the GPS time as time benchmark of the entire vehicle scanning system, through the on-board computer to unified control of each sensor. In this way, we could collect synchronous experimental region of space feature information [5]. The acquisition of POS positioning orientation system is the original position of on-board scanning system to measure information. When the GPS receiver satellite is failure, it can be collected by IMU inertial navigation system information, the time to make up for the failure data. When the POS data collection is ready, it calculates the GNSS/INS raw data collected in the survey by Inertial Explorer software. Through the POS system data accuracy calculation, we can get on-board scanning system information in the process of driving such as space position, speed and the attitude information.

2.2 Related Parameters of System Equipment

See Table 1.

Table 1. Related parameters of system equipment

The sensor	Brand models	The main parameters
Vehicular laser scanner	Riegl VQ-180	The scanner is a 2D laser scanner, it has a fast speed, high resolution, high accuracy and the advantages of multiple ports. Angle Measurement Resolution: 0.001°; Accuracy: 15 mm; Field of View: +60°/−40° = 100°; Max Effective Measurement Rate: 83000 measurements/s
GPS	Novatel ProPak-V3	Novatel ProPak - V3 GPS receiver, post-processing dynamic accuracy of 5 mm +/−1 PPM, original position output frequency and output frequency of 20 Hz, signal capture time of 0.5 s (L1), 1 s (L2), time accuracy of plus or minus 20 ns
Inertial navigation	SPAN-CPT	SPAN-CPT calculating accuracy under different patterns can be applied to different position demand, support including SBAS, l-band (Omnistar and CDGPS) and RTK differential a variety of ways

3 The Principle of Data Processing and System

In the vehicle-borne laser scanning system, the digital image collection system, GPS systems and inertial navigation systems are the spatial information described of each sensor in the three-dimensional feature space. Because of the differences in vehicle installation position and features of laser scanner, GPS and inertial navigation in the instrument itself, which makes the information that collected in the coordinate system differently [6]. It is often necessary to locate and merge the different view measured cloud data in a unified vehicle coordinate system. In order to get the cloudy data matching with a common principle, the transition between coordinate systems is necessary before data fusion. We call the process spatial registration.

3.1 Spatial Registration

In order to detail the space registration algorithms, this paper introduced five reference coordinate systems: scanner original coordinate, scanner space coordinate, vehicle coordinate, East-North-Up coordinate and Earth-Centered Earth-Fixed(ECEF) coordinate.

The Definition of the Reference Coordinate

(a) Scanner original coordinate
Scanner original coordinate system is a polar coordinate which is decided by the principle of laser scanner. The rationale of the measurement is to use range ρ and angle observation θ to calculate the 3D coordinate of space point.
(b) Scanner space coordinate
The vehicle scanner launcher is defined scanner space coordinate origin. The X-axis shows the direction perpendicular to vehicle traveling on the right side, while the Y-axis represents direction of the vehicle, and the Z-axis is the direction of zenith distance. We establish a right-handed coordinates called O-XYZ.
(c) Vehicle coordinate
The inertial navigation center is defined scanner space coordinate origin. The X-axis shows the direction perpendicular to vehicle traveling on the right side, while the Y-axis represents direction of the vehicle, and the Z-axis is the direction of zenith distance. We establish a right-handed coordinates called $O\text{-}X_cY_cZ_c$.
(d) East-North-Up coordinate
In this article the local level is based on carrier position as a coordinate system origin o, by measuring point in the normal direction of the ellipsoid as the z axis, by measuring point in the earth on the northern end of the meridian and the ground plane of the earth (north) as the y axis, intersection curve on the earth parallel to the intersection of a plane with the earth as east of the x axis, constitute a right hand $O\text{-}X_LY_LZ_L$ space rectangular coordinate system.
(e) Earth-Centered Earth-Fixed (ECEF) coordinate
Dim the center of the ellipsoid as ECEF Cartesian coordinate system origin, with initial meridian plane NGS and equatorial plane WAE of intersecting line for X axis and short axis to ellipsoid for the Z axis (north is positive), in the equatorial

plane and the direction of X axis orthogonal to Y axis, constitute a right hand $O\text{-}X_{ECEF}Y_{ECEF}Z_{ECEF}$ space rectangular coordinate system.

The Conversion Relation of Coordinate Systems

(a) The translation from Scanner original coordinate to the Scanner space coordinate
Scanners original coordinate system of polar coordinates (ρ, θ) can be measured by the instrument. The scanner space coordinates (X, Y, Z) can be obtained according to the following formula:

$$\begin{cases} X = \rho\cos\theta \\ Y = 0 \\ Z = \rho\sin\theta \end{cases} \tag{1}$$

(b) The translation from Scanner space coordinate to the Vehicle coordinate
Scanner space coordinate system O-XYZ transformation to Vehicle coordinate $O_c\text{-}X_cY_cZ_c$ requires transformation as follows:

$$\begin{bmatrix} X_C \\ Y_C \\ Z_C \end{bmatrix} = \begin{bmatrix} \Delta X \\ \Delta Y \\ \Delta Z \end{bmatrix} + R_c \begin{bmatrix} X \\ Y \\ Z \end{bmatrix} = \begin{bmatrix} \Delta X \\ \Delta Y \\ \Delta Z \end{bmatrix} + \begin{bmatrix} a_1 & b_1 & c_1 \\ a_2 & b_2 & c_2 \\ a_3 & b_3 & c_3 \end{bmatrix} \begin{bmatrix} X \\ Y \\ Z \end{bmatrix} \tag{2}$$

R_c is a rotation matrix, its parameters calculation according to the following formula:

$$\begin{cases} a_1 = \cos\varphi\cos\kappa - \sin\varphi\sin\omega\sin\kappa & b_1 = \cos\omega\sin\kappa & c_1 = \sin\varphi\cos\kappa + \cos\varphi\sin\omega\sin\kappa \\ a_2 = -\cos\varphi\sin\kappa - \sin\varphi\sin\omega\cos\kappa & b_2 = \cos\omega\cos\kappa & c_2 = -\sin\varphi\sin\kappa + \cos\varphi\sin\omega\cos\kappa. \\ a_3 = -\sin\varphi\cos\omega & b_3 = -\sin\omega & c_3 = \cos\varphi\cos\omega \end{cases} \tag{3}$$

In the formula, ΔX, ΔY, ΔZ, φ, ω, κ are the laser scanner coordinate translation and rotation parameters which relative to the vehicle coordinate system. They are measured by joint arm scanner in this paper. They can also be measured by other field measurement for calibration.

(c) The translation from Vehicle coordinate to the East-North-Up coordinate
The transformation between Vehicle coordinate to East-North-Up coordinate is based on three attitude angles matrix, the head angle, the pitch angle and the roll angle. If the coordinate of the point at East-North-Up coordinate is (X_L, Y_L, Z_L), it has the following transformation relationship:

$$\begin{bmatrix} X_L \\ Y_L \\ Z_L \end{bmatrix} = \boldsymbol{R_L} \begin{bmatrix} X_C \\ Y_C \\ Z_C \end{bmatrix} \tag{4}$$

R_L is a rotation matrix consisting of three attitude angles which are collected by inertial navigation.

$$\boldsymbol{R_L} = \boldsymbol{R_R} * \boldsymbol{R_P} * \boldsymbol{R_H} \tag{5}$$

$$R_R = \begin{bmatrix} \cos\alpha & 0 & -\sin\alpha \\ 0 & 1 & 0 \\ \sin\alpha & 0 & \cos\alpha \end{bmatrix}$$
$$R_P = \begin{bmatrix} 1 & 0 & 0 \\ 0 & \cos\beta & \sin\beta \\ 0 & -\sin\beta & \cos\beta \end{bmatrix} \tag{6}$$
$$R_H = \begin{bmatrix} \cos\gamma & -\sin\gamma & 0 \\ \sin\gamma & \cos\gamma & 0 \\ 0 & 0 & 1 \end{bmatrix}$$

$\boldsymbol{\alpha}$, $\boldsymbol{\beta}$, γ are present the head angle, the pitch angle and the roll angle collected by inertial navigation. Roll angle is around Y axis rotation. The right is positive displacement. Pitching angle is around the X axis rotation. Upward is positive displacement. Heading angle is around the X axis rotation. Clockwise is positive displacement.

(d) The translation from East-North-Up coordinate to the ECEF coordinate

$\mathbf{R_{ECEF}^{LL}}$ is the rotation matrix between East-North-Up coordinate and ECEF coordinate. B represents the latitude, L represents the longitude.

$$\mathrm{R_{ECEF}^{LL}} = \mathrm{R_{ECEF}^{LL}X} \cdot \mathrm{R_{ECEF}^{LL}Z} = \begin{bmatrix} -\sin(L) & \cos(L) & 0 \\ -\sin(B)\cos(L) & -\sin(B)\sin(L) & \cos(B) \\ \cos(B)\cos(L) & \cos(B)\sin(L) & \sin(B) \end{bmatrix} \tag{7}$$

The translation from East-North-Up coordinate to the ECEF coordinate is as follows:

$$\begin{bmatrix} X_{ECEF} \\ Y_{ECEF} \\ Z_{ECEF} \end{bmatrix} = \left(\boldsymbol{R_{ECEF}^{LL}}\right)^{-1} \begin{bmatrix} X_L \\ Y_L \\ Z_L \end{bmatrix} + \begin{bmatrix} \Delta X_{ECEF} \\ \Delta Y_{ECEF} \\ \Delta Z_{ECEF} \end{bmatrix} \tag{8}$$

3.2 Time Registration Algorithm

In this paper,the data of this study is collected when the laser scanner rate are 50 kHz, 100 kHz, 150 kHz, 200 kHz, and the data acquisition of the POS system frequency is 10 Hz. Firstly the GPS data must go through the settlement processing, using the observation data between the base station and the main antenna for single epoch post-processing differential settlement of the main antenna's center coordinates, and then based on the laser scanner and POS system data acquisition frequency recording the time and automatic interpolating 6 outer orientation elements. Finally, the data is applied to space registration and time registration. As interpolating part of the data points, we can use linear interpolation in linear trajectory. For broken line and other locus, we can use B-Spline curves or polynomial interpolation to process data.

4 Experiment Results

Based on the research method introduced in this paper, the software is designed by C# language in Windows 7 system. We select about 200 points randomly to make measurement error analysis. The position error is about 0.012 m. Figure 2 shows the results after synchronous processing.

Fig. 2. Vehicle-borne laser scanning (Vehicle speed is 10 km/h, the scanning frequency is 200 kHz)

5 Conclusion

On the basis of the existing theoretical research, this study design completed a set of three-dimensional spatial information acquisition for urban vehicle-mounted mobile laser scanning system. The system adopts the Riegl VQ - 180 laser scanner, Novatel GPS and SCAN-CPT inertial navigation system combination, guarantee the precision of scanning.

Acknowledgements. The project is supported by the National Natural Science Foundation of China (Grant No. 41301429); Beijing Municipal Natural Science Foundation (4144071); Scientific Plan Project of National administration of Surveying, Mapping and Geoinformation (2013CH-15); Open Research Foundation of Key Laboratory of Precise Engineering and Industry Surveying National Administration of Surveying, Mapping and Geoinformation (PF2012-1); Key Laboratory for Urban Geomatics of National Administration of Surveying, Mapping and Geoinformation (No. 20111223 N); Doctoral Research Fund of Beijing University of Civil Engineering and Architecture (101201605, 33/613004);

References

1. Sheng, Y., Zhang, K., Ye, C.: Spatial data acquisition and processing based on vehicle-borne mobile mapping system, pp. 1–12 (2008)
2. Youmei, H.: Study on the calibration of laser scanner and line scan camera based on the vehicle-borne mobile surveying system. Shandong University of Science and Technology (2011)
3. Lu, X., Li, Q., Feng, W.: Vehicle-borne urban information acquisition and 3D modeling system. Eng. J. Wuhan Univ. **36**(3), 76–80 (2003). doi:10.3969/j.issn.1671-8844.2003.03.017
4. Shi, B., Lu, X., Wang, D.: Space and time registration of vehicle-borne 3D measurement system based on muti-sensor fusion. Transducer Microsyst. Technol. **26**(9), 14–16, 19 (2007). doi:10.3969/j.issn.1000-9787.2007.09.005
5. Shen, Y., Li, L., Ruan, Y.: Mobile mapping technology by vehicle-borne lidar. Infrared Laser Eng. **38**(3), 437–440, 451 (2009). doi:10.3969/j.issn.1007-2276.2009.03.014
6. Kang, Y., Zhong, R., Wu, Y.: Research of calibrating vehicle laser scanner's external parameters. Infrared and Laser Engineering, pp. 453–457 (2008)
7. Wei, B., Zhang, A., Li, Y., Wang, Y.: Design and implementation of a vehicle-borne system of 3D data acquisition and processing. Chin. J. Stereology Image Anal. **13**(1), 30–33 (2008). doi:10.3969/j.issn.1007-1482.2008.01.007
8. Guo, B., Qu, X., Huang, X., Zhang, F., Li, Q.: Calibration and accuracy analysis of vehicle-borne mobile 3D measurement system. Laser Infrared **41**(11), 1205–1210 (2011). doi:10.3969/j.issn.1001-5078.2011.11.007
9. Diange, Y., Jian, B., Sifa, Z.: Research and realization of vehicle navigation system. Automot Technol. **1**, 1–4 (2005). doi:10.3969/j.issn.1000-3703.2005.01.001
10. Dayong, Z.: Research on System Consistency and Optimal States Estimation of LIDAR/INS Integrated Navigation System. Nation University of Defense Technology (2010). doi:10.7666/d.d138398

11. Stamos, I., Alien, PE.: 3-D model construction using range and image data. In: Proceedings of IEEE Conferenee on Computer Vision and Pattern Recognition, vol. 1, pp. 531–536 (2000)
12. Besl, P.J., Jain, R.C.: Segmentation through variable—order surface fitting. IEEE Trans. Pattern Anal. Mach. Intell. **10**(2), 167–192 (1988)
13. Fruh, C., Zakhor, A.: An automated method for largescale, ground—based city model acquisition. Int. J. Comput. Vision **60**(1), 5–24 (2004)
14. Joerger, M., Pervan, B.: Automous ground vehicle navigation using integrated GPS and laser-scanner measurement. IEEE/ION PLANS (2006)
15. Fekete, S.P., Klein, R., Nüchter, A.: Online searching with an autonomous robot. Comput. Geom. **34**, 14 (2006)
16. Guivant, J., Nebot, E., Baiker, S.: Autonomous navigation and map building using laser range sensors in outdoor applications. J. Rob. Syst. **17**(10), 565–583 (2000)

Simulation and Analysis of Backscattering Characteristics of Soil with Row Structure

Nan Yin[1(✉)], Jingwei Wang[2], and Shaobin Zhan[3]

[1] School of Surveying and Prospecting Engineering, Jilin Jianzhu University, No. 5088, Xincheng Street, Changchun, China
yinnanbaby@126.com

[2] China Power Engineering Consulting Group, Northeast Electric Power Design Institute, Changchun, China

[3] Shenzhen Institute of Information Technology, Shenzhen, China

Abstract. In order to retrieve soil moisture through active microwave methods, the quantitative relation between the backscattering coefficients and soil moisture should be established. The dependence of backscattering coefficients on soil parameters, radar parameters and row structure parameters was analyzed by using Ulaby model. It was found that M values (the sensitivity of backscattering coefficient to azimuth angle) in different polarization modes showed as HH > VV > VH. The M value is influenced by the variation of soil moisture and surface roughness in HH polarization, but that won't happen in VV and VH polarizations, so VV and VH polarizations are more suitable for retrieving soil moisture with row structure. The responding function is different for different incident angles, if the incident angles changes significantly in a SAR image, the effect must be considered in building inversion model of soil moisture.

Keywords: Backscattering coefficients · Row structure · Soil moisture · M value · Incident angle

1 Introduction

Most of the dry farmland (soybean, wheat and corn) in the north of China use ridge tillage method which is used frequently in many places around the world. A number of investigations have shown that a strong relationship exists between backscattering coefficients (or emissivity) and row structure parameters. Therefore the effect of periodic structure can't be ignored when using active or passive microwave method to retrieve soil moisture [1–5]. The backscattering behavior of plat random surface is mainly influenced by soil parameters and SAR parameters. But for the soil with row structure, the height of ridge, the space period of ridge and the azimuth angle can also effect the backscattering coefficients [6, 7].

Supported by Training plan of Guangdong Province outstanding young teachers in Higher Education Institutions (Grand No. YQ2013194) and Shenzhen strategic emerging industry development funds (JCYJ20140418100633634).

F. Bian and Y. Xie (Eds.): GRMSE 2015, CCIS 569, pp. 456–463, 2016.
DOI: 10.1007/978-3-662-49155-3_46

Row structure and direction effects have been verified and modeled [1, 5]. Daniel S used time frequency approach and a new random periodic surface scattering model, and used a polarimetric parameter that is quasi insensible to this phenomenon to perform the inversion [8]. The effect of random roughness associated with soil clods is never less than 2 dB, and the effect of periodic row structure can be as strong as 10 dB [9]. Yin N established an error correction model based on OH model, and removed the influence of a row pattern [10].

According to the above results, the empirical models depending on certain parameters are limited. The goal of this paper is to analyze the quantitative relation between the backscattering coefficients and the row structure parameters. The findings may have important implications for improving the soil moisture inversion accuracy.

2 Methodology

2.1 Ulaby Model

The soil surface includes random surface height superimposed on a much larger one-dimensional periodic surface height [1].

$$C(x, y) = s(x, y) + z(x, y) \tag{1}$$

where C(x, y) is the soil surface height variation; s(x, y) is random surface height; z(x, y) is periodic surface height. The periodic height variation only occurs in y-direction as shown in Fig. 1, the formula can be denoted by

$$z(x, y) = z(y) = z(y + nT) \tag{2}$$

where n is an integer; T is the row spacing of the periodic pattern in the y-direction.

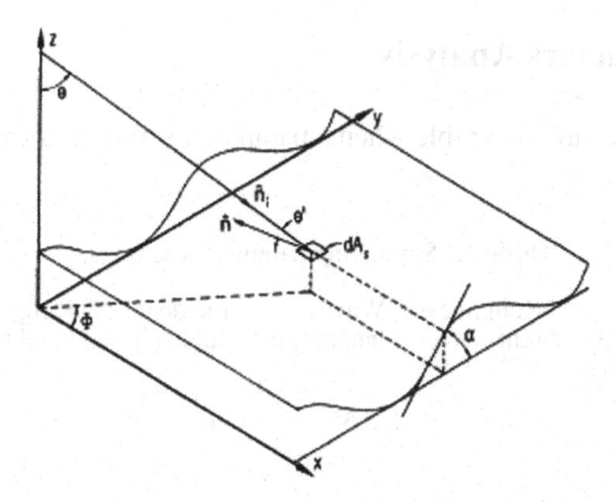

Fig. 1. Structure schematic of the periodic surface

Ulaby derived the scattering model applied to periodic surface, the derivation process won't be covered here.

$$\sigma_{ij}^{z}(\theta,\phi)=\frac{1}{T}\int_{y=0}^{T}\sigma_{ij}^{s}\left(\theta'\right)\sec\alpha\left(y\right)dy \tag{3}$$

where $\sigma_{ij}^{z}(\theta,\varphi)$ represents the scattering coefficient of periodic surface; $\sigma_{ij}^{s}\left(\theta'\right)$ is the scattering coefficient of the differential area; θ' is the local angle of incidence; φ is azimuth angle relevant to the look direction of the row; α is the angle whose tangent is equal to the slope of the surface z(y), z(y) is simulated by sine function according to previous research results; ij stands for the receive-transmit polarization of the antenna.

In this paper, $\sigma_{ij}^{s}\left(\theta'\right)$ is simulated by OH model [11] which is commonly used as a scattering model of random surface.

2.2 Look-Direction Modulation Ratio

The ratio of the backscattering coefficient observed in the direction perpendicular to row direction ($\phi = 90°$) to that observed in the direction parallel to row direction ($\phi = 0°$) is called the look-direction modulation ratio M [1].

$$M_{ij}(\theta)=\frac{\sigma_{ij}^{z}\left(\theta,90^{\circ}\right)}{\sigma_{ij}^{z}\left(\theta,0^{\circ}\right)} \tag{4}$$

M mainly measures the sensitivity of backscattering coefficient to azimuth angles, it can also be expressed in dB.

$$M_{ij}(\theta)=\sigma_{ij}^{z}\left(\theta,90^{\circ}\right)-\sigma_{ij}^{z}\left(\theta,0^{\circ}\right) \tag{5}$$

3 Influence Factors Analysis

The other parameters are invariable when a parameter serves as a variable as shown in Table 1.

Table 1. Simulating parameters settings.

Parameter	Soil moisture (cm^3/cm^3)	Roughness (cm)	Wave length (cm)	Incident angle (°)	Ridge height (cm)	Ridge space (cm)
Value	0.20	1.00	5.55	45	9	64

3.1 Row Structure Response

Ridge height (2A, A represents the amplitude of sine function which simulates the ridge structure) and ridge space (T) determine the geometrical structure of periodic surface. First, T is considered as a constant (T = 64 cm), the bigger ridge height, the stronger sensitivity of backscattering coefficient to azimuth angle is (as shown in Fig. 2(a)). Then A is considered as a constant (2A = 9 cm), the smaller ridge space, the stronger sensitivity of backscattering coefficient to azimuth angle is (as shown in Fig. 2(b)). By comparing the same color curves in Fig. 2(a, b), it's not difficult to find that the same A/T values correspond the same backscattering characteristics. The response rules of VV and VH polarization are same with the rule of HH polarization as shown in Fig. 2(c, d).

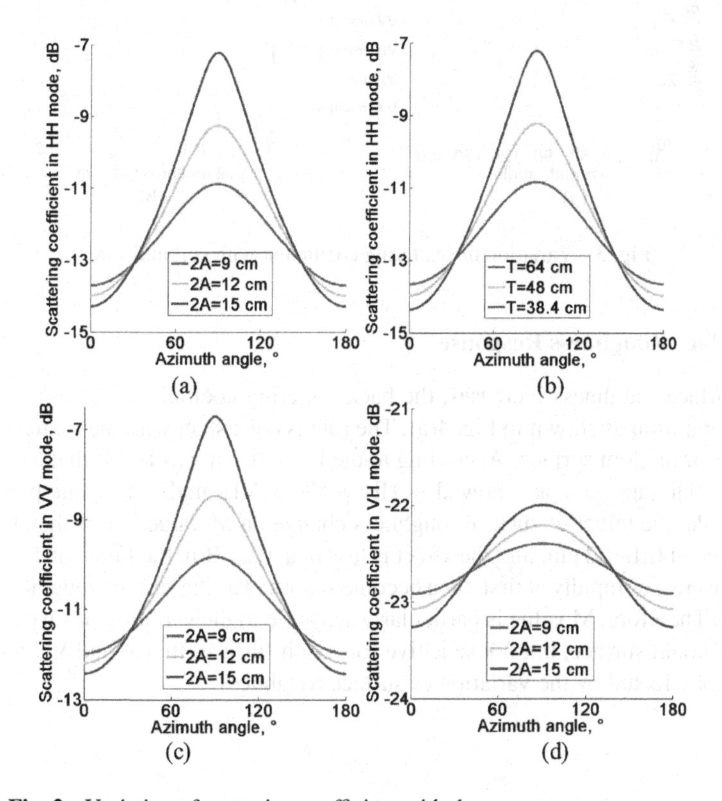

Fig. 2. Variation of scattering coefficient with the row structure parameters

3.2 Soil Moisture Response

As shown in Fig. 3(a), the greater soil moisture, the bigger backscattering coefficient, several responding curves corresponding different soil moisture are nearly parallel to each other. According to the Fig. 3(b), it was found that M values in different polarization modes showed as HH > VV > VH, under the same parameters. In HH mode, M value slowly increases at first, then become saturated with the increase of soil moisture. In the range of 0.04~0.1 cm^3/cm^3, the effect of soil moisture change on M value is less than 0.277 dB, and the effect is less than 0.157 dB in the range of 0.1~0.3 cm^3/cm^3. Therefore, the effect of soil moisture change on M value is very weak for dry soil, and the effect can be ignored for wet soil. In VV and VH modes, M value is not effected by the soil moisture change, M value equals 2.148 dB and 0.520 dB, respectively.

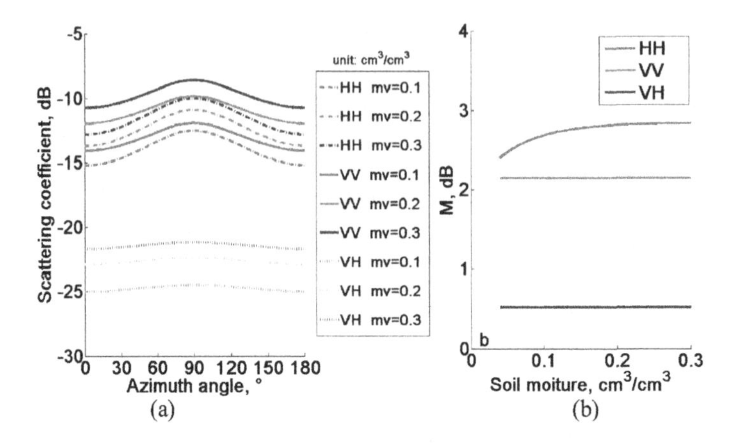

Fig. 3. Variation of scattering coefficient with soil moisture

3.3 Surface Roughness Response

As the surface roughness increases, the backscattering coefficient increases, and then reaches saturation as shown in Fig. 4(a). The rule is consistent with the scattering characteristics of random surface. According to the Fig. 4(b), it was found that M values in different polarization modes showed as HH > VV > VH, under the same parameters. In HH mode, the effect of surface roughness change on M value is less than 0.847 dB in the range of 0.1~1.3 cm, and the effect is less than 0.5 dB in the range of 1.3~6.0 cm. M value decreases rapidly at first, then become saturated as the surface roughness equals to 3.0 cm. Therefore, M value is particularly sensitive to the variation of surface roughness for smooth surface, but not sensitive for rough surface. In VV and VH modes, M value is not effected by the variation of surface roughness.

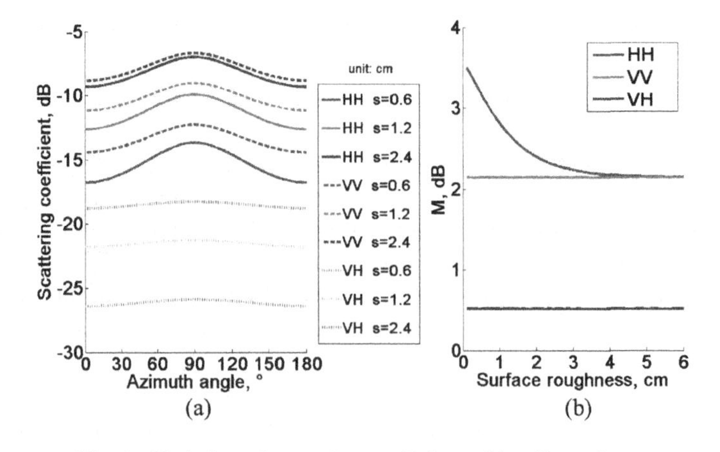

Fig. 4. Variation of scattering coefficient with soil roughness

3.4 Wavelength Response

As shown in Fig. 5(a), backscattering coefficient reaches the maximum in X wavelength, and reaches the minimum in L wavelength. Backscattering coefficient decreases with the longer wavelength. The rule is consistent with the scattering characteristics of random surface. As shown in Fig. 5(b), the M value is greater with the longer wavelength in HH mode. M value is constant in VV or VH mode. M values are showed as HH > VV > VH for any wavelength.

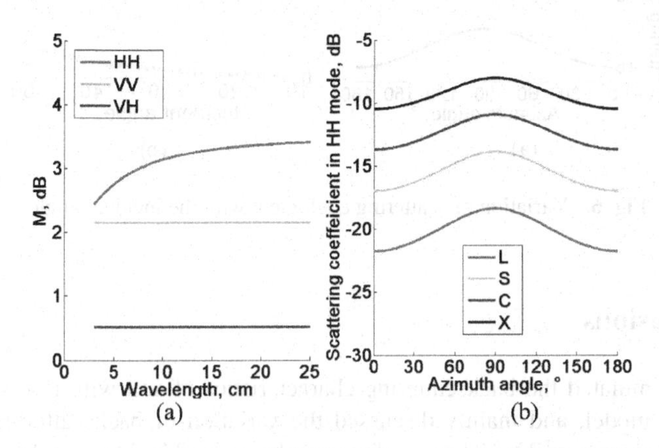

Fig. 5. Variation of scattering coefficient with the azimuth angle in different bands

3.5 Incident Angle Response

This paper only examines the range of incident angle from 10° to 50°. The responding curves for different incident angles are showed in Fig. 6(a). The shapes and amplitudes of curves are different for different incident angles, and may intersect. When A/T is equal to 0.07, the variation of M value with incident angle is showed in Fig. 6(b). In HH mode, M value is equal to 3.162 dB in the position of 10°, then increases gradually, reaches maximum 6.893 dB in the position of 23°, and then decreases gradually, reaches minimum 2.688 dB in the position of 50°. M values are showed as HH > VV > VH in any incident angle. In VV mode, M value reaches to maximum 6.138 dB in the position of 23°. In VH mode, M value increases gradually with the greater incident angle. In the range of 10° to 25°, M value is close to 0 dB, and reaches to maximum 0.72 dB in the position of 50°. Therefore, VH mode is not sensitive to the variation of azimuth angle from the incident angle of 10° to 50°.

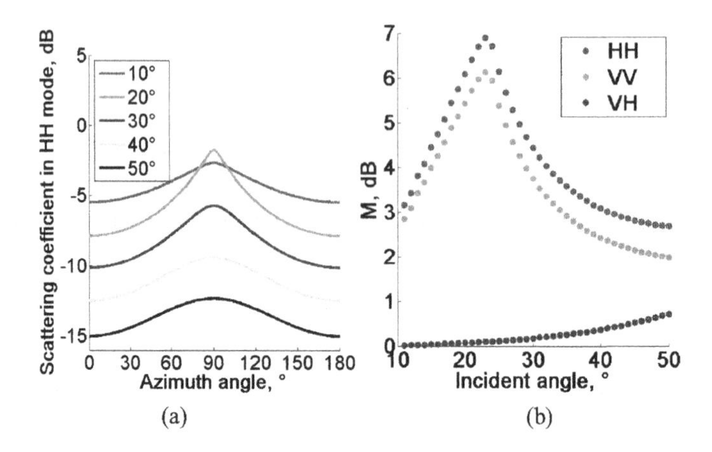

Fig. 6. Variation of scattering coefficient with the incident angle

4 Conclusions

This paper simulated the backscattering characteristics of soil with row structure by using Ulaby model, and mainly discussed the variation of backscattering with row structure parameters, soil moisture, surface roughness, and incident angle.

(1) The A/T value changes little generally in a given area, so the effect of A/T can be ignored. The direction of the ridge is random generally even in very small areas, therefore the variation of azimuth angle is considered to be the strongest affecting factors.
(2) It was found that M values in different polarization modes showed as HH > VV > VH, under the same parameters.
(3) M value is effected by soil moisture and surface roughness in HH mode. But it is always constant in VV and VH modes. When the soil moisture and surface roughness changes greatly in the study area, the M value changes obviously in HH mode, so VV and VH modes are more appropriate to retrieve soil moisture.
(4) The responding function is different for different incident angles, therefore, when the incident angle changes significantly in a SAR image, the effect must be considered in building inversion model of soil moisture.

References

1. Ulaby, F.T., Moore, R.K., Fung, A.K.: Microwave Remote Sensing: Active And passive, vol. III-Volume Scattering and Emission Theory, Advanced Systems and Applications. Artech House, Dedham (1986)
2. Li, Y.Q., Zhang, L.X., Jiang, L.M., Chai, L.N., Zhang, T.: A study on the effect of wheat row-structure on microwave emissivity using field experiment data. In: 2011 IEEE International Geoscience and Remote Sensing Symposium, pp. 779–782. IEEE Press, Vancouver (2011)

3. Wang, J.R., Newton, R.W., Rouse, J.W.: Passive microwave remote sensing of soil moisture: the effect of tilled row structure. IEEE Trans. Geosci. Remote Sens. **GE-18**, 296–302 (1980)
4. Promes, P.M., Jackson, T.J., O'Neill, P.E.: Significance of agricultural row structure on the microwave emissivity of soils. IEEE Trans. Geosci. Remote Sens. **26**, 580–589 (1988)
5. Zheng, X.M., Zhao, K., Zhang, S.W.: Results of soil moisture inversion from radiometer biased by periodic change of row structure on farmland. J. Remote Sens. **16**, 1310–1330 (2012). (in Chinese)
6. Ulaby, F.T., Kouyate, F., Fung, A.K., Sieber, A.J.: A backscatter model for a randomly perturbed periodic surface. IEEE Trans. Geosci. Remote Sens. **GE-20**, 518–528 (1982)
7. Engman, E.T., Wang, J.R.: Evaluating roughness models of radar backscatter. IEEE Trans. Geosci. Remote Sens. **GE-25**, 709–713 (1987)
8. Daniel, S., Allain, S., Ferro-Famil, L, Pottier, E: Surface parameter estimation over periodic surfaces using a time-frequency approach. In: 2008 IEEE International Geoscience and Remote Sensing Symposium, pp. 406–409. IEEE Press, Boston (2008)
9. Beaudoin, A., Le Toan, T., Gwyn, Q.H.J.: SAR observations and modeling of the C-Band backscatter variability due to multiscale geometry and soil moisture. IEEE Trans. Geosci. Remote Sens. **28**, 886–895 (1990)
10. Yin, N., Jiang, Q.G., Meng, Z.G., Li, Y.H.: Use of fully polarimetric RADARSAT-2 data to retrieve soil moisture of periodic surfaces. Trans. Chin. Soc. Agric. Eng. **29**, 72–79 (2013). (in Chinese)
11. Oh, Y.: Quantitative retrieval of soil moisture content and surface roughness from multipolarized radar observations of bare soil surfaces. IEEE Trans. Geosci. Remote Sens. **42**, 596–601 (2004)

High Resolution SAR Coherence and Optical Fused Images Applied in Land-Use Cover Classification

Liping Ai(✉), Lei Pang(✉), Hui Liu, Mengxin Sun, and Shuguang He

Beijing University of Civil Engineering and Architecture, 1st Exhibition Road, Beijing 100044, China
ailipingsl@163.com, panglei@bucea.edu.cn

Abstract. With sensitive to ground scatterers, SAR coherence image can be used for the detection of surface changes and the classification of land-use cover. From a new point of view, this paper synthetically used the change information of high resolution SAR coherence image, and spectral information from optical image, based on the PCA, to obtain the fusion image. And finally land-use and cover classification of the fusion image and test results prove that it is effective and provides a valuable reference.

Keywords: SAR · Coherence image · Image fusion · PCA analysis · Land-use cover classification

1 Introduction

Studies on urban land-use has always been a hot topic. The urban land-cover change trend is a base for rational land-use planning and sustainable development of the related policy. Better tools for land-cover classification are remote sensing technology. Remote sensing images can be used to monitor a wide range of urban land and can also be combined with other data analysing related issues. In recent years, high resolution SAR interferometry technique has more and more applications in such as surveying and mapping, hydrograph investment, urban monitoring, forest assessment and so on, with a satisfied stability and precise. However, most of the relative researchers pay more attention to extracting DEM or deformation information with InSAR or D-InSAR method [1–6]. For the SAR coherence image, there is few research involved and developed out [7]. At the same time, we also note that SAR coherence value has a very strong sensitivity to the change of ground objects' scattering characteristics, and, the value of the coherence coefficient between two SAR

Lei Pang, Beijing Natural Science Foundation (No. 8154043), Key Laboratory for Urban Geomatics of National Administration of Surveying, Mapping and Geoinformation (20131207NY) and Research Fund for the Doctoral Program of Beijing University of Civil Engineering and Architecture (Z12069). And this research work achieved in and supported by the Key Laboratory of Geo-Informatics of National Administration of Surveying, Mapping and Geoinformation (201327, Z13152).

F. Bian and Y. Xie (Eds.): GRMSE 2015, CCIS 569, pp. 464–470, 2016.
DOI: 10.1007/978-3-662-49155-3_47

SLC images contains the change information, especially for the buildings which often expressed an obvious coherence [8–10]. It can be used for the detection of surface changes, the classification of land-use cover and urban evolution. This paper proposes a new method, which will take into account the SAR coherence image to combine with high resolution optical image, utilizing the change and spectral information to find potential and furthermore building change and types based on the conventional land-use cover classification [20, 21].

2 LUCC Method on Coherence and Optical Fusion Image

For the conventional land-use and cover classification (LUCC), high resolution optical images become the appropriate dataset [19]. Yet obviously, this method can obtain the static state ground objects' classification, and, it is hard to find the change progress information. From a new point of view, we propose a method combined with the SAR coherence information to find the change progress information of buildings, in addition to the conventional supervised classification.

2.1 SAR Coherence Image

During InSAR data processing, the coherence is an essential and key size of measurement. Through the coherence value, the two complex SAR images can be estimated the degree of similarity and the quality of interferometric fringes. It represents the important characteristics of ground scatterers. And, its mathematical expression is as the following:

$$\rho = \frac{\sum \left| S_1(\mathrm{j},\ \mathrm{k}) \times S_2^*(\mathrm{j},\ \mathrm{k}) \right|}{\sqrt{\sum \left| S_1(\mathrm{j},\ \mathrm{k}) \times S_1^*(\mathrm{j},\ \mathrm{k}) \right|} \sqrt{\sum \left| S_2(\mathrm{j},\ \mathrm{k}) \times S_2^*(\mathrm{j},\ \mathrm{k}) \right|}} \tag{1}$$

Where * represents conjugate complex number; S1 and S2 are the single look complex SAR images; (j, k)∈ω, and ω is the rectangle window for coherence estimation operator. The value of coherence ρ is between 0 and 1, which 0 represents irrelevant and 1 represents completely relevant. If we stretch [0, 1] to [0, 255], the SAR coherence image will be obtained. Originating from two SLC SAR images' interferometric processing, the coherence image implies the change information of them, and is very sensitive on the change detection [11, 12]. This is an important advantage for finding the progress change information especially for buildings in SAR images.

2.2 Coherence and Optical Image Fusion

For the image fusion of SAR coherence and optical images, we analyse the actual effects and application fields for conventional methods of image fusion, and chose Principal Component Analysis (PCA) fusion method to implement the image fusion processing for coherence and optical images.

PCA algorithm is an effective method in statistical data analysis. Its main purpose is to reduce the dimensionality of complex data sets, in other words, to project the R-dimension space data to the M-dimension space, R ≥ M, conserving the most information and making them apt to be processed. During the data processing of PCA transform, it will seek one original point from the mean value at the new coordinate system, and achieve the maximum variance of high relevant multi-bands data through coordinate axis' rotation, to generate the unrelated output bands. One-dimensional principal component analysis can express in a mathematical way as (2):

$$F = \Phi^{*^T}(f - \mu) = A(f - \mu) \tag{2}$$

And inverse principal component analysis as (3):

$$(f - \mu) = \Phi^{*}F = A^{T}F \tag{3}$$

Where f is random vector set; μ is mathematical expectation of f; and $\Phi = [\Phi_1 \Phi_2 \ldots \Phi_n]$ represents eigenvectors of the covariance matrix of f.

Hereinto, the first principal component comprised with the maximum-variance data percentage, the second one comprised with the second maximum-variance data, and so forth, the last one will only exhibit noise characteristic because of the minimum-variance data. Moreover, derived from unrelated multi-bands data, PCA transform can generate more colours' combination images [13].

2.3 LUCC Method on Coherence and Optical Fusion Image

The land-use and cover classification information, which represents the land exploitation degree, usually has an important reference function in rational utilizing of land and urbanization progress. LUCC is the main research method of classification in this article. Using two different times for SLC image through coherence calculation to get a image of a coherence information. Image registration between the coherence and optical image can make them the same coordinate system. Then target a area with rich land-cover types and tailor it as the study area. After fusion [14–18] by using principal component analysis method, the fusion image contains the spectrum information of optical image that can make image colorful and easy to identify all kinds of ground objects and contains the information of change features reflected by coherence image. It is not available using a separate optical image for land-cover classification. Through supervised classification, one of the most commonly used classification methods, we can obtain a different classification result. From which, the progress change information of buildings will be obviously found out, and have more clear ground objects' edge. The flow chart about LUCC Method on Coherence and Optical Fusion Image is seen as follows (Fig. 1).

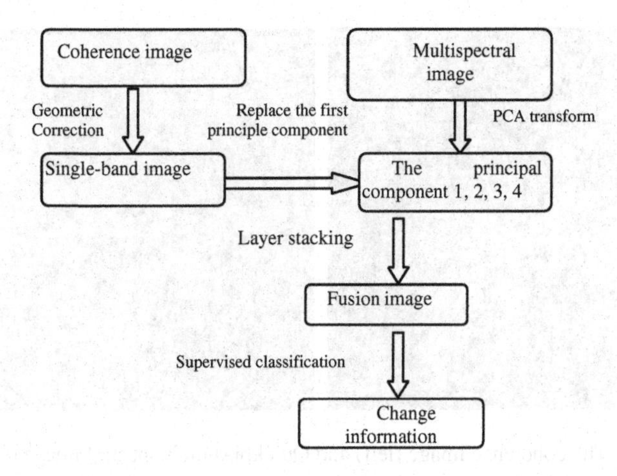

Fig. 1. Flow chart about LUCC method on coherence and optical fusion image.

3 Test Data and Results Analysis

The test data sets include the coherence image generated from interferometric data processing of two high resolution TerraSAR-X SLC images, 3 m resolution, in Nov. 2011 and Dec. 2013 respectively, and one Quickbird multispectral optical image in the same area, 2 m resolution, with 4 bands, in Beijing. The SAR coherence and Quickbird multispectral images just as the following see as Fig. 2. And, through data preprocessing, the test site cuts out near by the Forbidden City, where are mainly distributed by old Si-He-Yuan buildings, intermingled with trees, roads, and new constructed buildings. In addition, the obvious water body is the Beihai, Shichahai, and Zhongnanhai. The data preprocessing consists mainly of geometric correction, image resize, and image fusion. During the processing of PCA fusion, we involve 4 steps, that is:

A. Choose the multispectral data to execute PCA transform and 4 bands as the output bands number;

B. Use the SAR coherence image to replace the first principle component data and make bands synthesis operator. Before that, the image registration between the two data sets has been carried out to avoid information distortion.

C. Operate the PCA inverse transform to obtain the fusion image, see as Fig. 3(left).

D. Carry out the supervised classification, and choose 6 typical ground objects as the samples, such as water, trees, roads, permanent buildings, general buildings, and bare soils(which represent new-built buildings in urban areas), see as Fig. 3(right).

As shown in the above fusion image, the classification result has more distinct details and color classes than the original Quickbird multispectral image. Water, trees, roads, buildings are also be interpreted from it. We sampled the classification results of the actual situation and the classification accuracy are obtained(see as Table 1). Besides of those objects, the new-built buildings express a kind of light brown color, which is difficult to be find out in multispectral Quickbird image.

Fig. 2. The coherence image (left) and Quickbird multispectral image (right).

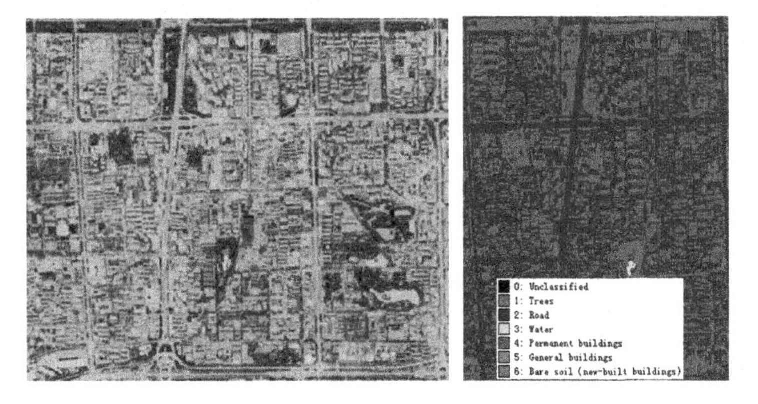

Fig. 3. The PCA fusion image (left) and supervised classification result (right).

For a furthermore demonstration, we compare 8 polygons with typical new-built buildings, in two areas, with those from WorldView-2 image, with 0.5 m resolution. The actual investigation proved that 6 of them are right new-built buildings, and the other wrong 2 polygons are derived from the reflection of buildings' walls and outsides decoration. The results have been shown in Fig. 4.

Table 1. Classification accuracy table.

Objects	Sample number	Correct classification number	Classification accuracy
trees	10	9	0.9
water	6	5	0.83
roads	6	5	0.83
buildings	10	8	0.8

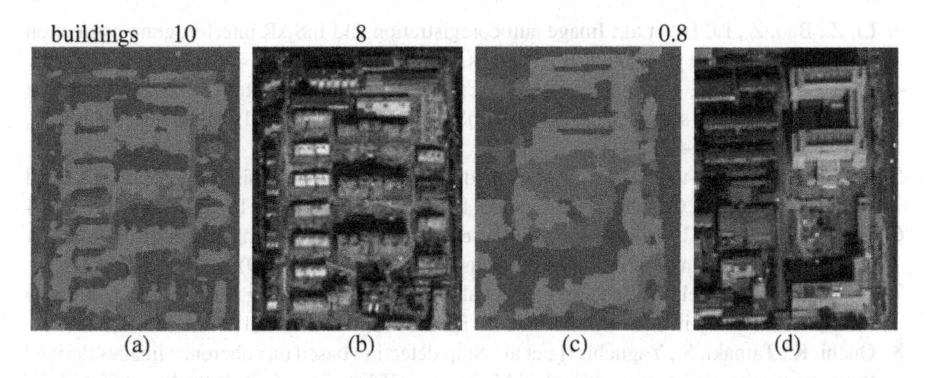

Fig. 4. The new-built buildings from the classification image (a and c) and the actual buildings from 0.5 m WorldView-2 image (b and d).

From this, it can be found that the PCA fusion image from SAR coherence and optical images has the potential to discriminate the general buildings and new-built ones, which are difficult to be found out with multispectral optical images. We can see that the under-construction land in the fusion image classification diagram of automatic interpretation effect is obvious, but there are also results that taking the general building mistakenly into unused land. Compared with optical image and field validation, error is partly caused by the wall reflection and construction of outer decoration. Therefore, land-cover classification method based on SAR coherence image is an effective and novel classification method, but further more research should be done for applying it to urban land-use planning.

4 Conclusions

From a new point of view, this paper synthetically use the change information of high resolution SAR coherence image, and spectral information from optical image, based on the PCA analysis, to obtain the fusion image. And finally through supervised classification and test results proved that it is an effective method to find out the under-construction buildings or projects with SAR coherence information. During the city urbanization, it can provide a valuable reference for detecting the under-construction buildings automatically. The further more research on it is waiting for development.

References

1. Lohman, R.B., Simons, M.: Some thoughts on the use of InSAR data to constrain models of surface deformation: noise structure and data downsampling. Geochem. Geophys. Geosyst. **6**(1) (2005). Q01007. doi:10.1029/2004GC000841
2. Thiele, A., Cadario, E., Schulz, K., et al.: Building recognition from multi-aspect high-resolution InSAR data in urban areas. IEEE Trans. Geosci. Remote Sens. **45**(11), 3583–3593 (2007)

3. Li, Z., Bao, Z., Li, H., et al.: Image autocoregistration and InSAR interferogram estimation using joint subspace projection. IEEE Trans. Geosci. Remote Sens. **44**(2), 288–297 (2006)
4. Hyde, P., Dubayah, R., Walker, W., et al.: Mapping forest structure for wildlife habitat analysis using multi-sensor (LiDAR, SAR/InSAR, ETM+, Quickbird) synergy. Remote Sens. Environ. **102**(1), 63–73 (2006)
5. Hooper, A.: A multi-temporal InSAR method incorporating both persistent scatterer and small baseline approaches. Geophys. Res. Lett. **35**(16) (2008). L16302. doi:10.1029/2008GL034654
6. Ferretti, A., Prati, C., Rocca, F.: Multibaseline InSAR DEM reconstruction: the wavelet approach. IEEE Trans. Geosci. Remote Sens. **37**(2), 705–715 (1999)
7. Marinelli, L., Michel, R., Beaudoin, A., et al.: Flood mapping using ERS tandem coherence image: a case study in Southern France. In: ESA SP, pp. 531–536 (1997)
8. Ouchi, K., Tamaki, S., Yaguchi, H., et al.: Ship detection based on coherence images derived from cross correlation of multilook SAR images. IEEE Geosci. Remote Sens. Lett. **1**(3), 184–187 (2004)
9. Moeremans, B., Dautrebande, S.: Soil moisture evaluation by means of multi-temporal ERS SAR PRI images and interferometric coherence. J. Hydrol. **234**(3), 162–169 (2000)
10. Gaveau, D.L.A., Balzter, H., Plummer, S.: Forest woody biomass classification with satellite-based radar coherence over 900 000 km^2 in Central Siberia. For. Ecol. Manag. **174**(1), 65–75 (2003)
11. Damini, A., Mantle, V., Davidson, G.: A new approach to coherent change detection in VideoSAR imagery using stack averaged coherence. In: 2013 IEEE Radar Conference (RADAR), pp. 1–5 (2013)
12. Dou, X.Y., Han, L.G., Wang, E.L., et al.: A fracture enhancement method based on the histogram equalization of eigenstructure-based coherence. Appl. Geophys. **11**(2), 179–185 (2014)
13. He, C., Liu, Q., Li, H., et al.: Multimodal medical image fusion based on IHS and PCA. Proc. Eng. **7**, 280–285 (2010)
14. Poulain, V., Inglada, J., Spigai, M., et al.: High-resolution optical and SAR image fusion for building database updating. IEEE Trans. Geosci. Remote Sens. **49**(8), 2900–2910 (2011)
15. Poulain, V., Inglada, J., Spigai, M., et al.: Fusion of high resolution optical and SAR images with vector data bases for change detection. In: 2009 IEEE International Geoscience and Remote Sensing Symposium IGARSS 2009, vol. 4, pp. IV-956–IV-959. IEEE (2009)
16. Zhang, J.: Multi-source remote sensing data fusion: status and trends. Int. J. Image Data Fusion **1**(1), 5–24 (2010)
17. McNairn, H., Champagne, C., Shang, J., et al.: Integration of optical and synthetic aperture radar (SAR) imagery for delivering operational annual crop inventories. ISPRS J. Photogramm. Remote Sens. **64**(5), 434–449 (2009)
18. Yang, S., Wang, M., Lu, Y.X., et al.: Fusion of multiparametric SAR images based on SW-nonsubsampled contourlet and PCNN. Sig. Process. **89**(12), 2596–2608 (2009)
19. Amarsaikhan, D., Blotevogel, H.H., Van Genderen, J.L., et al.: Fusing high-resolution SAR and optical imagery for improved urban land cover study and classification. Int. J. Image Data Fusion **1**(1), 83–97 (2010)
20. Weng, Q., Lu, D.: A sub-pixel analysis of urbanization effect on land surface temperature and its interplay with impervious surface and vegetation coverage in Indianapolis, United States. Int. J. Appl. Earth Obs. Geoinf. **10**(1), 68–83 (2008)
21. Snyder, W.C., Wan, Z., Zhang, Y., et al.: Classification-based emissivity for land surface temperature measurement from space. Int. J. Remote Sens. **19**(14), 2753–2774 (1998)

Urban Extraction Based on Multi-scale Building Information Extra-Segmentation and SAR Coherence Image

Mengxin Sun(✉), Lei Pang, Hui Liu, Xuedong Zhang, Liping Ai, and Shuguang He

Beijing University of Civil Engineering and Architecture, Beijing, China
1078570048@qq.com, panglei@bucea.edu.cn

Abstract. Urban building outline is not only a very important land cover type, but also the key of studying urban building. It is significance on urban construction planning and natural disasters monitoring. This paper discussed the method for building information extracting according to high-resolution SAR coherence image. Through theoretical analysis and test verifying, it had obtained a better effect for extracting the buildings' profiles or top figures from the coherence image, which often expressed as 'L'-style top structures from SAR SLC image. This method has been proved the further potential in extracting urban buildings' structure and profile information, and has some extent of reference means.

Keywords: Terrasar · Coherence image · Urban building · Multi-scale segmentation · Information extraction

1 Introduction

Urban building information is the key element of geospatial data, and it has an important role in planning and development of the city. In the remote sensing images, the majority of urban areas are building information. It provided favorable conditions to identify from building information. Contrasted with conventional optical imaging, SAR images have the characteristics of not affecting by climate and night, for the reason that different radar wavelengths can penetrate certain covering material and SAR image resolution have nothing to do with wavelength, height and operating range. Therefore, radar images have certain advantages in terms of building information extraction [1, 2].

There is no doubt for extracting DEM or deformation information has more applications in actual fields for InSAR or D-InSAR technique. For high resolution SAR system, people often pay more attention to the single or multi-look SAR complex but

L. Pang—Beijing Natural Science Foundation (No. 8154043), Key Laboratory for Urban Geomatics of National Administration of Surveying, Mapping and Geoinformation (20131207NY) and Research Fund for the Doctoral Program of Beijing University of Civil Engineering and Architecture (Z12069). And this research work achieved in and supported by the Key Laboratory of Geo-Informatics of National Administration of Surveying, Mapping and Geoinformation (201327, Z13152).

F. Bian and Y. Xie (Eds.): GRMSE 2015, CCIS 569, pp. 471–479, 2016.
DOI: 10.1007/978-3-662-49155-3_48

the coherence images. After the interferometric data processing, most of researchers note the interferometric phase, at the same time, the coherence image is thought less application because of the noise interference. Although, there are research results about extracting the urban borders utilizing the coherence image, interpreting the inclined building from SAR optimal coherent coefficient, extracting the settlement information from POLSAR polarization coherence coefficient, the actual application research on SAR coherence image are quite few [3–5]. This paper proposes a new point of view, which will utilize SAR coherence image to extract urban buildings' profile and structure information based on the multi-scale image segmentation method.

2 Urban Building Information Extraction Based on SAR Coherence Image

2.1 SAR Coherence Image Analysis

The interferometric coherence is a measure of relative stability for SAR image, coherence is the stability estimate of two radar images' interference. During InSAR data processing, the coherence is an essential and key size of measurement. Through the coherence value, the two complex SAR images can be estimated the degree of similarity and the quality of interferometric fringes. It represents the important characteristics of ground scatters. And its mathematical expression is as the following:

$$\gamma = \frac{\left|E\left[s_1 s_2^*\right]\right|}{\sqrt{E\left[|s_1|^2\right]E\left[|s_2|^2\right]}} \tag{1}$$

where * represents conjugate complex operator; S1 and S2 are the single look complex SAR images; The value of coherence is between 0 and 1, 0 represents irrelevant, and 1 represents completely relevant. If we stretch [0, 1] to [0, 255], the SAR coherence image will be obtained [11, 12].

For the coherence value, the influence factors include the decorrelation from spatial base, temporal interval, Doppler central offset, complex images registration, and systemic thermal noise. Originating from two SLC SAR images' interferometric processing, the coherence image implies the change information of them, and is very sensitive on the change detection. This is an important advantage of information extraction for urban buildings in SAR images.

2.2 Multi-scale Segmentation

Multi-scale segmentation using minimal heterogeneity tends to merge algorithm. This algorithm start from an arbitrary pixel. Firstly single pixel should be merged into smaller image objects, then smaller image objects merged into a large polygon objects. In the multi-scale segmentation process, variability objects becomes larger and larger, but its heterogeneity must be the least so that the whole the average of image heterogeneity threshold is the smallest. Therefore, the scale determines the segmentation size of ROI and the accuracy of building information extraction. To improve the accuracy extract feature information for different surface features, different threshold values and related properties are different scale segmentation feature. Finally, we combined with the actual characteristics of surface features to achieve efficient extraction of building information [6–10].

Scale parameter is an important parameter in image segmentation, which is examining the standard whether the pixels can be incorporated into the adjacent image of the object. Scale refers to the heterogeneity of the object segmentation threshold, which determines the minimum image generated polygon object level and size. In the multi-scale image segmentation, the size and number of polygons generated image objects are determined by the selected scale. The smaller segmentation scale value is, the smaller area and the greater number of a polygon in generated objects layer are. Appropriate scale shall be used to select the relevant image segmentation [13–15]. In this paper, image segmentation using a variety of information objects in the classification process carried out building information extraction. The software providing the tool of Rule Based Feature Extraction to accomplished the determination of segmentation scale size. Wherein, homogeneity index (V) is expressed as:

$$V = \frac{\sum_{i=1}^{n} \alpha_i v_i}{\sum_{i=1}^{n} \alpha_i} \tag{2}$$

Where, v_i is the Standard deviation of the object; α_i is the area of the object; n is the number of feature object to be segmented. Heterogeneity index ΔC_L is expression as:

$$\Delta C_L = \frac{1}{l} \sum_{i=1}^{n} l_{si|\overline{C_L} - \overline{C_{li}}|} \tag{3}$$

Where $|\overline{C_L} - \overline{C_{li}}|$ is the absolute value of the L-band single scale split objects and its neighborhood means difference; L represents an image where the band layer; l represents the length of the target object's border; n represents the number of adjacent objects; l_{si} represents the common length of the i-th adjacent object boundary; $\overline{C_L}$ represents the average value of the object in the L-band; C_{li} represents the average value of the i-th adjacent object in the L-band.

Index homogeneity and heterogeneity ratio is expressed as:

$$HD = \frac{V}{\Delta C_L} \tag{4}$$

After multi-scale image segmentation, we can get different scale segmentation image. The smaller of index homogeneity and heterogeneity ratio is, the better its segmentation are.

2.3 SAR Coherence Building Information Extraction Technology Processes

Experiments based on the edge segmentation method, this method is applicable to a significant edge in the identification of the target image, such as rivers and city buildings. Different regions of the edge information are the main basis for segmentation. Usually in the edges of image gray value changes huge, the method of gray value using the first-order and second-order differential operator to recognize edge is called parallel border technology. The whole band Lambda same extraction method is used in this paper to merger chunk and texture strong regional, such as trees, clouds, etc. The flow charts have been shown in Fig. 1:

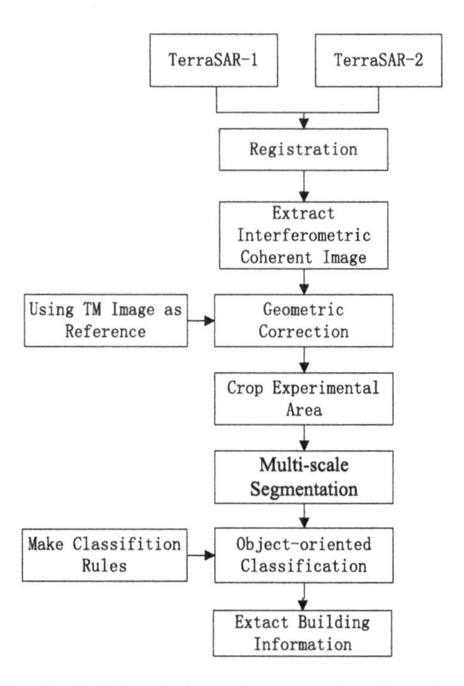

Fig. 1. Building information extraction flow chart

3 Test Data and Results Analysis

3.1 Experimental Area

The test data set include the coherence image generated from interferometric data processing of two high resolution TerraSAR-X SLC images, 3 m resolution, in 2013 respectively, in Beijing. The test area resized from the original coherence image is near by the Sports Park, 4 km^2 within the fifth wring of Beijing, where distributing dense and big buildings. The original SAR coherence, after-correction coherence image and resized images just as the following see as Fig. 2. It must be note that, during the

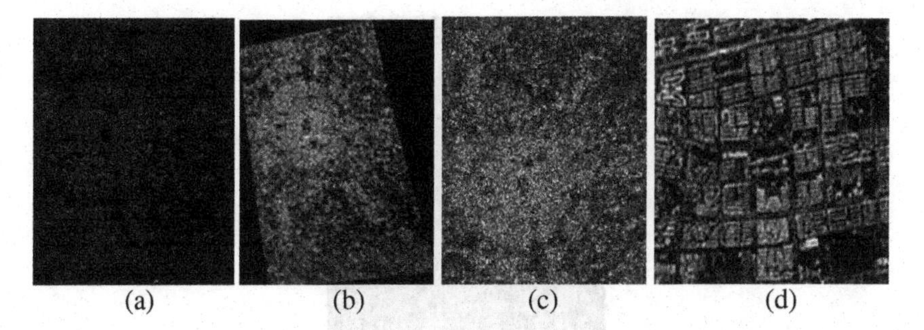

Fig. 2. Original image (a), the coherence image (b), after-correction coherence image (c), and resized coherence image (d)

geometric correction, a high resolution Worldview-2 image in the same area has been chosen as the base image, and after correction, the original coherence SAR image has a 180° rotation, putting up the similar ground objects distribution.

3.2 Experimental Process

Data Preprocessing. The data which is imported from TerraSAR SSC format should be transferred into ENVI grid and SLC format that we need. There is lots of speckle noise in SAR image in the processing of SAR image, Multi-view processing could be using to reduce the speckle noise in the SLC data.

The using of image smoothing can inhibit and reduce image speckle and noise. By multi-view processing the image radiometric resolution can be enhanced, but the spatial resolution will be reduced. It leads to the boundary of the buildings' roof structure not easy to distinguish, and effecting of coherence processing and multi-scale segmentation information extraction. Instead SLC data provides relatively clearly the boundaries of city buildings, the reflection of image are more properly makes it easier to distinguish the boundaries of buildings than multi-view image. Therefore, the experimental design do not be used the multiple-process of single-view complex images.

Extract the DEM Data. Though data preparation, we can get DEM, which is a common source of data. The use of reference DEM can eliminate terrain phase in the interference patterns. Reference images and the type of resampling in SRTM 3 Version2 tools was used to get DEM we need. Interpolation model, which is the most effective method of reconstruction of the unknown terrain, is suitable for small areas and unexpressed terrain architectural changes area. Beijing DEM has been shown in Fig. 3.

Obtain the Coherent Coefficient Images. The noise of coherent coefficient images is relatively more, so it is not widely used in SAR images. Coherent coefficient images in this article are obtained by two single-view complex images, it embodies the two coherent SAR image change information, and it is sensitive in change detection. The obtained coherent coefficient image with the original image such as shown in Fig. 4.

Fig. 3. Beijing DEM

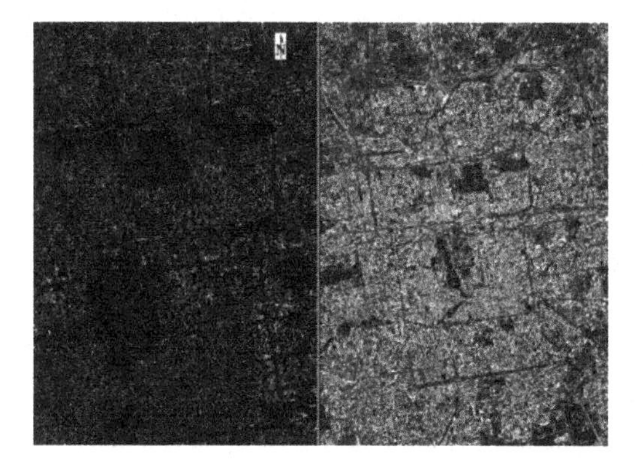

Fig. 4. Coefficient image and original image

Multi-scale Segmentation. In Multi-scale segmentation, high-scale segmentation images will be divided into a small number of patches, and low-scale segmentation images will be segmented more patches. Segmentation results to some extent determine the accuracy of the classification results. Though "preview" we can preview segmentation results, and select an ideal segmentation threshold will reflect edge features more clearly.

This article is used the method of based on edge detection, where the threshold value is set to 40, the combined threshold value is set to 90. Reasonable segmentation and merge thresholds can make building information extraction more accurate. In image segmentation, if the threshold is too low or too high, some features may be segmented fallacious.

Extraction Buildings Outline Based on Rules. In the classification rule interface, each category has a number of rules, and each rule have several property expressions to describe. Different objects are described by different rules, and its shape, size and reflectivity are also difference, such as water bodies, vegetation, buildings, etc. Rule-making as shown in Fig. 5.

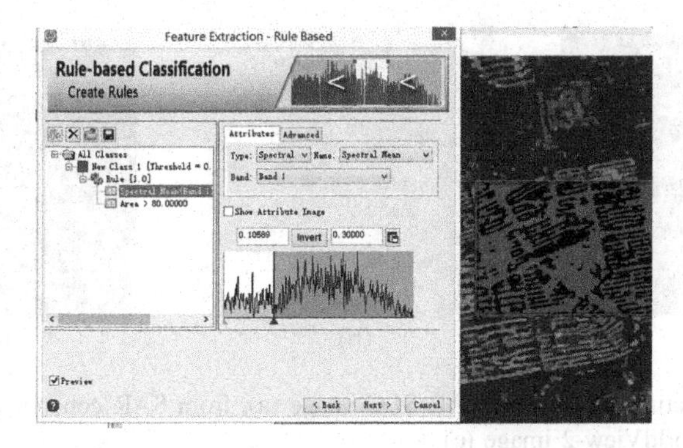

Fig. 5. Building information extraction rules

3.3 Object-oriented Classification Results and Analysis

Object-oriented classification technique is to combine the nearby pixels as one object to discriminate the interesting spectral factors. Its output information usually represent as high precise classification result or vector maps. Among the three methods of Object-oriented classification, the rule-based multi-scale segment method has been adopted to extract buildings' profile, including finding object and characteristics extracting the two parts. In the first part, the generation of an object depends on the relate segment and merge algorithms, which can combine the nearby pixels to an object through adjusting the threshold of relate algorithms.

The most important part is to choose the classification rules for buildings' information extracting. In this test, we work out the relate rules as following:

Rule 1: The spectral characteristic is the mean spectral value, with a threshold of 0.3–0.9, getting rid of the vegetables and water body.

Rule 2: The spatial information attribute is chosen as rectangular extent, with a threshold of 0.35–1, getting rid of roads and other non-building area.

Rule 3: The spatial information attribute is chosen as area, with a threshold of 60-maximum, getting rid of the small parking area in front of buildings and noise. The final buildings' information extraction results have been shown in Fig. 6.

Fig. 6. The buildings' profile information results which are shown in different output styles

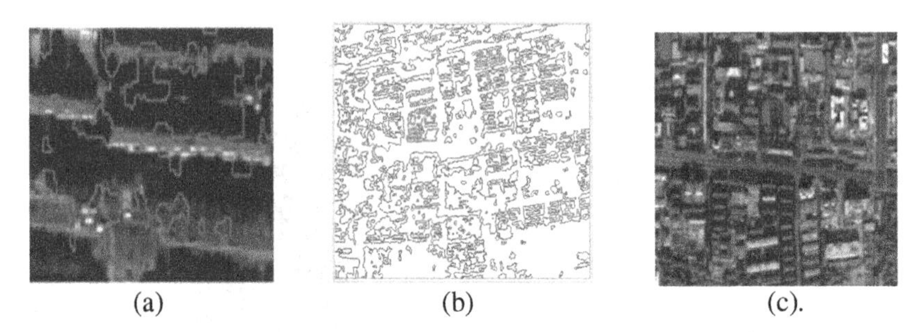
(a) (b) (c).

Fig. 7. The extracting result from SAR SLC image (a), from SAR coherence image (b), and from reference WorldView-2 image (c)

For a furthermore analysis, we take the high-resolution WorldView-2 image in the same area as a reference, to verify the actual effect of this test, see as Fig. 7. The other wrong 2 polygons are derived from the reflection of buildings' walls and outsides decoration.

From the above, it can be found that the buildings' information extracting image from SAR coherence has a more clear profile styles than that from SAR SLC image, although, the effect cannot equal to that from high-resolution WorldView-2 image yet, it still exhibits a valuable application method especially for urban buildings' structure information extracting.

4 Conclusions

This paper proposed to utilize the multi-scale segment method and SAR coherence image to extract urban buildings' profile or structure information, and finally through TerraSAR-X dataset realize it, achieving an effective and valuable reference method to find out more clear buildings information. This provided a new idea for ground objects' information extraction from SAR images.

References

1. Cagatay, N.D., Datcu, M.: Complex-valued Markov random field based feature extraction for InSAR Images. In: Proceedings of EUSAR 2014; 10th European Conference on Synthetic Aperture Radar; VDE, pp. 1–4 (2014)
2. Xu, H.P., Li, S., Feng, L.: Interferometric phase statistics and estimation accuracy of strong scatterer for InSAR. Chin. J. Electron. **21**(4), 740–744 (2012)
3. Zhang, W.Y., Liu, Q.C., Meng, X.J.: Urban boundary extraction based on SAR coherence image. J. Geomatics Spat. Inf. **5**, 56–59 (2014)
4. Zhang, L., Guo, H.D., Li, X.W.: Residents extraction of using POLSAR data exploration polarization dependency. J. Sens. Technol. Appl. **4**, 474–479 (2010)
5. Lin, Q., Chu, T., Zebker, H.A.: Pol-In SAR optimal coherence estimation and its application in imaging forest canopy. AGU Fall Meet. Abstr. **1**, 1736 (2012)

6. Zhang, X., Xiao, P., Song, X., et al.: Boundary-constrained multi-scale segmentation method for remote sensing images. ISPRS J. Photogram. Remote Sens. **78**, 15–25 (2013)
7. Johnson, B., Xie, Z.: Unsupervised image segmentation evaluation and refinement using a multi-scale approach. ISPRS J. Photogram. Remote Sens. **66**(4), 473–483 (2011)
8. Syed, A.H., Saber, E., Messinger, D.: Encoding of topological information in multi-scale remotely sensed data: applications to segmentation and object-based image analysis. In: International Conference on Geographic Object-based Image Analysis, Rio de Janeiro, Brazil, pp. 102–107 (2012)
9. Drăguţ, L., Csillik, O., Eisank, C., et al.: Automated parameterisation for multi-scale image segmentation on multiple layers. ISPRS J. Photogram. Remote Sens. **88**, 119–127 (2014)
10. Chaokui, L., Xiaojiao, D., Zhang, Q.: Multi-scale object-oriented building extraction method of Tai'an city from high resolution image. In: 3rd International Workshop on Earth Observation and Remote Sensing Applications (EORSA), IEEE, pp. 91–95 (2014)
11. Barrett, B., Whelan, P., Dwyer, N.: The use of C-and L-band repeat-pass interferometric SAR coherence for soil moisture change detection in vegetated areas. Open Remote Sens. J. **5**(1), 37–53 (2012)
12. Bickel, D.L.: SAR Image Effects on Coherence and Coherence Estimation. Sandia National Laboratories Report, SAND2014-0369 (2014)
13. Li, C., Yin, J., Bai, C., et al.: An object-oriented method for extracting city information based on high spatial resolution remote sensing images. Int. J. Adv. Comput. Technol. **3**(5), 80–88 (2011)
14. Zhang, C., Zhao, Y., Zhang, D., et al.: Application and evaluation of object-oriented technology in high-resolution remote sensing image classification. In: 2011 International Conference on Control, Automation and Systems Engineering (CASE), IEEE, pp. 1–4 (2011)
15. Chen, P., Wu, J., Liu, Y., et al.: Extraction Method for Earthquake-Collapsed Building Information Based on High-Resolution Remote Sensing. IOP Conference Series: Earth and Environmental Science, vol. 17(1), p. 012096. IOP Publishing, Bristol (2014)

Study of Greenhouse Remote Monitoring System Based on Webaccess

Xuepin Lyu, Yonggang Wang, Tongyu Xu, and Zhixia Zhang(✉)

College of Information and Electrical Engineering, Shenyang Agricultural University, Shenyang 110866, China
13591409759@163.com

Abstract. Based on Advantech Web access configuration software and a series of ADAM-2000 wireless sensor network modules and ADAM-4000 I/O modules, a distributed computer control system was developed under the need of greenhouse monitor and control. Using the hybrid control method of fuzzy and PID, the control script was written. The design of monitoring interface and management was finished. The test result shows that the system which runs stably achieved the remote monitoring of greenhouse.

Keywords: Webaccess · Fuzzy control · PID control

1 Introduction

From the previously using of the measuring instrument for greenhouse environment parameters, to now using computer technology to achieve automatic control, greenhouse environment control technology is developed rapidly. With the rapidly development of network technology, people put forward higher request to greenhouse monitoring. Network configuration software Web Access constitute a monitoring system based on computer network. In the system according to the allocation of each computer functions, respectively undertakes data acquisition, processing, display, storage, management, operation, alarm, report forms, trends, and other functions. The data transmitted between the various subsystems through the network, and connected to the various automation equipment through the computer subsystem, form a distributed control system (DCS), archive the monitoring of the whole system. This research adopts the web access configuration software and ADAM-2000 series wireless sensor module to form a monitoring network, using fuzzy PID controller to adjust and control the greenhouse environment, archive remote monitoring of greenhouse.

Zhixia Zhang is the corresponding author. Shenyang Agricultural University, 110866, Shenyang, Liaoning Province, China. E-mail: 13591409759@163.com. The Agricultural Technology Key Project of Shenyang (Grant No. F12-130-3-00), The Special Fund for Agro-scientific Research in the Public Interest (Grant No. 201503136), Key Scientific and Technological Project of Liaoning Province(Grant No. 2015103031)

F. Bian and Y. Xie (Eds.): GRMSE 2015, CCIS 569, pp. 480–486, 2016.
DOI: 10.1007/978-3-662-49155-3_49

2 The Design of the System Monitoring Program

The purpose of designing monitor system is that provided an appropriate environment for greenhouse crop growth and development. Based on the greenhouse environment parameter information inside or outside and the ideal setting value of environment parameter, by driving environment control equipment (fan, heater, insulation screen, spray, open a window, etc.), to regulate the greenhouse environmental factors (temperature, humidity, light intensity and co2 concentration, etc.), so that to meet the needs of the greenhouse crop growth and development. The whole system is divided into the following two parts: firstly, it is the system monitoring scheme. In accordance with the requirements of greenhouse production and distribution, formulate the corresponding monitoring scheme. You must to sure the environmental factors and the reasonable distribution of monitoring points that can accurately reflect environment condition in greenhouse, to provide data for greenhouse environment regulation. Secondly, that is the system control scheme. According to the ideal environment requirements for greenhouse development and the characteristics and performance of the greenhouse environment control equipment, control scheme is determined, to archive the local and remote, manual or automatic control.

3 Monitor and Control System Structure

The WebAccess are mainly consisted of the Project Node, the SCADA Nodes and the Clients. The monitoring node network are comprised of the monitoring nodes and the monitoring equipments. One or more monitoring nodes network together with project node make up the monitoring system. The different types of nodes perform different functions. Their functions are:

Project Node: working based on the principle of ASP (Active Server Pages), integrated Web Server and Access database, bear the function of the "project manager". Implementation of setting system parameter and storing data for the whole system, save all the figure of the project, a copy of the script, and other components, edit and create I/O points, alarm and graphics, providing the initial connection between the client and monitor node. Sending the edited result to the monitoring nodes by downloading. The user browse the dynamic running status of the monitoring nodes through the project node [1, 2].

Monitoring node network: monitor node connected to the control equipment and environmental monitoring equipment through the serial port on a computer, complete data acquisition, processing and control the transmission of information. Monitoring equipment is mainly composed of temperature sensor, humidity sensor, light intensity sensor, co2 sensor, analog input module and digital quantity input module, in implementation of data acquisition of analog quantity (temperature, humidity, light intensity and co2 concentration, etc.) and digital quantity (such as equipment running state), transfer data to computer monitor node through the RS485 bus and ADAM-2000 wireless sensor network (WSN), transmit collected greenhouse environmental information to the project node and the client through the network. Control equipment are mainly consisted of the analog output module, digital output module, relay and control equipment. Based on the environment parameters transmitting from monitoring equipment, combined with the suitable setting value of greenhouse parameters and control strategy, by using the ADAM-4000 series digital output modules and analog output modules through the RS485 bus, monitor node archive the control of start-stop and open degree for each control equipment.

Client: a computer or hand terminals running internet explorer, connecting the computer monitor node through TCP/IP protocol, transmitting the worked greenhouse environment information gathering by monitoring equipment to the browser, display in variety kinds of digital, animation, trends and so on, allowed the manager to adjust the setting value of environment parameters and the alert threshold based on permissions in the client [1, 2].

In the whole monitoring system, the project node in the system is the only one; The number of monitoring nodes is determined based on the distribution and the number of automation equipment in the system. In the whole control system, there is no limit to the number of monitoring nodes, the client is the interface accessed to the system for management, without the limit for number in the system, but given different control permissions in accordance with the identity of visitors.

4 Network Monitoring Nodes

4.1 Hardware Equipment and Architecture

Monitoring node network architecture is shown in figure 1.

In Fig. 1. Equipment is introduced in detail as follows.

ADAM-2031Z is the wireless sensor network terminal module that supporting 802.15.4 protocol, supporting two of AA lithium batteries, reaching up to 110 m of outdoor transmission, built-in temperature and humidity sensor, fulfilled the requirements of greenhouse measurement and control precision, saved the temperature and humidity. ADAM-2017PZ is the wireless sensor network terminal module that has six differential analog input channels, mainly used for collecting carbon dioxide concentrations, analog input signals such as light intensity. ADAM-4050 is a digital I/O module that has seven input channels mainly used for the start-stop state acquisition of the greenhouse control equipment, that has eight output channel mainly used for output the start-stop signals to the control equipment. ADAM-4024 is a four channels module for analog output, converted data into analog signals using the D/A, can achieve 3000VDC isolation transformer protection with the aid of optical isolation of the D/A components. ADAM-2520Z is a wireless sensor network coordinator gateway that supported the RS-422/485 and USB interface reaching up to 1000 m of outdoor range, mainly used for the connection between the wireless sensor terminals and the computer. ADAM-2510Z is the router of wireless sensor network (WSN) used for the expansion of wireless sensor network, connected the wireless sensor terminals and the coordinator.

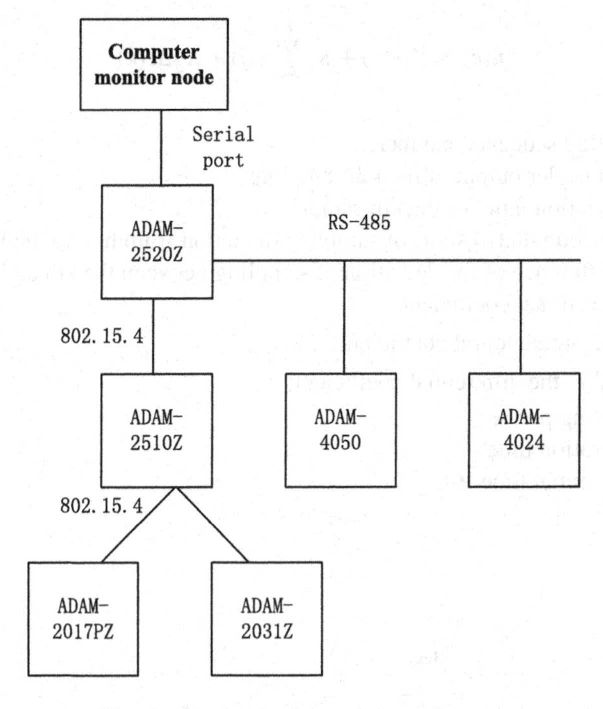

Fig. 1. Monitor node network architecture

4.2 Control Strategy and Software Implementation

PID control is a kind of the earliest, most widely used control method in automatic control. The essence of the PID is the calculative operations according to the function relation of the proportion, integral and differential based on the deviation of input, and then output the calculative results for the control. The general form of PID control is:

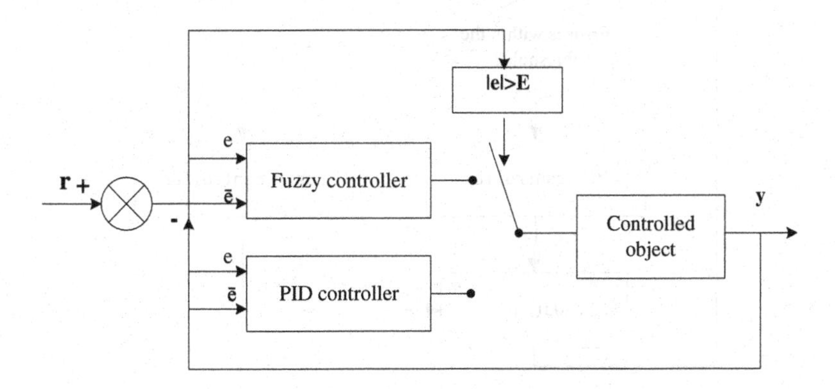

Fig. 2. Fuzzy PID hybrid control

$$u(k) = K_p e(k) + K_i \sum_{j=0}^{k} e(j) + K_d \Delta e(k)$$

k–the sampling sequence number;
$u(k)$–the controller output of the kth sampling;
$e(k)$–the deviation input of the kth sample;
$\sum e(k)$–the accumulated sums of sampling deviation from first to the kth;
$\Delta e(k)$–the difference of the deviation of sampling between the kth and the k−1th;
K_p–the proportional coefficient;
$K_i = K_p T / T_i$–the integral coefficient;
$K_d = K_p T_d / T$–the differential coefficient;
T–the sampling period;
T_i–the integration time;
T_d–the differential time [6].

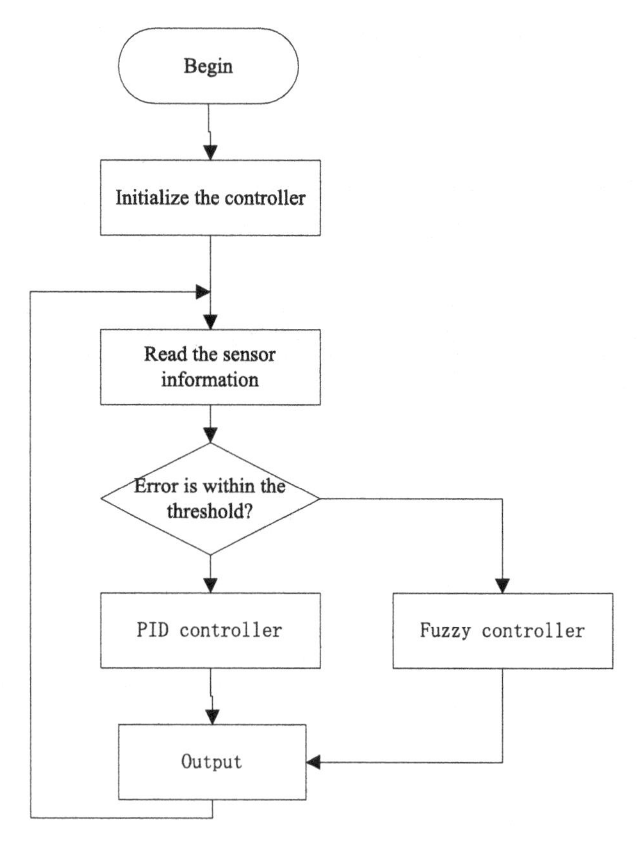

Fig. 3. Flow chart of control program

Fuzzy control based on fuzzy set theory, fuzzy language and fuzzy reasoning, is a nonlinear control method, through the computer to complete control activities that describing by people using natural language. Fuzzy control has many good properties as that does not need to know the mathematical model of the controlled object in advance and has fast response, small overshoot, short transition time, etc. Fuzzy control have higher adjust speed, good robustness but poor steady-state accuracy than PID control that can only achieve a rough control [11].

Comprehensive considering the advantages and disadvantages of the fuzzy control and PID control, this system adopts the method of the combination of the fuzzy control and the PID controller. When the error is bigger, fuzzy controller make rapidly system response. When the error is small, PID controller improve the control precision (Fig. 2).

Web access support three kinds script language of VB, JAVA, and TCL. This research adopts the TCL script language to write the control program. Program flow chart is as the following chart 3 (Fig. 3).

5 Project Node

5.1 The Human-Machine Interface

Web Access use the Microsoft's Internet information services (IIS) as a web Server, provided a friendly man-machine interface with principle of the ASP (Active Server Pages), Supported three kinds of the most popular bitmap file format(BMP, JPEG and GIF), can used this way to display the actual greenhouse frame through the background image, can also provided the structure diagram of the system and the visual image of the data acquisition equipment and control equipment in the form of superposition layer. Web Access also support to embed the Flash animation to the Web Access monitoring pages that can real-time dynamic display the temperature and humidity, light intensity and co2 concentration in greenhouse and the equipment working status and other information through the data I/O channels provided by the components.

5.2 Management Functions

Web access specify the administrator type for all users which determined that what type of pages the user can browse. Administrator types are divided into: super administrator, project manager, online manager, online operators and administrators. This system established two types of users. the super administrator (highest authority) log in system for project management, database setting, using information dialog to access all the points, modify project node properties, drawing graphics and browsing all of the monitoring ages. The online manager was limited to browse only the real-time monitoring data and the running status of the equipment.

Web access built-in alarm functions which can specify the alarm method, alarm type and alarm priority for the monitoring nodes. This system set up alarm functions for the greenhouse environment parameters and the running state of the equipment, which sent email to remind users when the greenhouse environment parameter exceeds the setting value and when the regulation equipment can not work.

Installed Web access will automatically configure a storage database (access database) as the default database which can also used ODBC interface to access other types of database. Greenhouse environment historical parameter information and equipment running information will be stored in a database which can be convenient for user to browsing history data in the form of web page report.

6 Conclusion

This study based on the shenyang science & technology bureau's research project (research and development of the greenhouse intelligent monitoring system based on internet of things) in agricultural science and technology, completed the development of remote monitoring system for greenhouse. Main work includes: aimed at the requirements of the greenhouse monitoring adopted the Webaccess configuration software, ADAM-2000 series of wireless sensor network modules(WSN), ADAM-4000 series I/O module and monitoring computer, completed the construction of the monitoring network and the setting of the monitoring node, proposed a hybrid fuzzy PID control method rely on comprehensive considering the advantages and disadvantages of the fuzzy control and PID control, coded the control script programming, designed the man-machine interface and the management functions. This system that can be formed fast, be convenient for man-machine interaction and solved better the problem of the greenhouse remote monitoring, has the certain promotion value.

References

1. Zhang, X., Lian, X., Yu, C., Gao, J.: Research and implementation of ground-source heat pump remote monitor system based on Webaccess. J. Meas. Control Technol. **28**(6), 51–53, 57 (2009)
2. Liu, J., Liu, L., Zou, W.: Energy monitoring and management system for solar sanitary hot water heater based on advantech WebAccess. J. Meas. Control Technol. **31**(1), 34–37 (2012)
3. Gong, G., Liu, J., Xiao, T., Jian, C.: Design of remote monitor and control system for sludge dewatering based on WebAccess. J. Instr. Tech. Sens. **2**, 52–54 (2013)
4. Hu, B., Ying, H.: Review of fuzzy PID control techniques and some important issues. J Acta Automatica Sin **27**(4), 567–584 (2001)
5. Wang, W., Zhang, J., Chai, T.: A survey of advanced PID parameter tuning methods. J. Acta Automatica Sin. **26**(3), 347–355 (2000)
6. Zhou, L., Zhao, G.: Application of fuzzy-PID control algorithm in uniform velocity temperature control system of resistance furnace. J. Chin. J. Sci. Instr. **29**(2), 405–409 (2008)
7. Qi, I., Wang, J.: A new type of fuzzy PID controller. J. Chin. J. Sci. Instr. **22**(z1), 72–74 (2001)
8. Li, H., Shi, G.: New design method of fuzzy and PID hybrid controller and its application. J. Autom. Instrum. **18**(4), 39–42 (2003)
9. Zhang, E., Shi, S., Weng, Z.: New type of parameters-varying PID control based on fuzzy rules. J. Shanghai Jiaotong Univ. **34**(5), 630–634 (2000)
10. Liu, H., Li, S., Cai, T.: Fuzzy hybrid controller based on fuzzy switching and its application. J. Control Decis. **18**(5), 615–618 (2003)
11. Li, Y., Du, S.: Advances of intelligent control algorithm of greenhouse environment in China. J. Trans. Chin. Soc. Agric. Eng. **20**(2), 267–272 (2004)

Typical Feature Scattering Analysis and Classification in Karst Mountain Plateau Based on Radarsat-2

Ping Wang[1,2], Zhongfa Zhou[1,2(✉)], and Juan Liao[1,2]

[1] Institute of Karst Guizhou Normal University, Guiyang 550001, Guizhou, China
fnwhwp@163.com

[2] National Remote Sensing Center Guizhou branch, Guiyang 550001, Guizhou, China
fa6897@163.com

Abstract. In order to explore the agricultural monitoring with microwave RS technology in karst plateau mountain by full-polarimetric radarsat-2 data, through the pretreatment of the radar data, using statistical analysis method to analyse feature's scattering characteristics for classification from the perspective of multi-temporal and multi-polarized SAR; using the experimental validation to validate classification accuracy. We can see the total accuracy is 80 % above. This study effectively improve the level of microwave RS technology on agriculture in karst area, which can provides scientific method on increasing income in modern tobacco agriculture.

Keywords: Karst plateau mountain area · Radarsat-2 · Backscattering coefficient · Full-polarization · Classification

1 Introduction

Guizhou Province is located in the southwestern China plateau mountain area, complex terrain, rainy days in the whole year. Using conventional visible light imaging is difficult to obtain complete information of surface features. The satellite borne SAR data has the advantages of full time, all-weather, high resolution and rich information, which can make up for the shortage of optical remote sensing [1–3]. Research shows that the first of using satellite borne SAR data for ground object recognition, we must determine the ability to distinguish between the representative parameters of ground objects [4, 5]. In recent years, many researches have different features of ground objects to different features. Therefore, the research on the different features of different ground objects and different features of different objects in different stages of the feature identification, provide technical support for the development of efficient agriculture.

2 General Situation and Data of Study Area

2.1 Research Area General Situation

The study area is located at the modern agricultural base of tobacco in Qingzhen Liuchang of Guizhou province. The base is located at 106°7′6″E-106°29′37″E,

F. Bian and Y. Xie (Eds.): GRMSE 2015, CCIS 569, pp. 487–495, 2016.
DOI: 10.1007/978-3-662-49155-3_50

26°24′5″N-26°45′45″N. The total area is 489 km^2. It has a humid subtropical monsoon climate, and the annual average temperature is 14 °C and the average annual precipitation is 1150.4 mm; Its soil mainly is yellow loam Soil; The pH is 5.5~6.5, and it's rich in organic matter content. The altitude is between the 600–1400 m, mainly on peak cluster depressions and valley. The land is in irregular shape and the typical feature are mainly forestlands, residential areas and crops.

2.2 Data Selection and Image Preprocessing

Data Selection In this study we use the RS and test data. It is synchronous between the monitoring period and the growing period of tobacco. The nominal spatial resolution is 8 m. Through establish the training field by using the 0.5 m resolution aerial photo and by using the method of field survey. To get the test data, first of all, we should establish each 10 training of every kind of ground feature. Because the land of the study area is broken, we choose the typical features including tobacco, residential areas, woodland, corn and paddy field to identify. Using GPS positioning, and at the same time gathering the growth parameters of crops which includes the length, width and water of its leaf, ridge spacing and plant spacing and so on. We could better classify and identify combining the information of radar image texture in later period (Fig. 1).

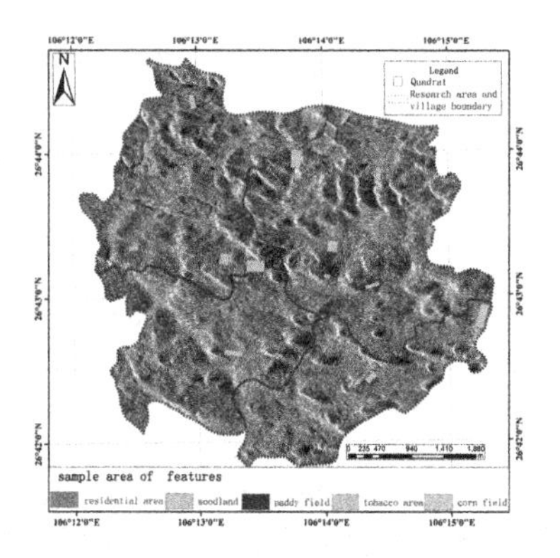

Fig. 1. SAR image data and the sample areas' outline

Data Preprocessing. Compared with optical images, SAR images have poor visual readability and are affected by the geometrical features, such as speckle noise and shadow or perspective shrinkage. Imaging radar transmits the microwave in a specific wavelength band through electromagnetic wave, receives the electromagnetic wave energy from the surface, with which to form coherent imaging. The scattering coefficient of the target is not entirely determined by the scattering coefficient of target. Due to the complex clutter, it comes about a speckle noise in the image, which is the biggest obstacle of SAR image quality [3, 6, 7]. Therefore, we need to preprocess the radar image to reduce the error. SAR image preprocessing mainly includes reading head file, multilook processing, filtering, geocoding and radiometric calibration, with which can increase the useful informations of ground objects.

3 Theory of Backscattering Characteristics Analysis on Ground Objects

The interaction between the electromagnetic wave and the objects itself determines the scattering characteristics of targets. The radar backscattering coefficient is sensitive to the dielectric properties of the object. Influenced by the cloud cover, optical and thermal channel data can not penetrate the vegetation, but in a certain extent, radar can do it [8–10]. Radar is the only reliable data source that can be obtained. The main scattering types of radar include surface, volume and hard target scattering. Different objects have different electromagnetic wave reflection and radiation characteristics, the same object in different bands and different polarization show different tones in the image. This study analyzed the scattering characteristics of the typical ground objects from the aspects of multi-temporal and multi-polarization. Then explored the recognition ability of the spaceborne radar data in karst area.

4 Analysis and Results

According to the backscattering characteristics of the different objects, five typical ground objects in the image which include residential area, corn, tobacco, woodland and paddy field were extracted.

4.1 Analysis from the Single Polarization and Multiple Phase

HH is sensitive to canopy water content, and the reflection is strong and stability. Affected by the terrain, the distribution of tobacco in the hillside and planted in the depression due to the surface roughness factor, the backscattering coefficient is also different (Fig. 2).

(1) Residents point: The residential buildings are mainly composed by buildings. Its structure makes the reflection enhanced. A corner reflector is made to enhance the polarization scattering, which is most obvious in HH. The point in the three periods showed higher backscatter coefficient.
(2) Corn and tobacco: For C band, in the near vertical and small incidence angle, the vegetation scattering is mainly affected by soil moisture content, density, dielectric constant of plant and plant type. Due to the growth period of tobacco and corn are same, it is difficult to distinction. However, it shows that in the root-extended and maturity period, the scattering characteristics have obvious differences, more easy to distinguish. And in the flourishing-growth period, the scattering coefficient of corn and tobacco is very large. However, the water content of the two plants in different stages is different., so the back scattering characteristics are different.
(3) Forest: Woodland is covered by shrubs or trees. Their trunks is tall, canopy is stable, surface rough degree is high [11]; HH is sensitive to canopy water content, reflection is strong and stable.

(4) Paddy field: Backscatter coefficient of paddy field is higher because the field water is enough, and complex dielectric constant is high. Rice seedling's shape is close to the vertical and the water surface is flat, which makes the HH polarization scattering strengthen [12, 13]. So the other fields can be well separated from it. In the multiple scattering process, the backscattering ability is greatly weakened. But in general, the HH polarization is higher than other.

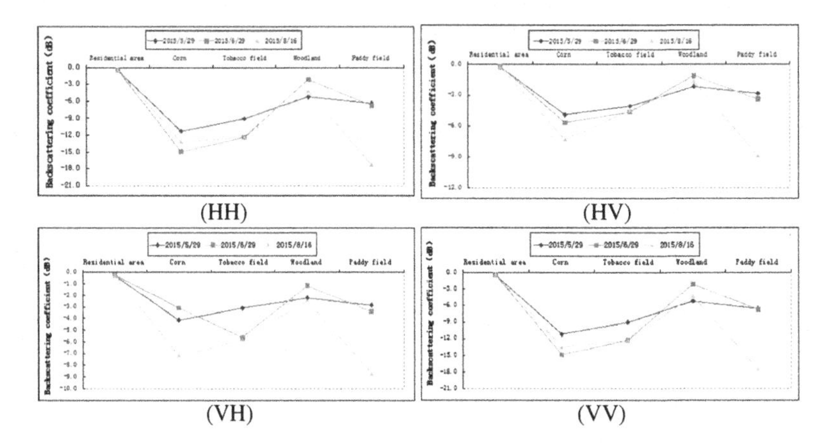

Fig. 2. Typical features scattering characteristics in single polarization and multi phase

4.2 Analysis from the Single Phase and Multiple Polarization

(1) During the period of the tobacco root-extended period (May), it is time that tobacco is transplanted into the field. It is in the seedling stage, the leaves are not fully extended and a large number of soil is bare. The backscattering characteristics are enhanced by soil.

(2) In the flourishing-growth period (June–July), leaf is long and wide, leaf water content is rich, HH is sensitive to water content of the canopy, which can increase the reflection of HH polarization; and its natural curve showing as a bow, the angle between the stem and the leaf becomes large. It also increases the scattering intensity of HH polarization. In the flourishing-growth period leaf is completely open, canopy density is high, plant canopy directly reduce the penetration of VV polarization.

(3) In the maturity period (July–October), tobacco, corn and other crops have entered into the mature period. Partially bare soil exposed. The backscattering properties of field mainly determined by tobacco leaf and stem. So it shows that HH > VV. In the two kinds of polarization modes, HH and VV, the backscattering coefficient is higher, and the two polarization modes have the same trend (Fig. 3).

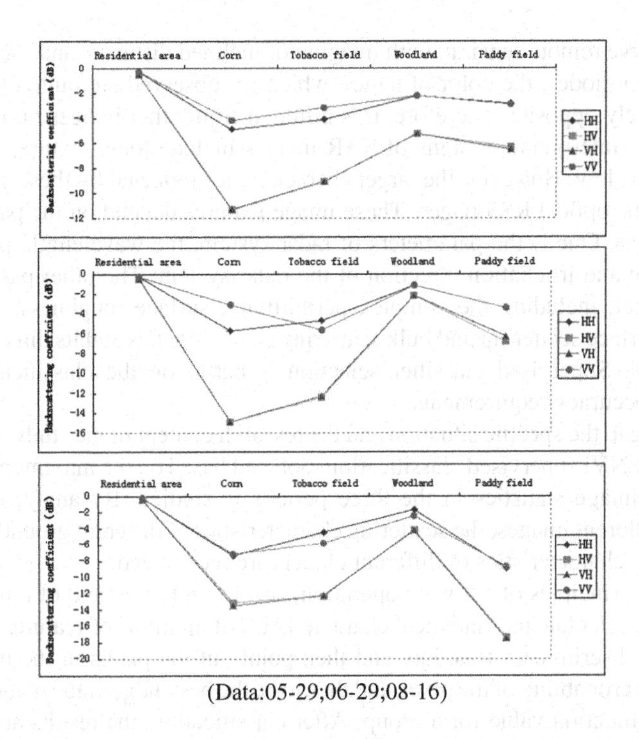

Fig. 3. Typical features scattering characteristics in single phase and multi polarization

The tobacco mature stage is also the mature period of rice, when rice leaf and stalk are basically vertical to the ground [14]. Through field investigation and aerial photograph verification, paddy field distribution is in the surrounding of residential areas. In the August SAR image, according to different backscatter intensity, different bands are shown to different colors. For rice, HH and HV are more sensitive to the incident angle. The HH polarization is shown in green band, red band and blue band respectively, the paddy and the tobacco field are both clearly distinguished [15, 16] (Fig. 4).

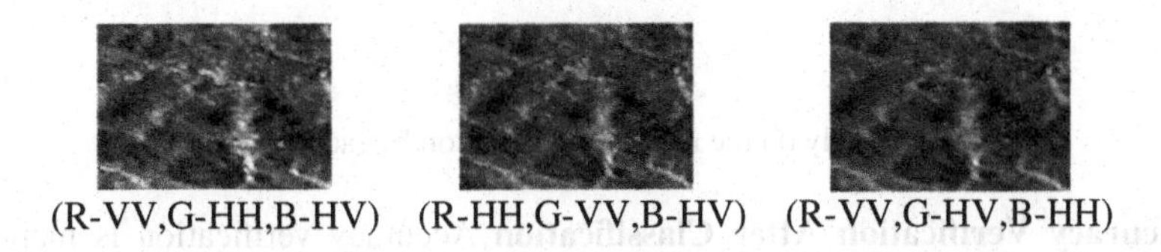

Fig. 4. False color composite image based on backscatter intensity in August 16th

4.3 Supervised Classification

Supervised Classification Method and Classification Results The imaging mechanism of radar sensor is unique, which is different from the visible and infrared images.

Since it is active remote sensing, with imaging of inclined distance, and it has a variety of polarization modes, the color of nature which we observed are quite different, and even completely opposite. Therefore, it is different from other images on image interpretation. The interpretation signs of SAR images include tone, texture, shape, size, pattern and shadow. However, the target characteristics reflected by these signs are not the same as the optical RS images. These image features depend on the parameters of the two aspects. One is the parameters of radar system, the wavelength, polarization, incident angle and irradiation direction of the radar system; The other parameters are from the target, including the complex permittivity, surface roughness, geometrical properties, surface scattering and bulk scattering characteristics and its directional characteristics [6]. Supervised classifier selection is based on the classification of the complexity, accuracy requirements.

According to the specific situation and the research content of the study area, mainly based on the ENVI supervised classification tool, and based on the maximum likelihood method, the image statistics in the three periods is studied. By analyzing different features in different images, the scattering characteristics of different ground objects are analyzed. The characteristics of different objects are recognized.

The basic principles of software operation, according to the field of typical ground training field, calculate the statistical characteristics of all kinds of features, establish a classification discriminant function, and then point out the pixels in the image. Then determine the probability of the pixel, and then to be the best judge value of the maximum discriminant function value for a group. After classification, the results are processed by Majority/Minority and Clump. Classification results are shown in Fig. 5.

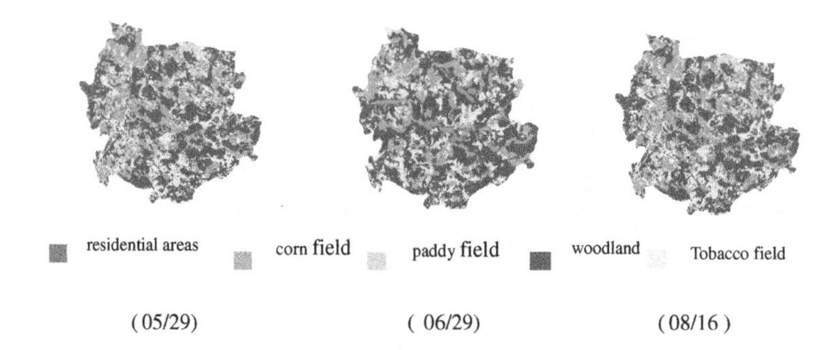

Fig. 5. Study on the image interpretation by radar in 2014

Accuracy Verification After Classification Accuracy verification is mainly to evaluate the classification results. Hypothesis, within quadrats is actually in tobacco fields were divided into tobacco fields, that is N_{TT}; plot is actually in tobacco fields were divided into non tobacco recorded as N_{TO}; plot is actually non tobacco fields were divided into non tobacco fields denoted by N_{OO}; plot is actually non tobacco is divided into tobacco recorded as $N_{OT,}$ (where T represents tobacco fields, O stands for non tobacco fields) [17]. Well, the total classification accuracy (P_A) can be expressed as formula (1).

$$P_A = (N_{TT} + N_{OO}) / (N_{TT} + N_{TO} + N_{OO} + N_{OT}) * 100\ \%. \quad (1)$$

It is seen from Table 1 that analysis scattering coefficient of typical features and the use of the maximum likelihood value in the study area classification, with which total accuracy of the classification in addition is to 84.18 %, the other two period are higher than 85 %, which is due to a large number of soil information in May. It is difficult to distinguish crops category. In June and August, crops have been basically grow into their own unique form, the overall accuracy has improved.

Table 1. Accuracy evaluation of typical features distinguish in 2014

Category date	N_{TT}	N_{TO}	N_{OO}	N_{OT}	Total classification accuracy (%)
05/29	104090.2	19563.3	510326.7	70409.0	84.18
06/29	105485.3	19575.4	510252.5	69076.0	87.40
08/16	104087.5	19521.6	510320.1	70460.0	87.23
Average	104554.3	19553.4	510299.7	69981.7	86.27

5 Conclusion and Discussion

By the classification results, the distribution of residential points in the classification results is not very high, due to the existence of a large number of residential points maked by the building of the corner reflector, and could enhance its reflection ability. Compared with other ground scattering ability, more easily to distinguish with other surface features [18]. Forest scattering characteristics is different between May and the other two period, which is due to that May is the late spring season, more arid, trees and shrubs layer is less flourishing than the other two, results in low water content of canopy. The distribution of paddy field in the three periods is obvious, because of its dominant surface in May, the scattering characteristics are caused by the surface scattering. In June, rice growth strong, scattering characteristics of paddy field is affected by interaction between rice and water. Because tobacco and maize are same as the dry land crops and the growth period are same, the like-polarized scattering ability of the two is higher than that of the cross polarization, and the scattering characteristics is semblable. But we can analyse the causes of the changes in the distribution of the results from different plant characteristics in different periods.

(1) During the period of May, the tobacco and corn are in the same stage, and the soil is bare. In May, mainly by soil moisture. At this time, the tobacco and maize leaves have been carried out, the leaves of corn are long and narrow, water content is lower than the tobacco leaves, the plant form is in horizontal structure, so the HH polarization scattering ability of corn is lower than that of tobacco.

(2) In June, the biggest difference between growing period of tobacco and corn is water content. Tobacco leaves in this period are rich in water content, large and thick; and corn leaves are narrow, water content less, so the HH polarization of corn is less than tobacco. Since this period of tobacco leaves have been launched, plant canopy was greatly easy to penetrate by VV polarization, and VV polarization is higher than that of tobacco. Due to the complexity of the plant, the scattering characteristics of the tobacco and corn were similar, and the classification accuracy of tobacco decreased.
(3) In August, corn at this time is to enter to mature period, the plant's water reduced, leaf drooped, color become yellow green from deep green gradually. During the same period, the tobacco in mature period, tobacco top was removed, the leaves began to curly yellow, which is mainly constituted by leaf, color is dark green. HH polarization scattering ability of corn is lower than that of tobacco. In the classification results showed tobacco classification accuracy decreased.

In a word, through the study area of Liuchang modern tobacco agriculture base in Guizhou, base on the microwave remote sensing identification research, explore the quantitative monitoring technic through microwave RS in Guizhou Karst Plateau Mountain. This microwave remote sensing monitoring technology is more mature, the classification results are more accurate. It is suitable for the complex terrain and climate conditions in Guizhou mountainous area. It can provide real-time, accurate and effective monitoring for agriculture plantation in Guizhou mountainous area and also provide technical support for tobacco industry in Southwest China even the whole country.

Acknowledgments. In this paper, the research was sponsored by the National Basic Research Program of China (973 Program, Project No. 2012CB723202), Guizhou province science and technology plan (Research and application of key technologies on SAR remote sensing platform monitoring and identification in Karst mountainous area) (Guizhou S&T Contract GY[2013]3062);The major application foundation research project of Guizhou Province (Guizhou S&T Contract JZ 2014-200201)

References

1. Wang, A., Zhou, D., Gong H.: A research on method of wetland vegetation identification and classification based on radar bakscatter characteristics. Remote Sens. Inf. (2), (2012)
2. Shu, N.: Principle of radar remote sensing, pp. 80–89. Surveying and Mapping Press, Beijing (1996)
3. Hua, G.: Study on maize growth monitoring and maize mapping based on full-polarization SAR data. Nanjing University of Information Science and Technology, Nanjing (2011)
4. Hua, G., Xiao, J., Huang, X.: Analysis of backscattering characteristics of corn based on fully polarimetric SAR data. Jiangsu Agri. Sci. **399**(3), 562–565 (2011)
5. Sui, L.: Active radar remote sensing, pp. 153–154. Surveying and Mapping Press, Beijing (2009)
6. Chen, J., Lin, H., Shao, Y.: Application of microwave remote sensing in rice growth monitoring, pp. 71–77. Science Press, Beijing (2010)

7. Zhang, B.: Methods for rice identification and sown area measure using ASAR data—a case study in Baoyang COUNTY of Jiangsu, pp. 27–31. Nanjing Agricultural University, Nanjing (2004)
8. Jia, M.: Microwave scattering characteristics of rice and its parameter inversion, pp. 55–59. University of Electronic Science and technology, Chengdu (2013)
9. Oliver, C.: Understanding synthetic aperture radar images, pp. 273–293. Electronics Industry Press, Beijing (2009)
10. Huadong, G.: Analysis of radar remote sensing image in China, pp. 1–35. Science Press, Beijing (1999)
11. Kurosu, T., Fujita, M., Chiba, K.: The identification of rice fields using multi-temporal ERS-1 C band SAR data. Int. J. Remote Sens. **18**(14), 2953–2955 (1997)
12. McNairn, H., Kross, A., Lapen, D., Caves, R., Shang, J.: Early season monitoring of corn and soybeans with TerraSAR-X and RADARSAT-2. Int. J. Appl. Earth Obs. Geoinf. **28**, 252–253 (2014)
13. Zhang, Y., Liu, X., Su, S., Wang, C.: Retrieving canopy height and density of paddy rice from Radarsat-2 images with a canopy scattering model[J]. Int. J. Appl. Earth Obs. Geoinf. **8**, 170–171 (2014)
14. Zhang, Y., Li, Y.: Key techniques of SAR image processing, pp. 340–347. Electronics Industry Press, Beijing (2014)
15. Zhao, C., Zhao, Z., Zhao, X.: Digital image processing and analysis, pp. 314–320. Tsinghua University Press, Beijing (2013)
16. Shao, Y., Liao, J., Fan, X.: Analysis of time domain backscattering characteristics of rice: a comparison of the results of radar satellite observation and model simulation. J. Remote Sens. **6**(6), 222–229 (2002)
17. Hu, J., Zhou, Z.: Multitemporal multipolarity difference map in mountain plateau of tobacco recognition. Jiangsu Agri. Sci. **43**(3), 388–391 (2015)
18. Jia, L., Zhou, Z., Li, B.: Application of high resolution SAR to tobacco growth modeling in Karst Mountainous Region. Chin. Tobacco Sci. **34**(5), 104–112 (2013)

Research on Data Calibration Model for the Working State Detection About Traffic Sound Barrier

Qi He[1], Zhengwei He[1,2(✉)], Dian He[3(✉)], Chaowei Hu[1], and Bo Wu[1]

[1] Wuhan University of Technology, Wuhan China
280360869@qq.com, wwwhzw@whut.edu.cn
[2] Hubei Inland Shipping Technology Key Laboratory, Wuhan 430063, China
[3] Wuhan Geomatic Institute, Wuhan China
115055030@qq.com

Abstract. Traffic sound barrier is an effective measure for suppressing road or railway traffic noise. The data measured from the inner acceleration sensor of sound barrier, is used for detecting the sound barrier's working state. But the data usually exists error, and will cause difficulties for maintaining the sound barriers specifically. Based on simulation and numerical integration on existing acceleration data, this paper get the sound barrier movement characteristic curve. And as integrated use of iterations average filter model, high pass filter model, delay correction model and amplitude correction model, a data calibration model is constructed to calibrate the data error. Finally, the model test experiments are implemented on the experimental data of traffic sound barrier state, and achieved good effects.

Keywords: Data calibration model · Traffic sound barrier · Acceleration sensor · Error correction · Iteration average filter · Model validation

1 Introduction

Sound barrier is an effective measure of a road or railway traffic noise suppression, and under normal conditions, the swing of the sound barrier in fluctuating wind pull under pressure effect should be within a certain range, and it needs to be repaired when above the normal range. Through detect the working status of the acceleration sensor inside the sound barrier detector, we can be targeted for repairing the sound barrier. In practice, the data of sensor measured is usually exist error [1]. Because of the errors, it will accumulate a lot of interference in use of numerical integration method for calculating the vibration displacement and will cause great inconvenience in repairing. Therefore, the measured data need to be corrected. But there is no literature to give a solution currently[1].

F. Bian and Y. Xie (Eds.): GRMSE 2015, CCIS 569, pp. 496–503, 2016.
DOI: 10.1007/978-3-662-49155-3_51

2 Data Correction Based on Iterative Average Filter Model

2.1 Random Error Elimination [2–3]

For the series of acceleration data obtained by, first, based on the physical formula of acceleration-velocity and acceleration-displacement and numerical integration to calculate the velocity and displacement, and make velocity-time and displacement-time curve

Taking into account the acceleration sensors in sound barrier detector record data by intensive sampling and have large amounts of data, using the trapezoidal method can achieve approximate integration. Trapezoidal method, were used the adjacent two data as the upper line and lower line of the trapezoid and the reciprocal of the sampling frequency as high trapezoid, the trapezoid area is calculated, and then accumulate the trapezoid area.

In general, acceleration data is very unstable and a large degree of dispersion, the curve obtained from it volatility with no regularity and there are many non-normal extreme points on it, which will lead to errors in the integral curve. So it is judged to be a part of measurement error from the acceleration sensor. It appears unavoidable and does not have the law, belongs to random errors.

So use the iteration average filtering method to eliminate errors, and the main methods is:

$$a_i' = \frac{a_i + a_{i+1} + \cdots + a_{i+N}}{4(i = 1, 2 \ldots M|N)}$$

Wherein a_i' is a filtered data, M is the total number of data.

For the selected N of the total number in an average calculation, when N is a large number, the signal have good smoothness, but the sensitivity is low; when the N value is smaller, the lower the degree of signal smoothing, but the higher the sensitivity. This paper selects $N = 4$, and a second filter, can eliminate abnormal extreme points but also to ensure sensitivity.

2.2 System Error Elimination

Ideally, the initial acceleration value of the sound barrier should be zero, but the acceleration data does not actually get to zero, that zero drift errors [4–6]. Due to the existence of the zero drift errors at any time, the zero drift acceleration sensor classified as systematic errors.

To eliminate errors caused by zero drift, we use a high-pass filter model. Based on high-pass filtering idea, the acceleration - time curve seen as a signal curve. The curve shows as low-frequency signal of a small perturbation when not added incentive. And plus performance incentives, the curve is the disturbance signal superimposed high-frequency excitation signal, the signal through a high-pass disturbance signal filter to filter out low frequency.

On the basis of the average iterative filtering model, we will fit all acceleration data with no incentive using least squares curve fitting that the fitting out of the curve is the low-frequency disturbance signal, and subtract the disturbance signal from the entire curve to eliminate the zero drift.

Specific algorithm is as follows:

- Using the least squares to seek fitting function of the disturbance signal [7–10]: For a given acceleration a_i^0, seeking polynomial function $p_n(\mathbf{t})$, and the square of should be minimized, that:

$$\sum_{i=0}^{m} r_i^2 = \sum_{i=0}^{m} \left[p_n(t_i) - a_i^0\right]^2 = \min$$

  Where t_i represents a time, a_i^0 indicates that a value of the acceleration is not energized, m represents the total number of the acceleration data recorded when the excitation not applied.
- The acceleration $a\left(t_i\right)$ at a time subtracting the time perturbation signal $a\left(t_i\right)$

$$a'\left(t_i\right) = a\left(t_i\right) | p_n\left(t_i\right) \quad (\mathrm{i} = 1, 2 \ldots \mathrm{M})$$

  Where $a'\left(t_i\right)$ represents the acceleration data after the correction for time t_i, $a\left(t_i\right)$ is the acceleration data not corrected for time t_i, M is the total number of the acceleration data.

The calculated velocity data computed by the acceleration data, which is less than 0 after shocking, is inconsistent with the actual situation. Suppose some components used in acceleration measurement instrument such as capacitors has a certain lag impact on the signal (especially negative signal), making acceleration rather long duration of action (or there are other circumstances makes acceleration lag), thus affecting the rate of return.

This paper use the lag correction model and introduce the correction factor λ, eliminate the lag effect by modifying the reverse effect of acceleration of time, the performance is the speed eventually return to zero. It is need to meet the following conditions:

$$\begin{cases} \int_{t_1}^{t_2} a(t_i)dt + \int_{t_2}^{t_3} a(t_i)dT = 0 \\ T = \lambda t \end{cases}$$

Where λ is the correction factor, T is the time after reduction, t_1, t_2, t_3 represent a particular abscissa where vibration acceleration curve intersection with the horizontal axis.

After the model for simulation of the improved model, from displacement map, we found that there will be a phenomenon that the deviation from horizontal axis at the end of it, and such displacement deviation concentrated in the end of the last movement, which then made the following improvements:

In the sound barrier acceleration images, acceleration amplitude correction parameters have great difference with different locations, and therefore this paper, using amplitude correction model, try to fix acceleration amplitude, and compare the results of the double integral acceleration to correct displacement deviation, thereby introducing pieces value of the correction factor β, must meet

$$s = \int\int_{t_1}^{t_2} \beta\, a(t_i)d^2t + \int\int_{t_2}^{t_3} \beta\, a(t_i)d^2T + \int\int_{t_4}^{t_5} \beta\, a(t_i)d^2T + \int\int_{t_5}^{t_6} \beta\, a(t_i)d^2t = 0$$

Where s is the final displacement, T is corrected time (T = λt)|$t_{\downarrow}1, t_2, t_3, t_4, t_5, t_6$ representing the intersection of acceleration curve with the horizontal axis (in a back and forth motion, for example).

3 Model Validation and Results Analysis

Shows the acceleration of the sampling data and the integration results in a particular experiment with the sound barrier under the normal state, when the end of the experiment, there is a significant deviation of the displacement speed and displacement of the sound barrier with the actual situation. Now there exist three groups of normal state of sound barrier experimental, the acceleration data is measured by acceleration detector with imitating sound barrier vibration, the sampling frequency of acceleration sensor is 1000 Hz, acceleration units is $g \cdot \frac{m}{s}$ (g is the acceleration of gravity). The data for the three groups: single-direction movement from point A to point B, denoted first experiment; return from C to D and then to C, denoted by the second experiment; from E to F, then by F to E, and then repeat remember to experiment three. Its initial velocity are all 0. As shown below.

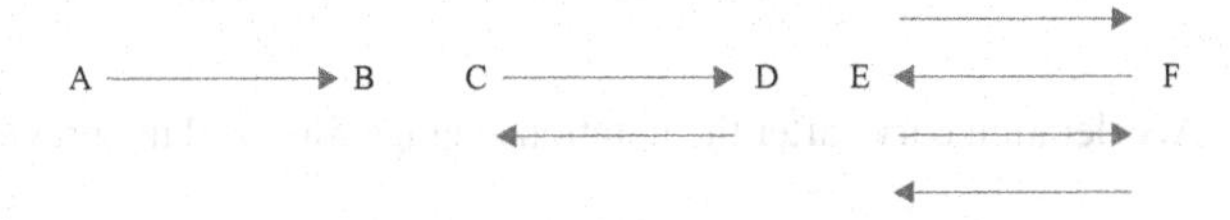

Theoretically single direction movement from point A to point B, the rate of change is increasing rapidly at first, slowly reach extremes, the speed drops from slow to fast until the speed drops to zero. But using the acceleration data to calculate the speed, the rate of decline is less than 0, inconsistent with the actual situation.

First, calculated velocity - time curve using the trapezoidal method, and then get the displacement - time curve. Contrasting Curve and the theoretical curve and do error analysis.

Experiment two acceleration - time curve as follows (Fig. 1):

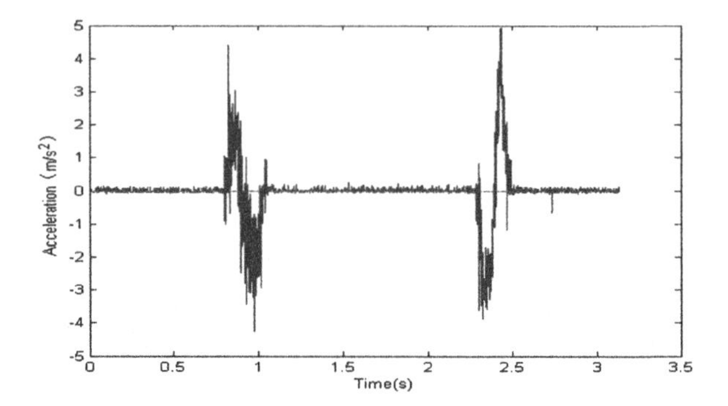

Fig. 1. The second experiment acceleration - time image

Then iterate average filter, the results obtained were as follows (experimental one example)

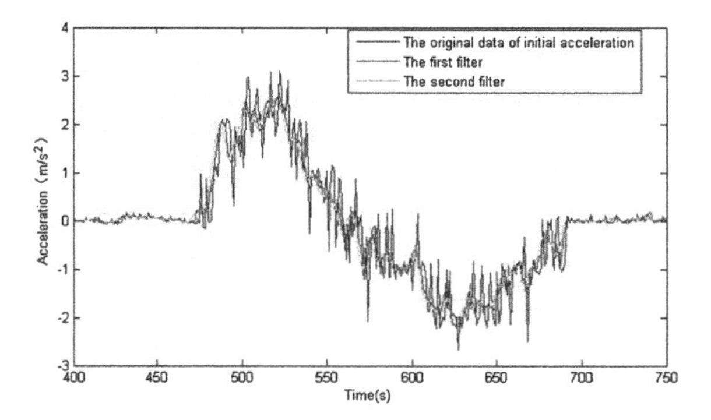

Fig. 2. Acceleration curve after the iteration average filter and the previous image

Then eliminate the zero drift error and the corrected acceleration signal with speed - time image as follows (in the second experiment as an example):

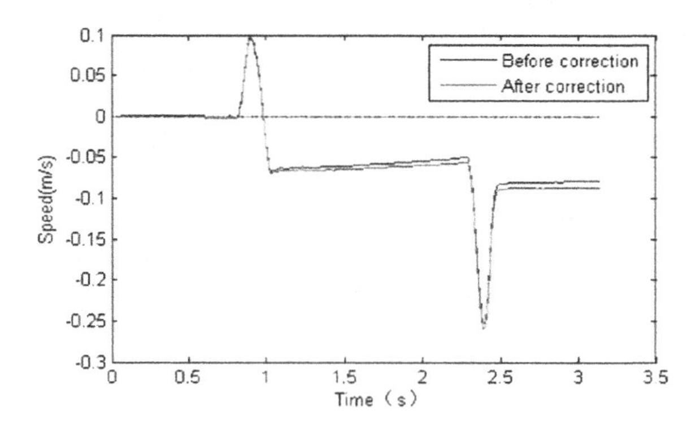

Fig. 3. Experiment two speeds - time zero drift correction image

By comparing the curve before with corrected curve, we find that the cumulative integration errors caused by acceleration zero drift have been eliminated substantially.

Then lag correction, at last, the experiment one calculated $\lambda = 0.809$; experiment two get from point C to D $\lambda = 0.605$, from point D to point C $\lambda = 0.879$; experiment three get from E to F point $\lambda = 0.825$ from point F to point E $\lambda = 0.754$, and the second from the point E to F $\lambda = 0.575$, and the second from the point E to point F, $\lambda = 0.745$. Comparison of the corrected and the before image below:

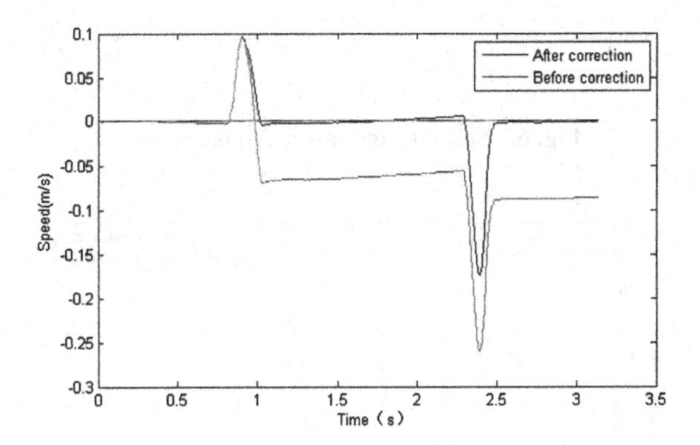

Fig. 4. Experiment two speeds – time

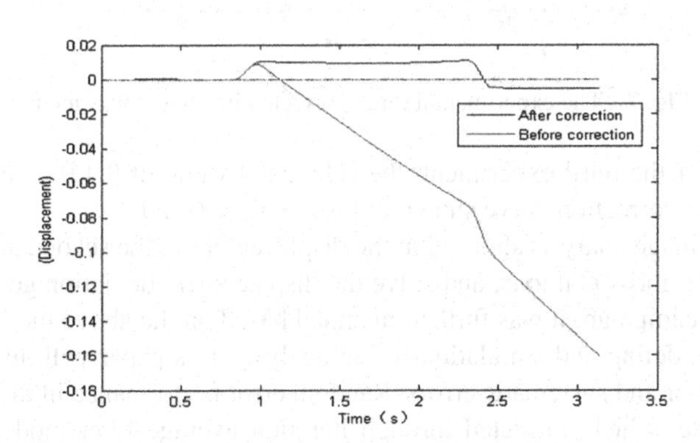

Fig. 5. Experiment two speeds - correction image displacement correction image

Finally, the amplitude correction. For the second experiment, based on the acceleration curve after filtering, using MATLAB to calculate the β value of 1.032, after amendments to calculate the corrected displacement - time curve and the uncorrected image. The contrast image below left:

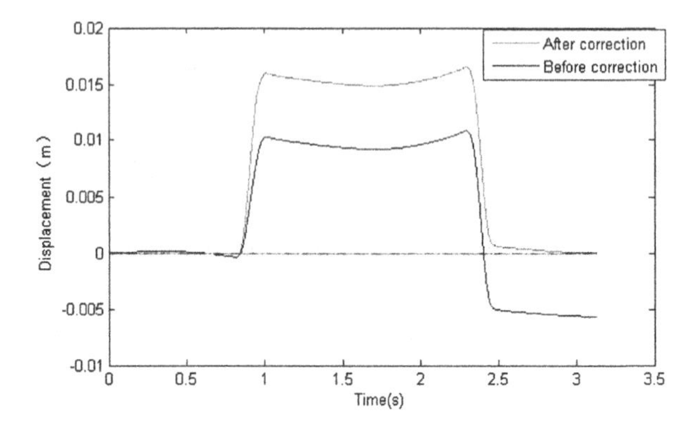

Fig. 6. Second experiment displacement

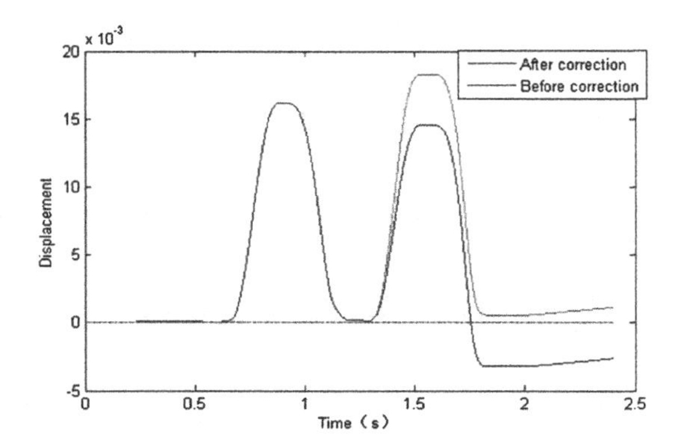

Fig. 7. The experimental three-corrected image displacement

Similarly for the third experiment, the calculated value of β 1.022, calculated the before and after correction curve shown in Figs. 3, 4, 5, 6 and 7.

Corrected image analysis shows that the displacement in the second and third trials end close to the horizontal axis, and solve the displacement deviation generated at the end of the last campaign, it was further amended based on the above model.

Through modeling and simulation fitting analysis, this paper will divide the error into random error and systematic errors. Random error performance in the acceleration data on discrete, which corrected through iteration average filter model; systematic errors are divided into two categories according their differences in characteristic: zero drift error and the lagged effects errors. Zero drift error corrected through a high-pass filter model, the lag effect of the error corrected by the amplitude correction model. After we made a bold assumption based range of applications of the data models and correction ideas on the promotion of the model. Specifically described as follows:

Iteration average filtering model is the basis of this article correction. It is widely used in handling the random errors, because of the mature technology and the good effect for random error correction;

High-pass filter model applied the idea of filter circuit which pass high-frequency and low-frequency resistance to achieve the elimination of the integral error caused by linear zero drift. But lacking of a common standard parameter to calibrate the characteristic curve, there has a concave trend in the upper and rear part of the displacement curve, and does not match the ideal situation, which blamed for the instability in the experimental state and the environment, or other causes such as the model is not perfect, it is necessary to improve the models from other angles.

Acknowledgements. This article is supported by "the Fundamental Research Funds for the Central Universities (WUT: 2014-IV-094)" and "China Western Project (NO. 108-45120691)". We thank for the supports.

References

1. Weeks acquaintance. Analyze the sound barrier displacement control. Shanxi Archit. **39**(02), 209–210 (2013)
2. Broad, Liang, F.: Acceleration signals and random noise elimination method for real-time trend term study. Electron. Des. Eng. **21**(14), 18–22 (2013)
3. Rui Tang, Y.: Introduction to measurement system error and random error. Econ. Technol. Cooperation Inf. (21), 149–149 (2008)
4. Shuang, Z., Zhou H., Xiao, X., Qin H., Kun, Q.: Based on non-local means filtering and high-pass filtering of the time domain Non-uniformity correction algorithm. Photons J. **43**(1), 153–156 (2014)
5. Zhong Bo, Z.: Pressure sensor zero drift solution analysis. Technol. Horiz. (34), 100 (2014)
6. Soares, Sun Timber, Li, G.: Relatively hot zero pressure sensor drift compensation of various calculation methods. Sens. Technol. **17**(3), 375–378 (2004)
7. Liu, W.: MATLAB programming and applications, 2nd edn. Higher Education Press, Beijing (2006)
8. Zhuo, J., Yongsheng, S.: MATLAB applications in mathematical modeling. Beijing University of Aeronautics and Astronautics Press, Beijing (2011)
9. Division, S.-K., Xi, S.J.: Mathematical modeling algorithms and applications. National Defense Industry Press, Beijing (2013)
10. Pang, Z.-H., Depression, J.: Predictive control algorithm based on error correction is reviewed. Chem. Autom. Instrum. **32**(2), 1–4 (2005)

Comparison and Fusion of Multispectral and Panchromatic IKONOS Images Using Different Algorithms

Dong Liang[1,2], Fan Yang[1], Jinling Zhao[1,2(✉)], Yan Zuo[1], and Ling Teng[1]

[1] Key Laboratory of Intelligent Computer and Signal Processing, Ministry of Education, Anhui University, Hefei 230039, China
aling0123@aling0123.com

[2] Anhui Engineering Laboratory of Agro-Ecological Big Data, Anhui University, Hefei 230601, China

Abstract. The fusion of multi-resolution remote sensing images has become a hot issue for enhancing the original images. In comparison with low- and middle-resolution remote sensing imagery, high spatial resolution images have competitive advantages in identifying fine spatial features of land cover features. In this study, an IKONOS image of Hefei, Anhui Province, was used to compare the fusion effects based on seven typical transform methods including the HSV (Hue-Saturation-Value), Brovey, Wavelet Transform (WT), Principal Component (PC) transform, Gram-Schmidt, Pan Sharpening, and Color Normalized (CN) transform. The spatial texture, spectral feature and classification accuracy were used to evaluate the fusion effects. The results showed that PC transform had a optimal performance; WT transform had better ability to keep spectral information; Pan Sharpening provided superior structure information and classification effect; Gramm-Schmidt transform had better spatial and spectral information but a general classification effect; CN transform maintained a good spectral information; and Brovey transform was the worst algorithm. In addition, the classification was also performed using the seven fused images and the accuracy was 85.03 %, 84.20 %, 90.26 %, 88.18 %, 85.48 %, 88.18 %, 85.48 %, respectively.

Keywords: Image fusion · IKONOS · High resolution imagery · Supervised classification · Remote sensing

1 Introduction

Remote sensing technology has been widely used in urban planning, road construction, land use survey, etc. Compared with the middle and low resolution satellite image (e.g., Landsat, MODIS and AVHRR), the high spatial resolution images (e.g., IKONOS and QuickBird) have the better identification abilities of identifying various land cover features. These images have high-resolution multispectral band and higher resolution panchromatic band.

At present, there are different fusion algorithms, in which the Brovey transform, wavelet transform (WT), principal components (PC) transform, Gram-Schmidt and Pan

F. Bian and Y. Xie (Eds.): GRMSE 2015, CCIS 569, pp. 504–513, 2016.
DOI: 10.1007/978-3-662-49155-3_52

Sharpening have been extensively used [1–9]. Each fusion algorithm has its advantages and disadvantages in keeping the spectral and spatial information. To evaluate the fusion methods, visual interpretation and quantitative evaluation are usually used. Hu et al. [1] found that the Pan Sharpening and WT transform were more suitable for the Pléiades image by comparing four fusion algorithms of principal component (PC) transform, high-pass filtering, WT, Ehlers and Pan Sharpening. He [2] showed that PC transform was particularly suitable for the fusion of IKONOS panchromatic and multispectral bands. Bie et al. [5] has shown that Gram-Schmidt was the best choice to merge the panchromatic and multispectral bands of Geoeye-1 image. Dong et al. [6] found that the adaptive wavelet packet fusion algorithm was suitable for the high-resolution satellite image of IKONOS. Wen [9] found the HPF fusion method works best on the IKONOS satellite image. Tan et al. [10] suggested that different image data needs different fusion methods. In general, Gram-Schmidt and Pan sharpening perform better in fusing the hyperspectral image, and traditional fusion methods can also have an advantage in the fusion of moderate resolution images.

To make use of these data more effectively and to overcome the shortage of a single data source, it is highly necessary to fuse the panchromatic and multispectral bands effectively for a certain remote sensing imagery. However, in previous studies. In order to find out a more suitable fusion algorithm to provide a technical support for mine environment monitoring, disaster assessment, urban planning. The IKONOS panchromatic and multispectral bands were used as the data source. The objective of this study is to evaluate the fusion effects using seven typical fusion algorithms. To quantitatively analyze each method, the spatial texture, spectral feature and classification accuracy were used.

2 Materials and Methods

2.1 Data Source and Preprocessing

The IKONOS satellite, successfully launched by Spacing Imaging Company on September 24, 1999, is the first high-resolution commercial satellite in the world, and the revisit cycle is about three days (Table 1). The satellite has strong data acquisition ability and more flexible attitude. Its successful launch means that the advent of the era of high resolution remote sensing, and provides a convenient way to get the latest geographic data access. The satellite has been widely applied to the mine environment monitoring, disaster evaluation, urban planning, forestry resource survey and monitoring of the land use and land cover. The panchromatic and multispectral bands of IKONOS (Fig. 1) in Hefei were used to compare the fusion effects of different methods, which was acquired on October 17, 2010. The experimental area is located in the southwestern corner of Hefei, Anhui Province, in which the primary land cover types mainly include the building, road, water, vegetation and others.

The original images have been corrected in radiation and geometry, and they have been assigned to a geographic reference of Universal Transverse Mercator (UTM) coordinate system. ENVI (The Environment for Visualizing Images) software was used to perform the data processing and image fusion. To fuse the two bands, accurate geometric correction must be carried out and the quadratic polynomial was used in our experiment. Here, the panchromatic band was used as the reference image, the original multispectral images were respectively geometrically corrected (Fig. 1). To calculate the statistics of fused images, all the bands of seven fused image were formed into one image by the "Layer Stacking" and the statistics was obtained by "Compute the Statistics" in ENVI.

Table 1. The main parameters of IKONOS satellite image.

Panchromatic band	Spectral range: 0.45–0.90 μm	Spatial resolution: 1 m
Multispectral band	Band 1 (blue): 0.45–0.52 μm Band 2 (green): 0.51–0.60 μm Band 3 (red): 0.63–0.70 μm Band 4 (near infrared): 0.76–0.85 μm	Spatial resolution: 4 m

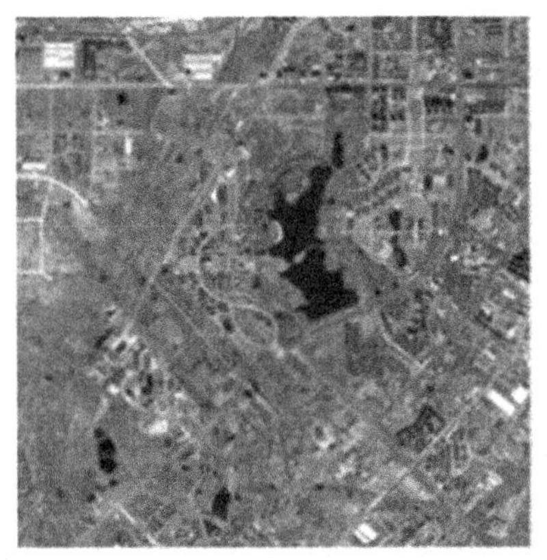

(a) The panchromatic image (b) The multispectral bands images

(c)the corrected multispectral image

Fig. 1. Original (a) panchromatic and (b) multispectral image, and (c) geometrically corrected multispectral image

2.2 Methodology

2.2.1 Fusion Methods

We chose seven fusion algorithms to compare the fusion effects, including the HSV (Hue-Saturation-Value), Brovey, Wavelet Transform (WT), Principal Component (PC) transform, Gram-Schmidt, Pan Sharpening, and Color Normalized (CN) transform. Specifically, the fusion methods of HSV, Brovey, PC transform, Gram-Schmidt and CN transform were carried out in ENVI. Conversely, the WT transform was performed using the wavelet packet programmed by IDL (Interactive Data Language) language. We chose the geometrically corrected multispectral images as the image A, and the panchromatic band was treated as the image B. Pan Sharpening works by the Pan Sharpening function in the ENVI ZOOM module. In comparison with the other algorithms, HSV and Brovey transform have been extensively used, and the WT, PC, Gram-Schmidt, Pan Sharpening and CN transform were just described in this study.

The fusion of WT is to filter the panchromatic image and multispectral image and merge them into one fused image (Fig. 2). Specifically, the WT is respectively applied in the panchromatic image and multispectral image.

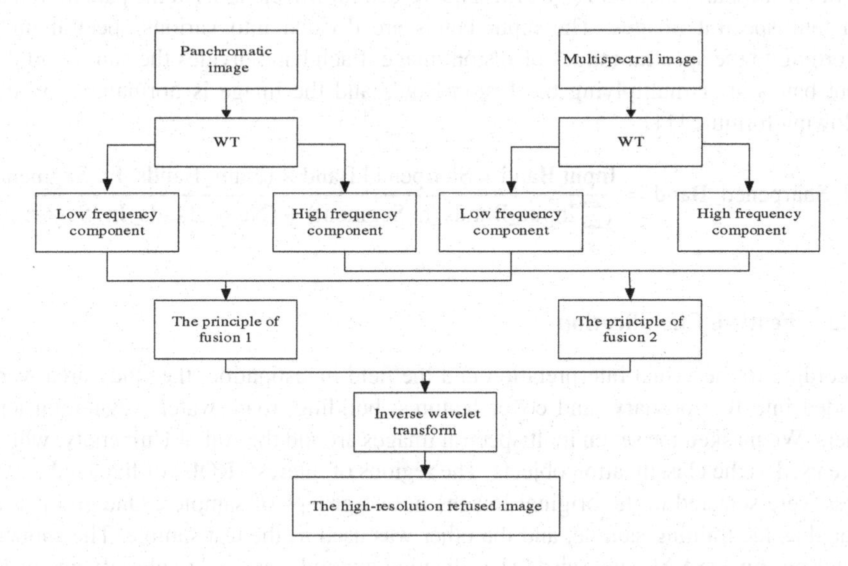

Fig. 2. Schematic diagram of fusing the panchromatic and multispectral images using the WT.

PC transform is the process of sharpening a multispectral image based on the panchromatic image, which is carried out in the multispectral image. We replaced the first principal component (PC1) band by the high resolution band matched to the PC1 to avoid spectral distortion information. Then, the inverse transformation was performed. Finally, the multispectral image was resampled to the high resolution by the nearest neighbor, bilinear interpolation and cubic convolution.

Gram-Schmidt transform is an orthogonalization transformation commonly used in linear algebra. The panchromatic band is firstly replaced by one band of the multispectral bands and is treated as the first band, and the multispectral bands are carried out by the Gram-Schmidt transform. Then, we replaced the first band of the fusion image by the panchromatic band and performed an inverse transformation to get the final fusion image.

Pan Sharpening is a fusion algorithm to merge the panchromatic image with the multispectral image. Both the two images should be simultaneously acquired at the same platform. The panchromatic image is applied to enhance the spatial resolution of multispectral image. It is assumed that the observation process of panchromatic image and multispectral image is simulated by combing the characteristics of two separate sensors. The expectation of the multispectral image is estimated by the prior knowledge. The method, increasing the spatial resolution and keeping the spectral information, has a good performance in automatically adjusting the panchromatic and multispectral images.

CN transform enhances the resolution of a multispectral image by the panchromatic image. It merges the panchromatic image and the selected band of the multispectral image, and the rest bands are exported directly. The spectral range of band is limited by the full width-half maximum (FWHM) and the central wavelengths of the panchromatic and multispectral images. The input bands are divided into various spectral units according to the spectral ranges of fusion image. Each band divides the sum of all the input bands after multiplying the fusion bands and the image is normalized by the following formula [11]:

$$\text{CN\_Sharpened\_Band} = \frac{\textbf{Input Band} \times \textbf{Sharpened Band} \times (\textbf{Num\_Bands\_In\_Segment})}{(\sum \textbf{Input\_Bands\_n\_Segment}) + (\textbf{Num\_Bands\_In\_Segment})}$$

2.2.2 Feature Classification

According to the visual interpretation and the field investigation, the study area were divided into five primary land cover features: building, road, water, vegetation and others. We masked the seven multispectral images around the Anhui University, which were used as the classification objects. The regions of interest (ROIs) of five land cover types were selected in the original images in two groups of samples. One group was treated as the training sample, and the other was used as the test sample. The support vector machine (SVM) supervised classification method was used to classify and evaluate the classification ability.

3 Analysis of Fusion Effect

3.1 Visual Interpretation

It could be found that the fused images have been obviously enhanced compared with original multispectral image. It was obvious that the spatial structure of land cover types were much clearer, and especially for the edges of trees, roads and buildings. In comparison with the spatial information, the images fused by the PC transform, Gram-Schmidt transform, Pan Sharpening and CN transform had higher resolution and their spatial textures was obviously enhanced. In comparison with the color, the fused images produced by the WT transform, PC transform, Gram-Schmidt transform, and Pan Sharpening were much close to the original multispectral image. Conversely, they could retain more spectral information, but there were more color differences for the HSV, Brovey and CN transform, which showed that the spectral information was lost and the fusion effects were relatively worse (Fig. 3).

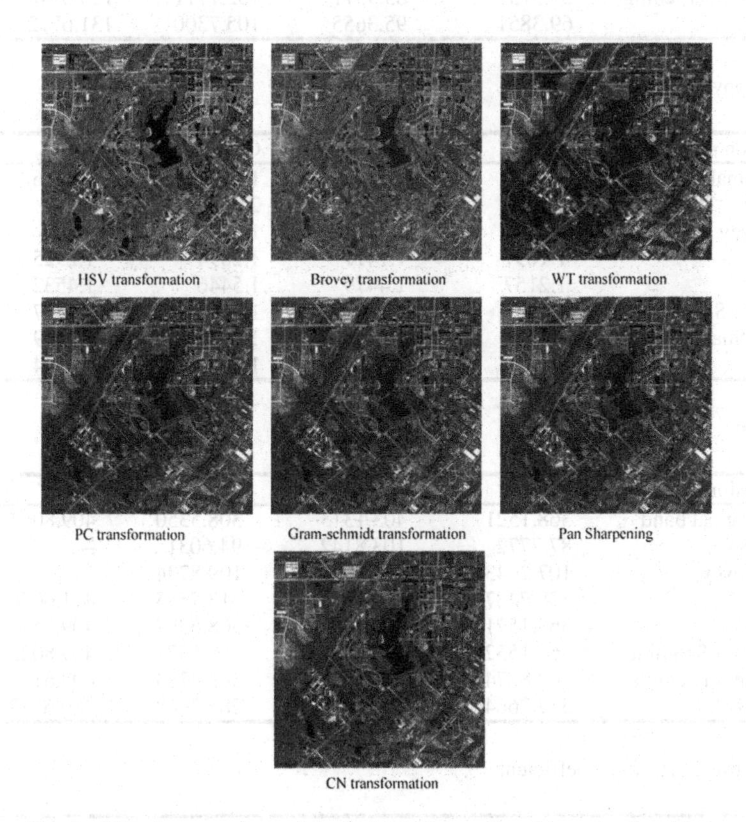

Fig. 3. Visual comparison of the fusion effects using the seven algorithms.

3.2 Quantitative Evaluation

We evaluated the fusion images from two aspects of spatial textures and spectral features. The standard deviation and entropy were used to analyze the spatial structure, and the mean and spectral correlation coefficient were used to analyze the spectral characteristics (Table 2).

Table 2. Comparison of the evaluation results using the seven fusion methods.

(a)Standard deviation

Fusion method	B	R	G	NIR
Original band	59.5993	88.0170	104.7277	117.6400
HSV	46.9825	53.8493	54.5093	—
Brovey	31.9684	33.6836	41.5944	—
WT	62.4195	89.2745	104.6567	115.8858
PC	56.0830	83.5573	100.8519	125.7375
Gram-Schmidt	55.1502	83.2745	101.5396	126.9424
Pan Sharpening	55.5751	83.9571	102.1711	125.0148
CN	69.3851	95.3655	105.7300	131.6772

(b) Entropy

Fusion method	B	R	G	NIR
Original band	0.9517	1.3048	1.4295	1.7996
HSV	2.1027	2.1017	2.0634	—
Brovey	0.9136	1.4160	1.5209	—
WT	1.1697	1.1411	1.2094	1.2925
PC	1.2157	1.4479	1.5446	1.4532
Gram-Schmidt	1.2581	1.4556	1.5380	1.4227
Pan Sharpening	1.4652	1.5354	1.4774	1.2609
CN	1.5014	1.4288	1.3076	1.3524

(c) Mean

Fusion method	B	R	G	NIR
Original band	368.1521	403.1313	308.5350	409.8004
HSV	87.7772	103.8142	94.6031	—
Brovey	107.2043	123.7877	109.8734	—
WT	358.7487	398.2811	313.7585	411.6078
PC	368.1571	403.1355	308.5392	409.7648
Gram-Schmidt	368.1532	403.1337	308.5371	409.8025
Pan Sharpening	367.8874	402.9655	308.4283	409.6188
CN	337.2664	369.6797	283.3468	380.6733

(d) Spectral correlation coefficient

Fusion method	B	R	G	NIR
Original band	1	1	1	1
HSV	0.6636	0.6497	0.7439	—
Brovey	0.7046	0.6413	0.7424	—
WT	0.8978	0.9278	0.9425	0.9294
PC	0.8887	0.8672	0.8675	0.9146
Gram-Schmidt	0.8789	0.8591	0.8640	0.9281
Pan Sharpening	0.8541	0.8326	0.8377	0.9171
CN	0.7463	0.8424	0.9231	0.9278

The spatial structure can be determined by the standard deviation and entropy. In comparison with the original multispectral image, the spatial information of the fused image by CN transform has been obviously enhanced. The fused images by Pan Sharpening, WT, PC or Gram-Schmidt transform changed a little, but the damping of fused images of HSV and Brovey became larger. Considering the entropy, the fused images by the HSV, Gram-Schmidt, PC, Brovey, CN and Pan Sharpening just changed a little, and the amplitude of the fused image by the WT was slightly reduced. We concluded that the fused images by the CN, Pan Sharpening, PC and the Gram-Schmidt could maintain good spatial structure, followed by the WT transform.

The spectral correlation coefficient could show the ability in keeping the spectral information of the fused image. As shown in Table 2, the spectral correlation coefficient of the image with the fusion of the WT transform is close to 1, and the spectral correlation coefficients were relatively lower of the fused images by the CN, Gram-Schmidt, PC transform. It just showed a general performance of the fused images by the Pan Sharpening, and they were the worse of the fused images by the HSV and Brovey transform. Considering the average gray, there were little changes in the fused images by the PC, Gram-Schmidt, Pan Sharpening and WT transform. It was slightly reduced in the fused images of the CN transform. Conversely, they were decreased sharply in the fused images of the HSV and Brovey transform. The WT and Gram-Schmidt transform had the best abilities in keeping spectrum, followed by the Pan Sharpening, and the abilities of HSV and Brovey in in keeping spectrum were the worst.

3.3 Analysis in Spatial and Spectral Structure

The analysis results in spatial structure and spectral information were consistent with the visual interpretation. We concluded that either of the Gram-Schmidt and PC transform was good, but it was difficult to determine the best. The fused images of the CN transform and Pan Sharpening could maintain the good spatial structure and the general spectral keeping ability, while the WT kept a good ability of keeping spectral information and a general ability of keeping spatial information. Conversely, the Brovey had a relatively band ability in keeping spatial structure and spectral information.

3.4 Comparison of Classification Accuracy

Seven classification maps were produced using the SVM method (Fig. 4). The classification accuracy of seven fused images were compared by the confusion matrix. It was concluded that the it was 85.03 %, 84.20 %, 90.26 %, 88.18 %, 85.48 %, 89.01 % and 84.17 %, respectively, for the HSV, Brovey, WT, PC, Gram-Schmidt, Pan Sharpening and CN transform.

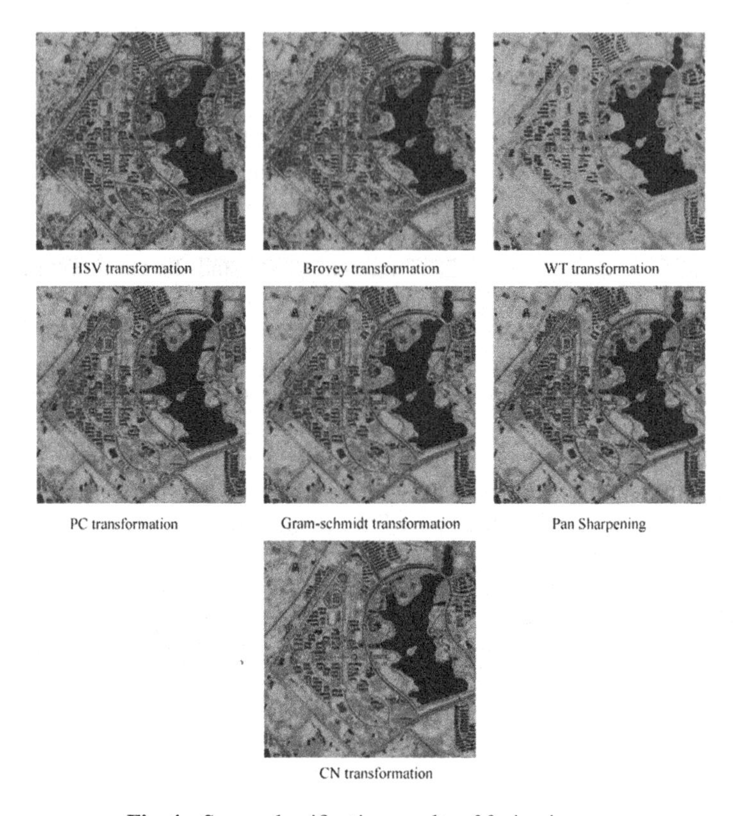

Fig. 4. Seven classification results of fusion images

It could be found that they showed a good performance in identifying the five land cover types using the seven fusion algorithms. The terrain contour in seven classification maps were all clear. Specifically, the water classification was the best; the road and other types was obviously clear but with a little mixing of the building and road. However, the shadow caused the misclassification of the water. The rank of the classification performance was the WT transform, Pan Sharpening, HSV, PC transform, Gram-Schmidt, CN and Brovey. Considering the spatial and spectral characteristics, the PC transform was the best. The WT transform had the better classification result and ability in keeping spectral information. Pan Sharpening had the better classification result and ability in keeping spatial structure. Gram-Schmidt had a relatively good ability in keeping spectral and spatial information as well as the general classification result. CN transform had good ability in maintaining the spectral information, the general ability in keeping the spatial information and the general classification effect. On the whole, HSV had the general ability in keeping spatial structure, spectral information and classification effect, while Brovey was the worst method.

4 Conclusion

The panchromatic and multispectral images of IKONOS were fused by the HSV, Brovey, WT transform, PC transform, Gram-Schmidt, Pan Sharpening and CN transform respectively. We compared three aspects of visual interpretation, quantitative evaluation and classification result. It can be concluded that the PC transform is the best; WT transform has the better ability of keeping spectral information and good classification; Pan Sharpening provides the superior spatial structure and classification effect; Gram-Schmidt transform has the better ability in keeping spatial and spectral information but the general classification effect; CN transform maintains good spectral information, which has a general ability of keeping spatial structure information and a classification effect; HSV transform has the general ability of keeping spatial information, spectral characteristics and classifying land cover features; and Brovey transform is the worst.

References

1. Hu, Y., Xi, X.H., Wang, C., Xiao, Y.: Study on fusion methods and quality assessment of Pléiades data. Remote Sens. Technol. Appl. **29**, 476–481 (2014)
2. He, H.P., He, G.J.: Comparison of image fusion algorithms for IKONOS high spatial resolution satellite image. Sci. Technol. Rev. **27**, 33–37 (2009)
3. Zhang, S., Zhao, C.S., Yang, G., Chen, Y.T., Zhao, J.M.: Study on algorithm of multi-spectral image and high resolution image. Remote Sens. Inf. **5**, 56–60 (2007)
4. Wang, Z.J., Ziou, D., Armenakis, C., Li, D.R., Li, Q.Q.: A comparative analysis of image fusion methods. IEEE Trans. Geo. Remote Sens. **43**, 1391–1402 (2005)
5. Bie, Q., He, L., Zhao, C.Y.: Study on vegetation information extraction based on object-oriented image analysis. Remote Sens. Technol. Appl. **29**, 164–171 (2014)
6. Dong, G.J., Zhang, Y.S., Dai, C.G.: Comparison of fusion algorithms of high resolution imagery. Opt. Technol. **32**, 827–830 (2007)
7. Yang, Z.X., Xu, J., He, X.F.: A novel method for merging Geoeye-1 panchromatic and multispectral images. Opto-Electron. Eng. **38**, 120–126 (2011)
8. Kim, Yong-Hyun, Kim, Yong-II, Kim, Youn-Soo: Fusion techniques comparison of GeoEye-1 imagery. Korean J. Remote Sens. **25**, 517–529 (2009)
9. Weng, Y.L., Tian, Q.J., Hui, F.M.: Research on the fusion effects of IKONOS high-resolution remote sensing image. J. Southeast. Univ. (Nat. Sci. Ed.) **34**, 274–277 (2004)
10. Tan, Y.S., Shen, Z.Q., Jia, C.Y., Wang, X.H., Deng, J.S.: The study on pixel-level data fusion for remote sensing images of medium and high resolution. Remote Sens. Technol. Appl. **22**, 536–542 (2007)
11. Deng, S.B.: ENVI Remote Sensing Image Processing Method. Sci. Publishing House, 118 (2010)

Discussion on Water Source Reservoir Dredging Effects on Water Quality—Case Study of Duihekou Reservoir

Shuang Zhao[1(✉)], Bin Huang[2], and Bin Guo[1]

[1] College of Economics and Management of Zhejiang Radio & Television University, Hangzhou 310030, Zhejiang, China
shuang_zhao1984@163.com
[2] Power China Huadong Engineering Corporation, Hangzhou 310014, Zhejiang, China

Abstract. To know whether the water reservoir dredging engineering affect the water source reservoir, Duihekou reservoir is taken as an example in this paper, and a test engineering of dredging at the end of the reservoir is done. First introduced the general situation of reservoir and the test dredging engineering, and combined with the characteristics of reservoir sediment, and on-site monitoring data, the impacts of water quality on the dredging operation of the reservoir, water intake and the test dredging work area were investigated and analyzed. Comprehensive routine monitoring data and special monitoring data, test dredging engineering did not have a significant impact on the water quality of Duihekou reservoir. The water quality of water intake and the test area was not significantly changed. And the upstream tended to improve, through precipitation and water purification treatment.

Keywords: Water source reservoir · Test dredging · Duihekou reservoir · Water environment influence assessment

1 Introduction

Due to the deposition of pollutants in the reservoir, the accumulation of sediment accumulation is increasing, which has a great influence on the water quality and the life of the reservoir. In order to ensure the water quality of the reservoir and the reservoir dredging, the dredging and water source protection engineering is very necessary. However, as the accumulation of COD, BOD, NH_3-N, TN and TP in the bottom of the reservoir, and the sediment releasing in dredging and water source protection engineering, there are also some risks of influence on water quality of the reservoir and the water supply guarantee rate. Therefore, whether the water reservoir dredging engineering affects water quality, also need a further research. Duihekou reservoir is taken as an example in this paper, and a test engineering of dredging at the end of the reservoir is done, to understand the effect of dredging engineering on the water source reservoir.

F. Bian and Y. Xie (Eds.): GRMSE 2015, CCIS 569, pp. 514–519, 2016.
DOI: 10.1007/978-3-662-49155-3_53

2 The Overview of Duihekou Reservoir and Its Test Dredging Engineering

2.1 The Necessity of Dredging and Water Source Protection Engineering

At present, Duihekou reservoir is the main water source of Deqing County. It is a comprehensive reservoir, with the functions of flood control, water supply, irrigation, power generation and so on. Since 1964, it has never been dredged. And the reservoir sedimentation is serious, and the water quality is also affected. Therefore, to restore its storage capacity, improve the flood control capacity and the guarantee rate of water supply, improve water quality, water conservation, improve ecological environment in the reservoir area, and promote the development of tourist economy, carrying out dredging and water source protection engineering is necessary.

2.2 The Overview of the Test Dredging Engineering and Its Implementation

According to the requirements of Deqing County people's government office file ([2006]82), in order to ensure the safety of water resources, a test dredging engineering was carried out at the end of the reservoir. And then if the test engineering was certified, a fully dredging engineering would be implemented. The test dredging engineering of Duihekou reservoir includes a 0.5 km^2 stock-yard. The actual dredging volume is 2954500 m^3. Using bucket chain dredger and self-propelled mud barge. And the stock-yard is 14.38 hm^2.

In this test dredging engineering, wetland type dredging dredge, and the dredging range and dredging depth are different with the test dredging engineering. The test dredging process, dredging machine and dredging the number of dredgers are consistent with test dredging engineering.

3 Reservoir Sediment Characteristics

According to the report of the geological survey of the reservoir in Deqing County, the reservoir area can be divided into three layers:

I: Containing gravel silty clay, with sand, roots and other impurities. The soil sample is rich in humus and has an effect on the water quality.

II: Gravel layer. Grain diameter is 10–15 cm. The maximum is over 30 cm. Roof height is 36.03–23.60 m. Sand content is 11–20 %. Average particle size of sand is 0.52 mm. Granularity modulus is 3.06–3.46. Mud content is 0.36–0.65 %. Uneven coefficient is 2.42–6.35.

III: Bedrock, whose color is meat red. Other area is the Ordovician shale and silty shale. The end of the reservoir area is Jurassic tuffs.

The dredging engineering mainly involves the first and second layer.

4 The Influence of Test Dredging on Water Quality

To master all kinds of environmental problems in the process of practical dredging, especially the influence degree of dredging process and mud piling on water quality of the reservoir. And for the feasibility of the dredging, and provide a scientific basis for the implementation of the engineering. The construction unit carried out a test dredging engineering in December 2007. The environmental impact of the test dredging engineering, the implementation of the main measures, effectiveness, are all reviewed and evaluated in this paper.

4.1 Construction Technology

For water transportation, dredging mud, sand, stone and muddy water is transported to the barge, by barge transported to the bank terminal for unloading. For underwater mud (sand, stone), dredging ship is used in this engineering, including the chain bucket and wetland dredger.

(1) Chain bucket dredger

The bucket is underwater mud, sand, stone dug to the dredging boat. This equipment is mainly used for deep-water and shallow water dredging.

Construction process: reservoir basin dredging layer, chain bucket dredger, mud, sand and stone mixture (slurry) in semi closed barge, barge self-propelled transport, bank quay, terrestrial mud, sand, stone separation.

(2) Wetland dredger

Wetland dredger dredges sand, stone and mud into the barge, transports them to the bank pier, screens and soon. It is mainly used in beach area, where the chain bucket dredger cannot be implemented.

Construction process: library sediment layer, the wetland dredger, mud, sand, stone mixture (slurry) into lighter barge self-propelled transport, bank quay, terrestrial mud, sand, stone separation.

4.2 Investigation and Analysis of Water Environment Impact of Test Dredging Engineering

To solve the environmental impact of test dredging engineering, construction unit commissioned Deqing County environmental monitoring station to conduct a number of monitoring. The results showed that the influence of test dredging engineering on water environment is mainly from the dredging operation.

(1) The influence of dredging on the whole reservoir

The routine monitoring data of Liudong bridge section, intermediate reservoir section and water intake section, from 2007 to 2010, see Figs. 1, 2 and 3. (Unit: pH is non dimensional, water temperature is °C, the other is mg/L).

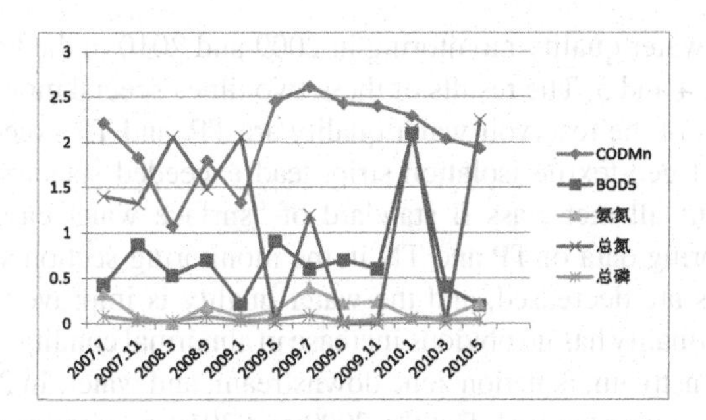

Fig. 1. Routine water quality monitoring results of surface water of Duihekou reservoir (Upstream, Liu Dong bridge section).

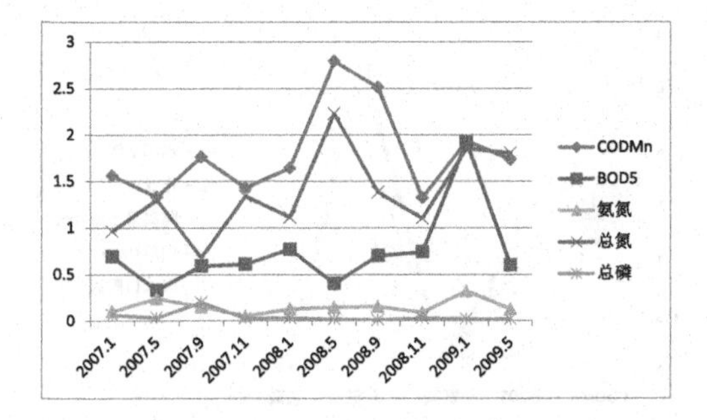

Fig. 2. Routine water quality monitoring results of surface water of Duihekou reservoir (Middle section of reservoir).

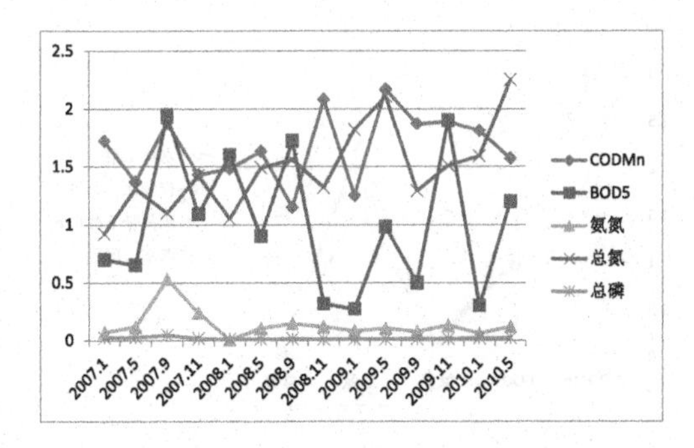

Fig. 3. Routine water quality monitoring results of surface water of Duihekou reservoir (Near the water intake).

In summary, digging process doesn't have an obvious effect on the water quality, and water intake any significant changes. Also according to the comparison and analysis, the possibility of reservoir water quality indicators occasionally anomaly causing by dredging was small.

(2) The influence of dredging on water quality of the test dredging area

The results of water quality monitoring in 2009 and 2010 in the test dredging area, are shown in Figs. 4 and 5. The results of these two times special monitoring show that the main problems of the reservoir water quality are TP, and TP exceeded. In 2009, in the downstream of geo-textile isolation strip, lead exceeded standard, the rest of the monitoring sections all met class II standard of "surface water environment quality standard". Monitoring data of TP and TN in the monitoring section show that TP and TN concentrations are decreased, and the water quality is improved. In the dredging process, the water quality has no obvious increase in abnormal conditions, by contrasting the upstream, downstream, isolation zone downstream, and water. In 2009, in the isolation zone, lead was over standard. But the 2009 and 2010 monitoring data showed that no one exceed standard. And combined with the monitoring results in Figs. 1, 2 and 3, the probability of lead anomaly causing by dredging was small.

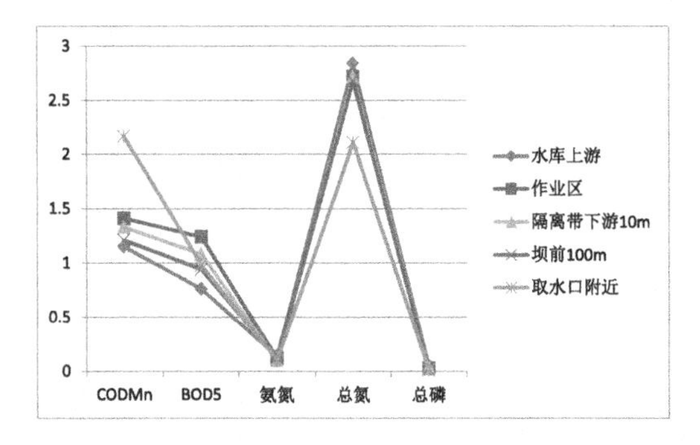

Fig. 4. Water quality monitoring results during the test dredging period in 2009.

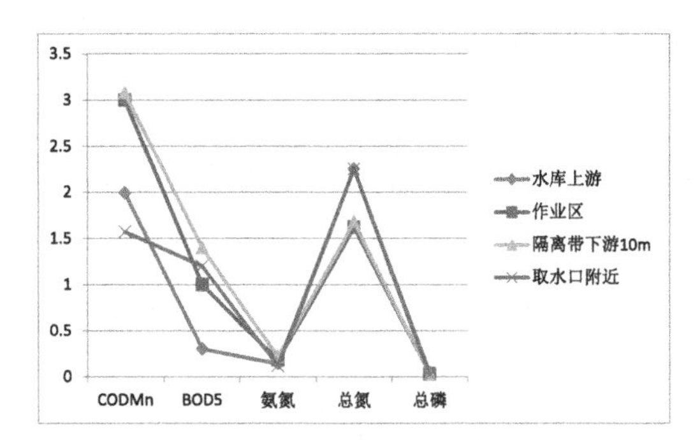

Fig. 5. Water quality monitoring results during the test dredging period in 2010.

In summary, in test dredging process, test dredging area and the surrounding water quality indexes rose in the certain range, but the impact range was generally small. Except for TP and TN, the other indexes basically met the class II standard of surface water environment quality standard. The test dredging engineering had little influence on the water quality of test dredging area and the surrounding.

5 Conclusion

Comprehensive routine monitoring data and special monitoring data, test dredging engineering did not have a significant impact on the water quality of Duihekou reservoir. The water quality of water intake and the test area was not significantly changed. And the upstream tended to improve, through precipitation and water purification treatment. Special monitoring data show that the construction waste water, after the sedimentation tank treatment compliance, reused and discharged into Yuying stream, has little effect on the water environment. But it still needs to do a further improvement on the drainage pipe network.

The Duihekou reservoir as a water head site reservoir, its test dredging engineering implementation and data monitoring, for the guidance of the similar water head site dredging work has an important role.

Acknowledgments. This work is supported by Natural Science Foundation Project of Zhejiang Province (LQ14G030020); Education Department Project of Zhejiang Province (Y201328633); Scientific research project of Zhejiang Radio and Television University (GRJ-13).

References

1. He, R.Y., Luo, J.L., Ni, H.W., et al.: Water quality valuation and pollution control measures for estuarine reservoirs. J. Zhejiang Hydrotechnics **2**, 1–3 (2010)
2. Luo, J.L., Yan, K.Y.: Reservoir dredging construction technology and disturbance source analysis. J. Zhejiang Hydrotechn. **2**, 1–3 (2010)
3. Xie, H.C., Wang, C.L., Chen, L.H.: On pollution controlling and protection of water environment in Duihekou Reservoir. J. Zhejiang Water Conservancy and Hydropower College 18, pp. 18-20 (2006)
4. Guo, Y.Z., Xue, X.: The environmental impact assessment of river dredging on agricultural soil on both sides of the river. J. Yangtze River. **33**, 34–35 (2002)
5. Cao, H.Q., Zhou, J.J.: Development and prospect of dredging at water conservancy works in China. J. Sediment. Res. **5**, 67–72 (2011)
6. Zhang, M.L., Yang, H., Lin, Z.S., et al.: Harmful substances residue and environmental pollution evaluation of Zhushan Bay, in Taihu Lake. J. China Monit. Sci. **31**, 852–857 (2011)
7. Liu, C.J., Liu, F.L.: Strengthen comprehensive improvement and protect water quality. J. Reservoir Fish. **26**, 73–74 (2006)
8. Chang, Q.J.: Status analysis of water resource in "south-north water diversion" middle route project in Ankang and water source area protection countermeasure. J. Anhui Agric. Sci. **37**, 7576–7578 (2009)
9. Liu, X.W., Xu, C.C., Shen, X.Z., et al.: Experimental study of drainage effect for different drainage measures on Yellow river dike body reinforced by desilting. J. Rock Soil Mech. **31**, 3563–3568 (2010)
10. An, C.H., Guo, X.Y. Li, Y.F., et al.: Study and practice on intake layout of the desilting experiment project in Xiaobeiganliu reach of the Yellow river. J. Sediment Res. 1, pp. 38–44 (2006)

Water Environment Prediction and Evaluation on Reservoir Dredging and Source Conservation Project—Case Study of Duihekou Reservoir

Shuang Zhao[1(✉)], Bin Huang[2], and Bin Guo[1]

[1] College of Economics and Management of Zhejiang Radio and Television University, Hangzhou 310030, Zhejiang, China
shuang_zhao1984@163.com
[2] Power China Huadong Engineering Corporation, Hangzhou 310014, Zhejiang, China

Abstract. In order to study the effect of dredging project on water supply safety of the reservoir, a case study of Duihekou reservoir is taken in this paper. On the basis of the characteristics of water environment, sediment and the surrounding pollution, analyze of the related activities of the project, predict the influence of the water head site dredging project on water environment, especially water intake and scenic area. The results show that chain bucket dredger only has a certain impact on the dredging area and nearby, the overall impact on the water quality of the reservoir is small, and the dredging project does not affect the water intake basically. Disturbance effect of wetland type dredging dredger is similar to chain bucket dredger.

Keywords: Dredging and protection water sources · Forecast EIA · Duihekou reservoir

1 Introduction

The reservoir not only has the function of flood control, but also water supply of a city. With the rapid expansion of the economy and the need of the construction of the new socialist countryside, Deqing County government passed the "Deqing County water supply planning (2000~2020)". Duihekou reservoir was defined and protected as the water source for the city water supply. And rely on the abundant water resources, to accelerate the process of integration of urban and rural water supply. However, the reservoir operates for more than 40 years. At present, there is a serious sediment deposition in it. And for many years, the storage of waste residues from agricultural production activities has seriously affected the water quality of reservoirs and flood control, water supply, irrigation and power generation efficiency. In recent years, a large number of planktonic algae bloom lead to the development of reservoir eutrophication. Reservoir water quality is declining. The implementation of reservoir basin dredging engineering has been very urgent and necessary.

F. Bian and Y. Xie (Eds.): GRMSE 2015, CCIS 569, pp. 520–524, 2016.
DOI: 10.1007/978-3-662-49155-3_54

2 Engineering Survey

Duihekou reservoir located in Duihekou village, Wukang town, Deqing County, Zhejiang province, Yu Yuyingxi upstream of Dongshaoxi anabranch. The reservoir built and water in July 1964 and in September 2003 the reservoir reinforcement engineering implemented. Reservoir watershed area upward dam site is 148.7 km^2, main dam height is 38.2 m, total capacity is 0.147 billion m^3, normal water storage is 50.2 m and corresponding capacity is 0.0805 billion m^3. This is a water conservancy engineering with the functions of flood control, water supply, irrigation, power generation, etc. To protect water quality in the reservoir area, ensure the normal play of the reservoir flood control and water supply efficiency, Deqing County People's government decided to implement Duihekou reservoir dredging source conservation engineering.

The implementation of the dredging engineering will greatly increase the storage capacity of the reservoir, guarantee and improve the reservoir function of flood control and water supply. It will ensure that the water supply requirement and water quality safety, and promote the sustainable and stable development of the local community and the surrounding region. In addition, silt and sand will be reused, as the local and nearby areas building materials and products, supplying the marketing construction market, for achieving financial balance of the engineering.

According to "the People's Republic of China environmental impact assessment law", "construction project environmental protection management regulations" and other relevant laws and regulations, the reservoir dredging engineering is a rebuilding ones. Based on the characteristics of water environment, sediment and the surrounding pollution in the engineering area, do an analysis of the characteristics of the engineering construction activities. Then, forecast the effect of the engineering activities on evaluation regional environment, especially landscape and famous scenery.

3 Forecast EIA of Operation Effect on Water Environment

Water environmental impact assessment scope is the reservoir area and the scope from the stock-yard to downstream 5 km and the focus is water intake. Forecast evaluation level year if 2010.

3.1 Selection of Evaluation Model

(1) Predictive factors: According to the digger pollutant source strength, can determine the predictors for suspension.

(2) Prediction model: Use persistent pollutant recommended by guidelines-Carla U Scherf model (windless lake library), i.e.

$$c_r = c_p - \left(c_p - c_{r_0}\right)\left(\frac{r}{r_0}\right)^{Q_p/\Phi H M_r}$$

Type:
r—The distance from forecast point to discharge port, m;
c_r—Pollutant concentration in forecast point, mg/L;
Q_p—Sewage flow, m^3/s;
C_p—Pollutant concentration of sewage, mg/L;
H—The average depth of the lake reservoir, m;
Φ—Mixed angle, radian. Φ can be determined by the shape of the lake and the water flow in the lake, take the Lake sewage discharge angle 2π, Straight shore π;
r_0—The distance from reference point to discharge port, m. You can choose a certain point enough far away from the discharge port. The impact of construction engineering on the water quality of the point is negligible;
C_{r_0}—The concentration of reference point, mg/L. Can take the status value at r_0 (Can be considered not be affected by C_p);
M_r—Radial mixing coefficient of Lake Reservoir, m^2/s.

3.2 Calculation Conditions and Parameters Selection

(1) Chain bucket dredger (in shallow and deep water operation area): According to the monitoring data of suspended solid concentration when chain bucket dredger constructing, the suspended solid source intensity is taken the value 23 mg/L of the bucket dredger 1 m vertical average concentration. In reference point, the concentration of suspended solids in the water intake port 6 mg/L. Take the distance from the dredging dredger to water intake 3000 m, as the distance value r_0 between forecast point and reference point. This pollutant characteristic of this engineering is similar to Mid-lake emission, so take the mixing angle 2π. The average depth of the deep water operating area is 6 m, and the shallow water operation area 16 m. In the area of the dredger, the average water depth is 10 m.

In consideration of the effect of dredging dredger construction disturbance on downstream water quality, the quantity of sewage can be considered as the water flow of pollutants mixed section caused by dredging disturbance. As it is difficult to define the size of the mixed section, considering the condition under adverse effect, make multi-year average flow as equivalent sewage quantity generated by dredging disturbance.

According to the monitoring results of suspended solid concentration when the dredger is dredging, and the relevant parameters of dredger construction, adjust the lake reservoir radial mixing coefficient, determine the lake reservoir radial mixing coefficient is 0.63 m^2/s.

(2) Wetland type dredging dredger (Reservoir bank): Wetland type dredging dredger is used on the reservoir bank. According to the operation effect characteristic, the operation mod of wetland type dredging dredger and chain bucket dredger is similar, and the disturbance effect is similar too. So the overall effect is close to the chain bucket dredger.

3.3 Calculation Result Analysis

(1) Chain bucket dredger (in shallow and deep water operation area): The forecast of suspended solids concentration change when the engineering is operating in shallow and deep water operation area, see Fig. 1.

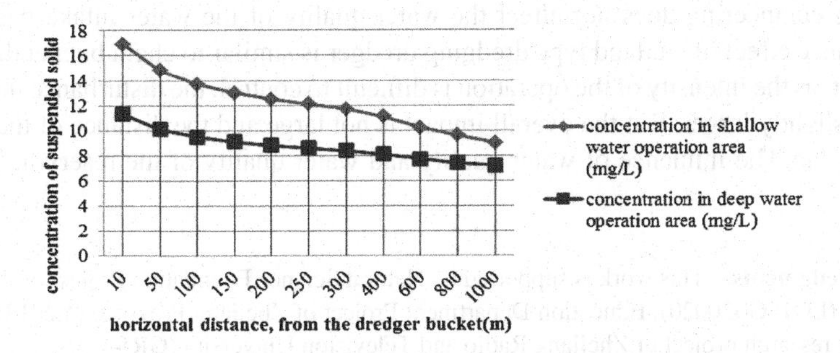

Fig. 1. The trend chart of suspended solids concentration when dredging engineering operation.

From Fig. 1, when the chain bucket dredger operating, the suspended solid concentration will decay rapidly, with the increase of the distance from the center point of the bucket. In shallow water area, suspended solids concentration falls below 10 mg/L, in the digging bucket center down to 700 m. In deep water area, suspended solids concentration falls below 10 mg/L, in the digging bucket center down to 100 m. Therefore, chain bucket dredger dredging only has a little impact the dredging area and nearby, and the overall impact on the water quality of the reservoir is small.

According to the construction plan, deep-water area, the shortest distance is about 400 m from dredging operation point to the water intake. At this time, the suspended material decay along the path, the concentration of the water intake is 8.1 mg/L. According to the monitoring data of the current water intake concentration, the suspended solids concentration in the water intake is 6 mg/L. In the case of not setting up earthwork cloth, the chain bucket dredger construction will have a certain impact on the intake water quality. According to the construction arrangement, when dredging, two line of earthwork cloth will be disposed in the downstream of the operating area. In the distance of radius 30 m from the water intake, arrange a line of earthwork cloth. At the line of 300 m upstream of the water intake, arrange another one. So, when the chain bucket dredger operating, with the filter of the two earthwork cloths, the dredging engineering does not affect the water quality of the water intake basically.

(2) Wetland type dredging dredger (Reservoir bank): Disturbance effect of wetland type dredging dredger is similar to chain bucket dredger. As the main influence water area is shallow water area, suspended solid concentration is down to 10 mg/L below, at downstream 700 m away from the center of digging bucket. The impact of wetland type dredging dredger on water quality is small totally. According to the construction plan, the shortest distant from working point to water intake is 3500 m, so there is basically no influence of the operation of wetland type dredging dredger on the water quality in water intake.

4 Conclusion

According to the prediction results of the mathematical model, chain bucket dredger only has a certain impact on the dredging area and nearby, the overall impact on the water quality of the reservoir is small. With the filter of the two earthwork cloths, the dredging engineering does not affect the water quality of the water intake basically. Disturbance effect of wetland type dredging dredger is similar to chain bucket dredger. However, as the intensity of the operation is difficult to control, the disturbance intensity may be slightly larger. But the overall impact is not large and the distance of the water intake is far. The influence of water quality and water quality of the reservoir is very small.

Acknowledgments. This work is supported by Natural Science Foundation Project of Zhejiang Province (LQ14G030020); Education Department Project of Zhejiang Province (Y201328633); Scientific research project of Zhejiang Radio and Television University (GRJ-13).

References

1. Luo, J.L., Yan, K.Y.: Reservoir dredging construction technology and disturbance source analysis. J. Zhejiang Hydrotechnics **2**, 1–3 (2010)
2. Guo, Y.Z., Xue, X.: The environmental impact assessment of river dredging on agricultural soil on both sides of the river. J. Yangtze River **33**, 34–35 (2002)
3. Cao, H.Q., Zhou, J.J.: Development and prospect of dredging at water conservancy works in China. J. Sediment. Res. **5**, 67–72 (2011)
4. Zhang, M.L., Yang, H., Lin, Z.S., et al.: Harmful substances residue and environmental pollution evaluation of zhushan bay, in Taihu Lake. J. China Environ. Sci. **31**, 852–857(2011)
5. Chang, Q.J.: Status analysis of water resource in "south-north water diversion" middle route project in Ankang and water source area protection countermeasure. J. Anhui Agric. Sci. **37**, 7576–7578 (2009)
6. An, C.H., Guo, X.Y., Li, Y.F., et al.: Study and practice on intake layout of the desilting experiment project in xiaobeiganliu reach of the yellow river. J. Sediment Res. **1**, 38–44 (2006)
7. Li, J., Li, L., Yang, M.F.: An analysis of the impact of different intake programs of reservoir on the water ecosystem of river downstream. J. China Rural Water Hydropower **3**, 21–25 (2011)
8. Jiang, F.G., Song, J., Dai, M.M., et al.: Research on water quality prediction of tongzi reservoir. J. China Rural Water Hydropower **2**, 27–31 (2015)
9. Liu, A.J., Kong, F.X., Wang, D.: Water quality risk assessment for sediment dredging operations, wulihu in Taihu Lake. J. Environ. Sci. **27**, 1946–1952 (2006)
10. Xu, Z.K., Huang, H.P., Wei B., et al.: Comprehensive evaluation method of flood risk based on multi-level gray system model—a case study in Poyang Lake Basin. J. South-to-North Water Transf. Water Sci. Tech. **13**, 20–23(2015)

Analyzing the Spatial Correlation Between Regional Economic Level and Water-Use Efficiency in Jiangsu Province

Lingling Zhang[1], Fengxia Yan[1], Zongzhi Wang[2(✉)], and Jiayao Shen[1]

[1] School of Public Administration, Hohai University, Nanjing, China
[2] Key Laboratory of Hydrology Water Resources and Hydraulic Engineering, Nanjing Hydraulic Research Institute State, Nanjing, China
wangzz77@163.com

Abstract. A spatial regression model between regional economy and water-use efficiency was built by using per capita GDP and water consumption of every ten thousand yuan of all the cities in Jiangsu province during 2009 to 2011. Based on spatial correlation analysis between economy and water-use efficiency in Jiangsu province through Moran's index, local Lisa figure and Moran scatter chart, the regional economy development and water-use efficiency in Jiangsu province are obviously spatial-clustered, decreasing progressively from southern, central to northern cities. When considering the relationship between per capita GDP and water use of per every ten thousand yuan, HH zone does not exist, LH zone includes Xuzhou, Huaian, Lianyungang, Suqian, Yancheng and Yangzhou cities, LL zone is Taizhou, and HL zone includes Nanjing, Wuxi, Changzhou, Suzhou and Zhenjiang cities. It could be concluded that spatial differences of regional economic developing level is the most significant factor to water-use efficiency.

Keywords: Regional economy · Water-use efficiency · Spatial correlation

1 Introduction

In order to achieve the sustainability of water resources and social economic development, the Chinese Government released the strictest water resources institution in which addresses the total volume control of water use, water use efficiency and water quality. Improving water-use efficiency means to conserve water resources, reduce waste, promote structural adjustment of water use, change the ways of water use and production. However water use efficiency is different in terms of different regions, and related with regional economic development. Regional is an open and complex system, and inter-regional interaction and mutual influence can lead to a high correlation between regions in many characters (Cui and Wei 1999). How to quantify the relationship between water use efficiency and economic development in different regions concerns the improvement of water use efficiency.

In the context of spatial difference of regional economy discussions have taken place in the literature on how to reveal the spatial difference of regional economy through

F. Bian and Y. Xie (Eds.): GRMSE 2015, CCIS 569, pp. 525–533, 2016.
DOI: 10.1007/978-3-662-49155-3_55

spatial econometric models (Barro and Sala 1991; Chen and Fleisher 1996; Giuseppe 2006; Sergio et al. 1999; Wei and Fan 2000; Long 2001; Shen and Zhai 2003; Ou and Gu 2004). Wang and Wu (2005) investigated that regional economic growth was not only dependent on its internal factors, but also increasingly the surrounding areas in an open regional economy. Li (2006) estimated the spatial difference of water use efficiency of China. The results showed a higher level of economic development were accompanied by a high rate of urbanization, the increasing proportion of non-agricultural industries and a reducing proportion of agricultural water consumption. So it can reduce the water consumption per ten thousand yuan GDP and improve outcome per water use unit. The above studies emphasize either the regional spatial difference of economy or one of water use efficiency, and less address both of them. This paper explores the regional spatial relationship difference of both economy and water use efficiency using the exploratory spatial data analysis in the 13 cities of Jiangsu province.

Exploratory spatial data analysis is a spatial data analysis method with identification function. The spatial autocorrelation coefficient is usually used to measure and test the space distribution of the adjacent objects and their attributes. It includes the positive spatial correlation with similar values and trend and the negative spatial correlation with the opposite values and trends. This paper uses the exploratory spatial data analysis method, combines, and builds the spatial correlation model of single factor and double factors with the data of regional economic development level and water-use efficiency to improve the efficiency of regional water by analyzing the spatial difference of regional economic. The specific analysis of Moran index, local Lisa I figure and Moran scatter diagram is used to explore the space relation of regional economy of Jiangsu province and water-use efficiency with Geoda spatial analysis software.

2 Study Area and Data Sources

Since 1992 Jiangsu province has kept a stable growth year by year. In 2011 year, the GDP was 54058.2 billion yuan, and per capita GDP was 68347 yuan with an increase of 6057 yuan than one of 2000 year.

Jiangsu province has 13 cities, and water use among the cities is obviously different. The acreages of the southern, northern and central sub-regions of Jiangsu province account for 28 %, 20 % and 52 %, however the water uses of the three sub-regions water consumption of southern Jiangsu region account for 41.8 %, 34.8 % and 23.4 %. Jiangsu province is rich in water resources, but scarce per capita water resources. Water shortage use efficiency and worse water environment has become constraints in promoting the economic development of Jiangsu (Gu 2013). Since 1949, Jiangsu province has implemented a balanced development and unbalanced development, regional common development, southern Jiangsu development, coastal development five development strategies in the regional economic development. However because of the difference of nature, society, economy, history, and other factors, economic differences of among the cities and among sub-regions are obvious (Zhu et al. 2004). With the development of regional economy, the demand of water resources increases gradually.

It is a necessary way to improve the utilization rate of water use to ease the water shortage (Sun et al. 2004).

The per capita GDP and water use per ten thousand yuan output value of 13 cities in Jiangsu province are used from year 2009 to 2011 in the following model. The GDP per capita data come from the statistical yearbook of Jiangsu province, and water use data of ten thousand Yuan output value is from the official reports on water resources in Jiangsu province. Combined with the administrative map of Jiangsu province, per capita GDP was taken as a variable indicator to measure regional economic differences, and ten thousand yuan output value of water consumption as measures of regional water use efficiency.

3 Methods

3.1 Global Spatial Autocorrelation

Global spatial autocorrelation indicates the characteristics of the spatial distribution of global location and attribute variables, and is used to test the spatial correlation or spatial heterogeneity across the adjacent regions in the whole research area. Moran's I is the commonly used indicator, and the formula is shown as following.

$$I = \frac{N \sum_{i=1}^{N} \sum_{j=1}^{N} w_{ij}(x_i - \bar{x})(x_j - \bar{x})}{\sum_{i=1}^{N} \sum_{j=1}^{N} w_{ij} \sum_{i=1}^{N} (x_i - \bar{x})^2} \tag{1}$$

where, N is the number of the space units; x_i and x_j indicate the attributes of regional i and j; $\bar{x}$ is the sample mean, $\bar{x} = \frac{1}{N} \sum_{i=1}^{N} x_i$; S2 is the sample variance, $S2 = \frac{1}{N} \sum_{i} (x_i - \bar{x})^2$; w_{ij} is the binary spatial weight (using adjacent standard, when adjacent regional i and $j = 1$, when regional i and j non-adjacent = 0).

$$w_{ij} = \begin{bmatrix} w_{11} & w_{12} & \cdots & w_{1N} \\ w_{21} & w_{22} & \cdots & w_{2N} \\ \vdots & \vdots & \ddots & \vdots \\ w_{N1} & w_{N2} & \cdots & w_{NN} \end{bmatrix} \tag{2}$$

Moran index generally ranges from −1 to 1. A Moran value greater than zero indicates the positive correlation, the Moran value of samples closer to 1 indicates that they have similar attributes and put them together. So the overall spatial difference is small. A value less than zero indicates the negative correlation, the Moran value of samples closer to −1 indicates that they are put together with different attributes, and they have the huge spatial difference. When Moran index I equals zero, it shows that the samples are independent each other, and randomly distributive in space. So there is no spatial autocorrelation among them.

3.2 Local Spatial Autocorrelation

Ansalin proposed a local Moran index (LMI), also known as LISA (Local indicator of Spatial Association). LISA is used to measure the effects of aggregation or discrete to each geographical unit as the center of a small area, and identify the spatial concentration (hot or cold spots) and outliers (Anselin 1995). Moran scatter chart and Lisa are used to explore the spatial distribution by the visualization of spatial relationship. For example, LMI of zone i can be expressed as follows.

$$I_i = \frac{(x_i - \bar{x})}{S^2} \sum_{j \neq i} w_{ij}(x_j - \bar{x}) \tag{3}$$

where I_i is the LMI.

4 Results

4.1 Spatial Data Analysis of Regional Economy in Jiangsu Province

The univariate Moran's index autocorrelation test of per capita GDP of 13 cities in Jiangsu province was made by software Geoda. Figure 1 showed that the Moran values of economic were 0.5811, 0.5979 and 0.6029 from year 2009 to 2011. Moran value is positive and greater than 0.5 under the Significant levels 0.01. So the 13 cities economic development demonstrated a strong positive correlation in space, Moran value was increasing year by year, and their mutual penetration degree and spillover effect were becoming more and more obvious.

The indicators of four quadrants in Fig. 1 were: high - high (HH), low - high (LH) and low - low (LL), high - low (HL). The indicators of the 13 cities focused mainly on the HH and the LL quadrants. Two cities located in the LH quadrant, and no city were in the fourth quadrant. So it showed that – HH and LL coexisted. Suzhou, Wuxi, Changzhou, Zhenjiang, Nanjing cities located in the – HH. Suzhou, Wuxi and Changzhou cities are an economic developed part of the Yangtze River Delta. Although the per capita GDP of Zhenjiang is slightly lower than other four cities located in HH quadrant, it is adjacent to Nanjing and Changzhou, and located in Shanghai-Nanjing industrial belt with transportation, ports, freight and other advantages, and the economy has made a rapid development in recent years. As Jiangsu's political, economic and cultural center, the economic development of Nanjing has been in a leading position in a long time. Taizhou and Nantong cities were in LH quadrant. They are located in the north of the Yangtze river without obvious advantages, and economic development has been slow. Yangzhou, Yancheng, Huaian, Xuzhou, Lianyungang and Suqian cities were in LL quadrant. Yangzhou is close to the economy strongly developed southern regions:

Suzhou and Wuxi and some other rich cities. The economy of its' northern cities were poor, so they also showed a low value.

Figure 2 shows that there is an obvious geographic difference of 13 cities' economy in Jiangsu province. It includes the developed south, the middle central and the less developed north in Jiangsu province. The gray zone in Fig. 2 indicates no significant, the red zone (HH) shows economy development levels are high, the blue zone (LL) indicates low levels of economic development, and light blue zone represents the low level of economic development and high level development. The aggregation of south and north of Jiangsu province is significantly in spatial, and the central Jiangsu is non-significant. It reveals that the regional economic development in Jiangsu province is uneven in spatial distribution and presents a big north-south differences. The cities like Suzhou and Nanjing in south are the priority development cities, focusing on investment and construction. So it plays an important role in the process of rapid economic development, and the north only get more development in recent years because they are put into a part of a national strategy (Yang and Zhu 2013).

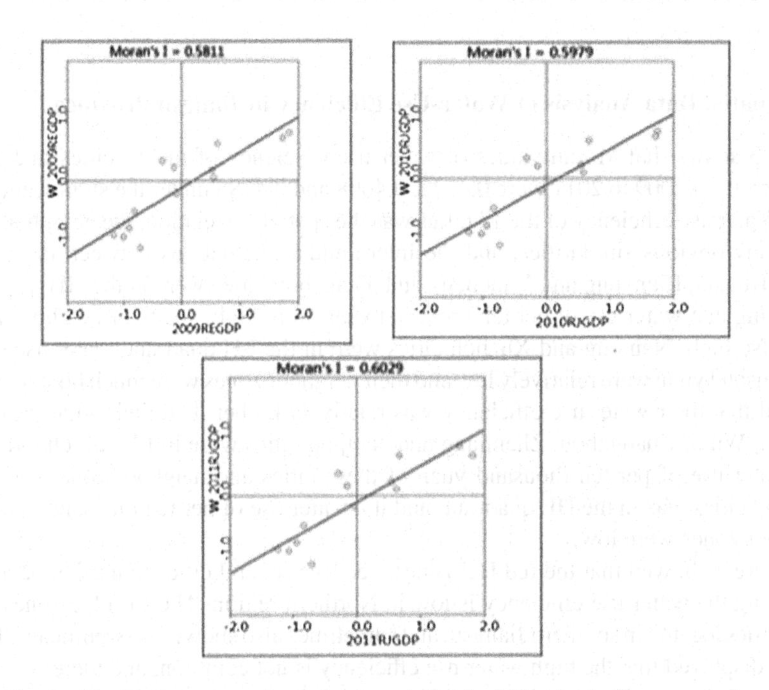

Fig. 1. The Moran scatter chart of economic level of 13 cities in Jiangsu province

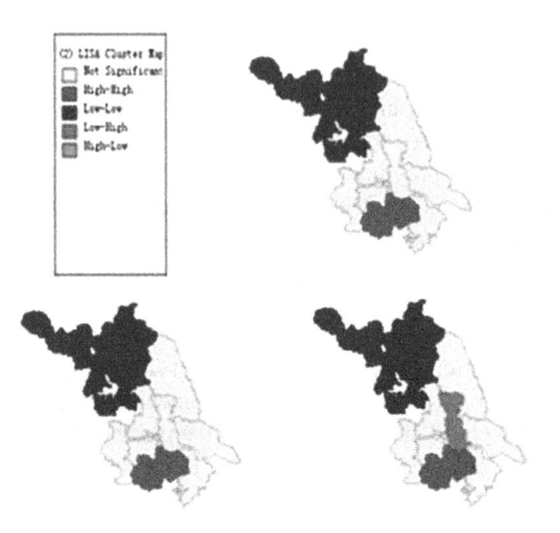

Fig. 2. The Lisa cluster of economic level of 13 cities in Jiangsu province (Color figure online)

4.2 Spatial Data Analysis of Water-Use Efficiency in Jiangsu Province

Figure 3 shows that Moran values of water use efficiency of the 13 cities in Jiangsu province from 2009 to 2011 were 0.4572, 0.4608 and 0.4188 under the significant level 0.01. Water use efficiency of the 13 cities was the spatial correlation, the neighbor cities had many obvious similarities, and the inter-annual fluctuations between them were small. Huaian, Lianyungang, Yancheng and Yangzhou cities were in the HH quadrant, indicating that water use of per ten thousand yuan were high, and their neighbor zones were also high. Nantong and Xuzhou cities were in the LH quadrant, water use of per ten thousand yuan were relatively low and their neighbor zones were much higher, which implied that their water use efficiency was relatively higher than their adjacent cities. Suzhou, Wuxi, Changzhou, Zhenjiang and Nanjing cities were in LL, which indicated that water use of per ten thousand yuan of these cities and neighbor zone were low. Taizhou cities was in the HL quadrant, and its' water use of per ten thousand yuan and neighbor zones were low.

Figure 4 showed that the red HH zones were the several cities in northern Jiangsu, indicating the water use efficiency is low in Northern region. The LL blue zones were a few cities located in southern Jiangsu, and sometimes also showed no significant. These results displayed that the high water use efficiency is not common, and there is enough space for water use. Xuzhou city located in northern Jiangsu was in LH zone, surrounded by red zone, and its water use efficiency was higher in contrast with other cities. Taizhou city located in the central of Jiangsu province was in the HL zone. Its water use efficiency was low and neighbor was high. The above analysis shows that the order of water use efficiency level is decreasing progressively from southern, central to northern of Jiangsu province.

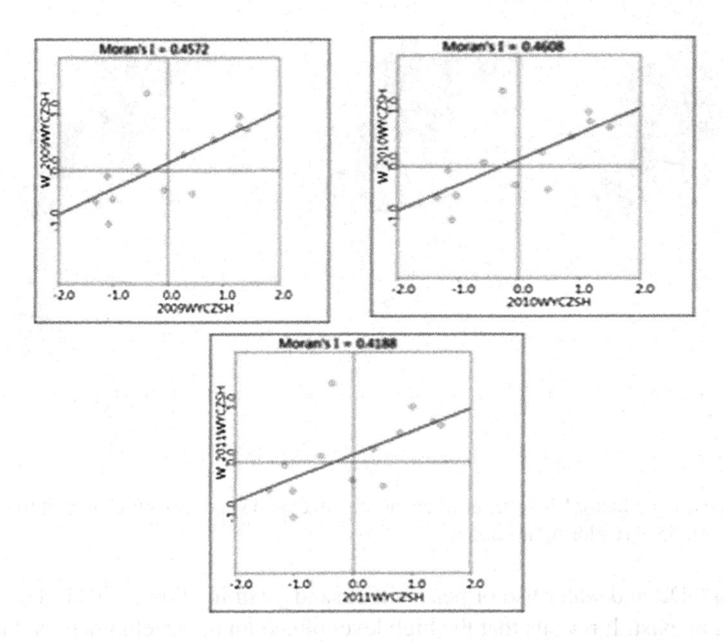

Fig. 3. The Moran scatter chart of water use efficiency of 13 cities in Jiangsu province

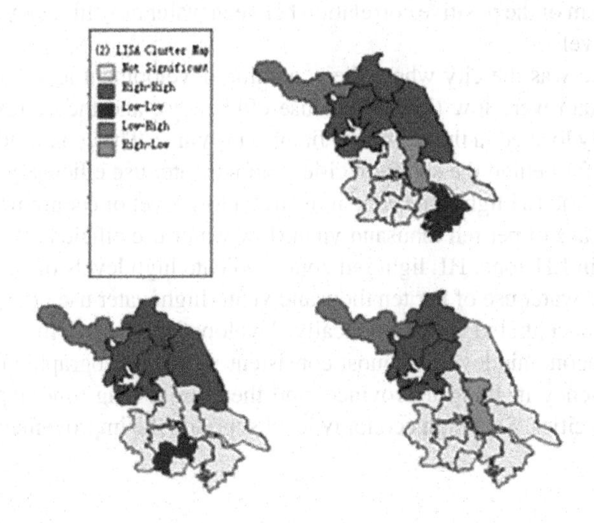

Fig. 4. The Lisa cluster of water use efficiency of 13 cities in Jiangsu province (Color figure online)

4.3 Two-Factor Association Analysis of Regional Economy and Water-Use Efficiency in Jiangsu Province

Figure 5 showed the spatial relationship between regional economy development level and water use efficiency of the 13 cities in Jiangsu province. The model data were from

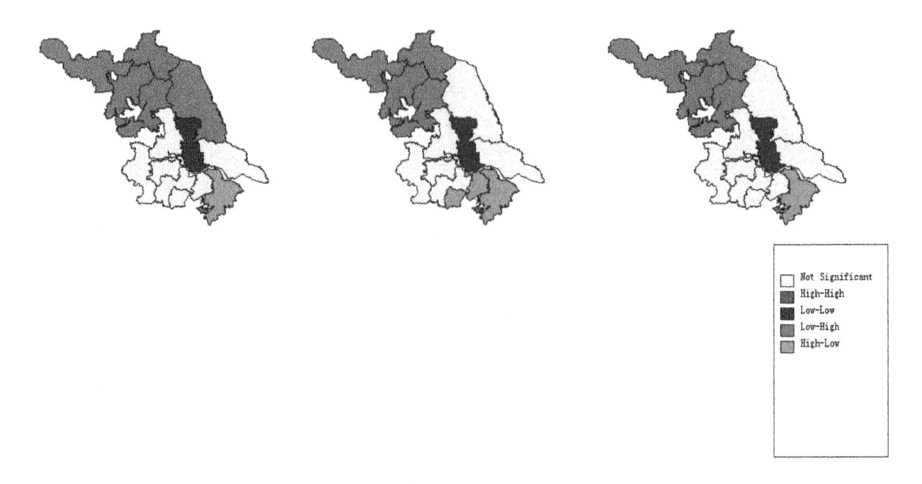

Fig. 5. The double factor Lisa cluster of economic level and water use efficiency of the 13 cities in Jiangsu province (Color figure online)

per capita GDP and water use of per ten thousand yuan in 2009 to 2011. The HH red zone did not exist. It reveals that the high level of economic development and the high water use of per ten thousand yuan (low water use efficiency) zones did not exist. This supports the claim of the positive correlation between water use efficiency and economic development level.

The LL zone was the city where the economic development and water use of per ten thousand yuan were low (high water use efficiency), and the corresponding area was Taizhou city located in the central of Jiangsu province. The economic development of the city lags far behind the southern cities and its water use efficiency is higher than northern cities. The LH light blue zone indicates a low level of economic development and high water use of per ten thousand yuan (low water use efficiency). There are five northern cities in LH zone. HL light red zones indicate high levels of economic development and low water use of per ten thousand yuan (high water use efficiency), mainly gathered and concentrated in economically developed southern cities. These results imply that the economic level is almost consistent with the geographic distribution of water use efficiency in Jiangsu Province, and there are a long time gap between the north and south cities. So a high economy level supports the improvement of water use efficiency.

5 Conclusions

Both economy level and water use efficiency display the significant spatial clustering in Jiangsu province from 2009 to 2011. There is a clear difference among the southern, central and northern cities, and the spatial correlation of the 13 cities is obvious. The water use efficiency of the southern cities is high because of high economic level, and the one of middle and northern cities is low because of the backward economy. The flow of production factors and human resource makes the gap of different cities bigger and bigger. So the economic level has been the most significant factor affecting the water use efficiency in a long time.

Acknowledgements. Support for this research was provided by a grant from the National Natural Science Foundation of China (No. 51109055, 51279223, 51479119, 51579064), Public-services Foundation of Ministry Water Resources of China (No. 201301003, 201201022), The National Soft Science Research Program (2014GXS4B047), and Social Science Foundation from the Jiangsu Province (No. 11GLA001).

References

Cui, G.H., Wei, Q.Q.: Regional Analysis and Planning, pp. 217–244. Higher Education Press, Beijing (1999)

Barro, R., Sala, I.M.: Convergence across states and regions. Brookings Pap. Econ. Act. **22**(1), 107–182 (1991)

Chen, J., Fleisher, B.M.: Regional income inequality and economic growth in China. J. Comp. Econ. **22**(2), 141–164 (1996)

Giuseppe, A.: Spatial Econometrics: Statistical Foundations and Applications to Regional Convergence, pp. 135–143. Springer, New York (2006)

Sergio, J., Rey, B.D., Montouri, U.S.: Regional income convergence: a spatial econometric perspective. Reg. Stud. **33**(2), 143–156 (1999)

Wei, Y.H., Fan, C.C.: Regional inequality in China: a case study of Jiangsu province. Prof. Geogr. **3**, 455–469 (2000)

Long, G.Y.: The political economy of intra-provincial disparities in post-reform China: a case of Jiangsu province. Geoforum **32**, 215–234 (2001)

Shen, Z.P., Zhai, R.X.: On the gap of economic development between north and south Jiangsu and their coordination. Econ. Geogr. **6**, 742–746 (2003)

Ou, X.J., Gu, C.L.: Quantitative analysis of regional economic polarization and dynamic mechanisms in Jiangsu province. Acta Geogr. Sinica **5**, 791–799 (2004)

Wang, Z., Wu, W.: An analysis to growth spillover cross regions in China. Geogr. Res. **24**(2), 243–252 (2005)

Li, S.X.: Study on the regional differences of the efficiency of water resources usage in China, vol. 5, p. 37. China University of Geo sciences, Wuhan (2006)

Wang, J.F.: Spatial Analysis. Science Press, Beijing (2006)

Gu, X.M.: Linking between water resources utilization and economic growth in Jiangsu province. China Environ. Sci. **32**(2), 351–358 (2013)

Zhu, C.G., Chou, F.D., Li, X.W.: Reconstruction on Jiangsu's development strategy of regional economy. J. Nanjing Univ. Finan. Econ. **2**, 17–22 (2004)

Sun, C.Z., Wang, Y., Li, H.X.: Influencing factors of water utilization efficiency in Liaoning province. J. Econ. Water Resour. **2**, 17–22 (2004)

Anselin, L.: Local indicators of spatial association-LISA. Geogr. Anal. **27**(2), 93–115 (1995)

Yang, L.J., Zhu, C.G.: The research in county economic divergence of Jiangsu province. J. Nanjing Univ. Finan. Econ. **3**, 7–12 (2013)

Shangguan, J.Z.: The analysis of the formation reason of regions economic development difference in Jiangsu province. Stat. Decis. **14**, 109–111 (2010)

Advanced Geospatial Model and Analysis for Understanding Ecological and Environmental Process

The Evaluation of Water Resources Sustainable Utilization in Kosi Basin Based on DPSIR Model

Bo Kong, Bing He(✉), Xi Nan, Wei Deng, and Ainong Li

Institute of Mountain Hazards and Environment, CAS, Chengdu, China
{kongbo,hebing}@imde.ac.cn

Abstract. Kosi basin located between the north Nepal and south Tibet of China has abundant water resources. Its recharge sources are glacier meltwater and precipitation in this area. However due to the topographical effects, the distribution of rainfall is uneven in space and time. In order to guarantee the harmonious development of social economy, it's necessary to evaluate the sustainable utilization of water resources. This paper adopts GIS spatial analysis methods to establish evaluation indexes of water resources sustainable based on DPSIR model, and determine the weights of each evaluation index by utilizing analytic hierarchy process. The results show that: (1) the proposed methods can reflect the development level of regional water resources system and its harmonious state with society and economy. (2) The comprehensive evaluation value changes are between 4.0 and 6.9. (3) The sustainable utilization level gradually increases from northwest to southeast. (4) The presented methods can quantitatively evaluate the sustainability of regional water resources.

Keywords: DPSIR · Water resources sustainable utilization · Kosi basin

1 Introduction

Since the United Nations Commission on Environment and Development put forward the concept of sustainable development in the publication entitled "our common future" in 1987, sustainable development strategy has become a common choice for the world [1]. The core issue is the rational utilization of natural resources for sustainable socio-economic development. As an important part of natural resources, water resources' sustainable utilization has become a research hotspot around the world [2].

Water resources sustainability relates to some factors, such as society, economy, environment, technology, culture and so on, all of which constitute a complex system. Therefore, many researchers take complex system science as methods, and introduce new theory and technology into sustainable research [3, 4]. Based on PSR (Pressure-state-Response) and DSR (Driving Force-State-Response) model, DPSIR model (Driving Force-pressure-State-Impact-Response) was proposed in 1993 [5]. DPSIR includes four factors (i.e. economic, social, environmental, policy). It not only shows that socio-economic development and human actions could impact the environment, but also that the state of environment can be feedback to social and economy development [6]. The feedback is constituted by solutions which are adopted to address

F. Bian and Y. Xie (Eds.): GRMSE 2015, CCIS 569, pp. 537–548, 2016.
DOI: 10.1007/978-3-662-49155-3_56

the changes of environment as well as its negative impact on the human environment [7]. DPSIR model describes a causal chain which is caused by the environmental problems [8].

This paper adopts GIS spatial analysis to construct evaluation index system based on DPSIR model and spatializes these indexes in order to reflect the spatial distribution in different regions, water conditions, levels of social development, socio-economic status. Due to the complex characteristics of basic data, we firstly normalize these data and unify the design principles in order to present the otherness at the same analysis level in various countries and territories. The 16 evaluation indexes are classified into five categories, including "driving force", "pressure", "state", "effect", and "response" [9, 10]. The purpose of this paper is to try to evaluate and analyze representative indexes for sustainable utilization of water resource evaluation based on GIS spatial analysis and DPSIR model.

2 Research Area

The research area is located in Kosi basin between southern part of Qinghai-Tibet plateau, China, Eastern Nepal and Bihar, India, and the area is 11.81×10^4 km^2. The North of Kosi Basin is located in Everest Nature Reserve, and is the major river in Everest area. Kosi has two tributaries in China, i.e. Pumqu River and Pocu River. Pumqu belongs to Ganges, and its source is Dasuopu Glacier in the north slope of Shishapangma, and finally flows into Nepal territory nearby Chen Tang [11] (Fig. 1).

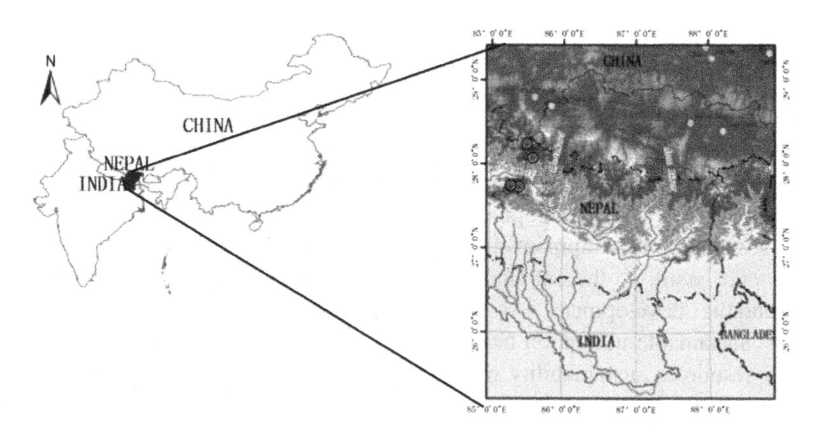

Fig. 1. The location of Kosi basin

Kosi basin has abundant water resources. The recharge sources are glacier melt-water and precipitation in this area. But due to the topographical influence, the distribution of rainfall is uneven in space and time. Currently, it faces two main threats, namely the increasing resources demand due to economic development and deterioration of water quality. Moreover, the utilization of water resources in this area has some problems, such as: (1) the low utilization ratio of water, and serious waste; (2) severe contamination of water; (3) vegetation damage, increasing soil erosion; (4) serious engineering water shortage; (5) water resources management system is weak. According to the connotation of sustainable utilization of water resources, evaluation index system for sustainable utilization not only reflected the characteristics of regional water resources and the status of development, utilization and management of water resource, but also the harmonious development levels between water resources system, society, economy and environment.

3 Methods

3.1 Establish Models

DPSIR model provides a preferable research approach for the analysis of sustainable utilization of water resources [6, 12], and can effectively integrate the water resources, environment, social and other relevant subjects. This model includes the socio-economic "drive force" caused by water resources utilization of Kosi, the "pressure" of sustainable utilization of water resources, the "Status" of environment and resources, the "effect" induced by the changes of "state", and the "response" solutions adopted by society to handle negative influence. The five aspects in DPSIR model is not only reflected the status of sustainable utilization of water resources in Kosi, and also is the necessary indexes in order to establish evaluation index system. Therefore, this paper will focus on these aspects.

3.2 Indexes Selection

The previous section has analyzed the water system based on the DPSIR framework, and determined the scope of influence factor corresponding with indexes in evaluation index system. Based on the investigation of water resources utilization status in Kosi basin, we have analyzed 22 papers related to sustainable utilization of water resources in recent years, and finally chose 16 indexes related to Kosi (Table 1).

Table 1. The evaluation indexes and definitions of sustainable utilization of water resources

Factor	Index	Definition	Unit
Driving force factor	Per capita water resources	Water resources quantity/total population	$M^3/P \cdot Y$
	Population growth rate	The natural population growth rate in a certain period/average population	%
	Population density	The population in 1 square kilometers	P/km^2
	GDP growth rate	(current year GDP-last year GDP)/last year GDP	%

(*Continued*)

Table 1. *(Continued)*

Factor	Index	Definition	Unit
Pressure factor	Drawing water coverage rate	Drawing water population usage/total population	%
	Groundwater utilization coverage rate	Groundwater population usage/total population	%
	Runoff depth	Average runoff depth for many years	mm
State factor	Produced water coefficient	Total water resources/total annual rainfall	%
	Produced water modulus	Total water resources/total area of each county	10 K m^3/km^2
	Guarantee rate of irrigation	Irrigation area/agricultural acreage	%
	Land reclamation rate	Agricultural acreage/area of land	%
	Area ratio of slope cropland (slope ≥25°)	Area of sloping fields(slope ≥25°)/agricultural acreage	%
Effect factor	Area ratio of impervious surface	City and rural construction land/county area	%
	Forest coverage rate	Forestry area/area of land	%
Response factor	Agricultural investment proportion	Agricultural investment/the same year GDP	%
	Utilization ratio of water resources	Area of water resources development project/water area	%

3.3 Weights Calculation and Evaluation Methods

This paper chooses the analytic hierarchy process (AHP) as the method to determine the index weights of resources support potential capability [13]. The index system in this paper is hierarchical structure. It includes three layers: goal layers, domain layers and index layers. The domain layers are divided into "driving force", "pressure", "state", "effect", and "response" in DPSIR model. The evaluation goal is the rank of Kosi water resources sustainable utilization.

The importance between two elements will be assigned values during the process of pairwise comparison. After pairwise comparison, it's necessary to compute the relative weight of N indexes in the principle layer. Finally, the largest eigenvalue and eigenvector could be computed by AHP method, and then we get the index weight [14]. The evaluation model for sustainable utilization of water resources is Eq. 1.

$$P = \sum_{i=1}^{n} (A_i B_i) \tag{1}$$

Where P is the sustainable utilization evaluation, A_i is evaluation index, and B_i is weight of each evaluation index.

4 Evaluation and Analysis

4.1 Index Spatialization

The data are mainly extracted from land-use, DEM, remote sensing image and socio-economic data. The land-use data and DEM are provided by the Institute of Mountain Hazards and Environment, CAS. The rainfall and temperature in recent 50 years are downloaded from University of Delaware, USA. We have purchased 2012 Regional Development Statistical Yearbook of Nepal, 2010 Statistical Yearbook of Shigatse, Tibet, and 2011 Statistical Yearbook of India.

Firstly, the statistical data of these countries and raster data are spatialized at county level by using ArcGIS spatial analysis tool. The spatialized indexes includes population

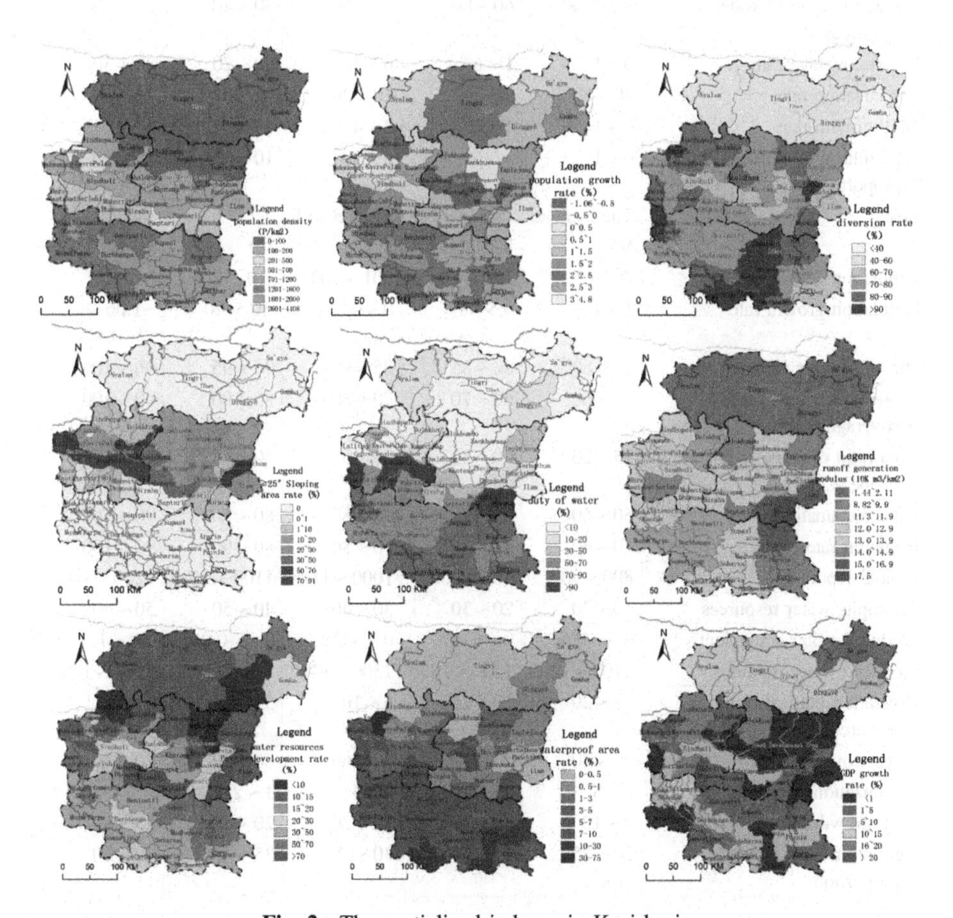

Fig. 2. The spatialized indexes in Kosi basin

Table 2. The normalized levels of water resources sustainable utilization evaluation indexes

Index	The evaluation standard from lower to higher				
	V1	V2	V3	V4	V5
Population density	1000 ~ 1200	801 ~ 1000	701 ~ 800	601 ~ 700	501 ~ 600
Population growth rate	4 ~ 4.8	3 ~ 4	2.5 ~ 3	2 ~ 2.5	1.5 ~ 2
Drawing water coverage rate	0 ~ 20	20 ~ 40	40 ~ 60	60 ~ 70	70 ~ 75
Groundwater utilization coverage rate	0 ~ 10	10 ~ 20	20 ~ 30	30 ~ 40	40 ~ 50
Area ratio of slope cropland (slope ≥25°)	0	0 ~ 1	1 ~ 5	5 ~ 10	10 ~ 15
Land reclamation rate	0 ~ 10	10 ~ 20	20 ~ 30	30 ~ 40	40 ~ 50
Irrigation rate	0 ~ 10	10 ~ 20	20 ~ 30	30 ~ 40	40 ~ 50
Runoff depth	0 ~ 100	100 ~ 200	200 ~ 400	400 ~ 600	600 ~ 800
Per capita water resources	0 ~ 2	2 ~ 4	4 ~ 6	6 ~ 8	8 ~ 10
Produced water coefficient	0 ~ 0.1	0.1 ~ 0.2	0.2 ~ 0.3	0.3 ~ 0.4	0.4 ~ 0.5
Produced water modulus	100 ~ 200	200 ~ 400	400 ~ 800	800 ~ 1000	1000 ~ 1100
Utilization ratio of water resources	80 ~ 100	60 ~ 80	40 ~ 60	30 ~ 40	20 ~ 30
Forest coverage rate	0 ~ 1	1 ~ 6	6 ~ 10	10 ~ 20	20 ~ 30
Urbanization rate	40 ~ 80	20 ~ 40	10 ~ 20	9 ~ 10	8 ~ 9
GDP growth rate	0 ~ 1	1 ~ 2	2 ~ 4	4 ~ 6	6 ~ 8
Agricultural investment proportion	0 ~ 1	1 ~ 6	6 ~ 10	10 ~ 15	15 ~ 20
Index	The evaluation standard from lower to higher				
	V6	V7	V8	V9	V10
Population density	401 ~ 500	301 ~ 400	201 ~ 300	101 ~ 200	0 ~ 100
Population growth rate	1 ~ 1.5	0.5 ~ 1	0 ~ 0.5	−0.5 ~ 0	−1.06 ~ −0.5
Drawing water coverage rate	75 ~ 80	80 ~ 85	85 ~ 90	90 ~ 95	96 ~ 100
Groundwater utilization coverage rate	50 ~ 60	60 ~ 70	70 ~ 80	80 ~ 90	90 ~ 100
Area ratio of slope cropland (slope ≥25°)	15 ~ 20	20 ~ 30	30 ~ 50	50 ~ 70	70 ~ 100
Land reclamation rate	50 ~ 60	60 ~ 70	70 ~ 80	80 ~ 90	90 ~ 100
Irrigation rate	50 ~ 60	60 ~ 70	70 ~ 80	80 ~ 90	90 ~ 100
Runoff depth	800 ~ 900	900 ~ 1000	1000 ~ 1100	1100 ~ 1200	1200 ~ 1400
Per capita water resources	10 ~ 20	20 ~ 30	30 ~ 40	40 ~ 50	50 ~ 60
Produced water coefficient	0.5 ~ 0.6	0.6 ~ 0.7	0.7 ~ 0.8	0.8 ~ 0.9	0.9 ~ 1
Produced water modulus	1100 ~ 1200	1200 ~ 1300	1300 ~ 1400	1400 ~ 1600	1600 ~ 1800
Utilization ratio of water resources	15 ~ 20	10 ~ 15	8 ~ 10	5 ~ 8	0 ~ 5
Forest coverage rate	30 ~ 40	40 ~ 50	50 ~ 60	60 ~ 70	70 ~ 80
Urbanization rate	6 ~ 8	4 ~ 6	2 ~ 4	1 ~ 2	0 ~ 1
GDP growth rate	8 ~ 10	10 ~ 15	15 ~ 20	20 ~ 25	25 ~ 30
Agricultural investment proportion	20 ~ 25	25 ~ 30	30 ~ 35	35 ~ 40	40 ~ 50

growth rate, population density, drawing water coverage rate, area ratio of slope cropland (slope ≥25°), guarantee rate of irrigation, produced water modulus, utilization ratio of water resources, area ratio of impervious surface, GDP growth rate and so on.

4.2 Weight Computation and Normalization

18 indexes will be normalized, and each index is divided into 10 levels (Table 2) which means high to low. Then, the weights obtained from AHP algorithm and normalized indexes value are substituted into the Eq. 1, and dimensionless value for each index and total evaluation value could be computed. Finally, Dimensionless value is divided into five levels from stronger to weaker (Table 3).

Table 3. Water resources supporting potential capability

Class	State	The meaning of index
I	Higher	Water resources can sustain, stable and adequate access. The support level for water utilization and socio-economic development in the current or future periods is stronger. Groundwater resources are abundant, the rainfall is frequent. Can recover soon in the event of emergencies (such as drought).
II	High	Water resources can sustainable and stable access. The support level for water utilization in the current or future periods is strong. Groundwater resources are rich, the rainfall is frequent. Be in a recoverable state in the event of emergencies (such as drought).
III	Moderate	Water resources can sustainable access. The support level for water utilization in the current or future is moderate. Groundwater resources are relatively abundant, the rainfall is relatively frequent. Have general recoverable capability in the event of emergencies (such as drought), and need the outside support.
IV	Low	Water resources can access. The support level for water utilization in the current or future periods is weaker. Groundwater resources are in shortage, less rainfall. Other water conservancy facilities can provide more assistance. The recoverable capability in the event of emergencies (such as drought) is weaker. It's need outside assistance (water) to operate.
V	Lower	The ability of accessing water resources is weak. The support level for water utilization in the current or future periods is weak. Lack of rainfall, runoff and groundwater resources. Can't repair in the event of large-scale drought events. The outside assistance is necessary otherwise it's difficult to operate.

This research adopts the analytic hierarchy process (AHP) algorithm to determine the evaluation index weights of resources support potential capability. The relative weight of driving force, pressure, state, effect and response factors are computed. The values for these factors respectively are 0.2447, 0.0367, 0.2798, 0.1212, and 0.0386 (Table 4).

Table 4. Weights estimation based on AHP method

Factor	Weight	Index	Weight
Drive factor	0.2447	The average per capita water availability	0.0402
		population growth rate	0.0453
		population density	0.1003
		GDP growth rate	0.0589
Pressure factor	0.1745	drawing water coverage rate	0.0151
		Groundwater utilization coverage rate	0.0216
		runoff depth	0.1378
Status factor	0.2798	Water production coefficient	0.0398
		Runoff modulus	0.0435
		probability of irrigation	0.0656
		Land reclamation rate	0.1031
		Slope cultivated land area than 25° or higher	0.0278
Impact factor	0.2624	Opaque surface area ratio	0.1408
		forest coverage rate	0.1216
Response factor	0.0386	Agricultural investment proportion	0.0118
		utilization ratio of water resources	0.0268

5 Results

For Kosi basin, the shorted supplied water resources quantity is the main natural limiting factors. Per capita water resource is one of the most important indexes for water resources. The less the population is and the more abundant water resources are, the higher water volume each person can have. The water storage in northern slope of Himalayas is more serious than that in the southern slope (Fig. 2). External driving force has a little change due to the effect of population growth rate and GDP growth rate in economic System. The population growth trend is negative in Nepal hill area, but it is positive in counties. The GDP and population is bi-directional growth in Shigatse. This is mainly factors which induces the domestic water rapidly increasing. The supplied water and domestic water is higher in densely population area, and the

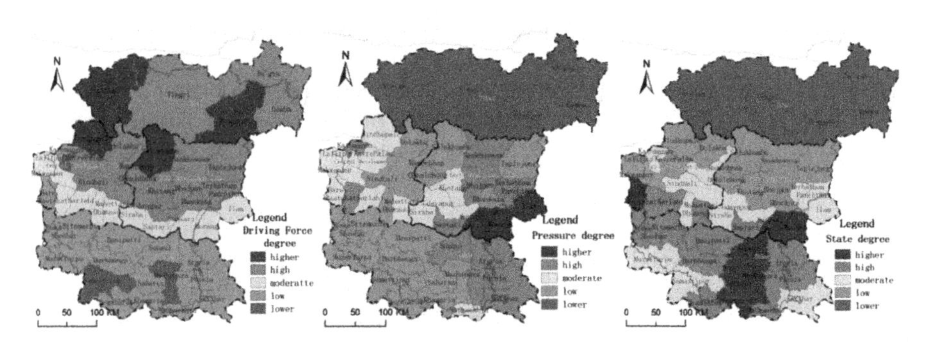

Fig. 3. The maps for driving forces, pressure, state, evaluation level

GDP is higher too (Fig. 3). But, due to short of driving force factors in sparsely population area, the water is less consumed. The internal and external driving force in Kosi basin from north to south is limited by natural and human disturbances. So, the driving force in this area is becoming increasingly scarce.

The pressure for water resources sustainable utilization induced by socio-economic driving forces is reflected in the competition of water resources utilization between the departments of social economy. The direct cause for the short of water resources in Kosi is the growth of social total water consumption (Fig. 2), the wasted water and pollution also aggravate the lack of water resources. The water resources pressure of Kosi in China is weak. This is due to the lower water consumption of social, agricultural, industrial and human life in Shigatse. The pressure index is very low, and the pressure value is between 0.1 and 1.0. Although the population density is large in Bihar, India, the water resource is abundant. For Nepal as it is located in main streams of Ganges. The population density in eastern development area is large, the water conservancy facilities is underdeveloped, the water shortage rate is higher in hill rural, and the pressure value is between 1.0 and 1.6 (Fig. 2).

"State" is the representation of water resource system under various pressures. The water resources in Kosi basin are uneven in time, and the distribution is space variation. The ratio of water resources utilization is lower. The distribution trend of runoff modulus is not completely consistent with water production coefficient. The water production coefficient is mainly influenced by rainfall, but the runoff modulus is related to the area of county. They are different in spatial state. However, the bigger the value of the two indexes, the more the water resources is abundant. The Irrigation guarantee rate, land reclamation rate, and area ratio of slope cropland are used to represent the efficiency of water resources utilization, ecological environment and management level (Fig. 2). The water resources scheduling and management level in Bihar India is obviously above the other two areas. The state value is between 1.0 and 2.2 in Bihar, and 0.3 ~ 1.0 in Shigatse due to the less human disturbance for environment and water resources (Fig. 3).

The change of water resources system state in Kosi is mainly affected by environment and socio-economic development. Ecological imbalance may cause natural disasters such as floods, droughts, mudslides, and landslides. The indexes of reflecting influences include forest cover rate, impervious area ratio (urbanization rate), and cultivated area. The forest mainly distributes in hills and plains of Nepal, and a small percentage in both Tibet China and plains of India. The rate of forest coverage is about 10.1 % (Table 4), and is one of the key factors influencing the distribution of index level. With high-speed urbanization all over the world, the distribution ratio in Nepal and India is relatively high, especially the regions near the Eastern Bihar and Kathmandu. From previous analysis, the forest, water, temperature and rainfall strongly affect the production and life in Nepal. The impact value for Nepal is between 1.4 ~ 2.4. But this value is between 0.5 ~ 1.4 in China and India.

The pressure induced by socio-economic development results in the change of water resources systems state. In return, the changed state will affect the structure, quality and quantity of water resource ecosystem, and ultimately impact human health. Therefore, in order to realize sustainable utilization of water resources, human must adjust their own behavior. The adjusting process is the response to the environment.

We use the water facilities investment ratio to represent the government support of these relevant countries. The investment of Nepal is significantly lower than the other two countries. We use the ratio of water resources development and utilization to reflect the effects of social policies. The development and utilization in plain area is significantly higher than that in the hills and highlands area. At present, under the support of China government, the response level is higher than India. That is to say, the development and investment of water resources in China has gained some effects. The ability of water conservation and water resources reasonable allocation in India is insufficient. The response index of Kosi basin is lower which is between 0.03 ~ 0.4 (Fig. 4).

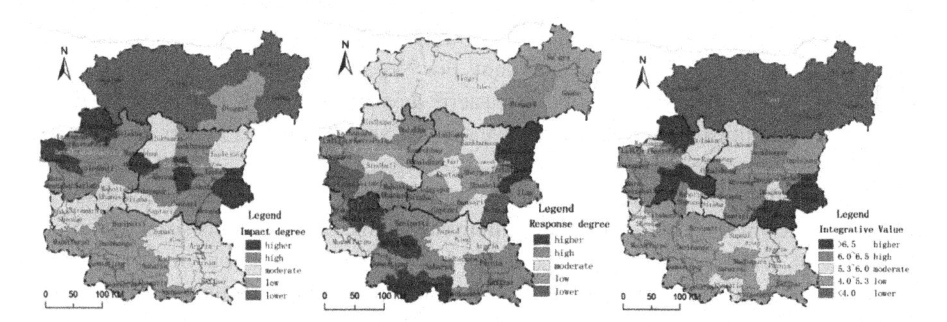

Fig. 4. The maps for influence, response levels and comprehensive evaluation

According to the analysis of drive force, pressure, state, effect and response, the sustainable utilization index of water resources in Shigatse is the lowest. The conditions of water facilities for domestic water, production water supply are poor and need to be improved. The integrated index value is between 3.2 and 3.9. The integrated value in Kosi basin is lowest, while the mean value of water resources sustainable utilization of China is changing between 2.0 to 3.0. So the value of sustainable utilization of water resources in this area is also in high value [15]. Due to intensive population and bigger agricultural acreage, the water conservancy facilities relatively lag in technology. The water resources sustainability is moderate level. The integrated value is 4.0 ~ 6.0. Because Nepal has abundant forest resources and rainfall, sparse population, moderate temperature and locates in Himalayas south slope, the water resources sustainability is from strong to strongest level comparing with other areas. The integrated values is 6.0 ~ 6.9 (Fig. 4). The water resources sustainable utilization level in Kosi basin is gradually increased from northwest to southeast.

6 Conclusions

At present, the comprehensive evaluation of DPSIR model mainly adopts sustainable integrated indexes to estimate. According to the definition of sustainable, the evaluation process is a comprehensive evaluation including drive force, development trend, harmony, equity. Some parts of this area need to improve. The value of evaluation level is up to 6.9 from 4.0. The sustainable utilization level is changing from moderate level to stronger. Generally speaking, due to the abundant water resources in Kosi basin, the state of sustainable utilization is developing to benign direction. This paper use spatial analysis based on DPSIR model to evaluate the five level of sustainable water resources utilization. The results show that the water resources sustainable utilization in Kosi basin is in good condition. Meanwhile, the evaluation results have a guidance meaning to the sustainable utilization of water resources in this area. The evaluation results obtained by the proposed method match the practical case, and testify the reliability and sensitivity of integrated method by utilizing GIS spatial analysis based on DSPIR model.

Acknowledgements. This research is supported by the National Natural Science Funds of China (grant no. 41301094), the Key Research Program of the Chinese Academy of Sciences (grant no. KZZD-EW-08-01, the lead strategic project of the Chinese Academy of Sciences (grant no. XDB03030507), the "one-three-five" directivity innovation project of IMHE (grant no. sds-135-1205-03), and the open fund for spatial information technology key lab bases of the Ministry of Land and Resources of China (grant no. KLGSIT2014-06).

References

1. Wang, W.Z., Guo, R.S., Huang, J.: Introduction to Local Sustainable Development, pp. 32–45. Commercial Press, Beijing (1999)
2. Liu, H., Geng, L.H., Chen, X.Y.: Indicators for evaluating sustainable utilization of regional water resources. Adv. Water Sci. **14**(3), 265–270 (2003)
3. Song, S.B.: Study of Indicators System and Assessment Methods for Regional Sustainable Utilization in Relation to Water Resources. Northwest Sci-tech University of Agriculture and Forestry Doctor's Degree (2003)
4. Cheng, G.M., He, S.H.: Progress on evaluating methods for sustainable utilization of water resources. J. Water Res. Water Eng. **17**(1), 52–56 (2006)
5. Smeets, E., Weterings, R.: Environmental Indicators: Typology and Overview, Technical Report No. 25. European Environmental Agency, Copenhagen (1999)
6. Stephen, C., Mangi, E.: Reef fisheries management in Kenya: Preliminary approach using the drive force-pressure-stat-impacts-response (DPSIR) scheme of indicators. Ocean Coast. Manage. 98–106 (2006)
7. Cao, H.J.: An initial study on DPSIR model. Environ. Sci. Technol. **28**(6), 110–111 (2005)
8. Caeiro, S., Mourao, I., Costa, M.H.: Application of the DPSIR. In: 7th AGILE Conference on Geographic Information Science, pp. 391–403 (2004)
9. Peter, H., Stephan, B., Tony, H.: Measuring Sustainable Development: Review of current Practice. International Institute for sustainable, Canada Cataloguing in Publication Data (1997)
10. George, C.Z., Maria, A.T.: Selecting agri-environmental indicators to facilitate monitoring and assessment of EU agri-environmental measures effectiveness. J. Environ. Manage. **70**, 315–321 (2004)
11. Liu, W.L.: Tibet's Natural and Ecological. Tibet People's Publishing Press, Lhasa (2007)
12. Reena, S.: International frameworks of environmental statistics and indicators. In: Inception Workshop on the Institutional Strengthening and Collection of Environment Statistics, vol. 4, p. 6 (2002)

13. Seaty, T.L.: The Analytic Hierarchy Process. University of Pittsburgh, Pittsburgh (1998)
14. Luo, Z.Q., Yang, S.L.: Comparative study on several scales in AHP. Syst. Eng. Theory Pract. **9**(9), 23–33 (2004)
15. Gao, H., Jin, H.: Evaluation of sustainable utilization for regional water resources based on AHP and fuzzy synthetic judgment—a case study for Jiangmen City in Guangdong Province. J. Water Res. Water Eng. **18**(3), 51–59 (2007)

A Feature Selection Method Based on Multi-objective Optimisation with Gravitational Search Algorithm

Bolou Bolou Dickson, Shengsheng Wang(✉), Ruyi Dong, and Changji Wen

College of Computer Science and Technology, Jilin University, Changchun 130012, People's Republic of China
boloubh@yahoo.com, wss@jlu.edu.cn

Abstract. The process of feature selection (FS) is a substantial task that has a significant effect in the performance of a given algorithm. The goal is to choose a subset of available features by eliminating the unnecessary features. This hybrid algorithm is in maximising the classification performance and minimising the number of features to achieve an outstanding performance through a less complex procedure. From the experiments, FSMOGSA was noted to be quite unparalleled in comparison with other methods in reducing the error rate, and maximising the general performance through irrelevant feature reduction.

Keywords: Gravitational search algorithm · Indexed non-dominated solutions · Pareto front · Multi-objective optimisation · Feature selection

1 Introduction

Today, the relevance of feature selection (FS) in machine learning cannot be under-rated. FS has considerable importance in real life applications such as medicine, astronomy, biology, to mention but a few. The goal is to choose a subset of available features by eliminating unnecessary features [1]. To obtain any desired result in using datasets, high dimensionality imposes learning difficulties by degradation of relevant information on the learned models. Real world datasets are entangled with many irrelevant and misleading features for this reason (FS) is adopted to eliminate such impediments. Furthermore, the objective of FS is to select a relevant subset of features say q, from a set of p features (q < p) in a given dataset. To extract sufficient information for example from an image set, it is appropriate to eliminate the features with no predictive information and avoid redundant features.

2 Related Works on Feature Selection

Efficient processing and retrieval of features rely on the number of relevant features extracted [2]. Hamdani et al. developed an entirely different method called, multi-objective feature selection algorithm using non-dominated sorting-based multi-objective GA II (NSGAII), however, it was not compared with any algorithm [3].

F. Bian and Y. Xie (Eds.): GRMSE 2015, CCIS 569, pp. 549–558, 2016.
DOI: 10.1007/978-3-662-49155-3_57

Our work is focused on Multi-Object feature selection with gravitational search algorithm (FSMOGSA) which is completely new in terms of feature selection. Tian et al. [4] proposed a work on multi-objective optimization of short-term hydrothermal scheduling using non-dominated sorting gravitational search algorithm with chaotic mutation. A. R. Bhowmik and A. K. Chakraborty, proposed, Solution of optimal power flow using non-dominated sorting multi-objective opposition based gravitational search algorithm (NSMOOGSA) [5]. And in 2013, Bing Xue et al. proposed PSO for feature selection and classification: a multi-objective approach and investigate two PSO-based multi-objective feature selection algorithms [6].

Our FSMOGSA is in maximising the classification performance and minimising the number of features to achieve an outstanding performance through a less complex method. It finds the non-dominated (Pareto fronts) solutions and groups such solutions into subsets of indexed non-dominated solutions.

2.1 Basic Gravitational Search Algorithm

Gravitational search algorithm was introduced in 2009 by Rashedi et al. [7], where the solutions of optimisation problems are regarded as agents. All agents attract one another in the solution space due to the force of gravity, lighter agents are attracted (converge) towards the heavier agents, known as the optimal solution based on the law of motion. Given a system of N agents the position of the ith agent is:

$$X_i = (x_i^1, \ldots, x_i^d, \ldots, x_i^n), \quad \text{for i} = 1, 2, \ldots, \text{N} \tag{1}$$

Where x_i^d is the position of the ith agent in the dth dimension and n is the dimension of the space.

$$F_{ij}^d(t) = G(t)\frac{M_{pi}(t) \times M_{aj}(t)}{R_{ij}(t) + \varepsilon}(x_j^d(t) - x_i^d(t)) \tag{2}$$

Where M_{aj} is called active gravitational mass, M_{pi} is passive gravitational mass, G (t) is gravitational constant at time t, ε is infinitesimally small value then $R_{ij}(t)$, is a Euclidean distance between masses i and j, $R_{ij}(t) = |X_i(t), X_j(t)|_2$.

Equation (3), is the force of the object i

$$F_i^d(t) = \sum_{j=1, j\neq i}^{N} rand_j F_{ij}^d(t), \tag{3}$$

$$\text{The acceleration of the ith object is}: a_i^d(t) = \frac{F_i^d(t)}{M_{ii}} \tag{4}$$

2.2 Velocity and Position of Particles

The successive velocity of a given object is obtained by the addition of its current velocity to its acceleration that is Eq. (5), and the current position of the object can be obtained by (6).

$$v_i^d(t+1) = rand_i^d v_i^d(t) + a_i^d(t) \tag{5}$$

$$x_i^d(t+1) = x_i^d(t) + v_i^d(t+1) \tag{6}$$

where $rand_i$ is a random number between 0 and 1, present velocity $v_i^d(t)$, next possible velocity $v_i^d(t+1)$, next possible position $x_i^d(t+1)$, present position $x_i^d(t)$, and acceleration $a_i^d(t)$ of the ith particle at time t.

3 Multi-objective Gravitational Search Algorithm and Pareto Front

This method operates based on the concept of dominance of set of optimal solutions called Pareto front. A given multi-objective optimisation entails maximisation or minimisation of multiple conflicting objective functions. From the training sets, any of the subsets possessing fewer features is presumed to achieve higher quality function value, as such the extrapolated features with the best fitness are chosen mostly out of such subsets of features. Reducing the number of irrelevant features will have a positive effect on the performance of the entire process. The minimisation is expressed below,

$$min, F(x) = [f_1(x), f_2(x), \ldots, f_M(x)] \tag{7}$$

$$g_i(x) \leq 0, \quad i = 1, 2, \ldots m \tag{8}$$

$$h_i(x) = 0, \quad i = 1, 2, \ldots l \tag{9}$$

where x is the vector of decision variables, $f_i(x)$ is a function of x, and k is the number of objective functions to be minimised, $g_i(x)$ and $h_i(x)$ are the constraint functions of the problem. Given any minimisation task, solution x_1 will dominate solution x_2 if both satisfy this condition:

$$\forall m \in [1, M], f_m(x_1) \leq f_m(x_2) \quad \text{and} \quad \ni n : f_n(x_1) < f_n(x_2) \tag{10}$$

For $m, n \in [1, 2, \ldots, M]$.

3.1 The Main Optimisation Process of (FSMOGSA)

If a given solution is not dominated by another set of solutions, then that solution is called a Pareto-optimal solution. A collection of all the sets of Pareto-optimal solutions yields the Pareto front, some basic principles of choosing dominant or non-dominated solutions in our algorithm is based on;

(a) The number of individuals a given individual dominates.
(b) The Pareto front an individual is located.
(c) The number of individuals that dominates a given individual solution.

Multi-objective tasks result if there is the necessity to make optimal decisions between two or more conflicting objectives in a solution space. Hence, we will make the equation of fitness of particles which is effective in a single objective to an equation adoptable for multi-objective as shown below:

$$M_i(t) = ||\varepsilon|| + \sum_{k=1}^{k} \left[m_i^k(t)\right]^2 / \sum_{j=1}^{N} \sum_{k=1}^{K} \left[m_j^k(t)\right]^2 \tag{11}$$

$$m_i^k(t) = \frac{fit_i^k(t) - worst^k(t)}{best^k(t) - worst^k(t)}, for\ k \in [1, k] \tag{12}$$

where $||\varepsilon||$ is an infinitesimally small error value, $m_i^k(t)$ is the normalised fitness value of the ith agent in the kth objective; $fit_i^k(t)$ is the fitness value of the ith agent in the kth objective; K is the number of objectives; $best^k(t)$ is the best fitness of all agents in the kth objective; $worst^k(t)$ is the worst fitness of all the agents in the kth objective (Fig. 1).

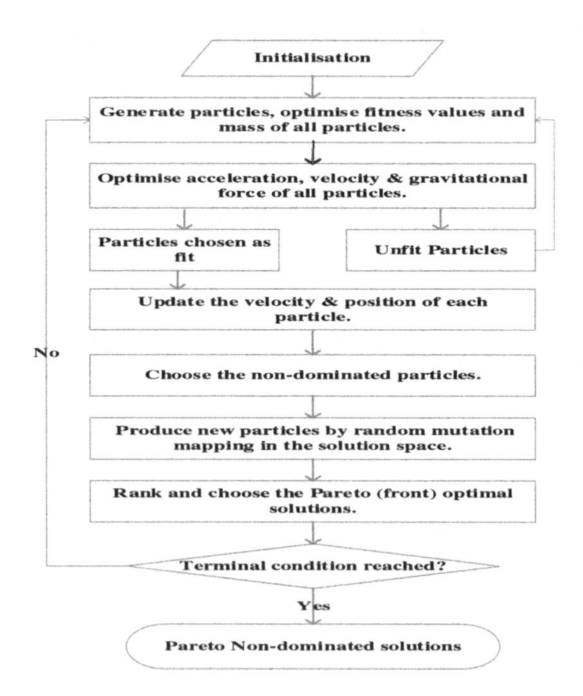

Fig. 1. Shows the FSMOGSA process.

The particles are initialised and the fitness values, velocity, acceleration and position of each particle is calculated and updated. Then, the non-dominated solutions are chosen which is followed by the random mutation to produce a new population for another optimisation.

3.1.1 The fitness function in Eq. (11) will improve the performance of the (FSMOGSA) algorithm by prudently minimising the convergence rate of agents in the process. The error factor $||\varepsilon||$ in the fitness equation stabilises the motion of the agents, it assumes infinitesimally small values within (0,1). The chosen features in the training set will be in categories of; False negative, False positive, True negative and True positive. After the fitness function had been utilised the error rate is further evaluated by Eq. (13) below:

$$ErrorRate(\psi) = \frac{Fn + Fp}{Fn + Fp + Tn + Tp} \tag{13}$$

For Fn is false negative, Fp is positive, Tn is true negative and Tp is true positive feature. This error rate can be adjusted to minimise error during feature selection.

3.1.2 Then the second purpose is reducing the number of features by choosing only very highly ranked features, redundant features are left out. This work is a multi-objective feature selection algorithm so a function other than Eq. (13) which will perform the dual purpose of the fitness function is used to minimise the classification error rate and as well guarantee minimisation of the number of features with high classification performance is adopted as in Eq. (14):

$$Fitness_{function} = \left\{ \frac{F_{selected}}{F_{All}} * \alpha + \frac{ER_{selected}}{ER_{All}} * (1 - \alpha) \right. \tag{14}$$

Where $F_{selected}$ is number of selected features, F_{All} is all the available features, α is a negligible constant within (0, 1), $ER_{selected}$ is classification error rate of the selected feature subset, ER_{All} is the classification error rate of all available features of the training set. A preponderantly negative occurrence in SI optimisation is stagnation, where the swarm agents get confined in local optimum.

3.2 Random Mutation to Generate New Agents

After every iteration period a mutation process is added to the population to randomly generate a new solution population. As a result of the unforeseen random factors, a random mapping process is employed to overcome premature convergence in FSMOGSA algorithm in the creation of new agents. Whenever a new agent dominates the current existing agent, the newly generated agent replaces the existing agent. In other words, it updates the masses and chooses the agent with heavier mass. Equations (15), (16) and (17) below are the mutation equations:

$$\zeta_i^d = [x_i^d(t) - x_{\min}^d(t)]/[x_{\max}^d(t) - x_{\min}^d(t)] \quad (15)$$

$$\eta_i^d = \lambda\zeta_i^d(1 - \zeta_i^d), \zeta_i^d \in [0, 1] \quad (16)$$

$$x \cdot c_i^d(t) = \eta_i^d[x_{\max}^d(t) - x_{\min}^d(t)] + x_{\min}^d(t) \quad (17)$$

where ζ_i^d represents the normalised position of the ith agent in the dth dimension; λ is a constant; η_i^d is the transformed value by random mutation; $xc_i^d(t)$ is the new position of the ith agent [8].

To determine the position of a particle undergoing mutation Eq. (17) is adopted. In the end of the mutation process, the steps to update the velocity and position of the offspring population is developed for ranking of the solutions then another optimisation process is carried out to select another set of Pareto solutions. The quantity ζ_i^d in Eq. (15) is randomly generated within the interval [0,1] as the process starts, λ is considered to be a constant in Eq. (16).

The mutation begins by choosing a particle say p_1, in a random pattern within the current population. Then from a given Pareto front, another two particles p_2 and p_3 are chosen lying within a bound. By Eq. (16) the mutation factor of particle p_1, is evaluated in the dth dimension from when d assumes the value 1 to n, then a newly mutated particle is produced. The next step is the substitution of p_1 with the newly mutated particle. When the mutation process is over the fitness values of the new population is evaluated, then, the error rate is also re-evaluated of the chosen features with Eqs. (13) and (14). Every abnormally copied code of a feature will result to a new feature (mutant), the process is not done orderly but randomly.

3.3 Indexed Non-dominated Solutions (Pareto Front Subsets)

A multi-objective method, unlike a single objective one searches for sets of optimally non-dominated solutions. The set of solutions are indexed (grouped) as non-dominated sets according to the various individual feature types. This idea helps to reduce extraneous features from the relevant features and improves the classification performance because the number of features is minimised.

Definition 3.3.1 Given different indexed feature sets, let F be a finite feature space, for every feature f in F, there is a set of features S_f such that a set of common features collected as G of S_f sets is known as an indexed collection of feature sets, that is a collection of feature sets indexed by F in different dimensions. The collection G is represented by $\{S_f\}_{f\in F}$, is the finite sets of indexed non-dominated feature sets represent Pareto fronts. Let G be the indexed sets of non-dominated solutions, so $\{S_f\}_{f\in F} = \left\{\{S_{f_1}\}_{f_1\in F}, \{S_{f_2}\}_{f_2\in F}, \ldots \{S_{f_n}\}_{f_n\in F}\right\}$ hence, $\{S_f\}_{f\in F} \subset G$.

Generally, for finite sets of non-dominated solutions, the **union** and **intersection**; $\bigcap_{i=1}^{n} G_i = \{x \in U : \exists i \in \{1,2,\ldots,n\}, : x \in G_i\}$ and $\bigcap_{i=1}^{n} G_i = \{x \in U : \forall i \in \{1,2,\ldots,n\}, : x \in G_i\}$.

The idea of indexed sets of non-dominated solutions is adopted here in two aspects:

(i) Disjoint Non-dominated Solutions (Pareto Fronts) For some indexed sets of non-dominated solutions, there are some indexed finite Pareto fronts $\{G_1, G_{2,\ldots,} G_n\}$, the arbitrary intersection of the Pareto fronts is; $G_1 \cap G_{2\ldots} \cap G_n = \emptyset$. It implies, $\bigcap_{f \in F} U_f = \{x | x \in U_f, \forall f \in F\} = \emptyset$, and called *"distinct or disjoint indexed non-dominated solutions"* in F which are ranked.

(ii) Connected or Intersecting Non-dominated Solutions (Pareto Fronts) If the solution subsets of non-dominated solutions have one or more common features in the indexed non-dominated solutions, then, the arbitrary intersection is non-empty. That is, $G_1 \cap G_{2\ldots} \cap G_n \neq \emptyset$, so $\bigcap_{f \in F} U_f = \{x | x \in U_f, \forall f \in F\} \neq \emptyset$, hence, called *"connected or intersecting indexed non-dominated solutions"*.

This segregates the non-dominated solutions and the irrelevant solutions are neglected.

The collection G is represented by $\{S_f\}_{f \in F}$, in this algorithm the finite sets of indexed non-dominated feature sets represent Pareto fronts.

3.4 K-Nearest Neighbor (K-NN) Classifier

The K-nearest neighbor (K-NN) classifier is employed here to evaluate our method as a result of its simplicity. The introduction of the K-nearest neighbor (K-NN) method in 1951 by Fix and Hodges has greatly contributed in the improvement of new algorithms. One reason of the (K-NN) algorithm is for the classification of new features after the random mutation. Due to the attributes and training samples obtained, the K-NN evaluates the classification performed by our FSMOGSA method. As a multi-objective task, the K-NN method consists of a supervised learning task where new indexed non-dominated solution sets are evaluated in the K-neighbourhood and classified based on the ranking in the solution space.

4 Experiments

The experiments were performed using some data sets from the UCI open access repository, in which three other algorithms were used to compare with the performance of our (FSMOGSA) method. Two of the algorithms are single objective method, that is, gravitational search algorithm (GSA) and Binary particles swarm optimisation (BPSO). The other is a multi-objective optimisation algorithm called Non-dominated solutions particle swarm optimisation feature selection (NSPSOFS), these three methods were used in the comparison with our proposed method. Precisely, four different data sets were used in the experiment to ascertain the efficiency of FSMOGSA. The experiment was implemented on a 32-bit windows 7 operating system, processor: Intel®, core™, Duo 3.00 GHz, RAM: 4 GB computer, with MATLAB (R2012a) suite (Table 1).

Table 1. Is the description of the four data sets

Data sets	Iris	Ionosphere	Vehicle	Wine
No. of instants	150	351	846	178
No. of classes	3	2	4	3
No. of features	4	34	18	13

All the four methods were tested on each of the four (4) data sets and the results were compared with one another. The outcome of each of the four algorithms for each data set shows the rate of error with regards to the number of features in the data set obtained.

Table 2, shows the percentage number of features and the error values of the four methods, the best results (error values) obtained in each data set are underlined in bold character. Every data set has different number of features as such the iterations are different in number. Obviously, the error values for FSMOGSA algorithm are the least. As the number of features increases the error value increases also. The Iris data set was left out here due to the few features in it.

Table 2. Is a performance Comparison of the error values with percentage number of features.

Wine_Dataset↓	**5%feature**	**15%**	**25%**	**35%**	**45%**	**55%**	**65%**	**75%**	**85%**	**95% ... %**
GSA	19.312	20.114	21.089	20.400	21.822	21.997				
BPSO	17.216	18.102	23.182	24.377	24.921	23.994				
NSPSOFS	13.882	14.876	14.993	15.734	15.871	16.822				
FSMOGSA	12.002	12.399	13.118	14.723	14.892	**15.177**				
Ionosphere_Dataset↓										
GSA	20.400	20.113	17.391	14.764	16.430	16.412	17.023	17.321	17.110	18.021
BPSO	24.032	20.132	17.943	12.730	12.023	11.899	11.071	14.132	16.234	16.942
NSPSOFS	11.115	11.172	11.093	10.987	10.123	10.141	10.112	10.023	9.897	8.098
FSMOGSA	12.976	11.132	10.398	9.341	9.076	**8.991**	9.112	9.354	9.310	
Vehicle_Dataset↓										
GSA	27.114	13.344	14.019	20.786	23.324	23.124	22.921			
BPSO	26.221	10.223	15.216	21.576	23.113	24.901	24.984			
NSPSOFS	22.223	16.912	17.445	18.139	18.939	18.175	17.945			
FSMOGSA	13.897	15.897	14.111	15.897	15.987	16.391	**17.012**			

Figure 2(a–d) show the graphical display of the experiment on each data set.

The FSMOGSA algorithm shows a great degree of stabilisation of the low error rate and minising the number of features to achieve our objectives. Since our multi-objective was to maximise performance by reducing the error rate and at the same time reducing the number of features, we used the error function in Eq. (14) to achieve these goals. While the generation of new particles increased the chances of obtaining optimal sets of non-dominated solutions, the indexed sets facilitate the ranking and choosing of the best solution sets. The experimental validation of (FSMOGSA) is a

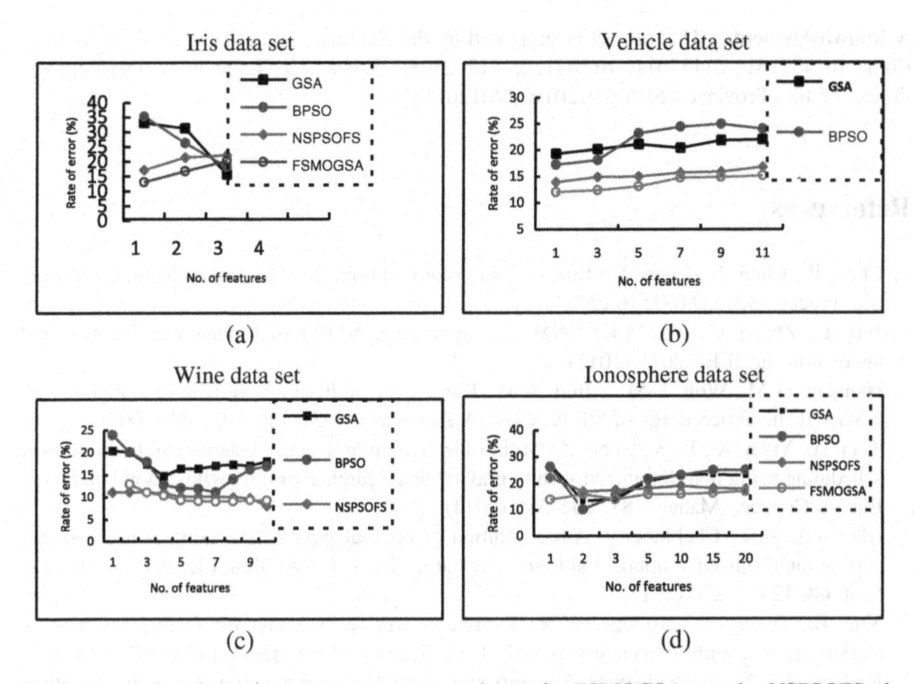

Fig. 2. (a) Is the Iris data set, the error rate was lower for FSMOGSA next by NSPSOFS, the features are fewer in this data set. For the data set; vehicle in (b), FSMOGSA still has the least error rate with reduced features. Then, the wine data set (c) the error rate is lower for NSPSOFS than our FSMOGSA but it is negligible and on the fourth data set (d), FSMOGSA showed the best result among the four algorithms employed on the Ionosphere data set.

good indication of the efficiency in its application to feature selection in a multi-objective task over most existing feature selection methods. Again, it indicates that FSMOGSA searches for the non-dominated solutions through the integration of minimal feature set numbers and classifier performance to yield optimised indexed non-dominated (Pareto fronts) solutions with high classification accuracy than the other three methods.

5 Conclusion

The experimental validation indicates that our method is a more efficient one The best results for the multi-objective algorithm's validation were obtained by our FSMOGSA algorithm, the next best result was NSPSOFS both of which are hybrid methods. This shows that the hybrid methods have a better performance than the regular methods. From the experiments, FSMOGSA was noted to be quite unparalleled in comparison with the other methods in reducing the error rate and maximising the general performance by minimising the irrelevant features. For the objective of efficient performance we suggest a future work in binary GSA hybrid method.

Acknowledgements. This project is supported by the National Natural Science Foundation of China (61472161, 61133011, 61303132, 61202308), Science & Technology Development Project of Jilin Province (20140101201JC, 201201131).

References

1. Chen, B., Chen, L., Chen, Y.: Efficient ant colony optimization for image feature selection. Sig. Process. **93**, 1566–1576 (2013)
2. Jing, L., Zhang, C., Ng, M.K.: SNMFCA: supervised NMF-based image classification and annotation. In: IEEE 2011 (2011)
3. Hamdani, T.M., Won, J.-M., Alimi, A.M., Karray, F.: Multi-objective feature selection with NSGA II. In: Proceedings of 8th ICANNGA Part I, vol. 4431, pp. 240–247 (2007)
4. Tian, H., Yuan, X., Ji, B., Chen, Z.: Multi-objective optimization of short-term hydrothermal scheduling using non-dominated sorting gravitational search algorithm with chaotic mutation. Energy Convers. Manage. **81**, 504–519 (2014)
5. Bhowmik, A.R., Chakraborty, A.K.: Solution of optimal power flow using non dominated sorting multi-objective opposition based gravitational search algorithm. Electr. Power Energy Syst. **64**, 1237–1250 (2015)
6. Xue, B., Zhang, M., Browne, W.N.: Particle swarm optimization for feature selection in classification: a multi-objective approach. IEEE Trans. Cybern. **43**(6), 1656–1671 (2013)
7. Rashedi, E., Nezamabadi-Pour, H., Saryazdi, S.: GSA: a gravitational search algorithm. Inform. Sci. **179**, 2232–2248 (2009)
8. Tian, H., Yuan, X., Ji, B., Chen, Z.: Multi-objective optimization of short-term hydrothermal scheduling using non-dominated sorting gravitational search algorithm with chaotic mutation. Energy Convers. Manage. **81**, 504–519 (2014)

Extracting Tempo-Spatial Features of Paddy Rice Using Time-Series MOD09A1-A Case Study in Hunan, China

Jinling Zhao, Linsheng Huang, and Dongyan Zhang(✉)

Key Laboratory of Intelligent Computing and Signal Processing, Ministry of Education, Anhui University, Hefei 230039, China
aling0123@163.com

Abstract. Time-series of 8-day composite MODIS surface reflectance product (MOD09A1) data were used to identify the spatial patterns of double-season early rice, single-season middle rice and double-season late rice of Hunan Province, China in 2010. Firstly, the available MODIS images of transplanting and heading stages were assured in accordance with the schedules of local traditional paddy fields tillage in 2010, and then time-series NDVI images were smoothed by Harmonic Analysis of Time Series (HANTS) algorithm to remove the noise and the atmospheric effects. Secondly, the spatial distribution and planting acreage of three types of rice were derived from combining the Enhanced Vegetation Index (EVI) and Land Surface Water Index (LSWI) according to the water background and variation characteristics of NDVI values at transplanting and heading stages. Finally, the accuracy evaluation was performed by the statistical data. The results showed that three types of rice were mainly distributed along the Dongting Lake Basin in Hunan Province and the relative errors were −10.99 %, 1.46 %, −5.87 %, respectively.

Keywords: Spatial pattern · Time-series MODIS · Paddy rice · Vegetation index

1 Introduction

Paddy rice is the most important grain crop for Hunan Province and its planting acreage and total yield also rank the top in China. It is of great significance to learn about the cropping system, planting intensity and spatial distribution by studying the tempo-spatial characteristics in such a typical study area. However, the tempo-spatial difference is greatly noticeable in different planting zones due to rice varieties, meteorological features, soil types, topography, relief as well as social development levels [1]. Consequently, it is vital to monitor the spatial patterns of paddy rice and estimate the acreage of fields as accurately as possible.

In recent years, remote sensing has been applied in identifying various kinds of rice information due to its wide swath, fast information acquisition, shorter revisiting cycle, and large amount of spatio-temporal information [2–5]. In comparison with other moderate resolution remote-sensing data, Landsat series images have been widely used

F. Bian and Y. Xie (Eds.): GRMSE 2015, CCIS 569, pp. 559–565, 2016.
DOI: 10.1007/978-3-662-49155-3_58

in monitoring growth, estimating planting area and other applications. Although the resolution of moderate images can reach to dozens of meters, their revisiting cycle is relatively longer. They cannot satisfy the demands of continually monitoring the fast change information of physiological and biochemical status of paddy rice at specific growth periods. Microwave and Radar data are also used in monitoring paddy rice because of its characteristics of penetrating cloud and frost, but their swath is usually very small. It can obviously improve the image quality by confusing multi-source images, strict geometric correction and more varieties of images are required. Conversely, high temporal resolution images can provide fruitful information in monitoring and identifying paddy rice [6, 7].

As a major producer of rice in China, the rice production in Hunan Province accounts for 80 % of its total grain production. Both the planting area and total yield hold the first place in China. Taking Hunan Province as the study area, this study is to find out the spatial distribution patterns of paddy rice based on time-series MODIS product. In order to achieve this goal, several spectral indices were calculated: normalized difference vegetation index (NDVI), normalized difference water index (NDWI), normalized difference soil index (NDSI), land surface water index, (LSWI), and enhanced vegetation index (EVI). According to the water background during the growing progress, Harmonic Analysis of Time Series (HANTS) software was used to smooth time-series NDVI. The identification results were evaluated by the field GPS positioning points and the statistical data.

2 Materials and Methods

2.1 Data Sources

Time-series MODIS/Terra data sets (8-day MOD09A1 V005) in 2010 were freely downloaded through the Land Process Distributed Active Archive Center (LPDAAC, http://reverb.echo.nasa.gov/reverb/). A total of three MODIS HDF tiles are required to cover the whole study area (Fig. 1). MOD09A1 provides Bands 1–7 at 500-m resolution in an 8-day gridded level-3 product in the Sinusoidal projection. There are hundreds of images for different MOD09A1 bands in a year, so batch processing must be performed to improve work efficiency. Here, Cygwin tools, which provide a Linux look and feel environment for Windows, were combined with MODIS Reprojection Tools (MRT) to change data format, transform projection and mosaic images to cover the whole Hunan Province.

In addition, the administrative boundary data of Hunan, a total of 64 GPS field survey data of paddy rice in 2010 were also acquired in Taoyuan County, Linli County and Hanshou County. These data are important to understand the real situation on the ground for paddy rice identification. High resolution imagery zoom-in-views from Google Earth can also provide visual information on various landscape features and types such as agriculture, forests and barren land. The statistical data of planting acreage in paddy rice by cities were also acquired from Hunan Province Statistical Bureau, which can be used to compare to the satellite-derived estimates at the city level.

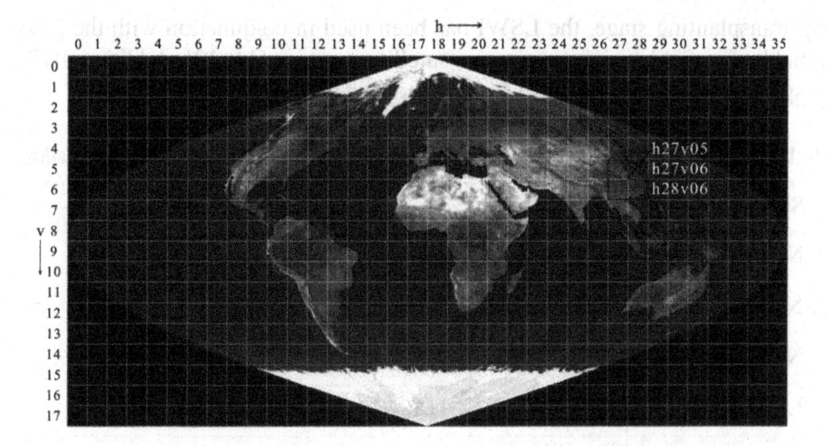

Fig. 1. Paths/rows of Hunan province located in the sinusoidal (SIN) projection.

2.2 Smoothing Time-Series MODIS NDVI Data

A HANTS algorithm for time series dataset analysis has been developed [8] and successfully applied in vegetation index analysis on the European continent [9]. The primary thought of HANTS is that the growth status of vegetation has extreme seasonal fluctuation as NDVI describes. It can be described by a series of low-frequency Sine functions with different phases, frequencies and amplitudes. Cloud and other intervention factors always randomly change and they can be made as the high-frequency noise. In order to smooth the time series NDVI images, the first step is to transfer observational samples from time domain to frequency domain; then, a new time series can be reconstructed using a reversed Fourier Transformation or synthesis, which can generate values of the variable at any desired time. A Fast Fourier Transformation (FFT) is often used in the Fourier Transformation process in order to save computational time. In order to smooth different time-series dataset, National Aerospace Laboratory (NLR) developed the NLR HANTS software package. Figure 2 shows the flow chart of smoothing time-series NDVI images.

2.3 Identification of Paddy Rice Fields

In comparison with other crops, a unique physical feature of paddy rice fields is that rice plants are grown on flooded soils. With the temporal dynamics of paddy rice fields, three main periods can be used: (1) the flooding and rice transplanting period; (2) the growing period (vegetative growth, reproductive, and ripening stages); and (3) the fallow period after harvest [10, 11]. At different growth stages, the mixture of surface water, soil and green rice plants show different combination forms. As a result, the dry land and the paddy field can be distinguished according to such a feature. In addition to the intervention of dry land, some other surface features which have similar reflectance characteristics must be also weeded out. Due to the unique spectral features of paddy rice fields during transplanting stage, the LSWI has been used in conjunction with the NDVI to map paddy rice fields using time-series satellite imagery [11, 12]. Additionally, some other spectral vegetation indices were also needed in this study (Table 1).

Table 1. Calculated spectral indices and corresponding expressions for identifying paddy rice.

Spectral index	Expression
NDVI	NDVI = $(R_{NIR} - R_{RED})/(R_{NIR} + R_{RED})$
NDWI	NDWI = $(R_{RED} - R_{SWIR})/(R_{RED} + R_{SWIR})$
NDSI	NDSI = $(R_{SWIR} - R_{NIR})/(R_{SWIR} + R_{NIR})$
LSWI	LSWI = $(R_{NIR} - R_{SWIR})/(R_{NIR} + R_{SWIR})$
EVI	EVI = $G^*(R_{NIR} - R_{RED})/(L + R_{NIR} + C_1 R_{RED} - C_2 R_{BLUE})$

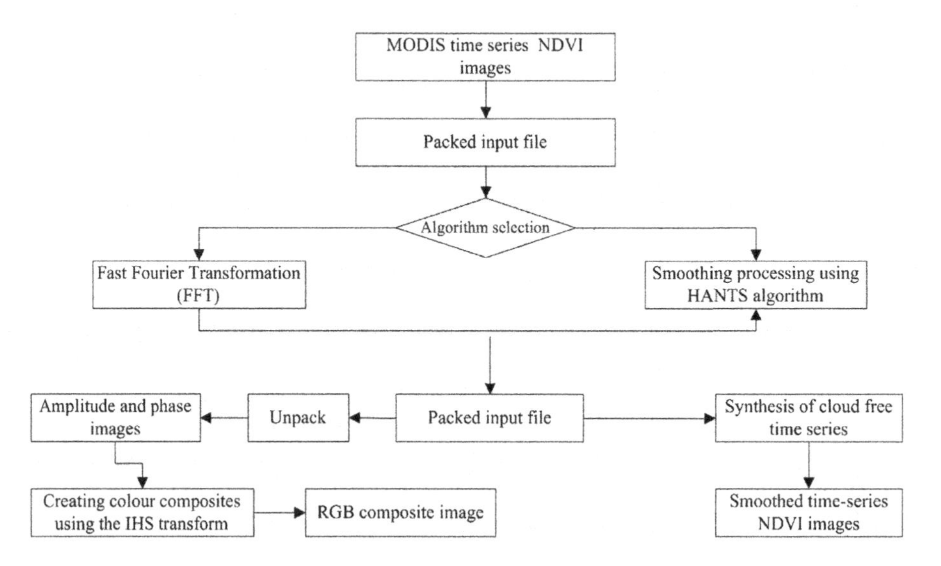

Fig. 2. Flow chart of smoothing time-series NDVI images by HANTS algorithm.

where L is a soil adjustment factor, and C_1 and C_2 are coefficients used to correct aerosol scattering in the red band by the use of the blue band. The R_{BLUE}, R_{RED}, R_{NIR}, R_{SWIR} represent reflectance at the blue (MODIS-Band3, 459–479 nm), red (MODIS-Band1, 620–670 nm), Near-Infrared (NIR, MODIS-Band2, 841–876 nm), and Short-Wave Infrared (SWIR, MODIS-Band6, 1628–1652 nm) wavelengths, respectively. In general, G = 2.5, C1 = 6.0, C2 = 7.5, and L = 1.

3 Results and Discussions

3.1 Mapping Paddy Rice Fields

Three maps were derived from time-series MOD09A1 products and other ancillary datasets (Fig. 3) and their identified acreage were 1152.3 kha, 1265.0 kha and 1309.2 kha, while they were 1294.6 kha, 1246.8 kha and 1390.6 kha according to the China Statistical Yearbook 2010. The relative errors were −10.99 %, 1.46 % and −5.87 %, where the positive value indicated that the calculated acreage was greater than the statistical values, while it was reverse for the negative value.

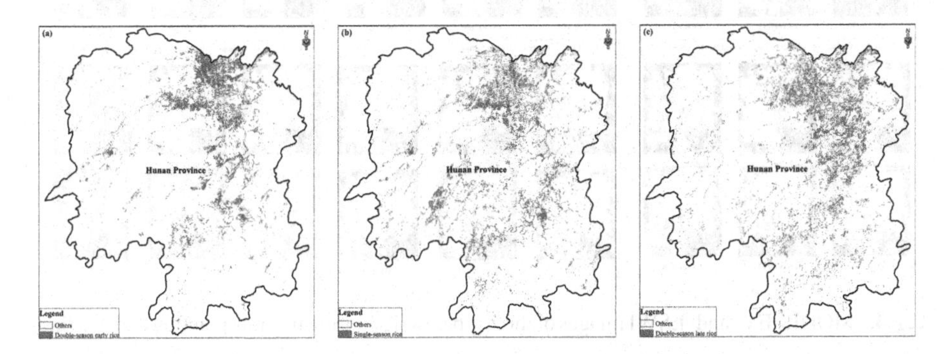

Fig. 3. Distribution of the early, middle and late rice of Hunan province in 2010.

We could know the spatial distribution of three types of rice: (1) the early rice in Hunan Province intensively located in the Dongting Lake Basin and basins in the mountain valleys, especially in the south plains of Dongting Lake Basin; (2) the single rice distributed in the northwest of Dongting Lake and in the western mountains; and (3) the late rice was mainly accumulated in the regions of the Dongting Lake Basin and the central plains, but the spatial distribution was relatively dispersed.

3.2 Analysis of Identification Errors

Two reasons can be used to analyze the deification errors of paddy rice fields using the time-series MODIS products: (1) the image qualities of MODIS products were worse due to the weather-related effects (e.g., cloud, fog, rainfall) at transplanting and heading stages, and sometimes most of the study area was even shaded by the thick clouds (e.g., 105 tile in Fig. 4A and 265 tile in Fig. 4F); (2) the spatial resolution of MODIS09A1 products is only 500 m, which could have a relatively good identification accuracy for the large areas of rice-planted regions, but it was worse for the dispersed and small paddy fields. In our study, in comparison with the early and late rice with the relative errors worse than 5.0 %, the identification accuracy was best with an only relative error of 1.46 % for the middle rice. The reason for this phenomenon was that the image qualities were worse for the early or late rice at transplanting or heading stage (e.g. Figure 4B and E), but conversely they were better for the middle rice at the two growth stages (e.g. Figure 4C and D). Consequently, the best image of the three images but not the average was used to identify the paddy rice fields.

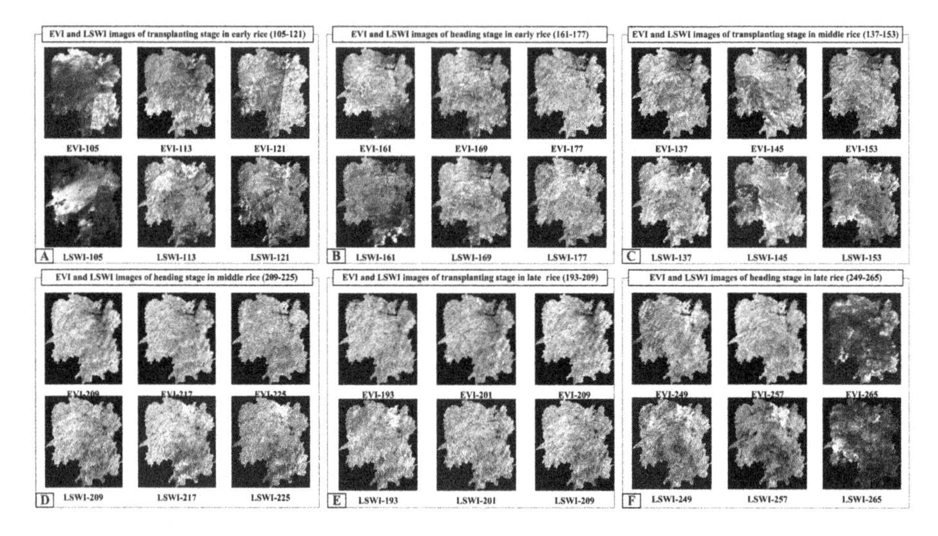

Fig. 4. MODIS-EVI and -LSWI images of three types of rice during the transplanting and heading stages.

4 Conclusion

A time series of MODIS 8-day composite products can be used to identify paddy rice fields, but the accuracy is usually affected due to the constraint of spatial resolution and available tiles at the two key growth stages of transplanting and heading. Fortunately, the spatial patterns of paddy rice can be well reflected based on time-series MODIS products. More ancillary data are usually needed to assist scholars in identifying the corresponding information of paddy rice on the regional scale. The reasons for under or over estimating the paddy rice acreage based on MODIS imagery can attribute to several factors: (1) failure of the moderate resolution data in identifying minor types like wastelands and settlements among the fields, especially on the small paddy rice fields; (2) similarities of some spectral indices (e.g., NDVI, LSWI) between paddy rice plants and some aquatic plants; (3) uncertainty occurred during the MODIS imagery reconstruction process; (4) the alternative planting of three types of rice caused the acreage errors for each rice; and (5) uncertainty existed in the statistical data for validating the identification accuracy.

Acknowledgements. This work was financially supported by Anhui Provincial Natural Science Foundation (grant no. 1408085QF126, 1308085QC58) and the Leadership Introduction Project of Academy and Technology of Anhui University (grant no. 10117700024).

References

1. Tennakoon, S.B., Murty, V.V.N., Eiumnoh, A.: Estimation of cropped area and grain yield of rice using remote sensing data. Int. J. Remote Sens. **13**, 427–439 (1992)
2. Ewe, H.T., Chuah, H.T., Ismail, A., Loh, K.F., Nasruddin, N.: Paddy crop monitoring using microwave remote sensing technique. Geocarto Int. **10**, 33–41 (1995)
3. Motohka, T., Nasahara, K.N., Miyata, A., Mano, M., Tsuchida, S.: Evaluation of optical satellite remote sensing for rice paddy phenology in monsoon Asia using a continuous in situ dataset. Int. J. Remote Sens. **30**, 4343–4357 (2009)
4. Li, W.G., Li, H., Zhao, L.H.: Estimating rice yield by HJ-1A satellite images. Rice Sci. **18**, 142–147 (2011)
5. Li, P., Feng, Z., Jiang, L., Liu, Y., Xiao, X.: Changes in rice cropping systems in the Poyang Lake Region, China during 2004–2010. J. Geogr. Sci. **22**, 653–668 (2012)
6. Xiao, X., Boles, S., Liu, J.Y., Zhang, D.F., Frolking, S., Li, C.S., Salas, W., Moore, B.: Mapping paddy rice agriculture in Southern China using multi-temporal MODIS images. Remote Sens. Environ. **95**, 480–492 (2005)
7. Lv, T.T., Liu, C.: Study on extraction of crop information using time-series MODIS data in the Chao Phraya Basin of Thailand. Adv. Space Res. **45**, 775–784 (2010)
8. Verhoef, W., Menenti, M., Azzali, S.: A color composite of NOAA-AVHRR-NDVI based on time series analysis (1981–1992). Int. J. Remote Sens. **17**, 231–235 (1996)
9. Roerink, G.J., Menenti, M.: Reconstructing cloud free NDVI composites using fourier analysis of time series. Int. J. Remote Sens. **21**, 1911–1917 (2000)
10. Le Toan, T., Ribbes, F., Wang, L., Floury, N., Ding, K., Kong, J., Fujita, M., Kurosu, T.: Rice crop mapping and monitoring using ERS-1 data based on experiment and modeling results. IEEE Trans. Geo. Remote Sens. **35**, 41–56 (1997)
11. Xiao, X., Boles, S., Frolking, S., Li, C.S., Babu, J.Y., Salas, W., Moore, B.: Mapping paddy rice agriculture in South and Southeast Asia using multi-temporal MODIS images. Remote Sens. Environ. **100**, 95–113 (2006)
12. Xiao, X., Boles, S., Frolking, S., Salas, W., Moore, B., Li, C., He, L., Zhao, R.: Observation of flooding and rice transplanting of paddy rice fields at the site to landscape scales in China using VEGETATION sensor data. Int. J. Remote Sens. **23**, 3009–3022 (2002)

A Spatio-Temporal Geocoding Model for Vector Data Integration

Xiaojing Yao(✉), Ling Peng, and Tianhe Chi

Institute of Remote Sensing and Digital Earth,
Chinese Academy of Sciences, Beijing 100101, China
yaoxj@radi.ac.cn

Abstract. Vector data integration is an important function in Urban Public Participation GIS Platform (UPPGP). Most current researches drill down the issue without considering two points: (1) the inner connections among different urban elements. (2) The temporal meaning of each object. The neglect of these points causes redundant storage and inefficient retrieval problems in smart city applications. In view of that, a spatio-temporal geocoding model for vector data integration is proposed in this paper. The model regards the urban entity element as the bridge between economic element and event element, so the task turns to find a way to uniquely identify the urban entities to avoid ambiguity and redundancy when entity objects connect with other type of objects during integration. Based on the object-oriented spatio-temporal data model, the entity object is constructed by type, space and time codes using concept lattice and regional GeoHash technologies. The method computes code similarity for each entity object to decide whether to put the object into storage. Experiments on the real UPPGP of Sino-Singapore Tianjin Eco-city show that it can avoid data redundancy and ambiguity effectively.

Keywords: Geocoding · Concept lattices · Ontology · GeoHash

1 Introduction

Big data issue is leading a new intelligent revolution in recent years. Geographic information systems (GIS) have sprout up continually with applications that include infrastructure maintenance, resource management, planning et al. But a lack of standards leads to a general inability for one GIS to interoperate with another [1]. It accelerates the production of Urban Public Participation GIS Platform (UPPGP). Since 2008, enterprise forerunners leading by IBM have formed a series of UPPGPs based on data integration and sharing [2]. The platform has many functions containing data extraction, cleaning and integration. Among these operations, integration for the vector data is one of the most important steps. It is a process of arranging the multi-source vector data in the geographical frame actually.

Multi-source vector data are different in terms of semantic meaning, spatio-temporal status, gathering approach, attribute structure et al. There are 3 integration levels of vector data currently: (1) Level 1 is a loose way of unifying the vector data in the same coordinate framework. This level refers to some basic pre-processes before deeper

F. Bian and Y. Xie (Eds.): GRMSE 2015, CCIS 569, pp. 566–577, 2016.
DOI: 10.1007/978-3-662-49155-3_59

integration, such as coordinate transformation and overlaying approaches [3, 4]. (2) Level 2 is to build relationships between individual objects in different data sets explicitly [5]. Compared with level 1, level 2 is a more compressed approach for integration. It is related with some data interoperation and information fusion technology, such as data warehouse and spatial "Extract-Transform-Load (ETL)" [1, 6, 7]. (3) The top level is the true integration level including two aspects: semantic integration and geometric integration. The former integration is related to the common attributes of the data objects. Based on the attributes belonged to different data, researchers proposed a valid method called "Geographic ontology" to identify the similarities and heterogeneities between geographic categories. Geographic ontology is a theory of abstracting entities from geographic knowledge, information and data. It makes up a system mixed with certain relationships by defining entities conceptually and formally [8]. Numerous attempts have been carried out to deal with geographic ontology integration. Among them, Kokla et al. (2001) and Kang et al. (2012) use ontology and concept lattice for geographical classification [9, 10], which provides gist for our researches. The latter integration refers to the geographic matching technology of vector data sharing the same name. Some studies have been proposed based on the type of the vector data [11] or the topology of objects [12].

The top level integration can compress several vector datasets into one really. However, most current studies concentrate on the semantic and geographic integration separately, and barely consider two points: (1) the connections among different urban elements. Urban data contain entity and event elements. Generally, there are always M: N relationships between entities and events. Taking building data for example, the management departments and the planning departments have distinct descriptions about the same building. Even more, different buildings may have the same event data also. The classical integration methods cannot describe the relationships of single building, building area, and different events happened on these individuals, because these data items have different spatial and time granularities. (2) The temporal meaning of each object. Most integration processes emphasize more on the spatial and sematic meaning than the temporality of single object. Actually, vector data from different departments always show the changes of objects along time axis. Without the time constraint, the data would lose their meanings. To sum up, the two points cause redundant storage and inefficiency retrieval in practical applications really.

Owing to that, by using geographic ontology classification and geocoding technologies, this paper proposes a spatio- temporal geocoding model for vector data, which is composed of type, space and time codes. It's a transition level between the existing level-2 and the level-3 integration, as well as a compensating model based on the content and logical connections of urban elements, and lets the disordered urban data expand infinitely and orderly in the geographical framework. A series of applications and experiments on the real UPPGP of Sino-Singapore Tianjin Eco-city show that the model can avoid data redundancy and ambiguity effectively.

The advantages of our model are summarized into threefold: first, it is excellent robustness. The granularity of property and spatial attributes of object will not break the coding framework. We can expand the code length to express more detailed information about the object. Second, it can avoid the redundant storage remarkably, because a geocoding method is proposed to compute the coding similarity for each vector object to decide its entering. Third, the code isn't an identity authentication only, since the code itself contains information about type, space and time of an entity. By this code, irrelevant instances can be filtered out firstly, which can improve the whole efficiency of data analysis.

The paper is organized as follows. We present the constitution and logic connection of urban elements firstly in Sect. 2. Section 3 presents the integrated coding framework of urban entity elements. Then we give the implement of the urban entity coding algorithm in Sect. 4. Section 5 presents applications and experiments to demonstrate the improvements brought by the coding method for the real UPPGP. Finally, the study is summarized in Sect. 6.

2 The Construction and Logic Connection of Urban Elements

Urban is an integrated and self-operating system by combining different social and nature spaces together. To describe the variety of urban data, urban planners divide the urban data into entity elements, economic elements and event elements. Entity elements are the general terms of artificial elements and the nature elements involved by human, such as urban zone, building, traffic, water conservancy, urban pipeline. Economic elements are financial organisms that have free behavior and independent responsibility, such as government, company and human. Event elements are the sum of the activities that entity and economic elements involved in, such as urban security, urban management, and urban operation [13]. The instance belonging to these elements are called "object" in the following.

There are two types of spatial-related events in urban: one represents the status of entity objects along the time axis, such as the natural damages of streetlights. The other represents the status of economic objects depending on entity objects along the time axis, such as workers report the broken streetlights. The main difference of the two is: the former is a description of entity objects without the participation of economic objects; the latter is a description of economic objects with the participation of entity objects. That is to say, the entity elements give space supplies for the event elements. From the previous analysis, entity element is the link between economic element and event element. They have complex dependencies in time and space scales. Figure 1 is an illustration about it.

(1) Spatial dependencies. Entity elements are the spatial carriers of event and economic elements. In other words, event elements and economic elements depend on entity elements in spatial terms. As shown in Fig. 1, event elements, such as macro economy, region planning and street management, spatially depend on entity elements such as administration region, community and street separately. Specially, these entity elements are hierarchical in spatial aspect. Similarly, economic elements, such as government, company and human, spatially depend on entity elements like building.
(2) Time dependencies. Entity element and economic element are not static but have time effect. The appearance of the time effect is expressed by event elements. For example, as shown in Fig. 1, the construction information, real-estate information

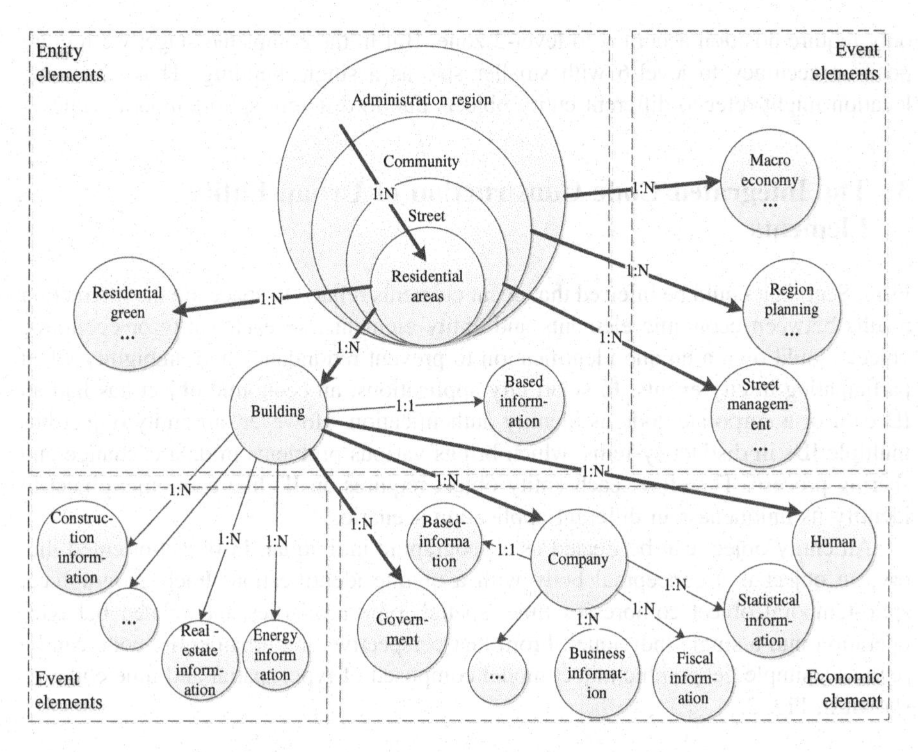

Fig. 1. The dependencies of urban elements

and energy information are events happened on the entity element of building along the time axis. The business information, fiscal information and statistical information are events happened on the economic element of company along the time axis.

(3) Dependencies among objects of these elements along sptio-temporal scales. An entity object consists of three basic characteristics at least: location, property and time status [14]. In another word, different temporal and spatial statuses map to different triples containing entity object, economic object and event object. The dependency can be expressed as follows.

$$O_{event} = (O_{time}, O_{location}, < O_{economy} > , O_{entity}) \tag{1}$$

where O_x denotes the corresponding object of type x. $O_{economy}$ is an optional item. For example, "planning results = (planning stage, level-2 area location, planning bureau, the north zone)" is a complete description of an event. In fact, all artificial entity objects in urban have similar experience timeline as "planning stage - design stage - implementation stage – completed stage - destruction stage". Each stage is an event node with spatial and temporal limitations. The target entity described by event nodes change from rough to fine in spatial scale with time flowing. In the planning stage, we only require position accuracy to level-2 zone. But in the completed stage, we require position accuracy to level-6 with smaller size as a single building. Thus, the same location might refer to different entity objects due to different sptio-temporal scales.

3 The Integrated Code Construction of Urban Entity Elements

From Sect. 2, it could be inferred that event elements related to space are the interaction results between economic elements and entity elements, so each entity or economic object should own a unique identification to prevent redundancy and ambiguity when participating in city events. In smart city applications, an economic object has had an ID card or a corporate code as identity authentication. However, an entity object has multiple IDs in distinct systems, which brings various problems in data exchange and sharing process. Therefore, each entity object requires an ID like economic object to identify its uniqueness in different application scenarios.

An entity object can be treated as a geographic individual. In object-oriented theory, an object is a conceptual body with a unique identification. Each geographical sptio-temporal object compresses time, spatial, type properties and related behavior operation into a single individual. From that perspective, the urban entity code can be seen as a simple geographic object model composed of type, spatial and time codes as shown in Fig. 2.

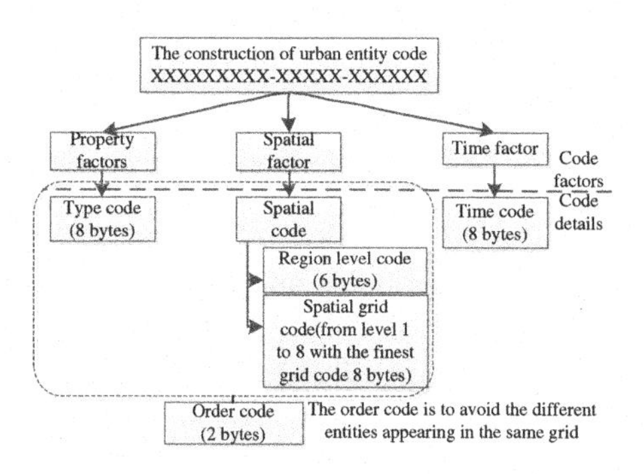

Fig. 2. The integrated coding framework of an urban entity object

From Fig. 2, it can be seen that type code, spatial code and time code correspond to property, space and time factors of an entity object respectively. Among them, type code and time code have consistent length once the initial works are executed before geocoding integration. But it's not the same as the spatial code. Spatial code consists of 2 parts: region level code and spatial grid code. The latter are a series of codes form level 1 to level 8 with the finest one 8 bytes. Its length varies with the spatial range of the object. To avoid different objects appearing in the same grid, the order code is necessary. More information will be presented in the next state.

4 Urban Entity Coding Algorithm

4.1 The Framework of Coding Integration

The UPPGP is a host platform for spatial data. When new vector data are entering, they are merged into the code database of the platform quickly. The vector data waiting for integrating should meet two conditions at least: (1) the data must have type attributes; (2) the geometric type (point, line or polygon) of the input data should map to the code dataset sharing the same type with the input data. The whole process contains three steps, followed by an illustration in Fig. 3(a). Any vector data can be merged into the data center by this method without changing the construct of the original data, so it can be easily used.

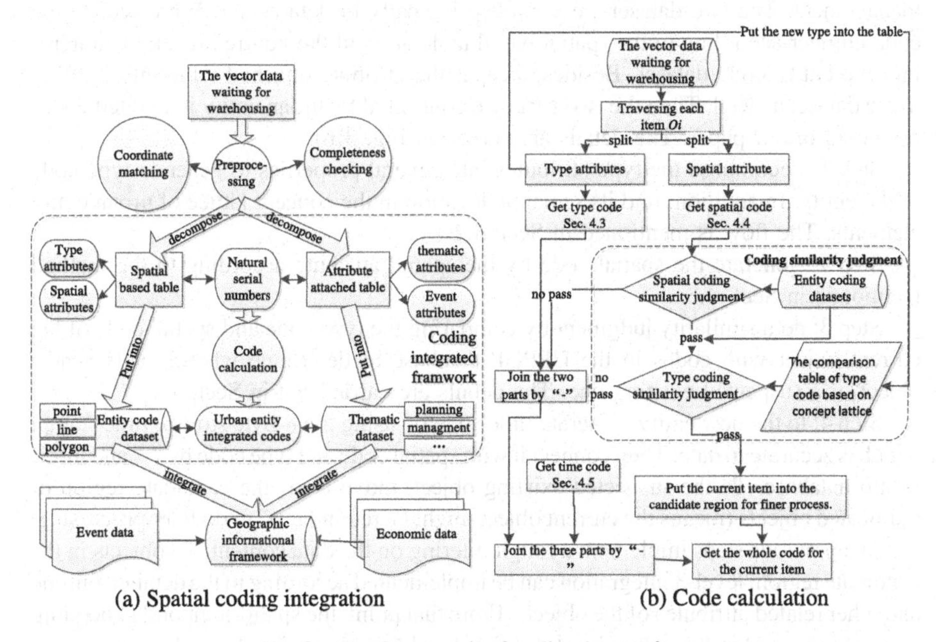

Fig. 3. The framework of coding integration

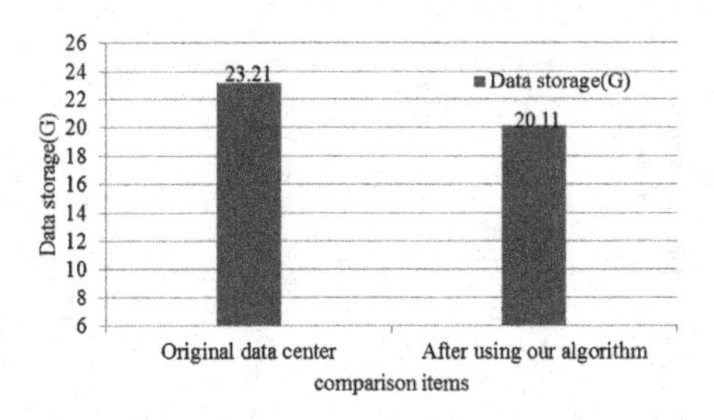

Fig. 4. Data storage improvement

First, pre-process the vector data. It is the level-1 integration proposed in Sect. 1. Rather than regurgitating, pre-process such as coordinate matching and data completeness checking before deeper coding process are discussed a lot in previous researches, so we don't present too much detail about that. **Second, decompose the data.** The process decomposes the data artificially into double parts: spatial based table and attribute attached table. The former only contains attributes that relate to the type and spatial issues of the data. The latter contains other attributes referring to the event and thematic issues of the data. The two parts are connected by the natural serial numbers or the numbers from their previous systems. **Third, calculate the integrated code and put the data into storage.** The vector data center of the UPPGP contains two parts: entity code dataset and thematic dataset. The former contains non-redundant vector data of point, line and polygon with unique codes. The latter contains 2-dimensional tables that related with different thematic domains such as planning and management. The two dataset are separated logically in data center. We calculate the code number according to the spatial based table and put the non-redundant data items into the entity code dataset. Besides, we put the attribute attached table into the thematic dataset as well. Then the two parts are connected by urban entity integrated codes presented in our paper. The details are shown in Fig. 3(b).

Step 1: decompose the type attributes into several properties to generate type code of the entity. In addition, find its semantic location in the concept lattice of urban entity elements. The flow is mentioned in Sect. 4.2.

Step 2: generate the spatial code by the spatial attribute according to the method mentioned in Sect. 4.3.

Step 3: do a similarity judgment by comparing the type code and spatial code of the current object with codes in the UPPGP database to determine whether it is a new object or a suspected existing one. The details are carried out in Sect. 4.4.

Step 4: to the new entity, generate time code by using the 8-byte storage time stamp, which is accurate to date. Then connect it with spatial code and type code by "-" and insert it into database. To the suspected existing object, move it to "the candidate region of duplicated objects (means the current object might be redundant)" to do finer processing.

The previous steps implement a rough filtering on the data content. To objects in the candidate region, level-3 integration can be implemented according to the detailed outline and other related attributes of the objects. From that point, the spatio-temporal geocoding method proposed in this paper is a transition level from level-2 to level-3 integration to make the computing complexity of the top integration much lower than before.

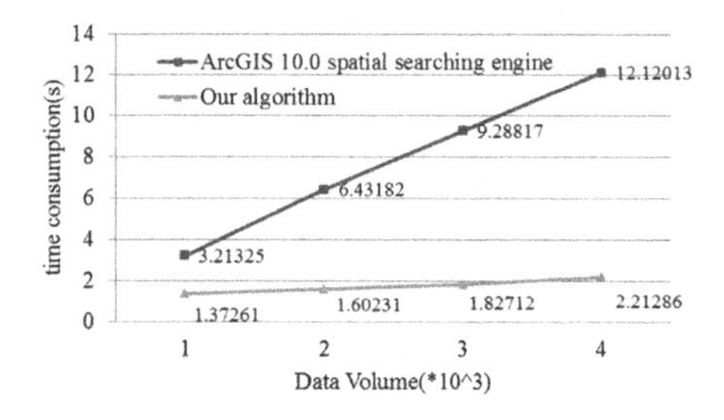

Fig. 5. Regional search improvement (street level)

4.2 Type Coding Algorithm Based on Concept Lattice

Type code expresses the essential attribute of the entity elements. In this section, we design a method to calculate the type code of urban entity element by building a property tree based on concept lattice and geographic ontology. This coding method can quantize the difference of massive entity elements. The processes are as follows.

First, collect the nature descriptions of urban geographical entity elements from the current standards to build the basic geographic concept dataset. The dataset will be enriched when new vector data are put into storage, since the type descriptions of the new data waiting for warehousing are also the source of the concept dataset.

Second, change the nature descriptions of each concept term into formal expressions. Each entity element has attributes that relate to materiality, reason, form, position, timeliness, and function [15]. Based on the attribute hierarchy tree, we denote each type as a list of formal words.

Third, Change each formal expression into binary code. If the current entity has the same attribute, the corresponding position is marked with "1"; otherwise it is marked with "0". It can be inferred that the granularity of the attributes effect the length of the type code. The longer the code is, the more certain the entity is. To avoid excess manual work, the numeric attributes like ">60 g" are ignored. But it is better to refine the attributes to improve the precision of similarity judgment. In addition, the granularity of type code is limited at the beginning usually.

Fourth, convert the binary code to 8-byte 64 hexadecimal code, with a cover of zeros in front if the code size is less than 8. The step can enlarge the capacity of the type code dataset in a limited storage circumstance.

At last, built the concept lattice for all entity type code and get the type code comparison table. The classification based on concept lattice [9] is used to the construct the code comparison table.

4.3 Spatial Coding Algorithm Based on Geocoding

Spatial code expresses the spatial characteristic of an entity object. It's an important part of the complete coding system. Current spatial coding methods include two types: regular grid geocoding and irregular grid geocoding. The former is based on the theory that the position of the point of interests (POIs) can be replaced by tiny areas. The method cuts the earth surface into regular and adjacent grids according to certain mathematical rules, and then gives each grid a unique code to represent the position of POIs in the grid. The "National Area Code (NAC)" developed by Canadian NAC company and GeoHash code proposed by Gustava Niemeyer both belong to this category [16]. The latter is represented by address geocoding method. It implements a conversion from descriptive geographic language (such as street or postal number) to spatial coordinates by building the coordinate correspondences between city administration units (like administrative area, street and house number) and their geometric center. The "Topologically Integrated Geographic Encoding and Referencing" [17] and "Postal Code Address Data" [18] both belong to that. The paper absorbs advantages of the two approaches. The processes are as follows.

First, calculate the Minimal Body Rectangular (MBR) for each entity object based on the spatial attribute. Second, to the scope larger than the street level (such as province or county level), use 6-byte postal code. To the scope smaller than the street level(such as residential area), consider not only postal codes but also regional GeoHash code [19] to cover tiny entities that are important but cannot be expressed by irregular grids, such as dustbins. Third, get the whole GeoHash codes arranged from level 1 to 8 combined with "-".

4.4 Coding Similarity Judgment

The coding similarity judgment is to distinguish the suspected redundant data items form the non-redundant ones. It contains two aspects:

(1) Spatial coding similarity judgment

From Sect. 4.3, we know that spatial code contains different size of indexed grids. It's a good idea to fully use the organized grid codes to get the similarity between two entity objects by calculating grid overlapping area in the same postal unit. The formula is:

$$Sim\_S(O_i, O_j) = \sum_{m=1}^{level} \frac{1}{2m} \left(\frac{GridSum(O_{im} \cap O_{jm})}{GridSum(O_{im})} + \frac{GridSum(O_{im} \cap O_{jm})}{GridSum(O_{jm})} \right) \quad (2)$$

where $Sim\_S(O_i, O_j)$ is the overlapping degree between the current object O_i waiting for warehousing and the target object O_j in the code dataset. *level* is the grid level. $GridSum(O_{im})$ is the grid sum of O_i in the current *level m*, $GridSum(O_{im} \cap O_{jm})$ is the cardinal number of the intersection of O_i and O_j in the current *level m*. If $Sim\_S(O_i, O_j) > \gamma$ (γ is the threshold), then the process goes to the next "type coding similarity judgment" step. Otherwise, the method combines the spatial code of the current entity with its type code and time code, and inserts the coded entity into database.

(2) Type coding similarity judgment

To entities passing the spatial coding similarity judgment, the coding method uses Eq. (3) [9] to calculate the type code similarity between the current object O_i and the target object O_j in the code dataset.

$$Sim\_T(O_i, O_j) = \omega_1 \left| \frac{O_i^{\triangleright} \cap O_j^{\triangleright}}{O_i^{\triangleright} \cup O_j^{\triangleright}} \right| + \frac{\omega_2}{DisMin(O_i, O_j)} \quad (3)$$

where $Sim\_T(O_i, O_j)$ is the type similarity between the current object O_i and the target object O_j. $O_i^{\triangleright}$ is the attribute set of O_i. $O_i^{\triangleright} \cup O_j^{\triangleright}$ is the intersection of O_i and O_j attribute set. $O_i^{\triangleright} \cup O_j^{\triangleright}$ is the union of O_i and O_j attribute set. $DisMin(O_i, O_j)$ is the minimum sum of edges from O_i to O_j in the attribute concept lattice. ω_1 and ω_2 are two adjustable parameters to meet the constraint of $\omega_1 + \omega_2 = 1$. If $Sim\_T(O_i, O_j) > \eta$ (η is the threshold), the current entity pass the judgment.

5 Applications and Experiments

The coding method has been successfully used in the UPPGP of Sino-Singapore Tianjin Eco-city already. It brings remarkable improvements to the data storage and searching efficiency for the UPPGP. The data storage improvement is shown in Fig. 4. In our platform, the volume of the data center is 23.21 G. After using our coding method, the data volume changes to 20.11 G. The Algorithm brings a 3.1 G storage decline for the vector data center. The searching improvement is shown in Fig. 5. Taking the block scale for example, we selected 4 cases of data volume with 10^3 intervals to test the efficiency of our method. The result shows that our method can save a lot of time compared with spatial searching engine of ArcGIS 10.0. We get similar results in other spatial scales as well. Furthermore, Multi-level searching based on this code is shown in Fig. 6.

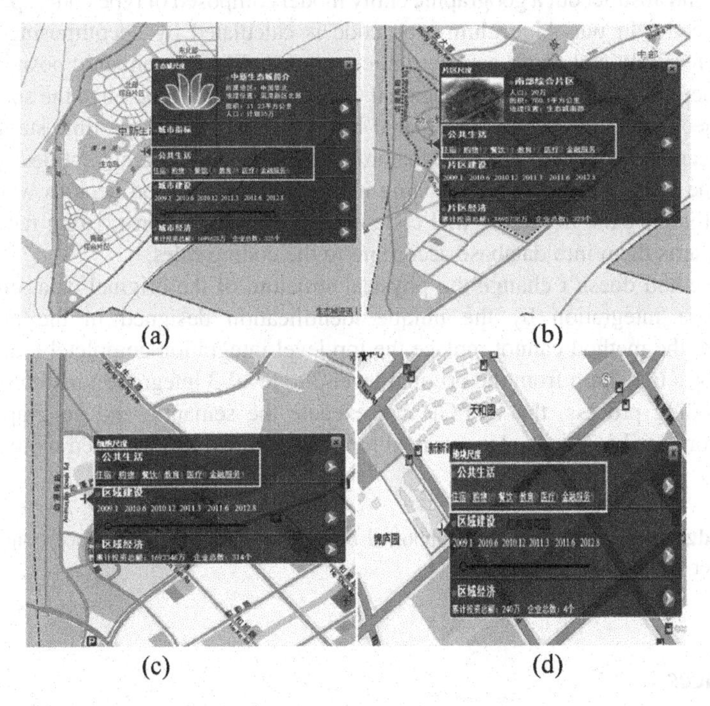

Fig. 6. Multi-level searching based on our code. (a) shows the function window on the global scale. It can be seen that 4 types of indicators related on the whole regional scale: urban comprehensive indicators, urban entity statistic indicators, urban construction indicators and macro-economic indicators. Similarly, (b) shows the function window referred to indicators on the district scale; (c) shows indicators on the block scale, and (d) shows indicators on the plot scale. Each spatial scale corresponds to different function windows.

Firstly, each spatial entity object is implemented the two-step judgment before storage. If it does not pass the test, the object will be given a unique code and inserted into the database. Otherwise, the duplicates will be put into the candidate region for finer process. By that way, the vector data are cleaned. Even though the method is taken as a rough filtration, it is really an effective way to avoid redundancy.

Secondly, we get the information about type, location and generated time of the target object from the code easily, and then do a rapid query based on the same part of the current entity code to find related ones. The matching and query tasks are similar to the coding process. However, the application case is only the tip of an iceberg.

6 Conclusion

From Sect. 5, it can be seen that the spatio-temporal geocoding method is an effective way to solve data redundancy and inefficient retrieval problems during the data integration. This method is based on the content and inner connections among urban elements. We ensure the key role of urban entity elements in connecting event elements and economic elements, and abstract out a geographic entity model composed of type code, spatial code and time code. In our Algorithm, type code is calculated by decomposing entities' attributes formally and using concept lattice to generalize the semantic position of the entity, space code is calculated by regional GeoHash code to generalize the space range of the object, and time code is calculated by recording the storage time stamp of the object. Meanwhile, we also consider the order code to avoid different entities appearing in the same grid. When doing geocoding integration, the method decides whether the outer data items exist in the original code dataset or not. To non-existing records, our method inserts them into database according to the coding rules.

The method doesn't change the physical structure of the original data, but implements loose integration by the unique identification designed in the paper. To emphasize, the method cannot replace the top level integration completely, but can be regarded as a transition from the original level-2 to level-3 integration methods. Due to this geocoding process, the UPPGP can execute the semantic and geographic integration more easily. That is also our next topic, which will be discussed more in future studies.

Acknowledgments. This work was supported by the National Natural Science Foundation of China under Grants 2011FU125Z24.

References

1. Uitermark, H.T.: Ontology-based geographic data set integration. Ph.D. Dissertation, Deventer, The Netherlands (2001)
2. Schönberger, V.M., Cukier, K.: Big Data: a Revolution That Will Transform How We Live, Work and Think. John Murray, England (2013)
3. Christen, P.: A survey of indexing techniques for scalable record linkage and deduplication. IEEE Trans. Knowl. Data Eng. **9**(24), 1537–1555 (2012)
4. Grayson, T.H.: Address Matching and Geocoding. Massachusetts Institute of Technology Department of Urban Studies and Planning, vol. 14 (2000)

5. Huh, Y., Yang, S., Ga, C., Yu, K., Shi, W.: Line segment confidence region-based string matching method for map conflation. ISPRS J. Photogrammetry Remote Sens. **78**, 69–84 (2013)
6. Kyung, M.J., Yom, J.H., Kim, S.Y.: Spatial data warehouse design and spatial OLAP implementation for decision making of geospatial data update. KSCE J. Civil Eng. **16**(6), 1023–1031 (2012)
7. Ok, G.H., Lee, D.W., You, B.S., Bae, H.Y.: A spatial data cubes with concept hierarchy on spatial data warehouse. Korea Inf. Process. Soc. **1**(13), 35–38 (2006)
8. Egenhofer, M., Mark, D.: Naive geography. In: Kuhn, W., Frank, A.U. (eds.) COSIT 1995. LNCS, vol. 988, pp. 1–15. Springer, Heidelberg (1995)
9. Kang, X.P., Li, D.Y., Wang, S.G.: Research on domain ontology in different granulations based on concept lattice. Knowl. Based Syst. **27**, 152–161 (2012)
10. Kokla, M., Kavouras, M.: Fusion of top-level and geopraphical domain ontologies based on context formation and complementarity. Int. J. Geogr. Inf. Sci. **15**(7), 679–687 (2001)
11. Filin, S., Doytsher, Y.: A linear mapping approach to map conflation: matching of polylines. Surveying Land Inf. Syst. **59**(2), 107–114 (1999)
12. Frank, A.U.: Qualitative spatial reasoning about distances and directions in geographic space. J. Vis. Lang. Comput. **3**(4), 343–371 (1992)
13. Shao, J., Yang, L.-N., Peng, L., Yao, X.-J., Zhao, X.-L.: Research of data resource management platform in smart city. In: Bian, F., Xie, Y. (eds.) GRMSE 2014. CCIS, vol. 482, pp. 14–22. Springer, Heidelberg (2015)
14. Butenuth, M., Gosseln, G., Tiedge, M., Heipke, C., Lipeck, U., Sester, M.: Integration of heterogeneous geospatial data in a federated database. ISPRS J. Photogrammetry Remote Sens. **62**(5), 328–346 (2007)
15. Wang, H., Li, L., Zhu, H.H.: The Key Research of National Fundamental Geographic Information Ontology (in Chinese). Science Press, Beijing (2011)
16. Parker, N.: A Look at NAC Geographic Directions Magazine, US (2004)
17. Census Bureau TIGER. http://www.census.gov/geo/maps-data/data/pdfs/tiger/tgrshp2013/TGRSHP2013_TechDoc.pdf
18. Canada Postal Guide-Addressing Guidelines. http://www.canadapost.ca/tools/pg/manual/PGaddress-e.pdf
19. Jing, A., Cheng, C.Q., Song, S.H., Chen, B.: Regional query of area data based on Geohash (in Chinese). Geogr. Geoinf. Sci. **29**(5), 31–35 (2013)

Research on the Relationship Among Urban Amenities, Talents and Urban Income in China: Structural Equation Model in Practice

Ting Wen[1,2], Jianming Cai[1(✉)], and Liou Xie[3]

[1] Institute of Geographic Sciences and Natural Resources Research, Chinese Academy of Sciences, Beijing 100101, China
tingwen111@126.com, Caijm@igsnrr.ac.cn
[2] University of Chinese Academy of Sciences, Beijing 100101, China
[3] State University of New York at Plattsburgh, Plattsburgh, NY 12901, USA
lxie001@plattsburgh.edu

Abstract. This paper aims at examining the relationship between urban amenities, talents concentration and urban income in Chinese cities. Structural equation model (SEM) is employed in investigating the hypothesis that urban amenities attract talents and thus can help improve urban income. The results draw three conclusions. Firstly, the variable of *Wholesale and retail trades*, *Real Estate, Culture, Sports and Entertainment* are significant in predicting talents concentration. Secondly, there is a strong correlation between talents and urban income performance. Lastly, talents play as a very crucial link between urban amenities and urban income. The implications are urban amenities should be paid more attention to at both the policy and planning levels.

Keywords: Talents · Urban amenities · Talents · Urban income · China

1 Introduction

Since 1950s, economic geographers have emphasized that cities need to attract more talents in order to improve its competitiveness [1, 2], because the productive forces created by talent accumulation could greatly help drive urban and regional growth. Traditionally, income disparity was considered as a main element for locational decision for talents. However, many researches demonstrated that urban amenities have been increasingly emerging as a new key factor, replacing income disparity, for attracting talents over the past few decades [3–6].

Studies conducted by Chinese scholars also show that people in urban areas in China are willing to pay more for urban amenities [7, 8]. In addition, urban amenities are considered as a key criterion for building livable city in China [9]. However, most of the existing research focuses only on the effect of urban amenities on urban residence living. Few of them look at the relationship between urban amenities and overall urban development. The purpose of this paper is thus to explore the influence of urban amenities on talents as well as urban income in Chinese cities.

F. Bian and Y. Xie (Eds.): GRMSE 2015, CCIS 569, pp. 578–585, 2016.
DOI: 10.1007/978-3-662-49155-3_60

Consumption amenity is increasingly being emphasized as an important urban amenity [6, 10, 11] as Clark [12] argues that the role of public facilities including schools, church, and social organizations are becoming less important while consumption amenities such as entertainment and recreational facilities are gaining importance. Considering the availability of data, we will use consumption sectors when evaluating urban amenities.

2 Methodology

2.1 Methods

Structural Equation Model (SEM) was developed based on Joreskog and Dag Sorbom's statistical theory in 1970s. It has many advantages in dealing with complex multiple causal relationships between variables. The general expression of SEM is as below:

$$\eta = \beta\eta + \Gamma\xi + \zeta \tag{1}$$

$$y = \Lambda y\eta + \varepsilon \tag{2}$$

$$x = \Lambda x\xi + \delta \tag{3}$$

Where η is vectors of unobserved criterion, and ξ is vectors of explanatory variables. β and Γ are matrices of coefficient parameters for η and Γ respectively. y is observed criteria and x is explanatory variables. Λy and Λx are regression matrices, and ε and δ are residual vectors.

Location Quotient (LQ) is normally used to test the specialization degree of a certain urban sector. We apply LQ model to access the performance of consumption sectors. The specialization degree of urban sector *j* in city *i* can be assessed by:

$$LQ_{ij} = \frac{L_{ij} / \sum_{j=1}^{m} L_{ij}}{\sum_{i=1}^{n} L_{ij} / \sum_{i=1}^{n} \sum_{j=1}^{m} L_{ij}} \tag{4}$$

Where Lij is the employment of sector j in city i, and $\sum_{j=1}^{m} L_{ij}$ is the overall employment of all sectors in city i. $\sum_{i=1}^{n} L_{ij}$ is the employment of sector j in all cities, and $\sum_{i=1}^{n} \sum_{j=1}^{m} L_{ij}$ is the overall employment of all sectors in all cities (m and n are the number of sectors and cities respectively).

2.2 Indexes and Data

The empirical analysis covers all 286 prefecture level cities in China except Lhasa (due to data unavailability) for 2010. All data is from China City Statistical Yearbook in 2011 and China's 6th National Population Census in 2010 and the description of variables is as Table 1 shows.

Table 1. Description of variables.

Variables	Description
Urban income	Gross domestic product
Talents	People with a bachelor or higher level degree, standardized by population aged 20 or above
Wholesale and retail trades	LQ of employment in wholesale and retail trades
Hotels and catering services	LQ of employment in hotels and catering services
Financial intermediation	LQ of employment in financial intermediation
Cultures, sports and entertainment	LQ of employment in culture, sports and entertainment
Real estate	LQ of employment in real estate

2.3 Structural Equation Model

The genetic model is established as Fig. 1 shows. In this way, the impacts of urban amenities on talent and urban income, and the effects of talent on urban income are tested separately.

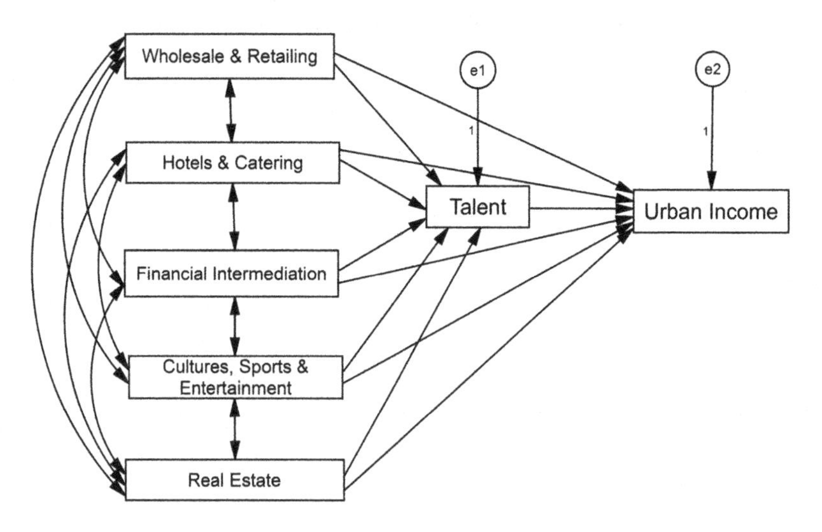

Fig. 1. Generic model

3 Correlation Between Urban Amenities, Talent and Urban Income

According to the output of the generic model run in AMOS, we cut the path when P > 0.05. The modified model is presented in Fig. 2 below. For the modified model, Chi-square is 1.656. Probability level is 0.437 > 0.05. It accepts null hypothesis meaning that the model is adapted to the sample. Besides, RMSEA = 0.000 < 0.05, AGFI = 0.980 > 0.900, GFI = 0.998 > 0.900, which also show that the model reaches the standard of adaptation.

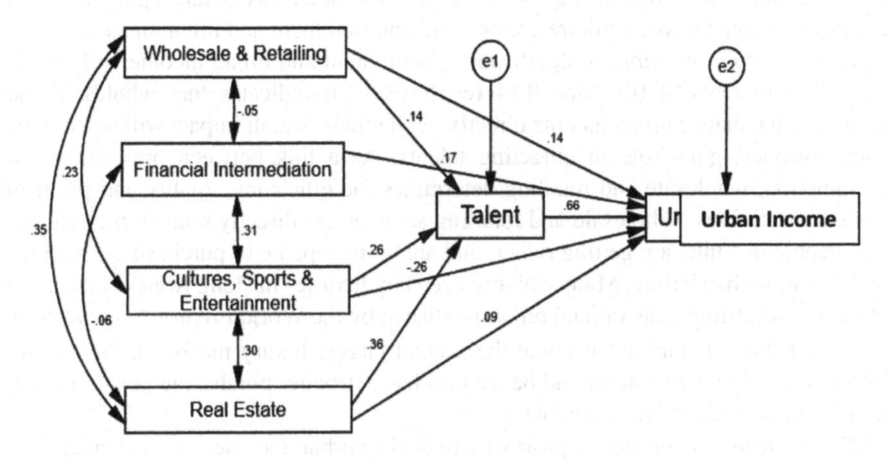

Fig. 2. Modified genetic model

In the modified model (Fig. 2), the factor of *Hotels and Catering Service* is removed because its influences on both talent and urban income are not significant. Although hotels and catering services have experienced rapid development in the past years in China, it seems that it contributes to neither talents concentration nor urban income. It seems to still mainly act as an attractive factor for tourists, but not for the decision of where to live and work.

The other four factors are all tested to be significant, of which *Real Estate* is the most significant one. Since the housing system reforms that started in 1998, real estate industry in China has been thriving. Investment in the real estate sector has been accumulating gradually, contributing significantly China's GDP. In 2008 the direct and indirect contribution of the real estate industry to the country's GDP amounted to 21.76 % [12]. Talents have more demands on real estate services than other people as there is a strong positive correlation between education background and income [13]. Talents are more capable of purchasing a house and improving their living conditions by moving into a better quality house in the same city, compared with the others. As a result, the concentration of talents requires more and better real estate service, which is another important sector in the real estate industry. The number of employment in real estate agent service represents the vitality of the real estate market. Additionally, people with higher education usually demand more and higher quality property management, which is also a part of real estate service.

The variable of *Cultures, Sports and Entertainment* is positively significant for talents with a standardized coefficient of 0.26, which is consistent with what we expect. It proves that the level of cultures, sports and entertainment is a significant predictor of talents concentration. Interestingly, although the correlation between *Cultures, Sports and Entertainment* and *Talent* is positive, its relationship with *Urban Income* is negatively significant. It seems that cultures, sports and entertainment activities are popular with talents, but they do not affect urban income directly. However, they can contribute to urban income through attracting talents, which also indicates that talent plays a crucial intermediary role between cultures, sports and entertainment and urban income.

Wholesale and retailing is significant to both talent and urban income with standardized coefficients of 0.14 and 0.14 respectively. It indicates that wholesale and retailing can enhance urban income directly, while their overall impact will be doubled when considering its role in attracting talents. As a link between production and consumption, wholesale and retailing determines the efficiency, quality and profit of economic activities. Wholesale and retailing services are directly related to consumption. People in China are getting richer, and are more capable of purchasing what they need and want than before. Many Chinese even buy luxuries in Hong Kong, Macau, and overseas. According to an official report published by the World Luxury Association in the end of 2010, China has become the second largest luxury market in the world[1]. Talents with higher education and better jobs have stronger purchasing power and are more willing to pay for new products.

Financial Intermediation is positively related to urban income with a standardized coefficient of 0.17. It is also insignificant for talents, which runs against our expectation. The level of financial intermediation does not lead to significant talents concentration although most people in the field of Financial are talents themselves holding "white collar" or "gold collar" jobs. As an important part of producer services, the financial sector contributes directly to urban income.

There is a strong correlation between *Talent* and *Urban Income* with a standardized coefficient of 0.66, indicating that China has transitioned from an industrial economy towards a more knowledge-oriented path. Talents function as a link between consumption amenities and urban income. In recent years, talent is been highly emphasized as the most important resource especially in the fields of innovation and high-tech. And China has put a lot of efforts on attracting talents as reflected in China's talent attraction plans since 2002. In 2006, talent attraction plan was included in the National (11th) Five-Year Plan for the first time. Special talents plans were established by the central and local governments including Changjiang Scholars Program (Changjiang xue zhe), Thousand Talents Program (Qian ren ji hua) and 100 Talents Program (Bai ren ji hua) to attract outstanding overseas researchers to work in China. Based on the results above, we suggest that the strategies of attracting talents can be diversified and the development of consumption amenities could be a positive factor for attracting talents.

---

[1] World Luxury Association. 2011. 2010–2011 World Luxury Association Annual Report. [Online] Available at: http://www.worldluxuryassociation.org/.

4 Evaluation of Urban Amenities in China

In order to further understand the impacts of urban amenities on talents, we evaluate the level of urban amenities of major cities in China based on their performance in the sectors of Wholesale and Retailing, Cultures Sports and Entertainment, and Real Estate. Figure 3 not only demonstrates the spatial distribution of urban amenities, but also shows the pattern of talent attractions of China in an illustrative way.

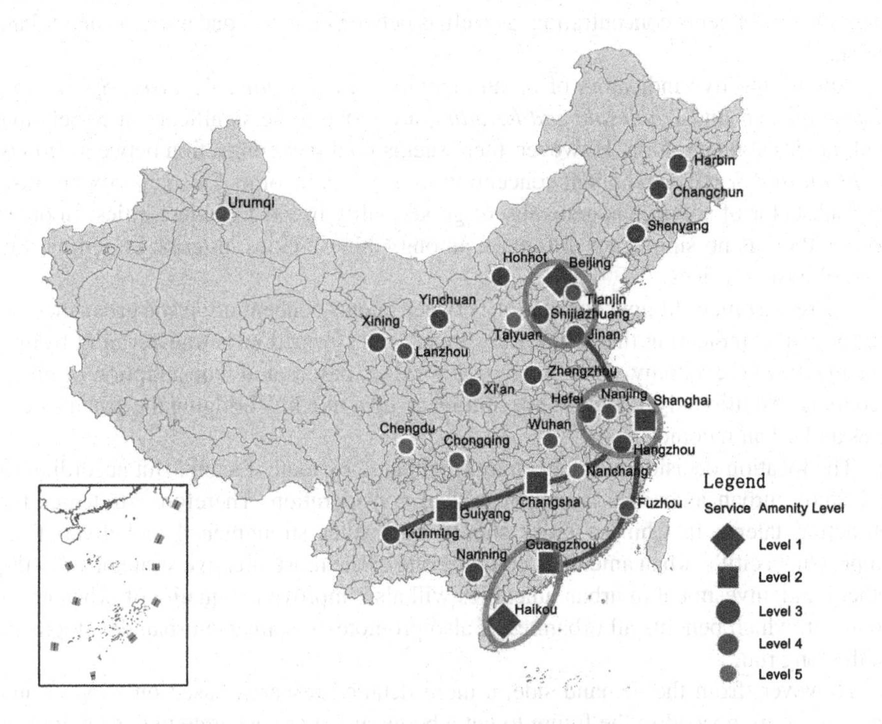

Fig. 3. Evaluation of amenities in major cities in China

Most of the cities with higher rank levels are in the east coast region, which form a south-north urban amenities corridor with three main nodes. The first node is Beijing-centered, located in the north including Beijing, Shijiazhuang, Tianjin and Jinan. The second node consists of Shanghai, Nanjing, Hangzhou and Hefei. And the third node locates in the southernmost of China and includes Guangzhou and Haikou. Besides, there is also an east-west corridor which incorporates the Shanghai-centered node, Changsha, Guiyang and Kunming. These two corridors form a shape of "bow and arrow". Interestingly, this "bow and arrow" layout of service amenities in cities is consistent with the spatial layout of economic levels of cities in China.

5 Conclusions and Discussions

The ongoing economic restructuring in China from traditional manufacturing industry and production-oriented economy to a more knowledge-based and consumption-oriented economy triggers and stimulates the importance of amenity as a significant factor in urban development. Talents play a key role in urban innovation and competitiveness, and they usually pay much more attentions on urban amenities. Out study applies SEM on empirical data shows that strong relationships exist between urban amenities and talents concentration, as well as between talents concentration and urban income.

Among the five indicators of urban amenities, *Real Estate; Cultures, Sports and Entertainment;* and *Wholesale and Retailing* are tested to be significant in correlation with talents concentration. However, there seems no direct connection between *Hotels and Catering Service* and talent concentration or urban income. This is partly because this subsector of services is generally of good quality in most Chinese cities. In other words, there is no significant difference among Chinese cities in terms of getting this kind of basic services.

There is a much stronger correlation between talents concentration and urban income performance, indicating that China has already transformed or is transforming from a manufacturing economy to a more knowledge based and consumption-oriented economy. Additionally, talents are becoming an effective link bridging the urban amenities and urban income.

The location decision of talents may be affected by many factors. But according to our study, urban amenities is one of critical consideration. Therefore, strategies for attracting talents in Chinese cities should be further strengthened and diversified. Improving a city's urban amenities could be one of the most effective strategies. On the other hand, investment in urban amenities will also improve the quality of urban environment, which benefits all urbanites. It also promotes sustainable urban development in the long run.

However, from the demand side, a more detailed research based on surveys and interviews are needed in the future to get a better and more accurate understanding of the priorities of urban talents, taking into consideration of their occupation, background and life cycle stage, etc. More importantly, as human needs and behavior are ever changing over time and space, future studies of consumption amenity or urban amenities should account for these changes.

References

1. Ullman, E.L.: Regional development and the geography of concentration. Pap. Reg. Sci. **4**, 179–198 (1958)
2. Jacobs, J.: The Death and Life of Great American Cities. Vintage, New York (1961)
3. Clark, T.N.: The City as an Entertainment Machine. Elsevier, Amsterdam and Boston (2004)
4. Florida, R.: The economic geography of talent. Ann. Assoc. Am. Geog. **92**, 743–755 (2002)
5. Florida, R., Mellander, C., Stolarick, K.: Inside the black box of regional development—human capital, the creative class and tolerance. J. Econ. Geogr. **8**, 615–649 (2008)

6. Glaeser, E.L., Kolko, J., Saiz, A.: Consumer city. J. Econ. Geogr. **1**, 27–50 (2001)
7. Zhou, J.K.: Regional differences of the impacts of urban amenities on house price and wages: an empirical test based on panel data of China's cities from 1999–2006. J. Finan. Econ. **35**, 80–91 (2009). (in Chinese)
8. Ding, W., Zheng, S., Guo, X.: Value of access to jobs and amenities: evidence from new residential properties in Beijing. Tsinghua Sci. Technol. **15**, 595–603 (2010). (in Chinese)
9. Zhang, W.: A Study of Livable Cities in China. Social Sciences Academic Press (China), Beijing (2006). (in Chinese)
10. Clark, T.N., Lloyd, R., Wong, K.K., Jain, P.: Amenities drive urban growth. J. Urban Aff. **24**, 493–515 (2002)
11. Rappaport, J.: Consumption amenities and city population density. Reg. Sci. Urban Econ. **38**, 533–552 (2008)
12. Liu, L.: What is the contribution of real estate to GDP? China Investment **10**, 17 (2007). (in Chinese)
13. Zhang, H.L., Li, M.: Research on relationship between education and income. Mod. Econ. Inf. **7**, 304 (2012). (in Chinese)

Driving Force Analysis of Cropland Loss in a Rapid Urbanizing Area—The Case of Beijing

Hongrun Ju[1,2(✉)], Lijun Zuo[1], and Zengxiang Zhang[1]

[1] Institute of Remote Sensing and Digital Earth, Chinese Academy of Sciences, Beijing 100101, China
jhr621@126.com
[2] University of Chinese Academy of Sciences, Beijing 100049, China

Abstract. Understanding the driving forces of cropland loss is important for land resource management and sustainable development. This paper aimed to identify the effects of physical and socioeconomic factors of cropland loss in a rapid urbanizing area–Beijing. Geographical detector was used to analyze the importance of drivers and the cropland loss intensity in space. Our results showed that physical factors were generally more influential than socioeconomic factors and their effects changed over time. Urban land was the most important factor during the late 1980s–2000, while woodland became the most influential one in 2000–2010 due to the Sloping Land Conversion Program. Also, the rural settlement in the surrounding area got more influential than urban land in the later period. At last, the cropland loss intensity showed clear but different relationships with most factors. These findings can offer government useful information to protect the cropland and thus maintain sustainable development.

Keywords: Geographical detector · Cropland loss · Driving force · Beijing · Rapid urbanizing

1 Introduction

Stable agricultural production, which relies on sufficient cropland maintenance, is the guarantee of food security and sustainable socioeconomic development [1]. Cropland change is an important issue for China, since the nation supplies food to 22 % of the world population with only 7 % of the world's cropland base [2]. Although the area of cropland is stable from the late 1980s in China, the area of traditional cropland with higher production actually decreases [3]. Cropland loss in China is especially serious in big cities where built-up land construction occupies large amounts of cropland. At the same time, with the implement of some environmental protection policies in China, the area of cropland also decreases in some ecologically fragile areas. Under such situations, it is essential to understand the driving forces of cropland loss deeply, which may provide important implications on cropland conservation of the nation.

Beijing, as the capital city of China, is experiencing rapid urban expansion along with serious cropland loss. More than 80 % of the expansion area of built-up land was converted from cropland between the late 1980s and 2010 [3]. Also, Beijing's cropland

F. Bian and Y. Xie (Eds.): GRMSE 2015, CCIS 569, pp. 586–596, 2016.
DOI: 10.1007/978-3-662-49155-3_61

loss was influenced by environmental protection policies. Beijing began to implement Sloping Land Conversion Program and Beijing and Tianjin Sandstorm Source Control Project in 2000 and 2002, respectively, which promoted the conversation of cropland to woodland in Beijing evidently. So Beijing was an optimal area to study the driving force of cropland loss in rapid urbanizing areas, whose cropland loss was affected by both urbanization and environmental protection policies.

Several studies have examined the spatial patterns and driving forces of cropland loss in rapid urbanization areas, for example, in Su-Xi-Chang region in Jiangsu [1], Shanghai [4, 5] and the delta of the Pearl River [6]. Current studies mostly get the driving forces using temporal statistical analysis such as bivariate regression and multiple linear regression. But relatively less attention has been paid to the temporal variation of the driving forces. The analysis of physical and socioeconomic factors as a whole in rapid urbanizing areas is also worth further exploration. In this study, we used a relatively new spatial statistical method called geographical detector to identify the spatial relationships and their changes between cropland loss and its drivers, which could help better understand the questions mentioned above.

This paper aims to examine the effects of physical and socioeconomic factors that leading to cropland loss and their changes during the late 1980s–2000 and 2000–2010 in Beijing, China. We would deal with the following questions: (1) What factors had dominant influence on cropland loss in Beijing during the late 1980s–2000 and 2000–2010, respectively? (2) How the effects of different factors changed over time? (3) Where did cropland lose more seriously?

2 Study Area, Drivers and Data

Beijing, the capital of the People's Republic of China, is located between 115.7°E-117.4°E and 39.4°N-41.6°N at the northern tip of the North China Plain, with a total area of about 16410 km^2. The average elevation of Beijing is 43.5 m, with Taihang Mountain and Yanshan Mountain surrounded in the west and north. Beijing has fourteen districts and two counties. The city has a monsoon-influenced humid continental climate with hot, humid summers and cold, dry winters. As the nation's political, cultural and educational center, Beijing has experienced rapid urbanization since the market reform initiated in 1978. Its population increased from 11.08 million in 2000 to 19.61 million in 2010, and the proportion of the urban population grew from 77.54 % to 85.96 % [7]. Along with the rapid urban expansion and population growth, the area of cropland in Beijing decreased by 27.4 % from 3704.85 km^2 at the end of 1980s to 2689.53 km^2 in 2010 [3]. The cropland is mainly located in the southeast of Beijing around the city center and in the plain area of the outer suburbs. The cropland loss process from the late 1980s to 2010 is shown in Fig. 1.

Factors leading to cropland loss are complicated. After a comprehensive literature review, we found three types of drivers have been typically considered in similar studies: physical factors, socioeconomic factors and policy factors. Physical factors included topography [8–12], climate [8] and neighborhood factors (e.g. urban land in the surrounding area or undeveloped land in the surrounding area) [8–13]. Socioeconomic

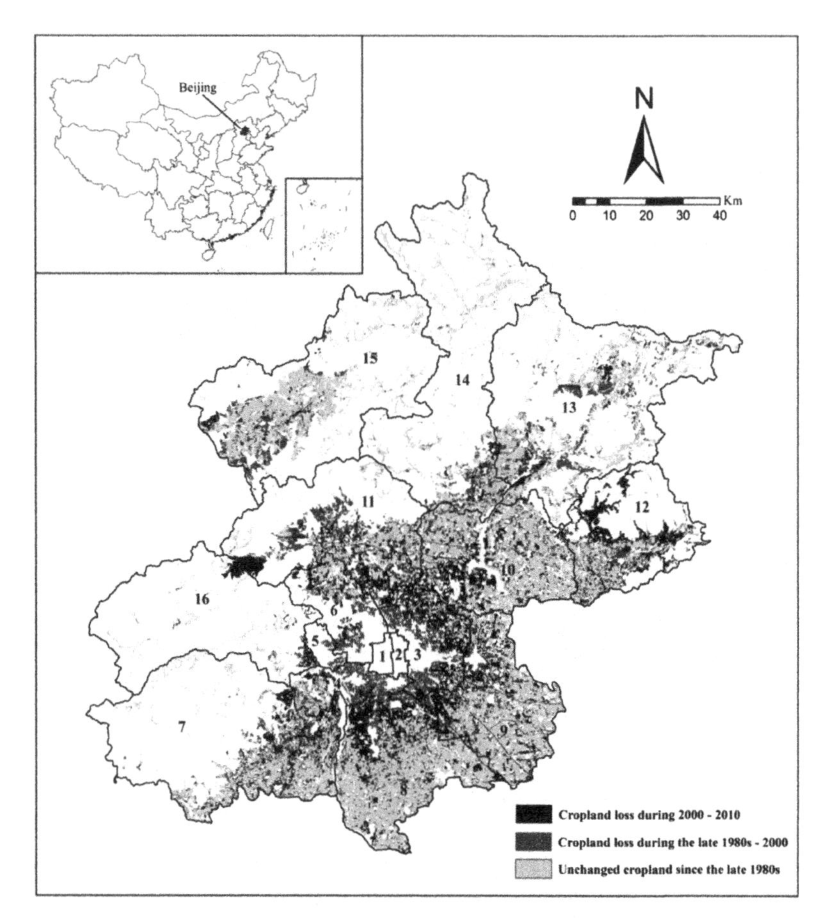

Fig. 1. The study area and its cropland loss process from the late 1980s to 2010. (1: Xicheng, 2: Dongcheng, 3: Chaoyang, 4: Fengtai, 5: Shijingshan, 6: Haidian, 7: Fangshan, 8: Daxing, 9: Tongzhou, 10: Shunyi, 11: Changping, 12: Pinggu, 13: Miyun, 14: Huairou, 15: Yanqing, 16: Mentougou).

factors included population [8–10, 12, 14, 15], economy [8, 14, 15], access to roads [8–12, 14], agricultural input intensity (e.g. tractor density and fertilizer use) [10, 11, 15] and so on. At the same time, policy factors such as cropland protection legislations and urban land policy were also considered in some research [8, 14]. Based on these studies, we selected eleven physical and socioeconomic factors that may influence the cropland loss during the two periods (Table 1). Policy factors were not included for the difficulty to express its spatial heterogeneity in this area. For period of the late 1980s–2000, factors of GDP and Prm_Indu were not included because of the lack of data in late 1980s.

Data used in this study include: (i) land use raster of Beijing of the late 1980s, 2000 and 2010 with six first level types (cropland, woodland, grassland, water bodies, built-up land and unused land). Cropland comprises two second level types: Paddy and dry land, while built-up land comprises three second level types: urban land, rural settlement

Table 1. List of selected factors in this study.

Category	Factors	Abbreviation
Physical factors	Elevation	–
	Slope	–
	Woodland in the surrounding area	Woodland
	Urban land in the surrounding area	Urban
	Rural settlement in the surrounding area	Rural_St
	Industry-traffic land in the surrounding area	Indu_Trf
Socioeconomic factors	Permanent population	Per_Pop
	Rural population	Rural_Pop
	Gross domestic product	GDP
	Proportion of primary industry in GDP	Prm_Indu
	Per capita income of rural residents	Rural_Icm

and industry-traffic land. The data are from national land use/cover database of China, which are mapping by digit human-computer interaction method based on multiple sources of remote sensing data (the Landsat Thematic Mapper, the China-Brazil Earth Resources Satellite and HJ-1A) [16]; (ii) digital elevation model (DEM) of 1980s from 1:250000 topographic database produced by National Fundamental Geographical Information System of China. Land use and DEM data are both raster files with a 100 m resolution; (iii) socioeconomic data of 16 districts and counties of Beijing including GDP, permanent population, rural population, per capita income of rural residents, proportion of primary industry in GDP [7, 17–19].

3 Method–Geographical Detector

Geographical detector is a spatial statistical method to test the consistency of spatial distribution between study objects and their potential driving factors. When the method is applied to cropland loss, we assume that the spatial distribution of cropland loss is similar to that of its potential factors. It consists of four factors: factor detector, risk detector, ecological detector and interaction detector. In order to answer the questions mentioned above, factor detectors is used to explore which factors are more important during each time period and their changes, while risk detector is used to answer where the cropland lost more seriously.

Figure 2 demonstrates the mechanism of geographical detector [20]. First, the study region A is divided with a grid system $G = \{g_i; i = 1, 2, \ldots, n\}$ and the area of cropland loss in every grid is calculated: $y_1, y_1 \ldots y_n$. $D = \{D_i; i = 1, 2, 3\}$ is the geographical stratum of potential factors that can be both continuous and categorical variables. Then the distribution of cropland loss is overlaid with the geographical stratum *D*. Every grid in system *G* will record the value of cropland loss area and the attribute of potential factors according to where it located. For factors of elevation and slope, the value of the grid is equal to the type with the largest proportion. The mean value and the dispersion variances over sub-regions D_i are denoted as $\bar{y}_{D,i}$ and $\sigma^2_{D,i}$ (i = 1, 2, 3), respectively. Let n be the total number of samples over the entire region *A* and let $n_{D,i}$ be the number of samples in sub-region D_i. The global variance of cropland loss in the region *A* is σ^2.

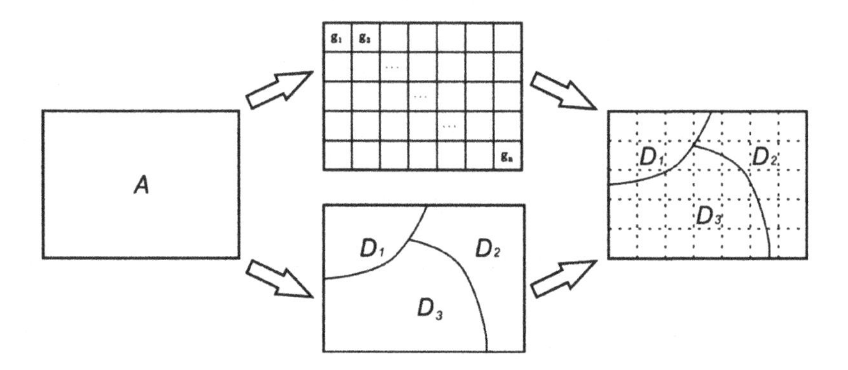

Fig. 2. Mechanism of geographical detector

The factor detector can quantitatively indicate the relative influence of different factors. In this study, the power determinant (PD) is defined as the difference between one and the ratio of accumulated dispersion variance of cropland loss area over each sub-region to that over the entire study region:

$$\text{PD} = 1 - \frac{1}{n\sigma^2}\sum\nolimits_{i=1}^{3} n_{D,i}\sigma^2_{D,i} \tag{1}$$

It means that if factor D is one determinant of cropland loss, the dispersion variance of cropland loss area of each sub-region is small, whereas the variance between sub-regions is large. The value of PD lies between 0 and 1. The larger the PD value is, the more influential the factor is. In this study, PD value represents the consistency of the spatial distribution between cropland loss and its factors.

The risk detector uses a t-test to compare the difference in average values between sub-regions of factor *D*. In this study, we only use the average value ($\bar{y}_{D,i}$) to calculate cropland loss intensity (I_d) which is the average percentage of cropland loss area of the grids in a sub-region D_i:

$$I_d = \frac{1}{S \cdot n_{D,i}}\sum\nolimits_{1}^{n_{D,i}} y_{D,i} \tag{2}$$

where $y_{D,i}$ denotes the cropland loss area of a grid in sub-region D_i, $n_{D,i}$ denotes the number of grids in the sub-region and S denotes the area of a grid of GD. With I_d values, it is more convenient to compare the effects of different sub-regions. The greater the I_d value is, the more dramatically the cropland loss in space.

4 Results and Discussion

The relative importance of each driver was listed in order of decreasing PD values, while the cropland loss intensities of each driver at different levels were shown in Figs. 3 and 4.

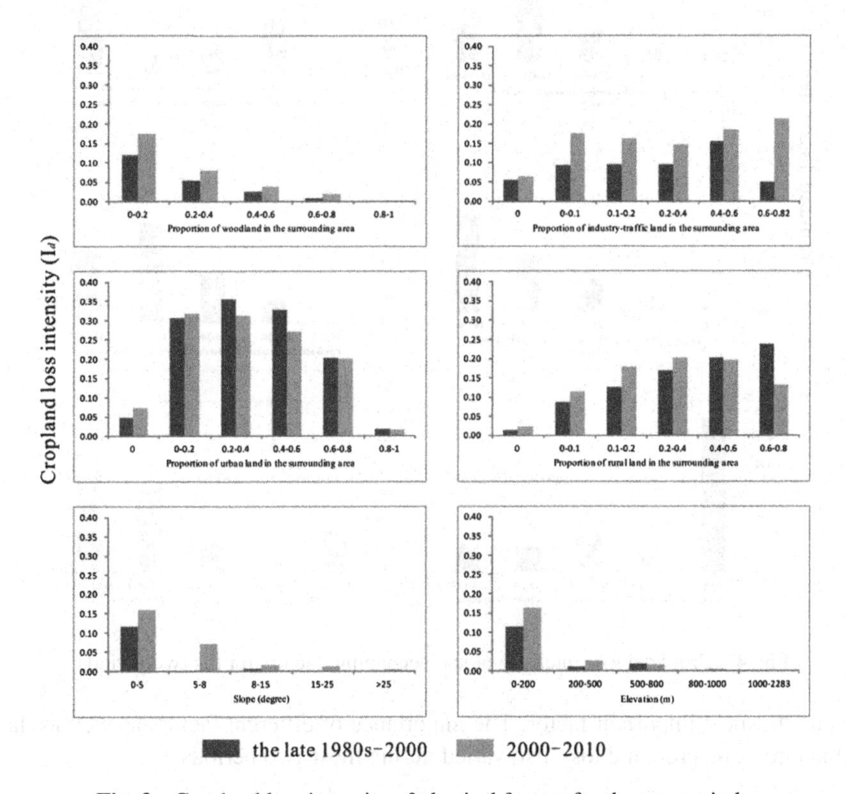

Fig. 3. Cropland loss intensity of physical factors for the two periods.

Period of late 1980s–2000: Slope (0.2071), Urban (0.2038), Woodland (0.1888), Elevation (0.1729), Rural_Icm (0.1693), Rural_St (0.1648), Per_Pop (0.1315), Rural_Pop (0.1048), Indu_Trf (0.0106).

Period of 2000–2010: Woodland (0.2683), Elevation (0.2396), Slope (0.2363), Rural_St (0.2031), Prm_Indu (0.1664), Urban (0.1372), P_Pop (0.1318), Rural_Pop (0.1198), GDP (0.1101), Indu_Trf (0.0837), Rural_Icm (0.0597).

The most influential factors during the late 1980s–2000 were slope and urban land in the surrounding area, while during 2000–2010, woodland in the surrounding area

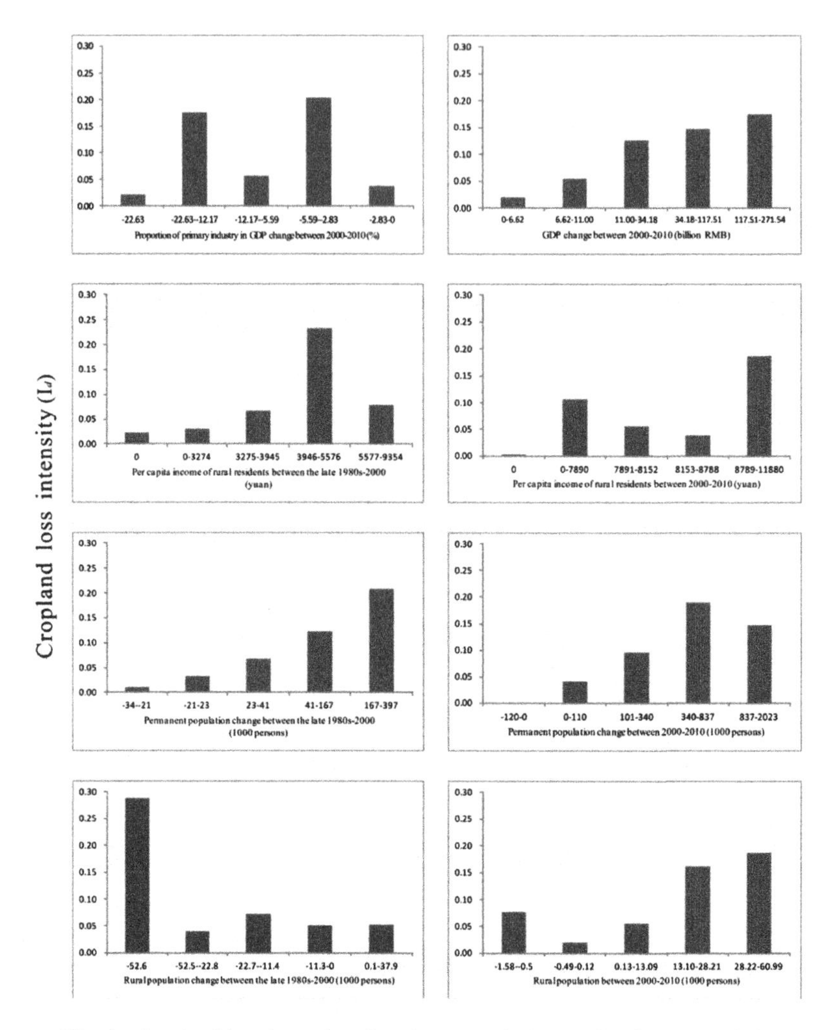

Fig. 4. Cropland loss intensity of socioeconomic factors for the two periods.

became the most important factor. The importance of different factors as well as their spatial effects on cropland loss also varied during these two periods.

4.1 Physical Factors

Physical factors showed significant effects on cropland loss in Beijing during the whole period.

The most obvious changed factor was woodland in the surrounding area which increased greatly as the most influential factor during the later period. It meant that the spatial distribution of cropland loss was more consistent with that of woodland after 2000. Similarly, the I_d value at different levels of woodland surrounding was larger during 2000–2010. This phenomenon was primarily due to the fact that Beijing began to implement the policy of China's Sloping Land Conversion Program in 2000. The program covered four districts and two counties in Beijing: Pinggu, Huairou, Changping, Mentougou, Fangshan, Miyun and Yanqing. Beijing completed the task of the program in 2004 and afforested 306.67 km^2 cropland [21]. The afforested farm land were mainly distributed in ecological fragile areas, low food production areas, desertified and sloping land and sides of main roads. It may be due to this policy, the PD and I_d value of woodland in the surrounding area, elevation and slope all increased (Fig. 4). Elevation and slope became more influential because the afforested cropland was mainly located in lower and less sloping area. Investigation showed that 69.37 % of the afforested cropland was located in places with slope less than 15° (the desertified cropland accounted for 78.45 %) [21]. At the same time, it was shown that cropland lost more dramatically where the woodland was sparse. This might be caused by two reasons. First, the returned cropland was mainly located in flat area where the woodland was sparse. Second, the land use types in Beijing mainly consisted of built-up land, cropland and woodland. So the less woodland might indicate a larger proportion of built-up land in the surrounding area which usually caused the cropland loss seriously.

As for other surrounding land use types, urban land and rural settlement also showed significantly effects during both periods. In the first period, urban land in the surrounding area was more influential than woodland and rural settlement, while its relative influence decreased during the later ten years. Its I_d value was generally higher than all the other factors reflecting its great effects on cropland loss intensity. It was interesting to find that rural settlement contributed more than urban land in causing cropland loss during 2000–2010. Also, the I_d value of urban land in the surrounding area at some levels decreased, while that of different levels of rural settlement surrounding was basically larger in 2000–2010. It indicated that the urban growth in Beijing was more sustainable in terms of the intensity of cropland loss and consistency of spatial distribution between urban land and cropland, but the rural settlement which was fragmented in space actually occupied the cropland more seriously than before. There was another phenomenon that worth noting. The largest I_d value of both the urban land and rural settlement in the surrounding showed a trend of moving to less developed areas. For example, the largest intensity point of rural settlement moved from 0.6–0.8 level to 0.2–0.4 level during the two periods. It reflected a trend that the less developed area might occupy cropland in its surrounding more seriously in future. Industry-traffic land in the surrounding area was the least influential physical factor. Its PD and I_d value both increased during the two periods. The I_d value did not show large difference between different proportions of industry-traffic land in the surrounding area.

Elevation and slope were also important factors in causing cropland loss and they both influenced the I_d value negatively. The I_d value was much higher where elevation was below 200 m and slope was less than 5°. The reason might be that the cropland in Beijing was mainly distributed in plain areas in the southeast. In addition, built-up land expansion which occupied most cropland mainly concentrated in these regions, for the lower development cost in flatter areas [22].

4.2 Socioeconomic Factors

Compared with physical factors, socioeconomic factors had a relatively low influence.

Per capita income of rural residents (Rural_Icm) was the most important socioeconomic factor during the late 1980s–2000. It indicated the well spatial distribution consistency between Rural_Icm and cropland loss. But during 2000–2010, the spatial distribution consistency between Rural_Icm and cropland loss decreased greatly. The cropland loss intensity did not show a clear relationship with this factor. During 2000–2010, proportion of primary industry in GDP was influential but its I_d value did not show a regular pattern with in space. By contrast, GDP had a positive effect on I_d value. Places with more GDP growth tended to lose cropland more dramatically. It indicated that economy growth in Beijing was at the cost of cropland loss to some extent.

The influence of permanent population remained generally unchanged during the two periods. From the late 1980s to 2000, the I_d value was well positively correlated with the permanent population change, indicating the more the population grew, the more dramatically the cropland lost. It was because population in Beijing increased, but they required more urban land instead of cropland. This phenomenon was similar in the later ten years, however, when the population increased dramatically of 837–2023 thousand persons, the I_d value decreased. Further analyzing these high population increased area, they were Chaoyang, Haidian, Fengtai and Changping Districts where urban land was concentrated. This was consistent with the finding that the large intensity occurred where there were less proportion of urban land surrounding in 2000–2010.

Factor of rural population became a bit more influential within these years. It was interesting to find that the rural population change showed different relationships with the I_d value during the two decades. During the first decade when rural population decreased in Beijing, the largest loss intensity happened where the rural population dropped most, while in the later decade, the I_d value showed positive relationship with rural population growth. This might be explained by the fact that people living in the rural area depended less than agriculture to earn a living in Beijing. The rural population increased by 32.07 % from 3.798 million in 1996 to 5.016 million in 2006 and the structure of rural population also changed a lot in Beijing. Nearly one third of the rural population in Beijing was from other cities and this figure reached 58.7 % in the city functional expansion area (Chaoyang, Haidian and Fengtai Districts). However, rural population employed in the primary industry decreased from 855,000 in 1996 to 657,000 in 2006. There were 79.3 % of the employed rural population being involved in the secondary and tertiary industry in 2006, while only 61 % in 1996 [23]. These phenomena indicated that increased rural population in the later period had a higher requirement of rural settlements instead of cropland. It also explained the rural settlement in the surrounding area becoming a more important factor than urban land during 2000–2010.

It should be noted that the relative importance of different factors obtained by factor detector only reflected the spatial consistency between cropland loss and potential factors. So the higher relative importance was not equal to a large amount area of lost cropland. Although woodland and rural settlement in the surrounding area outweighed the influence of urban land in the later period, cropland in Beijing was still primarily converted into urban land in terms of area.

5 Conclusion

Based on multi-temporal LUCC data getten from satellite images, we observed a serious loss of cropland in Beijing from the late 1980s to 2010. Physical and socioeconomic factors significantly affected the cropland loss in Beijing. Based on the spatial statistic results of geographical detector, we could draw the following conclusions:

(1) During the late 1980s–2000, slope and urban land in the surrounding area are the dominant factors in causing cropland loss in Beijing with the relative importance of 0.2071 and 0.2038, respectively. By contrast, in period 2000–2010, woodland in the surrounding area was the most influential factor with the relative importance of 0.2683.
(2) The relative importance of different factors varied over time. Woodland in the surrounding area became the most influential factor after 2000, reflecting the great interference of the policy of China's Sloping Land Conversion Program. Moreover, we found that the influence of rural settlement outweighed urban land in the surrounding area during the later period, indicating a relative sustainable development of the urban land but less controlled rural settlement expansion in Beijing.
(3) The cropland loss intensity was influenced by various factors combined. Places with urban land and rural settlement concentrated were more likely to experience serious cropland losing, but less developed area tended to lose cropland more dramatically in the future. Also, the cropland loss intensity was much higher where the elevation was lower than 200 m and the slope was less than 5°. As for socioeconomic factors, GDP and population both showed a positive relationship with cropland loss intensity. However, rural population showed different relationships with cropland loss intensity during the two periods, indicating a changed land requirement and working structure of rural residents in Beijing.

Based on the findings of this study, it was evident that the urbanization and environmental policy had a significant influence on the cropland loss in Beijing. So we suggested that effective cropland protection management combined with urban planning should be implemented to control the cropland loss in Beijing. Moreover, government should take action to control the rural settlement expansion to protect the cropland in rural areas in the future.

References

1. Zhou, X., Han, J., Meng, X., Yi, B., Cao, W., Huang, L., Xiang, W.: Comprehensive analysis of spatio-temporal dynamic patterns and driving mechanisms of cropland loss in a rapidly urbanizing Area. Re. Sci. **36**(06), 1191–1202 (2014). (in Chinese)
2. Zuo, L., Wang, X., Zhang, Z., Zhao, X., Liu, F., Yi, L., Liu, B.: Developing grain production policy in terms of multiple cropping systems in China. Land Use Policy **40**, 140–146 (2014)
3. Zhang, Z., Zhao, X., Wang, X.: Remote Sensing Monitoring of Landuse in China. Star Map Press, Beijing (2012). (in Chinese)
4. Han, J., Hayashi, Y., Cao, X., Imura, H.: Evaluating land-use change in rapidly urbanizing China: case study of Shanghai. J. Urban. Plann. Dev. **135**(4), 166–171 (2009)

5. Meng, F., Bao, W., Shan, B.: Research on dynamic changes and driving forces of Shanghai cultivated land. J. Shan. Jian. Univ. **22**(5), 403–407 (2007). (in Chinese)
6. Yeh, A.G.O., Li, X.: Economic development and agricultural land loss in the Pearl River Delta, China. Habitat. Int. **23**(3), 373–390 (1999)
7. Beijing Municipal Statistical Bureau: Beijing Statistical Yearbook 2011. China Statistics Press, Beijing (2011). (in Chinese)
8. Liu, X., Wang, J., Liu, M., Meng, B.: Spatial heterogeneity of the driving forces of cropland change in China. Sci. China Series D Earth Sci. **48**(12), 2231–2240 (2005)
9. Lakes, T., Müller, D., Krüger, C.: Cropland change in southern Romania: a comparison of logistic regressions and artificial neural networks. Landscape Ecol. **24**(9), 1195–1206 (2009)
10. Müller, D., Kuemmerle, T., Rusu, M., Griffiths, P.: Lost in transition: determinants of post-socialist cropland abandonment in Romania. J. Land Use Sci. **4**(1–2), 109–129 (2009)
11. Müller, D., Leitão, P.J., Sikor, T.: Comparing the determinants of cropland abandonment in Albania and Romania using boosted regression trees. Agr. Syst. **117**, 66–77 (2013)
12. Prishchepov, A.V., Müller, D., Dubinin, M., Baumann, M., Radeloff, V.C.: Determinants of agricultural land abandonment in post-Soviet European Russia. Land Use Policy **30**(1), 873–884 (2013)
13. Tan, M., Li, X., Xie, H., Lu, C.: Urban land expansion and arable land loss in China—a case study of Beijing–Tianjin–Hebei region. Land Use Policy **22**(3), 187–196 (2005)
14. Xie, Y., Mei, Y., Guangjin, T., Xuerong, X.: Socio-economic driving forces of arable land conversion: a case study of Wuxian City, China. Global Environ. Chang. **15**(3), 238–252 (2005)
15. Hazell, P., Wood, S.: Drivers of change in global agriculture. Philos. T. R. Soc. B **363**(1491), 495–515 (2008)
16. Zhang, Z., Wang, X., Zhao, X., Liu, B., Yi, L., Zuo, L., Wen, Q., Liu, F., Xu, J., Hu, S.: A 2010 update of national land use/cover database of China at 1: 100000 scale using medium spatial resolution satellite images. Remote Sens. Environ. **149**, 142–154 (2014)
17. Beijing Municipal Statistical Bureau: Beijing Statistical Yearbook 2000. China Statistics Press, Beijing (2001). (in Chinese)
18. Beijing Municipal Statistical Bureau: Beijing Statistical Yearbook 1988. China Statistics Press, Beijing (1988). (in Chinese)
19. Beijing Municipal Statistical Bureau, National Bureau of Statistics Survey Office in Beijing: The 30th Anniversary of Reform and Opening up. China Statistics Press, Beijing (2008). (in Chinese)
20. Wang, J.F., Li, X.H., Christakos, G., Liao, Y.L., Zhang, T., Gu, X., Zheng, X.Y.: Geographical detectors-based health risk assessment and its application in the neural tube defects study of the Heshun Region, China. Int. J. Geogr. Inf. Sci. **24**(1), 107–127 (2010)
21. Lu, J.: Analysis and consolidation strategies of the results of the "Returning Farm Land to Forest" project in Beijing. Forest Resour. Manage. **5**, 44–49 (2009). (in Chinese)
22. Li, X., Zhou, W., Ouyang, Z.: Forty years of urban expansion in Beijing: what is the relative importance of physical, socioeconomic, and neighborhood factors? Appl. Geogr. **38**, 1–10 (2013)
23. Beijing Office of Second National Agricultural Census in China. http://www.bjstats.gov.cn/nypc/pcdt/pcxw/200804/t20080416_110328.htm. (in Chinese)

Primary Research on Geo-Informatic Tupu for Crime Spatio-Temporal Analysis

Dong Cai[1(✉)], YanMing Chen[2], and Chao Gao[3]

[1] Public Security Management Department, Jiangsu Police Institute, Nanjing, China
jickie_cd@126.com

[2] School of Geographic and Oceanographic Sciences, Nanjing University, Nanjing, China

[3] Key Laboratory of Geographic Information Technology of Public Security Ministry, Changzhou City Public Security Bureau, Changzhou, China

Abstract. Geo-Informatic Tupu is a complex spatio-temporal analysis method. Its detailed, simple image analysis and expression ways can be better meet the crime spatio-temporal analysis needs. This paper summarizes the research background and current situation of crime spatio-temporal analysis, and discusses the significance and content of this research. And also this paper puts forward own ideas about research ways, which is in order to provide a new method for method references and decision supports in the crime spatio-temporal analysis practices.

Keywords: Geo-Informatic tupu · Crime spatio-temporal analysis · Review

1 Introduction

First, crime spatio-temporal characteristics require constant prolongation of crime spatio-temporal information analysis technology. As a part of society, the occurrence of criminal activities is an inevitable social phenomenon, whose change with time and space rules of typical. A growing body suggests there is an inevitable relationship between the crime phenomenon and the geographical environment. From the city spatial anticrime point of view, crime spatio-temporal analysis will become an important means of policing activities understanding, analysis and prediction. The current crime spatio-temporal analysis technology is developing rapidly, so the old style of pin map has been taken place with crime mapping technology. Advantage of crime spatio-temporal analysis has a profound deeply impact on the crime analysis efficiency.

Second, crime spatio-temporal analysis provides a methodology for spatial information compound analysis. The crime spatial distribution is a complex temporal evolution processing, the inversion, simulation and prediction research of this is a composite study of an integrated time and space, which needs an effective support of effective temporal and spatial analysis theory. The geographical information Tupu theory is based on inheriting the traditional China research results, which is on the basis of temporal 3S technology, information network and other contemporary advanced technology. As a showing way of related information processing and display, the geographical information Tupu provides theoretical basis for crime spatio-temporal

F. Bian and Y. Xie (Eds.): GRMSE 2015, CCIS 569, pp. 597–602, 2016.
DOI: 10.1007/978-3-662-49155-3_62

analysis. The related information includes variation characteristics and time sequence of phenomena spatial structure.

Third, the investment in the ministry of public security and the ministry on the police geographic information technology provide guarantee for the crime spatio-temporal analysis work. In July 13, 2012, in the guidance and support of the ministry of public security and Jiangsu province public security division, Changzhou City Public Security Bureau, Nanjing Normal University, Jiangsu Police Institute, founder International Software Co., Ltd. and other units jointly established the Jiangsu Provincial Public Security Bureau police Spatial Information Technology Key Laboratory. The Ministry public security PGIS project office members attend to laboratory construction and operation. In December 28, 2013, Changzhou City Public Security Bureau and Nanjing Normal University establish "Key Laboratory of the Ministry of public security police geographic information technology" together. The construction of Laboratory solves the practical business application problems such as grassroots work and investigation and case solving, which provides researchers with a good exchange platform. One of the key research directions in the laboratory is crime spatial analysis and crime mapping, which puts forward higher requirements on crime spatio-temporal composite analysis.

2 Research Review at Home and Abroad

The existing crime spatio-temporal analysis technical usually uses crime mapping technology. On the one hand, crime mapping technology has accurate and number requirements of crime data, which has some difficulties in the data acquisition and application. On the other hand, the crime spatio-temporal analysis based on crime mapping technology mainly is the embodiment of the results, which is lack of process embodiment and causes attention.

Thus, Geo-Informatic Tupu is a complex spatio-temporal analysis method. Its detailed, simple image analysis and expression ways can be better meet the needs of crime spatial analysis, which will not only reduce the crime data requirements, but also can promote the understanding of crime spatio-temporal analysis process.

Even so, crime spatio-temporal analysis based on Geo-Informatic Tupu is still in the blank at home and abroad. There is a huge space in this study. This research can contribute to the understanding of crime spatio-temporal process, which has the characteristics of image concise. At the same time, as a new kind method of spatio-temporal variation analysis theory and expression method, the application of Geo-Informatic Tupu in crime spatio-temporal analysis can xpand and enrich the theory and method system.

3 Research Significance and Application Value

From the theoretical point of view, the present crime spatio-temporal analysis mainly includes method research, model study, and researching typical area, all of this should be breakthrough and upgrade. Then Geo-Informatic Tupu is pulled into crime spatio-temporal analysis. Geo-Informatic Tupu analysis provides a theoretical basis for crime spatio-temporal analysis and a new idea for the study of crime spatio-temporal analysis. At present, the present Geo-Informatic Tupu research mainly focuses on physical geography and human geography, and less involve in criminal geography. The application of Geo-Informatic Tupu in crime spatio-temporal analysis, can not only expand the application field of Geo-Informatic Tupu, further enrich the theory of Geo-Informatic Tupu, but also open up new ideas for the study of crime spatio-temporal analysis.

From the practice point of view, Geo-Informatic crime Tupu is a concept extension into certain fields, and also a practical research of Geo-Informatic crime Tupu research. The research content, method and extraction process of crime spatio-temporal analysis, and the discussion about crime spatio-temporal analysis based on Geo-Informatic Tupu in this paper can provide similar studies with a prior knowledge and experience. The results of crime spatio-temporal analysis can carry out spatial distribution of crime and achieve optimization of spatial anticrime. All this can promote the transition from traditional policing mode to the new police model, which has important practical significance.

The more direct application value embodied in the study is the method reference and decision support in crime spatio-temporal analysis practice, which can be available in many relevant administrative departments, such as the information department, intelligence department, and emergency department.

4 The Main Research Contents

Taking the remote sensing image, statistics data and crime data as the data source, using Geo-Informatic Tupu theory as the support, this paper researches the composition, generation method and application of Geo-Informatic crime Tupu. And also this paper studies on driving force which affecting the crime spatio-temporal analysis. The research process can be described as following: firstly, crime spatio-temporal information must be collected, from which the time and spatial information Tupu can be construct. Basing on the information Tupu, the overall local characteristics can be carried out by crime spatio-temporal analysis. And then driving force of Geo-Informatic crime Tupu is researched in order to cognize the time evolution of criminal activities from the spatial graphics point. On the basis of the criminal activities time evolution, spatial differential features and crime driving mechanism are deeply studies. Specifically, the main contents of the research include as follows:

First, Geo-Informatic crime Tupu is consist of crime time evolution Tupu and spatial characteristics Tupu. All these Tupu foundation is based on the analysis of Geo-Informatic crime Tupu, which is only on the condition of crime spatio-temporal analysis.

Second research is crime time series of whole evolution analysis. By the crime information extraction, we can analysis the crime time evolution, crime time transfer evolution, and also the overall characteristics evolution by means of a variety of indices to measure.

Third research is criminal spatial distribution analysis. The difference of spatial distribution characters in every area, every position and every direction can be attracted by spatial index and crime spatio-temporal analysis model. Then the overall crime spatial distribution characters such as the overall, local, and detailed features can be summarized, which will come to conclusion about the leading direction, expansion mode and phase morphology.

Forth research is the driving force related with Geo-Informatic crime Tupu. The crime spatial distribution has stage, diversity and complexity characteristics, which is under the comprehensive effect of internal, external social forces, expansion in the different stages. In this paper, the combination of qualitative and quantitative analysis is used to analysis the correlation between city development and Geo-Informatic crime Tupu.

Crime spatio-temporal analysis mainly focus on three aspects which includes temporal, spatial and the driving force. The details are as follows:

First is characteristics extraction and analysis of crime temporal Tupu. With all the crime statistics data as data source, multi-temporal crime information metastasis boundary is extracted, and the crime temporal Tupu is built in the certain time. Then taking the crime temporal Tupu in each time point as Tupu unit, the crime temporal Tupu can be analysis overall.

Second is crime characteristic Tupu and spatial difference analysis. On the basis of the overall crime temporal Tupu analysis, The local spatial difference Tupu and detail spatial difference Tupu can be extracted by the analysis based on the 1 hectares unit. Then the detail and local spatial characteristic can be analysis, which can conclude the spatial distribution characteristic and transfer mode in the different stage of crime.

Third is related effect analysis of driving force on crime time division. The crime temporal and spatial distribution is not only a single driving force result, but also many factors related influence result. The aspects of driving force related effects on regional spatial and temporal information of crime from the natural, economic, social, historical and cultural, in order to help the relevant policies and measures to promote crime prevention and control system, the development of the region to the direction of rationalization.

5 The Research Ideas and Methods

5.1 Research Ideas

The research on spatial information Tupu takes crime statistics data as the main data source, and uses GIS spatial statistical analysis methods and spatial analysis to study on Geo-Informatic Tupu. The study step can be described as the three following steps: crime spatio-temporal information extraction, time evolution of crime distribution, spatial differentiation analysis, and associated driving force effects research. Research general framework and technical route are shown in Fig. 1.

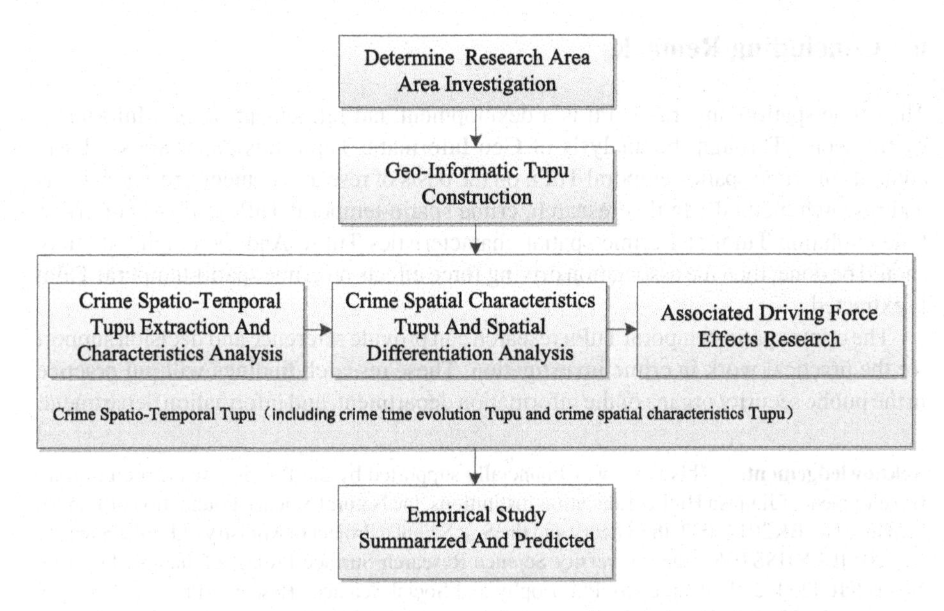

Fig. 1. Research general framework

5.2 Research Methods

(1) Literature research method. Through checking books and reference materials and computer network technology, the study can learn from the latest research achievements and utilization. Then the summarizing analysis can be done by the monograph books, research reports, journals and newspapers and other relevant information etc. which carries on the thorough discussion to the crime temporal-spatial information.
(2) Comparative method. To study the theory of crime spatio-temporal analysis, some kind of association between the spatio-temporal characteristics and criminal behavior can be summarized. The theory of crime spatio-temporal analysis technology is widely used in Europe and the United States, and there are some successful cases. With the idea of science and technology strengthening police, comparison method can be used to provide deeply background support for this dissertation.
(3) The pilot study. The pilot study is selecting the appropriate cities as pilot while studying. The study closely cooperates with the city public security bureau and police geographic information technology laboratory, Geo-Informatic Tupu analysis technology is put into practice. And the theory becomes perfection in the practice, which the conclusion of the study has application value and promotion effect.

6 Concluding Remarks

The crime spatio-temporal TuPu is a development and refinement of Geo-Informatic Tupu theory. Through the analysis of Geo-Informatic Tupu, this paper araises basic contents of crime spatio-temporal TuPu on the basis of research contents, research ideas and research methods. In this research, crime spatio-temporal TuPu is consis of crime time evolution TuPu and crime spatial characteristics TuPu. And the empirical study should be done, then the association driving force effects on crime spatio-temporal TuPu is extracted.

The crime spatio-temporal TuPu research can provide reference and decision support for the practical work in crime investigation. These research findings will put practice in the public security organs of the information department, and information department.

Acknowledgements. This work was financially supported by the Priority Academic Program Development of Jiangsu Higher Education Institutions, the Natural Science Foundation of Jiangsu Province No. BK20141033, the Theory and the Soft Science Project of Ministry of Public Security No. 2014LLYJJSST061, the University Science Research Surface Project of Jiangsu Province No. 13KJB420002, the University Philosophy and Social Sciences Research Project of Jiangsu Province No. 2014SJD194 and the Research Project of Jiangsu Police Institute No. 13Q18.

References

1. Chen, W., Wu, J.P.: Research review on temporal and spatial distribution of foreign criminal. World Reg. Stud. **22**(2), 151–158 (2013)
2. Herold, M., Couclelis, H., Clarke, K.C.: The role of spatial metrics in the analysis and modeling of urban land use change. Comput. Environ. Urban Syst. **29**, 369–399 (2005)
3. Irwin, E., Geoghegan, J.: Theory, data, methods: developing spatially-explicit economic models of land use change. Agric. Ecosyst. Environ. **85**, 7–24 (2001)
4. Liu, W., Seto, K.C., Sun, Z.: An ART-MMAP neural network based spatio-temporal data mining approach to predict urban growth[OB/OL] (2006). http://www.isprs.org/commission8/workshop_urban
5. White, R., Engelen, G.: The use of constrained cellular automata for high-resolution modeling of urban landuse dynamics. Environ. Plann. B: Plann. Des. **24**, 323–343 (1997)
6. Pijanowski, B.C., Pithadia, S., Shellito, B.A., et al.: Calibrating a neural network-based urban change model for two metropolitan areas of the upper midwest of the United States. Int. J. Geogr. Inf. Sci. **19**(2), 197–215 (2005)
7. Ridd, M., Liu, J.: A comparison of four algorithms for change detection in an urban environment. Remote Sens. Environ. **63**, 95–100 (1998)
8. Schneider, L.C., Pontius, R.G.: Modeling land-use change in the Ipswich Watershed, Massachusetts, USA. Agric. Ecosyst. Environ. **85**, 83–94 (2001)
9. Batty, M.: Agents, cells, and cities: new representational models for simulating multi scale urban dynamics. Environ. Plann. A **37**(8), 1373–1394 (2005)
10. Verburg, P.H., Veldkamp, A., Bouma, J.: Land use change under conditions of high population pressure: the case of Java. Glob. Environ. Change **9**, 303–312 (1999)
11. White, R., Engelen, G., Uljee, I., et al.: Developing an urban land use simulator for European Cities. In: Fullerton, K. (ed.) Proceedings of the 5th EC GIS Workshop: GIS of Tomorrow, European Commission Joint Research Centre, pp. 179–190 (2000)

Possible Influences of Land Use and Cover Change on Vegetation Cover: A Case Study in China's Yongding River Basin

Hong Wang, Kaikai Xu, Xiaobing Li(✉), Honghai Liu, and Dengkai Chi

State Key Laboratory of Earth Surface Processes and Resource Ecology, College of Resources Science and Technology, Beijing Normal University, Beijing, China
bnuxbli@163.com

Abstract. Understanding the influence of land uses on vegetation cover can guide both the restoration and the historical reconstruction of vegetation communities. In this study, we developed a linear decomposition model for vegetation cover, with the model's endmembers determined using a pixel purity index. Using remote sensing images from 1978, 1987, 2000, and 2005, we calculated normalized-difference vegetation index (NDVI) values, and used the linear decomposition model to estimate the changes in vegetation cover in China's Yongding River Basin due to climate change. We performed regressions among NDVI values, climate factors, and slope, and reconstructed NDVI (thus, vegetation cover). Based on classification of the remote sensing images from the four dates, we analyzed the influence of land use and cover change on vegetation cover under two climate scenarios: (1) The climate in the previous period remains unchanged, so the precipitation and temperature in 1978, 1987, and 2000 replaced the values in 1987, 2000, and 2005, respectively. In this scenario, the decrease in vegetation cover from 1978 to 1987 in most of the basin ranged from 20 to 60 %. The vegetation cover in most of the basin increased from 1987 to 2000. From 2000 to 2005, vegetation cover decreased in northern parts of the basin but increased in the middle, eastern, and southern parts. (2) The 1978 precipitation and temperature values remain unchanged. In this scenario, vegetation cover increased by 40 to 60 % from 1987 to 2000 in the eastern part of the basin and by 0 to 40 % in the middle and western parts. From 2000 to 2005, vegetation cover decreased in most of the basin, but increased in some middle and eastern areas.

Keywords: Land use and cover change · Vegetation cover · Linear decomposition model · Yongding River · China

1 Introduction

Coverage of the ground surface by vegetation is an important parameter in global and regional climate models, but also provides important basic data for describing an ecosystem. Changes in terrestrial ecosystems are therefore an important component of

F. Bian and Y. Xie (Eds.): GRMSE 2015, CCIS 569, pp. 603–610, 2016.
DOI: 10.1007/978-3-662-49155-3_63

research on global change, and the relationship between vegetation cover and environmental changes is a complex and active area of research [1, 2].

To investigate the influence of land use changes on vegetation cover, researchers first obtain data on the land use and cover types in the study area, then estimate vegetation cover using a model built from the remote sensing data that relates these data to the vegetation data. A relationship between land use and cover change and vegetation cover for different land use and cover types can then be determined. Previous research has shown that land use and cover changes can influence the vegetation cover [3, 4], but there has been little research in this area in China. Thus, we chose a representative study area in China and investigated the influence of land use and cover changes on vegetation cover over a long period to provide reference data that could support efforts to protect the regional ecological environment.

2 Overview of the Study Area

The Yongding River Basin (Fig. 1) is located between longitudes 111°35′E and 117° 12′E, and between latitudes 41°30′N and 38°35′N. There is sufficient sunshine for good vegetation growth, with mean annual hours of sunshine ranging from 2800 to 3000, except in Shanyin and Hunyuan counties, where annual sunshine totals less than 2800 h. The vegetation is classified as a warm grassland, with shrub grassland and small areas of forest present in some areas.

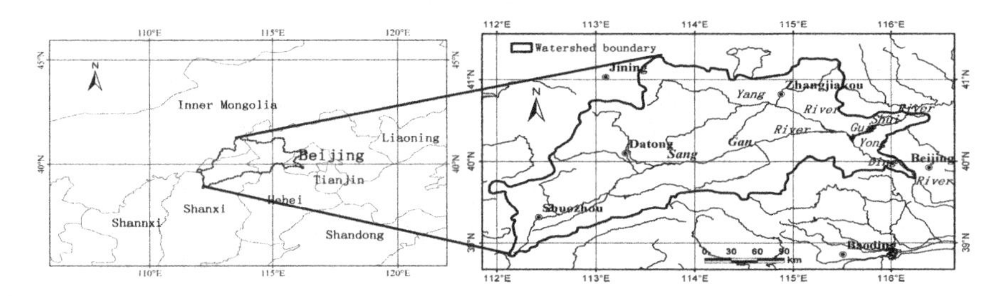

Fig. 1. Location of the study area.

3 Data

For our analysis, we chose the following remote sensing images of the study area: Landsat TM images from 1987 and 2005, Landsat ETM+ images from 2000, and Landsat MSS images from 1978. We performed the necessary geometric corrections, radiation calibrations, atmospheric corrections, and image mosaic of the remote sensing images. We then classified the images into land use and cover type maps by combining unsupervised and supervised classification, and verified the accuracy of the image classification using field data from 230 sampling points for different land use and cover types. The overall accuracy was 81.1 % and the kappa coefficient was 0.71, indicating good overall classification accuracy. For the climate data, we used monthly precipitation and average temperature data from 1978, 1987, 2000, and 2005 provided by the State Meteorological Administration.

4 Model for the Estimation of Vegetation Cover

4.1 Linear Decomposition Model of Vegetation Cover

Because of the low resolution of the remote sensing images, each pixel may composed of several different land use and cover types. All components of each pixel contribute to the overall information for that pixel that is obtained by the sensor. However, each pixel can be decomposed into its components by means of a combination of linear and nonlinear models, which can then be used to estimate vegetation cover.

The information represented by each pixel can be decomposed into two parts: one that represents vegetation information and another that represents soil information. The information contained by a pixel can be represented by the following formula:

$$I_{\text{pixel}} = I_{\text{veg}} + I_{\text{soil}} \tag{1}$$

where I_{pixel}, I_{veg}, and I_{soil} represent the information for the pixel as a whole, for the vegetation component, and for the soil component, respectively. We can then define fc_{veg} and fc_{soil} as the vegetation and soil cover components, respectively. The sum of these two components must add up to 1 (i.e., to 100 % of the information content for that pixel):

$$fc_{\text{veg}} + fc_{\text{soil}} = 1 \tag{2}$$

If the area represented by a pixel is totally covered by vegetation, $I_{\text{pixel}} = I_{\text{veg}}$ and $I_{\text{soil}} = 0$, and fc_{veg} acts as a weight coefficient for the vegetation information in the pixel. The vegetation components of a mixed pixel can then be represented by the following formula:

$$I_{\text{veg}} = I_{\text{g}} \times fc_{\text{veg}} \tag{3}$$

where I_{g} represents the remote sensing information when a pixel is totally covered by vegetation. If the area represented by a pixel is totally covered by soil, $I_{\text{pixel}} = I_{\text{soil}}$ and $I_{\text{veg}} = 0$, and fc_{soil} acts as a weight coefficient for the soil information in the pixel. The soil components of a mixed pixel can then be represented by the following formula:

$$I_{\text{soil}} = I_{\text{s}} \times \left(1 - fc_{\text{veg}}\right) \tag{4}$$

where I_{s} represents the remote sensing information when a pixel is totally covered by soil. We can derive the following formula from Eqs. (3) and (4):

$$I_{\text{pixel}} = [I_{\text{g}} \times fc_{\text{veg}}] + [I_{\text{s}} \times (1 - fc_{\text{veg}})] \tag{5}$$

Solving for the vegetation component, we get:

$$fc_{\text{veg}} = \left(I_{\text{pixel}} - I_s\right) / \left(I_g - I_s\right) \quad (6)$$

Because the vegetation index can reflect the growth status and distribution of the vegetation, we chose NDVI to calculate the vegetation cover:

$$fc_{\text{veg}} = \left(NDVI_{\text{pixel}} - NDVI_s\right) / \left(NDVI_g - NDVI_s\right) \quad (7)$$

where $\text{NDVI}_{\text{pixel}}$ represents the NDVI for the whole pixel, and $NDVI_g$ and $NDVI_s$ represent the NDVI values when the area represented by the pixel is totally covered by vegetation and totally covered by soil, respectively.

NDVI_s is influenced by the atmosphere and by changes in surface soil moisture, and will therefore change over time. Because NDVI_s is also influenced by surface roughness, soil type, and other factors, it will have obvious spatial variability. Because of differences in vegetation types, seasonal changes in vegetation cover, and "noise" due to surface soil moisture, snow, the presence of dried leaves, and other factors, NDVI_g will also change temporally and spatially [5, 6]. Thus, the values of NDVI_g and NDVI_s at different locations within an image will be influenced by many factors, and it is difficult to determine these factors.

4.2 Determination of Spectral Endmembers

Over MNF conversion, remote sensing image chooses wave band without noise over MNF conversion to calculate pixel purity index (PPI). An appropriate threshold value is set based on the PPI value and is used to identify relatively pure pixels in the image. After the location of a pure pixel is determined, the land use and cover type corresponding to the pure pixel cannot be determined. The pixels corresponding to high PPI values are identified, and the typical land types for these pixels are determined together with a scatterplot of wavelength band as a function of the MNF change. After the endmembers for vegetation and soil have been determined, NDVI values corresponding to the locations of pure pixels can be determined based on the NDVI image for the study area.

5 Influence of Land Use and Cover in the Yongding River Basin on Vegetation Cover

By means of kriging interpolation, we interpolated the monthly precipitation and temperature data from April to July in 1978, 1987, 2000, and 2005 to produce gridded data with the same spatial resolution as the remote sensing images, and used the resulting precipitation and temperature data in our subsequent analyses. In each of the four years, we used the total precipitation from April to July to represent precipitation during the growing season and the mean temperature from April to July to represent the mean temperature during the growing season.

5.1 The Influence of Land Use and Cover Change on Vegetation Cover

(1) Influence of Land Use and Cover Change on Vegetation Cover in Scenario 1

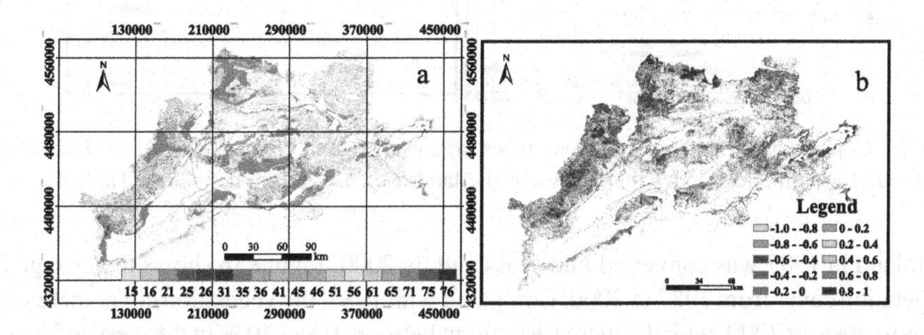

Fig. 2. Conversion of land use and cover types (a) and change in vegetation cover (b) from 1978 to 1987. The numerical codes that represent the land use changes are defined in Table 1.

Table 1. Numerical codes used to define the conversions among different land use types. The actual conversions are shown in Figs. 8 to 10.

21 Residential and industrial land to farmland	15 Farmland to forest	16 Farmland to grassland
31 Roads to farmland	25 Residential and industrial land to forest	26 Residential and industrial land to grassland
41 Water body to farmland	35 Roads to forest	36 Roads to grassland
51 Forest to farmland	45 Water body to forest	46 Water body to grassland
61 Grassland to farmland	65 Grassland to forest	56 Forest to grassland
71 Unused land to farmland	75 Unused land to forest	76 Unused land to grassland

Table 1 defines the code numbers used to represent each type of land use transition discussed in the remainder of this section. Figure 2a shows the pattern of conversion among different land use and vegetation cover types from 1978 to 1987 under Scenario 1. Because the vegetation cover changes to 0 after a vegetation type is converted into a non-vegetation type, we will only show the conversions among other land use types. From 1978 to 1987, the forest in most of the basin was converted into grassland, and large areas of forest were also converted into farmland. In the west, some farmland was converted into forest and grassland, and there were few conversions among other types. Figure 2b shows the vegetation cover change from 1978 to 1987. In most area of the basin, vegetation cover decreased, with the magnitude of the decrease ranging from 20 to 60 %. The decrease was between 0 and 20 % in some northern and northwestern areas, but was greater than 60 % in a few areas. In parts of the middle of the basin, the vegetation cover increased by 20 to 40 %.

Figure 3a shows the spatial distribution of the conversions among land use and cover types from 1987 to 2000. The forest land in most areas was converted into grassland, and large areas of grassland in the northwest were converted into farmland. The periphery of

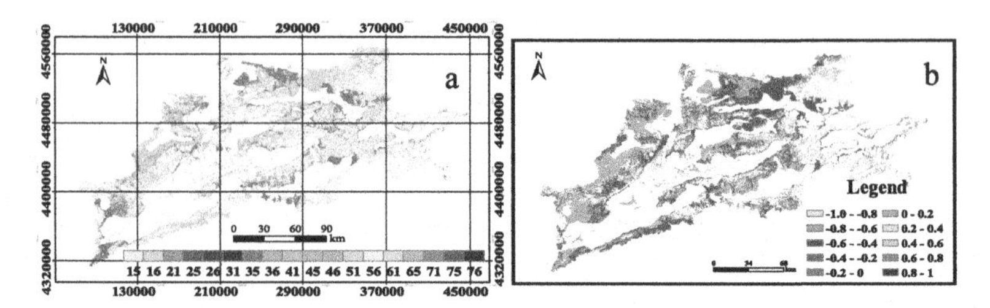

Fig. 3. Conversion of land use and cover types (a) and change in vegetation cover (b) from 1987 to 2000. The numerical codes that represent the land use changes are defined in Table 1.

farmland in 1987 was converted into grassland by 2000. Figure 3b shows the change in vegetation cover from 1987 to 2000. Compared with 1987, the vegetation cover increased in most areas in 2000, with the increase ranging between 0 and 40 % in the west and in the middle areas of the basin and between 40 and 60 % in parts of the east. It decreased in small areas throughout the basin, and particularly in the east.

Figure 4a shows the spatial distribution of the conversions among different land use and cover types from 2000 to 2005. The area in which conversions occurred was less than in the two previous periods. Most of the grassland in the north and some of the grassland in the middle of the basin were converted into farmland. Influenced by a government policy of converting farmland into forest, the grassland in some areas was converted into forest, but some forest was also converted into grassland. Figure 4b shows the change in vegetation cover from 2000 to 2005. The vegetation cover decreased in many parts of the basin, particularly in the northern half, with the decrease ranging between 20 and 40 % in parts of the north and between 40 and 60 % in parts of the northeast. In the middle of the basin, the east, and the south, vegetation cover increased by up to 40 %.

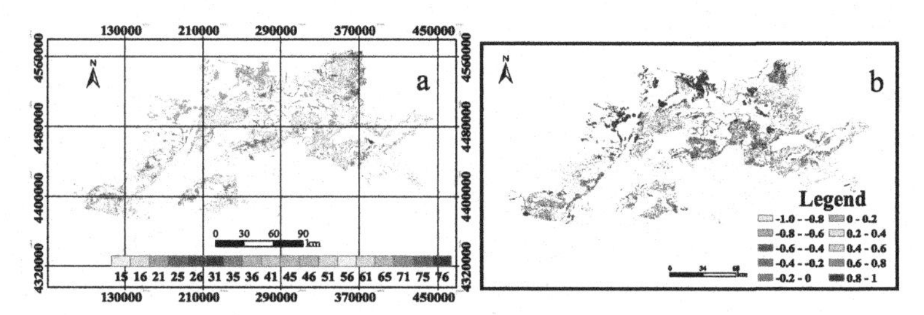

Fig. 4. Conversion of land use and cover types (a) and change in vegetation cover (b) from 2000 to 2005. The numerical codes that represent the land use changes are defined in Table 1.

(2) Influence of Land Use and Cover Change on Vegetation Cover in Scenario 2
Figs. 2, 3 and 4 show the spatial distribution of conversions among land use and cover types in the different periods. (These are the same in Scenario 2 as they were in Scenario 1 because these are based on land use and cover type data, not our model's predictions.) Figure 5 shows the change in vegetation cover from 1987 to 2000 and from 2000 to 2005. Vegetation cover mainly increased from 1987 to

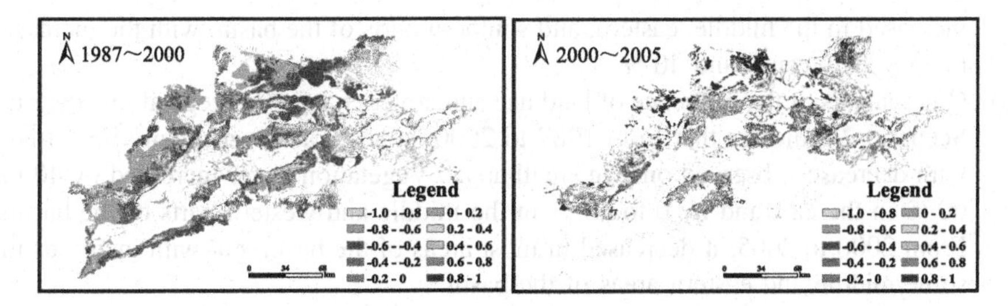

Fig. 5. Change in vegetation cover from 1987 to 2000 and from 2000 to 2005 with NDVI in all four years calculated using the temperature and precipitation data from 1978.

2000, but decreased in a few small areas. Vegetation cover increased by 40 to 60 % in parts of the east and by up to 40 % in parts of the middle of the basin and the west. From 2000 to 2005, vegetation cover decreased in most areas, with the decrease ranging from 40 to 60 % in the north and from 20 to 40 % in middle part of the basin and the south. However, vegetation cover increased by up to 20 % in some parts of the middle of the basin and the east.

6 Conclusion and Discussion

Based on a linear pixel decomposition model, we calculated the vegetation cover in the Yongding River Basin in 1978, 1987, 2000, and 2005 using Landsat MSS, ETM+, and TM images. We then developed a regression model with precipitation, temperature, and slope as the independent variables and used the model to reconstruct the NDVI values in two climate change scenarios: one in which the temperature and precipitation in 1978, 1987, and 2000 were used to predict the values in 1987, 2000, and 2005, respectively, and another in which the 1978 climate data were used to predict NDVI in all years. Under these scenarios, we investigated the influence of land use and cover change on vegetation cover. We reached the following conclusions:

(1) The vegetation cover in 2005 was verified using field-measured vegetation cover data, and revealed that our model had acceptable accuracy. We found a significant linear correlation between vegetation cover in the basin and the vegetation cover estimated using our model. Therefore, the linear decomposition model could be used to estimate vegetation cover in the basin.
(2) Despite these variations among the years, the vegetation cover generally tended to be high in the east and low in the west.
(3) This result demonstrates that the regression model is sufficiently reliable to be used to reconstruct NDVI values of vegetation in the basin.
(4) Our analysis of the influence of land use and cover changes on vegetation cover in Scenario 1 indicates that from 1978 to 1987, the vegetation cover decreased in most areas of the basin, with the magnitude of the decrease ranging between 20 and 60 %. Compared with the situation in 1987, the vegetation cover in most of the basin increased in 2000. From 2000 to 2005, it decreased in the north but

increased in the middle, eastern, and southern parts of the basin, with the increase ranging between 0 and 40 %.

(5) Our analysis of the influence of land use and cover changes on vegetation cover in Scenario 2 indicates that from 1987 to 2000, vegetation cover mainly increased, with decreases observed only in small areas. Vegetation cover increased by 40 to 60 % in the east and by 0 to 40 % in the middle and western parts of the basin. From 2000 to 2005, it decreased in most areas of the basin, but with increases in some middle and eastern areas of the basin.

Our reconstruction of NDVI under the two climate scenarios appears to be reliable, indicating the usefulness of this method for the study area. The linear regression model based on precipitation, temperature, and slope as independent variables was able to reconstruct NDVI in the basin. Although the overall precision was acceptable, some NDVI values revealed large estimation errors, and this introduced errors in the estimation of vegetation cover in the two climate scenarios. As a result, a better estimation method for NDVI must be developed in future research to increase the accuracy of the reconstruction and consequently improve the simulation accuracy.

Acknowledgements. This work was supported by the National Natural Science Foundation of China (No. 41471350), the National Key Basic Research Program of China (2014CB138803), the Funds for Creative Research Groups of China (41321001) and State Key Laboratory of Earth Surface Processes and Resource Ecology.

References

1. Nunes, C., Auge, J.I. (eds.): Land-Use and Land-Cover Change (LUCC): Implementation Strategy (1999)
2. Pan, Y.Z., Li, X.B., He, C.Y.: Research on comprehensive land cover classification in China: based on NOAA/AVHRR and Holdridge PE index. Quat. Sci. **20**(3), 270–281 (2000). (in Chinese)
3. Xiao, J.F., Moody, A.: A comparison of methods for estimating fractional green vegetation cover within a desert-to-upland transition zone in central New Mexico, USA. Remote Sens. Environ. **98**, 237–250 (2005)
4. Todd, M.S., John, D.A., Kelly, K.C., Chris, A.W.: Determining land surface fractional cover from NDVI and rainfall time series for a savanna ecosystem. Remote Sens. Environ. **82**, 376–388 (2002)
5. Gutman, G., Ignatov, A.: The derivation of the green vegetation fraction from NOAA/AVHRR data for use in numerical weather prediction models. Int. J. Remote Sens. **19**(8), 1533–1543 (1998)
6. Ma, J.H., Liu, D.D.: On the application of dimidiate pixel model to inversion of vegetation coverage in land-use investigation. Bull. Surveying Mapp. **4**, 13–16 (2006). (in Chinese)

A Novel Method to Downscale Daily Wind Statistics to Hourly Wind Data for Wind Erosion Modelling

Zhongling Guo[1(✉)], Chunping Chang[1], and Rende Wang[2]

[1] Hebei Key Laboratory of Environmental Change and Ecological Construction, College of Resource and Environmental Sciences, Hebei Normal University, Shijiazhuang 050024, People's Republic of China
changchunping@126.com,
gzldhr@mail.hebtu.edu.cn,
[2] Institute of Geographical Science, Hebei Academy of Sciences, Shijiazhuang 050011, People's Republic of China
wangrende10@163.com

Abstract. Wind is the principal driver in some wind erosion models. The hourly wind speed data were generally required for precisely wind erosion modeling. In this study, a novel method to generate hourly wind speed data from daily wind statistics (daily average and maximum wind speeds together or daily average wind speed only) was established. Two typical windy locations (Lubbock and Big Spring, Texas, USA) with measured hourly wind speed data were used to validate the downscaling method. The results showed that the overall agreement between observed and simulated cumulative wind speed probability distributions appears excellent, especially for the wind speeds greater than 5 m s^{-1} range (erosive wind speed). The results further revealed that the values of daily average erosive wind power density (AWPD) calculated from generated wind speeds fit the counterparts computed from measured wind speeds well with high models' efficiency (Nash-Sutcliffe coefficient).

Keywords: Downscaling method · Daily wind statistics · Hourly wind speed · Erosive wind power density · Wind erosion modeling

1 Introduction

Wind erosion play an important role in shaping the Earth's surface. The need to estimate soil erosion yields many wind erosion models [1, 2]. For example, wind data with high temporal resolution is fundamentally important in precisely wind erosion modeling. For example, the Wind Erosion Prediction System (WEPS) and Revised Wind Erosion Equation (RWEQ) generally require hourly wind series [3, 4]. Van Donk et al. [5] demonstrated that four-wind data per day (measured at LT 0200, 0800, 1600, 2000) are suitable for use in WEPS while Guo et al. [6] revealed that the same type of wind data can be used to evaluate wind erosion potential in RWEQ. However, hourly wind data or four-wind data per day are not always available for some locations. Meteorological observations with daily wind statistics may only be available for some sites [5, 7].

F. Bian and Y. Xie (Eds.): GRMSE 2015, CCIS 569, pp. 611–619, 2016.
DOI: 10.1007/978-3-662-49155-3_64

One naturally wonder whether one can directly use daily average wind data to estimate period wind erosion for WEPS and RWEQ. Namikas et al. [8] have shown that the intervals of wind data affect shear velocity estimates and possibly further influence predicting aeolian transport sediment flux. Larsén and Mann [9] have shown that the wind speed with long averaging time can remove substantial extreme gusts. Some of the gusts may be above threshold and crucial to evaluate wind erosivity. Guo et al. [10] have shown that the periods of wind speed can significantly under-estimate values of wind erosivity, and may further under-predict soil loss. These studies indicate that the type of wind data have significant impacts on wind erosion prediction. Thus, we cannot directly use wind data averaged over a day in WEPS and RWEQ.

One also might ask whether one can convert longer-period (such as a day) wind speed data to hourly wind speed data. For accuracy estimates of wind erosivity, the wind speeds greater than threshold (erosive wind speeds) are critical. In practice, many studies have been performed to make converting between different wind data types. A gust factor (or Durst Curve) is generally used to predict the extreme gust for a period [11, 12]. Dynamical and statistical methods are used to downscale monthly or daily wind data for global climate simulations [13, 14]. But these conversion method mentioned above are unable to evaluate some gusts, which may be above threshold and critical to wind erosion modeling. Several stochastic or deterministic wind generation methods have been developed for describing diurnal pattern of wind speed [4, 15–18]. Yet these simulation equations need historical wind statistics, which are obtained from hourly or more detail wind data.

Where only daily wind statistics (such as daily average wind data) are available, the challenge for application of the wind erosion models requiring hourly wind data, such as WEPS and RWEQ, to various sites still remains. The purpose of this study is to develop a method of downscaling the daily wind statistics to hourly wind series for wind erosion modeling.

2 Materials and Method

For some sites, only some daily wind statistics are saved. Here we discussed two general case: (1) the daily average, maximum wind speed are available, how to use the two daily wind statistics to predict the magnitude of hourly wind within a day. The diurnal variation for the magnitude of hourly wind speeds can be calculated by:

$$W_n = W_{ave} + \frac{1}{\pi} W_{\max} COS(\frac{n \cdot \pi}{12}) \tag{1}$$

where W_n is wind speed at hour of n, W_{ave} is the daily average wind speed, W_{max} is the daily maximum wind speed. 24 wind speeds were reproduced for a day using Eq. 1 (n varies from 0 to 23); (2) In some cases, only daily average wind speed data are available, the hourly wind speeds variation can be adjusted as:

$$W_n = W_{ave} + \frac{1}{2} W_{ave} COS(\frac{n \cdot \pi}{12}) \tag{2}$$

The other steps are as same as that mentioned above. Accordingly, we can also obtain 24 wind speeds from Eq. 2. Occasionally, the value of W_n generated by Eq. 1 or 2 may be less than zero, in this case, the W_n was assumed as 0 (calm).

The measured wind data were used to validate the downscaling methods. The hourly wind data are obtained from Lubbock and Big Spring, Texas, USA. The two sampling sites are located on the level plains of the Llano Estacado, a region known for its windy conditions and associated wind erosion problems [19]. For Lubbock, wind speed was measured at a height of 2 m using a propeller-type anemometer. The sampling period extended over a period of 9 years from January 1, 2001 to December 31, 2009. During this time, only 2 days of wind data were lost. For Big Spring, wind speed was measured at a height of 3 m. The sampling period is from January 1, 2001 to December 31, 2007, and 51 days of wind data were not used because some hourly wind data were not recorded in these days. The 3 m wind data of Big Spring were adjusted to a 2 m height using Elliot's method [20]. The daily wind statistics (daily average and maximum wind speeds) dataset are extracted from the hourly average wind speed data.

Compared with wind speed, the wind erosivity describing the potential of wind to generate sediment transport is more closely related to soil loss rate by wind [21]. In this paper, a widely used wind erosivity formula defined as the erosive wind power density (WPD) [3, 5, 22, 23] was chosen to further evaluate how well wind is generated by Eqs. 1 and 2. The WPD is calculated by [5]:

$$WPD = \frac{1}{2}\rho(U - U_t)U^2 \quad (3)$$

where WPD is erosive wind power density (W m^{-2}), U is wind speed (m s^{-1}) and U_t is threshold wind speed (m s^{-1}). When wind speed is below the threshold wind speed, WPD is zero and there is no sediment transport. The average WPD is computed based on wind statistics [5]. For each estimation period, the average WPD can be calculated by:

$$AWPD = \int_{U_t}^{U_{max}} P(U)\ WPD\ dU \quad (4)$$

where AWPD is the average WPD (W m^{-2}), Umax is the maximum wind speed (m s^{-1}), and P(U) is the wind speed probability for the estimation period. Here the threshold wind speed (U_t) and air density (ρ) were assumed as fixed values of 5 m s^{-1} and 1.293 kg m^{-3}, respectively.

The Nash-Sutcliffe coefficient [24] was used to access the two equations' efficiency for daily AWPD estimation.

$$NSC = 1 - \frac{\sum_{i=1}^{n}(O_i - P_i)^2}{\sum_{i=1}^{n}(O_i - O_m)^2} \quad (5)$$

where NSC is Nash-Sutcliffe coefficient, O_m is the mean of the observed values, O_i is the observed values and P_i is the predicted values.

3 Results and Discussion

The hourly wind speed data of Lubbock and Big Spring were generated using Eqs. 1 and 2, respectively. For Lubbock, the average or median for the measured and generated hourly wind speed data change slightly while the maximum varies from 13.95 to 14.44 m s^{-1} (Table 1). For Big Spring, the average or median of the daily AWPD shows the similar trends as that of Lubbock, but the maximum ranges from 11.71 to 13.02 m s^{-1} (Table 1). The measured cumulative distribution of hourly wind speeds for Lubbock and Big Spring were compared with the distribution generated by Eqs. 1 and 2 (Fig. 1), respectively. Although the cumulative distribution of Lubbock (Fig. 1a) is different from that of Big Spring (Fig. 1b), the overall agreement between measured

Table 1. Statistical characteristic of the observed and predicted hourly wind speeds for Lubbock and Big Spring.

Site	Statistical characteristic	Wind Speed (ms^{-1})		
		Observed	Predicted by Equation 1	Predicted by Equation 2
Lubbock	Average	3.36	3.36	3.36
	Median	3.17	3.04	2.97
	Maximum	14.44	13.37	13.95
	Minimum	0.00	0.00	0.10
Big Spring	Average	3.15	3.15	3.15
	Median	2.99	2.90	2.82
	Maximum	13.02	11.95	11.71
	Minimum	0.00	0.00	0.03

and simulated cumulative wind speed probability distribution appears satisfactory, with a slight under-prediction in less than 2 m s^{-1} wind speed and a slight over-prediction in 2 to 4.5 m s^{-1} wind speed range by the generation equations for the two sites (Fig. 1). Figure 1 further illustrates that the performance of Eq. 1 is little better than that of Eq. 2 for reproducing hourly wind speeds. The threshold wind speed varies with surface conditions, the threshold is generally assumed as 5 m s^{-1} at the height of 2 m with a erodible field surface condition [4]. For the wind speeds greater than 5 m s^{-1} (erosive wind speed), the agreement appears excellent (Fig. 1). The results indicated that the two downscaling models are able to generate hourly wind speeds with high accuracy.

The core module of the wind generators in WEPS and RWEQ is the cumulative wind speed probability distribution (Fig. 1) that is obtained from measured hourly wind speeds [4, 5]. Wind speed data can be generated from the cumulative distribution curve using linear interpolation in WEPS [5], while the important coefficients (c and k for Weibull distribution) for the wind generator in RWEQ are derived from the cumulative distribution curve [4]. The Eqs. 1 and 2 are capable of reproducing hourly wind speeds magnitude, which further produces the cumulative wind speed probability distribution curve with satisfactory performance (Fig. 1), implying that the cumulative distribution curve for generated hourly wind data by Eq. 1 or 2 (Fig. 1) can also be used in the wind generators for simulating hourly wind speed data when only daily wind statistics data are available for a site.

Daily AWPD values for measured and generated wind speed data were computed using Eq. 4. For Lubbock, a total of 1489 days during the 3285 day sampling period had non-zero values of daily AWPD calculated from measured wind data. For Big Spring, the daily AWPD values of 1234 days in the 2555 day sampling period are more than zero. Thus these non-zero values days were used to evaluate the performance of the Eqs. 1 and 2, respectively (Table 2). Comparisons between observed and predicted daily AWPD values are presented in Figs. 2 and 3. For Lubbock, the NSC are 0.92 and 0.81 for Eqs. 1 and 2, respectively (Fig. 2). For Big Spring, the NSC are 0.88 and 0.74 for Eqs. 1 and 2, respectively (Fig. 3). The predicted daily AWPD fit observed counterparts very well (Figs. 2, 3). Good agreement is shown in Fig. 3 due to the good performance of the downscaling methods for generating hourly wind speeds (Fig. 1). For the two downscaling methods, the Eq. 1 requires more inputs (daily average and maximum wind speed), therefore, its performance is better than that of the Eq. 2, which only needs daily average wind speed (Figs. 2, 3). Overall, the performance of the downscaling methods are also satisfactory for the wind erosivity (AWPD) estimation. The downscaling methods may span the gap between daily wind statistics and hourly

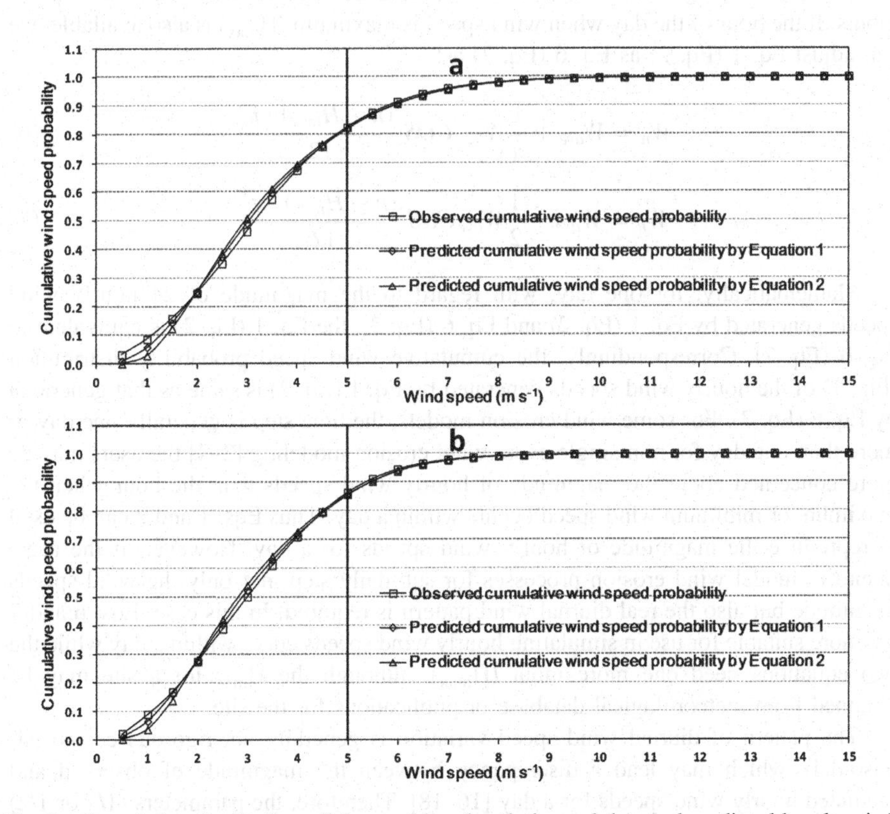

Fig. 1. Cumulative distribution of observed hourly wind speed data and predicted hourly wind speed data using Eqs. 1 and 2 for (a) Lubbock, and (b) Big Spring.

Table 2. Statistical characteristic of the daily AWPD values computed from observed and predicted hourly wind speeds for Lubbock and Big Spring.

Site	Statistical characteristic	AWPD (average erosive wind power density, W m^{-2})		
		Observed	Predicted by Equation 1	Equation 2
Lubbock	Average	16.46	16.78	16.96
	Median	5.47	5.78	5.38
	Maximum	355.44	363.02	400.91
	Minimum	0	0	0
Big Spring	Average	9.81	9.89	10.23
	Median	3.04	3.21	3.22
	Maximum	241.33	215.35	203.51
	Minimum	0	0	0

wind data. With these methods, we can estimate the wind erosivity from some rough wind data, such as daily wind dataset.

The downscaling methods for daily wind statistics can be flexible for the type of inputs. If the hour of the day when wind speed is maximum (H_{max}) is also available, we can adjust Eq. 1 (Eq. 2) as Eq. 6 (Eq. 7) as:

$$W_n = W_{ave} + \frac{1}{\pi} W_{\max} COS(\frac{(n - H_{\max}) \cdot \pi}{12}) \quad (6)$$

$$W_n = W_{ave} + \frac{1}{2} W_{ave} COS(\frac{(n - H_{\max}) \cdot \pi}{12}) \quad (7)$$

Mathematically, for one day, with regard to the magnitude of 24 hourly wind speeds generated by Eq. 1 (Eq. 2) and Eq. 6 (Eq. 7), the Eq. 1 (Eq. 2) is equivalent to Eq. 6 (Eq. 7). Correspondingly, the cumulative wind speed probability distribution (Fig. 1) of the hourly wind speeds generated by Eq. 1 (Eq. 2) is same as that generated by Eq. 6 (Eq. 7). For some wind erosion models, the time step is generally one day or more than one day for non-single event wind erosion modeling [2–4] the users may be more concerned about the magnitude of hourly wind speeds than the hour when the maximum or minimum wind speed occurs within a day. Thus Eqs. 1 and 2 can be used to reproduce the magnitude of hourly wind speeds for a day. However, if the users intend to model wind erosion processes for sub-daily step, not only the wind speeds magnitude but also the real diurnal wind pattern is required. In this case, Eqs. 6 and 7 are more suitable for use in simulating hourly wind speeds curve within a day while the two equations need one more input (H_{max}), although the H_{max} for a site may be obtained from meteorological database or publications for the site.

The pattern of diurnal wind speed variation is generally not rigorous cosine (sinusoidal), which may lead a discrepancy between the magnitude of observed and predicted hourly wind speeds for a day [16–18]. Therefore, the parameters ($1/\pi$ or $1/2$) of the second term in the Eq. 1 (Eq. 6) and Eq. 2 (Eq. 7) may have different values for

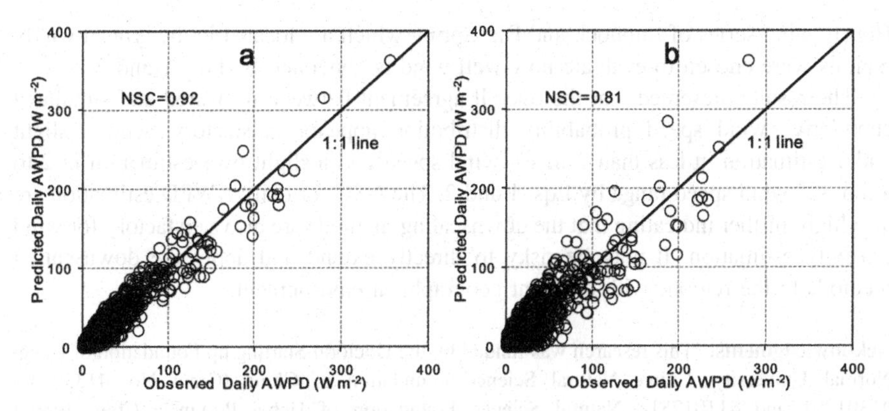

Fig. 2. Observed daily AWPD values calculated from measured hourly wind speeds for Lubbock VS (a) predicted daily AWPD values calculated from the generated hourly wind speeds by Eq. 1, and (b) predicted daily AWPD values calculated from the generated hourly wind speeds by Eq. 2. Here the daily wind statistics used in the Eqs. 1 and 2 are extracted from measured hourly wind data of Lubbock. AWPD is the average erosive wind power density, NSC is the Nash-Sutcliffe coefficient.

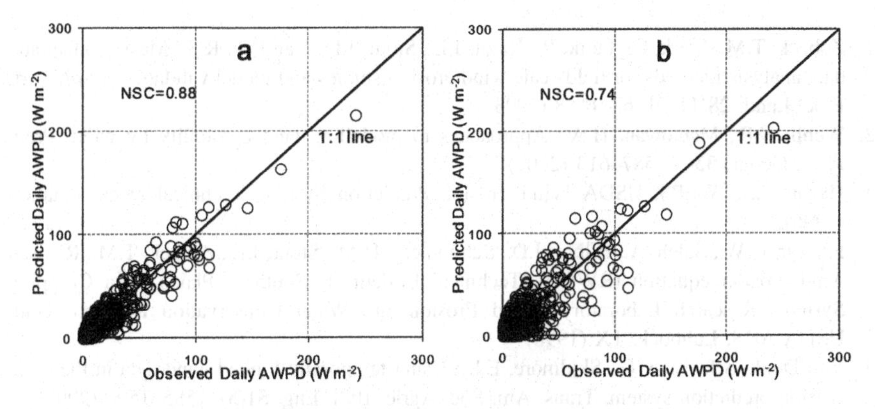

Fig. 3. Same as Fig. 2, but for Big Spring.

different geographical regions. Accordingly, it may be risky to use the four formulas to downscale daily wind data for other sites with different wind fluctuations. The results allow researchers to explore new equations based on the formulations of the Eq. 1 (Eq. 6) and Eq. 2 (Eq. 7) for downscaling daily wind data. Further studies are needed to validate the downscaling methods using other sites wind data.

4 Conclusion

In this study, we established a simple method to reproduce hourly wind speed data from daily average and maximum wind speeds together or daily wind speed data only.

Hourly wind series of Lubbock and Big Spring which are located in the typical windy regions were chosen to evaluate how well wind is generated by Eqs. 1 and 2.

The results presented that the overall agreement between measured and simulated cumulative wind speed probability distribution appears satisfactory, with a slight under-estimation in less than 2 m s^{-1} wind speed and a slight over-estimation in 2 to 4.5 m s^{-1} wind speed range by Eqs. 1 and 2. The NSC for daily AWPD estimation are also high, further indicating that the downscaling methods are also satisfactory for wind erosivity estimation. It may be risky to directly extend and apply the downscaling methods to the regions with different geographical environment.

Acknowledgments. This research was funded by the Doctoral Starting up Foundation of Hebei Normal University, China, Natural Science Foundation of China (Grant No. 41330746, 41301291 and 41101251), Natural Science Foundation of Hebei Province, China (Grant No. D2014205063 and D2013302034), and the key project on soft sciences of Hebei province, China (Grant No.13454213D). The authors would like to thank J.E. Stout and T.M. Zobeck for supplying the field data used in this paper.

References

1. Zobeck, T.M., Sterk, G., Funk, R., Rajot, J.L., Stout, J.E., Van Pelt, R.S.: Measurement and data analysis methods for field-scale wind erosion studies and model validation. Earth Surf. Proc. Land. **28**(11), 1163–1188 (2003)
2. Webb, N.P., McGowan, H.A.: Approaches to modelling land erodibility by wind. Prog. Phys. Geogr. **33**(5), 587–613 (2009)
3. Hagen, L.J.: WEPS, USDA Wind Erosion Prediction System. Technical Documentation (1996)
4. Fryrear, D.W., Saleh, A., Bilbro, J.D., Schomberg, H.M., Stout, J.E., Zobeck, T.M.: Revised wind erosion equation (RWEQ). Technical Bulletin 1. Southern Plains Area Cropping Systems Research Laboratory, Wind Erosion and Water Conservation Research Unit, USDA-ARS, Lubbock, TX (1998)
5. Van Donk, S.J., Liao, C., Skidmore, E.L.: Using temporally limited wind data in the wind erosion prediction system. Trans. Am. Soc. Agric. Biol. Eng. **51**(5), 1585–1590 (2008)
6. Guo, Z., Zobeck, T.M., Zhang, K., Li, F.: Estimating potential wind erosion of agricultural lands in northern China using the revised wind erosion equation (RWEQ) and geographic information systems. J. Soil Water Conserv. **68**(1), 13–21 (2013)
7. Liu, B., Qu, J., Wagner, L.E.: Building Chinese wind data for wind erosion prediction system using surrogate US data. J. Soil Water Conserv. **68**(4), 104A–107A (2013)
8. Namikas, S.L., Bauer, B.O., Sherman, D.J.: Influence of averaging interval on shear velocity estimates for aeolian transport modelling. Geomorphology **53**, 235–246 (2003)
9. Larsén, X.G., Mann, J.: The effects of disjunct sampling and averaging time on maximum mean wind speeds. J. Wind Eng. Ind. Aerodyn. **94**, 581–602 (2006)
10. Guo, Z., Zobeck, T.M., Stout, J.E., Zhang, K.: The effect of wind averaging time on wind erosivity estimation. Earth Surf. Proc. Land. **37**(7), 797–802 (2012)
11. Durst, C.S.: Wind speeds over short periods of time. Meteorol. Mag. **89**, 181–186 (1960)
12. Deacon, E.L.: Wind gust speed: averaging time relationship. Aust. Meteorol. Mag. **52**, 11–14 (1965)

13. Sailor, D.J., Smith, M., Hart, M.: Climate change implications for wind power resources in the Northwest United States. Renewable Energy **33**(11), 2393–2406 (2008)
14. Pavlik, D., Söhl, D., Pluntke, T., Mykhnovych, A., Bernhofer, C.: Dynamic downscaling of global climate projections for Eastern Europe with a horizontal resolution of 7 km. Environ. Earth Sci. **65**(5), 1475–1482 (2012)
15. Peterson, T.C., Parton, W.J.: Diurnal variations of wind speeds at the short-grass prairie site–a model. Agric. Meteorol. **28**, 365–374 (1983)
16. Skidmore, E.L., Tatarko, J.: Stochastic wind simulation for erosion modeling. Trans. Am. Soc. Agric. Eng. **33**(6), 1893–1899 (1990)
17. Ephrath, J.E., Goudriaan, J., Marani, A.: Modelling diurnal patterns of air temperature, radiation, wind speed, and relative humidity by equations from daily characteristics. Agric. Syst. **51**(4), 377–393 (1996)
18. Donatelli, M., Bellocchi, G., Habyarimana, E., Confalonieri, R., Micale, F.: An extensible model library for generating wind speed data. Comput. Electron. Agric. **69**, 165–170 (2009)
19. Stout, J.E.: Diurnal patterns of blowing sand. Earth Surf. Proc. Land. **35**, 314–318 (2010)
20. Elliot, D.L.: Adjustment and analysis of data for regional wind energy assessments. Paper presented at the Workshop on Wind Climate, Asheville, NC, 12–13 November (1979)
21. Shao, Y.P.: Physics and Modelling of Wind Erosion. Kluwer Academic Publishers, Dordrecht (2008)
22. Lettau, H.H., Lettau, K. (eds) Experimental and micrometeorological studies of dune migration. In: Exploring the World's Driest Climate. Institute of Environmental Studies, University of Wisconsin, Madison, pp. 110–147 (1978)
23. Greeley, R., Iversen, J.D.: Wind as a Geological Process: On Earth, Mars, Venus and Titan. Cambridge University, New York (1985)
24. Nash, J.E., Sutcliffe, J.V.: River flow forecasting through conceptual models part I-A discussion of principles. J. Hydrol. **10**, 282–290 (1970)

The Identification to the Palm Color Spots Based on Improved HSV Model

Bing Kang, Fu Liu(✉), and Shoukun Jiang

College of Communications Engineering,
Jilin University, Changchun 130000, China
{kangbing, liufu}@jlu.edu.cn

Abstract. Palm color [1] is of important significance for the clinical auxiliary diagnosis [2]. The extraction [3] of palm color feature used HSV colormodel [4]. The interval of three color components are extracted and enlarged after experimental analysis. To reduce the processing complexity, each component has been quantized according to color density. Extract the color spots used watershed segmentation algorithm based on control tags and the spots have been divided into several categories. It provides high performance with recognition rate, and reduces the running time, which can identify the palm color spots quickly.

Keywords: HSV color model · Image processing · Feature extraction [5]

1 Introduction

Palm color reflects people's health [6]. Healthy palm color is uniform light red; diseased palms will appear a variety of significant abnormal color spots, the cancer patient's palm will appear brown, white or dark red spots [7]. Identification of color spots on palms can help people discover disease as early as possible. Therefore, using it to provide reasonable health information timely do have a sense of social applications [8]. So far, there are few studies on palm color spots recognition. In 2003, Professor Wang conducted the first attempt to classify the cancer and non-cancer palm images [9], but this algorithm is of high computational complexity and large amount of computation, not easy to achieve detection quickly. This paper does researches on the features of color spots, using its distribution characteristics in the HSV color space, and proposes a monitoring method of H-S-Gray features color spots recognition.

2 Distribution of Palm Color Spots [10]

For identifying the color spots quickly and accurately, it is very important to select the appropriate color model [11]. In order to make the process of color image processing similar to expert medical consultation, it is HSV color model appropriate human visual perception when dealing with palmprint images.

Color parameters in the HSV color model [12] are Hue(H), Saturation(S) and Value (V). Palm color distributes within a very small range in RGB color space [13], therefore

F. Bian and Y. Xie (Eds.): GRMSE 2015, CCIS 569, pp. 620–628, 2016.
DOI: 10.1007/978-3-662-49155-3_65

it isn't suitable for palmprint extraction. HSV color model is mainly user-oriented design, which is suitable for color perception of people. Each component in HSV space is independent, and is suitable for computer processing separately. Hence the method using HSV color model can reduce the complexity and accelerate the speed of image processing.

HSV model is a three-dimensional model, evolved from RGB color model. When k1 = max{r, g, b}, k2 = min{r, g, b} and r, g and b are greater than 0 but less than 1, the conversion from RGB color model to HSV color model can be accomplished using the formulas as follows:

$$v = k_1 \tag{1}$$

$$h = \begin{cases} 0^\circ & if \quad k_1 = k_2 \\ 60^\circ \times (g-b)/(k_1-k_2) + 0^\circ & if \quad k_1 = r \quad and \quad g \geq b \\ 60^\circ \times (g-b)/(k_1-k_2) + 360^\circ & if \quad k_1 = r \quad and \quad g < b \\ 60^\circ \times (b-r)/(k_1-k_2) + 120^\circ & if \quad k_1 = g \\ 60^\circ \times (r-g)/(k_1-k_2) + 240^\circ & if \quad k_1 = r \end{cases} \tag{2}$$

$$s = \begin{cases} 0 & if(k_1 = 0) \\ (k_1-k_2)/k_1 & if(k_1 \neq 0) \end{cases} \tag{3}$$

H is the main characteristic component to detect the color spots on palm, because H is less susceptible to the influence of light intensity; S represents color saturation, and its value is from 0 to 1; V represents color brightness, and its value is from 0 to 1. Collect a lot of palmprint images [14], remove the background and convert them to the HSV color space, then make statistical histogram of three components (H, S, and V) as shown in Fig. 1.

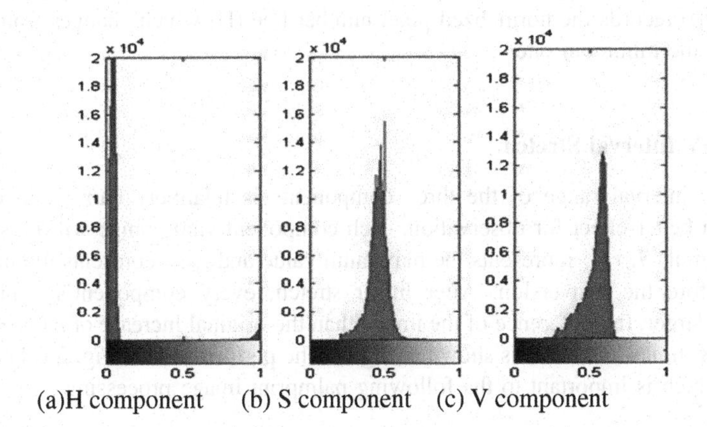

Fig. 1. H-S-V color component histogram

(1) The value of S, V is within consecutive intervals respectively: value distribution of S is in the continuous interval [0.2, 0.7]; value distribution of V is in the continuous interval [0.3, 0.72].
(2) The value of H distributes in the two discrete intervals: [0, 0.2], [0.8, 1].

3 H-S-Gray Features of Palm Color Spots

3.1 H Interval Translation and Combination

In order to combine both intermittent intervals into one continuous interval, use formula to translate H value and it is between [0.2, 0.6] after translation.

$$s(x,y) = T(r) = \begin{cases} r(x,y) + 0.4 & 0 \leq r(x,y) \leq 0.2 \\ r(x,y) - 0.6 & 0.8 \leq r(x,y) \leq 1.0 \end{cases} \quad (4)$$

s(*x*,*y*) represents the brightness after conversion, *r*(*x*,*y*) represents the brightness before conversion.

For the best effect, the image's histogram needs to be translated. Translation rule is to obtain its minimum value according to the histogram superimposed using the formulas as follows:

$$M = \min\left(\sum_{i=m}^{m+1} count(i)\right) \quad 0 \leq m \leq 255 \quad (5)$$

$$i = \begin{cases} i & if \quad 0 \leq i \leq 255 \\ i - 256 & if \quad i \geq 256 \end{cases} \quad (6)$$

count(*i*) records the normalized pixel number hue (H) which changes from 0 to 1, and *M* is the minimum one.

3.2 HSV Interval Stretch

Since the interval range of the three components is relatively narrow, in order to achieve a better effect for observation, each component value range will be stretched using formula 7. r_{max} represents the maximum value and r_{min} represents the minimum value before the conversion. After linear stretch, every component's value range becomes larger, the difference of the image than the original increase of k (as shown in formula 8) times, the effect is shown in Fig. 1. The performance in Fig. 2 is better than Fig. 1, which is important to the following palmprint image processing.

$$s(x,y) = T(r) = r(x,y) \times \frac{1}{r_{max} - r_{min}} \quad (7)$$

$$k = 1/(r_{max} - r_{min}) \quad (8)$$

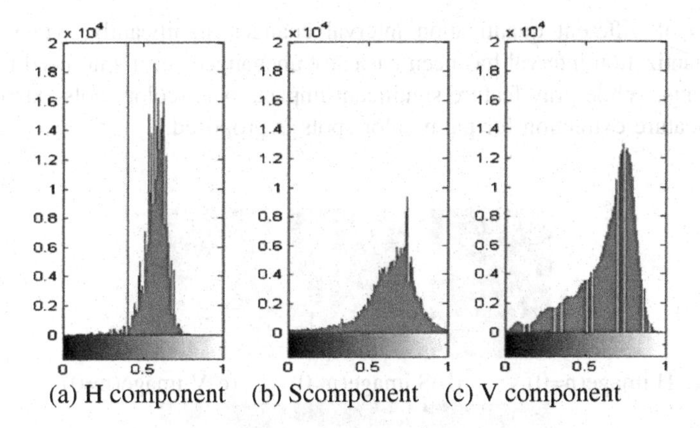

(a) H component (b) Scomponent (c) V component

Fig. 2. Stretched H, S, V color component histogram

3.3 HSV Interval Quantization

The colors that human eyes can distinguish are much less than colors in HSV color model, so it is necessary to quantize [15] the HSV color model to simplify the number of colors to increase the operating speed and efficiency. To minimize color distortion and not change the original color of the contrast, this paper has taken the isometric quantization interval. Here the length and number of quantization interval are *k* and n, the quantized value of each component can be obtained by the formulas 9 and 10.

$$k = 1/n \tag{9}$$

$$s(x,y) = k \times [r(x,y)/k] \tag{10}$$

s(*x*,*y*) stands for the new value and *r*(*x*,*y*) stands for the original value, the operator [] represents the integer arithmetic. Through experimental data analysis, when n is 20, it takes good quantization effect. Figure 4 is n = 20, 10 and 0 for each component (Fig. 3).

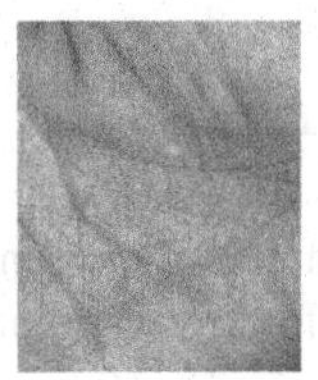

Fig. 3. Palm image

Figure 4(a–c) are the H, S and V component original images, (d), (e) and (f) are the corresponding component images after quantization (n = 20), (g), (h) and (i) are the corresponding component images after quantization (n = 10). When n is too large, the color of part region will be distorted, it is not conducive to identify the color spots on palms. Analysis of Fig. 4, in HSV color space, palm spot characteristic in H and S components of different quantization interval changes significantly, while in V of different quantization interval has been basically unchanged, so it's no need to use the V components. While gray feature significant impacts palm color spots extraction, so H-S-Gray feature extraction for palm color spots is proposed.

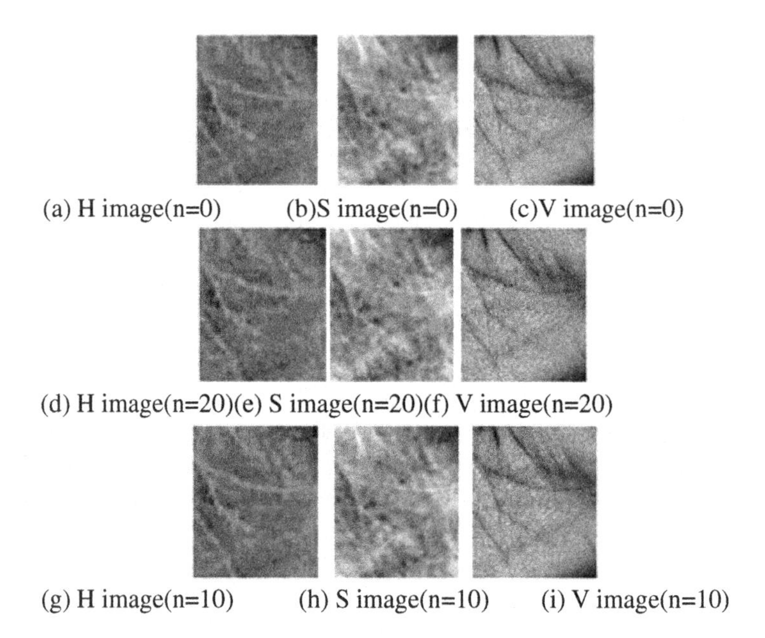

Fig. 4. Different quantization interval images of H, S and V

3.4 Watershed Segmentation Algorithm

Assuming the range of image gradient value is $D = Z^2$, gradient value I is $[0, 1, \cdots, N] \quad 0 \leq N \leq 255$, definition 1 as follows:

$$I = \begin{cases} D \subset Z^2 \to \{0, 1, \cdots, N\} \\ p \to I(p) \end{cases} \tag{11}$$

Definition 1: In communication path between q and p, there are $l + 1$ points $(p_0, p_1, \cdots p_{l-1}, p_l), p_0 = p, p_l = q$, for every point, $\forall i \in [1, l], (p_{i-1}, p_i) \in G$, G is a communication path.

Definition 2: Polecell *(M)* of height (*h*) of image (*I*) is a communication region constituted by one or a plurality of gradient values of the pixels. Every point in *M* make a communication path(*L*) with a pixel (gradient value less than *h*), so there must be one point whose gradient value higher than *h*. The formula is shown as follows:

$$\begin{aligned} &\forall p \in L, \forall q \notin L, I(q) < I(p) \\ &\forall L = (p_0, p_1, \cdots p_{l-1}, p_l), p_0 = p, p_l = q \\ &\exists i \in [1, l], I(p_i) > I(p_0) \end{aligned} \tag{12}$$

In the formula, p and q are two points, I is gradient value, and L is communication path.

Definition 3: Geodesic distance: Point a and b are in communication path (A), the communication path belongs to A completely and is the shortest path.

$$d_A(a, b) = \inf\{l(P)\} \tag{13}$$

P is a communication path between a and b and belongs to A. In the formula, d stands for distance.

Definition 4: Geodesic influence zone: Assuming A contains B, B is composed of $B_1, B_2, \cdots, B_k$, $B_1, B_2, \cdots, B_k$ are disconnected region with each other. Define geodesic influence zone of B is $iz_A(B_i)$.

$$iz_A(B_i) = \{p \in A, \forall j[1, k] \; and \quad j \neq i, d_A(p, B_i) < d_A(p, B_j)\} \tag{14}$$

If we use the watershed segmentation algorithm for image segmentation directly, it is usual to produce more serious over-segmentation phenomenon, in order to control the over-segmentation, we use a watershed segmentation algorithm based on control tags to segment spots [16]. In Fig. 5(a) is a color palm image, (b) is the extracted spots

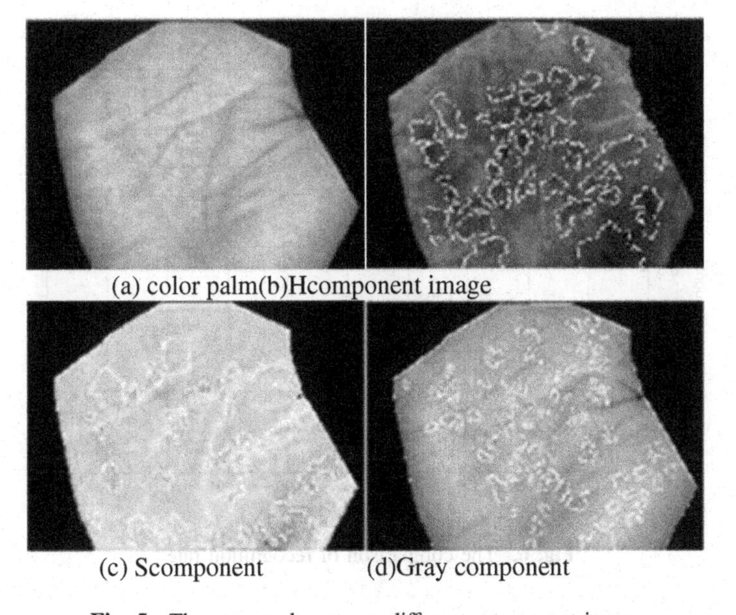

Fig. 5. The extracted spots on different component image

Table 1. Spots color data

n	h	s	gray	n	h	s	gray
1	0.2772	0.3976	102.8396	23	0.2726	0.0685	151.9878
2	0.0011	0.1376	142.7692	24	0.6192	0.3369	121.2536
3	0.3353	0.2969	114.4231	25	0.6140	0.7077	93.6108
4	0.6469	0.8672	112.6296	26	0.6195	0.7186	120.0814
5	0.6937	0.8812	110.0212	27	0.6445	0.7026	130.3751
6	0.6368	0.9965	99.35000	28	0.7384	0.4953	126.7238
7	0.4052	0.6481	118.2409	29	0.7451	0.5437	153.1752
8	0.4311	0.5233	112.0158	30	0.6538	1.0000	108.5522
9	0.4103	0.2689	132.0925	31	0.6752	0.9989	98.4457
10	0.3674	0.3663	114.2545	32	0.6616	0.9945	93.5818
11	0.6068	0.6180	107.7848	33	0.7028	0.8335	102.4404
12	0.6479	0.9485	110.0932	34	0.6344	0.8528	97.0595
13	0.6859	0.9047	109.4058	35	0.7385	0.7318	117.3972
14	0.6268	0.3997	210.1595	36	0.5729	0.8831	80.66090
15	0.5708	0.6134	179.7918	37	0.7090	0.7887	98.35600
16	0.6106	0.4646	194.8241	38	0.6322	0.7193	104.6517
17	0.5724	0.6148	180.3913	39	0.5980	0.7756	101.1048
18	0.5621	0.6031	165.2969	40	0.6268	0.8179	102.4740
19	0.5503	0.5357	168.9005	41	0.5529	0.7604	83.3614
20	0.6432	0.3927	150.9147	42	0.5966	0.7039	115.7219
21	0.5251	0.6194	119.5629	43	0.4467	0.5307	123.1703
22	0.4636	0.5168	118.3917	44	0.5263	0.5401	116.1521

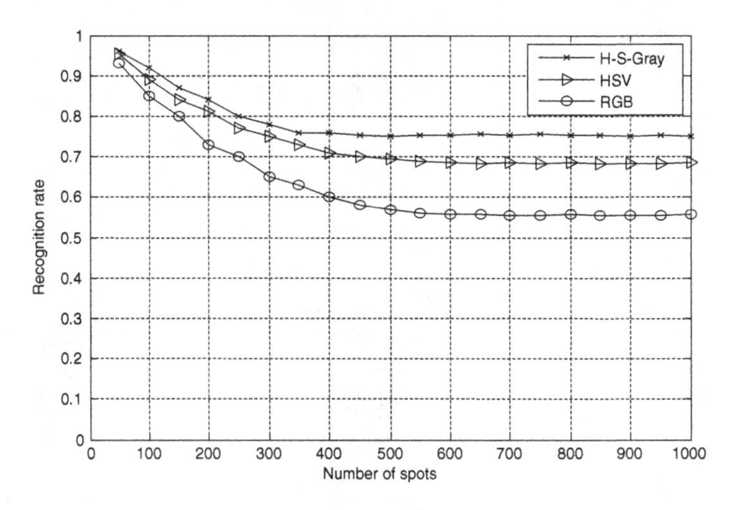

Fig. 6. The comparison of recognition rate

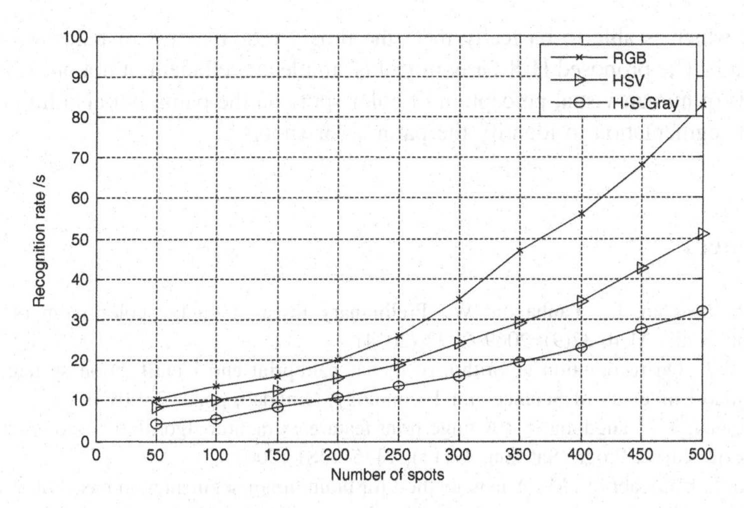

Fig. 7. The comparison of running time

on the H component, (c) is the extracted spots on the S component and (d) is the extracted spots on the Gray component.

4 Conclusions

In this paper, the pixel area of the spots between 20 and 90 are selected to be analyzed. Respectively, use the RGB color model, HSV color model and the H-S-Gray model proposed in this paper to extract each component image, use the watershed segmentation algorithm based on control tags to identify the palm color spots from the various components of the images, Table 1 is part of spot color data.

From the Fig. 6, the final recognition rate of RGB color model, HSV color model and H-S-Gray is 53 %, 68 % and 75 %. Obviously, H-S-Gray's performance is better than the first two. And with the increasing number the performance is better.

In Fig. 7, from up to down they are followed by RGB model, HSV model and H-S-Gray model. The running time of RGB model is 10.25 s, that of HSV model is 8.33 s, and the H-S-Gray is 4.25 s when the number of spots is 50. Contrasted to RGB model and HSV model, the running time of H-S-Gray is 41.5 % and 51 % of RGB and HSV. When the number becomes larger, the performance become more obvious.

This paper mainly studies the fast identification of palm color spots. Because palmprint color is rich, and spots species is relatively concentrated, this paper has stretched HSV color space interval to expand the color range of the palm spots, and according to color characteristics the of abnormal spots, quantifies the color interval and improves the extraction efficiency of the color components; uses watershed segmentation algorithm based on control tags to segment H, S and Gray quantified component image respectively, and it turns to be efficient to identify spots; compares the H-S-Gray model algorithm with the RGB and HSV color model algorithm, it is obvious that recognition rate significantly improves and the running time significantly reduces, which is able to basically meet the timely requirements of palm color spots recognition. The proposed H-S-Gray model is simple and efficient, it not only can meet the needs of human visual perception of color spots on the palm, but also improve the speed of segmentation to identify the palm color spots.

References

1. Zhongshenglan, L., Yanhaixia, W.: Preliminary research on the color palm of healthy people. Chin. Med. **32**(9), 2069–2072 (2014)
2. Bai, Q.J.: On recognition algorithm of thenar palmprint and identification system design. Qingdao University of Science and Technology, Qingdao, pp. 1–3 (2013)
3. Huaqiang, Y., Yangdong, Y.: A fingerprint feature extraction algorithm based on curvature of bezier curve. Progr. Nat. Sci. **17**(11), 1376–1381 (2007)
4. Maiti, I., Chakraborty, M.: A new method for brain tumor segmentation based on watershed and edge detection algorithms in HSV colour model. In: 2012 National Conference on Computing and Communication Systems, NCCCS 2012, pp. 192–196 (2012)
5. Lü, L.-T., Zhou, X.-J., Yang, Y.X.: Feature extraction algorithm for palm bioimpedance spectroscopy based on wavelet transform. In: 2nd International Conference on Intelligent Materials, Applied Mechanics and Design Science, IMAMD 2013, pp. 1576–1579 (2013)
6. Wang, C.X.: Palmprint diagnosis treatment. Northern Arts Publishing Press, Guangzhou (2007)
7. Zhang, D.: Automated Biometrics technologies and Systems, pp. 111–134. Kluwer Academic Publishers, Dordrecht (2000)
8. Xuning, Z.: Palmprint diagnosis expert system based on neural network. Appl. Res. Comput. **18**(2), 4–6 (2001)
9. Wang, K.-Q., Liu, L., Wu, X.-Q., Zhang, D.-P.: Multi-central dynamic clustering algorithm classifying colour palm images for cancer diagnosis. Comput. Sci. **30**(3), 24 (2003)
10. Hairuddin, M.A., Md, T.N., Baki, S.R.S.: Overview of image processing approach for nutrient deficiencies detection in Elaeis Guineensis. In: International Conference on System Engineering and Technology, ICSET 2011, pp. 116–120 (2011)
11. Zhang, X.: Common color space and its conversions in color image project. Comput. Eng. Des. **29**(5), 1210–1212 (2008)
12. Gonzalez, R.C., Woods, R.E.: Digital Image Processing, 3rd edn., pp. 61–88. PixelSoft Inc, Los Altos (2001)
13. Cui, C., Zhu, M.: Real-time human face detection and tracking based on HSV models pace of skin color. J. Fu Zhou Univ. (Nat. Sci.) **34**(6), 11 (2006)
14. Wu, X.-Q.: Cancer diagnosis based on the palm colour features. Computer Assembly in the Application and Research (2005)
15. Yu, Y.C.: ISC with color quantization based on HEVC. Popular Sci. Technol. **14**(11), 11–13 (2014)
16. Hieu, T., Worring, M.: Watersnakes: energy-driven watershed segmentation. IEEE Trans. Pattern Anal. Mach. Intell. **25**, 330–342 (2003)

Coupling Model of Land Subsidence Considering Both Effects of Building Load and Groundwater Exploitation

Bin Liu(✉), Jianping Yue, Jing Li, and Shun Yue

Hohai University School of Earth Science and Engineering, Nanjing 210098, China
957761759@qq.com

Abstract. Among the existing study of land subsidence, there is almost no consideration about the effects of building load and groundwater exploitation at the same time. Based on the consideration of these issues, this study designs the physical model test to simulate the subsidence effected by the building load and analysis the tendency of subsidence from the soil mechanics. Lastly, taking the typical geological structure of floodplain as an example, a coupling model is established considering both effects of building load and groundwater exploitation, its high accuracy of fitting and prediction shows that the coupling model can be used to simulate the land subsidence.

Keywords: Groundwater exploitation · Building load · Land subsidence

1 Introduction

Land subsidence is a geological phenomenon characterized by a decrease in surface elevation due to the compression of loose surface soil. It is a major geologic hazard for urban modernization. The triggering factors of land subsidence include natural and human factors. Excessive extraction of groundwater and concentrated building load were found to be the main causes of land subsidence [1, 2].

Existing applied research using numerical simulation of subsidence has considered groundwater extraction or loads on the surface to be one cause of land subsidence, but few studies have explored the influences of both groundwater extraction and building load on land subsidence [3, 4]. In this context, this research was conducted to investigate the land subsidence caused by both groundwater extraction and concentrated building load [8–10]. Furthermore, a coupling model of a typical floodplain structure under the influences of groundwater extraction and concentrated building load was designed to simulate and analyze the characteristics of land subsidence under different conditions. The results would serve as a scientific basis for effective control of land subsidence and sustainable development of cities.

2 Land Subsidence Induced by Concentrated Building Load

By conducting a subsidence simulation experiment using the physical model, the horizontal influence area of the concentrated building load on the land subsidence was

F. Bian and Y. Xie (Eds.): GRMSE 2015, CCIS 569, pp. 629–637, 2016.
DOI: 10.1007/978-3-662-49155-3_66

obtained, as shown in Fig. 1(a). The contour map shows that the length and width of this influence area are about two times those of the building zone. Therefore, this area was divided into sections to generate a $4a \times 4b$ grid centered on the load M for the calculation of nodal loads. Assuming that $M(0,0)$ was a building load in the study area and P was a subsidence observation point near it, the influence area of M in the first quadrant (Fig. 1(b)) was analyzed. To make the analysis easier, the building load was distributed to the four nodes of the grid [5–7].

$$M_{i,j} = \frac{1}{2^{i+j}} \cdot \frac{MS}{\sqrt{i^2a^2 + j^2b^2}} \tag{1}$$

Where M is the building load; $M_{i,j}$ is the nodal load of M distributed by distance. $i,j = 0, 1, 2$, but the i and j can not be zero at the same time; $S = \sqrt{a^2 + b^2}$, a and b are the length and width of the building.

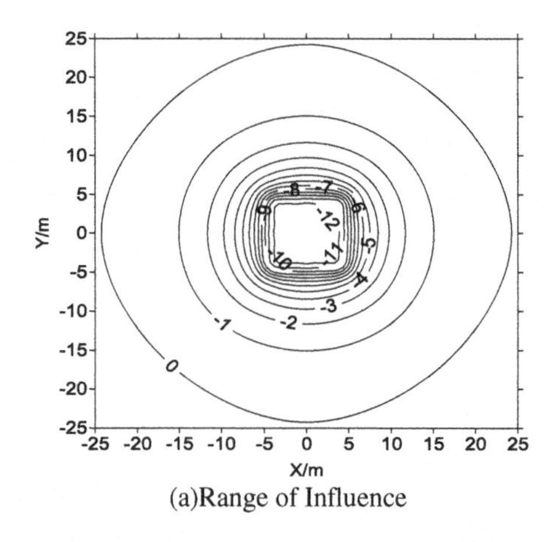

(a)Range of Influence

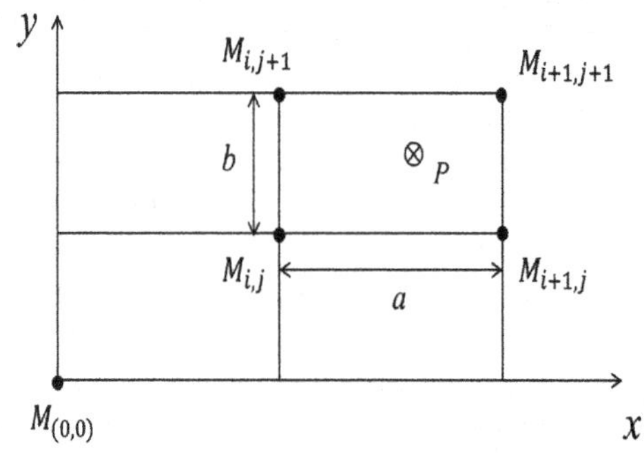

(b)Distribution of nodal load

Fig. 1. Distribution of load

According to the theory of elasticity, when a concentrated force F is applied on a semi-infinite elastic medium, the vertical strain at its centroid $M(x, y, z)$ is expressed as:

$$\xi_z = \frac{1}{E}(\sigma_z - \mu(\sigma_x + \sigma_y)) \tag{2}$$

Where σ_x, σ_y and σ_z are the positive stress in the three axis directions of spatial coordinate system, μ is the poisson ratio, E is the deformation modulus of soil, so the positive stress in the z direction can be got by Boussinesq formulas.

$$\sigma_z = \frac{3F}{2\pi} \times \frac{z^3}{R^5} \tag{3}$$

Where R is the distance between M and F.

In semi-infinite elastic ground, if only the vertical normal stress σ_z is considered, i.e. $\sigma_x = \sigma_y = 0$, then the amount of land subsidence caused by the concentrated building load, denoted as S_d, can be obtained by integrating the strain in the z-direction:

$$S_d = \int \xi_z dz = \int \sigma_z dz = \frac{F(1+\mu)}{2\pi E}\left[\frac{z^2}{R^3} + \frac{2(1-\mu)}{R}\right] \tag{4}$$

When the value of z is zero, the land subsidence caused by building load can be got.

$$\begin{aligned} S_d &= \sum_{i=1}^{n} \frac{F(1-\mu^2)}{\pi E R_i} \\ &= \sum_{i=1}^{n} \frac{GM_i(1-\mu^2)}{\pi E R_i} \end{aligned} \tag{5}$$

Where G is the acceleration of gravity; n is the number of all the nodal loads; R_i is the distance between nodal load and M_i.

3 Land Subsidence Induced by Groundwater Extraction

3.1 Relationship Between the Changes in Groundwater Level and the Land Subsidence

The analysis is based on the following assumptions: the soil layer is homogeneous, the stress to it is uniformly distributed along its height, and the particles and pore water in the soil are incompressible. The level of groundwater before extraction is denoted as h; the total stress and the effective stress of the saturated soil are represented by p and σ respectively, and the pore water pressure is represented by U_w. Before groundwater extraction, their relationship can be expressed as:

$$p = h\gamma_o + (\eta - h)\gamma_f \tag{6}$$

$$U_w = (\eta - h)\gamma_w \tag{7}$$

$$\sigma = p - U_w = h\gamma_0 + (\eta - h)(\gamma_f - \gamma_w) \tag{8}$$

where γ_0 is the bulk density of the soil above the groundwater level, γ_f is the bulk density of the saturated soil below the groundwater level, and γ_w represents the bulk density of pore water.

After the groundwater above the level of η is extracted, the pore water pressure is borne solely by the soil skeleton, and the effective stress σ' is then given by:

$$\sigma' = h\gamma_0 + (\eta - h)\gamma_f \tag{9}$$

The increment in effective stress is obtained:

$$\Delta\sigma = \sigma - \sigma' = (\eta - h)\gamma_w \tag{10}$$

From a geometrical perspective, the vertical displacement of a point on the surface resulting from the drop in groundwater level, denoted as S_w, is equal to the amount of compression of the soil, Δh. When the groundwater level declines to H, the amount of subsidence relative to the land surface can be obtained by the following integration:

$$\begin{aligned} S_w &= \int_h^H \frac{\Delta e}{1+e_0} d\eta = \int_h^H \frac{\alpha_v \Delta\sigma}{1+e_0} d\eta = \int_h^H \frac{\alpha_v(\eta - h)\gamma_w}{1+e_0} d\eta \\ &= \frac{\alpha_v\gamma_w}{2(1+e_0)}H^2 - \frac{\alpha_v\gamma_w h}{1+e_0}H + \frac{\alpha_v\gamma_w}{2(1+e_0)}h^2 \\ &= a_0 + a_1 H + a_2 H^2 \end{aligned} \tag{11}$$

where Δe represents the change in the soil's porosity; is the soil's initial porosity; is the soil's coefficient of compressibility; and a_0, a_1 and a_2 are constants.

3.2 Relationship Between the Consolidation of Soft Soil and the Land Subsidence

The hysteretic consolidation of cohesive soil is taken into consideration. Then solving Terzaghi's consolidation differential Eq. (11) gives:

$$\frac{\partial u}{\partial t} = C_v \frac{\partial^2 u}{\partial t^2} \tag{12}$$

$$C_v = \frac{K}{\gamma_w m_v} = \frac{(1+e)}{\gamma_w \alpha_v} K \tag{13}$$

Where, u is the pore water pressure; C_v is the coefficient of consolidation; K is the coefficient of flow; m_v is the compressibility of volume; γ_w is the test weight of water; e is the void ratio; α_v is the compressibility of soil.

As the soil's deformation depends on its effective stress, the degree of consolidation can be represented by the ratio of the amount of compression at a point to the total compression of the soil after the consolidation stops.

$$U_t = S_t / S_\infty \tag{14}$$

Where S_∞ is the final land subsidence; S_t is the compression of soil during a period of time; U_t is the degree of consolidation of soil.

The total subsidence can be calculated using the layer-wise summation method. If the excess pore water pressure at the average depth shows a linear distribution, the average degree of consolidation can be calculated using this formula:

$$U_t = 1 - \frac{32}{\pi^3} \cdot \frac{\frac{\pi}{2}\alpha_v - \alpha_v - 1}{1 + \alpha_v} \cdot e^{-\frac{\pi^2}{4}T_v} \tag{15}$$

Where T_v is the time factor.

Then, the land subsidence caused by consolidation of soil can be got.

$$\begin{aligned} S_t &= S_\infty U_t \\ &= C(1 - De^{-dt}) \\ &= c_0 - c_1 e^{c_2 t} \end{aligned} \tag{16}$$

Where $C = S_\infty$; $D = \frac{32}{\pi} \cdot \frac{\frac{\pi}{2} - \alpha_v + 1}{1+\alpha_v}$; $d = \frac{\pi^2 C_v}{4H^2}$; c_0, c_1 and c_2 are the constants.

4 Subsidence Coupling Model Under the Influences of Groundwater Extraction and Surface Load

Formula (16) indicates that the soil consolidation-induced subsidence is associated with the time dependent component t, and the subsidence of an aquifer is a quadratic function of H. The total subsidence caused by the building load can be calculated using Formula (5). To design a coupling model under the influences of groundwater extraction and concentrated building load, it is necessary to associate the load-induced subsidence and the aquifer's subsidence with the time dependent component t in Formula (16). The method is detailed below:

Formula (11) is the function for calculating an aquifer's subsidence. Associating H in this function with the time dependent component t yields the change in the groundwater level (relative to the initial level) at the time t.

Formula (5) calculates the land surface's total displacement in the z-direction caused by the building load from the perspective of soil mechanics. To associate this displacement with the time dependent component t, different weights are then assigned to different times during the observation of subsidence and the load-induced subsidence at the time t, S_d', is then given by:

$$S_d' = W_i S_d \tag{17}$$

where W_i is the corresponding weight coefficient for the time dependent component t. Due to the consolidation and compression of the soil, the influence of concentrated building load on land subsidence gradually reduces. Therefore, relative large weights are assigned to the early stage of subsidence caused by the surface load.

The value of W_i is set as $\frac{2^{t-i}}{2^t-1}$; So $\sum_{i=1}^{t} W_i = 1$; t is the aging components which is measured by month.

After associating Formulas (11), (16), and (17) with the time dependent component t, the function for calculating the total subsidence is yielded:

$$\begin{aligned} S &= S_w + S_d' + S_t \\ &= a_0 + a_1 H + a_2 H^2 + W_i \sum_{i=1}^{n} \frac{GM_i(1-\mu^2)}{\pi E R_i} + c_1 e^{c_2 t} \end{aligned} \tag{18}$$

Where $\boldsymbol{H}$ is the change to the first phase of the groundwater level; $\boldsymbol{n}$ is the number of all the nodal loads; $\boldsymbol{R_i}$ is the distance between nodal load and $\boldsymbol{M_i}$.

5 Case Study

The subsidence coupling model under the influences of both the surface loads and groundwater extraction, as expressed by Formula (18), was used to predict land subsidence occurring in a region. During one-month period of observation, data about the land subsidence and groundwater level changes occurring at 64 stages were collected at an observation point in this area. Figure 2 shows the collected data.

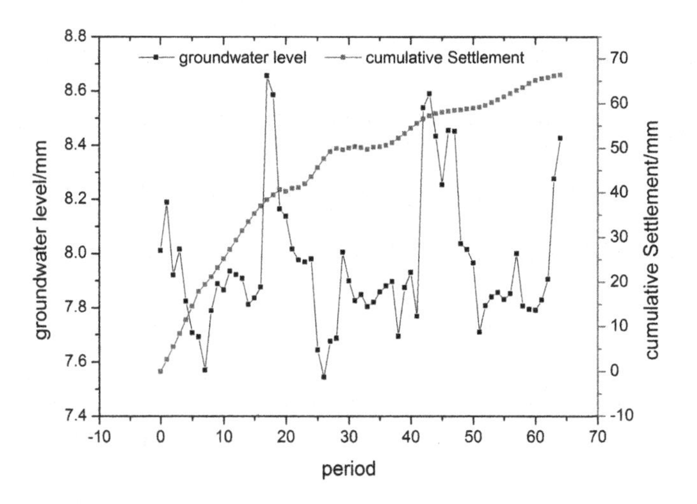

Fig. 2. Change in land subsidence and groundwater levels

To make the subsequent calculation and analysis easier, the data were divided into two groups: the historical data of the early 50 stages for fitting and the data of the later 14 stages for prediction and verification. The results of fitting and prediction are listed in Table 1.

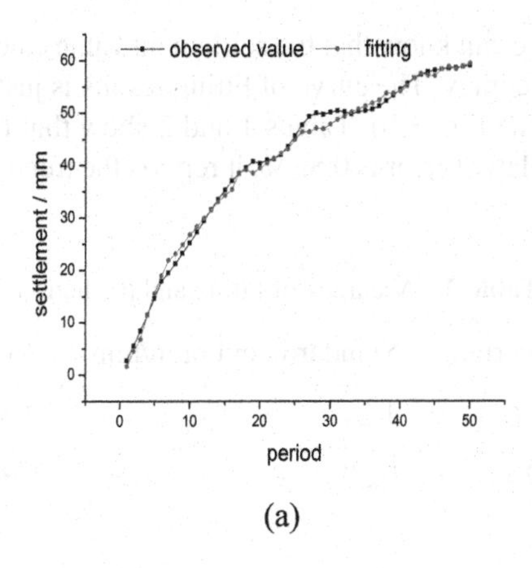

(a)

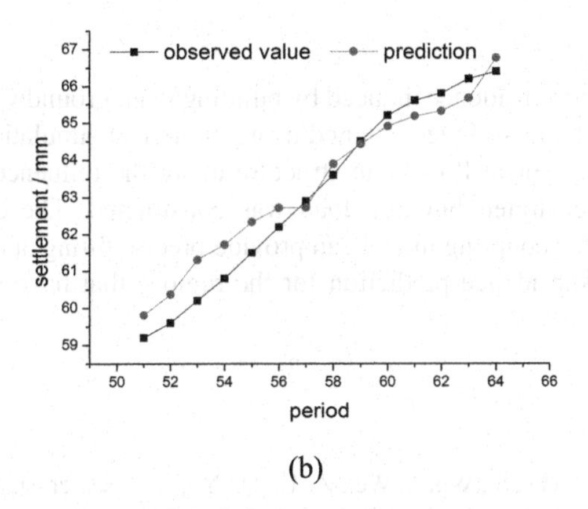

(b)

Fig. 3. Result of fitting and prediction

Table 1. Prediction of cumulative settlement

Period	1	2	3	4	5	6	7
Observed value	59.23	59.64	60.25	60.86	61.52	62.22	62.90
Predictive value	59.81	60.31	61.30	61.72	62.33	62.71	62.73
Residuals	−0.58	−0.67	−1.05	−0.86	−0.81	−0.49	0.17
Period	8	9	10	11	12	13	14
Observed value	63.64	64.51	65.28	65.61	65.82	66.27	66.43
Predictive value	63.89	64.43	64.91	65.18	65.32	65.67	66.76
Residuals	−0.25	0.08	0.37	0.43	0.50	0.60	−0.33

From the Fig. 3(a), we can know that the settlement value calculated by the Formula (18) has a high fitting accuracy. The curve of fitting results is just the same as the curve of original settlement. The Fig. 3(b), Tables 1 and 2 show that the biggest residuals is 1 mm and the average relative error is 0.83 %, it reports the trend of the land subsidence with a high accuracy.

Table 2. Accuracy of fitting and prediction

Result	Periods	Standard deviation/mm	Average relative error/%
Result of fitting	50	1.24	2.34
Result of prediction	14	0.59	0.83

6 Conclusion

In this paper, the land subsidence induced by building load, groundwater level changes, and consolidation of soft soil were studied using numerical simulations. A subsidence coupling model of a typical floodplain structure under the influences of groundwater extraction and concentrated building load was constructed. The calculation results demonstrated that this coupling model can provide precise fitting and prediction and is thus applicable to subsidence prediction for the regions that have similar geological conditions.

References

1. Xue, Y., Zhang, Y., Ye, S., Wu, J., Wei, Z., Li, Q., Yu, J.: Research on the problems of land subsidence in China. Geol. J. China Univ. **2**, 153–160 (2006)
2. Jianping, Y., Fang, L.: Research advances of monitoring and controlling technology for urban land subsidence. Bull. Surveying Mapp. **03**, 1–4 (2008)
3. Qinfen, L., Zheng, F., Hanmei, W.: A mathematical model and forecast of groundwater workable reserves for Shanghai. Shanghai Geol. **02**, 36–43 (2000)
4. Liu, H.: The study on the land subsidence with the effect of high-rise buildings in Tianjin Binhai New Area, Chang'an University (2010)
5. Heng, S., Jianping, Y.: On mathematical model of urban land subsidence based on building load. Bull. Surveying Mapp. **04**, 15–17 (2013)
6. Yuxin, J., Yan, G., Guangxin, L.: Analysis on the land subsidence induced by city construction. Geotech. Eng. Tech. **02**, 78–82 (2007)
7. Jie, Y., Gao, Y., Li, G.: Statistical pattern of building spacing and its influence additional stress. Ind. Constr. **S1**, 62–65+61 (2010)
8. Shengzhong, W., Pengfei, F.: Land subsidence computational theories and methods. J. Taiyuan Univ. Technol. **02**, 162–166 (2000)
9. Zhao, H., Qian, H., Li, Y., Peng, J.: Land subsidence model under dual effects of groundwater pumping and construction loading. J. Earth Sci. Environ **01**, 57–59 (2008)

10. Demin, D., Fengshan, Ma., Yamin, Z., Jie, W., Jie, G.: Characteristics of land subsidence due to both high-rise building and exploitation of groundwater in urban area. J. Eng. Geol. **03**, 433–439 (2011)
11. Lei, W.: Application of Terzaghi's 1D consolidation theory in research of the urban land subsidence, Jilin University (2005)

Study on Risk Assessment Framework for Snowmelt Flood and Hydro-Network Extraction from Watersheds

Dong Liu, Shaobo Zhong(✉), and Quanyi Huang

Department of Engineering Physics/Institute of Public Safety Research, Tsinghua University, Beijing 100084, China
zhongshaobo@tsinghua.edu.cn

Abstract. Disasters caused by snowmelt flood in mountains and high latitudes (e.g. Xinjiang, China) happen seriously every year for the lack of effective means for accurate prediction, early warning and risk assessment at present. As snowmelt flood involves meteorology, hydrology, disaster science and other disciplines, multidisciplinary research is demanded largely. This paper comprehensively considers the integration of snow-covered area monitoring, snow-depth retrieval, snowmelt runoff, hydrology and the assessment of disaster risk, proposes a new risk assessment framework for snowmelt flood. And then the extraction of hydro-network with Arc Hydro Tools is described taking Juntanghu Basin, Xinjiang Autonomous Region, China where snowmelt flood happens frequently as a case. Hydro-Network is a data model for following hydrology simulation.

Keywords: Snowmelt flood · Risk assessment · Hydro-Network

1 Introduction

Due to the economic development, global climate change and other factors, various natural disasters have caused ever-increasing losses, among which the flood disaster threatens human's daily life and property safety directly because of its strong burst, high frequency, serious damage degree, intensive damage range and other characteristics. In addition, the seasonal snowmelt flood in China's western mountains and high latitudes, caused by sudden rise of temperature in spring, also affects local economic development and residents' personal safety seriously. Statistical data displays that in 1966, 1971, 1977, 1985, 1993, 2005 and 2010, severe spring floods occurred in Wusu, Manasi, Hutubi and other Northern areas in Xinjiang, China [1]. The most serious snowmelt flood happening in Xinjiang in March, 1988, ever since the founding of the People's Republic of China, destroyed 130 km of the 312 National Highway, 181 km of canals, 70 hydraulic structures, 9600 hm^2 of farmland and three reservoirs because of the failure in flood discharge, resulting in direct economic losses of billions of Yuan [2].

As flood disaster is common and serious, domestic and foreign scholars have carried out plentiful flood disaster risk assessment. Li [3] introduced fuzzy mathematics method of information diffusion into the flood disaster risk analysis based on the system theory

F. Bian and Y. Xie (Eds.): GRMSE 2015, CCIS 569, pp. 638–651, 2016.
DOI: 10.1007/978-3-662-49155-3_67

and fuzzy mathematics theory, established a theoretical framework and methods for the flood disaster risk assessment and carried out the research aiming at small samples of flood risk assessment system. Apel et al. [4] developed a stochastic flood risk model consisting of simplified model components associated with the components of the flood process chain, parameterized these model components based on the results of the complex deterministic models and used them for the risk and uncertainty analysis in a Monte Carlo framework. Nonetheless, there are few researches in the risk assessment of snowmelt flood disaster at present, because of its scarcity and particularity and the lack of plentiful relevant data, resulting in less models and systems applied in forecasting and early warning directly, and then causing the short of effective prediction. Studying on integration techniques aiming at snowmelt process, hydrological simulation and disaster risk assessment and establishing effective forecasting, early warning and disaster response system can provide technical support for the emergency management personnel to respond effectively to the flood disaster and then realize the resource utilization of flood, promoting regional economic development.

Firstly, this paper presents a risk assessment framework for the snowmelt flood disaster based on the integration of methods in snow-covered area monitoring and snow-depth retrieval and models of snowmelt and hydrology, the combination of meteorological, hydrological and disaster theory and the consideration of regional characteristics, meteorological, hydrological and topographic data and social economic background data of different areas. Every part of the framework is described in detail. Secondly, aiming at the extraction of hydro-network prepared for sequent hydrological simulation, this paper illustrates the extraction process of hydro-network and carries out this method in extracting the hydro-network of study area (i.e., Juntanghu River Basin) with Arc Hydro Tools. Finally, the risk assessment framework and basic research work of the snowmelt flood risk assessment are summarized.

2 Risk Assessment Framework for Snowmelt Flood

This paper proposes a method and framework of risk assessment for snowmelt flood based on the research and analysis of present risk assessment methods for flood as shown in Fig. 1.

Firstly, investigation and survey of the regional characteristics of the study area should be carried out to collect and gather topography, meteorological, hydrological and other relevant data. On a basis of this, the topographical, meteorological and hydrological characteristics of the study area can be analyzed and summarized to select the appropriate methods or models for monitoring of snow-covered area, retrieval of snow depth, simulation of snow melting and hydrology. As a result, the data of snow-covered area and snow depth can be calculated by above methods and the data of snowmelt runoff can be simulated by snowmelt runoff model based on the input data, including regional terrain, meteorological and hydrological data.

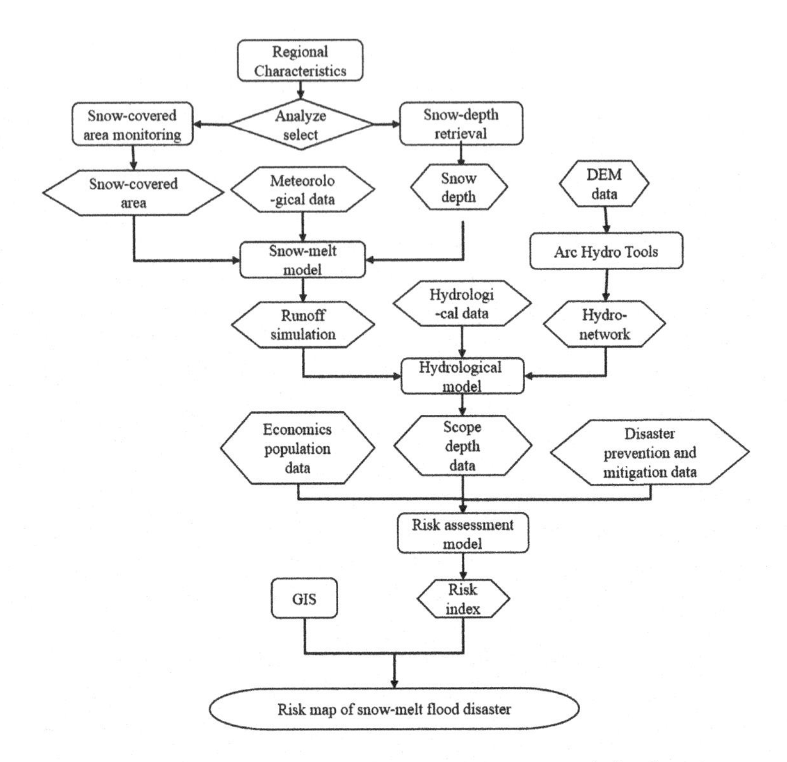

Fig. 1. Risk assessment framework for snowmelt flood

Secondly, the hydro-network of study area will be extracted based on DEM (digital elevation model) data by an ArcGIS-based extension: Arc Hydro Tools and the flooding submerged scope and depth of the water is supposed to acquire by applying hydrological model, combing simulative runoff and hydrological data.

Finally, with the integration of the above scope and depth data and regional background data, including the corresponding natural geography, social economy, and cultural education, etc., the impact of flood disaster is predicted and the corresponding risk map of the flood disaster is generated, and then the risk assessment of snowmelt flood disaster can be achieved.

2.1 Extraction of Snow Information

Meteorological and hydrological stations distribute sparsely in most study areas, which leads to relatively scarce snow information available including snow depth, snow-covered area, snow density, etc. Therefore, an alternative approach is using remote sensing data as the main data source, with those station data regarded as reference data.

Snow-covered area monitoring refers mainly to the extraction of snow-covered area information from remote sensing data, whose key process lies in the interpretation and recognition of snow cover, i.e., the recognition of pixels of snow. Current methods of identification are abundant, mainly including threshold-pixel statistic, normalized difference snow index method, decision tree, multi-spectral image classification, mixed-pixel decomposition method, etc., with the former two methods widely used [5]. Atmosphere has great influence on albedo and brightness temperature measured by remote sensing in the practical application of threshold method, which leads to difficulty to determine the uniform threshold, therefore two thresholds are needed to debug with a certain step in the specific study area to select the best threshold by visual confirmation finally [5, 6]. This method is simple and convenient, but the accuracy of the snow-covered area is dependent on the selection of the threshold. Normalized difference snow index (NDSI) works effectively in distinguishing thick clouds from snow, however often fails to distinguish thin cloud from snow, usually requiring the cloud-mask method to removing interference of cloud from remote sensing image [6]. There are also many other constraints in the application of NDSI, for example the need of adjusting the threshold to increase judgment basis according to different underlying surface of snow and introducing into the normalized difference vegetation index (NDVI) to improve accuracy, however the interpretation accuracy is very high, especially in the areas with thick clouds and plenty types of underlying surface [6, 7].

Snow depth varies in discontinuous distribution in a study area influenced by the amount of snow, slope degree, slope aspect and many other factors, resulting in much complexity of retrieval of snow depth. One of the extraction methods by means of the remote sensing data is based on optical remote sensing using visible light and near infrared band data to extract snow depth, mainly depending on the correlation analysis between snow depth and reflection spectrum of snow. In this method, the optimal combination of image band is selected to establish the regression equation between snow depth data in observation point and the combination of image band, by which regression coefficients can be computed to build the retrieval model finally. Optical remote sensing has strong application foundation as it generally covers a wide range and has a high spatial resolution and obvious texture features, but it also suffers a lager effect of cloud and doesn't have a good retrieval of deep snow. Another is based on the brightness temperature of microwave remote sensing image to extract the snow depth. As the microwave brightness temperature of snow decreases with the increase of snow depth, the corresponding relation between brightness temperature and snow depth can be established to extract snow depth. Microwave remote sensing has relatively high cost, large effects on brightness temperature data caused by snow grain size, liquid water content and underlying surface type and limitation of deep-snow retrieval only, but it will not be affected by weather conditions and works well for the extraction of night and cloud cover data [8].

2.2 Snowmelt Runoff Model

Snowmelt runoff model is a tool quantitatively describing snow melting process and simulating snowmelt runoff, and developed from experiential model establishing input and output relationship between meteorological factors and snowmelt runoff to conceptual model considering freezing, thawing and evapotranspiration process, which is represented by degree-day model based on the index of air temperature, and then to physical model considering physical mechanism, which is represented by the energy balance model. Snowmelt model also experienced from the single point model regarding the regional snow cover as an integral whole to the distributed model taking spatial heterogeneity of snow cover into account [9]. A variety of snowmelt models have been developed till now, as each model has its own advantages and disadvantages and applicable scope, it is very important to select an appropriate model in carrying out the research in a specific area. Here are two of the most widely used models named SRM and SWAT as an example to show the choice of a model.

SRM (Snowmelt Runoff Model) is a runoff model recommended by the WMO (World Meteorological Organization), calculating concentration of runoff with runoff coefficient, extinction coefficient and other conceptions of hydrology, and has been widely used in more than 100 basins of more than 30 countries and regions [10]. The flow is calculated by the following formula:

$$Q_{n+1} = \left[c_{S_n} a_n \left(T_n + \Delta T_n\right) S_n + c_{R_n} P_n\right] \left(A \cdot \frac{10000}{86400}\right) \cdot \left(1 - k_{n+1}\right) + Q_n k_{n+1} \quad (1)$$

Where Q is average daily flow, c is runoff coefficient, c_s is snowmelt runoff coefficient, c_R is rainfall runoff coefficient, a is degree-day factor, T is degree-days, S is snow-covered area percentage, P is precipitation, A is watershed zoning area, k is recession coefficient, n is the number of days, 10000/86400 is unit conversion coefficient, T, S and P are all model variable and the others are model parameters [11].

The snowmelt module of Soil and Water Assessment Tool (SWAT) adopts algorithms similar to sine equation, assuming the potential snowmelt rate varies from minimum value (for instance, this value is acquired on December 21) to maximum value (for instance, this value is acquired on June 21) following the laws of sine function to calculate snowmelt runoff by the following equation calculation:

$$SNO_{mlt} = B_{mlt} \cdot SNO_{cov} \left[\frac{T_{snow} + T_{max}}{2} - T_{mlt}\right] \quad (2)$$

Where SNO_{mlt} is snowmelt water, B_{mlt} is snowmelt factor, SNO_{cov} is score of snow-covered area, T_{snow} is the temperature of snow cover, T_{max} is daily maximum temperature, T_{mlt} is temperature threshold of snow melt [12, 13].

A comparison of SRM and SWAT is shown in Table 1:

Table 1. The comparison of SRM and SWAT.

Name	SRM	SWAT
Categories	Experiential, distributed	Experiential, distributed
Input variable	Daily temperature and precipitation, snow coverage rate	
Input parameters	Runoff coefficient, degree-day factor, recession coefficient, etc	Daily temperature, precipitation and relative humidity, etc
Output parameters	Daily and seasonal runoff, etc	Daily and seasonal runoff, etc
Sensitive parameters	Temperature, degree-day factor, etc	Temperature, temperature hysteresis factor of snow, etc
Simulation method	Degree-day factor algorithm	Degree-day factor algorithm
Characteristics	Altitude-zone, recession coefficient representing flow characteristics, single runoff coefficient considering dissipation like infiltration, evapotranspiration etc	Physical mechanism, adopting representative basic unit, leaving out actual spatial distribution, snow-depth threshold estimating snow-covered area, etc
Advantages	Simple structure, easily used, less demand for data, high simulating precision, etc	Conventional data, continuous long-time, detailed and efficient simulating, less effect of underlying surface, etc
Disadvantages	Different parameters with same effect, sketchy simulation, less carefully considering terrain and meteorological elements	High data accuracy, different data standards in domestic and overseas, data conversion required
Main error sources	Massive simplification of physical processes like evapotranspiration, etc., scarce station data, great daily change of temperature	Low accuracy of meteorological data, unclear physical process like infiltration, evaporation snowmelt, etc. lots of effect of seasonally frozen ground
Applicable conditions and scope	Successfully applied in different geographical and climate conditions (from wet to semi-arid areas, from 305 to 7690 m height, etc.), mainly suitable for alpine regions with sparse meteorological and hydrological stations	Mainly suitable for humid and semi-humid regions and plains with rich rainfall, low simulation precision in arid and semi-arid and alpine region with rare rain
Key demand	Accurate input of model parameters	Accurate and detailed data

As the table above shows, SRM and SWAT are both experiential models paying less attention to physical details of melting process relative to physical models demanding much more accurate and detailed data to simulate. These experiential models are preferentially selected for the study in the mountainous of northwest China with frequent flood disasters on account of scare meteorological and hydrological stations without conditions of describing fine physical processes, possessing good precision in simulation and forecasting runoff flow with low data requirements.

According to the comparison of models above, suitable model can be selected clearly and conveniently on the consideration of geographic, meteorological and hydrological characteristics of the study area, the acquisition of data to simulate and the simulation purposes. At the same time, a fast and relatively accurate simulating of snowmelt runoff is demanded first other than the higher accuracy from a contingency perspective in risk assessment of snowmelt flood disaster, therefore corresponding simplified processing based on the actual situation is indispensable in the selection of model parameters, input data and application aspects. The parameters can be properly ignored sometimes, such as the neglect of physical detailed processes like infiltration, intercepting, evaporation, etc. when paying no attention on the effect of precipitation on snow melting in scare rainfall area and regarding the glacier area as a part of snow-covered area. At other times, constant can replace the parameter, including referring to the experience of snowmelt models, such as specifying the critical temperature as 3 at the beginning and 0.75 at the end of the snowmelt season, and according to some results announced by World Meteorological Organization to value, such as specifying the watershed lag as 3 h and snow density as 0.3 g/cm [14]. And also some research results of study regions or similar areas made by scholars in the past can be relied on to specify values needed, such as applying the research results of watershed nearby directly in the acquisition of runoff coefficient [14, 15]. The above simplification treatment would sacrifice simulation accuracy to some extent, but as long as the accuracy achieves the level demanded by risk assessment and emergency management, it would greatly accelerate the speed of obtaining simulation results and improve the efficiency of risk assessment in the case of ensuring the relative accuracy of risk assessment.

2.3 Risk Assessment

Snowmelt flood disaster system is a complex system composed by hazard-bearing body, disaster-causing factor and disaster-inducing environment and affected by nature and human social factors, therefore the risk assessment of snowmelt flood disaster should consider both the temporal and spatial distribution of the disaster and the economic situation of the area flood may occur in. Simulating snowmelt runoff by snowmelt model, extracting the hydro-network of study area by Arc Hydro Tools and simulating flood submerged area and depth not only consider regional topographic, hydrologic, etc. environment features, but also reflect the snowmelt flood risk of different depth and scope. Therefore the size of overwhelmed scope and flood depth which comprehensively reflects the features of disaster-inducing factor and disaster-inducing environment can be regarded as the risk assessment standard of comprehensive evaluation for the two above factors [12, 16]. Then a risk evaluation index system [16] can be established by following steps: first, the size of snowmelt water depth and hazard-bearing body related factors should be selected as the main factors generally reflecting the risk of snowmelt flood disaster. Second, the above factors will be quantified and classified to obtain the corresponding factor index, basing on which the risk index can be acquired then through simple mathematical calculations. Finally, the risk index is divided into risk indicators which will be displayed on GIS by different colors. Basing on the above analysis and the features of flood disasters, the flood factor, economic factor and disaster reduction factor have been preliminary considered, with the mathematical calculation method to obtain risk index following:

$$\text{Risk Index} = \text{Water} - \text{Depth Factor Index} \times \text{Economic Factor Index}/\text{Disaster Reduction Factor Index.} \tag{3}$$

The water-depth factor index is determined by flood submerging scope and depth, the greater the depth, the higher the index. The economic factor considers economic and social situation of the area flood may occur in, including population, cultivated land, GDP, etc. Combining snowmelt flood disaster characteristics, the population and GDP per area of flood-threatened areas are selected as a measure of economic factor, the larger the population and the GDP per area, the higher the index. Social disaster reduction ability and flood control engineering facilities, etc. are involved in the disaster reduction factor, with the facilities including flood discharge project, flood storage project and flood diversion project, etc. Basing on the above consideration and combining characteristics of disaster and hazard-prone region, quantification processing result of flood-storage standards of reservoirs with high flood-storage ability is preliminary considered, the higher the standards, the higher the index [1, 2, 16].

3 Extraction of Hydro-Network

As shown in Fig. 1, Hydro-Network is an important prerequisite for submerging analysis of flood. And extracting hydro-network accurately and effectively has a great effect on the risk assessment of flood disaster [17, 18]. In this section, we will present how to extract hydro-network through an ArcGIS Extension: Arc Hydro Tools taking Juntanghu Basin, Xinjiang Autonomous Region, China as a study area.

3.1 Study Area

The Juntanghu basin located in Hutubi County, Changji State, Xinjiang Province, China, was selected as a typical study area in this paper. This closed basin possesses an elevation ranges from 1000 to 1500 m, except the source located in 3400 m and a main river named Juntanghu River originating from the northern slope of Tianshan Mountains, which is about 45.20 km from the source to the Hongshan Reservoir which retains the river in the mountain-pass and have a catchment area 833.57 km^2 above the Hongshan reservoir. Snow appears in mountains of this basin since mid- September and accumulates to the maximum in January of the following year as the decreasing of air and ground temperature, then it begins to melt in mid to late February as air and ground temperature starts to bottom out and melt in large area in early March, when the snowmelt runoff is really easy to form a snowmelt flood. This basin is not only small and basically closed and have typical basin characteristics, in line with the purpose of this study and also where snowmelt flood occurred frequently in recent years, which is the reason for being chosen as the study area [20] (Fig. 2).

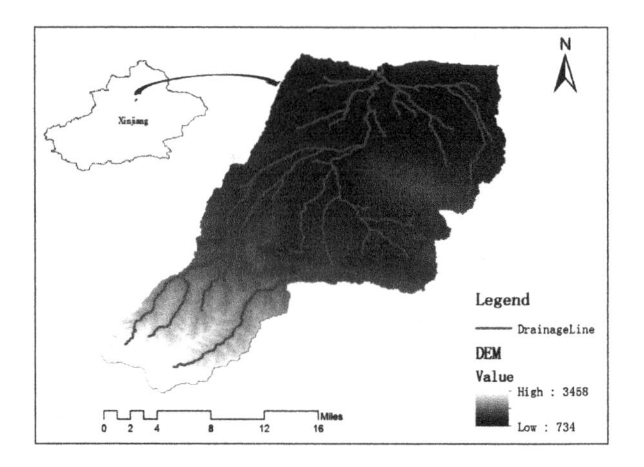

Fig. 2. Study area map

In this case, ASTER GDEM with a resolution of 30 m, which is a product of METI and NASA is selected as the digital elevation data.

3.2 Arc Hydro Tools

In 1999, ESRI (Environmental Systems Research Institute, Inc.) joint Water Resources Research Center, University of Texas began to design Arc Hydro data model, an ArcGIS-based system used to build hydrological information systems. Arc Hydro integrates geographic spatial data and time-series data of water resources, and is always used for hydrological analysis and simulation combining hydrological models [19]. It contains two key elements: Arc Hydro Data Model and Arc Hydro Tools. These two elements and generic programming framework provide a basic database design methodology and a set of analytical tools to enhance the field of water resources [19]. So we can make full use of Arc Hydro Tools based on DEM data to extract hydro-network for the flood disaster assessment.

Watershed delineation using Arc Hydro Tools contains a series of steps as shown in Fig. 3, and the following is a brief description of main steps applied in research areas.

3.3 Extraction Process

DEM Reconditioning. As the difference of elevation is relatively small between adjacent grids of DEM data of plain areas, it is likely to create deviation like parallel or straight rivers from the real drainage by drainage modelling directly. To solve this problem, Agree DEM method is applied by setting parameters like Stream buffer, Smooth drop\raise, Sharp drop\raise, etc. and combing the actual river data to adjust the surface elevation of DEM in Arc Hydro to reduce this bias [21].

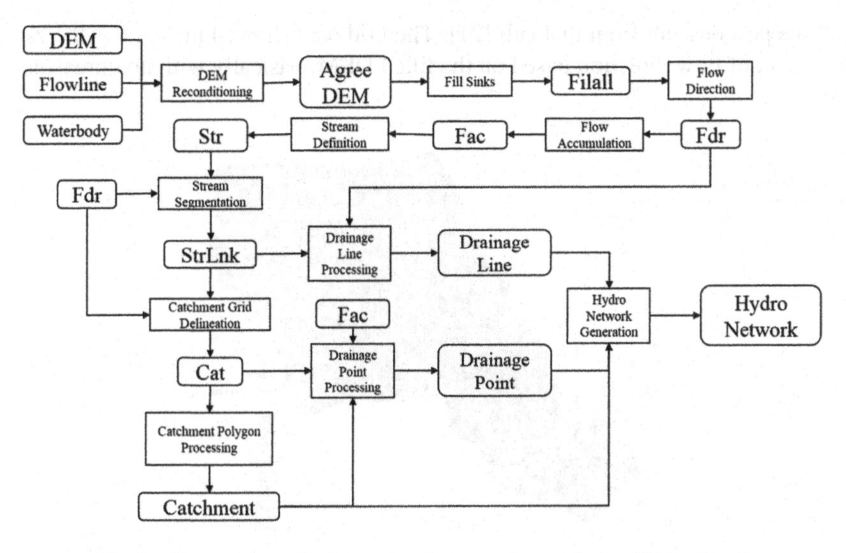

Fig. 3. Flow chart of hydro-network extraction with Arc Hydro Tools

Fill Sinks. One of the hot issues of watershed features extraction is filling sinks. As sinks are where flow direction is not reasonable, they can be determined with the help of flow direction and will be filled mainly through changing their elevation by interpolation function. What needs to notice is that not all sinks are created by data error as some of these are true reflection of the surface morphology, therefore a filling threshold should be reasonably set based on the calculation of sinks depth before filling sinks, which is 10 point here. The filled DEM created by this function is shown in Fig. 4.

Fig. 4. Fill Sinks Map, left is raw DEM, right is DEM with filled sinks.

Flow Direction. As flow direction analysis is the premise of the following steps including the analysis of drainage, the division of watershed, etc., it is very important in the extraction of hydro-network. Arc Hydro computes flow direction and stores a value representing the steepest direction in each grid cell based on the given raster data by D8 method. The values in the cells of the flow direction grid indicate the direction of the steepest descent from that cell [21]. The grid data showed in Fig. 5 is the calculation result of flow direction based on the filled DEM, basically with no unreasonable places.

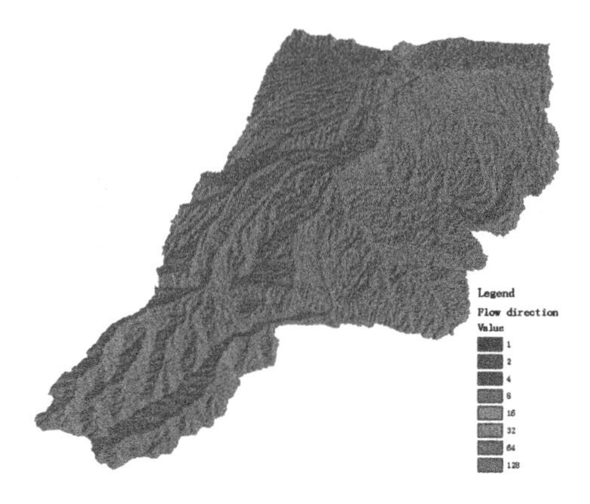

Fig. 5. Flow directions.

Stream Definition. Stream Definition is one of the key steps in the analysis of hydro-network, which computes a stream grid based on a flow accumulation grid and a user specified threshold through setting the grid value to 0 or 1 to definite the stream [21]. The setting of the threshold is directly related to the intensive degree of hydro-network which represents the fine degree of analyzing, the smaller the threshold, the more intensive the hydro-network and the longer the drainage line. The default threshold of Arc Hydro is 1 % of the maximum flow accumulation value. Figure 6 shows the hydro-network extracted from filled DEM and Fig. 7 shows the contrast between hydro-networks extracted with different thresholds.

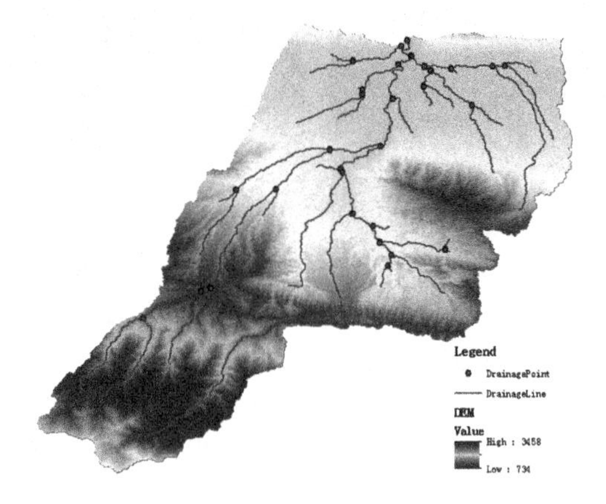

Fig. 6. Hydro-network extracted from filled DEM.

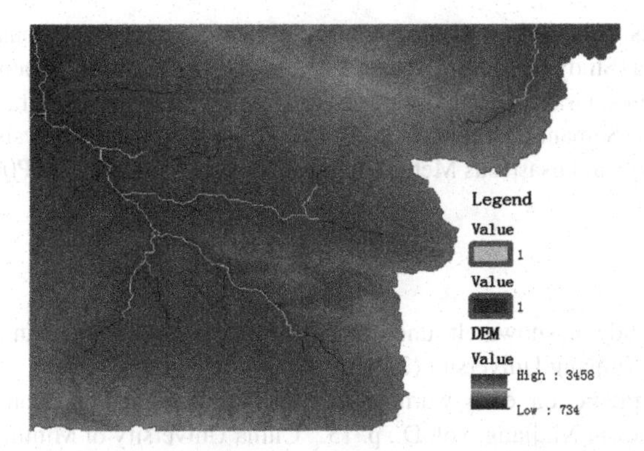

Fig. 7. Contrast between hydro-networks extracted with different thresholds (yellow lines represent higher threshold, purple lines represent smaller threshold) (Color figure online).

Arc Hydro has been relatively mature till now after gradual improvement and its extraction contents have become a major input data for most of the hydrological models, which is widely used in flood control, disaster mitigation, water resources development and utilization, etc [22]. The extraction of hydro-network with Arc Hydro provides a premise for the simulation of water depth and flood inundated area in risk assessment of snowmelt flood disaster.

4 Conclusion

Snowmelt flood takes place frequently and causes serious consequences every year at home, including severe personnel casualty and billions of Yuan of direct economic losses, which is the result of lacking enough efficient prevention means. It has a great significance to study on the risk assessment of snowmelt flood disaster. Studying on integration technique aiming at snowmelt process, hydrological simulation and disaster risk assessment and establishing effective forecasting, early warning and disaster response system can provide technical support for the emergency management personnel to respond effectively to the flood disaster.

This paper presents a snowmelt flood risk assessment framework based on the data of meteorology, hydrology, topography and social economic background and the integration of snow information extraction, snowmelt runoff model, hydrological simulation, risk assessment and other techniques, which combines meteorology, hydrology and disaster theory. According to the risk assessment framework, Hydro-Network extraction, one of the prerequisite tasks in the framework has been carried out with Arc Hydro Tools: an ArcGIS Extension, which lays a good foundation for the following hydrological analysis. On the basis of the work presented in this paper, further technology integration and system implementation will be carried out to establish the risk assessment system of snowmelt flood, which can not only support snowmelt flood disaster prevention and response, but also has an important realistic significance in making full use of the flood resources and promoting regional economic development.

Acknowledgements. The authors would like to thank the support of the National Natural Science Foundation of China (Study on Pre-qualification Theory and Method for Influences of Disastrous Meteorological Events, Grant No. 91224004) and the youth talent plan program of Beijing City College (Study on Semantic Information Retrieval of Decision Analysis of Emergency Management for Typical Disastrous Meteorological Events, Grant No. YETP0117).

References

1. Fang, S.F.: A study on snowmelt runoff forecasting and its uncertainty analysis in Xinjiang, vol. D, p. 193. Xinjiang University (2010)
2. Liu, Z.H.: The prediction, early-warning for snowmelt flood and decision support based on '3S' technologies in Xinjiang, vol. D., p. 152. China University of Mining and Technology (2009)
3. Li, Q.: The research and improvement of risk analysis and evaluation method on flood disaster, vol. D, p. 112. Huazhong University of Science And Technology (2012)
4. Apel, H., Thieken, A.H., Merz, B., Blöschl, G.U.N.: Flood risk assessment and associated uncertainty. Nat. Hazards Earth Syst. Sci. **4**, 295–308 (2004)
5. Tachiiri, K., Shinoda, M., Klinkenberg, B., Morinaga, Y.: Assessing Mongolian snow disaster risk using livestock and satellite data. J. Arid Environ. **72**, 2251–2263 (2008)
6. Tian, G.L., Xiao, D.H., Cai, X.W., Xing, Y.L., Wen, L.L., Zheng, G.G., Ji, Z.R.: An application of MODIS data to snow cover monitoring in a pastoral area: a case study in Northern Xinjiang. China. Remote Sens. Environ. **112**, 1514–1526 (2008)
7. Zhang, F., Guan, H.J., Xu, C.H.: Study on the methods of snow information extraction in upstream of the Manas River Basin on TM image. Plateau Mt. Meteorol. Res. **1**, 69–73 (2011)
8. Liu, Y.Q.: Key technologies and realization of early warning DSS for snowmelt flood in Xinjiang, vol. D, p. 182. Xinjiang University (2007)
9. Bin, C.J.: Improvement of passive microwave snow algorithm over Qinghai-Tibet Plateau, vol. M, p. 69. Shijiazhuang University of Economics (2013)
10. Martinec, J.: Snowmelt - runoff model for stream flow forecasts. Nord. Hydrol. **6**, 145–154 (1975)
11. Gao, J.: Spatiotemporal distribution of snow and snowmelt modeling in Alpine Regions, vol. D, p. 122. Tsinghua University (2011)
12. Youssef, A.M., Pradhan, B., Hassan, A.M.: Flash flood risk estimation along the St. Katherine road, southern Sinai, Egypt using GIS based morphometry and satellite imagery. Environ. Earth Sci. **62**, 611–623 (2011)
13. Li, H., Lei, X.Y., Jin, S.: Simulation about the daily runoff of the montane mixedreplenishment of ice and snowmelt river based on the SWAT model. J. Irrig. Drainage **6**, 105–108 (2010)
14. Wang, J. Ma, M,G, Paolo, F.: Simulating snowmelt runoff in mountainous watershed of Italy using GIS and remote sensing data. Journal of Glaciology and Geocryology 436–441 (2001)
15. Liu, W., Li, Z.L., Li, K.B.: Study on Tashikuergan River Basin with SRM. Technical Supervision in Water Resources 43–46 (2007)
16. Tan, X.M., Zhang, W.B., Ma, J.M., Su, Z.C.: Research on regional assessment of flood risk and regionalization mapping in China. J. China Inst. Water Resour. Hydropower Res. **1**, 54–64(2004)
17. Liu, J.F., Li, J., Liu, J., Cao, R.Y.: Integrated GIS/AHP-based flood risk assessment: a case study of Huaihe River Basin in China. J. Nat. Disasters **17**, 110–114 (2008)
18. Zhang, J., Song, L., Feng, F., Gong, H.: Hydrologic information extraction for flood disaster risk assessment in Pearl River Basin and Luan River Basin, China, pp. 1–4. IEEE (2011)

19. Tools, A.H.: 1.3-Overview [EB/OL]. ESRI Press (2009)
20. Wei, Z.C.: The study on snow melt process simulation and snow characteristic analysis, vol. M, p. 60. Xinjiang University (2010)
21. Tools, A.H.: 1.3-Tutorial [EB/OL]. ESRI Press (2009)
22. Zhu, S.R.: Introduction of ERSI Arc Hydro data model. Geomatics Spat. Inf. Technol. **5**, 87–90 (2006)

Risk Perception and Adaptive Demands of Climate Change in Metropolis: A Case Study of Shanghai

Lei Song(✉)

Department of Economics and Business, China Executive Leadership, Academy (Pudong), Shanghai 201204, People's Republic of China
alei0829@163.com

Abstract. This research put emphasis on the specific characteristics of climate risk as well as the impact factors of vulnerabilities in metropolis. Furthermore, grounded on the approach of participatory analysis, this research carried out the case studies in Shanghai, which reflected how different the stakeholders had the perception of climate risk and the demands of adaptation. The participatory stakeholder analysis in Shanghai depicted the different adaption demands. The research outcomes discussed the force driving factors of metropolitan climate risk including coastal geographic location, the population vulnerabilities (aging and high-density population), ecological vulnerabilities (the degraded ecological bearing capacities and resilience) and the vulnerabilities of economic structures. All in all, the metropolitan areas should build more incremental adaptive capabilities. In Shanghai' case, it should be highlighted of adaptive capabilities building in communicates and eco-logical environments.

Keywords: Climate risk perception · Metropolis · Adaptive capacity · Participatory stakeholder analysis · Shanghai · Vulnerability

1 Introduction

There are arguments about the definition of climate change risks, [1–4] which interwove a buck of impact factors including the frequency of hazard, socioeconomic conditions, population exposure, environmental vulnerability, and sanitation vulnerability [5, 6]. In this search, the definition of climate risk and vulnerability are cited from the IPCC [4] and Ellis [7], which claim that Climate change risks refer to adverse effects of climate change including climate variability and extremes. The direct climate change risks result from the natural hazard including heat, floods, droughts, storms and sea-level rising. Compared with that, the indirect of climate change risks are caused by the vulnerability of ecological and social systems in climate change such as crop production reduction, property loss and injuries of people. So the climate risks can be defined as a function of two main parameters: (1) the frequency of climate hazards and (2) vulnerability. Regarding vulnerability, some researches integrates physical vulnerability (P.V), exposure to hazards, and socioeconomic vulnerability [8]. For instance, Heijmans [9] finds that disaster response agencies are increasingly using 'the concept of "vulnerability" to analyze processes that lead to disasters and to identify responses', but at the same time,

F. Bian and Y. Xie (Eds.): GRMSE 2015, CCIS 569, pp. 652–662, 2016.
DOI: 10.1007/978-3-662-49155-3_68

'agencies use the concept in the way that best fits their practice- in other words, focusing on physical and economic vulnerability. Safi [1] argued that vulnerability evolved from being centered on the physical vulnerability and became a comprehensive concept, within which both P.V. and socioeconomic conditions of threatened individuals or communities are intertwined. In this research, we still use the definition of vulnerability stated by IPCC [2] and Smit and Pilifosova [10]. IPCC report [2] describes vulnerability as a function of the character, magnitude, and rate of climate variation to which a system is exposed, its sensitivity, and its adaptive capacity. Smit (2003) argued that the vulnerability of a given community or system to climate change is specific to particular stresses or stimuli at particular locations and periods of time. There are two main factors: (1) exposure and (2) adaptive capacity. Just as aforementioned, climate risk may be represented as model (1).

$$\begin{aligned} \text{RISK} &= f\{\text{HAZARD, VULNERABILITY}\} \\ &= f\{\text{HAZARD, EXPOSURE, ADAPTIVE CAPACITY}\} \end{aligned} \tag{1}$$

From this formula to see, climate change risk results from three components. One is relating to natural hazard. Hazard may be represented as the long-term climate variation involving intense rainfall, rising sea level and the occurrences of extreme temperature, as well as the shot-term climate change such as the increased frequency of drought, flooding, typhoon, sandstorm, heat waves, acid rain and other environmental change.

The second factor refers to the exposure of regional system to hazard, particularly the exposure of population and socioeconomic condition [11]. The exposure refers to the probability or incidence of hazardous conditions relative to the presence of humans at a particular location at a particular time [12]. So the parameter of exposure is determined by the specific location, and the sensitivity of population and socioeconomic condition to hazard. For instance, the location near to floodplain or coastal zone may be easily harmed by flood or typhoon. The elderly, children, and people with disability are more sensitive to risks such as tornadoes, tsunamis, and earthquakes, where they have more difficulty in escaping and adapting to them. Furthermore, some industries such as agriculture, tourism, forestry and fishing are more sensitive than others to some hazards. Thereby, the region with high proportion of these sensitive industries may suffer with property loss [1, 13, 14].

Another factor is adaptive capacity, which refers to the ability of a system to cope with or survive from hazards. The adaptive capacity of a system reflects on the climate-proofing infrastructures and carrying capacity of environment. For instance, increasing forest coverage rate or dam reinforce may enhance the ecological adaptive capacity to flood. Adaptive capacity also involves the policy measures and institutional organization. The high perception of climate variation may increase the abilities of coping with hazard when it happened. Furthermore, some measures for raising awareness of climate risks such as setting up the system of meteorological disaster early warning, or providing self-help training and education for community would have great effort on preventing disasters and reducing damages.

In light of the conception of climate risks discussed above, climate risks have evident regional characteristics. The climate risks in metropolis may have more adverse impacts on population health as well as the social-economic system because of the agglomeration development. So this paper firstly focuses on the special characteristics of climate change risk in metropolis together with a case study of climate risk analysis in Shanghai. Four main vulnerabilities have been depicted in this part, involving the frequency of climate hazard happening, population factor, natural environment and economic structure. Generally, the aim of adaptation to climate change is to reduce vulnerability and increase resilience to impacts [10]. So promoting adaption capacity is the other task in this research. Adaptation capacities in a region depend on a variety of factors, including economic development condition, engineering measures such as building defensive infrastructures and other political measures such as public education or hazard insurance policies. Furthermore, the perceptions and action willingness of stakeholders to climate risks, especially the policy-makers in this region, have a significant role in building adaptation capacities. Thus ground on the climate risks analysis in Shanghai, this paper tried to analysis the different perceptions and adaptive demands among stakeholders in Shanghai so as to find the retrofit strategies for adaption to climate risks.

2 The Characteristics of Climate Change Risks in Shanghai

2.1 Main Climate Hazard

Shanghai is located midway along China's highly populated and generally prosperous Eastern seaboard. The city itself sits on mainly flat land at the confluence of the Yangtze and Huangpu Rivers, on the south-east edge of the Yangtze delta. The coastal lines of Shanghai is about 172 km, where is affiliated to Pacific Rim with multiple climate disasters especially rainstorm. In the Grounded on the meteorological data and disaster statistics from the Shanghai Meteorological Bureau, the main natural hazards in Shanghai point to typhoon, rainstorm and strong wind (Seen in Table 1). In these climate hazards, flood, typhoon and thunder happen in Shanghai most frequently than other climate hazards. However, the typhoon brings the most losses of economic losses, especially agricultural industry, resulting in economic losses about annual 1390.2 million$ and 8529 hm^2 areas of crops affected. Flood is the second worst hazard which cause the annual economic losses of 361.7 million$ and 6501.9 hm^2 areas of crops affected. The casualties mainly result from thunder and lightning, 3 or 4 people would die in this kind of climate hazard in every year, and about 2 or 3 people may be led to death by typhoon or tornado. The statistic data from Table 1 explains that the happing frequency of a certain hazard does not have linear relation with economic losses or casualty. The potential reason may be related to the adaption capacities or the perception of stakeholders to this kind of climate hazard, which will be discussed in detail later.

Table 1. Annual average losses resulted from different disaster in Shanghai

Category	Frequency (times)	Economic losses (million$)	Affected crops (hm^2)	Death population
Typhoon	143	1390.2	8529.0	2.4
Flood	163	361.7	6501.9	1.1
Strong wind	90	96.7	530.7	1.8
Tornado	48	68.4	756.5	2.0
thunder and lightning	142	23.4	231.8	3.4
Dense fog	35	0	0	1.3

**Shanghai Climate Center, Special assessment of climate change in Yangtze River delta urban agglomerations, 2012*

2.2 Ecological Environment Factors

Shanghai is an economic powerhouse in China that is already one of the world's major business cities. As the largest economic center in China, this city has kept sprawling rapidly with increasing consumption of energy and natural resources. Therefore, the intensive land utilization and high consumption of ecological sources bring great challenges to the ecological environment capacity, especially the ecological service functions of water resources, cultivated land and ecological diversity have decayed gravely [5]. Since 2003 the municipal government has paid more concern about increasing the ecological resiliency, which results in the ecological service value in the downtown of Shanghai restoring by 19.45 %. However, compared with the ecosystem before 1943, the ecological service value has dropped by 88 % [15]. Taken the water resource as the example, the rivers in Shanghai have been landfilled to increase the available land areas. So the water system has declined quickly in Shanghai. During the period from 1990 to 2009, the channel density had dropped from 6.5 km/km^2 to 3.4 km/km^2, which had fallen about 67 %. Especially, the small channels whose slop range from 200 to 1000 m dispersed most fast, which have accounted for 60 % of the whole dispersed rivers [16]. Similar to the water system, the increasing demand of land resources also triggers the irrational land utilization, which intensifies the deterioration of environment capacity. The data has shown that the temperature would elevate by 0.91°C resulting from heat island while the population in Shanghai grow per 1 million people.

2.3 Population Factors

The two factors of population have impacts on the climate risks, one related to the density; the other on the population structure. Shanghai is the city with the most population density in China. Until 2010, the density of population in Shanghai has swollen to 3632 people per kilometer square, increasing 40.3 % compared with the density of 2588 people per kilometer square in 2000. In addition, fifty percent of population has crowded in the downtown of Shanghai, which only account for one-tenth areas of the whole Shanghai [17]. This kind of uneven population distribution brings more and more pressure to the central town not only in the infrastructure service but also the resource supplies, which intense the risk of population exposures to abnormal climate. Furthermore, the aging population structure also takes disadvantage to the exposure of climate risks. Generally, the poor or the age arranging from seventeen to sixty refer to the vulnerable population to climate risks such as heat wave, frozen and flooding. In Shanghai the vulnerable population has accounted for 33.8 percent of the whole population. The aging people over 80 years have kept growth in recent ten years. In China the social bonding in communities are weak. If the development of health care system cannot tag along with the trend of population structure change and provide enough social sources to care this vulnerable group, more and more population in the big city will be exposure in the threat of climate risks. Compared with the aging population, since 2002 the poverty has fallen and the adaptive capacities of the poverty to climate risk have been improved by increasing social insurance payments. Until 2007 the poverty group has reached 0.34 million people and the poverty rate is only 2.84 %.

3 Risk Perception of Different Stakeholders and Adaptive Capacities Analysis

Adaptive capacity is the ability or potential of a system to respond successfully to climate variability and change, and includes adjustments in both behavior and in resources and adaptive technologies. [2] adaptive capacity is influenced not only by economic development and technology, but also by social factors such as social capital, social networks, values, perceptions, customs, and governance structures. Smith and Pilifosova (2003) argued that even though some region own rich resources or stay in a good economic condition, the climate vulnerability is higher than the poverty areas [10]. The main causes rely on the local poor cognition and deficient adaption measures. This study relies on a participatory action research approach to probe the impact of perception and adaptive demand on the design and implementation of effective adaptation strategies. So in the research, some approaches such as semi-structured interviews, scoring and ranking, will be applied to analyze the perception and action willingness of stakeholders.

The sampling design of this interview involved 53 participants, all of whom were the crucial policy-makers or actors in the implementation of local adaptation strategies. The participants would be divided into three groups. The first group refers to municipal officials from different sectors including departments of urban planning, water management, transportation, health care, forestry and disaster management. The second group belongs to meteorological experts who are engaged in independent researchers or consultants in meteorological agencies. The third group involves the community residents who mainly come from the downtown of Shanghai (Seen in Table 2).

Table 2. Number of interviewees and stakeholder groups represented

	Governmental, officials	The community residents	Academic, independent expert	Total
Interview participants	17	12	6	35
Scoring participants	14		4	18
Total	31	12	10	53

3.1 Risk Reception of Different Stakeholders to Climate Risk

The aim of this interview refers to understand their direct feeling or cognition about the climate risk through semi-interview with governmental officials, the meteorological experts and the residents in shanghai. Therefore, the interview question was designed as "what kinds of extreme climate events impressed you happened in Shanghai. The participants from three groups all perceived typhoon, rainstorm and heat wave as the main climate risks, which marches with the frequency of these climate hazards happened in Shanghai (Table 1). However, there were some divergences about the impacts caused by these hazards. Firstly, the meteorological experts paid more concern on the climate hazard itself, involving rainstorm, heat wave, flooding, frozen, draught and lightning. As the professionals, the direct perceptions from meteorological experts are more objective and specific. For instance, one of the most impressed extreme climate event to the experts was the rainstorm happened in August 1977, which cost the losses about 28 million dollars and 2 people dead. The experts cared more about exact extreme climate events rather than the impacts and damage resulting from the climate events. Secondly, the governmental officials concerned more about the impacts of climate risk on economic and social development. Taken the rainstorm as an example, the interviewed officials were impressed by the hazard impact on transportation operation, the crop losses and casualties resulting from the rainstorm. So the governmental officials care more about indirect climate risks such as waterlogging, infrastructure destroy, environmental impacts, economic losses and public health care. Thirdly, the perception of residents derived from their living experiences rather than the professional acknowledge. The residents, as rational-economic men, cared more about the short-term interest and their own losses in hazards. Usually they did not consider certain climate events as risks until their life or properties are threatened. Furthermore, one outcome from the interview with residents has drawn the attention. Though the residents are the important stakeholders in the climate adaptation, 76 % of the residents involved in this survey stated that even they do what they can for the adaption to climate risk, but it is useless or does not make a difference because their efforts are trifle compared with the systemic adaptive strategies. So the residents felt that the climate change was "governmental tasks and none of my business". From the outcome of the survey to analysis, the residents in Shanghai have a low perception of climate risks, which not only bring dilemmas to adaptive action but also increasing the vulnerability of population in climate hazards.

3.2 The Analysis of Adaptive Demand in Shanghai

Climate adaptation refers to anticipating the adverse effects of climate change and taking appropriate action to prevent or minimize the damage they can cause, or taking advantage of opportunities that may arise [6]. Due to the different perception in climate risks, even in the same region, stakeholders would have distinct opinions in the adaptation demand. The adaptation demand may be multiple, but when the sources are limited, adaptation action should be carried out to satisfy the most emergent demands. In this study, the officials coming from different departments and the experts from meteorological agencies were required to indicate the urgency of climate vulnerabilities and rank their levels. The results have been shown in the Tables 3 and 4. In the opinion of meteorological experts, Typhoon is the most dangerous climate risk, to which the agriculture productions are very vulnerable as well as energy supply and causalities. Furthermore, the experts believed that rainstorm is the other risk which results in the urban waterlogging. Overall, the agriculture, urban drainage and energy supply are vulnerable to the climate risk. Compared with meteorological experts' opinion, the governmental officials deemed that the agriculture only account for little percent of GDP development in Shanghai, so they did not put the emphasis of adaptation on agricultural development, but treated rainstorm as the most emergency, which usually bring damage and threaten to transportation operation and energy supply. Typhoon is another primary climate risk in the official's perception, and they agreed the impact of typhoon on agriculture and energy supply. The heat wave is treated as the third climate risk in Shanghai not only to the experts but also governmental officials. But the expert argued that its impacts mainly focus on the energy supply, and the officials concerned its impact on agriculture. Though the threat of heat wave to population health has been noticed by both groups, they did not believe that the vulnerability pf population to heat wave is a serious problem.

Table 3. Evaluation results of climate vulnerabilities in Shanghai meteorological agency

effect / cause	A1 20	A2 12	A3 11	A4 10	A5 8	score
typhoon 9	1 1	16 7	7 7	3 14	8 11	40
heat wave 6	26 20	0 0	0 2	6 3	3 3	28
rainstorm 6	0 0	15 14	11 2	1 4	8 6	36
score	27	31	18	10	36	

Table 4. Evaluation results of climate vulnerabilities in Shanghai municipal departments

	Rainstorm	Typhoon	Heat wave	Scores
Transportation	5	0	1	16
Energy supply	3	2	0	13
Losses of crops	0	2	3	7
Environment	0	1	1	2
Constructions	0	0	0	1
Casualties	0	1	1	2

**Institute for Urban and Environmental Studies Chinese Academy of Social Sciences, The investigation report of climate risks in Shanghai and its adaptive strategies, 2011*

Compared with the perception of the officials and meteorological experts in Shanghai, the participants from residents group felt through their direct living experiences that heat wave and rainstorm may bring more damage to their property or lives. However, regarding to the adaptive capacities for these two climate hazards, they felt satisfactions with the efforts of governmental adaption actions in minimizing the damage from rainstorm, but sounded the alert for preventing heat waves. The responders from resident groups reflected that they are easy to be stuck in the heat wave due to three action dilemmas. Firstly, the heat wave has turned to be an emerging climate disaster as the global temperature kept rising in recent years. But the residents in Shanghai still have not enough awareness of its hazard, so when they, especially the children or the old, take heatstroke or other disease caused by heat wave, the residents lack the related knowledge and experiences for timely medical aid. Secondly, the power and water supply for households still stay in the tension state, even though some factories and shopping centers in Shanghai have been restricted in the consumption of utilities on the hottest days of the summer. The power supply in Shanghai depends on the import of electricity from the other cities in Yangtze River Delta, where the energy tensions are also highlighted during the hot weather. Third dilemma is related to the medical care system. The residents in the interview described that it was difficult to call the ambulance or other medical aid services in the hottest days. The high temperature alert did not stir up enough attention to the medical care system, so the shortage of the medical reserves might delay the medical aid, which can be drawn a lesson from the heat wave disaster in Chicago during the July 1995.

Grounded on the distinct perception of potential climate risks, the stakeholders have different demands for engineering measures, technological innovation, policies or other adaptation measures for preventing the most threatening climate risks. The meteorological experts considered that weather forecast should involve the meteorological impact analysis in the health care, transportation, agriculture and other urban infrastructure operations so as to increase the capacities of anticipating potential damages and planning disaster prevention. Compared with the meteorological experts, the sectoral departments in Shanghai keep more interest and concerns in defense work for rainstorm flooding, which had hit shanghai frequently in the history and resulted in serious damage.

For instance, the urban drainage system engineering is one of the most important infrastructure development projects in Shanghai. Until 2010, the drainages have been built for 11488 km. Furthermore, since 2008 the drainages have been dredged for twice per year so as to improve the water discharge capacity. In the decade the drainage ability in Shanghai has increased obviously. Furthermore, during the interview, the governmental officials considered that the disaster warning system should be taken advantage to break an early alert for the vulnerabilities transportation. They deemed that rapid and unreasonable underground space uses brought more burdens and vulnerabilities to underground pipe system, which was prone to Waterlogging disasters. The digital management system for underground pipe network integrate with weather forecast information, which both are an important components in the urban disaster warning system, would minimize the impacts of climate risks such as rainstorm on the urban drainage systems

4 Conclusions

This research depicted the specific characteristic of climate risks in four aspects, which involved hazard itself and the vulnerability of population, economy and environmental capacity. Firstly, the most metropolis areas mainly lie in the coastal zones and are sensitive to the climate change. The increasing climate hazard in these locations such as the rising sea level, the rainstorm, and heat wave are threatening economic and social develop. Secondly, the aging population and the increasing population density in metropolitan areas have exposed more and more the old and vulnerable people to the climate hazard such as heat wave and flooding. The vulnerability of population has been outstanding in metropolitan climate risks. In addition, the climate hazard interweaved with the pressure of urbanization development has aggravated the environmental deterioration and the shortage of natural resources, as well as the degeneration of ecological rehabilitation capacity. Finally, the economic vulnerability to climate change has also been analyzed in this research.

Furthermore, this research probed in the risk perceptions and adaptive demands of different stakeholders in Shanghai through interview investigations. The perception level of stakeholders to climate risks determine if the adaption strategies can be planned exactly and the adaptive demands have a role in the priorities of various adaption actions. In this research, the outcomes of interviews showed that the main climate hazards such as typhoon, rainstorm and heat wave have been perceived by the stakeholders involving governmental officials, meteorological experts and the public. However, the adaptive demands among the stakeholders were distinct. The meteorological experts and local officials concerned more about the adaption on vulnerable transportation operation, urban flooding, agriculture development and energy securities in Shanghai. Compared with the former stakeholders, the public argued that the most emergent adaption action should be put into heat wave, which have turned to be a serious disaster in the decade and threatened the health of vulnerable people. So this research suggests that a disaster prevention system for heat wave should be set up as the priority adaption action, as well as the healthy services and insurances should be improved further.

In addition, from the analysis of the interview outcomes to see, both officials and metrological experts have noticed the necessity of taking adaptation measures for the climate change risks. Especially the vulnerabilities of urban flooding, energy supply, transportation and agriculture development have been put into the emphasis of adaptation action. However, the adaptation to environmental vulnerability in Shanghai has been ignored. For instance, even the vast investment has been contributed to the retrofit engineering of urban drainage so as to relief the risks of urban flooding, urban river channel keep dropping, which play a role of self-rehabilitation to digest the flooding.

Acknowledgments. We thank reviewers and the Editor for the critical comments which improved an initial version of the paper. We also thank Dr. Pan Jiahua who brought more advices for the research and Shanghai Meteorological Service Center who provides abundant metrological data to support this research. Furthermore, the study was funded by Chinese National Social Science Foundation Project (Project No.: 11CJL055), and by China Postdoctoral Science Foundation Funded Project (Project No.: 2014T70163).

References

1. Safi, A.S.: Rural Nevada and climate change: vulnerability, beliefs, and risk perception. Risk Anal. **32**, 1041–1059 (2012)
2. IPCC: Climate Change 2007: Impacts, Adaptation and Vulnerability: Working Group II Contribution to the Fourth Assessment Report of the IPCC Intergovernmental Panel on Climate Change, vol. 4: pp 13–17. Cambridge University Press, Cambridge (2007)
3. Wisner, B.: At Risk: Natural Hazards, People's Vulnerability and Disasters. Psychology Press, East Sussex (2004)
4. Adger, W.N.: Vulnerability. Glob. Environ. Change **16**, 268–281 (2006)
5. Bruneau, M.: A framework to quantitatively assess and enhance the seismic resilience of communities. Earthq. Spectra **19**, 733–752 (2003)
6. Smit, B., Wandel, J.: Adaptation, adaptive capacity and vulnerability. Glob. Environ. Change **16**, 282–292 (2006)
7. Ellis, C.J.: A risk-based model of climate change threat: hazard, exposure, and vulnerability in the ecology of lichen epiphytes. Botany **91**, 1–11 (2012)
8. Xiangrong, W.: Global Climate Change and Vulnerability Assessment of Estuary City—A Case Study of Shanghai. Science Press, Beijing (2010)
9. Heijmans, A.: From Vulnerability to Empowerment. Mapping Vulnerability: Disasters, Development and People. Earthscan, London (2004)
10. Smit, B., Pilifosova, O.: Adaptation to climate change in the context of sustainable development and equity. Sustain. Dev. **8**, 9–16 (2003)
11. Burg, J.: Measuring populations' vulnerabilities for famine and food security interventions: the case of Ethiopia's Chronic Vulnerability Index. Disasters **32**, 609–630 (2008)
12. McLeman, R., Smit, B.: Vulnerability to climate change hazards and risks: crop and flood insurance. Can. Geogr. **50**, 217–226 (2006)
13. Jotzo, F. et al.: Fulfilling Australia's International Climate Finance Commitments: Which Sources of Financing are Promising and How Much Could They Raise? Centre for Climate Economics and Policy, Crawford School of Economics and Government, ANU (2011)

14. Dobes, L.: Adaptation to Climate Change: Formulating Policy under Uncertainty: Centre for Climate Economics and Policy, Crawford School of Economics and Government, ANU (2012)
15. Jiang, C.: Impact assessment of the land use change in Shanghai based on ecosystem service value. China Environ. Sci. **4**, 95–100 (2009)
16. Qixin, X., Shiyuan, X.: Study on water-environment effects and counter measures for high-speed urbanization process in Shanghai. World Reg. Stud. **12**, 54–59 (2003)
17. Xizhe, P.: Population increasing and urban public security. http://theory.people.com.cn

Spatial Estimation of Mean Annual Precipitation (1951–2012) in Mainland China Based on Collaborative Kriging Interpolation

Fushen Zhang, Shaobo Zhong(✉), Zhitao Yang, Chao Sun, and Quanyi Huang

Department of Engineering Physics/Institute of Public Safety Research, Tsinghua University, Beijing 100084, China
zhongshaobo@tsinghua.edu.cn

Abstract. Spatially explicit distribution of mean annual precipitation are required in the quantitative research on several water-related issues. The difference of distribution of precipitation has complicated reasons, one of them being the spatial correlation between multivariate meteorological factors. In this study, collaborative kriging interpolation (CKI) was used to estimate the spatial distribution of mean annual precipitation in China. Precipitation data from 756 meteorological stations were used, and spatial correlations between seven meteorological factors were analyzed, including annual precipitation, average barometric pressure, average wind speed, average temperature, average water pressure, average relative humidity, and annual average sunshine hours. The estimation results were assessed by means of cross-validation with the mean error (ME), mean absolute error (MAE), and root mean square error (RMSE). The results indicated that adding the spatial correlation analysis between multivariate meteorological factors can help improve the prediction performance.

Keywords: Spatial distribution · Spatial correlations · Multivariate meteorological factors · Collaborative kriging interpolation

1 Introduction

Spatial estimation of precipitation over any region is a basic parameter in the study of climatic change, flood disasters caused by heavy rain, and several other water-related problems. Although precipitation is typically measured using ground-based meteorological stations, interpolating these discrete observations as a continuous map is helpful for many water-related studies [1, 2].

Aiming at the problem of spatial estimation for precipitation, many scholars carry on the exploration regarding this by using different methods, such as inverse distance weighting [3], linear or nonlinear regression [4], geographically weighted regression [5], artificial neural networks [6–8], Kriging method [9].

However, ground-based meteorological stations currently unable to interpret the explicit distribution information of precipitation, partly because of the sparsity of meteorological stations [10], and partly because of spatial correlations between

F. Bian and Y. Xie (Eds.): GRMSE 2015, CCIS 569, pp. 663–672, 2016.
DOI: 10.1007/978-3-662-49155-3_69

meteorological factors. Many researchers have confirmed that more accurate estimation results can be acquired by integrating relationship information between precipitation and ancillary observation data. Thus, multivariate spatial interpolation techniques as an necessary tool are applied in precipitation mapping, which estimate the values at unmeasured locations based on data collection locations.

In this paper, annual precipitation data from 756 meteorological stations were used, and spatial correlations between seven meteorological factors were analyzed, including mean annual precipitation, average barometric pressure, average wind speed, average temperature, average water vapor pressure, average relative humidity, average annual sunshine durations. Besides, two spatial prediction methods (i.e., CKI, normal kriging interpolation) were implemented with different calculation parameters. A comparison of the results obtained by these two methods was assessed by using cross-validation with the mean error (ME), mean absolute error (MAE), and root mean square error (RMSE).

2 Materials and Method

2.1 Flowchart of Study Method

The flowchart of the study method used in this paper is composed of three parts, i.e., the study area and sampling data, statistical analysis and variograms, and collaborative kriging interpolation technique (Fig. 1). Collaborative kriging interpolation technique was used to determine the explicit spatial mapping of mean annual precipitation, and six meteorological factors (including average barometric pressure, average wind speed, average temperature, average water vapor pressure, average relative humidity, and average annual sunshine duration) were used as the co-variables.

2.2 Study Area and Sampling Data

The data set used in this work contains 756 meteorological stations, consisting of 60-year-average (1951–2012) values of each meteorological factor per station along with their geographical coordinates [10]. We initially prepared covariates of longitude, latitude, mean annual precipitation, average barometric pressure, average wind speed, average temperature, average water vapor pressure, average relative humidity, average annual sunshine durations. Annual precipitation was selected as the main estimation variable in the data sets, other prepared covariates are selected as the reference variables. Some missing data during the period were excluded to ensure the quality of the observation data. Finally, the meaningful data from 756 meteorological stations were used in this study (See Fig. 2).

2.3 Statistical Analysis and Variograms

Density scaled histograms use regularly spaced bins between specified minimum and maximum values. The histogram has the meaning of a probability distribution function, and the sum of the displayed rectangular surfaces is equal to 1. Kernel density estimates

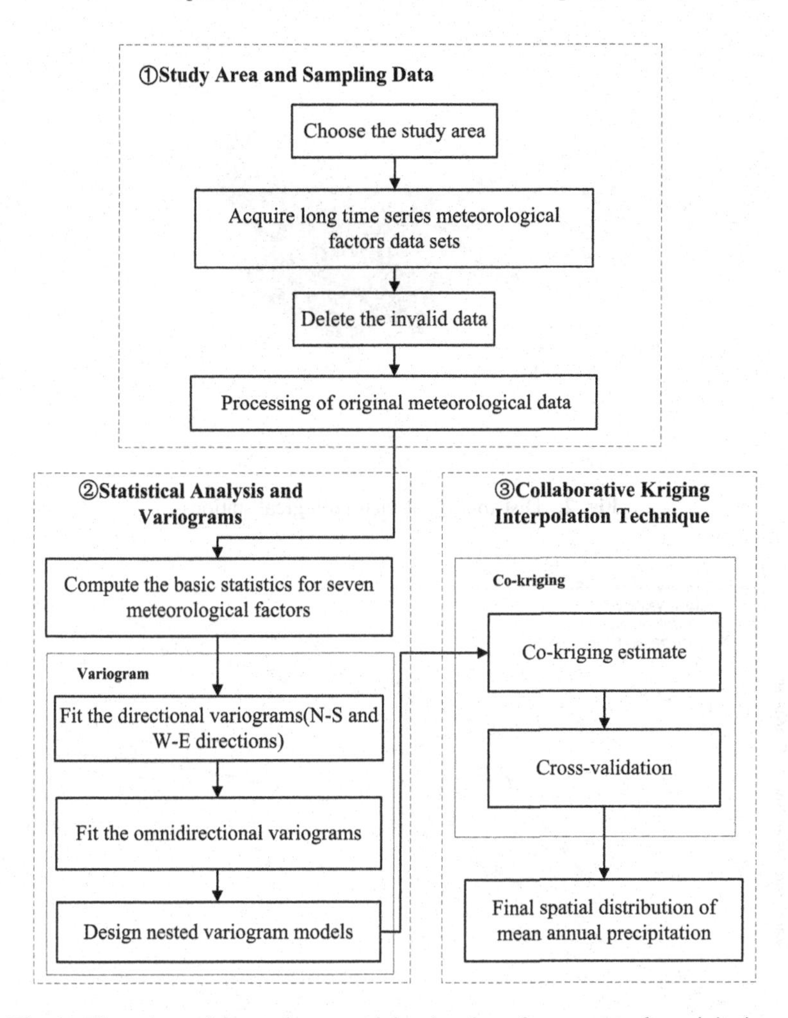

Fig. 1. Flow chart of the study on spatial estimation of mean annual precipitation.

evaluated from −700 to 3500 by step of 100 (Annual precipitation), 43 to 130 by step of 2 (Average barometric pressure), −3 to 15 by step of 0.1 (Average wind speed), −17 to 40 by step of 0.5 (Average temperature), −700 to 3500 by step of 10 (Average water pressure), 15 to 110 by step of 1 (Average relative humidity) and 200 to 4200 by step of 110 (Annual average sunshine hours), respectively.

Spatial correlation deal with how variance and covariance depend on the distance between seven meteorological factors. Variograms and cross-variograms are the prerequisite for analysis, and it can enhance the accuracy of the estimated results. To characterize the spatial structure of the studied meteorological factors, the construction of variogram models was essential.

We used a four-step procedure to fit a nested variogram model.

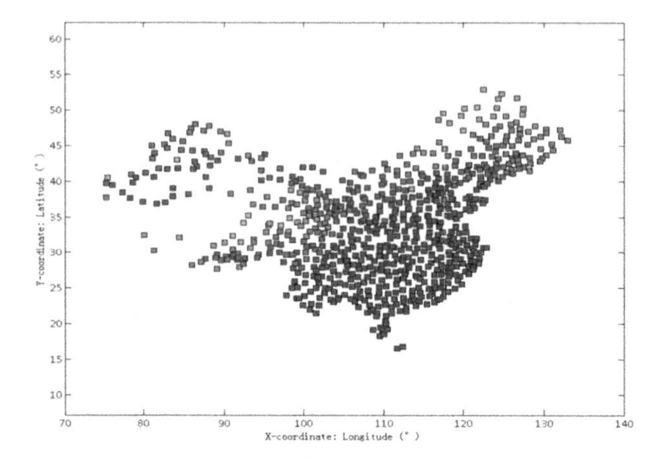

Fig. 2. Distribution of meteorological stations.

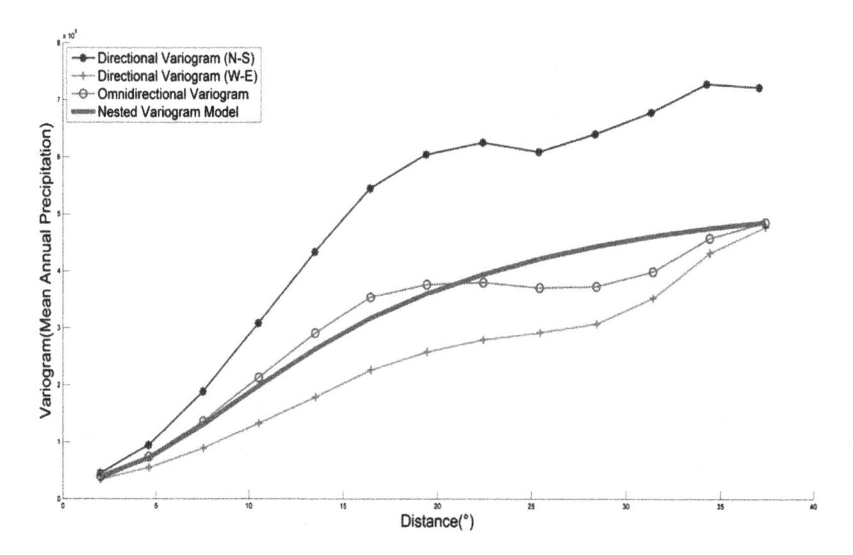

Fig. 3. Experimental and theoretical variograms/cross-variograms.

Step 1: We compute the directional variograms for the mean annual precipitation along the N-S and the W-E directions, using an angular tolerance of 90°, and distances classes are 3°.
Step 2: We compute the omnidirectional variograms for the mean annual precipitation, distances classes are the same than above.
Step 3: We plot the nested variogram models (See Fig. 3). Nested models are defined as linear combinations of several basic models composed of one nugget effect and two standard Gaussian model (See Eq. 1).

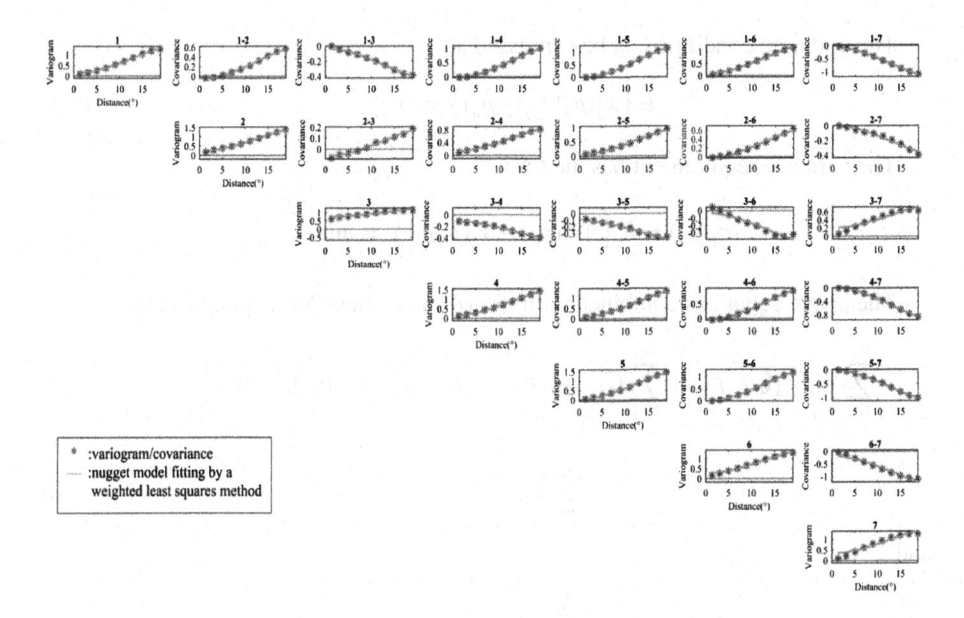

Fig. 4. The whole set of variograms and cross-variograms. Variables one through seven represent variables related to mean annual precipitation, average barometric pressure, average wind speed, average temperature, average water vapor pressure, average relative humidity, average annual sunshine durations, respectively.

$$r(h) = \begin{cases} 3\times10^{4}, h = 0 \\ 3\times10^{4} + 2.2\times10^{5}(1 - e^{-\frac{3h^2}{20^2}}) + 2.6\times10^{5}(1 - e^{-\frac{3h^2}{42^2}}),\ 0 < h \le 20 \\ 3\times10^{4} + 2.2\times10^{5} + 2.6\times10^{5}(1 - e^{-\frac{3h^2}{42^2}}),\ 20 < h \le 42 \\ 3\times10^{4} + 2.2\times10^{5} + 2.6\times10^{5}, h \ge 42 \end{cases} \quad (1)$$

Step 4: By using the weighted least squares method and the same nested model above, the whole set of variograms and cross variograms were fitted (See Fig. 4).

2.4 Collaborative Kriging Interpolation Technique

Integrating spatial correlation information between seven meteorological factors and estimating spatial distribution of mean annual precipitation can be done by using a multivariate extension of the normal kriging technique, known as collaborative kriging interpolation [11], which can be expressed as

$$x_2^*(p_0) = \sum_{i=1}^{n_1} \lambda_{1i} x_1(p_{1i}) + \sum_{j=1}^{n_2} \lambda_{2j} x_2(p_{2j}) \quad (2)$$

where λ_{1i} and λ_{2j} are the weights associated with x_1 and x_2; n_1 and n_2 are the number of neighbors of x_1 and x_2 involved in the estimation of the position p_0, respectively.

The unbiased condition can be expressed as

$$E\left\{x_2^*(p_0) - x_2(p_0)\right\} = 0 \tag{3}$$

The variance requirement has the form

$$\sigma_{ck}^2(p_0) = E\left\{\left[x_2^*(p_0) - x_2(p_0)\right]^2\right\} = min \tag{4}$$

The above equations combined with (5)–(8), yield the CKI technique [11]:

$$\sum_{i=1}^{n_1} \lambda_{1i} C_{11}(p_{1i}, p_{1k}) + \sum_{j=1}^{n_2} C_{12}(p_{1k}, p_{2j}) - \mu_1 = C_{21}(p_0, p_{1k}),\ \ k = 1, n_1 \tag{5}$$

$$\sum_{i=1}^{n_1} \lambda_{1i} = 0 \tag{6}$$

and

$$\sum_{j=1}^{n_2} \lambda_{2j} = 1 \tag{7}$$

and the minimum estimation variance,

$$\sigma_{ck}^2(p_0) = C_{22}(0) + \mu_2 - \sum_{i=1}^{n_1} \lambda_{1i} C_{21}(p_0, p_{1i}) - \sum_{j=1}^{n_2} \lambda_{2j} C_{22}(p_{2j}, p_0) \tag{8}$$

The interpolation results of two different methods, i.e., CKI and normal kriging, were assessed and compared by using cross-validation with the mean error (ME), mean absolute error (MAE), and root mean square error (RMSE). And its basic idea consists of re-estimating mean annual precipitation at the locations of meteorological stations after removing.

3 Plots and Results

Figure 3 shows the variograms for mean annual precipitation.

Figure 4 shows the whole set of variograms and cross-variograms.

Figure 5 shows the interpolation results by using different maximum hard station number, and Fig. 6 shows the interpolation results by using different lag distance.

Overall, the spatial distribution of mean annual precipitation generally decreases from south to north and increase from west to east, which is consistent with the actual situation.

Table 1 shows the assessment result of the interpolator performance. The result show that, after parameter optimization and considering the statistical correlation between multivariate meteorological factors, CKI result is more reliable. Table 1 shows the assessment result of the interpolator performance.

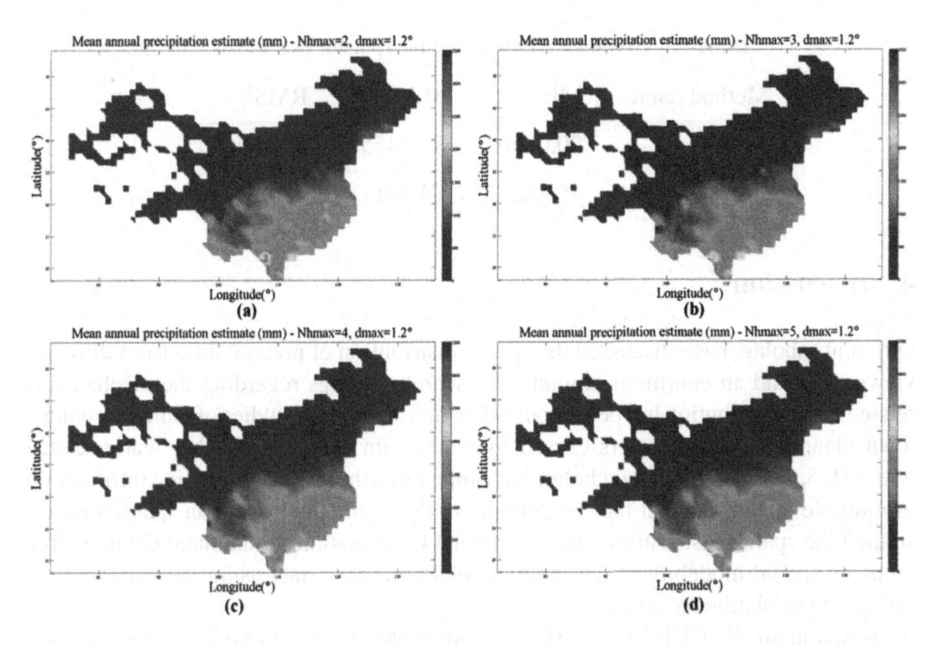

Fig. 5. The interpolation results by using different maximum hard station number.

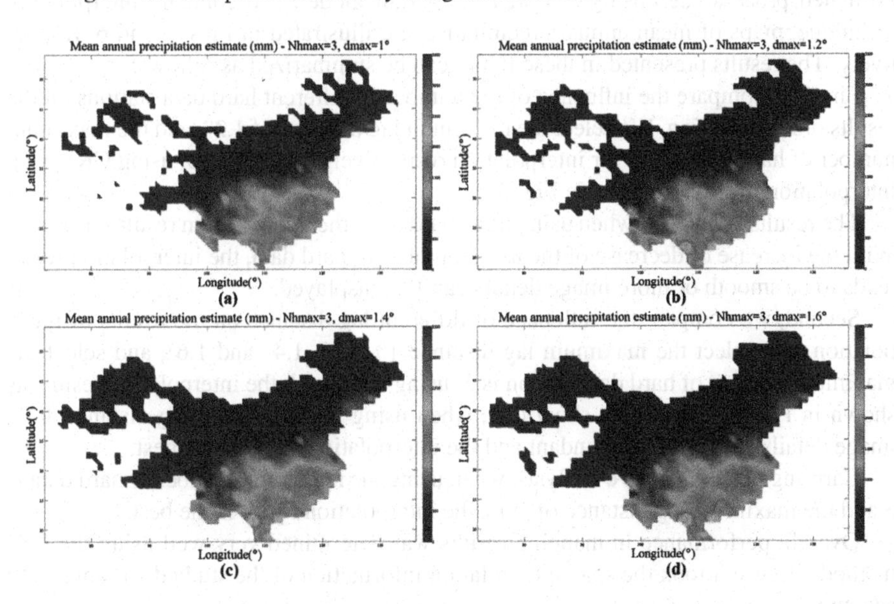

Fig. 6. The interpolation results by using different lag distance.

Table 1. Assessment result of the interpolator performance.

Method name	ME	MAE	RMSE
Kriging	20.0890	75.9455	133.8063
CKI	15.4027	74.4011	126.7393

4 Discussion

Different scholars have discussed the spatial distribution of precipitation from different viewpoints, and an enormous amount of research findings regarding the mechanisms related to the distribution has been reported, which are used in studies of climatic change, town planning, flood disasters caused by heavy rain and several other water-related issues [1, 3, 7]. However, few scholars have integrated the spatial correlation information of multiple meteorological factors to improve the estimation precision. Therefore, we studied the spatial distribution of mean annual precipitation in mainland China with a focus on spatial modeling of the sampled meteorological data using the collaborative kriging interpolation technique.

Based upon the CKI framework, the variogram model can be integrated into the estimated process (i.e., using a nested variogram model). To compare interpolation technique, maps of mean annual precipitation are illustrated in Figs. 5 and 6, respectively. The results presented in these figure can be summarized as follows:

First, we compare the influence of the number of different hard data stations on the results of interpolation. We select the maximum lag distance of 1.2°, and the maximum number of hard data using for interpolation respectively are 2, 3, 4, 5, using CKI, each interpolation results as shown in Fig. 5.

The results show that, when using three hard data, the interpolation result is the best. With the increase of decrease of the using number of hard data, the interpolation result tends to be smooth or more image details can't be displayed.

Second, we compare the influence of different lag distance on the results of interpolation. We select the maximum lag distance 1°, 1.2°, 1.4° and 1.6°, and select the maximum number of hard data station is 3, using CKI, each the interpolation results as shown in Fig. 6. The results show that, when using the maximum lag distance 1.4°, image details are the most abundant and the interpolation result is the best.

Carrying on comparative analysis, when using the maximum number of hard data is 3 and the maximum lag distance of 1.4°, the interpolation result is the best.

Overall, performance in mapping results was determined very well using the CKI method because it took the spatial correlation information of the studied variables into account.

This study has limitations that should be addressed. First, the versatility of the study approach has yet to be verified. Some of the problems associated with CKI may be overcome by extending the study to other research areas and studying additional types of meteorological factors. Second, we plan to obtain more field-based meteorological data using mobile weather stations to improve the accuracy of spatial estimation.

5 Conclusions

This paper study on spatial estimation of mean annual precipitation based on CKI technique by integrating statistical correlation information between multivariate meteorological factors. Cross-validation tests were used to compare the estimation performances between two interpolation techniques. From the cross-validation results, CKI seems to give the more accurate results. Further, the fusion of statistical correlation information between multivariate meteorological factors and other regional variables such as topographic factors will help to improve more accurate interpolation result.

Acknowledgments. The authors would like to thank the support of the National Natural Science Foundation of China (Study on Pre-qualification Theory and Method for Influences of Disastrous Meteorological Events, Grant No. 91224004) and the youth talent plan program of Beijing City College (Study on Semantic Information Retrieval of Decision Analysis of Emergency Management for Typical Disastrous Meteorological Events, Grant No. YETP0117).

References

1. Gao, C., Zhang, Z., Zhai, J., Qing, L., Mengting, Y.: Research on meteorological thresholds of drought and flood disaster: a case study in the Huai River Basin, China. Stoch. Env. Res. Risk Assess. **29**(1), 157–167 (2015)
2. Wang, S., Jiang, F., Ding, Y.: Spatial coherence of variations in seasonal extreme precipitation events over Northwest Arid Region. China Int. J. Climatol. **35**(15), 4642–4654 (2015)
3. Chen, C., Zhao, N., Yue, T., Guo, J.: A generalization of inverse distance weighting method via kernel regression and its application to surface modeling. Arab. J. Geosci. **8**(9), 6623–6633 (2014)
4. Stojković, M., Ilić, A., Prohaska, S., Plavšić, J.: Multi-temporal analysis of mean annual and seasonal stream flow trends, including periodicity and multiple non-linear regression. Water Resour. Manage. **28**(12), 4319–4335 (2014)
5. Taghipour Javi, S., Malekmohammadi, B., Mokhtari, H.: Application of geographically weighted regression model to analysis of spatiotemporal varying relationships between groundwater quantity and land use changes (case study: Khanmirza Plain, Iran). Environ. Monit. Assess. **186**(5), 3123–3138 (2014)
6. Nastos, P.T., Paliatsos, A.G., Koukouletsos, K.V., Larissi, I.K., Moustris, K.P.: Artificial neural networks modeling for forecasting the maximum daily total precipitation at Athens, Greece. Atmos. Res. **144**, 141–150 (2014)
7. Dahamsheh, A., Aksoy, H.: Markov chain-incorporated artificial neural network models for forecasting monthly precipitation in arid regions. Arab. J. Sci. Eng. **39**(4), 2513–2524 (2014)
8. Young, C., Liu, W.: Prediction and modelling of rainfall–runoff during typhoon events using a physically-based and artificial neural network hybrid model. Hydrol. Sci. J. (2014) 141217125340005
9. Estimating spatially downscaled rainfall by regression kriging using: TRMM precipitation and elevation in Zhejiang Province, southeast China. Int. J. Remote Sens. **35**(22), 7775–7794 (2014)

10. Ding, Y.: China Meteorological Disaster Authority (Comprehensive Volume). China Meteorological Press, Beijing (2008)
11. Sideris, I.V., Gabella, M., Erdin, R., Germann, U.: Real-time radar-rain-gauge merging using spatio-temporal co-kriging with external drift in the alpine terrain of Switzerland. Q. J. R. Meteorol. Soc. **140**(680), 1097–1111 (2014)

Assessment of Spatio-Temporal Vegetation Productivity Pattern Based on MODIS-NDVI and Geo-Correlation Analysis

Zhenyu Wang[1], Gaole Li[1], Yiru Dai[1], Zhibo Wang[1,2], and Zongyao Sha[1(✉)]

[1] International School of Software, Wuhan University, Wuhan 430079, China
zongyaosha@whu.edu.cn

[2] Software College, East China Institute of Technology, Nanchang 330013, China

Abstract. This paper explored the spatio-temporal patterns of vegetation productivity based on MODIS-NDVI and spatial auto-correlation analysis in the grassland of Inner Mongolia, China during 2011–2013. Two statistics indices, i.e., spatial auto-correlation and semi variance function, were applied in the analysis. The results showed that: (1) at regional scale, the NDVI presented a positive spatial auto-correlation, while at local scope NDVI showed high-high auto-correlation in the eastern part of the study region where the vegetation cover was relatively best. In contrast, NDVI displayed low-low auto-correlation in the western area where the vegetation cover was poor. (2) During 2011 to 2013, the structural factors explained 70 % of the total spatial variations impacting the vegetation cover, and the annual precipitation also played a significant role in the spatial variation of vegetation cover.

Keywords: NDVI · Inner Mongolia · Spatial auto-correlation · Semi variance function

1 Related Work

Grassland vegetation plays an important role in the global ecology and understanding the spatial patterns of vegetation distribution can contribute to sustainable development of grassland management. As a proxy for vegetation productivity, NDVI has been produced routinely from the imagery acquired by various remote sensing platforms. As a matter of fact, NDVI has become the best indicator in spatial distribution density and vegetation growth status [1]. Spatial auto-correlation is one of the most important methods to analyze the similarity between a certain geographic phenomena or properties in two adjacent cells, and it is also an effective mean to explore spatial relationship in economic, natural and social studies [2].

Many studies have made a great effort on the dynamic changes of vegetation based on NDVI. Further, vegetation growth is closely relevant to climatic conditions, such as precipitation and average temperature. It has been found that NDVI is significantly correlated to climatic factors [3]. For instance, Li found the link between NDVI variation

F. Bian and Y. Xie (Eds.): GRMSE 2015, CCIS 569, pp. 673–681, 2016.
DOI: 10.1007/978-3-662-49155-3_70

and climate factors in the Red River Valley [4]. NVDI variation has very clear differences in different climatic conditions, land uses and time periods.

Spatial auto-correlation analysis has already been well and widely used in the field of landscape ecology, demography and economics. For example, by means of spatial statistics, Zhang et al. studied vegetation cover in Mongolian Plateau and concluded that the vegetation of stable state on the whole and showed globally positive spatial auto-correlation [5]. Besides, Zhao et al. used MODIS-NDVI image data from 2000 to 2006 in Jilin Province and concluded that the vegetation ecosystem had been restored but vegetation cover variation had a clear spatial difference [6]. Wang et al. reached the conclusion that vegetation cover in both Shandong Peninsula and Liaodong Peninsula showed strongly positive global spatial auto-correlation [7].

Inner Mongolia was ever a vast prairie while it has been desertification in some regions, which is a calamity for the residents and the ecosystem. This paper explored the space auto-correlation features and clustering patterns of vegetation productivity using MODIS-NDVI, and climate data. The results can provide more scientific foundation for the implementation of ecological protection and some references for analyzing the external factors' effect to the vegetation productivity.

2 Data Source and Process

2.1 Data Source

The data used in this experiment consists of three parts, MODIS-NDVI, meteorological data, and vegetation type cover. MODIS-NDVI is at a resolution of 250 m and generated globally each 16 days, comes from LP DAAC/NASA (The Land Processes Distributed Active Archive Center). We downloaded 12 scenes of during June to September each year from 2011 to 2013. NDVI is calculated as,

$$\mathrm{NDVI} = (\mathrm{NIR} - \mathrm{R})/\mathrm{NDVI}\,(\mathrm{NIR} + \mathrm{R}\,) \tag{1}$$

NIR represents the reflectivity in near-infrared wavelength. R represents the reflectivity in infrared wavelength. Meteorological data, i.e., monthly precipitation and average temperature measured by 48 meteorological stations in Inner Mongolia from 2011 to 2013, was provided by China Meteorological Data Sharing Service System. Vegetation type cover comes from national survey of grassland resources in 1980.

2.2 Spatial Auto-Correlation

Spatial auto-correlation is an important indicator to test and analyze whether the property of a factor is significantly correlative with that of its adjacent point. There are two kinds of classification, positive correlation and negative correlation.

Spatial Weight Matrix shows the spatial layout of objects in different space, such as adjacency, topological relationships, etc. Therefore, to represent the adjacent relationship among some positions, we usually define a binary and symmetric spatial weight matrix, W, whose form is as follow:

$$W = \begin{bmatrix} W_{11} & W_{12} & \cdots & W_{1n} \\ W_{21} & W_{22} & \cdots & W_{2n} \\ \cdots & \cdots & \cdots & \cdots \\ W_{n1} & W_{n2} & \cdots & W_{nn} \end{bmatrix} \quad (2)$$

N represents the amount of the spatial unit. W_{ij} represents the relationship between region i and j.

General indicators of spatial auto-correlation is used to measure the spatial distribution features of unit property in the entire region and the degree of association of adjacent units. To explore spatial pattern of NDVI in Inner Mongolia, the paper uses Moran's I index which is calculated by the following formula:

$$I = \frac{n}{\sum_{i=1}^{n}\sum_{j=1}^{n} W_{ij}} \times \frac{\sum_{i=1}^{n}\sum_{j=1}^{n} W_{ij}(x_i - \bar{x})(x_j - \bar{x})}{\sum_{i=1}^{n} (x_j - \bar{x})^2} \quad (3)$$

where x_i and x_j represent the value of variable x in adjacent pairing spatial units. $\bar{x}$ represents the average value of x. W_{ij} represents the adjacent weight. n represents the amount of spatial units. The Coefficient of Moran's I is between −1 and 1. While it is more than 0, it shows the vegetation cover trends to clustering status and has good integrity and if it is less than 0, it shows the spatial distribution trends to fragmentation.

Local indicators of spatial auto-correlation is mainly used to analyze the heterogeneity of the various spatial units. So it can improve the instability of General indicators of spatial auto-correlation. Its formula is

$$I' = n\frac{(x_i - \bar{x})}{\sum_{i=1}^{n} (x_i - \bar{x})^2} \sum_{j=1}^{n} (x_j - \bar{x}) \quad (4)$$

Moran scatterplot, usually used to explore the instability of local space, has four kinds of relationship form. High-high and low-low represent strongly spatial positive correlation while high-low and low-high suggest strongly spatial negative correlation.

2.3 Semi Variance Function

Heterogeneity, one of the main reasons of spatial pattern's forming, is the variability of a certain property in the space. It can be quantitatively described by semi variance function, which defines the distribution feature of the property by measuring their variability degree and the distance between them and the formula is:

$$\gamma(h) = \frac{1}{2N(h)} \sum_{i=1}^{N(h)} \left[Z(x_i) - Z(x_i + h)\right]^2 \quad (5)$$

Z(x) represents regional variable and is NDVI in this paper.

After using the above formula, we can require some estimates which can be used to generate theoretical model, such as linear model, the Gaussian model, exponential model and so on. We select exponential model in this paper and the formula is as follow:

$$\gamma(h) = C_0 + C(1 - e^{-\frac{h}{a}}) \tag{6}$$

where C_0 represents nugge, a represents range, and h represents the lag distance.

According to geostatistics principle, we can use $C_0/(C_0 + C)$ to study the degree of the random factors' influence in the variability of the whole space. If the value is high, it shows random factors have more important impact, otherwise, structural factors do.

3 Results

3.1 Temporal and Spatial Property of NDVI

Based on the MODIS-NDVI, higher vegetation productivity was found in the east and lower vegetation productivity in the west of the study region. The spatial difference of vegetation in Inner Mongolia is obvious. In 2011, statistics shows that average NDVI is 0.42 in the growing season and the proportion of the region with NDVI > 0.6 occupied 24.2 %. In 2012, average value of NDVI reduced to 0.21, showing very poor vegetation growth. While in 2013, the NDVI > 0.6 showed an increase.

Figure 1 showed the absolute difference between the years. It can be seen that compared to 2011, the NDVI in 2012 decreased a lot on average, especially in the eastern part. Compared to 2012, the NDVI in 2013 showed increasing trend.

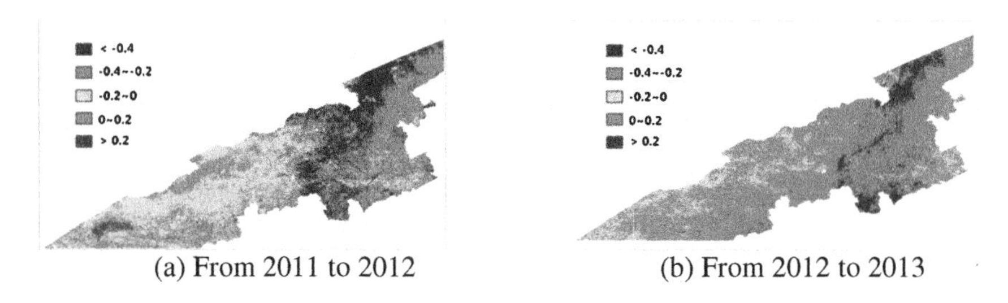

(a) From 2011 to 2012 (b) From 2012 to 2013

Fig. 1. The changes of growing season NDVI between adjacent years

3.2 General Indicators of Spatial Auto-Correlation

The Moran's I index of Inner Mongolia vegetation region in 2011 to 2013 is 0.9254, 0.8542 and 0.8900 respectively. It is obvious that NDVI in Inner Mongolia vegetation region has significantly spatial positive correlation, or the vegetation cover in this region is of clustering pattern (Fig. 2).

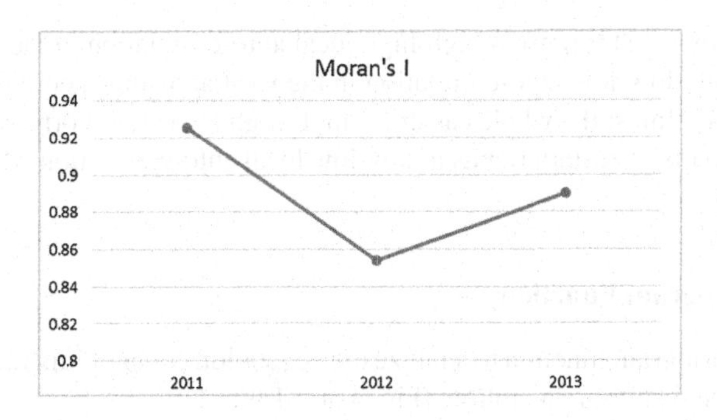

Fig. 2. Moran's I changes from 2011 to 2013

3.3 Local Indicators of Spatial Auto-Correlation

Global indicators provide a good reflection to spatial clustering status, but they cannot show the detail of each local region. Therefore, local indicators of spatial auto-correlation were also used in this study.

Figure 3 indicated the clustering character of vegetation productivity based on local indicator. The red region represents high-high clustering, the blue region represents low-low clustering which means NDVI of both is relatively low, the light blue region represents low-high clustering which means NDVI of this unit is lower than that of surrounding units and the light pink region represents high-low clustering which means NDVI of this unit is higher than that of surrounding units. White means non-significant.

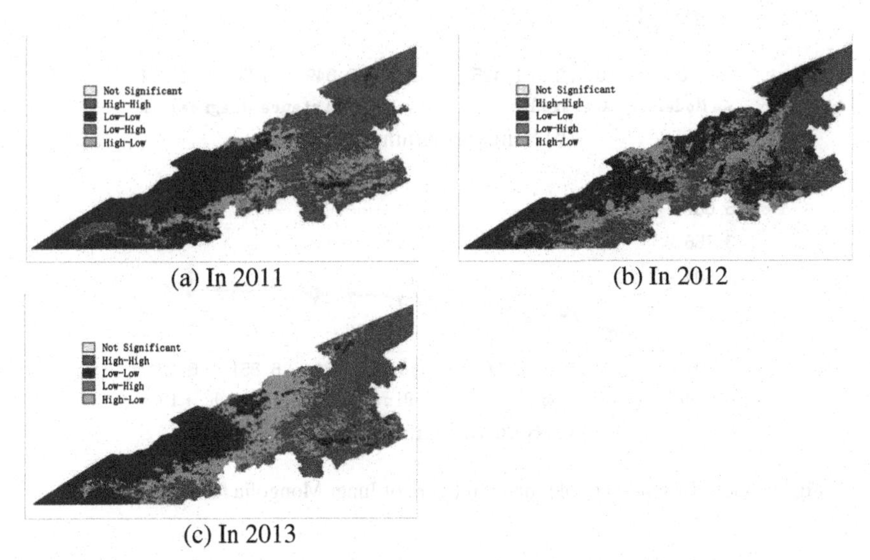

Fig. 3. LISA Diagram of Inner Mongolia vegetation region from 2011 to 2013

Figure 3 shows that it present high-high local auto-correlation in the east and some sporadic area and low-low auto-correlation in the west and some sporadic area. As we can see, in 2011, almost the whole eastern is high-high local auto-correlation, however, in 2012, some parts of eastern trends to low-low local auto-correlation while it began to restore in 2013.

3.4 Semivariogram Function

We use Semivariogram function to analyze the vegetation cover of Inner Mongolia from 2011–2013. The results are as follow (Fig. 4 and Table 1):

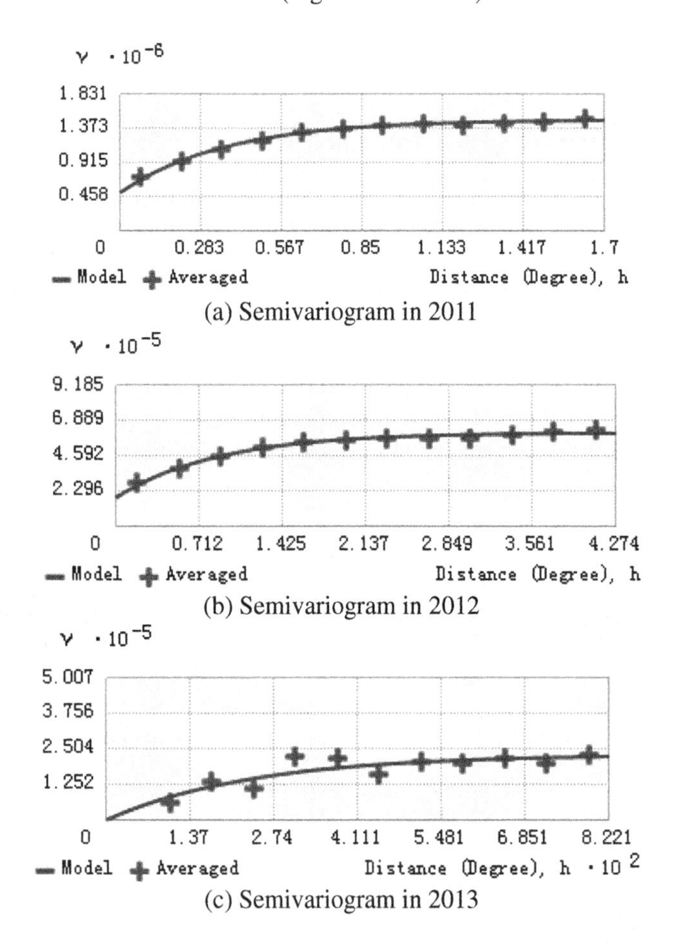

(a) Semivariogram in 2011

(b) Semivariogram in 2012

(c) Semivariogram in 2013

Fig. 4. General semi variable function figure of Inner Mongolia from 2011 to 2013

Table 1. NDVI spatial variability from 2011 to 2013

YEAR	$C_0(*10^3)$	$C_0 + C(*10^3)$	$C_0/(C_0 + C)$	A_0(m)	Fitting model
2011	509.19	1487.66	0.34	278.5	Exponential model
2012	184.02	608.56	0.30	650	Exponential model
2013	58.28	214.172	0.27	177.5	Exponential model

Sill (C0 + C) represents the highest variation degree that the function can reach. The spatial variation showed decreasing trend from 2011 to 2013. Moreover, the nugget also decreased in the period, meaning the spatial variation caused by random factors showed decreasing tendency. Besides, the range in 2012 is 650 m while that in 2011 and 2013 is 278.5 m and 177.5 m respectively, suggesting that the spatial auto-correlation of 2011 and 2013 is more significant than that in 2012.

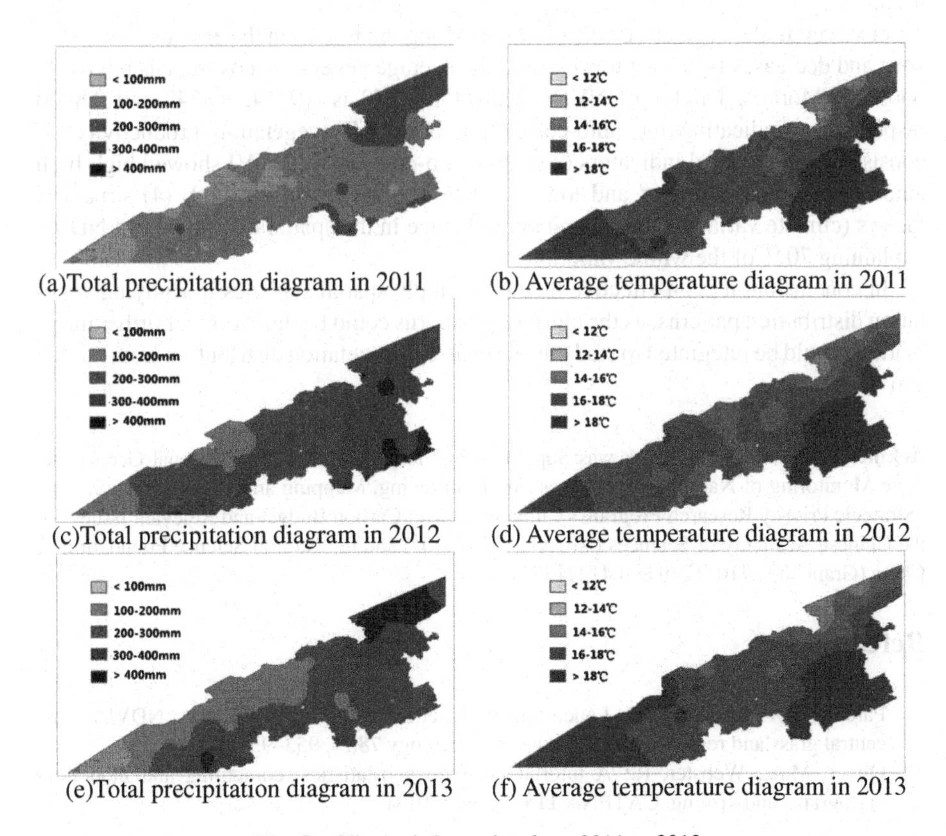

(a)Total precipitation diagram in 2011 (b) Average temperature diagram in 2011
(c)Total precipitation diagram in 2012 (d) Average temperature diagram in 2012
(e)Total precipitation diagram in 2013 (f) Average temperature diagram in 2013

Fig. 5. Climate information from 2011 to 2013

3.5 Variance in Climate Conditions and Elevation

To explain the vegetation growth and the variance of NDVI, the average temperature and precipitation were investigated further in Inner Mongolia from 2011 to 2013 (Fig. 5).

The average annual precipitation of the 3 years is 231.66 mm and the average temperature is 17.3 °C in the growing season. Spatially, the precipitation increased and average temperature decreased from southwest to northeast generally.

It shows high-high local auto-correlation in the northeast where the vegetation cover status is relatively good with precipitation greater than 300 mm and temperature between 12 °C and 18 °C. Besides, it shows low-low local auto-correlation in the northeast where the vegetation cover status is relatively poor with precipitation less than 150 mm and temperature between 15 °C and 20 °C.

4 Conclusion

By means of spatial statistics and geostatistics, MODIS-NDVI was used to investigate the spatio-temporal pattern in the Inner Mongolia vegetation during 2010 to 2013. Moran's I spatial auto-correlation analysis and semivariogram function analysis were used.

(1) Overall, the vegetation cover in Inner Mongolia is high in the east and low in the west and decreases from east to west, and the average vegetation coverage is relatively poor. (2) Moran's I index of NVDI in 2011 to 2013 is 0.9254, 0.8542 and 0.8900 respectively, indicating the auto-correlation of NDVI (vegetation productivity) is consistent. (3) As local indicators of spatial auto-correlation, NDVI showed high-high auto-correlation in the east and low-low auto-correlation in the west. (4) structural factors (climate variables) had dominant influence in the spatial distribution of NDVI, explaining 70 % of the whole variation.

In conclusion, it is an effective method to apply spatial statistics in studying vegetation distribution patterns, as the clustering patterns could be discovered. Further information should be integrated to study the influence of vegetation distribution from human activities.

Acknowledgments. This work was Supported by "Key Laboratory for National Geographic State Monitoring of National Administration of Surveying, Mapping and Geoinformation", the "Strategic Priority Research Program - Climate Change: Carbon Budget and Relevant Issues" of the Chinese Academy of Sciences (No. XDA05050402) and the Natural Science Foundation of China (Grant Nos. 41071249 and 41371371).

References

1. Paruelo, J.M., Epstein, H.E., Lauenroth, W.K., et al.: ANPP estimates from NDVI for the central grassland region of the United States. Ecology **78**(3), 953–958 (1997)
2. Oliver, M.A., Webster, R.: A tutorial guide to geostatistics: computing and modelling variograms and kriging. CATENA **113**, 56–69 (2014)

3. Jaime Gómez-Hernández, J., Horta, A., Jeanée, N.: Geostatistics for environmental applications. Spat. Stat. **5**, 1–2 (2013)
4. Li, Y., He, D.: The spatial and temporal variation of NDVI and its relationships to the climatic factors in Red River Basin. J. Mt. Sci. **3**, 333–340 (2009)
5. Zhang, X., Hu, Y., Zhuang, D., et al.: NDVI spatial pattern and its differentiation on the Mongolian Plateau. Geograph. Res. **28**(1), 10–18 (2009)
6. Zhao, C., Shu, H., Xuan, G.: Temporal-spatial analysis of vegetation change based on MODIS images. Sci. Surveying Mapp. **35**(5), 173–175 (2010)
7. Wang, X., Jiang, D., Ma, D.: Spatial auto-correlation analysis of vegetation cover using MODIS NDVI time series data-regional comparison of Shandong Peninsula and Liaodong Peninsula. J. Arid Land Resour. Environ. **27**(10), 24 (2013)
8. Crabtree, R., Potter, C., Mullen, R., et al.: A modeling and spatio-temporal analysis framework for monitoring environmental change using NPP as an ecosystem indicator. Remote Sens. Environ. **113**, 1486–1496 (2009)
9. Lin, G.-F., Chen, L.-H.: A spatial interpolation method based on radial basis function networks incorporating a semivariogram model. J. Hydrol. **288**, 288–298 (2004)
10. Aragao, L.Y., Shimabukuro, S.F., et al.: Landscape pattern and spatial variability of leaf area index in Eastern Amazonia. For. Ecol. Manag. **211**(3), 240–256 (2005)
11. Zeng, L., Levy, G.: Space and time aliasing structure in monthly mean polar-orbiting satellite data. J. Geophys. Res.: Atmos. (1984–2012) **100**(D3), 5133–5142 (1995)
12. Overmars, K.P., De Koning, G.H.J., Veldkamp, A.: Spatial autocorrelation in multi-scale land use models. Ecol. Model. **164**(2), 257–270 (2003)

Area Errors Between Grid Imagery Boundaries and Vector Actual Boundaries Identifying Waterbodies from Remote Sensing Imagery

Zhaofei Liu(✉) and Zhijun Yao(✉)

Institute of Geographic Sciences and Natural Resources Research, Chinese Academy of Sciences, Beijing 100101, China
{zfliu,yaozj}@igsnrr.ac.cn

Abstract. In identifying water bodies from remote sensing imagery, a mismatch between grid data boundary and vector boundary has always existed but was seldom studied. Therefore, area errors between grid imagery boundaries and vector real boundaries are the subject of this study. A solution based on the sub-pixel classification method was developed to analyse these errors. A case study from Lake Manasarowar in China showed that the area error proposed in this study is larger than that from different interpretation methods. It was concluded that uncertainties from mixed boundary pixels were greater than that from different methods for identifying lake area using the remote sensing imagery in the study area. Overall, area error analyses for grid imagery boundaries and vector real boundaries are necessary for identifying water bodies from remote sensing imagery. It is also useful for the interpretation of other continuous bodies, such as glaciers.

Keywords: Area error analyses · Sub-pixel classification · Grid imagery boundary · Vector actual boundary · NDWI

1 Introduction

Remote sensing has been widely applied to waterbodies monitoring because of it could quickly interpret waterbodies information. There are several methods for identifying waterbodies from remote sensing imagery. Dozier [1] developed a normalized difference snow index (NDSI), which is calculated by (Green-MIR)/(Green+MIR), Green and NIR mean visible green and mid-infrared reflected spectrums respectively. This index is based on the difference between strong reflection of visible radiation and near total absorption of middle infrared wavelengths by snow [2]. It is effective in distinguishing snow from similarly bright soil, vegetation and rock, as well as from clouds [3]. McFeeters [4] developed the normalized difference water index (NDWI) method. The NDWI index is calculated by (Green-NIR)/(Green+NIR), in which NIR mean near infrared reflected spectrum. It could maximize the reflectance of water features in the green band, (2) minimize the low reflectance of water features in the NIR band and (3) take advantage of the high reflectance of terrestrial vegetation and soil features in the NIR band. Xu [5] proposed a modified

F. Bian and Y. Xie (Eds.): GRMSE 2015, CCIS 569, pp. 682–692, 2016.
DOI: 10.1007/978-3-662-49155-3_71

NDWI, which was calculated as same as the NDSI but for identifying open water. There were several assessments for evaluating performance of these NDWIs on identifying waterbodies from remote sensing imagery [6–9]. A threshold of a ratio image of Red/MIR (TM band3/TM band4) or NIR/MIR (TM band4/TM band5) was also used by many studies to delineate glacier outlines from imagery [10–15].

Although there has been much improving spatial and temporal resolutions for remote sensing imagery, the accuracy assessment is still necessary in the interpretation of remote sensing imagery [16, 17]. Calculation of uncertainties from multi-satellite imagery was developed by William [18], and widely used in glaciers mapping from imagery [19–21]. But it is not suitable for the single imagery accuracy assessment. The accuracy assessment for single imagery needs a referenced data, which is usually taken as manually interpretation higher spatial resolution imagery [22] or topographic map [23], GPS results [24].

However, errors between grid imagery boundaries and vector actual boundary were seldom mentioned. Actually, it is always a mismatch between grid data boundary and vector boundary. There is always a fractional area of waterbodies in any boundary grid. It could be estimated by the interpretation of imagery into the sub-pixel components. The interpretation of sub-pixel components from mixed pixels is often referenced as 'soft' classification method [25]. The corresponding 'hard' classification method identifies only one entity for each pixel.

There were several methods for the interpretation of imagery into sub-pixel, such as the fuzzy classification [26], the maximum likelihood classification [27], the Bayesian classification [28], and the linear spectral mixture (LSM) classification [29, 30]. The LSM method differs from many other methods that it links image spectra to laboratory or field spectral reflectance of materials [31]. In other words, it is a physically based image processing method [32].

The object of this paper is to propose area errors between grid imagery boundaries and vector actual boundaries in identifying waterbodies from remote sensed imagery. Lake area is selected as a case study to evaluate these errors.

2 Materials and Methods

2.1 Descriptions of Area Errors Between Grid Imagery Boundaries and Vector Real Boundaries

In the interpretation of remote sensing imagery, area errors always existed between grid and vector boundaries, because boundaries identified from imagery are grid, whereas real boundaries are vector. That difference is described in Fig. 1, which takes a lake as an example. As shown in Fig. 1, the imagery grid boundary is identified by water body pixels. It is obviously different from the real lake (vector) boundary. Area errors between them are affected by mixed pixels, which are crossed by the real lake boundary. In generally, the lake area was directly calculated by water body pixels, whereas mixed pixels were neglected. Although sub-pixel classification methods considered mixed pixels, and the accuracy metrics derived from them missed information on the spatial distribution of error [33]. In this study, area errors between grid imagery boundaries and vector real boundaries are estimated for each mixed pixels. Therefore, the spatial distribution of errors can be assessed.

2.2 Study Area and Datasets

The study area, Lake Manasarowar (also called Manasarovar), was generally recognized as the highest body of fresh water in the world [34]. The mean altitude of the study area was more than 4,600 meters above sea level.

Multispectral imagery of Landsat Thematic Mapper (TM) was obtained from http://glovis.usgs.gov/. An image was acquired over Lake Manasarowar on 21 October 2004.

Ground control points for the boundary of Lake Manasarowar (Fig. 2) were collected using the Magellan Triton 400 handheld GPS. Built-in signal augmentation reception provides an accuracy of 3 m.

2.3 Image Pre-processing

The correction and calibration of images, which was done to achieve as faithful a representation of the Earth surface as possible, was a fundamental consideration for all applications [35]. First, the raw pixel values of the image (known as digital numbers, or DN) were converted into true measures of reflective power (spectral radiance). Then, the radiance image data were atmospherically corrected to obtain the spectral reflectance image. Finally, image reflectance values were used for the following application. Details for this image processing method can be found in the Landsat 7 Science Data Users Handbook (http://landsathandbook.gsfc.nasa.gov/).

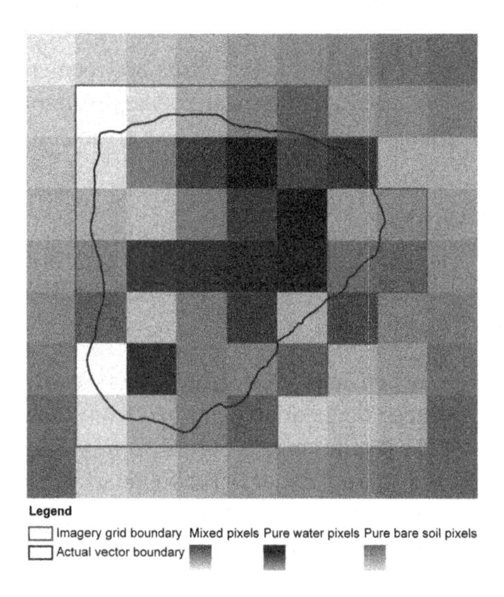

Fig. 1. Descriptions for area errors between the grid imagery boundary and the vector real boundary, with a lake as an example.

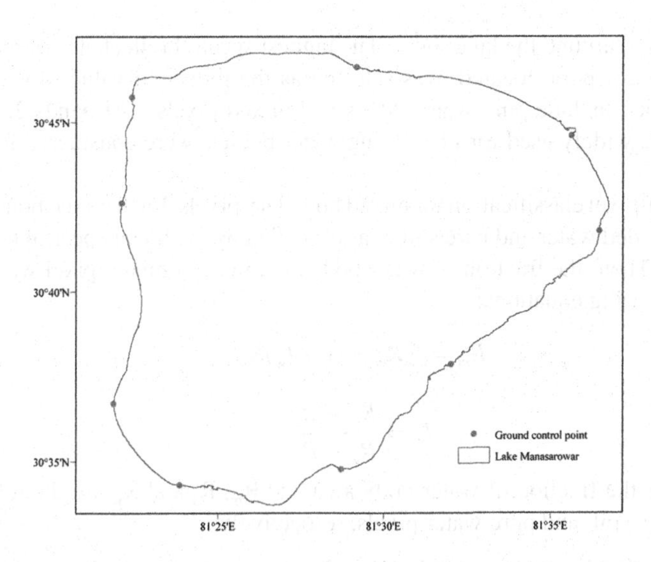

Fig. 2. The location of Lake Manasarowar and the ground control points.

2.4 Sub-pixel Classification

The LSM classification from an imagery band can be expressed as follows [28, 29]:

$$R_s = \sum_{i=1}^{n} f_i R_i + \varepsilon. \quad (1)$$

where R_s is the sum reflectance of the mixed spectrum in band i, and f_i and R_i are the fraction and the reflectance of the i_{th} components, respectively, and ε is the residual. Because the expectation of ε is zero [30], the residual for a specific band is neglected in this study.

2.5 Error Analyses

There were three steps for analysing area errors between grid imagery boundaries and vector real boundaries. Using Lake Manasarowar as an example:

Step 1. Identifying grid water body boundary from imagery. Spectral knowledge is important in extracting remote sensing imagery information. Although spectral reflectance of water is affected by many factors (such as depth and total dissolved solids), it is generally much lower than that of bare soil. For example, the reflectance of water and bare soil from TM band 5 is 0.18% and 2.98%, respectively, based on ground control points at Lake Manasarowar. Therefore, it was easy to differentiate water bodies from bare soil based on the threshold of spectral reflectance, because there was only bare soil around the lake shore. The highest spectral reflectance of pixels, which ground control point located, was selected as the threshold value to identify water pixels, which include pure water pixels and mixed pixels. TM bands 2, 3, 4 and 5, which were widely used for identifying water bodies, were considered for the interpretation.

Step 2. Sub-pixel classification for mixed boundary pixels. Each mixed boundary pixel, which included water and bare soil, was classified by the linear spectral mixture classification. Then, the fraction of water body area in each mixed pixel was calculated by the following equations:

$$R_m = f_w R_w + (1 - f_w) R_s. \tag{2}$$

$$f_w = \frac{R_m - R_s}{R_w - R_s}. \tag{3}$$

where f_w is the fraction of water body area and R_m, R_s and R_w are the reflectance of mixed, pure soil, and pure water pixels, respectively.

As shown in Fig. 3, for each mixed pixel, pure water pixels were searched around it at a radius of $30\sqrt{2}$ m. The radius was not identified until pure water pixel(s) were searched. The lowest reflectance of these pure water pixel(s) was applied to Eq. (3). This reflectance was the same as that for pure soil pixels.

Step 3. Area error analyses for identifying water bodies from imagery. Area error in identifying the grid boundary from imagery compared with the vector real boundary was calculated by the following equation:

$$E = \sum_{i=1}^{n} (1 - f_{wi}) P_a. \tag{4}$$

where f_{wi} is the fraction of water body area at the ith mixed boundary pixel, P_a is the pixel area, and n is the number of mixed boundary pixels.

Uncertainties from different methods of identifying water bodies from imagery were also estimated. The NDWI method, which has been widely applied in identifying water bodies from imagery, was selected. Three NDWIs, including (Green – NIR)/(Green + NIR) [4], (Green – MIR)/(Green + MIR) [5] and (NIR – MIR)/(NIR + MIR) [36], were calculated from multispectral TM imagery. Green, NIR, and MIR describe TM bands 2, 4 and 5, respectively.

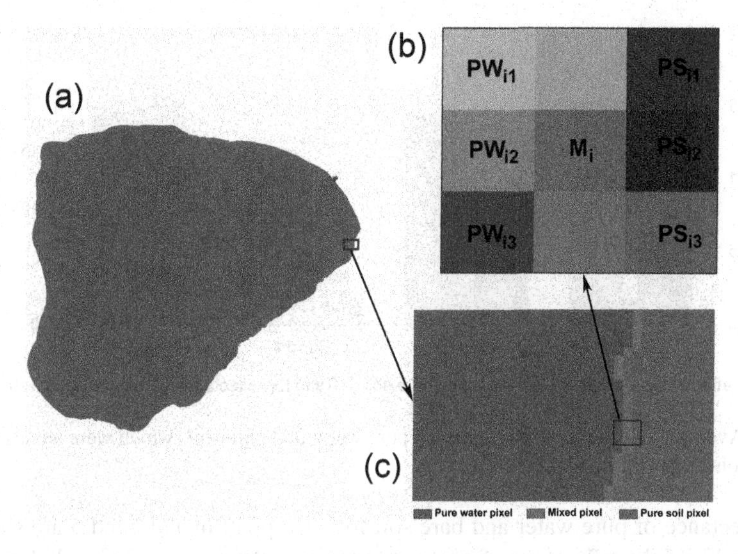

Fig. 3. Sub-pixel classification for mixed boundary pixels (Mi is the ith mixed pixel; PWij and PSij are the jth pure water and soil pixels around Mi, respectively).

3 Results

Based on the spectral reflectance threshold method, grid water body boundaries were identified from bands 2, 3, 4, and 5. The method of identifying mixed boundary pixels from each band is shown in Table 1. The interpretation showed the best performance in TM band 5. It was similar to the results of delineating water bodies from TM imagery at the Murray-Darling Basin [37]. Reflectance of pure water and bare soil pixels in TM band 5, which were adjacent to the ground control points, were also calculated in this study (Fig. 4). It can be seen in Fig. 4 that the ratio for reflectance of bare soil and water in band 5 was considerably larger than that in other bands. The reflectance of water was an order of magnitude lower than that of bare soil in band 5. In other words, the difference in reflectance of bare soil and water was greater in band 5 than in other bands. Therefore, TM band 5 was more suitable for identifying water bodies and bare soil at Lake Manasarowar.

Table 1. Identifying mixed boundary pixels from each TM band.

TM bands	Threshold value for mixed pixel	Correct ground control points number	Total ground control points number
2	0.141	6	9
3	0.125	8	9
4	0.091	6	9
5	0.019	9	9

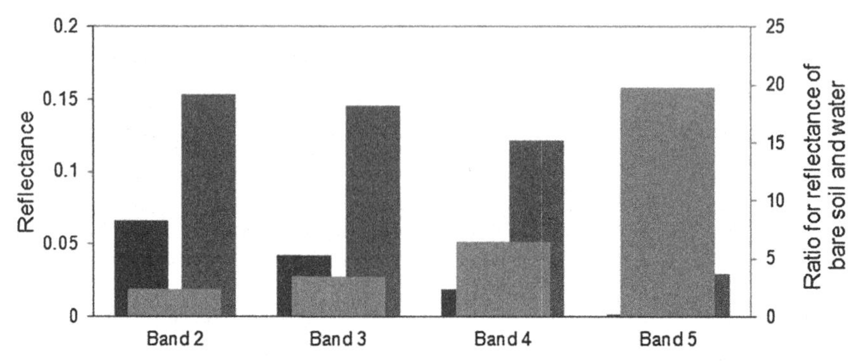

Fig. 4. Average reflectance of all pixels for pure water and bare soil (which were adjacent to the ground control points) in TM band 5.

Reflectance of pure water and bare soil for each pixel in TM band 5 are shown in Fig. 5. In Fig. 5 the reflectance of pure water and mixed pixels are obviously lower than that of pure soil pixels. That finding means the threshold classification method used in this study is reasonable to differentiate water pixels (pure water and mixed pixels) from pure soil pixels.

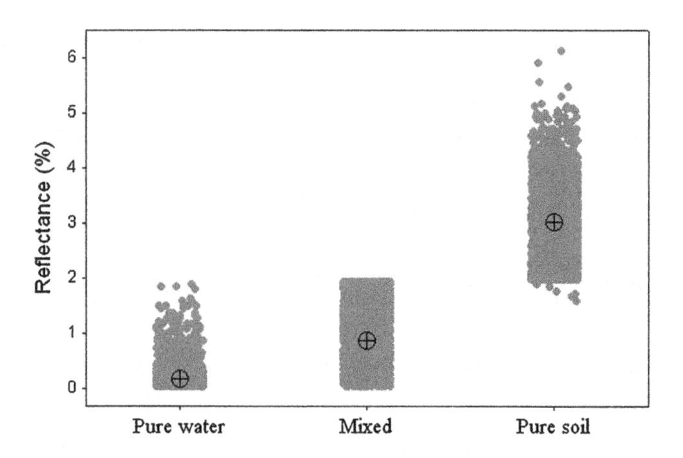

Fig. 5. Reflectance of pure water and bare soil for each pixel in TM band 5.

The grid boundary of Lake Manasarowar, which was identified from band 5, is shown in Fig. 2a. Results of area error analyses showed 3,476 mixed boundary pixels. The area error, which was between the grid imagery boundary and the vector real boundary, was 1.3 km^2. It was equal to the bare soil area in mixed pixels. Spatial distribution of error (fraction of bare soil area) for each mixed boundary pixel is shown in Fig. 6. The fraction of bare soil area in mixed boundary pixels was mainly located at 0.25 to 0.58, with mean value of 0.43 (Fig. 7).

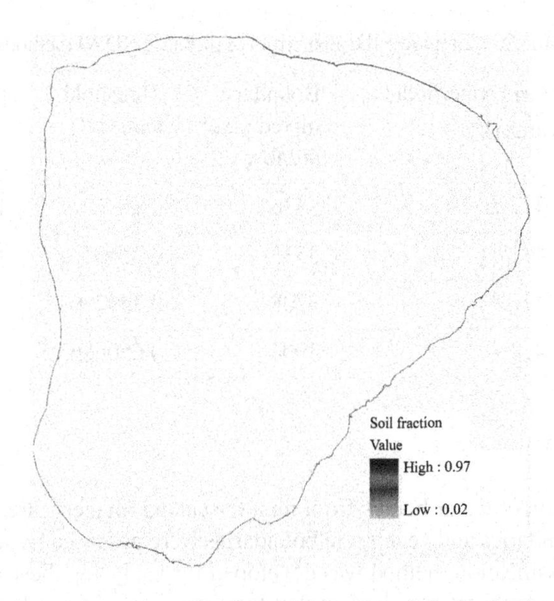

Fig. 6. Spatial distribution of error (fraction of bare soil area) for each mixed boundary pixel.

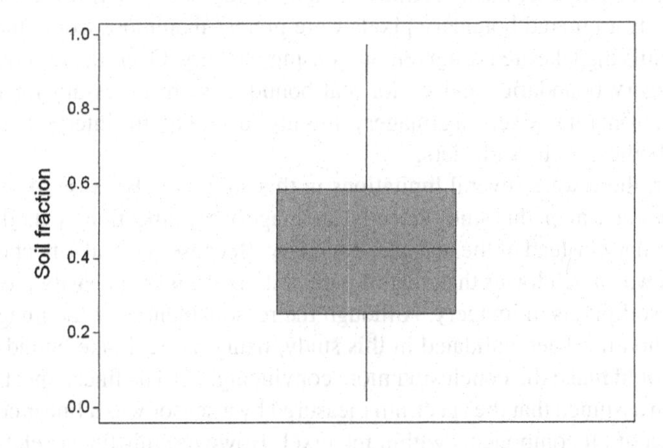

Fig. 7. Box plot for fraction of bare soil area at mixed boundary pixels.

The grid area of Lake Manasarowar identified from three NDWIs is shown in Table 2. The area difference among the three NDWIs is 1.0 km^2. It is lower than the error between the grid imagery boundary and the vector real boundary (1.3 km^2). In other words, uncertainties from mixed boundary pixels are greater than that from other methods for identifying lake area using imagery. Therefore, analyses of area errors between grid imagery boundaries and vector real boundaries are important and necessary for identifying water bodies from remote sensing imagery.

Table 2. Comparing the area errors of different NDWI methods.

Methods	Grid waterbodies area (km^2)	Boundary mixed pixel number	Threshold value	Errors (km^2)
Band 5	412.1	3476		1.3
NDWI-24	411.4	3533	0.2146129	1.0
NDWI-25	411.8	3708	0.7592704	
NDWI-45	412.4	3904	0.6506863	

4 Discussion

In the identification of water bodies from remote sensing imagery, area errors between grid imagery boundaries and vector real boundaries were proposed. A solution based on the sub-pixel classification method was developed for analysing these errors. The case study from Lake Manasarowar showed that the area error proposed in this study was larger than that from other interpretation methods (three NDWIs). It was concluded that uncertainties from mixed boundary pixels were greater than those from other methods used for identifying lake area using remote sensing imagery. Overall, area error analyses for grid imagery boundaries and vector real boundaries are necessary for identifying water bodies from remote sensing imagery. It is also useful for the interpretation of other continuous bodies, such as glaciers.

However, there were several limitations in this study. (1) Because of the lack of a real lake boundary map, this study selected an imagery classification method to identify the lake boundary instead of the real lake boundary. Because spectral reflectance of pure water pixels was much lower than that of pure soil pixels, it could easily identify water pixels from soil pixels in imagery. Although the reasonableness of the imagery classification method has been validated in this study, using the real lake boundary map, if available, would make the conclusion more convincing. (2) The linear spectral mixture classification assumed that the spectrum measured by a sensor was a linear combination of the spectra of all components within the pixel. However, this linear relation seldom completely or accurately reproduces a real spectrum. Other sub-pixel analyses methods might provide optional solutions. (3) Ground control points were limit in this study because of a terrible plateau environment (mean elevation was above 4,600 meters above sea level). It was necessary to get ground control points as often as possible, to sufficiently represent spectral reflectance of entities.

Acknowledgements. This study was supported and funded by the International Science & Technology Cooperation Program of China (2013DFA91700), National Key Technology Support Program of China (2012BAC06B02) and National Natural Science Foundation of China (41201035).

References

1. Dozier, J.: Snow-reflectance from Landsat-4 thematic mapper. IEEE Trans. Geosci. Remote **GE-22**(3), 323–328 (1984)
2. Hall, D.K., Riggs, G.A., Salomonson, V.V.: Development of methods for mapping global snow cover using moderate resolution imaging spectroradiometer data. Remote Sens. Environ. **54**, 127–140 (1995)
3. Dozier, J.: Spectral signature of alpine snow cover from the landsat thematic mapper. Remote Sens. Environ. **28**, 9–22 (1989)
4. McFeeters, S.K.: The use of the normalized difference water index (NDWI) in the delineation of open water features. Int. J. Remote Sens. **17**, 1425–1432 (1996)
5. Xu, H.: Modification of normalised difference water index (NDWI) to enhance open water features in remotely sensed imagery. Int. J. Remote Sens. **27**, 3025–3033 (2006)
6. Campos, J.C., Sillero, N., Brito, J.C.: Normalized difference water indexes have dissimilar performances in detecting seasonal and permanent water in the Sahara-Sahel transition zone. J. Hydrol. **464**, 438–446 (2012)
7. Sun, F.D., Sun, W.X., Chen, J., Gong, P.: Comparison and improvement of methods for identifying waterbodies in remotely sensed imagery. Int. J. Remote Sens. **33**(21), 6854–6875 (2012)
8. Li, W., Du, Z., Ling, F., Zhou, D., Wang, H., Gui, Y., Sun, B., Zhang, X.: A comparison of land surface water mapping using the normalized difference water index from TM, ETM+ and ALI. Remote Sens. **5**, 5530–5549 (2013)
9. McFeeters, S.K.: Using the normalized difference water index (ndwi) within a geographic information system to detect swimming pools for mosquito abatement: a practical approach. Remote Sens. **5**, 3544–3561 (2013)
10. Bayr, K.J., Dorothy, K.H., William, M.K.: Observations on glaciers in the eastern Austrian Alps using satellite data. Int. J. Remote Sens. **15**(9), 1733–1742 (1994)
11. Rott, H.: Thematic studies in alpine areas by means of polarmetric SAR and optical imagery. Adv. Space Res. **14**(3), 217–226 (1994)
12. Paul, F.: Evaluation of different methods for glacier mapping using Landsat TM. In: Proceedings of EARSeL-SIG-Workshop Land Ice and Snow, pp. 239–245, Dresden, Germany (2000)
13. Paul, F., Kääb, A.: Perspectives on the production of a glacier inventory from multispectral satellite data in Arctic Canada: Cumberland Peninsula, Baffin Island. Ann. Glaciol. **42**, 59–66 (2005)
14. Andreassen, L.M., Paul, F., Kääb, A., Hausberg, J.E.: The new Landsat-derived glacier inventory for Jotunheimen, Norway, and deduced glacier changes since the 1930s. Cryosphere **2**(2), 131–145 (2008)
15. Bris, R.L., Paul, F., Frey, H., Bolch, T.: A new satellite-derived glacier inventory for western Alaska. Ann. Glaciol. **52**(59), 135–143 (2011)
16. Congalton, R.G.: A review of assessing the accuracy of classifications of remotely sensed data. Remote Sens. Environ. **37**, 35–46 (1991)
17. Gao, J., Liu, Y.S.: Applications of remote sensing, GIS and GPS in glaciology: a review. Prog. Phys. Geog. **25**(4), 520–540 (2001)
18. Williams, R.S.J., Hall, D.K., Sigurdsson, O., Chien, J.Y.L.: Comparison of satellite-derived with ground-based measurements of the fluctuations of the margins of Vatnajfkull, Iceland, 1973-92. Ann. Glaciol. **24**, 72–80 (1997)
19. Silverio, W., Jaquet, J.M.: Glacial cover mapping (1987–1996) of the Cordillera Blanca (Peru) using satellite imagery. Remote Sens. Environ. **95**, 342–350 (2005)

20. Hall, D.K., Bayr, K., Schfner, W., Bindschadler, R.A., Chien, Y.L.: Consideration of the errors inherent in mapping historical glacier positions in Austria from ground and space (1893-2001). Remote Sens. Environ. **86**, 566–577 (2003)
21. Hendriks, J.P.M., Helsinki, P.P.: Semi-automatic glacier delineation from Landsat imagery from Landsat imagery over Hintereisferner in the Austrian Alps. Z. Gletsch.kd. Glazialgeol. **41**, 55–75 (2007)
22. Paul, F., Kääb, A., Maisch, M., Kellenberger, T., Haeberli, W.: The new remote-sensing-derived Swiss glacier inventory: I. Methods. Ann. Glaciol. **34**(1), 355–361 (2002)
23. Albert, T.H.: Evaluation of remote sensing techniques for ice-area classification applied to the tropical Quelccaya ice cap. Peru. Polar Geogr. **26**(3), 210–226 (2002)
24. Frezzotti, M., Capra, A., Vittuari, L.: Comparison between glacier ice velocities inferred from GPS and sequential satellite images. Ann. Glaciol. **27**, 54–60 (1998)
25. Bastin, L.: Comparison of fuzzy c-means classification, linear mixture modelling and MLC probabilities as tools for unmixing coarse pixels. Int. J. Remote Sens. **18**(17), 3629–3648 (1997)
26. Wang, F.: Fuzzy supervised classification of remote sensing images. IEEE Trans. Geosci. Remote **28**, 194–201 (1990)
27. Bolstad, P.V., Lillesand, T.M.: Rapid maximum likelihood classification. Photogramm. Eng. Remote Sens. **57**, 67–74 (1991)
28. Gorte, B., Stein, A.: Bayesian classification and class area estimation of satellite images using stratification. IEEE Trans. Geosci. Remote **36**(3), 803–812 (1998)
29. Roberts, D.A., Smith, M.O., Adams, J.B.: Green vegetation, non-photosynthetic vegetation, and soils in AVIRIS data. Remote Sens. Environ. **44**, 255–269 (1993)
30. Settle, J.J., Drake, N.A.: Linear mixing and the estimation of ground cover proportions. Int. J. Remote Sens. **14**(6), 1159–1177 (1993)
31. Adams, J.B., Sabol, D.E., Kapos, V., Filho, R.A., Roberts, D.A., Smith, M.O., Gillespie, A.R.: Classification of multispectral images based on fractions of endmembers: application to land cover change in the Brazilian Amazon. Remote Sens. Environ. **52**, 137–154 (1995)
32. Weng, Q.H., Lu, D.S.: A sub-pixel analysis of urbanization effect on land surface temperature and its interplay with impervious surface and vegetation coverage in Indianapolis, United States. Int. J. Appl. Earth Obs. Geoinf. **10**(1), 68–83 (2008)
33. Foody, G.M.: Status of land cover classification accuracy assessment. Remote Sens. Environ. **80**, 185–201 (2002)
34. Encyclopædia Britannica. http://www.britannica.com/EBchecked/topic/363541/Lake-Mapam
35. Eastman, J.R.: Introduction to Remote Sensing and Image Processing, Guide to GIS and Image Processing, vol. 1, pp. 27–28. Clark Labs, Worcester (2001)
36. Gao, B.C.: NDWI-A normalized difference water index for remote sensing of vegetation liquid water from space. Remote Sens. Environ. **58**(3), 257–266 (1996)
37. Fracier, P.S., Page, K.J.: Water body detection and delineation with Landsat TM data. Photogramm. Eng. Remote Sens. **66**(12), 1461–1467 (2000)

Watershed-Scale Phosphorus Balance Evaluation Using a Mass Balance Method

Shiyu Li[1,2], Bin Liu[1,2], Changliang Yang[1,2(✉)], Guiming Chen[2], Linna Yuan[1], Yangyu Song[2,3], Xiaomei Li[2,3], Deshou Cun[2,3], and Shuang Hu[2,3]

[1] Engineering Technology Institute, Yunnan University, Kunming 650091, China
lishiyukm@163.com, lisy@ynu.edu.cn, YANGCL227@163.com

[2] School of Ecology and Environmental Sciences, Yunnan University, Kunming 650091, China

[3] School of Urban Construction and Management, Yunnan University, Kunming 650091, China

Abstract. It is crucial for assessing the eutrophication risk of lake by analyzing the phosphorus (P) balance of lake watershed quantitively. A mass balance method was used to calculate P balance of both Yangzonghai lake watershed and the lake itself in one year. The imported P load was 725.1 t in 2010, while the exported P load was 317.3 t, which indicated that 56.2 % (407.8 t yr^{-1}) of P was retained in the lake watershed. Such a high retention load implied that the lake, which was mesotrophic, was under great pressure of further eutrophication. Among all the input pathways, the largest P input contributor was fertilizer, contributing 679.0 t P and accounting for 93.6 % of input P, followed by atmospheric deposition (44.7 t P, 6.2 %). Plant product (264.6 t, 83.4 %) was the largest P output contributor, followed by animal products (50.2 t, 15.8 %).

Keywords: Phosphorus load · Eutrophication · Watershed · Phosphorus balance

1 Introduction

Lake eutrophication is still one of the main water pollution problems in China [1]. Watershed development often increases lake nutrient inputs, leading to lake eutrophication [2]. Excessive lake P input is a major driver of algal blooms in lakes [3, 4]. Phosphorus (P) balance of lake watershed is important to determine if the lake is threatened by eutrophication. When there was more P imported into the watershed than exported, the redundant P retained in the watershed will flow into the lake and may finally result in lake eutrophication, or it will flow into lake through soil erosion [5], increasing the potential of lake eutrophication.

Yangzonghai Lake, a deep-water lake, which located in southwest China is working as a drinking water supply (before 2008) besides providing agricultural and industrial water. Due to nutrient overload, Yangzonghai Lake turns to be mesotrophic currently, which has a potential to threat the safety of drinking water, this is tightly

F. Bian and Y. Xie (Eds.): GRMSE 2015, CCIS 569, pp. 693–706, 2016.
DOI: 10.1007/978-3-662-49155-3_72

related to 60,000 people's health [6]. There are few previous studies reported the lake total P balance. However, it is important to draw up P controlling measures through deeply understanding of the P balance, especially the P flux of each input pathways.

Scholars have made much effort to study P input and output from lake watersheds. P inputs are presented in the form of atmospheric deposition, fertilizer, and other products. P outputs are presented in the form of harvested crops, livestock and its products, and the lake outlet [2, 7–9]. In this study, we identified the P input and output ways of the Yangzonghai Lake on watershed scale, and calculated the P balance of the watershed through mass balance methods, which will provide helpful information to control the lake eutrophication.

2 Materials and Methods

2.1 Study Area

Yangzonghai Lake (102°5′–103°02′E, 24°51′-24°58′N), with an average water area of 3,160 ha and catchment of 19, 200 ha. Has an elevation of 1,769.9 m. The volume of the lake is about 5.69×10^8 m^3 [10] and the main soil type is latosol. There are three inflow streams and one channel called Baiyi River outflow from the watershed, but only one outflow river called Tangchi River which is controlled by a sluice gate in the north (Fig. 1). There are 58,743 people living in the catchment (in 2010), most of whom are farmers. Agricultural activities are the main anthropogenic factors in the catchments, and there are only four industrial enterprises (Fig. 1, Brilliant Resort & Spa, Spring City Golf, Yangzong Thermal Power Plant and Yunnan Aluminum Factory) in which the

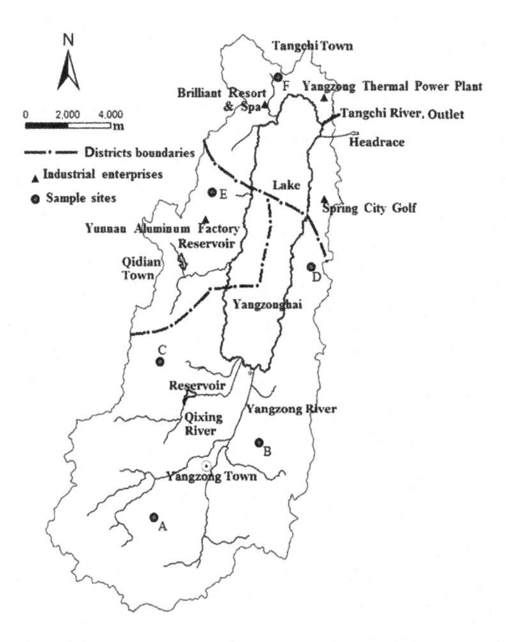

Fig. 1. Map of the Yangzonghai Lake catchment with sampling sites.

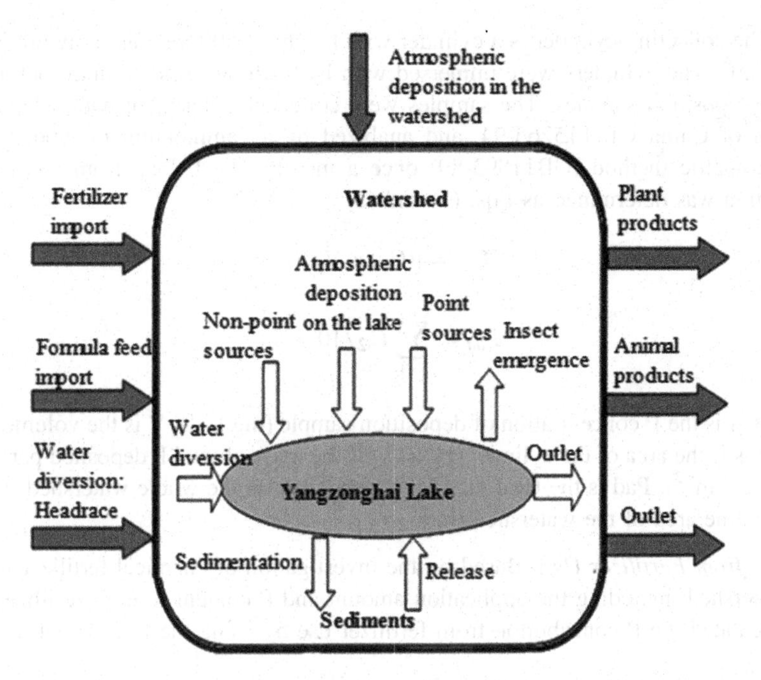

Fig. 2. Concept model of phosphorus balance in the Yangzonghai Lake watershed. In the diagram, solid arrows indicate the phosphorus transfer between the watershed boundaries, and hollow arrows indicates the phosphorus transfer between the lake subsystem boundaries.

sewage discharges into wastewater treatment plants and the treated sewages is discharged into Tangchi River and outflow from the watershed, not into the lake. For convenience, we divided the watershed into three districts: Yangzonghai District, Tangchi District, and Qidian District.

2.2 Methodology

We used a mass balance method [8, 11, 12] in this study to calculate P balance of the watershed which was regarded as a whole in this study and the lake itself in one year (2010). The watershed boundary is the natural catchment area, not including the watershed form where water is channeled. The concept model of the watershed is shown in Fig. 2. In 2010, the P input/output pathways of the watershed in Fig. 2 were studied. The specific methods are described below:

P Input Pathways on Watershed Scale. As Fig. 2 shows, the input pathways taken into consideration in the study included atmospheric deposition, fertilizer use, water diversion, formula feed input.

P Input Through Atmospheric Deposition. In order to calculate the P load from atmospheric deposition, we selected 6 sites (Fig. 1A–F) in the watershed and placed three dust collecting cylinders in each site to collect the total deposition samples for one year. The collecting cylinder is a cylinder with height of 500 mm and a mouth area of 0.0143 m^2. The cylinders were immersed with hydrochloric acid solution before collecting deposition samples. The samples were collected in terms of national standard method of China GB/T15265-94, and analyzed by an ammonium molybdate spectrophotometric method (GB11893-89) once a month. The P load from atmospheric deposition was determined as Eqs. (1) and (2):

$$C_{i2} = [(C_{i1} \times V)]/s. \quad (1)$$

$$P_{ad} = \sum_{i=1}^{n} C_{i2}/10 \times S. \quad (2)$$

where C_{i1} is the P concentration of deposition sample (mg L^{-1}), V is the volume of the sample, s is the area of the cylinder (m^2), C_{i2} is the mass of total P deposited per square meter (mg m^{-2}), Pad is the total mass of P deposited on the whole watershed (t yr^{-1}), and S is the area of the watershed (ha).

P Input from Fertilizer Use. Based on the investigation of chemical fertilizer used in the watershed, including the application amount and P content of each fertilizer type, *we* calculated the P contribution from fertilizer use by using the Eq. (3) below:

$$F_P = \sum_{i=1}^{n} Q_i \times C_{iP} \times 100. \quad (3)$$

where FP is the P load from chemical fertilizer, Q_i is the application amount of type i fertilizer, and C_{iP} is The P content of type i fertilizer.

P Input Through Water Diversion. The water flow from water diversion channels was continuously monitored by a hydrometric station, and the water quality was analyzed monthly by the local environmental monitoring institution, making it easy to obtain the P input load from water diversion.

P Output Pathways on Watershed Scale. As Fig. 2 shows, the output pathways identified in the study included plant products, animal products, outlet.

P Output Through Plant Products. P exported from one plant product was calculated by the output amount (data provided by the local government) multiplied by the P content of the product. We calculated the total P mass exported from plant products (including food products) by adding the P output of each product. In this study, the P content of the products was determined in terms of the recommended analyzing method of the China Ministry of Agriculture (NY/T1018-2006, the number of replications for each plant products ≥6). The plant products consumed by humans who lived in the watershed are not included.

P Output Through Animal Products. The P exported from one animal product was calculated by the output amount (data provided by the local government) multiplied by the P content of the product. We calculated the total P mass output from animal products by adding the P output of each animal product. In this study, the P content of the main animal products was determined according to the recommended analysis method of the China Ministry of Agriculture (NY/T1018-2006, the number of replications for each animal product ≥6). The animal products consumed by people who lived on the watershed were excluded.

P Output Through Outlet. The water from the outlet was continuously measured by a hydrometric station, and the effluent water quality was monitored monthly by the local environmental monitoring institution. We calculated data each month by multiplying the water flow of each month with the water total P concentration of the month, and obtained the total P output from the outlet by adding all of the data of 12 months.

P Input and Output of the Lake. We identified the P inflows and outflows pathways of the lake and calculated the P flux of each pathways (Fig. 2). The input pathways included sediment release, water diversion, atmospheric deposition, point pollution source input, and non-point pollution source input (Fig. 2). The outflows pathways included outlet, sedimentation and insect emergence [13] (Fig. 2). Similar methods were used to calculate P fluxes of lake through water diversion, atmospheric deposition, outlet. We calculated P load which flowed to the lake in terms of different pollutant types. Point source and non-point source P load was quoted from Yang's research [14]. The P atmospheric deposition was calculated by using Eq. 2, in which S means the surface area of the lake (i.e. S = 3,160 ha). Methods about how to calculate P fluxes through anaerobic sediment release and sedimentation were described below:

P Imported to Lake Through Anaerobic Sediment Release. The P flux from sediment was measured by an experiment carried out in the laboratory (n = 15). Sediments and overlying water were collected from the bottom of the lake (liquid: solid ratio of 20:1, i.e. 4000 mL water: 200 g sediments). N_2 was aerated into a closed glass box to keep environment anaerobic (Dissolved Oxygen <3.0), making sure that it is similar to the bottom environment of lake. The net P release fluxes ($P_{release}$, t) experiments were carried out from February 1, 2010 to January 31, 2011. The P concentrations of overlying water was tested in the 1st day ($C_{background}$, mg L^{-1}) and then measured once each 15 days, details about analyzing method of P concentration are deccirbed above. The equations we used are described below as Eqs. (4) and (5):

$$C_{Change} = C_{max} - C_{background}. \tag{4}$$

$$P_{release} = \left(C_{Change} \times V_{water}\right) / \ S_{interface} \times S_{lake} \times 10^{-9}. \tag{5}$$

where C_{Change} is changes of P concentrations in the overlying water between the maximum P concentration and the background P concentration (mg L^{-1}); C_{max} is the maximum P concentration value during the whole experiment (mg L^{-1}); V_{water} is the volume of overlying water in experiment (L); $S_{interface}$ is the area of the interface in simulated device (m^2); S_{lake} is the area of lake (m^2).

Most of the basin is a rural area, the residents on the basin lived a self-sufficient life. According to the information provided by the local government, food imported from outside the basin was negligible and unrecorded; human flows out of the basin and flows in the basin were substantially balanced in 2010; there was little formula feed input from outside the watershed. Thus, the P input from formula feed could be ignored in this balance calculation of Yangzonghai Lake watershed in 2010.

P Exported from Lake Through Sedimentation. The P exported via sedimentation of the lake was calculated by using a mass balance method: inputs - outputs = change in storage of lake water. The formula of P exported from sedimentation (Qs) is as Eq. 6:

$$Q_s = Q_{in} - Q_{outlet} - Q_{IM} - \Delta P. \tag{6}$$

where Q_{in} is the sum of the P load entering into the lake in 2010, Qoutlet is the P load exported from the outlet, and Q_{IM} is the P load exported from insect emergence (to the lake subsystem, the P export pathways are just from sedimentation, outlet, and insect emergence, Fig. 2). ΔP is the P increment in the lake water, which is the difference of P in lake water between December 2009 and December 2010. The water samples from each site were collected at different depths using a water pump (the first at 0.5 m below the surface, then every 1 m down the water column, and the last sample at 0.5 m from the bottom).

Due to the lake was polluted by arsenic in 2008 and was treated with a ferric flocculants ($FeCl_3$) from October 2009 to December 2011, see Liu et al. [10] in details. The P flux of sedimentation has been accelerated by the remediation of arsenic pollution. The Fe-induced sedimentation of P was calculated by establishing relationship model between the amount of arsenic sedimentation and P sedimentation in the overlying water when added the flocculants ($FeCl_3$). The model was established by an experiment carried out in the laboratory (n = 102) in August 30–31, 2012. Two parts of overlying water (8 Barrels × 20 L for each part) were taken from the bottom of the lake in August 30, 2012 in southern and northern of the lake, respectively. Measured the concentration of P and arsenic for each part, then the water samples were poured into 102 glass bottles (3 L for each bottle). Added different amount of the flocculants ($FeCl_3$) into the bottles, sampled and measured the concentration of P and arsenic in the water in the next day and calculated the amount of arsenic and P sedimentation at the same bottle, and gained 102 arsenic and phosphorus corresponding data. Then we established the relationship model. The amount of arsenic removal from the lake in 2010 was 48.43 t (the data estimated by the local environmental monitoring institution). We calculated the amount of P sedimentation induced by iron based on the model and the data of arsenic removal from the lake in 2010. The P concentration analyzed method see above mentioned. The concentrations of arsenic were determined using a 933AFS atomic fluorescence spectrometer (Titan Instruments, Beijing, China; detection limit = 0.5 $\mu g\ L^{-1}$).

Statistical Analysis. SPSS software version 16.0 was used to analyze the data. All comparisons were carried out using analysis of variance (ANOWA, when sample sizes were the same) and General Linear Model (GLM, when sample sizes were different). A significance level of 0.05 was used in the analysis.

3 Results

3.1 P Input of the Yangzonghai Lake Watershed

P Input through Atmospheric Deposition. The total atmospheric deposition P input was 44.7 t in 2010 (44.7 t yr^{-1}). The area weighted flux was 2.3 kg ha^{-1} yr^{-1}. The average monthly flux was 3.7 t. The atmospheric deposition was relatively high in the beginning of the rainy season (Fig. 3).

P Input through Fertilizer Use. According to the investigation, the main P fertilizer applied to the watershed was $Ca(H_2PO_4)_2 \cdot CaHPO_4$, a compound fertilizer (N:P:K, 10:7:8) and a fertilizer specially used on golf courses (Table 1). About 679.0 t P was imported into the watershed in 2010 by fertilizer use.

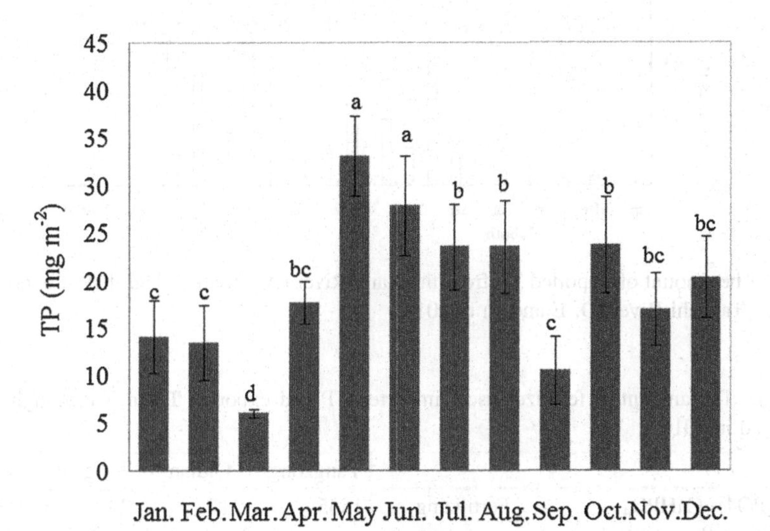

Fig. 3. Trend of phosphorus atmospheric deposition. The different letters displaying the quantity of phosphorus deposition between two months indicated a significant difference (i.e. $p < 0.05$), according to the multiple-comparison tests carried out. Vertical bars are standard deviations (n = 9).

P Input through Water Diversion. The water quality of the diversion water outside the watershed varied seasonally. The worst water quality appeared in September (0.331 mg L^{-1}, Fig. 4A). According to monitoring data, 5,676,566 m^3 of water was from the Baiyi River water diversion in 2010 (Fig. 4B), and it contained 1.4 t P (Fig. 4C), which flowed into the lake. In addition, P from anaerobic sediment source was 1.9 t yr^{-1}.

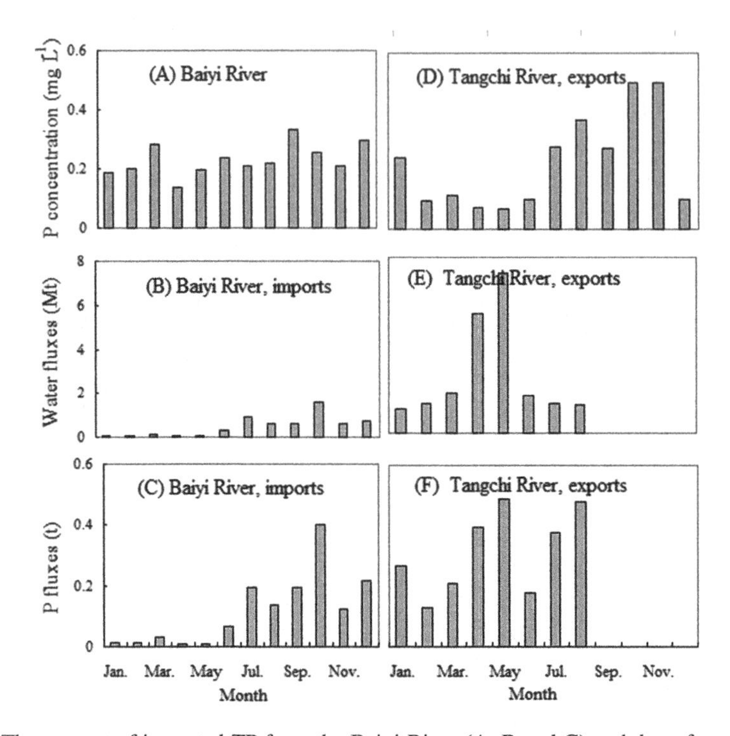

Fig. 4. The amount of imported TP from the Baiyi River (A, B and C) and that of exported TP from the Tangchi River (D, E and F) in 2010.

Table 1. The amount of fertilizer used, imported TP and exported TP in Yangzonghai Lake watershed in 2010.

District		Yangzong	Qidian	Tangchi	Total
$Ca(H_2PO4)_2{\cdot}CaHPO_4$ fertilizer, P content: 25.128 %	Fertilizing amount (t)	1565	17	233	1815
	P (t)	393.2	4.3	58.5	456.1
Compound fertilizer (N:P:K, 10:7:8) P content: 7 %	Fertilizing amount (t)	2582	28	436	2976
	P (t)	180.7	2	30.5	213.2
Golf fertilizer, P content: 15 %	Fertilizing amount (t)	–	–	65	65
	P (t)	–	–	9.7	9.7
P imported in total (t)		573.9	6.3	98.8	679
P-output from plant products (t)		203.24	55.04	6.35	264.63
P-output from animal products (t)		23.26	23.08	3.87	50.21
P exported in total (t)		226.5	78.12	10.22	314.84

3.2 P Output from the Yangzonghai Lake Watershed

P Output through Plant Products. The amount of P exported from the watershed by plant products (including food products) was 264.6 t in 2010 (Table 1). Among all the output pathways, plant products contributed the most (264.6 t yr^{-1}). Among plant products, timber exported 199.4 t yr^{-1} and accounted for 72.6 %, rice exported 38.6 t yr^{-1} and accounted for 14 %, corn exported 19.5 t yr^{-1} and accounted for 7.1 %, and other plant products accounted for 6.3 %.

P Output through Animal Products. The amount of P exported from the watershed by animal products (meat, egg, milk, etc.) was 50.2 t in 2010 (Table 1). Animal products ranked second in all output pathways (50.2 t yr^{-1}). Among animal products, pork exported 31.1 t yr^{-1} and accounted for 59.6 %, eggs exported 7.0 t yr^{-1} and accounted for 13.4 %, poultry exported 5.2 t yr^{-1} and accounted for 10 %, milk exported 3.7 t yr^{-1} and accounted for 6.3 %.

P Output through Outlet. There was 21,389,702 m^3 of water flowed out form the outlet and 2.5 t P in 2010. Variation of P concentration in water, water fluxes and P fluxes of the outlet could be observed in this period of time (Fig. 4D–F).

P Input and Output of the Lake. P retention in lake was 23.4 t at the end of 2009, and was 18.0 t at the end of 2010, so the increment was −5.4 t. By calculation with the mass balance method using Eq. 4, we found that P exported from sedimentation was 44.5 t in 2010. The P budget of the lake subsystem is shown in Fig. 5. As for the lake, non-point sources contributed the largest P input, accounting for 77.7 % of the total P, and sedimentation contributed the largest P output, accounting for 94.6 % of the total P.

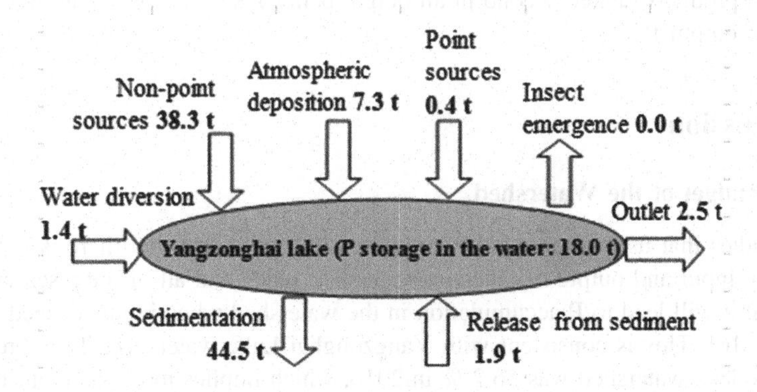

Fig. 5. P budget of the Yangzonghai Lake subsystem in 2010

When adding flocculants ($FeCl_3$) into the lake, the concentrations of P and arsenic were decreased simultaneously ($y = 5.9464x + 0.0006$, $R^2 = 0.9946$, $n = 102$, $p < 0.05$). The Fe-induced sedimentation of P was 7.9 t yr^{-1} based on the relationship model between the amount of P sedimentation and arsenic sedimentation in the water when

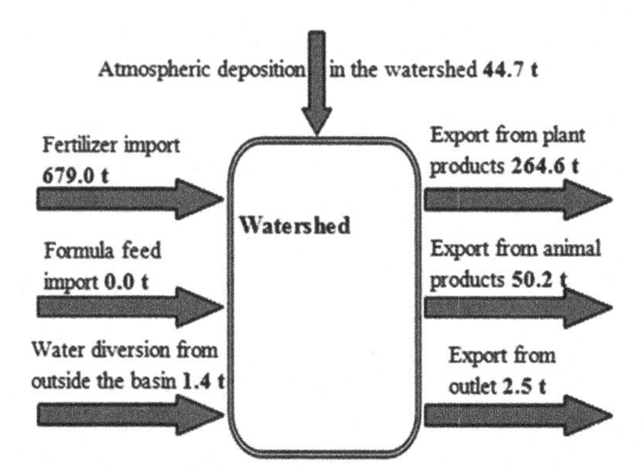

Fig. 6. P budget of Yangzonghai Lake watershed

adding different amount of $FeCl_3$ ($y = 6.1557x + 9E^{-05}$, $R^2 = 0.9946$, $n = 102$, $p < 0.05$) and the amount of arsenic removal from the lake in 2010 (48.43 t).

3.3 P Budget of the Watershed

The imported P load was 725.1 t, and the exported P load was 317.3 t in 2010, which indicates that 407.8 t P was retained in the watershed in this period of time (Fig. 6). Among all the input pathways, fertilizer use ranked first, contributing 679.0 t P and accounting for 93.6 % of input P. Then followed by atmospheric deposition, which contributed 44.7 t P, accounting for 6.2 % of input P. For output pathways, plant products ranked first, exporting 264.6 t P and accounting for 83.4 % of output P. Animal products ranked second in all output pathways (50.2 t yr^{-1}), accounting for 15.8 % of output P.

4 Discussion

4.1 P Budget of the Watershed

Many studies that used a mass balance method have revealed that there is imbalance between P input and output of lakes on watershed scale, and all of the researchs have found that it will lead to P accumulation in the watershed when inputs exceed outputs [2, 7, 15, 16]. This is consistent with Yangzonghai Lake watershed. The P retention rate in the lake watershed was 56.2 % in 2010, which implies more than half of the P inputs was retained in the watershed. The retention rate is comparable with the upper Potomac River basin (60 %, [7]) and Lake Okeechobee (80 %, [16]).

The long term accumulation of P in the watershed will finally result in lake eutrophication [2]. Yangzonghai Lake is a deep lake with an 11.1 year-long water renewal period [17]. The P concentration in the lake has been rising up in the last six years [6], which means that the risk of eutrophication is increasing. Therefore, watershed governors should seek to draw up strategies which can keep the balance of inputs and outputs (or outputs > inputs) in the watershed. Besides reducing fertilizer use, it is necessary to construct wetlands in the riparian zone to decontaminate nutrients in surface runoff on Yangzonghai Lake watershed.

4.2 P Input Fluxes of Watershed and Lake

The average P inflow of the watershed was 37.8 kg ha^{-1} yr^{-1}, and that of Yangzonghai Lake was 15.6 kg ha^{-1} yr^{-1}, which was 2 times higher than that of the eleven recreational lakes in Minnesota State of the United States (0.32–6.0 kg P ha^{-1} yr^{-1}, [2]). Non-point source pollution from agriculture caused river and lake pollution problems in United States, and the control of P from agriculture is still the key aspect in American water pollution treatment [18]. Pollution from agriculture is also responsible for much of the water pollution and drinking water shortages in China. In Yangzonghai Lake watershed, the main P inflow was from fertilizer application, which contributed 679.0 t and accounted for 93.4 % of the total in 2010. In terms of the lake P budget, fertilizer application contributed 65.1 % of the total P input. Cultivated land account for 27.7 % of the watershed, as the crop farming ecosystems, the P budget strongly relied on P fertilizer use [19]. Therefore, it is the key countermeasure to control the fertilization intensity in the watershed to keep the lake from further eutrophication, which can be done by improving fertilizer use efficiency or changing land use from planting nutrient exhaustive crops to poor resistance crops.

Atmospheric deposition is also a P input pathway which cannot be ignored. The Yangzonghai Lake watershed was 44.7 t yr^{-1}, accounting for 6.2 % of the total in 2010. In the studies of the Lake Simcoe in Canada, Winter et al. [20] found that P from atmospheric deposition accounted for 23–56 %. Yang et al. [21] studied the P contributed by atmospheric deposition in Tai Lake, China, and found that the P load from atmospheric deposition accounted for 46.2 % of the total load.

The P deposition in the Yangzonghai Lake watershed was in the medium level compared with other watersheds which had been reported in China. The monthly average P deposition load in the Yangzonghai Lake watershed in 2010 was 0.192 kg ha^{-1} $month^{-1}$, which was lower than that in Xingyun Lake watershed (0.370 kg ha^{-1} $month^{-1}$, [22]) and in the Tai Lake watershed, East China, in 2007 (0.230 kg ha^{-1} $month^{-1}$, [21]), but it was three times higher than that in the Lake Fuxian watershed (0.066 kg ha^{-1} $month^{-1}$, [22]), a nearby lake in the same province. The coal-fired power plant beside Yangzonghai Lake perhaps accounted for the significant difference.

The P from lake sediment release was 1.9 t yr^{-1}, contributing only 3.9 % to the P input (49.3 t yr^{-1}). This indicates that sediment dredging would not solve lake internal problems related to eutrophication, the results are similar to Shahe Reservoir's [1]. Input flux of P from water diversion was 1.4 t yr^{-1} in 2010, much smaller than other inflow fluxes. But taking consideration of its high P concentration (between 0.135 - 0.331 mg L^{-1}), the influence of water diversion should not be neglected. The imported P load could result in lake P concentration increasing 0.0028 mg L^{-1}. Thus, it is quite necessary to purify the channeled water before it enters into the lake.

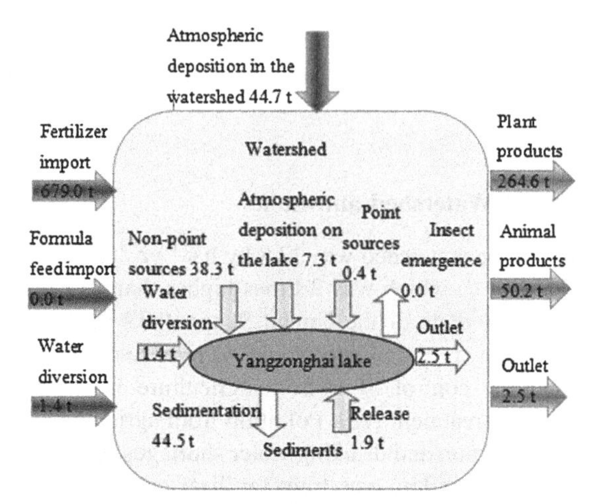

Fig. 7. The amount of P flux for each way in watershed-scale and lake-scale of Yangzonghai Lake

5 Conclusions

The amount of P flux for each way in watershed-scale and lake-scale of Yangzonghai Lake see Fig. 7, and P fluxes between inputs and outputs in Yangzonghai Lake watershed were imbalanced; 56.2 % (407.8 t) of input P was retained in the watershed. Such a high retention rate implies that the lake, which was on mesotrophic level, has a great pressure of further eutrophication. Among all the countermeasures to control the eutrophication of the lake, reducing the fertilizer application should be put first.

Acknowledgments. This research was financially supported by the National Natural Science Council, P.R. China (Project numbers 51168047, 31300418). We gratefully acknowledge Prof. Jie Chang and Prof. Ying Ge of Zhejiang University for their valuable comments that improved our manuscript. We thank Luke Driskell for language improving.

Conflicts of Interest

State any potential conflicts of interest here or "The authors declare no conflict of interest".

References

1. Zhang, H., Coudrier, B.P., Wen, S., Müller, B., Shan, B.: Budget and Fate of phosphorus and trace metals in a heavily loaded shallow reservoir (Shahe, Beijing City). CLEAN–Soil Air Water **43**, 210–216 (2015)
2. Schussler, J., Baker, L.A., Chester-Jones, H.: Whole-system phosphorus balances as a practical tool for lake management. Ecol. Eng. **29**, 294–304 (2007)

3. Graneli, E., Weberg, M., Salomon, P.S.: Harmful algal blooms of allelopathic microalgal species: the role of eutrophication. Harmful Algae **8**, 94–102 (2008)
4. Yuan, H., Liu, E., Pan, W., Geng, Q., An, S.: Species and characteristics of organic phosphorus in surface sediments of northwest region of Taihu Lake, Eastern China. CLEAN–Soil Air Water **42**, 1518–1525 (2014)
5. Fraterrigo, J.M., Downing, J.A.: The influence of land use on lake nutrients varies with watershed transport capacity. Ecosystems **11**, 1021–1034 (2008)
6. Yang, C.L., Li, S.Y., Yuan, L.N., Liu, W.H., Liu, R.B., Li, Z.Y.: The eutrophication process in a plateau deepwater lake ecosystem: response to the changes of anthropogenic disturbances in the watershed. In: International Conference on Biomedical Engineering and Biotechnology (iCBEB), vol. 3, pp. 1815–1821 (2012)
7. Jaworski, N.A., Groffman, P.M., Keller, A.A., Prager, J.C.: A watershed nitrogen and phosphorus balance: the upper Potomac River basin. Estuaries **15**, 83–95 (1992)
8. Bennett, E.M., Reed-Andersen, T., Houser, J.N., Gabriel, J.R., Carpenter, S.R.: A phosphorus budget for the Lake Mendota watershed. Ecosystems **2**, 69–75 (1999)
9. La Jeunesse, I., Elliott, M.: Anthropogenic regulation of the phosphorus balance in the Thau catchment–coastal lagoon system (Mediterraean Sea, France) over 24 years. Mar. Pollut. Bull. **48**, 679–687 (2004)
10. Liu, R.B., Yang, C.L., Li, S.Y., Sun, P.S., Shen, S.L., Li, Z.Y.: Arsenic mobility in the arsenic-contaminated Yangzonghai Lake in China. Ecotoxicol. Environ. Safety **107**, 321–327 (2014)
11. Schauser, I., Chorus, I.: Water and phosphorus mass balance of Lake Tegel and Schlachtensee–a modelling approach. Water Res. **43**, 1788–1800 (2009)
12. Wu, H., Zhang, J., Li, C., Fan, J., Zou, Y.: Mass balance study on phosphorus removal in constructed wetland microcosms treating polluted river water. CLEAN–Soil Air Water **41**, 844–850 (2013)
13. Havens, K., Fukushima, T., Xie, P., Iwakuma, T., James, R., Takamura, N., Hanazato, T., Yamamoto, T.: Nutrient dynamics and the eutrophication of shallow lakes Kasumigaura (Japan), Donghu (PR China), and Okeechobee (USA). Environ. Pollut. **111**, 263–272 (2001)
14. Yang, C.L., Li, S.Y., Liu, R.B., Li, Z.Y., Liu, K.: Analysis on external load of Nitrogen and Phosphorus in Yangzonghai Lake. Shanghai Environ. Sci. **33**, 47–52 (2014)
15. Lowrance, R.R., Leonard, R.A., Asmussen, L.E., Todd, R.L.: Nutrient budgets for agricultural watersheds in the southeastern coastal plain. Ecology **66**, 287–296 (1985)
16. Fluck, R., Fonyo, C., Flaig, E.: Land-use-based phosphorus balances for Lake Okeechobee, Florida, drainage basins. Appl. Eng. Agric. **8**, 813–820 (1992)
17. Yang, C.L., Chen, J., He, B., Wang, H.: Origin analysis of overproof of total phosphorus and its environmental capacity in Yangzonghai Lake. Yunnan Environ. Sci. **27**, 44–46 (2008)
18. Kröger, R., Dunne, E., Novak, J., King, K., McLellan, E., Smith, D., Strock, J., Boomer, K., Tomer, M., Noe, G.: Downstream approaches to phosphorus management in agricultural landscapes: regional applicability and use. Sci. Total Environ. **442**, 263–274 (2013)
19. Senthilkumar, K., Nesme, T., Mollier, A., Pellerin, S.: Regional-scale phosphorus flows and budgets within France: the importance of agricultural production systems. Nutr. Cycl. Agroecosyst. **92**, 145–159 (2012)
20. Winter, J.G., Dillon, P.J., Futter, M.N., Nicholls, K.H., Scheider, W.A., Scott, L.D.: Total phosphorus budgets and nitrogen loads: Lake Simcoe, Ontario (1990 to 1998). J. Great Lakes Res. **28**, 301–314 (2002)

21. Yang, L.Y., Qin, B.Q., Hu, W.P., Luo, L.C., Song, Y.Z.: The atmospheric deposition of nitrogen and phosphorus nutrients in Taihu Lake. Oceanologia Et Limnologia Sinica **38**, 104 (2007)
22. Jin, X., Lu, Y., Jin, S.Q., Liu, H., Li, Y.X.: Research on amount of total phosphorus entering into Fuxian Lake and Xingyun Lake through Dry and Wet deposition caused by surrounding phosphorus chemical factories. Yunnan Environ. Sci. **29**, 39–42 (2010)

Modelling the Risk of Highly Pathogenic Avian Influenza H5N1 in Wild Birds and Poultry of China

Ping Zhang[1(✉)] and Peter M. Atkinson[2]

[1] Geo-Exploration Science and Technology College, JiLin University, Changchun, China
zp@jlu.edu.cn
[2] School of Geography and Environment, University of Southampton, Southampton, UK

Abstract. This paper applied an integrated spatial regression model to explore the associations between ten environmental variables and the highly pathogenic avian influenza (HPAI). A subtype H5N1 cases in wild birds and poultry in China, and to predict the spatial distribution of HPAI H5N1 relative risk. Here a generalized linear mixed model (GLMM) incorporated with a variogram model through its random effects item, used as the spatial regression model. Four environmental variables were found to have significant effects, including annual mean temperature, poultry density, distance to lakes and wetlands, and distance to bird migration routes. The Root Mean Square Error of arbitrary 15 sample data was 11.56. Further, the high predicted relative risk areas of HPAI H5N1 were mainly in the Northwest, Middle, Southwest and Southeast part of China. With its simple structure and good prediction ability, this spatial regression model was very promising for predicting the risk of other disease.

Keywords: Spatial regression model · Generalized linear mixed model · Variogram model · Risk · Avian influenza

1 Introduction

Since late 2003, highly pathogenic avian influenza (HPAI) outbreaks caused by infection with the H5N1 virus have led to the death of millions of poultry and tens of thousands of wild birds. As of February 8, 2012, 42 laboratory-confirmed human infections have occurred in China [1]. Although HPAI H5N1 has taken place in a limited number of provinces in China, spreading might occur at any time due to movement of domestic birds, migration of wild birds, and interaction of both. This ongoing H5N1 avian influenza epidemic in China poses risks to animals as well as human health, and will be elevated by the potential cross-species transmission to humans and subsequent re-assortment of avian and human influenza viruses in co-infected individuals [2]. Thus, it is urgent and important to model the risk of the H5N1 infection in China. Modeling may help to detect areas of unusually high and low risks so that actions may be taken in advance to allow better resource allocation for prevention and control.

F. Bian and Y. Xie (Eds.): GRMSE 2015, CCIS 569, pp. 707–721, 2016.
DOI: 10.1007/978-3-662-49155-3_73

So far, studies aiming to identify HPAI H5N1 risk factors and predict risk have been undertaken in many countries where the disease was introduced, such as Thailand [3–5], Vietnam [6], Indonesia [7], Bangladesh [8], the U.S. [9], the Netherlands [10], Romania [11] and Southern Africa [12]. Only three studies tried to model the risk of HPAI H5N1 in China [13–15]. Despite this research effort, a central goal still exists: to understand the factors favoring the continuing reoccurrence of the virus [16]. Specifically, little is known about the agro-ecological conditions associated with highly pathogenic avian influenza H5N1 virus spread and persistence [4, 13].

In this research, risk refers to the likelihood distribution for the number of cases of avian influenza in a particular area. The goal of modeling the risk of avian influenza is to examine spatial variation in risk in terms of number of cases for a given country or region. There are few examples applying spatial modeling methods to predict avian influenza risks [17]. Logistic regression models have been used widely [11, 13, 15, 16, 18]. One may implement logistic regression to characterize the statistical association between avian influenza cases or outbreaks and environmental covariates. However, considering risk modeling in a spatial context, particularly in the case that the areas are small, one would expect "residual" dependence between counts in areas that are geographically close, due to unmeasured risk factors or errors in the data that have spatial structure [19]. In such cases, simple logistic regression modeling is insufficient.

To account for spatial dependence in the residuals, a model involving spatial autocorrelation may be fitted [20], such as a spatial regression model [9, 21, 22], and Bayesian geostatistical logistic regression model [23]. Ignoring autocorrelation may lead to the erroneous conclusion that a variable is significant in explaining avian influenza cases when the variable is in fact insignificant [9]. Such spatial regression models have the advantage that both environmental covariates and spatial autocorrelation can be estimated and full posterior distributions can be produced to quantify uncertainties in the parameters of interest [23]. It is only in recent years that researchers have begun to apply spatial regression models to avian influenza risk [9]. However, they have not yet been applied to avian influenza risk in China.

Compared with previous studies, this research brings two improvements. First, we modeled the risk of H5N1 in poultry and wild birds, while others studies examine the risk in poultry only [15]. Surveillance for HPAI in wild birds will help to predict the spread of the avian influenza virus, and is an important component of a comprehensive surveillance program [24]. Therefore, it is included here. As the first attempt to model the distribution of HPAI H5N1 risk in China, Fang et al. [13] predicted areas at high risk in ecological areas that would not support the maintenance and transmission of the virus (e.g., the extremely large desert regions of Inner Mongolia, Tibet and Xinjiang autonomous regions) [15]. Moreover, the research conducted by Cao et al. [14] was mainly for risk analysis, rather than for risk modeling. Secondly, a generalized linear mixed model combined with variogram modeling was used for risk modeling here to account for spatial dependence. These two new aspects shed light on the risk of HPAI H5N1 in wild birds and poultry in China.

2 Methods

2.1 Data and Test for Spatial Dependence

Data on the number of cases of the HPAI H5N1 in wild birds and poultry in China reported from January 2004 to March 2011 were provided by OIE, a world organization for animal health [25]. Basic geographic data were provided by the Data Sharing Infrastructure of Earth System Science [26]. During the period January 2004 to March 2011, three main epidemic waves occurred in the number of HPAI H5N1 cases in China (Fig. 1). Some periodicity is evident. The H5N1 outbreaks in poultry during wave I were mainly distributed in central and South China, while outbreaks in wild birds only appeared in south (Fig. 2a). During wave II, the outbreaks in wild birds expanded from Southern to Western and North-Eastern China; and outbreaks in poultry moved to the North-Western and Northern part of China, while outbreaks in central China decreased significantly (Fig. 2b). During wave III, the HPAI H5N1 outbreaks in both poultry and wild birds decreased further. Outbreaks among wild birds were distributed in the West and South of China only, and those among poultry were mainly distributed in Xinjiang, Tibet and Guangdong Province (Fig. 2c). The HPAI H5N1 outbreaks in wild birds were distributed along the bird migration flyways, especially along the eastern one (Western Pacific Route) and the western one (Middle-Asia India Route) (Fig. 2).

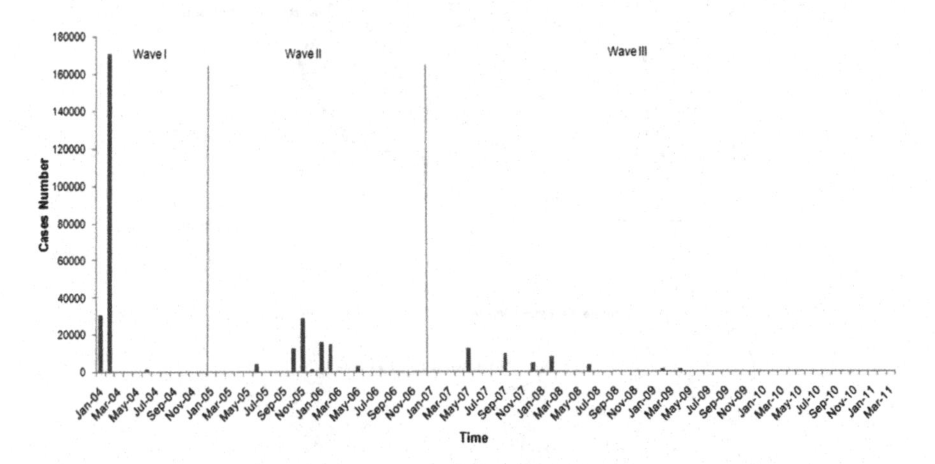

Fig. 1. Temporal distribution of monthly HPAI H5N1 cases numbers reported in China

Spatial dependence is the propensity for nearby locations to influence each other and to possess similar attributes [27, 28]. It is necessary to test for spatial dependence in the model residuals. If spatial dependence exists in the model residuals then it needs to be considered in the model for predicting HPAI H5N1 cases in different geographical areas. To test for spatial dependence in the model of HPAI H5N1 cases in wild birds and poultry in China, Ripley's K function and Moran's I statistic as the statistical measures of spatial dependence for point locations were used here [29].

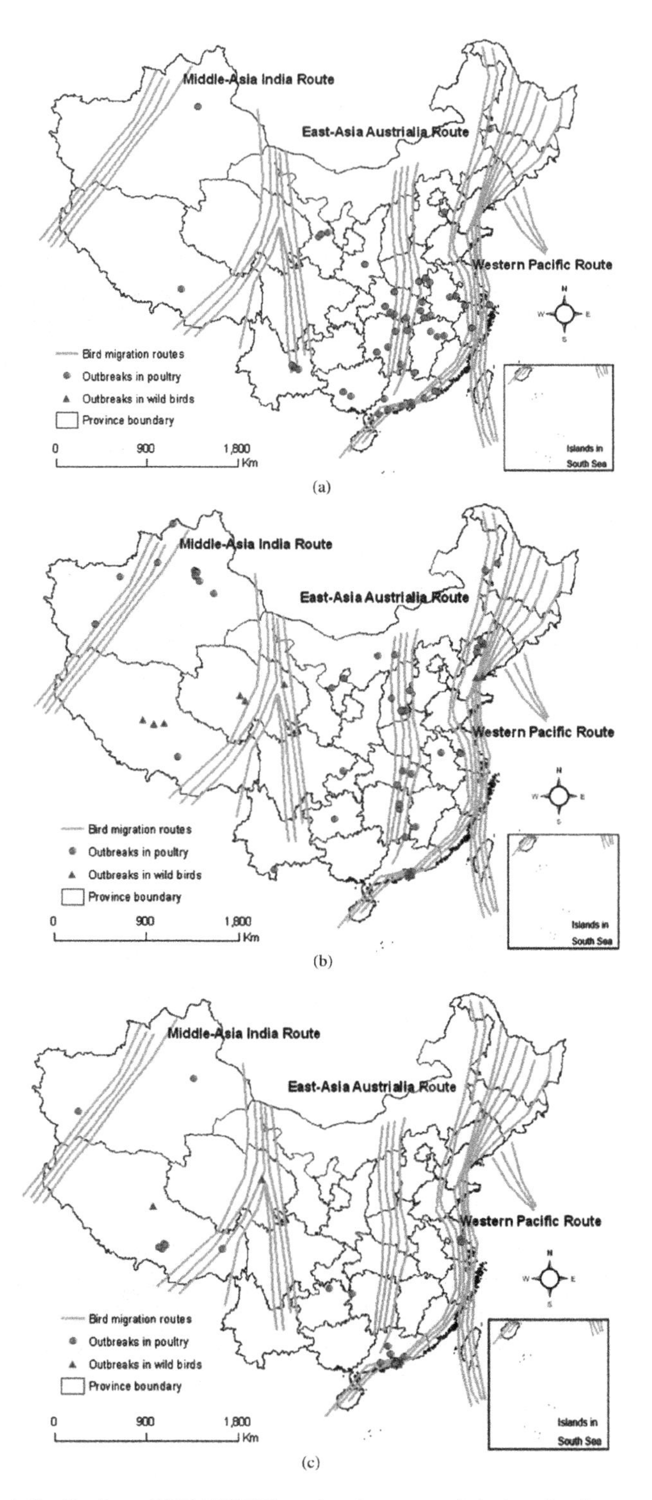

Fig. 2. Spatial distribution of HPAI H5N1 outbreaks reported during the three main epidemic waves in China. (a) Wave I: 01 January of 2004–30 December of 2004; (b) Wave II: 01 January of 2005–30 December of 2006; (c) Wave III: 01 January of 2007–31 March of 2011

Table 1. Global spatial autocorrelation by Moran's *I* statistic on the incidence rate of HPAI H5N1 in poultry of China

Study period	Moran's *I* statistic	P-value	Pattern
January 2004–December 2004	0.03	0.02	Clustered
January 2005–December 2005	0.05	0.01	Clustered
January 2011–March 2011	0.43	0	Clustered

Usually, *L*-function $L(d)$ is used instead of the *K*-function to test for autocorrelation in a spatial point distribution [29]. Figure 3 shows that the observed value of $L(d)$ for the HPAI H5N1 cases in wild birds and poultry of China between 2004 and 2011 was outside of the two envelope bounds (min and max), which is the confidence interval of Monte Carlo test. The value of $L(d)$ increases with the increase in distance of separation from 10 km to 1200 km. It shows that the spatial distribution of the HPAI H5N1 cases in wild birds and poultry of China 2004–2011 was clustered. The value of $L(d)$ increases rapidly at the distance between 11 km and 24 km and then it reaches its plateau slowly (Fig. 3). This might be related to the poultry activity radius of 11–24 km and the wild bird activity radius of greater than 24 km in China.

In addition, spatial autocorrelation analysis has also been performed on the incidence rate of HPAI H5N1 in poultry by Moran's *I* statistic (Table 1). It shows that the Moran's *I* statistics on the incidence rate of HPAI H5N1 in poultry for the year of 2004, 2005 and 2011 are significantly greater than zero. It means that the HPAI H5N1 incidence in poultry is positively spatially autocorrelated (Table 1). Complete wild bird population data at county level in China are unavailable. Therefore, spatial autocorrelation analysis was not performed on the incidence rate of HPAI H5N1 in wild birds. From the above test

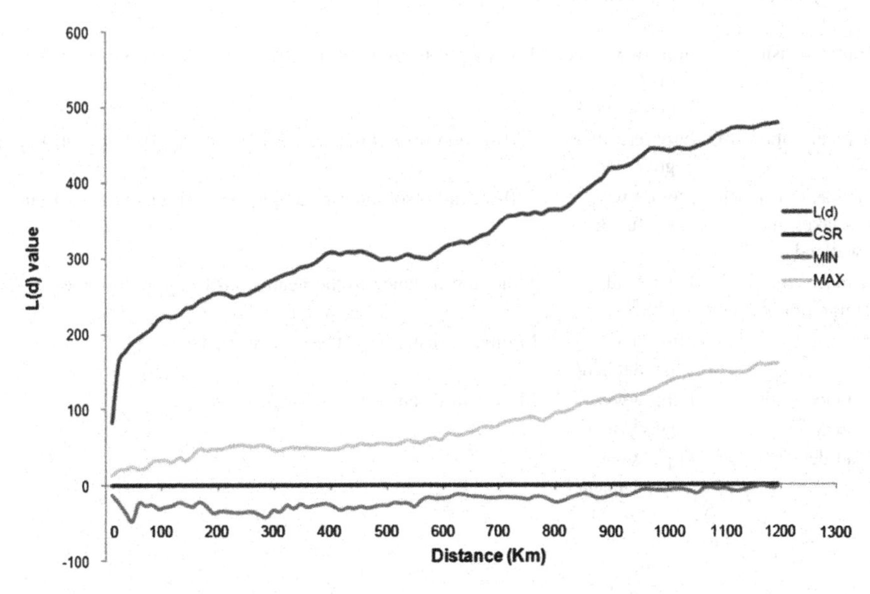

Fig. 3. Ripley's *K* function for the HPAI H5N1 cases in wild birds and poultry

results, it can be found that spatial dependence does exist on the HPAI H5N1 cases in wild birds and poultry in China from 2004 to 2011. Therefore, it is important to include spatial component in the model to predict the HPAI H5N1 cases in various geographic areas. And it is suitable to apply spatial regression model in this research.

2.2 Environmental Covariates

Ten environmental covariates were considered as risk factors for HPAI H5N1 cases in wild birds and poultry in China. The covariates include human population density, annual mean temperature, annual precipitation, poultry density, mean elevation, Euclidean distance to lakes and wetland, minimum distance to the nearest bird migration route, minimum distance to the nearest road, minimum distance to the nearest city and road density (Table 2).

Table 2. Environmental variables associated with the HPAI H5N1 cases in wild birds and poultry of China

Environmental variable	Source	Manipulation
Human population density	http://www.ornl.gov/sci/landscan/	30-second resolution; resample cell size (10 km × 10 km)
Annual mean temperature (°C)	http://www.worldclim.org/	30-second resolution; resample cell size (10 km × 10 km)
Annual precipitation (mm)	http://www.worldclim.org/	30-second resolution; resample cell size (10 km × 10 km)
Poultry density	http://www.fao.org/geonetwork	180-second resolution; resample cell size (10 km × 10 km)
Mean elevation (m)	http://eros.usgs.gov/	30-second resolution; resample cell size (10 km × 10 km)
Euclidean distance to lakes and wetland	http://www.wwfus.org/	30-second resolution; resample cell size (10 km × 10 km)
Distance to bird migration routes	Fang et al. (2008)	Minimum distance to the nearest bird migration route
Distance to roads	http://www.geodata.cn/	Minimum distance to the nearest road
Distance to the cities	http://www.geodata.cn/	Minimum distance to the nearest city
Road density	http://www.geodata.cn/	

Human population density was chosen as one of the risk factors because it was found to be associated with HPAI H5N1 in several studies conducted in countries with different agro-ecological conditions such as Thailand, Bangladesh, Vietnam, Romania and China [6, 8, 15, 16, 30, 31]. Human population density may indicate higher levels of trading activity [18]. Disclosure of the HPAI H5N1 cases in Thailand poultry markets in 2006 and 2007 suggested that the HPAI virus had continued to spread among poultry through trade activities despite the presence of control measures [32].

Annual mean temperature and annual precipitation were chosen as climatic factors here. Some researchers found that a sudden drop in temperature occurred shortly before HPAI H5N1 outbreaks among birds in the Eurasian regions in 2005 and 2006 [33]. Climate change and subsequent immune-suppression may have allowed the H5N1 virus to proliferate more efficiently in birds which have already been carrying the virus, thereby, hastening the inter-species spread of the virus and the deaths of wild birds [33]. HPAI H5N1 virus transmitted by migratory birds could be spread during the migration period, and the speed of such a spread may be elevated in particularly cold winter [34]. Lower levels of moisture and precipitation may affect the availability of food resources and, thereby, influence the distribution of wild birds [18].

Previous research has shown an association between poultry density and HPAI H5N1 outbreaks [3, 15, 16, 30]. In China, the HPAI H5N1 cases were not always positively related to poultry density [13, 15]. Chickens in areas with high population densities are usually bred in industrialized farms with good animal husbandry practices and proper vaccination [35]. Mean elevation was chosen as an environmental factor here because some research has reported an increased HPAI H5N1 risk in lowland and river delta areas [6, 16, 36]. Also it has been demonstrated HPAI H5N1 outbreaks in South Asia and China to be significantly associated with elevation [15, 16]. Normally suitable habitats are concentrated in the lowland, and then elevation influences the availability of food resources and shelter for waterfowl, which are natural hosts for the HPAI H5N1 virus [18]. Water bodies and wetlands have been found to be significantly associated with HPAI H5N1 outbreaks in China, India and Bangladesh [13, 37, 38]. Lakes and wetlands are important for migratory (and local) waterfowl and provide potential, suitable habitats [18].

Some researchers have found that infected wild birds can carry the avian influenza virus for long distances during migration [39]. Wild bird migration is important for avian influenza virus transmission [13]. Usually migratory birds cannot fly the full distance to their annual migratory destination at once. Instead, they usually interrupt their migration to rest and refuel [40]. Avian influenza virus may be spread between wild and domestic birds when migratory birds search for food, water and shelter [13]. Proximity to cities and proximity to roads were included as risk factors here since they relate to poultry trade and movement which may facilitate the mechanical spread of the HPAI H5N1 virus [13]. During long distance transportation, a variety of birds and animals from various origins are caged on top of each other, possibly providing an easy cross-infection route for avian influenza. Moreover, many open live poultry markets are established along or near roads, which may further increase the chance of avian influenza virus transmission [13].

2.3 Spatial Regression Model

The statistical model represents the number of avian influenza cases per geographical unit as a Poisson-distributed random variable, which is appropriate for analyzing disease cases in which some geographic units have many cases but most units have few or no cases [9, 21]. Since not accounting for spatial autocorrelation when predicting the number of HPAI H5N1 cases may lead to the erroneous conclusion that an environmental variable is significant when it is in fact non-significant [9] a spatial regression model was used here to predict the number of HPAI H5N1 cases per geographical area.

Spatial regression models are regression models with a term to account for spatial dependence, which is assumed to arise from some unobservable latent variable(s) that are spatially correlated [41]. There are a lot of forms for spatial regression models. Here a generalized linear mixed model (GLMM) incorporating a variogram model was used to explore the statistical association between HPAI H5N1 cases in wild birds and poultry and environmental factors, to quantify the relative importance of the main environmental factors, and to predict the number of HPAI H5N1 cases in geographical areas [9, 21].

The key elements of a classical linear model are (i) the observations are independent, (ii) the mean of the observation is a linear function of some covariates, and (iii) the variance of the observation is a constant [42]. The extension to generalized linear models (GLM) consists of modification of (ii) and (iii) above; by (ii)' the mean of the observation is associated with a linear function of some covariates through a link function; and (iii)' the variance of the observation is a function of the mean [42]. Generalized linear mixed models (GLMM) are natural extensions of GLM and linear mixed models that allow for additional components of variability due to latent random effects [43].

The Poisson log-linear mixed model was used in this research. The Poisson distribution is often used to model responses that are counts [42]. Suppose that, given the random effects α, the counts $y_1 \ldots y_n$ are conditionally independent such that

$$y_i|\alpha \sim Poisson(\lambda_i) \tag{1}$$

$$\log(\lambda_i) = x_i'\beta + z_j'\alpha \tag{2}$$

where x_i' and z_j' are known vectors, β is a vector of unknown parameters (the fixed effects), and λ_i is the expected number of occurrences during the given interval [42].

Here, glmmPQL() in the MASS package of R was used to run a GLMM [44, 45]. The Poisson log-linear model with a random intercept estimated through the PQL method can be written as:

$$\ln(\lambda_i) = \beta_0 + x_i\beta_i + b_i \tag{3}$$

where λ_i is the number of avian influenza cases, β_0 and β_i are the unknown parameters for the fixed effects, x_i are the environmental covariates, b_i are the random effects with distribution assumption:

$$b_i \sim N(0, \sigma^2) \tag{4}$$

Formula (4) means that the random effects of b_i are normally distributed with a mean of 0 and a variance of σ^2.

A GLMM was chosen here because it is not only one of the fundamental tools in the analysis of longitudinal data in epidemiology [44, 45], but also it allowed for a spatial correlation structure through its random effects term [9]. The random effects term of b_i is similar to the residual (error) term in classical linear models. Thus, the GLMM incorporates spatial autocorrelation in the residuals through its random effects term. Here, variogram modeling of the spatial autocorrelation in the residuals r_i was used, where:

$$r_i \sim N(\mu, \sigma^2) \tag{5}$$

$$\sigma^2 = I\sigma_1^2 + F\sigma_2^2 \tag{6}$$

$$F = \exp(-d_{ij}/\rho) \tag{7}$$

where, the residuals r_i of the GLMM (formula (3))are distributed normally with mean μ and variance σ^2, σ_1^2 is the nugget of the residuals' semi-variance, σ_2^2 is the sill, ρ is the range, d_{ij} is the lag distance, I is the adjusted coefficient.

The final model used to predict the number of HPAI H5N1 cases per geographical area can be written as:

$$\ln(\lambda_i) = \beta_0 + x_i\beta_i + r_i \tag{8}$$

where λ_i, β_0, x_i, β_i and r_i are as in formulas (1)–(7).

3 Results

From Table 3, four environmental covariates are significant in predicting the HPAI H5N1 risks in wild birds and poultry in China between 2004 and 2011. The significant covariates are annual mean temperature, poultry density, distance to lakes and wetlands and distance to bird migration routes. In particular, the estimated coefficient for poultry density is 0.00. This means that HPAI H5N1 cases in wild birds and poultry is negatively correlated with poultry density. It might be partially contributed by the fact that

Table 3. Effects of environmental variables on the HPAI H5N1 cases in wild birds and poultry in China 2004–2011

Variable (X_i)	Estimated coefficient (β_i)	Std. error	DF	T-value	P-value
(Intercept)	6.88	0.34	78	19.98	0
Annual mean temperature	0.01	0	12	2.18	0.05
Poultry density	0	0	12	−2.38	0.03

poultry are normally fed in industrialized farms where poultry density is high, and where poultry have been vaccinated and well managed. On the contrary, poultry density is low in rural villages where poultry usually are fed in open backyards without having been vaccinated. This would imply a rural-urban divide such that poultry fed in backyards in rural areas are more likely to become infected.

There were 111 sample data altogether in this research, 96 of which were chosen randomly and used in the GLMM, while the remaining 15 data were used for validation. After doing regression, the 96 points' residuals have been done variogram modeling to test their spatial autocorrelation. It clearly shows that the curve of semi-variance rising up steadily with lag distance increasing from 0 up to about 100 km, and then it keeps level when lag distance is greater than 100 km (Fig. 4). It indicates that residuals have spatial autocorrelation when lag distance between 0 and 100 km, and spatial autocorrelation doesn't exist when lag distance greater than 100 km.

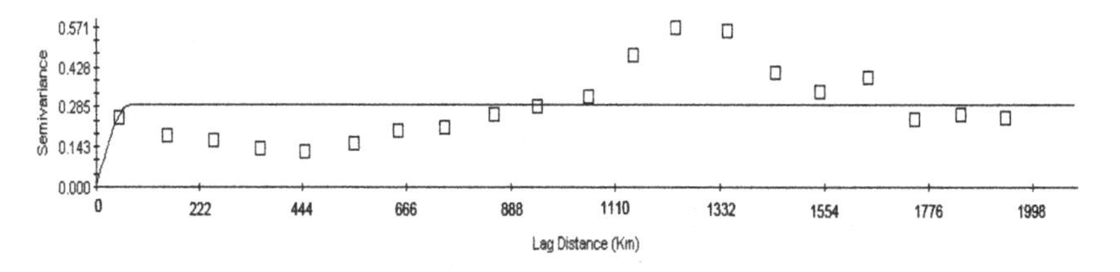

Fig. 4. Isotropic variogram of fitted spherical model for the 96 sample data's residuals

Root Mean Square Error (RMSE) has been used here to do the validation. The RMSE of the left 15 sample data is 11.56 cases per 10 km × 10 km pixel when the adjusted coefficient of *I* in formula (6) is 3.5. This means that its prediction results are desirable to utilize the spatial regression model which is GLMM incorporating with variogram modeling in this research. Model validation statistics revealed that the final spatial regression model has good predictive ability for HPAI H5N1 cases in geographical areas.

Moreover, relative risks of the HPAI H5N1 in wild birds and poultry in China were divided according to the predicted number of HPAI H5N1 cases in geographical areas (Fig. 5). Risk maps generated from the model shows a heterogeneous distribution and importantly risk of HPAI H5N1 in wild birds and poultry in China was found to highly varied across all regions. The highest predicted relative risk of HPAI H5N1 in wild birds and poultry mainly occurs in Northwest, Central and Southwest China, which are very near the Middle-Asia India bird migration route (Fig. 5). Another high predicted relative risk area occurs in Southeast China which is near the Western Pacific bird migration route (Fig. 5). It implies that wild birds and bird migration may play an important role in HPAI H5N1 virus spreading in the wild birds and poultry in China.

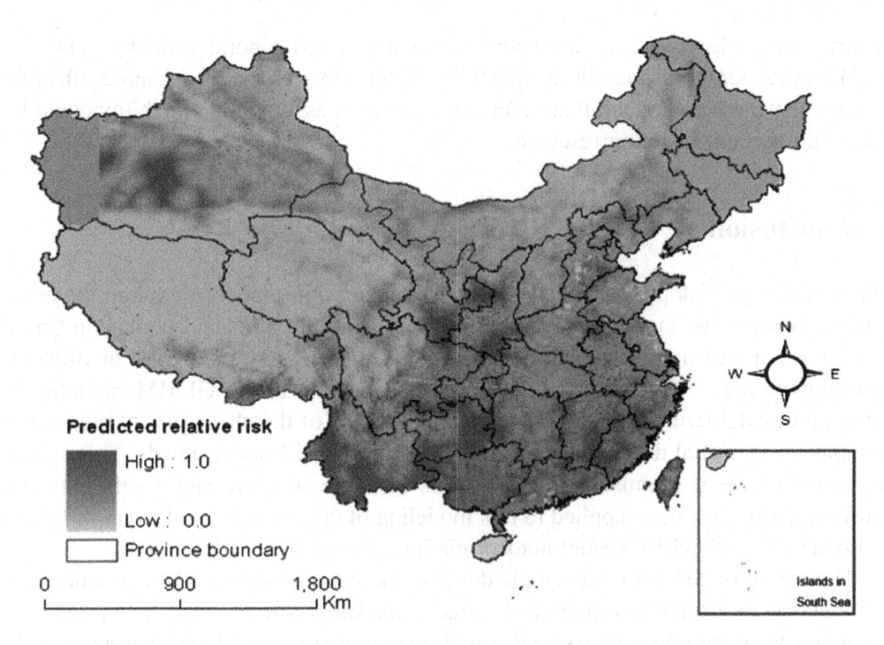

Fig. 5. Predicted relative risk of the HPAI H5N1 in wild birds and poultry in China 2004–2011

4 Discussion

Risk modeling treats the entire transmission cycle as a black box, and focuses on the spatial position and environmental characteristics of sites where humans or poultry contract the disease [46]. As such, independent testing and repeated challenging of models to be predictive and general are central to this application of risk modeling [17]. In this research, GLMM incorporating variogram modeling was used to predict the number of HPAI H5N1 cases in geographical areas. The model was limited by the data available, which were themselves limited to those events that have occurred in the last decade. Given more data, it is possible that a greater number of environmental covariates may be seen to affect the number of HPAI H5N1 cases in wild birds and poultry in China, and more sample data would be available to use in validation. A traditional problem with risk distribution maps predicted by statistical models, based on linking the presence/absence of a disease or species to a series of predictors, is that they often lose much of their predictive power when extrapolated outside of the spatial range of the training data, which makes external validation difficult [16]. This problem exists in the present research also and, therefore, caution is warranted when extrapolating the results beyond the present spatial and temporal domains.

Another problem that cannot be avoided is how to choose environmental covariates and how to represent their effects on the number of HPAI H5N1 cases in wild birds and poultry, because the processes including environmental factors influencing the spread of HPAI H5N1 virus are not clearly understood [13]. Here, we have chosen ten well defined environmental covariates, but the analysis may benefit from addition of other covariates and other representations (e.g., the role of live bird markets, cropping intensity etc.). Moreover, environmental covariates have temporal variability and this could lead to temporal variability in HPAI H5N1 virus cases. Thus, more effective environmental covariates and their interactions, as well as temporal variability could be taken into account in future research.

5 Conclusion

This research has adopted an integrated spatial regression model to explore the associations between the number of HPAI H5N1 cases in wild birds and poultry in China and ten environmental covariates such as to predict HPAI H5N1 risk in different geographical areas. This spatial regression model comprises a GLMM including a variogram model term allowing a quantitative analysis of the effects of environmental covariates and spatial dependence in the HPAI H5N1 incidence residuals. This spatial regression model is promising because it has a simple structure and good predictive capability. Thus, it can be applied to risk modeling of other subtypes of avian influenza and other diseases where spatial autocorrelation persists in model residuals.

The spatial regression model applied to risk modeling of HPAI H5N1 in wild birds and poultry in China has produced some interesting results. Four environmental covariates were significantly associated with the number of HPAI H5N1 cases in wild birds and poultry. These four covariates were annual mean temperature, poultry density, distance to lakes and wetlands, and distance to bird migration routes. Predicted high risk areas were identified in Northwest, Central, Southwest and Southeast China. These high risk areas fall within two bird migration routes: the Middle-Asia India Route and the Western Pacific Route. This implies that wild birds and bird migration may play an important role in outbreaks of HPAI H5N1 in China. Further research should be undertaken to explore further these findings, with the possible goal of targeting these geographical regions for future surveillance and control.

References

1. World Health Organization, WHO (2012). http://www.who.int/en/
2. Ferguson, N.M., Fraser, C., Donnelly, C.A., Ghani, A.C., Anderson, R.M.: Public health risk from the avian H5N1 influenza epidemic. Science **304**, 968–969 (2004)
3. Gilbert, M., Chaitaweesub, P., Parakamawongsa, T., Premashtira, S., Tiensin, T., Kalpravidh, W., Wagner, H., Slingenbergh, J.: Free-grazing ducks and highly pathogenic avian influenza. Thailand Emerg. Infect. Dis. **12**, 227–234 (2006)
4. Gilbert, M., Xiao, X., Chaitaweesub, P., Kalpravidh, W., Premashthira, S., Boles, S., Slingenbergh, J.: Avian influenza, domestic ducks and rice agriculture in Thailand. Agric. Ecosyst. Environ. **119**, 409–415 (2007)
5. Paul, M., Wongnarkpet, S., Gasqui, P., Poolkhet, C., Thongratsakul, S., Ducrot, C., Roger, F.: Risk factors for highly pathogenic avian influenza (HPAI) H5N1 infection in backyard chicken farms. Thailand Acta. Trop. **118**, 209–216 (2011)

6. Pfeiffer, D.U., Minh, P.Q., Martin, V., Epprecht, M., Otte, M.J.: An analysis of the spatial and temporal patterns of highly pathogenic avian influenza occurrence in Vietnam using national surveillance data. Vet. J. **174**, 302–309 (2007)
7. Yupiana, Y., de Vlas, S.J., Adnan, N.M., Richardus, J.H.: Risk factors of poultry outbreaks and human cases of H5N1 avian influenza virus infection in West Java Province. Indonesia Int. J. Infect. Dis. **14**, e800–e805 (2010)
8. Loth, L., Gilbert, M., Osmani, M.G., Kalam, A.M., Xiao, X.: Risk factors and clusters of highly pathogenic avian influenza H5N1 outbreaks in Bangladesh. Prev. Vet. Med. **96**, 104–113 (2010)
9. Fuller, T.L., Saatchi, S.S., Curd, E.E., Toffelmier, E., Thomassen, H.A., Buermann, W., DeSante, D.F., Nott, M.P., Saracco, J.F., Ralph, C.J., Alexander, J.D., Pollinger, J.P., Smith, T.B.: Mapping the risk of avian influenza in wild birds in the US. BMC Infect. Dis. **10**, 187 (2010)
10. Boender, G.J., Hagenaars, T.J., Bouma, A., Nodelijk, G., Elbers, A.R.W., de Jong, M.C.M., Boven, M.V.: Risk maps for the spread of highly pathogenic avian influenza in poultry. PLoS Comput. Biol. **3**(4), e71 (2007)
11. Ward, M.P., Maftei, D., Apostu, C., Suru, A.: Environmental and anthropogenic risk factors for highly pathogenic avian influenza subtype H5N1 outbreaks in Rominia, 2005–2006. Vet. Res. Commun. **32**(8), 627–634 (2008)
12. Cumming, G.S., Hockey, P.A.R., Bruinzeel, L.W., Plessis, M.A.D.: Wild bird movements and avian influenza risk mapping in Southern Africa. Ecol. Soc. **13**(2), 26 (2008)
13. Fang, L., Vlas, S.J., Liang, S., Looman, C.W.N., Gong, P., Xu, B., Yan, L., Yang, H., Richardus, J.H., Cao, W.: Environmental factors contributing to the spread of H5N1 avian influenza in mainland China. PLoS ONE **3**(5), e2268 (2008)
14. Cao, C., Xu, M., Chang, C., Xue, Y., Zhong, S., Fang, L., Cao, W., Zhang, H., Gao, M., He, Q., Zhao, J., Chen, W., Zheng, S., Li, X.: Risk analysis for the highly pathogenic avian influenza in mainland China using meta-modeling. Chinese Sci. Bull. **55**(36), 4168–4178 (2010)
15. Martin, V., Pfeiffer, D.U., Zhou, X., Xiao, X., Prosser, D.J., Guo, F., Gilbert, M.: Spatial distribution and risk factors of highly pathogenic avian influenza (HPAI) H5N1 in China. PLoS Pathog. **7**(3), e1001308 (2011)
16. Gilbert, M., Xiao, X., Pfeiffer, D.U., Epprecht, M., Boles, S., Czarnecki, C., Chaitaweesub, P., Kalpravidh, W., Minh, P.Q., Otte, M.J., Martin, V., Slingenbergh, J.: Mapping H5N1 highly pathogenic avian influenza risk in Southeast Asia. PNAS **105**(12), 4769–4774 (2008)
17. Peterson, A.T., Williams, R.A.J.: Risk mapping of highly pathogenic avian influenza distribution and spread. Ecol. Soc. **13**(2), 15 (2008)
18. Si, Y., Wang, T., Skidmore, A.K., de Boer, W.F., Li, L., Prins, H.H.T.: Environmental factors influencing the spread of the highly pathogenic avian influenza H5N1 virus in wild birds in Europe. Ecol. Soc. **15**(3), 26 (2010)
19. Wakefield, J.: Disease mapping and spatial regression with count data. BioStat. **8**(2), 158–183 (2007)
20. Rezaeian, M., Dunn, G., Leger, S.S., Appleby, L.: Geographical epidemiology, spatial analysis and geographical information systems; a multidisciplinary glossary. J. Epidemiol. Commun. H **61**, 98–102 (2007)
21. Kleinschmidt, I., Sharp, B.L., Clarke, G.P.Y., Curtis, B., Fraser, C.: Use of generalized linear mixed models in the spatial analysis of small-area malaria incidence rates in KwaZulu Natal. South Africa Am. J. Epidemiol. **153**(12), 1213–1221 (2001)
22. Kazembe, L.N.: Spatial modelling and risk factors of malaria incidence in Northern Malawi. Acta Trop. **102**, 126–137 (2007)

23. Reid, H., Haque, U., Clements, A.C.A., Tatem, A.J., Vallely, A., Ahmed, S.M., Islam, A., Haque, R.: Mapping malaria risk in Bangladesh using Bayesian geostatistical model. Am. J. Trop. Med. Hyg. **83**(4), 861–867 (2010)
24. Zepeda, C.: Highly pathogenic avian influenza in domestic poultry and wild birds: a risk analysis framework. J. Wildl. Dis. **43**(3), S51–S54 (2007)
25. World Organization for Animal Health, OIE (2012). http://www.oie.int/
26. Data Sharing Infrastructure of Earth System Science (2012). http://www.geodata.cn/
27. Anselin, L.: What is special about spatial data? Alternative perspectives on spatial data analysis. Technical report 89-4, National Center for Geographic Information and Analysis, Santa Barbara, CA (1989)
28. Goodchild, M.F.: Geographical information science. Int. J. Geogr. Inf. Sci. **6**(1), 31–45 (1992)
29. Ripley, B.D.: Chapter 8 Mapped point patterns. Spatial Statistics, pp. 144–190. Wiley, Chichester (1981)
30. Tiensin, T., Ahmed, S.S.U., Rojanasthien, S., Songerm, T., Ratanakorn, P., Chaichoun, K., Kalpravidh, W., Wongkasemjit, S., Patchimasiri, T., Chanachai, K., Thanapongtham, W., Chotinan, S., Stegeman, A., Nielen, M.: Ecologic risk factor investigation of clusters of avian influenza A (H5N1) virus infection in Thailand. J. Infect. Dis. **199**, 1735–1743 (2009)
31. Paul, M., Tavornpanich, S., Abrial, D., Gasqui, P., Charras-Garrido, M., Thanapongtharm, W., Xiao, X., Gilbert, M., Roger, F., Ducrot, C.: Anthropogenic factors and the risk of highly pathogenic avian influenza H5N1: prospects from a spatial-based model. Vet. Res. **41**, 28 (2010)
32. Amonsin, A., Choatrakol, C., Lapkuntod, J., Tantilertcharoen, R., Thanawongnuwech, R., Suradhat, S., Suwannakarn, K., Theamboonlers, A., Poovorawan, Y.: Influenza virus (H5N1) in live bird markets and food markets. Thailand. Emerg. Infect. Dis. **14**(11), 1739–1742 (2008)
33. Liu, C., Lin, S., Chen, Y., Lin, K.C.M., Wu, T.S.J., King, C.C.: Temperature drops and the onset of severe avian influenza A H5N1 virus outbreaks. PLoS ONE **2**, e191 (2007)
34. Keller, I., Korner, F., Jenni, N.L.: Within-winter movements: a common phenomenon in the Common Pochard Aythya farina. J. Ornithol. **150**, 483–494 (2009)
35. World Health Organization: Direct and indirect factors facilitating the spread of the avian influenza virus (2006). http://www.searo.who.int/LinkFiles/PublicationsandDocuments factors
36. Williams, R.A.J., Peterson, A.T.: Ecology and geography of avian influenza (HPAI H5N1) transmission in the Middle East and Northeastern Africa. Int. J. Health. Geogr. **8**, 47 (2009)
37. Adhikari, D., Chettri, A., Barik, S.K.: Modelling the ecology and distribution of highly pathogenic avian influenza (H5N1) in the Indian subcontinent. Curr. Sci. **97**(1), 73–79 (2009)
38. Biswas, P.K., Christensen, J.P., Ahmed, S.S.U., Das, A., Rahman, M.H., Barua, H., Giasuddin, M., Hannan, A.S.M.A., Habib, M.A., Debnath, N.C.: Risk for infection with highly pathogenic avian influenza virus (H5N1) in backyard chickens. Bangladesh Emerg. Infect. Dis. **15**(12), 1931–1936 (2009)
39. Olsen, B., Munster, V.J., Wallensten, A., Waldenström, J., Osterhaus, A.D.M.E., Fouchier, R.A.M.: Global patterns of influenza a virus in wild birds. Science **312**, 384–388 (2006)
40. Alerstam, T., Lindstrom, A.: Optimal bird migration: the relative importance of time, energy and safety. In: Gwinner, E. (ed.) Bird Migration: Physiology and ecophysiology, pp. 331–351. Springer, Berlin (1990)
41. Altman, M., Gill, J., McDonald, M.P., LeSage, J.P.: Spatial regression models (Chap. 9). In: Numerical Issues in Statistical Computing for the Social Scientist. Wiley, Hoboken (2004)

42. Jiang, J.M.: Chapter 3 Generalized linear mixed models: Part I. Linear and Generalized Linear Mixed Models and Their Applications. Springer Series in Statistics, pp. 119–162. Springer, Berlin (2007)
43. Zhu, H.T., Lee, S.Y.: Analysis of generalized linear mixed models via a stochastic approximation algorithm with Markov chain. Stat. Comput. **12**, 175–183 (2002)
44. Gan, X.: Generalized linear mixed models. J. Kunming Univ. Sci. Technol. **32**(4), 107–113 (2007)
45. Dean, C.B., Nielsen, J.D.: Generalized linear mixed models: a review and some extensions. Lifetime Data Anal. **13**, 497–512 (2007)
46. Peterson, A.T.: Ecological niche modelling and spatial patterns of disease transmission. Emerg. Infect. Dis. **12**, 1822–1826 (2006)

Study on Regional Cultivated Land Resource Sustainable Utilization Evaluation— A Case Study of Chengdu Plain

Fashuai Qin[1], Meixiu Zhou[2], Ruoheng Tian[2], Tingting Huang[2], Weizhong Zeng[2], and Chengyi Huang[2(✉)]

[1] College of Water Conservancy and Hydropower Engineering, Sichuan Agricultural University, Yaan 625014, China
[2] College of Management, Sichuan Agricultural University, Chengdu 611130, China
ahuang_6@sohu.com

Abstract. Cultivated land resource is not only the material foundation of survival and development of human, but also the essential factors guaranteeing the national food security strategy, maintaining the social stability and ensuring the sustainable development of eco-environment. The evaluation of sustainable utilization of cultivated land resource is a significant topic of regional land utilization research. In this paper, current situation and problems in cultivated land resource is analyzed for the research region centered on Chengdu Plain focusing on the urban-rural development and new rural construction based on the basic theories, including the sustainable development theory, human-land relation theory, control theory, etc. and the evaluation index system of sustainable cultivated land resource utilization is constructed for evaluating the state and variation trend of sustainable utilization of cultivated land resource from 2002 to 2010. According to the results, over the past decade, the cultivated land resource of Chengdu Plain was in sustainable, displaying the fluctuation characteristics. Although from 2003 to 2008, the comprehensive evaluation of the sustainable utilization of cultivated land resource increased, the overall sustainability was low, and the situation was not optimistic.

Keywords: Cultivated land resource · Sustainable utilization · Evaluation · Chengdu plain

1 Introduction

Cultivated land resource consists of paddy field, dry land, vegetable plot, etc. as a valuable and rare natural resource [1]. With the acceleration of socio-economic development and urbanization process, the area of cities and towns expands constantly, substantial cultivated land resource are occupied and human-land relationship is relatively tense, for instance, the cultivated lands decrease; cultivated land quality degenerates, the pollution aggravates day by day; the cultivated land resource utilization is not adapted to the urban socio-economic development and industrial distribution; the cultivated land resource varies greatly, and the regional distribution is uneven [1, 2].

F. Bian and Y. Xie (Eds.): GRMSE 2015, CCIS 569, pp. 722–732, 2016.
DOI: 10.1007/978-3-662-49155-3_74

However, cultivated land resource is not only the material foundation of survival and development of human, but also the essential factors guaranteeing the national food security strategy, maintaining the social stability and ensuring the sustainable development of eco-environment. Consequently, how to realize the sustainable utilization of cultivated lands has already become the top priority of the sustainable development of China's socio-economic society, while the launch of sustainable utilization evaluation is a significant step of exploring the impact factor of sustainable utilization of cultivated lands and realizing the sustainable utilization of cultivated lands [3].

In recent years, domestic and foreign scholars have conducted substantial research work targeting at related problems from different perspectives. Hubbard and Flores-endoza [4] discussed the sustainable utilization of cultivated lands according to the correlation between the crop yield, per unit yield, cultivated area and climatic changes. Marshall [5] analyzed the sustainable development of American agriculture by combining related test results of 'international geosphere – biosphere'. Berroteran and Zinck. [6] launched national scale studies on the sustainable utilization of cultivated land resource in Venezuela. Lefroy et al. [7] launched related studies based on the investigation into farmers for the sustainable utilization of cultivated land resource in Vietnam, Indonesia and Thailand. Chinese scholars Liang et al. [8] conducted studies on the dynamic changes of social sustainability, economic sustainability and ecological sustainability of cultivated lands with the energy analysis method. Wen et al. 9] proposed the basic principles of sustainable utilization of cultivated land resource, 'soil improvement condition, soil formation type and soil features, as well as complexity of soil improvement technology' through the soil investigations on Gansu, Qinghai and Xinjiang regions. Wang et al. [10] calculated the sustainable development capacity and load factor of cultivated lands and construction lands in Lulong County, which represented the cultivated land and construction land scale that could be hold by the sustainable development of regional economy, and discussed the potential transformational relation between the cultivated lands and construction lands in the research area. Tang and Ning [11] studied the sustainable utilization of cultivated land in Zhejiang, Jiangxi, etc. From the perspective of microcosmic body and concluded that there was a close relationship between the land utilization behaviors and sustainable utilization of cultivated land resource. Yang [12] applied the comprehensive evaluation model and studied the sustainable utilization of cultivated land resource in Qianjiang District of Chongqing. Liu [13] constructed the evaluation index system with the overall improvement of ecological rationality, economic feasibility and social acceptability as research objective, found out the key barriers impacting the sustainable utilization of land resource in Dalian, and predicted and conducted regressive analysis of the cultivated land variation trend in the future few years. Yang et al. [14] proposed the maintenance of dynamic equilibrium in total cultivated lands, protection and gradual improvement of cultivated land resource quality and implementation method, and proposed the strategic approaches and countermeasures guaranteeing the sustainable utilization of cultivated land resource by targeting at the practical conditions in the research area, with Xichou County as an example. The previous studies had provided abundant fundamental data and methods for this paper. Established on the fundamental theories, such as sustainable development theory, human-land relationship theory and control theory, etc. study on the sustainable utilization of cultivated land resource is launched in the research region with Chengdu Plain as the core area, which mainly focusing on the overall rural and urban development, as well as new rural construction demonstration area, so that it can provide theoretical references for the study on the scientific optimization of the structural functions, sustainable management and related work of the cultivated land resources.

2 General Situation of the Research Area

Chengdu Plain locates in the core of Chuanxi Plain, covering the core area of Chuanxi Plain in the south of Guanghan and Shifang [2, 15]. The land is flat, and the soil is fertile in the region. The soil is mainly dominated by the paddy soil and purple soil. Currently, the cultivated land resource is 478069 hm^2, accounting for 35.90 % of the total area, known as the land of abundance. The region located in the area of subtropical moist monsoon climate, the annual average temperature ranges between 15.2 and 16.5 °C, the heat resource is relatively abundant, and the annual sunshine duration lasts for about 1100 h. In the region, the precipitation is relatively abundant, and the average annual precipitation is about 900 to 1300 mm. In July and August, it rains the most, and the precipitation is above 200 mm [16, 17]. Chengdu Plain has always been the significant political and cultural center in southwest China. In 2010, the total population reached 12.756 million. Chengdu Plain is also the significant production base of food, oil plants, vegetables of Sichuan and even China. In 2010, the grain total output was 3310800 tons, the oil plant output was 261600 tons, and the vegetables were 5596400 tons. Meanwhile, as one of the pilot regions for comprehensive reforms, Chengdu Plain was economically developed, with high urbanization rate and abundant rural labor force. In 2010, the per capita gross domestic product was 47163 Yuan. In the GDP constitute of other regions, the primary industry, secondary industry and tertiary industry was 33.541 billion, 266.357 billion and 286.294 billion Yuan respectively [8].

3 Construction of Evaluation Index System for the Sustainable Utilization of Cultivated Land Resource in Chengdu Plain

3.1 Factor Selection and Determination of the Optimal Value

It shall follow the setting principle, scientific, systematic and operable principles of the index system. On the basis of referring to the latest domestic and foreign research achievements of related problems concerning the utilization of cultivated land resource, and by combining the characteristics of the cultivated land resource system and practical utilization of cultivated land resource utilization, related factors impacting the sustainable development of cultivated land resource are divided, and factors are selected from three aspects, including the nature, economy and society. Furthermore, each index is selected and condensed, and the evaluation index system of the sustainable utilization of the cultivated land resource was concluded for Chengdu Plain (Table 1).

The weight of evaluation index is mainly determined with the Delphi method, which mainly empowers each type of index system and its subordinate index factors, and the grade is eventually determined according to the comparison of importance degree. Later, the feature vector of the matrix obtained is applied for determining the contribution of the substratum target to the super-stratum target. As a result, the evaluation result of the fundamental indexes to the overall target can be obtained, and eventually, the empowerment result of the index to the total target is eventually determined (Table 1).

Table 1. The evaluation index system of the sustainable utilization of the cultivated land resource was concluded for Chengdu Plain

Control subject	Assessing target	Index description	Control order parameter	Weight	Threshold
Cultivated land resource sustainable utilization	Social factor	Quantity of resources	Farmland acreage per capita X_{11} (hm^2 per capita)	0.2796	0.46
		Structure	The proportion of farmland in the total land area X_{12} (%)	0.2532	40
		The proportion of urban population in total population	Urbanization rate X_{13} (%)	0.2219	70
		Cultivated land use &Employment	The proportion of first industry ' practitioners in total GDP practitioners X_{14} (%)	0.2453	41
	Natural factor	Quality of resources	Soil quality index X_{21} (0–1)	0.3112	1
		Environmental quality	Soil pollution index X_{22} (0–1)	0.2265	1
		Change Rates	Annual reduction rate of farmland X_{23} (%)	0.2521	5
		Cultivated land use &Landscape	Shannon uniformity index's patch number (SHEI)X_{24} (0–1)	0.2102	0.2
	Economical factor	Economic development	GDP per capita X_{31}(Ten thousand yuan per capita)	0.2216	5
		Income level	Growth rate of farmers' average income X_{32} (0–1)	0.2457	0.3
		Industrial structure	The proportion of first industry ' output in total GDP X_{33}(%)	0.2739	10
		Output of resources	Production value per unit area X_{34} (Ten thousand yuan per hm^2)	0.2588	7

3.2 Standardization of the Index Value

Since there are differences in the dimension, order of magnitudes, positive and negative orientation, it is quite necessary to conduct standardization processing for the initial data, so as to reflect the significance of each index effectively. The greater the index value is, the better the sustainable utilization of cultivated land resource will be. On that basis, the positive and negative index method is employed, namely:

$$\text{Positive index: } X'_{ij} = (X_{ij} - \min\{X_j\})/(\max\{X_j\} - \min\{X_j\}) \quad (1)$$

$$\text{Negative index : } X'_{ij} = (\max\{X_j\} - X_{ij}) / (\max\{X_j\} - \min\{X_j\}) \quad (2)$$

In the equation, X'_{ij} is the standardized data, $\min\{X_j\}$ is the minimum j index in the past years, while $\max\{X_j\}$ is the maximum j index in the past years.

3.3 Comprehensive Evaluation Model of the Sustainable Utilization of Cultivated Land Resource

The sustainable utilization evaluation of cultivated land resource is actually the evaluation of mutual coordination between sub-systems of the cultivated resource land system. Within certain regions, the cultivated land resource utilization may respond to the regional economic development, social progress and resource and environment protection, thus to promote the coordination and optimization of cultivated land resource utilization, and the social, economic and resource environment, enhance the mutual constraint and coordination of numerous factor in each sub-system of the cultivated land resource, and eventually realize the sustainable utilization of cultivated land resource. Therefore, comprehensive coordination degree shall be selected for evaluating the sustainable utilization of cultivated land resource, and the evaluation function is:

$$T_S = \sum_{i=1}^{m} W_i * X'_{ij}$$

$$T_Z = \sum_{i=1}^{m} W_i * X'_{ij}$$

$$T_E = \sum_{i=1}^{m} W_i * X'_{ij}$$

The secondary comprehensive assessment function is:

$$T = \sum T * W$$

In the above function, W is the weight of each evaluation index, T is the development level of the sustainable utilization of cultivated land resource, T_S is the sustainable degree between the cultivated land resource utilization and closely related social factors, T_Z is the sustainable degree between the cultivated land resource utilization and closely related natural factor, T_E impacts the sustainable state between the cultivated land resource utilization and closely related economic factor. In different evaluation interval, there are differences in the coordination degree of T, and it mainly corresponds to the sustainable utilization of cultivated lands.

3.4 Determination of the Sustainable Utilization Standard of Cultivated Land Resource

Focusing on the sustainable utilization evaluation of cultivated land resource, the sustainable varying interval is divided as follows by combining the expert opinions in related fields (as shown in Table 2).

Table 2. Determination of the sustainable varying interval and utilization standard of cultivated land resource

C	State of sustainable utilization	Meaning of order degree
$C \geq 0.9$	Highly sustainable utilization	The best sustainable utilization relationship between cultivated resource protection, social economic sustainable utilization, Economic rationality and Social acceptability show orderly state
$0.7 \leq C < 0.9$	Relatively sustainable utilization	Good sustainable utilization relationship between cultivated resource protection, social economic sustainable utilization, Economic rationality and Social acceptability show relatively orderly state
$0.5 \leq C < 0.7$	Basically sustainable utilization	Basically sustainable utilization relationship between cultivated resource protection, social economic sustainable utilization, Economic rationality and Social acceptability show basically ordered
$C < 0.4$	unsustainable utilization	Unsustainable utilization relationship between cultivated resource protection, social economic sustainable utilization, Economic rationality and Social acceptability show exhibits disordered state

Note:C refers to the value of farmland resources sustainable utilization degree

4 Evaluation and Result Analysis of Sustainable Utilization of Cultivated Land Resource in Chengdu Plain

4.1 Comprehensive Evaluation of Sustainable Utilization of Cultivated Land Resource in Chengdu

Targeting at the practical utilization of cultivated land resource, related calculation equation in 2.2 and 2.3 is applied for the dynamic evaluation of the sustainable utilization of cultivated land resource in Chengdu Plain from 2002 to 2010. The related evaluation result of sustainable utilization of cultivated land resource in Chengdu Plain was concluded (Table 3).

Table 3. The related evaluation result of sustainable utilization of cultivated land resource in Chengdu Plain

Year	2002	2003	2004	2005	2006	2007	2008	2009	2010
Evaluation result of social factors	0.53	0.36	0.36	0.43	0.46	0.55	0.56	0.47	0.47
Evaluation result of natural factors	0.71	0.65	0.77	0.75	0.71	0.64	0.57	0.47	0.47
Evaluation result of economical factor	0.27	0.30	0.35	0.41	0.43	0.56	0.63	0.66	0.73
Comprehensive evaluation result	0.52	0.46	0.52	0.55	0.55	0.59	0.59	0.53	0.55

4.2 Analysis of the Sustainable Utilization of Cultivated Land Resource in Chengdu Plain

(1) Comprehensive variation trend analysis for the social factor sub-system of cultivated land resource in Chengdu Plain

In Fig. 1, from 2002 to 2010, the comprehensive variation trend of the social factor sub-system of cultivated land resource in Chengdu Plain mainly ranged between 0.4 and 0.5, and the influence of social factors on the sustainable utilization of cultivated lands in research area occurs in sustainable state, in which, in 2003 and 2004, owing to the comprehensive impact of the constant acceleration of urbanization process, decreasing per capita cultivated land caused by the urban expansion, etc. on the sustainability and stability of cultivated land resource utilization, the figure fluctuates suddenly. However, with the intervention of government into the rural labor transfer, the gradual standardization of cultivated land resource, as well as the increasingly remarkable integration effect of the urban and rural economy, the impact of social factors on the sustainable utilization of cultivated land resource occurs in positive effect, and the entire system occurs in sustainable state.

(2) Comprehensive variation trend analysis of the natural factor sub-system index of the sustainable development of cultivated land resource in Chengdu

In Fig. 2, from 2002 to 2003, the comprehensive value of the natural factor sub-system of sustainable utilization of the cultivated land resource in Chengdu Plain occurs in declining trend and state. The natural factors have a huge impact on the sustainable utilization of cultivated land resource in Chengdu Plain, and the figure

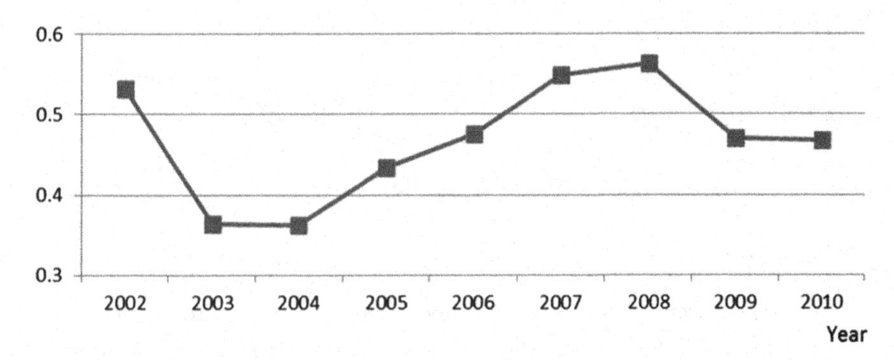

Fig. 1. The comprehensive variation trend of the social factor sub-system of cultivated land resource in Chengdu Plain during 2002 to 2010

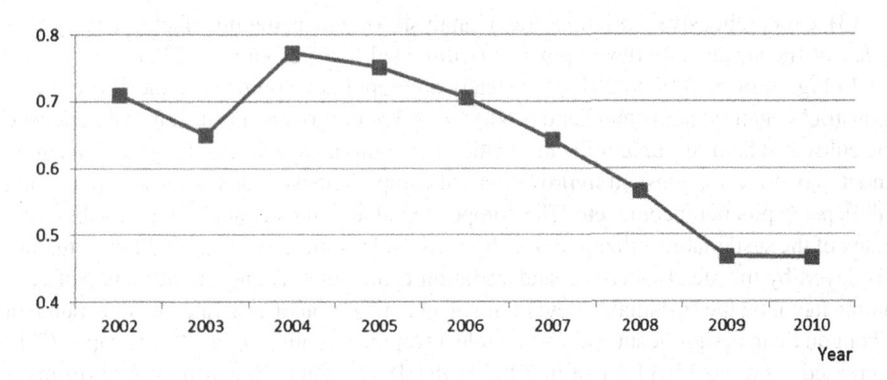

Fig. 2. The comprehensive value of the natural factor sub-system of sustainable utilization of the cultivated land resource in Chengdu Plain during 2002 to 2010

gradually turned to be the 'basically sustainable' state in 2007 and 2008 from the 'relatively sustainable' state in 2002, till the 'unsustainable state' in 2009 and 2010. As for the causes, Chengdu Plain, as the main production area of China's agriculture, the cultivated agriculture enjoys a long history, and it is superior in the climate conditions dominated by the warm and humid sub-tropical climate. The soil, mainly the purple soil and purple paddy soil, is quite fertile, with high resource endowment [19]. In recent years, with the substantial occupation of cultivated land resource, the resource decreases year by year, but due to the rigid demand of food and agricultural product demand, the agricultural production has to rely on the dense cultivation, which shall adopt the fertilizers and pesticides to guarantee the output. However, it would result in a series of resource and environment problems, such as declining soil fertility, degenerating quality of cultivated land and soil environment pollution, etc. [2]. The production and development of these problems would severely restrain and impact the healthy, orderly and sustainable development of cultivated land resource in Chengdu Plain.

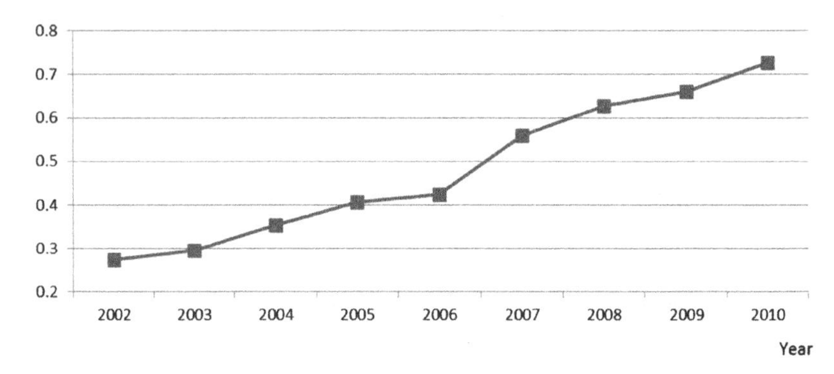

Fig. 3. The comprehensive value of the economical factor sub-system of sustainable utilization of the cultivated land resource in Chengdu Plain during 2002 to 2010

(3) Comprehensive variation trend analysis of the economic factor sub-system index of the sustainable development of cultivated land resource in Chengdu

In Fig. 3, from 2002 to 2010, the steady and rapid development of social economy, industrial structural adjustment and productive force improvement substantially enhance the cultivated land resource utilization efficiency, output capacity, and output efficiency, and it may drive the constant improvement of comprehensive indexes, such as per capita GDP, per capita net income, etc. The composite value of the economic factor subsystem index of the sustainable utilization of cultivated land resource in Chengdu Plain is mainly displayed by the steady increase and variation characteristic, and the influence of economic factor on the sustainable development and utilization of cultivated land resource in Chengdu Plain is significant and sustainable. Over the past nine years, the per capita GDP increased from the 15611 Yuan in 2002 to the 48312 Yuan, the proportion of primary industry output in GDP decreased from the 8.94 % in 2002 to the 5.14 % in 2010, the unit area output of the cultivated resource increased from the 30500 Yuan/hectare in 2002 to the 64900 Yuan/hectare in 2010, and the rural per capita net income increased form 3332.85 Yuan/year to 8205.63 Yuan. Under the comprehensive function of numerous influence factors, the economic factor poses a positive impact on the sustainable utilization of the cultivated land resource in Chengdu Plain.

(4) Comprehensive variation trend analysis of the sustainable utilization of cultivated land resource

In Fig. 4, it reflects the basic state and variation trend of the sustainable utilization of cultivated land resource in Chengdu Plain. In the past nine years, the comprehensive evaluation of sustainable utilization of cultivated land resource in Chengdu Plain ranged between 0.45 and 0.6, in basically sustainable state, and fluctuating variation feature. In which, from 2002 to 2003, the composite value of the natural factor sub-system and social factor sub-system declines at the same time, which gave rise to the sustainable declining trend of the cultivated land resource system, and the entire system transformed from the 'basically sustainable' state to the 'unsustainable' state; from 2003 to 2008, the composite value of the economic factor sub-system and social factor sub-system increased constantly, while the composite value of the natural factor sub-system declined constantly. Due to its impact, the composite value of sustainable

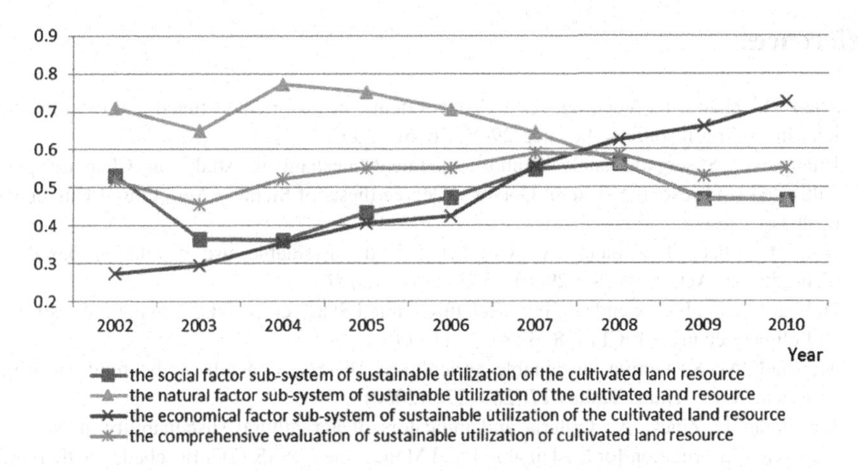

Fig. 4. The basic state and variation trend of the sustainable utilization of cultivated land resource in Chengdu Plain

utilization of cultivated land resource increased constantly, in fluctuating increasing trend; from 2008 to 2010, due to the dual influence of the composite value of the natural factor sub-system and social factor sub-system, the comprehensive evaluation of sustainable utilization of the cultivated land resource occurred in declining trend, and the overall system was on the edge of the 'unsustainable' state, and the situation was not quite optimistic.

5 Conclusion and Discussion

In this paper, the sustainable utilization of cultivated land resource in Chengdu Plain is analyzed comprehensively. In recent ten years, the overall cultivated land resource utilization is in sustainable state and dynamic fluctuation trend, but is close to unsustainable edge. The sustainable utilization of cultivated land resource is quite severe, and it is mainly caused by the acceleration of economic development and industrialization, constant growth of population, severe degeneration of cultivated land quality, aggravating pollution, lack of effective connection between the overall primary land utilization planning, urban planning, and industrial development planning, and low resource utilization efficiency. Therefore, during the land utilization process in the future, attention shall be paid to the protection and proper utilization of cultivated land resource, improvement of cultivated land resource utilization rate, control of population growth, and improvement of sustainable utilization level of the cultivated land resource.

Acknowledgments. This research was supported by "Cultivated land resource multi-source information intelligent management & Synergetic analysis and assessment system development" research fund, "Study on regional cultivated land resource multi-source information intelligent management and sharing key techniques of application service platform" research program and Social sciences special research funds of Sichuan agricultural university (2014).

References

1. Zhao, H., Zhang, F., Yueqing, X.: Urban residents. needs-oriented functional of cultivated land in Beijing resources. Science **29**(1), 56–62 (2007)
2. Huang, C.: Sysergetic analysis and adjustment mechanisms study on Chengdu plain cultivated land resource system. Doctor's Degree thesis of Sichuan Agricultural University, 6 (2011)
3. Niu, H., Zhao, T., Zhang, A.: Cultivated land sustainable use evaluation based on niche-fitness. Acta Ecol. Sin.**29**(10), 5335–5543 (2009)
4. Hubbard, K.G., Flores-endoza, E.J.: Relating United States cropland use to natural resources and climate change. J. Clim. **8**, 329–335 (1995)
5. Marshall, A.: Sustaining sustainable agriculture: The rise and fall of the fund for Rural America. Agric. Hum. Values **17**, 267–277 (2000)
6. Berroteran J., Zinck, A.: Criteria and indicators of agricultural sustainability at National level. Geo-Information for Sustainable Land Management ISSS/ITC, Enschede, Netherlands (1997)
7. Lefroy R.D.B., Bechstedt, H.D., Rais, M.: Indicators for sustainable and management based on farmer surveys in Vietnam, Indonesia and Thailand. Ag III Ecosystem and Environment
8. Liang, S., Zhang, C., Li, S.: Analysis on cultivated land sustainable performance value of Xi'an city. Ecol. Econ. **2**, 263–266 (2012)
9. Wen, Z., Shi, Y., Huang, R.: Xinjiang Soil Geography. Science Press, Beijing (1965)
10. Wang, X., Men, M., Wang, S.: Calculation on land resource sustainable development capacity and analysis on potential transferring relationship base on ecological footprint method. Acta Ecol. Sin. **30**(14), 3772–3783 (2010)
11. Tang, M., Ning, A.: Farmer's land use behavior impact on the sustainable utilization of cultivated land in the process of urbanization – Take zhejiangJiangxi as an example. J. Agric. Econ. **5**, 68–69 (2011)
12. Yang, Y.: Study on regional cultivated land resource sustainable utilization evaluation—a case study of Qianjiang District in Chongqing Doctor's Degree thesis of Southwestern University, 5 (2006)
13. Liu, Y.: Municipal land resources sustainable utilization evaluation and prospect of cultivated land quantity analysis in Dalian City Master's Degree thesis of Liaoning Normal University, 5 (2005)
14. Yang, Y., Yang, Y., Chen, Y.: Study on sustainable method of cultivated land utilization in Xichou County. Environ. Sci. Tribune **27**(4), 61–63 (2008)
15. Tan, T., Wang, C., Li, B.: Pollution and evaluation of PB in soil in Chengdu plain. Resour. Environ. Yangtze Basin **14**(1), 71–75 (2005)
16. Zhang, B., Yi, H., Jingwen, X.: Characteristic analysis of agricultural climate in Chengdu plain in Autumns. J. Chengdu Inst. Meteorol. **14**(4), 322–328 (1999)
17. Yi, H., Zhu, K., Li, Y.: Study on the climate characteristics and its change in recent 40 years in the midwest of Chengdu Plain. J. Chengdu Inst. Meteorol. **19**(2), 223–231 (2004)
18. Sichuan Province Statistical Yearbook. China Statistical Press, 8 (2011)
19. Huang, C.: Study on the supply and demand of cultivated land in Sichuan province. Master's Degree thesis of Sichuan Agricultural University, 5 (2007)

Value Evaluation and Analysis of Space Characteristics on Linear Cultural Heritage Corridor Ancient Puer Tea Horse Road

Hui Li(✉), Qiucen Duan, Zhanjing Zeng, Xiaoling Tan, and Guoyan Li

College of Urban Construction and Administration, Yunnan University, Cuihubei Road 2, Kunming 650091, China
{ydlihui,kmguoyanli}@126.com

Abstract. The ancient Puer tea horse road is one of significant nodes in the ancient Yunnan-Tibet tea horse road. Now By way of site investigation and GIS information technology, using the analytic hierarchy process (AHP) we has carried on the value evaluation and classification, and using GIS software combined with of the adjacent index model has carried on the analysis of space characteristics. The main results are as follows: We have completed the resources composition and value assessment of the linear cultural heritage corridor. And have determined heritages identification and login, and have completed the heritage spatial characteristics analysis. The spatial distribution of heritage is two typical characteristics: between heritage point the spatial agglomeration degree is high and there are significant grade differences. The distance between heritage point and the ancient road within 10 km covers 80 % of the number of heritage, it provide the basis for the width of the protecting corridor.

Keywords: The ancient tea horse road · Linear cultural heritage · Heritage corridor · Analysis of space characteristics · Puer

1 Introduction

Foreign research progress: From the green channel, heritage corridor, cultural route to the linear cultural heritage [1]. The United States is the first for practice research in this field. Specified Illinois and Michigan canal national heritage corridor by American Assembly in 1984 that is the United States also the world's first national heritage corridor [2]. In addition, Europe, Canada, Mexico, Japan and other developed countries have similar research, its protect means are about the same.

Overseas for linear cultural heritage research are mainly following three aspects: First of all, the economic impact for tourism. Secondly, the research for assessment of heritage corridor, Third, according to the development of heritage, heritage product marketing and restricting factors have presented relevant suggestions for promotion value of the corridor.

The domestic research progress: From the perspective of the protection planning and utilization, Tongqian Zou et al. have discussed the case of the protection to planning

F. Bian and Y. Xie (Eds.): GRMSE 2015, CCIS 569, pp. 733–740, 2016.
DOI: 10.1007/978-3-662-49155-3_75

design, planning guideline and management system [3]. Jingwei Wang et al. have discussed heritage tourism development [4]. Kongjian Yu et al. have proposed establishment of the linear cultural heritage network method based on the national level [5].

Related research on the ancient tea-horse road. Yihong Mu, Baoya Chen, Xun Li, Yongtao Xu, Xiaosong Wang, Lin Li have investigated "The ancient tea-horse road", and have co-authored "the Yunnan Tibet Sichuan "big triangle" culture exploring" in 1992, for the first time put forward the "The ancient tea-horse road" name. After that, the ancient tea-horse road received widespread attention. The current study of "The ancient tea-horse road" is focused on the theoretical connotation, historical and cultural development, protection and exploitation and utilization such as three aspects. The largest research literature concentrate on ancient history and change, the development of ancient road settlement and the historic blocks. On the macro level to the exploration of the ancient tea horse road remains to be further [6].

2 Research Methods

Literature research combined with field investigation. Literature research is mainly to collect Puer city ten areas related local chronicles, historical data and maps, such as a large number of literature, and a region of city master plan (2011–2030), a region of Simao ancient tea horse road scenic area overall plan (2012–2030), building regulations and text, cultural heritage and other related information, and clear up in detail the above information. Multiple site investigation, field scientific research and interviews, have collected in the name of the linear cultural heritage, spatial location, surrounding environment, culture, customs and other important information, have provided the basis for heritage point determination and assessment.

2.1 Multidisciplinary Crossover Method

We have combined with the landscape ecology, landscape design, geography, human geography, ecology, urban planning, heritage protection, archaeological objects, and geographic information systems theory and method.

2.2 Analytic Hierarchy Process (AHP) and Delphi Method

Linear cultural heritage value evaluation is a key link in the heritage corridor protection program. Using analytic hierarchy process (AHP) and Delphi method to complete establishment of the linear cultural heritage value assessment system of Puer ancient tea horse road, and then have evaluated and graded to heritage.

2.3 Spatial Analysis

To enter space attribute of heritage into the GIS database, through operation for the GIS spatial analysis in combination with the adjacent index model, then have obtained the space distribution features of heritage, and further have researched relationship between the heritage points and between the heritage points and the ancient road.

2.4 Technical Route

Determining "source" (heritage corridor resources) → evaluating of heritage value → analyzing of characteristics of space.

3 Puer Ancient Tea Horse Road Situation

3.1 Natural Geographical Background

The study area Puer city is located in the southwest of Yunnan province, has area of 45385 square kilometers and a population of about 2.37 million. The Puer city is located in the Hengduan mountains stretches out to the south area, Lanchang, Honghe and Nu drainage rivers arrange alternate and winding down to south. Ailaoshan, Wuliangshan, Nushan three big mountains from south to north end to end set-up, and the mountains and rivers alternate arrangement, close in the north, and out to southwest and southeast. Regional vertical three dimensional terrain lead to 3d climate significantly, and domestic distribution of five different climate types. Puer is the region with the most abundant diversity of biological resources in Yunnan province, have higher plants of 352 families, 5600 species, accounting for 40 % of the province, forest land area of 14193 sq.km, it enjoy "the forest kingdom", "the pearl of the vegetation" reputation.

3.2 Cultural Background

Puer culture form is varied and co-exist on the honeycomb culture development. About the origin of "ancient tea horse road", academic circles at present, there are a large number of scholars believe that "ancient tea horse road" come into the world after "tea horse frontier trade" in Tang and Song. In a strict sense, with the breadth of human migration and supplies trading frequency is more and more big that is the real reason of "ancient tea horse road" origin.

3.3 Ethnic Background

Puer city has a total of more than 40 nationalities, every minority in the process of life and production, cultural blend and form their own unique customs and etiquette, diet, clothing, literature and art and characteristics of the local-style dwelling building, and hand down a large number of valuable material culture and non-material cultural heritage for along the line Puer ancient tea horse road.

3.4 Religious Background

In Puer religious sort is more, can be roughly divided into two types, one is Puer native produce, such as natural worship, ancestor worship, animal and plant worship and so on, two is by the abroad or the central region artificial propagation religion into the Puer and which has a far-reaching influence on Puer nationalities such as Taoism, Buddhism, Islam, Christianity and so on.

Puer ancient tea horse road belongs to important nodes of the Yunnan Tibet ancient tea horse road network, that have the following route: tea horse north road, tea horse south road, tea horse east road, tea horse west road.

4 Results

4.1 Corridor Resources Composition Analysis

Puer ancient tea horse road linear cultural heritage includes material cultural heritage and intangible cultural heritage, the relationship between heritage and ancient tea horse road is divided into function, history and spatial correlation. Heritage that don't have one of these three correlation not included this study of heritage corridor. According to research from the initial potential heritage point have screened potential material cultural heritage 272, intangible cultural heritage 294. Then the heritage point after again research and geography information technology inspection, it is concluded that the heritage list. A total of 222 heritage point, among them, the material culture heritage 104, non-material cultural heritage 118, and then the material cultural heritage points have input into the GIS database.

4.2 Material Cultural Heritage Resources Constituting

Along the ancient tea horse road distributing of a large number of precious heritage, in this study have divided into the material culture heritage of ancient relics ancient buildings, ancient tombs, ancient ruins, cave temples, stone carving and important historic and modern representative buildings, a total of seven categories and 34 subclass.

4.3 Heritage Type Composition Analysis

Through heritage type, location, protection level, to protect the status composition analysis have obtained: In the Puer ancient tea horse road the type of material cultural heritage is rich. Among them, number of proportion the ancient buildings, ancient ruins, ruins of the ancient road are larger. Heritage history old, large time span, from the New Stone Age to the period of the republic of China, in the majority with Ming and Qing dynasties and modern period. Heritage protection levels have national, provincial, municipal, county and unprotected level five categories, among them, unprotected level hold the largest, and protection status is divided into five grades, generic and better is the largest. From the distribution of different regional heritage, Simao and Ninger have most varied, followed by Ching, Mojiang and Lanchang, Kengtung, Zhenyuan, Cimo center, Jiangcheng and Menglian most single. Relative to the history time Lanchang has most species distribution, the least is the Jingcheng. Heritage in Jiangcheng and Menglian are best preserved area, Simao, Kengtung, Cimo etc. because of its distribution type more and save state worse are key reserves, save state in Ninger is good.

4.4 Intangible Cultural Heritage Resources

Intangible cultural heritage includes mainly the following contents: oral tradition and form, including the language as the medium of intangible cultural heritage, the performing arts, social practices, rituals and festivals; about the knowledge and practice of nature and the universe, Traditional handicrafts.

4.5 Heritage Type Composition Analysis

Puer ancient tea horse road intangible type is very rich and vast, and its majority have local characteristics, Puer ancient tea horse road type distribution of intangible cultural heritage folk-custom is the most, 36.4 % of the total, traditional dances and traditional art in second place, the traditional sports, recreation and acrobatics and traditional medicine at least distribution. From the perspective of the distribution area and type of heritage, various types of intangible distribution around the region. For heritage typicality and representative sample area with intangible resources more such as Simao, Ninger, Ching, Menglian that can be through the establishment cultural partition for protection. Due to the intangible cultural heritage resources are "living state", "mobile" heritage, investigation for intangible cultural heritage resources is unlikely to be as accurately as material cultural heritage of space positioning, only know its distribution area and scope.

4.6 Cultural Heritage Value Assessment and Grading

Heritage value evaluation is the key step in the heritage corridor construction, and according to the evaluation results, determine the level of heritage protection. Historical, artistic and scientific value, the three value of is the height generalization and scientific norms of the linear cultural heritage value, which covers basic the core cultural value system. The material and non-material culture heritage in Puer ancient tea horse road embodies the uniqueness and diversity of the linear cultural heritage, history, science, art, economy and society value, have built the evaluation system from five aspects, assessment method have adopt analytic hierarchy process (AHP).

Heritage the historical value highest (31.55 %), it reflect importance the history properties of material and cultural heritage, the second is the art and economic value. In factor layer, importance weights of heritage in residents affective commitment (16.80 %) rank first that reflects the protection material cultural heritage in residents affective commitment, must be protection the local residents interests. In addition, the old degree of history weight (12.66) is higher, also reflects that heritage longer history has more protection value.

For 104 material cultural heritage, the first rank score above 3.98 have total 35 points. ancient ruins 16, ancient ruins 1, ancient building 8, cave temples and stone carvings 3, important historic and modern representative building 2, ancient settlements 5. The second rank score between 2.60 to 3.98, a total of 42 heritage. ancient ruins 6, ancient ruins 10, ancient building 8, cave temples and stone carvings 6, important historic and modern representative building 11, ancient settlements 1. The third rank score below

2.60, a total of 27 heritage. ancient ruins 10, ancient tombs 2, ancient buildings 6, cave temples and stone carvings 5, important historic and modern representative building 4.

From the point of distribution areas, Simao, Ninger top level heritage point distribution are maximum, In other parts the top heritage point number is less. Number of secondary heritage point number distribution is Simao, followed by Ching, Mojiang, other cities distribution is uniform, number of level 3 point distribution areas is still the most in Simao, less in other parts of the distribution. Embodies importance about Simao as Puer ancient tea horse road origin region, the most distribution of heritage, heritage value is highest. From the type, the largest value is the ancient road heritage, in turn, is the ancient villages, ancient architecture, ancient ruins, important historic and modern representative building, cave temples and stone carvings and ancient tombs.

Intangible cultural heritage of historical value weights are still the highest (26.8 %) that reflecting the importance of the non-material cultural heritage of historical attribute, the weight of social value (20.15 %) reflects the intangible cultural heritage the important degree of residents affective commitment, Residents' acceptance of heritage in general are more likely to come down, in addition, the historical inheritance status (10.02 %) and the intact degree of the heritage status quo preserved (9.25 %), two indicators of weight also is relatively high, it reflects importance of the non-material cultural heritage inheritance and our protecting state.

For 118 intangible cultural heritage, the first rank score of 4.00 above, a total of 38 heritage. Among them, folk literature 6, traditional music 2, traditional dance 8, traditional drama 3, traditional skills 9, folk 10. The second rank score between 2.90 to 4.00, a total of 46 heritage. folk literature 2, traditional music 2, traditional dance 7, traditional operas 1, and Quyi 2, traditional arts 4, traditional skills of 8, traditional medicine 2, folk 18. The third rank score below 2.90, a total of 34 heritage. traditional music 1, traditional dance 7, Quyi 2, traditional sports, recreational and acrobatics 2, traditional art 2, traditional craft 5, folk 15. According to non-material cultural heritage value level form case have obtained, heritage highest total number of grade is Menglian, the least is Mojiang. Among them, the first level heritage distribution of the largest number of is Simao, the least is Mojiang. The second level, including heritage the largest number of distribution is the MengLian, minimum Mojiang. In the third grade, heritage largest number of distribution is Menglian, the least is Zhenyuan. In terms of heritage type distribution folk heritage number is far more than other heritage type, and the primary, secondary, tertiary heritage value with folk distribution is also most, followed by the traditional arts and traditional dance, amount of traditional sports, recreational and acrobatics and traditional medicine is the least.

4.7 Analysis of Characteristics of Heritage Space

Through to analyze the heritage point distribution characteristics and the characteristics of linear distribution, namely after analysis spatial relations between point and point and between points and lines can obtain results: first, there are high spatial agglomeration degree between heritage points, and there are an obvious distribution level difference. For the degree of distribution between two points, between 80 % of the heritage point the straight line distance less than or equal to 10 km. Only suitable for living and the transportation convenient produced heritage point, so overall there are gathered characteristics between heritage points, but there are a few heritage point exist alone. Second, the distance between heritage point and the ancient road within 10 km cover 80 % of number of heritage point. According to the results of the analysis, in Puer tea horse road corridor width of 8~10 km range is more suitable.

5 Conclusion and Discuss

5.1 Conclusion

Combed the Historical Context. The ancient Puer tea horse road is one of significant nodes in the ancient Yunnan-Tibet tea horse road. The historical evolution period of the linear cultural heritage ancient Puer tea horse road can be divided into: before the Tang dynasty and the Tang dynasty, Song Yuan and Ming Dynasties, Qing dynasty and the Republic of China three stages. And further have obtained evolution pattern formed by the different historical period of heritage.

The Discrimination and Login of Heritage. Heritage discriminant and login is also the determination of the "source" in the protection system, that is sure studied material and non-material cultural heritage. Finally after a quadratic discriminant login heritage list, a total of 222 heritage, including material culture heritage there are 104, and non-material cultural heritage has 118.

Corridor Resources Composition Analysis. Material culture heritage has the following characteristics: (1) Heritage type connotation is rich that is divided into 7 categories, 34 subclasses. (2) The history of the heritage so long have wide time span, in the majority with heritage of Ming Qing dynasty and modern period. (3) Heritage protection levels have national, provincial, municipal, county and unprotected level five categories, among them, there is no protection level accounted for the largest. To protect the status quo is divided into five grades, general and better are most.

Intangible Cultural Heritage Resources Composition Analysis. Intangible cultural heritage type are rich and vast, and include ten types all, and most have regional characteristics.

From the distribution area and type of heritage, the types of distribution of intangible cultural heritage are various. Amount of heritage no protection level most, protection situation is not optimistic.

The Linear Cultural Heritage Value Evaluation and Classification Research. The value of the linear cultural heritage include science, history, art, economy and social value. The heritage points are divided into three value of primary, secondary and tertiary level.

Heritage Space Research for Identifying Characteristics. Between heritage point the spatial agglomeration degree is high and there are significant grade differences. The distance between heritage point and the ancient road within 10 km covers 80 % of the number of heritage, it provide the basis for the width of the protecting corridor.

5.2 Discuss

This article was studied only on the construction of heritage corridor from the macroscopic aspect, for the protection and management work about the micro scale ancient villages, ancient ruins and so on, needs to be further deepen and perfect.

Due to the intangible cultural heritage is the living state of culture and information collection difficult, has a wide distribution area, sampling difficulties, bring the basic data of some evaluation factors in the evaluation system are insufficient precise. Study is only the initial stage of protection work in this article.

Acknowledgement. The authors would like to express appreciations to support by the National Natural Foundation of China (No. 51468064). The authors would like to express appreciations to colleagues in our laboratory for their valuable comments and other helps.

References

1. National Park System: National Heritage Areas Contact Information (2009)
2. Lanning, D.M.: Regional History and Public Planning Potential in Delaware Navigation Canal National Heritage Corridor. State University of New York (1994)
3. Zou, T.Q., Wang, Z.Y., Zheng, C.H.: Preliminary exploration development and protected mode on the China linear cultural heritage. World Heritage **4**, 106–109 (2010)
4. Wang, J.W., Han, B.N.: Linear cultural heritage tourism development potential evaluation and empirical study. J. Yunnan Normal Univ. **20**(5), 120–126 (2008)
5. Yu, K.J., Xi, X.S., Li, D.H., Li, H.L., Liu, K.: Building China national linear cultural heritage network. Hum. Geogr. **3**, 11–16 (2009)
6. Wang, L.P.: Yunnan-Tibet ancient tea horse road linear heritage area protection research. Geogr. Geogr. Inf. Sci. **28**(3), 101–105 (2012)

Spatial Pattern of Projectile Direction of Crater Rays on the Moon

Jiao Wang[1,2], Chengdu Zhou[1(✉)], Weiming Cheng[1], and Zengpo Zhou[1]

[1] State Key Laboratory of Resources and Environmental Information System, Institute of Geographic Sciences and Natural Resources Research, Chinese Academy of Sciences, Beijing 100101, China
{wjiao,zhouch,chengwm,zhouzp}@lreis.ac.cn
[2] University of Chinese Academy of Sciences, Beijing 100049, China

Abstract. Crater rays are among the most prominent lunar features that are radial or subradial to fresh impact craters. According to the latest more than 800 rayed impact craters with asymmetric ejecta in a latitude zone from 75°N to 75°S using Chang'E imageries, the spatial pattern of projectile direction of crater rays on the moon, namely the nearside-farside, leading side-trailing side and latitudinal variations, have been analyzed. The results suggest that: (1) the projectiles the rayed craters created mainly come from northeast direction while the number of projectiles coming from southwest direction is small; (2) in the nearside, projectiles from north direction can account for nearly 15 % of the whole projectiles. Yet in the farside, the number of projectiles makes up no more than 10 % of the total. On the other hand, the discrepancy is characterized by smaller frequency in the east direction on the trailing side while that is in higher frequency on the leading side direction; (3) with the latitude increases, the projectiles concentrate coming from one or two directions. The spatial variation suggests that the asteroid belt can well explain the dominant direction of projectiles concentrates in northeast direction and rayed craters on the Moon are formed mainly by near-Earth asteroids rather than comets with higher encounter velocities.

Keywords: Spatial pattern · Rayed carters · Projectile direction · Chang'E · GIS

1 Introduction

Lunar crater rays are those obvious bright streaks of material that we can see extending radially away from many impact craters on the Moon, Mercury, and large, icy Galilean satellites [1]. They are filamentous, high-albedo features that are generally narrow in relation to the crater radius and often extend many crater radii from their parent craters. They are often discontinuous, and feathery ray elements exhibit no noticeable topographic relief. Rays from a given lunar crater often cross a variety of lunar terrains including maria, terrae, and mountain ranges [2]. Crater rays are generally considered

F. Bian and Y. Xie (Eds.): GRMSE 2015, CCIS 569, pp. 741–750, 2016.
DOI: 10.1007/978-3-662-49155-3_76

to be formed when jets of rock flour are thrown out just as an impacting body forming the central crater begins to penetrate the lunar surface for the vast majority of lunar craters are of impact origin [3]. For planetary impactors approaching from random directions, virtually all impact craters on the Moon results from nonvertical impacts, which directly leads to the appearance of the rays' ejecta blanket varying with impact angle. Our research is focus on the spatial pattern of crater rays on the moon. Understanding how the crater rays distributing on the lunar surface with different projectile direction characteristics because of distinctly nonvertical impacts, or oblique impacts, is, therefore, critical to understanding planetary impact crater formation and to using the cratering record to study the geologic history of a planet.

The rayed crater is the parent feature containing crater rays. McEwen et al. [4] have identified 96 rayed craters larger than 10 km in diameter within a latitude zone from 60°N to 60°S on the far side. Grier et al. [5] have investigated rayed craters with diameters larger than 20 km globally using OMAT images. Morota and Furumoto [6] have identified 222 rayed craters larger than 5 km in diameter mainly on the far side of the Moon excluding the lunar mare areas and the South Pole-Aitken basin. Werner and Medvedev [7] have considered a significant subset of the young craters discussed by [4–6], which are as homogenous as possible. With these rayed crater catalogs, major research on the formation, the composition, the ejecta patterns, the age of lunar rays have long been done. Kadono et al. [8] performed impact experiments with granular targets to reveal the formation process of crater rays. For the composition of crater rays, both recent and ongoing researches have concluded that there are essentially two types of lunar rays: (1) immature rays that are bright because of the presence of fresh, high-albedo materials, which contrast sharply with adjacent mature surfaces and (2) mature compositional rays that are bright because of compositional contrasts, usually highlands materials from the source crater deposited above darker mare. Some rays may be a combination of the two types [2, 4, 5]. Other researchers surveyed the impact crater populations of the moon and found the ejecta patterns of crater rays could be classified in to four groups according to declining zenith angle of incidence [9, 10]. In most cases, crater rays are very useful chronologic markers. For example, the Copernican Era, was defined by the radiometric dating techniques applied to glassy materials recovered from what is believed to be ejecta from the rayed crater Copernicus and spectral studies of the optical maturity of lunar soils/ray material have been used in an attempt to further divide it and constrain age differences between rayed craters that possess sets of optically immature rays [4, 5, 11, 12]. Despite a lot of researches have been done on crater rays, the spatial pattern of crater rays are not well understood and analyzed.

In this work, we survey the rayed craters with asymmetrical ejecta and characterize how the shape of crater rays vary with impact angle using topography data derived from stereo imagery. We also attempt to investigate the spatial variation in the projectile direction characteristics of crater rays on the lunar surface to unravel potential spatial asymmetries.

2 Data Source

The data used in this paper include imageries from CCD stereo camera onboard ChangE-1 (CE-1) and ChangE-2 (CE-2). Their spatial resolution are 120 m and 50 m respectively [13, 14]. Thus, our observed crater diameter range include considerably smaller crater diameters than in former rayed crater catalogs. Positively identified craters are marked by their centre points and scaled by their crater diameters in a Geographic Information System (GIS) system. A compiled a global catalog of 1933 rayed craters is registered with craters as small as 200 m in diameter in Fig. 1. In this study, 857 craters among the 1933 craters are identified as rayed craters with asymmetrical crater ejecta, which are the main object in this study; the other 1076 are regarded as rayed craters with homogenous crater ejecta. The incidence angle of these 857 rayed craters are almost less than 45° and the ejecta blanket become asymmetric. For clarity and simplicity, we define abbreviated forms of direction to identify projectile direction. No more than three letters represent a fixed azimuth for the projectile direction. The scheme is illustrated in Fig. 2. Also each projectile direction is illustrated by a typical rayed crater, as shown in Fig. 3. Since images in high latitude regions have high phase angles, they represent illuminated topography rather than the albedo differences [15]. Therefore, the regions poleward of 75°latitude are ignored from our study area.

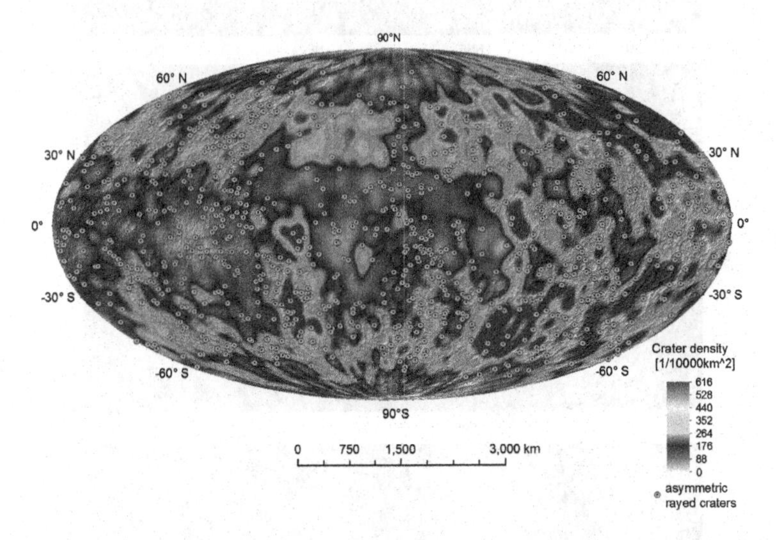

Fig. 1. Map of Moon showing spatial distribution of asymmetric rayed craters. Background shows crater density calculated from Chang'E-1 global crater catalog by Wang et al. [16]

We examine the completeness of catalog by comparing the cumulative distribution plots. A comparison of the cumulative size-frequency distributions for the rayed craters identified in this study and McEwen et al. [4] are shown in Fig. 4. McEwen et al. [4] report 96 rayed craters larger than 10 km in diameter in an area of $1.31 \times 10^7 km^2$ on the far side while we identify 857 rayed craters larger than 200 m in an area of $3.79 \times 10^7 km^2$. The crater densities obtained in the two catalogs are within a difference

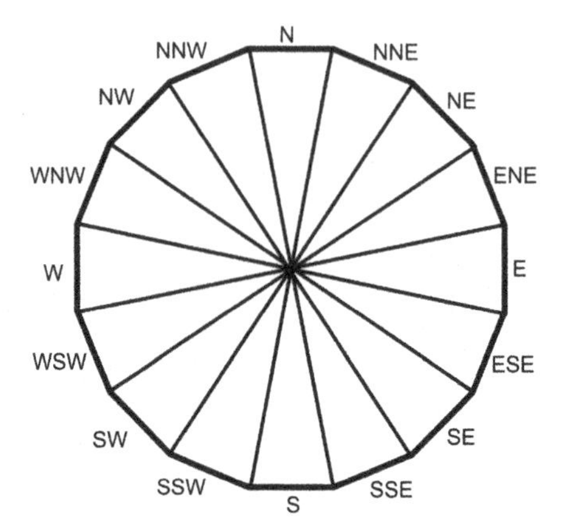

Fig. 2. The scheme for identifying projectile direction

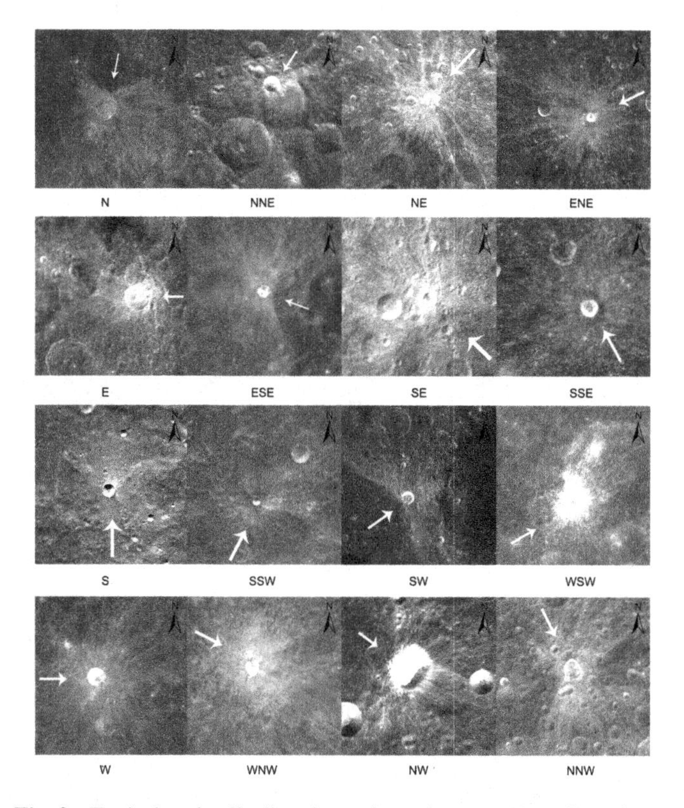

Fig. 3. Typical projectile directions of rayed craters (Data from CE-1)

larger than 65 %. The crater distribution measured by McEwen et al. [4] is in general steeper in comparison to 1Ga isochrones, the same to the inhomogeneous rayed-crater population described here. This is mainly brought about by a lack of large rayed craters in the limbs of the near side. The crater frequency drops below the isochrone at a diameter around 5 km, we can assign this deviation to the transition between simple and complex craters. The general features of the size-frequency distributions for craters larger than 45 km well agree with each other. Therefore, we conclude that our identification for rayed crater is sufficiently performed for craters larger than 45 km.

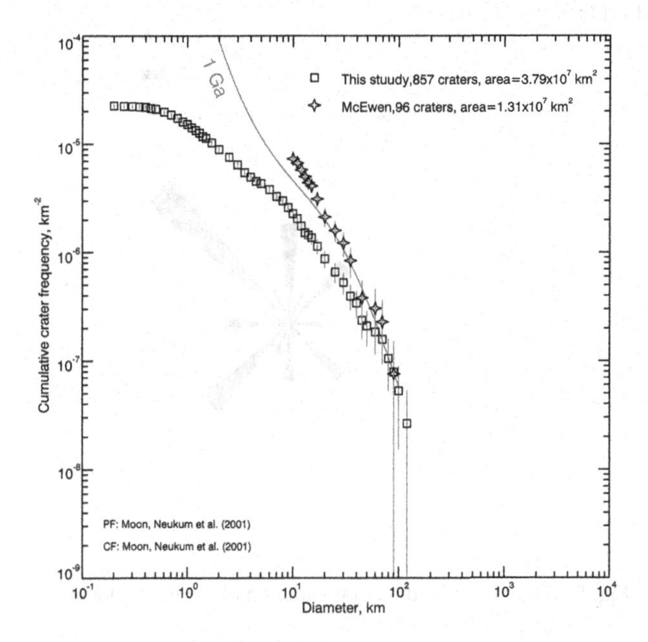

Fig. 4. The crater size-frequency distribution of the rayed-crater population in comparison with isochrones for 1Ga, calculated using the crater-production function and the cratering chronology function given by Neukum et al. [17]. For comparison with earlier studies, the farside lunar rayed crater distribution by McEwen et al. [4] is plotted

3 Spatial Pattern of Projectile Direction of Crater Rays

The study of the projectile direction aims to explore the origins of the projectiles. Due to the Moon is in synchronous rotation with Earth, it always shows the same face to the Earth. So the direction variations between nearside and farside, leading side and trailing side, latitudinal bands have special significance to the study.

3.1 Global Dependence

Figure 5 shows the projectile direction characteristics of rayed craters on the Moon. The figure clearly shows that projectiles for creating rayed craters from northeast direction are much more than from other directions. The projectiles occupy 21.24 % of the whole projectiles. The second largest number of projectiles comes from southeast direction, occupying 12.84 % of the whole projectiles. These two directions cross at an angle of 90°. The least number of projectiles is from west southeast direction, occupying 0.47 % of the whole projectiles. Calculating from the north and south direction, the number of projectiles from north direction is 1.71 times higher than that from south direction. Calculating from the east and west direction, the number of projectile from east direction is 2.03 times higher than those that from west direction. The projectile number difference among every direction is so large that it cannot be a chance fluctuation from an even probability of cratering.

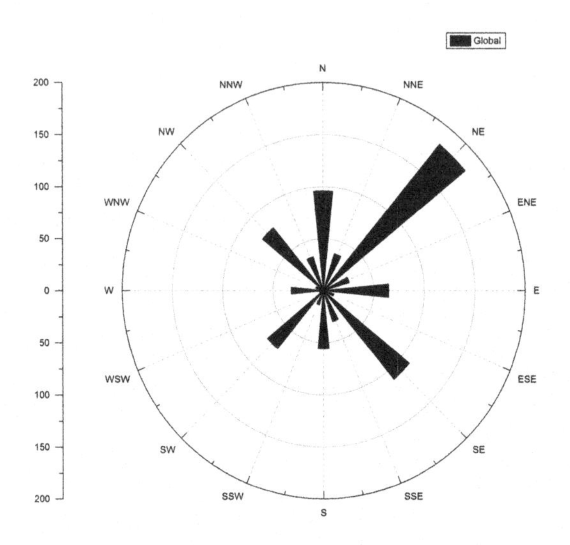

Fig. 5. The projectile direction variation of the lunar global area

3.2 Nearside-Farside Dependence

Figure 6 shows the projectile direction variation of the nearside and farside of the Moon. The figure clearly shows that the direction pattern of the nearside is similar to the farside one. The projectiles were concentrated from northeast direction, occupying more than 20 % of the whole projectiles. However, relatively fewer projectiles are from the west southwest direction. Through comparison, some slight differences exist between the two patterns. In the nearside, projectiles from north direction can account for nearly 15 % of the whole projectiles. Yet in the farside, the number of projectiles makes up no more than 10 % of the total. The projectiles from southeast direction in the farside of the Moon are a little more than corresponding direction in the nearside.

3.3 Leading Side-Trailing Side Dependence

In comparing the frequency of the two units, we find the leading side has a little more rayed craters than that of the trailing side which is similar to the pattern predicted by Shoemaker and Wolfe [18]; Horedt and Neukum [19]; Zahnle et al. [20]. They predicted that the leading hemisphere of synchronously rotating satellites is expected to be cratered at higher rate on the trailing side hemisphere. The asymmetric rayed craters appear in the leading side is 10 % more than those appear in the trailing leading side. Figure 7 shows the projectile direction variation of the leading side and trailing side. The pattern

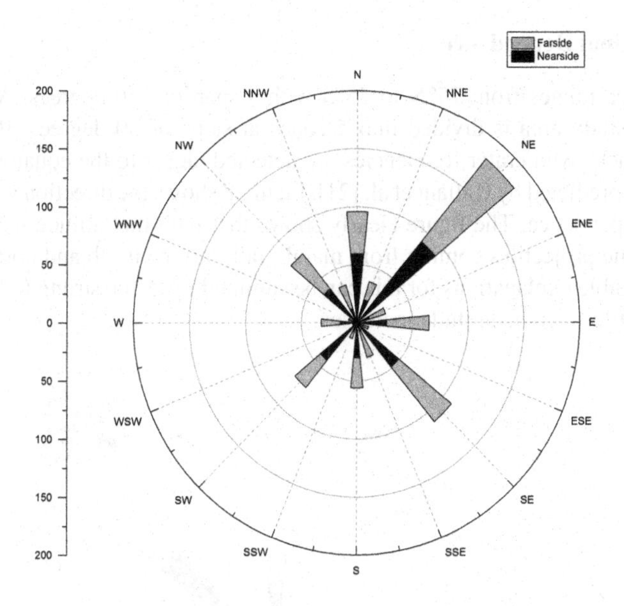

Fig. 6. The projectile direction variation of the nearside and farside of the Moon

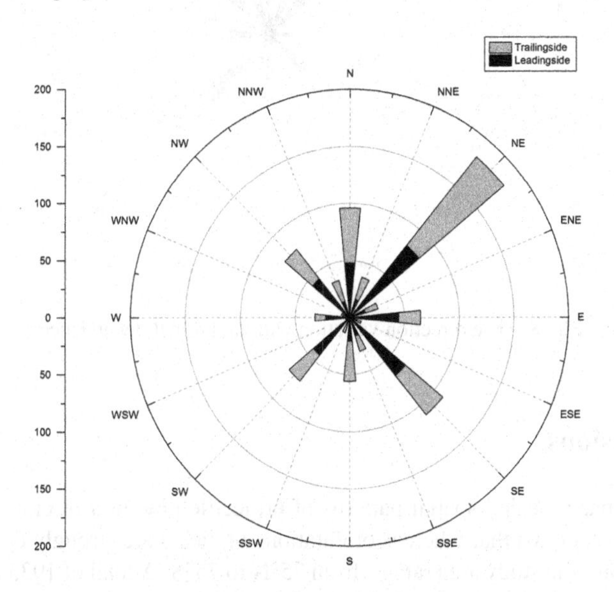

Fig. 7. The projectile direction variation of the leading side and trailing side of the Moon

indicates that both sides have higher frequency in northeast direction. And on the vertical direction (northwest and southeast direction), the frequency is high too. The discrepancy is characterized by smaller frequency in the east direction on the trailing side while that is in higher frequency on the leading side direction. The difference can reach 6 %.

3.4 Latitudinal Dependence

The study area ranges from 75°N to 75°S with a span of 150 degrees. According to latitude, the study area is divided into 5 zones at steps of 30 degrees. By statistical analysis, higher rayed crater frequencies are detected closer to the equator than to the pole which is predicted by Gallant et al. [21]. Figure 8 shows the direction variation with latitudinal dependence. The figure clearly shows that with the latitude increasing, the direction of the projectiles coming from mainly fall into the north and northeast direction. The possible explanations for this can be summed up as the strong Coriolis effects on the projectile with the latitude increases.

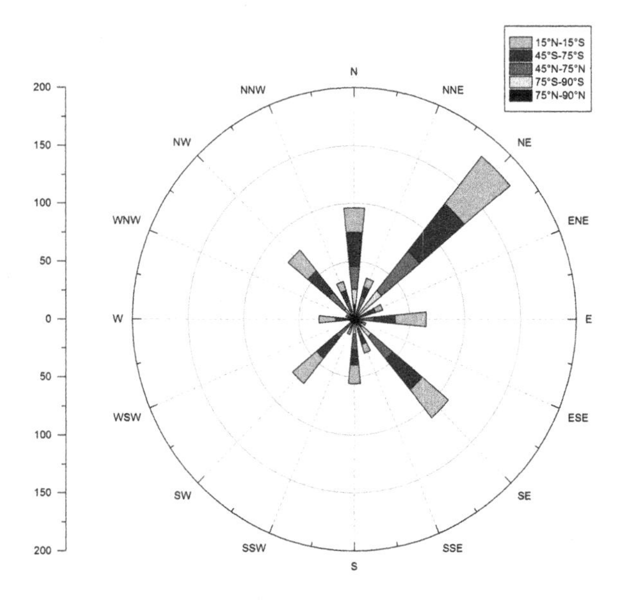

Fig. 8. The direction variation with latitudinal dependence

4 Conclusions

Our analysis has revealed spatial patterns of projectile direction of crater rays on the moon. We have shown that whereas orientations of ejecta are strongly correlated with spatial location. The study area ranges from 75°N to 75°S. A total of 1933 rayed craters are indentified, but only 857 craters with asymmetric ejecta are considered to identifying their projectile directions. The results show the projectiles which rayed craters created mainly come from northeast direction. And on the vertical direction (northwest and southeast direction), the frequency is high too. However, the number of projectiles coming from west southwest direction is small. In the nearside, projectiles from north direction can account for nearly 15 % of the whole projectiles. Yet in the farside, the number of projectiles makes up no more than 10 % of the total. On the other hand, the discrepancy is characterized by smaller frequency in the east direction on the trailing side while that is in higher frequency on the leading side direction. With the latitude increase, the projectiles concentrates coming from one or two directions. In the north polar regions, the projectiles mainly coming from the north direction, the second is southeast direction. In the south polar regions, the projectiles mainly coming from northeast direction, reaching 27 % of the total projectiles. Near the equator regions, the directions of projectiles are more diversified than the pole regions, but the dominant direction remains northeast.

The phenomenon of spatial asymmetries introduced by the particular orbital configuration of a synchronously rotating satellite has been previously noted [22–24]. Our analysis corroborates this point of view, but extents it to a global scale with firm statistics. Judging by density of rayed craters on the Moon, spatial asymmetries can be detected easily. Higher crater density is found closer to the equator than to the pole, which is consistent with the expectation by Halliday et al. [25]. It could be a result of the predicted cratering rate distribution: a combination of the crater-forming projectile flux and the micrometeorite bombardment. The latter acts on the maturation of the rayed deposits and is, following the predicted pattern, strongest at equatorial latitudes. However, the trailing side has less rayed craters than the leading side. A likely explanation for the difference is that the leading hemisphere of synchronously rotating satellites is expected to be cratered at a higher rate than on the trailing hemisphere, which results from the orbital velocity of the satellite is large compared to the space velocity of the impactor. As statistics of the projectiles' direction, they show perspicuous asymmetry. This asymmetry is unlikely caused by comets for the higher velocity of impactors. And the asteroid belt, which lies between Mars and Jupiter, orbits the Sun eastward day and night. This is a good explanation for the dominant direction of projectiles concentrates in northeast direction. So we can conclude that the rayed craters are formed by asteroids and the ones formed by direct comet impacts are minor.

Acknowledgements. The CE data are presented by the National Astronomical Observation and Chinese Academy of Sciences. This work is supported by National Natural Science Foundation of China (Grant No. 41171332) and Self-innovative Projects of Institute of Geographic Sciences and Natural Resources Research, CAS (Grant No. 201001005). The authors also thank anonymous reviewers for constructive and insightful reviews of the manuscript.

References

1. Tornabene, L.L., Moersch, J.E., McSween, H.Y., McEwen, A.S., Piatek, J.L., Milam, K.A., Christensen, P.R.: Identification of large (2-10 Km) rayed craters on mars in THEMIS thermal infrared images: implications for possible martian meteorite source regions. J. Geophys. Res. **111**, 1–25 (2006)
2. Hawke, B.R., Blewett, D.T., lucey, P.G., Smith, G.A., Bell III, J.F., Campbell, B.A., Robinson, M.S.: The origin of lunar crater rays. Icarus **170**, 1–16 (2004)
3. Baldwin, R.B.: The Measure of the Moon. The University of Chicago Press, Chicago (1963)
4. McEwen, A.S., Moore, J.M., Shoemaker, E.M.: The phanerozoic impact cratering rate: evidence from the farside of the moon. J. Geophys. Res. **102**, 9231–9242 (1997)
5. Grier, J.A., McEwen, A.S., Lucey, P.G., Milazzo, M., Strom, R.G.: Optical maturity of ejecta from large rayed lunar craters. J. Geophys. Res. **106**, 32847–32862 (2001)

6. Morota, T., Furumoto, M.: Asymmetrical distribution of rayed craters on the moon. Earth Planet Sci. Lett. **206**, 315–323 (2003)
7. Werner, S.C., Medvedev, S.: The lunar rayed-crater population-characteristics of the spatial distribution and ray retention. Earth Planet. Sci. Lett. **295**, 147–158 (2010)
8. Adono, T., Suzuki, A.I., Wada, K., Mitani, N.K., Yamamoto, S., Arakawa, M., Sugita, S., Haruyama, J., Nakamura, A.M.: Crater-ray formation by impact-induced ejecta particles. Icarus **250**, 215–221 (2015)
9. Herrick, R.R., Forsberg-Taylor, N.K.: The shape and appearance of craters formed by oblique impact on the moon and venus. Meteorit. Planet. Sci. **38**, 1551–1578 (2003)
10. Gault, D.E., Wedekind, J.A.: Experimental studies of oblique impact. In: Lunar and Planetary Science Conference, Houston, Texas, USA 9 (1978)
11. Wentworth, S.J., McKay, D.S., Lindstrom, D.J., Basu, A., Martinez, R.R., Bogard, D.D., Garrison, D.H.: Apollo 12 ropy glasses revisited. Meteoritics **29**, 323–333 (1994)
12. Ucey, P.G., Blewett, D.T., Taylor, G.J., Hawke, B.: Imaging of lunar surface maturity. J. Geophys. Res. **105**, 20377–20386 (2000)
13. Li, C.L., Liu, J.J., Ren, X., Mou, L.L., Zou, Y.L., Zhang, H.B., Lü, C., Liu, J.Z., Zuo, W., Su, Y., Wen, W.B., Bian, W., Zhao, B.C., Yang, J.F., Zou, X.D., Wang, M., Xu, C., Kong, D.Q., Wang, X.Q., Wang, F., Geng, L., Zhang, Z.B., Zheng, L., Zhu, X.Y., Li, J.D., OuYang, Z.Y.: The global image of the moon obtained by the Chang'E-1: data processing and lunar cartography. Sci. China Earth Sci. **53**, 1091–1102 (2010)
14. Liu, J.J., Ren, X., Tan, X., Li, C.L.: Lunar image data pre-processing and quality evaluation of CCD stereo camera on Chang'E-2. Geomatics Inf. Sci. Wuhan Univ. **38**, 186–190 (2013). (in Chinese)
15. Ito, T., Malhotra, R.: Asymmetric impacts of near-earth asteroids on the moon. Astron. Astrophys. **519**, 2–9 (2010)
16. Wang, J., Cheng, W.M., Zhou, C.H.: A Chang'E-1 global catalog of lunar impact craters. Planet. Space Sci. **112**, 42–45 (2015)
17. Neukum, G., Ivanov, B.A., Hartmann, W.K.: Cratering records in the inner solar system in relation to the lunar reference system. Space Sci. Rev. **96**, 55–86 (2001)
18. Shoemaker, E.M., Wolfe, R.F.: Cratering time scales for the Galilean satellites. In: Satellites of Jupiter, Tucson, AZ, USA 1, pp. 277–339 (1982)
19. Redt, G.P., Neukum, G.: Cratering rate over the surface of a synchronous satellite. Icarus **60**, 710–717 (1984)
20. Zahnle, K., Dones, L., Levison, H.F.: Cratering rates on the Galilean satellites. Icarus **136**, 202–222 (1998)
21. Gallant, J., Gladman, B., Ćuk, M.: Current bombardment of the earth-moon system: emphasis on cratering asymmetries. Icarus **202**, 371–382 (2009)
22. Craddock, R.A., Howard, A.D.: The case for rainfall on a warm, wet early Mars. J. Geophys. Res. **107**, 1–33 (2002)
23. Irwin, R.P., Craddock, R.A., Howard, A.D., Flemming, H.L.: Topographic influences on development of martian valley networks. J. Geophys. Res. **116**, 1–18 (2011)
24. Barnhart, C.J., Howard, A.D., Moore, J.M.: The influence of cratered slopes on late-noachian valley network formation. In: Lunar and Planetary Science Conference, Woodlands, Texas, USA 42, 1983 (2011)
25. Halliday, I.: The variation in the frequency of meteorite impart with geographic latitude. Meteoritics **2**, 271–278 (1964)

Research on Regional Environmental Pollution Analysis Technology Based on Atmospheric Numerical Model in Shenyang City

Xiaofei Shi[1(✉)], Yunfeng Ma[1], Qi Wang[1], Kunyu Gao[2], and Xu Liu[1]

[1] College of Energy and Environment, Shenyang Aerospace University, No. 37 Daoyi South Avenue, Shenbei New District, Shenyang 110136, China
shixiaofei0707@163.com

[2] Liaoning Province Environmental Monitor Centre, No. 37 Daoyi South Avenue, Shenbei New District, Shenyang 110136, China

Abstract. Shenyang City is one of the largest and important industrial cities in the northeast in China which plays an important role in the development of regional economy. With the increasing level of urbanization and industrialization, the air pollution problem has also been frequently appeared in the public sight. In order to better understand the distribution and time-varying of the pollutants in this area. A coupled model of WRF/CALPUFF was chosen to simulate the changing trend of sulfur oxides from 30 January 00 h UTC to 1 January 00 h UTC(winter episode) in 2014. By comparing the two datasets of measured and simulated, a conclusion that the effects of terrain characteristic and wind speed and direction on SO_2 are better reproduced by the model than the effects of temperature. Also, the result shows that the WRF/CALPUFF model can reproduce satisfactorily meteorology fields and SO_2 changing trend.

Keywords: Shenyang City · Air pollution problem · WRF/CALPUFF model · Sulfur oxides

1 Introduction

In recent years, the phenomenon of haze and $PM_{2.5}$ pollution has widely appeared in the Shenyang City especially in winter episode. There are two factors to explain this cause. One is the effect of weather fields, during winter the burning of fossil fuels, including civil heating and industrial production, produces a large amount of gas which can lead the appearance of inversion layer, it is not in favor of the diffusion of air pollutants; the other is the effect of terrain height of this area, there is a series of hills in the southeast part of Shenyang Area, while in this season the wind direction is mainly north and northwest, which can prevent the spread of pollutants. As a heavy industrial city with so many people in winter season, the total suspend particulate (TSP) and sulfur dioxide (SO_2) have become the main pollutants which have been closely associated with urban air quality problem [1].

Due to the complexity of containment concentrations, the development of air quality models to assist monitoring has become so necessary in application. Many studies, in fact, have emphasized that the California Puff (CALUFF) modeling system

F. Bian and Y. Xie (Eds.): GRMSE 2015, CCIS 569, pp. 751–759, 2016.
DOI: 10.1007/978-3-662-49155-3_77

has proved to a powerful tool in predicting and evaluating the plume dispersion [2–5]. Also in China, Zou Xudong et al. used CALPUFF model to simulate the PM10 concentration distribution of the typical source in Shenyang City [6]. In research of CALPUFF coupled with WRF research, Bo Xing et al. studied the effect of air pollution of the thermal power enterprises in Beijing, Tianjin and Hebei regions by the coupled model WRF/CALPUFF [7].

This study was carried out to model the dispersion of SO_2 emission from about 100 sources from the plant in and around the Shenyang City area by employing the coupled model WRF/CALPUFF. The WRF model was introduced to generate the meteorological data (surface and upper air meteorological fields).

2 Methods

Shenyang City is the capital of Liaoning Province which lies in the northeast of China, where is also a famous heavy industrial base. The area of the city has reach 3495 km^2 with population about 7.28 million. In this research, about 100 emission sources were collected in and near the area. See Fig. 1:

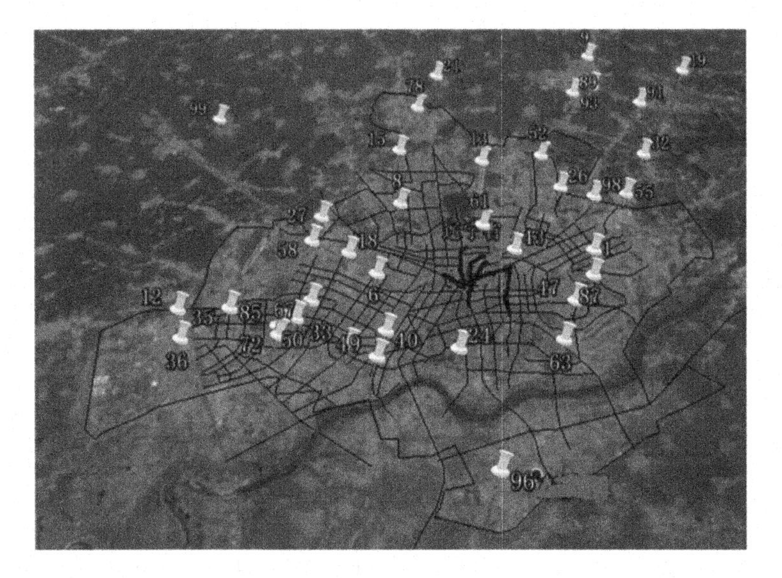

Fig. 1. Emission sources locations

From Table 1, the related information as follow:

2.1 Modeling Domain and Initialization

2.1.1 WRF Domains and Initialization

The modelling domains are shown in Fig. 2(a):

Table 1. Emission source information

Name	X coordinate (km)	Y coordinate (km)	Stack height (m)	Stack diam. (m)	Emission rate (g/s)
Point 1	541.24	4629.42	45	4.16	593.62
Point 2	529.03	4727.56	58	5.20	511.12
Point 3	508.35	4590.57	42	3.28	421.16
……					

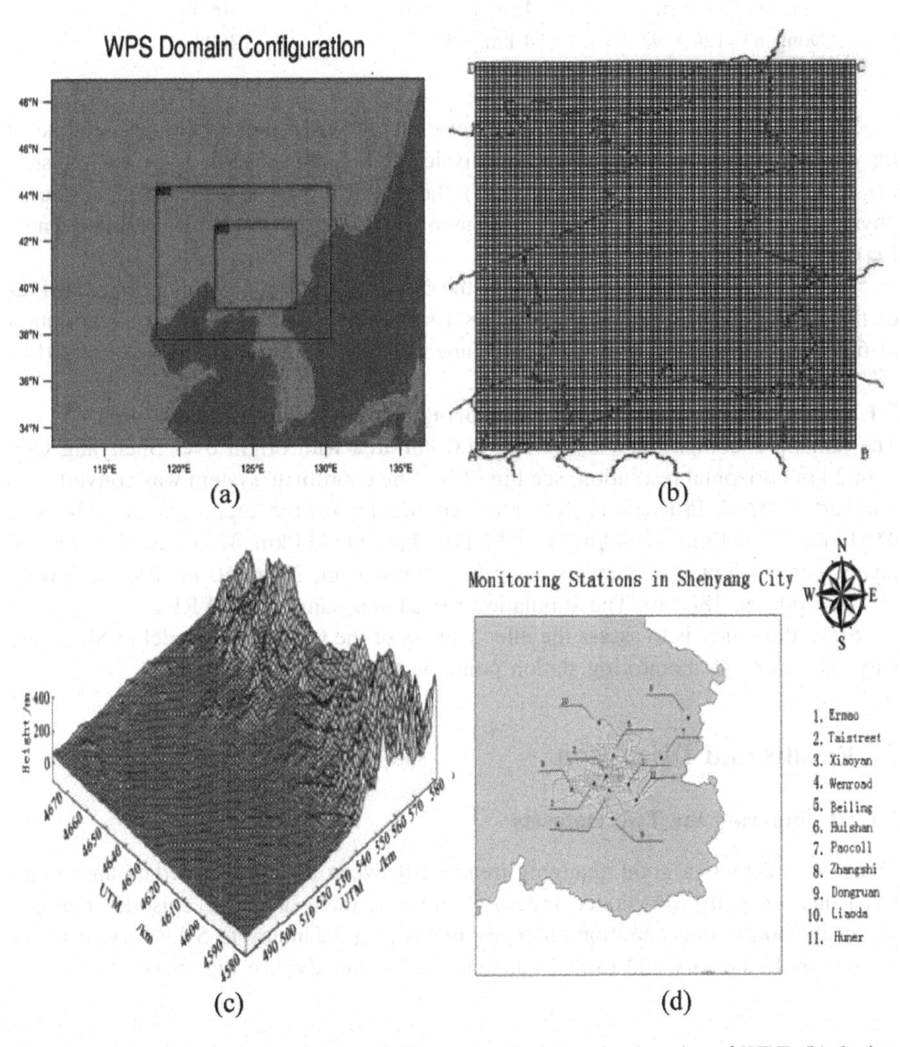

Fig. 2. Domains and Terrain of the modelling area. (a) shows the domains of WRF, (b) depicts the grid cells of CALPUFF, (c) shows the terrain height of Shenyang City area, (d) locates the Monitoring Stations in Shenyang City

The mother domain (D1) is centred at 41°N,121°E with 36 km spatial resolution. The first nested domain (D2) covers the whole Liaoning Provinence with 12 km spatial resolution. The innenrmost domain consists of 124 columns and 97 rows of 4 × 4 km^2 grid cells. Detail information can be seen in Table 2:

Table 2. WRF domains' information

Domain	Grid cells (X × Y)	Spatial resolution (Δ X, Δ Y)	Time steps (Δ t)
Domain1	65 × 49	36 km, 36 km	3600 s
Domain2	91 × 61	12 km, 12 km	3600 s
Domain3	124 × 97	4 km, 4 km	3600 s

Several main physical options were selected (1) WSM 6-class graupel scheme in mp_physics, (2) rrtm scheme in ra_lw_physice, (3) Dudhia scheme in ra_sw_physics, (4) YSU scheme in bl_pbl_physics, (5) thermal diffusion scheme in sf_surface_-physics, (6) MM5 Monin-Obukhov scheme in sf_sfclay_physics, (7) Kain-Fritsch (new Eta) scheme in cu_physics and so on.

The WRF simulation was driven by the National Centres for Environmental Prediction Global Tropospheric Analyses 1 × 1 spatial resolution and temporal resolution of 6 h from 30 January 00 h UTC to 1 January 00 h UTC (winter episode) in 2014.

2.1.2 CALPUFF Domain and Monitoring Station Points Information

The domain encompasses a 200 km × 200 km area with origin over Shenyang City with 2 km horizontal resolution, see Fig. 2(b). The coordinate system was converted to Lambert Conical Projection grid, each coordinate of the area was A (434 km, 4564 km), B (634 km, 4564 km), C (634,4764 km), D (434 km, 4764 km). The vertical layers were separated in 8 layers, each height was 0 m, 20 m, 50 m, 100 m, 200 m, 500 m, 1000 m, 1800 m. The simulation period was same as in WRF.

Since this study is to assess the effectiveness of the CALPUFF model in Shenyang City. There are 11 monitoring station points in the area, see Fig. 2(d).

3 Results and Discussion

3.1 Comparing the Two Datasets

Figure 3(a) depicts a good changing trending between the simulated data and monitored data on daily SO_2 concentration. Moreover, there are three episodes that can reflect the hourly concentration changing trend from 3 January to 5 January, from 11 January to 13 January and from 16 January to 18 January, see Fig. 3(b,c,d)

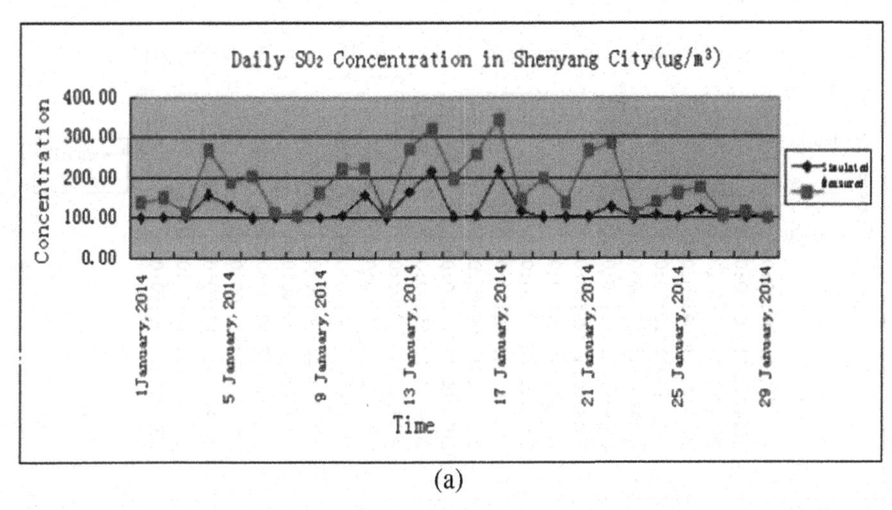

(a)

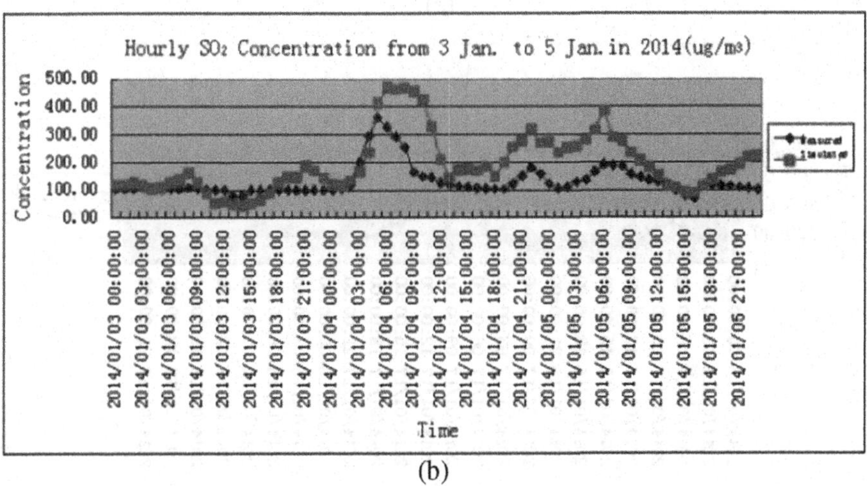

(b)

Fig. 3. The comparison of measured data and simulated data. (a) shows the daily SO_2 concentration in Shenyang City. (b) (c) (d) depict hourly SO_2 concentration during three periods, 3 Jan. to 5 Jan. in 2014, 11 Jan. to 13 Jan. in 2014 and 16 Jan. to 18 Jan. in 2014 correspondingly

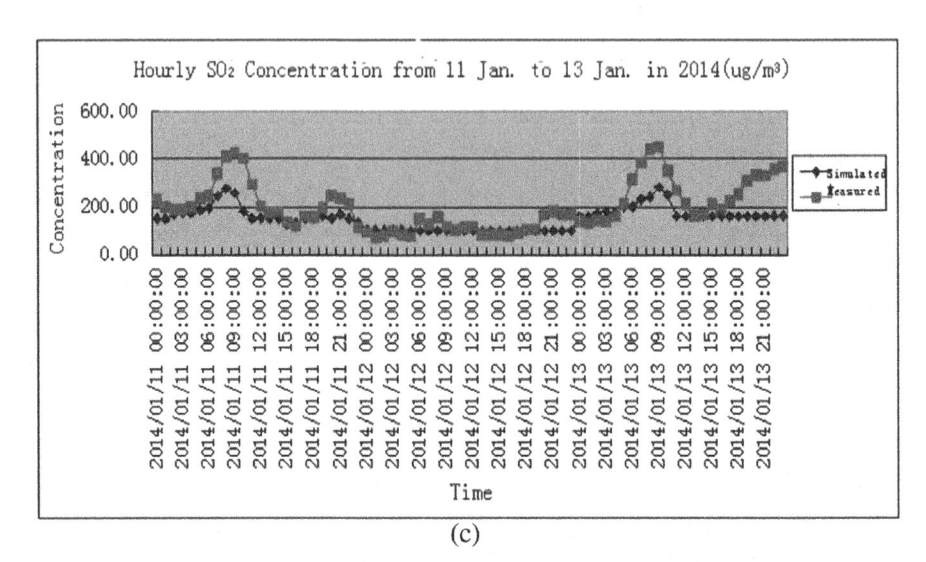

(c)

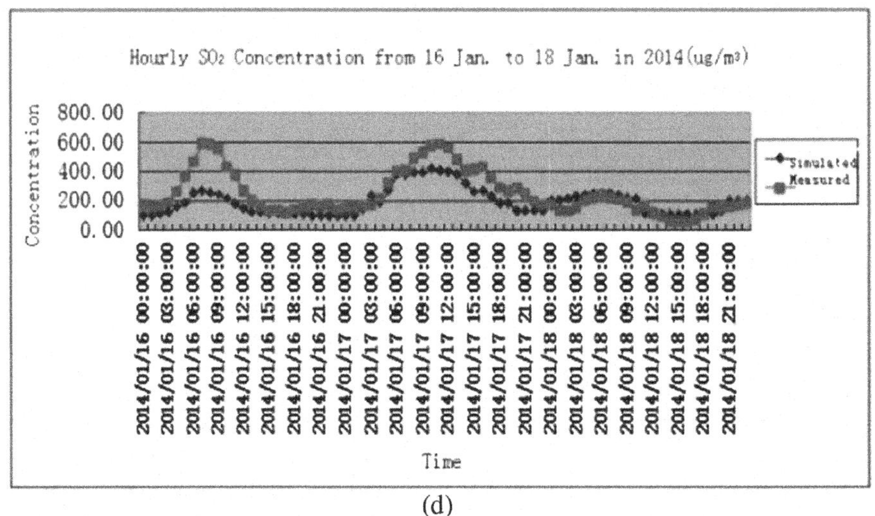

(d)

Fig. 3. (*continued*)

In total, by comparing the whole simulation period, we can get the Table 3:

Table 3. Monthly statistical relationship

Period	Monitoring item	Simulated (ug/m3)	Measured (ug/m3)	Correlation coefficient R
1 Jan. to 30 Jan. in 2014	Daily SO_2 mean	119.22	184.41	0.76

3.2 Visualization the Wind and Concentration Dispersion

In order to better analysis the effect of wind and terrain on the dispersion sulfur oxide in the area of Shenyang City. The last period, 16 January to 18 January, was chosen. Figure 4 showed the wind rose of 17 January (see Fig. 4(a)) and 16–18 January (see Fig. 4(b)). The main directions of the area are northwest and north with the speed reaching the maximum 10 m/s.

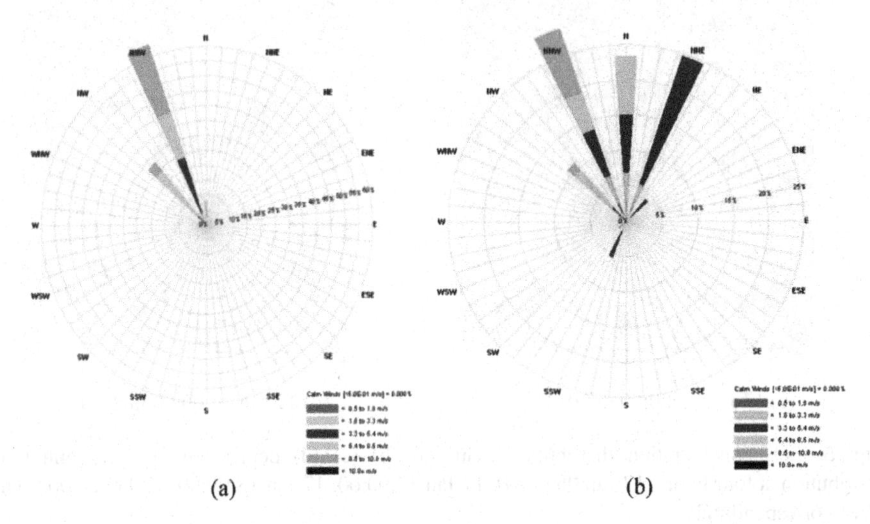

Fig. 4. Wind rose map. (a) shows the wind rose map on 17, Jan. 2014, (b) depicts the wind rose map during 16 Jan. to 18 Jan. in 2014

The concentration distribution on 17 January was shown in Fig. 5.

As is clearly showed in the above four pictures, the wind direction plays an important role on the dispersion of air pollutants. But considering the terrain height in Fig. 2(c), the shifting maximum from Fig. 5(c) to (d) is easily acceptable.

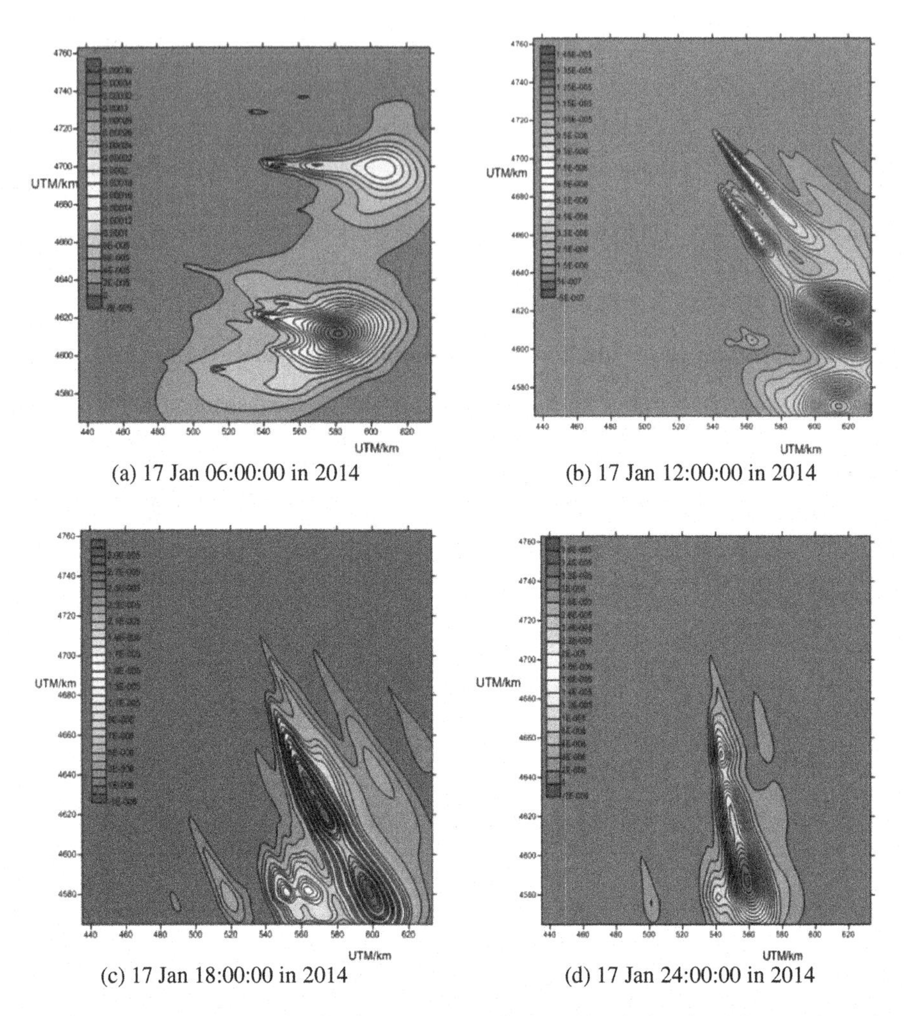

(a) 17 Jan 06:00:00 in 2014

(b) 17 Jan 12:00:00 in 2014

(c) 17 Jan 18:00:00 in 2014

(d) 17 Jan 24:00:00 in 2014

Fig. 5. SO_2 concentration distribution map. (a) (b) (c) (d) depict the SO_2 concentration distribution at four hours, 17 Jan 06:00:00, 17 Jan 12:00:00, 17 Jan 18:00:00, 17 Jan 24:00:00 in 2014 correspondingly.

4 Conclusion

This study represented a comparison of simulation concentration of SO_2 using WRF/CALPUFF model with the monitoring data in Shenyang City. About 100 point sources in and near the area were taken to simulate the surface concentration of SO_2 from 1 January to 30 January in 2014. It was found that:

(1) The coupled WRF/CALPUFF model performed well in this area, the correlation coefficient had reached 0.76 in the episode;
(2) Also the WRF model generated a good prognostic meteorological field to CALPUFF;

(3) The wind direction and terrain height had a great influence on the dispersion of atmospheric pollutants and the most easily polluted area was in southwest of this area.

References

1. Sahin, U., Ucan, O.N., Bayat, C., Oztorun, N.: Modeling of SO_2 distribution in istanbul using a artificial neural networks. Environ. Model. Assess. **10**(2), 135–142 (2005)
2. Choi, Y.J., Fernando, H.J.: Simulation of smoke plumes from agricultural burns: application to the San Luis/Rio Colorado air-shed along the U.S./Mexico border. Sci. Total Environ. **388** (1–3), 278–289 (2007)
3. Elbir, T.: Comparison of model predictions with the data of an urban air quality monitoring network in Izmir. Turkey Atmos. Environ. **37**(15), 2149–2157 (2003)
4. Fisher, A.L., Parsons, M.C., Roberts, S.E., Shea, P.J., Khan, F.L., Husain, T.: Long-term SO_2 dispersion modeling over coastal region. Environ. Technol. **24**(4), 399–409 (2003)
5. Tuaycharoen, P., Wongwises, P., Aran, R.P., Satayopas, B.: Nitrogen Oxide (NO_X) dispersion model for Khanom power plant area. In: International Conference on Environmental Research and Technology (ICER 2008), Penang, Malaysia (2008)
6. Zou, X., Yang, H., Zhang, Y., et al.: Distribution simulation analysis of PM_{10} concentration from typical sources of Shenyang in winter. Chin. J. Environ. Eng. **4**(4), 881–886 (2010)
7. Bo, X., Wang, G., Wen, R., et al.: Air pollutant effect of the thermal power plants in Beijing-Tianjin-Hebei region. Chin. Environ. Sci. **35**(2), 364–373 (2015)

The Influencing Factors of Biochar's Characteristics and the Development of Carbonization Equipments: A Review

Lingzhi Xia, Yonggang Wang, Jun Meng, Wenfu Chen, and Zhixia Zhang(✉)

Shenyang Agricultural University, Shenyang 110866, Liaoning, China
349802215@qq.com, syzzx7@163.com

Abstract. The biochar is one of the products of biomass pyrolysis. Because of its aromatic structure it has the potential for long-term carbon sequestration. The biochar also has a great effect on the improvement for the soil fertility, and it incorporation in soils influences soil structure, density, texture, porosity, and particle size distribution. In this paper, the research status of the influencing factors of biochar`s characteristics and the development of the carbonization equipment are summarized, which indicates researches for biochar in future should focus on systematic comparison and summary of the influencing factors of biochar`s characteristics, which made by the same species growing in different regions and different species, and the combination of different factors for the effect of biochar. The future carbonization equipment should be designed with higher efficiency, stability and automation.

Keywords: Pyrolysis · Biochar · Influencing factors · Carbonization equipment

1 Introduction

People living in the Amazon River Basin in Brazil use a special fertilizer for a long time, which has a strong ability to recover the soil fertility. T he locals call it 'dark earth of the Indians' [1]. The apparent high agronomic fertility, ordinarily existing in tropic, has attracted interests, which let people see the hope of developing agricultural land without cutting down forests [2]. Modern studies have shown that the black soil was made in 2500 to 6000 years ago by people living in the Amazon River Basin, who used animal`s feces or bones, fish, and plant waste for the materials of the black soil [3].

Produced through the high temperature pyrolysis of wood, grass, corn stalks or other crops waste in the absence of oxygen—incomplete combustion, biochar is not a general charcoal but a charcoal in rich carbon [4]. Biochar has a long history and the oldest method for fabrication of biochar is lighting the heap of weeds, straw, twigs, leaves or others,then covering them by a thin layer of mud, and eventually through the incomplete

Y. Wang—China Postdoctoral Science Foundation (Grant No. 2015M571328).
Z. Zhang—The Special Fund for Agro-scientific Research in the Public Interest (Grant No. 201503136), Key Scientific and Technological Project of Liaoning Province (Grant No. 2015103031).

F. Bian and Y. Xie (Eds.): GRMSE 2015, CCIS 569, pp. 760–769, 2016.
DOI: 10.1007/978-3-662-49155-3_78

combustion of materials in a low oxygen environment,the biochar is formed [5]. Climate increasingly has become one of the most far-reaching global environmental issues [6]. The molecular structure of biochars shows a high degree of chemical and microbial stability,and burying it in the ground can fix carbon in soil in hundreds years, which will slow down the greenhouse effect [7]. This binding also can enhance the availability of macro-nutrients such as elements of nitrogen and phosphorus. And the porous structure of biochar is very easy to gather nutrients and beneficial microorganisms and influences the binding of important nutritive cations and anions, so as to make the soil more fertile and conducive to the growth of plants [6].

In modern times, researchers from all the world have invented many advanced equipments for the production of biochar, and whatever the efficiency of production or the quality of biochar are greatly improved. International Biochar Initiative (IBI) noted that the biochar has the value of agricultural applications and environmental benefits [4]. Based on recent observations and simulations in the field of biochar, this paper conclude the influencing factors of biochar`s characteristics and the carbonization equipments in order to provide the reference for the researches studying in biochar.

2 The Influencing Factors of Biochar`s Characteristics

The extent of macromolecular structure, polymerization and the specific surface area (caused by porous structure of the biochar) of biochar is influenced by the type of raw material (vegetation type,size or pretreatment for material), catalysts, the duration and temperature and heating rate of pyrolysis in the absence of oxygen [3]. With larger specific surface area, it show higher affinity for pollutants, particularly planar aromatic compounds, which will maintain mineral elements and fertilize soil [8].

Most of the research scholars believe that the inert gas in the pyrolysis reaction has no effect on the properties of the carbon. For example, Borrego`study [9] in the pyrolysis of different materials,such as rice husk, forest residues, sawdust with different atmosphere (nitrogen and carbon dioxide) founded that whatever the shape, specific surface area or the characteristics of the biochar are similar with each other. Increasing air pressure can reduce the activation energy required for biomass pyrolysis, and increase the rate of pyrolysis reaction. But higher pressure could prolong the gas retention time, which will increase the probability of the reaction of the secondary decomposition, so that the production of biochar reduce. Biochar can be used as combustion or smelting metal. The release of its heat will be increased by the growth of calorific value. And the pyrolysis pressure has a certain effect, but not in others, on the heating value of the product, and Nader found [10] that in the condition of 137.9 Pa (air pressure) wheat straw carbon will obtain the maximum value (about 24670 J/g). Thus, we can conclude that the four main influencing factors are the temperature and heating rate of pyrolysis, duration, catalyst and the pretreatment of raw materials.

2.1 The Temperature and Heating Rate of Pyrolysis

At temperatures lower than 300°C and low heating rates, the production of biochar is highest because of the dominance of the dehydration step of the cellulose, forming anhydrocellulose, its more stable form [11]. The characteristics and production of biochar are greatly influenced by the temperature and heating rate in the process of pyrolysis. Dehydration is the fourth degradation step in pyrolysis, occurring after depolymerization, hydrolysis and oxidation and preceding decarboxylation [6].

Kuhlbusch [12] discussed the biochar`s formation coefficient as that part of the charcoal which is produced by vegetation fires and which is biologically not decomposable. This coefficient increases linearly with lower burning efficiency, which depends on factors such as heat temperature, the heating rate or inherent moisture in the biomass. As Chun Y`s research [13], if the temperature increases from 300 °C to 700 °C, the specific surface area of the biochars made by straw added from 116 m^2/g to the 363 m^2/g and the yield of biochar decrease sharply. And Isaac [14] indicated temperature and heating rates are both considered as main factors for producing biochars, which are smokeless in quality and harder in texture across biomass species. Nocak`s research [15] points out that biochar with low temperature has more effect on the increase of soil`s pH (composition with biochar) than with high temperature.

Thus, we can conclude that the higher the pyrolysis temperature, the lower yield of biochar, but the high temperature can optimize aromatic structure (increasing biochar`s lifetime), specific surface area (availability of macro-nutrients, increasing in electrical conductivity and cation exchanging capacity) and porosity (suitable for the growth of microorganisms). Although the increase of heating rate can decrease the yield of carbon, the pore structure of biochars can be increased. The most suitable temperature and heating rate are eventually determined by the raw material.

2.2 Duration

With the extension of the duration, the yield of biomass increased, but too long reaction time would reduce the volatile components of biochar and increase the gaseous content. Rahul Sinha`s research [16] of pyrolysis of linseed seed indicated the secondary decomposition reaction of biomass is conducive to the growth of gaseous productions, and the longer duration, the more chance for the secondary decomposition reaction occurring. And Wu [17] had the same conclusion with Rahul. she thought, with the extension of the duration, the content of alkaline groups of biochar (beneficial to the supplement of soil chemical elements [10]) showed an increasing trend while production decreased.

2.3 Catalysts

Generally the ion of alkali metals can promote the decomposition of biomass in a low temperature, but it will decrease the reaction rate and increase the yields of solid products [18]. Tsung-Ying Lin et al. [19] studied on the characteristics of the slow-pyrolysis reaction of bagasse and wood chips with different catalysts. The conclusion is that, with the increasing amount of the iron-ion, the yield of bio-oil and gaseous products increased, but the yield of biochar decreased. Li J et al. [20] in their study once mentioned: catalysts lower the activation energy for pyrolysis of biomass by providing a new reaction pathway. Simultaneously, the catalysts speed up the primary and the secondary decomposition reactions of biomass. Therefore, the carbonization of biomass with a suitable application of catalyst could take place at a lower temperature with fewer residues leaving. And Vamvuka D. [21] experimentalized the thermal gravimetric test for the biomass already removed metal element, which showed that calcium, magnesium, potassium and silicon can decrease the reaction rate and the temperature of carbonization but have less effect on the its kinetic parameters. Yang [22] has the same conclusion with Vamvuka D, and moreover some elements of metal could promote the formation of carbonyl compounds and carbon dioxide, but not the carbon monoxide and methane. The analysis of the products of carbonization with the raw material already washed by Fahmi [23] showed washing raw material could promote the formation of biochar and increased the content of acetaldehyde in biochar while decreased the content of levoglucose. The results of Tan Hong`s research [24] on the biomass carbonization with addition of chloride-ion showed that the element of potassium can reduce the production of bio-oil but increase the production of biochars and gases. So we can conclude most of metal-ion not only speeds up primary and secondary decomposition reactions of pyrolysis, but also makes the process of carbonization with a lower temperature and fewer residues leaving.

2.4 The Pretreatment of Raw Materials

The pretreatment of raw materials plays a very important role affecting biochar`s characteristics. There are kinds of pretreatment methods such as the particle size of raw materials, immersion in solution or drying dispose. LIU [25] found that the size of the raw materials influences the products. While the particle size decreased, the yield of bio-oil increased. Tstephanidis [26] found pickling is conducive to enhancing the yield of sugar (especially levoglucosan) and at the same time the content of acids, ketones and phenols in bio-oil decreased. Wang used the dilute acid solution to immerse pine woods for pretreatment [27]. The results show that pretreatment have effects on production and components of biochar, and especially in the pretreatment of vitrioling (the concentration of vitrioling is 1%), the yield and quality of bio-oil (high pH, high calorific value and low water content) is the best.

Carrier got similar results [28]. He found that using ethanol and acetone to immerse biomass can change the direction of the cellulose fibers, and it can improve the yield of pyrolysis products. Biswas [29] used willow sawdust having been done the pretreatment of steam explosion as raw material, the results showed that the contents of ash and alkali metals in the raw materials decreased. While the carbon content increases, the oxygen content decreases so the heating value increases. XU [30] observed that the pretreatment of steam explosion for wool fiber residue is conducive to the increase of the production of biochars, and the reason may be that the processing of pretreatment remove some loose components of raw materials. Through the study of pretreatment of baking the rice straw and cotton stalk.

In summary, we find the pretreatment should not be ignored. General physical methods (grinding, drying) in a certain range affect the production of biochar, while the pretreatment of solution soaking or steam explosion can affect the its element content and properties (specific surface area, the size of internal hole, etc.), and baking can enhance the carbon content in biochars, reduce moisture content and oxygen content (influencing biochar`s heat value).

3 The Carbonization Equipment

At present, in the efforts of the national research personnel, the carbonization equipment has been more advanced [31]. Modern carbonization equipment should have the following characteristics: (1) The pyrolysis temperature is controllable and the insulation capability of the furnace is strong. (2) The air tightness of the furnace is in high performance. (3) The equipments can ease to expand production as well as correct the faults, the process of manufacture is convenient, and the cost of maintenance is low.

3.1 Kiln

Charcoal craft has a long history, and the traditional kilns mainly are earth kilns or brick kilns. In the aspects of temperature control and exhaust pipes, New-type kilns have great changes, in which the main structure include a sealing cover, furnace and the bottom grate, the device for gas-liquid condensation, separation and recycling equipment. The materials of furnace body are low alloy steel and refractory materials, the degree of mechanization, the quality of product and the adaptability of which is higher than old-type. Gas-liquid products in the process of producing biochars can be recycled [32].

WANG researched on a spontaneous combustion furnace, called the open type fast pyrolysis carbonization kiln [33]. This furnace adopts the type of top ignition and internal combustion to control the carbonization process. When the temperature controlled in furnace is up to 190 °C with the absence of oxygen, the flame can gradually goes into the carbonization chamber, then a variety of raw materials in kiln could be carbonized, and simultaneously clean, high calorific value of the combustible gases generate.

A new type carbonization kiln called BA-I, which has good heat insulation performance, was designed by the Institute of forestry in Japan [34], which use the local bamboos, mulberries as raw materials of carbonization. The furnace body and cover are designed for the double sealing structure in heat insulation materials, and the connecting parts of the furnace body and cover is sealed with sand to make the heat difficult to leak, where the furnace has good performance of insulation, and the small change of furnace temperature and ventilation volume, thus the yield of biochar is large because of avoiding the excessive self burning of raw materials.

3.2 Fixed-Bed Reactor

With the further research on the fixed-bed reactor, the technology of fixed-bed carbonization has become increasingly mature.

Aiming at the problems of carbonization equipments, such as relatively low productivity, large energy consumption and poor in raw material adaptability, CONG [35] managed to design internal heating continuous type biomass carbonization furnace, as shown in Fig. 1. It can continuously carbonize biomass particles, peanut shell and corncob, realizing the continuous production of biochar, while kilns almost are discontinuous-working type. Material compaction device can realize the effective compaction and leveling of the material in the furnace, which provides a stable carbonization environment for the material. The perturbator can destroy the material compaction layer, improve the carbonization quality, and reduce the slag clustering phenomenon of the material in the heating preservation area. It used the heat of internal raw materials slowly burning in the absence of oxygen for the heat source to pyrolysis and carbonization, and the use of hot air to dry the raw material will reducing energy consumption. Its productivity is 108 kg/h with thirty percent productivity of biochars, and the power consumption is 15 kW·h·T^{-1}.

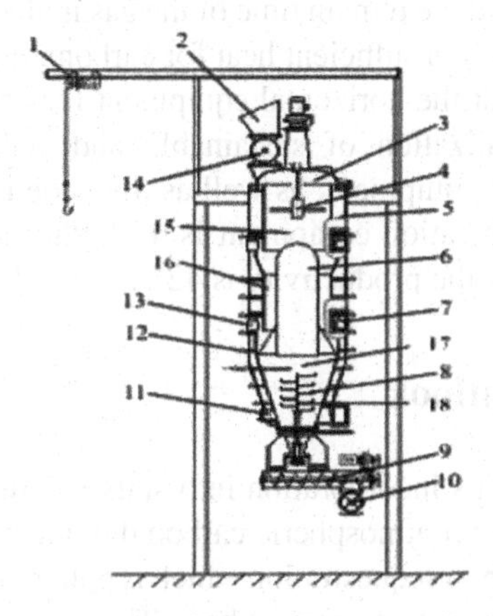

Fig. 1. Schematic diagram of internal heating continuous type biomass carbonization furnace. (1. Crane; 2. Hopper; 3. Furnace lid; 4. Compactor; 5. Furnace; 6. Dryer; 7. Air inlet; 8. Perturbator; 9. Screw; 10. Rotary feeder for biochar discharging; 11. Furnace door; 12. Detection of air inlet; 13. Air duct; 14. Rotary feeder for materials feeding; 15. Material preparation area; 16. Material drying area; 17. Pyrolysis zone; 18. Carbonization zone)

3.3 Moving Bed Reactor

The furnace of moving bed reactor is a moving platform, the production of which is continuous. At present, the production of most of the fixed bed reactor is intermittent. Many researchers from all over the world greatly pay attention on this reactor, and a typical and advanced example is given here.

YUAN managed to manufacture a horizontal continuous biomass carbonization equipment, which does not use external heating source (at the beginning of pyrolysis, the external heat source is needed) [36]. It utilizes high temperature gases produced during biomass pyrolysis to dry raw materials and heat biomass, as shown in Fig. 2.

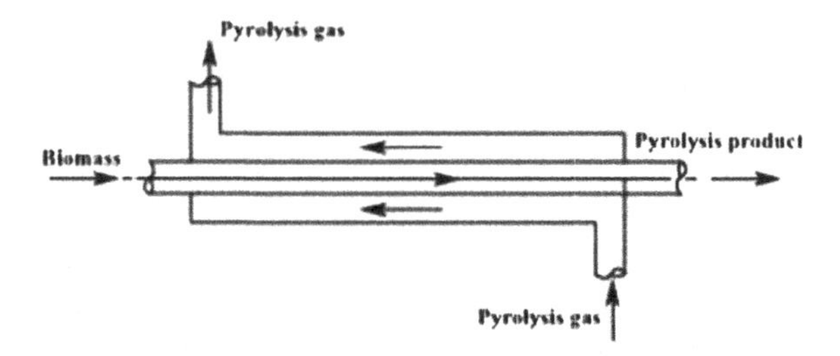

Fig. 2. Schematic diagram of the horizontal continuous biomass carbonization equipment

The high temperature gases flow in the reactor outer sleeve, and the outer cylinder has baffles and fins to extend the remain time of the gas in flue, which increases the heat transfer coefficient to achieve a sufficient heat for carbonizing the biomass. The results of experiment indicates that the horizontal equipment for continuous carbonization of biomass achieves the realization of sustainable and stable operation, reasonable designing of carbonization equipment, as well as the wide range of applicability. The productivity of this carbonization equipment is 45 kg/h, the terminal temperature of carbonization is 500°C and the productivity is 42%.

4 Summary and Outlook

Production of biochar and its incorporation into soils is a relatively novel concept for setting up a long-term sink for atmospheric carbon dioxide storage. Biochar provides a stable and inert form of carbon sequestration which is potentially long-term and substantial, with a low risk of return into the atmosphere. The use of biochar as a source of soil carbon sequestration appears to have important adaptational potential, providing economical solutions to its production and incorporation can be found. The biochar`s characteristics are greatly influenced by the type of raw material (vegetation type,size or pretreatment for material), catalysts, the duration and temperature and heating rate of pyrolysis.

At present, there are no systematic comparison and summary of the biochar `s characteristics made by same species from different region or different species in other papers. And there is also a lack of the results of the various combination of influencing factors, while current researches almost focus on what a single influencing factor affects. Therefore, the author believes that the future research should be aimed as follows:

(1) We should systematically research on the comparison of the influencing factors of carbonization of the homogeneous biomass gained from different fields and different kinds of biomass to obtain parameters, which will be in order for the industrial production to provide the basic data.
(2) What kinds of biochars can be produced by different combinations of the influencing factors, and what is the application of this kind of biochar should be given out to provide us the data that we can produce the biochar we want.

Carbonization is an immemorial but emerging industry. From the original kiln to modern various carbonization equipment, the technologies of carbonization are almost in the same principle but in different types. Every equipment has its advantages and disadvantages, the scientific and technological content of which are quick increase. According to conclude investigation and analysis a large number of data, the author`s outlook of carbonization equipment is as follows:

(1) In future, carbonization equipment should have the technologies of the better resistant of high temperature, the better sealing, safety pre-warning, and anti-explosion. These carbonization reactors should be designed for working continuously and the poly generation of biochar, bio-oil and gaseous products.
(2) Higher mechanization and automation.

The production and application of biochars both have important economic, social and environmental benefits, but in the early stage of the development of biochar industry, establishing the public cognition and market cultivation need a lengthy process. Therefore, from the production of biochar to the application of biochar, the current government should carry out policies to give the necessary financial support.

References

1. Harder, B.: Smoldered-Earth policy: created by ancient amazonian natives, fertile, dark soils retain abundant carbon. Sci. News **169**, 133–143 (2006)
2. Lehmann, J.: A handful of carbon. Nature **447**, 143–144 (2007)
3. Marris, E.: Putting the carbon back: black is the new green. Nature **442**, 624–626 (2006)
4. Xu, G., Lü, Y., Sun, J., et al.: Recent advances in biochar applications in agricultural soils: benefits and environmental implications. CLEAN-Soil Air Water **40**, 1093–1098 (2012)
5. Lehmann, J., Da Silva, J.P., Steiner, C., et al.: Nutrient availability and leaching in an archaeological anthrosol and a ferralsol of the central amazon basin: fertiliser, manure and charcoal amendments. Plant Soils **249**, 343–357 (2006)
6. Liu, F., Guo, M.: Comparison of the characteristics of hydrothermal carbons derived from holocellulose and crude biomass. Mater. Sci. **50**, 1624–1631 (2015)
7. Wang, D., Min, Y., Youhai, Yu.: Facile synthesis of wheat bran-derived honeycomb like hierarchical carbon for advanced symmetric super capacitor applications. Solid State Electrochem. **19**, 577–584 (2015)
8. Schwarzenbach, R., Gschwend, P., Imboden, D.: Sorption processes involving organic matter. In: Environmental Organic Chemistry, 2nd edn., pp. 275–330. Wiley-Interscience, New York (2002)
9. Borrego, A.G., Garavaglia, L., Kalkreuth, W.D.: Characteristics of high heating rate biomass chars prepared under N2 and CO2 atmospheres. Int. Coal Geol. **77**, 409–415 (2009)
10. Nader, M., Pulikesi, M., Thilakavathi, M., et al.: Analysis of bio-oil, bio-gas, and biochar from pressurized pyrolysis of wheat straw using a tubular reactor. Energy Fuels **23**, 2736–2742 (2009)
11. Vigouroux, R.Z.: Pyrolysis of Biomass Thesis. Royal Institute of Technology, Stockholm (2001)
12. Kuhlbusch, T.A.J.: Black carbon as a product of savanna fires in southern Africa: a sink of biospheric carbon. In: Proceedings of the 1993 AGU Fall Meeting, San Francisco, CA, USA, December 1993

13. Chun, Y., Sheng, G.Y., Cary, T.C.: Compositions and sportive properties of crop residue-derived chars. Environ. Sci. Technol. **38**, 4649–4655 (2004)
14. Isaac, F.T., Armando, G.: Effect of temperature on biochar product yield from selected lignocellulosic biomass in a pyrolysis process. Waste Biomass Valor. **3**, 311–318 (2012)
15. Novak, J.M., Frederick, J.R.: Rebuilding organic carbon contents in coastal plain soils using conservation tillage systems. Soil Sci. Soc. America J. **73**, 622–629 (2009)
16. Rahul, S., Sachin, K.: Production of bio-fuel and biochar by thermal pyrolysis of linseed seed. Biomass Conv. Bioref. **3**, 327–335 (2013)
17. Hongxiang, W., Zengli, Z., Wei, Z., et al.: Effects of alkali/alkaline earth metals on pyrolysis characteristics of cellulose. Trans. CSAE **28**, 215–220 (2012)
18. Ottino, J.M., Khakhar, D.V.: Mixing and segregation of granular materials. Annu. Rev. Fluid Mech. **32**, 55–91 (2000)
19. Lin, T.-Y., Kuo, C.-P.: Study of products yield of bagasse and sawdust via slow pyrolysis and iron-catalyze. J. Anal. Appl. Pyrol. **96**, 203–209 (2012)
20. Li, J., Yan, R., Xiao, B., Liang, D.T., Lee, D.H.: Preparation of nano-NiO particles and evaluation of their catalytic activity in pyrolyzing biomass components. Energy Fuels **22**, 16–23 (2008)
21. Vamvuka, D., Troulinos, S., Kastanaki, E.: The effect of mineralmatter on the physical and chemical activation of low rank coaland biomass materials. Fuel **85**, 1763–1771 (2006)
22. Changyan, Y., Jianzhong, Y., Xuesong, L.: Influence of K^+ and Ca^{2+} on the mechanism of biomass pyrolysis. Acta Energiae Solaris Sin. **27**, 496–502 (2006). (in Chinese with English abstract)
23. Fahmi, R., Bridgwater, A.V., Donnison, I., et al.: The effect of ligninand inorganic species in biomass on pyrolysis oil yields. Qual. Stab. J. Fuel **87**, 1230–1240 (2008)
24. Hong, T., Shurong, W., Zhongyang, L., et al.: Influence of metallic salt on biomass flash pyrolysis characteristics. J. Eng. Thermophys. **26**, 742–744 (2005). (in Chinese with English abstract)
25. Yunquan, L.I.U., Minnan, L.O.N.G.: Rapid pyrolysis of several different biomass. Chem. Ind. Eng. Prog. **29**, 126–132 (2010)
26. Tstephanidis, S., Nitsos, C., Kalogianis, K., et al.: Catalytic upgrading of lignocellulosic biomass pyrolysis vapours: effect of hydrothermal pre-treatment of biomass. Catal. Today **167**, 37–45 (2011)
27. Wang, H., Srinivasanr, Y.U., et al.: Effect of acid, alkali, and steam explosion pretreatments on characteristics of bio-oil produced from pinewood. Energy Fuels **25**, 3758–3764 (2011)
28. Carrier, M., Neomagus, H.W., Görgens, J., et al.: Influence of chemical pretreatment on the internal structure and reactivity of pyrolysis chars produced from sugar cane bagasse. Energy Fuels **26**, 4497–4506 (2012)
29. Biswas, A.K., Yang, W., Blasiak, W.: Steam pretreatment of salix to upgrade biomass fuel for wood pellet production. Fuel Process. Technol. **92**, 1711–1717 (2011)
30. Xu, W., Ke, G., Wu, J., et al.: Modification of wool fiber using steam explosion. Eur. Polym. J. **42**, 2168–2173 (2006)
31. Zhu, B., Wang, X., Chen, Y., Yang, H., Chen, H.: Experiment of agricultural straw baking properties. Chem. Ind. Eng. Prog. **29**, 120–125 (2010)
32. Novak, J.M., Busscher, W.J., Laird, D.L., et al.: Impact of biochar amendment on ertility of a southeastern coastal plain soil. Soil Sci. **174**, 105–112 (2009)
33. Wang, Y., Wang, H., Wang, X.: The Open Type Fast Pyrolysis Carbonization Kiln: China, 200610048274.3[P], 17 June 2009
34. Ma, Y.: Introduce a mobile carbonization furnace. J. Chem. Ind. Forest Prod. **4**, 24–26 (1993). (in Chinese with English abstract)

35. Hongbin, C., Lixin, Z., Zonglu, Y., Haibo, M., Yanwen, Y.: Development of internal heating continuous type biomass carbonization equipment. Acta Energiae Solaris Sin. **8**, 1526–1535 (2014)
36. Yanwen, Y., Yishui, T., Lixin, Z., Zonglu, Y.: Design and manufacture of horizontal continuous biomass carbonization equipment. Trans. Chin. Soc. Agric. Eng. **13**, 203–209 (2014)

Dynamic Nonlinear Relationships between Carbon Emission Allowance and Reduction Credit Markets-Based on the IRF-DCC Model

Jingjing Jiang[1], Bin Ye[2(✉)], Dejun Xie[1], and Lixin Miao[2(✉)]

[1] School of Financial Mathematics and Engineering, South University of Science and Technology of China, Shenzhen 518055, China

[2] Research Center on Modern Logistics, Graduate School at Shenzhen, Tsinghua University, Shenzhen 518055, China

ye.bin@sz.tsinghua.edu.cn, lxmiao@tsinghua.edu.cn

Abstract. Emission trading and market mechanism have increasingly become crucial policy measures to promote sustainable development. Coupled with carbon emission reduction credit trading, carbon emission allowance trading under the cap-and-trade scheme is also steadily developing in China. To learn from the EU ETS's experience, the paper applied the IRF-DCC model to explore the dynamic nonlinear relations between EUA and CER markets. Empirical results indicate that EUA and CER are dynamically and conditionally correlated both in the spot and future markets. Correlations of spot volatilities are highly instable and market dependent while correlations of future volatilities are relatively stable and independent.

Keywords: Carbon emission allowance · Carbon reduction credit · IRF-DCC · Carbon market · EU ETS

1 Introduction

Carbon emissions from China have exceeded the sum of those from the Unite States and the Europe Union in 2013 [1]. Meanwhile, China's carbon emissions per capita have outstripped that of the EU, being more than 7.2 tons. The government of China is actively promoting the application of market mechanisms in the field of energy-saving and emission-reduction. China is the largest supplier of carbon emission reduction credit in the global CDM market. The "China's National Interim Measures for voluntary greenhouses emission reduction" has been enacted since June 13, 2012, to mobilize the conscious participation in the entire society's carbon emission mitigation activities. In recent years, the cap-and-trade carbon emission trading scheme and the related carbon allowance trading market are also growing quickly around the world. China has started seven regional cap-and-trade carbon emissions trading pilots during the period of 2013-2014 [2], and explores to construct a national carbon allowance trading market during the period of 2016-2020. Carbon emission allowance and

F. Bian and Y. Xie (Eds.): GRMSE 2015, CCIS 569, pp. 770–777, 2016.
DOI: 10.1007/978-3-662-49155-3_79

reduction credit trading markets are becoming two crucial measures to realize the low-cost carbon emission reduction in China.

The European Union Emissions Trading Scheme (EU ETS) was launched in 2005 and creates the most influential carbon emission trading market in the world. Under the scheme, the European Union Allowance (EUA) is traded as default carbon emission right asset for the regulated enterprise to fulfill its compulsory carbon emission reduction obligation. Besides, the Certified Emission Reduction (CER) is introduced as carbon offset credits to provide flexibility for the regulated enterprise to lower its emission reduction cost. The linkage of carbon emission allowance and reduction credit markets has become more and more important as carbon derivatives and structured financial instruments springing out in recent years. As a result, it is of great significance for the development of China's structured carbon market to study the relationship between allowance and credit carbon markets under the EU ETS.

2 Literature Overview

In the literature, most researches focus on investigating the relationships of basic carbon products and their derivatives. Milunovich and Joyeux [3] applied the granger causality tests to detect price discovery in the first Phase EUA allowance carbon market and find that its spot and future prices presented bilateral information transmission and jointly contributed to price discovery. Uhrig-Homburg and Wagner [4] analyzed the dynamic connection between EUA spot and future markets, and uncovered a long-run relationship between the observed allowance futures prices and the theoretical allowance future prices derived from allowance spot prices by the cost-of-carry model. Daskalakis *et al.* [5] developed a two-factor equilibrium model for pricing option on future based on the jump-diffusion price process, and empirically conclude that the framework they propose behaves better than the Black formula in capturing the correlation between allowance option and future prices.

Besides, some researches begin to pay their attention to the linkage and relationship between the EUA and CER carbon markets. Barrieu and Fehr [6], Mizrach [7] applied the co-integration and granger causality tests, VAR and VECM models to explore the one-order relationships between allowance and credit carbon markets under the EU ETS. The results indicated that the co-integration and causally influence between the prices of EUA and CER carbon markets. However, the existing researches paid little attention to the interdependence and interaction in term of second-order price volatility between EUA and CER market. Based on the above literature overview, this paper sheds light on developing econometric models to further explore the nonlinear second moment relationships between EUA and CER carbon markets.

3 Method and Model

This article dedicates to detecting and modeling the time-varying conditional volatility relations between EUA and CER carbon markets. To this end, the paper starts from several statistical tests to get some basic understanding of the log return volatilities and their interactions. Then based on these preliminary analyses, a two-step procedure is proposed to build the econometric models. In the first step, the Impulse Response Function (IRF) is constructed to describe the reactions of system to shocks over time. In the second step, the residuals generated from the first step are used to estimate the Dynamic Conditional Correlation (DCC) model, in order to uncover dynamic conditional volatility correlations between the carbon allowance and credit markets.

3.1 Statistical Tests

In order to detect the volatility characteristics of EUA and CER log return rates and their interplay, several statistical tests are conducted in this paper. They consist of:

(1) Unit root test to investigate whether the log return rate series act as stationary process. (2) Ljung-Box Q and Q^2 tests to detect effects of autocorrelation and heteroskedasticity in single series. (3) Correlation coefficient analysis to measure how well the studied series vary jointly. (4) Cointegration test to investigate whether or not there exist long-term equilibrium relationship between EUA and CER. (5) Granger causality test to detect the short-term causality.

3.2 Impulse Response Function

IRF provides an effective instrument to analyse causality and policy effectiveness, which tracks the impact of any variable on others in the system. Therefore, the VAR model is selected as mean-filter to tackle the autocorrelation and cross-correlation in EUA and CER log returns, together with IRF to analyse the dynamic reactions of system to shocks. The VAR (k) model with k-order maximized lag can be expressed as Eq. (1). Here Y_t is a N*1 vector of daily return or loss rate at t time and its element $y_{i,t}$ is the corresponding observation of the i-th variable. ω is a N*1 vector of constants, $\prod_i$ is a N*N regression coefficient matrix for the i-th order lagged Y_{t-i}, and can be used to analyze the mean autocorrelation and interdependence in the system. The last item ε_t is a N*1 innovation of random error generated in the VAR filtering process, which meets the requirement of zero mean and has a N*N conditional variance-covariance matrix H_t.

$$Y_t = w + \sum_{i=1}^{k} \prod\nolimits_i Y_{t-i} + \varepsilon_t,\ \varepsilon_t \in \mathrm{N}\ (0, H_t) \tag{1}$$

Given the above VAR model, the IRF model can be expressed as Eq. (2).The Φ_{i} represents the MA coefficients owning the ability to measure the impulse response. More specifically, its element $\Phi_{mn,i}$ represents the response of variable m to an unit impulse in variable n occurring i-th period ago.

$$Y_t = w + \sum_{i=1}^{k} \prod\nolimits_i Y_{t-i} + \varepsilon_t = w + \Phi(B)\varepsilon_t = w + \sum_{i=0}^{\infty} \Phi_i \varepsilon_{t-i} \tag{2}$$

3.3 DCC Model

Based on the decomposition of the conditional variance-covariance matrix into conditional standardized deviations and correlations, DCC model has advantage to provide intuitive interpretation of time-varying volatility correlations between variables. In general, a two-step process is proposed to develop DCC model. Firstly, the univariate GARCH (*p*, *q*) equations are calculated. Then the DCC is estimated by using the standardized residuals generated from the above univariate equations. This paper does not detail the first step, but pay more attention to the second step of estimating DCC model. Assuming the conditional correlations are time-dependent, the DCC model can be expressed as Eq. (3), where H_t and R_t refer to the dynamic variance-covariance and conditional correlation matrices respectively.

$$H_t = D_t R_t D_t \tag{3}$$

As to the specification of conditional correlations, this paper follows the structure introduced by Engle [8], as illustrated by Eq. (4).

$$R_t = Q_t^{*-1} Q_t Q_t^{*-1};\ Q_t = (1-\alpha-\beta)Q_c + \alpha\varepsilon_{t-1}\varepsilon_{t-1}^{'} + \beta Q_{t-1} \tag{4}$$

Here the time-dependent R_t is given as the first equation, in which Q_t comes to be the determinant. In the second equation, Q_t is a symmetric N*N positive definite matrix representing the standardized conditional correlations, Q_c is the constant conditional correlation matrix in absence of dynamics, and ε_{t-1} refers to the 1-order lagged standardized residuals generated by the univariate GARCH process. In addition, α is a positive and β is a non-negative DCC scalar parameter satisfying $\alpha + \beta < 1$. According to the framework, α and β illustrate the impact on current conditional volatility correlations of last shocks and past correlations respectively. Therefore, if α and β are statistically significant, this model rejects the invariant conditional correlations and can be used to describe and generate the dynamic conditional correlations of price volatilities in different markets.

4 Results and Discussions

The paper article focuses on the dynamical nonlinear relations of EUA and CER spot and future under the second phase of the EU ETS. According to the liquid concern, EUA and CER spot and future samples (EUAS, CERS, EUAF, CERF) are respectively extracted from the daily closing market indexes provided by the BlueNext Exchange and the ECX Exchange.

4.1 Volatility Statistical Properties

Tables 1 and 2 summarize results of several statistical tests. (1) Results of unit root tests indicate that all the log return series are stationary. And they are auto-correlated and heteroscedastic, indicated by the significantly high values of Ljung-Box Q and Q^2

statistics at different time lags. (2) EUA and CER are positively linked in the sense of linear statistics, with highly significant 0.87 and 0.86 of Pearson coefficients respectively. (3) Johansen test strongly suggests bi-direction cointegration between EUA and CER (4) Unilateral granger causality from EUA to CER is detected.

Table 1. Summary results of statistical tests.

	EUAS	CERS	EUAF	CERF
Jarque-Bera	812.664	930.504	2077.278	2064.436
ADF t-Statistic	−28.2864***	−28.5157***	−24.7037***	−24.9301***
Q(5)	13.993**	8.7117	22.615***	22.961***
Q(10)	19.178**	16.367*	26.628***	33.362***
$Q^2(5)$	109.05***	304.67***	96.258***	183.51***
$Q^2(10)$	197.36***	443.65***	140.89***	209.38***

Table 2. Summary results of relationship tests.

Correlation test	Pearson correlation coefficient	
EUA-CER spot	0.8726	
EUA-CER future	0.8566	
Johansen cointegration test	Trace statistic	Prob.
None CE in EUA and CER spots	322.8966***	0.0001
At most 1 CE in EUA and CER spots	151.8522***	0.0000
None CE in EUA and CER futures	327.9000***	0.0001
At most 1 CE in EUA and CER futures	146.7964***	0.0000
Granger test	F-statistics	Prob.
EUAS does not granger cause CERS	1.9402*	0.0853
CERS does not granger cause EUAS	1.8052	0.1092
EUAF does not granger cause CERF	5.5313***	0.0000
CERF does not granger cause EUAF	1.6970	0.1328

4.2 DCC Estimations and Conditional Volatility Correlations

The paper estimates DCC (1, 1) models for spot and future carbon markets respectively using MVGARCH.ZIP package in MATLAB. Empirical results are presented in Table 3.

The significance of DCC parameters suggests that correlations of EUA and CER volatilities are time-dependent in both the spot and future markets. More specifically, parameters a and b are estimated to be positive and their sum is about 0.93 and 0.88 respectively, which indicate that correlations of EUA and CER volatilities are positively and prevalently impacted by the lagged shocks and correlations. Besides, it should be noted that the sum of $\alpha + \beta$ for future is smaller than that for spot. As a result,

Table 3. Estimations of DCC coefficients and statistics of conditional volatility correlations.

	EUAS and CERS	EUAF and CERF
α	0.1052***	0.0392**
β	0.8387***	0.8566***
Mean	0.8497	0.8176
Maximum	0.9514	0.9037
Minimum	0.2666	0.6765
Std. dev	0.0851	0.0308
Log likelihood	4883.5	4533.1
AIC	−12.68	−11.75
SC	−12.20	−11.32

impact of lagged information on EUA and CER volatility correlations in the future market is weaker than that in the spot market. The relations of EUA and CER future volatilities may be more stable than those of their spot volatilities.

The paper proceeds to analyze the evolutions of conditional correlations between EUA and CER volatilities. In the spot markets, the correlation is slightly larger than 0.8 at the beginning and walks between 0.5 and 0.9 in the early stage (2008.08-2009.05). In the middle stage (2009.05-2011.03), the linkage of EUAS and CERS volatility is stronger with their conditional correlation stably leveling off within 0.8 and 0.95. But in the last stage (2011.03-2012.03), the correlation slumps to 0.26 and then jumpily evolves between 0.26 and 0.95. More importantly, a valuable relation is discovered between volatility correlation and volatility size (denoted by the squared log return). The volatility correlation of EUAS and CERS jump up and down in the first and last stage when their squared log returns shift very much, but it remains strong and steady when their squared log returns evolve at low level. As a result, relationships of EUAS and CERS volatility are highly market dependent in the spot market. A close linkage can be detected when market is peaceful. However, the linkage reduces quickly when market steps into the episodes of turbulence, which may be caused by heterogeneous reactions carried by heterogeneous participants in EUAS and CERS traders. For example in the early stage, many EU companies sell off their EUA spots to obtain positive cash flow in order to fight financial crisis. It seriously disturbs the normal relations between EUA and CER spots until May 2009 that companies gradually reduced their heavy sell of EUA in the spot market. Judging from these, hedge strategies may dysfunction in the turbulent market situations, which diversify EUA and CER spot positions based on the mean of their volatility correlations. Traders should adjust strategies according to market situations and built positions according to dynamical relations of EUA and CER volatilities in the spot carbon market [9].

As to the future market, conditional correlation of EUAF and CERF volatilities begins at 0.80 and moderately walks between 0.68-0.90 in the whole period. Moreover, their correlations evolve in a relatively independent process, in which no significant relation can be found between volatility correlation and size. Consistent with many financial and commodity markets, it demonstrates that information process is more efficient and trade liquidity is higher in the future market. As EUA and CER future volatilities are highly related in the whole period, strategies based on their volatility correlation may be effective no matter how market condition is.

5 Conclusions

EUA market is the most famous mandatory carbon allowance market while CER market is the largest voluntary carbon credit market. To study the relationship between EUA and CER markets is of great significance for the development of China's structured carbon market. The paper develops an integrated IRF-DCC model to analyze the relationships of EUA and CER markets under the second phase EU ETS. The main findings include:

(1) Returns of EUA and CER are closely related both in the spot and future markets. In spot market, returns of EUA and CER are co-integrated and any deviations from long-run equilibrium are corrected by their error correction mechanism at similar speed. EUA and CER spots evolve in the same direction and EUA spot plays a leading role in the dynamics of information reaction. In futures market, reactions of EUA future to deviations gradually adjust the system toward their long-run equilibrium, while reactions of CER future to deviations are explosive. It can be inferred that EUA future generates more influence on the evolution of returns in the future market. Both future returns of EUA and CER react to their own lagged returns, while the reaction of CER to its own lagged returns in the future market is weaker than that in the spot market.

(2) Volatilities of EUA and CER are dynamically correlated in both spot and future markets. Current correlations of EUA and CER volatilities are positively impacted by their lagged shocks and correlations. In spot market, correlations of EUA and CER volatilities are highly instable and market dependent, jumping between 0.26 and 0.95. Volatilities of EUA and CER spots are closely linked when market is peaceful whereas the linkage reduces quickly when market steps into violent shakes. It may be caused by heterogeneous reactions to information shocks carried by heterogeneous participants in EUA and CER spot trades. Judging these, hedge strategies based on the mean of EUA and CER volatility correlations may dysfunction in the turbulent market situations. Spot traders should adjust their strategies according to changeable market situations and built positions according to dynamical volatility correlations of EUA and CER spots. In future market, correlations of EUA and CER volatility are relatively stable and independent, walking between 0.68 and 0.90. Consistent with many financial and commodity markets, it demonstrates that information process is more efficient and liquidity is higher in future market. As EUA and CER future volatilities are highly related in the whole period, strategies based on their volatility correlation may be effective no matter how market condition is.

Acknowledgments. This work was supported by The China Postdoctoral Science Foundation (Grant No. 2014M560993) and The Natural Science Foundation of Guang Dong Province, China (Grant No. 2014A030310404).

References

1. Ye, B., Jiang, J., Miao, L., et al.: Sustainable energy options for a low carbon demonstration city project in Shenzhen, China. J. Renew. Sustain. Energy **7**(2), 23117–23122 (2015)
2. Jiang, J.J., Ye, B., Ma, X.M.: The construction of Shenzhen's carbon emission trading scheme. Ener. Pol. **23**, 256–267 (2014)
3. Milunovich, G., Joyeux, R.: Pricing efficiency and arbitrage in the EU-ETS carbon futures market. J. Invest. Stra. **2**, 23–25 (2007)
4. Uhrig-Homburg, M., Wagner, M.: Futures price dynamics of CO_2 emission allowances: an empirical analysis of the trial period. J. Deri. **17**, 73–88 (2009)
5. Daskalakis, G., Psychoyios, D., Markellos, R.N.: Modeling CO_2 emission allowance prices and derivatives: evidence from the european trading scheme. J. Bank. Fina. **33**, 1230–1241 (2009)
6. Barrieu, P., Fehr, M.: Integrated EUA and CER Price Modeling and Application for Spread Option Pricing. Centre for Climate Change Economics and Policy, Working Paper No. 50, Washington, D.C. (2011)
7. Mizrach, B.: Integration of the global carbon markets. Ener. Econ. **34**, 335–349 (2012)
8. Engle, R.F.: Dynamic conditional correlation: a simple class of multivariate generalized autoregressive conditional heteroskedasticity models. J. Bus. Econ. Stat. **20**, 339–350 (2002)
9. Jie, T., Bin, Y., Qiang, L., et al.: Economic analysis of photovoltaic electricity supply for an electric vehicle fleet in Shenzhen, China. Int. J. Sustain. Transp. **8**(3), 202–224 (2014)

Three Dimensional Scene Modeling Based on SketchUp, Tiling and ArcGIS

Shuai Liu, Lingli Zhao(✉), Junsheng Li, and Wei Xiong

School of Engineering, Honghe University, Mengzi 661100, China
{liushuai_csu,zll_csu}@126.com

Abstract. The paper proposed a kind of 3Dscene modeling considering of SketchUp software, tiling map and ArcGIS for three-dimensional data management, query and spatial analysis. The experiment shows that the approach is much validated, and something useful is obtained.

Keywords: 3D modeling · Sketchup · Tiling map · ArcGIS

1 Introduction

With the development of three-dimensional visualization technology, the emergence of a large number of 3D visualization software and development platforms, there has been a lot of three-dimensional visualization software and development platform. Foreign representative are ArcGIS, MapInfo, IMAGINE Virtual GIS, Skyline, GoogleEarth and so on; domestic representative are MAPGIS, SuperMap, GeoStar, IMAGISClassic, VRMap etc.

The difference to complete and design three-dimensional visualization software platform is mainly the use of underlying technologies, which has characteristics of high speed and three-dimensional strong analysis, and has become an option for three-dimensional visualization systems for applications ranging from urban planning, underground exploration, surface simulation, environmental monitoring, aerospace and other fields.

By means of the software packages ERDAS Imagine 9.1, ESRI ArcGIS 9.2, 3D Nature Visual Nature Studio 3 (VNS), Digi-Art 3DZ Extreme and Avaron Tucan 7.2 the glacier conditions during the “Little Ice Age” (+/− 1850) and the following two dates were reconstructed. Subsequently, several derivates of these data sets were generated [1].

The Digital City Kyoto Experimentation Forum was launched [2]. The forum includes several universities, local authorities, leading computer companies, local newspaper companies, historical temples, as well as photographers, programmers, students, volunteers and so on. James D, etc. developed the package BoreIS based on ESRI’s ArcScene3D platform for the management of underground drilling data visualization, query and analysis. Bo Song embedded development using COM technology and

L. Zhao—PhD, interest covers 3D modelling, data integration and data mining.

F. Bian and Y. Xie (Eds.): GRMSE 2015, CCIS 569, pp. 778–786, 2016.
DOI: 10.1007/978-3-662-49155-3_80

OpenGL based on ArcGIS three-dimensional environment to build editing component toolbar which had geographic features such as editor, interface and simplified interactive. Zhu Yinghao used OpenGL technology in Visual C ++ environment to integrate urban landscape visualization and MapInfo software, developing urban landscape visualization system to implement three-dimensional visualization of digital topographic map [4, 5].

The paper proposed a kind of 3Dscene modeling considering of SketchUp software, tiling map [6, 7] and ArcGIS for three-dimensional data management, query and spatial analysis.

2 Tiling and Image Maps

2.1 Tile Cache

Figure 1 shows that the actual image area of each layer is the same range to reduce the number of tiles (Fig. 2 shows the example), reduce the number of tiles and scale is to zoom in. The number of tiles between adjacent layers is reduced to the original 1/4, the scale down is to zoom out, and the number of tiles adjacent layers increased to four times to the original.

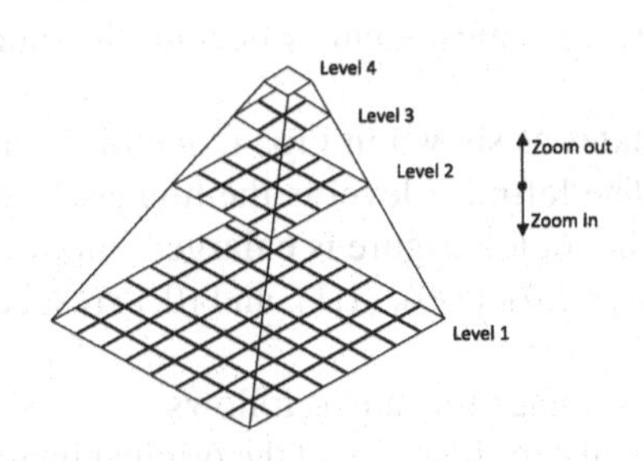

Fig. 1. Schematic model of tile pyramid

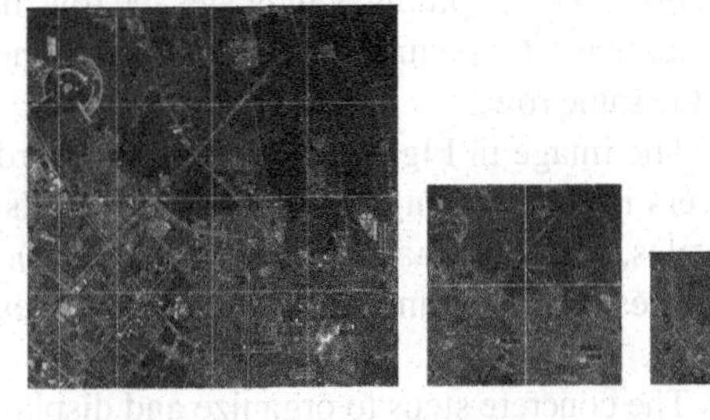

Fig. 2. The layers generation of tile map example

The construction of tile maps follows pyramid principle, and the steps are as follows:

First, determine the zoom level of the map and select the largest scale image of that layer as the first layer of the pyramid as shown in Fig. 2. It is divided into the same size (e.g. 256×256 pixels) square image tiles block shown in Fig. 2 after the blue line separator forming the first layer of the tile matrix which encodes order starting from the top left of the map, from left to right and top to bottom. Then, every 2×2 pixel is composited as a pixel method of generating the next image layer on the basis of the first layer image on the tile, splitting it to form the new layer matrix as shown in the middle of Fig. 2, using the same method to generate the next tile matrix layer as shown in the right of Fig. 2. Finally, the cycle continues until the first N layer to constitute the entire tile pyramid.

You can automatic load maps as you need simply loading the current tile, thus reducing the number of images to improve loading speed, and also saving computer memory.

2.2 Offline Image Map Production

Offline image is one that has been downloaded to the local computer, you can browse using image maps no network. Use image downloading software, For example: Google satellite image map download, download the regional map. The map data downloaded by the software contains image tile data at all levels, which represent different scales images, you can judge from the name of the folder belonging to the level t. Each folder contains multiple raster image shown in Fig. 3.

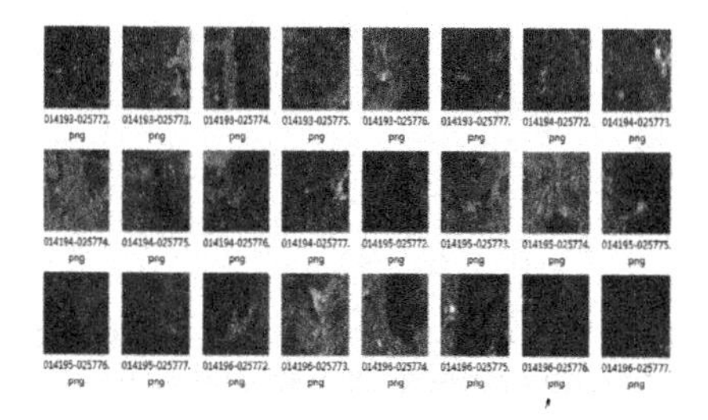

Fig. 3. Original raster image

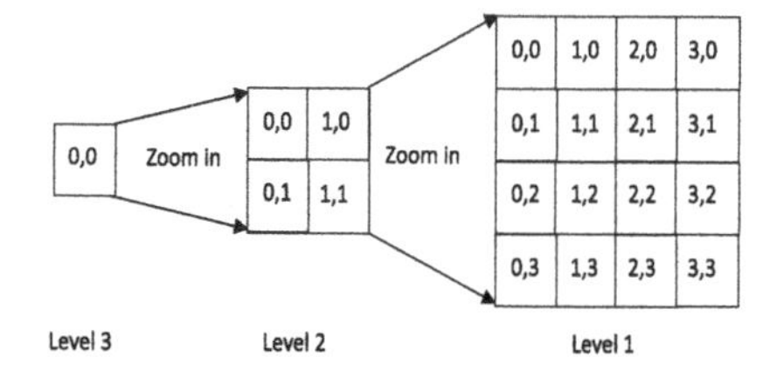

Fig. 4. Tile coordinate encoding of different layers

The grid image in the form of xy naming (file suffix is unified png) shown in Fig. 3, x represents the column, y indicates the row, namely that represents the column of x and the number of y picture, x represents the same as the same column, y denotes the same as the same row.

The image in Fig. 2 is set for local coordinates as shown in Fig. 4, and different layers is representing different zoom levels. The leftmost level is the highest level of tiles, tile coordinates is encoded as (0,0). The middle figure is enlarged into four pictures, and the numbers of tile coordinates are (0,0), (1,0), (1,1), and (0,1) respectively.

The concrete steps to organize and display the image tile, are as follows:

Firstly, the area of images is setting for the local coordinates, and the original image number “column number - line number” becomes ranks number relatively according to the encoding in Fig. 4.

Secondly, the raster image is stored by hierarchical according to each level of tile map file, setting up and naming the column number in the same folder, for example, the first layer needs to establish four folders of 0,1,2 and 3 in Fig. 4, Layer 2 needs to establish two folders of 0 and1, layer 3 simply creates a file of folder 0;

Thirdly, analysis the ranks of image files and the images in the same column are stored in the file column fold moving the image file to the appropriate folder in the file list, and renaming the line numbers (file extension unchanged).

Finally, zoom out map according to a comparative image possible by excursion and move by the ranks.

Specific processes are shown in Fig. 5. After the above processing, each layer set is composed by a folder of multiple columns, the folder name stands for column number shown in Fig. 6. The tile image in each folder is named in their line number shown in Fig. 7. The location of any one tile image is constituted by the three kinds of information, namely line number information (file name), the column number information (parent folder name), and belongs to the layer (zoom level).

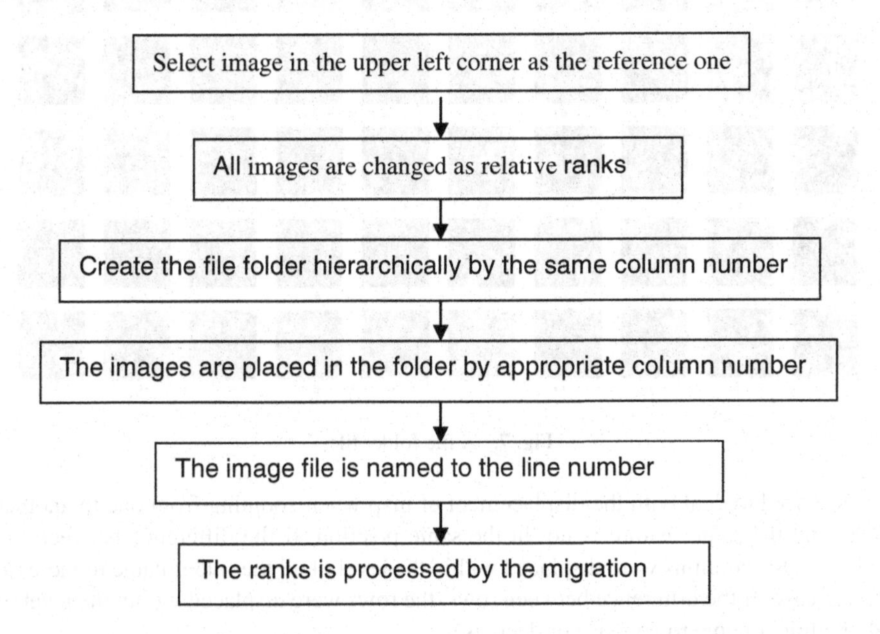

Fig. 5. The flow chart processing of tile document

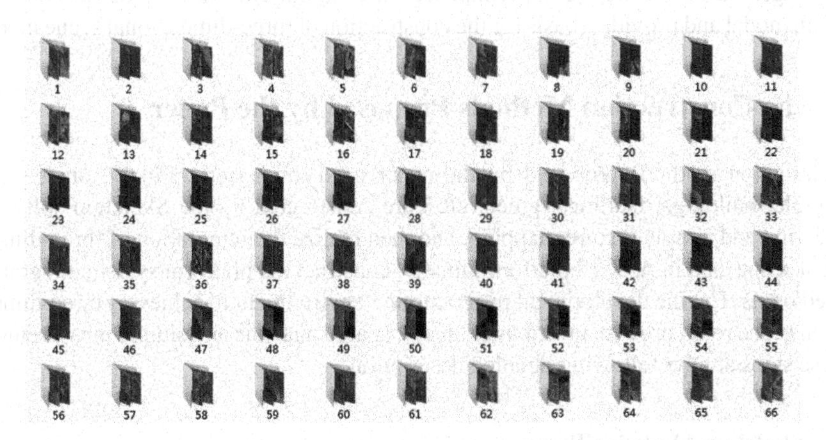

Fig. 6. Column folder of the layer

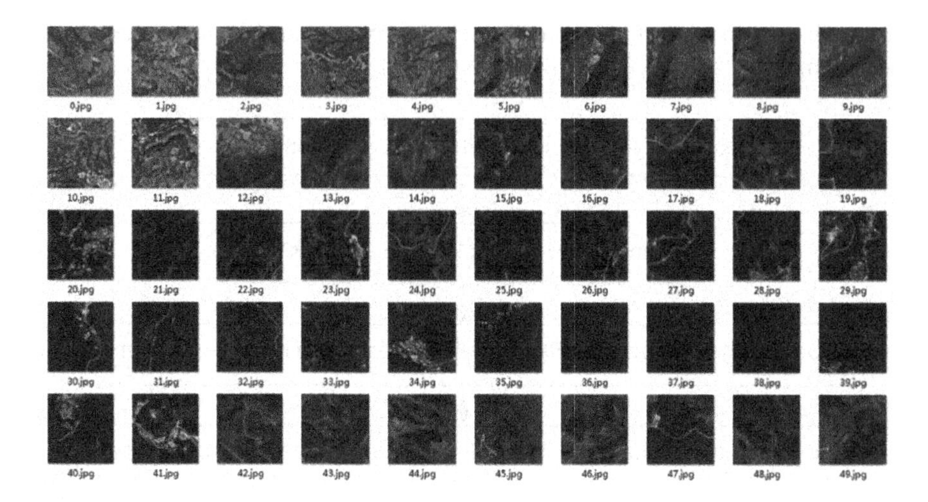

Fig. 7. A file folder tile

We need to deal with the displacement of map when zooming from one to another level and the same feature is not in the same position. If the difference between the columns, the columns were displaced within the level of all the folder name to increase or decrease; If the difference between rows, the rows were displaced within the level of all the folder name to increase or decrease.

Images based on tile are superimposed on a digital elevation model to construct terrain model and provide a basis for the construction of three-dimensional scene model.

3 The Construction Methods Proposed by the Paper

Construction methods proposed by the paper need three steps. Firstly, objects (for example: buildings, facilities, green, etc.) are constructed by the Sketchup software, rendering and simple texture mapping, and then realize the integration of three-dimensional scene data in ArcGIS platform, finally, construct 3D platform system integration based on ArcEngine development programming environment to achieve a three-dimensional scene roam browse, information query, spatial analysis and simple measurement. These steps are the following detailed description.

3.1 Sketchup Modeling Steps

To model three-dimensional objects is a fundamental step in the entire three-dimensional visualization software by SketchUp software, and construction of three-dimensional model should be 1: 1 ratio. The exported data can be compatible in the ArcGIS. Here are the steps by SketchUp, and the modeling flow char is shown in Fig. 8.

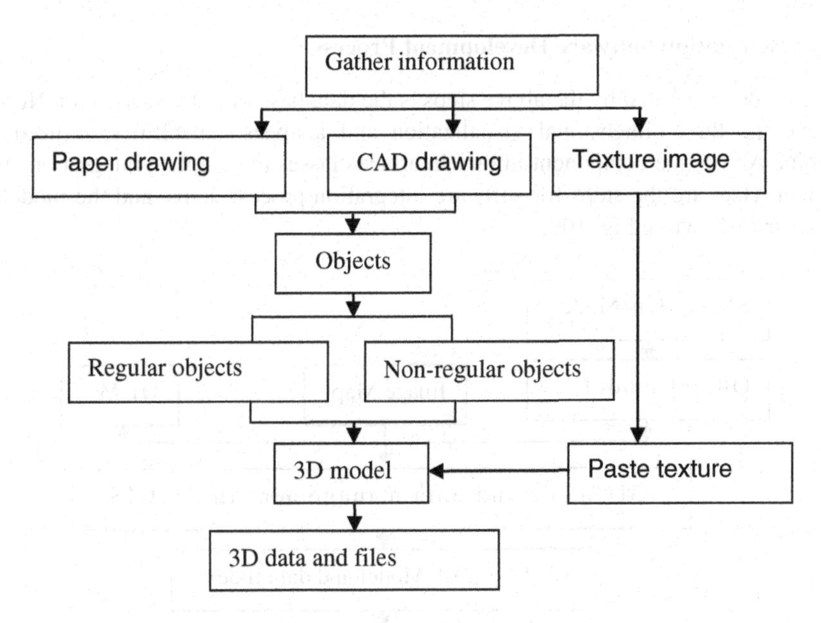

Fig. 8. The flow chart of SketchUp modeling

3.2 ArcGIS Modeling Steps

ArcGIS scene is generally divided into three layers: floating layers, superimposed layers and elevation layers. How to make a three-dimensional object model and terrain model integrate into the scene is the focus for this step. Here are the steps for ArcGIS modeling steps, and the modeling flow char is shown in Fig. 9.

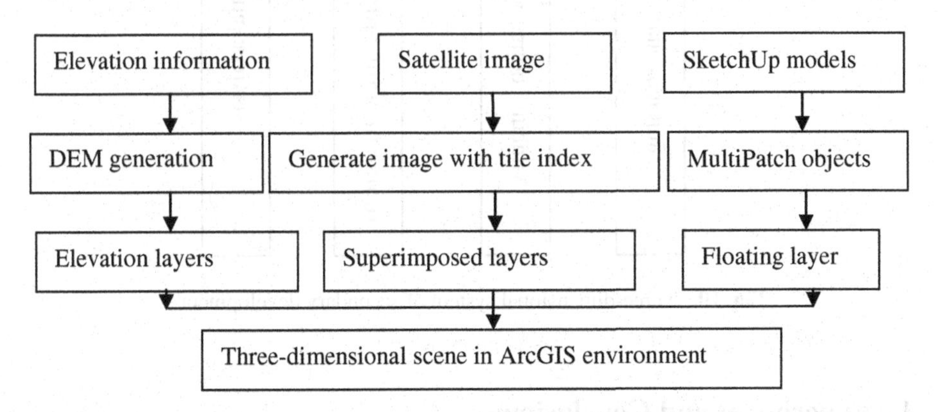

Fig. 9. The flow chart of 3D scene modeling in ArcGIS

3.3 Visualization Software Development Process

Scene model generated by the above steps is the data base of 3D visualization How to achieve the three-dimensional visualization and analysis capabilities, it needs to combine ArcEngine component library interface classes to complete integration environment. Here are the steps for software integration process steps, and the modeling flow char is shown in Fig. 10.

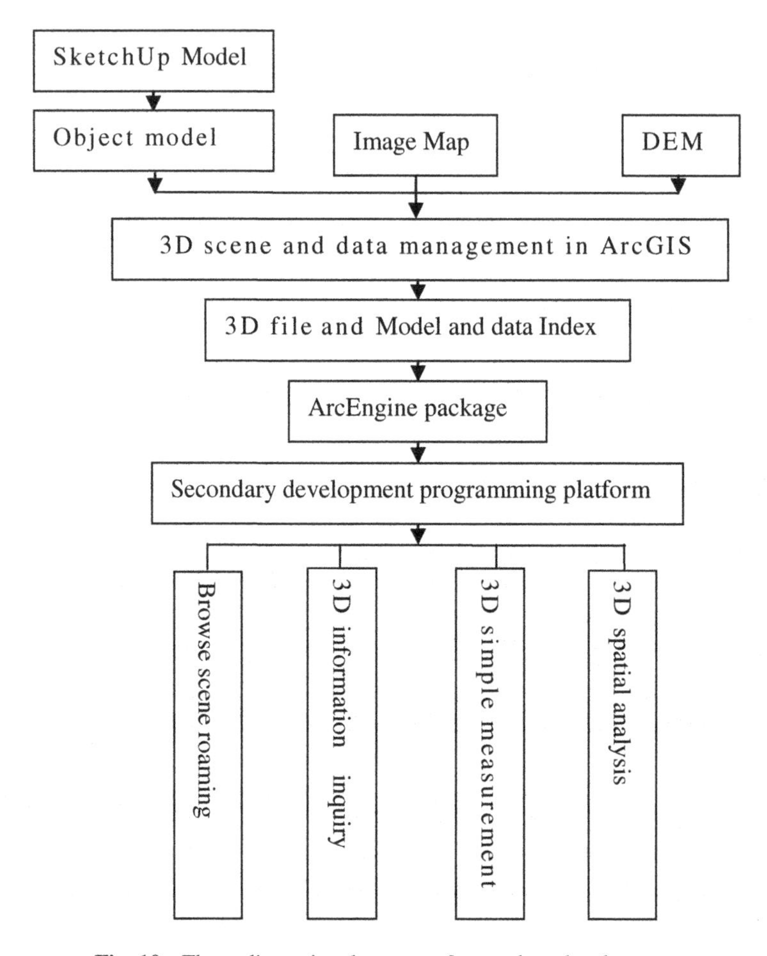

Fig. 10. Three-dimensional system of secondary development

4 Experiment and Conclusions

In the example to construct a school campus visualization system adopted the modeling method proposed in this paper. Figure 11 is the main interface visualization systems. Tile image is superimposed on elevation model constitutes three-dimensional scene, and three-dimensional building models are placed on the scene. Figure 12 is flight roaming of the campus scene.

Fig. 11. Three-dimensional system main interface

Fig. 12. Flight roaming

This paper proposed a kind of 3D integrated method based on tile-based imaging, sketch modeling and ArcGIS, and the results demonstrate the effectiveness and feasibility of the method.

Acknowledgements. This research is supported by the National Natural Science Foundation (No. 41201418, 41301442), and Pecuniary aid of Yunnan Province basic research for application (2013fz127).

References

1. Bruhm, K., Buchroithne, M., Hetze, B.: True-3D Visualization of Glacier Retreat in the Dachstein Massif. Cross-Media Hard-and Softcopy Displays, Austria
2. Ishida, T.: Activities and technologies in digital city Kyoto. In: Besselaar, P., Koizumi, S. (eds.) Digital Cities 2003. LNCS, vol. 3081, pp. 166–187. Springer, Heidelberg (2005)
3. Zhang, L.: Key Technologies for 3D Digital Earth Development. Chinese Academy of Sciences (2004). (in Chinese)
4. McCarthy, J.D., Graniero, P.A.: A GIS-based borehole data management and 3D visualization system. Comput. Geosci. **32**(10), 1699–1708 (2006)
5. Song, B., He, Y., He, X., et al.: Study on interactive editing method in 3D GIS environrnent, 12–14 October 2008
6. Liu, Z., Pierce, M.E., Fox, G.C., et al.: Implementing a caching and tiling map server: a web 2.0 case study. In: International Symposium on Collaborative Technologies and Systems, CTS 2007, pp. 247–256. IEEE (2007)
7. Ji, H., Wong, W.H.: TileMap: create chromosomal map of tiling array hybridizations. Bioinformatics **21**(18), 3629–3636 (2005)

Analysis and Calculation of Ecological Footprint in Shaanxi Province from 2000 to 2010

Guozhang Cen(✉) and Xuelu Liu

College of Resources and Environmental Science,
Gausu Agricultural University, Lanzhou 730070, China
{cengz, liuxl}@gsau.edu.cn

Abstract. In this paper, the ecological footprint in Shaanxi Province during the period from 2000 to 2010 is calculated and analyzed. The analysis is based on the theory and methodology of ecological print, with natural resource production and consumption, population and land use data from the FAO database and Shaanxi Statistical Yearbooks. Ecological deficit was demonstrated that from 2000 to 2010, indicating an unsustainable ecological development in Shaanxi province.

Keywords: Ecological footprint · Ecological capacity · Ecological deficit · Sustainable development

1 Introduction

Since the United Nations Conference on Environment and Development (UNCED) in 1992, in-depth studies on sustainable development have been done internationally. Consequently, the quantitative evaluation methods for sustainable development have also become the leading edge and hotspot of researches. Some intuitive and operable index systems and quantitative evaluation methods for sustainability have emerged, for example the Green Gross Domestic Product, the National Wealth index by the World Bank, the Index of Sustainable Economic Welfare model, and the Barometer of Sustainability model. The aim of these index systems and models is to make sustainable development measurable, by converting sustainability to specific indicators to assess whether human being is living within the carrying capacity of ecological system. However, there are certain limitations with these studies and the progress has been slow. The key of sustainable development evaluation is to understand the utilization of nature by human and quantitatively measure whether human demand exceeds nature's re-production capacity. Ecological footprint serves such purpose being a method quantitatively assessing human being's utilization of natural resources and nature's functions of providing human being with support services. In the 1990s, Canadian ecologist and economist Rees raised the concept of ecological footprint. Later on, Wackernagel completed this theory and method. While proposing the concept, indicators and methodology of ecological footprint, they applied the indicators to calculate the ecological footprints of 52 countries and regions. Ecological footprint analysis was

F. Bian and Y. Xie (Eds.): GRMSE 2015, CCIS 569, pp. 787–794, 2016.
DOI: 10.1007/978-3-662-49155-3_81

introduced into China in 1999 and quickly applied to analyses and studies of some provinces, cities and regions' sustainable development as a new theoretical method. The creation of this index system provided a simple framework for the assessment of the natural resource usage in a region, a country or even the world. By measuring the gap between human's demand for nature's ecological services and the ecological services that nature can provide, we can thoroughly understand human's utilization of the ecological system, and compare human's consumption of nature and nature's carrying capacity in a regional, nationwide, or global scale. Therefore, ecological footprint analysis is an advantageous quantitative analysis index for assessing the influence of human activities on natural environment. In recent years, because of its scientific and complete theoretical basis and concise and unified index system, ecological footprint analysis has been applied widely and efficiently in domestic and foreign studies. In this study, we applied the principle and method of ecological footprint to analyze the dynamic evolution pattern of Shaanxi's ecological footprint and ecological capacity during 2000–2010, and calculate the per capita ecological deficit of Shaanxi. The problems in Shaanxi's sustainable development are analyzed in this paper in order to provide reference for the sustainable development within the carrying capacity of the resources and environment.

2 Study Area

2.1 Geography

Located in the central part of China mainland, Shananxi borders Shaanxi to the east with the Yellow River in between, Gansu and Ningxia to the west, Inner Mongolia to the north, and Sichuan, Chongqing, Hubei and Hunan to the south. The province spans across a south-north length of 870 km from 105°29′E to 115°15′E, and an east-west width of 510 km from 31°42′N to 39°35′N, with a total area of 205,800 km^2. Overall, it has a terrain higher in the south and north parts and lower in the middle. In the northern part of Shaanxi is the Shanbei Plateau, which consists of mainly the Loess Plateau, with an altitude of 900–1500 m, and an area accounting for 45 % of the province's total. In the southern part is the Qinba Mountains with an altitude of 1500–300 m and an area accounting for 36 % of the province's total. In the middle is the Guanzhong Plain, which is a typical valley plain with an altitude of 320–650 m, and this region accounts for 19 % of the province's total area.

The Annual mean temperature of Shaanxi is around 11.6 °C. On average, the annual precipitation is 653 mm, and there's about 150 to 270 frost-free days per year. The province can be divided into three regions from south to north according to the climate, including Qinba Mountains with north subtropical humid monsoon climate, Guanzhong Plain and Shanbei Plateau with temperate semi-humid semi-arid monsoon climate, and the blown-sand region with temperature monsoon climate. Currently, 136 kinds of mineral resources have been discovered in Shaanxi, and the proved reserves of 57 kinds rank top 10 in China. In total, there are 11 provincially administrated cities/districts and 107 counties/cities/districts in the province.

2.2 Demographics and Economy

As of 2010, Shaanxi province hosts a population of 37.35 million, including 17.07 million of urban population and 20.28 million of rural population. The Guanzhong region is the center of population distribution. In the Shanbei Plateau region, the population mostly distributes in the mainstream of the Yellow River and the downstream of its branches. In southern Shaanxi, the population concentrated in valleys and plains in distributions of dots and strips. In the mountain region, the scarce population scatters in low density. The 107 counties/cities/districts of the province include 32 in Guanzhong Plain region and 25 in Weibei Loess Plateau region, thus these two regions have the densest populations of the province.

According to statistical records of 2010, the province owns cropland of 4.05 million hectare, which accounts for 19.68 % of the province's total area. The areas of pasture and forestry are 3.064 and 10.354 million hectares, respectively, accounting for 14.89 % and 50.31 % of the total area. The land use in Shaanxi is characterized by its diversity, and dominated by agricultural use. In different regions, land use exhibits distinct difference. In Guanzhong region, there is more cropland, while in southern Shaanxi there is more forestry, and in northern Shaanxi there is more pasture.

3 Data Source and Methodology

3.1 Data Source

Data used in this study are from the *Shaanxi Statistical Yearbook*, and the global average yield data of products are from the *Analysis of* Ecological *Footprints of Shaanxi Province in 2001* and the *Analysis of Ecological Footprints of Shaanxi Province in 2003*.

3.2 Methodology

Calculation of Ecological Footprint. The per capita ecological footprint for a particular product can be calculated with the following formula.

$$A_i = C_i/Y_i = (P_i + I_i - E_i)/(Y_i * N)(i = 1, 2, 3, \ldots, 6) \quad (1)$$

Where, i marks the product, A_i is the per capita biological area (hm^2) for the i-th product, C_i is the average tonnage of this product consumed per person, Y_i is the global average annual yield (kg/hm^2) of this product produced by corresponding biologically productive land, P_i is the annual production of the product, I_i and E_i are the annual import and export of this product, respectively, and N is the population.

Calculation of Ecological Capacity. The per capita ecological capacity is calculated with the following formula.

$$ec = \sum \left(a_j \times C_j \times y_j\right) (j = 1, 2, 3, \ldots, 6) \quad (2)$$

Where, a_j is the average area of biologically productive land per person, C_j is the equivalence factor, and y_j is the yield factor.

The yield factor is calculated with the following formula.

$$y_j = p_j / P_j (j = 1, 2, 3, \ldots, 6) \quad (3)$$

Where, y_j is the yield factor of the j-th type of biologically productive land, p_j is the average productivity of this land type in Shaanxi, and P_j is the world average productivity of this land type.

The equivalence factor is calculated with the following formula.

$$C_j = d_j / D (j = 1, 2, 3, \ldots 6) \quad (4)$$

Where, C_j is the equivalence factor of the j-th type of biologically productive land, d_j is the global average productivity of this land type, and D is the global average productivity of all biologically productive land.

4 Results and Analysis

4.1 Dynamic Characteristics of the Ecological Footprint in Shaanxi

The calculation of the ecological footprint in Shaanxi is detailed with the calculation for 2010 as an example. We selected 12 products for the calculation of ecological footprint of cropland, including rice, wheat, corn, beans, sorghum, oilseeds, rapeseed, cotton, fiber, tobacco, vegetables, and pork. For the calculation of average ecological footprint of forestry, due to the lack of data, we only selected tung seed, walnut, chestnut and pepper, and the results are relatively low comparing with the data in other literatures. For the ecological footprint of pasture, we selected eggs, beef, lamb, dairy, and hair. For fisheries, consumption of aquatic products was used for calculation. For construction land, residential area, industrial land, and transportation land were included. And for the calculation for energy land, coal, oil, natural gas and electricity were selected.

It can be seen that the ecological footprint of Shaanxi from 2000 to 2010 exhibits a clear ascending trend. Over the 11 years, the per capita ecological footprint increased from 0.9125 to 2.2855. The increase was rapid in the first 4 years and smooth during 2005–2006. Then, after a rapid increase in 2007, the increase became smooth again during 2008–2009. The increase of ecological footprint from 2009 to 2010 was the most intense in the recent decade.

4.2 Dynamic Characteristics of the Ecological Capacity of Shaanxi

Based on the natural resources data of Shaanxi over the years (Table 2), the ecological capacity was calculated as presented in Table 3. The yield factor was identified according to the ratio of each land type's average productivity to global average productivity. Due to the difficulty in acquiring data, the average productivity data are in national scale. For some of the years, the areas of water were calculated based on available data of other years. The area of construction land is the sum of residential area, industrial land and transportation land.

Table 1. Per capita ecological footprints in Shaanxi from 2000 to 2010 [hm^2]

	Cropland	Forestry	Pasture	Fisheries	Construction land	Energy land	Total
2000	0.3756	0.0012	0.1837	0.0586	0.0200	0.2733	0.9125
2001	0.3765	0.0009	0.1914	0.0586	0.0201	0.3162	0.9636
2002	0.4035	0.0014	0.2202	0.0603	0.0203	0.3584	1.0640
2003	0.4183	0.0014	0.2545	0.0624	0.0204	0.4062	1.1633
2004	0.4643	0.0018	0.2764	0.0639	0.0205	0.5640	1.3909
2005	0.4922	0.0018	0.2981	0.0684	0.0206	0.7211	1.6022
2006	0.4029	0.0018	0.2382	0.0446	0.0207	0.9261	1.6342
2007	0.3776	0.0019	0.2597	0.0464	0.0207	1.0267	1.7330
2008	0.4338	0.0025	0.2867	0.0479	0.0207	1.1535	1.9450
2009	0.4111	0.0029	0.2705	0.0513	0.0208	1.2490	2.0057
2010	0.4290	0.0026	0.2639	0.0557	0.0208	1.5135	2.2855

Note: All consumption data in Table 1 are from *Shaanxi Statistical Yearbook.*

It can be seen that the ecological capacity of Shaanxi from 2000 to 2010 exhibits a clear descending trend. Over the 11 years, the per capita ecological capacity decreased from 0.9803 to 0.8880. As shown in Fig. 2, the capacity decreased drastically during 2000–2005, especially 2001–2003. During 2005–2010, the ecological capacity decr eased in a slower manner. The decrease of ecological capacity indicates the continuously deteriorating ecological environment of Shaanxi.

The increase of ecological footprint is also closely related to the year-by-year decrease of cropland and pasture and the increase of construction land. Although it is suggested in Table 2 that the area of forestry increased, the increase was insufficient to compensate the increasing ecological footprints. In addition, the decrease of cropland and pasture worsened the ecological capacity decrease. The year-by-year increase of ecological footprints and the year-by-year decrease of ecological capacity resulted in the rapid increase of ecological deficit. This indicates the grim condition of ecological development, and the root cause is that the excessive use of natural resource and energy has exceeded the capacity of the ecological system. It can be seen from 4 that the areas of cropland, pasture and water have been reducing over the years, although in slow rates, still affecting the ecological changes in Shaanxi. In the meantime, the per capita ecological footprint for energy land has increased from the 0.2733 of 2000 to the 1.5135 of 2010, with an over 5 times of increment. Apparently, the consumption of energy has

Table 2. Natural resources of 2000–2010 in Shaanxi [104 hm^2]

Year	Cropland	Pasture	Forestry	Fisheries	Construction land
2000	480.0	317.9	962.6	40.3	73.0
2001	468.5	320.3	969.3	40.2	73.4
2002	450.6	321.5	984.3	40.1	74.4
2003	424.2	316.0	1011.9	39.9	74.9
2004	415.4	313.4	1020.3	39.9	75.5
2005	408.9	311.7	1028.5	39.9	75.9
2006	405.8	307.1	1034.7	39.9	76.5
2007	404.9	306.6	1035.4	39.9	76.8
2008	404.9	306.6	1035.4	39.9	76.8
2009	405.0	306.4	1035.4	39.9	77.6
2010	405.0	306.4	1035.4	39.9	77.6

Table 3. Per capita ecological capacity [hm^2]

Year	Cropland	Pasture	Forestry	Fisheries	Construction land	Total
2000	0.6123	0.0083	0.2644	0.0022	0.0931	0.9803
2001	0.5961	0.0083	0.2656	0.0022	0.0934	0.9656
2002	0.5719	0.0083	0.2691	0.0022	0.0944	0.9459
2003	0.5370	0.0082	0.2758	0.0022	0.0948	0.9180
2004	0.5245	0.0081	0.2775	0.0022	0.0953	0.9076
2005	0.5151	0.0080	0.2790	0.0022	0.0956	0.8999
2006	0.5099	0.0079	0.2800	0.0022	0.0961	0.8961
2007	0.5075	0.0079	0.2795	0.0022	0.0963	0.8933
2008	0.5062	0.0078	0.2788	0.0021	0.0960	0.8909
2009	0.5051	0.0078	0.2781	0.0021	0.0968	0.8899
2010	0.5040	0.0078	0.2775	0.0021	0.0966	0.8880

Note: Since the ecological capacity of energy land is 0, it is not listed in Table 4.

Table 4. The equivalence factors and yield factors [4]

	Cropland	Pasture	Forestry	Fisheries	Construction land
Equivalence Factor	2.8	0.5	1.1	0.2	2.8
Yield Factor	1.66	0.19	0.91	1	1.66

been rapidly increasing during the 11 years. Therefore, there is a close relation between the ecological deficit and the excessive consumption of energy. Moreover, the increase of population, the decrease of cropland, pasture and water, and the increase of construction land all resulted in the increase of per capita ecological footprints. Accompanied by the decrease of ecological capacity, the ecological deficits keep growing. Therefore, we can conclude that from 2000 to 2010, Shaanxi was in an unbalanced and unsustainable mode of development, and such condition is getting worse.

During the calculation of annual per capita ecological footprints, we didn't consider the consumption of fruits and wood, thus the footprints calculated for cropland and forestry are smaller than actual numbers. Meanwhile, the areas of fisheries for some of the years were deduced from available data of other years during the calculation of ecological capacity. These factors would cause deviations from the actual ecological deficit/surplus. However, the deviation does not significantly impact the determination of whether the ecological development in Shaanxi is sustainable.

5 Discussion

The calculations and analyses presented in this paper demonstrated that the economic social development mode of Shaanxi is unsustainable and the ecological environment is in a dangerous status. To maintain the sustainable development of this region, it is necessary to reduce ecological deficits, reduce ecological footprint and improve ecological capacity. Existing studies showed that the more economically developed a region is, the greater its ecological footprints. With economic development, the ecological footprint in Shaanxi is likely to continue increasing. In order to keep the pressure on nature from exceeding the threshold of the ecological environment and maintain sustainable development, it is crucial to reduce the ecological deficits. For such situation, we propose three types of countermeasures based on Wackernagel model, including adopting high-tech to improve the productivity of unit natural system, utilizing available resource reserves efficiently, and controlling the population in order to reduce consumption and change people's production and living and consumption styles in order to establish resource-saving social production and consumption system.

These three measures are fundamental for reducing ecological deficits, and they can be applied for any regions, however different regions should come up with more practical solutions and measures based on their own situations. The ecological condition under the original economic pattern of Shaanxi can no longer satisfy the demands of the rapidly developing society and economy. To ensure sustainable development, more practical solutions should be formulated for Shaanxi on the basis of these three ecological deficit-reducing measures. Based on existing studies, five measures can be adopted in Shaanxi. First, use high-tech to improve the productivity of unit natural ecological system. With the help of high-tech, we can prevent the per capita ecological capacity from decreasing drastically because of population increase and cropland decrease. Second, use available resources efficiently. Currently, China's energy consumption of unit product is about 50 % to 70 % higher than the ones of developed countries, and the energy consumption of unit GDP is 318 times of global average. Third, control the population and reduce per capita consumption, change people's production, living and consuming styles, and build a resource-saving social production and consumption system. Limited resources decided that the ecological capacity is also limited, and the increasing population and enjoyment-pursuing lifestyles will continuously increase ecological footprints. Fourth, enforce the current resource protection measures and reduce cropland loss. The decrease of cropland in Shaanxi over the past half century was also an important reason for the increase of ecological deficits. In particular, in the loess hilly-gully region of northern Shaanxi, the

Weibei gully region of Loess Plateau, and the hilly region of southern Shaanxi, soil erosion has been the primary reason for cropland decrease. Fifth, develop circular economy, find ways to recycle and reuse wastes and intermediate products in order to tap the potential of resources and energy. For Shaanxi, a province with large population and limited resources, implementing these measures will help reduce the pressure human being pose on nature, reduce ecological deficits and gradually realize sustainable development.

References

1. Xie, H.Y., Wang, L.L., Chen, X.S., et al.: Modification and Application of Ecological Footprint Evaluation Models. Chemical Industry Press, Beijing (2008)
2. Kang, W.X., Liu, K., Zhang, S.F., et al.: Ecological Environment Protection in Shaanxi. Shaanxi People's Publishing House, Beijing (2006)
3. Zhang, X.N.: Dynamic measurement and analysis of ecological footprint – study of sustainable development of Shaanxi Province. J. Xi'an Univ. Post Telecommun. **13**, 49–53 (2008)
4. Wu, J.J., Cai, L., Zhang, Q.H., et al.: Analysis and assessment of ecological footprint in Shaanxi Province from 1993 to 2004. J. Shaanxi Normal Univ. (Nat. Sci. Ed.) **34**, 104–108 (2006)
5. Zhang, Q.F., Wu, F.Q., Tian, D., et al.: Analysis of ecological footprint of Shaanxi Province in 2003. Agric. Res. Arid Areas **25**, 35–40 (2007)
6. Li, M.Y., Jiang, H.: Defects of hypothesis of ecological footprint model, scientific and technological management of land and resources, pp. 72–75 (2005)
7. Yang, K.Z., Yang, Y., Chen, J.: Ecological footprint analysis: concept, method and cases. Adv. Earth Sci. **15**, 630–636 (2000)
8. Zhao, X.G., Xiao, L., Lan, Y.X.: Dynamics of ecological footprint and ecological capacity of Shaanxi. Scientia Agricultura Sinica **38**, 746–749 (2005)
9. Yang, Q.X., Hu, Y.Y, et al.: Ecological deficit and its causes. Soc. Sci. Yunnan, 54–59 (2007)
10. Wang, J.H., Ren, Z.Y., Su, Y.L., et al.: Evaluation of ecological carrying capacity based on ecological footprint model in Xi'an during 1997-2009. Agric. Res. Arid Areas **30**, 224–260 (2012)
11. Mathis, W., William, E.R.: Our Ecological Footprint: Reducing Human Impact on the Earth. New Society Publishers, Gabriola Island (1996)
12. Gossling, S., Hansson, C.B., et al.: Ecological footprint analysis as a tool to assess tourism sustainability. Ecol. Econ. **43**, 199–211 (2002)

Research on Collision Risk Model in Free Flight Based on Position Error

Zhaoning Zhang and Ruijun Shi(✉)

College of Air Traffic Management, Civil Aviation University of China, Tianjin, China
srjworkmail@163.com

Abstract. The problem of collision risk in free flight with a certain safe distance is studied. In free flight, the collision risk is closely related to the aircraft position error. With the stochastic characteristics, we set the position error to satisfy a three-dimensional Gaussian distribution, and established the position error model of the aircraft in the horizontal and the vertical direction. Determine the location and the time when the two aircrafts are closest by the nominal route, and calculate the probability that the aircraft appears at any point near the theoretical position by the position error model. Because every collision occurs after conflict, we calculated the conflict probability first, and then modeled the collision risk given safe distance in free flight. Moreover, the algorithm is presented to use this collision risk model to obtain the minimum safe distance with a safety target level. The results of case study show the feasibility of this model.

Keywords: Free flight · Position error · Conflict probability · Collision risk · The minimum safe distance

1 Introduction

In free flight, pilots can freely choose the flight path, speed and direction, which can save time, reduce fuel consumption and increase traffic flow and is an effective way to solve the problem of traffic congestion. Currently, free flight has been listed as the core work of the environment of FANS (Future Air Navigation System) brought by ICAO, and has been identified by FAA as the core concepts of its modernization program. In order to ensure flight safety in free flight, it is needed to study the risk of collision between aircraft for safety assessment.

Currently, in order to study collision risk assessment in free flight, scholars established collision risk model from different angles. The earliest model is the Reich collision risk model, from which the concept of collision slab is brought up [1]. Thereafter some scholars have improved this model, and established models based on conflict region [2–4] or models based on the position error [5]. Moreover, some scholars build collision risk assessment model with stochastic simulation method [11, 12] or stochastic dynamic colored Petri net [13].

This paper is aimed to study the problem of aircraft collision risk given a certain safe distance. Because pilots are free to choose the flight path, it cannot be known whether the route will be cross or parallel in advance, so the model based on conflict zone or

F. Bian and Y. Xie (Eds.): GRMSE 2015, CCIS 569, pp. 795–803, 2016.
DOI: 10.1007/978-3-662-49155-3_82

based on position error are more appropriate. However, the former, although considering the relationship between conflict and collision, is not accuracy in estimating the weight coefficient and the latter, although considering the position error, can't be used to know the relationship among conflict probability, collision probability and the safe distance.

In this paper, it is planned to establish a mixture model which has the advantages of the two models mentioned above, while avoiding their drawbacks. The 3D position error probability model is improved. Firstly, establish the elliptic cylinder collision slab according to the position error of aircrafts in the direction of in the horizontal and vertical direction, and the correlation of errors in different directions. Secondly, calculate the probability that the aircraft appears at any point near the theoretical position by the position error model, and then obtain the conflict probability. Finally, according with the condition probability relationship between conflict and collision, establish the collision risk model, and from the model calculate the minimum safe distance.

2 Modeling of Position Error of Aircrafts

In the horizontal plane, the factors that generate position error are different between the direction x that the aircraft flying along and the direction y that the aircraft flying across, so reference to the literature [8], position error is orthogonally decomposed to two components, of which one is along direction x, and the other is y. Assuming the error along x complies $E_1 \sim N(u_1, \sigma_1^2)$, in which u_1 is the theoretical optimal value of aircraft and the standard deviation increases linearly with time $\sigma_1^2 \sim r_1^2(\Delta t)^2$. The error perpendicular to the flying direction complies $E_2 \sim N(u_2, \sigma_2^2)$, and the standard deviation $\sigma_2^2 \sim \min\{r_2^2 s^2(t), \sigma_{2c}^2\}$, which increases linearly with distance that the aircraft fly over but in the end tends to a constant σ_{2c}.

In the vertical plane, because the aircraft get the height information mainly from the measurement system, the main source of height error is measuring system error [1], so we assume that E_3, the position error of vertical height, satisfy Gaussian distribution $E_3 \sim N(u_3, \sigma_3^2)$. Thus it is reasonable to assume that the aircraft position error subjects to a three-dimensional Gaussian distribution. The horizontal projection of forecasted flight path is show in Fig. 1.

Take t_0 as the initial time. The initial coordinates of aircraft A, B are (x_{10}, y_{10}, z_{10}) and (x_{20}, y_{20}, z_{20}). After a short moment Δt, the theoretical coordinates of aircraft A, B are $(x_1(t), y_1(t), z_1(t))$ and $(x_2(t), y_2(t), z_2(t))$, and the speed of A, B is V_1 and V_2. θ_i is the angle between the aircraft A and the positive direction of axis X. Then

$$\begin{cases} x_i(t) = x_{i0} + V_i t \cos\theta_i \\ y_i(t) = y_{i0} + V_i t \sin\theta_i, \quad i = 1, 2 \\ z_i(t) = z_{i0} \end{cases} \tag{1}$$

The joint probability density function of the three-dimensional Gaussian distribution is

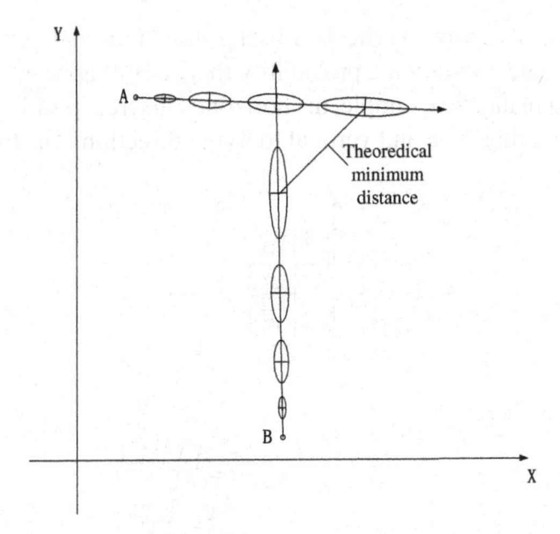

Fig. 1. The horizontal projection of forecasted flight path

$$f_x = \frac{1}{\left(\sqrt{2\pi}\right)^3 |\Sigma|^{\frac{1}{2}}} \exp\left[-\frac{1}{2}(x-u)^T \sum^{-1} (x-u)\right] \tag{2}$$

In which the mean value is Eq. (3), and the covariance matrix is as Eq. (4).

$$u_i = \begin{bmatrix} u_{i1} \\ u_{i2} \\ u_{i3} \end{bmatrix} = \begin{bmatrix} x_{i0} + V_i t \cos\theta_i \\ y_{i0} + V_i t \sin\theta_i \\ z_{i0} \end{bmatrix} \tag{3}$$

$$\sum = \begin{bmatrix} r_{i1}^2 (\Delta t)^2 & 0 & 0 \\ 0 & \min\left\{r_{i2}^2 s^2(t), \sigma_{i2c}^2\right\} & 0 \\ 0 & 0 & \sigma_{i3}^2 \end{bmatrix} \tag{4}$$

3 Establishment of Collision Risk Model

Each collision accidents will experience from conflict to collision, so the collision probability closely relates to conflict, so does the minimum safe distance. In order to obtain collision risk between aircraft, firstly find the probability of conflict in any time t, then calculate the collision probability, and finally model the collision risk of a time period. Additionally, calculate the minimum safe distance given a target level of safe through this model.

3.1 Calculation of Conflict Probability

In order to get the collision probability at any time, take the mean value of position as the center, and differentiate the region that the aircraft may appear into a number of small spaces, in the flying direction, vertical to flying direction, and vertical to horizontal direction. And ultimately obtain the probability that aircraft appears at each point.

In the horizontal plane, for example, the horizontal discrete position region of aircraft A and B in the flying direction and vertical to flying direction is as Fig. 2 shown.

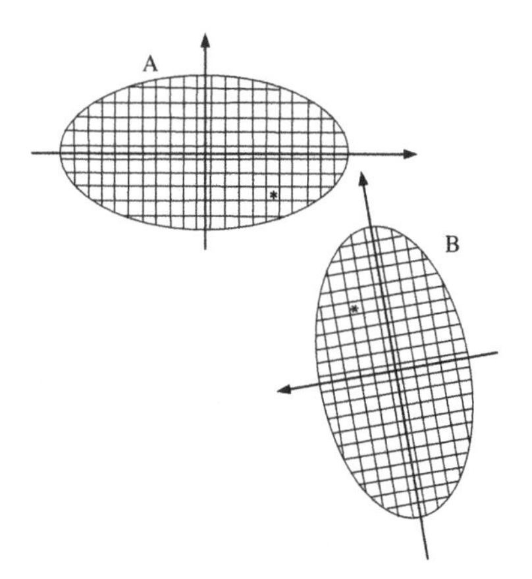

Fig. 2. The horizontal discrete position region of aircraft A and B

Assuming that divide the region into m pieces in flying direction, k pieces in vertical to flying direction, and l pieces in vertical to horizontal plane. To obtain the probability of the center of each of the small box, it is needed to integrate the small cuboid, and when $m \to \infty, k \to \infty, l \to \infty$, the probability that the aircraft appears in any point can be calculated.

$$P_t = \iiint\limits_{V \to 0} f_x dv = \iiint\limits_{V \to 0} \frac{1}{\left(\sqrt{2\pi}\right)^3 |\Sigma|^{\frac{1}{2}}} \exp\left[-\frac{1}{2}(x-u)^T \sum^{-1} (x-u)\right] dv \quad (5)$$

During flight, if not the position error, the probability of collision of two aircraft reached a maximum at the closest point, so at the time that the two get closest, by optimization, the worst case about the position—the shortest distance between the two aircraft—can be found.

In the three-dimensional space, it is assumed that the center position vectors of the aircrafts are $\overrightarrow{r_{10}}$ and $\overrightarrow{r_{20}}$, the factual position vectors are $\overrightarrow{r_1}$ and $\overrightarrow{r_2}$, the velocity vectors are $\overrightarrow{v_1}$ and $\overrightarrow{v_2}$, and the distance between them is $\rho = |\vec{\rho}| = |\overrightarrow{r_1} - \overrightarrow{r_2}|$. Let the mean value plus a stochastic error, i.e. $\overrightarrow{r_1} = \overrightarrow{r_{10}} + \overrightarrow{w_1}, \overrightarrow{r_2} = \overrightarrow{r_{20}} + \overrightarrow{w_2}$. The initial time $t_0 = 0$. At the moment t, the relative position vector of the aircrafts is $\overrightarrow{\rho_t} = \overrightarrow{r_1(t) - r_2(t)} = \overrightarrow{r_1} + \overrightarrow{v_1}t - \overrightarrow{r_2} - \overrightarrow{v_2}t = \vec{\rho} + \overrightarrow{v_{relative}}t$, and the square of the relative distance is $\rho_t^2 = \vec{\rho}_t \cdot \vec{\rho}_t$. Derivative with respect to time, make the derivation 0, and then the shortest distance between the aircrafts is obtained, as well the time

$$t_{PCA} = -\frac{\vec{\rho} \cdot \overrightarrow{v_{relative}}}{\overrightarrow{v_{relative}} \cdot \overrightarrow{v_{relative}}} \tag{6}$$

In this case, the relative position vector is perpendicular to the relative velocity vector. The relative movement of the aircraft is shown in Fig. 3.

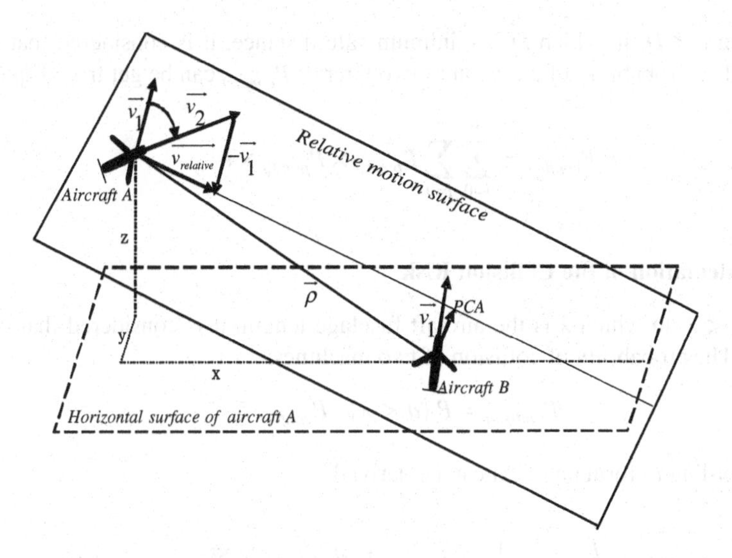

Fig. 3. The relative movement of the two aircrafts

Then at t_{PCA}, the position of the two is

$$\overrightarrow{r_i(t_{PCA})} = r_{i0} + v_i \cdot t_{PCA} + w_i, i = 1, 2 \tag{7}$$

And the distance between the two:

$$d_{PCA} = \left|\overrightarrow{\rho(t_{PCA})}\right| = \left|\vec{\rho} + \overrightarrow{v_{relative}} \cdot t_{PCA}\right| \tag{8}$$

Take (x_1, y_1) and (x_2, y_2) as the coordinate of center position. Taking stochastic factor into account, the boundary of position is approximate to

$$\begin{cases} \frac{(x-x_i)^2}{a^2} + \frac{(y-y_i)^2}{b^2} = 1, \quad i = 1, 2 \\ -c^2 \le z \le c^2 \end{cases} \tag{9}$$

In which a, b is respectively the long semi-axis and the short semi-axis of the elliptic boundary. c is the border in the vertical direction. The distance between the aircrafts can be expressed as $d = \sqrt{(x_x - x_2)^2 + (y_1 - y_2)^2}$.

Two aircrafts may appear at any point of the position error area, whose center is theoretical coordinate of aircrafts. The initial position of the aircraft A, B is respectively A_0, B_0. Take η as the iteration step length, and let $B_1 = B_0 + \eta$, $B_2 = B_1 + \eta$,……

$B_i = B_{i-1} + \eta \ldots\ldots$; Take η as another iteration step length, and let $A_1 = A_0 + \eta'$, $A_2 = A_1 + \eta', \ldots\ldots A_j = A_{j-1} + \eta' \ldots\ldots$, construct function ε satisfying:

$$\varepsilon = \begin{cases} 0 & d_{ij} > D \\ 1 & d_{ij} \leq D \end{cases} \tag{10}$$

When $d < D$, in which D is minimum safe distance, it is considered that conflict occurs. The probability of collision of two aircraft $P_{conflict}$ can be get from Eq. (11).

$$P_{conflict} = \sum_{i=0}^{m} \sum_{j=0}^{l} P_{plot1,i} \times P_{plot2,j} \times \varepsilon \tag{11}$$

3.2 Calculation of the Collision Risk

When $d < \lambda$, in which λ is the aircraft fuselage length, it is considered that collision occurs. The probability of collision of two machines:

$$P_{collison} = P\{d < \lambda\} \cdot P_{conflict} \tag{12}$$

According to formula (2), it can be derived:

$$P\{d < \lambda\} = \int_{-\lambda}^{\lambda} \frac{1}{\left(\sqrt{2\pi}\right)^3 |\Sigma|^{\frac{1}{2}}} \exp\left[-\frac{1}{2}(x-u)^T \sum^{-1} (x-u)\right] dx \tag{13}$$

Generally speaking, it is believed that one accident can be seen as two collisions. The collision risk CR of 10^7 flight hours can be expressed as Eq. (14).

$$CR = 2 \cdot NP \cdot P_{collision} \frac{\Delta t'}{10^7} \tag{14}$$

Here NP is aircraft sorties, sorties /h; $\Delta t'$ is the time that the two planes stay in the airspace.

3.3 Calculation of the Minimum Safe Distance

According to the collision risk model established above, we can obtain the probability of conflict and collision risk at a given safe distance. Then calculation of the minimum safe distance given a target safety level is the inverse problem of the model. But it is hardly solving the minimum safe distance directly from the function is almost impossible. However, the safe distance can be carried out by iterative calculations. Take length of the aircraft fuselage as the initial value of the iteration, set a certain step, and set the target safety level as termination condition. The algorithm is shown in Fig. 4, in which ξ is calculation accuracy, p_t is the target safety level, γ is the step length, and D_0 is the initial value of minimum safe distance.

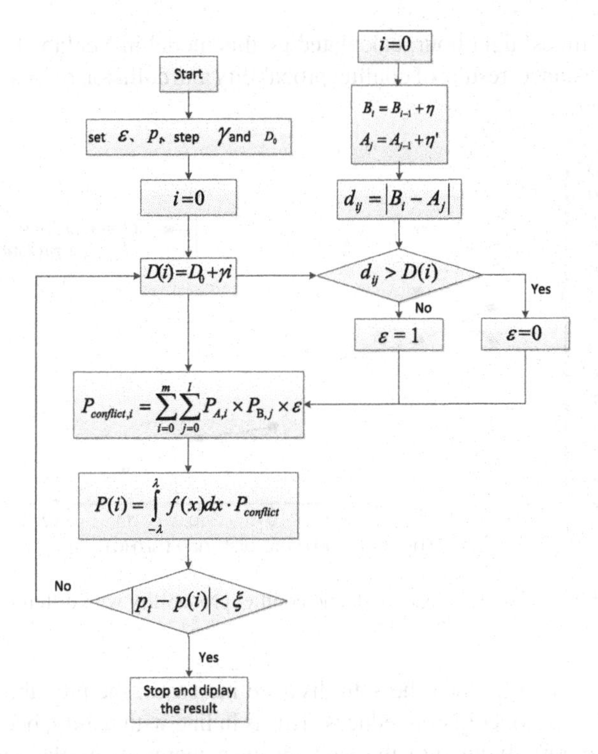

Fig. 4. Algorithm flow chart of minimum safe distance

At the time that the two aircrafts are closest, calculate the distance between two any point of the error position area by differentiation, and determine whether the current distance is less than the safe distance and calculate the conflict probability. Then calculate the collision risk by the collision risk model established above, and compare the result with p_t. If the former is larger, return to D_i, and plus a step with D_i. Stop iteration when the result of collision risk close to p_t and meet the accuracy requirements, and then displays the results.

4 Case Study

Because free flight is not implemented, reference to related data in non-free-flight, which refers to the way implemented currently. Assuming that in a specific airspace of free flight, at the initial time t = 0, the aircraft A and aircraft B enter the airspace, with the respective direction 0°, 150°(or angle between them is 150°), the speed 400 knots, 500 knots, span 40 meters, fuselage length 40 meters, and the relative position vector (−35 nm, −25 nm, 0). Take RNP6, RCP6, RSP6, communication delay 3 s, human factors operational delay 3 s. the variance of position error, which is caused by CNS, in the longitudinal direction, the lateral and vertical direction are respectively 20, 10, 10 [7]. The average height measured error of aircrafts is ±80ft [19]. If the minimum safe distance is 8 nmile, the conflict probability is 1.51×10^{-4}, and the collision risk is

4.25144×10^{-9} times/flight hour, calculated by this model in Matlab. Taking a certain minimum safe distance, results of conflict probability and collision risk can be calculated as the Fig. 5 shows.

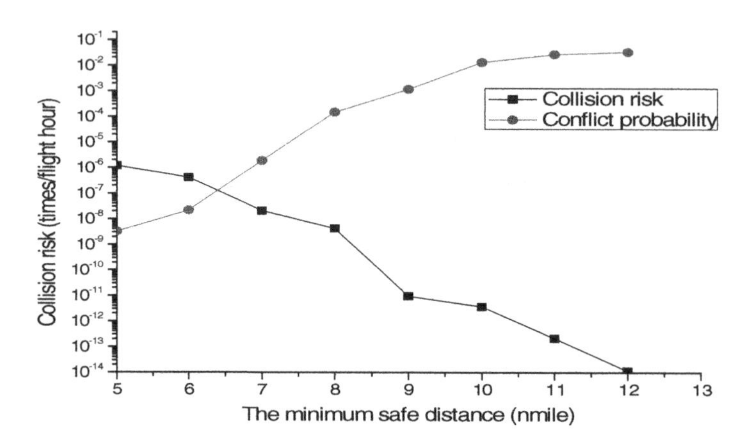

Fig. 5. Calculation results of collision risk and conflict probability with different minimum safe distance

As the Fig. 5 shows, when the safe distance increases, the probability of conflict increases, but the risk of collision reduces. This is in line with reality, because the probability of an aircraft flying into the safe distance increases as the safe increasing, however, the collision happens after aircraft entering the safe distance, due to the safe distance increases, so the risk of collision will reduce. Selecting an appropriate safe distance can not only guarantee full use of airspace, but also meet the security needs. Here, under the target level 1.5×10^{-8} of ICAO safety regulations, and according to the collision risk model built in this paper, it is reasonable to prescribe the safe distance 8nm in this situation and it is reasonable.

5 Conclusions

According to the idea of the position error model and the conflict area model, the collision risk model in free flight is built. The model incorporates the advantages of the conventional position error model and the conflict area model, and take into account the relationship between collision and conflict is considered, as well the distribution of the horizontal and vertical directions aircraft position error and the error correlation. Firstly, the position-error model is built; secondly, elliptic cylindrical collision slab is built; finally, the collision risk model is built. The model can be used to analyze the relationship between conflict probability, collision risk and safe distance, and also offers algorithm to calculate the minimum safe distance. Example results have proved the feasibility and rationality of the model, providing a theoretical basis for the safety assessment under free flight.

References

1. Zhang, Z., Wang, L., Li, D.: Flight Separation Safety Assessment Introduction. Science Press, Beijing (2009)
2. Zhang, Z., Zuo, J., Fei, L.: Study on free flight collision risk in free flight based on conflict. Int. Refereed J. Eng. Sci. **5**(2), 1–8 (2013)
3. Zhang, Z., Sun, C., Zhou, P.: Safety separation assessment in free flight based on conflict area. J. Comput. **7**(10), 2488–2495 (2012)
4. Peng, D., Li, Y.: Flying target track generation and conflict detection algorithm. China Saf. Sci. J. **23**(10), 110–113 (2013)
5. Zhaoning, Z., Gao, J.: Risk assessment model of stochastic differential equations in free flight considering the three -dimensional error distribution correlation. Sci. Technol. Eng. **13**(24), 7282–7286 (2013)
6. Zhang, Z., Zhou, P.: Collision risk model in free flight based on stochastic differential equations. J. Civil Aviat. Univ. China **30**(3), 1–5 (2012)
7. Zhang, Z., Wang, Y.: Study of collision risk in free flight based on fuzzy stochastic differential equations. China Safety Sci. J. **22**(10), 14–18 (2012)
8. Zhang, Z., Zuo, J.: Study of collision risk in free flight based on Brownian motion. China Safety Sci. J. **22**(8), 44–47 (2012)
9. Cai, M., Zhang, Z., Wang, L.: Study of collision risk in free flight environment. Aeronautical Comput. Tech. **41**(1), 51–56 (2011)
10. Ming, C., Zhaoning, Z., Lili, W.: Study of collision risk in free flight based on fuzzy fault tree analysis method. Aeronautical Comput. Tech. **3**(41), 22–26, 30 (2011)
11. Blom, H.A.P., Jaroslav Krystul, G.J., (Bert), B.: A particle system for safety verification of free flight in air traffic. In: Proceedings of the 45th IEEE Conference on Decision and Control. Manchester Grand Hyatt Hotel, San Diego, CA, USA (2006)
12. Everdij, M.H.C., Blom, H.A.P., et al.: Modelling lateral separation and separation for airborne separation assurance using petri nets. Simulation **83**(5), 401–414 (2007)
13. Roger, S., Rick, C., Rajeev, T., et al.: A reduced aircraft separation risk assessment model. In: AIAA Guidance, Navigation and Control Conference, New Orleans, pp. 1–16 (2003)

3D Planning and Design of Fire Disaster Evacuation Path in Buildings

Huixian Jiang[1,2](✉)

[1] Institute of Geography, Fujian Normal University, Fuzhou 350007, China
Jhxl55@163.com
[2] College of Geographical Sciences, Fujian Normal University, Fuzhou 350007, China

Abstract. This paper established a 3D evacuation path model aimed at the internal area of buildings through a case study on fire emergency in large-scale buildings. Besides, the optimum evacuation path was planned. By combining analytic hierarchy process with expert scoring method, factors of influencing path in planning were analyzed and corresponding weights were gained as coefficients of distance attribute in the path. Besides, the actual path was worked out via A* algorithm, so as to obtain the 3D evacuation path model. With very important theoretical significance for safe personnel evacuation when fire disasters are handled, this path model will help the government enhance its ability of disposing and preventing fire disasters and other emergencies.

Keywords: 3D evacuation path · Influence factor · Fire disaster · Large-scale building

1 Introduction

Owing to urban development and rapid growth of population, large-scale public buildings increase sharply, including large-scale shopping centers and high-rise office buildings. The structures of these building are complicated and involve high population density, so they are bound to trigger a series of hidden dangers. During emergencies (like fire disaster, earthquake and leakage of hazardous articles), safety of personnel in the building will be threatened inevitably [1]. As a result, the scheme of safe and efficient evacuation becomes especially important. According to the past fire accidents in buildings, mass death and casualty accidents are caused by improper personnel evacuation behaviors in most cases [2]. At present, China is still at a stage of qualitative analysis in fire safety design of buildings. However, developed countries like America and Japan stipulate that dynamic simulation must be carried out for personnel evacuation behaviors and smoke flow of fire in large-scale buildings. Therefore, it is very necessary to set up evacuation models suitable for the practical situations of China according to specific circumstances through dynamic simulation. Besides, safety performance of buildings in fire should be reasonably evaluated as per the models and methods of safe evacuation guidance and management must be proposed.

F. Bian and Y. Xie (Eds.): GRMSE 2015, CCIS 569, pp. 804–815, 2016.
DOI: 10.1007/978-3-662-49155-3_83

By directing at the increasingly severe difficulties of evacuation during fire emergencies in complex buildings, this paper constructed a model of inner spatial data in buildings aimed at emergency evacuation by taking fire emergency in large-scale buildings as an example [3, 4]. Meanwhile, the influence on the evacuation path was analyzed from five aspects: characteristics of personnel structure, products of fire disaster, building structure, and safe evacuation and management. In addition, environmental risk assessment model of escape path in large-scale public buildings was established, to describe the escape probability on each road in the escape network as well as the escape time consumed on the path during building fire. Moreover, in order to optimize the network of the shortest escape time based on escape probability, the optimal dynamic model of 3D escape path in the building was established. For interior 3D space in the building, the shortest path is planned to provide theoretical foundation for planning of crowd escape and evacuation path after fire disaster happens [5]. The achievement can efficiently provide data verification and decision aid supports for virtual practice and emergency rescue during sudden fire accidents in large-scale urban buildings. At the same time, China is able to perfect the emergency management system of unexpected public incidents and risk management system of catastrophic disasters by referring to this achievement.

2 Selection of Test Site and Collection and Processing of Relevant Data

By a case study of the typical example of China's real estate business – Wanda Plaza, this paper selected the main body of the urban complex, Cangshan Wanda Plaza as the test site. 3D data model of the building body was constructed, the planning algorithm of evacuation path was analyzed, and the simulation system of emergency evacuation process in complex buildings based on path model was preliminarily designed and realized.

Cangshan Wanda Plaza is an urban complex with a total area of about 650 000 m^2 developed by Wanda Group in Fuzhou City. It is located in Jinshan Plate of Cangshan District in Fuzhou City, next to Pushang Avenue in the south and adjacent to Jinzhou Road in the west; West Ring Road is 1 km away from it in the east. The land area is 126 300 m^2 and the total construction area is 644 000 m^2. In which the ground area is 490 100 m^2 and the underground area is 156 500 m^2. Completed in Dec. 2011, the entire project is composed of complex region (Zone A for short) in the south, service apartment region (residence) (Zone B for short) in the north, and SOHO office area (Zone C for short) in the east.

Acquisition of spatial data is the foundation of 3D modeling and visualization. Multiple methods including field measurement and remote-sensing image overlay were used to acquire data about the research area in this paper. The data mainly cover data of plane (2D vector data such as control polygon of building bottom), data of height (mainly height of the building), and data of digital elevation model (used for ground modeling). Besides, attribute data of ground features were needed during spatial query. The process of data acquisition is as follows: quantize the chart about internal structure of the building; determine the positions of function division areas, fire-fighting equipment and emergency exits in the building, and provide the actual photos about fire evacuation taken at the site. By combining remote-sensing imagery interpretation with ground survey, planar data of the building, linear data (inner passage and surrounding roads), and point data (important rescue organizations including fire brigade and hospital as well as refuge and evacuation points) were acquired. The following steps are mainly involved: ① Data about core areas (inside the building): Quantize the chart about the internal structure of the building; determine the positions of function division areas, fire-fighting equipment and emergency exits in the building, and provide relevant photos taken at the site; investigate the crowd situation and work and rest schedule. As for the establishment method, field investigation was conducted in the research area and field operation was carried out via measuring instruments like laser range finder. As a result, various attributes of spatial elements, such as distance and land area, were obtained. Laser range finder was used to collect the spatial distance between elements in Wanda Plaza, such as gate, store, elevator and emergency exit. Moreover, data of height were acquired. Vector map of the entire Wanda Plaza was drawn, involving three layers: point, line and plane. The layer of point includes positions of key points like store, evacuation exit and stairway. The layer of line covers abstract elements like major pedestrian path and emergency staircase in the market and office building; topological link relation was established between the layer of point and the layer of line. The layer of plane contains planar contour of all buildings in the entire plaza. ② Data in buffer area (within a certain scope outside the building): Establish a digital elevation model based on the quantized topographic map of 1:500, and acquire data about high-resolution images of satellite remote sensing based on aerial image or QUICKBIRD; acquire planar data of the building, linear data (roads), and point data (fire brigade, hospital, important organizations, and refuge and evacuation points) by combining remote-sensing imagery interpretation with ground survey, and provide relevant photos taken at the site; investigate data like distribution of permanent resident population.

3 Planning of 3D Evacuation Path in Buildings

The foundation for construction of 3D dynamic path in buildings is to establish a 3D network data set. By creating the 3D environment in buildings and calculating weights of various influence factors, the shortest 3D path was obtained and visualized. In path analysis, selection of weight factors directly affects the analysis result of optimal path. On the basis of fully considering inherent features of the building and the state of escaping personnel, path analysis was made according to spatial positions of the nodes (position of crowd and store). Attributes of the nodes are different (for instance, differences can be discovered in space, floor and distance to ancillary facilities), so different influence factors of the path were selected. At the same time, different weights were dynamically set according to the characteristics of population distribution in the building structure such as variations of quantity and behavior in geographical positions, the trend of fire propagation, and the specific time of fire occurrence. Realization process [6, 7]: ① Generate the topological graph of planar path in specific floors according to the plane graph of different floors, and the link up the path diagrams of adjacent floors at the vertical direction through stairway nodes; for lower floors, provide the path to window nodes according to the height of scaling ladder and the open degree outside the building. ② On the basis of fully considering inherent features of the building (such as floor height, floor and room distribution) and real-time trend of fire (such as fire behavior, temperature and smoke scope), determine the weights of different path factors. ③ Solve the optimal escape path. The path weight will change with variation of the real-time conditions, so this model is a dynamic path model meeting the practical situations, different from the previous static path models.

3.1 Determination of Factors

In the analysis of path planning, selection of weight factors directly affects the analysis result of optimal path. In this paper, influence factors of evacuation and escape were divided into two categories: static factor and dynamic factor. Static factor refers to the existing structure of the building layout as well as the property elements; dynamic factor means elements required by supervision or evacuation with change of the situations after the fire disaster happens. The values of the influence factors were determined by combining analytic hierarchy process with expert scoring method. Influence factors of the path in planning were analyzed to gain corresponding weights as coefficients of distance attribute in the path. Thus the actual path was worked out via the follow-up path planning algorithm. Analytic hierarchy process (AHP) is able to solve qualitative and quantitative problems well. Complex decision was made for the information according to few materials through pairwise comparison by setting up a clear hierarchical structure. Therefore, the judgment matrix was established layer by layer, and then the weight of judgment matrix was solved. Finally, comprehensive weight of the factors was calculated and ranking was conducted [8].

3.1.1 Static Factor

Static Factor as shown in Fig. 1.

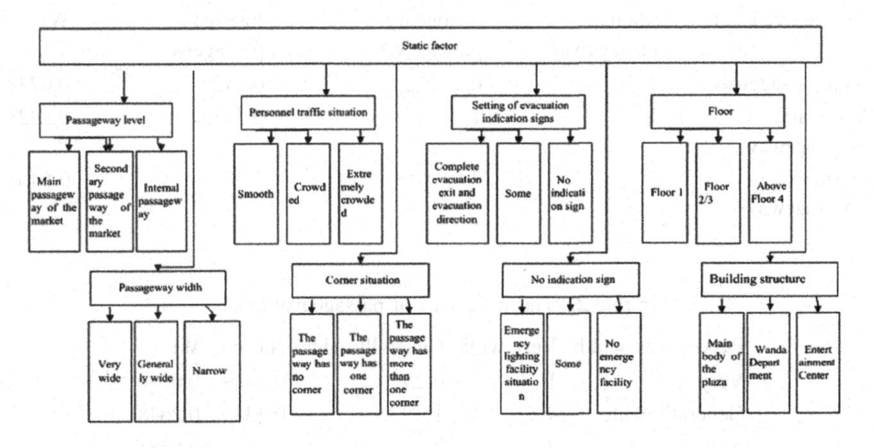

Fig. 1. Static factor of evacuation path planning

(1) Passageway level

This includes area of the store and height of the storey. According to the practice, when area of the store is large, the optimal path can be figured out easily due to the wide view and the evacuation time is short. Besides, a large space can provide a wide horizon, so the stampede rate is low. The weight wi of passageway level is gained via the judgment matrix, as shown in Table 1.

(2) Passageway width

The width of passageway is one of the major factors that affect evacuation, and it directly influences the evacuation time. Generally speaking, the width of passageway is in direct proportion to evacuation time. In another word, the longer the passageway is, the more time it will take. The width of passageway is also an important factor of affecting evacuation, especially under congestion. The larger the width of passageway is, the faster the evacuation speed will be. Effective width of the passageway, width after deducting boundary dimensions, was adopted in this project. The weight wi of passageway level was gained via the judgment matrix, as shown in Table 2.

(3) Setting of evacuation indication signs

The following contents are also factors of affecting path planning: whether distribution of evacuation indication facilities is reasonable; whether the sign is clear; whether indication signs are set in the main channel and corner; whether firefighting equipment is complete; whether the regional distribution is reasonable, including fire hydrant and auto-induction fire sprinkler. Planning of fire compartment, reasonable setting of evacuation facility and good condition of the facility also play an important role in evacuation path selection. Therefore, the parameter about setting and service situations of evacuation facilities should be considered in the model. The weight wi of passageway level was gained via the judgment matrix, as shown in Table 3.

Table 1. The weight wi of passageway level

Passageway level	Main passageway	Secondary passageway	Internal passageway	Wi
Main passageway	1	0.2	0.1429	0.0738
Secondary passageway	5	1	0.3333	0.2828
Internal passageway	7	3	1	0.6434

Table 2. The weight wi of passageway level

Passageway width	Very wide	Generally wide	Narrow	Wi
Very wide	1	0.25	0.1429	0.0796
Generally wide	4	1	0.3333	0.2648
Narrow	7	3	1	0.6555

(4) Building structure

With differences in the system layout, number of exits, type of exits, walkway width, and evacuation distance of the building structure, the evacuation requirements also vary. Different calculation methods of evacuation analog should be adopted for different evacuation paths. According to the functions in evacuation, the entire building structure can be divided into four types: room structure, lobby structure, stairway structure, and elevator structure. The weight wi of passageway level was gained via the judgment matrix, as shown in Table 4.

(5) Personnel traffic situation

The weight wi of passageway level was gained via the judgment matrix, as shown in Table 5.

Table 3. The weight wi of passageway level

Setting of evacuation indication signs	Complete evacuation exit	Some	No indication sign	Wi
Complete evacuation exit	1	0.2	0.1429	0.0738
Some	5	1	0.3333	0.2828
No indication sign	7	3	1	0.6434

Table 4. The weight wi of passageway level

Passageway attribute	Main body of the plaza	Wanda Department	Entertainment Center	Wi
Main body of the plaza	1	0.2	0.1429	0.0738
Wanda Department	5	1	0.3333	0.2828
Entertainment Center	7	3	1	0.6434

Table 5. The weight wi of passageway level

Personnel situation	Smooth	Crowded	Extremely crowded	Wi
Smooth	1	0.2	0.1429	0.0738
Crowded	5	1	0.3333	0.2828
Extremely crowded	7	3	1	0.6434

(6) Corner situation

The weight wi of passageway level was gained via the judgment matrix, as shown in Table 6.

Table 6. The weight wi of passageway level

Corner situation	No corner	One corner	More than one corner	Wi
no corner	1	0.2	0.1429	0.0738
one corner	5	1	0.3333	0.2828
more than one corner	7	3	1	0.6434

(7) The weight wi of passageway level was gained via the judgment matrix, as shown in Table 7.

Table 7. The weight wi of passageway level

Emergency lighting facility situation	Complete emergency lighting facility	Some	No emergency facility	Wi
Complete emergency lighting facility	1	0.2	0.1429	0.0738
Some	5	1	0.3333	0.2828
No emergency facility	7	3	1	0.6434

(8) Floor

The weight wi of passageway level was gained via the judgment matrix, as shown in Table 8.

Table 8. The weight wi of passageway level

Floor	Floor 1	Floor 2/3	Above Floor 4	Wi
Floor 1	1	0.2	0.1429	0.0738
Floor 2/3	5	1	0.3333	0.2828
Above Floor 4	7	3	1	0.6434

3.1.2 Dynamic Factor

Dynamic Factor as shown in Fig. 2.

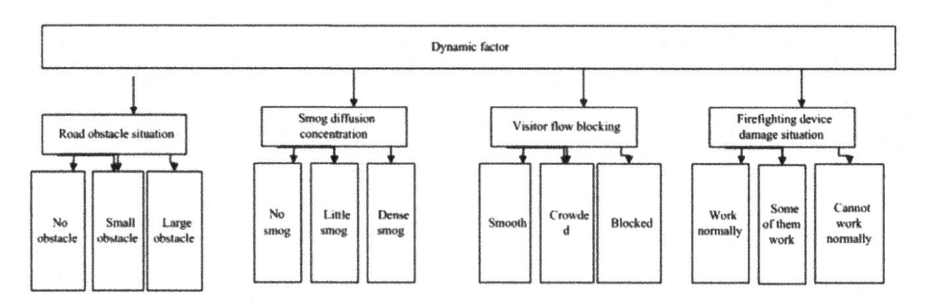

Fig. 2. Dynamic factor of evacuation path planning

(1) Road obstacle situation

The weight wi of passageway level was gained via the judgment matrix, as shown in Table 9.

(2) Smog diffusion (visibility)

The weight wi of passageway level was gained via the judgment matrix, as shown in Table 10.

(3) Visitor flow

Due to the difference of room functions in the building, the personnel density also varies. Related to number of people in the evacuation exit, personnel density decides the evacuation time and evacuation speed. Therefore, the factor of personnel density should be considered when evacuation model is established. The weight wi of passageway level was gained via the judgment matrix, as shown in Table 11.

(4) Firefighting device damage situation

The weight wi of passageway level was gained via the judgment matrix, as shown in Table 12.

Table 9. The weight wi of passageway level

Road obstacle situation	No obstacle	Small obstacl	Small obstacle	Wi
No obstacle	1	0.2	0.1429	0.0738
Small obstacle	5	1	0.3333	0.2828
Small obstacle	7	3	1	0.6434

Table 10. The weight wi of Smog diffusion

Smog diffusion concentration	No smog	Little smog	Dense smog	Wi
No smog	1	0.2	0.1429	0.0738
Little smog	5	1	0.3333	0.2828
Dense smog	7	3	1	0.6434

Table 11. The weight wi of Visitor flow

Visitor flow blocking	Smooth	Crowded	Blocked	Wi
Smooth	1	0.2	0.1429	0.0738
Crowded	5	1	0.3333	0.2828
Blocked	7	3	1	0.6434

Table 12. The weight wi of Firefighting device damage situation

Firefighting device damage situation	Work normally	Some of them work	Cannot work normally	Wi
Work normally	1	0.2	0.1429	0.0738
Some of them work	5	1	0.3333	0.2828
Cannot work normally	7	3	1	0.6434

3.2 Construction of 3D Dynamic Path in Buildings

The acquired 3D data about the inner space of the building were abstracted, and analysis was made for the topological space relation, sequence space relation and metric space relation among spatial entities distributing on various floor planes. Based on this, database model of the inner space was established via digital information simulation. Planar path diagram of relevant floors was generated for plane graphs of various floors through the algorithm of image thinning and image feature point extraction based on mathematical morphology according to the established database model of the inner space. Then path diagrams of the adjacent floor were linked up at the vertical direction through stairway nodes, to construct a complete 3D path diagram inside the building. In terms of lower floors, path to window nodes should be provided according to the height of scaling ladder and the open degree outside the building. Inner space of the building involves point, line, plane, surface and body. Doors, windows, stairways, emergency exits and ancillary facilities can be abstracted into nodes; corridors, passageways and pipeline facilities can be abstracted into lines. These spatial entities distributed on various floors have topological space relation (adjacency, containing and intersection), sequence space relation (ordering) and metric space relation (distance). The database model of the inner space established via simulation for digital information by analyzing these space relations can objectively and standardly express the background state of the 3D space structure in buildings. Besides, it can further bring about an immersive visual experience to users with the help of 3D modeling technology.

3.3 Analysis on the Planning Algorithm of Evacuation Path

By summarizing various influence factors, planning of the evacuation path is decided by static factor (S) and dynamic factor (D) together. Static factor is determined by eight factors including passageway level, passageway width, situation of personnel traffic, corner situation, setting of evacuation indication signs, setting of emergency lighting, floor, and passageway attribute (as shown in Table 13); dynamic factor is determined by four factors including situation of road obstacle, smog diffusion (visibility), visitor flow blocking, and situation of firefighting device damage (as shown in Table 13). Selection of after-fire evacuation path in complicated buildings is a function constituted by the above factors: R = f(S,D). The core of emergency response application model is to select the optimal path. By combining with previous algorithms, this paper adopted weighted A* algorithm in which several weights were added. The algorithm not only worked out the shortest path from all stores in the market to various evacuation exits but also dynamically added and avoided the obstacles. Finally, the statistical result of evacuation path was output.

Step 1: Enter the evacuation network constructed G(E,V), the position of the fire O, the evacuation path P (the optimal path from the classroom S_i to the evacuation position D_j), the set of the classroom S_i (i = 0,1,2,3……n), the number of the students in each classroom F_i, the set of the evacuation position D_j $(j = 0, 1, 2, 3\ldots\ldots n)$, the total population evacuated $F = \Phi$, the flow statistics in the

Table 13. The weight wi

Factor	Wi
Passageway width	0.3252
Personnel traffic situation	0.2262
Passageway level	0.156
Passageway attribute	0.1072
Corner situation	0.0733
Floor	0.04
Emergency lighting facility situation	0.0378
Setting of evacuation indication signs	0.0343
Visitor flow blocking	0.5902
Road obstacle situation	0.2583
Smog diffusion concentration	0.1023
Firefighting device damage situation	0.0492

evacuation position $FD_j = \Phi$, the set of the evacuation cost on each path (it can be the length of the path, the time consumed, the comprehensive weight, etc.) $C_{ij} = \Phi$, and the total cost during the evacuation process $BC = \Phi$;
Step 2: Calculate the path C_{kj} with the lowest consumption from S_k to various exits D_j via A* algorithm, and let the calculation result $T = \min\{Ckj \mid k \in i\}$; $BC = BC \cup T$. At the same time, work out the estimated visitor flow N_k in store S_k; $F = F \cup F_k$, $FD_j = F \cup FD_j$.
Step 3: Calculate the round buffer area centering on the disaster point and add this buffer area into the layer of network analysis as evacuation obstacle.
Step 4: Obtain the store node S_i from top floors to lower floors successively and calculate the path C_{ij} with the lowest consumption from S_i $(i \neq k)$ to various exits D_j via a* algorithm. In the calculation process, if obstacles are encountered, the barrier should be bypassed. Let the calculation result $T = \min\{C_{ij} \mid i \neq k\}$. Work out the number of people N_i in store $F = F \cup F_i$ and $FD_j = F \cup FD_j$.
Step 5: Repeat Step 4 till people in all stores are evacuated.
Step 6: Output the total consumption of evacuation BC, number of people that have been evacuated F, and accepted number of people in evacuation FD_j.

3.4 Example of Simulation Application

Analysis was made for various static and dynamic influence factors of influencing evacuation and escape from the building after emergencies like fire disaster happen in a complex building. Besides, corresponding calculation was carried out. Guidance for the path choice was offered to people in different positions of the building, and decision analysis was made for the process of emergency evacuation in the entire building and even surrounding areas. Besides, the result was visually displayed at corresponding terminals. The whole path model included F1-6 of the commercial region in Wanda Plaza as well as F7-22 of the office building. The plane structures of F1-6 are

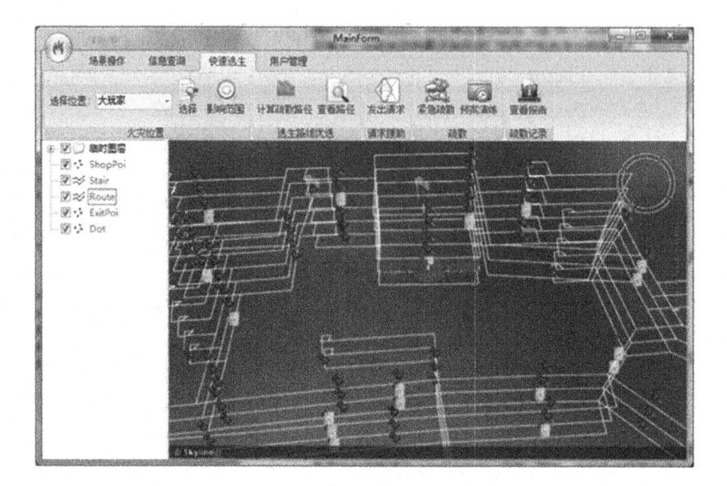

Fig. 3. Presents the three-dimensional path model of the entire region

independent and superposed to some extent; all floors of the office building have the same structure. Figure 3 3D path model of the entire region.

4 Conclusion

Sudden fire accident has already become a potential menace of urban public security in China, and emergency rescue in large-scale buildings and outdoor evacuation are big problems in emergency management of urban disaster prevention and reduction. By directing at the increasingly severe difficulties of evacuation during fire emergencies in complicated buildings, this study makes it possible to evacuate and rescue people orderly and effectively under simple operation by digitizing and modeling the issue of disaster emergency response. Selection of the optimal path aimed at emergency evacuation in buildings is an issue of planning the shortest path based on 3D topological structure of inner building space. However, the actual situations are complex and variable, so the factors to be considered are far more complicated when compared with the classic shortest path algorithm. From the angle of characteristics and specific application of the algorithm, it has important significance to study the algorithm of planning the shortest path in buildings. However, personnel evacuation is quite a complicated problem, and people with different psychological qualities, educational degrees and living habits will select different evacuation paths. In addition, the system just blindly designs the path of the optimal weight at present for the points representing the stores, without considering visitor flow restriction at the exit. However in practical situation, congestion might happen in some exits, while other exits can be unoccupied. In this way, resource waste will be caused and meanwhile stampede can also happen. Therefore, more attention will be paid to factors like human behaviors and mental activities in the follow-up studies. Meanwhile, the evacuation algorithm will be improved by adding crowded points into the sub-layer of point obstacle in path analysis as obstacle points.

Acknowledgments. This research is supported by Socially useful program of Fujian Province (No. 2015R1101029-6); Colleges and universities in Fujian Province Research on the teaching reform (No. JAS14677).

References

1. Luis, A.-A., Tralhao, L., Santos, L.: A multi-objective approach to locate emergency shelters and identify evacuation routes in urban areas. Geogr. Anal. **49**, 9–29 (2009)
2. Xianzhi, L., Tingshen, Z., Jingjin, T.: Study on decision support system of emergency rescue based on GIS. In: International Symposium on Emergency Management 2009 (ISEM 2009), pp. 535–538, 22–24 December 2009
3. Xiao, X., Daming, Z., Rong, X.: Based on 3D GIS campus fire emergency rescue decision support system research. Henan Sci. **19**(1), 88–91 (2011)
4. Jiaqi, D.: Research on Disaster Response Education Mechanisms of College Student. East China Normal University Master's thesis (2009)
5. Yang, Y.: The Research and Design on Functional Adaptation of Campus as a Shelter for Earthquake. Southwest Jiaotong University Master's thesis (2007)
6. Bo, W., Yuan-xiang, L., Xing, X., et al.: Particle swarm optimization algorithm based on stable strategy. Comput. Sci. **38**, 221–223 (2011)
7. Blum, A.L., Furst, M.L.: Fast planning through planning graphic analysis. Artif. Intell. (1997)
8. Lozano-Perez, T., Wesley, M.: An algorithm for planning collision—free paths among polyhedral obstacles. Commun. ACM **22**(5), 436–450 (1979)

Adaptability Analysis of Cosmic-Ray Neutron Method to Monitoring Soil Moisture in Desert Steppe

Zhiguo Pang[1,2(✉)], Jingya Cai[1,2], Jun'e Fu[1,2], Wenlong Song[1,2], and Yizhu Lu[1,2]

[1] China Institute of Water Resources and Hydropower Research, Fuxing Road. A-1, Beijing 100038, China
pangzg@iwhr.com, jingyacai@126.com

[2] Research Center on Flood and Drought Reduction of the Ministry of Water Resources, Fuxing Road. A-1, Beijing 100038, China

Abstract. In order to research the adaptability of cosmic-ray neutron method in soil moisture measurement and serve the management and decision of animal husbandry, continuous monitoring was conducted by the Cosmic-Ray Sensing probe (CRS) in desert steppe. By comparing with measuring results of the Time Domain Reflectometry (TDR), the consistency of soil moisture measured by CRS and TDR and the sensitivity in response of soil moisture with both methods to rainfall were researched. Results show that a good consistency of soil moisture measured by CRS and TDR is achieved and that TDR isn't sensitive to rainfall, while CRS can not only respond quickly to rainfall but also reveal clearly corresponding changes of soil moisture led by different rainfalls. It's concluded that cosmic-ray neutron method can measure soil moisture and reflect its dynamic change accurately in desert steppe and provide decision basis for the modern animal husbandry management.

Keywords: Soil moisture · CRS · TDR · Rainfall

1 Introduction

The common methods to measure soil moisture are divided into two types consisting of point scale measurements and area scale measurements. Point scale measurements include oven drying method, dielectric property method (such as Time Domain Reflectometry (TDR) and Standing-Wave Ratio), neutron method, tensiometer method and so on. Precision of point scale measurements is generally high. Among them, oven drying method, the most accurate standard method, is often applied to detect and correct the measuring results of other methods. But point scale measurements are only suitable for measurement of small scale soil moisture with poor spatial representative. Area scale measurement is mainly to use remote sensing to retrieve soil moisture at present. Remote sensing makes it possible to measure large scale soil moisture with the advantages of multisource, multi-temporal and multi-polarization. But because of the influence of various factors such as the texture

F. Bian and Y. Xie (Eds.): GRMSE 2015, CCIS 569, pp. 816–824, 2016.
DOI: 10.1007/978-3-662-49155-3_84

and bulk density of soil, vegetation cover and land slope, the accuracy and validation of remote sensing products are all along one part of problem for remote sensing to retrieve soil moisture quantitatively.

In 1940s, there has been a new approach, the cosmic-ray neutron method, to measure soil moisture. This method calculates the average soil moisture of a certain area by measuring the background fast neutrons emitted naturally from soil that are created by secondary Cosmic-Ray fluxes coming in through the atmosphere. Its measurement range is between those of point scale measurements and remote sensing approaches filling the gap of current methodologies. This approach can not only measure soil moisture continuously and non-invasively but also provide a means of effective validation for remote sensing approaches in pixel scale. So it's widely applied to fields like soil moisture measurement, drought monitoring, agricultural irrigation and slope stability analysis. Chrisman *et al.* [1] maps soil moisture over large areas using cosmic-ray rover to calibrate and validate satellite soil moisture data products. Almeida *et al.* [2] uses cosmic-ray neutron counts, capacitance probes network measuring soil moisture and a multiple adaptive neutron-fuzzy inference system to obtain reliable estimates of soil moisture in the top portion of soil during wet periods. Baroni *et al.* [3] proposes a soil moisture scaling approach to estimate directly the correct soil moisture from cosmic-ray neutron counts avoiding the need to introduce one correction for each hydrogen contribution and to estimate indirectly all the related time-varying hydrogen pools. Xujun Han [4], Zhongli Zhu [5] apply this approach at heterogeneous farmland of the Heihe Watershed and prove that cosmic-ray neutron method is capable of measuring the real soil moisture at the intermediate spatial scaling. Because fast neutron counts are influenced by many factors such as atmospheric pressure and water vapor, above- and below-ground biomass, litter layer, intercepted water in the canopy, soil organic matter and lattice water of the soil minerals [3] and researches about most of factors is not yet mature, the accuracy measurement of neutron counts is interfered to some extent and the rapid promotion of cosmic-ray neutron method is hindered.

Soil moisture plays a decisive function during the grassland degradation and its reverse process. The Cosmic-Ray Sensing probe (CRS) is used to monitor soil moisture nearly four months continuously choosing typical desert steppe. By correcting the main influence factors (atmospheric pressure and water vapor) of the research area, the consistency of soil moisture measured by CRS and TDR is analyzed and the sensitivity in response of soil moisture with both methods to rainfall is compared to research its adaptability to desert steppe and provide technical support for management of steppe animal husbandry.

2 Study Area

The study area is situated in the experimental base of Institute Of Water Resources for Pastoral Area, MWR, belonging to Xilamuren Town, Baotou City, Inner Mongolia Autonomous Region. It covers an area of 150 ha with a center geographic coordinate of 41°22′N, 111°12′E within Wulanchabu desert steppe region. Its geomorphology is low and rolling hill in the north foot of Yinshan Mountain. The terrain of north is higher than that of south and east and west higher than middle with a maximum altitude of 1690.3 m and minimum 1585.0 m. It has temperate semi-arid continental monsoon climate with 284 mm precipitation, 2305 mm evaporation, 2.5 °C temperature and 4.5 m/s wind speed (all are annual average). Vegetation is distributed homogenously. The constructive species is Stipakrylovii and the dominant species is Leymuschinensis. The study area is not affected by livestock foraging and human activity and land surface conditions are relatively uniform, providing favorable environment conditions for adaptability research of cosmic-ray neutron method to measure soil moisture in desert steppe.

3 Data and Methodology

3.1 Measurement Range

The lateral extent of the measurement is the region within which 86 % of the horizontally counted neutrons originate [6]. It's a circular region with CRS as the center. It's not influenced by soil moisture but inversely proportional to atmospheric pressure [7]. The relationship [8] of radius and pressure is:

$$R = R_0 * \frac{P_0}{P}. \tag{1}$$

where R_0 is the radius with referenced pressure P_0 and is 300 m [9] under the standard pressure (1013.25 hpa); R is the radius with actual pressure P. The actual measurement radius of study area is about 360 m according to annual average air pressure (840 hpa) acquired from automatic weather station data.

The depth of measurement is the region within which 86 % of the vertically counted neutrons originate [6]. The whole measurement range is a cylinder from surface to a certain depth (12~70 cm). The effective depth, related to soil moisture, is commonly calculated as what Franz *et al.* [10, 11] proposes:

$$Z = \frac{5.8}{\frac{\rho_b}{\rho_w}\tau + \theta_m + 0.0829}. \tag{2}$$

where Z is effective depth of the soil layer (cm); ρ_b is the bulk density of soil (g/m^3); ρ_w is density of liquid water with a default of 1 g/cm^3, τ is the weight fraction of lattice water in the mineral grains and bound wate 0~0.05 [8], ignored in this letter), θ_m is the weight soil moisture (kg/kg). During the measurement period, there are 25 points (Fig. 1, point) to sample several times in horizontal measurement range of CRS. All the soil samples are dried and weighted to calculate the average soil moisture (the arithmetic mean) and the bulk density of soil in the range. The actual measurement depth of the study area is calculated as 26 cm approximately.

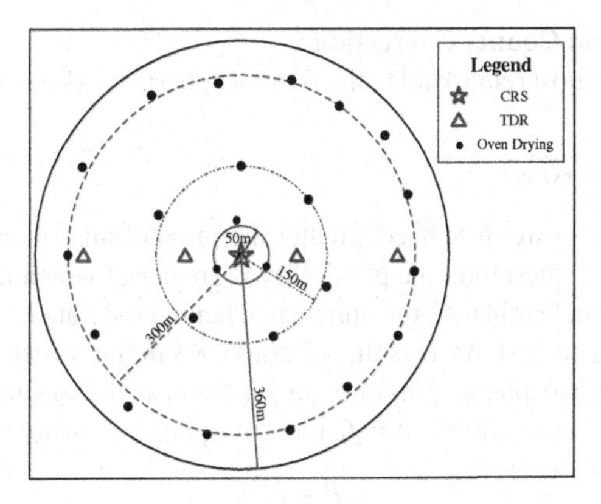

Fig. 1. Positions of CRS, TDR and soil samples

3.2 Data Acquisition

During 01/08/2014–19/11/2014, one CRS (CRS1000B) is implemented to monitoring soil moisture in study area continuously. The frequency of data collection is set as 1 time/hour to keep consistent with that of TDR.

Four TDR sample points (Fig. 1, triangle) are distributed within the measurement range of CRS. The soil is divided as 3 layers according to depth, i.e., 0~10 cm, 10~20 cm, 20~30 cm. There is one probe in each layer respectively. Likewise, the arithmetic mean is made as the average soil moisture in the range measured by TDR. The result is converted to weight soil moisture to keep consistent with that of CRS according to formula (3):

$$\theta_m = \frac{\theta_v}{\rho_b}. \quad (3)$$

where θ_v is volume soil moisture (cm^3/cm^3).

At the same time, rainfall sensor monitors synchronously in the automatic weather station to compare the sensitivity in response of soil moisture with CRS and TDR to rainfall.

3.3 Data Processing

3.3.1 Data Screening

CRS data should be screened before retrieving soil moisture to eliminate data acquired when CRS is under abnormal operation and ensure the quality of data. Under abnormal operation, counting data are with an interval of 1 h, counts from adjoining times differ by less than 20 %, the relative humidity is less than 80 % inside the probe box and the battery voltage is greater than 11.8 V [7].

3.3.2 Fast Neutron Counts Correction

The factors with the most remarkable effects, atmospheric pressure and water vapor, are analyzed.

(1) *Atmospheric Pressure*

Higher the air pressure in a place, greater the mass of air column above and more atmospheric particles, therefore, the possibility is greater of which fast neutrons generated colliding with nuclear in the atmosphere and being moderated, leading to reduction of neutrons reaching ground. As a result, neutrons CRS monitors decrease with the same soil moisture. So atmospheric pressure may influence the accuracy of cosmic-ray neutron method. The correction factor f_P for atmospheric pressure is defined as

$$f_P = \exp\frac{P - \mathrm{P}_0}{\mathrm{L}}. \tag{4}$$

where P_0 is the average air pressure during the measurement period (mb); L is the mass attenuation length for high-energy neutrons that varies progressively between ~ 128 g/cm^2 at high latitudes and 142 g/cm^2 at the equator (g/cm^2) [7].

(2) *Atmospheric Water Vapor*

Rosolem *et al.* [12] verify by MCNPX model that the influence of water vapor on neutron counts near the ground may be as much as 12 %, relting in an error of 0.1 m3/m3 in soil moisture. A correction factor CWV for water vapor is proposed as

$$\mathrm{CWV} = 1 + 0.0054 * (\rho - \rho_0) \tag{5}$$

where ρ is the actual density of water vapor at the measurement time (g/m^3); ρ_0 is the referenced density of water vapor (usually chosen as 0).

Above all, the neutron counts after correction for atmospheric pressure and water vapor is

$$\mathrm{N} = \mathrm{N}_{\mathrm{raw}} * \mathrm{f}_{\mathrm{P}} * \mathrm{CWV}. \tag{6}$$

where N is the corrected neutron counts; $\mathrm{N}_{\mathrm{raw}}$ is the neutron counts monitored by CRS.

3.3.3 Soil Moisture Calculation

Desilets *et al.* [13] establish the relationship between corrected fast neutron counts and soil moisture using MCNPX model:

$$\theta_{\mathrm{m}}(\mathrm{N}) = \frac{\alpha_0}{{}^{\mathrm{N}}/_{\mathrm{N}_0} - \alpha_1} - \alpha_2. \tag{7}$$

where $\theta_m(N)$ is the weight soil moisture corresponding to neutron counts N; $\alpha_0, \alpha_1, \alpha_2$ are coefficients and $\alpha_0 = 0.0808$, $\alpha_1 = 0.372$, $\alpha_2 = 0.115$ under the condition of soil moisture more than 0.02 (it's usually met); N_0 represents the neutron counts when no water is in soil in the study area and is calculated by formula (7) with the mean of corrected neutron counts and the average soil moisture with oven drying method.

4 Comparison and Analysis of CRS and TDR Results

4.1 Consistency

The hourly soil moisture time series using CRS and TDR are showed in Fig. 2. As a whole, soil moisture acquired with both CRS and TDR are relatively consistent. Soil moisture during the trial period using CRS distribute within the range of 0.05~0.25 kg/kg and the average is 0.11 kg/kg. Those of TDR are within the range of 0.06~0.16 kg/kg and the average is 0.10 kg/kg. Except that results of both methods are of large difference during 12~19 in September (the maximum is over 0.05 kg/kg and the minimum is more than 0.02 kg/kg), they are well consistent. The

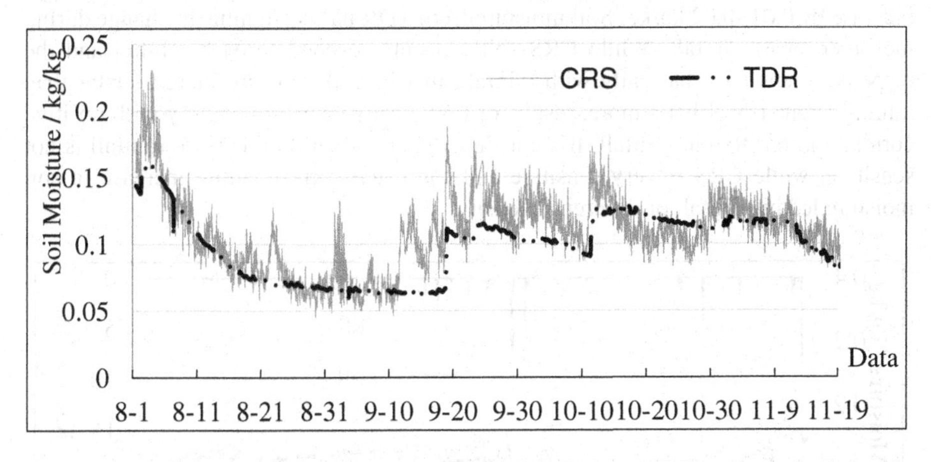

Fig. 2. Comparison of hourly soil moisture measured by CRS and TDR respectively during the trial period

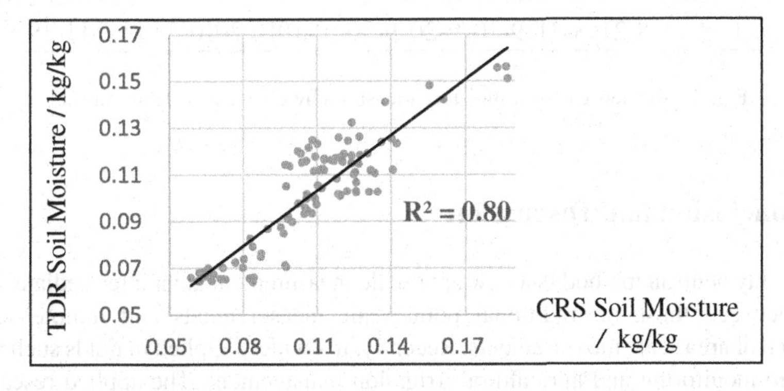

Fig. 3. Comparison of average daily soil moisture measured by CRS and TDR in the trial period except for 12 ~ 19 in September

results with an absolute errors less than 0.01 kg/kg and a relative error less than 10 % account both for 55.45 % and the results with a relative error more than 20 % is only 11.88 %. The coefficient of determination R2 is as much as 0.80 (Fig. 3) and the root-mean-square error RMSE is 0.0135 kg/kg. So the consistency of long-term soil moisture results using CRS and TDR is quite high and soil moisture retrieved with CRS is able to satisfy the requirement of production practice.

4.2 Sensitivity of Response to Rainfall

The sensitivity in response of soil moisture with both methods to rainfall is compared with the data from rainfall sensor of automatic weather station (Fig. 4). The result shows that among the rainfall more than thirty times during rhe trail period, TDR responds only to the heaviest three rainfalls, including two with the hourly maximum rainfall being 8.03 mm and 6.13 mm in 2 August and 18 September respectively and one with the hourly rainfall of continuous 3 h being all about 1 mm in 18 September, and soil moisture increase by 0.01~0.02 kg/kg. Soil moisture from TDR has no significant change during the other small rainfall. While CRS soil moisture respond to each rainfall and the response keeps pace basically with rainfall, that is, soil moisture increase later than rainfall about 1 h. What's more, results of CRS change correspondingly with positive correlation to different rainfall. It's concluded that response of TDR to rainfall is not sensitive, while CRS is very sensitive and can catch corresponding increase in soil moisture led by rainfall of different intensity.

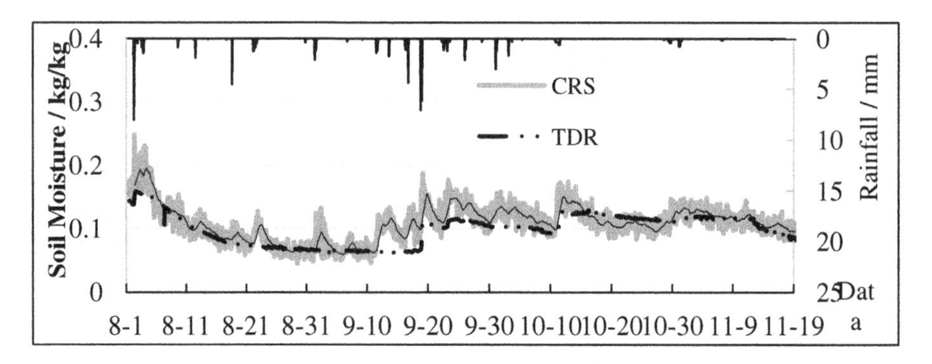

Fig. 4. Response of soil moisture measured by CRS and TDR to rainfall

5 Conclusion and Discussion

Cosmic-ray neutron method is a new area scale measurement at an intermediate spatial scale between those of traditional point scale measurements and remote sensing approach of area scale measurements. Recently, it's widely applied in fields such as soil moisture monitoring and agricultural irrigation management. The applied research is carried out for CRS to measure soil moisture in typical desert steppe. The result shows that the consistency of soil moisture with CRS corrected for atmospheric pressure and water vapor and TDR is relatively high. R2 is 0.80 and RMSE is only 0.0135 kg/kg. For the sensitivity of response to rainfall, TDR makes response only to soil moisture increase led by heavy rainfall and is not sensitive for that of small rainfall event. While CRS is very sensitive in response to rainfall and measuring results are able to reveal clearly soil moisture increase resulting from different rainfall. It can be concluded that cosmic-ray neutron method measures dynamic change of soil moisture precisely in desert steppe and provides basis for management and decision of animal husbandry. This research considers only two remarkable influence factors, i.e., atmospheric pressure and water vapor and ignore the other insignificant factors according to actual situation of study area. For regions influenced greatly by other factors, the correction of influence factors will be undoubtedly emphasis and difficulty of cosmic-ray neutron method to measure soil moisture. Further research can promote further generalization for this approach.

Acknowledgments. This paper was funded by the "948" plan of Ministry of Water Resources, P.R. China (No. 201401).

References

1. Chrisman, B., Zreda, M.: Quantifying mesoscale soil moisture with the cosmic-ray rover. J. Hydrol. Earth Syst. Sci. **17**, 5097–5108 (2013)
2. Almeida, A.C., Dutta, R., Franz, T.E.: Combining cosmic-ray neutron and capacitance sensors and fuzzy inference to spatially quantify soil moisture distribution. IEEE Sens. J. **14**(10), 3465–3472 (2014)
3. Baroni, G., Oswald, S.E.: A scaling approach for the assessment of biomass changes and rainfall interception using cosmic-ray neutron sensing. J. Hydrol. (2015). doi:http://dx.doi.org/10.1016/j.jhydrol.2015.03.053
4. Han, X.J., Jin, R., Li, X., et al.: Soil moisture estimation using cosmic-ray soil moisture sensing at heterogeneous farmland. J IEEE Geosci. Remote Sens. Lett. **11**(9), 1659–1663 (2014)
5. Zhu, Z.L., Tan, L., Gao, S.G., et al.: Observation on soil moisture of irrigation cropland by cosmic-ray probe. J IEEE Geosci. Remote Sens. Lett. **12**(3), 472–476 (2015)
6. Zreda, M., Desilets, D., Ferré, T.P.A., et al.: Measuring soil moisture content non-invasively at intermediate spatial scale using cosmic-ray neutrons. J. Geophys. Res. Lett. **35**, L21402 (2008)
7. Zreda, M., Shuttleworth, W.J., Zeng, X.: COSMOS: The COsmic-ray Soil Moisture Observing System. J. Hydrol. Earth Syst. Sci. Discuss. **9**, 4505–4551 (2012)
8. Jiao, Q.S., Zhu, Z.L., Liu, S.M., et al.: Research and application of cosmic-ray neutron method to measuring soil moisture at farmland. J. Adv. Earth Sci. **28**(10), 1136–1143 (2013). 焦其顺, 朱忠礼, 刘绍民, 等. 宇宙射线中子法在农田土壤水分测量中的研究与应用[J].地球科学进展, **28**(10), 1136–1143 (2013)
9. Desilets, D., Zreda, M.: Footprint diameter for a cosmic-ray soil moisture probe: theory and monte carlo simulations. J. Water Resour. Res. (2013). doi:10.1029/wrcr.20187
10. Franz, T.E., Zreda, M., Ferré, T.P.A., et al.: Measurement depth of the cosmic-ray soil moisture probe affected by hydrogen from various sources. J. Water Resour. Res. **48**, W08515 (2012)

11. Franz, T.E., Zreda, M., Rosolem, R.: Field validation of a cosmic-ray neutron sensor using a distributed sensor network. J. Vadose Zone (2012). doi:10.2136/vzj2012.0046
12. Rosolem, R., Shuttleworth, W.J., Zreda, M., et al.: The effect of atmospheric water vapor on neutron count in the cosmic-ray soil moisture observing system. J. Hydrometeorology **14**, 1659–1671 (2013)
13. Desilets, D., Zreda, M., Ferré, T.P.A.: Nature's neutron probe: land surface hydrology at an elusive scale with cosmic rays. J. Water Resour. Res. **46**, W11505 (2010)

Research on Land Use-Cover Change of Wuhan City Based on Remote Sensing and GIS

Fan Bai(✉)

Laboratory for Information Engineering in Surveying, Mapping and Remote Sensing, Wuhan University, Wuhan, Hubei, China
fateatfish@hotmail.com

Abstract. Since the 1990s, the research of Land use - cover change (LUCC) has become an important field of global environment change. This problem becomes especially important in China because of its huge population and relatively small area. During the process of urbanization, land-use changed dramatically in suburban area of large cities in the central area of China such as Wuhan. With the classification interpretation of the TM/ETM remote sensing images of Wuhan city in 2000 and 2007, this paper comprehensively analyzes land use dynamic characteristics of Wuhan city during the 8 years by using GIS technology. This study reveals that during the 8 years, the arable land declined fast and the confliction between the human beings and land resources becomes more severely, as the continued expansion of urban construction land, part of local section is not highly developed and it also has an impact on the ecological environment.

Keywords: Land use-cover change (LUCC) · Remote sensing · GIS · Wuhan city

1 Introduction

Land resources, as one of the most important natural resources, are the foundation of where people can have social and economic activities and rehabilitate, and are also the material basis for Human survival and sustainable development.

Since the twentieth century, with the rapid population growth and the development of economic, the demand of the human being for a variety of production increased dramatically which significantly deepened the dependence on land-use in depth and width. Meanwhile, industrialization and urbanization had a profound impact on land-use, too. The development of human civilization and technology made the human way of land-use diversity, complexity and technicality. This advancement improved the scope and extent of land available constantly, but at the same time, some irrational behavior such as over-exploitation and land reclamation around the lake caused serious and negative effects of land itself. Arable land declined dramatically; large numbers of agricultural land turned into construction land. The confliction between the human being and land resources increases prominently. Meanwhile, as the human living condition is continuously demanded for improvement, the pressure on earth resources and environment are increasing dramatically. This leads to many problems of resource, ecological

F. Bian and Y. Xie (Eds.): GRMSE 2015, CCIS 569, pp. 825–833, 2016.
DOI: 10.1007/978-3-662-49155-3_85

and environmental, such as landslides, dust storms, desertification, soil erosion, acid rain, greenhouse effect, which are all closely related to land-use.

2 Review of Relevant Literature

The Land use-cover change (LUCC) is an important component of the global environmental change. Since the 1990s, LUCC has become the main area of global environmental change researches, and is also the focus of resource science, geography, remote sensing and many other subjects.

In 1991, a special committee was formed by International Geosphere-Biosphere Programme (IGBP) and International Human Dimensions Programme (IHDP) to investigate the feasibility of natural and social scientists joining together to study LUCC, which was one of the problems of global change. Based on the conclusions of the special committee, IGBP and IHDP decided to establish a joint "Core Project Planning Committee"(CPPC), and published a report entitled "Joint Land-use and Global Land-use-Cover Change". The report highlighted a series of tasks to guide the research for three main focuses to meet the long-term goal of LUCC ultimately. In 1995, the two organizations jointly issued a "Land Use-Cover Change Research Plan" [1] as one of the core projects, making the LUCC research focus and cutting-edge issue of natural and social sciences. "Land Use-Cover Change Implementation Strategy" published in 1999 further proposed that the process of land-use, land use-cover change of the human response, integrated global and regional model as the subject of study. The article pointed out that the two prominent features of land use- cover change are comprehensive and regional, and it stressed the research should be linked to sustainable development problems such as regional land degradation, water and poverty issues. It called on scientists in each area (human ecologists, land economists, remote sensing/GIS experts, LUCC model experts and comprehensive evaluation focused experts) to strengthen the cooperation [1, 2]. With the deepening of global change research, since 1993, many international organizations and countries which were deeply concerned about global environmental change have also started their own LICC research project. For example, in 1994, United Nations Environment Programme (UNEP) started "Evaluation and simulation of land-cover" project; in 1995, International Institute for Applied Systems Analysis started "European and North land-use- cover change manipulation (LACM)" research; "Land-use for Environmental Conservation Research " started by Center for Global Environmental Research of Japan's National Academy of Sciences. IGBP and IHDP considered and-cover to be the earth's land surface and near-surface layer of the natural state, including vegetation, soil, glaciers, lakes, wetlands and various buildings, etc., it is the natural and the result of human activities; land-use referred to human development and utilization of land resources with the destination of all activities, including agricultural land, industrial land, transportation land, residential land-use, etc. [5]. Land-use change is constantly accelerating change of land-cover, which together constituted the dual socio-economic and natural properties of land resources.

3 Method of Research Analysis

This paper combined with remote sensing technology and geographic information system (GIS) technology using the TM/ETM remote sensing images from Wuhan City in 2000 and 2007. Integrating the theoretical basis of land-use classification, basic principles and methods using qualitative and quantitative analysis of Methods, this paper analyses and studies the land-use change situation in Wuhan City in the nearly eight years. Figure 1 is the workflow chart of this study. ERDAS is used to classify the TM/ETM images of Wuhan City in 2000 and 2007 from land-use. Then a remote sensing interpretation is run to derive the classified maps of land-use. A detection of land-use change and extract the land-use change information is then applied to find land use changes.

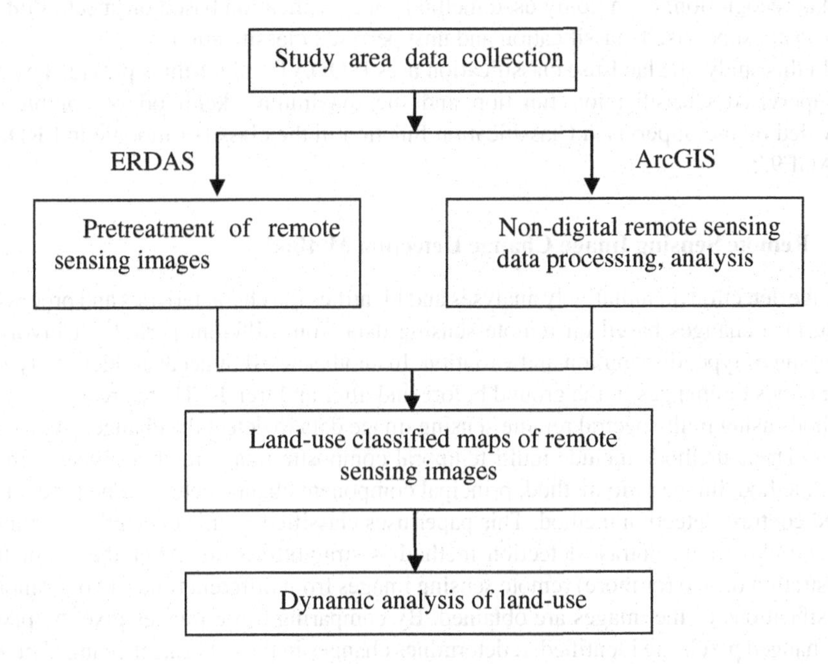

Fig. 1. Workflow chart

3.1 Selection and Establishment of Land-Use Classification System

One of the most common classification systems in the world is Anderson classification. It divides the land-use type into seven categories including urban areas, built-up areas, agricultural lands, grasslands, woodlands, wetlands and bare areas. The classification system has been modified for various purposes since then [3].

In China, according to the 1984 "Land-use Investigation Technical Specification", the national land-use classification system has two classes, a unified coding order. It has 8 classes including arable land, orchard, woodland, traffic land, water, residential and industrial land, grassland, unused land and 46 secondary classes.

Based on the criteria above, this research develops a classification according to the characteristics of the experimental region and the study purposes as described in Sect. 4.3.

3.2 Land-Use Classification of Remote Sensing Images

Image classification is the first step for remote sensing image change detection. Remote sensing image classification includes computer automatic classification and manual visual interpretation. Manual visual interpretation is a natural identification of human intelligence, and computer classification is the use of computer technology to artificial human recognition. Commonly used methods of classification based on spectral information are supervised classification and unsupervised classification.

In this study, the land-use classification uses ISODATA algorithms provided by the Unsupervised Classification function and the maximum likelihood of completion provided by the Supervised Classification function in the classifier module in ERDAS IMAGE9.2.

3.3 Remote Sensing Image Change Detection Method

Change detection quantitatively analyses and identifies the characteristics and processes of surface changes based on remote sensing data from different periods. It involves changing of type, distribution and variation. In another word, it needs to identify types, boundaries of changes in the ground before and after and trends. There are a variety of methods using multispectral remote sensing image data to detect the changes of ground types. These methods include multi-temporal composite image method, image difference method, image ratio method, principal component change detection method, classified contrast detection method. This paper uses classified contrast detection method.

The classified contrast detection method is straightforward. After the geometric registration of two (or more) remote sensing images from different times, two (or more) classifications of the images are obtained. By comparing those images pixel by pixel, the changed pixels are identified. It determines changes in the patterns of changed pixels based on the change detection matrix. The advantage of this method is that it not only determines the spatial extent of change, but also provides information on the changing nature, such as the change of types. The disadvantage is that, on one hand it needs to classify the images twice, on the other hand, the accuracy of change analysis depends on the accuracy of image classification. The reliability of image classification could seriously affect the accuracy of change detection. However, in the actual selection of methods, classified contrast detection method is still the most commonly used method.

4 Presentation and Explanation of Findings

4.1 Study Area: Wuhan City

Wuhan is the capital city of Hubei Province of the People's Republic of China, and is one of the 15 sub-provincial cities. Wuhan located at east longitude 113° 41′–115° 05′, latitude 29° 58′–31° 22′ in the eastern Jianghan Plain. It has a total area of 8494 km^2, including urban and suburban area of 1171.70 km^2, the main city urban area of 684 km^2. The built-up area has 500 km^2.

The terrain of Wuhan is featured by hills and undulating plains, lying north to south, with a flat central. Wuhan has a wide distribution of red soil and yellow brown soil, a wide range of land suitability and a high rate of utilization. According to the national "Land-use Investigation Technical Specification", the city land is divided into eight classes and 44 secondary classes. The city's land-use types show an obvious distribution of spheres: central city is mainly based on vegetables and food production; suburban area grow grain, cotton, oil production as the mainstay, and also have economic crops and animal husbandry; hilly areas near suburban are based on horticulture and planting.

Currently, Wuhan City land-use has some primary problems featured by the following aspects:

First, as the arable land declining fast, the contradiction between human being and land resources is growing more seriously.
Second, the production potential of agricultural land has not been fully realized.
Third, as the continued expansion of urban construction land, part of local section is not highly developed.
Fourth, the irrational uses of land in some areas have an impact on the ecological environment.
Fifth, construction of rural settlements is scattered and the index of construction land per capita is high.

Overall, because of the high development status of the city's land, reserve land resources are relatively insufficient. The potential use of land resources is to rationalize the major land-use structure, and tap the potential of existing land for construction to improve the level of various types of land-use.

4.2 Experimental Data

Based on the needs and the actual situation of the study area, the following data were collected by this paper: 2 phase of the U.S. resources Lands at TM/ETM+ images (ETM + 2000, TM2007) of Wuhan City, 1:1 million land-use map of Wuhan in 2000 and digital map of administrative divisions of Wuhan City.

Remote Sensing Images. These remote sensing images collected in this paper are all from http://landsat.datamirror.csdb.cn [4], the site is maintained and operated by the Chinese Academy of Sciences Computer Network Information Center. The remote sensing images of Wuhan in 2000 are spliced from 2000 Lands at/ETM (123-38) and

ETM (123-39) using the same phase. The remote sensing images of Wuhan in 2007 are spliced from 2007 Lands at/ETM (123-38) and ETM (123-39) using the same phase.

Non-remote Sensing Data. The land-use information includes 1:1 million land-use map of Wuhan in 2000; digital map of administrative divisions of Wuhan City; map information of Wuhan City related to topography and environment. The digital map of administrative divisions is mainly used in extracting the target area after the remote sensing images were spliced. The land-use map and topography map are mainly used as a reference for the supervision of remote sensing image classification and the accuracy assessment of classification.

4.3 Category Design

Limited by the topography of Wuhan and the resolution of remote sensing images, woodland and grassland will be classified as one class as shown in Table 1.

Table 1. Land-use classification system of Wuhan City

Codes	1	2	3	4	5
Name	Water	Urban and rural construction land	Unused land	Arable land	Woodland (Including grass-land)

4.4 Land-Use Classification

The land-use classification maps and classification results of Wuhan in 2000 and 2007 are showed in Figs. 2, 3, 4 and Table 2.

Table 2. The area and percentage of different land-use type

Type	2000		2007	
	Area (m^2)	Ratio	Area (m^2)	Ratio
1	934605900	11.2011 %	729545400	8.7435 %
2	2878245000	34.4952 %	3480798600	41.7167 %
3	22722300	0.2723 %	10563300	0.1266 %
4	3705081300	44.4047 %	3554797500	42.6036 %
5	803238300	9.6267 %	568188000	6.8096 %
Total	8343892800	100 %	8343892800	100 %

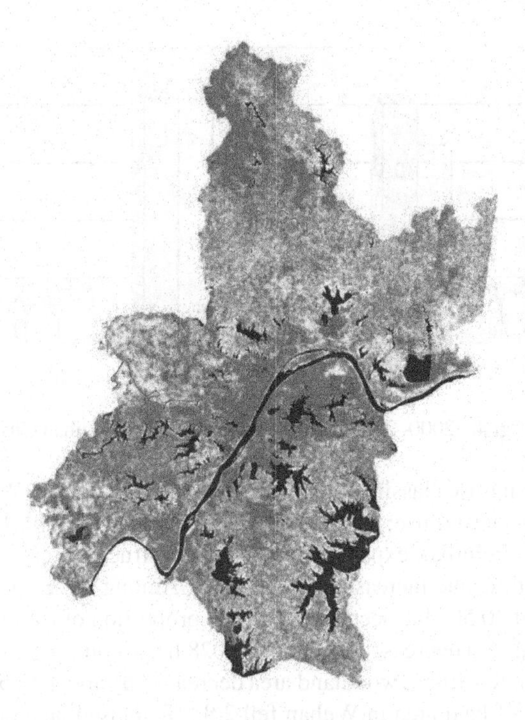

Fig. 2. The land-use classification result of Wuhan City in 2000

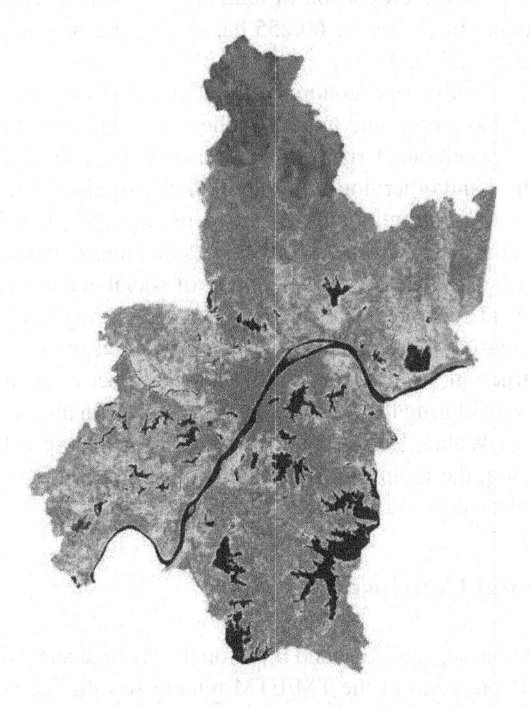

Fig. 3. The land-use classification result of Wuhan City in 2007

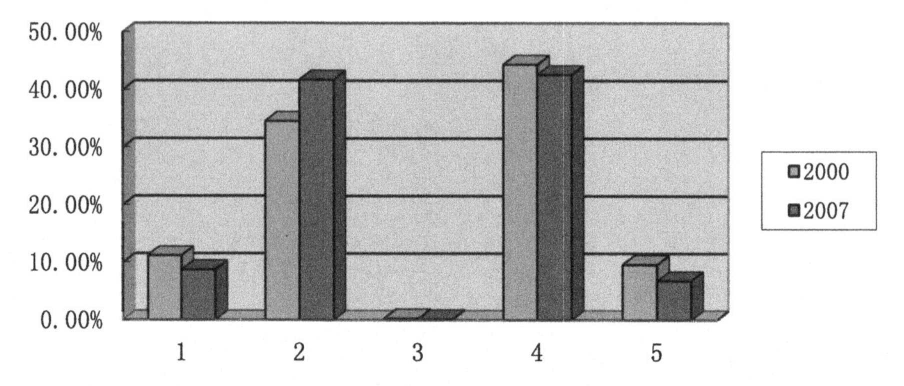

Fig. 4. 2000–2007 land-use type changes in Wuhan City

According to land-use classification results, from 2000 to 2007, Wuhan experienced a significant change in land-use: the size of water, unused land, arable land and woodland has been reduced; the urban expansion and road construction make the area of urban and rural construction land increased substantially. Among those above, the water area decreased by about 20,506 ha, accounting for the proportion of land area in Wuhan fell 2.46 %; arable land area decreased by about 15,028 ha, accounting for the proportion of land area in Wuhan fell 1.8 %; woodland area decreased by about 23,505 ha, accounting for the proportion of land area in Wuhan fell 2.82 %; unused land decreased by about 1216 ha, accounting for the proportion of land area in Wuhan fell 0.14 %; urban and rural construction land increased by 60,255 ha, accounting for the proportion of land area in Wuhan increased 7.22 %.

Since 2000, along with rapid economic growth, the process of industrialization and urbanization, which has a profound impact on the rate and trends of land-use change in Wuhan, continues to accelerate. Urbanization refers to a process that the rural population moved into industrial and other non-rural population, dispersed rural population gathered into urban area and the continuous generation of new cities, which lead to the urban population growth, the increase of number of cities, the improvement and formation of urban systems. It leads to the rapid development of social economy, but also makes a lot of agricultural land occupied, especially in paddy fields, dry land and farmland, and it is also the main reason of the increasing of the urban construction land. At the same time, wetland destruction, water and soil pollution and other environmental issues also became more seriously during the process of urbanization. On the other hand, the industrial development of Wuhan leads the development of township enterprises along the Yangtze River region, the regional requirements of rapid economic development also makes a large number of rural farmland occupied.

5 Summary and Conclusions

This paper uses remote sensing, GIS and traditional statistical analysis methods to have a classification interpretation of the TM/ETM remote sensing data of Wuhan in 2000 and 2007, and analyzes land-use quantity changes, the extent of land use and other dynamic characteristics of Wuhan in 8 years, and then reveals the law of change in land-use of Wuhan.

The experiments of this paper reveals that, as the arable land declining fast, the contradiction between human being and land resources is growing larger and larger. The production potential of agricultural land has not been fully realized. As the continued expansion of urban construction land, part of local section is not highly developed. The irrational uses of land in some areas have an impact on the ecological environment; construction of rural settlements is scattered and the index of construction land per capita is high.

References

1. Turner II, B.L., Skole, D., Sanderson, S., et al.: Land-use and land-cover change science/research plan. IGBP Report No. 35 and IHDP Report No. 7. IGBP, Stockholm (1995)
2. Aspinall, R.: Modelling land use change with generalized linear models—a multi-model analysis of change between 1860 and 2000 in Gallatin Valley, Montana. J. Environ. Manage. **72**(1–2), 91–103 (2004)
3. Anderson, J.R., Hardy, E.E., Roach, J.T., Witner, R.E.: A land use and land cover classification system for use with remote sensor data. USGS Professional, vol. 946 (1976)
4. http://landsat.datamirror.csdb.cn/
5. Tuner II, B.L., Meyer, W.B., Skole, D.L.: Global land use/land cover change: towards an integrated program of study. Ambio **23**(1), 91–95 (1994)

Applications of Geo-Informatics in Resource Management and Sustainable Ecosystem

An Application System for Calculation, Evaluation and Analysis of Municipal Water Resources Based on GIS

Zhencai Cui[1(✉)], Xiaoli Gao[1], Jiyu Yu[1], Weiqun Cui[1], and Xiangli Yin[2]

[1] Shandong Water Polytechnic, Rizhao, Shandong Province, China
cuizc@sina.com
[2] Rizhao Reservoir Management Bureau, Rizhao, Shandong Province, China

Abstract. In order to promote the construction of water conservancy modernization, an architecture of municipal water resources application system based on GIS, along with the calculation model and functional structure of surface water resources management subsystem and groundwater resources management subsystem, that are the core components within the application system, were discussed throughout this paper. The application system has been used in several small-medium size cities in Shandong Province, and has achieved good economic and social benefits.

Keywords: GIS · Municipal water resources · MIS · Public class libraries · Information mining

1 Introduction

The large scope of water shortage has affected the industrial and agricultural production and the quality of people's daily life. At the same time, environmental harm by widespread behavior of blind exploitation and unconscionable utilization of water resources exacerbates the crisis of water resources further. On the other hand, management technologies of municipal water resources are not adaptations to the modern water conservancy demands and the strictest water resources management in many small-medium size cities. To some degree, this situation has restricted the development of social economy [6].

Therefore, it is significant to try to break the traditional extensive management model, and to build a municipal water resources management information system which is an important part of water conservancy modernization management. The municipal water resources application system can make those related departments know water condition well and rapidly, and supply scientific basis for decision making.

This work is sponsored by the Special Water Resources Projects Aided by Special Fun of Water Resources Department of Shandong Province (No. sdw200709027) and the Provincial Water Conservancy Science Research and Technology Promotion Project of Shandong Province (No. SDSLKY201304).

F. Bian and Y. Xie (Eds.): GRMSE 2015, CCIS 569, pp. 837–845, 2016.
DOI: 10.1007/978-3-662-49155-3_86

With the rapid development of GIS (Geographic Information System) technology, the municipal water resources management system caused people attention and concern gradually, and some outstanding systems have been successfully applied in some cities. In 2009, Hebei Province carried out the construction of the provincial water resources management system. In 2014, Jiangsu Province has finished building the Province-Municipality-County tri-level application platform for water resources management and the fundamental water resource's data center.

However, water resources system has characteristics such as a large amount of data, complexity of data type and format. For this reason, we designed and implemented municipal water resources MIS (Management Information System) based on advanced technologies of GIS and information mining. This system can integrate the spatial data and attribute data organically, and manage them uniformly [2]. It can also analyze spatial data, inquire by map and data bidirectional, provide various charts and report forms automatically, provide spatial and static synthetic information services for study, management and decision of water resources, regional geology and eco-environment, etc.

2 Architecture of Application System

The architecture of water resources application system based on GIS includes the following subsystems: Surface water resources management subsystem, groundwater resources management subsystem, water permission management subsystem, water fee collection and use of management subsystem, water quality monitoring management subsystem, water resources annual report management subsystem, data management and maintenance subsystem, and system management and maintenance subsystem. The core functions are embodied in surface water resources management subsystem and groundwater resources management subsystem.

2.1 Surface Water Resources Management Subsystem

Surface water resources management subsystem is based on electronic map, space database, and the layer of the administrative regions. It can complete the analysis of surface water resources of administrative regions or 4th-level water resource regions (mainly includes analysis of statistical characteristics of annual precipitation, analysis of statistical characteristics of annual runoff depth and calculation of surface water resources, etc.) and the result can be queried, displayed or printed on the forms of thematic map, histogram, pie chart or data report form, etc.

The subsystem mainly realizes the following three functions: The analysis of statistical characteristics, the calculation of resources, the operations of map and spatial information. Each kind of function contains several functional modules, which complete many different sub-functions.

The three functions work through more than 20 classes, interfaces and other factors. The core module is named DXSMainForm, which contains hundreds of attributes and methods, is responsible for map loading, management of map elements, map zooming, map panning, generating bird's eye view, and many other kinds of spatial operations.

The functional structure of DXSMainForm module (partial) as showed in Fig. 1.

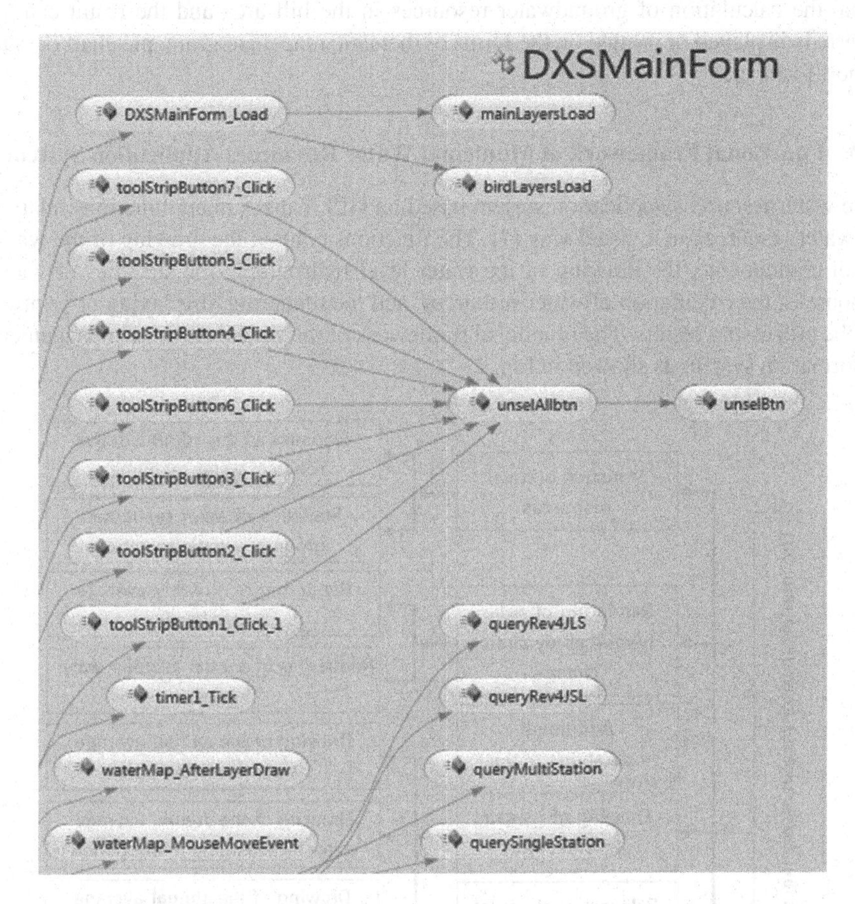

Fig. 1. The functional structure of DXSMainForm module (partial).

In this subsystem, with the help of GIS technology, spatial operation and non spatial operation can be unified in the electronic map, so as to achieve the seamless integration of spatial information and non spatial information.

2.2 Groundwater Resources Management Subsystem

Similar to the surface water resources management subsystem, the groundwater resources management subsystem is based on the electronic map and accessed through the plain groundwater resources' division, hill area, administrative area, groundwater, spatial layer data and corresponding attribute data.

The subsystem can complete the drawing of the groundwater level hydrograph, the drawing of the groundwater level contour line, the statistics of groundwater resources (mainly includes maximum, minimum and extremes ratio) [1], the analysis of the development and utilization of groundwater resources, the analysis of the groundwater level regression between different wells, the calculation of groundwater resources in the plain area, the calculation of groundwater resources in the hill area and the result can be queried, displayed or printed on the forms of thematic map, histogram, pie chart or data report form, etc.

2.3 Functional Framework of Municipal Water Resources Application System

The water resources application system based on GIS realizes many functions relative to water resources in a visual way [7]. The functions contain the drawing of the water level contour line, the drawing of the water level hydrograph, the statistics of water resources, the calculation of water resources, and the rendering, displaying or printing of the processing results. The functional framework of the water resources management information system as showed in Fig. 2.

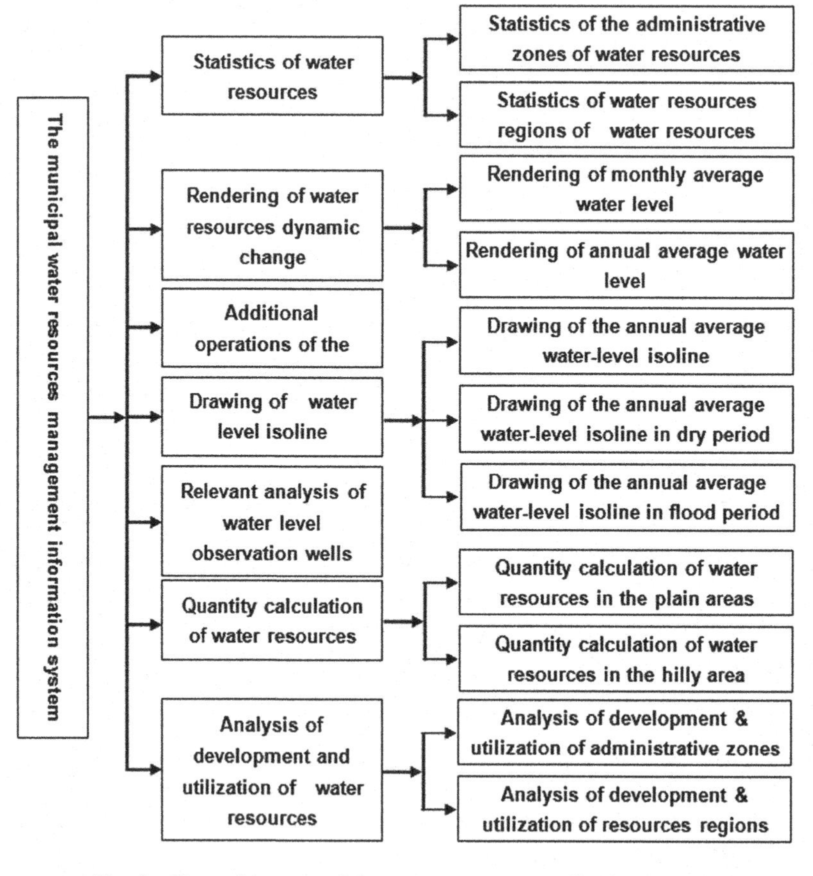

Fig. 2. The architecture of the water resources application system.

3 Core Technologies in the Application System

In municipal water resources application system, various mathematical models, many mathematical methods such as the arithmetic average method, or the Tyson polygon method, are employed in calculating various hydrological parameters [3].

3.1 Statistical Analysis of Annual Precipitation Series

The Calculation of annual average precipitation by using the arithmetic average method, the Tyson polygon method or the isohyetal method, resulting in annual precipitation series [1].

1. **The extremal statistic of annual precipitation.**
   The extremal statistic of annual precipitation includes maximal value $P_{\max}$, minimal value $P_{\min}$ and extremes ratio $j = \frac{P_{\max}}{P_{\min}}$.
2. **The parameters statistical of annual precipitation.**

$$\bar{P} = \frac{1}{n}\sum_{i=1}^{n} P_i \tag{1}$$

$$C_V = \sqrt{\frac{\sum_{i=1}^{n}(k_i - 1)^2}{n-1}} \tag{2}$$

$$C_s \approx \frac{\sum_{i=1}^{n}(K_i - 1)^3}{(n-3)\,C_v^3}, \quad k_i = \frac{P_i}{\bar{P}} \tag{3}$$

3. **The calculation of annual precipitation under different guarantee rate.**
   There are various kinds of different guarantee rates are used in practical applications, varying greatly from 5 % to 75 %. For example, annual precipitation under 50 % guarantee rate and annual precipitation under 75 % guarantee rate are shown below.

$$\begin{aligned} P_{50\%} &= \bar{P}(\Phi_{50\%} C_V + 1) \\ P_{75\%} &= \bar{P}(\Phi_{75\%} C_V + 1) \end{aligned} \tag{4}$$

3.2 Statistical Analysis of Annual Runoff Series

There are many kinds of statistical parameters are used in practical applications.

1. **The extremal statistic of annual runoff.**
   The extremal statistic of annual runoff includes years of maximal value $R_{\max}$, years of minimal value $R_{\min}$ and extremes ratio $j = \frac{R_{\max}}{R_{\min}}$.

2. **The parameters of annual runoff.**

$$\bar{R} = \frac{1}{n}\sum_{i=1}^{n} R_i \tag{5}$$

$$C_V = \sqrt{\frac{\sum_{i=1}^{n}(k_i - 1)^2}{n-1}} \tag{6}$$

$$C_s \approx \frac{\sum_{i=1}^{n}(K_i - 1)^3}{(n-3)\,C_v^3}, \quad k_i = \frac{R_i}{\bar{R}} \tag{7}$$

3. **Annual runoff depth under different guarantee rate.**
   There are many kinds of different guarantee rates are used in practical applications, varying greatly from 5 % to 75 %. Three kinds of these guarantee rates are shown below.

$$\begin{aligned} R_{20\%} &= \bar{R}(\Phi_{20\%} C_V + 1) \\ R_{50\%} &= \bar{R}(\Phi_{50\%} C_V + 1) \\ R_{75\%} &= \bar{R}(\Phi_{75\%} C_V + 1) \end{aligned} \tag{8}$$

4. **Annual runoff under different guarantee rate.**
   Corresponding to the three kinds of guarantee rates in the example above, there are three kinds of annual runoff, as follows.

$$\begin{aligned} W_{20\%} &= 0.1FR_{20\%} \\ W_{50\%} &= 0.1FR_{50\%} \\ W_{75\%} &= 0.1FR_{75\%} \end{aligned} \tag{9}$$

3.3 Information Mining Based on Spatial Features and Parameters of Water Resources

In order to realize the query functionality between map and text mutually, and to achieve the information mining technology based on spatial features and parameters of water resources, The associated mechanism of spatial information data and attribute data needs to be built. At the same time, to create the attribute tables through SQL statements, thus map data and text data can be accessed and manipulated mutually through the connected attribute [4].

The interface of map rendering as showed in Fig. 3.

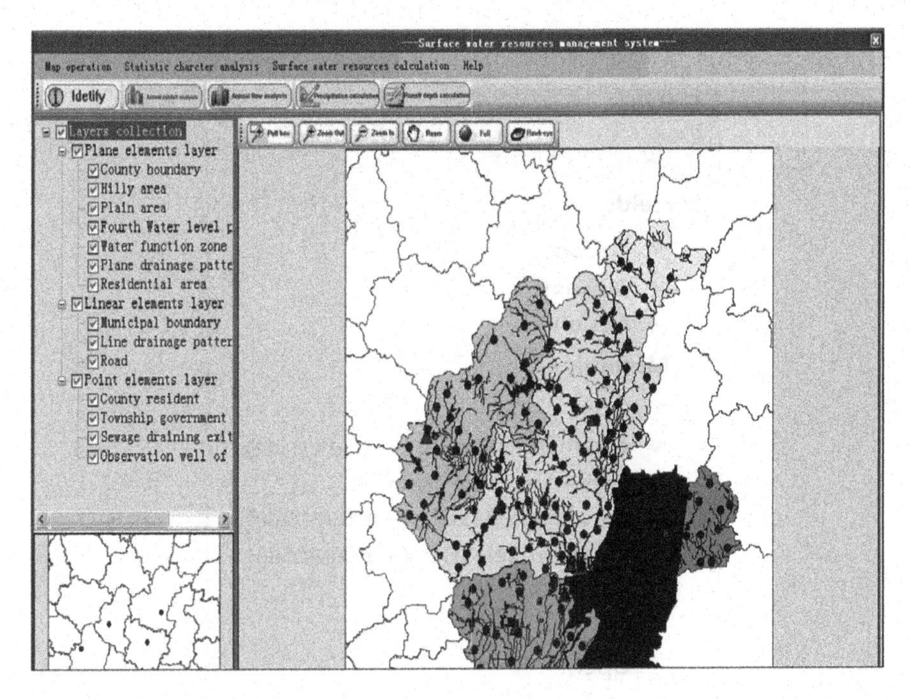

Fig. 3. The interface of map rendering.

3.4 Public Class Libraries for Database Operations

The analysis of statistical characteristics module, and the calculation of resources module, are two central functional modules of surface water resources management subsystem. Their crucial functions are to calculate various statistical indicators, or hydrographic parameters, to display and render the results through accessing the attribute database, with an efficient algorithm.

In surface water resources management subsystem, almost every functional module will access water resource data in application system's databases [5]. In order to facilitate the database operations and to achieve code reuse, application system compiles those frequently-used database operations (such as insert, query, backup, etc.) into a set of routines, and integrates these functional routines into a class file named DBLib.

The class diagram of DBLib as showed in Fig. 4.

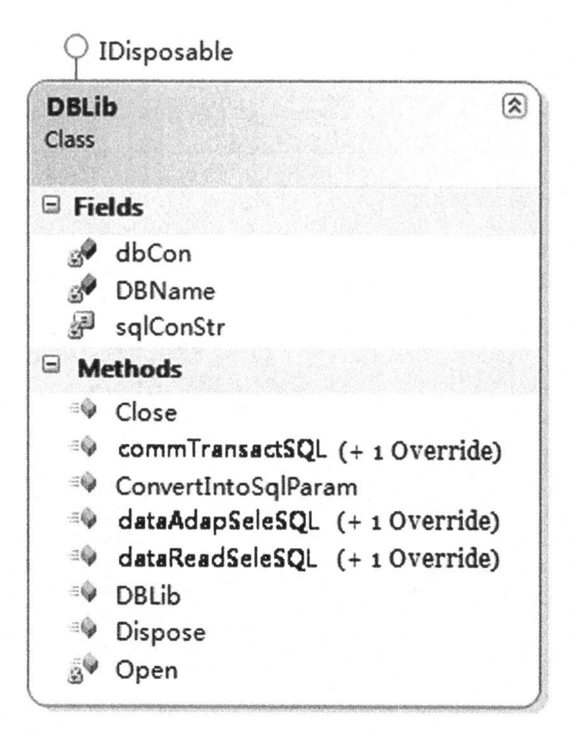

Fig. 4. The class diagram of DBLib (partial).

4 Conclusion

The municipal water resources MIS has realized the integrated management of hydrology, water resources, ecological environment, etc. It can complete the drawing of water level contour line and water level hydrograph in a visual way. It can complete the statistics of water resources (mainly includes maximum, minimum and extremes ratio), the analysis of the development and utilization of water resources, the analysis of water level regression between different wells, the calculation of water resources in the plain regions, the calculation of water resources in the hilly regions and the result can be queried, displayed or printed on the forms of thematic map, histogram, pie chart or data report form, etc.

The following advantages or technical innovations are our distinctive achievements gained in the system.

The UI (user interface) is friendly and unified because all operations of the system are based on GIS and maps. The query operation between map and text mutually based on information mining technology is flexible and efficient.
An algorithm for auto-generating of groundwater isoline based on TIN and the same color weighted minimum distance is proposed, which makes the spatial features integrating with the parameters of groundwater resources. Lots of experimental results

illustrated that the hydraulic characteristics of the contour lines produced by the algorithm are in agreement with the actual situation.

The municipal water resources MIS has served in Rizhao Municipal Water Conservancy Bureau and Linyi Municipal Water Conservancy Bureau, and has achieved good economic benefit and social effectiveness. The results show that this application system has a remarkable popularization value and application prospect.

References

1. Deng, Z., Wang, J.: Application of GIS to the information system of regional water resource. Hydrogeol. Eng. Geol. **12**(1), 106–108 (2006)
2. Elbir, T.: A GIS based decision support system for estimation, visualization and analysis of air pollution for large Turkish Cities. Atmos. Environ. **38**(2), 83–87 (2004)
3. Salem, B.B.: Application of GIS to biodiversity monitoring. Arid Environ. **54**(3), 91–114 (2003)
4. Li, Y., Cui, Z.: Algorithm for creating contour of underground water based on MapObjects. Geospatial Inf. **7**(5), 84–86 (2009)
5. Zhencai, C., Jiyu, Y., Naibo, J., Weiqun, C., Xiangli, Y.: Realization and development of municipal groundwater resources MIS Based on GIS. In: 6th International Conference on Intelligent Human-Machine Systems and Cybernetics, vol. 2, pp. 350–353. IEEE Computer Society CPS Press, USA (2014)
6. Yang, X., Chen, S., Tao, T.: Application on Com GIS in water resource management system. Hydrogeol. Eng. Geol. **5**(1), 46–49 (2007)
7. Pamanikabud, P., Tansatcha, M.: Geographical information system for traffic noise analysis and forecasting with the appearance of barriers. Environ. Model. Softw. **18**(1), 959–973 (2003)

Water Environment Early Warning System in Tongzhou District

Xiuju Zhang(✉), Jiahuan Li, Wenrong Zhao, and Kaisen Ding

College of Hydrology and Water Resources, Hohai University, Nanjing 210098, People's Republic of China
xjzh03@sina.com

Abstract. Water environment early warning is an effective mean to predict and evaluate the impacts of droughts or water pollutions and provide a basis for decision-making to relieve the damage, which will be beneficial to social security and economic development. Taking Tongzhou District, Nantong City as the research area, this paper established a water environment early warning system based on Microsoft Visual Basic 6.0, the main functions of which include gradual changing and sudden water environment early warning modules. Then the paper simulated an industrial wastewater leakage accident in Tongzhou District and calculated the influence time, influence scale and warning level. The result demonstrates that this early warning system is practicable and can be applied to provide technical supports for relevant departments to take emergency measures.

Keywords: Tongzhou District · Gradual changing water environment early warning · Sudden water environment early warning · Emergency measures

1 Introduction

With the rapid economic development and improvement of people's living standards, large amounts of industrial wastewater and domestic sewage are discharged into rivers, which has made the water environment badly polluted and has been one of the major factors that restrict the national economic development of China. The main water pollution accidents in China can be divided into two categories. One is gradual changing water pollution, such as persistent droughts [1]; the other is sudden water pollution [2], such as industrial accidental emissions, ship oil spills and so on. Comparatively, people pay more attention to the studies of sudden water pollution accidents [3, 4], while researches on the gradual changing water pollution accidents are still not enough. On the whole, the current research in the field of water environment early warning system is relatively weak [5].

Early warning system was originally applied in the studies of natural disasters. White [6] established a land flood early warning system utilizing risk management and decision theory in the 1970s. By taking advantage of early warning system, water quality of some rivers which were severely polluted in 1950s and 1960s, such as the Chicago River, the Thames River, the Rhine River, the Ruhr River, the Ohio River and Mississippi River, have been greatly improved [7–10]. Currently, the Rhine Biological

F. Bian and Y. Xie (Eds.): GRMSE 2015, CCIS 569, pp. 846–854, 2016.
DOI: 10.1007/978-3-662-49155-3_87

Early Warning System (RBEWS) and the Danube Accident Early Warning System (DAEWS), which are relatively mature early warning systems abroad, have played an important role in regional water pollution prevention and regulation [11, 12]. As for China, in recent years, China has established water quality early warning systems in the Guijiang River, the Han River, the Liao River, the Yellow River and the Yangtze River, among which the Han River water quality early warning system and Dalian early warning system for severe water pollution accidents are relatively successful and can rapidly respond to water pollution accidents to minimize the loss.

In conclusion, the existing researches mainly focus on some aspects of water environment early warning, such as monitoring and forecasting sudden water pollution accidents, or simply on the early warning systems of water quality. However, how to establish a practicable water environment early warning system reflecting gradual changing and sudden water pollution, still need to be further studied.

This paper takes Tongzhou District, Nantong City as the study area. It locates in the southeast of Jiangsu Province, off the west coast of the Yellow Sea and the north side of the Yangtze River. The local per capita water resources is only a quarter of China's per capita level, and 65 % of annual precipitation concentrates in the wet season. Moreover, its capacity of water conveyance and storage only depend on the criss-crossed rivers, which results in the poor capacity of water regulation and storage. It mainly relies on the Yangtze River to alleviate water shortage contradiction. However, in recent years, several droughts and sudden water pollution accidents have occurred in the Yangtze River basin, which caused great influence on water taking and water supply security of Tongzhou District. In addition, currently large amounts of industrial wastewater and domestic sewage are discharged into the rivers, which have badly affected the local water environmental quality and social sustainable development. Consequently, it is necessary to study the water environment early warning system of Tongzhou District.

According to the actual situation of Tongzhou District, this paper developed a water environment early warning system based on Microsoft Visual Basic 6.0. It designed a convenient and practicable user interface to classify the degree and possible impacts of water pollution accidents and provide technical supports for responsible authorities to formulate emergency plans and take measures.

2 Water Environment Early Warning System in Tongzhou District

2.1 System Functions

The water environment early warning system can comprehensively analyze the water environment state and predict the warning alarms and the results, including single indexes and comprehensive indexes, will be showed in the form of a group of alarm lights. Also, the exact score will be calculated to judge the condition of regional environmental sustainable development. It consists of five modules; the function of each part is shown as follows:

(1) General situation management module

An introduction of Tongzhou District is showed on the visual interface of VB, including regional general situation, river situation and water function zone situation, which helps the visitors learn more about its geographical location, hydrology and meteorology, social economy, location of water function zones and so on.

(2) Water environment evaluation module

Not only can this module assess the changes of water quality in each water function zone over time, but also inquiry the distribution rule of water quality timely, and then generates the distribution diagrams of the principal water quality categories graded by the major pollution indicators in selected monitoring sections, classifies the water quality categories, and calculates excessive factors and excessive multiples.

(3) Sewage outlets management module

The main contents of this module include, inquiring the information of the sewage outlets of all the rivers, calculating the surplus water environment capacity of the water function zones, and the amount of pollutants that should be reduced. Also this module can simulate the concentration of pollutants at the water intake section of downstream after being degraded and diffused, and calculate the influence distance that the pollutants can reach.

(4) Gradual changing water environment early warning module

Based on the analysis of precipitation and water demand of Tongzhou District and combined with the data of water quantity, water level and water quality, this module can identify the state of local water resources and the pollution degree of water bodies. Then it predicts the change tendency and carries out water environment early warning.

(5) Sudden water environment early warning module

This module includes two aspects, qualitative analysis and quantitative calculation. It evaluates the alert situation comprehensively and grades the warning level, by calculating the influence scale, influence time and influence degree of the accidents to the social economy, water body and ecological environment respectively.

2.2 Gradual Changing Water Environment Early Warning

The program of gradual changing early warning system can be divided into two parts. One is to evaluate the possibilities of the latent droughts caused by the continuous poor flow and low river level according to the monitoring data. And the other is to assess the changes of the water quality and predict its trends, and then, the system classifies the water quality early warning level and responds to it (see Fig. 1).

Persistent Drought Early Warning. (1) Indices of drought evaluation

The three indicators used to judge the drought degree and the changes of water quantity in the study area, are precipitation anomaly percentage, continuous non-precipitation days, and river levels. The specific formula is as follows:

$$D_p = \frac{P - \overline{P}}{\overline{P}} \times 100\%. \tag{1}$$

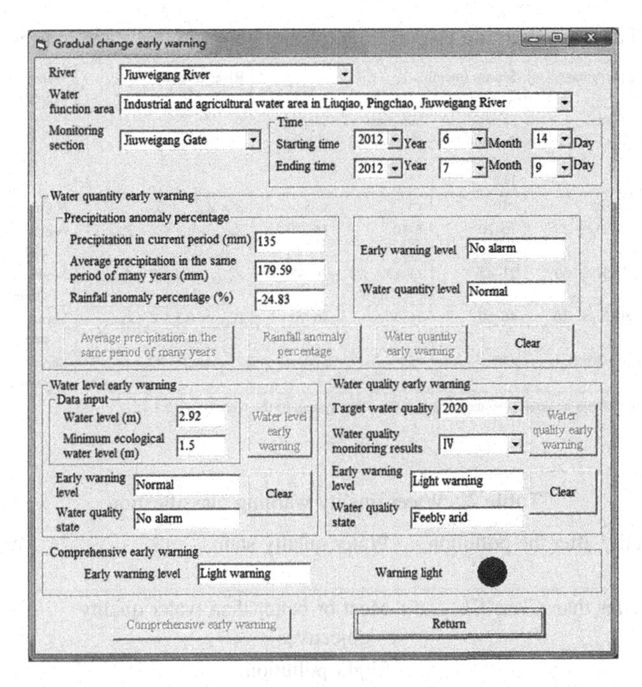

Fig. 1. Gradual changing water environment early warning

Where D_p = precipitation anomaly percentage (%); P = rainfall in calculation interval (mm); $\overline{P}$ = average rainfall in the same period (mm).

(2) Warning classification of droughts

According to the three indexes above, early warning will be issued when the warning level meets a certain degree and the weather continues to be cloudy or sunny in follow-up period. The classification of droughts with the indexes is shown in the Table 1 (Standard of Classification for Drought Severity (2008) [13]).

Water Quality Early Warning. According to the water quality monitoring results in "Water Quality Bulletin of Tongzhou District" or actual class of water quality tested, an alarm will be issued if the water quality does not meet requirements of each water function zone in Tongzhou District. ① For centralized drinking water source areas, the warning should be alarmed immediately once water quality changes more than one class for two consecutive months and the requirements of the water environment are not reached; ② For industrial and agricultural water consumption zones, if water quality changes more than two classes for two consecutive months and the water quality can not meet the requirements, early warning work should be performed immediately. The classification standard of water quality warning level is established based on water quality goals of water function zones in Tongzhou District (see Table 2).

Table 1. Warning classification of droughts with the three indices

Precipitation anomaly percentage D_P		Season (month)			River level Z	Water degree	Warning level
Monthly	Seasonal	Spring; Autumn (3–5; 9–11)	Summer (6–8)	Winter (12–2)			
$D_P > -40$	$D_P > -25$	<10	<5	<15	$Z \geq Z_{min,j} + 0.4$	Normal	No alarm
$-60 < D_P \leq -40$	$-50 < D_P \leq -25$	10–20	5–10	15–25	$Z_{min,j} + 0.1 \leq Z < Z_{min,j} + 0.4$	Feebly arid	Light warning
$-80 < D_P \leq -60$	$-70 < D_P \leq -50$	21–45	11–15	26–45	$Z_{min,j} - 0.2 \leq Z < Z_{min,j} + 0.1$	Moderate drought	Moderate warning
$-95 < D_P \leq -80$	$-80 < D_P \leq -70$	46–60	16–30	46–70	$Z_{min,j} - 0.5 \leq Z < Z_{min,j} - 0.2$	Severe drought	Serious warning
$D_P \leq -95$	$D_P \leq -80$	>60	>30	>70	$Z < Z_{min,j} - 0.5$	Super drought	

Note: where $Z_{min,j}$ = the lowest ecological water level of NO.j water system in the table (j = 1,2,3,4).

Table 2. Water quality warning classification

The water quality after the pollution accident	Water quality status	Warning level
Class III or better than Class III	Meet or better than water quality objectives	No alarm
Class IV	Slight pollution	Light warning
Class V	Intermediate pollution	Moderate warning
Inferior to Class V	Serious contamination	Serious warning

Note: where the classification of water quality is according to Environmental Quality Standards for Surface water (2002) [14].

2.3 Sudden Water Environment Early Warning

(1) Basic data sources maintenance

This part mainly includes basic data and all kinds of warning standards of sudden water pollution accidents, such as water quality standards of water function zones, classification threshold standards of early warning indexes and standards of early warning levels.

(2) Calculation and analysis

Before calculating the comprehensive evaluation indexes, it is necessary to judge whether the pollutants can be degraded. The calculation can be divided into two parts: ① Qualitative analysis: the qualitative indexes include accident type, scale of affected water body, categories of affected water function zones and damage degree to ecological environment, which can be manually chosen on the early warning system interface according to the actual situation of water pollutions to evaluate the accidents more exactly. ② Quantitative calculation: including choosing the pollution indexes, water function zones and the downstream monitoring cross sections and calculating the influence degree, influence time and influence scope respectively.

(3) Early warning analysis

The main function of the early warning analysis module is to depict and analyze the water pollution accidents comprehensively, predict and evaluate the influence and then issue the warning alarm timely. Based on the comprehensive evaluation indexes calculated above, the module grades the accidents according to standards of early warning level, and the results will be shown in the form of a group of alarm lights to judge the severity, urgency and controllability of water pollution accidents.

2.4 Mathematical Model

This system applies the following formulas to calculate the time that the pollutants take to reach the downstream water intakes, predict the change of pollutants concentration of pollutants over time, and the time the water intakes should stop and resume obtaining water.

① The sudden water pollution is usually considered as instantaneous point source pollution, where the one-dimensional and two-dimensional unsteady models are generally used to calculate the pollutants concentration. When the distance between the accident site and the calculated section is longer than the transverse mixing length, i.e. $x > L_m$, one-dimensional model should be adopted, otherwise two-dimensional model should be adopted to predict the water quality.

In straight rivers, when the pollution source is in the center of the river, the formula used to calculate the lateral distance is as follow:

$$L_m = \frac{0.1 \cdot u \cdot B^2}{E_y}. \tag{2}$$

When the pollution source is on the shore, the formula is as follow:

$$L_m = \frac{0.4 \cdot u \cdot B^2}{E_y}. \tag{3}$$

Where L_m = the transverse mixing length of pollutants (m); u = the mean velocity of river flow (m/s); B = the mean width of river channel (m); E_y = the lateral dispersion coefficient (m^2/s).

② Two-dimensional unsteady water quality model

$$\mathrm{C}(x, y, t) = \frac{M}{H 4\pi t \sqrt{E_x E_y}} \exp\left[\frac{-(x - ut)^2}{4E_x t} - \frac{y^2}{4E_y t} - 2Kt\right] + C_h. \tag{4}$$

Where M = the leakage of pollutants (g); H = the depth of the river channel section (m); t = the prediction time (s); C_h = the initial concentration of pollutants in river (mg/L); E_x = the longitudinal diffusion coefficient (m^2/s); K = the degradation coefficient of pollutants (l/s); x = the vertical distance between discharge point to water intake (m); y = the transverse distance from discharge point to water intake (m); The meaning of the rest symbols is as above.

③ One-dimensional unsteady water quality model

$$C(x,t) = \frac{M}{A\sqrt{4\pi E_x t}} \exp\left[\frac{-(x-ut)^2}{4E_x t}\right] + C_h. \quad (5)$$

Where A = the cross sectional area of rivers (m^2).

3 Application Examples

It is necessary to test the rationality and feasibility of the early warning system to ensure the system is practical and effective. Therefore, it will apply the early warning system in an assumed industrial wastewater leakage accident which occurs at 9:00 one day in September, in the center of Jiuweigang River, Pingchao Town of Tongzhou District, to examine its rationality and feasibility. The parameters of this water pollution accident are as follows:

The discharge flow of the pollutants at leaks Q is 0.5 m^3/s; The discharge time t is 30 min; The river width of the accident point B is 285 m; The mean water depth H is 3.3 m; The mean flow velocity of the section u is 0.18 m/s; The distance from the leakage point to the shore d is 178 m; The water function zone the leakage point located is industrial and agricultural water consumption zone in Liuqiao, Pingchao Town, Tongzhou District. The distance between the water intake and the leakage point is 4660 m; the distance from the water intake to the shore is 5 m. The main pollution indexes of the wastewater are NH_3-N, COD and BOD_5, and the concentrations of them are 1.5 mg/L, 30 mg/L and 5 mg/L respectively, and the initial concentrations are 1 mg/L, 20 mg/L and 4 mg/L.

The procedures for simulating this case include two parts, qualitative analysis and quantitative calculations. The former one is mainly to analyze the accident type, the class of affected river, the categories of affected water function area and the impacts on the ecological environment. The quantitative calculation mainly contains calculating the influence degree, influence time and influence scale. Entering the monitoring data, the system will emulate and predict the possible influences of this case. For this case, the influence time of this accident is 24.5 h, the influence distance is 4969.82 m, i.e. the impacts on the downstream water intake will last for 1 d 30 min, during which the water intake should stop obtaining water from the river.

After calculating the integrated indexes of water pollution accidents by qualitative analysis and quantitative calculation, it selects other primary indexes evaluation options on the sudden warning interface considering the actual situation, to evaluate the impacts of this accident. The calculation result shows that the level of this accident is Class III, and warning light is yellow, the result is shown in Fig. 2.

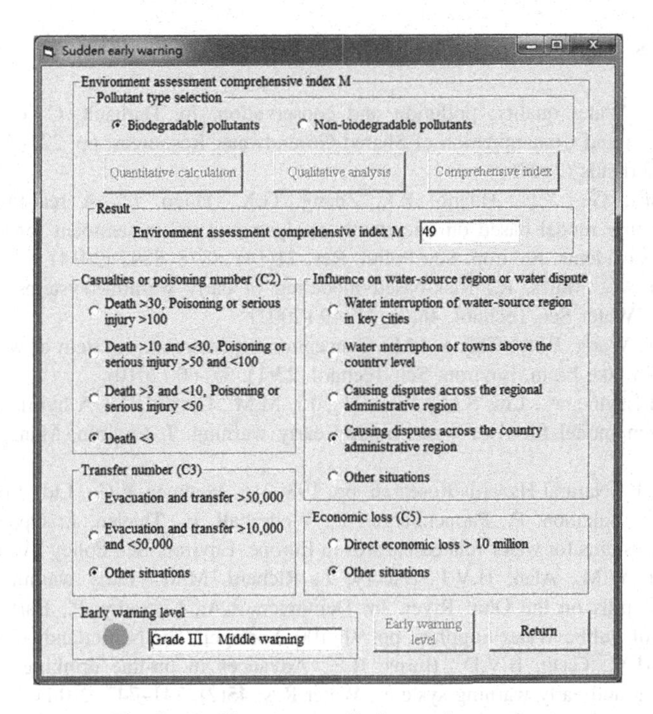

Fig. 2. Sudden water pollution early warning

4 Conclusions

In view of the risk characteristics and environmental risks receptors of water pollution accidents in Tongzhou District, this paper established a water environment early warning system based on Microsoft Visual Basic 6.0 taking both gradual changing early warning and sudden early warning into account. Through qualitative analysis and the quantitative calculations, this early warning system can evaluate and predict the security condition of the water environment, which enables the system to further analyze and assess the possible impacts of water pollution incidents, and then classify the water pollution of the accidents. To test the rationality and feasibility of the system, this paper simulated an industrial wastewater leakage accident in Jiuweigang, Pingchao Town, Tongzhou District comprehensively, and then alarmed the warning timely. It demonstrates that this system could analyze the water pollution accidents comprehensively and alarm the warning timely. And the case shows that the system is effective and practicable and can provide technical supports for the follow-up emergency monitoring, emergency control and risk assessment, and contributes to improve the condition of water environment and reduce the economic loss.

Acknowledgments. This research was supported by the projects initiated by the Department of Science and Technology of Jiangsu Provincial Water Resources Department. The authors would like to thank the help of Hydrological Bureau of Nantong City and Tongzhou District Water Conservancy Bureau.

References

1. Yetiş, Ü.: Water quality, pollution and conservation. In: Darnault, C.J.G. (ed.) Over-exploitation and Contamination of Shared Groundwater Resources, pp. 227–256. Springer, The Netherlands (2008)
2. Hou, D.B., Ge, X.F., Huang, P.J., Zhang, G.X., Hugo, L.: A real-time, dynamic early-warning model based on uncertainty analysis and risk assessment for sudden water pollution accidents. Environ. Sci. Pollut. Res. **21**(14), 8878–8892 (2014)
3. Grayman, W., Males, R.: Risk-based modeling of early warning systems for pollution accidents. Water Sci. Technol. **46**(3), 41–49 (2002)
4. Zhang, F., Wang, W.L., Cheng, J.M.: Conception on emergency system of water pollution accident in lake basin. Environ. Sci. Technol. **23**(1), 57–60 (2010)
5. Burchard-Levine, A., Liu, S.M., Vince, F., Li, M.M., Ostfeld, A.: A hybrid evolutionary data driven model for river water quality early warning. J. Environ. Manag. **143**, 8–16 (2014)
6. White, G.F.: Natural Hazards Research, pp. 193–216. Methuen & Co. Ltd., London (1973)
7. Alfieri, L., Salamon, P., Pappenberger, F., Wetterhall, F., Thielen, J.: Operational early warning systems for water-related hazards in Europe. Environ. Sci. Policy **21**, 35–49 (2012)
8. Grayman, W.M., Alan, H.V.J., Vicory, J., Richard, M.M.: Early warning system for chemical spills on the Ohio River. In: Deininger, R.A., Literathy, P., Bartram, J. (eds.) Security of Public Water Supplies, pp. 91–100. Springer, The Netherlands (2000)
9. Storey, M.V., Gaag, B.V.D., Burns, B.P.: Advances in on-line drinking water quality monitoring and early warning systems. Water Res. **45**(2), 741–747 (2011)
10. Yu, F.C., Fang, G.H., Shen, R.: Study on comprehensive early warning of drinking water sources for the Gucheng Lake in China. Environ. Earth Sci. **72**(9), 3401–3408 (2014)
11. Puzicha, H.: Evaluation and avoidance of false alarm by controlling Rhine water with continuously working biotests. Water Sci. Technol. **29**(3), 207–209 (1994)
12. Pintér, G.G.: The Danube accident emergency warning system. Water Sci. Technol. **40**(10), 27–33 (1999)
13. Standard of Classification for Drought Severity, Water conservancy standard in China, SL 424-2008 (2008)
14. Environmental Quality Standards for Surface Water, National standard, GB 3838-2002 (2002)

Research on Soil Heat Balance Theory of Ground Coupled Heat Pump System

Chao Lü[1(✉)], Feng Yu[1], Maoyu Zheng[2], and Jiachen Zhong[1]

[1] Zhejiang Sci-Tech University, Hangzhou 310018, People's Republic of China
lvchao-929@163.com
[2] Harbin Institute of Technology, Harbin 150090, People's Republic of China

Abstract. Ground coupled heat pump (GCHP) is an efficient air-conditioning technology, which is good to sustainable development of energy use. In order to make GCHP system operate efficiently in long term, soil needs to be maintained heat balance in an annual cycle. If necessary, auxiliary heat or cold source is also needed. A soil heat balance theory was proposed, and it contained heat balance equation and annual heat balance rate of soil. Soil heat balance point and floating range were derived, and the upper and lower limit of the range correspond to cold and hot critical region, respectively. And heat balance point 80 % and floating range 65 %~95 % were obtained according to experimental data, which can be used as indicators to measure soil heat balance. The soil heat balance theory can be used to investigate the regional applicability of a GCHP system with auxiliary heat or cold source, and design an efficient system which can maintain soil heat balance.

Keywords: Ground coupled heat pump · Soil heat balance · Annual heat balance rate of soil · Auxiliary heat source

1 Introduction

The sustainable development of energy use is one of the themes of the world. Ground source heat pump (GSHP) is a kind of high efficiency and energy saving technology using shallow geothermal energy as heat source or heat sink. It can be divided into three types: ground coupled heat pump (GCHP), surface water heat pump (SWHP) and ground water heat pump (GWHP). And the environmental GCHP technology is considered as one of the most promising air-conditioning technology nowadays.

It needs to keep the soil heat balance annually to make the GCHP system operate long-term and efficiently. Especially for the area where the heating and cooling loads of buildings are imbalanced, it is very important to supply heat or cold (auxiliary heat source or cold source) to the soil. Some scholars called this kind of system hybrid ground source heat pump (HGSHP) system.

At present, the research on GCHP and its hybrid systems is a hot research problem in the field of building energy saving and heating, ventilating and air conditioning. Domestic and foreign scholars have studied it extensively and deeply, mainly concentrated in the theory and model of ground heat exchanger, simulation of underground temperature field,

F. Bian and Y. Xie (Eds.): GRMSE 2015, CCIS 569, pp. 855–861, 2016.
DOI: 10.1007/978-3-662-49155-3_88

selection of backfill materials, development of design and simulation software, heat and cold storage technology, equipment matching of the system, operation mode conversion, thermodynamic and economic analysis of the system [1–5]. As for the soil heat balance problem, although it has also been studied to some extent, the total research depth and the actual operation still needs to be further strengthened.

China is vast in territory, so the cold and warm degree in different climate zones has a big difference. The region where soil heat balance can be maintained only rely on the GCHP system is very limited. For the northern region where the heating load is greater than the cooling load, auxiliary heat source is needed to supply heat to the soil, such as the use of solar energy, i.e., solar-ground coupled heat pump (SGCHP) system; for the southern region where the cooling load is greater than the heating load, auxiliary cold source is needed to supply cold to the soil, such as the use of cooling tower or cold storage device. Whether the GCHP system itself or adding auxiliary heat and cold source, it needs to further research the soil heat balance theory and reasonably design the system, to keep the soil heat balance annually and improve the system's efficiency. In the following, an actual system is used as an example to mainly analyze soil heat balance problem of the GCHP system adding auxiliary heat source (SGCHP system).

2 System Description

References [6, 7] have detailedly introduced a demonstration project of solar-ground coupled heat pump for heating and cooling in severe cold zone. The composition and operation principle of the system was presented, and the practical heating and cooling results, the variation of soil temperature and heat utilization of the system were also analyzed using the field data.

The system was built in Harbin (45°45′N, 126°39′E), China. It mainly consists of 4 subsystems, i.e., solar collection system, heat pump unit, underground heat exchange system and radiant floor heating and cooling system. Figure 1 shows the schematic diagram of the system. The system stored solar energy in spring, summer and autumn into the soil using solar collectors and ground heat exchangers (GHEs), and extracted the heat for heating the building using GCHP in winter. And solar energy can also be used for direct heating in the winter sunny days. This realized the use of solar energy in the whole year. In addition, heat extraction in winter decreased the soil temperature, so the soil can be used as natural cold source in summer for direct cooling. In this paper, this system is named ground coupled heat pump with solar seasonal heat storage (GCHPSSHS) system.

GCHPSSHS system selected the following 4 operating modes: Mode 1 - solar heat storage in the soil; Mode 2 - solar direct heating; Mode 3 - GCHP heating; Mode 4 - soil cold source direct cooling. The system was tested for 3 years with annual cycles of heat storage, heating and cooling experiments.

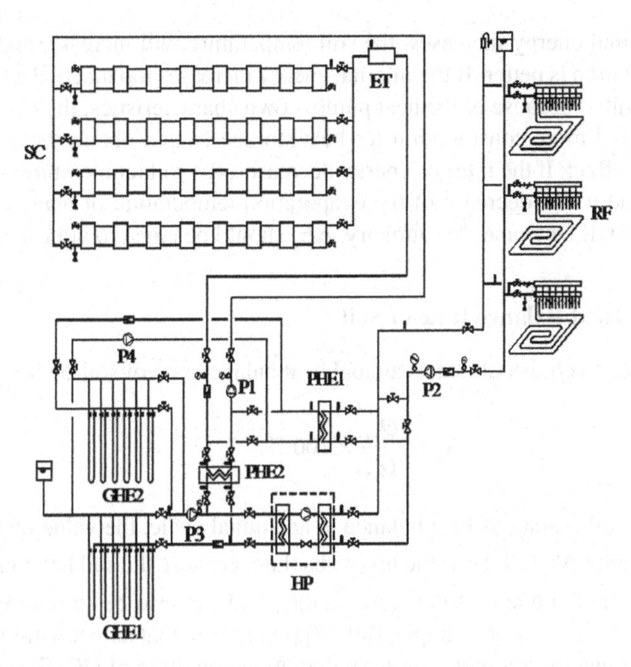

Fig. 1. Schematic diagram of the GCHPSSHS system. (SC: solar collectors; GHE1, GHE2: ground heat exchangers; HP: heat pump; RF: radiant floor; PHE1, PHE2: plate heat exchangers; ET: expansion tank; P1, P2, P3, P4: circulating pumps)

3 Soil Heat Balance Theory

3.1 Soil Heat Balance Equation

Since the environment's thermal effect on the soil changes in an annual cycle, the total thermal effect can be treated as zero. So the main impact to soil heat balance is heat storage and heat extraction. The soil around the GHEs is analyzed as a control volume, the heat balance equation is:

$$Q_{in} - Q_{di} - Q_{ex} + Q_{as} = \Delta E_g. \quad (1)$$

Where, Q_{in} is the total heat input to the soil, including solar heat storage Q_{st} and the heat exhaust when cooling Q_{co}, i.e., $Q_{in} = Q_{st} + Q_{co}$, GJ; Q_{ex} is the heat extraction of the heat pump, GJ; Q_{di} is the heat loss of heat storage, i.e., the heat flowing from the control volume to the outside, GJ; Q_{as} is the heat compensation of heat extraction, i.e., the heat flowing from the outside to the control volume, GJ; ΔE_g is the internal energy change of the control volume, GJ.

The so-called soil heat balance annually means soil temperature remains unchanged after a year of heat injection and extraction, i.e., ΔE_g is almost zero.

If the internal energy increases, the soil temperature will increase, and the heating effect of heat pump is better. If the internal energy increases too much, the COP will not increase unlimited because of the heat pump's own characteristics; the heat loss will be also greater, and more consumption for heat storage cannot obtain the corresponding better heating effect. If the internal energy decreases, the soil temperature will decrease, which will lead to the decrease of the evaporation temperature of heat pump, and the heating effect will be worse. In summary, ΔE_g should be equal to almost zero.

3.2 Annual Heat Balance Rate of Soil

The ratio of Q_{ex} to Q_{in} is defined as annual heat balance rate of soil Δ_g, i.e.:

$$\Delta_g = \frac{Q_{ex}}{Q_{in}} \times 100\,\%. \tag{2}$$

When the soil maintains heat balance in an annual cycle, the value of Δ_g is the soil heat balance point Δ_g^0. But the value is not absolute, because the soil has a natural ability to recover. A small amount of float around the point can also be considered as the soil maintains heat balance. For example, Ref. [8] pointed out that a float within 20 % would not lead to the deterioration of long-term operating condition of GCHP system. So the floating range of annual heat balance rate is 80 $\%\Delta_g^0 \sim 120\,\%\Delta_g^0$.

In the cold critical region, heating load/heat extraction is very large, and cooling load is very small or even none. When all the time in the whole year except the heating season is used for heat storage, Δ_g can just meet the upper limit of soil heat balance 120 $\%\Delta_g^0$. When $Q_{co} = 0$, $Q_{in} = Q_{st} + Q_{co}$ changes into $Q_{in} = Q_{st}$. Then Eq. 2 changes to the expression of annual heat balance rate of soil for the cold critical region:

$$\Delta_g^c = \frac{Q_{ex}}{Q_{st}} \times 100\,\% = 120\,\%\Delta_g^0. \tag{3}$$

In the hot critical region, cooling load/heat exhaust is large enough, solar heat storage is not needed to maintain soil heat balance, and Δ_g even reaches the lower limit 80 $\%\Delta_g^0$. When $Q_{st} = 0$, $Q_{in} = Q_{st} + Q_{co}$ changes into $Q_{in} = Q_{co}$. Then Eq. 2 changes to the expression of annual heat balance rate of soil for the hot critical region:

$$\Delta_g^h = \frac{Q_{ex}}{Q_{co}} \times 100\,\% = 80\,\%\Delta_g^0. \tag{4}$$

When the GCHPSSHS system meets the heating and cooling load requirements of the building, there are the following formulas:

$$Q_h = Q_{hp} + Q_{so} = Q_{ex} + W_{hp} + Q_{so}. \tag{5}$$

$$Q_c = Q_{co}. \tag{6}$$

Where, Q_h is the heating load of the building, GJ; Q_c is the cooling load of the building, GJ; Q_{hp} is the heat pump heating quantity, GJ; Q_{so} is the solar heat quantity, GJ; W_{hp} is the power consumption of heat pump, GJ.

Put Eq. 5 and Eq. 6 into Eq. 3 and Eq. 4 respectively, then

$$\Delta_g^c = \frac{Q_h - Q_{so} - W_{hp}}{Q_{st}} \times 100\ \% = 120\ \%\Delta_g^0. \tag{7}$$

$$\Delta_g^h = \frac{Q_h - Q_{so} - W_{hp}}{Q_c} \times 100\ \% = 80\ \%\Delta_g^0. \tag{8}$$

The GCHPSSHS system is only applicable to the region among the cold and hot critical region. It is no longer applicable in the region colder than the cold critical area, or the efficiency is very low. It is also no longer applicable in the region hotter than the hot critical area, and cooling tower or other auxiliary cold source (heat dissipation equipment) is needed.

Annual heat balance rate of soil Δ_g can be used as a measurable index of soil heat balance. And it can be used in the following two aspects: applicable region of the GCHPSSHS system can be obtained according to the scope of Δ_g; for one region, the scope of Δ_g is used to design an efficient system which can maintain soil heat balance.

4 Example Calculation

Reference [7] obtained that the soil maintained heat balance in an annual cycle through analyzing soil temperature change of the GCHPSSHS system in an annual cycle. So the actual heat injection to the soil should be roughly equal to the heat extraction from the soil in the long term. As the heat storage period is quite long, there must be some heat loss. Table 1 shows the heat quantities of the system in the 3 operating years. It can be seen that the total Q_{ex} and Q_{in} of 3 years were 177.33 GJ and 220.54 GJ, respectively. At this time, Δ_g of the 3 years were 73 %, 90.4 % and 78.6 %, respectively. And the overall Δ_g of the 3 years was 80.4 %. So 80 % can be used as heat balance point Δ_g^0. Then the float of 20 % should base on 80 %, namely, the absolute amount is 16 %. So 15 % can be chosen as a floating amount. So the floating range of Δ_g is 65 %~95 %, that is, the cold and hot critical regions corresponding to 95 % and 65 %, respectively.

Table 1. Heat quantities of the system in the 3 operating years [GJ]

Heat quantities	First year	Second year	Third year
Q_{ex}	55.65	63.25	58.43
Q_{st}	70.76	64.83	69.02
Q_{co}	5.52	5.10	5.31
Q_{in}	76.28	69.93	74.33

5 Soil Heat Balance Theory for GCHP System Adding Auxiliary Cold Source

The above theory is used to analyze the soil heat balance problem for GCHP system adding auxiliary heat source. If this theory is extended to GCHP system adding auxiliary cold source, the principle is the same, and the heat balance equations are also consistent, but the meanings of some parameters change.

For Eq. 1, Q_{in} only includes the heat exhaust when cooling Q_{co}; and Q_{ex} not only contains the heat extraction of the heat pump, but also contains the cool storage (or the auxiliary heat dissipation).

In the cold critical region, the heating load/heat extraction is large enough. Cool storage is not needed to maintain soil heat balance, and Δ_g even reaches the upper limit 120 $\%\Delta_g^0$. In the hot critical region, the cooling load/ heat exhaust is very large, but the heating load is very small or even none. When all the time in the whole year except the cooling season is used for cool storage, Δ_g can just meet the lower limit of soil heat balance 80 $\%\Delta_g^0$.

6 Conclusions

This paper researched on GCHP and its auxiliary heat source system, and obtained some conclusions about soil heat balance as follows:

(1) In order to make GCHP system operate efficiently in long term, soil needs to be maintained heat balance in an annual cycle. If necessary, auxiliary heat source or cold source is also needed.

(2) A soil heat balance theory was proposed, heat balance equation was given, and annual heat balance rate of soil Δ_g was also defined. Soil heat balance point Δ_g^0 and floating range 80 $\%\Delta_g^0$ ~ 120 $\%\Delta_g^0$ were derived, whose upper limit 120 $\%\Delta_g^0$ and lower limit 80 $\%\Delta_g^0$ correspond to cold and hot critical region, respectively. And heat balance point 80 % and floating range 65 %~95 % were obtained according to experimental data, which can be used as an index to measure soil heat balance.

(3) Annual heat balance rate of soil Δ_g can be used in the following two aspects: applicable region of the GCHPSSHS system can be obtained according to the scope of Δ_g; for one region, the scope of Δ_g is used to design an efficient system which can maintain soil heat balance.

(4) To the GCHP systems adding auxiliary cold and heat source, the principles are the same, and the heat balance equations are also consistent, but the meanings of some parameters change.

Acknowledgements. This work is supported by Program for Innovative Research Team of Zhejiang Sci-Tech University, and Science Foundation of Zhejiang Sci-Tech University (ZSTU) under Grant No. 1205828-Y.

References

1. Chiasson, A.D., Spitler, J.D., Ree, S.J., Smith, M.D.: A model for simulating the performance of a pavement heating system as a supplemental heat rejecter with closed-loop ground-source heat pump systems. ASME J. Sol. Energy Eng. **122**, 183–191 (2000)
2. Yang, W., Shi, M.: Study on hybrid ground source heat pump system. Build. Energy Environ. **25**, 20–26 (2006)
3. Ma, Z., Lü, Y.: Design And Application Of Ground Source Heat Pump System. China Machine Press, Beijing (2006)
4. Kjellsson, E., Hellström, G., Perers, B.: Optimization of systems with the combination of ground-source heat pump and solar collectors in dwellings. Energy **35**, 2667–2673 (2010)
5. Lubis, L.I., Kanoglu, M., Dincer, I., Rosen, M.A.: Thermodynamic analysis of a hybrid geothermal heat pump system. Geothermics **40**, 233–238 (2011)
6. Lü, C., Zheng, M.: Effect analysis on heat storage in soil, heating and cooling for SGCHPS. J. Harbin Inst. Technol. **43**, 104–108 (2011)
7. Lü, C., Zheng, M.: Soil heat balance and heat utilization of system for SGCHPS. J. Harbin Inst. Technol. **44**, 106–111 (2012)
8. He, X., Liu, X.: Problems needed attention for GSHP application in north region. Low Temp. Archit. Technol. **26**, 85–86 (2004)

Mapping Forest Composition in China: GIS Design and Implementation

Ruren Li[1], Xinyue Ye[2(✉)], Ruixiu Wang[3], Mark Leipnik[4], and Bing She[5]

[1] College of Transportation Engineering, Shenyang Jianzhu University, Shenyang 110168, Liaoning, China
rurenli@163.com

[2] Department of Geography, Kent State University, Kent, OH 44242, USA
xye5@kent.edu

[3] Esri China Information Technology Co. Ltd., Beijing 100007, China
2241165161@qq.com

[4] Department of Geography and Geology, Sam Houston State University, Huntsville, TX 77340, USA
GEO_MRL@shsu.edu

[5] State Key Laboratory of Information Engineering in Surveying, Mapping and Remote Sensing, Wuhan University, 129 Luoyu Road, Wuhan 430079, Hubei, China
coolnanjizhou@163.com

Abstract. Geographic Information System (GIS)-based mapping and decision support systems use in forestry in China is still relatively limited, especially the large-scale mapping and specialized software development. However this application area has witnessed a growth in its market from various customers. This research has designed and implemented a new Forest Form Mapping System for China's Forest Resources, as well as a platform for research, education and decision support to achieve the goal of sustainable environmental management. The project has demonstrated how GPS data and data from the second national forest survey of China can be integrated, visualized, and reported in a.NET and ArcEngine-based system. Given its flexible architecture and user friendly interface, it is suitable for a variety of applications in forest resource management.

Keywords: GIS · Spatial data integration · Forest resources · Forest form mapping

1 Introduction

The world's first true geographic information system (GIS), the Canadian GIS, which was initiated in the middle 1960's, had a significant forest mapping and map overlay component [1]. Most American, Canadian and European commercial timber companies and the U.S. Forest Service have for decades relied on GIS and related geospatial technologies [2]. Many international efforts particularly in rainforest preservation rely of GIS as an integral part of forest management. Germany, Finland, Canada and the USA are probably at the forefront of efforts to use GIS and related technologies [3–5]. Typical

F. Bian and Y. Xie (Eds.): GRMSE 2015, CCIS 569, pp. 862–871, 2016.
DOI: 10.1007/978-3-662-49155-3_89

forestry applications include inventory, timber harvest optimization, protection of watersheds, inventory and protection of endangered species, management of access both haul and recreational road design and location, preservation of scenic resources through geo-visualization, fire management particularly fire spread modeling and more generally protection of areas like rainforests [6, 7]. Forest resource are the foundation of woodlands and vegetated wild lands generally and support all the living organism in them including many human populations that live in close proximity to forests [8]. Conservation and sustainable management of forests is the foundation of forest development, and forests and timber are an indispensable natural resource which can promote economic growth [9]. With the rapid development of our social productivity and the continuous improvement of science, technology and human civilization, the management and maintenance of forest resources also will unceasingly tend to become an information technology application area in China as it has already done in many advanced countries [10].

China has the fifth largest forest area in the world. In 2010, forest area was 1.34 million ha, which accounts for 3.9 % of total land area. Unfortunately, due to the 1.3 billion population of China, its per-capita forest area was only ranked 119th in the world. The average per-capita forest area of the world was 0.6 ha, and developing countries and developed countries possess 0.5 ha and 1.07 ha respectively. China has 9.789 billion cubic meters of timber, which is the eighth highest in the world, and this accounts for 2.55 % of world total harvestable timber biomass. Because China is a large country with a huge population and economic base, forests are important and the relatively small amount of forest land per capita must be carefully nurtured. That underscores the importance of the type of logical and advanced forest resource management fostered by the use of GIS and geospatial technologies.

2 Literature Review

In China, forestry mapping software has developed to a moderate level, and also has accumulated a lot of information and experience on the part of expert users, but it cannot meet all the needs of forest resources mapping. This paper focuses on a new advanced forest resources form mapping system aimed at addressing the problems cited above, and the implementation of the forest form mapping system on a large scale project. Specifically the project involved creation of a GIS-based forest form map at a scale of 1:2000. This approach can accurately position the stand (sub-lot) boundary related information of specific tree species, determine the area and volume of timber associated with that species, while roughly determining the positional information of single tree. Thus, it fully satisfies modern precision forestry needs, it reduces computational workload and reduces and aids field work, thus enhancing the efficiency of map generation and data management.

This system was developed under Esri's object oriented ArcGIS second generation development platform - ArcEngine, and the develop environment is dot net, the development language used was C Sharp. Developers can use ArcEngine to embed the GIS functionality into the existing application, such as Microsoft office applications, including the Word and Excel products. ArcEngine provides four development modes: COM, .net bar, Java and C++, developers can choose any development environment which supports these APIs for system development. The functions of the forest form mapping GIS software constructed by using Arc Engine include: management of vector and raster data types; support shape file (.shp) and GDB or TIF formats; read/write MXD files; have a data display and map viewer; import aerial photos or satellite images; have tracking and editing of mapping elements (such as point, line (arc), polyline, and polygon feature classes). The ArcGIS software suite has thousands of functions and scores of extensions that too numerous to detail but which greatly exceed those available in previously used forest mapping software in China [11].

3 Design and Implementation

Since forest resources mapping software should support large scale mapping and user needs, it should be designed according to the following principles: First, integrity of systems is necessary. The functions of the map drawing platform should be comprehensive, and meet the needs of users as much as possible. With the functions of spatial data and attribute data input, editing, display and query, as well as stock and thematic map generation, stock symbol creation and modification, map plotting and dissemination including over the web and to mobile devices and in GPS and surveying environments. Second, it should meet the needs of for stock production and operation, but also for professional cartographers. Third, it needs to have symbol libraries specific to forest mapping symbols in use in China. These symbols are in accordance with those used in existing 1:2000 scale maps, with different symbols for different tree species common in China. Fourth, in terms of system efficiency, the GIS design must consider the connection to large spatial databases, the management of attribute data and interoperability. Fifth, in respect to system usability, not all cartographers are forestry professionals or computer experts. Therefore, it is important to design easy and intuitive user interfaces, try to simplify interface functions for use by professionals, and make the software more convenient to use and operate. Lastly, the system must meet preset national standards and norms. These include standard forestry symbols used in forest stock mapping and conformance with surveying and national map accuracy standards. These standards specifically include: The Chinese National stock map schema, forest resources classification and codes for forest type, name and codes for state farm, the forest resources non-spatial data standard and the major technical regulations of forest resource planning, design and survey.

Using a set of standard charts and tools, the basic method of structural system design is based on the logic functioning of the system design and data flow. It is assisted, by using a data flow chart and a data dictionary. The system is divided into multiple modules which have clear functions, limited independence and are easy to implement; thus the complex system design is transformed into several simple modular design steps. The ultimate goal of the project is to create a forest resources stock mapping system in order to provide a convenient tool for large scale forest mapping. That supports functions of data input, editing, management and report/statistical output. The system design mainly follows the principles of practicability and standardization. The system design should be simple and able to solve practical problems. Additionally a requirement is that the data format and output needs to conform to existing standardized practices so that dissemination and utilization of the data can be maximized. The system design structure schematic is shown as Fig. 1. The data used in the system is mainly obtained in DEM, DOM and DLG data format. The data is typically aerial imagery and forest resource inventory data. The map database stores forestry graphics and forestry symbols, the attribute database stores attributes derived from information gathered in the field. The user database stores information established by users relevant to stock mapping. The Design layer mainly consists of the graphical user interface and related connections between ArcEngine and C#. The presentation layer consists of modules designed to support forestry data query, display, processing, editing, plotting and output.

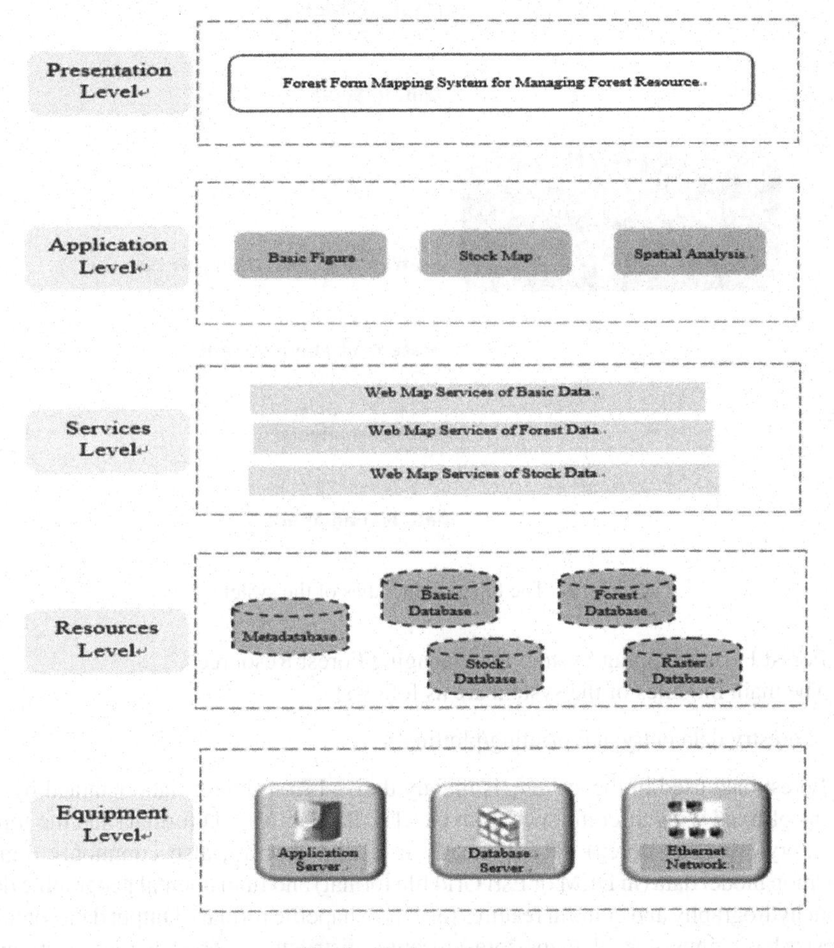

Fig. 1. Design structure.

System module functional design directly affects the use of software system. The design of the forest resources stock mapping software system is in strict accordance with the Ministry of Forestry standards because all forest lands are government controlled in China. The system's functional modules are: forestry data entry and plotting/printing, derived raster image management derived from scanned stock maps, basic map operation, forestry graph editor and output processing, and structured query language (SQL) query based determination of the absolute position of trees, spatial analysis functionality, forestry statistics management and output, stock map and special charts production, and help menu. The functional modules of the system are shown in Fig. 2.

Fig. 2. The function modules of the system

Forest Form Mapping System for Managing Forest Resource
The main modules of the system are as follows:

(1) Forestry data entry and printing/plotting.

Forest data used in the system, is mainly derived from image data obtained by the means of fixed wing aircraft (such as in GeoTif, IMG or MrSID format) and the forest inventory (second survey) data (typically in a .shp format), also commonly digital elevation model data (in DEM or Esri Grid file format) and fundamental geographic data about hydrography and cultural features (also is a shapefile format). Output data contains forestry basic figures, stock maps, forest resource distribution, forest districts, stationing figures, and special charts that reflect special investigations; and bar chart or pie chart that is generated from a SQL based query of statistical results and analysis.

(2) The basic operations of the stock map.

The basic operation of the stock map module includes the basic operations of the forestry layer and the printing of forest maps. It mainly includes amplification, narrowing, shifting and measurement (including distance measurement, area measurement, measurement of elements), scale and coordinate system display; layer display, hide, remove, the reclamation and attribute browsing with inquiry, attribute data manipulation.

(3) The editing and processing of the stock map.

The map editing function is designed to support drawing and editing the map, symbolizing the features in a layer, and executing commands such as transforming of symbols, selections of elements, deleting, modifying, copying, pasting, cutting, undo and redo steps in the editing process.

(4) Spatial analysis.

This mainly consists of a buffer zone generation function, superimposing analysis (map overlay), statistical analysis, and other spatial analysis functions. Based on what type of feature class (point, line and polygon) is involved, the buffer zone analysis function operates in three ways. The analysis is subdivided into point and polygon superposition analysis, line and polygon superposition analysis and polygon and polygon superposition analysis. Statistical analysis is based on forestry related properties, and can generate pie charts, bar charts, figures, etc.

(5) Statement management and output.

The reports generated by the system include statistics of land area, land area by type, the summarized age of trees and stumpage, stumpage accumulation, tree by tree stock statistics, shrub land area, timber type, timber volume statistics, etc.

(6) Stock map and thematic mapping

This module mainly completes sub-lo and, compartment level maps. It also supports generation of many kinds of special charts and maps. Special charts mainly includes single value special charts, which makes all objects in the layer the same style by using different settings of the fill color, line color, and line weight and line style in that layer. The software system supports templates which have preset industry standard illustrations, map titles, scale, north arrow (compass), graticule, or kilometer nets, etc.

In the discussion above of the functional modules, we introduced the main functions of the system. Here we only list the primary functions used by typical users. Symbol production tools consist of a symbol manager, symbol editor, and symbol selector. The user chooses symbol parameter settings by using the symbol editor. The symbol selector allows users to browse symbol libraries so that users can select symbols designed for specific purposes in the application system. The symbol selector is used in the period when the system is running. It allows users to add or remove symbol libraries. It is also the location where a new symbol library is created or added from the web and where individual map symbols are added, removed and revised. Although certain symbols contained in ArcGIS can be modified, resized, and moved to meet the requirements for topographic or terrain mapping, the majority of the existing symbols in the system need for of stock mapping cannot be obtained from the available symbol libraries in ArcGIS. Instead, we custom tailor some special symbols according to the required symbols adopted by the Ministry of Forestry. An example of the typical stock map along with the system main interface is shown in Fig. 3 below.

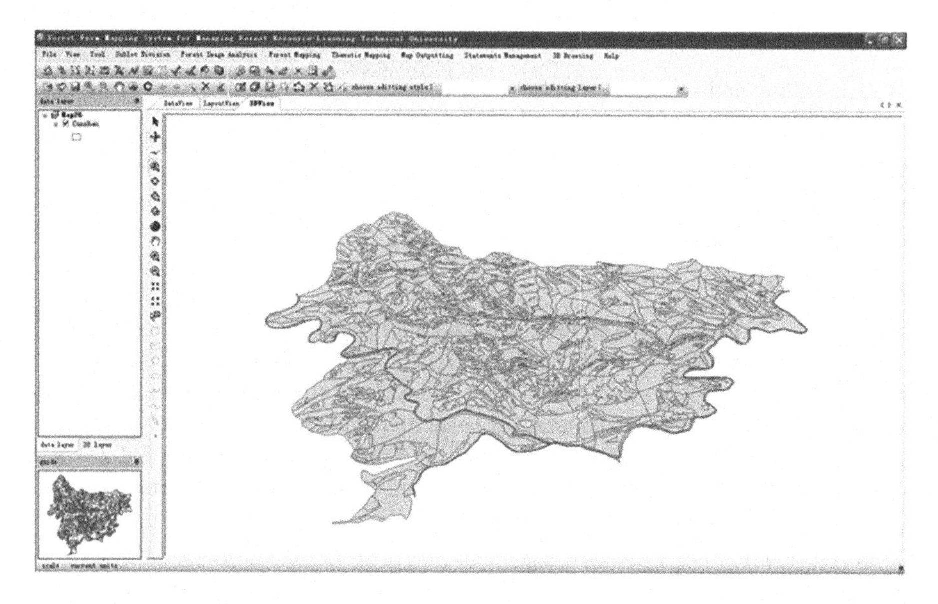

Fig. 3. An example of the typical stock map

These custom designed Chinese forestry mapping symbols as stored in special symbol library files in a *.style format that is supported by ArcMap, but what ArcEngine uses are the files in a *.server style format. So we must first create then transform the symbols.

In addition to labeling conventional geographical features (such as roads, hydrography, and human settlements) in the stock map, the paramount thing is the labeling (annotation) of forestry related information. Most of this information is noted by placing a call out box or text box inside a forest stand or compartment polygon with a label that reflects forest_id-area. The forest_id is directly extracted from the database, while the area information can be calculated by the spatial analysis capabilities of the system and added to the forest polygon ID tag. For larger features a text box is used for smaller features a call out box with a arrow pointing to the feature is used.

In summary, the system design adopts several important principals. Firstly, functions need to be as complete and comprehensive as possible with the ability to support database updates, display, query, mapping, import/export, etc. Hence, the needs of forest product production and management can be addressed to the greatest extent possible. Secondly, this software must be a tool which specifically targets forest resource management. It was designed with components that were based on a survey of actual users and is directed toward solving the real world issues and challenges faced by forestry management agencies. Thirdly, system efficiency should be optimized by integrating attribute data and spatial data, as well as minimizing data redundancy. Fourthly, the system needs to be relatively easy to use, and technical complexity should be minimized. This is because most of the users are the level of the basic administrative unit of forestry management agencies such as district offices and the workers are not well trained in computer techniques with high or even diverse educational levels. This means that the design of a convenient and intuitive graphical user interface is essential to achieve the smooth operation of the system and the adoption of the technology by a growing number of satisfied users. Last but not least, this GIS platform will need to conform to a series of laws, policies and regulations regarding forest resource management in China.

4 Conclusions

Forestry applications of GIS have a long history, with widespread use in many diverse applications areas from protecting rainforests and endangered species to timber harvest optimization to wild fire modeling to management of scenic and recreational resources in over 100 countries [12]. There is a fast-growing and pressing need to provide users in China with an integrated tool for mapping and spatial analysis in support of various tasks involved with managing forest resources.

We have reached several important conclusions: Firstly, we have determined that forest mapping and spatial analysis using GIS and related geospatial technologies like RS and GPS has reached a high level in many developed countries with important forest resources such as Canada, the USA, Germany and Finland. We have also concluded that existing forest mapping although now starting to be computerized is far behind world class standards. Secondly, we have concluded that an important part of forest mapping in China is design and creation of forestry symbol sets. The key of drawing the stock map is correct use of the various kinds of forestry symbols long used by the Ministry of Forestry. These contain information on issues such as sub-lot boundaries, compartment boundaries, various boundary line types, text and sub-lot notes and compartment notes, etc. We used ArcMap software to make the simpler custom symbols, and wrote special code to make other more complicated custom symbols. Thirdly, we present a discussion of the realization of the forest resources stock mapping software.

There is a need for ongoing research and additional system development to improve the system in respect to several important points: firstly the system has a complete set of forestry symbols, but it does not allow users to modify or add new forestry symbols by themselves; if the users need to add or modify symbols, they can only do this through technical support. Secondly, the statistics output function of the system does not connect with the database. Therefore, the monitoring of statement output is not accomplished in real-time. This step is under development and is the next improvement under development for the system.

The forest stock mapping GIS software system has been popular with users in the forest profession in China and is being increasingly adopted there. Once widely in use it will be able to effectively promote precision forestry management, digitization of forest related data, attribute data base design and entry, and the scientific construction of a modern decision support tool [11]. Compared with other previously used software systems used in forestry in China, the GIS based approach possesses important advantages such as integration of multiple co-registered spatially accurate and reliable data sources, efficient data editing functions; strong spatial analysis capabilities and rapid and standardized stock map output functionality.

China's forest resources are not abundant compared with its vast size and population; Nevertheless tens of thousands of forest administrative departments manage irreplaceable state-owned forest resources that provide timber and watersheds and habitat for many diverse plant and animal communities some unique to China such as the giant bamboo forest giant panda habitat. It is a vital issue in forestry how to use and protect these resources [12, 13]. The construction of large scale digital forest maps and an associated GIS based decision support system is an important part of this mission.

Acknowledgments. This work has been supported by State High-Tech Development Plan 863 Program (No. 2008AA121305-4), Key Project of Chinese Ministry of Education (No. 13JJD790008).

References and Notes

1. U.S. Forest Service. New I-tree software, (used in the United States, Canada, and 108 countries) so far (2012). http://www.fs.fed.us/news/2012/releases/10/valuesoft.shtml Accessed 30 May 2014
2. Williams, M.A., Baker, W.L.: Variability of historical forest structure and fire across ponderosa pine landscapes of the Coconino Plateau and south rim of Grand Canyon National Park, Arizona, USA. Landscape Ecol. **28**, 297–310 (2013)
3. Iverson, L.R., Dale, M.E., Scott, C.T., Prasad, A.: A GIS-derived integrated moisture index to predict forest composition and productivity of Ohio forests (USA). Landscape Ecol. **12**, 331–348 (1997)
4. Lehtomäki, J., Tomppo, E., Kuokkanen, P., Hanski, I., Moilanen, A.: Applying spatial conservation prioritization software and high-resolution GIS data to a national-scale study in forest conservation. Forest Ecol. Manage. **258**, 2439–2449 (2009)
5. Mustonen, J., Anttila, P., Lehtonen, A., Tuominen, S.: In method for estimating forest biomass potentials for Central Finland with biomass maps and GIS analysis. In: Precision Forestry Symposium Stellenbosch, South Africa, 1–3 March 2010
6. Sacchelli, S., De Meo, I., Paletto, A.: Bioenergy production and forest multifunctionality: a trade-off analysis using multiscale GIS model in a case study in Italy. Appl. Energy **104**, 10–20 (2013)
7. Yu, X.W., Zhang, W.G., Yang, Y.C., Lei, Z.Y., Zhang, X.: A GIS based assistant information system for forest fire prevention direction. Appl. Mech. Mater. **303**, 2215–2218 (2013)
8. Hock, B., Blomqvist, L., Hall, P., Jack, M., Möller, B., Wakelin, S.: Understanding forest-derived biomass supply with GIS modelling. J. Spat. Sci. **57**, 213–232 (2012)
9. Liu, X.F., Geng, Y., He, Z.M., Zhang, J.H., Zhang, L., Chen, Z.X., Li, D.Y.: Design and realization of forest fire monitoring system based on GIS in Henan Province, China. Adv. Mater. Res. **610**, 3665–3669 (2013)

10. Xie, H., Kung, C.-C., Zhao, Y.: Spatial disparities of regional forest land change based on ESDA and GIS at the county level in Beijing-Tianjin-Hebei area. Front. Earth Sci. **6**, 445–452 (2012)
11. Bu, R., He, H.S., Hu, Y., Chang, Y., Larsen, D.R.: Using the LANDIS model to evaluate forest harvesting and planting strategies under possible warming climates in Northeastern China. Forest Ecol. Manage. **254**, 407–419 (2008)
12. Ohmann, J.L., Gregory, M.J.: Predictive mapping of forest composition and structure with direct gradient analysis and nearest-neighbor imputation in coastal Oregon, USA. Can. J. Forest Res. **32**, 725–741 (2002)
13. Ehlers, S., Grafström, A., Nyström, K., Olsson, H., Ståhl, G.: Data assimilation in stand-level forest inventories. Can. J. Forest Res. **43**, 1104–1113 (2013)

Deformation Monitoring of Bridge Structures by Ground-Based SAR Interferometry

Zhiwei Qiu[1,2](✉), Jianping Yue[2], Xueqin Wang[2], and Shun Yue[2]

[1] Key Laboratory of Precise Engineering and Industry Surveying of National Administration of Surveying, Mapping and Geoinformation, Wuhan 430072, China
qiuzhiwei-2008@163.com
[2] Earth Science and Engineering, Hohai University, Xikang Road 1, Nanjing 210098, China

Abstract. In this paper, a ground-based SAR interferometry technology was used to monitor major engineering. This technology has been recognized as a powerful tool for terrain monitoring and structural change detecting. Deformation monitoring for Bridge has been a hot issue among them. According to GBSAR interferometry principle and characteristics of IBIS system, the authors analysis the error sources of deformation monitoring, and experimentally extract atmospheric phase which should removed based on permanent scatterer analysis. Atmospheric disturbance effect analysis is discussed in this paper, and an atmospheric correction method is proposed to remove atmospheric effect, then the effective displacement can be retrieved. Results from this approach have been compared with that from traditional method in this campaign, GBInSAR technology can be exploited successfully in deformation monitoring for major projects with high accuracy [1–3].

Keywords: Atmospheric effect · GBSAR interferometry · IBIS system · Permanent scatterer (PS)

1 Introduction

Ground-based SAR system can receive the images for illuminated area by active imaging capability. Synthetic aperture radar and step-frequency continuous wave technology are employed to improve the special resolution in azimuth and rang direction. Interferometry technique can improve the accuracy on a fine resolution map of the scene. Typical applications are: landslide [4–7], glaciers [8], the dam [9], building [10] and bridges [11] and earthquake, volcanic hazard assessment, land subsidence monitoring [12]. GBInSAR with high precision, low cost, special continuity is a new approach for monitoring; its high sampling frequency is sensitive to the movements of architectural structures such as bridges and dams in real time. This technique overcomes deficiencies belonged to space-borne SAR such as spatial and temporal decorrelation, low resolution.

GBInSAR technology has been largely used in interference applications in recent years, technology defects such as atmospheric effects, temporal and special decorrelation

F. Bian and Y. Xie (Eds.): GRMSE 2015, CCIS 569, pp. 872–881, 2016.
DOI: 10.1007/978-3-662-49155-3_90

become more prominent. In 2004, Guido Luzi et al. analyzed the decorrelation caused by baseline or other factors combined with GBInSAR monitoring data for landslide somewhere, and proposed atmospheric effects can be eliminated through the PS points. In 2005, Linhsia Noterini pointed out the true deformation values were submerged in phase errors caused by atmospheric disturbance for valley monitoring data, then the effective displacement could be extracted by meteorological correction for GBInSAR data with PS technique [17]. In 2006, Massimiliano Pieraccini et al. used GBInSAR technology to monitor the landslide somewhere unstable, eliminated the effect caused by temporal and meteorological decorrelation through PS technique, and verified the effectiveness of PS in GBInSAR.

2 Atmospheric Disturbance Analysis

As the displacement can be extracted from interference phase received by Interferometry SAR, the phase quality affected by decorrelation is most important factor for deformation monitoring. The decorrelation of space borne radar is due to: a. geometric decorrelation caused by baseline; b. decorrelation caused by doppler shift; c. temporal decorrelation caused by revisiting period or temperature, meteorological condition. Although the measurement principle of GBInSAR is same with space-borne SAR in essence, the first and second one can be ignored because there is no baseline for GBSAR. Therefore, the temporal decorrelation caused by atmospheric disturbance is the key factor to improve phase quality for GBInSAR [12, 13].

Atmospheric refractivity n changing with temporal and special distribution is the source of phase error due to atmospheric disturbance. Firstly, we can suppose radar frequency as f, the echo phase of the target located with a distance r from radar antenna center can be expressed as:

$$\varphi(t) = \frac{4\pi f}{c} \int n(r,t)dr \tag{2.1}$$

In Eq. 3.1, c is the propagation speed of radar wave in vacuum, refractivity n is the function of time t and range r. If the range R between static target and radar is constant, refractivity n only has relation with time t, the echo phase difference between times t_1 and t_2 can be expressed as:

$$\Delta\varphi = \varphi(t_2) - \varphi(t_1) = \frac{4\pi fR}{c}[n(t_2) - n(t_1)] \tag{2.2}$$

The phase difference $\Delta\varphi$ is interference phase contributed by atmospheric disturbance. If the atmospheric conditions in time t_1 and t_2 are identical totally, the phase difference will be zero. However, the atmospheric condition is definitely not the same in different time. Therefore, this error would not be zero and should be corrected.

GBInSAR signal once travels in troposphere, refractivity n is the function of degree Kelvin T, atmospheric pressure P and humidity H. The refractivity is very close to 1 as usual, so the refraction N always takes place of the refractivity in electromagnetic wave propagation. The relation between them can be expressed as:

$$N = (n - 1) \times 10^{6} \tag{2.3}$$

Tropospheric delay can be divided into two components, dry N_{dry} and wet N_{wet}. The refraction N can stated as:

$$N = N(P, T, H) = N_{dry} + N_{wet} = 0.2589\frac{P_d}{T} + (71.7 + \frac{3.744 \times 10^5}{T})\frac{e}{T} \tag{2.4}$$

In Eq. 3.4, P is total atmospheric pressure, P_d is dry atmospheric pressure, e is water vapor pressure, the unit of them is mbar. The relation between them is $P_d = P - e$, the water vapor pressure can be expressed as (e_{sat} is standard vapor pressure saturation):

$$e(T, H) = \frac{H}{100} \cdot e_{sat}(T) = \frac{H}{100} \cdot 6.1016 \times 10^{(\frac{7.5T}{T+237.3})} \tag{2.5}$$

We can see that dry delay component has relation to total atmospheric pressure P and degree Kelvin T, and wet delay component has relation to water vapor pressure e and degree Kelvin T. As the water vapor distributions in troposphere vary in time and space greatly, wet delay component is main factor of atmospheric effect.

Therefore, phase difference caused by atmospheric effect is a function composed of range R and atmospheric refractivity difference Δn_{atm}. Supposed the central frequency is 2 GHz, the error curves are shown in Fig. 1. The error caused by atmospheric effect is up to millimeter level when the atmospheric refractive difference increases with volume 10^{-5}, and the displacement is proportional to range, so we should take measures on the atmospheric effect correction.

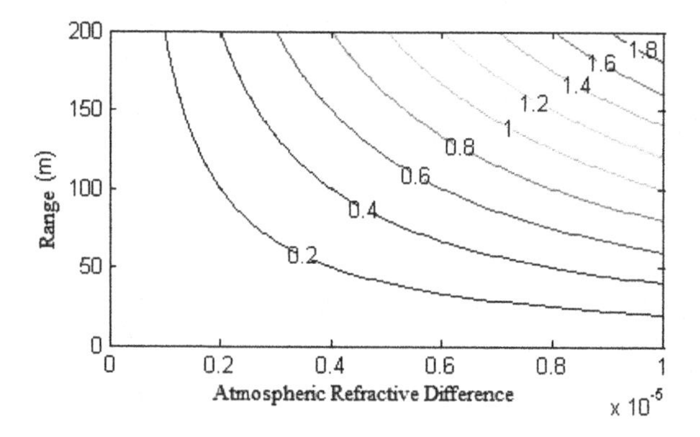

Fig. 1. Atmospheric disturbance effect for GBsar.

3 Artificial Control Point Atmospheric Correction Method

Stable artificial targets was placed in illuminated area before measurement, artificial control point correction method can eliminate the atmospheric disturbance by analyzing interference phase of these control point targets. Spain's L. Pipia and South Korea's H. Lee et al. arranged a stable metal reflector as artificial control points in observation region. The atmospheric effect were restricted by these points, the errors were rectified well. The stable artificial target can be regarded as a special PS point, and the parameters of the point can be accurately measured, processed simply. Therefore this method of atmospheric effects correction will be accomplished in our experiment, and then we will analyze this method.

Supposed $\varphi_{atm}(r,t)$ stands for phase error caused by atmospheric effect, the interference phase can be expressed as Eq. 3.6 without considering the influence due to the vibration of scatterers and the noise caused by system.

$$\varphi(r,t) = \varphi_{dis}(r,t) + \varphi_{atm}(r,t) \tag{3.1}$$

In Eq. 3.6, $\varphi_{dis}(r,t)$ means interference phase due to displacement, this value would be zero for static target. If the illuminated region is small scale, this phase model is linear:

$$\varphi_{atm}(r,t) = a \cdot r \tag{3.2}$$

The phase $\varphi_{dis}(r,t)$ is zero for artificial target without movements, $\hat{a}$ is the estimated value of coefficient a can be expressed as:

$$\hat{a} = \varphi_0 / r \tag{3.3}$$

The average value of a few control points can be calculated to improve the estimation accuracy as usual, other monitoring targets will be corrected with this value, $\varphi_{corr}(r,t)$ interference phase after correction is:

$$\varphi_{corr}(r,t) = \varphi(r,t) - \hat{a}(t) \cdot r \tag{3.4}$$

When the monitoring region is large scale and complex, the interference phase due to atmospheric effect which is nonlinear with the range r can be expressed by quadratic function:

$$\varphi_{corr}(r,t) = \varphi(r,t) - \hat{a}_1(t) \cdot r - \hat{a}_2(t) \cdot r^2 \tag{3.5}$$

The equation above can be solved by two stable artificial targets at least, the coefficients a_1 and a_2 will be calculated for atmosphere correction. Finally, the phase after correction is:

$$\varphi_{corr}(r,t) = \varphi(r,t) - \hat{a}_1(t) \cdot r - \hat{a}_2(t) \cdot r^2 \tag{3.6}$$

3.1 Atmospheric Correction Experiment

This experiment platform is IBIS system; artificial targets observed are reflector with three corners. Artificial control point atmospheric correction method is employed to eliminate atmospheric disturbance here.

The target distance is within 100 m, the frequency center is 16.9 GHz, and wavelength is 17.6 mm. The range resolution is 0.5 m, and the dynamic monitoring measurement accuracy is 0.01 mm, static monitoring measurement precision is 0.1 mm. Other system parameters of IBIS are shown in Table 1.

Table 1. System parameters of IBIS.

Parameter name	Parameter value
Target distance	1–100 m
Bandwidth	200 MHz
Central frequency	17.0 GHz
Range resolution	0.5 m
Measurement accuracy	0.01 mm/0.1 mm
Max samples	200 Hz

Test site is Stone City along the Qinhuai River in Nanjing as shown in Fig. 2, two corner reflectors CR1 and CR2 was arranged along the bank of this river. 24 IBIS measurements were received during the time from 10:00 to 17:00, the reflector CR2 was static all the time and the other had displacement with a few millimeters recorded by venire caliper in this experiment.

Figure 3 describes the range profile of IBIS data, the X axis means pixels (pixel resolution is 0.5 m), and Y axis means the reflectivity. The accurate distance of corner reflectors and the piers are measured automatically by Georobot, the location of these points are demonstrated in Fig. 3.

Fig. 2. Atmosphere disturbance experiment scene correction and corner reflector.

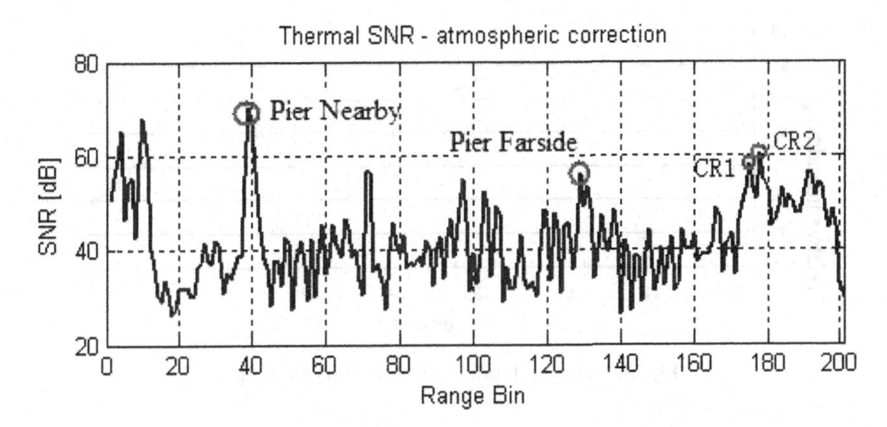

Fig. 3. The range profile of IBIS data.

The atmospheric variation for illuminated area during data acquisition is drawn in Figs. 4 and 5, the atmosphere pressure of the scene was kept within 1006 hPa to 1010 hPa, the temperature varied from 25 °C to 30 °C, and relative humidity was from 39 % to 47 %. This measuring system has strong stability, the phase error caused by system and frequency shift can be ignored. As the corner reflectors are stable scatterers, phase error due to scattering characteristic is also negligible. Therefore, it can be considered that the interference phase is mainly caused by the atmospheric effect for stable artificial point target CR2 and the piers, but the interference phase for CR1 is influenced by the target displacement and atmospheric effect.

For noise reduction, the displacements of each target during observation period are averaged, and the processed displacements for piers nearby and farside are drawn according to the monitoring period in Fig. 6. Because the piers are stable and unshakable, the displacements measured for piers can be recognized as the phase error caused by atmospheric effects which is proportional to the distance, pier farside has greater variations than nearby obviously.

Since the scene is relatively small, the atmospheric effect factor a can be estimated from Eq. 3.7. The "distant pier" and "pier nearby" can be seen as the stable control

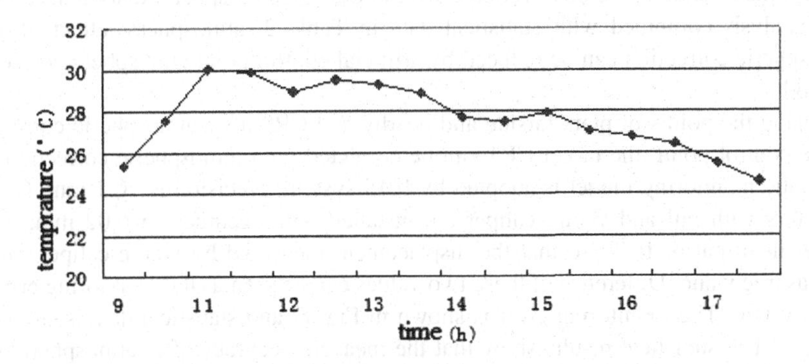

Fig. 4. Temperature variation of the scene.

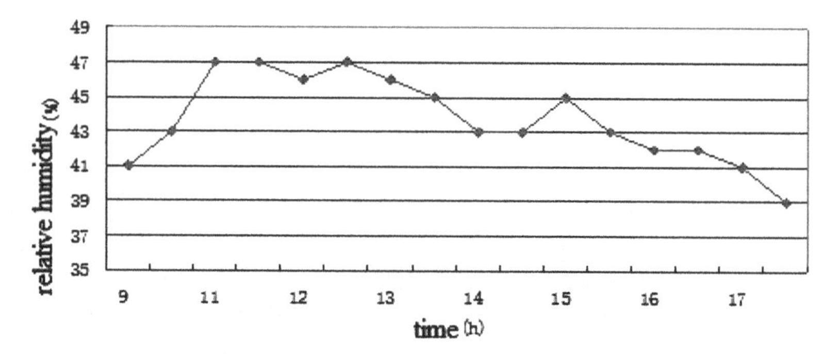

Fig. 5. Relative humidity variation of the scene.

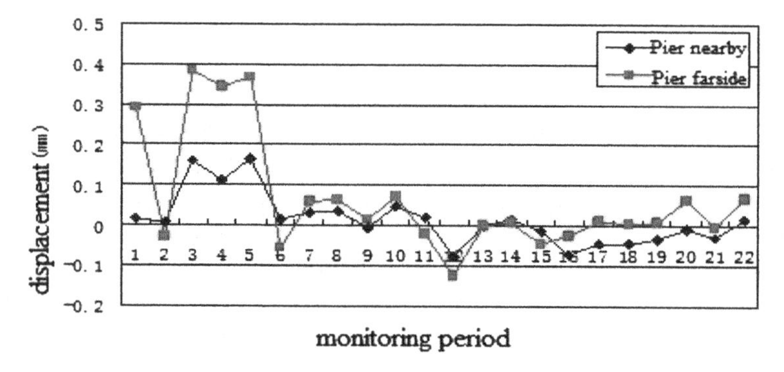

Fig. 6. Displacement diagram for pier nearby and farside.

point to estimate atmospheric influence factor. After taking the average of these two estimations as a regional atmospheric influence factor, coarse phase received by IBIS can be rectified by Eq. 3.9. After atmospheric influence factors have been finished, the displacements before and after atmospheric correction for CR2 with the time series can be given in Fig. 7.

Since the corner reflector CR2 is static, and the displacement should be zero, the displacement value is the observation error induced by atmospheric disturbance. After data analysis combined with statistical data in Table 2, atmospheric effect through atmospheric correction can be reduced by artificial control point atmosphere correction method.

Using the points of piers farside and nearby and CR2 after atmospheric correction as the control point, the target CR1 can be corrected with atmospheric correction.

Static monitoring model is adopted by IBIS system, precision is ±0.1 mm. Corner reflectors with rail and venire caliper are installed with accuracy of 0.02 mm, much higher than that of IBIS, so that the displacement measured by venire caliper can be seen as true value. Difference of these two values can be seen as the monitoring error of IBIS system. The monitoring error is shown in Fig. 8, and statistical data is shown in Table 3. The statistical results show that the measure accuracy after atmospheric correction is improved greatly, and the max error could be controlled within 1 mm.

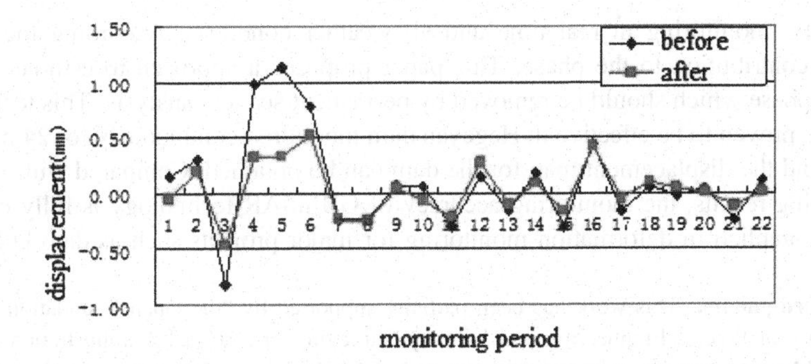

Fig. 7. Displacement with atmospheric correction before and after for CR2.

Table 2. The statistical error of CR2 with atmospheric correction before and after.

	MEAN	RMSE	MAX
BEFORE	0.31 mm	±0.32 mm	1.13 mm
AFTER	0.19 mm	±0.15 mm	0.52 mm

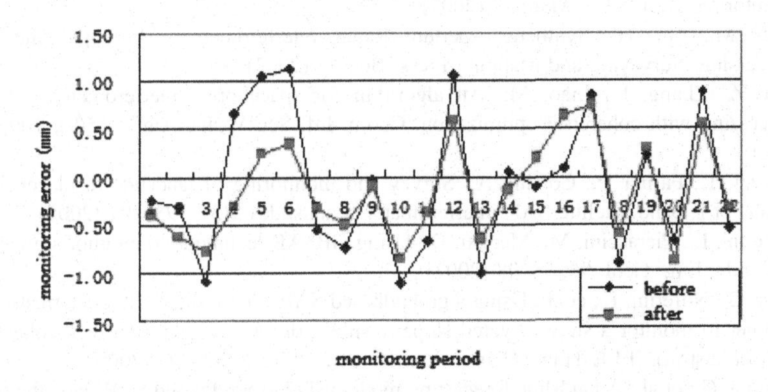

Fig. 8. Displacement atmospheric correction before and after for CR1.

Table 3. The statistical error of CR1 atmospheric correction before and after.

	MEAN	RMSE	MAX
BEFORE	0.54 mm	±0.52 mm	1.12 mm
AFTER	0.46 mm	±0.21 mm	0.73 mm

4 Conclusion

GBInSAR technology is a useful tool for extracting deformation by virtue of radar interferometry technique, which has been applied in deformation monitoring for major projects such as dam, bridge and slope. Compared with traditional method, GBInSAR is a kind of potential monitoring technology with high precision, 2D imaging capability, monitoring in real time and all weather condition. According the atmospheric contribution to the phase, This paper proposes a approach to estimate atmospheric phase which should be removed by permanent scatters analysis. This technique has been proven to be effective in Hegeyan dam monitored continuously for 24 h in this paper, and the displacement map for the dam can be obtained. Compared with vertical monitoring results, the monitoring accuracy of GBInSAR technology is fully capable of being applied in deformation monitoring for major projects such as dam [14, 15].

Acknowledgments. This work has been partially supported by "the Open Foundation of Key Laboratory of Precise Engineering and Industry Surveying of National Administration of Surveying, Mapping and Geoinformation," (Grant No. PF2013-2), and also been funded by "The Scientific Research Innovation Projects of Ordinary University Graduate Students in Jiangsu Province," (Grant No. KYLX_0498).

References

1. Yue, J., Fang, L., Li, N.: Research advances of theory and technology in deformation monitoring. Bull. Surv. Mapp. **7** (2007)
2. Liao, M., Lin, H.: Synthetic Aperture Radar Interferometry— Principle and Signal Processing. Surveying and Mapping Press, Beijing (2003)
3. Qiu, Z., Zhang, L., Liao, M.: An algorithm for spaceborne interferometric sar signal processing with coherence optimization. Geom. Inf. Sci. Wuhan Univ. **35**(9), 1065–1068 (2010)
4. Strozzi, T., Farina, P., Corsini, A.: Survey and monitoring of landslide displacements by means of L band satellite SAR interferometry. Landslides **2**(3), 193–201 (2005)
5. Noferini, L., Pieraccini, M., Mecatti, D.: Using GBSAR technique to monitor slow moving landslide. Eng. Geol. **95**, 88–98 (2007)
6. Luzi, G., Noferini, L., et al.: Using a groundbased SAR interferometer and a terrestrial laser scanner to monitor a snow covered slope: results from an experimental data collection in Tyrol (Austria). IEEE Trans. Geosci. Remote Sens. **47**(2), 382–393 (2009)
7. Herrera, G., et al.: A landslide forecasting model using ground based SAR data: the portalet case study. Eng. Geol. **105**(3/4), 220–230 (2009)
8. Luzi, G., et al.: Monitoring of an alpine glacier by means of ground based SAR interferometry. Geosci. Remote Sens. Lett. IEEE **4**(3), 495–499 (2007)
9. Mario, A., Giulia, B., Alberto, G.: Measurement of dam deformations by terrestrial interferometric techniques. In: Congress of the International Society for Photogrammetry and Remote Sensing in Beijing, ISPRS, pp. 133–139 (2008)
10. Tarchi, D., Rudolf, H.: Remote monitoring of buildings using a ground-based SAR: application to cultural heritage survey. Remote Sens. **21**(18), 3545–3551 (2000)
11. Dei, D., Pieraccini, M., et al.: Detection of vertical bending and torsional movements of a bridge using a coherent radar. NDT E Int. **6**, 741–747 (2009)
12. Pipia, L., Fabregas, X., Aguasca, A., Lopez-Martinez, C., Mallorqui, J., Mora, O.: A subsidence monitoring project using a polarimetric GB-SAR sensor. In: Workshop POLinSAR, vol. 1, pp. 22–26 (2007)
13. Leva, D., Nico, G., Tarchi, D., et al.: Temporal analysis of a landslide by means of a ground-based SAR interferometer. IEEE Trans. Geosci. Remote Sens. **41**(4), 745–752 (2003)

14. Takahashi, K., et al.: Continuous observation of natural disaster affected areas using ground-based SAR interferometry. Appl. Earth Obs. Remote Sens. **6**(3), 1–8 (2013)
15. Bernardini, G., et al.: Dynamic monitoring of civil engineering structures by microwave interferometer. In: Conceptual Approach to Structural Design Venice, vol. 6 (2007)

The Fishing Ground Analysis and Forecasting Information System for Chinese Oceanic Fisheries

Weifeng Zhou(✉), Xuezhong Chen, Xuesen Cui(✉), Wei Fan(✉), Shenglong Yang, Fenghua Tang, Xiumei Fan, Chengjun Hua, Yumei Wu, Heng Zhang, and Shengmao Zhang

Key Laboratory of Fisheries Resources Remote Sensing and Information Technology, East China Sea Fisheries Research Institute, Chinese Academy of Fishery Sciences, Jungong Roud 300, Shanghai 200090, China

{zhouwf,xuezhong,cuixuesen,fanw,yangsl,tangfh,fanxm,huacj,wuym, zhanghl,zhangsm}@ecsf.ac.cn

Abstract. Chinese oceanic fisheries have already covered seven main ocean area, including East Pacific, West Pacific, Middle Atlantic, North Pacific, Southeast Pacific, Southwest Atlantic and Indian Ocean. Based on the ocean environment data obtained from remote sensing and fishing historical data, the fishing ground analysis and forecasting information system has been constructed for Chinese oceanic fisheries. This paper briefly described the concerned remote sensing data, system architecture, client and models. Then a case application for tuna of the Indian Ocean was given at the end of the paper. The whole system has already in operation as making and delivering the forecasting information product every week.

Keywords: Fishing ground forecasting · Ocean environment · Remote sensing

1 Introduction

Ocean is the fish living environment, influencing fish actions considerably. Satellites can see a large area of the ocean at one time as they orbit Earth to get the information about the ocean. The acquisition and interpretation of oceanic environmental information have the great significance for the rapid location of fishing grounds. Remote sensing techniques show great potential to support global fisheries management and the exploitation of pelagic species. Both short- and long-term environmental variations are often reflected in satellite-derived oceanographic variables such as ocean temperature and primary productivity. Joint analysis of satellite and historical fishing catch data can be used to identify habitat changes and their impact on migration, size or recruitment of a particular fish stock, helping to regulate maximum catches and to preserve fishing resources. With such support of satellite-derived information, fisheries management decisions can be made to protect target species from possible overfishing [1].

Co-corresponding author. Funded by the Key Technologies R&D Program of China (No. 2013BAD13B06).

F. Bian and Y. Xie (Eds.): GRMSE 2015, CCIS 569, pp. 882–889, 2016.
DOI: 10.1007/978-3-662-49155-3_91

The fishing ground analysis and forecasting information system has been constructed to provide on-line analysis and interactive process. It is an integrated platform of analysis of fishing grounds environment and fishing ground probability forecasts. The system has involved over 15-year span of the world's major fishing grounds' environmental data and production data. It has the capabilities to achieve a variety of environmental data and fishery data, query, to playback, to overlay, and to visualize these data via the Internet for data transmission in near real-time.

2 Main Oceanic Fishery Area

From 1985 to now, with a near-30-year development, Chinese oceanic fisheries have already covered seven main ocean area, including East Pacific, West Pacific, Middle Atlantic, North Pacific, Southeast Pacific, Southwest Atlantic and Indian Ocean (Fig. 1).

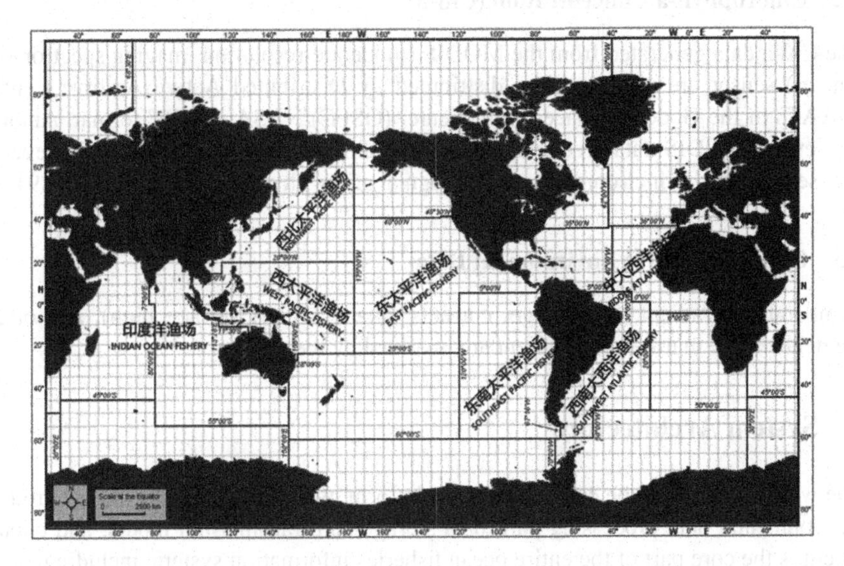

Fig. 1. Distribution map of Chinese pelagic operation fishing ground

3 Remote Sensing Data

Many satellites and remote sensors provide data on oceanographic parameters as standard products. Sea surface temperature and Chlorophyll-a concentration are the main two environmental data which were considered in the system, and also known as the critical parameters to the marine fishery ecosystem.

3.1 Sea Surface Temperature (SST)

Ocean temperature often determines the scope of most fish to survive. Sea surface temperature (SST) is an important geophysical parameter. In the past, SST could only be measured by ships and buoys, whose ranges were limited. Satellite technology has improved upon our ability to measure SST by allowing frequent and global coverage. On a more local scale, SST can be used operationally to assess eddies, fronts and upwelling for marine navigation and to track biological productivity. Eddies, fronts and upwelling are important marine environment features, which can affect fish distribution significantly.

SST data were generated from the information collected by the Advanced Very High Resolution Radiometer (AVHRR) sensor on board the National Oceanic and Atmospheric Administration (NOAA) satellites. This data set is produced and distributed by the Physical Oceanography Distributed Active Archive Center (PODAAC) of the Jet Propulsion Laboratory (JPL)/National Aeronautics and Space Administration (NASA) in the Hierarchical Data Format (HDF). SST data were in a regular grid of 9 × 9 km.

3.2 Chlorophyll-a Concentration (Chl-a)

The Chl-a data generated from the MODIS (moderate-resolution imaging spectroradiometer) sensor are produced and distributed by Distributed Active Archive Center (DAAC) of the Goddard Space Flight Center (GSFC)/NASA in HDF format. Among the several MODIS data products, available in different processing levels, the level 3 data set was used, which corresponds to a regular grid with spatial resolution of 9 × 9 km.

3.3 Other Derivative Environment Data

Temperature gradient, SST range, currents data, SSH(sea surface high), and other elements of the marine environment are involved by the system.

4 System Architecture

The whole system can be considered as a platform, which is composed of background environmental data processing part, data storage and transmission model, and client. Client is the core part of the entire ocean fisheries information system, including automatic data update, visualization of ocean environment data, fisheries production statistics and its visualization, and management of four fishing grounds forecasting models. Figure 2 is the graph of system architecture.

4.1 Background Environmental Data Processing Model

Background environmental data processing model is responsible for processing the ocean environment data, including sea surface temperature and Chlorophyll-a concentration, which are in the Hierarchical Data Format (HDF). Depending presetting area, the data in one week will be combined and cut. It also products vector or stream files from ocean current data in shapefile format. Background data processing model has processed all the various dimensions of data.

4.2 Data Storage and Transmission Model

In order to improve the efficiency of storage and management, to promote to be convenient for follow-up program development, the platform system defined a unified data format, stored in binary file form. This data file consists of two parts. See Fig. 3.

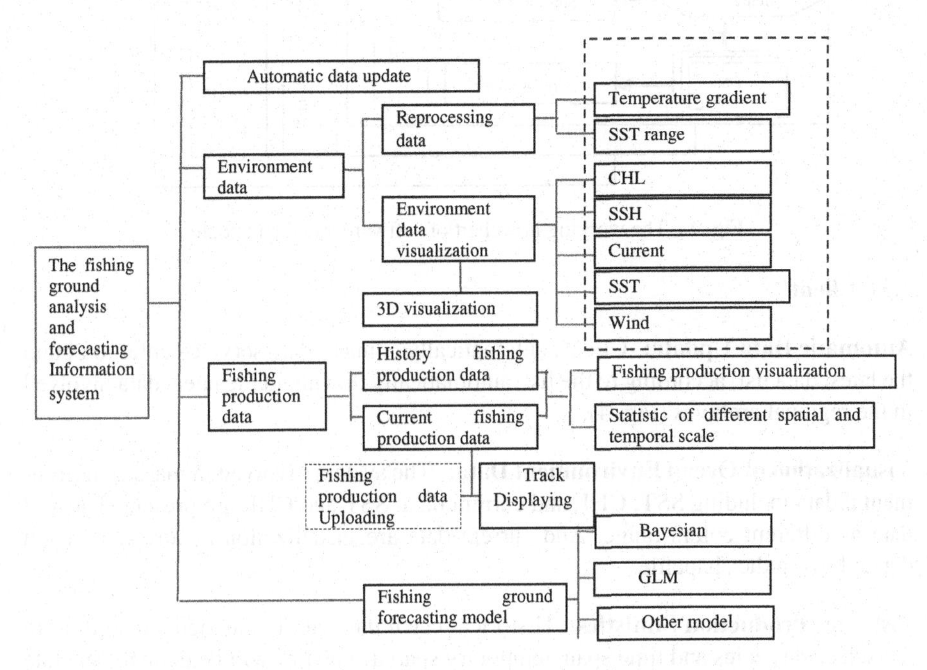

Fig. 2. The architecture of the fishing ground analysis and forecasting information system

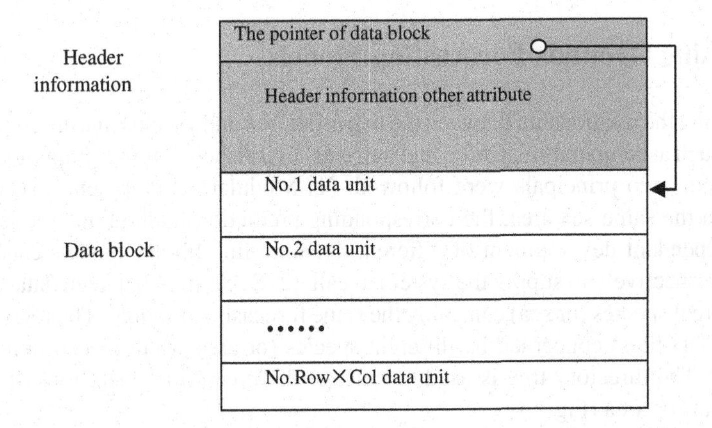

Fig. 3. Normalized grid data storage format of the fishing ground analysis and forecasting information system

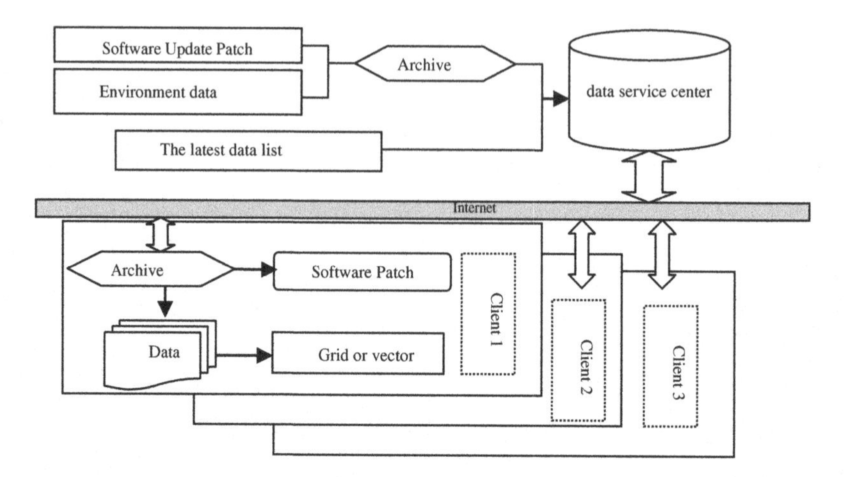

Fig. 4. The working flowchart of software and data update

4.3 Client

Automatic Data Update: Clients automatically connect data service center to obtain the latest data list, according to the list, automatically download the latest data involved in the regional archives (Fig. 4).

Visualization of Ocean Environment Data: The system involved in marine environmental data including SST, CHL and current data. SST and CHL are presented as grid data as different colors range. And current data are visualization as arrow or steam depending on the shapefile.

Fisheries Production Statistics: Historical catch data can be queried. According to the selected species and time span, temporary spatial statistics will be done to generate thematic pie chart 3.

5 Fishing Grounds Forecasting Models

Considering the relationship between the fish migration and various marine environment factors, such as temperature, Chl-a, and currents, four fisheries forecasting models have established. Two principals were followed: (1) Modular development. For the same species in the same sea area, the corresponding prediction method may be more than one. Independent development of different dynamic link library models can be independent respectively to supply the system to call. (2) Separation between data and code. The different species (or sea) can enjoy the same forecasting module. The relevant database (or data files) concerned by different species (or area) were placed in a different directory. The directory tree is used to manage the prediction operations of different species data or area (Fig. 5).

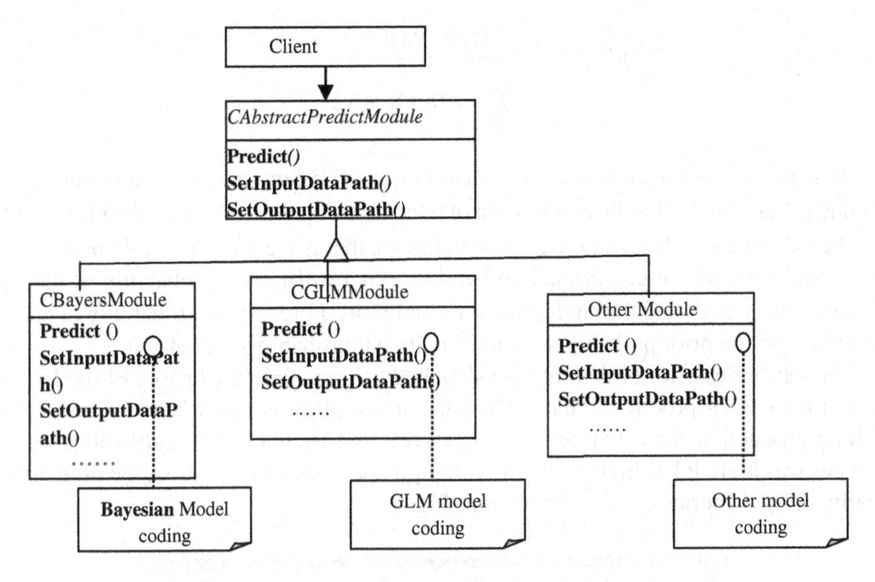

Fig. 5. Class structure of fishing ground forecasting model

6 Application Case: Tuna of the Indian Ocean

The Indian Ocean, having about 20 % of the global tuna production, is the second largest proportion of principal tuna market in the world [2] (Fig. 6).

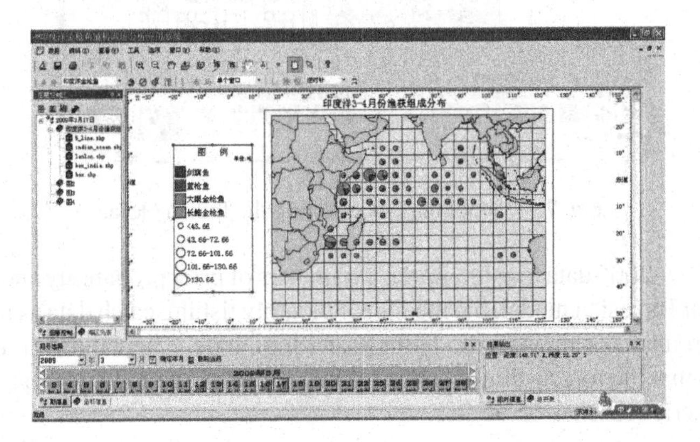

Fig. 6. Thematic pie chart of catch data of bigeye tune in the Indian Ocean

6.1 Bayesian Probabilistic Forecasting Model

Based on the framework of Bayesian Probability model, the prediction model has been set up for the Indian Ocean tuna longline fisheries [4].

$$p(h_0/e) = \frac{p(e/h_0) \times p(h_0)}{\sum_{i=0}^{1} p(e/h_i) \times p(h_i)} \quad (1)$$

Which, h_0 is assumed to be true situation, where fishing area can be defined as a fishing ground, and h_i for the assumption of none true situation. $P(h_0/e)$ is the probability for the fishing grounds under the given conditions, that is the posterior probability. $P(e/h_0)$ is the environment conditional probability. $p(h_0)$ is the prior probability of fishing grounds not considering the environmental conditions. Forecast the probability of major fisheries and the prior probability by its conditional probability calculated.

Tuna fisheries prior probability is calculated using mainly the history of the Indian Ocean tuna catch production data. The basic assumption is that where the history of fishing production (or catch per unit effort, referred to as CPUE) and fishing yields catches was high, it has high probability of high abundance distribution, the high probability of the fisheries grounds formation (Fig. 7).

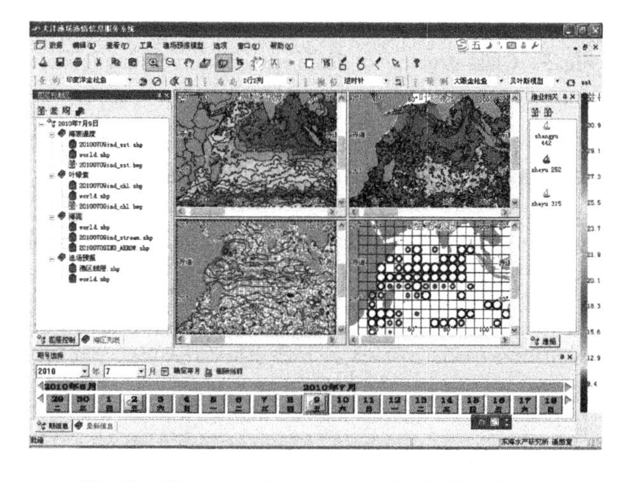

Fig. 7. Client interface covering the Indian Ocean

Using 1997–2006 data to estimate the parameters of prior probability and conditional probability of Bayesian model, the median of monthly fishing catch data as the threshold of fishing grounds for the specific latitude and longitude. The validation result of the model show that the forecasting accuracy of the model for the Indian Ocean bigeye tuna longline fisheries reaches 65.96 %.

References

1. Zagaglia, C.R., Lorenzzetti, J.A., Stech, J.L.: Remote sensing data and longline catches of yellowfin tuna (*Thunnus albacares*) in the equatorial Atlantic. Remote Sens. Environ. **93**(1–2), 267–281 (2004)

2. FAO: Review of the status of the world fishery resources: marine fisheries. FAO Fisheries Circular No. 920 FIRM/C920, Rome (1997)
3. Wei, F., Xue-zhong, C., Xin-qiang, S.: Tuna fishing grounds prediction model based on Bayes probability. J. Fish. Sci. China **13**(3), 426–431 (2006)
4. Zhou, W., Fan, W., Cui, X., Yang, S., et al.: Fishing ground forecasting of Bigeye Tuna in the Indian Ocean based on bayesian probability model. Fish. Inf. Strategy **27**(3), 214–218 (2012)

Research on Labeling Land Use Management Information for Land Use Change Survey Based on "3S" Technology

Chenchen Guo(✉), Shenmin Wang, Xiaohua Shao, Qifang Ma, Xinyue Zhang, and Wenqi Liu

College of Geography and Remote Sensing, Nanjing University of Information Science and Technology, Nanjing 210044, China
975533895@qq.com

Abstract. Many scholars have done a lot of research on "3S" technology in application of land change. But theoretical research for land management information system of labeling is less. In this paper, based on the "3S" technology as the support, researched construction land management information sheathed annotation of Nanjing area, and analyzed the study results. The conclusion shows: The 3S technology application in the register of land management information has a great advantage, to a great extent; improve the working efficiency and the accuracy of the result data.

Keywords: 3S technology · Land use change survey · Nested annotation · Nanjing

1 Introduction

Land use survey is on the basis of national land survey, according to the change of the urban and rural land use situation and the ownership, to investigate the cadastral alteration and land use change of towns and villages, and summarize on a regular basis. Investigation of land use is the basis of land resources, asset management, the survey data of land use is the basis of national economic plans, macro-control policies and the preparation of land use planning [1, 2]. Our country carried out the "Investigation on the current situation of land use in China" (the first land detailed survey) from 1984 to 1996, obtained the number of land resources, distribution, utilization and the status of the land ownership. Since that, our country has carried out the investigation of land change across the country every year. Due to the first round of land survey and land change survey conducted yearly is still in the traditional artificial visual interpretation and simple measuring method, the accuracy of the survey data is hard to guarantee. The second national land survey (second tone) completed from 2007 to 2009, integrated use of Remote Sensing technology (RS), geographical Information System (GIS), Global

College Students Practice and Innovation Training Program of 2014(Study on the theory and practice of regional land smart use, 201410300047).

F. Bian and Y. Xie (Eds.): GRMSE 2015, CCIS 569, pp. 890–900, 2016.
DOI: 10.1007/978-3-662-49155-3_92

Positioning System (GPS), database and network communication technology, the combination of inside and outside the industry to the investigation method, obtained the land types, area, the ownership and distribution information. Since 2010, our country take the annual "One Map" project construction combined with the land change survey, to achieve nationwide remote sensing monitoring, and strengthen the scientific nature of the dynamic monitoring.

RS technology can obtain real-time terrain information to identify the land which land use changed; GPS technology provides all-weather, continuous, high precision positioning; GIS technology, based on the powerful function of spatial data processing and analysis, can be used to manage and storage data dynamically. "3S" technology integrated by these three, grasping the feature of land resource utilization accurately, realizing the dynamic monitoring of land resources, having the incomparable superiority to traditional survey method, is widely used at the beginning of the land change survey.

But at present, the application of the "3S" technology in the land change is mainly in the change of the plot, the verification and the database change, remote sensing monitoring pattern set alignment annotation is less. Therefore this paper take Nanjing area as an example, explore the feasibility of applying remote sensing monitoring images to set of synthetic annotation, in order to provide guidance in land change survey in land management information fast and accurate annotation.

2 Research Methods and Principles

2.1 Research Methods

This paper mainly adopts the method of "3S" technology. The "3S" technology is a general designation of remote sensing technology, geographic information system and global positioning system. The RS is a science and technology which detecting target under the condition of non-contact target object feature by remote sensors, obtains the reflection, radiation and scattering of electromagnetic wave information to extract, judge, process and analyze the information [3]. Global positioning system is a satellite navigation and positioning system, established by the United States, using the system, users can realize the all-weather, continuous, real-time three-dimensional navigation, location and velocity in the global scope; In addition to the use of the system, the user can also carry out high precision time transfer and high precision positioning [4]. Geographic information system, which supported by computer hardware and software system, is a data acquisition, storage, management, operation, analysis, display and description of technical system, for the whole or part of the earth surface (including the atmosphere) space about the geographic distribution [5, 6].

"3S" integration technology, GIS, GPS, RS has a natural advantage of complementary. In the study of land change survey, RS is a major means of information gathering, used to detect the land use change in the plot; GPS is mainly used for real-time, rapid verification of suspected change of land, and to give the coordinates and other properties; GIS is a new integrated system, which is based on the comprehensive processing and management of the multi-source spatial-temporal land use data.

2.2 Research Idea

Utilizing the characteristics that remote sensing image can reflect the characteristics of the spectral to produce digital orthophoto map (DOM). Extract the construction land information in the graph information from this year's DOM and last years'. Extract the information of land use change by computer or the man-machine interactive interpretation. Change information is represented in the form of tables and sketch maps. Field investigation under the technical guidance and the accurate positioning of GPS, confirm the change pattern of the type, size, ownership, verify features width and fragmentary surface features (including the omission of small patches) measurements, forming standard change map of land use status [7]; On the basis of the ground investigation, with the help of GIS software, realized the coordinate transformation, graphics editing, information query, and the generation of the change survey record form. Use ArcGIS software overlaying the change data and the record reproduction of the construction land to complete construction land information set of annotations of this year. The technical route is shown in Fig. 1.

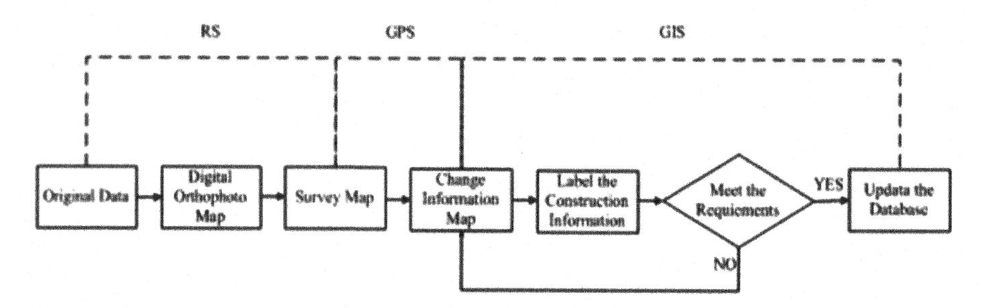

Fig. 1. Research idea figure

3 Research Process

Using "3S" technology to complete the register of land management information, mainly is the use of satellite remote sensing images reflect the characteristics of spectral features, playing its advantages in the identification of land category, extracting the change information on the basis of the existing land use database; Utilizing global positioning system to investigate and measure the changing information; Utilizing geographic information system to establish the land use database and realize the construction land information set of annotations, to achieve unified management of spatial data and attribute data, to guarantee the updating of the database.

3.1 Production of Digital Orthophoto Map

Firstly, according to the basic information to select and correct the corresponding land use remote sensing data, and make digital orthophoto map [8]. Due to the camera can not be ensured absolute horizontal in photography instant, the resulting image, receiving from aerial photography or satellite sensors, with a certain inclination angle, and the scale of each part of the image is not consistent [9]. In addition, the camera imaging is the central projection, topographic relief will cause projection difference in the images. The maps we using are orthographic projection, in order to make the image map of orthographic projection characteristics, it is necessary to take rectification. Then splic and finish the ortho image, which after correction of various kinds of deformation, to obtain digital orthophoto map [10]. DOM has the geometric accuracy and texture features, measurement in the digital orthophoto map is equivalent to measurement on the surface of the earth, so the accuracy of DOM affect the accuracy of the entire database directly.

In production of DOM actual operation, when choosing images of different years in one area, use the same remote sensing satellite data, and ensure on time as far as possible close to, in order to reduce the unnecessary errors caused by differences in the data source. Utilizing ground control points and digital elevation model to rectify inclination angle, correct projection difference and resample the image to be ortho images. Splicing and inlaying ortho images before the color balance treatment, and then obtain DOM of Nanjing area after cutting by a certain range.

3.2 Discovery and Extraction of Change Information

The discovery and extraction of change information is an important step in the investigation, and the accuracy of information determines whether the follow-up work can achieve the desired effect [11]. The principle of extracting new construction land is: extracting plots which there is no building in the former phase images, while the later image has been pushed filling or has formed building (structures). Including: the former phase image is covered with vegetation or non obvious characteristic of land use construction, while there is obvious construction trace(such as foundations, buildings, etc.) in the later image; The former phase image is covered with vegetation or non obvious characteristic of land use construction, while there is construction push fill trace in the later image; There is construction push fill characteristic in the former image, while there is obvious construction trace in the later image.

Traditional change detection method is putting remote sensing images, land use status data and the land change survey data together to form spatial overlay layer, then discovering changed map patches with human-computer interactive method. This method not only has heavy workload, but also has poor reliability. With RS software, it is convenient to discover the changes by extracting construction land from the former image and the later one by choosing representative training samples, highlighting the changing information by overlay analysis training samples. When extracting the change information, extract the linear features firstly, and then is the change plots. It is necessary to check spots one by one to avoid omissions. For uncertain suspected, the spots should also be recorded for field investigation to verify.

3.3 Field Verification of the Change Information

Produce the survey map and query information in the field with global positioning system. Survey new plots of graphics structure, and input the attribute information. The "Investigation Star", based on the integration of GPS and PDA (Personal Digital Assistant), is a successful example of the application of the "3S" integration technology to land change survey. The "Investigation Star" operating system is composed of two parts: hardware, which including field data capture system and vehicle base station, and software, which including data-collecting software, data-processing software and the base station server software [12, 13].

When surveying in the wild, opening pre-processed image in the work space and set GPS parameters, choosing real-time difference mode, the GPS will receive GPS satellite data, and complete difference processing. With the navigation function, the system will display the location in the corresponding image, and then can be accurately positioned to the need to check the map. Draw and edit the spots, the change of which have be confirmed, boundary, land use types, the ownership and other properties on the base map, as the basis of updating land survey database [14].

3.4 The Updating of Land Use Database

Transmit data from PDA to PC and complete the change of land use database by utilizing GIS software. Edit graphic information, gotten from field data acquisition, and draw boundary in the changing spots layer to form the file of changing spots. The file contains all the changing spots and attribute; Draw linear feature and input corresponding attribute in the new linear feature layer; Draw sporadic feature and input corresponding attribute in the new sporadic feature layer. With the support of GIS platform, modify and edit graph layer to establish a complete topological relationship, to correct topological errors and attribute errors.

Mark the new construction land type and other information in the database, according to the field construction and approval conditions. Mark the file number of those with legal permitting procedure. The flow chart of marking construction land management information can be established based on the above analysis (Fig. 2).

4 Empirical Analysis

4.1 The Generality of the Study Region

Nanjing is located in the lower reaches of the Yangtze River, southwest of Jiangsu province. The geographical coordinates is between latitude 31°14″ to 32°37″, east longitude 118 °22″ to 119°14″. Nanjing, with a total area of 6582.31 km^2, is an important center city of the Yangtze River Delta, and is the important portal of driving the development of the central and western regions [15]. With the development mode of Nanjing city from single center to polycentric, Nanjing construction land shows a trend of accelerated expansion. The geographical location of the study area is shown in Fig. 3.

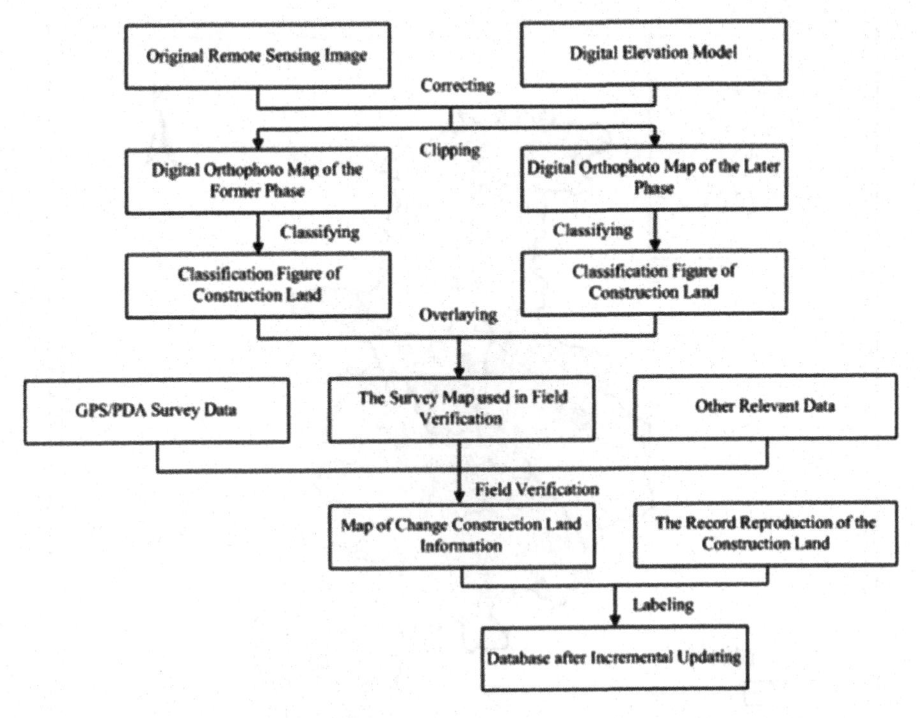

Fig. 2. Construction land management information labeling process

4.2 Data Source

Research data including SPOT5 images of 2012 and 2013, the spatial resolution is 2.5 m, the record reproduction of the construction land of 2012 and 2013, and the land-use map of 2012. Data is mainly obtained from surveying and mapping departments.

4.3 The Research Process

(1) Firstly, process orthophoto, and identify the figure of the change of construction land. By selecting control points in the research area, utilize ERDAS IMAGINE IMAGINE orthoBase module to make digital orthophoto map of two phases, and extract the information of construction land by supervised classification. Complete information interpretation of changing construction land, By examining the anastomosis degree of boundary and the consistency of construction land, record the results in the check list(see Table 1), and mark the change situation on the on the corresponding land-use maps. Do special markers for changes in the remote sensing image are not clear spot, investigate and confirm these spots in the later field investigation.

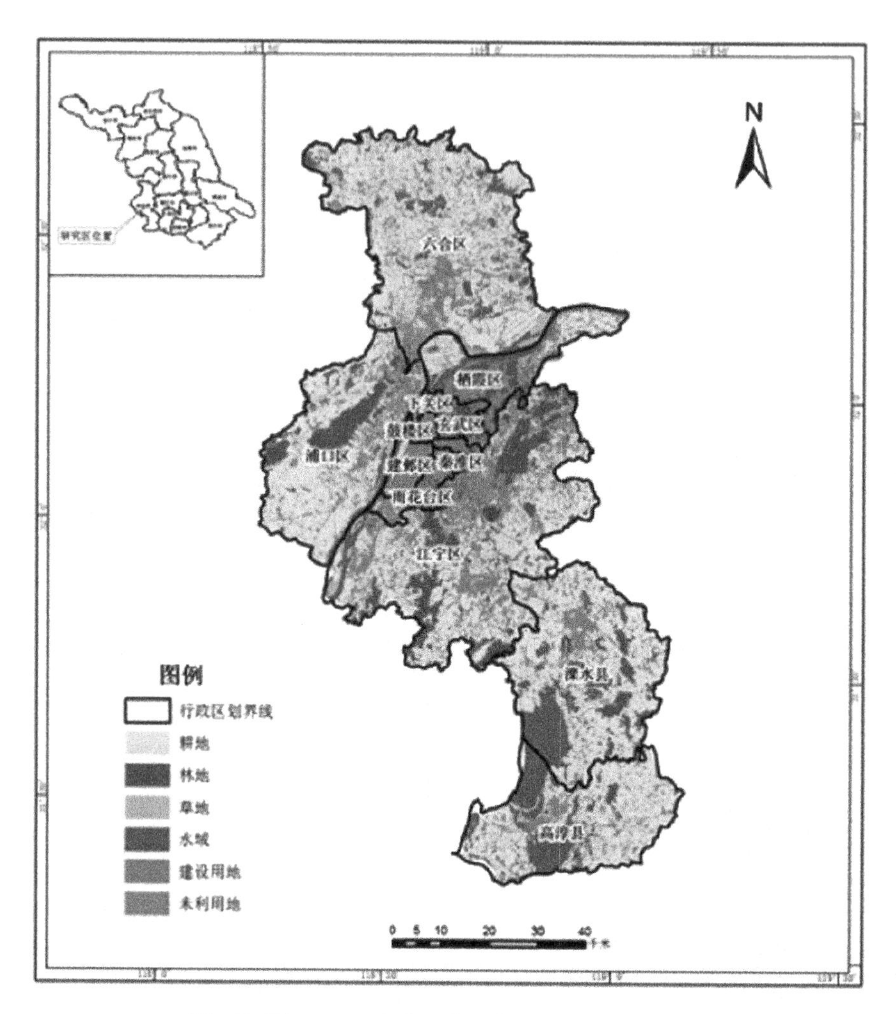

Fig. 3. Diagrammatic location sketch of research area

Table 1. Remote sensing monitoring patch information inspection record form of Nanjing area

The interpretation results in the industry	Serial number	1	2	3
	Location	Luhe District	Luhe District	Jiangning District
	Figure number	12836	12845	18140
	Land category before changed	Drought land	Paddy field	Drought land, paddy field
	Land category after changed	Village	Highway land	Village
	Area (m^2)	899.87	17885.6	84125.51

(*Continued*)

Table 1. (*Continued*)

Field survey results	Changed or not	Yes	Yes	Yes
	Actual land category	Village	Highway land	Village
	Ownership	Village collective	Village collective	Village collective
	Remarks			

(2) Carry out field verification according to the specific location of the spots, for the changes of the situation is not clear picture spot. Input the uncertain change area as well as other relevant data into PDA, as the basis of field change survey data. Collect the location information by GPS positioning function, carry on the real-time editor, and edit attribute information of the changes.

(3) Transfer the data from PDA to the PC, edit and modify the construction information, collecting by the foreign industry, then output into SHP format data. For the new construction land this year, according to the field construction and approval, mark the new construction land type and other information in the database. Mark the file number of those with legal permitting procedure. Put data pattern and construction land record map of 2013 superimposed, fusion or comminute the spots according to concrete conditions, modify the corresponding attribute, and form new construction land layer. Figure 4 is construction spots after confirmation.

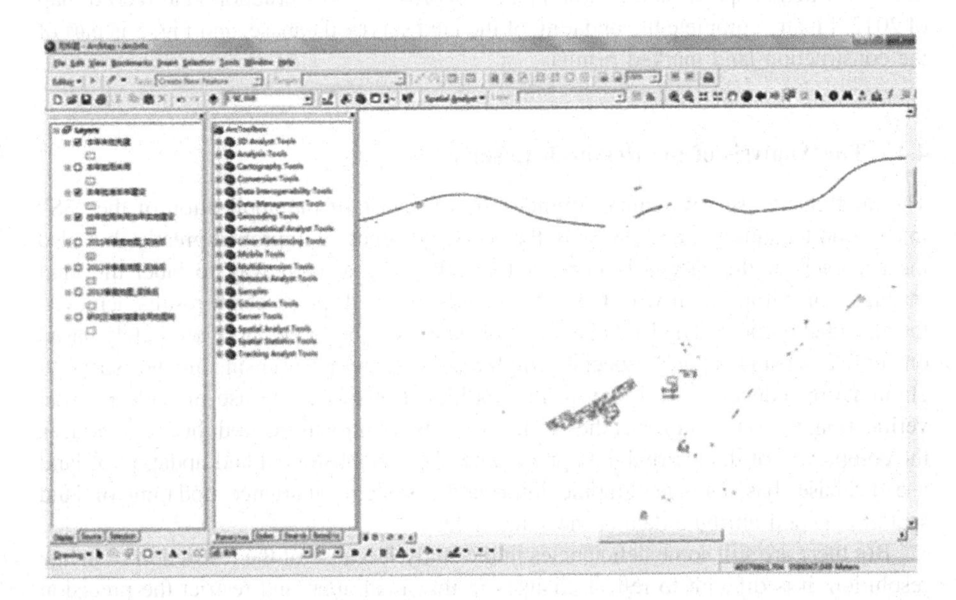

Fig. 4. New construction land

(4) Using the vector superposition of the GIS function, put construction spots, after confirmation, and construction land record map of 2012 and 2013 superimposing respectively. Specific labeling requirements reference the regulations of Ministry of Land and Resources:

① The construction approved this year: The land spots, which have been building block and with the approval procedures, change to the construction site according to the situation, and label "the construction approved this year" (B).

② Approved while not utilized this year: The land spots, which haven't been building block and with the approval procedures, change to the construction site according to the approval, and label "approved while not utilized this year" (P).

③ The construction hasn't been approved: The land spots, which have been building block without the approval procedures, change to the construction site according to the situation, and label "the construction without approval procedures" (W).

④ The construction approved in previous years: The land spots, which have been building block this year with the approval procedures of previous years, change to the construction site according to the situation, and label "the construction approved in previous years" (PJ) [16].

Therefore, label "the construction approved this year" (B) to the new construction spots, of which boundary accords with construction land record map of 2013; Label "the construction without approval procedures" (W) to the part beyond the record map; Label the parts, which on the record map beyond the new construction, "approved while not utilized this year" (P); Label "the construction approved this year" (PJ) to the new construction spots, of which boundary accords with construction land record map of 2012. Finally complete the updating of the land survey database, and Fig. 5 is part of the construction land marked results.

4.4 The Analysis of the Research Results

The land change survey realizes digitalization because of the application of the "3S" integration technology, and not only the work efficiency is improved greatly, but also the accuracy of the survey is improved greatly. It is a good way to label the land management information with the "3S" technology, and the research results achieved the expected purpose. The land change information can be found more accurately based on remote sensing image, especially high-resolution remote sensing image, so as to eliminate or reduce the omission in the traditional method; The use of GPS external verification, not only improves the accuracy of the traditional method, but also reduces the complexity of the internal data processing; The establishment and updating of land use database, based on geographic information system, guarantee updating of land spatial data and attribute data at the same time.

But there are still some deficiencies in the study: because of the restriction of image resolution, it is difficult to reflect changes in the small area, and restrict the precision and accuracy of survey data; The two interpretations of the two phase remote sensing

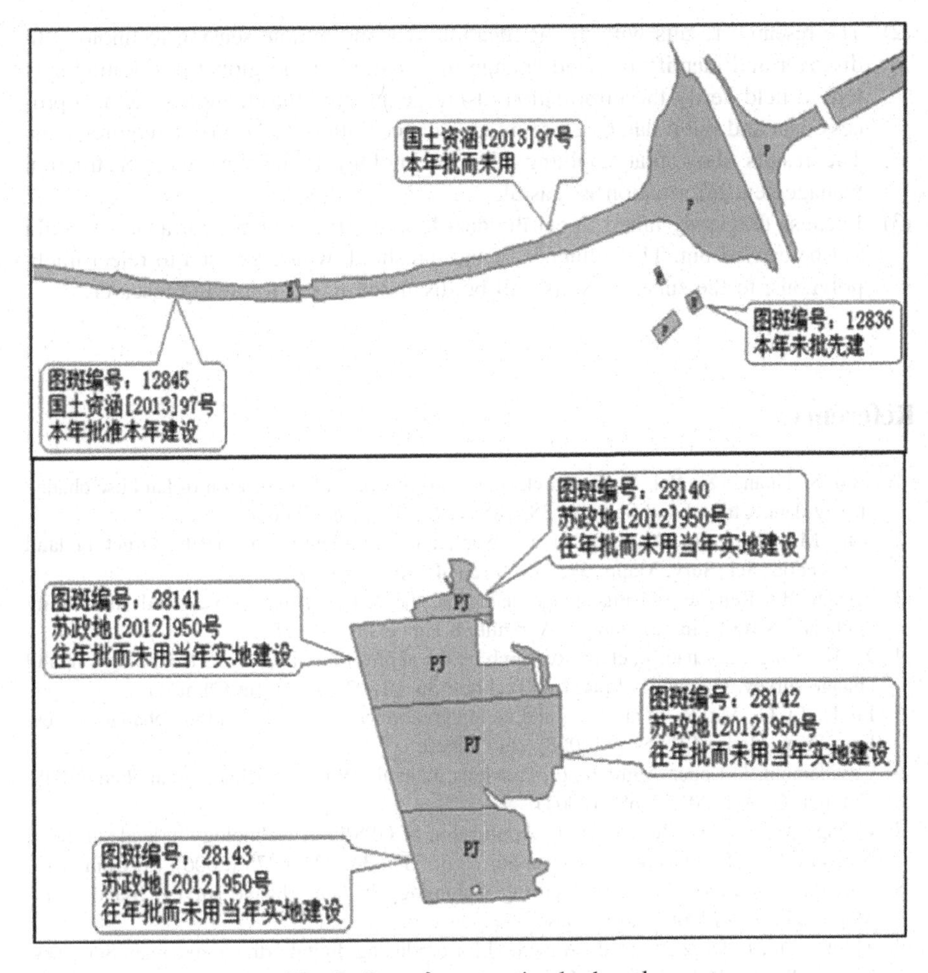

Fig. 5. Part of construction land mark

images will produce a certain classification error, which will result in second error. The low precision will generate massive finely polygons when overlay the change boundary, which will cause adverse effect to the last result. These deficiencies, which affect the accuracy and workload, will be solved in the further research.

5 Conclusions and Discussions

(1) In this paper, based on the "3S" technology as the support, research Nanjing area of construction land management information annotation, and analyze the study results. Despite massive exclusion and artificial error correction in the process of the research, the final results are in accordance with change survey requirements.

(2) The research results with the application of using remote sensing technology to discover and identify the land change information, using global positioning system to field verify the information, using geographic information system to process data and build database are in accordance with change survey requirements. The results show that applying "3S" technology to labeling the construction management information is feasible.
(3) Because this study failed to get the data before 2012, so the error analysis could not be carried out. The influence of the threshold, which be used to reject finely polygons, to the survey results will be discussed in the following research.

References

1. Guo, S., Huang, X., Bai, X.: Research status and prospect of application of land use change survey data. China Land Science. **38**, 188–199 (2013). (in Chinese)
2. Yan, M., Zhang, L., Yan, Q., et al.: Application of identifying specific target in land verification. Sci. Surv. Mapp. **39**, 71–75 (2014). (in Chinese)
3. Zhang, H.: Remote sensing image using in the land change survey and monitoring applications (in Chinese). Jiangxi Agriculture University (2012)
4. Xi, K., Yang, X., Zhao, J., et al.: Research on the application of PDA+3S technology in land change survey. Geomatics Spat. Inf. Technol. **36**, 91–92 (2013). (in Chinese)
5. Fu, L., Li, G., Ma, R., et al.: "3S" and its integration technology in land use change survey. Eng. Surv. Mapp. **17**, 55–59 (2008). (in Chinese)
6. Casademont, J., Lopez-Aguilera, E., Paradells, J., et al.: Wireless technology applied to GIS. Comput. Geosci. **30**, 67–682 (2004)
7. Chang, Q., Wei, D., Zhu, G., et al.: Application of GPS/PDA technology in land survey in Nanyang City, Henan Province. J. Anhui Agric. Sci. **34**, 627–6275 (2006). (in Chinese)
8. Cai, F.: Application of remote sensing technology in land change survey. Stand. Surv. Mapp. **25**, 1–14 (2009). (in Chinese)
9. Ma, D., Cui, J., Ding, N., et al.: A method for producing digital ortho image map. Sci. Surv. Mapp. **38**, 188–199 (2013). (in Chinese)
10. Guan, Y., Zhou, S., Lu, T.: The production of digital ortho image map based on IMAGINE ERDAD. Bull. Surv. Mapp. **12**, 31–33 (2015). (in Chinese)
11. Walter, V.: Object-based classification of remote sensing data for change detection. ISPRS J. Photogram. Remote Sens. **58**, 22–238 (2003)
12. Chen, D., Liu, Y., Yu, H., et al.: Research on the key technologies of GPS-PDA application in land survey. Sci. Surv. Mapp. **35**, 137–139 (2010). (in Chinese)
13. Li, X., Li, W., Ma, L., et al.: Research on the application of GPS-PDA in land change survey based on "3S" Technology. Sci. Surv. Mapp. **33**, 20–210 (2008). (in Chinese)
14. Clegg, P., Bruciatelli, L., Domingos, F., et al.: Digital geological mapping with tablet PC and PDA: a comparison. Comput. Geosci. **32**, 1682–1689 (2006)
15. Chen, J., Gao, J., Xu, M., et al.: Characteristics and mechanism of urban construction land expansion in Nanjing City. Geogr. Res. **33**, 42–438 (2014). (in Chinese)
16. Ministry of Land and Resources: Notification of the 2012 national land change survey and remote sensing monitoring. Nat. Land Resour. Inf. **24**, 1–32 (2012). (in Chinese)

Design and Application of Green Building Based on BIM

Deyue Xing and Jun Tao(✉)

School of Mathematics and Computer Science, Jianghan University, Wuhan Hubei, China
985342098@qq.com, martintao2006@163.com

Abstract. BIM technology is a kind of brand-new concept which involves a series of innovation and change from the planning, design theory to construction, maintenance technology, it is the development trend of information construction. BIM technology research will have important application value and broad application prospects for building life-cycle management, it has the extremely vital significance to improve the level of the construction design, construction, and operation of scientific and technological level, and promote the construction industry level of modernization and informatization.

Keywords: BIM · Green building · 3D modeling · Design and application

1 Introduction

BIM (Building Information Modeling) is regarded relevant information and data in the architectural engineering project as the basis of the model to establish the model of building and simulate real information of the building through the simulation of digital information. The characteristics of this technology are: Visualization, Coordination, Simulation, and Optimization and to create graphics.

The green building is in a life-cycle, the maximum saving resources, energy, land, water, materials, so that to protect environment and reduce pollution, also for providing people with healthy applicable and efficient use of space, the building matches perfectly with the nature and embodying the harmony between architecture and environment, human and human, human and nature. The construction of green-building is to green construction, in order to achieve environmental protection energy saving and to examine the traditional construction technology with the vision of sustainable development. This is a systems engineering which includes ecological-design (Eco-design), Eco-construction, Eco-restoration and so on.

The traditional Building-Project is a long process from the beginning though the planning and design stage, Engineering construction stage, Completion inspection and acceptance to delivery. There are many uncertain factors in this process and in order to overcome this shortcomings and disadvantages, using the BIM technology to design, construct, and operations management of the architectural engineering project. All kinds of Architectural-information are managed by a whole and throughout the whole life-cycle of the architectural process. Use computer technology to build a model of architectural information, for the construction space geometry information, the function of

F. Bian and Y. Xie (Eds.): GRMSE 2015, CCIS 569, pp. 901–907, 2016.
DOI: 10.1007/978-3-662-49155-3_93

building space information, and the construction management and equipment information to precede data integration and integrated management.

The application of the BIM technology, will bring huge benefit for the development of the architecture industry, and make the architectural planning and design, construction of the building engineering, operations management, the quality of the whole project and the efficiency of resource management significantly improved.

2 Features and Values of BIM

BIM technology which is based on 3d technology and integrates Engineering data model of all kinds of information related to construction project is a digital expression about physical facility and functional characters of construction project. A perfect information model can link the data, process and resource of different stage of the construction project's life cycle, which is a complete description and can be used by all participants of the construction project. BIM technology with single engineering data can solve the consistency and global Shared problems of distributed and heterogeneous engineering data, and it also supports the creating, managing and sharing of the dynamic engineering data of the construction project's life cycle.

2.1 General Features of BIM

Completeness of Model's Information. In addition to use 3D Geometrical information and topological relation to describe the engineer object, it also include the total engineer information description: such as object name structure type building materials and engineering properties; construction procedure, schedule, cost, quality, human resource, machine and material resource; engineering safety performance, material durability; engineering logic relation between objects etc.

Relevance of Information about the Model. The objects of information model are interrelated and can be recognized. System can make a statistics and analysis according to model information to create corresponding graphics and documents. If one object of the model was changed, all the relevant objects would change to maintain the completeness and vigorous of the model.

The Consistency of the Model's Information. The model and information are consistent in the different stages of building life-cycle. This eliminates the need for the user to duplicate the entry of the same information and can automatically evolution not to create a new model in the different stages of model. It can have a simple extend and modify and avoid the error of inconsistent information.

2.2 Values of BIM

The state supports the application of BIM—technology research results, and achieves fruitful research results. The relative achievements of software are demonstration applied to practical engineering application; its further development will greatly promote the application of BIM technology process.

To make the industry standards of BIM—technology, IFC (Industry Foundation Classes) is a kind of BIM technology standards, which is used to express and exchange data, it is the guarantee and foundation of the application of BIM—Technology to make all kinds of the relevant standards.

The actual demand of construction information, Architectural enterprise is the pillar of the national economic, taking on the heavy responsibility of country capital construction. At present, our country is carrying on the word's largest capital construction, increasing the scale of the project, especially large projects is emerging in endlessly, and due to the complexity of structure form is more an more, making enterprise and project faced with a huge investment risk, technology risk and management risk. However, their current management modes and the means of information cannot meet the needs of modernization construction.

3 Green Building and BIM

Speaking from the life-cycle of construction, BIM technology and green building focus on the building's whole life-cycle which extends up to the building material's production and raw material and extend down to end of life the building demolition, recovery and utilization.

From the perspective of information management, BIM technology is the best carrier of building information, using the vector can be realize the centralized management of building information in various stages of construction, according to the requirement of it. The modeling of BIM technology and the concept of target is the integrity of information, accuracy, consistency and controllability that will be used to control the cost, ensure the quality and improve efficiency. Green building also requires all kinds of building materials, equipment, and system information which is complete, accurate and controllable so that save fund more, material, energy, land, water and try to reduce emissions of carbon dioxide. Relatively speaking, BIM is similar to a kind of implementation approach, and Green building is the expectation objective. Despite of the different angles, both of them that are direction and objective are consistent.

From the perspective of research methods to see, BIM technology with model is regarding as an information carrier. Green-building which requires the simulation and forecast is based on model. The model of technology will be the perfect combined with the simulation of the Green-building. If the BIM model is only regarded as the carrier of information, rather than use for the top of this emulation, it is only a kind of three-dimensional statistical database and just extensive utilization of the information, losing the advantage of calculation basis and the simulation of carrier. Significance will be discounted. For the Green-building, if we don't have a uniform and reliable information model as the carrier, the performance simulation is built on the basis of all kinds of simplified model, the simplified model of the difference people set up difference will lead to the difference between the simulation results. It is only a relative reference.

The core of the BIM technology is to build a model, it can carry the architectural information with precise and complete under the strict control policy, and to avoid parts of redundancy, conflict and fuzzy, then make the model use and transfer in the stages of architectural. In general, the research content of BIM technology is to set up a good-model, use and pass it. The working process of the BIM technology is to finish the circulation and transmission of "modeling—using—passing—processing—(continue modeling)—reusing—passing again".

BIM-technology and Green-building are to achieve complementarities, the application of BIM technology which can change the traditional construction management ideas to lead the construction of information technology to a higher level, and will greatly improve the informatization level of management with BIM.

4 Design and Application of Green Building

4.1 Design of Green Building Based on BIM

The application of BIM technology is in conformity with the development direction of science and technology of construction and national policy guidance, BIM technology is applied to the engineering construction to strengthen the management of engineering control, increase the professional service ability, and improve the quality of products and the development efficiency. The feature is in line with the green building design idea.

The Stage of Solution. According to the scheme of information, BIM-model is established and applied to the design of visual communication. The Three-Dimensional visual performance is more accurate, richer information than traditional 2D drawings. It is so easy to understand and communication that it can improve the communication efficiency. BIM model also can be imported to the related performance analysis software in order to avoid repeat modeling and provide the reasonable reference and judgment for the owners' optimization scheme.

The Stage of Preliminary Design. According to the preliminary design of architectural drawings, a new expansion model is established. BIM can assist the project company to further confirm the design of architectural space and relationship between each of system for the design to proceed a preliminary inspection and to proceed a complete collision check for each of specialties. The inspection reports and the improving suggestions would be submitted to the project companies and design institute which had gotten the construction drawings for updating the composite model and helping to optimization design method. It is so necessary for construction works to avoid some mistakes that can reduce the waste after changes.

The Stage of Constriction Drawing Design. BIM model which based on the construction drawings was a technology for information integration at the design stage of a project. It can provide all kinds of summary information for deeply subsequent adjustment of the design in detail. Updating the model for major engineering adjustment and small and medium-sized projects can provide a platform of data for information integration and working nodes. As is proved, BIM technology which applies in the construction drawing will contribute to each of the stakeholders of engineering go to deepen adjustment, construction investigation, cost estimates and make measured decisions.

The Coordination of Design. Assisting the project company to coordinate each related design which included civil engineering curtain wall, steel structure, internal decoration for common area, according to the related drawings, the designer would put the drawings into the model to design and check up, and issues found in this phase submitted to the project company and related department of consultant to update for improving the quality of project design service.

The coordination of multiple design of design institute will be a test of building relative to a unit of construction drawing design before the actual construction. BIM technology for each of party information integration and testing will contribute to control the foreseeable risks and coordinate the function of each of department. The application of BIM technology will have an good effect on conforming the construction schedule and cost—control.

4.2 Application Analysis of Green Building Based on BIM

The performance indicators of Construction projects have been basically determined in the early development, which including landscape visibility, sunshine, wind environment, thermal environment, and sound environment and so on. But because of the lack of proper technology, general project is very difficult to have the time and cost to analyze and simulate above indicators in many solutions. And BIM technology provides the possibility to popularize and apply for green building performance analysis.

The Simulation of Outdoor Wind Environment. To improve comfort of surrounding residential buildings area, improve the quality of residential environment by adjusting the planning building layout, landscape planting arrangement, improving residential wind flow field distribution, reduce the eddy current and hysteresis phenomenon; Under the condition of high winds, Analyze which areas could be triggered safe problems by funneling.

The Simulation of Natural Lighting. To analyze the indoor natural lighting effect of related design schemes, and improve it by adjusting the building layout, finishing materials, visible light transmittance of palisade structure, and according to the lighting effect to adjust indoor layout arrangement.

The Simulation of Indoor Natural Ventilation. To analyze related design scheme, improve indoor flow field distribution by adjusting the location and size of air vents and architectural layout, and improve indoor comfort conditions by guiding the indoor airflow organization to ventilate effectively.

The Simulation Analysis of Residential Thermal Environment. To simulate and analyze the heat island effect of residential area, weaken the heat island effect in the way of optimizing single architectural design and group layout and strengthening evanescence.

The Simulation Analysis of Building Environmental Noise. The advantage of computer simulations of the acoustic environment is that it can predict the acoustic quality of construction, as well as the feasibility of the projections for architectural acoustics retrofit scheme after setting up the geometric model through the change of the material and the interior room in a short period of time.

The application of Green-building technology and green products will greatly improve the comfort of room products and the environmental performance, which provides higher quality green life for the owners. In the process of the practice of architecture and technology, through the BIM technology constantly being adjusted and perfected is more and more strict for determining the "best" of design and engineering management which is also the concept of Green-building—creating more environmental products for the society and more urban ecological beauty.

5 Conclusion

The characteristic of BIM technology in a nutshell is an approach that can apply to the entire design, construction, and management of the digital method, which supports the integrated management environment of construction, and can make significantly improve efficiency and reduce risk in the whole process.

Because of the building information model needs to support the integrated management environment, so it is a composite structure which includes a data model and behavior model. It also contains the behavior model is related to management in addition to contain the data model is related to geometrical figure and data. It is an extremely important way to give meaning to the data to simulate the behavior of the real world and to make the two points combine by correlation. For example, simulated the condition of the architectural stress and the situation of heat transfer in the envelope, certainly, the simulation of behavior is closely related to the quality of information.

The application of BIM technology, it can support various kinds of information which is high-quality, high-reliability, high-coordination, high degree of integration to apply consecutively and real-time application, BIM reaches the goal of substantially reducing cost and greatly improving the quality and efficiency of both the design and the whole project.

Acknowledgment. Supported by research creation fund of Graduate School of Jianghan University (C12), and supported by research fund on humanities and social sciences of Hubei Provincial Education Department (14G169, 14G172), and supported by educational research fund of Hubei Provincial Department of Education (H2014269), and supported by S&T research fund of Hubei Provincial Department of housing and urban rural construction (Hubei Construction File [2014] No. 54), and supported by college educational research fund for graduate education of Jianghan University (C14), and supported by research fund on humanities and social sciences of Wuhan Research School of Jianghan University (jhdxwyy20141109, jhdxwyy20141109), and supported by college educational research fund of Wuhan Educational Bureau (2014086).

References

Yen, S.-J., Yang, J.-K., Kao, K.-Y., et al.: Bitboard knowledge base system and elegant search architectures. Knowl.-Based Syst. **34**, 43–54 (2012). (Special Issue)

Wu, I.-C., Lin, H.-H., Lin, P.-H., et al.: Job-level proof-number search. Comput. Games **6515**, 11–22 (2011)

Jun, T.: 3D modeling of small object based on the projector-camera system. Kybernetes **41**(9), 1269–1276 (2012)

Yen, S.-J., Yang, J.-K.: Two-stage monte carlo tree search. IEEE Trans. Comput. Intell. AI Games **3**(2), 100–118 (2011)

Wu, I.-C., Lin, P.-H.: Relevance-zone-oriented proof search. IEEE Trans. Comput. Intell. AI Games **2**(3), 191–207 (2010)

Jun, T.: Development and application of functionally gradient materials. In: International Conference on Industrial Control and Electronics Engineering, pp. 1022–1025 (2012)

Qiao, Z., Yang, M., Wang, Z.: Technologies Analysis of Computer Game. In: Achievements in Engineering Materials, Energy, Management and Control Based on Information Technology, pp. 679–682 (2011)

Xu, C.-M., Ma, Z.M., Tao, J.-J., Xu, X.-H.: Enhancements of proof number search. In: Proceedings of 21st Chinese Control and Decision Conference, vol. 1–6, pp. 4525–4529 (2009)

Jun, T.: Design and visualization of optical feedback laser based on computer vision. In: International Conference on Industrial Control and Electronics Engineering, pp. 1030–1032 (2012)

Lin, Y.-S., Wu, I.-C., Yen, S.-J.: TAAI 2011 computer-game tournaments. ICGA J. **34**(4), 248–250 (2011)

Yoshizoe, K., Kishimoto, A., Mueller, M.: Lambda depth-first proof number search and its application to go. In: 20th International Joint Conference on Artificial Intelligence, pp. 2404–2409 (2007)

Jun, T.: Face reconstruction based on camera-projector system. In: International Conference on Industrial Control and Electronics Engineering, pp. 1026–1029 (2012)

Tao, J.-J., Xu, C.-M., Han, K.: Construction of opening book with its application. In: Proceedings of the 21st Chinese Control and Decision Conference, vol. 1–6, pp. 4530–4534 (2009)

Wu, I.-C., Lin, H.-H., Sun, D.-J.: Job-level proof number search. IEEE Trans. Comput. Intell. AI Games **5**(1), 44–56 (2013)

Saito, J.-T., Winands, M.H.M., van den Herik, H.J.: Randomized Parallel Proof-Number Search. Adv. Comput. Games **6048**, 75–87 (2010)

Research on the Typical GIS Service Chain Application in Land Resource Management

Yanjun Wang[1,2(✉)]

[1] National-Local Joint Engineering Laboratory of Geospatial Information Technology, Hunan University of Science and Technology, Taoyuan Road, Xiangtan 411201, China
[2] Hunan Province Engineering Laboratory of Geospatial Information, Hunan University of Science and Technology, Taoyuan Road, Xiangtan 411201, China
wongyanjun@163.com

Abstract. Service-oriented architecture and web service technology provide a method for geospatial information sharing and service chain building. From the development of geographical information system, this paper explores the concept and theory of geospatial information service, focuses on the key technologies in service chain, such as registration and discovery of spatial information service, demand expansion customized UDDI and service composition method. By designing service chain reference model, services registry and catalog, this paper builds a house demolition and land expropriation processing service portfolio in the city railway line planning. The geospatial information service chain provides practical application reference, which can be applied to the construction of spatial information sharing and professional application.

Keywords: SOA · Geospatial information service · Web feature service · Web processing service · Service chain

1 Introduction

Geographic Information System (GIS) is unceasing developed with the development of computer information technology, which formed distributed network processing of Web GIS based on network technology, but due to that system construction method is different, heterogeneous data, is not conducive to the cross platform interoperability [1]. In geo spatial framework of the construction of digital cities to promote, the spatial data processing analysis of the growing demand, GIS development of open interoperability of geographic information platform, the application of geographic information has also been deeply to land, planning, public security, water conservancy and other economic and social fields, and toward a more flexible and efficient and intelligent development [2, 3]. Based on SOA (service-oriented architecture, service oriented architecture (SOA) geographic information public service platform, in the next generation Internet, networking, cloud computing and intelligent of the observation sensor network rapid development background, application of different functional units of the service module of permutation and combination, to form advanced spatial analysis service chain, to avoid the traditional data update, operation and maintenance and upgrade transformation

F. Bian and Y. Xie (Eds.): GRMSE 2015, CCIS 569, pp. 908–916, 2016.
DOI: 10.1007/978-3-662-49155-3_94

brought about by the repeated construction and waste of resources and other issues, it is important research topic in current geographic information public platform and wisdom urban infrastructure construction.

There are problems such as heterogeneous and semantic heterogeneity, which need to improve the interoperability and reuse of the spatial analysis model, so as to realize the application of distributed geographic information processing [4, 5]. OGC (open geospatial consortium, open geospatial consortium) of interoperability standards WMS (web map service, web map service), WFS (Web feature service, Web feature service, WCS (WEB coverage service, web coverage service) and WPS (web processing service, geographical network processing services, to facilitate the realization of the application of space information service under the distributed network environment, this paper based on SOA and OGC standards, space research information service registration, discovery and composition technology, realizing the geospatial information chain construction and typical applications.

2 The Principle of Geospatial Information Chain

Geospatial information chain is a sequence of spatial information service, necessary conditions for each service, a service is a service execution, and service chain have served to find, composition and execution ability, no need to manage maintenance of basic data and services can on-demand find and bind to space information service, executive senior complex spatial analysis tasks [6, 7]. Compared to the web service technology into the field of spatial information, combined with the formation of SOA Service GIS [8] with the traditional component GIS, easy to under the environment of network data acquisition, integration and sharing, is conducive to the rapid construction of application system; and Web GIS can flexibly to overcome geographical processing models and semantic heterogeneous problems. The effective use and free combination of geospatial information chain to the distributed spatial information service is the developing direction of the geographic information service.

Rich data and functional services is the basis of the construction of geographic information chain, how to find the required services, the data and the single function of the atomic service effective combination is the focus of research. OGC summarizes the classification system of spatial information service, and puts forward three kinds of service chain: transparent chain, semi-transparent chain and opaque chain, and [9] has been discussed in detail. Chain research of geo spatial information related to spatial information service registration and discovery, according to need to customize the UDDI (Universal Description, discovery and integration, universal description, discovery, and integration), service composition and workflow engine key technology, results mainly concentrated in spatial information directory service, based on the workflow service chain process description and BPEL4WS (BPEL for web services, web service business process execution language) service chain architecture, to achieve business process customization, management and execution of geospatial information chain automation [10].

This paper explore multi-source and multi-scale geospatial information chain framework (see Fig. 1), the distributed dynamic service composition on-demand integration to complete a specific task and to verify that geospatial information chain application in digital city geographic information sharing service platform.

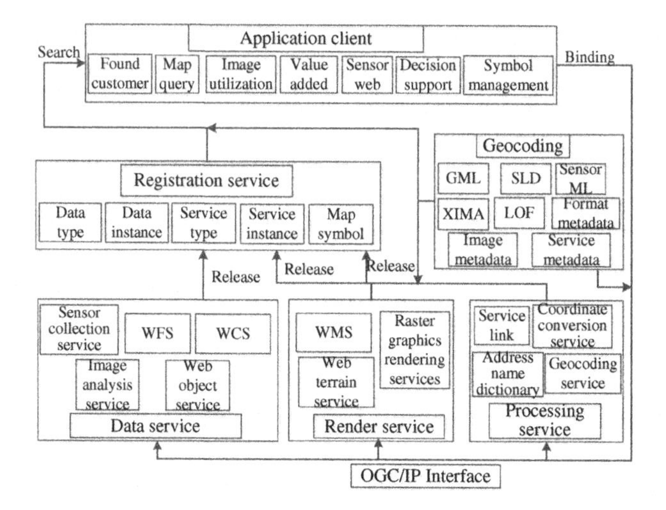

Fig. 1. Geographic spatial information chain framework.

3 Key Technology of Geospatial Information Chain

3.1 Spatial Information Service Registration and Discovery

The registration and discovery mechanism of spatial information service is mainly to facilitate the search and discovery of services, from the distributed network environment of complex services to efficiently and accurately find the required services. Registration of spatial information service is a service metadata of spatial data and processing functions to be registered with a UDDI, generally use two ways: one is to establish a private directory service registration center, the second is using common UDDI registration center. The former is the OGC registry, to relatively complex spatial information service for describing objects, the use of special communication protocol and interface, poor compatibility with the general registration center, hinder information interoperability; the latter is based on XML and HTTP storing spatial data and information service, based on ebXML registry information model, the spatial information service customization register contents, better meet the unique requirements of spatial information. Therefore, both the design space information service registration center, combining the advantages of private directory service register center and common UDDI, in the standard network communication and request protocol based on fully reflect spatial information service in the field of characteristics.

Spatial information service registration center to meta data services as registration information, support online publishing, finding and binding, realize the structure shown in Fig. 2(a). Registration center directory service to store metadata service, provide service discovery and management functions to achieve the retrieval and the registration interface. The UDDI extension data structure and access to services and the enhancement function to realize the unity of register center of spatial information service private and general, there are two ways to realize: (1) spatial information service registered to OGC registry. The private registry is added to the common registry as the node of the UDDI, and the access protocol specified in OGC is found and invoked by the space information service. (2) the OGC private information service is registered to the UDDI, and the user directly through general UDDI queries and access has been registered for spatial information services or non spatial information services.

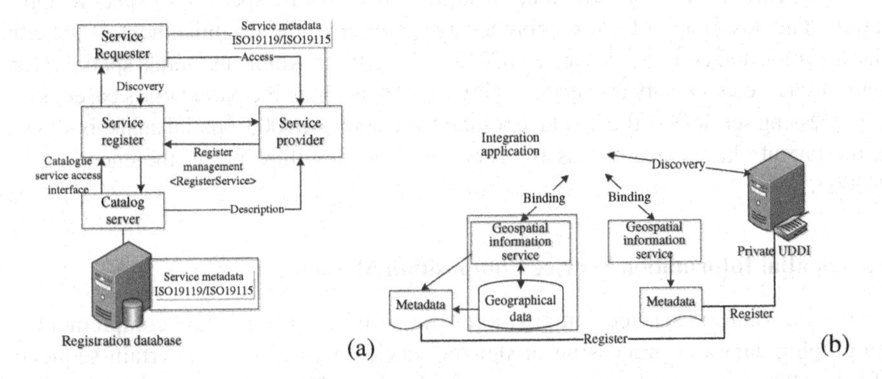

Fig. 2. (a) is OGC service registration implementation structure, (b) is the framework structures for the registration and discovery of spatial information services.

3.2 Expand the Custom Description UDDI as Needed

The existing UDDI provides three core classification systems: the North American industrial classification system, the general standard product and the service product and the geographic region classification method. Describe the UDDI although increased geographic position classification and geographical network classification method, but only according to the regional division, and can not reflect the characteristics of the spatial information service, resulting in lower recall ratio and precision ratio. Using the geographic area classification method, the UDDI can be extended to describe the data service and the automatic classification of the geographic processing service.

Private and general registration center has the same structure of service registration and discovery method, essence is the metadata of spatial information service issued to private registry, need to consider and data closely coupling and loose coupling with the data service in Fig. 2(b), the former should not only registered service metadata, metadata and data. On-demand custom extensions realize service registration and discovery work flow is: (1) service provider classified information released registered to a central directory, and register the validation service; (2) the use of spatial information service classification method to classify the service; (3) service users search service, catalogue center found and classification of spatial information service is called verification.

If it is successful, the classified information contained in the service entity is marked as a result of the check, and returns the result set of the spatial information service.

According to specific data service and application characteristics, spatial data is according to the type and to scale classification method, combined with ISO and OGC spatial information service classification, spatial information service can be points to describe the services, registration services, data services and processing services four categories, for each classification design, define the name, hierarchy and classification and coding rules, storage for XML document for directory for call center. The classification of spatial information services by OGC and ISO can be found that the former is from the perspective of information, the latter is from the perspective of specific application. The advantage of the comprehensive reference, the establishment of specific classification and coding rules are as follows: using three Arabic numerals spatial information service layer, subclass node coding for expansion in the parent class based, such as processing services in the first layer of the three coding to 003, spatial analysis service in the parent class space processing services under the third place, the encoding for 003003.

3.3 Spatial Information Service Composition Method

Spatial information service composition is the number of functional complementary geographic data and processing model for service according to a certain sequence of permutation and combination cooperatively. Complex spatial analysis tasks, often combined with workflow management. Workflow is a business process that can be performed automatically or semi automatically, so that the spatial information service can be transmitted and executed according to a series of process rules, documents or tasks.

Spatial information service workflow, management of spatial information and non spatial information and its process, formed by the execution of tasks and data, the logical relations constitute the space stream processing control of the process, before the event of the output flow as the input stream. BPEL programming language based on XML description of business processes and interaction, each activity in the process by the web service concrete realization, two types of business processes include: executable processes, defined within a specific task and the corresponding processing interface, the entire process can be execution engine; the abstract process, a detailed definition of the public message exchange method, but there is no internal details, cannot be the direct execution engine.

The workflow reference model, the abstraction of the specific business processes, said interaction of workflow definition of component interface, including process definition tools, workflow engine, workflow execution service, the calling application, client application and monitoring and management tools. The relationship between spatial information service workflow modeling the specific business process formalization, composed of processes, activities and sub process three entity and application components, first order, integration and reference etc. description of the process. Petri net is represented by the elements of status and status change is composed of network information flow model, workflow according to the nature of the Petri net [11], in order to define, parallel, alternative and four kinds of process logic cycle. At the same time, the definition and bifurcation, merged with, or bifurcation and or with four structural module, and bifurcation, and merging said parallel process logic or bifurcation, or a combination of said a selection process logic.

The spatial information service node, the logical relationship between the workflow reference model based on the definition of service, service composition, order form, parallel selection and circulation of four kinds of logical structure, as indicated in Fig. 3, detailed implementation process is as follows: (1) the sequence structure, composition, a process of order execution in accordance with the 2 (before and after;) parallel structure, composed of branching process concurrent execution at the same time, and by the bifurcation and with composition; (3) structure, with each other between mutually exclusive relationship, and add constraints to select one or more from, or by the bifurcation, or merger; (4) circular structure a process of change, can be repeated several times in the implementation of the given conditions, until the results meet the conditions.

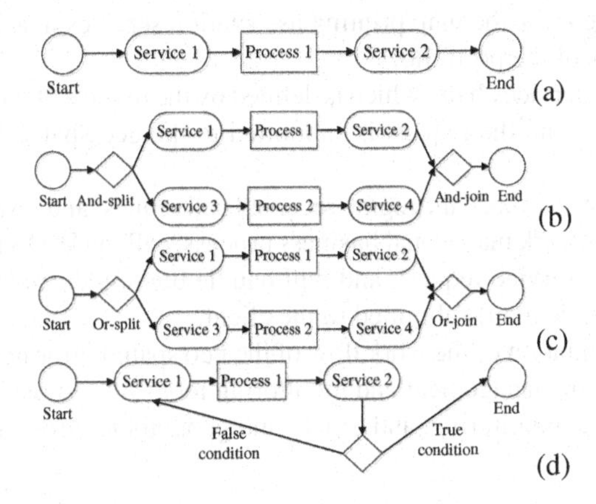

Fig. 3. Logical structure of geographical spatial information chain model. (a) is sequence structure, (b) is parallel structure, (c) is selection structure, and (d) is circular structure.

4 The Typical Application of GIS Service China

4.1 Geospatial Information Chain Reference Model

Geospatial information chain will multiple unit services are available according to certain rules of procedure, combined to form complex functions and can perform the service, to meet the demand request information and knowledge, spatial information value-added services. The application of the reference model of workflow management of geospatial information chain, spatial information service workflow modeling, combination, dissemination, implementation and monitoring, combined with the SOA architecture and workflow reference model, improved the design of the geospatial information chain reference model, mainly from the following 5 parts: (1) the service chain engine, responsible for generating and executing process examples; (2) the requester calls the service chain, service chain and client application; (3) the definition of service chain, service chain is responsible for generating business processes; (4) the service chain management and monitoring, the implementation is responsible for monitoring the entire service chain process in the life cycle; (5) spatial information service, provide data and processing functions for service chain.

Geo spatial information chain reference model, which provides a model support for the combination of spatial information services. Analysis of urban planning and construction of railway network map visualization business as an example, the purpose of which is to output a full map containing a map, scale line, the compass and the legend, in the client, you first need to call getMap requests return a map to be printed, and then call the GetLegendGraphic request to obtain information on the map legend, the thematic map affiliated information output. The execution process corresponding to the service chain reference model:

(1) in accordance with the order structure, parallel structure, select the structure or the structure of the loop, the map printing association services in accordance with the business needs of the portfolio;
(2) a network map service chain, which is defined by the request of the requesting party, is called step 1, and the request is submitted to the geo spatial information chain engine;
(3) geospatial information chain engine received call request and instantiation step one to define the network map visual business process, call the WMS requests and other non processing service request, and will handle the implementation of the results of intact (thematic map) submitted to the client.
(4) in the implementation of the work flow of the geo spatial information chain engine, the service chain management and monitoring tools are responsible for the implementation of the monitoring, and timely and geographic information chain engine interaction.

4.2 Typical Geospatial Information Chain Implementation

Urban planning and construction of railway analysis service chain, function realization is retrieval is located in an urban area along the railway line within 50 meters of all housing and land use data, and the statistical analysis of demolition result information. Chain service realization principle: first in the map by clicking, marquee or buffer query to obtain the factors of geographical planning and analysis; using WFS getFeature interface getFeatureIntersects method. The obtained above elements and administrative region do intersect query and gets urban result data.

In particular, the Intersection service will be invoked to obtain the city polygon elements and the railway layer data to do the intersection query, to obtain a certain area within the railway line. Call the buffer service to get the 50 meter range buffer zone within a certain area, access to the buffer polygon A. Then call the Intersection service will A and the district residents to data to do the intersection query, access to the railway within the range of 50 m of housing data, is the result of the analysis. Call the intersection service a and the land use data layer do intersect query and get into railway within 50 m of land data, the data of the land in accordance with the type of land of output statistics of, for the land acquisition and analysis results. Service chain of buffer a and retrieval to be housing units results are shown in Fig. 4(a), in which the shadow region; of the demolition of the service chain to be levied land classification and area statistical results is shown in Fig. 4(b), geospatial information chain analysis can be a step execution, a variety of results at a glance.

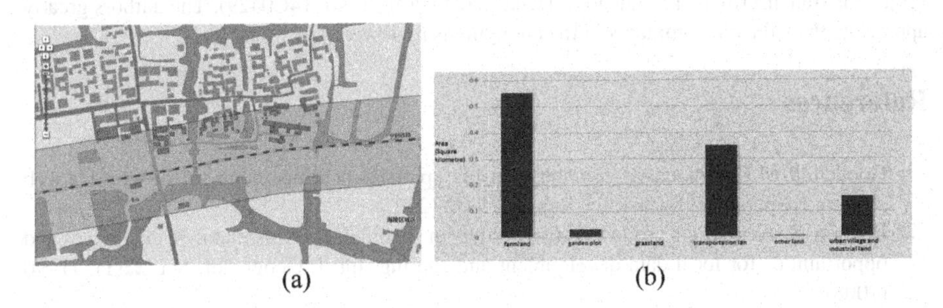

(a) (b)

Fig. 4. The left (a) is housing demolition analysis results, the right (b) is analysis of land expropriation and requisition statistics.

5 Conclusion

Service oriented architecture provides a new architecture and concept model for the data sharing and sharing of geographic information system. Service oriented architecture will release all kinds of network resources in the form of service specification, provide the standard call interface, realize the separation of business logic and implementation process. In this paper the spatial information service from SOA, the network data and geoprocessing model as atomic services are unified registration, release and call, based on the workflow reference model design the logical structure of the combination of spatial information service. In practical applications, there are a large number of mutual independent geographic information service, how to effectively retrieve and arrange the required services and feedback results still need to be further studied.

SOA for geospatial information chain provides architecture support, web service technology for geospatial information chain provides the concrete realization technology and OGC specifications for a variety of services provides the norm basis of geospatial information chain and workflow engine provides geospatial information chain workflow definition and control. The geospatial information chain construction and application as the research content, focused on the principle and key technology of geographic spatial information chain, makes beneficial attempt of spatial information sharing and the value-added services, design the geospatial information chain reference model and typical cases in the application practice validation. Geospatial information chain is the new tendency of the development of geographic information under the distributed network environment, is the important support of the digital city and the wisdom of urban construction, in order to achieve spatial information on-demand services and take the initiative to push. Geospatial information chain, the construction of open and interoperable geographic information infrastructure, which can provide a solid foundation for the application of geographic information in the network.

Acknowledgments. The work described in this article was substantially supported by the National Natural Science Foundation of China (Project No. 41271390), Scientific Research Project of Hunan Provincial Education Department (Project No. 14C0429). The authors greatly appreciate the valuable comments of the anonymous reviewers.

References

1. Goodchild, M.F.: Citizens as voluntary sensors: spatial data infrastructure in the world of web 2.0. Int. J. Spat. Data Infrastruct. Res. **2**, 24–32 (2007)
2. Elwood, S.: Grassroots groups as stakeholders in spatial data infrastructures: challenges and opportunities for local data development and sharing. Int. J. Geogr. Inf. Sci. **22**(1), 71–90 (2008)
3. Peng, Z.R.: A proposed framework for feature-level geospatial data sharing: a case study for transportation network data. Int. J. Geogr. Inf. Sci. **19**(4), 459–481 (2005)
4. Yang, C., Raskin, R.: Introduction to distributed geographic information processing research. Int. J. Geogr. Inf. Sci. **23**(5), 553–560 (2009)
5. Zhang, C., Li, W.: The roles of web feature and web map services in real-time geospatial data sharing for time-critical applications. Cartography Geogr. Inf. Sci. **32**(4), 269–283 (2005)
6. Xu, K., Wu, H., Gong, J.: Research on mechanism of geographic information service supported by multi-tier heterogeneous geospatial database. Geomatics Inf. Sci. Wuhan Univ. **33**(4), 402–404 (2008)
7. Yang, H., Sheng, Y., Wen, Y., et al.: Distributed geographic models sharing method based on web services. Geomatics Inf. Sci. Wuhan Univ. **34**(2), 142–145 (2009)
8. Shi, Y., Li, L., Zhang, L.: Framework and its core contents research of ubiquitous geographic information. Geomatics Inf. Sci. Wuhan Univ. **34**(2), 0150–0153 (2009)
9. Guo, R., Liu, J., Peng, Z., et al.: Technologies connotation and developing characteristics of open geospatial information platform. Acta Geodaetica Cartogr. Sin. **41**(3), 323–326 (2012)
10. Fan, X., Ren, Y., Yang, C., et al.: Research on scalable cluster-based cloud GIS platform. Appl. Res. Comput. **29**(10), 3736–3739 (2012)
11. Yang, C.-W., Michael, G., Huang, Q.-Y., et al.: Spatial cloud computing: how can the geospatial sciences use and help shape cloud computing. Int. J. Digit. Earth **4**(4), 305–329 (2011)

The Design and Implementation of Geographic Information Storage System Based on the Cloud Platform

Zhibo Wang[1,2], Kuai Hu[1], Ying Li[1], Gaole Li[1], Tianrun Sun[1], Weiping Zhu[1], and Xiaohui Cui[1(✉)]

[1] International School of Software, Wuhan University, Wuhan 430079, China
xcui@whu.edu.cn

[2] Software College, East China Institute of Technology, Nanchang 330013, China

Abstract. Faced with the status in big data era of information explosion, the conventional geographic information system can no longer meet the demands of storage and processing with a sea of data. The cloud computing as with ultra-large-scale server and high versatility and reliability of the new generation service model, just to meet the future needs of GIS. The paper design and implement the Geographic Information Storage System with Hadoop. The thesis has the following components: The brief analysis of the Hadoop implementation of the advantages of geographic information systems and the shortcomings of conventional GIS. Analysing the process of image processing and storage in Hadoop and conventional GIS to explore the suitability of the geographic information storage system built in Hadoop. The efficient image storage on HDFS cutting and generate image pyramid data for a large graphic.

Keywords: Cloud computing · Geographic information storage · GDAL · Hadoop · HDFS · Mapreduce

1 Introduction

The rapid growth in the total amount about information of human society in the past decade, not only increased the difficulty of mass data storage management and analysis process, but also brought new opportunities for the application and development of information technology. Remote sensing technology is one of the prominent technologies. At present, the change of the remote sensing digital image technology promotes the increase of the acquisition approach of GIS data, which, under the further derivation, shows a trend of development with large scale, multi-temporal, multi-resolution and other characteristics.

Faced with the multitude of GIS data, it has become a core issue about how to store, analysis and process it efficiently. As an emerging Internet service model, the cloud computing provides a realistic and effective solution for the above problems in storing and processing large amounts of data because of its strong parallel-computing storage capacity and many other advantages. Therefore it is an appropriate combination, which

F. Bian and Y. Xie (Eds.): GRMSE 2015, CCIS 569, pp. 917–928, 2016.
DOI: 10.1007/978-3-662-49155-3_95

is very necessary, of cloud computing technology to storage and analysis processing of geographic information with an optimized resource management.

Early GIS data was operated on a single machine and managed as a file unit. In 1970s, due to the further development of computer cartography, people found the link between the image data stream and the database and used the database system to manage remote sensing data successfully. And in 1990s, with the popularity of the Internet, there was a further innovation, from centralized storage into a distributed and networked one, in storing geographic information data. GIS also turned from a system in concept to a business model which is known as centering on "service" now, such as Ethernet-based ultra-large-scale cluster mass storage system.

As for the storage and use of technology in considerable remote sensing data, many experts, researchers and companies start their ways to innovate. Like HPSS produced by IBM, United States Department of Energy and Lawrence Livermore Lab; RASCHAL system operated by NASA; Massive data storage system named BW1 K (or Whale 100) system, which was used in 2008 Peking Olympic Games.

In summary, they can be divided into two categories. One is an emerging system based on distributed systems and large-scale network storage technology; the other one is the conventional system which manages the remote sensing image storage based on a relational database and file system. The conventional technology is still active in the practical application. Although it is mature enough, it has many shortcomings, such as the speed of access to massive data is not fast enough, access methods is not flexible and it cannot process unstructured data. Further, its storage capacity is fully constrained by the DBMS (Data Base Management System). However, with the utility of storage technology and virtual storage system, the emerging distributed system just has its uses. It can be well adapted to a variety of complex image data structure and support operation and maintenance of data efficiently so that it will enable the whole system to be practical and scalable.

The rest of the paper is organized as follows:

Section 2 introduce the research procedure about the specific implementation.

Section 3 describes methods used in this research procedure.

Section 4 shows the specific implementation and their experimental results.

Section 5 concludes the paper.

2 Related Work

As a practical field which has much utilization in our daily life, the applications of GIS have been widely researched by many researchers all over the world. For example, Chuvieco E, Congalton R G have a deep introduction about the application of remote sensing and GIS to forest fire hazard mapping [1]; Zhilin Li and Tao Cheng has ever studied in a multi-scale approach for spatio-temporal outlier detection [2]; Huanping Wu has research on an application of GIS in Meteorology which concerning about the spatial analysis and data sharing [3]; Yu Hai-long, Wu Lun, Liu Yu, Li Da-jun and Liu Li-ping finish a study of GIS-based model based on Web service [4]; Ole Einar Tveito use GIS to manage an application of spatial interpolation of Climatological and

Meteorological elements [5]. Meanwhile, Hartwig Dobesch, Pierre Dumolard and Lzabela Dyras apply GIS to the spatial interpolation for climate data [6];

There are some applied studies about Hadoop as well. For instance, Boss G, Malladi P, and Quan D, has a concise introduction about Cloud computing [7]; Yonggang Wang, and Sheng Wang has studied in topic concerning research and implementation on spatial data storage and operation based on Hadoop Platform [8]; Li XueFeng, CHENG ChengQi, GONG JianYa, and GUAN Li make a review of data storage and management technologies for massive remote sensing data [9]; Yang HC, Dasdan A, Hsiao RL, and Parker DS conduct a research on Map-Reduce-Merge that is about simplified relational data processing on large clusters [10]; The team of Divyakant does its work to design and implement an elastic data infrastructure for cloud-scale location services [11];

Besides, there are some works applying processing work of GIS into cloud computing platform. For example, firstly, a group leading by Qunying Huang has their perspective answer to the question: how can the geospatial sciences use and help shape cloud computing? [12]; Rama Naga Durga Rao Khaja and Venkateswara Rao Kota. Almeer start a work in hybrid Cloud framework for object storage based Geo-Spatial remote sensing Data processing [13]; Mohamed H. Almeer researched on cloud Hadoop Map Reduce for remote sensing image analysis [14]; Yang Chi, Zhang Xuyun, Zhong Changmin, Liu Chang, Pei Jian, Ramamohanarao, Kotagiri, and Chen Jinjun have a team work in spatio-temporal compression based approach for efficient big data processing on Cloud [15]; In addition, Junfeng KANG, Zhenhong, and Xiaosheng Li conducts a study about the framework of remote sensing image map service on Hadoop [16].

3 Research Procedure

Because of the large amount, high calculation intensity, harsh demanding to memory and hard drive and other reasons of the remote sensing image data, the conventional computing model cannot cope with its storage and computing. At the same time, cloud computing is a cross-age product which can process data in virtual parallelization of a super large scale. Therefore, we will be bold to merge them, aimed at solving the bottleneck problem of large data image storage and processing in an independent computer. Based on Hadoop system, this research uses MapReduce, parallel programming model, to solve the problem of image processing, while using HDFS, Hadoop Distributed File System, to store the image data across multiple computers.

In order to confirm the feasibility and efficiency of data storage and processing on cloud platform. We analyze and summarize the flow of mainstream conventional remote sensing processing model in China. On the basis of these, we innovate for a cloud version under the Hadoop cluster: store the files in HDFS, and process the images by MapReduce method. Then compare the flows and results between these two versions. After that, we do some research about block storage for image and generating image pyramid under Hadoop system, for layering storage and efficiently preview images, which is Compatible with different software.

4 Approach

4.1 Partition of Geographic Information Data Storage

In general, volumes of image data in geographic information system are large, so we just process a part of the image rather than a whole remote sensing image whose size is several GB or even dozens of GB. If the whole image is read during each process, huge amount of time will be used for I/O and the limit for memory will not allow to directly read a large image of dozens of GB in one time. So in consideration of the bottleneck of hardware and the pursuit of the most efficient process, we should partition image processing preferentially.

The file system in Hadoop (HDFS) also has the concept of block. HDFS splits the file with 64 MB per block, then each block is allocated from a single computer to the whole cluster as a single storage unit. Because of the inadaptability of Hadoop for storing a set of small files, when we are splitting the image, we need to avoid the blocks which are too small. In general, the block size is greater than 64 M, the storage unit in Hadoop.

Using GDAL library to manipulate the image. In our research, we implement two methods for image block: dividing with constant block number and dividing with constant block size. After GDAL reading the image, it can directly obtain image data set, band, image width, and drive for corresponding image format. When divide with constant block number, user inputs the number of dividing weight side M and the number of dividing height side N, then the image will be divided into M*N identical (border-blocks may be greater than or less than other blocks) blocks.

For instance, if we divide a 52000 * 20000 three-band satellite image into 6*2 blocks (Fig. 1), we will get blocks whose size is 8666*10000. But the last two blocks are greater than others, 8670*10000, because of indivisible.

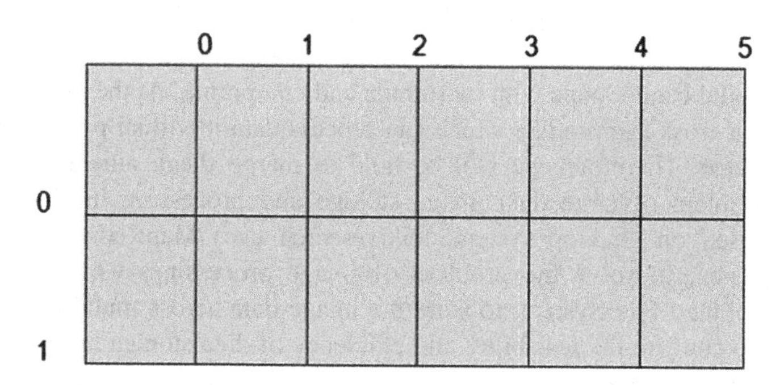

Fig. 1.

When we divide an image with constant size of each block, user input the weight and height of each block, then the image will be divided into several identical blocks.

For instance, if we divide a 20000*20000 single-band satellite image into blocks whose size is 5500*7500, the height of the last row is 5000 rather than 7500; the weight of last column is 3500 rather than 5500 (Fig. 2). These two kinds of blocks divided by different methods will be labeled according to the row and column since 0. When divided

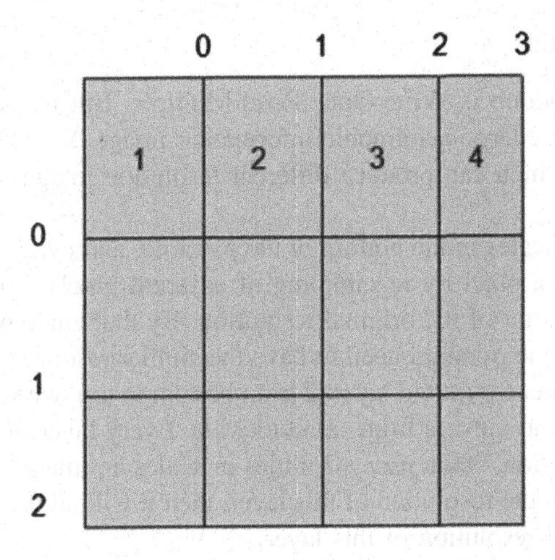

Fig. 2.

images have been sent to HDFS, the images will be named as image name + row number + column number. Image above (photo.tiff) shows the first block on the top left corner is named as photo.0.0.tiff, in a similar way, the second block is named as photo.0.1.tiff, and the third is named as photo.0.2.tiff. All of these blocks will be stored in a same folder, and the original remote sensing image will be sent to this folder.

As for the methods of dividing blocks, because the size of block is always greater than the number of block, the gap between the blocks in last row and column which is divided by constant block number and the other blocks is far less than the gap made by dividing with constant block size. As a result, except the situation where it requires a specific block size, using dividing the image with constant block number can get much uniform blocks.

The flow chart (Fig. 3):

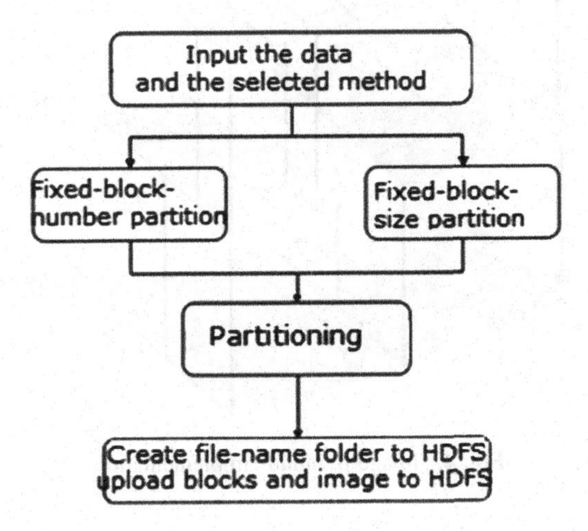

Fig. 3. Flow chart of partition

4.2 Image Pyramids

The core idea of Hadoop is Write-Once Read-Multiple. But in fact, it will take long time to read visually a large geographic information image. So it is necessary to build an image pyramid which can produce different resolution images according to user's requirement.

Original image locates in the bottom of the pyramid, and every four adjacent pixels are incorporated for a pixel by re-sampling of adjacent pixels. So that gets an upper image file with a quarter of the original resolution. By that analogy, we get the image located at the top of the pyramid need to have the similar resolution on the computer's display. Finally, we get a pyramid layered from bottom to top, whose quantity of data is from huge to small and view is from small to wide. Every layer of the pyramid has its corresponding resolution, when user magnifies or scales an image, if the current visual range does not match the resolution of this layer, then it will adjust to display an image with similar pyramid resolution of this layer.

The Pyramid is an effective way for storing grid file decreasing step by step. It can display the appropriate resolution image in the corresponding area by simple image coordinates query and resolution calculation. General computer's resolution is much lower than Satellite images', so using the pyramid file can avoid loading the whole original image when the user open the image. Instead, display small area's high resolution image when user magnifies the part of the whole image so that reduce the time of loading and preview.

Our research implements the building of 2*2 model pyramid. (Figure 4)

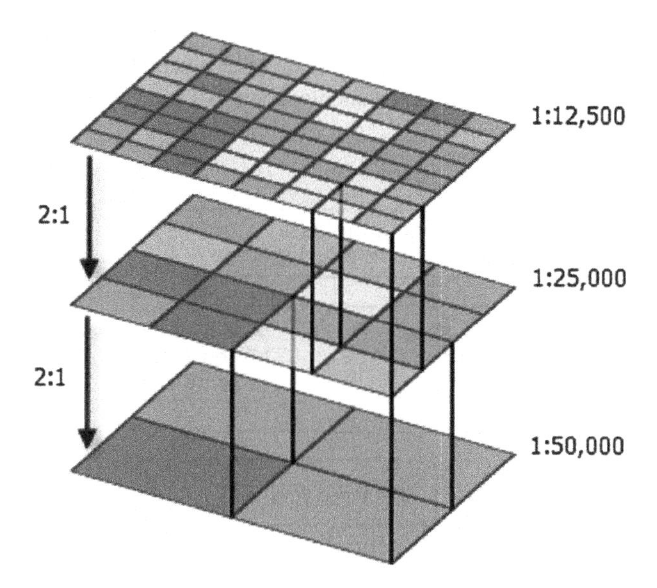

Fig. 4. Image pyramids implementation

The 2*2 model means each pixel on upper layer is generated by re-sampling with two pixel height and two pixel weight on below layer. In this graph, the top 64 images are the lowest level of pyramids, also are the images with the highest resolution. The middle 16 images' resolution reduces to original image's 1/4, but the view range of each grid equals to four grid on the above layer. The next layer can use such analogy. Then we can know, each grid has the same resolution, so when user's display resolution is constant, the corresponding layer can be chosen according to the needed view range.

In general, we can use the bilinear interpolation method, the nearest neighbor method and cubic convolution method to implement re-sampling. Cubic convolution method has the highest accuracy. In our research, we use GDAL and it chooses cubic convolution method to build the image pyramid. In GDAL, BuildOverviews (String, int [], ProgressCallback) can help us to build the pyramid. It resamples the image according to the value of int [], then confirm the minimum standard pixel value. Take the superposition calculation of the pixel values can get corresponding set of sampling rate. Finally, obtain Driver object of the original image when generate the pyramid file, then get the file format for judging whether it generates an pyramid inside files or an outside pyramid files.

5 Experiment

We implement our experiment in Hadoop environment to optimize our resource management. Based on Openstack platforms, we could manage the sources much better, for it could combine the specific resource pools of separate machines into one integrated virtual machine and, in the same way, divide a super computer into several virtual machines. For instance, we could divide a computer with 32 GB memory and 520 GB HDD into 4 virtual machines with 8 GB memory and 130 GB HDD. In this way, we could easily build several machines in same configurations to meet the need of our experiments.

5.1 Hadoop Configuration Details and Graphics

Hadoop Cluster consists of 7 machines which has the configuration mentioned above (Table 1).

There are one Master Machine and left 6 are Slave Machines.

Table 1. Cluster configuration

Configuration	Details
Mainboard	MSI Z97 MPOWER MAX AC(Intel Z97/LGA 1150)
CPU	Intel Core i7-4790 k 22 nano-processor
Memory	16 GB
HDD	Seagate 1 TB ST1000NM0033 7200 128 M SATA 6 Gb/s Enterprise HDD
Operation system	Centos6.6
Hadoop version	2.2.0 fully distributed cluster

5.2 Test About Feasibility and Efficiency of Data Storage and Processing on Cloud Platform

According to the analysis in the previous part, we manage one machine with same configuration as the standalone version. And the cluster version consists of 1 Master with 6 Slaves, corresponding to 1 control node and 6 compute nodes in Openstack.

Detail about attributes of the Grid Image:

GEOTIFF Type, 54000*54000 pixels, 5.4 GB single-band;

After 3*3 partition: 18000*18000 pixels, 618 MB single-band;

Considering about reducing the errors, we take three 54000*54000 pixels images to make process for four times. And the result is showing in Table 2 and Fig. 5:

Table 2. Image process summary

Zoom	Time of standalone version	Time of Hadoop version	Time of splicing image in standalone	Ratio of efficiency	Ratio of speed increase in standalone version	Ratio of speed increase in Hadoop version
160 times zoom	126 s	8 s	45 s	15.57 s		
40 times zoom	134 s	11 s	45 s	12.18 s	6.3 %	37.5 %
10 times zoom	167 s	69 s	46 s	2.42 s	24.6 %	527 %

From this possibility demo result, we could figure out that paralleled way to process in Hadoop is much faster than it's in conventional GIS process like standalone. For the standalone taking almost 45 s to splice the partition images, the effect in the small-size image could be even more obvious with the Hadoop which avoid the process to merge the partition images. Like the ratio of efficiency is 15.75 in 160 times zoom. It could infer that the efficiency has dropped down with the time for re-sampling getting smaller. The first reason for this circumstance is that the ratio of 45 splicing time to total time has decreased. Besides, we found Hadoop have sensitivity to the rapid increase of the intermediary data because of its remarkable amount of the intermediary product during the matrix multiplication. Conventional method like standalone stores the data into memory directly and does the matrix multiplication. However, the key-value pairs as the output of the MapReduce method Map are stored in the local hard disk, which means it would cost a lot for method Reduce to read them through I/O. Furthermore, the size of the dense Matrix from the re-sampling has increased with the time of zoom getting smaller. This is because the Hadoop should upload the matrix

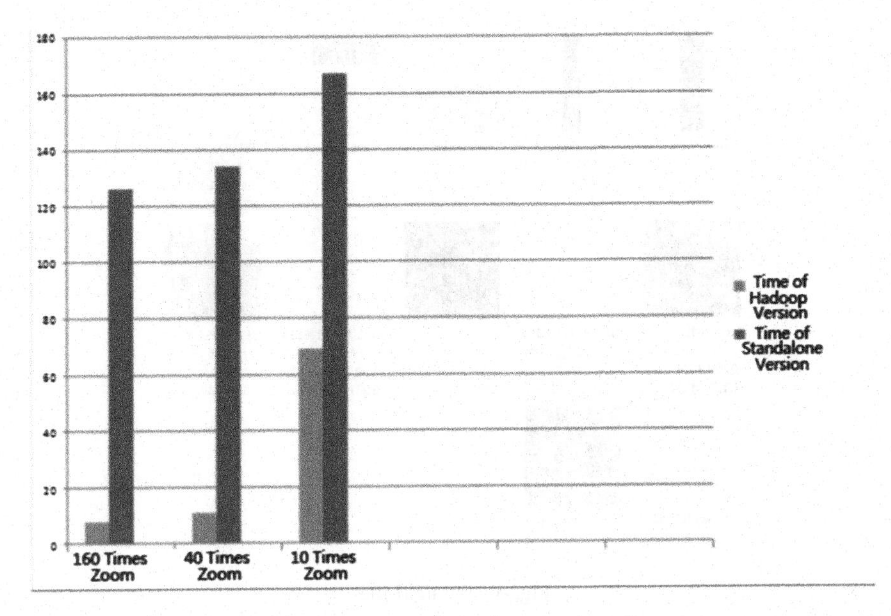

Fig. 5. Time of image processing bar chart

files to HDFS so that it enables matrix to do the parallel processing, which would create extra time for file transmission as well.

In conclusion, comparing to conventional frameworks of Geo-information Storage system, it is more efficient to create it based on Hadoop in paralleled way with equal configurations. In this way, based on Openstack, we could have a better resource management about considering the way of Hadoop to satisfy our requirement, which indicates that we could, if possible, manage large enough resources to do one specific task. However, one thing needs our attention is that the cursory algorithm would make too much intermediary values for systems to be efficient, for they take much time concerning I/O transmission. Though the transmission of file to HDFS is inevitable, we could reduce the intermediary data to alleviate the extra time cost of the HDFS transmission.

5.3 Pyramid Efficiency Comparison

We use three 54000*54000 grid images to do the test for reading speed of pyramid files. After getting.aux files from the previous algorithm, we could process them by the specific software like ArcGIS or Erdas. Like Fig. 6 showing:

We made three grid images into pyramid files with various layers and used ArcGIS to get average time to reduce the errors. The results are in Table 3.

In experiment, we choose remote images in different magnitudes to compare the experimental algorithm with ArcGIS concerning creating Pyramid in Table 4.

Experimental algorithm have so obvious advantage in reading image comparing to file without pyramids that it has equal effects on the file with pyramid created in

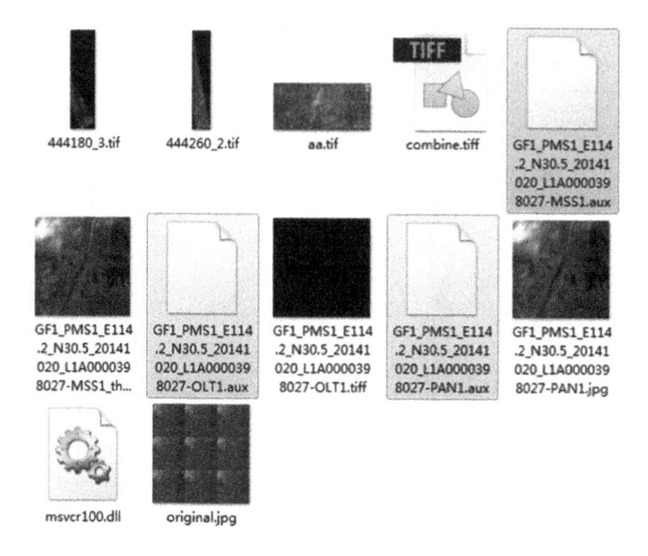

Fig. 6. Pyramid file

Table 3. Reading time

Without pyramid	Experimental algorithm	ArgGis
25.3 s	1.63 s	1.10 s
23.4 s	1.66 s	1.30 s
25.7 s	1.46 s	0.80 s

Table 4. Creating pyramids time

	Experimental algorithm	ArgGis
100 M	5 s	11 s
600 M	7 s	23 s
6 GB	122 s	170 s

ArcGIS. When considering about creating pyramid, experimental algorithm could finish work in a shorter time, especially for bulky image. So experimental algorithm could not only use in image reading, but also batch in creating pyramid files.

6 Conclusion

In this study, it achieves the design and implementation of geographic information storage system based on the cloud platform. The paper records the test to simulate the conventional image storage and processing of GIS in a single machine and the image manipulation of the version with MapReduce and HDFS based on Hadoop cluster, which proves the feasibility and advantages of building geographic information systems on Hadoop. Finally, it achieves an efficient image partition and storage on HDFS and the combination across the cluster hardware by the distributed file system, which fully shows the advantage, high performance and throughput, of Hadoop. And the oversized images are stored by Pyramid files, which greatly reduce the speed of visual reading. Taking all content into consideration, we could conclude that distributed way in Hadoop provides us a optimization of resource management that enable us to process images much more efficiently.

Acknowledgement. This research is supported in part by National Nature Science Foundation of China No. 61440054, Fundamental Research Funds for the Central Universities of China No. 216274213, and Nature Science Foundation of Hubei, China No. 2014CFA048. Outstanding Academic Talents Startup Funds of Wuhan University, No. 216-410100003.S

References

1. Chuvieco, E., Congalton, R.G.: Application of remote sensing and geographic information systems to forest fire hazard mapping. Remote Sens. Environ. **29**(2), 147–159 (1989)
2. Li, Z., Cheng, T.: A multiscale approach for spatiotemporal outlier detection. Trans. GIS **10**(2), 253–263 (2006)
3. Wu, H.: Application of GIS in Meteorology. Meteorol. Mon. (2010)
4. Yu, H.-L., Wu, L., Liu, Y., Li, D.-J., Liu, L.-P.: A study of integration between GIS and GIS-based model based on web services. Acta Geod. et Cartographic Sin. **35**(2), 153–159 (2006)
5. Tveito, O.E.: Applications of Spatial Interpolation of Climatological and Meteorological Elements by the Use of Geographical Information Systems (2006)
6. Dobesch, H., Dumolard, P., Dyras, L.: Spatial Interpolation for Climate Data: The Use of GIS in Climatology and Meteorology. Wiley-ISTE, Hoboken (2007)
7. Boss, G., Malladi, P., Quan, D., et al.: Cloud computing. IBM white paper, 1369 (2007)
8. Wang, Y., Wang, S.: Research and implementation on spatial data storage and operation based on Hadoop platform. In: 2010 Second IITA International Conference on Geoscience and Remote Sensing, pp. 275–279
9. Li, X.F., Cheng, C.Q., Gong, J.Y., Guan, L.: Review of data storage and management technologies for massive remote sensing data. Sci. China (Technol. Sci.) **54**(12), 3220–3232 (2011)
10. Yang, H.C., Dasdan, A., Hsiao, R.L., Parker, D.S.: Map-Reduce-Merge: simplified relational data processing on large clusters. In: Proceedings of the 2007 ACM SIGMOD Int'l Conference on Management of Data, pp. 1029–1040. ACM Press, New York (2007)
11. Agrawal, D., Abbadi, A.E., Nishimura, S., Das, S.: Design and implementation of an elastic data infrastructure for cloud-scale location services. Distrib. Parallel Databases 31(2), 289–319 (2013)
12. Huang, Q., Nebert, D., Raskin, R., Xu, Y., Bambacus, M., Fay, D., Yang, C., Goodchild, M.: Spatial cloud computing: how can the geospatial sciences use and help shape cloud computing? Int. J. Digit. Earth **4**(4), 305–329 (2011)
13. Khaja, R.N.D.R., Kota, V.R., Almeer: Hybrid cloud framework for object storage based geo spatial remote sensing data processing. Int. J. Eng. Sci. Res. Technol. **3**(12), 449–454 (2014)
14. Almeer, M.H.: Cloud Hadoop map reduce for remote sensing image analysis. J. Emerg. Trends Comput. Inf. Sci. **3**(4), 637–644 (2012)

15. Yang, C., Zhang, X., Zhong, C., Liu, C., Pei, J., Ramamohanarao, K., Chen, J.: A spatiotemporal compression based approach for efficient big data processing on cloud. J. Comput. Syst. Sci. **80**(8), 1563–1583 (2014). ISSN:0022-0000, doi:10.1016/j.jcss.2014.04.022
16. Kang, J., Zhenhong, Li, X.: The framework of remote sensing image map service on Hadoop. In: Information Technology and Computer Science—Proceedings of 2012 National Conference on Information Technology and Computer Science, 16 November 2012

Using Fuzzy Synthetic Evaluation in Site Selection of Post-earthquake Reconstruction Based on Remotely Sensed Imagery and Geographic Information

Li Peng(✉), Wunian Yang, and Hanhu Liu

Key Laboratory of Geoscience Information Technology of Ministry of Land and Resources of P.R. China, Chengdu University of Technology, Chengdu 610059, China
{pengli,Wunian.Yang,Hanhu.Liu}@cdut.edu.cn

Abstract. Site selection of post-earthquake reconstruction needs to process spatial data, and economic and human factors. There are both certainties and fuzzy uncertainties. Also quantitative data and qualitative data coexist due to different data types and precision. A reasonable location decision requires a comprehensive understanding and full use of these data. This paper introduced the Fuzzy Synthetic Evaluation (FSE) approach to the assessment of candidate areas acquired from the geological safety analysis of the study area. The proposed evaluation indices mainly consisted of natural resources, ecological protection, human protection, urban construction, transportation and development potential. The valid results show that the proposed method can be applied to site selection of post-earthquake reconstruction.

Keywords: Earthquake · Site selection · RS · GIS · Fuzzy synthetic evaluation

1 Introduction

There has been an increase in the numbers of earthquakes in world-wide locations in recent years. The strong earthquakes have brought high threatens to lives and properties of residents there. In 2008, the 8.0 Richter scale earthquake attacked Wenchuan county of Sichuan province in China. It caused havoc in lots of towns, where the public services and traffic lines were destroyed, and countless houses were broken down. The earthquakes have damaged the living environment. Also the secondary geological disasters triggered by aftershocks and heavy rainfalls raise serious risks to post-earthquake reconstruction tasks. Thus how to make the scientific and rational decision for post-disaster relocation is to be solved. The spatial decision making in site selection will handle both spatial data and attribute data, including structured, unstructured and semi-structured data. The constraints of site selection consist of certainties and uncertainties, as well as dynamic and fuzzy features. Both quantitative factors and qualitative factors coexist due to different precise of data, where some criteria are difficult to be quantified accurately. Hence the site selection of post-earthquake reconstruction

F. Bian and Y. Xie (Eds.): GRMSE 2015, CCIS 569, pp. 929–937, 2016.
DOI: 10.1007/978-3-662-49155-3_96

presents as a complex multi-objective synthetic decision making considering qualitative and quantitative factors.

At present, some mathematical models based on quantitative spatial analysis technique are used in site selection. The models using Principal Component Analysis [1] or Analytic Hierarchy Process [2], and the models using buffer and overlay analysis of Geographic Information System (GIS) [3, 4] were employed to select sites for immigration. Also, [5, 6] developed the spatial analysis of GIS to locate factories and industrial lands, and [7, 8] applied FSE method to determine bank ATM locations.

The FSE is the decision making process in fuzzy environment, using fuzzy theory to develop the synthetic assessments in a number of uncertainties [9]. The FSE is able to take into account a variety of factors and quantify the uncertainties, which has been proved to be an effective multi-factor decision making approach in practices. This paper introduced FSE to the decision making in site selection of post-earthquake reconstruction with the attempt to make full use of quantitative and qualitative data to obtain a rational assessment upon certain and fuzzy factors.

2 Study Area and Data

The study area is Miansi town at Wenchuan county in China. It is damaged in 5.12 earthquakes and the subsequent secondary geological disasters. Wenchuan is located at the southeastern Aba Tibetan & Qiang Autonomous Prefecture from 30°45′N-31°43′N to 102°51′E-103°44′E, where the Qiang people live. Miansi town is in the southwest of Wenchuan, 18 km from the count, with an administrative region of 25,140 hm^2 and population of 8606 people (2007 statistics) consisting of 14 administrative villages and 44 village groups. It is located at a dry valley in the upper Minjiang River, featured as the mountain valley landform at an average altitude of about 1,400 m, where the State Road 213 passes through.

The study data are as follows.

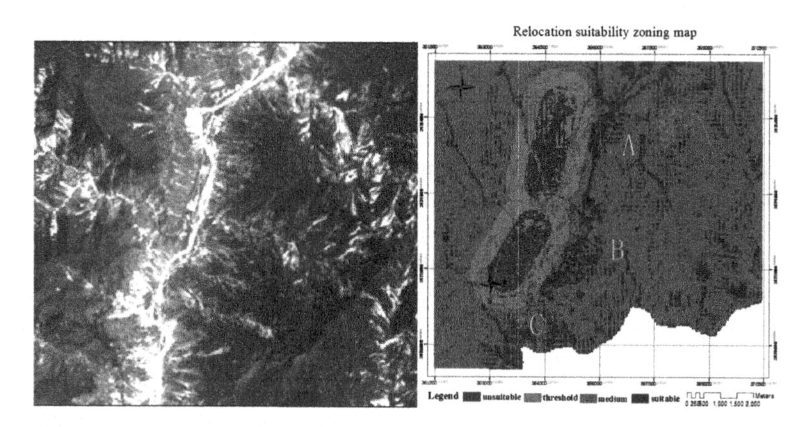

Fig. 1. (a) Airborne image of study area; (b) Relocation suitability zoning map.

(1) *Remotely Sensed Imagery.* The airborne color infrared images were taken by the Ministry of Land and Resources on May 18th, with spatial resolution of 1.0 meter. Figure 1a shows the study area. Other data include geological, landform and mountain disaster data (such as distribution of landslips, landslides and debris-flows) acquired by means of remote sensing technology and field survey.
(2) *Human and Economic Data.* Basic information includes natural and human environment, natural resources distribution, urban construction, infrastructure construction, and transportation data.
(3) *Distribution of the Candidate Areas.* GIS spatial analysis was performed on geological safety, terrain, water distribution factors to obtain the relocation suitability zoning map (Fig. 1b) [10]. Figure 1b shows three candidate regions: area A from Sanguanmiao Village to Miansi School, area B from Gaodianzi village to Xindian, and area C from Sandaoguai to Daping. Area A is partially overlapped with original town center.

3 Methods

3.1 Method Outline

Base on the preliminary relocation suitability zoning map (Fig. 1b) obtained by previous studies [10], this study acquires human, economic data and expert knowledge, and then performs FSE in the three candidate areas to obtain the optimal location scheme. The research method (Fig. 2) is divided into three steps: evaluation factors selection and index system construction, spatial database construction, and fuzzy synthetic evaluation.

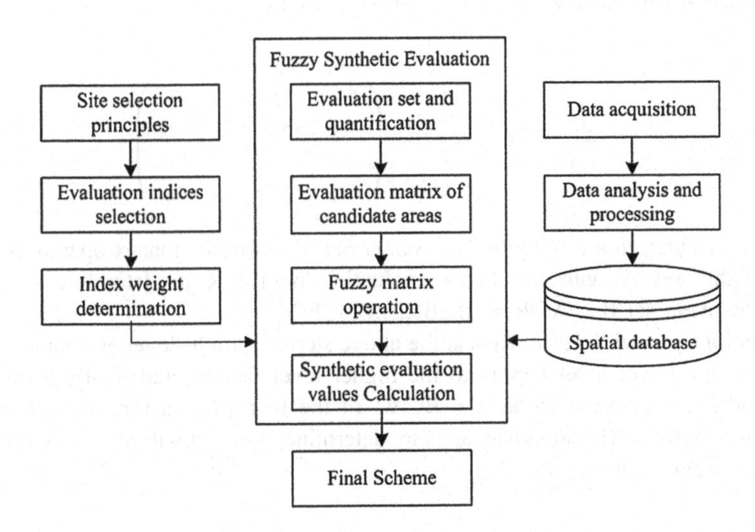

Fig. 2. Research method outline.

(1) *Evaluation Factors Selection and Index System Construction.* Determine the site selection principles, and then consider natural, human and economic characteristics of the study area to select the appropriate factors for the evaluation index system.
(2) *Spatial Database Construction.* Acquire various data related to the selected factors, including qualitative and quantitative data, certain and uncertain data, and then analyze, classify and save these data to build the spatial database.
(3) *Fuzzy Synthetic Evaluation.* Build the fuzzy synthetic evaluation model, and then integrate expert knowledge to assess the candidate areas to obtain location results.

3.2 Fuzzy Synthetic Evaluation Model

The Fuzzy Synthetic Evaluation models are built on the fuzzy theory. According to the different definitions of operations, the basic model can be further developed into five application models [9]. Considering that each individual constraint contributes to the evaluation, this paper build FSE model as follows.

(1) *Produce Factor Set.* According to the principles of site selection, create factor set $U = \{U_1, U_2, \ldots, U_m\}$. The individual factors can be further divided into the second-level factors until they can no longer be subdivided.
(2) *Acquire Weights Set.* Determine the weights of the individual factors $A = (A_1, A_2, \ldots, A_m) \in F(U)$, here $\sum A_i = 1$.
(3) *Determine Evaluation Set.* Determine the evaluation set $V = \{V_1, V_2, \ldots, V_n\}$ and quantify the evaluation values, where the evaluations are not absolutely positive or negative.
(4) *Evaluate Individual Factors.* Perform assessment on individual factors to obtain the evaluation matrix $R = (r_{ij})_{m \times n} \in F(U \times V)$, as (1).

$$R = \left(r_{ij}\right) = \begin{array}{c} \\ U_1 \\ U_2 \\ \cdots \\ U_m \end{array} \begin{array}{c} \begin{array}{cccc} V_1 & V_2 & \cdots & V_n \end{array} \\ \begin{pmatrix} r_{11} & r_{12} & \cdots & r_{1n} \\ r_{21} & r_{22} & \cdots & r_{2n} \\ \cdots & \cdots & \cdots & \cdots \\ r_{m1} & r_{m2} & \cdots & r_{mn} \end{pmatrix} \end{array} \quad (1)$$

(5) *Matrix Operation to Obtain Assessment Set.* Perform the matrix operation on the weights set A with the fuzzy evaluation matrix R to determine the fuzzy assessment set $B = A \bigcirc R = (B_1, B_2, \ldots, B_n) \in F(V)$.
(6) *Obtain Optimal Scheme.* Repeat the above steps for multi-level evaluation factors from the lower level factors to the higher level factors, and finally acquire the synthetic evaluation value $E = B \bigcirc V^T$ of the first-level factors of U. Calculate value E for each candidate area to determine the optimal scheme having the maximum value E.

4 Site Selection

The principles of site selection, the evaluation index system, and the implementation of fuzzy synthetic evaluation for the candidate areas are discussed here.

4.1 Site Selection Principles

Guided by the "Regulations on Post-Wenchuan Earthquake Rehabilitation and Reconstruction [11]," the principles of site selection are determined as follows.

(1) *Principle of Safety*. According to the requirements of the Wenchuan earthquake immigration, the post-disaster reconstruction should give top priority to urban safety [12]. Reconstruction should consider a variety of geological disasters and their active periods. The Holocene active faults must be avoided effectively.
(2) *Principle of Rational Exploitation of Resources.* The main natural resources enable development of mountain towns cover land resources, mineral resources, forest resources and water resources. Reasonable allocation of resources allows progresses in agriculture, tourism and other industries. Meanwhile, over-exploitation of resources should be prevented or minimized.
(3) *Principle of Ecological Protection.* Because of the complicated interactions between urban system and its natural environment, relocation must highlight the protection of ecosystem and natural landscape, especially water resources of upper rivers, primitive forests and wetland ecosystem, which provide high ecological values. At the same time, site selection will allow the restoration of the broken ecosystem.
(4) *Principle of Actions on Local Conditions.* According to the actual situation in earthquake regions and effects of surrounding towns, the relocation should take full account of ethnic factors. The reconstruction planning will help to maintain and develop characteristics and traditions of the native population.
(5) *Principle of Sustainable Development.* Site selection should consider the bearing capacity of the natural environment to develop environment-compatible industries. Moreover, site selection should fit into the future rational use of land to support the sustainable development of the reconstructed cities.

4.2 Evaluation Index System

Safety is the primary principle for site selection of post-earthquake reconstruction in mountain towns. It is necessary to analyze natural environmental factors such as geological disaster distribution, terrain and water distribution. The research group has performed spatial analysis upon indices including distribution of faults and geological disasters, ground surface slope, elevation, land coverage and water distribution to calculate the basic relocation suitability of the study area [10]. Consequently, this paper focuses on constraints mainly related to natural resources, ecological protection and future development. The selected indices are listed in Table 1. Their weights are acquired by means of the Delphi method.

Table 1. Evaluation indices and weights.

First-level indices and weights	Second-level indices and weights
Natural resources U_1 0.30	Land resources U_{11} 0.27 Mineral resources U_{12} 0.16 Forest resources U_{13} 0.19 Water resources U_{14} 0.38
Ecological protection U_2 0.21	Environmental protection U_{21} 0.75 Ecosystem restoration U_{22} 0.25
Human protection U_3 0.12	Ethnic customs U_{31} 0.31 Ethnic language U_{32} 0.46 Cultural heritage U_{33} 0.23
Urban construction U_4 0.10	Public infrastructure U_{41} 0.40 Building area U_{42} 0.60
Transportation U_5 0.19	External transport facilities U_{51} 0.67 Road accessibility U_{52} 0.33
Development potential U_6 0.08	Economic influence of neighbor towns U_{61} 0.19 Traffic development U_{62} 0.33 Urbanization U_{63} 0.48

(1) *Natural Resources.* This index indicates the status quo of the distribution, the configuration and the development of land resources, mineral resources, forest resources and water resources in candidate areas, which limits or supports the development of agriculture, industries and cities.
(2) *Ecological Protection.* It includes environmental protection and ecosystem restoration. Location should avoid breaking the eco-environment as far as reasonably possible. Also, location should support the restoration engineering of the ecosystem which was impaired grievously during the earthquakes and the subsequent secondary geological disasters.
(3) *Human Protection.* Qiang people live in Miansi town. Thus the relocation must consider the protection for their traditional customs, their language, and the cultural heritage tangible or intangible. Therefore, the relocation should avoid the incompatibility raised by the conflicts between the new human/geographical environment and the ethnic characteristics of native population.
(4) *Urban Construction.* It is the status quo of urban construction, including the second-level indices of public infrastructure and building area. The public infrastructure provides social services through municipal, culture, education and medical facilities. The building area is defined as the total area of the usable buildings at present.
(5) *Transportation.* This index contains two sub-indices, i.e. external transport facilities and road accessibility. The road accessibility is decided by the internal road network of a town.

(6) *Development Potential*. The development potentials of a town are influenced by economics of neighbor towns, internal and external traffic development, and its urbanization process, where the index of traffic development plays an important role in the development of a mountain town. The index of urbanization indicates the spatial expansibility of a town when considering its future economic development.

4.3 Fuzzy Synthetic Evaluation

The above FSE model was performed in the three candidate areas A, B and C (Fig. 1b). Let evaluation set as $V = \{V_1, V_2, \ldots, V_n\} = \{\text{Excellent, Good, Fair, Poor}\} = \{9, 7, 5, 3\}$. Then, based on the human and economic background information, the expert knowledge was used to assess the second-level individual factors to obtain the evaluation matrix table (Table 2).

Then, the matrix operation was performed on the weights set A with the fuzzy evaluation matrix R to determine the fuzzy assessment set of the second-level indices (U_{11}, U_{12}, U_{13}, U_{14}) of area A as follows.

$$B_{A1} = A_1 \circ R_{A1} = (0.27 \quad 0.16 \quad 0.19 \quad 0.38)\begin{pmatrix} 0.3 & 0.4 & 0.3 & 0.0 \\ 0.2 & 0.3 & 0.4 & 0.1 \\ 0.1 & 0.4 & 0.3 & 0.2 \\ 0.3 & 0.5 & 0.2 & 0.0 \end{pmatrix} = (0.246 \quad 0.422 \quad 0.278 \quad 0.054)$$

Similarly, the fuzzy assessment sets of other second-level indices of area A were calculated as follows.

$$B_{A2} = A_2 \circ R_{A2} = (0.375 \quad 0.475 \quad 0.125 \quad 0.025)$$
$$B_{A3} = A_3 \circ R_{A3} = (0.531 \quad 0.331 \quad 0.092 \quad 0.046)$$
$$B_{A4} = A_4 \circ R_{A4} = (0.640 \quad 0.360 \quad 0.000 \quad 0.000)$$
$$B_{A5} = A_5 \circ R_{A5} = (0.333 \quad 0.433 \quad 0.234 \quad 0.000)$$
$$B_{A6} = A_6 \circ R_{A6} = (0.200 \quad 0.404 \quad 0.348 \quad 0.048)$$

After that, the fuzzy assessment set of the first-level indices (U_1, U_2, U_3, U_4, U_5, U_6) of area A was calculated as follows.

$$B_A = A \circ R_A = (0.30 \quad 0.21 \quad 0.12 \quad 0.10 \quad 0.19 \quad 0.08)(B_{A1} \quad B_{A2} \quad B_{A3} \quad B_{A4} \quad B_{A5} \quad B_{A6})^T$$
$$= (0.360 \quad 0.417 \quad 0.193 \quad 0.031)$$

Finally, the synthetic evaluation value of area A was calculated as $E_A = B_A O V^T = 7.21$. Similarly, the synthetic evaluation value of area B was calculated as $E_B = B_B O V^T = 7.28$, and that of area C as $E_C = B_C O V^T = 6.63$. Here, area B has the highest value among the three areas. Area A is slightly lower than area B. Area C has the lowest value. As a result, area B, A can be selected as the first, second candidate area respectively, on which the in-situ surveys should be focused.

Table 2. Second-level indices fuzzy evaluation table of candidate areas. Excellent as E, Good as G, Fair as F, Poor as P.

Indices	Area A				Area B				Area C			
	E	G	F	P	E	G	F	P	E	G	F	P
U_{11}	0.3	0.4	0.3	0.0	0.3	0.5	0.2	0.0	0.0	0.3	0.5	0.2
U_{12}	0.2	0.3	0.4	0.1	0.2	0.4	0.3	0.1	0.1	0.4	0.4	0.1
U_{13}	0.1	0.4	0.3	0.2	0.2	0.5	0.3	0.0	0.2	0.5	0.2	0.1
U_{14}	0.3	0.5	0.2	0.0	0.2	0.5	0.3	0.0	0.2	0.6	0.2	0.0
U_{21}	0.4	0.5	0.1	0.0	0.5	0.4	0.1	0.0	0.6	0.4	0.0	0.0
U_{22}	0.3	0.4	0.2	0.1	0.4	0.4	0.2	0.0	0.2	0.4	0.3	0.1
U_{31}	0.6	0.4	0.0	0.0	0.6	0.4	0.0	0.0	0.6	0.4	0.0	0.0
U_{32}	0.7	0.3	0.0	0.0	0.7	0.3	0.0	0.0	0.7	0.3	0.0	0.0
U_{33}	0.1	0.3	0.4	0.2	0.2	0.4	0.3	0.1	0.1	0.4	0.4	0.1
U_{41}	0.7	0.3	0.0	0.0	0.3	0.4	0.3	0.0	0.1	0.3	0.4	0.2
U_{42}	0.6	0.4	0.0	0.0	0.5	0.4	0.1	0.0	0.0	0.3	0.5	0.2
U_{51}	0.3	0.4	0.3	0.0	0.3	0.4	0.3	0.0	0.3	0.4	0.3	0.0
U_{52}	0.4	0.5	0.1	0.0	0.3	0.5	0.2	0.0	0.0	0.4	0.4	0.2
U_{61}	0.2	0.5	0.3	0.0	0.2	0.5	0.3	0.0	0.2	0.5	0.3	0.0
U_{62}	0.2	0.5	0.3	0.0	0.1	0.3	0.5	0.1	0.0	0.2	0.5	0.3
U_{63}	0.2	0.3	0.4	0.1	0.6	0.4	0.0	0.0	0.0	0.2	0.5	0.3

5 Conclusions

Spatial decision making in site selection of post-earthquake reconstruction involves data processing on certainties and uncertainties. There are large differences in data precision, and there are quantitative and qualitative data. This paper applied the Fuzzy Synthetic Evaluation approach to assessing the candidate areas. The FSE model was employed to quantify qualitative factors and uncertainties. The valid results can provide information for reconstruction site selection planning.

It is worth noting that the reasonableness of the index system and weights depends on the expert knowledge. Also, for the same index system, different groups of experts may produce different results. Therefore, the spatial decision support system should be adopted in future practices. The constant developments of the knowledge base are supposed to further improve the rational and practical results.

Acknowledgments. This project is supported by the National Natural Science Foundation of China (Grant No. 41071265).

References

1. Tang, X.M., Zhou, W.C.: A study on the site selection model of mountain town: take Wusham county as example. J. Mt. Sci. **19**, 135–140 (2001)
2. Li, S.: The research of the method of geo-disaster immigration location in Xinpxian country.master dissertation. China University of Geosciences, Beijing (2007, unpublished)

3. Yang, W.N., Hu, G.C., Liu, H.H., Peng, L.: New site selection for the immigrant transit of the post-disaster reconstruction of Wenchuan earthquake regions: takes Yanmen district of Wenchuan as an example. In: Proceedings of SPIE the 2nd International Conference on Earth Observation for Global Changes (EOGC 2009), vol. 7471, pp. 747107-747107-7 (2009)
4. Peng, L., Yang, W.N., Pan, P.F.: Applying remote sensing and spacial analysis technology to locating transitional shelters in earthquake areas. In: Proceedings of the 2nd International Conference on Information Science and Engineering (ICISE 2010), vol. 9, pp. 6463–6466 (2010)
5. Gao, J.S., Guan, Z.Q.: Implementation of site locating strategies based on RS and GIS. Geomatics Inf. Sci. Wuhan Univ. **30**, 779–781 (2005)
6. Zhou, L.J., Zhang, S.H., Zang, S.Y.: The application of spatial analyst in sites selection: a case of Daqing city in the Ha-Da-Qi industrial corridor. Areal Res. Dev. **26**, 125–128 (2007)
7. Wu, J.Z., Hua, X.H., Zhou, Q.J.: Spatial decision support system for location selection based on fuzzy synthetic evaluation. Geospatial Inf. **3**, 45–47 (2005)
8. Li, W., Zhou, T.G., Zhang, W.: The application of spatial analysis of GIS and fuzzy synthetic evaluation in site location of ATM network. Sci. Surveying Mapp. **33**, 229–231 (2008)
9. Xie, X.L.: Decision-making models based on fuzzy synthetic evaluation. Stat. Decis. **21**, 57–58 (2005)
10. Peng, L., Yang, W.N., Liu, H.H., Shao, H.Y., Pan, P.F.: Study on spatial decision-making of site selection for earthquake-induced migration using remote sensing and GIS technology. J. Southwest Univ. **33**, 96–103 (2011)
11. The State Council of the People's Republic of China. Regulations on post-Wenchuan earthquake rehabilitation and reconstruction. Gazette of the State Council of the People's Republic of China, vol. 17, pp. 4–12 (2008)
12. Guan, S.Y., Sun, W.L.: Research on geological security of new Beichuan county seat after 5.12 Wenchuan earthquake. Geotech. Invest. Surveying **10**, 24–28 (2009)

The Utilization Status Research of Clay Resources Using RS and GIS, in Shanghai Area

Haiqing Wang(✉), Ling Chen, and Wei Chen

China Aero Geophysical Survey and Remote Sensing Center for Land and Resources, Beijing 100083, China
whq0705@126.com

Abstract. Shanghai is one height developed area in China, the clay resources is consumed rapidly, so it has important significance to do the utilization status research of clay resources using RS and GIS. The utilization status of clay resources could be known according to the production status of brickfield. In this article the remote sensing images which respectively gotten in 2011 and 2014 were used, and remote sensing interpretation interior were combined with field surveying. This study showed that one brickfield could be confirmed through identify three landmarks: brick kiln, brick dump, billet dump. There were 66 brickfields were confirmed in the remote sensing images of 2011, and 41 brickfields were confirmed in the remote sensing images of 2014. Some scientific decision-making support was provided by this research for mines management department in government, to maintain the normal development and using order.

Keywords: Shanghai · Clay · Utilization status · Remote sensing · Brickfield

1 Introduction

Mineral resources are important material foundation for social and economic development. Since twenty-first Century, resource consumption has been significantly increased, the contradiction between limited resources and social demand is growing [1], most mineral products prices rose sharply [2]. Face the China's mineral resources characteristics "big shortage, rare and not lack of" [3], mineral resources should be use reasonably. So, it is very important to grasp the utilization of mineral resources.

From 2003, Ministry of Land and Resources start a pilot study about remote sensing survey and monitoring of key mines, experience was accumulated for carrying out work in all country, a number of academic papers were published [4–10]. Since 2006, China Geological Survey has launched "remote sensing survey and monitoring of mineral resources", "remote sensing interpretation of the national mineral satellite", "comprehensive survey of mine environment", "mineral resources development environment remote sensing monitoring" and other projects. With the project work carried out, relevant technical methods were improved gradually, not only a large number of academic papers were published [11–19], but also work guide was published in 2009 [20], proceedings [21] and monograph [22, 23] were published in 2011 and 2014, the industry

F. Bian and Y. Xie (Eds.): GRMSE 2015, CCIS 569, pp. 938–945, 2016.
DOI: 10.1007/978-3-662-49155-3_97

standard was awarded also [24]. In short, the technical method of this investigation about mineral resources utilization status is gradually improved.

Shanghai is a highly developed area of our country, urban construction is developing rapidly, and the clay resource consumption is very large, so it is very important to study the utilization of clay resource in Shanghai area. Clay resources is mainly used for firing brick, the processing site is brickfield, so the utilization status of clay resources could be known according to the production status of brickfield. In this article the remote sensing images which respectively gotten in 2011 and 2014 were used, Arcgis platform was used also, and remote sensing interpretation interior were combined with field surveying. Based on these, the interpretation keys of brickfields and their utilization status was studied. There were 66 brickfields were confirmed in the remote sensing images of 2011, and 41 brickfields were confirmed in the remote sensing images of 2014. The objective and practical situation of utilization status of clay resources in Shanghai area was gotten pressingly through this research. And some scientific decision-making support was provided by this research for mines management department in government, to maintain the normal development and using order, and to attack the rule-breaking behavior of mining development and using.

2 Remote Sensing Images

Remote sensing images which respectively gotten in 2011 and 2014 were used for this utilization status research of clay resources in Shanghai area. The remote sensing

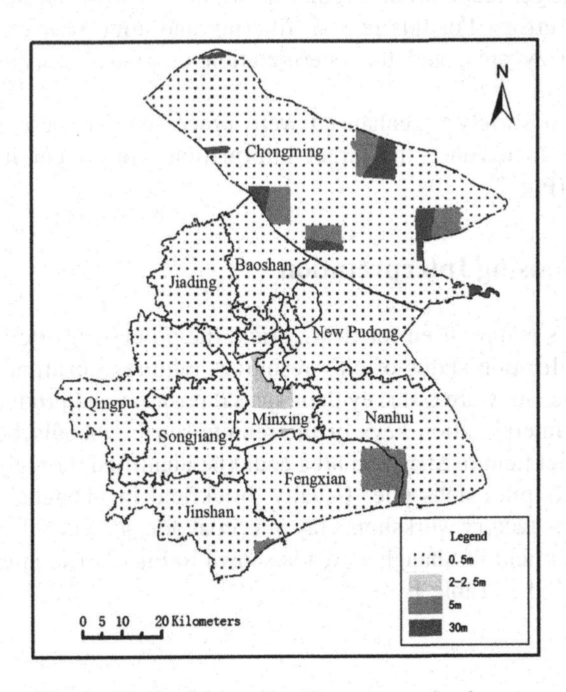

Fig. 1. The basic situation of remote sensing images

images gotten in 2011 are 8138 km^2 in total, among these, spatial resolution of 0.5 m is 7556 km^2, spatial resolution of 2 m and 2.5 m are 96 km^2, spatial resolution of 5 m is 305 km^2, and spatial resolution of 30 m is 181 km^2 (Fig. 1). The remote sensing images gotten in 2014 are 8138 km^2 in total, among these, spatial resolution of 0.5 m is 1659 km^2, spatial resolution of 1 m is 1546 km^2, and spatial resolution of 2 m is 4933 km^2.

In the remote sensing images gotten in 2011, spatial resolution of 0.5 m is composed of Worldview-1, Worldview-2, and GeoEye-1, spatial resolution of 2 m is composed of YG-2, spatial resolution of 2.5 m is composed of Spot5, spatial resolution of 5 m is composed of YG-8, spatial resolution of 30 m is composed of HJ-1. In the remote sensing images gotten in 2014, spatial resolution of 0.5 m is composed of Worldview-2, spatial resolution of 1 m is composed of YG-1, spatial resolution of 2 m is composed of 02C and TH-1.

3 Remote Sensing Image Processing

The quality of satellite remote sensing image is affected by weather conditions, cloud and fog (fuzzy area) is the main factor causing poor quality and poor application effect. The existence of fuzzy area in remote sensing image, while increasing the brightness of the surface, reduce the contrast between the features, not only to remote sensing image processing greatly inconvenience, but also to follow-up image classification, target recognition, ground information extraction accuracy decline, and even cause errors. For this reason, filtering and atmospheric correction were processed in fuzzy area, and the interpretation of remote sensing images been enhanced.

In addition, a variety of enhanced were processed for these remote sensing image, in order to enhance the target information, convenient interpretation of remote sensing (Fig. 2).

4 Remote Sensing Interpretation

Clay resources is mainly used for firing brick, the processing site is brickfield, so study on the utilization status of clay resources requires quantitative research on brickfields in the study area, survey their spatial position and production status.

First of all, interpretation keys of the remote sensing should be set up. Study shows that, a brickfield in Shanghai area could be identified through three features: brick kiln, bricks pile, slabs pile. Usually, brickfield in Shanghai area has some common features: factory workshop, clay pile, coal pile, gangue pile, transport ships etc. Typical brickfield in Shanghai area is shown in Fig. 3. The interpretation keys can be summarized as Table 1.

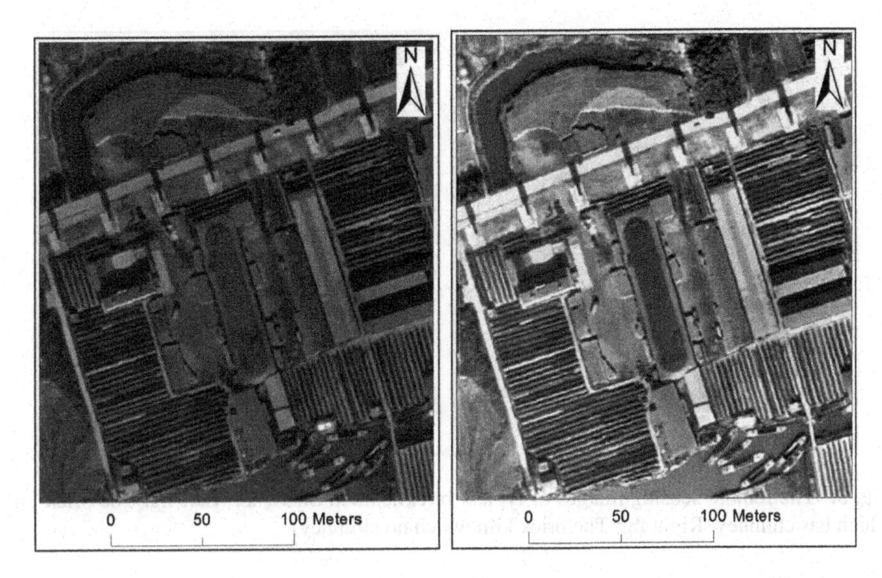

Fig. 2. The effect comparison of images enhancement (Left fig: The image before enhanced, Right fig: The image after enhanced)

Table 1. The interpretation keys of brickfield in Shanghai

No.	Feature	Shape	Color	Other
1	Brick kiln with chimney	Long moment	Brick red	Could find the towering chimney and chimney shadow
2	Brick kiln without chimney	Long strip, two ends with circular arc	Blue	Could not find the towering chimney
3	Slabs pile	Long strip of neat arrangement	Yellow, light gray, dark gray	The largest area of this brickfield
4	Bricks pile	Approximate moment	Brick red	General distribution in the brick kiln around
5	Factory workshop	Moment	Blue, brick red, gray	
6	Clay pile	Irregular shape	Soil yellow, light gray	
7	Coal pile	Irregular shape	Black	
8	Gangue pile	Irregular shape	Gray	

Based on these interpretation keys, using Arcgis platform, there were 66 brickfields were confirmed in the remote sensing images of 2011, and 41 brickfields were confirmed in the remote sensing images of 2014. So, there were 25 brickfields were closed between 2011 and 2014 (Fig. 4, Table 2).

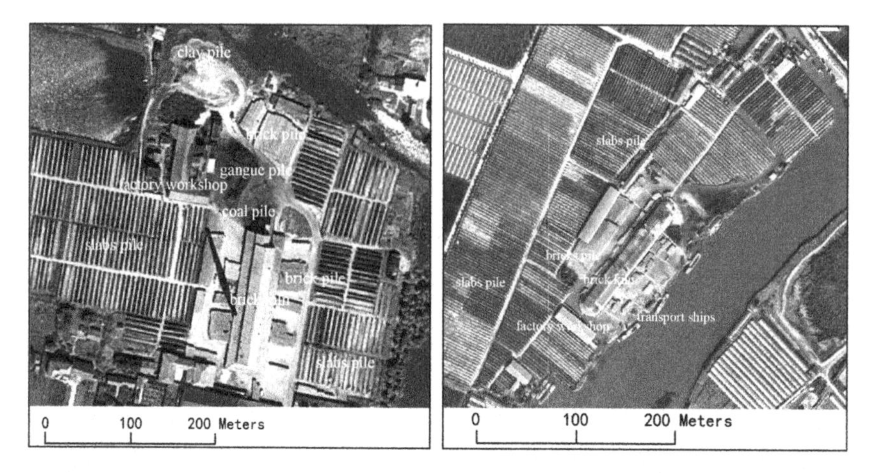

Fig. 3. The remote sensing images of typical brickfields in Shanghai (Left fig: The brick kiln which has chimney, Right fig: The brick kiln which no chimney)

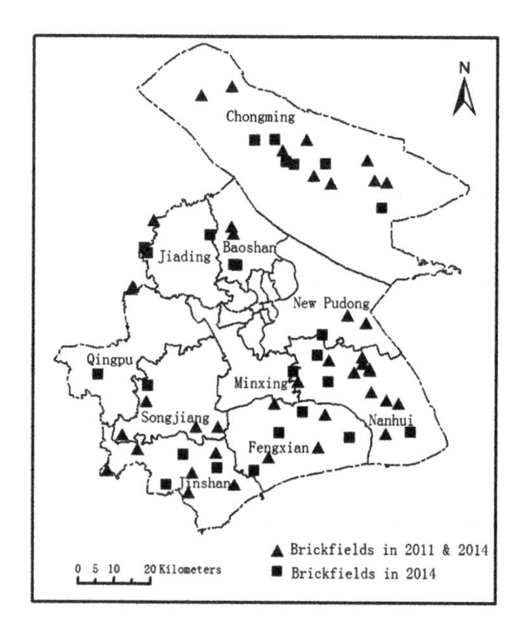

Fig. 4. The distribution map brickfields in Shanghai

Table 2. The brickfields quantity in every county

No.	Country	2011	2014	Field surveying
1	Minxing	1	0	1
2	Baoshan	5	2	2
3	Jiading	3	1	3
4	New Pudong	3	2	3
5	Jinshan	9	6	2
6	Songjiang	5	4	4
7	Qingpu	3	2	3
8	Nanhui	14	11	12
9	Fengxian	8	4	3
10	Chongming	15	9	7
Total		66	41	40

5 Field Surveying

In order to ensure the accuracy of interpretation, one part of the brickfields were surveyed in field. A total of 40 brickfields, accounting for 61 % of the total were surveyed in field.

Field work showed that, the 40 brickfields been confirmed in the remote sensing images, were completely correct, the interpretation accuracy of 100 %. Illustrate the interpretation keys, established in this research, are correct and feasible, so brickfield could be confirmed through interpretation of brick kiln, slabs pile and brick pile.

Field work found that, all brick products in Shanghai area are hollow brick, did not see any solid brick products, reflecting the practical measures to build a resource-saving society of government in Shanghai. And many brickfields had been closed, or even abandoned, especially in Chongming County.

The actual production situation of brickfields in Shanghai area was know through the interpretation of remote sensing and field work, and the objective and practical situation of utilization status of clay resources in Shanghai area was gotten indirectly.

The results show that brickfields in Shanghai area have some distribution characteristics: ① It shows negative correlation between degree of economic development and brickfields distribution in Shanghai area, all brickfields are located in suburbs which far away from towns, this could be shown in Fig. 4. ② All brickfields in Shanghai distributed beside rivers edge. The main reason is that, there are a lot of river in Shanghai area, so the water transportation developed, cheap water transportation provides convenient conditions for clay and other raw materials input and brick and other products output.

6 Conclusion

From this utilization status research of clay resources in Shanghai area, there were 66 brickfields were confirmed in the remote sensing images of 2011, and 41 brickfields were confirmed in the remote sensing images of 2014, there were 25 brickfields were closed between 2011 and 2014. The objective and practical situation of utilization status of clay resources in Shanghai area was gotten pressingly through this research. And some scientific decision-making support was provided by this research for mines management department in government, to maintain the normal development and using order, and to attack the rule-breaking behavior of mining development and using.

The specific interpretation keys, which established through this research, for brickfields in Shanghai area, and the distribution characteristics of brickfields in Shanghai area, lay a solid foundation for future investigation and monitoring, and have important reference value to the similar and related work in the neighboring area.

Acknowledgements. This work was financially supported by the China Geological Survey Foundations (1212011120027, 12120115062801, 12120113100100).

References

1. Ministry of land and resources of the People's Republic of China.: Mineral Resources report for China. Geological Press, Beijing (2011)
2. Information Center of Ministry of Land and Resources of People's Republic of China, ed.: Anual Review of World Mineral Resources from 2009 to 2010. Geological Press, Beijing (2011)
3. Jia, W.L., Chen, J.B., Hu, D.W.: The Main Mineral Supply and Demand Situation Analysis for 2009 in China. China Land Press, Beijing (2011)
4. Wang, X.H., Nie, H.F., Yang, Q.H., et al.: The different monitoring effects of QuickBird and Spot-5 data in mine exploitation. Remote Sens. Land Resour. **16**(1), 15–18, 80–81 (2004)
5. Wang, X.H., Nie, H.F., Yang, Q. H., et al.: Establishment of dynamic monitoring for mineral resources development status using remote sensing. In: National Land and Resources and Environment Remote Sensing Technology Exchange, Wuhan, China (2004)
6. Wang, J., Li, C.Z.: Using high-resolution satellite data to carry out dynamic monitoring, taking Shanxi Jincheng area for example. In: National Land and Resources and Environment Remote Sensing Technology Exchange, Wuhan, China (2004)
7. Wang, X.H., Nie, H.F., Li, C.Z., et al.: The application characteristics of different remote sensing data sources to the investigation of the mining situation and environment of mines. In: The Fifteenth Session of the National Remote Sensing Symposium, Guiyang, China (2005)
8. Wang, X.H., Nie, H.F., Li, C.Z., et al.: The application of characteristics of different remote sensing data sources to the investigation of the mining situation and environment ofmines. Remote Sens. Land Resour. **18**(2), 69–71 (2006)
9. Li, C.Z., Nie, H.F., Wang, J., et al.: A remote sensing study of characteristics of Geological disaster in a mine. Remote Sens. Land Resour. **17**(1), 45–48, 78 (2005)
10. Liu, Q., Nie, H.F., Lv, J.T., et al.: The application of GIS to the remote sensing dynamic monitoring of mine exploitation. Remote Sens. Land Resour. **17**(1), 61–65 (2005)

11. Nie, H.F., Yang, J.Z., Wang, X.H., et al.: The problems in the remote sensing monitoring technology for the exploration of mineral resources and the countermeasures. Remote Sens. Land Resour. **19**(4), 11–13 (2007)
12. Jing, Q.Q., Zhang, Z., Wang, X.: Collecting method of coal refuse distribution information based on Aster remote sensing images. Coal Sci. Technol. **36**(5), 93–96 (2008)
13. Zhou, Y.J., Zhu, Z.J.: The quality and capability of the Rapid-eye 3A product for discerning ground objects in the mine. Remote Sens. Land Resour. **21**(2), 62–65, 70 (2009)
14. Wang, H.Q.: A comparison and selection of the methods for mine geological environment assessment based on GIS and RS. Remote Sens. Land Resour. **22**(3), 92–96 (2010)
15. Wang, H.Q., Chen, L.: Evaluation of mine geological environment in LingLong-LaiXi area. Resour. Ind. **13**(3), 72–76 (2011)
16. Wang, H.Q., Chen, L.: Investigation of current surface subsidence situation using remote sensing images, in JiNing coal mine concentration area, Shandong Province. Chin. J. Geol. Hazard Control **22**(1), 87–93 (2011)
17. An, Z.H., Wang, X.H., Dai, L., et al.: An investigation research on exploitation of Bayan Obo based on high-resolution satellite remote sensing data. Geol. Explor. **47**(3), 462–468 (2011)
18. Wang, H.Q., Zhou, Y.J., Chen, L., et al.: The dynamic monitoring on mining collapsing around Xinglongzhuang coal mine based on remote sensing images. Adv. Mater. Res. **726-731**, 4625–4630 (2013)
19. Wang, H.Q.: Mining subsidence monitoring around Longgu coal mine based on remote sensing. Adv. Mater. Res. **1010–1012**, 489–495 (2014)
20. Yang, J.Z., Qin, X.W., Nie, H.F., et al.: Mining Remote Sensing Monitoring Guide. China Land Press, Beijing (2009)
21. Qin, X.W., Yang, J.Z., Kang, G.F., et al.: Mining Remote Sensing Monitoring Techniques. Surveying and Mapping Press, Beijing (2011)
22. Yang, J.Z., Qin, X.W., Zhang, Z., et al.: Theory and Practice of Mining Remote Sensing Monitoring. Geological Press, Beijing (2011)
23. Yang, J.Z., Qin, X.W., Nie, H.F., et al.: Remote Sensing Monitoring of Mine in China. Surveying and Mapping Press, Beijing (2014)
24. China Geological Survey: Technical Requirements for Development of Mineral Resources Remote Sensing Monitoring (DD 2001–06), Beijing (2011)

Estimation of Tobacco Yield Based on the Radarsat-2 in Karst Mountainous Area

Xiaotao Sun[1,2], Zhongfa Zhou[1,2(✉)], Yong Fu[1,2], Juan Liao[1,2], and Ping Wang[1,2]

[1] Institute of Karst Guizhou Normal University, Guiyang 550001, Guizhou, China
sxt66666666@sina.com, fa6897@163.com
[2] National Remote Sensing Center Guizhou Branch, Guiyang 550001, Guizhou, China

Abstract. In order to realize the quantitative monitoring of modern tobacco cultivation in karst mountainous area, This paper take the Liuchang modern tobacco agriculture base in Qingzhen City Guizhou as the study area, According to the tobacco planting characteristics in karst mountainous area, with Radarsat-2 data source, choose radar image of tobacco in different growth period. After image preprocessing, calculated the tobacco scattering coefficient in different periods with different polarization, combined the field investigation data, utilized the coupling relationship of different backscattering of polarization tobacco radar data and the corresponding period fresh tobacco of weight establish three polynomial estimation model to estimate the yield of tobacco in different growth periods and verified the accuracy of the modelby the measured data. The results shown that: the overall accuracy of tobacco vigorous, mature period estimation models were respectively above 97 % and 94 %. Establish three polynomial estimation model with SAR backscatter coefficient and tobacco fresh weight can reflect the weight of tobacco well and meet the tobacco yield of estimation requirements in karst mountainous area, provide technical support for modern tobacco monitor widely.

Keywords: Karst · Radarsat-2 · Backscattering coefficient · Estimate model · Tobacco

1 Introduction

Agriculture is the source of the basic livelihood of human society, the social division of labor and other sectors of the national economy to become the basis for the further development of the independent production sector and the existence and development of all non production sectors [1]. The information of the area and the output of the crops is the important basis for the national food policy and economic development plan [2]. Therefore, it is plays a more and more important role to monitor and estimate crop products. In the early 80's of 20th Century, Domestic and international has been in the optical remote sensing data of crop yield estimation achieved remarkable results [3]. Tobacco is an important economic crop, which has important significance for

F. Bian and Y. Xie (Eds.): GRMSE 2015, CCIS 569, pp. 946–955, 2016.
DOI: 10.1007/978-3-662-49155-3_98

quantitative monitoring of tobacco. However, the mountain is located in the subtropical monsoon climate zone in karst area, Guizhou Province, and it is difficult to obtain the optical remote sensing data; In addition, the Guizhou mountain area of karst, a large area of the exposed surface, rugged terrain, land fragmentation, making tobacco growing areas are not concentrated, The terrain of the tobacco planting demonstration area is complex and diverse and tobacco agricultural production is difficult to intensive; In tobacco growth monitoring and yield estimation of tobacco has also shown great difficulties, it is difficult to realize inversion of crop. So, on the basis of previous research, considering the unique advantages of SAR imaging, for estimation of tobacco yield.

Synthetic aperture radar (SAR) is a kind of high resolution microwave sensor, as a kind of active sensor without limitation of illumination and weather conditions, with a wide range of observations, cycle short, data timeliness and strong, all-time and all-weather of observational characteristics can through the earth's surface and vegetation to get underground information [4–7]. The ability of synthetic aperture radar can penetrate the cloud and soil, the field data and the basic geographic environment data of the tobacco growing areas in the tobacco growing areas are acquired by the technology of microwave remote sensing, then monitor the growth process and analysis of the growth and the establishment of tobacco yield estimation model. Tobacco is one of the most important economic crops in China that Its growth has long growth cycle, Influence factors, Aspects of complex and other characteristics, Tobacco planting is a capital intensive, Labor intensive, Technology intensive industry. Therefore, the quantitative monitoring of tobacco planting SAR in Karst mountain area of Guizhou province, Explores the application of in Guizhou Plateau Mountain microwave remote sensing of agricultural monitoring technical route, realize the precise regulation and control of the crop production.

2 Study Area

Guizhou Qingzhen Liuchang modern tobacco agriculture base unit, between the study area between 106°7'6"E ~ 106°, 29'37"E, 26°24'5"N ~ 26°45'45"N, 489 km^2 with the total area, base units under the jurisdiction of the Liuchang, Liwo, Hongfeng Lake three township (town), should smoke land 81420 acres, tobacco planting areas mainly concentrated in the Liuchang, Liwo two towns, the planting area accounts for the base unit 90.26 % [8]. In the study area of central and Eastern belongs to shallow, low mountain and hilly area, average altitude 1327 m. The rest of the region for shallow hill to middle low mountain canyon transition zone, with an average altitude of high 1295 m. Subtropical monsoon humid climate, the annual average temperature of 14°C. Annual relative humidity is 77 %, frost free period about 270 days, the average annual sunshine time is 1100 ~ 1150 h, 1180 mm annual precipitation, soil mainly to the sand soil, yellow soil; pH 5.5–6.5, was slightly acidic, rich in organic matter, soil and climatic conditions were favorable to the growth of Flue-cured Tobacco, the main cultivation of K326, Yunyan 85&87 and Jiangnan No.3 etc.

3 Data Selection and Acquisition

3.1 SAR Data

Radarsat-2 satellite is a single equipped with sensors for C band high resolution commercial satellite radar and microwave C band is mainly utilized for monitoring crop classification [9]. The study selected the full polarization, high resolution Radarsat-2 image data in May, June, and August 2014a which covered the study area. The basic information of the data is show in Table 1.

Table 1. Parameters of microwave remote sensing image

No.	1	2	3
Acquisition date	2014-5-29	2014-6-29	2014-8-16
Polarization mode	Full polarization	Full polarization	Full polarization
Incident angle	36.60°	30.42°	30.42°
Pattern	FQ11	FQ11	FQ11
Spatial resolution	8 m	8 m	8 m
Tobacco growth period	Group period	Prosperous growth period	Mature period
Processing level	SLC	SLC	SLC

3.2 Field Measurement Data Acquisition

Through established samples and recorded the detail growth information of tobacco, and the time of the sampling is consistent with the remote sensing image to achieve the best monitoring results. The sample area selected which the contiguous tobacco plant area more than 60 hm^2. According to the radar data pixel size, established 41plots of 16 m*16 m, in order to ensure the positioning accuracy of GPS positioning of each kind of the four point and the center point to match with the radar image. Basis the growth characteristics of tobacco, the data collection period is divided into the Resettling Growth Stage, the prosperous Stage and the mature stage. Acquisition of tobacco growth parameters including leaf length, leaf width, leaf dry weight, row spacing, row spacing, plant height, leaf number, crop growth analysis, which lay a foundation for radar remote sensing data further inversion tobacco leaf weight.

4 Data Processing and Model Establishment

4.1 Data Processing

Radarsat-2 data preprocessing, Mainly include: Multi view processing, Noise filtering, Geometric precision correction, Radiometric calibration and Geographic Encoding.

Ecautilize of the different filter algorithm is different, iltering and processing of different window and filtering method for the original radar image. He research has tried the Gamma/Gaussian filter, as well as Median, Lee, Frost filter windows of 5*5, 7*7, 9*9, 11*11. Contrast these filters, And the filtering results of different windows, finally select Gamma/Gaussian filter [10]. Utilized 1:10000 topographic map, two polynomial model were adopted to correct the images (Ortho rectification). After the image of encoding and the absolute radiometric calibration, the DN values of the image represented the backscattering coefficient.

Radiometric calibration of Radarsat-2 image, mainly according to the product of the Sigma set up the standard to find the table file, the achieve process is shown in the formula (1).

$$DN = (p_{real}^2 + p_{img}^2)/A \tag{1}$$

In this type, A for gain, read from the Sigma table file; The value of each pixel of the SLC image is represented by a complex number, the preal is fact part and the pimg is the imaginary part.

The calculation formula of the absolute radiometric calibration shown as follows [11]:

$$\sigma_{dB}^0 = 10 * \log\left(ks * |DN|^2\right) + \log_{10}(\sin\theta_{loc}) \tag{2}$$

In formula 2, the σ_{db}^0 represents image backscattering coefficient; The DN value represents the gray value of pixels; The θ_{loc} representing the local angle of incidence, Can be obtained by GIM file conversion; The Ks as an absolute constant. After calibrated, the radar image pixel value is the backward scattering coefficient. Partial datas were shown in Table 2.

Table 2. The partial backscattering coefficient and fresh weight of different polarization modes of tobacco the prosperous stage and maturity stage

Number	Exuberant growth period				
	Backscattering coefficient				Fresh weight (kg)
	HH	VV	HV	VH	
1	−6.39	−6.13	−12.27	−12.43	356.52
2	−5.83	−6.18	−11.41	−11.12	345.03
3	−7.46	−7.16	−11.33	−12.57	366.98
4	−6.04	−5.67	−10.6	−10.97	343.39
5	−6.46	−6.32	−10.47	−10.85	332.74
6	−4.57	−4.77	−9.64	−9.52	316.59
7	−6.72	−6.55	−11.74	−12.11	351.91
8	−5.13	−5.38	−8.61	−8.89	321.29

(Continued)

Table 2. (*Continued*)

Number	Exuberant growth period				
	Backscattering coefficient				Fresh weight (kg)
	HH	VV	HV	VH	
9	−5.24	−5.59	−8.74	−8.54	314.92
10	−5.85	−6.04	−12.39	−12.67	342.76
Number	Maturity period				
	Backscattering coefficient				Fresh weight (kg)
	HH	VV	HV	VH	
1	−5.92	−6.23	−10.16	−10.45	110.37
2	−7.63	−7.31	−13.23	−13.44	139.17
3	−5.08	−4.98	−9.61	−10.02	104.6
4	−4.23	−4.62	−9.03	−9.25	88.71
5	−4.09	−4.37	−9.39	−9.68	89.8
6	−6.25	−6.65	−11.13	−11.27	127.18
7	−6.83	−7.17	−10.35	−10.72	118.9
8	−7.63	−7.84	−12.02	−12.43	132.88
9	−5.82	−5.61	−9.32	−9.76	104.75
10	−6.39	−6.55	−10.99	−11.33	124.85

4.2 Estimation Model

Due to the rosette leaves small, the growth of tobacco center mainly concentrated in underground roots, above ground parts of tobacco leaves short and can not be directly used for estimation, so there is no do yield estimation model analysis. According to the data of the prosperous Stage and the mature stage tobacco field data, using SPSS19.0 software to analyze the correlation between the 30 sample plots of the sample plots, the data of the mature stage and the backscattering coefficient, the two regression analysis was carried out by using the three polynomial model, the two time polynomial model and one yuan linear regression model. The simulation results of different models show that the fitting accuracy of the three degree polynomial model based on the SAR backscattering coefficient is the highest, see Figs. 1 and 2.

Through analysis of Figs. 1 and 2 can be drawn:

(1) Prosperous stage, In the HH polarization mode linearity (R^2) is highest, followed by is VH polarization, VV and HV are close. Under the HV polarization mode, the fitting model of the HH mode is the best. Mainly becautilize of the strong polarization of HH, also the HH polarization is sensitive to the change of water content in the canopy [12]. Tobacco growing season in the study area of the rainy season, tobacco leaf water content is high; At the same time tobacco canopy mainly to leaf oriented, leaves spread rapidly, the rapid expansion of the total leaf area, full of smoke and the entire, broad leaves enhances the HH polarized reflection, weakened the penetration ability of VV polarization; And the leaf angle increased the scattering of HH polarization. And becautilize the cross polarized echo is scattered in all directions, the back scattered echo of the sensor is small, so the VH and HV are lower. Therefore, the linear

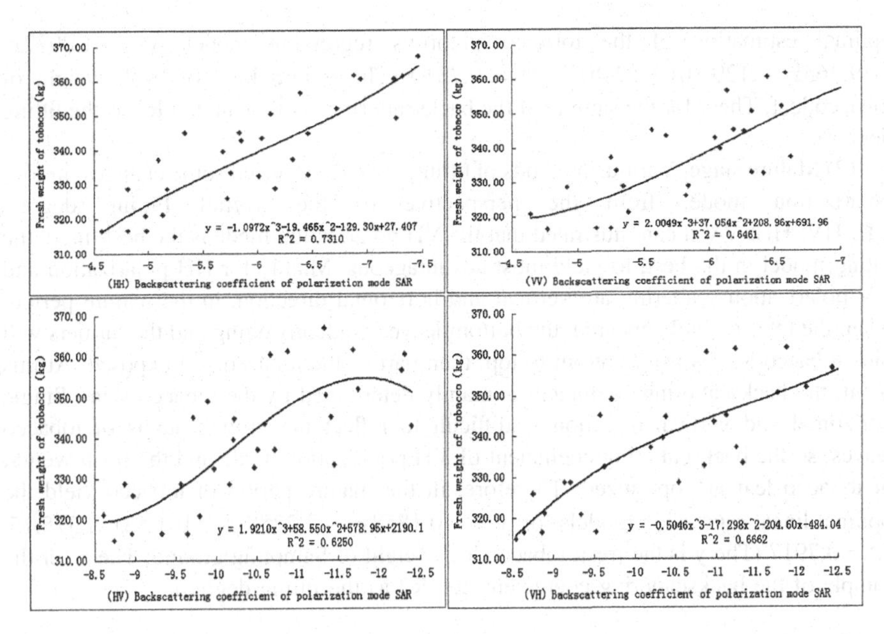

Fig. 1. Relationship between fresh weight of tobacco and backscattering coefficient in different polarization modes (Prosperous stage)

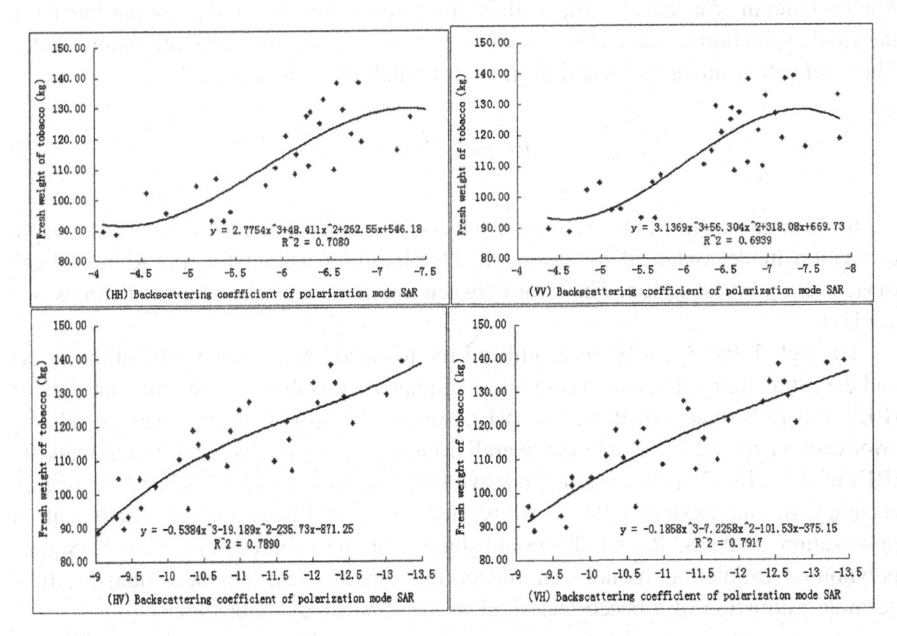

Fig. 2. Relationship between fresh weight of tobacco and backscattering coefficient in different polarization modes (Mature stage)

optimal estimation of the tobacco vigorous regression model: $y = -1.0972x^3 - 19.465x^2 - 129.30x + 27.407$, $R^2 = 0.7310$ (The y is a kind of fresh weight of tobacco leaf; The x for the sample of the backscattering coefficient, the R2 as the fitting degree).

(2) Mature stage, Various methods of fitting degrees occurred some changes, in four polarization mode, from the perspective of the overall Fitting degree: VH>HV>HH>VV it can illustrated that the VH polarization mode is the best fitted, the fitting model in the best, to yield most advantageous. Mainly for HH polarization and VV polarization scattering are vertical and horizontal direction, in the mature period, when the tobacco bud appeared, the bottom leaves gradually aging and the farmers will pick tobacco leaves from bottom to top, then part of the tobacco soil exposed. At this point, the backscattering coefficient is mainly determined by the tobacco stem. Single horizontal and vertical direction is difficult to reflect the growth status of tobacco leaves, so the backscattering coefficient of VH polarization mode and the fresh weight of tobacco leaf are optimized. Therefore, in the mature period of tobacco yield the optimal linear regression models for: $y = -0.1858x^3 - 7.2258x^2 - 101.53x - 375.15$, $R^2 = 0.7917$ (The y is the fresh tobacco leaf weight of Sampling ground; The x for the sample of the backscattering coefficient, the R2 as the fitting degree).

4.3 The Accuracy of Estimation Model

Mainly random selected 11 samples data in the study area which did not participate in the yield estimation model as the verification of the accuracy of yield estimation model. The verification method of yield estimation model are as follow [13]:

$$RE = \frac{|S_i - O_i|}{O_i} \tag{3}$$

In formula 3, the RE for relative error coefficient (%), S_i for tobacco leaf weight estimation model obtained by inversion, O_i Measured weight for any kind of fresh tobacco from investigation. The error between the model and the measured values, see Table 3.

Through Table 3 can be seen, utilized the tobacco growth yield estimation model and the estimation value with a coefficient of relative error between the measured value (RE). Where Strong growth period (RE) between 0.44–6.23 %, the average relative error coefficient is 2.77 %, and the overall accuracy of the model is 97 %. On maturity (RE) in 0.34–13.47 %, the average relative error coefficient is 5.92 %, and the overall accuracy of the model is 94 %. Illustrated that the Backscattering coefficient of polarization mode SAR and the establishment of the tobacco leaf 3 times Square polynomial estimation model can reflect the tobacco leaf biomass status, satisfy accurate estimation of tobacco leaf yield in the karst mountainous area.

Table 3. The tobacco yield estimation model accuracy evaluation

Sample number	Exuberant growth period				
	Backscattering coefficient (HH)	Yield value (kg)	Measured value (kg)	Absolute error (kg)	Relative error (%)
1	−5.33	329.73	345.03	15.3	4.43 %
2	−5.75	335.91	333.4	2.51	0.75 %
3	−5.72	335.48	326	9.48	2.91 %
4	−5.68	334.91	339.6	4.69	1.38 %
5	−6.16	341.75	328.79	12.96	3.94 %
6	−5.09	325.93	337.18	11.25	3.34 %
7	−6.01	339.6	332.46	7.14	2.15 %
8	−6.94	353.99	360.47	6.48	1.80 %
9	−7.26	360.02	349.06	10.96	3.14 %
10	−5.17	327.23	328.66	1.43	0.44 %
11	−5.97	339.04	361.56	22.52	6.23 %
Average				9.52	2.77 %
Sample number	Maturity period				
	Backscattering coefficient (VH)	Yield value (kg)	Measured value (kg)	Absolute error (kg)	Relative error (%)
1	−13.02	131.94	129.27	2.67	2.07 %
2	−10.28	106.82	95.74	11.08	11.57 %
3	−11.48	119.23	109.77	9.46	8.62 %
4	−12.44	127.35	128.6	1.25	0.97 %
5	−11.68	121.02	116.06	4.96	4.27 %
6	−10.87	113.34	118.63	5.29	4.46 %
7	−11.99	123.67	138.12	14.45	10.46 %
8	−9.29	93.42	93.1	0.32	0.34 %
9	−10.82	112.82	108.34	4.48	4.14 %
10	−11.72	121.37	106.96	14.41	13.47 %
11	−12.32	126.39	132.72	6.33	4.77 %
Average				6.79	5.92 %

5 Conclusions and Discussion

The national modern tobacco agriculture base unit as the study area in Qingzhen city of Guizhou province. According to the tobacco planting characteristics in Guizhou plateau mountainous area, Taking the Radarsat-2 as the data source, adopted the tobacco fresh weight and the corresponding period of the scattering characteristics of radar images to constructed tobacco yield estimation model. Study shown that the long-term simulation precision is higher than the mature period. On the one hand, becautilize the growth center of the tobacco growing center of the root of the earth is transferred to the ground, Rhizome quickly grow taller and thicker, leaves quickly expanded, the photosynthetic product accumulation, growing tobacco with luxuriant foliage, rich in water content. On the other hand, due to the mature period, the emergence of the flower bud, some leaves are farmers picking, the soil begins to bare; Also due to tobacco leaf picking progress is not the same, the small sample in tobacco schedule later. The mature period of existence part in vigorous growing period of tobacco.

The fragment of cultivated land, serious crop planting flower phenomenon, General intercropping and complex planting structure in the mountainous region of Guizhou Province, the coupling relationship between the linear regression and the three degree polynomial model, which can be effectively utilized for the establishment of the scattering coefficient and the fresh weight of tobacco, was effectively utilized for SAR; through the accuracy verification, the overall accuracy of the model of the tobacco growing period and maturity period was 97 % and 94 % respectively. The technology of microwave remote sensing monitoring is more mature, estimation results are more accurate, suitable for the Guizhou mountainous area complex topography and climate conditions, it can provide technical support for the realization of real time and precision in the tobacco growing areas in Southwest China and the national tobacco industry.

Acknowledgment. In this paper, the research was sponsored by the National Basic Research Program of China (973 Program, Project No. 2012CB723202); Guizhou province science and technology plan (Research and application of key technologies on SAR remote sensing platform monitoring and identification in Karst mountainous area) (Guizhou S&T Contract GY[2013] 3062); The major application foundation research project of Guizhou Province (Guizhou S&T Contract JZ 2014-200201); the Major science and technology in guizhou special plan (Guizhou S&T Special Plan 2013-6024).

References

1. Liao, J., Zhou, Z.F., Li, B., et al.: Analysis of data from high - resolution SAR data of tobacco at high altitude area. Chin. Tob. Sci. **35**(6), 74–79 (2014)
2. Yuan, W.Q., Zhou, G.X., Shu, Q.J., et al.: Based on high resolution satellite remote sensing data of winter wheat yield estimation. J. Agric. Eng. **25**(7), 118–123 (2009)
3. Li, Y., Peng, S., Qi, F.L.: Radarsat SNB SAR data in a large area of rice yield estimation. Earth Sci. Prog. **18**(1), 109–115 (2003)
4. Luckman, A.J.: Correction of SAR imagery for variation in pixel scattering area cautilized by topography. IEEE Trans. Geosci. Remote Sens. **36**(1), 344–350 (1998)
5. Bamler, R., Breit, H.: Experience with ERS-1 SAR signal processing at the german PAF. Geosci. Remote Sens. Symp. **2**(1), 1353–1355 (1992)
6. Graziano, M.D., D'Errico, M., Razzano, E.: Constellation analysis of an integrated AIS remote sensing spaceborne system for ship detection. Adv. Space Res. **50**(3), 351–362 (2012)
7. Atteia, G., Collins, M.J.: On the utilize of compact polarimetry SAR for ship detection. ISPRS J. Photogrammetry Remote Sens. **80**, 1–9 (2013)
8. Fu, Y., Zhou, Z.F., Jia, L.H.: Based on SAR technology in karst mountainous region of Guizhou tobacco yield estimation model. Hubei Agric. Sci. **53**(9), 2156–2159 (2014)
9. Chen, J.S., Lin, H., Shao, Y.: Study on Application of Microwave Remote Sensing of Agriculture, Rice Growth Monitoring 2010. Science Press, Beijing (2010)

10. ISO 11562 Geometrical product specification (GPS)-suface Texture: method-Metrological characteristics of phase correct Filters (1996)
11. Zhou, Z.F., Hu, J.C., Wang, J.: Identification of. based on single phase dual polarized TerraSAR-X data in mountain areas Hubei. Agric. Sci. **53**(23), 5851–5854 (2014)
12. Yang, S.B., Li, B.B., Shen, S., et al.: Based on ASAR data of rice remote sensing monitoring ENVISAT. Jiangsu Agric. J. **24**(1), 33–38 (2008)
13. Jia LH, Zhou ZF, Li B.: High resolution SAR modeling application in Karst mountain tobacco growth study. China Tob. Sci. **10**, **34**(5), 104–112 (2013)

Environmental Sustainability Assessment for Countries Involved in OBOR Initiatives Based on Planetary Boundary Theory

Yijing Li(✉)

One Belt One Road (OBOR) Research Center, China Executive Leadership Academy, No. 99 Qiancheng Road, Pudong District, Shanghai, China
liyijing@celap.org.cn

Abstract. Based on planetary boundary theory, this paper has analyzed the national biodiversity capability (BC) and environmental footprint (EF) among countries involved in the OBOR Initiatives; calculated their respective biocapability deficit or reserves (DR); and evaluated their environmental sustainability (ESPI) upon the humanity's development requirements (HDI). It aims to assess the environmental supporting capability from a regional broad view in realizing the success of OBOR strategy, and further to provide references for regional sustainable development. The methodology could also be suggestive in regional development studies and sustainability evaluation from a environmental friendly perspective.

Keywords: Planetary boundary · Ecological footprint · "One Belt, One Road" (OBOR) Initiatives · Environmental sustainability

1 Introduction and Theoretical Background

The "One Belt, One Road" (OBOR) Initiatives was firstly proposed by Chinese President XI Jinping in September 2013. It includes The Silk Road Economic Belt and the 21st Century Maritime Silk Road initiatives, and consists of a network of railways, highways and other forms of infrastructure, as well as oil and gas pipelines, power grids, Internet networks and aviation routes in the Eurasian area. Since the practice of OBOR strategy is still at the starting stage, so far the majority of relevant researches and discussions are focusing on its strategic importance, the expected impacts to regional development, economic and trade cooperation, cultural exchanges, etc., as well as some anticipating suggestions on practical realization of the strategy. For example, Feng (2014), Huo (2014) and Gao (2014) discussed that the OBOR strategic initiatives is helpful to promote regional cooperation at various levels, to quicken the pace for China's opening-up, and to propel the prosperity and development in countries getting involved; Han and Zou (2014) and Qian (2014) proposed the potential issues on regional cooperation might incur during the trade cooperation between China and Western Asian countries, and the resources cooperation between China and the Middle East, arriving at the conclusion that the economy and trade cooperation among regions could be enforced upon the

F. Bian and Y. Xie (Eds.): GRMSE 2015, CCIS 569, pp. 956–962, 2016.
DOI: 10.1007/978-3-662-49155-3_99

OBOR strategy; Cai (2014) required cultural exchange and cooperation among countries getting involved to realize a real multi-dimensional cooperation upon the OBOR strategy.

However, research so far rarely covers the environmental issues upon OBOR strategy, which do actually important topics need to be analyzed urgently. Because no matter of the desserts and gobi along the Silk Road Economic Belt, the exploitation of marine resources along the 21st Century Maritime Silk Road, or the obstructions from national trades environmental barriers, environment and resources are essential guarantee for the success of OBOR initiatives. It is crucial to solve all kinds of potential environmental conflicts during the propelling of OBOR strategy, and is of paramount importance to construct the OBOR into a road to peace, green and harmony, and to realize coordinated development among economy, society and environment.

One way of quantifying the resources usage and human pressure on the natural environment is to calculate humanity's "environmental footprint". In the previous decades, footprint calculations on various environmental elements have been utilized as measures for environmental sustainability against humanity's development, such as the carbon footprint, water footprint, ecological footprint, etc. Hoekstra and Wiedmann (2014) proposed that environmental sustainability depends on the size and spatiotemporal characteristics of humanity's footprint relative to Earth's carrying capacity, and requires that footprints remain below their maximum sustainable level—the planetary boundaries Rockström et al. (2009). So the index for environmental footprints (EF) has been proposed, to measure how much of the available capacity within the planetary boundaries is already consumed, and their results indicated that current environmental footprint is not sustainable upon the Earth's limited natural resources and assimilation capacity. So this paper tried to measure the environmental sustainability of countries involved in OBOR strategy, using their respective EF against the corresponding natural resources capability, in order to provide scientific references for the realization of sustainable development among these countries and regions, and further to make suggestions on the propelling of OBOR strategy from an ecological perspective.

2 Data and Methodology

In this paper, the environmental footprint is the summarization of 6 types of footprints, which are cropland footprint, grazing land footprint, forest land footprint, carbon footprint, fishing ground footprint and built up land footprint. It utilizes the *ESPI* index to measure the environmental pressure for sustainable development for a country/region; hence the reversed *ESPI* index could be interpreted as the reflection of regional environmental sustainability. It is calculated by Eq. (1).

$$ESPI_i = \sum_i NEF_{ij} \cdot \omega_{ij} = \sum_i NEF_{ij} \cdot \frac{EF_i}{BC_i} \qquad (1)$$

where $ESPI_i$ is the environmental pressure for sustainable development for country/region i, with NEF_{ij} as the environmental deficit for footprint j in country i, and ω_{ij} is the weight for footprint *j* in country *i*. The weight is actually the quotient for environmental footprint in country *i* (EF_i) divided by its biodiversity capability BC_i. The NEF_{ij} could be calculated by Eq. (2) below:

$$NEF_{ij} = \frac{EF_{ij} - EF_{jmin}}{EF_{jmax} - EF_{jmin}} \quad (2)$$

where EF_{jmin} represents for the minimum value environmental footprint *j* in country *i*, and the EF_{jmax} refers to the maximum value. For special occasions that environmental capability for resource *j* is 0 in country *i*, the corresponding weight is replaced by the overall weight for country *i*.

2.1 Data Source

There are 65 countries/regions getting involved in the OBOR strategy, and data utilized in this paper are mainly from Global Footprint Network (GFN) report (2014). It has summarized the environmental footprint data in 2011 for 232 countries, with 63 OBOR countries' data listed (data for Maldives and Palestine are blank). So the *EF* and *BC* for required countries could be derived from the report, in addition, the human development index (*HDI*) is also used to measure the humanity development. Other data on local population, economic and social development (e.g. GDP) are derived from the World Bank WDI database. Upon checking the required items, it is found that some items are missing for Cambodia and Oman, so there are finally 61 countries been discussed.

2.2 Methodology

As indicated previously, environmental footprint (EF) is the indicator for humanity's utilization of resources, the higher value it is, the greater people use domestic resources; biodiversity capability (BC) reversely represents the "planetary boundary" of specific country, and a larger value indicating higher resources support for development. Based on the index ESPI, another index, domestic biocapability deficit or reserve (DR), has been calculated to measure the sustainable development potentiality upon the respective planetary boundary in each target country. It is calculated by subtracting the biodiversity capability (BC) value by the environmental footprint (EF), if the utilization of resources hasn't exceeded its planetary boundary, then the value is larger than 0 and there is a reserve of domestic biocapability; otherwise, the minus value indicates that there is a deficit, and the more the worse.

(a) The HDI-EF could reflect the humanity's development pressure on resources upon the corresponding socio-economic development, and it is been deemed that if HDI in country *i* is larger than 0.67, then the country is at a high human development level.
(b) ESPI measures regional environmental sustainability, and HDI-ESPI reflects the environmental sustainability upon the humanity's development requirements.

3 Results

3.1 Natural Resources Utilization Efficiency

It can be read from Fig. 1 that, (1) regional biodiversity capability has been marked in red, the darker the higher: Singapore is the lowest, while Mogolia with the highest natural resources support for development. If divide the 61 countries by BC into 5 quartile categories, then China falls into the 2nd category, which indicating a lower BC. In general, except for the European countries, countries involved in the Silk Economic Road have higher BC values than those within the 21st maritime Silk Road. (2) The environmental footprint index reflects the domestic humanity's utilization of resources, and the lower is, the more friendly to environment. From the results, Timor-Leste and Afghanistan have the lowest EF values, while Kuwait and United Arab Emirates have highest humanity's footprint on using environmental resources. It is found that China's utilization of resources already reached the middle value the 3rd quartile category, which could be taken as overuse so far. The countries in the western part of OBOR strategy exhibit more footprints on environmental resources usage.

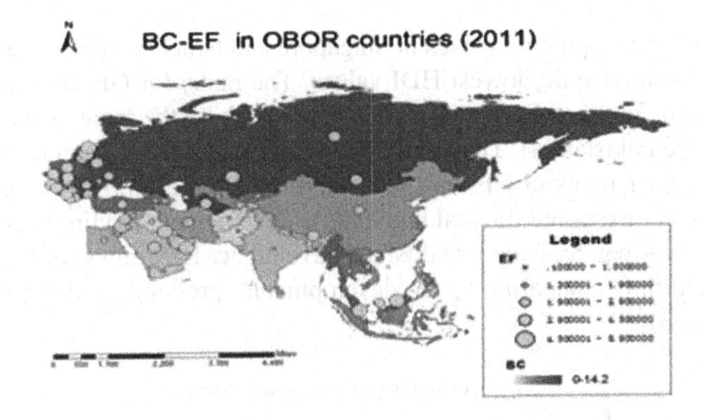

Fig. 1. Biodiversity capability and environment footprint in 2011 for the 61 OBOR countries (Color figure online)

The biocapability deficit or reserve value (DR) has been created as depicted in Fig. 2, the darker it is, the higher biocapability it has. It can be read that Kuwait and United Arab Emirates saw severest biocapability deficit, while Estonia and Mongolia have the highest resources reserves. Among the 61 OBOR countries, only 10 of them still have biocapability reserves, and others all saw resources deficit upon their footprints on environment resources. China is falling into the 2nd quartile, which indicating a severer resources deficit approaching, and an urgent requirement for sustainable development. Taking into consideration of all types of resources as well as the economic development structures in different countries, it can be arrived that OBOR countries in the middle are facing server deficit problems than those along the regional boundaries, which are with higher biocapability reserves.

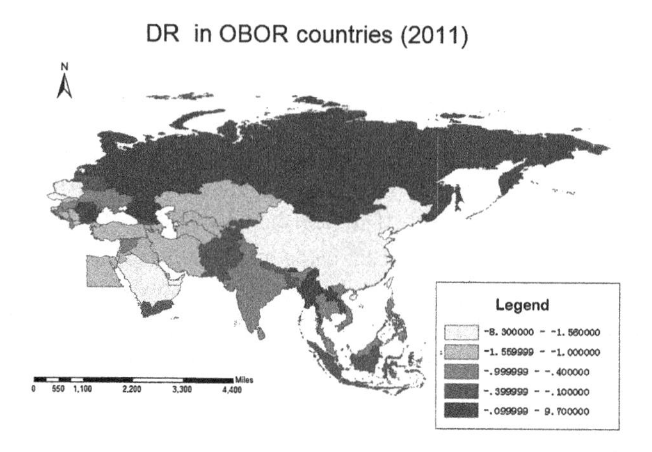

Fig. 2. Biocapability deficit or reserve in 2011 for the 61 OBOR countries (Color figure online)

3.2 Humanity's Development Pressure on Resources

The humanity's development indexes in Singapore and Israel are highest, while Afghanistan and Yemen have the lowest HDI values. The ranks for OBOR countries show different styles by indexes. For example, 3 countries exhibit the same ranks for HDI and EF, and they are Pakistan (4), Thailand (25) and Lebanon (38); 27 countries, including China, have higher ranks of EF than their HDI, which means their actual resources utilization already exceeded the real humanity's development requirements; while the other 31 countries, e.g. Singapore and Israel, have higher HDI ranks for their less environmental footprints comparing to the developments' pressures and requirements on resources (Fig. 3).

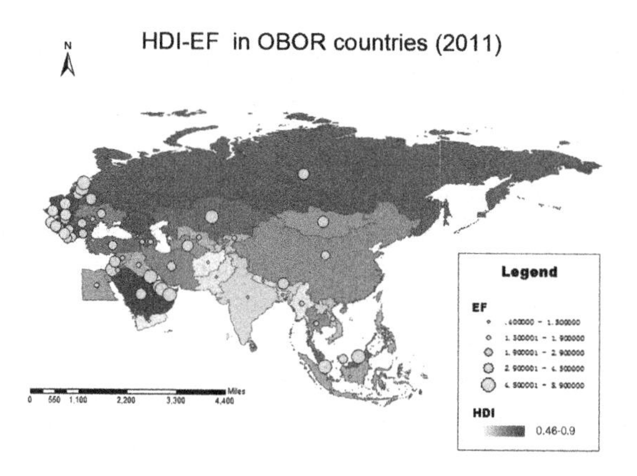

Fig. 3. Humanity's development index v.s. Environmental footprint among OBOR countries

3.3 Environmental Sustainability upon Humanity's Development

The ESPI index reflects domestic environmental sustainability pressures, and as shown in Fig. 4, regions in darker red carry more pressures hence have lower environmental sustainability (e.g. Singapore and Kuwait), while regions in lighter red exhibit higher environmental sustainability (e.g. Timor-Leste and Afghanistan). So upon the analysis, China's rank as 40/61 indicating an urgent attention should be paid to sustainable development. Taking the current development status into consideration, 5 countries, Bosnia and Herzegovina, Macedonia TFYR, Uzbekistan, Singapore, and Qatar, have the same ranks for HDI and ESPI, seeing an relatively equilibrium between their humanity's development and the environmental sustainability; 27 countries, including China exhibit higher ESPI ranks and demanding for more attention to sustainable development upon current humanity's requirement; the other 29 countries have higher HDI ranks than their ESPI ranks, showing a relatively higher potentiality on environmental sustainable development based on current developing requirements to resources.

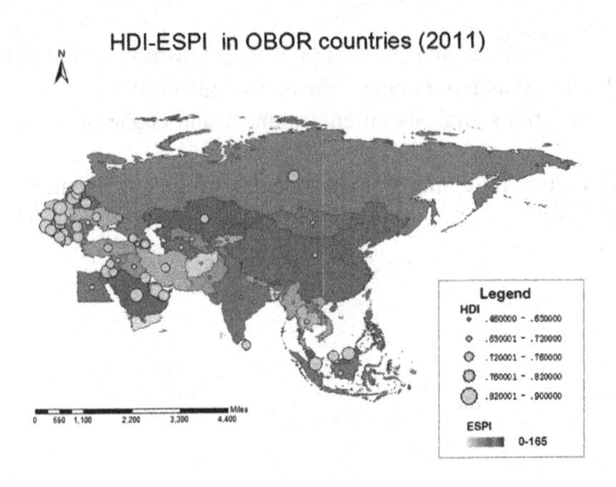

Fig. 4. Environmental sustainability v.s. Humanity's development among OBOR countries (Color figure online)

4 Conclusions and Discussions

The 61 countries involving in "One Belt One Road" Initiatives exhibit varied environmental resources capability, humanity's development requirements, and environmental footprints upon their respective developing modes; hence they present different environmental sustainability, with some of them already suffered severe natural resources deficit, while others still have resources reserves. Although environmental sustainability cannot simply be taken as the indicator for environmental quality, it reflects the environmental support potentiality for future development based on current humanity's development mode (HDI). So in order to realize the coordinated development of relevant countries, and to make the success of OBOR strategy, it is necessary to evaluate the regional environmental support capability from a overall view beforehand, and to propel regional sustainable development according to equilibrium utilization and allocation of biocapability.

References

Rockström, J., et al.: Nature **461**, 472–475 (2009)
Hoekstra, A.Y., Wiedmann, T.O.: Humanity's unsustainable environmental footprint. Science **344**, 1114–1117 (2014)
Borucke, M., et al.: Ecol. Indic. **24**, 518–533 (2013)
Feng, Z.X.: The strategic meanings of "One Belt One Road" initiatives. Guangming Daily, 20 October 2014
Huo, J.G.: Profound significance of "One Belt One Road" initiatives. People's Forum **5** (2014)
Gao, H.C.: Deepen economic and commercial cooperation, create new glorious era. People's Daily, 2 July 2014
Han, Y.H., Zou, J.H.: China and West-Asia countries' trade cooperation and outlook against the "One Belt One Road" initiatives. Int. Trades **8** (2014)
Qian, X.M.: Pivots for "One Belt One Road" initiatives: China and the middle east cooperation on energy resources. Arabia World Stud. **3** (2014)
Cai, W.: Culture construction in the first place for "One Belt One Road" initiatives. Qiushi **9** (2014)
Zhang, H.B.: Environmental and the International Relationship: Rational Thinking of Global Environmental Issues. Shanghai People's Press, Shanghai (2008)
Jiang, X.: Structural functions analysis on environment and economic conflicts. Ecol. Econ. **2** (2006)
Chen, T.: Literature review on China's environmental struggles. Henan Univ. J. (Soc. Sci.) **1** (2014)

Remarkable Ecological Restoration Due to Integrated Socio-Economic Policies in the Loess Plateau

Wenlong Song[1,2,3(✉)], Shengtian Yang[3], Jingxuan Lu[1,2], Zhiguo Pang[1,2], Xuefeng Wang[1], Wei Qu[1,2], June Fu[1,2], Xiaoyan Liu[4], Yizhu Lu[1,2], Yanan Tan[1], and Jingyi Han[1]

[1] China Institute of Water Resources and Hydropower Research, Fuxing Road. A-1, Beijing 100038, China
songwl@iwhr.com
[2] Research Center on Flood and Drought Reduction of the Ministry of Water Resources, Fuxing Road. A-1, Beijing 100038, China
[3] State Key Laboratory of Remote Sensing Science, Jointly Sponsored by Beijing Normal University and the Institute of Remote Sensing Applications of the Chinese Academy of Sciences, Beijing 100875, People's Republic of China
[4] Yellow River Conservancy Commission of the Ministry of Water Resources, Zhengzhou 450004, People's Republic of China

Abstract. Remarkable changes of vegetation cover in the Loess Plateau occurred in the latest 20 years. Land use/land cover (LULC) data of 1985, 1999 and 2009 were collected based on Landsat MSS, TM and MODIS remote sensing data. Results show that the widely implemented ecological restoration policies obtained a great success, achieving remarkable vegetation recovery and soil conservation in the Loess Plateau.

Keywords: The Chinese Loess Plateau · Ecological restoration · Economic and social policies · Remote sensing

1 Introduction

The fertile, yet highly erodible soil of the Loess Plateau in the Yellow River basin has contributed considerably to the development of Chinese civilization. However, widespread environmental degradation from soil loss and associated flooding has also inflicted misery on its people [1]. Soil loss from the Loess Plateau, which accounts for nearly 90 % of the total sediment yield, has always been a most serious obstacle to the sustainable development in the Yellow River basin [2].

The Loess Plateau has the largest and deepest loess distribution in the world. Historically, the region's vegetation was plentiful with large areas of shrub and grassland 2,300 years ago [3]. Over the centuries, population growth, agricultural development, wars and so forth, have destroyed much of its natural vegetation [4]. But, since the founding of the People's Republic of China, the government has implemented remarkable soil erosion control measures in reversing the environmental degradation,

F. Bian and Y. Xie (Eds.): GRMSE 2015, CCIS 569, pp. 963–971, 2016.
DOI: 10.1007/978-3-662-49155-3_100

while benefitting local economies and aiding the poor in using their land. In spite of these achievements, criticisms to Loess protection and control policies continued to ignite a debate until the end of 20th century. From this point, ecological restoration policies continued to be widely implemented with great success. Results of these policies are remarkable, with trends of soil erosion and land desertification be completely reversed in the Loess Plateau.

2 Study Area

The Loess Plateau is located in a region of 32°-41°N,107°-114°E in the northwest China, including Shanxi province, Gansu province, Ningxia province and other parts. It is an economically depressed region with a total area of 0.64 million km^2 and a total population of 110 million. The altitude is 1000 m to 2000 m, covered with loess of 50–80 m thick. It is the largest and deepest loess land in the world. Moreover, it has a dry climate with temporally concentrated 466 mm annual precipitation and sparse vegetation cover. The major regions under control and protection measures in the Loess Plateau chosen for the study area is shown in Fig. 1, with Shanxi, Henan provinces and the Hetao irrigation district excluded.

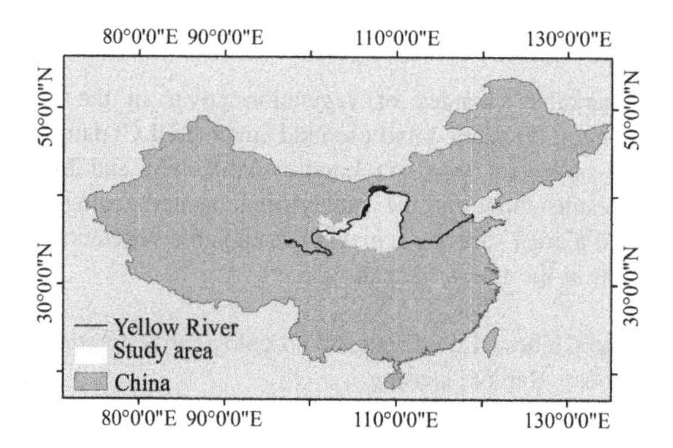

Fig. 1. Location of the study area

3 Data and Methodology

3.1 Data

(1) *Meteorological Data*

Annual mean precipitation and temperature from 1951 to 2010 were collected from 71 weather stations in the Loess Plateau.

(2) *Hydrological Data*

Runoff and sediment yield of the Yellow River from 1956 to 2010 were collected from the Tongguan hydrological station.

(3) *Remote Sensing Data*

Change and recovery of land use/land cover (LULC) data from the Loess Plateau in 1985, 1999 and 2009 were interpreted based on Landsat MSS, TM and MODIS remote sensing data. These data were then classified into six main types including forest, bush land, grass land, cultivated land, sandy and barren land and other types.

3.2 Methodology

3.2.1 Integrated Dynamic Degree of Land Use

The integrated dynamic degree of land use is an index which depicts regional differences in change rate of the land use type and reflects the comprehensive effect of human activities on the change of land use type in river basin. The mathematical model is:

$$IK_i = \frac{S_{it2} - S_{it1}}{S_{it1}} \times \frac{1}{t_2 t_1} \times 100\% \tag{1}$$

Where IK_i is the integrated dynamic degree of land use i in study area from t_1 to t_2; $\Delta S_{i,j} = S_{it2} - S_{it1}$ is the total area of land use type i that is converted to other land use type during the study period; S_i is the total area of land use type i.

3.2.2 Single Dynamic Degree of Land Use

The single dynamic degree of land use is an index which portrays the change rate and amplitude of different land use type in a certain period, and reflects the effect of human activities on the single land use type. The mathematical expression is:

$$SK_i = \frac{S_{it2} - S_{it1}}{S_{it1}} \times \frac{1}{t_2 t_1} \times 100\% \tag{2}$$

Where SK_i is the dynamic degree of land use type i from t_1 to t_2; S_{it1} and S_{it2} denote the area of land use type i at t_1 and t_2 respectively.

4 Result and Discussion

4.1 LULC in the Loess Plateau

The LULC changes between 1985 (Fig. 2a) and 1999 (Fig. 2b) show that forest and grassland areas decreased 1.0 % and 6.0 % respectively, while cultivated and barren land areas increased 0.6 % and 1.5 % respectively. In addition, bush land and other land areas increased 2.1 % and 2.8 % respectively. Meanwhile, the cultivated land dispersed in spatial structure and extended towards more northern arid region. It also shows that some grass lands in the Maowusu sandy area degraded into desert and forests in the Ziwu Mountain region deteriorated markedly. Although a large number of ecological protection and restoration measures were implemented, they did not achieve stable and satisfying effects in the Loess Plateau until the end of the 20th century.

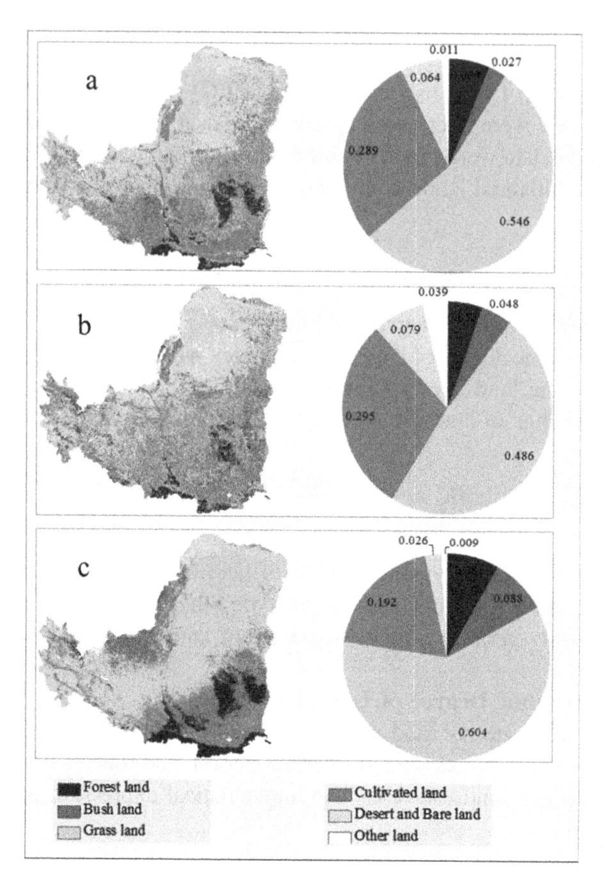

Fig. 2. Changes and recovery of land use/land cover (LULC) in the Loess Plateau. LULC and their percentages for each land type in 1985, 1999 and 2009 are shown in a, b and c respectively.

The LULC changes observed between 1999 (Fig. 2b) and 2009 (Fig. 2c) illustrate that areas of forest, bush land and grass land increased 2.8 %, 4.0 % and 11.8 % respectively while cultivated land, barren land and other land areas decreased 10.3 %, 5.3 % and 3.0 % respectively. The grass land ratio has risen to 60 %. Cultivated land is distributed in the southeastern region of the study area, mainly in river valley plains and the plateau areas with an annual rainfall over 500 mm. Regions with an annual rainfall below 500 mm were mostly covered by grass land. Forested areas in the Ziwu Mountain region and Qingling Mountain region recovered well, with barren land in the Maowusu sandy area decreased as a result of increased grass cover. In summary, the vegetation coverage ratio increased quickly.

4.2 Land Use Changes in the Beiluo River Basin as a Response to Human Activities

A case study of the Beiluo River Basin is presented as an example. Study results (Table 1) are as follows. (1) the dynamic degree of integrated land use in the Beiluo River Basin increased from 0.61 in 1976-1998 to 6.66 in 1998–2010; Farm land and grass land area reduced, and the reduce speed increased to 26.20 % and 23.33 % from

Table 1. Land use changes in the Beiluo River Basin since 1976 (unit:km^2)

Year	Cultivated land	Forest land	Grass land	Water	Land of urban and rural	Unused land	River basin
1976	8895	7293	10296	109	333	4	26931
1998	8717	7726	10020	109	355	4	26931
2010	6433	12336	7682	109	367	4	26931
1976–1998	−178	433	−277	−0.21	0.22	0	/
1998–2010	−2284	4610	−2338	−12	0.12	0	/

2.00 % and 2.69 %. Besides, the area of forest land and the land for urban and rural industrial or residents increased. The former increased from 5.93 % to 59.68 %, and the increase speed reduced from 6.59 % to 3.52 %; (2) the transfer direction of land use type from 1976–1998 and 1998–2010 tend to present a characteristics that the cultivated land and grass land mainly converted into wood land and small number of cultivated land into the land for urban and rural industrial or residents; (3) The comprehensive change index of land use degree in counties of the river basin expanded to 27 ~ 4 from −2 ~ 1. It proved that the influence of human activities on natural environment of the Beiluo River Basin comes to increase, with main features of the mutual conversion among cultivated land, grass land, forest land and the land for urban and rural industrial or residents. However, the degree of forest land area increase caused by human activities is far greater than that of cultivated land and grass land reduction, as well as the land increase for urban and rural industrial or residents.

4.3 Effect of Ecological Restoration Policies

4.3.1 Ecological Restoration Policies

From the founding of the People's Republic of China in 1949 to the end of the 1970s when economic reforms begun, China had been trying to transform itself into a modern, socialist, industrialized country, which resulted in a fast population growth along with unsustainable farming practices causing serious ecological problems to the Loess Plateau region.

Since the 1980s, a series of soil conservation policies have been implemented with progressive economic development and the gradual establishment of the market economy system. The principal strategic policy of ecological restoration has been the integrated watershed management policy [5]. To that end, new technology was used to develop a comprehensive approach to control the Loess soil erosion and desertification including: (1) overall planning using system's theory and computer technology for treatment measures; (2) engineering measures including hillside terracing, level steps, level ditches, fish-scale pits, plant barriers and so forth for landslide mitigation as well as gully head protection, silt-trap dams, check-dams for gully treatment, mountain torrent drainage and water storage projects; (3) vegetation protection measures effective for soil conservation, including economic incentives to farmers for reforestation, sustainable agricultural practices and cultivation of crops on suitable slopes and (4) innovative farming methods including soil tillage and crop cultivation technology. Economic development throughout the 1990s strengthened ecological protection and control measures through economic incentive programs such as auctioning the use rights of barren land in an exchange for restoration of the ecological environment in the watershed area.

Benefited from economic development and China's Grand Western Development strategy, a series of social policies were implemented, including giving fiscal support to agriculture, rural areas and farmers [6]. In 2000 experimental tax reforms were carried out to lower the tax burden on farmers. In 2006 the agricultural land tax was abolished and subsidies were given to farmers to narrow the widening income gap between rural and urban residents. Reforms in property rights law led to economic incentives for rural residents to reseed and protect the forest. A rural social security system was also established to increase the income level and quality of life for farmers. All of these social and economic policies have vigorously advanced ecological restoration in the Loess Plateau [7]. At the same time, the ecological restoration project Returning Farmland to Forests was carried out in a pilot zone in 1999 and has been widely implemented since 2003 [8]. International cooperation and financial support has played an especially important role in ecological restoration in the Loess Plateau. The World Bank supplied a $300 million loan supporting a two phases soil conservation project in the Loess Plateau, the first phase from 1994 to 2001 and the second phase from 1999 to 2005 [9].

The main conservation and restoration measures in the Loess Plateau initiated during this period are described as follows: (1) conversion of crop land to forest and grass land, prohibiting cultivation of sloping fields over 25 degrees; (2) closing of the mountains to accelerate reforestation, grazing prohibition and raising livestock indoors to recover the degraded grass and shrub land avoiding excessive grazing; (3) providing fund and grain allowance of 970 Yuan RMB per acre in the first 8 years and 550 Yuan RMB per acre for peasants after 8 years; (4) readjustment of the rural industrial structure to develop local industries such as animal husbandry and fruit cultivation; (5) enforcement of ownership and benefit rights of the land under treatment (the right of land use obtained by farmers can be extended to 70 years and is transferable); (6) adequate capital investments necessary to enhance engineering and vegetation restoration measures and (7) legal guarantees for policy implementation and environmental protection. The new Law of the People`s Republic of China on Water and Soil Conservation prescribes penalties for illegal activities related to soil conservation and the Regulations on Returning Farmland to Forests serve as the legal guarantee for the ecological restoration strategy. The Chinese government has invested more than 240 billion Yuan RMB in projects responsible for converting cropland to forest and grassland in the Loess Plateau.

4.3.2 Effect of Ecological Restoration Policies

The variation of LULC revealed considerable improper land development and occupation while control measures were implemented. As a result, some ecosystems in the region suffered. The change in runoff and sediment yield of the Yellow River (Fig. 3B) indicates an increase in soil conservation due to engineering measures. However, these effects were not stable. Consequently, new ecological restoration policies were developed to protect and prevent additional environmental damage.

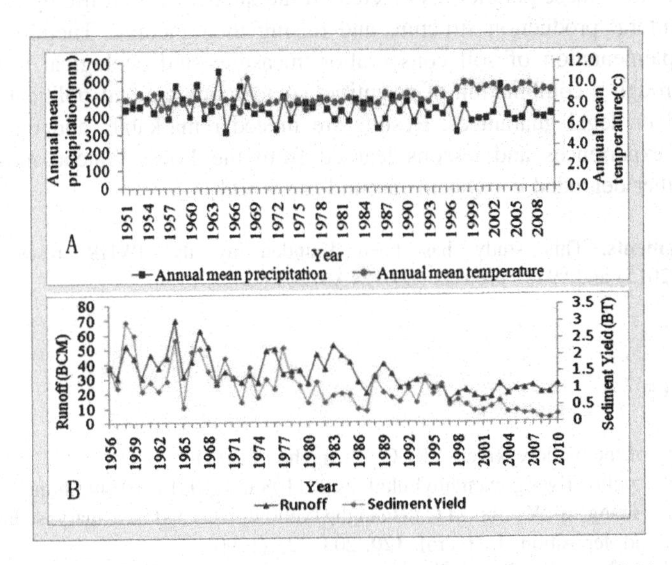

Fig. 3. Key climate factors in the Loess Plateau and the runoff and sediment yield of the Yellow River over the past 60 years. **(A)** Annual mean precipitation and temperature from 1951 to 2010 collected from 71 weather stations in the Loess Plateau. **(B)** Runoff and sediment yield of the Yellow River from 1956 to 2010 collected from the Tongguan hydrological station.

The LULC results indicate that the spatial land use pattern is consistent with the natural conditions such as rainfall distribution. Therefore, reciprocity between humans and the natural environment has been established. It is clear that runoff and sediment yield of the Yellow River has decreased drastically (Fig. 3B); the runoff value has decreased from an average annual 45 billion cubic meters (BCM) to about 20 BCM since 2000 while the sediment yield has decreased from an average annual 1.1 Billion Ton (BT) to 0.3 BT since 2000. Meanwhile, rainfall and temperature (Fig. 3A) have not changed drastically. Therefore, a policy implementing elements of prevention and control has proved scientifically effective, a result that is both exciting and hopeful [10].

5 Conclusion and Discussion

Ecological restoration in the Loess Plateau has steadily evolved. Control technologies have been developed from traditional forms to comprehensive and modern forms. The soil erosion control and prevention measures have benefitted the ecosystem as well as the economy and rural communities. Although prevention and control measures began as government mandated policies, market-based approaches were quickly incorporated due to the economic reform. With soil erosion under control, prevention is now the first priority along with intensified control measures.

During the process of soil erosion reversal, economic development and social policies were instrumental in spreading environmental protection awareness to farmers. At the same time, these policies also increased the quality of rural life by transforming their agricultural production structure and raising their income. Therefore, the successful implementation of soil conservation measures and ecological restoration is possible through a combination of scientific, socio-economic and political considerations based on legal guarantees. Results are indeed remarkable, and the ecological restoration experiences and lessons learned from the Loess Plateau may assist in restoring other degraded ecosystems around the world.

Acknowledgments. This study has been founded by the IWHR research projects (JZ0145B042015) and "948" projects (JZ0124A012014).

References

1. A Survey of the Yellow River. YRCC, Zhengzhou (2011)
2. http://www.yellowriver.gov.cn/hhyl/hhgk/zs/201108/t20110814_103443.html
3. Zhang, J., Huang, W.W., Shi, M.C.: Huanghe (yellow river) and its estuary: sediment origin, transport and deposition. J. Hydrol. **120**, 203–223 (1990)
4. Wang, S.C.: On ancient Loess Plateau vegetation. Geogr. Res. **9**, 72–79 (1990)
5. Li, R., Yang, W.Z., Li, B.C.: Research and Future Prospects for the Loess Plateau of China, pp. 140–143. Science Press, Beijing (2006)
6. In the Chinese Loess Plateau, the goals of ecological restoration have been to conserve water and soil, to improve the ecological environment, and to develop local economies and increase the income of peasants. Integrated watershed management means that a comprehensive control system is formed through the rational arrangements of land use such as agriculture, forestry, animal husbandry and side-line production; the layout of soil conservation measures such as agronomic measures, forest-grass measures and engineering measures and the combination of gully and slope control, based on the overall plan and at a watershed scale
7. Wen, J.B.: The way of Chinese development in agriculture and rural areas. Qiushi **2**, 203–223 (2012)
8. Liu, N.: Adhere to the line of Chinese water and soil conservation and ecological restoration. Qiushi **8**, 56–58 (2011)
9. Regulations on Returning Farmland to Forests. CPG, Beijing (2003)
10. http://www.mlr.gov.cn/zwgk/flfg/tdglflfg/200508/t20050819_69550.htm

11. Restoring China's Loess Plateau. The Word Bank, Washington, D.C. (2007)
12. http://www.worldbank.org/en/news/2007/03/15/restoring-chinas-loess-plateau
13. Planning Framework of Chinese Loess Plateau for Comprehensive Control (2010–2030). NDRC, Beijing (2010). http://www.gov.cn/gzdt/att/att/site1/20110117/001e3741a2cc0e9e318c01.pdf

Study on Burial History and Mesozoic Hydrocarbon Accumulation of Sugan Lake Depression on the Northern Margin of Qaidam Basin

Jiarui Fan[1(✉)], Guang Yang[1], and Hong Lu[2]

[1] College of Earth Sciences, Jilin University, Changchun 130061, China
346334197@qq.com
[2] Center for Computer Fundamental Education, Jilin University, Changchun 130061, China

Abstract. Mesozoic oil-gas exploration of Sugan lake depression in northern margin of Qaidam Basin is low relatively. In order to understand the controlling factors and enrichment rules of reservoir formation in this depression, this paper based on analysis of single well buried history, by using thermometry of fluid inclusions in reservoirs and the fourier transform infrared spectrograph, to determine the formation time and stage of reservoir. We restore burial history and original stratigraphic thicknesses, which is 1440 m and 319 m respectively. The mean homogeneous temperature value of saline fluid inclusion coexisted with oil-bearing fluid inclusion in Middle-Jurassic Dameigou Group reservoir is 72°C. Combined with single well burial history and temperature history, the formation time of reservoir is 100 Ma, which formed in tectonic return period of Early Cretaceous. Organic inclusions at different depths have the same characteristics of fourier transform infrared spectrograph, which means the accumulation periods are uniform.

Keywords: Sugan lake depression · Sucan1 well · Single burial history · Reservoir fluid inclusions · Homogeneous temperature · Accumulation time · Accumulation stage

1 Introduction

Qaidam Basin consists of three first-order tectonic units, know as north basin uplift, west basin uplift and Sanhu depression, Sugan lake depression is located in northwest of Qaidam Basin, it belonged to a second-order tectonic unit of north basin uplift of Qaidam Basin (Fig. 1). The joint of Aerjin Mount and Danghenan Mount lied to the north of Sugan Lake depression, in the north there are Dasaishiteng Mount and Xiaosaishiteng Mount, the southeast part of the depression reached the north side of Tuergendaban Mount, it covered an area of about 7000 km^2, two parts of Mesozoic was exposed, the Middle East part was the research target, which covered an area of 323 km^2,it was a Mesozoic remnant basin formed in late Cretaceous [1].

F. Bian and Y. Xie (Eds.): GRMSE 2015, CCIS 569, pp. 972–984, 2016.
DOI: 10.1007/978-3-662-49155-3_101

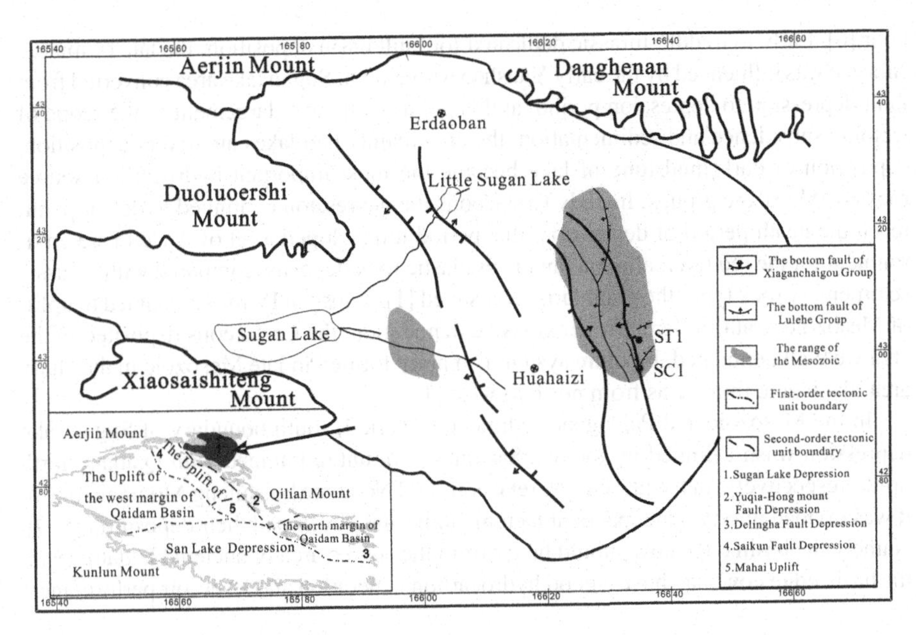

Fig. 1. The location map of sugan lake depression

Qaidam Basin has been explored for half a century, a huge mass of data of geological characteristics and distribution of oil and gas had been accumulated [2, 3], but data of accumulation time and stage of reservoir in Sugan Lake depression in northwest of Qaidam basin were insufficient [4, 5], research record for Mesozoic oil and gas accumulation issue is still missing. At present, the study has already reached two wells, SC 1 well and ST 1 well. Mesozoic strata drilling in oil and gas, according to well logging interpretation of reservoir, but the testing result showed as dry layer, the reason can be included that, previous studies considered it was caused by strong heterogeneity and poor physical property of the reservoir. In this paper, based on the recovery of single well burial history, in method of thermometry of fluid inclusion in reservoirs and the fourier transform infrared spectrograph technology, it reveals the Mesozoic reservoir formation time and stage for the first time, indicating that there are no breakthrough in prospecting oil and gas, because Mesozoic exploration well located in the north of fracture reservoir and formed in the early time of late Cretaceous. Experienced late Cretaceous and Miocene of two tectonic return movement, the reservoir near fracture layer was destroyed, thus no oil and gas can be explored in this area.

2 Tectonic and Sedimentary Evolution

Regional tectonic evolution study showed that, Sugan lake depression tectonic evolution was mainly influenced by India - Eurasian plate collision, the Aerjin strike-slip fracture and Qilian mount uplift, Jurassic -Cretaceous is the Continental rift stage, belong to the model of the strike slip type in the basement,and fault lateral seal performance is relatively poor [6]. Early to middle Jurassic dedicated for fault basin deposition, the late of middle Jurassic was influenced by the early Yanshan Movement, Sugan lake area converted from fault depression to depression period, at this point, with the enlargement of the scope of tectonic subsidence and sedimentation, the area mainly had lakes and rivers deposition, deposition of dark mudstone of lake became the most important hydrocarbon source rocks of Mesozoic groups. In early Cretaceous the depression continued to develop, but reach the death period of depression, this period it developed a set of dry climate environment for the red river, alluvial fan facies clastic rock deposits, compared with Jurassic sedimentary rock types the granularity coarsened [1]. Yanshan IV Movement led to uplift of Mesozoic strata, suffered from extensive denudation, the Cretaceous deposited in the area were all denudated. Fracture system is mainly formed in late Mesozoic in the study area, the whole move was from north to west [1].

In the Mesozoic and Paleogene sedimentary period, south boundary of Sugan lake depression wasn't limited by today's Saishiteng Mount, combined with Qaidam north uplift, respectively, in the late early Cretaceous and Neogene Saishiteng Mountain uplift, it was divided into two independent tectonic units and developed their deposition [8, 9]. Namely the basin edge now should be far from the source area of ancient Qaidam basin, the basin edge now may have a good hydrocarbon generation and reservoir performance.

3 Burial History and Thermal Evolution History

3.1 Restoration of Burial History

Sedimentary basin burial history is the foundation of the oil and gas basin analysis and an important part of oil and gas evaluation. Back stripping technology is applied in this article for the recovery of burial history, the theory basis of back stripping technology was the principle of heavy backlog, namely change the formation thickness into the change of porosity, on the basis of the hypothesis formation skeleton thickness, by establishing the relationship fomula of porosity depth to calculate strata thickness [10], resulting to the relationship between buried depth and geological time of all layers.

3.2 Recovery of Erosion Thickness

In the process of basin sedimentary recovery, when having uplift burial history, it's necessary to know its largest ancient buried depth and denudation thickness at the beginning of rise. Calculation of quantity of denudation had developed into many ways till now, such as method of Mirror reflectance, Apatite fission track, acoustic time difference, deposition rate, Fluctuation analysis method, etc. [11]. Currently, Sugan lake depression exploration degree is low, only two drill reached to Mesozoic exploration Wells, including SC1 well whose drilling maximum depth of 2878 m, full hole acoustic time logging data was full, it was suitable for recovery of denudation thickness of acoustic time difference method, according to denudation thickness calculation, burial history on the single well can be recovered.

Acoustic time difference method is used to restore denudation thickness, basically has the following three steps: (1) the strata denudation; (2) to extract effective mudstone sonic time difference value, make acoustic time - depth scatter plot, determine whether acoustic time difference method can be applied to calculate denudation amount; (3) calculate denudation thickness.

SC 1 Well located in the eastern part of Sugan lake depression, regional geology and drilling reveal that there is a unconformity the between Xiayoushashan Group to Hongshuigou Group. Late Yanshan Movement in Mesozoic resulted to extensive denudation, only remained two pieces of Mesozoic strata (Fig. 1), formed regional unconformity contact of the Paleogene groups and Hongshuigou Group (J_3), the uplift of the deposition eroded all the Quanyagou Group in Cretaceous, at the same time it caused erosion at the top Hongshui Group.

First to analysis the applicability on choosing the amount of acoustic time difference method to calculate denudation, and using the size of the old strata compaction curve slope and new layer compaction curve slope to determine whether could use acoustic time difference method to calculate erosion thickness [12]. When the old strata compaction curve slope is greater than the new formation compaction curve slope, it showed that under the denudation, plane formation compaction law is not damaged, we can use acoustic time difference method to restore the denudation thickness (Fig. 2).

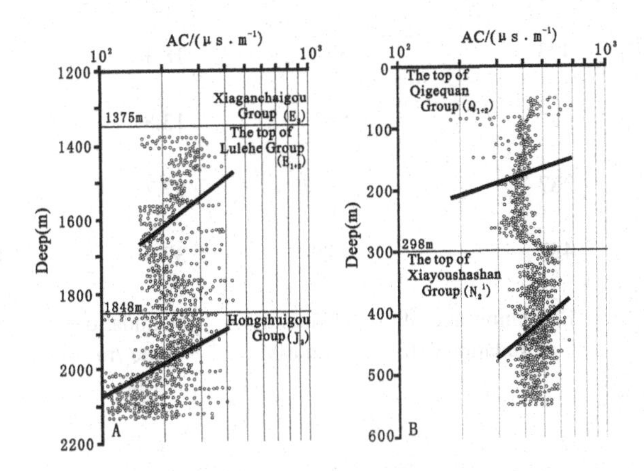

Fig. 2. Acoustic time - depth fitting graphs of SC1 well

Application of acoustic time difference method to calculate SC 1 well Mesozoic top denudation thickness is 1440 m (Table 1). But the erosion thickness is summation of Hongshuigou Group in upper Jurassic erosion amount and Quanyagou Group in the lower Cretaceous, need calculate respectively to the two groups. Since Quanyagou Group is eroded all, so the denudation amount of Quanyagou Group is equal to the thickness of sedimentary during this period. Deposition rate method is used for the thickness calculation of Quanyagou Group, namely sedimentary thickness = time by deposition rate; the determination of deposition time calculated using formula (1)

$$Te = [HeT_{i+1} + (H_i + He)T_i]/(H_i + 2He) \quad (1)$$

Te—beginning age of denudation (Ma); He—deposition thickness (m); T_{i+1}—Geological Age of Lower Boundary (Ma); H_i—Residual thickness of formation (m); Ti - geological age of upper boundary of eroded strata (Ma). Calculate the beginning age of the denudation of 105.5 Ma, denudation of 13.7 Ma. Because of the Quanyagou Group in lower Cretaceous was all denudation, unable to restore the deposition rate of this well, using this group's sedimentary rate of the adjacent Yuqia - Hongshan fault

Table 1. The denudation thickness data and recovery burial history data of SC1 well

Geological age	Thickness/m	Recovery thickness/m	Deposition rate /m $\cdot Ma^{-1}$	Deposition time /Ma
Qigequan group (Q_{1+2})	298	298	164.5	1.81
Xiayoushashan group (N_2^1)	248	578	250.2	2.31
Shangganchaigou group (N_1)	390	561	31.7	17.7
Xiaganchaigou group (E_3)	439	583	53.6	10.87
Lulehe Group (E_{1+2})	473	572	18.1	31.6
Quanyagou group (K_1)	–	552	13.8	40
Hongshuigou group (J_3)	287	1377	87.7	15.7
Dameigou group (J_{1-2})	487	545	37.8	14.4
Geological age	Beginning age of denudation /Ma	Denudation time/Ma	Denudation rate /m $\cdot Ma^{-1}$	Denudation thickness/m
Qigequan group (Q_{1+2})				
Xiayoushashan group (N_2^1)	3.01	1.21	263.6	319
Shangganchaigou group (N_1)	–	–	–	–
Xiaganchaigou group (E_3)	–	–	–	–
Lulehe group (E_{1+2})				
Quanyagou group (K_1)	105.5	13.7	40.3	552
Hongshuigou group (J_3)	91.8	22	40.3	888
Dameigou group (J_{1-2})	–	–	–	–

depression L1 well to calculate and result is 13.8 m/Ma (Table 1). Using (1) calculate the erosion amount of Quanyagou Group is 552 m, with the total denudation thickness minus this value, it is concluded that erosion thickness of Quanyagou Group is 888 m.

Late Miocene to Pliocene, depression has a long hiatus, lack of Shizigou Group. Acoustic time difference method is used to calculated the denudation amount of Xiayoushashan Group is 319 m (Table 1).

3.3 Burial History Analysis

With Yanshan Movement well-developed Sugan lake depression in the late of early Cretaceous and middle Pliocene exist two large-scale uplift and denudation (Fig. 3). When it was middle Jurassic, the Sugan lake depression began to enter fault depression period, the late shift to depression, sedimentary scope expanding gradually during this period, the deposition rate of 37.8 m/Ma. Late Jurassic, the Sugan lake depression into relatively quickly, the sedimentation deposition rate during this period is nearly 88 m/Ma (Table 1), sedimentary thickness is 1377 m. At this point, the largest Jura - buried depth is 2000 m. Sag to continue settlement at the beginning of the cretaceous, but significantly lower deposition rate, in the late early cretaceous in Jurassic buried as deep as 2200 m at the bottom. Late Yanshan IV Movement in early Cretaceous caused the rise of the differences between the Qaidam basin, making Sugan lake depression just deposits are of cretaceous strata in the start up, according to the uplift and erosion process continues to the end of the Cretaceous period, the formation of a total thickness of 1440 m. Cretaceous strata denudation of the tectonic movement will deposit is exhausted (552 m), also make the most of late Jurassic strata suffered denudation, erosion thickness is 888 m. From Eocene to Oligocene (E_{1+2}) (N_1) end, the deposition rate in the period of the change is not particularly evident. Early Eocene Lulehe Group period (E_{1+2}), sag in tectonic movement is relatively stable period, for the slow subsidence, sedimentation rate of only 18.1 m/Ma; To the Xiaganchaigou Group the deposition rate is relatively higher, the deposition rate at 53.6 m/Ma, this may be related to an event at this time on Aerjin mountain, is the deposition of the Aerjin fault strong activity corresponding [13] (Table 1). Shangganchaigou Group's deposition rate slows to 31.7 m/Ma. Into the Miocene (N_2^1), Himalayan movement II caused the strong tectonic activity, the sag early performance first, rapid subsidence middle-late stage in the rapid uplift. This period is the fastest phase deposition rate Sugan lake depression, SC1 well block deposition rate is 250 m/Ma, at this point, the sag of the main source rocks of Jurassic buried depth reaches the maximum value of about 2700 m since Mesozoic. After the rise time is shorter, less than 2 Ma, although denudation rate reached 266 m/Ma, but denudation amount is only 319 m. Enter the Pleistocene, deposition rate is 164.5 m/Ma, SC 1 well block is nearly 300 m thick sedimentary strata in nearly 2 Ma.

3.4 Restoration of Thermal History

In the production of oil, gas, transport and accumulation in the process of geothermal is universal control condition. In order to determine reservoir formation time, we recover the thermal -burial history for the single well (Fig. 3). The geothermal gradient in the study area choose the Jurassic geothermal gradient 3.27 °C/100 m, geothermal gradient in Paleogene choose 3.04 °C/100 m, since the Neogene geothermal gradient choose 2.40 °C/100 m [14]. Previous study of ancient climate of Qaidam basin, the Mesozoic belong to the temperate climate zone [15], Cenozoic belong to semi-arid - drought environment, therefore, the surface of the average annual temperature of Cenozoic before take 20 °C, the annual average temperature of 4 °C after the Cenozoic groups surface [16].

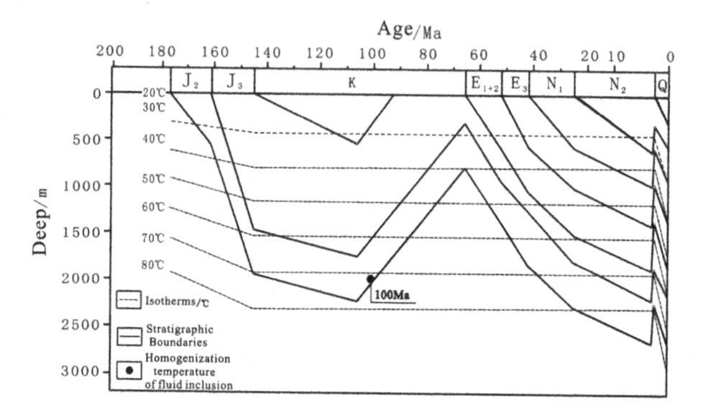

Fig. 3. Burial history of SC1 well

4 Accumulation Stage and Time of the Reservoir

Hydrocarbon accumulation is the process of reservoir formation and distribution of research one of the core content. Determine the method of hydrocarbon accumulation time is more, such as trap forming time method, hydrocarbon generation history method, reservoir fluid inclusion temperature method, saturation pressure method for oil and gas reservoir method, reservoir geochemistry method, isotope dating method, oil/water interface tracing method, organic petrology method [12], etc. Reservoir fluid inclusion temperature method is the most commonly used method. Hydrocarbon accumulation period to capture the mineral crystal growth at the boundary of fluid, the inclusions in the vast majority of cases (except a few strong damage caused by the burst and leakage), will not disappear because of the inheritance of oil and gas activity superposition reconstruction of the late, in the formation of surviving in oil and gas, transport, gathering, each phase of the inclusions of rare, is rich in oil and gas accumulation information, an advantaged tool to keep track of basin fluid flow.

Application of reservoir fluid inclusion technology and the fourier transform infrared spectrograph analysis to study Sugan lake depression of Mesozoic oil and gas accumulation time and stage. Reservoir fluid inclusion fluorescence characteristics observed, uniform temperature measurement use the Olympus BX-51 fluorescent light microscope in the Fluid laboratory of the Earth Science College in Jilin University and Linkam THMS-600 stages (UK), the precision of the machine is 0.1°C, the indoor temperature is 24°C, humidity is 40 %. Organic inclusions composition test, use the Nicolet type 6700 micro - Fourier transform infrared spectrometer, done by Beijing Nuclear Industry and Geology Institute.

4.1 Hydrocarbon Accumulation Time

Test of reservoir inclusions in samples taken from ST 1 well in Jurassic Dameigou Group (2555 ~ 2644.9 m). It contained 17 inclusions slices totally, and under fluorescent light could see organic inclusions in 8 slices.

Microscopic Characteristics of Organic Inclusions. Hydrocarbon inclusions are mainly found in gas liquid two phase inclusions, which located in the cement, and shape with rectangular to circular. The size of Inclusions individual is small, and distribution is uniform, mainly in 2 ~ 3 μm. Gas liquid ratio is between 5 ~ 10 %, and mainly 10 %, the fluorescent color is yellow and blue and white (Fig. 4).

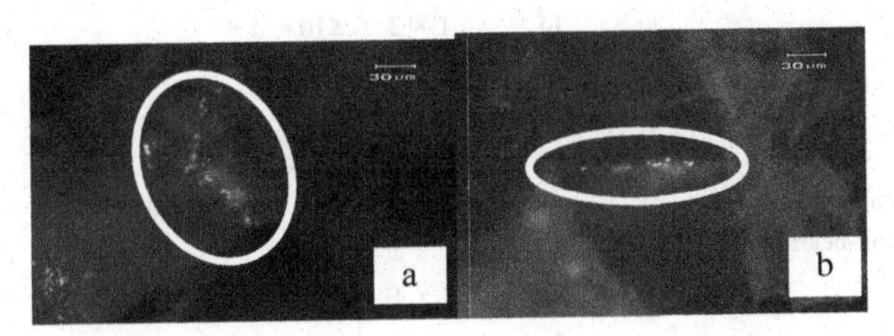

(Sample depth is 2622.0m ; a: in the cement; b: in the crack)

Fig. 4. Distribution under microscope of fluid inclusions of ST1 well (Color figure online)

Thermodynamic Characteristics of Saline Inclusions. Saline fluid inclusion coexisted with oil-bearing fluid inclusion is given priority to with round shape, followed by rectangular and triangular, distributed small groups, partially isolated distribution. The size of saline inclusion is very small, less than 5 μm, basic between 2 ~ 3 μm, are gas-liquid two-phase, gas liquid ratio is mainly between 10–15 %, with 10 % is given priority to, the individual is 5 %. For temperature measurement of saline inclusion, are mainly distributed in the cement, there were 35 measuring points, of which 31 samples were in cement, and the rest 4 samples were in quartz overgrowth boundary (Table 2).

The homogenization temperature distribution of the saline inclusion is between 55 ~ 85°C, there are two main peak values, one is 65 ~ 70°C, and another is 75 ~ 80°C (Fig. 5), the main peaks are 32 % and 28 % respectively. Because the two main peak temperatures are adjacent, it can be thought of as a main peak value, and uniform

Table 2. The temperature data of saline fluid inclusion coexisted with oil-bearing fluid inclusion

Depth/m	Fluid type	Symbiotic type	Number	Size/ μm	Gas liquid ratio /%	Homogeneous phase	Homogenization temperature /°C
2622.0	Inclusion coexisted with oil-bearing	Oil inclusion	1	1 × 2	≤10	Liquid	77.6
			2	1 × 2	≤10		70.8
			3	1 × 3	≤5		77.5
			4	1 × 2	≤10		88.0
			5	1 × 2	≤10		78.5
2625.0	Inclusion coexisted with oil-bearing	Oil inclusion	1	1 × 2	≤10	Liquid	75.8
			2	1 × 2	≤10		76.2
			3	1.5 × 2	≤10		89.6
			4	1 × 2	≤5		70.0
			5	1 × 2	≤10		75.4
			6	2 × 2.5	≤10		76.3
			7	1.5 × 2	≤10		78.9
			8	1 × 1	≤10		77.8
			9	1.5 × 2	≤10		59.6
			10	1.5 × 2	≤10		61.0
			11	1 × 2	≤10		62.3
			12	1 × 2	≤10		62.8
			13	1 × 2	≤10		63.8
			14	1 × 2	≤10		67.2
			15	1 × 1	≤10		65.2
			16	1 × 2	≤10		66.8
			17	1.5 × 2	≤15		50.7
2638.0	Inclusion coexisted with oil-bearing	Oil inclusion	1	2 × 2.5	≤10	Liquid	65.2
			2	1 × 2	≤10		65.8
			3	1 × 1	≤10		73.1
			4	1 × 2	≤10		73.8
			5	1.5 × 2	≤15		70.3
			6	1.5 × 2	≤10		67.3
			7	1 × 1	≤10		67.0
			8	1 × 1	≤10		66.8
2644.0	Inclusion coexisted with oil-bearing	Oil inclusion	1	1 × 2	≤10	Liquid	68.9
			2	1.5 × 2	≤10		74.6
			3	1 × 2	≤10		66.9
			4	1 × 2	≤5		78.6
			5	2 × 2.5	≤5		83.5

temperature is between 65 ~ 80°C with a total of 26 temperature measurement data, the average is 72°C.Then the value is voted to the Fig. 3, and the corresponding hydrocarbon accumulation time is about 104 Ma, which formed in tectonic return period of Early Cretaceous, and after the reservoir formed, and then experienced strong tectonic activities of Himalayan Movement II, and the reservoir suffered damage. This is why there was oil sand indication but no yield of oil in the SC1 well.

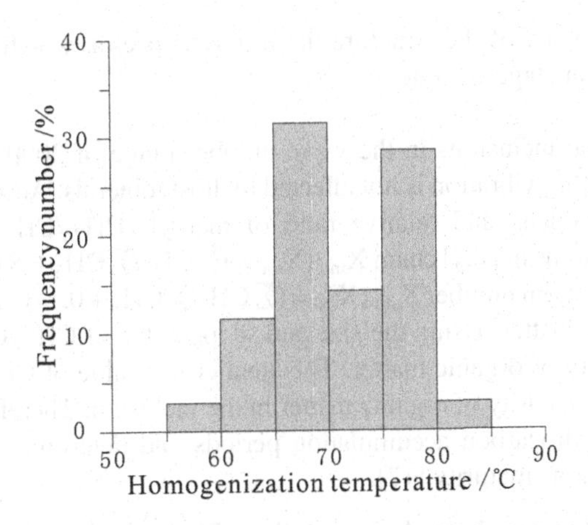

Fig. 5. Histogram of reservoir inclusions' homogenization temperature of ST1 well

4.2 Hydrocarbon Accumulation Period of Time

Infrared Spectrum Characteristics. Different infrared absorption peak intensity represents the relative abundance of the different chemical structure, the abundance ratio reflects the structure characteristics of the organic matter. Microscopy, infrared spectroscopy is the use of Fourier transform infrared spectrometer Instrument with a microscope area of micro samples, through a microscope observation of the appearance of the sample under test form or microstructure, physical and chemical test samples

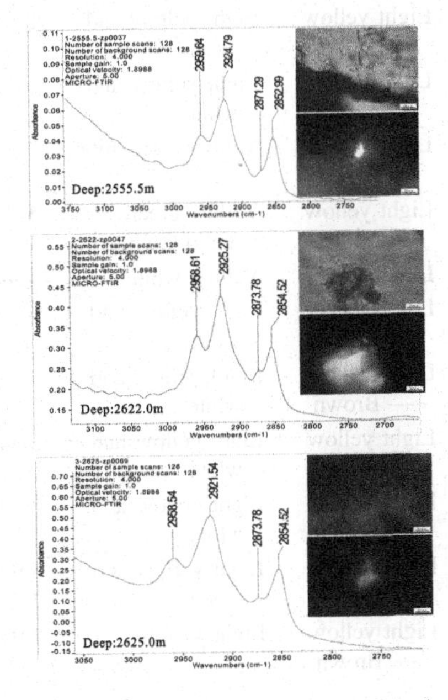

Fig. 6. The fourier transform infrared spectrograph of ST1 well

directly to specific parts of the structure, its infrared spectra, which divide the hydrocarbon accumulation stage of time.

Reservoir of organic inclusions in the wave number range of 3000 ~ 2800 cm - 1 fat hydrocarbon stretching vibration is not affected by host minerals. According to the wave number range methylene and relative ratio of methyl (CH_{2a}/CH_{3a}), organic matter number of carbon atoms in alkyl chain X_{inc} ($X_{inc} = (\sum CH_2/\sum CH_3\text{-}0.8)/0.09$), are straight chain alkane carbon atom number X_{std} ($X_{std} = (\sum CH_2/\sum CH_3 + 0.1)/0.27$) can analyze the structure of organic matter. Using the size and scope value of CH_{2a}/CH_{3a}, X_{inc}, X_{std} to evaluate the maturity of organic matter. The smaller the value of CH_{2a}/CH_{3a}, X_{inc} and X_{std}, the higher the maturity of organic matter in the inclusion. Therefore be used CH_{2a}/CH_{3a}, X_{inc}, X_{std} hydrocarbon accumulation periods and determine the parameters of hydrocarbon inclusions maturity [17].

Determine the Hydrocarbon Accumulation Stage and Maturity of Organic Inclusions. Study of reservoir single organic inclusion components can provide the direct evidence for oil and gas accumulation research. Organic inclusions component

Table 3. The fourier transform infrared spectrograph data of ST1 wel

Well number	Group	Depth/m	Characteristic of oil-gas inclusion		AREA [$\sum CH_2$]/ AREA [$\sum CH_3$]	X_{inc}	X_{std}
			Single polarized color	Fluorescent color			
ST1	J_2	2555.5	Light yellow	Light yellow and white	3.822	33.573	14.524
			Light yellow	Light yellow and green	3.89	34.329	14.776
			Light yellow	Light yellow and white	5.413	51.257	20.419
		2622.0	Light yellow —Brown	Yellow and white	3.857	33.971	14.657
			Light yellow — Brown	Yellow green 、 Yellow white	5.121	48.016	19.339
			Light yellow	Yellow white	4.999	46.658	18.886
			Light yellow — Brown	Light yellow and white	4.744	43.823	17.941
			Light yellow — Brown	Light yellow and white	3.382	28.687	12.896
			Light yellow — Brown	Light yellow and white	3.342	28.249	12.75
		2625.0	Light yellow — Brown	Light yellow and white	3.694	32.156	14.052
			Light yellow — Brown	Light yellow and green	5.609	53.431	21.144
			Light yellow — Brown	Light yellow	4.084	36.487	15.496

infrared spectral analysis detected ST 1 well 3 slide, 12 points. The samples are taken from three depth (2555.5 m, 2622 m, 2625 m, Fig. 6). For hydrocarbon reservoir fluid inclusion type liquid phase and gas-liquid two-phase inclusion. According to the results of the spectrum value calculation shows that the depth of three hydrocarbon inclusions infrared spectrum characteristics, specifications for one stage of oil and gas accumulation, and reservoir inclusions in uniform temperature distribution value a main analysis results are identical with each other. CH2a/CH3a value distribution between 3.34 ~ 5.61, X_{inc}, X_{std} value respectively between 32 ~ 51, 12 ~ 21 (Table 3), showed that methyl relative abundance of hydrocarbon chain is short, high maturity.

5 Conclusion

To Sugan lake depression's single-well burial history analysis as the foundation, mudstone acoustic time difference method is used to calculate the SC1 well's denudation thickness, since the Jurassic, main suffered late Cretaceous and the late Neogene strata two uplift and denudation, denudation thickness of 1440 m and 319 m respectively. The back strip technology to restore the original formation thickness and burial history of SC1 well. Application of thermometry of fluid inclusion in reservoirs and the fourier transform infrared spectrograph technology to determine the formation time and stage of middle Jurassic reservoir. The inclusions coexisted with oil-bearing's homogenization temperature peak is between 65 ~ 80 °C, and the average is 72 °C, be determined by the temperature reservoir formation time is 100 Ma, and formed in tectonic return period of Early Cretaceous; Using the fourier transform infrared spectrograph data of oil and gas inclusions to calculate the parameters (CH_{2a}/CH_{3a}, X_{inc}, X_{std}) characteristics, meaning that accumulation is in the same stage.

Acknowledgments. We thank Professor Guangxi Ou of the Beijing Nuclear Industry and Geology Institute for analyses of fluid inclusions.

National Natural Science Foundation of China (41472101);

China Petroleum Qinghai Oilfield Company Fund Project (QHKT/JL-03-013).

References

1. Xiao, A., Chen, Z., Yang, S.: The study of late cretaceous paleo-structural characteristics in northern Qaidam Basin. Earth Sci. Front. (China Univ. Geosci. Beijing; Peking Univ.) **12**(4), 451–457 (2005)
2. Fu, S.: Key controlling factors of oil and gas accumulation in the western Qaidam Basin and its implications for favorable exploration direction. Acta Sedimentol. Sin. **28**(2), 373–379 (2010)
3. Fu, S.: Natural gas exploration in Qaidam Basin. China Pet. Explor. **19**(4), 2–10 (2014)
4. Qi, W., Yang, G., Gao, X.: Formation condition of petroleum and exploration suggest in sugan lake depression. Qinghai Pet. **25**(3), 7–11 (2007)
5. Zhang, Y., Yang, G., Song, B.: Petroleum explored potential in the peripheral basins of Qaidam Basin. Qinghai Pet. Explor. **23**(4), 1–5 (2010)

6. Zhang, X., Sun, Y., Ma, H.: Faulted structure characteristics and its tightness evaluation in the northern margin of Qaidam Basin. Pet. Geol. Oilfield Dev. Daqing **30**(6), 43–45 (2011)
7. Tang, L., Jin, Z., Zhang, M.: Tectonic evolution and oil (gas) pool-forming stage in Northern Qaidam Basin. Pet. Explor. Dev. **27**(2), 36–39 (2000)
8. Wang, M., Hu, W., Peng, D.: Oil and gas potential of Jurassic strata in Northern margin of Qaidam Basin. Pet. Explor. Dev. **24**(5), 20–24 (1997)
9. Fang, X., Zhang, Y.: Cenozoic sediments and tectonic evolution in the western Qaidam Basin. Geol. Explor. **50**(1), 28–36 (2014)
10. Yuan, Y., Zheng, H., Tu, W.: Methods of eroded strata thickness restoration in sedimentary basins. Pet. Geol. Exp. **30**(6), 636–642 (2008)
11. Fu, X., Li, Z., Lu, S.: Recovering denudation thickness by interval transit-time. Pet. Geol. Oilfield Dev. Daqing **27**(1), 35–37 (2000)
12. Mu, Z., Chen, Z., Lu, Y.: The recovery of Mesozoic formation erosion thickness in the north margin of Qaidam Basin. Pet. Explor. Dev. **27**(1), 35–37 (2000)
13. Liu, H., Wu, G., Yang, M.: Sedimentary features of the Cenozoic in the western Qaidam basin response to strike-slipping of the Altun fault. Chin. J. Geol. **41**(2), 344–354 (2006)
14. Zhao, D., Li, J., Hu, Y.: Distribution of clay minerals in the reservoir strata of Qaidam Basin and its affecting factors. Geol. Geochem. **29**(4), 34–40 (2001)
15. Deng, S.: Paleoclimatic implications of main fossil plants of the Mesozoic. J. Palaeogeogr. **9** (6), 559–574 (2007)
16. Li, H., Ma, Y., Wang, Y.: Influence of climate warming on plant phenology in Qinhai plateau. J. Appl. Meteorol. Sci. **21**(4), 500–504 (2010)
17. Zou, Y., Yu, X., Li, S.: Study on hydrocarbon reservoir formation period using microscope-infrared spectroscopy method. Pet. Geol. Oilfield Dev. Daqing **24**(3), 33–34 (2005)

Author Index

Printed in the United States
By Bookmasters